Semicircle **10** 	$A = \dfrac{\pi r^2}{2}$ $\bar{y} = \dfrac{4r}{3\pi}$	$I_x = \dfrac{(9\pi^2 - 64)r^4}{72\pi} \approx 0.1098r^4$ $I_y = \dfrac{\pi r^4}{8} \qquad I_{xy} = 0 \qquad I_{BB} = \dfrac{\pi r^4}{8}$
Quarter Circle **11** 	$A = \dfrac{\pi r^2}{4}$ $\bar{x} = \bar{y} = \dfrac{4r}{3\pi}$	$I_x = I_y = \dfrac{\pi r^4}{16}$ $I_{xy} = \dfrac{r^4}{8}$ $I_{BB} = \dfrac{(9\pi^2 - 64)r^4}{144\pi} \approx 0.05488r^4$
Circular Sector **13** 	$A = \alpha r^2$ $\bar{x} = r \sin \alpha$ $\bar{y} = \dfrac{2r \sin \alpha}{3\alpha}$	$I_x = \dfrac{r^4}{4}(\alpha + \sin \alpha \cos \alpha)$ $I_y = \dfrac{r^4}{4}(\alpha - \sin \alpha \cos \alpha)$ $I_{xy} = 0 \qquad I_P = \dfrac{\alpha r^4}{2}$
Ellipse **16** 	$A = \pi ab$	$I_x = \dfrac{\pi ab^3}{4}$ $I_y = \dfrac{\pi ba^3}{4}$ $I_{xy} = 0 \qquad I_P = \dfrac{\pi ab}{4}(b^2 + a^2)$

STATICS AND MECHANICS OF MATERIALS

Barry J. Goodno

Georgia Institute of Technology

James M. Gere

Professor Emeritus, Stanford University

CENGAGE

Australia · Brazil · Mexico · Singapore · United Kingdom · United States

**Statics and Mechanics of Materials,
First Edition
Barry J. Goodno and James M. Gere**

Product Director, Global Engineering:
Timothy L. Anderson

Senior Content Developer: Mona Zeftel

Product Assistant: Teresa Versaggi

Marketing Manager: Kristin Stine

Senior Content Project Manager: John E. Fuller

Senior Art Director: Michelle Kunkler

Cover Designer: Jen Wahi

Internal Designer: Lou Ann Thesing

Cover Image: Image Source/Getty Images

Production Service: RPK Editorial Services, Inc.

Compositor: SPi Global

Intellectual Property
  Analyst: Christine Myaskovsky
  Project Manager: Sarah Shainwald

Text and Image Permissions Researcher:
Kristiina Paul

Manufacturing Planner: Doug Wilke

Library of Congress Control Number: 2017948554

Student Edition:
ISBN: 978-1-133-36440-5

Loose-leaf Edition:
ISBN: 978-1-337-63076-4

Cengage
20 Channel Center Street
Boston, MA 02210
USA

Cengage is a leading provider of customized learning solutions with employees residing in nearly 40 different countries and sales in more than 125 countries around the world. Find your local representative at **www.cengage.com.**

Cengage products are represented in Canada by
Nelson Education Ltd.

To learn more about Cengage platforms and services, visit
www.cengage.com.

To register or access your online learning solution or purchase materials for your course, visit **www.cengagebrain.com.**

Printed in the United States of America
Print number: 01   Print Year: 2017

CONTENTS

Barry J. Goodno

Barry John Goodno is Professor of Civil and Environmental Engineering at Georgia Institute of Technology. He joined the Georgia Tech faculty in 1974. He was an Evans Scholar and received a B.S. in Civil Engineering from the University of Wisconsin, Madison, Wisconsin, in 1970. He received M.S. and Ph.D. degrees in Structural Engineering from Stanford University, Stanford, California, in 1971 and 1975, respectively. He holds a professional engineering license (PE) in Georgia, is a Distinguished Member of ASCE and an Inaugural Fellow of SEI, and has held numerous leadership positions within ASCE. He is a past president of the ASCE Structural Engineering Institute (SEI) Board of Governors and is also a member of the Engineering Mechanics Institute (EMI) of ASCE. He is past-chair of the ASCE-SEI Technical Activities Division (TAD) Executive Committee, and past-chair of the ASCE-SEI Awards Committee. In 2002, Dr. Goodno received the SEI *Dennis L. Tewksbury Award* for outstanding service to ASCE-SEI. He received the departmental award for *Leadership in Use of Technology* in 2013 for his pioneering use of lecture capture technologies in undergraduate statics and mechanics of materials courses at Georgia Tech. He is a member of the Earthquake Engineering Research Institute (EERI) and has held several leadership positions within the NSF-funded Mid-America Earthquake Center (MAE), directing the MAE Memphis Test Bed Project. Dr. Goodno has carried out research, taught graduate courses and published extensively in the areas of earthquake engineering and structural dynamics during his tenure at Georgia Tech.

Dr. Goodno is an active cyclist, retired soccer coach and referee, and a retired marathon runner. Like co-author and mentor James Gere, he has completed numerous marathons including qualifying for and running the Boston Marathon in 1987.

© Barry Goodno

James M. Gere

James M. Gere (1925-2008) earned his undergraduate and master's degree in Civil Engineering from the Rensselaer Polytechnic Institute in 1949 and 1951, respectively. He worked as an instructor and later as a Research Associate for Rensselaer. He was awarded one of the first NSF Fellowships, and chose to study at Stanford. He received his Ph.D. in 1954 and was offered a faculty position in Civil Engineering, beginning a 34-year career of engaging his students in challenging topics in mechanics, and structural and earthquake engineering. He served as Department Chair and Associate Dean of Engineering and in 1974 co-founded the John A. Blume Earthquake Engineering Center at Stanford. In 1980, Jim Gere also became the founding head of the Stanford Committee on Earthquake Preparedness. That same year, he was invited as one of the first foreigners to study the earthquake-devastated city of Tangshan, China. Jim retired from Stanford in 1988 but continued to be an active and most valuable member of the Stanford community.

Courtesy of James and Janice Gere Family Trust

Jim Gere was known for his outgoing manner, his cheerful personality and wonderful smile, his athleticism, and his skill as an educator in Civil Engineering. He authored nine textbooks on various engineering subjects starting in 1972 with *Mechanics of Materials*, a text that was inspired by his teacher and mentor Stephan P. Timoshenko. His other well-known textbooks, used in engineering courses around the world, include: *Theory of Elastic Stability*, co-authored with S. Timoshenko; *Matrix Analysis of Framed Structures* and *Matrix Algebra for Engineers*, both co-authored with W. Weaver; *Moment Distribution*; *Earthquake Tables: Structural and Construction Design Manual*, co-authored with H. Krawinkler; and *Terra Non Firma: Understanding and Preparing for Earthquakes*, co-authored with H. Shah.

In 1986 he hiked to the base camp of Mount Everest, saving the life of a companion on the trip. James was an active runner and completed the Boston Marathon at age 48, in a time of 3:13. James Gere will be long remembered by all who knew him as a considerate and loving man whose upbeat good humor made aspects of daily life or work easier to bear.

Statics and Mechanics of Materials are basic engineering subjects that must be understood by anyone concerned with the strength and physical performance of structures, whether those structures are man-made or natural. At the college level, Statics is usually taught during the sophomore or junior year and is a prerequisite for the follow-on course in Mechanics of Materials. Both courses are required for most students majoring in mechanical, structural, civil, biomedical, petroleum, nuclear, aeronautical, and aerospace engineering. In addition, many students from such diverse fields as materials science, industrial engineering, architecture, and agricultural engineering also find it useful to study mechanics of materials.

Statics and Mechanics of Materials

In many university engineering programs today, both statics and mechanics of materials are taught in large sections of students from the many engineering disciplines. Instructors for the various parallel sections must cover the same material, and all of the major topics must be presented so that students are well prepared for the more advanced courses required by their specific degree programs. An essential prerequisite for success in a first course in mechanics of materials is a strong foundation in statics, which includes not only understanding fundamental concepts but also proficiency in applying the laws of static equilibrium to solutions of both two- and three-dimensional problems.

Part 1 consists of six chapters that cover basic concepts and applications in statics: forces, moments and resultants; equilibrium of bodies based on a properly drawn free-body diagram; application to trusses, frames and machines; centroids and moments of inertia; and internal forces and moments in bars, shafts, beams and frames.

In Part 2, stresses, strains, deformations and displacements of these elements are examined in detail. Numerous examples and end of-chapter problems are included to help students review the analysis of plane and space trusses, shafts in torsion, beams and plane and space frames, and to reinforce basic concepts learned in the prerequisite course.

Many instructors like to present the basic theory of say, beam bending, and then use real world examples to motivate student interest in the subject of beam flexure, beam design, etc. In many cases, structures on campus offer easy access to beams, frames, and bolted connections that can be dissected in lecture or in homework problems, to find reactions at supports, forces and moments in members and stresses in connections. In addition, study of causes of failures in structures and components also offers the opportunity for students to begin the process of learning from actual designs and past engineering mistakes. A number of the example problems and also the end-of-chapter problems are based upon actual components or structures and are accompanied by photographs so that the student can see the real world problem alongside the simplified mechanics model and free-body diagrams used in its analysis.

An increasing number of universities are using rich media lecture (and/or classroom) capture software (such as Panopto and Kaltura CaptureSpace) in their large undergraduate courses in mathematics, physics, and engineering. The *many photos and enhanced graphics* in this text are designed to support this enhanced lecture mode.

Key Features

The main topics covered in this book are: forces, moments, couples and resultants; development of simplified models of structures and components; equilibrium of these components; dry friction; centroids, center of mass and moments of inertia; internal forces and moments in bars, shafts, beams, frames and machines; and analysis and design of structural members subjected to tension, compression, torsion, and bending. Other important topics are the transformations of stress and strain, combined loadings and combined stress, deflections of beams, and stability of columns. Some additional specialized topics include bending of beams of two materials (or composite beams), bending of unsymmetric beams, maximum stresses in beams, and statically indeterminate beams.

Each chapter begins with a Chapter Overview highlighting the major topics covered in that chapter and closes with a Chapter Summary and Review in which the key points as well as major mathematical formulas in the chapter are listed for quick review. Each chapter also opens with a photograph of a component or structure that illustrates the key concepts discussed in the chapter.

Notable Features

Some of the notable features of this text, developed to meet the needs of modern courses in statics and mechanics of materials, are:

- **Problem-Solving Approach**—All examples in the text are presented in a Four-Step Problem-Solving Approach, which is patterned after that presented by R. Serway and J. Jewett in *Principles of Physics*, 5e, Cengage Learning, 2013. This structured format helps students to formulate simplified mathematical models of structures and components based on the relevant theory involved; to develop a correct free-body diagram(s) for use in analysis; to refine their problem-solving skills; and to improve their understanding of the main concepts illustrated in the example.

- **Chapter Overview and Chapter Summary and Review sections**—The *Chapter Overview* and *Chapter Summary* sections include *key equations* and *figures* presented in each chapter. These summary sections serve as a convenient review for students of key topics and equations presented in each chapter.

- **Emphasis on underlying fundamental concepts** such as equilibrium, constitutive, and strain-displacement/ compatibility equations in problem solutions. Example problem and end-of-chapter problem solutions have been designed to emphasize an orderly process of explicitly writing out the equilibrium, constitutive and strain-displacement/ compatibility equations before attempting a solution.

- **Extensive number of end-of-chapter problems**—The text includes more than 450 end-of-chapter problems in statics and over 900 problems in mechanics of materials. All end-of-chapter problems are grouped as **Introductory** or **Representative** and are arranged in order of increasing difficulty. Many problems in Part 1 are analyzed to find support reactions or internal forces, and then are examined in more detail in Part 2 to find deformations and stresses at key points.
- **Consistent notation and analysis approach**—A consistent set of notation is used throughout the text as the discussion proceeds from static analysis of *rigid* bodies at rest to the mechanics of *deformable* bodies and the stresses and strains associated with their internal distortions.

Importance of Example Problems

- Examples are presented throughout the book to illustrate the theoretical concepts and show how those concepts may be used in practical situations. All examples are presented in the **Four-Step Problem-Solving Approach** format so that the basic concepts as well as the key steps in setting up and solving each problem are clearly understood. Photographs have been added showing actual engineering structures or components to reinforce the tie between theory and application. Each example begins with a clear statement of the problem and then presents a simplified analytical model and the associated free-body diagrams to aid students in understanding and applying the relevant theory in engineering analysis of the system. In most cases, the examples are worked out in symbolic terms so as to better illustrate the ideas, and then numeric values of key parameters are substituted in the final part of the analysis step. In selected examples throughout the text, graphical display of results (e.g., stresses in beams) has been added to enhance the student's understanding of the problem results.

Example

FIGURE (a)

FIGURE (b)

Free-body diagram of truss model

In many cases, the problem involves the analysis of a real physical structure, such as this truss structure (a) representing part of the fuselage of a model air plane. Begin by sketching the portion of the structure of interest showing members, supports, dimensions and loadings. This **Conceptualization** step in the analysis often leads to a free-body diagram (b).

The next step is to simplify the problem, list known data and identify all unknowns, and make necessary assumptions to create a suitable model for analysis. This is the **Categorize** step.

Write the governing equations, then use appropriate mathematical and computational techniques to solve the equations and obtain results, either in the form of mathematical formulas or numerical values. The **Analysis** step leads to support reaction and member forces in the truss.

List the major steps in your analysis procedure so that it is easy to review or check at a later time.

Solution:

The solution involves the following steps:

1. *Conceptualize* [*hypothesize, sketch*]: First sketch a free-body diagram of the entire truss model (Figure (b)). Only known applied forces at C and unknown reaction forces at A and B are shown and then used in an equilibrium analysis to find the reactions.

2. *Categorize* [*simplify, classify*]: Overall equilibrium requires that the force components in x and y directions and the moment about the z axis must sum to zero; this leads to reaction force components A_x, A_y, and B_y. The truss is statically determinate (*unknowns*: $m + r = 5 + 3 = 8$, *knowns*: $2j = 8$) so all member forces can be obtained using the *method of joints*. . . .

3. *Analyze* [*evaluate; select relevant equations, carry out mathematical solution*]: First find the lengths of members AC and BC, which are needed to compute distances to lines of action of forces.

   Law of sines to find member lengths a and b: Use known angles θ_A, θ_B, and θ_C and $c = 10$ ft to find lengths a and b:

   $$b = c\frac{\sin(\theta_B)}{\sin(\theta_C)} = (10\text{ ft})\frac{\sin(40°)}{\sin(80°)} = 6.527\text{ ft},$$

   $$a = c\frac{\sin(\theta_A)}{\sin(\theta_C)} = (10\text{ ft})\frac{\sin(60°)}{\sin(80°)} = 8.794\text{ ft}$$

   Check that computed lengths a and b give length c by using the law of cosines:

   $$c = \sqrt{(6.527\text{ ft})^2 + (8.794\text{ ft})^2 - 2(6.527\text{ ft})(8.794\text{ ft})\cos(80°)} = 10\text{ ft}$$

4. *Finalize* [*conclude; examine answer—does it make sense? Are units correct? How does it compare to similar problem solutions?*]: There are $2j = 8$ equilibrium equations for the simple plane truss considered above and, using the *method of joints*, these are obtained by applying $\Sigma F_x = 0$ and $\Sigma F_y = 0$ at each joint in succession. A computer solution of these simultaneous equations leads to the three reaction forces and five member forces. The *method of sections* is an efficient way to find selected member forces.

List the major steps in the **Finalize** step, review the solution to make sure that it is presented in a clear fashion so that it can be easily reviewed and checked by others. Are the expressions and numerical values obtained reasonable? Do they agree with your initial expectations?

Problems

In all mechanics courses, solving problems is an important part of the learning process. This textbook offers more than 1350 problems, many with multiple parts, for homework assignments and classroom discussions. The problems are placed at the end of each chapter so that they are easy to find and don't break up the presentation of the main subject matter. Also, problems are generally arranged in order of increasing difficulty, thus alerting students to the time necessary for solution. Answers to all problems are listed near the back of the book.

Considerable effort has been spent in checking and proofreading the text so as to eliminate errors. If you happen to find one, no matter how trivial, please notify me by e-mail (*bgoodno@ce.gatech.edu*). We will correct any errors in the next printing of the book.

Units

Both the International System of Units (SI) and the U.S. Customary System (USCS) are used in the examples and problems. Discussions of both systems and a table of conversion factors are given in Chapter 1. For problems involving numerical solutions, odd-numbered problems are in USCS units and even-numbered problems are in SI units. This convention makes it easy to know in advance which system of units is being used in any particular problem. In addition, tables containing properties of plane areas, properties of different materials, and properties of structural-steel shapes in both USCS and SI units may be found in Appendix C so that solution of beam analysis and examples and end-of-chapter problems can be carried out in either USCS or SI units.

Supplements

Instructor Resources

An **Instructor's Solutions Manual** includes solutions to all end-of-chapter problems with Mathcad solutions available for some problems. The Manual includes appropriate free-body diagrams as well as the detailed steps required in the full solution. The digital ISM is accessible to instructors via the secure, password-protected Instructor Resource Center at http://login.cengage .com. The Instructor Resource Center also contains a full set of **Lecture Note PowerPoints** and an **Image Bank** of figures and tables from the book.

Student Resources

FE Exam Review Problems are available to students online. This supplement contains 61 statics problems and 106 mechanics of materials review problems and solutions, which cover the major topics presented in the text and are representative of those likely to appear on an FE exam. Each of the problems is presented in the FE Exam format and is intended to serve as a useful guide to the student in preparing for this important examination.

Many students take the *Fundamentals of Engineering Examination* upon graduation, the first step on their path to registration as a Professional Engineer. Most of these problems are in SI units, which is the system of units used on the FE Exam itself and require use of an engineering calculator to carry out the solution. The student must select from four available answers, only one of which is the correct answer. Go to http://www.cengagebrain.com to find the FE Exam Review Problems and the resources below, which are available on the student website for this book:

- *Answers to the FE Exam Review Problems*
- *Detailed Solutions for Each Problem*

Custom Options for Statics and Mechanics of Materials

Would you prefer to easily create your own personalized text, choosing the elements that best serve your course's unique learning objectives?

Cengage's Compose platform provides the full range of Cengage content, allowing you to create exactly the textbook you need. The Compose website lets you quickly review materials to select everything you need for your text. You can even seamlessly add your own materials, like exercises, notes, and handouts! Easily assemble a new print or eBook and then preview it on our site.

Cengage offers the easiest and fastest way to create custom learning materials that are as unique as your course is. To learn more about customizing your book with Compose, visit **compose.cengage.com**, or contact your Cengage Learning Consultant.

S.P. Timoshenko (1878–1972) and J.M. Gere (1925–2008)

Many readers of this book will recognize the name of Stephen P. Timoshenko—probably the most famous name in the field of applied mechanics. Timoshenko is generally recognized as the world's most outstanding pioneer in applied mechanics. He contributed many new ideas and concepts and became famous for both his scholarship and his teaching. Through his numerous textbooks he made a profound change in the teaching of mechanics wherever mechanics is taught. Timoshenko was both teacher and mentor to James Gere and provided the motivation for the mechanics of materials part of this text, authored by James M. Gere and published in 1972. The second and each subsequent mechanics of materials part of this book were written by James Gere over the course of his long and distinguished tenure as author, educator, and researcher at Stanford University. James Gere started as a doctoral student at Stanford in 1952 and retired from Stanford as a professor in 1988 having authored nine well-known and respected text books on mechanics, and structural and earthquake engineering. He remained active at Stanford as Professor Emeritus until his death in January of 2008.

Acknowledgments

To acknowledge everyone who contributed to this book in some manner is clearly impossible, but I owe a major debt to my former Stanford teachers, especially my mentor and friend, and co-author James M. Gere.

I am grateful to my many colleagues teaching both Statics and Mechanics of Materials at various institutions throughout the world who have provided feedback and constructive criticism about the text; for all those anonymous reviews, my thanks. Their advice has resulted in significant improvements in both content and pedagogy.

My appreciation and thanks also go to the reviewers who provided specific comments for this book as well as the others who did not wish to be acknowledged in the preface.

Erian Armanios, *University of Texas at Arlington*

Aaron S. Budge, *Minnesota State University, Mankato*

Virginia Ferguson, *University of Colorado, Boulder*

James Giancaspro, *University of Miami*

Andrew Grimm, *Marietta College*

Paul Heyliger, *Colorado State University*

Eric Kasper, *California Polytechnic State University, San Luis Obispo*

Brian Kohn, *University of Minnesota*

Richard Kunz, *Mercer University*

David Lattanzi, *George Mason University*

Prasad Marehalli, *Stevens Institute of Technology*

Gustavo Molina, *Georgia Southern University*

Sangram Redkar, *Arizona State University*

Suzannah Sandrik, *University of Wisconsin—Madison*

Morteza A.M. Torkamani, *University of Pittsburgh*

Jin Zhihe, *University of Maine*

I wish to also acknowledge my Engineering and Mechanics colleagues at the Georgia Institute of Technology, many of whom provided valuable advice on various aspects of the revisions and additions leading to this text. It is a privilege to work with all of these educators and to learn from them in almost daily interactions and discussions about structural engineering and mechanics in the context of research and higher education. I wish to extend my thanks to my many current and former students who have helped to shape this text. Finally, I would like to acknowledge the excellent work of Dr. Edwin Lim, who suggested new problems and also carefully checked the solutions of many of the examples and end of chapter problems.

I wish to acknowledge and thank the Global Engineering team at Cengage Learning for their dedication to this new book:

Timothy Anderson, Product Director; Mona Zeftel, Senior Content Developer; Kristin Stine, Marketing Manager; Teresa Versaggi, Product Assistant; and Rose Kernan of RPK Editorial Services, Inc.

They have skillfully guided every aspect of this text's development and production to successful completion.

I am deeply appreciative of the patience and encouragement provided by my family, especially my wife, Lana, throughout this project.

Finally, I am very pleased to continue this endeavor begun so many years ago by my mentor and friend, Jim Gere. I am committed to its continued excellence and welcome all comments and suggestions. Please feel free to provide me with your critical input at *bgoodno@ce.gatech.edu.*

Barry J. Goodno
Atlanta, Georgia

Statics and Mechanics of Materials is also available with **MindTap**, Cengage's digital learning experience. The textbook's carefully-crafted pedagogy and exercises are made even more effective by an interactive, customizable eBook accompanied by automatically graded assessments and a full suite of study tools.

MindTap gives you complete control of your course—to provide engaging content, to challenge every individual, and to prepare students for professional practice. Adopting MindTap cuts your prep time and lets you teach more effectively with videos, assessments, algorithmically generated problem sets, and more. Built-in metrics provide insight into engagement, identify topics needing extra instruction, and let you instantly communicate with your students. Finally, every MindTap adoption includes support from our dedicated, personalized team. We'll help you set up your course, tailor it to your specifications, and stand by to offer support to you and your students whenever you need us.

< CHAPTER 7: TENSION, COMPRESSION, AND SHEAR

Chapter 7: Tension, Compression, and Shear
Introduction to Mechanics of Materials - Normal Stress and Strain - Mechanical Properties · Plasticity, and Creep - Linear Elasticity, Hooke's Law, and Poisson's Ratio - Shear Stress and and Allowable Loads - Design for Axial Loads and Direct Shear - Chapter Summary and Rev

Chapter 7 Step-by-Step Tutorials

Chapter 7 Videos

Chapter 7 Quiz
Take this quiz to see what you've learned in Chapter 7.
No Submissions COUNTS TOWARD GRADE

Chapter 7 Problem Set
Solve this set of problems designed to help you master mechanics of materials challenges. see stepped through solutions for each problem part.
No Submissions COUNTS TOWARD GRADE

Chapter 7 Reflective Questions
These questions provide you with an opportunity to reflect on how you did in learning the
No Submissions COUNTS TOWARD GRADE

Chapter 7 Drop Box
Use this drop box to submit any other assignments your instructor has assigned to you.

How Does MindTap Benefit Instructors?

- Customize and personalize your course by integrating your own content into the **MindTap Reader** (like lecture notes, audio and video recordings, or problem sets to download) or pull from sources such as RSS feeds, YouTube videos, websites, and more.

- Save grading time by leveraging MindTap's **automatically graded assignments and quizzes**, including **algorithmically generated problem sets**. These problems include immediate, specific feedback, so students know exactly where they need more practice.

- The **Message Center** helps you to quickly and easily contact students directly from MindTap. Messages are communicated immediately by email, social media, or even text message.

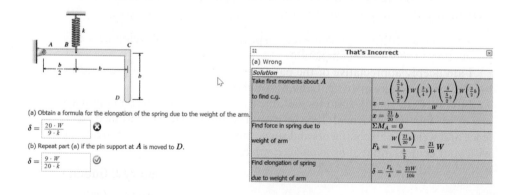

(a) Obtain a formula for the elongation of the spring due to the weight of the arm.

$\delta = \dfrac{20 \cdot W}{9 \cdot k}$ ✗

(b) Repeat part (a) if the pin support at A is moved to D.

$\delta = \dfrac{9 \cdot W}{20 \cdot k}$ ✓

That's Incorrect

(a) Wrong

Solution

Take first moments about A to find c.g.

$$x = \frac{\left(\frac{\frac{3}{2}b}{\frac{5}{2}b}\right) W \left(\frac{3}{4}b\right) + \left(\frac{b}{\frac{5}{2}b}\right) W \left(\frac{5}{2}b\right)}{W}$$

$$x = \frac{21}{20} b$$

Find force in spring due to weight of arm

$\Sigma M_A = 0$

$$F_k = \frac{W\left(\frac{21}{20}b\right)}{\frac{b}{2}} = \frac{21}{10} W$$

Find elongation of spring due to weight of arm

$$\delta = \frac{F_k}{k} = \frac{21W}{10k}$$

- **StudyHub** is an all-in-one studying destination that allows you to deliver important information and empowers your students to personalize their experience. Instructors can choose to annotate the text with **notes** and **highlights**, share content from the MindTap Reader, and create custom **flashcards** to help their students focus and succeed.
- The **Progress App** lets you know exactly how your students are doing (and where they are struggling) with live analytics. You can see overall class engagement levels and drill down into individual student performance, enabling you to identify topics needing extra instruction and instantly communicate with struggling students to speed progress.

How Does MindTap Benefit Your Students?

- The **MindTap Mobile App** includes the entire eBook accompanied by flashcards, quizzes, and course alerts to help students understand core concepts, achieve better grades, and prepare for their future courses.
- **Flashcards** are pre-populated to provide a jump start on studying, and students and instructors can also create customized cards as they move through the course.
- The **Progress App** allows students to monitor their individual grades, as well as their level compared to the class average. This not only helps them stay on track in the course but also motivates them to do more, and ultimately to do better.
- The **StudyHub** is a single-destination studying tool that empowers students to personalize their experience. They can quickly and easily access all notes and highlights marked in the MindTap Reader, locate bookmarked pages, review notes and Flashcards shared by their instructor, and create custom study guides.
- The **MindTap Reader** includes the abilities to have the content read aloud, to print from the digital textbook, and to take notes and highlights directly in the text while also capturing them within the linked **StudyHub App**.

For more information about MindTap for Engineering, or to schedule a demonstration, please call (800) 354-9706 or email higheredcs@cengage.com. For instructors outside the United States, please visit http://www.cengage.com/contact/ to locate your regional office.

A area

A_f, A_w area of flange; area of web

a, b, c dimensions, distances

C centroid, compressive force, constant of integration

c distance from neutral axis to outer surface of a beam

D diameter

d diameter, dimension, distance

E modulus of elasticity

e eccentricity, dimension, distance, unit volume change (dilatation)

F force

f shear flow, shape factor for plastic bending, flexibility, frequency (Hz)

f_T torsional flexibility of a bar

G modulus of elasticity in shear

g acceleration of gravity

H height, distance, horizontal force or reaction, horsepower

h height, dimensions

I moment of inertia (or second moment) of a plane area

I_x, I_y, I_z moments of inertia with respect to x, y, and z axes

I_{x1}, I_{y1} moments of inertia with respect to x_1 and y_1 axes (rotated axes)

I_{xy} product of inertia with respect to xy axes

I_{x1y1} product of inertia with respect to x_1y_1 axes (rotated axes)

I_P polar moment of inertia

I_1, I_2 principal moments of inertia

J torsion constant

K stress-concentration factor, bulk modulus of elasticity, effective length factor for a column

k spring constant, stiffness, symbol for $\sqrt{P/EI}$

k_T torsional stiffness of a bar

L length, distance

L_E effective length of a column

$\ln, \log$ natural logarithm (base e); common logarithm (base 10)

M bending moment, couple, mass

m moment per unit length, mass per unit length

N axial force

n factor of safety, integer, revolutions per minute (rpm)

O origin of coordinates

O' center of curvature

P force, concentrated load, power

P_{allow} allowable load (or working load)

P_{cr} critical load for a column

p pressure (force per unit area)

Q force, concentrated load, first moment of a plane area

q intensity of distributed load (force per unit distance)

R reaction, radius

r radius, radius of gyration $(r = \sqrt{I/A})$

S section modulus of the cross section of a beam, shear center

s distance, distance along a curve

T tensile force, twisting couple or torque, temperature

t thickness, time, intensity of torque (torque per unit distance)

t_f, t_w thickness of flange; thickness of web

u_r, u_t modulus of resistance; modulus of toughness

V shear force, volume, vertical force or reaction

v deflection of a beam, velocity

v', v'', etc. dv/dx, d^2v/dx^2, etc.

W force, weight, work

w load per unit of area (force per unit area)

x, y, z rectangular axes (origin at point O)

x_c, y_c, z_c rectangular axes (origin at centroid C)

$\bar{x}, \bar{y}, \bar{z}$ coordinates of centroid

α angle, coefficient of thermal expansion, nondimensional ratio

β angle, nondimensional ratio, spring constant, stiffness

β_R rotational stiffness of a spring

γ shear strain, weight density (weight per unit volume)

$\gamma_{xy}, \gamma_{yz}, \gamma_{zx}$ shear strains in xy, yz, and zx planes

γ_{x1y1} shear strain with respect to $x_1 y_1$ axes (rotated axes)

γ_θ shear strain for inclined axes

δ deflection of a beam, displacement, elongation of a bar or spring

ΔT temperature differential

ε normal strain

$\varepsilon_x, \varepsilon_y, \varepsilon_z$ normal strains in x, y, and z directions

$\varepsilon_{x1}, \varepsilon_{y1}$ normal strains in x_1 and y_1 directions (rotated axes)

ε_θ normal strain for inclined axes

$\varepsilon_1, \varepsilon_2, \varepsilon_3$ principal normal strains

ε' lateral strain in uniaxial stress

ε_T thermal strain

ε_Y yield strain

θ angle, angle of rotation of beam axis, rate of twist of a bar in torsion (angle of twist per unit length)

θ_p angle to a principal plane or to a principal axis

θ_s angle to a plane of maximum shear stress

κ curvature ($\kappa = 1/\rho$)

λ distance, curvature shortening

ν Poisson's ratio

ρ radius, radius of curvature ($\rho = 1/\kappa$), radial distance in polar coordinates, mass density (mass per unit volume)

σ normal stress

$\sigma_x, \sigma_y, \sigma_z$ normal stresses on planes perpendicular to x, y, and z axes

σ_{x1}, σ_{y1} normal stresses on planes perpendicular to $x_1 y_1$ axes (rotated axes)

σ_θ normal stress on an inclined plane

$\sigma_1, \sigma_2, \sigma_3$ principal normal stresses

σ_{allow} allowable stress (or working stress)

σ_{cr} critical stress for a column ($\sigma_{\text{cr}} = P_{\text{cr}}/A$)

σ_{pl} proportional-limit stress

σ_r residual stress

σ_T thermal stress

σ_U, σ_Y ultimate stress; yield stress

τ shear stress

$\tau_{xy}, \tau_{yz}, \tau_{zx}$ shear stresses on planes perpendicular to the x, y, and z axes and acting parallel to the y, z, and x axes

τ_{x1y1} shear stress on a plane perpendicular to the x_1 axis and acting parallel to the y_1 axis (rotated axes)

τ_θ shear stress on an inclined plane

τ_{allow} allowable stress (or working stress) in shear

τ_U, τ_Y ultimate stress in shear; yield stress in shear

ϕ angle, angle of twist of a bar in torsion

ψ angle, angle of rotation

ω angular velocity, angular frequency ($\omega = 2\pi f$)

GREEK ALPHABET

A	α	Alpha	N	ν	Nu
B	β	Beta	Ξ	ξ	Xi
Γ	γ	Gamma	O	o	Omicron
Δ	δ	Delta	Π	π	Pi
E	ε	Epsilon	P	ρ	Rho
Z	ζ	Zeta	Σ	σ	Sigma
H	η	Eta	T	τ	Tau
Θ	θ	Theta	Y	υ	Upsilon
I	ι	Iota	Φ	ϕ	Phi
K	κ	Kappa	X	χ	Chi
Λ	λ	Lambda	Ψ	ψ	Psi
M	μ	Mu	Ω	ω	Omega

Introduction

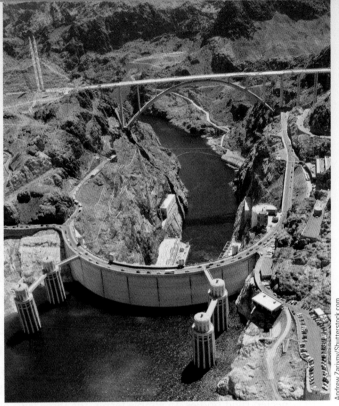

Andrew Zarivny/Shutterstock.com

Engineering mechanics principles are used in the analysis and design of structures such as bridges and dams that are subjected to large and variable forces during their lifetime (Mike O'Callaghan-Pat Tillman Memorial Bridge (below Hoover Dam)).

CHAPTER OBJECTIVES

- Introduce *fundamental concepts* in statics such as its origins, governing laws, and their expression in mathematics.

- Study *basic quantities* such as *mass, length, time*, and *force* as well as *units* to be used (both USCS and SI) to define these quantities.

- Review basic operations in *vector algebra*.

- Discuss *rounding* of numbers and the issue of *significant digits* to be retained in solution of *numerical problems* in statics.

- Present a *four-step problem-solving approach* for carrying out both numerical and *symbolic* solutions to problems.

Chapter Outline

1.1 Fundamental Concepts: Rigid and Deformable Bodies

Engineering mechanics is the science that deals with bodies at rest or in motion. Bodies at rest are considered in the branch of mechanics known as *statics*, while bodies in motion are studied in the branch known as *dynamics*. Classical mechanics originated with Isaac Newton's laws of motion[1] and involves considerable use of mathematical models to implement its basic theories. Study and mastery of basic mechanics is an essential prerequisite for an understanding of mechanics of materials, fluid mechanics, structural analysis and design, machine design, flight vehicle design, robotics and control, vibrations, structural dynamics, and so on.

Part 1 of the text focuses on bodies at rest that may be considered *rigid* during analysis, while part 2 provides an assessment of internal distortions of *deformable* bodies, including the variation of stresses, strains, and displacements throughout the body. In *statics*, bodies are idealized as rigid because the dimensions of the body are irrelevant to describing the position as well as the effects of any forces acting upon them. You often have to disassemble the body into its component parts to reveal any action–reaction forces acting between those parts, but the parts themselves remain rigid. Your goal is to find the forces required to maintain the equilibrium of the overall body. Once these *external* forces are known, you can then dissect the body to investigate *internal* interaction forces and associated stresses, strains, and deformations.

1.2 Newton's Laws of Motion and Gravitation

Newton's laws[2] of motion and gravitation explain the motion of particles and form the basis of the statics and dynamics of rigid bodies as well as interconnected rigid bodies. In simple terms, the three laws of motion are

First Law A particle remains at rest or continues to move at a constant speed along a straight line if no unbalanced forces act on the particle.

Second Law A particle accelerates in the direction of the vector sum of the forces acting on it, and the acceleration is proportional to the sum of the forces.

Third Law Action and reaction forces act between bodies in contact, and these forces are equal in magnitude, opposite in direction, and collinear.

The first law describes the force *equilibrium* of bodies that is at the heart of *statics* of particles and systems of particles. Here the particle's acceleration is zero. In the second law, the acceleration is nonzero, and the vector sum of forces F is equal to the mass m of the particle times the acceleration a, that is, $F = ma$. Both force and acceleration are vector quantities. Vectors F and a have

---

[1]René Dugas, *A History of Mechanics*, Dover Publications (2011).

[2]Sir Isaac Newton (1642–1727) was an English mathematician, physicist, and astronomer. He invented calculus and discovered the laws of motion and gravitation.

the same direction, and the magnitudes of **F** and **ma** must be equal. The second law is used extensively in the study of the *dynamics* of particles.

The third law is concerned with action–reaction force pairs that exist between bodies in contact (*internal forces*) or between bodies and the external world at points of support (*reaction forces*) of the body. An assessment of equilibrium only can be made once bodies in contact are separated and the force of the body acting upon the body of interest is clearly displayed. A **free-body diagram** is a display of a body with all the forces shown upon it. Constructing the free-body diagram is an essential step in the study of the equilibrium of rigid bodies in statics.

Newton's Law of Gravitation   The **weight** of an object is the **force of gravity** acting on that object and is measured in newtons (N). Since the force of gravity depends upon altitude and position on the earth, weight is not an invariant property of a body. The weight of a body as measured by a spring scale is affected not only by the gravitational pull of the earth but also by the centrifugal effects associated with the rotation of the earth.

As a consequence, there are two kinds of weight: **absolute weight** and **apparent weight**. Absolute weight is based upon the force of gravity alone, while apparent weight includes the effects of rotation. Newton formulated the *law of gravitation*, which can be used to compute the absolute weight of a body as

$$F = G\frac{m_1 m_2}{r^2} \qquad\qquad (1\text{-}1)$$

where

F = force (in lb or N) of mutual attraction between two bodies

m_1 and m_2 = mass of each particle (in slugs or kg)

G = universal gravitation constant (in U.S. units, 3.439×10^{-8} ft^4/(lb-s^4) and in SI units, 6.673×10^{-11} m^3/(kg·s^2))

r = distance (in ft or m) between the centers of the two bodies.

Apparent weight is always less than absolute weight (except at the poles). Apparent weight, which is the weight as measured with a spring scale, is the weight used in everyday life; absolute weight is used in astroengineering and certain kinds of scientific work. In this book, the term "weight" will always mean "apparent weight."

The **acceleration of gravity**, denoted by the letter g, is directly proportional to the force of gravity; therefore, it depends upon position. In contrast, **mass** is a measure of the amount of material in a body and does not change with location.

The fundamental relationship between weight, mass, and acceleration of gravity can be obtained from Newton's second law, which becomes

$$W = mg \qquad\qquad (1\text{-}2)$$

In this equation,

W = weight in newtons (N)

m = mass in kilograms (kg)

g = acceleration of gravity in meters per second squared (m/s^2) if SI units are used.

Equation (1-2) shows that *a body having a mass of one kilogram has a weight in newtons numerically equal to g*. The values of the weight W and the acceleration g depend upon many factors, including latitude and elevation. However, for scientific calculations, a standard international value of g has been established as

$$g = 9.806650 \text{ m/s}^2 \qquad \text{(1-3)}$$

The recommended value of g for ordinary engineering purposes on or near the surface of the earth is

$$g = 9.81 \text{ m/s}^2 \qquad \text{(1-4)}$$

Thus, a body having a mass of one kilogram has a weight of 9.81 newtons in SI units. In U.S. customary units, weight W is in pounds (lb) when mass m is in slugs, and acceleration g is taken with sufficient accuracy in statics as 32.2 ft/s^2.

1.3 Vectors and Scalars

Scalar   A scalar is a quantity with *magnitude* only. Scalars include such items as mass, volume, density, time, and speed. A scalar is shown as V using a regular font.

Vector   A *vector* is a quantity that has both *magnitude* and *direction*. Examples of vector quantities are force, displacement, velocity, and acceleration. A vector is shown using a bold italic font (V). The magnitude of a vector is a scalar quantity and can be shown as V or $|V|$. Note that speed, which is a scalar, is the magnitude of velocity, which is a vector.

A vector is usually represented in diagrams with an arrow pointing in the direction of the vector and having some inclination angle, θ, with respect to some reference line (Fig. 1-1).

The length of the arrow is a visual indication of the magnitude of the vector. The vector can also have a sign associated with it so that V (or $+V$) points in the positive direction while $-V$ has equal magnitude but points in the opposite direction. A vector tied to a specific point along a line is a *fixed* vector (Fig. 1-2a), while a vector positioned anywhere along a line is a *sliding* vector (Fig. 1-1). If a vector is not associated with a particular line or a point, it is a *free* vector (Fig. 1-2b).

FIGURE 1-1

Sliding vector

FIGURE 1-2

(a, b) Fixed and free vectors

(a)   (b)

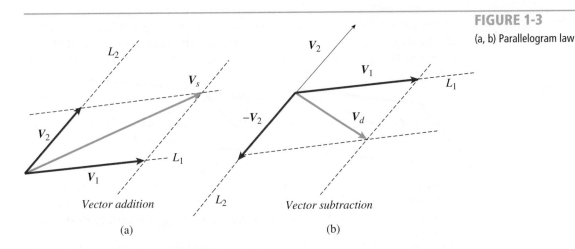

FIGURE 1-3

(a, b) Parallelogram law

Some Basic Vector Operations

Parallelogram Law: The *sum* of two vectors can be represented by the vector equation

$$V_s = V_1 + V_2 = V_2 + V_1 \tag{1-5}$$

Vector addition is represented graphically in Fig. 1-3a using two *free vectors* V_1 and V_2 whose sum is the resultant vector V_s. The order of vector addition does not affect the final result. Resultant vector V is shown to be the diagonal of the parallelogram formed from free vectors V_1 and V_2. The *difference* between two vectors is obtained using vector subtraction as

$$V_d = V_1 - V_2 = V_1 + (-V_2) \tag{1-6}$$

Vector subtraction is accomplished by reversing the direction of vector V_2 and then using the parallelogram law to find the vector difference V_d, as shown in Fig. 1-3b.

Note that vector addition ($V_s = V_1 + V_2$) and vector subtraction ($V_d = V_1 - V_2$) are not the same as scalar addition ($V_s = |V_1| + |V_2| = V_1 + V_2$) and scalar subtraction ($V_d = |V_1| - |V_2| = V_1 - V_2$).

Triangle Law: An alternative way to represent vector addition and subtraction is with the triangle law. Vector V_2 can be moved to the head of vector V_1 (since both are *free vectors*) to find the resultant vector (V_s or V_d for addition and subtraction, respectively) as shown in Figs. 1-4a and b.

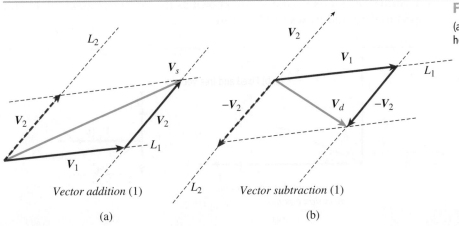

FIGURE 1-4

(a, b) Triangle law: move vector V_2 to head of V_1

FIGURE 1-5

(a,b) Triangle law: move vector V_1 to head of V_2

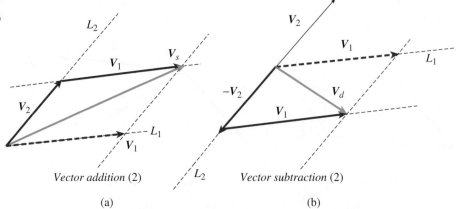

Vector addition (2)

(a)

Vector subtraction (2)

(b)

The same result is obtained if vector V_1 is moved instead to the head of vector V_2 (Figs. 1-5a and b).

Vector Components: In the discussion above, vectors V_1 and V_2 are the *components* of the *resultant* vector V_s. Vectors V_1 and V_2 act along lines L_1 and L_2, respectively, and resultant vector V_s is obtained using either the parallelogram law (Fig. 1-3) or the triangle law (Fig. 1-4). It is often convenient to use a Cartesian coordinate system with vectors V_1 and V_2 replaced by perpendicular vectors V_x and V_y (in two dimensions) which are directed along coordinate axes x and y (Fig. 1-6a).

Vectors V_x and V_y are the *rectangular components* of resultant vector V. Using vector addition [see Eq. (1-5)]:

$$V_s = V_x + V_y = V_y + V_x \tag{1-7}$$

and vector subtraction [see Eq. (1-6) and Fig. 1-6b] gives

$$V_d = V_x - V_y = V_x + \left(-V_y\right) \tag{1-8}$$

Note that the orientation of V with respect to the x and y axes can be defined using angles θ_x and θ_y where

$$\theta_x = \tan^{-1}\frac{V_y}{V_x} = \cos^{-1}\frac{V_x}{|V|}$$
$$\theta_y = \cos^{-1}\frac{V_y}{|V|} \tag{1-9}$$

The cosines of angles θ_x and θ_y are known as the *direction cosines* of V with respect to the x and y axes (see Fig. 1-6).

FIGURE 1-6

(a,b) Rectangular components (2D), unit vectors

2D rectangular components, direction cosines

(a)

(b)

FIGURE 1-7

Rectangular components (3D), unit vectors

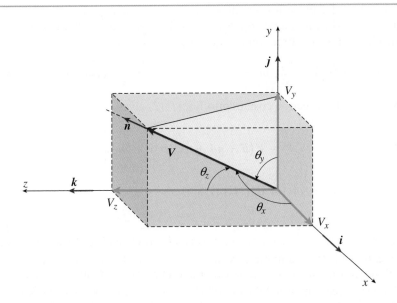

In three dimensions, vector V has three components (Fig. 1-7), one along each coordinate axis (x, y, and z). Vector addition is carried out as

$$V = V_x + V_y + V_z \qquad \textbf{(1-10)}$$

Vector V is the resultant vector in three dimensions, and a graphical display using the parallelogram law in three-dimensional space results in a parallelepiped (a three-dimensional figure formed by six parallelograms). In a three-dimensional Cartesian coordinate system, the parallelepiped becomes a rectangular parallelepiped with V as the diagonal and vector components V_x, V_y, and V_z as the length, width, and height (Fig. 1-7).

Unit Vectors: Vector V can also be expressed mathematically as the product of vector magnitude $V = |V|$ and unit vector $\boldsymbol{n}$ as

$$V = |V|\,\boldsymbol{n} = V\boldsymbol{n} \qquad \textbf{(1-11)}$$

Unit vector $\boldsymbol{n}$ has a magnitude of one, is dimensionless, and has the same direction as V. Unit vectors along x, y, and z axis directions are defined as $\boldsymbol{i}$, $\boldsymbol{j}$, and $\boldsymbol{k}$ (see Fig. 1-7) so that vector V can be expressed in terms of its rectangular components, in three dimensions, by expanding Eq. (1-11) as

$$V = V_x\boldsymbol{i} + V_y\boldsymbol{j} + V_z\boldsymbol{k} = \begin{pmatrix} V_x \\ V_y \\ V_z \end{pmatrix} \qquad \textbf{(1-12)}$$

where $V_x = V_x\boldsymbol{i}$, $V_y = V_y\boldsymbol{j}$ and $V_z = V_z\boldsymbol{k}$. Vectors are often represented as column vectors [as in Eq. (1-12)] in computer software.

It follows from the Pythagorean theorem that the magnitude of V is

$$V = \sqrt{V_x^2 + V_y^2 + V_z^2} \qquad \textbf{(1-13)}$$

In general, the *direction cosines* of a vector V are the cosines of the angles between the vector and the three coordinate axes. This follows from Eq. (1-9), which gives the direction cosines for a vector in two dimensions. From Figure 1-7 and using angles θ_x, θ_y, and θ_z between vector V and each of the coordinate axes, the direction cosines of V can be defined as

$$l = \cos\left(\theta_x\right) = \frac{V_x}{V}, \ m = \cos(\theta_y) = \frac{V_y}{V}, \text{ and } n = \cos\left(\theta_z\right) = \frac{V_z}{V} \quad \textbf{(1-14)}$$

Substituting the direction cosines into Eq. (1-12) gives

$$V = V\left(li + mj + nk\right) = V \begin{pmatrix} l \\ m \\ n \end{pmatrix} \quad \textbf{(1-15)}$$

It follows from Eqs. (1-13) and (1-14) that

$$l^2 + m^2 + n^2 = \frac{V_x^2 + V_y^2 + V_z^2}{V^2} = 1 \quad \textbf{(1-16)}$$

Position Vectors: Sometimes you know the coordinates of points A and B, which lie on the line of action of a vector V (Fig. 1-8).

Using a three-dimensional rectangular Cartesian coordinate system, you can define position vectors r_{OA} and r_{OB} (Fig. 1-8), which run from a reference point O (with coordinates $(0,0,0)$) to points A (coordinates x_A, y_A, z_A) and B (coordinates x_B, y_B, z_B) along the line AB as

$$r_{OA} = \left(x_A\right)i + \left(y_A\right)j + \left(z_A\right)k \text{ and } r_{OB} = \left(x_B\right)i + \left(y_B\right)j + \left(z_B\right)k \quad \textbf{(1-17)}$$

Vector addition [Eq. (1-5)] gives

$$r_{OA} + d_{AB} = r_{OB} \quad \textbf{(1-18)}$$

where d_{AB} is a distance vector from point A to point B. After substituting Eq. (1-17) in Eq. (1-18) and solving for d_{AB}, the distance vector from A to B can be expressed in terms of the coordinates of points A and B as

$$d_{AB} = r_{OB} - r_{OA} = \left(x_B - x_A\right)i + \left(y_B - y_A\right)j + \left(z_B - z_A\right)k \quad \textbf{(1-19)}$$

The magnitude of the distance vector is then:

$$d_{AB} = \left|d_{AB}\right| = \sqrt{\left(x_B - x_A\right)^2 + \left(y_B - y_A\right)^2 + \left(z_B - z_A\right)^2} \quad \textbf{(1-20)}$$

From this, you can define a unit vector along the line AB (see Fig. 1-8) as

$$n_{AB} = \frac{d_{AB}}{d_{AB}} = \frac{\left(x_B - x_A\right)i + \left(y_B - y_A\right)j + \left(z_B - z_A\right)k}{\sqrt{\left(x_B - x_A\right)^2 + \left(y_B - y_A\right)^2 + \left(z_B - z_A\right)^2}} \quad \textbf{(1-21)}$$

FIGURE 1-8

Position vectors r_{OA}, r_{OB} and distance vector d_{AB}

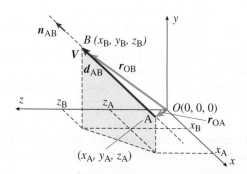

Finally, you can describe a vector V along line AB as the product of its magnitude V and unit vector n_{AB} as

$$V = Vn_{AB} \qquad (1\text{-}22)$$

Note that vector operations described above are summarized in Appendix A.

1.4 Systems of Units and Conversion Factors

Measurement systems have been a necessity since people first began to build and barter, and every ancient culture developed some sort of measurement system to serve its needs. Standardization of units took place gradually over the centuries, often through royal edicts. Development of the **British Imperial System** from earlier measurement standards began in the 13th century and was well established by the 18th century. The British system spread to many parts of the world, including the United States, through commerce and colonization. In the United States, the system gradually evolved into the **U.S. Customary System (USCS)**.

The concept of the **metric system** originated in France about 300 years ago and was formalized in the 1790s at the time of the French Revolution. France mandated the use of the metric system in 1840, and since then most other countries have done the same. In 1866, the United States Congress legalized the metric system without making it compulsory.

A new system of units was created when the metric system underwent a major revision in the 1950s. Officially adopted in 1960 and named the **International System of Units** (Système International d'Unités), this newer system is commonly referred to as **SI**. Although some SI units are the same as in the old metric system, SI has many new features and simplifications. Thus, SI is an improved metric system.

Length, *time*, *mass*, and *force* are the basic concepts of mechanics for which units of measurement are needed. However, only three of these quantities are independent since all four of them are related by Newton's second law of motion (Sec. 1.2), which states that $F = ma$, in which F is the force acting on a particle, m is the mass of the particle, and a is its acceleration. Since acceleration has units of length divided by time squared, all four quantities are involved in the second law.

SI is classified as an **absolute system of units** because measurements of the three fundamental quantities are independent of the locations at which the measurements are made; that is, the measurements do not depend upon the effects of gravity. Therefore, the SI units for *length*, *time*, and *mass* may be used anywhere on earth, in space, or even on another planet. This is one of the reasons why SI has always been preferred for scientific work.

The British Imperial System and the U.S. Customary System are based upon *length*, *time*, and *force* as the fundamental quantities with mass being derived from the second law. Therefore, the unit of mass is expressed in terms of the units of length, time, and force. The unit of force is defined as the force required to give a certain standard mass an acceleration equal to the acceleration of gravity, which means that the unit of force varies with location and altitude. For this reason, these systems are called **gravitational systems of units**. Such systems were the first to evolve, probably because weight is such a readily discernible property and because variations in gravitational attraction were not noticeable. It is clear, however, that in the modern technological world an absolute system is preferable.

SI Units

The International System of Units has seven **base units** from which all other units are derived. The base units of importance in mechanics are the meter (m) for *length*, second (s) for *time*, and kilogram (kg) for *mass*. Other SI base units pertain to temperature, electric current, amount of substance, and luminous intensity.

The **meter** was originally defined as one ten-millionth of the distance from the North Pole to the equator. Later, this distance was converted to a physical standard, and for many years the standard for the meter was the distance between two marks on a platinum-iridium bar stored at the headquarters of the International Bureau of Weights and Measures (Bureau International des Poids et Mesures) in Sèvres, a suburb on the western edge of Paris, France.

Because of the inaccuracies inherent in the use of a physical bar as a standard, the definition of the meter was changed in 1983 to the length of the path traveled by light in a vacuum during a time interval of 1/299792458 of a second [the reciprocal of the speed of light in a vacuum (299,792,458 meters per second)]. The advantages of this "natural" standard are that it is not subject to physical damage and is reproducible at laboratories anywhere in the world.

The **second** was originally defined as 1/86400 of a mean solar day (24 hours equals 86,400 seconds). However, since 1967, a highly accurate atomic clock has set the standard, and a second is now defined to be the duration of 9,192,631,770 periods of the radiation corresponding to the transition between the two hyperfine levels of the ground state of the cesium-133 atom. (Most engineers would probably prefer the original definition over the new one, which hasn't noticeably changed the second but is necessary because the earth's rotation rate is gradually slowing down.)

Of the seven base units in SI, the **kilogram** is the only one that is still defined by a physical object. Since the mass of an object only can be determined by comparing it experimentally with the mass of some other object, a physical standard is needed. For this purpose, a one-kilogram cylinder of platinum-iridium, called the International Prototype Kilogram (IPK), is kept by the International Bureau of Weights and Measures at Sèvres. (At the present time, attempts are being made to define the kilogram in terms of a fundamental constant, such as the Avogadro number, thus removing the need for a physical object.)

Other units used in mechanics, called **derived units**, are expressed in terms of the base units of meter, second, and kilogram. For instance, the unit of **force** is the **newton**, which is defined as the force required to impart an acceleration of one meter per second squared to a mass of one kilogram. From Newton's second law, you can derive the unit of force in terms of base units:

$$1 \text{ newton} = (1 \text{ kilogram})(1 \text{ meter per second squared})$$

Thus, the newton (N) is given in terms of base units by the formula

$$1\,\text{N} = 1\,\text{kg}\cdot\text{m/s}^2 \tag{1-23}$$

Note that a small apple weighs approximately one newton.

The names, symbols, and formulas for SI units of importance in mechanics are listed in Table 1-1. Some of the derived units have special names, such as newton, joule, hertz, watt, and pascal. These units are named for notable persons in science and engineering and have symbols (N, J, Hz, W, and Pa) that are capitalized, although the unit names themselves are written in lowercase letters. Other derived units have no special names (for example, the units of

Table 1-1

Principal Quantities used in Mechanics

Quantity	International System (SI)			U.S. Customary System (USCS)		
	Unit	Symbol	Formula	Unit	Symbol	Formula
Force	newton	N	$kg \cdot m/s^2$	pound	lb	(base unit)
Force per unit length (Intensity of force)	newton per meter		N/m	pound per foot		lb/ft
Length	meter	m	(base unit)	foot	ft	(base unit)
Mass	kilogram	kg	(base unit)	slug		$lb\text{-}s^2/ft$
Moment of a force; torque	newton meter		$N \cdot m$	pound-foot		lb-ft
Moment of inertia (area)	meter to fourth power		m^4	inch to fourth power		$in.^4$
Moment of inertia (mass)	kilogram meter squared		$kg \cdot m^2$	slug foot squared		$slug\text{-}ft^2$
Pressure	pascal	Pa	N/m^2	pound per square foot	psf	lb/ft^2
Time	second	s	(base unit)	second	s	(base unit)

acceleration, area, and density) and must be expressed in terms of base units and other derived units.

Loads on structures, whether due to gravity or other actions, are usually expressed in force units, such as newtons, newtons per meter, or pascals (newtons per square meter). Examples of such loads are a concentrated load of 25 kN acting on an axle, a uniformly distributed load of intensity 800 N/m acting on a small beam, and air pressure of intensity 2.1 kPa acting on an airplane wing. However, if the load acting on a structure is produced by gravity acting on a mass, then that load may be expressed in mass units (kilograms, kilograms per meter, or kilograms per square meter). The usual procedure in such cases is to convert the load to force units by multiplying by the acceleration of gravity ($g = 9.81 \, m/s^2$).

U.S. Customary Units

The units of measurement traditionally used in the United States have never been made mandatory by the government; hence they are called "customary" units. In this system, the **base units** of relevance to mechanics are the foot (ft) for *length*, second (s) for *time*, and pound (lb) for *force* (see Table 1-1). The **foot** is defined as

$$1 \, ft = 0.3048 \, m \, (exactly) \tag{1-24}$$

The **second** is the same as in SI and is described in the preceding section.

The **pound** is defined as the **force** that gives to a certain standard *mass* an acceleration equal to the acceleration of gravity. In other words, the pound is the weight of the standard *mass*, which is defined as 0.45359237 kg (exactly). The weight of this amount of mass is

$$W = (0.45359237 \text{ kg})(9.806650 \text{ m/s}^2) = 4.448222 \text{ N}$$

in which the standard international value of g is used. Thus, the pound is defined as

$$1 \text{ lb} = 4.448222 \text{ N} \tag{1-25}$$

which shows that the pound (like the foot) is actually defined in terms of SI units.

The unit of **mass** in USCS, called the **slug**, is a derived unit defined as the mass that is accelerated one foot per second squared when acted upon by a force of one pound. Writing Newton's second law in the form $m = F/a$, you get

$$1 \text{ slug} = \frac{1 \text{ pound}}{1 \text{ ft/s}^2}$$

which shows that the slug is expressed in terms of base units by the formula

$$1 \text{ slug} = 1 \text{ lb-s}^2/\text{ft} \tag{1-26}$$

To obtain the mass of an object of known weight, use the second law in the form

$$m = \frac{W}{g} \tag{1-27}$$

where m is the mass in slugs, W is the weight in pounds, and g is the acceleration of gravity in feet per second squared.

The value of g depends upon the location, but in calculations where location is not relevant, the standard international value of g may be used:

$$g = 32.1740 \text{ ft/s}^2 \tag{1-28}$$

For ordinary purposes, the recommended value is

$$g = 32.2 \text{ ft/s}^2 \tag{1-29}$$

Therefore an object having a mass of 1 slug weighs 32.2 pounds at the earth's surface.

Conversions between Units

Quantities given in either USCS or SI units can be converted quickly to the other system by using the **conversion factors** listed in Table 1-2.

If the given quantity is expressed in USCS units, it can be converted to SI units by *multiplying* by the conversion factor. For example, if the force applied to a body is 15 lb and you want to convert this quantity to SI units, Table 1-2 show that a force of 1 lb converts to 4.44822 N [see also Eq. (1-25)]. Therefore, the conversion of the given value is

$$(15.0 \text{ lb})(4.44822) = 66.723 \text{ N} = 66.7 \text{ N}$$

Because the original value is given to three significant digits, you must round the final result to three significant digits (see Section 1.5 for a discussion of significant digits). Note that the conversion factor of 4.44822 has units of newtons divided by pounds; therefore, the equation is dimensionally correct.

U.S. Customary Unit		Times Conversion Factor		Equals SI Unit	
		Accurate	Practical		
Force					
pound	lb	4.44822	4.45	newton (kg·m/s^2)	N
kip (1000 pounds)	k	4.44822	4.45	kilonewton	kN
Force per unit length					
pound per foot	lb/ft	14.5939	14.6	newton per meter	N/m
pound per inch	lb/in.	175.127	175	newton per meter	N/m
kip per foot	k/ft	14.5939	14.6	kilonewton per meter	kN/m
kip per inch	k/in.	175.127	175	kilonewton per meter	kN/m
Length					
foot	ft	0.3048*	0.305	meter	m
inch	in.	25.4*	25.4	millimeter	mm
mile	mi	1.609344*	1.61	kilometer	km
Mass					
slug	$\text{lb-s}^2\text{/ft}$	14.5939	14.6	kilogram	kg
Moment of a force; torque					
pound-foot	lb-ft	1.35582	1.36	newton meter	N·m
pound-inch	lb-in.	0.112985	0.113	newton meter	N·m
kip-foot	k-ft	1.35582	1.36	kilonewton meter	kN·m
kip-inch	k-in.	0.112985	0.113	kilonewton meter	kN·m
Moment of inertia (area)					
inch to fourth power	in.^4	416,231	416,000	millimeter to fourth power	mm^4
inch to fourth power	in.^4	$0.416{,}231 \times 10^{-6}$	0.416×10^{-6}	meter to fourth power	m^4
Moment of inertia (mass)					
slug foot squared	slug-ft^2	1.35582	1.36	kilogram meter squared	kg·m^2

Table 1-2

Conversions between U.S. Customary and SI Units for Selected Quantities

*An asterisk denotes an *exact* conversion factor

Note: To convert from SI units to USCS units, *divide* by the conversion factor

To reverse the conversion process (that is, to convert from SI units to USCS units), the quantity in SI units is *divided* by the conversion factor. For instance, suppose that the mass of a body is given as 100 kg. Then the mass in USCS units is

$$\frac{100\,\text{kg}}{14.5939} = 6.85\,\text{slug}$$

in which the term 14.5939 is the conversion factor for mass.

1.5 Accuracy, Approximations, and Significant Figures

Statics and mechanics of materials are problem-solving courses. You are expected to understand and be able to apply the underlying theoretical concepts to the solution of problems—many of which have practical engineering significance. Proper representation of the intermediate steps as well as the final solution is an important aspect of the overall solution process.

Types of Problems

Some of the homework problems in this book require symbolic solutions and others require numerical solutions. In the case of **symbolic problems** (also called *analytical, algebraic,* or *literal problems*), the data are supplied in the form of symbols for the various quantities, such as P for force, M for moment, and L for length. Such problems are solved in terms of algebraic variables, and the results are expressed as formulas or mathematical expressions. Symbolic problems usually do not involve numerical calculations, except when numerical data are substituted into the final symbolic result in order to obtain a numerical value. However, this final substitution of numerical data should not obscure the fact that the problem was solved in symbolic terms.

In contrast, **numerical problems** are those in which the data are given in the form of numbers (with appropriate units); for example, a force might be given as 12 kN, a length as 3 m, and a dimension as 150 mm. The solution of a numerical problem is carried out by performing calculations from the beginning, and the results, both intermediate and final, are in the form of numbers.

An advantage of a numerical problem is that the magnitudes of all quantities are evident at every stage of the solution, thereby providing an opportunity to observe whether the calculations are producing reasonable results. Also, a numerical solution makes it possible to keep the magnitudes of quantities within prescribed limits. Symbolic problems have several advantages too. Because the results are algebraic formulas or expressions, you can see immediately how the variables affect the answers. For instance, if a load appears to the first power in the numerator of the final result, you know that doubling the load will double the result. Equally important is the fact that a symbolic solution shows what variables do *not* affect the result. For instance, a certain quantity may cancel out of the solution, a fact that might not even be noticed in a numerical solution. Furthermore, a symbolic solution makes it convenient to check the *dimensional homogeneity* of all terms in the solution. And most important, a symbolic solution provides a general formula that is applicable to many different problems, each with a different set of numerical data. In contrast, a numerical solution is good for only one set of circumstances, and a completely new solution is required if the data are changed. Of course, symbolic solutions are not feasible when the formulas become too complex to manipulate; when that happens, a numerical solution is required.

In more advanced work in mechanics, problem solving requires the use of **numerical methods**. This term refers to a wide variety of computational methods, including standard mathematical procedures (such as numerical integration and numerical solution of differential equations) and advanced methods of analysis (such as the finite-element method). Computer programs for these methods are readily available. More specialized computer programs are also available for performing routine tasks, such as finding vector algebra solutions or solving simultaneous equilibrium equations. However, when studying mechanics, this book concentrates on the concepts rather than on the use of particular computer programs.

Dimensional Homogeneity

As noted previously, the basic concepts in mechanics are *length, time, mass*, and *force*. Each of these physical quantities has a **dimension**, that is, a generalized unit of measurement.

Every equation—whether in numeric form or symbolic form—must be **dimensionally homogeneous**, that is, the dimensions of all terms in the equation must be the same. To check the dimensional correctness of an equation, disregard numerical magnitudes and write only the dimensions of each quantity in the equation. The resulting equation must have identical dimensions in all terms. The four dimensions of length, time, mass, and force can be denoted by the symbols L, T, M, and F, respectively.

As an example, consider Newton's law of gravitation [Eq. (1-1)]:

$$F = G\frac{m_1 m_2}{r^2}$$

The corresponding dimensional equation is obtained by replacing each quantity by its dimensions; thus, the force F is replaced by the dimension F, the gravitational constant G is replaced by $L^3/M \cdot T^2$, the product of masses m_1 and m_2 is replaced by M^2, and the square of distance r is replaced by L^2. Therefore, the dimensional equation is

$$F = (L^3/MT^2)(M^2)/L^2 = ML/T^2$$

When simplified, this equation reduces to the dimensional equation F = F, as expected [see Eq. (1-23) and Table 1-1 in which force in SI is represented as $N = kg \cdot m/s^2$].

Dimensional equations can be written either in generalized terms using the LTMF notation or in terms of the actual units being used in the problem. For instance, if you are making calculations using the preceding gravitational law equation [Eq. (1-1)] with USCS units, you can write the dimensional equation as

$$lb = \left(ft^3/slug\text{-}s^2\right)\left(slug^2\right)/ft^2 = ft\text{-}slug/s^2 = ft\left(lb\text{-}s^2/ft\right)/s^2 = lb$$

which reduces to lb = lb and is dimensionally correct. Frequent checks for dimensional homogeneity (or *consistency of units*) help to eliminate errors when performing derivations and calculations.

Significant Digits, Numerical Accuracy, and Rounding

Engineering calculations are performed by calculators and computers that operate with great precision. For instance, some computers routinely perform calculations with more than 25 digits in every numerical value, and output values with 10 or more digits are available in even the most inexpensive hand-held calculators. It is important to realize that the accuracy of the results obtained from an engineering analysis is determined not only by the calculations but also by factors such as the accuracy of the given data, the approximations inherent in the analytical models, and the validity of the assumptions used in the theories. In many engineering situations, these considerations mean that the results are valid to only two or three significant digits.

To make clear the accuracy of a given numerical value, it is common practice to use **significant digits**. A significant digit is a digit from 1 to 9 or any zero not used to show the position of the decimal point; for instance, the numbers 417, 8.29, 7.30, and 0.00254 each have three significant digits. However, the number of significant digits in a number such as 29,000 is not apparent. It may have two significant digits, with the three zeros serving only to locate the decimal point, or it may have three, four, or five significant digits if one or more of the zeros is valid. By using powers of ten, the accuracy of a number such as the mass of the earth can be made clearer. When written as 5.973×10^{24} kg or 0.5973×10^{25} kg, the number is understood to have four significant digits; when written as 5.97×10^{24} kg or 0.597×10^{25} kg, it has three significant digits. When a number is obtained by calculation, its accuracy depends upon the accuracy of the numbers used in performing the calculations. A rule of thumb for **multiplication and division** is: The number of significant digits in the calculated result is the same as the least number of significant digits in any of the numbers used in the calculation.

Although the use of significant digits provides a handy way to deal with the matter of **numerical accuracy**, you should recognize that significant digits are not valid indicators of accuracy. Consider the numbers 999 and 101. Three significant digits in the number 999 correspond to an accuracy of 1/999, or 0.1%, whereas the same number of significant digits in the number 101 corresponds to an accuracy of only 1/101, or 1.0%. This disparity in accuracy can be reduced by always using one additional significant digit for numbers beginning with the digit 1. Thus, four significant digits in the number 101.1 give about the same accuracy as three significant digits in the number 999. Generally follow the rule that *final* numerical results beginning with the digits 2 through 9 should be recorded to three significant digits, and those beginning with the digit 1 should be recorded to four significant digits. However, to preserve numerical accuracy and avoid round-off errors during the calculation process, the results of *intermediate* calculations are usually recorded with additional digits. Note that many of the numbers entering into our calculations are exact, for example, the number π or fractions such as 1/2. Exact numbers are significant to an infinite number of digits and therefore have no role in determining the accuracy of a calculated result.

Finally, note that the process of discarding the insignificant digits and keeping only the significant ones is called **rounding**. To round a number to three significant digits, apply the following rules:

a. If the fourth digit is less than 5, leave the first three digits unchanged and drop all succeeding digits or replace them with zeros. For example, 37.44 rounds to 37.4 and 673,289 rounds to 673,000.

b. If the fourth digit is greater than or equal to 5 and is followed by at least one digit other than zero, then the third digit is increased by 1 and all following digits are dropped or replaced by zeros. For example, 26.37 rounds to 26.4 and 3.245002 rounds to 3.25.

c. Finally, if the fourth digit is 5 and any following digits are zeros, then the third digit is unchanged if it is an even number and increased by 1 if it is an odd number, and the 5 is replaced by a zero. (Trailing and leading zeros are retained only if they are needed to locate the decimal point.) This process is usually described as "rounding to the even digit." Since the occurrence of even and odd digits is more or less random, the use of this rule means that numbers are rounded upward about as often as downward, thereby reducing the chances of accumulating round-off errors.

These rules for rounding to three significant digits apply in the same general manner when rounding to any other number of significant digits.

1.6 Using a Four-Step Problem-Solving Approach

The study of mechanics divides naturally into two parts: first, *understanding* the general concepts and principles, and second, *applying* those concepts and principles to physical situations. You can gain an understanding of the general concepts by studying the discussions and derivations presented in this book. You can gain skill only by solving problems on your own. Of course, these two aspects of mechanics are closely related, and many experts in mechanics will argue that *you do not really understand the concepts if you cannot apply them*. It is easy to recite the principles, but applying them to real situations requires an in-depth understanding. Problem solving gives meaning to the concepts and also provides an opportunity to gain experience and develop judgment.

The more widespread adoption of the *flipped classroom* concept in college engineering programs (and beyond) in recent times can be attributed in part to recognition of the importance of problem solving in basic engineering courses. Students watch short pre-recorded lectures on fundamental concepts *before* coming to the classroom and then use class time as a computational lab, solving homework-style problems under the guidance of the instructor. In the process, students can develop and refine their problem-solving skills and are better prepared to retain and be able to apply fundamental concepts of mechanics in more advanced courses.

A major objective of this text is to assist you in developing a *structured solution process* for problems in statics and mechanics of materials. This process is referred to as a *four-step problem-solving approach* and is used in all example problems in the text. The four-step problem-solving approach involves the following four steps:

1. **Conceptualize** [*hypothesize, sketch*]: List all relevant data and draw a sketch showing all applied forces, support/boundary conditions and interactions between adjacent bodies. Development and refinement of the *free-body diagram* is an essential part of this step.

2. **Categorize** [*simplify, classify*]: Identify the unknowns in the problem and make any necessary assumptions to simplify the problem and streamline the solution process.

3. **Analyze** [*evaluate; select relevant equations, carry out mathematical solution*]: Apply appropriate theories, set up the necessary equations for the chosen mathematical model, and then solve for the unknowns.

4. **Finalize** [*conclude; examine answer—Does it make sense? Are units correct? How does it compare to similar problem solutions?*]: Study the answers, compare them to those for similar problems you have solved in the past, and test the robustness of the solution by varying key parameters to see how the results change (perhaps even plot the main result as a function of that parameter to investigate the sensitivity of the answer).

You are encouraged to study the four-step problem-solving approach presented in the example problems and then apply it to homework and in-class laboratory problems. This structured systematic approach will also be useful during examinations.

Example 1-1

FIGURE 1-9

(a–d) Man on surface of earth, at Dead Sea, at Mt. Everest, on space shuttle

A man of mass 75.0 kg would like to compare his mass and weight at various locations on the earth and above the earth.

(a) Find his mass (slugs) and weight (lb and N) if he is on the surface of the earth at sea level.

(b) Using the law of gravitation, find his weight (lb and N) at the Dead Sea (elevation: −1360 ft), on top of Mount Everest (+29,035 ft), and on the space shuttle in orbit (at +375 miles).

Solution:

Use the following four-step problem-solving approach.

Part (a): Find mass and weight in both USCS and SI units if on surface of the earth (Fig. 1-9a).

1. *Conceptualize* [*hypothesize, sketch*]: His elevation is zero on the surface of the earth at sea level, so it is appropriate to use $W = mg$ to compute his weight. Use conversion factors (Table 1-2) to convert between USCS and SI units.

2. *Categorize* [*simplify, classify*]: Only the mass in SI units is given: $m = 75.0$ kg. For the acceleration of gravity, use $g = 32.2 \, \text{ft/s}^2$ and $g = 9.81 \, \text{m/s}^2$ as appropriate. All data is given to three significant figures.

3. *Analyze* [*evaluate; select relevant equations, carry out mathematical solution*]: Based on given data, his weight will be expressed to three significant digits. The calculations in SI are

$$W_1 = mg = (75.0 \text{ kg})(9.81 \text{ m/s}^2) = 735.75 \text{ N} \approx 736 \text{ N}$$

In USCS, begin by converting the mass in kg to mass in slugs:

$$m = 75.0 \text{ kg}\left(\frac{1 \text{ slug}}{14.5939 \text{ kg}}\right) = 5.139 \text{ slugs} \approx 5.14 \text{ slugs}$$

Hence, the weight in USCS units is $W_2 = (5.14 \text{ slugs})(32.2 \text{ ft/s}^2) = 165.508 \text{ lb} \approx 166 \text{ lb}$.

Finally, convert the weight in USCS units to SI. Use the conversion factor for force in lb to force in N (see Table 1-2) to obtain

$$W_2 = (165.5 \text{ lb})(4.44822 \text{ N/lb}) = 736.18 \text{ N} \approx 736 \text{ N}$$

4. *Finalize* [*conclude; examine answer—Does it make sense? Are units correct? How does it compare to similar problem solutions?*]: The man's weight on the surface of the earth, computed using the formula $W = mg$ has the same numerical value (i.e., $W_1 = W_2$) to three significant digits. Note that values for weights W were rounded to three significant digits using the rules stated in Section 1.5.

Part (b): Using the law of gravitation, find his weight (lb and N) at the Dead Sea (elevation: -1360 ft), on top of Mount Everest ($+29{,}035$ ft), and on the space shuttle in orbit (at $+375$ miles).

1. *Conceptualize*: The mutual attraction force F between the man and the earth is equal to the man's weight, W. Now specify the distance between the mass centers of the man and the earth (distance r) for each of the three locations of interest. The mass of the man is given as $m = 75.0$ kg. To make weight (W) calculations using the law of gravitation [Eq. (1-1)], you will also need numerical values for the mass of the earth (m_e) and the gravitational constant (G).

2. *Categorize*: The mutual attraction force F between the man and the earth depends on constants G, m_e, m, and r_e. Assume that the mass of the earth (m_e), radius of the earth (r_e), and gravitational constant (G) are

$$m_e = 5.976 \times 10^{24} \text{ kg}, \; r_e = 6372 \text{ km}, \; G = 6.674 \times 10^{-11} \; \frac{\text{m}^3}{\text{kg}\cdot\text{s}^2}$$

Modifying Eq. (1-1) gives weight W as a function of elevation h at each location of interest:

$$W(h) = G\frac{m_e m}{\left(r_e + h\right)^2} \qquad\qquad \textbf{(1-30)}$$

On the surface of the earth, $h = 0$. At the other three locations, elevation h can be computed to three significant digits using conversion factors in Table 1-2 as

$$\textbf{Dead Sea: } h = \left(-1360\,\text{ft}\right)\left(0.3048\,\text{m/ft}\right)\left(1\,\text{km}/1000\,\text{m}\right) = -0.415\,\text{km}$$

$$\textbf{Mt. Everest: } h = \left(29{,}035\,\text{ft}\right)\left(0.3048\,\text{m/ft}\right)\left(1\,\text{km}/1000\,\text{m}\right) = 8.85\,\text{km}$$

$$\textbf{In space: } h = \left(375\,\text{mi}\right)\left(1.609344\,\text{km/mi}\right) = 604\,\text{km}$$

3. *Analyze*: First apply Eq. (1-30) using each of the values for elevation h to obtain:

Dead Sea:

$$W\left(-0.415\,\text{km}\right) = \left(6.674 \times 10^{-11}\;\frac{\text{m}^3}{\text{kg}\cdot\text{s}^2}\right)\left[\frac{\left(5.976 \times 10^{24}\,\text{kg}\right)\left(75.0\,\text{kg}\right)}{\left(6372 \times 10^3\,\text{m} - 415\,\text{m}\right)^2}\right]$$

$$= 736.8\,\text{kg}\cdot\text{m/s}^2 \approx 737\,\text{N}$$

Similarly, for the other two locations:

$$\textbf{Mt. Everest:} \qquad W\left(8.85\,\text{km}\right) = 734.7\,\text{N} \approx 735\,\text{N}$$

$$\textbf{In space:} \qquad W\left(604\,\text{km}\right) = 614.7\,\text{N} \approx 615\,\text{N}$$

Finally, divide by the conversion factor in Table 1-2 (i.e., 4.44822 N/lb) to convert the weights from newtons to pounds. The resulting weights (to three significant figures) are 166 lb (Dead Sea), 165 lb (Mt. Everest), and 138 lb (in space).

4. *Finalize*: The computed results are reasonable and have the correct units, either N or lb. As expected [based on Eq. (1-1)], his weight increases at the Dead Sea and decreases as he moves into space away from the center of the earth.

As an alternative to converting the final weights from SI to USCS units, start with numerical values for masses m and m_e, earth radius r_e, elevation h, and gravitational constant G—in USCS units—and then compute the weights W in pounds. (You are encouraged to do this and then compare the results to values reported previously.)

Finally, if a computer is available, the functional form of Eq. (1-30) makes plotting the weight W versus elevation h a straightforward exercise; the resulting plot is shown in Fig. 1-10.

FIGURE 1-10

Weight W (N) of man with mass of 75.0 kg versus elevation h (m)

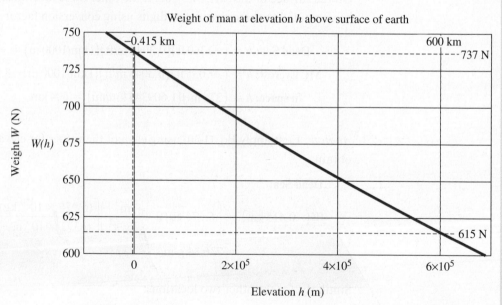

Example 1-2

At some point during positioning of a moveable overhead camera used for close-up viewing of field action at sporting events (Fig. 1-11), force vectors F_1 and F_2 are directed at angles $\alpha = 22°$ and $\beta = 40°$, as shown in Fig. 1-12. At this instant, the magnitudes of F_1 and F_2 are known to be 124 N and 150 N, respectively.

(a) Find the Cartesian components of vectors F_1 and F_2.

(b) Find unit vectors n_s and n_d along vectors F_s and F_d, respectively, in terms of their Cartesian components.

FIGURE 1-11

(a, b) Field camera and position vector to field camera

(a)

(b)

(c) Find a position vector r_{OP} from reference point O at the corner of the field to the position of the camera at point P (see Fig. 1-11), and a unit vector n_{OP} along line OP. Assume that the Cartesian coordinates of point P at this instant are (40, 15, 50) m.

Solution:

Use the following four-step problem-solving approach to find Cartesian components of force vectors F_1 and F_2, as well as associated unit and position vectors.

1. *Conceptualize*: First, draw vectors F_1 and F_2 in an *x-y* coordinate system whose origin is at common point P (Fig. 1-13). Label known angles α and β.

FIGURE 1-12

(a,b) Force vectors on overhead field camera

(a)   (b)

FIGURE 1-13

Force vectors in *x-y* coordinate system

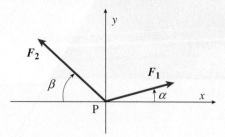

2. *Categorize*: Next use the *parallelogram law* to show vector sum F_s and and the *triangle law* to show vector difference F_d (Fig. 1-14). Finally, mark *x* and *y* components of each vector on the drawing (Fig. 1-15).

3. *Analyze*: Use Figs. 1-15a and b to guide you in solving for Cartesian components and unit and position vectors as follows.

FIGURE 1-14

(a, b) Parallelogram and triangle laws to find F_s and F_d

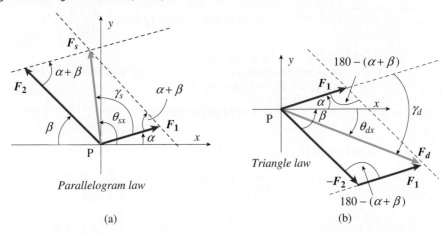

Parallelogram law

(a)

Triangle law

(b)

FIGURE 1-15

(a, b) Cartesian components of F_s and F_d

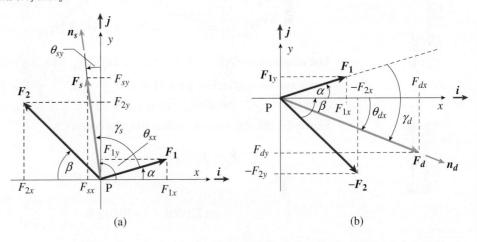

(a)

(b)

Part (a): Find the Cartesian components of vectors F_1 and F_2 using magnitudes 124 N and 150 N, respectively, and angles $\alpha = 22°$ and $\beta = 40°$.

The Cartesian components of vectors F_1 and F_2 are obtained using simple geometry (see Fig. 1-15a):

$$F_{1x} = F_1 \cos(\alpha) = 114.9 \text{ N}, \quad F_{1y} = F_1 \sin(\alpha) = 46.5 \text{ N}$$

$$F_{2x} = -F_2 \cos(\beta) = -114.9 \text{ N}, \quad F_{2y} = F_2 \sin(\beta) = 96.4 \text{ N}$$

so vector F_1 can be written in terms of its x and y components as

$$F_1 = F_{1x}i + F_{1y}j$$

$$= (114.9)i + (46.5)j \text{ N} = \begin{pmatrix} 114.9 \\ 46.5 \end{pmatrix} \text{ N}$$

and vector F_2 can be written as

$$F_2 = F_{2x}i + F_{2y}j$$

$$= (-114.9)i + (96.4)j \text{ N} = \begin{pmatrix} -114.9 \\ 96.4 \end{pmatrix} \text{ N}$$

The *sum* of the two vectors is

$$F_s = F_1 + F_2$$
$$= (F_{1x} + F_{2x})i + (F_{1y} + F_{2y})j$$

$$= (0)i + (142.9)j \text{ N} = \begin{pmatrix} 0 \\ 142.9 \end{pmatrix} \text{ N}$$

and the magnitude of F_s is computed as

$$|F_s| = \sqrt{(F_{1x} + F_{2x})^2 + (F_{1y} + F_{2y})^2} = \sqrt{F_{sx}^2 + F_{sy}^2}$$

$$|F_s| = \sqrt{(0)^2 + (142.9)^2} \text{ N} = 142.9 \text{ N}$$

The orientation of vector F_s is given by angle θ_{sx}, which is measured from the x axis and computed using Eq. (1-9) as $\theta_{sx} = \cos^{-1}\left(\dfrac{F_{sx}}{F_s}\right) = 90°$ measured CCW from the x axis, so angle $\theta_{sy} = 0$ (Fig. 1-15a).

Similarly, the *difference* between the two vectors is

$$F_d = F_1 - F_2$$
$$= (F_{1x} - F_{2x})i + (F_{1y} - F_{2y})j$$

$$= (229.9)i + (-50.0)j \text{ N} = \begin{pmatrix} 229.9 \\ -50.0 \end{pmatrix} \text{ N}$$

and the magnitude of F_d is

$$|F_d| = \sqrt{(F_{1x} - F_{2x})^2 + (F_{1y} - F_{2y})^2} = \sqrt{F_{dx}^2 + F_{dy}^2}$$

$$|F_d| = \sqrt{(229.9)^2 + (-50.0)^2} \text{ N} = 235 \text{ N}$$

The orientation of vector F_d is given by angle θ_{dx}, which is measured from the x axis and computed using Eq. (1-9) as $\theta_{dx} = \cos^{-1}\left(\dfrac{F_{dx}}{F_d}\right) = 12.3°$ measured CW from the x axis.

Part (b): Find unit vectors n_s and n_d along vectors F_s and F_d, respectively, in terms of their Cartesian components.

Unit vectors n_s and n_d are shown in Figs. 1-15a and b and can be obtained using Eq. (1-11). The resulting dimensionless unit vectors in Cartesian coordinates are

$$n_s = \frac{F_s}{F_s} = \frac{(0)i + (142.9)j}{142.9}$$

$$= (0)i + (1.0)j = \begin{pmatrix} 0 \\ 1.0 \end{pmatrix} \qquad \blacktriangleleft$$

$$n_d = \frac{F_d}{F_d} = \frac{(229.9)i + (-50.0)j}{235}$$

$$= (0.977)i + (-0.212)j = \begin{pmatrix} 0.977 \\ -0.212 \end{pmatrix} \qquad \blacktriangleleft$$

These are unit vectors, so $|n_s| = |n_d| = 1$.

Part (c): A position vector r_{OP} from reference point O at the corner of the field to some position of the camera at point P (see Fig. 1-11) and a unit vector n_{OP} along line OP. Assume that the Cartesian coordinates of point P at this instant are (40, 15, 50) m.

Point O is the origin of the coordinate system at the corner of the field, so it has coordinates (0, 0, 0) and point P has coordinates (40, 15, 50) m. You can find the position vector from O to P using Eq. (1-17):

$$r_{OP} = (x_P - x_O)i + (y_P - y_O)j + (z_P - z_O)k \text{ m}$$
$$= (40 - 0)i + (15 - 0)j + (50 - 0)k \text{ m}$$

$$r_{OP} = (40)i + (15)j + (50)k \text{ m} = \begin{pmatrix} 40 \\ 15 \\ 50 \end{pmatrix} \text{ m}$$

From Eq. (1-20), the magnitude of the position vector is

$$r_{OP} = |r_{OP}| = \sqrt{(x_P - x_O)^2 + (y_P - y_O)^2 + (z_P - z_O)^2}$$
$$= \sqrt{(40)^2 + (15)^2 + (50)^2} \text{ m} = 65.765 \text{ m}$$

Finally, a unit vector n_{OP} along line OP is computed using Eq. (1-21):

$$n_{OP} = \frac{r_{OP}}{r_{OP}} = \left(\frac{40}{65.765}\right)i + \left(\frac{15}{65.765}\right)j + \left(\frac{50}{65.765}\right)k$$

$$n_{OP} = (0.608)i + (0.228)j + (0.76)k = \begin{pmatrix} 0.608 \\ 0.228 \\ 0.76 \end{pmatrix} \qquad \blacktriangleleft$$

Note that $|n_{OP}| = 1$ as expected.

4. *Finalize*: In many cases, a Cartesian reference frame is most convenient for vector operations such as those carried out in this example. Sketches drawn roughly to scale (as those in Figs. 1-13, 1-14, and 1-15) are valuable aids when making geometric calculations.

Example 1-3

Two forces are applied to an automobile parking brake at point P (see Fig. 1-16). Assume that force magnitudes are $F_1 = 13$ lb and $F_2 = 15$ lb.

(a) Find the magnitude and orientation of the vector *sum* F_s of vectors F_1 and F_2.

(b) Find the magnitude and orientation of the vector *difference* F_d between vectors F_1 and F_2.

FIGURE 1-16

(a,b) Forces applied to automobile parking brake

(a)   (b)

Solution:

Part (a): Use the following four-step problem-solving approach to find the magnitude and orientation of the vector *sum* F_s of vectors F_1 and F_2; use the *parallelogram law*.

FIGURE 1-17

Force vectors applied at point P

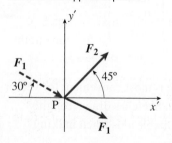

1. *Conceptualize*: First make a sketch (roughly to scale) of vectors F_1 and F_2 showing the vectors starting from the same point (at point P where their lines of action cross, as shown in Fig. 1-17). Slide vector F_1 along its line of action so that its tail (not its head) is at point P.

2. *Categorize*: The vector sum F_s can be represented as a *parallelogram* like the shaded area in Fig. 1-18. If vector F_1 is not moved along its line of action so that its tail is aligned with the tail of vector F_2 at point P, you will obtain an incorrect parallelogram and an incorrect resultant, as shown in Fig. 1-18.

FIGURE 1-18

Parallelogram law for addition

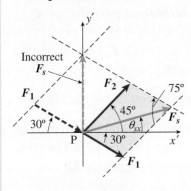

FIGURE 1-18

Parallelogram law for addition

FIGURE 1-19

Incorrect use of triangle law in vector addition

FIGURE 1-20

Force vectors for vector subtraction

3. *Analyze:* Use Eq. (1-5) to obtain vector sum F_s: $F_s = F_1 + F_2$. Vector magnitudes are given as

$F_1 = |F_1| = 13$ lb, $F_2 = |F_2| = 15$ lb. The magnitude of F_s can be obtained using the *law of cosines* (Appendix A):

$$F_s = |F_s| = \sqrt{F_1^2 + F_2^2 - 2F_1 F_2 \cos(180° - 75°)}$$

$$F_s = \sqrt{(13 \text{ lb})^2 + (15 \text{ lb})^2 - 2(13 \text{ lb})(15 \text{ lb}) \cos(180° - 75°)}$$

$$= 22.25 \text{ lb}$$

The direction or orientation of F_s can be computed using the *law of sines* (Appendix A):

$$\frac{F_s}{\sin(180° - 75°)} = \frac{F_2}{\sin(30° + \theta_{sx})}$$

$$\text{so } \theta_{sx} = \sin^{-1}\left(\frac{F_2}{F_s}\sin(105°)\right) - 30°$$

$$= 10.64°$$

Vector F_s has a magnitude of 22.25 lb and is oriented at a counterclockwise (CCW) angle θ_{sx} of 10.64° from the $+x'$ axis, as shown in Fig. 1-18.

4. *Finalize:* The graphical display of results and the orientation angle θ_{sx} computed using trigonometry appear to be in agreement. Note that vector addition ($F_s = F_1 + F_2$) is not the same as scalar addition ($F_1 + F_2$); the magnitude F_s of vector sum F_s cannot be obtained by summing the magnitudes of component vectors F_1 and F_2. The correct expression for the magnitude F_s is 22.25 lb as

$$|F_s| = F_s = 22.25 \text{ lb} \neq F_1 + F_2 = 28 \text{ lb}.$$

Note that you must slide vector F_1 along its line of action so that its tail is at point P to get a correct solution for vector sum F_s. Failure to do so results in a vector F_s of correct magnitude but its line of action is not through point P as required (see Fig. 1-19).

Part (b): Use the following four-step problem-solving approach to find the magnitude and orientation of vector *difference* F_d between vectors F_1 and F_2; use the *triangle law*.

1. *Conceptualize:* To find vector difference F_d, first make a sketch of vectors F_1 and $-F_2$ showing the vectors starting from the same point P (see Fig. 1-20).

2. *Categorize:* The vector difference F_d is represented as a *triangle*, as shown in Fig. 1-21. The tail of vector $-F_2$ is attached to the head of vector F_1 (Fig. 1-21), or the tail of F_1 is attached to the head of $-F_2$ to find F_d.

FIGURE 1-21

Triangle law for vector subtraction

3. *Analyze*: The vector difference F_d is obtained using Eq. (1-6): $F_d = F_1 - F_2$. Vector magnitudes are $|F_1| = F_1 = 13$ lb, $|F_2| = F_2 = 15$ lb. The magnitude of F_d is obtained using the *law of cosines* (Appendix A):

$$F_d = |F_d| = \sqrt{F_1^2 + F_2^2 - 2F_1 F_2 \cos(75°)}$$

$$F_d = \sqrt{(13\,\text{lb})^2 + (15\,\text{lb})^2 - 2(13\,\text{lb})(15\,\text{lb})\cos(75°)} = 17.12\,\text{lb}$$

The magnitude of the vector difference is 17.12 lb and the direction of F_d is quite different from that of F_s (see Figs. 1-18 and 1-21). The orientation of F_d is computed using the *law of sines* (Appendix A):

$$\frac{F_d}{\sin(75°)} = \frac{F_2}{\sin(\gamma_d)} \quad \text{so} \quad \gamma_d = \sin^{-1}\left(\frac{F_2}{F_d}\sin(75°)\right) = 57.8°$$

and

$$\theta_{dx} = \gamma_d + 30° = 87.8°$$

Vector F_d has a magnitude of 17.12 lb and is oriented at a clockwise (CW) angle θ_{dx} of almost 88° from the $+x'$ axis, as shown in Fig. 1-21.

4. *Finalize*: The graphical display of results (drawn to appropriate scale) and the orientation angle θ_{dx} computed using trigonometry appear to be in agreement. Note that vector subtraction ($F_d = F_1 - F_2$) is not the same as scalar subtraction ($F_1 - F_2$); the magnitude F_d of vector difference F_d cannot be obtained by taking the difference between the magnitudes of vectors F_1 and F_2. The correct expression for the magnitude F_d is 17.12 lb as $|F_d| = F_d = 17.12\,\text{lb} \neq F_1 - F_2 = -2.0\,\text{lb}$.

CHAPTER SUMMARY AND REVIEW

Some of the major concepts presented in this chapter are given here.

1. **Newton's third law** is concerned with action–reaction force pairs that may exist between bodies in contact (*internal forces*) or between bodies and the external world at points of support (*reaction forces*) of the body.

2. The display of a body or part of a body, with all forces shown upon it, is known as a **free-body diagram**.

3. The **absolute weight** of a body can be computed using the law of gravitation, as illustrated in Example 1-1. You will compute the **apparent weight** W of bodies, $W = mg$, where m is the mass in kilograms or slugs and g is the acceleration of gravity: $32.2\,\text{ft/s}^2$ in USCS units or $9.81\,\text{m/s}^2$ in SI units.

4. A **vector** is a quantity that has both magnitude and direction. Examples of vector quantities are force, displacement, velocity, and acceleration. A **scalar** on the other hand has only magnitude. Scalars include such items as mass, volume, density, time, and speed.

5. Vector addition and subtraction can be carried out using either the **parallelogram law** or **triangle law**.

Parallelogram law

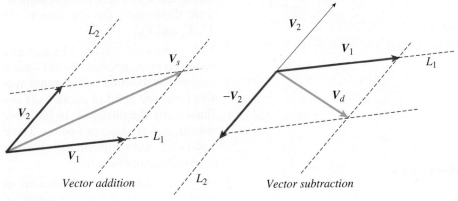

Vector addition   Vector subtraction

Triangle law; move vector **V_2** to head of **V_1**

Vector addition   Vector subtraction

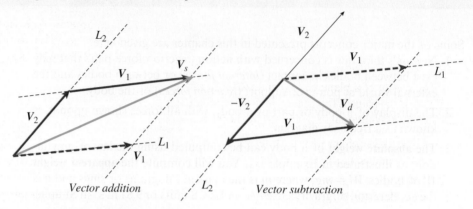

Vector addition   Vector subtraction

Rectagular components (2D) and unit vectors i and j

Rectagular components (3D) and unit vectors i, j, k

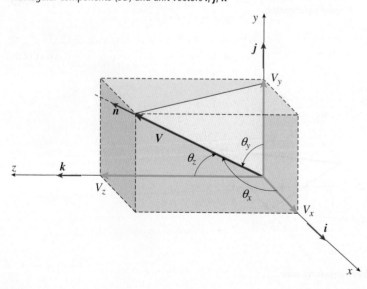

6. In many instances, it is convenient to use a Cartesian coordinate system with vector V_1 replaced by mutually perpendicular vector components V_x and V_y (in two dimensions) that are directed along coordinate axes x and y, respectively. In three dimensions, the components are V_x, V_y and V_z.

7. **Unit vectors** are obtained by dividing each vector component by the vector magnitude. **Position** and **distance vectors** from one point to another are defined using the difference in the x-y-z coordinates of those two points. These and other vector operations are summarized in Appendix A.

8. Basic and derived quantities used in mechanics were defined in terms of both the U.S. Customary (**USCS**) and **SI** systems (see Table 1-1). These quantities are force, mass, length, and time. **Conversion factors** used to translate quantities from one system to another are listed in Table 1-2.

9. **Round** numerical answers to three significant digits but use additional digits when doing intermediate calculations to preserve numerical accuracy and avoid round-off errors.

10. Find both **symbolic** and **numerical solutions** in statics using a **four-step problem-solving approach** that is illustrated in all example problems throughout the text.

PROBLEMS

1.2 Newton's Laws of Motion and Gravitation

Introductory Problems

1.2-1 The weight of a plane parked at an airport gate is 155,500 lb. If the plane is flying at an altitude of 35,000 ft, calculate the gravitational force exerted on the plane.

   (a) Using the law of gravitation, where $m_{earth} = 4.0925 \times 10^{23}$ slugs, $r_{earth} = 3959$ miles.

   (b) Using $W = Mg$, where $m =$ mass and $g =$ gravitational acceleration.

1.2-2 Find the gravitational force between the earth and the following bodies in space. The mass of each body is given below. R is the distance between the centers of the two bodies.

   (a) Moon ($R = 378,000$ km, $m_{Earth} = 5.9726 \times 10^{24}$ kg, $m_{moon} = 7.3477 \times 10^{22}$ kg)

   (b) Jupiter ($R = 588.5$ million km, $m_{Earth} = 5.9726 \times 10^{24}$ kg, $m_{Jupiter} = 1.8983 \times 10^{27}$ kg)

   (c) Mars ($R = 55.7$ million km, $m_{Earth} = 5.9726 \times 10^{24}$ kg, $m_{Mars} = 6.4174 \times 10^{23}$ kg)

   (d) Saturn ($R = 1.195$ billion km, $m_{Earth} = 5.9726 \times 10^{24}$ kg, $m_{Saturn} = 5.6836 \times 10^{26}$ kg)

Which body has the largest gravitational interaction (kN) with earth?

Representative Problems

1.2-3 What is the *weight* of a 180 lb man walking on the moon? Assume that the moon is 238,900 miles from the surface of the earth and has mass of 5.035×10^{21} slugs. Let the gravitational constant $G = 3.439 \times 10^{-8}$ ft^4/(lb-s^4). The moon has a radius of 1080 miles.

PROBLEM 1.2-3

1.2-4 Consider the interaction between the earth and the objects (see figure). Which object has the larger gravitational force? (Assume that $m_{earth} = 5.9726 \times 10^{24}$ kg).

$m_1 = 50$ kg

$m_2 = 150$ kg

$R_1 = 50$ km

$R_2 = 25$ km

PROBLEM 1.2-4

1.3 Scalars and Vectors

Introductory Problems

1.3-1 Find the direction cosines of the vector $V = 10i + 15j$. Also, find a unit vector along the line of action of V.

1.3-2 Find both the sum (V_s) and difference (V_d) of vectors $V_1 = 10i + 15j$ and $V_2 = 8i - 12j$. Confirm your results using the *parallelogram law*. Repeat using the *triangle law*.

1.3-3 Find both the sum (V_s) and difference (V_d) of vectors $V_1 = -10i + 15j$ and $V_2 = 8i - 12j$. Confirm your results using the *parallelogram law*. Repeat using the *triangle law*.

1.3-4 The *sum* of vectors V_1 and V_2 is $(4i - 8j + 7k)$. A unit vector along V_1 is $(0.635i + 0.773j)$. *Find orthogonal* vectors V_1 and V_2.

1.3-5 Vectors V_1 and V_2 are orthogonal. If V_1 is $(5i + 7j - 3k)$ and $V_2 = (7i + V_y j + 14k)$, *find* component V_y in vector V_2.

Representative Problems

1.3-6 Force F (see figure) is applied at C and is directed toward A. Its magnitude is $F = 18$ N. Find a unit vector along CA then find vector F. Also, find *distance vectors* from B to C (d_{BC}) and from B to D (d_{BD}).

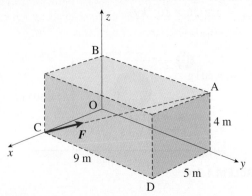

PROBLEM 1.3-6

1.3-7 Find the *direction cosines* of vector F in Prob. 1.3-6.

1.3-8 Find an expression for a force F, applied at the origin, which makes equal angles with the *x-y-z* axes.

1.3-9 Two forces are applied at point P to the automobile parking brake from Example 1-3 (see figure). Force magnitudes are $F_1 = 13$ lb and $F_2 = 15$ lb. Find the Cartesian components of vectors F_1 and F_2.

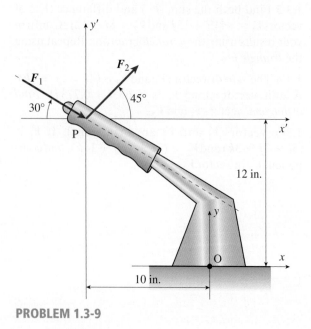

PROBLEM 1.3-9

1.3-10 A force defined by vector $F = 60i - 120j + 30k$ kN acts on a body. Find the angles between vector force F and the *x*, *y*, and *z* axes.

1.3-11 Two forces are applied at point P to the automobile parking brake from Example 1-3 (see figure). Force magnitudes are $F_1 = 13$ lb and $F_2 = 15$ lb.

(a) Find unit vectors n_s and n_d along vector sum F_s and vector difference F_d, respectively, in terms of their Cartesian components.

(b) Find a position vector r_{OP} from reference point O at the base of the hand brake to the point of application of the force vectors at point P (see figure) and a unit vector n_{OP} along line OP. The Cartesian coordinates of point O are (0, 0) and of point P are ($x = -10$ in., $y = 12$ in.).

PROBLEM 1.3-11

1.3-12 Consider the moveable overhead camera from Example 1-2 (see Fig. a). Force vectors F_1 and F_2 are directed at angles $\alpha = 22°$ and $\beta = 40°$ as shown (Fig. b), and the magnitudes of F_1 and F_2 are known to be 124 N and 150 N, respectively.

(a) Find the magnitude and orientation of vector sum F_s of vectors F_1 and F_2 using the *parallelogram law.*

(b) Find the magnitude and orientation of vector difference F_d between vectors F_1 and F_2 using the *triangle law.*

(a)

(b)

PROBLEM 1.3-12

1.4 Systems of Units and Conversion Factors

Introductory Problems

1.4-1 A car weighs 3500 lb. Find its *mass* in both slugs and kilograms.

1.4-2 A bicycle rider has a mass of 58 kg, and her bicycle has mass of 14 kg. Find the combined mass of the bicycle and rider in slugs; then find the combined weight in lbs.

Dudarev Mikhail/Shutterstock.com

PROBLEM 1.4-2

1.4-3 A right circular cone has height $h = 2.37$ in. and radius $r = 1.295$ in. Using the rule for significant digits, find its volume V. Also, show that the equation for volume is dimensionally homogeneous.

PROBLEM 1.4-3

Representative Problems

1.4-4 In SI units, the elastic modulus of a material (E) is expressed in pascals (Pa, N/m^2). The elastic modulus of an aluminum alloy is 10,600 ksi and its unit weight (γ) is 175 lb/ft^3. Find E in terms of Pa, kPa, MPa, and GPa. Also, find γ in terms of N/m^3 and kN/m^3. Finally, find the mass (in both slugs and kg) of an aluminum-alloy sphere with radius of 3 in.

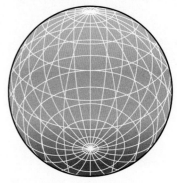

PROBLEM 1.4-4

1.4-5 A man lifts an object that weighs 15 lb. What is the *mass* of the object in slugs and also in kilograms?

PROBLEM 1.4-5

1.5 Accuracy, Approximations and Significant Figures

Introductory Problems

1.5-1 Find the percent error associated with use of the small angle approximation for trigonometric functions sine, cosine, and tangent for the following angles: (a) 5°; (b) 10°; (c) 15°; and (d) 20°.

Representative Problems

1.5-2 What is the percent error resulting from use of the small angle approximation in evaluating the following expression:

$$R(\alpha) = \frac{\sin(2\alpha)\cos(2\alpha)}{\cos(\alpha) + \dfrac{1}{2}\cos(2\alpha)}$$

Use the four angles given in Prob. 1.5-1.

1.5-3 Using the convention for significant digits, find the results of the following computations: $P + Q$; $P - Q$; $\dfrac{P}{Q}$; $(P + Q)^2$; and $\sqrt{-P + Q}$. Assume that $P = 2.728$ and $Q = 7.93$.

Forces, Moments, Resultants

sainthorant daniel/Shutterstock.com

Large forces are developed in the cables that support the masts on this sailboat.

CHAPTER OBJECTIVES

- Learn the action of *force* and *moment* vectors on bodies such as structures and machines used in engineering applications.

- Use *triangle, parallelogram,* and *polygon laws* to find the *resultant* of a system of forces or to resolve a vector into its *components* in selected directions or along specified axes.

- Define the turning effect of a force about a point, which is a vector quantity known as a *moment*.

- Define a *couple*, which consists of two parallel forces that are equal in magnitude but opposite in sense, so act in opposite directions along separate parallel lines of action.

- Understand how to replace a system of forces and couples with the *simplest possible force-couple system* that has the same overall effect on the rigid body under study.

Chapter Outline

Forces and their effects on structures and machines are at the heart of statics and mechanics of materials. Forces and the turning effect of forces, called moments, are both vector quantities. First, the basic properties of vectors are reviewed and then applied to force and moment vectors. You can use basic vector operations in the solution of a number of practical engineering problems. Begin by examining force vectors acting at a point and later look at forces acting on rigid bodies.

2.1 Forces: 2D, 3D

Forces are vectors; they have magnitude, line of action, direction, and point of application. The arrowhead indicates the sense of the vector. Two-dimensional forces and force systems usually can be handled using scalar operations, but three-dimensional forces often require vector algebra to determine their effects at various points in a structure.

Types of Forces

Various kinds of forces act on engineering structures. Assume that all forces are applied statically (or can be treated as such), so time-dependent forces such as wind, blast, and earthquake are not considered. In many instances, you can treat the effect of forces using scalar quantities and algebraic equations. For three-dimensional models, vector operations are often required, so you can build upon the introductory concepts about vectors from Section 1.3 to carry out these analyses, beginning with the properties and effects of a single force.

Consider a small truck crane used to hoist an inclined rectangular steel plate of constant thickness (Fig. 2-1). The plate has weight W, which can be treated as an *external applied* force located at the center of gravity of the plate. Force W is a vector with magnitude $W(|W| = W)$, direction (downward), and point of application (center of gravity, c.g., of the plate). The weight of the plate is transferred through clevis and pin attachments at each top corner to two cables suspended from the overhead crane. The forces in the pins, clevis attachments, and plate at each corner are examples of *internal forces* that result in localized stresses and deformations in the plate.

FIGURE 2-1

Small truck crane lifting inclined
rectangular steel plate

FIGURE 2-2

Isolate part of structure: plate and cables

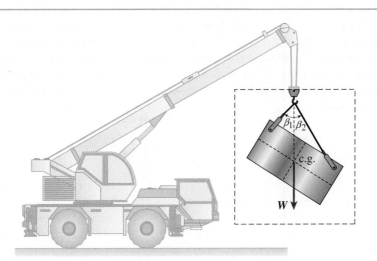

Our focus here is the *analysis* of the cable-clevis-pin lift system for the plate. Begin by isolating the portion of the structure containing the cables and plate. Only the cable forces (F_1, F_2) and the weight of the plate (W) are needed, so cut through the crane hook just above the plate (see Fig. 2-2). You want to find the resultant vector R associated with force vectors F_1 and F_2. You can ignore all parts of the system above and to the left of the point at which the crane hook is attached to cables 1 and 2.

For now, you are only interested in the forces in the cables, so cut through both cables to expose the *internal* cable forces. The resulting sketch (Fig. 2-3a) shows both the external force W and the internal cable forces F_1 and F_2. To find the resultant vector R from the two cable forces using vector addition, ignore stresses or deformations in the plate in the region of the pin and clevis. Also ignore gravity force W for now so that you can focus on the cable forces. You are interested in the magnitudes and directions of forces F_1 and F_2. Therefore, use the *principle of transmissibility* to slide the two cable force vectors to the point at which their lines of action cross. The principle of transmissibility states that forces F_1 and F_2 can be moved anywhere along their lines of action if you are looking for their resultant overall *external* effects on the rigid body (the plate) upon which they act.

FIGURE 2-3

(a, b) Resultant R from parallelogram law

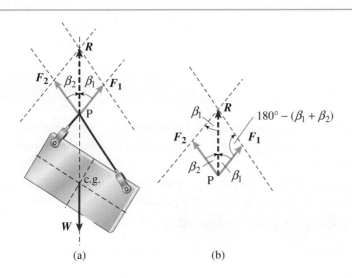

(a)   (b)

FIGURE 2-4

(a, b) Resultant **R** from triangle law

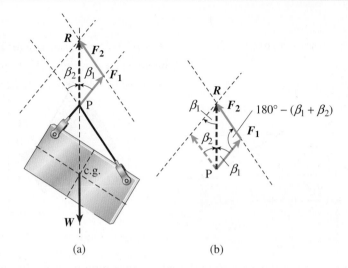

Note that forces F_1 and F_2 are both *coplanar* (they lie in the same plane) and *concurrent* (their lines of action intersect at common point P). Treat them as sliding vectors, move each of them to common point P, and then use the *parallelogram law* to find resultant force **R** (Fig. 2-3) as

$$R = F_1 + F_2 \tag{2-1}$$

Use the *triangle law* to find **R** to get the same result (Fig. 2-4).

Apply the law of cosines in Fig. 2-3b to find the magnitude of resultant **R** in terms of the magnitudes of cable forces F_1 and F_2 as

$$R = |R| = \sqrt{F_1^2 + F_2^2 - 2F_1F_2 \cos(180° - (\beta_1 + \beta_2))} \tag{2-2}$$

Alternatively, the law of sines in Fig. 2-4b gives

$$\frac{R}{\sin(180° - (\beta_1 + \beta_2))} = \frac{F_1}{\sin(\beta_2)} = \frac{F_2}{\sin(\beta_1)} \tag{2-3}$$

so the magnitude R of resultant **R** can be found as

$$R = F_1 \left(\frac{\sin(180° - (\beta_1 + \beta_2))}{\sin(\beta_2)} \right) = F_2 \left(\frac{\sin(180° - (\beta_1 + \beta_2))}{\sin(\beta_1)} \right) \tag{2-4}$$

Example 2-1

A continuous cable with known tension force T is attached to a horizontal frame member at B and C to support a sign structure (see Fig. 2-5). The cable passes over a small frictionless pulley at D.

(a) Find the magnitude and orientation of a single equivalent vector R_D that is the *sum* of the two cable forces T at D; use the *parallelogram law*.

(b) Find the projection of R_D onto the line DB.

FIGURE 2-5

Sign structure supported by cable BDC

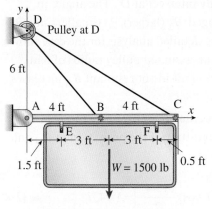

FIGURE 2-6

Cable force vectors intersect at D

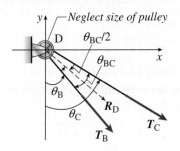

FIGURE 2-7

Resultant vector R_D

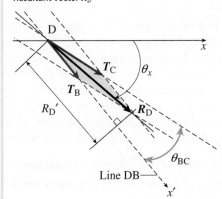

Solution:

Part (a): Use the four-step problem-solving approach to find the magnitude and orientation of a single equivalent vector R_D that is the *sum* of the two cable forces T at D; use the *parallelogram law*.

1. *Conceptualize* [*hypothesize, sketch*]: First isolate the part of the structure of interest here: the region around the pulley at D. Make a careful sketch (roughly to scale; see Fig. 2-6) of vectors T_B and T_C acting along lines DB and DC, respectively. Assume that the vectors start from the same point D (since the pulley is small) where their lines of action cross.

2. *Categorize* [*simplify, classify*]: Assume that the pulley size can be neglected so cable force vectors meet at point D. The vector sum or resultant R_D can be represented using a *parallelogram* like that in Fig. 2-7.

3. *Analyze* [*evaluate; select relevant equations, carry out mathematical solution*]: The vector sum is: $R_D = T_B + T_C$. The magnitude of R_D is computed using the *law of cosines*. First find the angle θ_{BC} between the lines DB and DC (see Fig. 2-7):

$$\theta_{BC} = \theta_C - \theta_B = \tan^{-1}\left[\frac{8}{6}\right] - \tan^{-1}\left[\frac{4}{6}\right] \quad \text{(2-5)}$$
$$= 53.13° - 33.69° = 19.44°$$

Use $|T_B| = |T_C| = T$ and the law of cosines to find magnitude of R_D:

$$R_D = |R_D| = \sqrt{T^2 + T^2 - 2T^2 \cos(180° - \theta_{BC})} = 1.971T \quad \text{(2-6)}$$

The direction or orientation of R_D is a CW angle θ_x, as measured from the x axis:

$$\theta_x = 90° - \theta_C + \frac{\theta_{BC}}{2} = 46.6° \quad \text{(2-7)}$$

So vector R_D has a magnitude of $1.971T$ and is directed at a CW angle θ of 46.6° from the $+x$ axis, as shown in Fig. 2-7.

4. *Finalize* [*conclude; examine answer—Does it make sense? Are units correct? How does it compare to similar problem solutions?*]: The graphical display of results (drawn approximately to scale in Fig. 2-7) and the orientation angle θ_x computed using trigonometry [Eq. (2-7)] are in agreement.

If the pulley is not "small," the pulley cannot be represented by a single point at D, and its dimensions must be

FIGURE 2-8

Cable force vectors intersect at D′ due to size of pulley

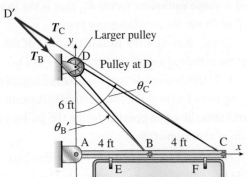

considered. The geometry changes to that shown in Fig. 2-8 and the lines of action of the cable forces T_B and T_C actually intersect at D′. The angles in Fig. 2-6 are changed: θ_B (larger), θ_C (smaller), and θ_{BC} (smaller). A more detailed analysis for the case shown in Fig. 2-8, with an assumed pulley radius of 5 in., reveals that the magnitude of resultant R_D increases by approximately 1%.

Part (b): Find the projection $R_D{}'$ of R_D onto the line DB.

3. *Analyze*: The magnitude of the projection of R_D onto line DB (Fig. 2-7) is

$$R_D' = \left|R_D\right|\cos\left[\frac{\theta_{BC}}{2}\right] = 1.971T\cos\left[\frac{19.44°}{2}\right] = 1.943T \qquad \text{(2-8)}$$

This is considerably larger than the magnitude of component vector T along line DB. If you take components of T_B and T_C along line DB (Fig. 2-7), you find from trigonometry that:

$$R_D' = T + T\cos(\theta_{BC}) = 1.943T \qquad \text{(2-9)}$$

4. *Finalize*: Therefore, the *projection* of a vector along an axis is only equal to the *component* of the vector along that axis if the axes used are perpendicular to one another.

Components of Force: 2D

For many engineering problems, it is more convenient to use a rectangular Cartesian coordinate system when dealing with coplanar concurrent forces acting at a point. The force parallelogram used to find the resultant becomes a rectangle and the force triangle is a right triangle. In Cartesian coordinates, the projections of vectors along the orthogonal axes are the vector components along those axes. For example, the resultant R of the two forces F_1 and F_2 supporting a steel plate can be written in terms of its Cartesian components as (see Fig. 2-9)

$$R = F_1 + F_2 = (F_{1x} + F_{2x})i + (F_{1y} + F_{2y})j$$
$$= (R_x)i + (R_y)j \qquad \text{(2-10)}$$

where

$$F_{1x} = F_1\sin(\beta_1), \ F_{2x} = -F_2\sin(\beta_2), \ F_{1y} = F_1\cos(\beta_1), \text{ and}$$
$$F_{2y} = F_2\cos(\beta_2) \qquad \text{(2-11)}$$

FIGURE 2-9

Cartesian components of resultant R

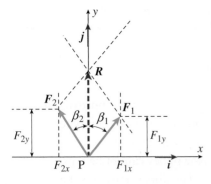

Force vectors R, F_1, and F_2 also can be written in a vector column format as

$$R = \begin{pmatrix} R_x \\ R_y \end{pmatrix} \qquad F_1 = \begin{pmatrix} F_{1x} \\ F_{1y} \end{pmatrix} = \begin{pmatrix} F_1 \sin(\beta_1) \\ F_1 \cos(\beta_1) \end{pmatrix}$$

$$F_2 = \begin{pmatrix} F_{2x} \\ F_{2y} \end{pmatrix} = \begin{pmatrix} -F_2 \sin(\beta_2) \\ F_2 \cos(\beta_2) \end{pmatrix} \tag{2-12}$$

FIGURE 2-10

Cartesian components of cable forces in Example 2-1

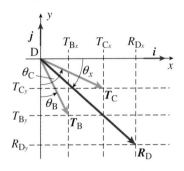

Rectangular components can be used to relate cable forces T_B and T_C to resultant R_D for the sign supported by the cable-pulley system in Example 2-1. From Fig. 2-10, resultant R_D is

$$R_D = (R_{Dx})i + (R_{Dy})j = T_B + T_C = (T_{Bx} + T_{Cx})i + (T_{By} + T_{Cy})j \tag{2-13}$$

in which $R_{Dx} = T_{Bx} + T_{Cx}$ and $R_{Dy} = T_{By} + T_{Cy}$.

The Cartesian components of the three vectors written in vector format are

$$R_D = \begin{pmatrix} R_{Dx} \\ R_{Dy} \end{pmatrix} \qquad T_B = \begin{pmatrix} T_{Bx} \\ T_{By} \end{pmatrix} = \begin{pmatrix} T_B \sin(\theta_B) \\ -T_B \cos(\theta_B) \end{pmatrix}$$

$$T_C = \begin{pmatrix} T_{Cx} \\ T_{Cy} \end{pmatrix} = \begin{pmatrix} T_C \sin(\theta_C) \\ -T_C \cos(\theta_C) \end{pmatrix} \tag{2-14}$$

Using numerical values from Example 2-1 ($\theta_B = 53.13°$, $\theta_C = 33.69°$ and $|T_B| = |T_C| = T$), Eq. (2-14) can be rewritten as

$$R_D \begin{pmatrix} 1.355 \\ -1.432 \end{pmatrix} T \qquad T_B = \begin{pmatrix} T_{Bx} \\ T_{By} \end{pmatrix} = \begin{pmatrix} 0.555 \\ -0.832 \end{pmatrix} T$$

$$T_C = \begin{pmatrix} T_{Cx} \\ T_{Cy} \end{pmatrix} = \begin{pmatrix} 0.8 \\ -0.6 \end{pmatrix} T \tag{2-15}$$

Using Cartesian components, the magnitude of R_D is then confirmed as

$$R_D = |R_D| = \sqrt{R_{Dx}^2 + R_{Dy}^2} = \sqrt{(1.355T)^2 + (-1.432T)^2}$$

$$= \sqrt{3.887T^2} = 1.971T \tag{2-16}$$

The direction cosine of R_D (CW angle θ_x in Fig. 2-10) is computed using Cartesian components as

$$\theta_x = \tan^{-1}\left(\frac{|R_{Dy}|}{R_{Dx}}\right) = \tan^{-1}\left(\frac{|-1.432T|}{1.355T}\right) = 46.6° \tag{2-17}$$

As a final example of Cartesian rectangular components in two dimensions, consider three vector forces (T_1, T_2, and T_3) acting on an eye bolt attached to a wall (Fig. 2-11). The resultant of the cable forces is vector R where $R = T_1 + T_2 + T_3$. The Cartesian components of R are R_x and R_y. Force component R_x is controlled by the allowable *tensile* stress in the bolt, and component R_y depends on the allowable *shear* stress across the bolt. In this case, an acceptable value of the resultant force magnitude R is $R = \sqrt{R_x^2 + R_y^2}$.

FIGURE 2-11

Eye bolt with three cable forces

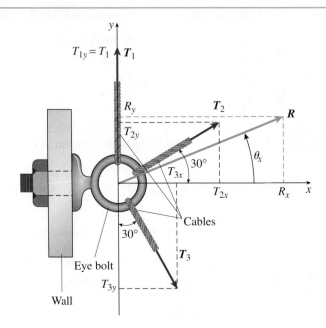

The direction angle θ_x (Fig. 2-11) for force vector $\boldsymbol{R}$ is $\theta_x = \tan^{-1}\left[\dfrac{R_y}{R_x}\right]$. Next find acceptable values for the three cable forces. This is a challenge because there is not enough information to find a unique solution. Equate rectangular components as in Eq. (2-13) to obtain

$$R_x = T_{2x} + T_{3x} = T_2\cos(30°) + T_3\sin(30) = 0.866T_2 + 0.5T_3$$
$$R_y = T_1 + T_{2y} + T_{3y} = T_1 + T_2\sin(30°) - T_3\cos(30°)$$
$$= T_1 + 0.5T_2 - 0.866T_3 \tag{2-18}$$

These two equations contain three unknown forces (T_1, T_2, and T_3), so a unique solution is not possible.

As an alternative description of the problem, suppose that you know cable force vectors $\boldsymbol{T}_1$, $\boldsymbol{T}_2$, and $\boldsymbol{T}_3$ in advance and must find resultant $\boldsymbol{R}$ and its components R_x and R_y. In this approach, magnitudes R_x and R_y are needed to *design* the bolt support system (by selecting a large enough bolt so that both tensile and shear stresses in the bolt are below allowable values). The resultant force components R_x and R_y are needed if the bolt is to be sized properly. Using vector addition, resultant $\boldsymbol{R}$ is

$$\boldsymbol{R} = \boldsymbol{T}_1 + \boldsymbol{T}_2 + \boldsymbol{T}_3 = R_x\boldsymbol{i} + R_y\boldsymbol{j} = (T_{2x} + T_{3x})\boldsymbol{i} + (T_1 + T_{2y} + T_{3y})\boldsymbol{j} \tag{2-19}$$

Components R_x and R_y are defined in more detail in Eq. (2-18).

You can also use successive applications of the *parallelogram* law to find resultant vector $\boldsymbol{R}$ from cable force vectors $\boldsymbol{T}_1$, $\boldsymbol{T}_2$, and $\boldsymbol{T}_3$, as illustrated in Fig. 2-12. First, find the resultant $\boldsymbol{R}_{12}$ of vectors $\boldsymbol{T}_1$ and $\boldsymbol{T}_2$ (Fig. 2-12a) and then add vectors $\boldsymbol{R}_{12}$ and $\boldsymbol{T}_3$ to find $\boldsymbol{R}_{123} = \boldsymbol{R}$ (Fig. 2-12b). The vector addition is

$$\boldsymbol{R}_{12} = \boldsymbol{T}_1 + \boldsymbol{T}_2 \quad \text{and} \quad \boldsymbol{R} = \boldsymbol{R}_{123} = \boldsymbol{R}_{12} + \boldsymbol{T}_3 \tag{2-20}$$

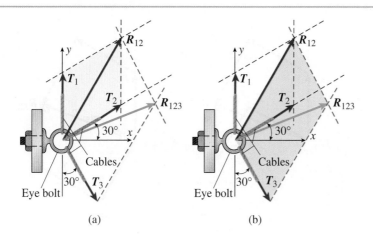

FIGURE 2-12

(a,b) Successive applications of the parallelogram law

It appears that use of Cartesian coordinates and rectangular components in Eq. (2-19) might provide a simpler approach for this problem.

Last, to complete finding the resultant of the three cable forces, note that an extension of the *triangle* law to a *polygon* law can be used to find resultant **R**, as in Fig. 2-13. Resultant **R** is not dependent on the order in which you carry out the vector addition or subtraction so, the resultant is

$$R = T_1 + T_2 + T_3 = T_2 + T_3 + T_1 = T_3 + T_2 + T_1 \qquad \textbf{(2-21)}$$

This is represented graphically by showing that alternate force polygons take us from components T_1, T_2, and T_3 to resultant **R** (see Fig. 2-14) but along different paths with the same result: $R = R_{123} = R_{231} = R_{321}$. Care must be exercised in carrying out the trigonometric solutions to avoid errors. Use of rectangular components (Fig. 2-12) appears to be the preferred solution approach.

FIGURE 2-13

Polygon law for forces in bolt-cable system

FIGURE 2-14

(a,b) Alternate force polygons

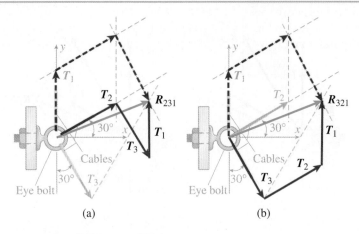

(a)   (b)

Components of Force: 3D

Use of the parallelogram and triangle or polygon laws in three dimensions can be a challenge for many problems. In most cases, a three-dimensional Cartesian coordinate system is the best solution approach. For example, consider three cables used to secure a cargo ship to a dock (Fig. 2-15).

Assume that the cable force magnitudes F_1, F_2, and F_3 are known; let $F_1 = F$, $F_2 = 1.25F$, and $F_3 = 1.5F$ where F is some known value of cable tension force. Also assume that the cables are straight and that the coordinates of the single tie-down point on the dock and the three attachment points on the ship are known (Fig. 2-16). You want to find the single vector R at the tie-down point on the dock that is the sum of cable force vectors F_1, F_2, and F_3 and then find the Cartesian components of R as

$$R = F_1 + F_2 + F_3 = R_x i + R_y j + R_z k \tag{2-22}$$

Accurate values of these forces for a variety of loadings are essential to design of the eye bolt connection.

Each of the cable force vectors F_1, F_2, and F_3 are represented by its components. For example, force vector F_1 is

$$F_1 = F_{1x} i + F_{1y} j + F_{1z} k \tag{2-23}$$

Equating like components in Eqs. (2-22) and (2-23) gives

$$R_x = F_{1x} + F_{2x} + F_{3x}, \quad R_y = F_{1y} + F_{2y} + F_{3y}, \quad R_z = F_{1z} + F_{2z} + F_{3z} \tag{2-24}$$

FIGURE 2-15

Cables secure cargo ship to dock

Using the coordinates of point O and those of each cable attachment point on the cargo ship (given in Fig. 2-16), you can create unit vectors along each of the cable lines from Eq. (1-21). For example, the unit vector along cable 1 is

FIGURE 2-16

Eye bolt on dock with three cable forces

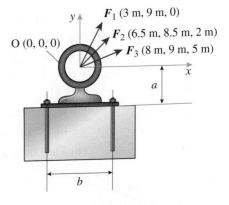

$$n_{O1} = \frac{r_{O1}}{r_{O1}} = \frac{(x_1 - x_O)i + (y_1 - y_O)j + (z_1 - z_O)k}{\sqrt{(x_1 - x_O)^2 + (y_1 - y_O)^2 + (z_1 - z_O)^2}}$$

$$= \frac{(3\text{ m} - 0)i + (9\text{ m} - 0)j}{\sqrt{(3\text{ m} - 0)^2 + (9\text{ m} - 0)^2}} = \left(\frac{3}{\sqrt{90}}\right)i + \left(\frac{9}{\sqrt{90}}\right)j \quad \textbf{(2-25)}$$

The resulting unit vectors for all three cables are

$$n_{O1} = 0.316i + 0.949j, \quad n_{O2} = 0.597i + 0.781j + 0.184k,$$
$$n_{O3} = 0.614i + 0.69j + 0.383k \quad \textbf{(2-26)}$$

Each of the cable force vectors now can be written as the given force magnitude (i.e., $F_1 = F, F_2 = 1.25F, F_3 = 1.5F$) times the corresponding unit vector. The results are

$$F_1 = F(0.316i + 0.949j), \quad F_2 = (1.25F)(0.597i + 0.781j + 0.184k),$$
$$F_3 = (1.5F)(0.614i + 0.69j + 0.383k) \quad \textbf{(2-27)}$$

or

$$F_1 = F(0.316i + 0.949j), \quad F_2 = F(0.746i + 0.976j + 0.23k),$$
$$F_3 = F(0.92i + 1.035j + 0.575k) \quad \textbf{(2-28)}$$

Find the components of the resultant vector R from Eqs. (2-24) and (2-28). The components of R, in terms of reference force magnitude F, are

$$R = F(1.983i + 2.96j + 0.805k) \quad \textbf{(2-29)}$$

From Eq. (1-13), the magnitude of R is

$$R = |R| = \sqrt{R_x^2 + R_y^2 + R_z^2} = 3.65F \quad \textbf{(2-30)}$$

FIGURE 2-17

Resultant vector R and its Cartesian components

Finally, compute the direction of resultant vector R using its direction cosines. From Eq. (1-14), the direction cosines are

$$l = \cos(\theta_x) = \frac{R_x}{R}, \quad m = \cos(\theta_y) = \frac{R_y}{R}, \quad n = \cos(\theta_z) = \frac{R_z}{R} \quad \textbf{(2-31)}$$

Substitute components of R from Eq. (2-29) and the magnitude of R from Eq. (2-30) into Eq. (2-31) to get

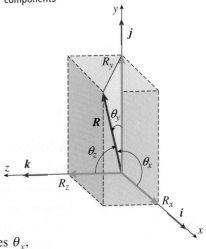

$$l = \cos(\theta_x) = 0.543, \quad m = \cos(\theta_y) = 0.81, \quad n = \cos(\theta_z) = 0.22 \quad \textbf{(2-32)}$$

Note that Eq. (1-16) is satisfied ($l^2 + m^2 + n^2 = 1$). Solving for angles θ_x, θ_y, and θ_z between vector R and each of the coordinate axes gives

$$\theta_x = \cos^{-1}(l) = 57.1°, \quad \theta_y = \cos^{-1}(m) = 35.9°, \quad \theta_z = \cos^{-1}(n) = 77.3° \quad \textbf{(2-33)}$$

The resultant vector R and its Cartesian components are depicted in Fig. 2-17.

From Eq. (1-22), you can describe resultant vector R [Eq. (2-29)] in more compact form by using a unit vector n_R along the direction of R as

$$R = Rn_R \quad \textbf{(2-34)}$$

where

$$n_R = \frac{R}{|R|} = \frac{F(1.983i + 2.96j + 0.805k)}{3.65F} = 0.543i + 0.81j + 0.22k \quad \textbf{(2-35)}$$

Example 2-2

Two separate cables AC and BC support a sign structure attached to a building (see Fig. 2-18). The sign is also supported by a pin (ball and socket) support at O and a restraint at D—in the z direction only—to stabilize the sign against rotation about the x axis. The tension force in cable AC is known to be 5.41 kN and that in cable BC is 4.48 kN.

(a) Determine the magnitude and orientation of resultant vector **R** that is the *sum* of the two cable forces at C; use the *parallelogram law*.

(b) Find the Cartesian components of vector **R**.

FIGURE 2-18

Sign structure supported by cables AC and BC

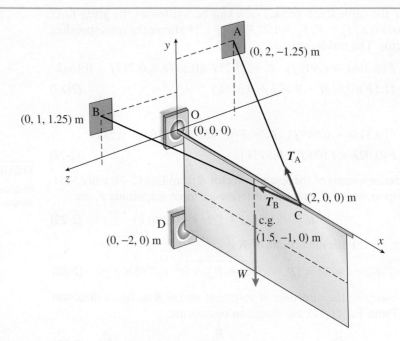

Solution:

Part (a): Use the four-step problem-solving approach to find the magnitude and orientation of resultant vector R that is the *sum* of the two cable forces at C; use the *parallelogram law*.

1. *Conceptualize*: First focus on the part of the structure of interest: the region around the cable attachment point at C where the lines of action of the two cable force vectors intersect. Make a careful sketch (see Fig. 2-18) of vectors T_A and T_B, both originating at C, acting along lines AC and BC, respectively.

2. *Categorize*: The vector sum or resultant **R** of cable force vectors T_A and T_B can be represented using a *parallelogram* like that in Fig. 2-19. Draw two lines: one parallel to line AC through the tip of vector T_B, and the other parallel to line BC through the tip of vector T_A. These two lines lie in the inclined plane ACB, so the resultant vector **R** also lies in this plane (Fig. 2-20).

FIGURE 2-19

Resultant vector **R** using the parallelogram law

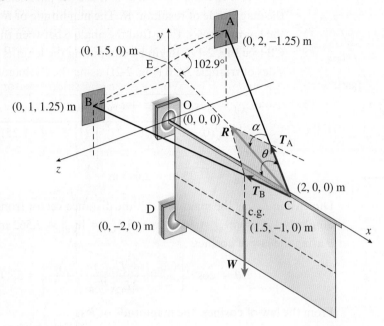

FIGURE 2-20

Vectors lie in plane ACB

3. *Analyze*: The vector sum is $R = T_A + T_B$. Given the coordinates of joints A, B, and C and the magnitudes of the cable forces, $T_A = 5.41$ kN and $T_B = 4.48$ kN, use the parallelogram law to find the magnitude of resultant R. The magnitude of R is computed using the *law of cosines*. First find the angle θ between the lines CA and CB (see Fig. 2-19). Begin by computing the lengths of each of the sides of triangle ACB (Fig. 2-21) using the Pythagorean theorem:

$$L_{AB} = \sqrt{(x_B - x_A)^2 + (y_B - y_A)^2 + (z_B - z_A)^2}$$
$$= \sqrt{(0-0)^2 + (1-2)^2 + (1.25-(-1.25))^2} \text{ m}$$

or   **(2-36)**

$$L_{AB} = \sqrt{7.25} \text{ m} = 2.693 \text{ m}$$

Dimension L_{AB} is the magnitude of the distance vector from A to B, r_{AB}. Similarly $L_{CA} = |r_{CA}| = 3.092$ m and $L_{CB} = |r_{CB}| = 2.562$ m. Now find angle θ using the law of cosines:

$$\theta = \cos^{-1}\left[\frac{(L_{CA}^2 + L_{CB}^2) - L_{AB}^2}{2L_{CA}L_{CB}}\right] = 55.93° \qquad \textbf{(2-37)}$$

From the law of cosines, the magnitude of R is

$$R = |R| = \sqrt{T_A^2 + T_B^2 - 2T_A T_B \cos(180° - \theta)}$$

or

$$R = \sqrt{(5.41 \text{ kN})^2 + (4.48 \text{ kN})^2 - 2(5.41 \text{ kN})(4.48 \text{ kN})\cos(180° - 55.93°)}$$
$$= 8.75 \text{ kN} \qquad \textbf{(2-38)}$$

The orientation of R can be determined by finding the angle α between vector T_A and vector R in the ACB plane (Fig. 2-21). Use the law of sines to find:

$$\alpha = \sin^{-1}\left[\frac{T_B}{|R|}\sin(180° - \theta)\right] = 25.1° \qquad \textbf{(2-39)}$$

Last, the angle β between R and the $(-x)$ axis (Fig. 2-20) is

$$\beta = \tan^{-1}\left(\frac{1.5 \text{ m}}{2 \text{ m}}\right) = 36.87° \qquad \textbf{(2-40)}$$

So, with angles α and β defined, the orientation of R is now known.

4. *Finalize*: Line AB crosses the y axis at its midpoint (see point E in Fig. 2-19). If you compute angle α in Figs. 2-20 and 2-21 using the geometry of triangle ACB and the law of cosines, you can see that this is equal to angle α in Eq. (2-39), which defines the orientation of vector R. Apparently resultant vector R has a line of action which passes through the y axis at point E (Fig. 2-19). This means that the Cartesian component R_z is, zero which you can confirm in part (b).

Part (b): Find the Cartesian components of vector R.

1, 2. *Conceptualize Categorize*: Find vector R as the sum of cable force vectors T_A and T_B then compare results to those obtained in part (a) based on the parallelogram law. First find position vectors and then unit vectors along lines CA and CB as

$$n_{CA} = \frac{r_{CA}}{r_{CA}} = \frac{-2}{3.092}i + \frac{2}{3.092}j - \frac{1.25}{3.092}k = -0.647i + 0.647j - 0.404k$$

(2-41)

$$n_{CB} = \frac{r_{CB}}{r_{CB}} = \frac{-2}{2.562}i + \frac{1}{2.562}j - \frac{1.25}{2.562}k = -0.781i + 0.39j + 0.488k$$

3. *Analyze*: The resulting cable force vectors are

$$T_A = T_A n_{CA} = (5.41 \text{ kN})(-0.647i + 0.647j - 0.404k) = \begin{bmatrix} -3.499 \\ 3.499 \\ -2.187 \end{bmatrix} \text{kN}$$

(2-42)

and

$$T_B = T_B n_{CB} = (4.48 \text{ kN})(-0.781i + 0.39j - 0.488k) = \begin{bmatrix} -3.498 \\ 1.749 \\ 2.186 \end{bmatrix} \text{kN}$$

Sum cable force vectors in Eq. (2-42) to find resultant vector R:

$$R = T_A + T_B = (-7.0i + 5.25j) \text{ kN} \qquad \longleftarrow \textbf{(2-43)}$$

The Cartesian components are $R_x = -7.0$ kN, $R_y = 5.25$ kN (see Fig. 2-23); component R_z is equal to zero as noted in the conclusion of part (a). The magnitude of vector R is

FIGURE 2-23

Cartesian components and direction cosines

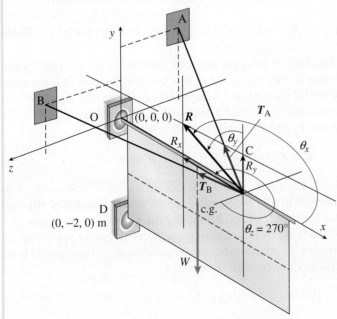

$$R = |R| = \sqrt{R_x^2 + R_y^2}$$
$$= \sqrt{(-7.0)^2 + 5.25^2} \text{ kN}$$
$$= \sqrt{76.5625} \text{ kN} = 8.75 \text{ kN} \qquad \textbf{(2-44)}$$

which is in agreement with Eq. (2-38) in part (a). Finally, the direction cosines of R define its orientation (Fig. 2-23) in the Cartesian reference frame:

$$\theta_x = \cos^{-1}\left(\frac{R_x}{R}\right) = 143.1°,$$

$$\theta_y = \cos^{-1}\left(\frac{R_y}{R}\right) = 53.1°,$$

$$\theta_z = \cos^{-1}\left(\frac{R_z}{R}\right) = 270° \qquad \textbf{(2-45)}$$

4. *Finalize*: Angle β in Eq. (2-40) can be computed from angle θ_x ($\beta = 180° - \theta_x = 36.87°$), so results from parts (a) and (b) are in agreement.

Resultant of a System of Forces

Determination of resultant forces is a fundamental exercise in statics. Statics is concerned with equilibrium of rigid bodies, a condition that is satisfied in part if the resultant of all forces acting on the body is equal to zero.

Thus far, this chapter has covered only systems of concurrent forces looking at both two-dimensional and three-dimensional force systems. By definition, the lines of action of these forces meet at a point, so there are no moments about the point of concurrency. Since there are no moments, the effect of the system of concurrent forces on either a particle or a rigid body is the same.

In most cases, you want to reduce the system of concurrent forces to a resultant force. The resultant force replaces the original system of forces since the resultant force is the simplest force combination with the same overall effect on the particle or rigid body. The resultant R is applied at the point of concurrency and can be obtained by simple vector addition of the individual forces, F, as

$$R = F_1 + F_2 + F_3 = \sum_{i=1}^{3} F_i \tag{2-46}$$

A *polygon law graphical approach* was used to find the resultant force for a bolt-cable system in Figs. 2-13 and 2-14, and then a *vector summation approach* was used for the ship tie-down example leading to the addition of Cartesian vectors to find the resultant [Eq. (2-29)]. With a Cartesian coordinate system, the magnitudes of the x, y, and z components of the resultant are equal to the sums of the corresponding components of the individual forces [see Eqs. (2-18) and (2-24)].

$$R_x = \Sigma F_x, \quad R_y = \Sigma F_y, \quad R_z = \Sigma F_z \tag{2-47}$$

From Eq. (1-13), the magnitude of resultant R can be expressed as

$$R = \sqrt{R_x^2 + R_y^2 + R_z^2} = \sqrt{(\Sigma F_x)^2 + (\Sigma F_y)^2 + (\Sigma F_z)^2} \tag{2-48}$$

Thus far, this chapter has only covered systems of concurrent forces, so *force equilibrium* is only discussed in this section. To look at force equilibrium for the aerial camera in Example 1-2, consider the weight of the camera, W, as an additional force acting in Fig. 2-24. Recall that F_s is the sum of cable forces F_1 and F_2, so the overall resultant R is

$$R = F_s - W = (F_1 + F_2) - W = 0 \tag{2-49}$$

You can also apply Eq. (2-49) to the truck crane example in Section 2.1 to find force equilibrium in that problem. To do so, add the weight of the steel plate to the force diagrams in Figs. 2-3b and 2-4b.

Example 2-3 adds a three-dimensional support structure for the aerial camera in Example 1-2 and examines *force equilibrium of a concurrent force system* that includes the weight of the camera.

FIGURE 2-24

Aerial camera from Example 1-2

Example 2-3

A moveable overhead camera, used for close-up viewing of action at sporting events (Fig. 2-25), is supported by four cables anchored on poles A, B, C, and D (Fig. 2-26). By adjusting the lengths of cables 1 through 4, the camera can be positioned at any point over the field. One such position is shown in Fig. 2-26. The four cable force vectors (F_1 to F_4) and a vector whose magnitude is the weight of the camera, $W = 28$ lb, are concurrent at point P. The forces in cables 1 through 3 (expressed in terms of camera weight W) are known to be $1.424W$, $2.261W$, $2.261W$, respectively; the force in cable 4 is unknown. A Cartesian coordinate system with origin at O is shown in Fig. 2-26. Find these vectors:

FIGURE 2-25

Aerial field camera

(a) Distance vectors and unit vectors from point of concurrency P to the top of each cable support pole (A through D).

(b) Expressions for force vectors F_1 to F_4 and W.

(c) Resultant R using force vectors F_1 to F_4 and W.

(d) Force vector F_4 if the resultant at point P must be zero.

FIGURE 2-26

Force vectors at Point P on overhead field camera

Solution:

Part (a): Use the four-step problem-solving approach to find distance vectors and unit vectors from the point of concurrency P to the top of each cable support pole (A through D).

1. *Conceptualize*: You have coordinates of all points (A through D and P; see Fig. 2-26). Vectors F_1 to F_4 and gravity force W can be treated as *sliding vectors* and moved to point P where their lines of action cross.

2. *Categorize*: Distance and unit vectors are found from joint coordinates at each end of a line.

3. *Analyze*: The distance vectors from P to the top of each pole are

$$r_{PA} = (-10 - 50)i + (25 - 15)j + (-10 - 50)k \text{ yds} = \begin{bmatrix} -60 \\ 10 \\ -60 \end{bmatrix} \text{yds}$$

$$r_{PB} = (80 - 50)i + (25 - 15)j + (-10 - 50)k \text{ yds} = \begin{bmatrix} 30 \\ 10 \\ -60 \end{bmatrix} \text{yds}$$

$$\qquad\qquad\qquad\qquad\qquad\qquad\qquad\qquad\qquad\qquad\qquad \Longleftarrow \textbf{(2-50)}$$

$$r_{PC} = (80 - 50)i + (25 - 15)j + (110 - 50)k \text{ yds} = \begin{bmatrix} 30 \\ 10 \\ 60 \end{bmatrix} \text{yds}$$

$$r_{PD} = (-10 - 50)i + (25 - 15)j + (110 - 50)k \text{ yds} = \begin{bmatrix} -60 \\ 10 \\ 60 \end{bmatrix} \text{yds}$$

where first a row and then a column format is used to display each vector. Magnitudes of the four distance vectors are

$$|r_{PA}| = |r_{PD}| = 85.44 \text{ yds}, \quad |r_{PB}| = |r_{PC}| = 67.82 \text{ yds} \qquad \textbf{(2-51)}$$

The four unit vectors from point of concurrency P to the top of each cable support pole (A through D) are

$$n_{PA} = \frac{r_{PA}}{r_{PA}} = (-0.702)i + (0.117)j + (-0.702)k$$

$$n_{PB} = \frac{r_{PB}}{r_{PB}} = (0.442)i + (0.147)j + (-0.885)k$$

$$n_{PC} = \frac{r_{PC}}{r_{PC}} = (0.442)i + (0.147)j + (0.885)k \qquad \Longleftarrow \textbf{(2-52)}$$

$$n_{PD} = \frac{r_{PD}}{r_{PD}} = (-0.702)i + (0.117)j + (0.702)k$$

4. *Finalize*: The cable support system is symmetric about an *x-y* plane passing through point P. As a result, cable length PA ($|r_{PA}|$) is equal to length PD and length PB is equal to length PC. Another check on results is that each of the unit vectors is of unit length; for example

$$n_{PD} = |n_{PD}| = \sqrt{(-0.702)^2 + (0.117)^2 + (0.702)^2} = 1.0 \qquad \textbf{(2-53)}$$

Part (b): Find expressions for force vectors F_1 to F_4 and W.

3. *Analyze*: Write each force vector as the product of its magnitude and the unit vector [Eq. (2-52)] along its line of action:

$$F_1 = F_1 n_{PA} = (1.424)(28 \text{ lb})((-0.702)i + (0.117)j + (-0.702)k) = \begin{bmatrix} -28 \\ 4.666 \\ -28 \end{bmatrix} \text{lb}$$

$$F_2 = F_2 n_{PB} = (2.261)(28 \text{ lb})((0.442)i + (0.147)j + (-0.885)k) = \begin{bmatrix} 28 \\ 9.334 \\ -56 \end{bmatrix} \text{lb}$$

$$F_3 = F_3 n_{PC} = (2.261)(28 \text{ lb})((0.442)i + (0.147)j + (0.885)k) = \begin{bmatrix} 28 \\ 9.334 \\ 56 \end{bmatrix} \text{lb}$$

$$F_4 = F_4 n_{PD} = F_{4x}i + F_{4y}j + F_{4z}k = \begin{bmatrix} F_{4x} \\ F_{4y} \\ F_{4z} \end{bmatrix}$$

$$W = W(-j) = -28j \text{ lb} = \begin{bmatrix} 0 \\ -28 \\ 0 \end{bmatrix} \text{lb} \qquad \textbf{(2-54)}$$

The magnitude of force vector F_4 is unknown at this point, so numerical values of its components cannot be listed here.

Part (c): Find resultant R using force vectors F_1 to F_4 and W.

2. *Categorize*: Now consider the weight of the camera, W, as part of the summation of forces at P in accordance with Eqs. (2-46) and (2-49). The camera weight is a body force that can be treated as a downward applied force at the center of gravity of the camera. Weight W was ignored in previous examples, but now all forces at point P have been accounted for. The resultant of all forces at the point of concurrency P is

$$R = (F_1 + F_2 + F_3 + F_4) - W \qquad \textbf{(2-55)}$$

3. *Analyze*: Substituting Eqs. (2-54) into Eq. (2-55) gives:

$$R = \begin{bmatrix} -28 + 28 + 28 + F_{4x} \\ 4.666 + 9.334 + 9.334 + F_{4y} - 28 \\ -28 - 56 + 56 + F_{4z} \end{bmatrix} \text{lb} = \begin{bmatrix} 28 + F_{4x} \\ F_{4y} - 4.666 \\ -28 + F_{4z} \end{bmatrix} \text{lb} \qquad \textbf{(2-56)}$$

or

$$R = (28 + F_{4x})i + (F_{4y} - 4.666)j + (-28 + F_{4z})k \text{ lb} = \begin{bmatrix} R_x \\ R_y \\ R_z \end{bmatrix} \quad \text{(2-57)}$$

Part (d): Find force vector F_4 if the resultant at point P must be zero.

2. *Categorize:* If the balance of forces at point P is of interest, the resultant R must be equal to zero in accordance with Eq. (2-49). This means that each of the components of R (R_x, R_y, and R_z) must be equal to zero. By enforcing this condition of *force equilibrium*, you can find the required Cartesian components of force vector F_4 from Eq. (2-56) as

$$F_4 = F_{4x}i + F_{4y}j + F_{4z}k = (-28)i + (4.666)j + (28)k \text{ lb} \quad \text{(2-58)}$$

3. *Analyze:* The magnitude of cable force F_4 is

$$F_4 = |F_4| = \sqrt{(-28)^2 + (4.666)^2 + (28)^2} \text{ lb} = \sqrt{1590} \text{ lb} = 39.9 \text{ lb} \quad \text{(2-59)}$$

and, after substituting for camera weight $W = 28$ lb, the magnitudes of the other forces at concurrency point P are

$$F_1 = 1.424 \ W = 1.424(28 \text{ lb}) = 39.9 \text{ lb} \quad \text{(2-60)}$$
$$F_2 = F_3 = 2.261 \ W = 2.261(28 \text{ lb}) = 63.3 \text{ lb}$$

4. *Finalize:* The fact that cable forces $F_1 = F_4$ and $F_2 = F_3$ is consistent with the fact that both the cable structure and the overall force system are symmetric about an *x-y* plane through point P.

FIGURE 2-27

Vectors *F* and *P*

Force Component Along a Line

Vector operations in statics often involve finding: (1) the component of a vector along (and/or perpendicular to) a line (or specified direction) and (2) the angle between two vectors. In both cases, use the *dot* or *scalar* or *inner product* of two vectors. The dot product of two vectors, *F* and *P*, is defined as the product of their magnitudes (*F* times *P*) times the cosine of the angle (θ) between them (Fig. 2-27):

$$F \cdot P = FP \cos(\theta) = (F \cos(\theta))(P) = F_P P$$

or

$$F \cdot P = P \cdot F = F(P \cos(\theta)) = FP_F \quad \text{(2-61)}$$

where $F_P = F \cos(\theta)$ is the orthogonal projection of *F* along the line of action of *P*, and $P_F = P \cos(\theta)$ is the orthogonal projection of *P* along the line of action of *F*. The dot product is a scalar and the order of the vectors is irrelevant. Note that the dot product of two vectors *F* and *P* [Eq. (2-61)] applies even if their lines of action do not intersect.

FIGURE 2-28

(a,b) Vector projections

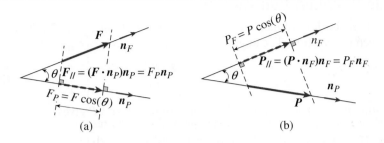

(a) (b)

To express the projection of vector F on the line of action of vector P as a vector, first multiply the scalar product by a unit vector n_P along the line of action of P (Fig. 2-28a):

$$(F \cdot n_P)n_P = F_P n_P = F_{//} \tag{2-62}$$

Likewise, to express the projection of vector P on the line of action of vector F as a vector, multiply the scalar product by a unit vector n_F along the line of action of F (Fig. 2-28b):

$$(P \cdot n_F)n_F = P_F n_F = P_{//} \tag{2-63}$$

If the dot product involves a vector and a unit vector [such as in Eqs. (2-62) and (2-63)], the result is the vector projection along (or *parallel to*) the direction defined by unit vector n_P [in Eq. (2-62)] or the direction defined by n_F [in Eq. (2-63)]. Call these vectors $F_{//}$ and $P_{//}$, respectively where the two slashes in each subscript refer to the vector orthogonal projection along (or *parallel to*) the direction defined by the unit vector. The difference between the original vector and the parallel projection is the *perpendicular* component (Fig. 2-29):

$$F_\perp = F - F_{//}, \quad P_\perp = P - P_{//} \tag{2-64}$$

If the vectors are written in terms of their Cartesian components

$$F = F_x i + F_y j + F_z k, \quad P = P_x i + P_y j + P_z k \tag{2-65}$$

the dot product of F and P is

$$F \cdot P = P \cdot F = F_x P_x + F_y P_y + F_z P_z \tag{2-66}$$

because, for the mutually perpendicular unit vectors i, j, and k along the x, y, and z directions, each has unit length, so

$$i \cdot i = j \cdot j = k \cdot k = 1 \tag{2-67}$$

and

$$i \cdot j = i \cdot k = j \cdot i = j \cdot k = k \cdot i = k \cdot j = 0 \tag{2-68}$$

so $F \cdot i = F_x,\ F \cdot j = F_y,$ and $F \cdot k = F_z.$

FIGURE 2-29

(a,b) Parallel and perpendicular vector components

(a) (b)

Angle Between Two Vectors

You can use the dot product definition [Eq. (2-61) and Fig. 2-27] to find the angle θ between two vectors. The angle θ between known vectors $\boldsymbol{F}$ and $\boldsymbol{P}$ is

$$\theta = \cos^{-1}\left[\frac{\boldsymbol{F}\cdot\boldsymbol{P}}{F\,P}\right] \tag{2-69}$$

Using unit vectors along the lines of action of $\boldsymbol{F}$ and $\boldsymbol{P}$ gives:

$$\theta = \cos^{-1}\left[\frac{\boldsymbol{F}\cdot\boldsymbol{P}}{F\,P}\right] = \cos^{-1}(\boldsymbol{n}_F\cdot\boldsymbol{n}_P) = \cos^{-1}(\boldsymbol{n}_P\cdot\boldsymbol{n}_F) \tag{2-70}$$

Equation (2-69) is especially useful for in 3D problems like in Example 2-2. To avoid the geometric calculations in the use of the law of cosines for inclined plane ACB to find angle θ between cable force vectors $\boldsymbol{T}_\mathrm{A}$ and $\boldsymbol{T}_\mathrm{B}$ [Eq. (2-37) in Example 2-2], use $\theta = \cos^{-1}[\boldsymbol{n}_\mathrm{CA}\cdot\boldsymbol{n}_\mathrm{CB}] = \cos^{-1}(0.56) = 55.93°$ where unit vectors $\boldsymbol{n}_\mathrm{CA}$ and $\boldsymbol{n}_\mathrm{CB}$ are given in Eq. (2-41). Likewise, angle α between resultant $\boldsymbol{R}$ and components $\boldsymbol{T}_\mathrm{A}$ and $\boldsymbol{T}_\mathrm{B}$ is more easily found using $\alpha = \cos^{-1}[\boldsymbol{R}\cdot\boldsymbol{n}_\mathrm{CA}\,/\,R] = 25.1°$ rather than the law of sines [Eq. (2-39) in Example 2-2].

If unit vectors $\boldsymbol{n}_F$ and $\boldsymbol{n}_P$ in Eq. (2-70) are the unit vectors $\boldsymbol{i}, \boldsymbol{j}$, and $\boldsymbol{k}$ along the Cartesian coordinate axes x, y, and z, then, as expected, angle $\theta = 0°$ when the conditions stated in Eq. (2-67) are satisfied, and $\theta = 90°$ for conditions defined in Eq. (2-68).

Example 2-4

Two force vectors $\boldsymbol{F}$ and $\boldsymbol{P}$ are applied at point O (Fig. 2-30). The magnitudes of each force vector are $F = 5$ kN and $P = 13$ kN.

(a) Determine the dot product of vectors $\boldsymbol{F}$ and $\boldsymbol{P}$.

(b) Find the components of each vector along (parallel to) and perpendicular to the other vector.

Solution:

Part (a): Use the four-step problem-solving approach to find the dot product of vectors $\boldsymbol{F}$ and $\boldsymbol{P}$.

FIGURE 2-30

Two force vectors $\boldsymbol{F}$ and $\boldsymbol{P}$

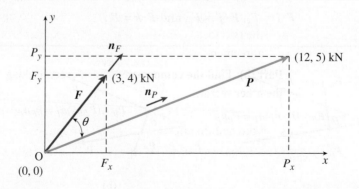

1. *Conceptualize*: Write each vector in terms of its Cartesian components. Find unit vectors $\boldsymbol{n}_F$ and $\boldsymbol{n}_P$ along the line of action of each force vector:

$$\boldsymbol{F} = F_x \boldsymbol{i} + F_y \boldsymbol{j} = 3\boldsymbol{i} + 4\boldsymbol{j} \text{ kN} = \begin{bmatrix} 3 \\ 4 \end{bmatrix} \text{kN},$$

$$\boldsymbol{P} = P_x \boldsymbol{i} + P_y \boldsymbol{j} = 12\boldsymbol{i} + 5\boldsymbol{j} \text{ kN} = \begin{bmatrix} 12 \\ 5 \end{bmatrix} \text{kN} \tag{2-71}$$

$$\boldsymbol{n}_F = \frac{\boldsymbol{F}}{F} = \frac{3\boldsymbol{i} + 4\boldsymbol{j} \text{ kN}}{5 \text{ kN}} = 0.6\boldsymbol{i} + 0.8\boldsymbol{j},$$

$$\boldsymbol{n}_P = \frac{\boldsymbol{P}}{P} = \frac{12\boldsymbol{i} + 5\boldsymbol{j} \text{ kN}}{13 \text{ kN}} = 0.923\boldsymbol{i} + 0.385\boldsymbol{j} \tag{2-72}$$

2. *Categorize*: Angle θ between the two vectors (Fig. 2-30) is easily found using a dot product as in Eq. (2-69) or (2-70), or from geometry in Fig. 2-30:

$$\theta = \tan^{-1}\left(\frac{4}{3}\right) - \tan^{-1}\left(\frac{5}{12}\right) = 30.51° \tag{2-73}$$

or, from Eq. (2-70):

$$\theta = \cos^{-1}\left[\frac{\boldsymbol{F} \cdot \boldsymbol{P}}{F\,P}\right] = \cos^{-1}(\boldsymbol{n}_F \cdot \boldsymbol{n}_P) = \cos^{-1}(\boldsymbol{n}_P \cdot \boldsymbol{n}_F) \tag{2-74}$$

3. *Analyze*: The dot product of force vectors $\boldsymbol{F}$ and $\boldsymbol{P}$ is [Eq. (2-66)]:

$$\boldsymbol{F} \cdot \boldsymbol{P} = F_x P_x + F_y P_y = [(3)(12) + (4)(5)] \text{ kN}^2 = (36 + 20) \text{ kN}^2 = 56 \text{ kN}^2 \quad \longleftarrow \tag{2-75}$$

Substitute Eq. (2-75) into Eq. (2-74):

$$\theta = \cos^{-1}\left[\frac{\boldsymbol{F} \cdot \boldsymbol{P}}{F\,P}\right] = \cos^{-1}\left[\frac{56}{(5)(13)}\right] = \cos^{-1}\left[\frac{56}{65}\right] = \cos^{-1}(0.862) = 30.51° \tag{2-76}$$

Finally, confirm the dot product result in Eq. (2-75) using angle θ and scalar calculations:

$$\boldsymbol{F} \cdot \boldsymbol{P} = FP\cos(\theta) = (5)(13) \text{ kN}^2 \cos(30.51°) = 65 \text{ kN}^2(0.862) = 56 \text{ kN}^2 \tag{2-77}$$

4. *Finalize*: You computed the dot product using two different equations and obtained the same scalar result. You also computed the angle θ between the two force vectors using two different approaches, and the results are in agreement.

Part (b): Find the components of each vector along (parallel to) and perpendicular to the other vector.

1. *Conceptualize*: Find vectors $\boldsymbol{F}_{//}$ and $\boldsymbol{F}_{\perp}$, and also $\boldsymbol{P}_{//}$ and $\boldsymbol{P}_{\perp}$. These vectors are sketched (to appropriate scale) in Fig. 2-31.

2. *Categorize*: Use the dot product followed by vector subtraction. See Eq. (2-64).

FIGURE 2-31

Parallel and perpendicular vector components

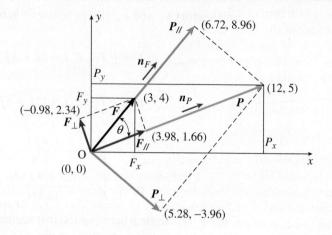

3. *Analyze*: First use Eqs. (2-62) and (2-63) to find the parallel components $F_{/\!/}$ and $P_{/\!/}$ (vector F along P and vector P along F, respectively):

$$F_{/\!/} = (F \cdot n_P)n_P = [(3i + 4j) \text{ kN} \cdot (0.923i + 0.385j)](0.923i + 0.385j)$$
$$= [3(0.923) + 4(0.385)](0.923i + 0.385j) \text{ kN} = (4.309)(0.923i + 0.385j) \text{ kN}$$
$$= 3.98i + 1.66j \text{ kN}$$

$$P_{/\!/} = (P \cdot n_F)n_F = [(12i + 5j) \text{ kN} \cdot (0.6i + 0.8j)](0.6i + 0.8j)$$
$$= [12(0.6) + 5(0.8)](0.6i + 0.8j) \text{ kN} = (11.2)(0.6i + 0.8j) \text{ kN}$$
$$= 6.72i + 8.96j \text{ kN} \qquad \qquad \blacktriangleleft \text{(2-78)}$$

Next, use vector subtraction [Eq. (2-64) and Fig. 2-29] to find the perpendicular components $F_\perp$ and $P_\perp$ (vector F component perpendicular to P and vector P component perpendicular to F, respectively):

$$F_\perp = F - F_{/\!/} = (3i + 4j) \text{ kN} - (3.98i + 1.66j) \text{ kN} = (-0.98)i + (2.34)j \text{ kN}$$
$$P_\perp = P - P_{/\!/} = (12i + 5j) \text{ kN} - (6.72i + 8.96j) \text{ kN} = (5.28)i + (-3.96)j \text{ kN}$$
$\qquad \qquad \blacktriangleleft \text{(2-79)}$

4. *Finalize*: Parallel and perpendicular components are plotted to scale in Fig. 2-31. For parallel components, the angle between associated vectors should be zero, and for perpendicular vectors, it should be 90°. As a check on these computations, confirm that $F_{/\!/}$ is along P (and that $P_\perp$ is perpendicular to F) and that $P_{/\!/}$ is along F (and that $F_\perp$ is perpendicular to P) by using Eq. (2-74) to show that:

$$\cos^{-1}\left[\frac{P \cdot F_{/\!/}}{PF_{/\!/}}\right] = 0°, \quad \cos^{-1}\left[\frac{F \cdot P_\perp}{FP_\perp}\right] = 90°$$

$$\cos^{-1}\left[\frac{F \cdot P_{/\!/}}{FP_{/\!/}}\right] = 0°, \quad \cos^{-1}\left[\frac{P \cdot F_\perp}{PF_\perp}\right] = 90°$$

(2-80)

2.2 Moments and Couples: 2D, 3D

Forces may exert a turning effect on a rigid body, causing it to rotate about an axis. The product of the magnitude F of the force and the perpendicular distance d from the axis to the line of action of the force is called a *moment*. For example, to tighten or loosen a lug bolt on a wheel (Fig. 2-32), use a lug wrench to apply a force F at some distance d, creating moment M. In general, moment is a vector quantity, and its magnitude is $M = Fd$. The units of moment are typically pound-feet (lb-ft; USCS) or newton-meters (N · m, SI).

Moment About a Point

In two dimensions, the moment M is usually defined with respect to, or about, some point [such as point O (Fig. 2-32)]. The magnitude of the moment is M, and the moment is taken about an axis through point O that is normal to the plane formed by force F, moment arm d, and reference point O. Actually, moment M is a *vector* normal to the plane, as shown in Fig. 2-33a. If force F is applied at A with line of action L_2, moment M is a *sliding* vector along the C–C axis (shown as an arrow with two heads) whose sense is dependent upon the direction that F tends to rotate the body. The C–C axis is the moment axis in the figure, and the sense of M is determined by the *right-hand rule*: Point the fingers of your right hand in the direction of force F, and curl your fingers about the moment axis through the point of interest (point O); your thumb then points in the direction of the associated moment vector (Fig. 2-33b). In Fig. 2-33, moment M is CCW and the moment vector is shown as a positive moment in this sign convention. Moment M also can be represented in two dimensions as the curved dashed arrow about axis C–C shown in Fig. 2-33a.

Cross Product   In many instances, it is more convenient to use vector operations to describe the moment of a force about a point. Define the moment M to be the cross product of a position vector r_{OA} from point O to some point A along line L_2 and force F as

$$M = r_{OA} \times F \qquad\qquad \textbf{(2-81)}$$

The magnitude M of moment M is computed using scalar operations as

$$M = F\, r_{OA} \sin(\alpha_A) = Fd \qquad\qquad \textbf{(2-82)}$$

This expression is consistent with the initial definition of moment shown in Fig. 2-32.

   Any point along the line of action of force F (such as alternate point B in Fig. 2-33) can be used, so an equivalent expression for moment M is

$$M = r_{OB} \times F \qquad\qquad \textbf{(2-83)}$$

where r_{OB} is a position vector from point O to some other point B that also lies on line of action L_2. Observe that moment arm distance d is unchanged and is now computed as $d = r_{OB} \sin(\alpha_B)$. So the magnitude of the moment obtained from either Eq. (2-81) or Eq. (2-83) is the same.

   Finally, note that the order of the terms in the cross-product expressions [Eqs. (2-81) and (2-83)] is important because $F \times r$ produces a moment vector of opposite sign to that obtained using $r \times F$.

   In two dimensions, the moment axis is normal to the plane and all forces lie in the plane containing the reference point about which you wish to compute moments. For these types of problems, a scalar approach [Eq. (2-82)] is often sufficient, as illustrated in Example 2-5.

FIGURE 2-32

Moment M of force F about point O

Jaggat Rashidi/Shutterstock.com

FIGURE 2-33

(a) Moment vector M in two dimensions; (b) right-hand rule

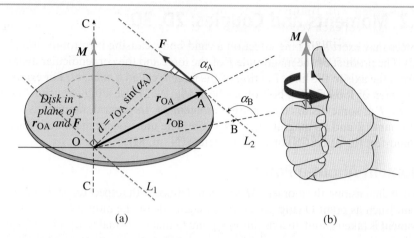

(a)   (b)

FIGURE 2-34

Force and moment vectors on 3D body

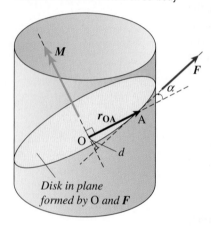

In three-dimensional problems, however, a vector solution approach [Eqs. (2-81) and (2-83)] is usually required because finding perpendicular distance d between the moment axis or reference point O and the line of action of a force can be a challenging task. However, Eq. (2-62) can be used to find $r_{OA//}$, and Eq. (2-64) is used to find $r_{OA\perp}$ where $d = |r_{OA\perp}| = r_{OA\perp}$; this option is illustrated when Example 2-4 is extended and in Example 2-5. Note also that $d = r_{OB\perp}$.

Now consider a body acted upon by a force F having a line of action through point A (Fig. 2-34). Use Eq. (2-81) to compute the moment M of force F about a moment axis through point O. Point O and force vector F form an inclined plane in Fig. 2-34, and moment vector M is perpendicular to that plane. Position vector r_{OA} extends from point O to any point along the line of action of force vector F, and moment arm d is the perpendicular distance from O to the line of action of F. Assume that the three-dimensional Cartesian components of force F and position vector r_{OA} are known (Fig. 2-35):

$$F = F_x i + F_y j + F_z k \quad \text{and} \quad r_{OA} = r_{OAx} i + r_{OAy} j + r_{OAz} k \quad \textbf{(2-84)}$$

The cross product is evaluated using Eq. (2-81) in determinant form as

$$M = r_{OA} \times F = \begin{vmatrix} i & j & k \\ r_{OAx} & r_{OAy} & r_{OAz} \\ F_x & F_y & F_z \end{vmatrix} \quad \textbf{(2-85)}$$

FIGURE 2-35

Cartesian components of position and force vectors

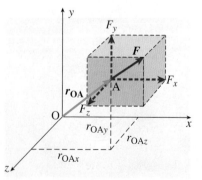

The expressions in Eq. (2-85) are obtained by determinant expansion by minors using the cofactors of the first row:

$$M = \begin{vmatrix} r_{OAy} & r_{OAz} \\ F_y & F_z \end{vmatrix} i - \begin{vmatrix} r_{OAx} & r_{OAz} \\ F_x & F_z \end{vmatrix} j + \begin{vmatrix} r_{OAx} & r_{OAy} \\ F_x & F_y \end{vmatrix} k \quad \textbf{(2-86)}$$

The resulting expression for M is

$$M = (r_{OAy}F_z - r_{OAz}F_y)i + (r_{OAz}F_x - r_{OAx}F_z)j + (r_{OAx}F_y - r_{OAy}F_x)k \quad \textbf{(2-87)}$$

The individual terms in parentheses in the determinant expansion in Eq. (2-87) can be confirmed by taking moments about the x, y, and z axes (force component times distance component, one at a time) in Fig. 2-35.

Example 2-4 is continued below so that you can gain more experience with the cross product and moments about a point. This example is concerned with finding the dot product of two vectors F and P and the components of each vector along (parallel to) and perpendicular to the other vector. Now, of interest are: (1) the cross product of vectors F and P; (2) the moment about some point A, of vectors F and P; and (3) perpendicular distances d from point A to the lines of action of both F and P. Finally, you will use scalar operations to confirm the moment magnitude for a specific point A.

Example 2-4 extended

Two force vectors F and P are applied at point O (Figs. 2-30 and 2-36). The magnitudes of each force vector are $F = 5$ kN and $P = 13$ kN.

(c) Find the cross product of vectors F and P (Fig. 2-36).

(d) Find moment M_A of vectors F and P about some point A (Fig. 2-37).

(e) Find perpendicular distances from point A to the lines of action of both F and P; then use scalar operations to confirm the magnitude M_A for a specific point A.

Solution:

Part (c): Find the cross product of vectors F and P.

3. *Analyze*: Force vectors F and P are (see Fig. 2-36):

$$F = 3i + 4j \text{ kN} = \begin{bmatrix} 3 \\ 4 \end{bmatrix} \text{ kN and } P = 12i + 5j \text{ kN} = \begin{bmatrix} 12 \\ 5 \end{bmatrix} \text{ kN}$$

To use the determinant expansion in Eq. (2-86) to evaluate the cross products of these two vectors, first expand both vectors to 3×1 to include a zero component in the z-direction, so $F_z = P_z = 0$. The determinant expansion to compute $P \times F$ is

$$P \times F = \begin{vmatrix} i & j & k \\ P_x & P_y & 0 \\ F_x & F_y & 0 \end{vmatrix} = (P_x F_y - P_y F_x)k = [(12)(4)-(5)(3)]k \text{ kN}^2 = 33k \text{ kN}^2 \quad \text{◀ (2-88)}$$

Cross product $P \times F$ is a CCW vector normal to the plane formed by P and F in Fig. 2-36. The magnitude of the cross product is $|P \times F| = PF\sin\theta$. A negative result (CW in Fig. 2-36) is obtained if the vectors are reordered:

$$F \times P = -(P \times F) = \begin{vmatrix} i & j & k \\ F_x & F_y & 0 \\ P_x & P_y & 0 \end{vmatrix} = (F_x P_y - F_y P_x)k$$

$$= [(3)(5) - (4)(12)]k \text{ kN}^2 = -33k \text{ kN}^2 \quad \text{◀ (2-89)}$$

4. *Finalize*: This confirms that the sequence of the vectors in the cross product expression is important.

Part (d): Find moment M_A of vectors F and P about some point A (Fig. 2-37).

1, 2. *Conceptualize, Categorize*: Point O is a convenient reference point to use for the position vector because it is on the line of action of both force vectors F

FIGURE 2-36

Force vectors *F* and *P* from
Example 2-4

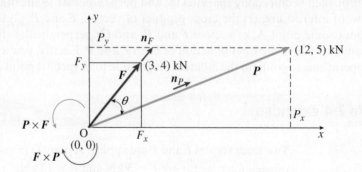

and *P*, so use position vector r_{AO}. Moment M_A is the sum of the moments of the individual forces so, by the distributive law for cross products,

$$M_A = r_{AO} \times P + r_{AO} \times F = r_{AO} \times (P + F) = r_{AO} \times R \qquad \textbf{(2-90)}$$

First define position vector $r_{AO} = (-x_A)i + (-y_A)j$. Resultant *R* is the sum of the two force vectors and can be represented graphically using the parallelogram law for vector addition:

$$R = F + P = (3 + 12)i + (4 + 5)j \ \text{kN} = 15i + 9j \ \text{kN} \qquad \textbf{(2-91)}$$

3. *Analyze*: It follows that moment M_A is obtained using a determinant expression as

$$M_A = \begin{vmatrix} i & j & k \\ -x_A & -y_A & 0 \\ R_x & R_y & 0 \end{vmatrix} = \begin{vmatrix} i & j & k \\ -x_A & -y_A & 0 \\ 15 \ \text{kN} & 9 \ \text{kN} & 0 \end{vmatrix} = (15y_A - 9x_A)k \ \text{kN·m} \quad \Longleftarrow \textbf{(2-92)}$$

4. *Finalize*: Moment M_A is a vector normal to the plane containing point A and vector resultant *R* and may be CCW (+) or CW (−), depending on the location of point A in Fig. 2-37. Now ask: Under what conditions would moment $M_A = 0$? An obvious answer is: $M_A = 0$ if point A about which you are finding the moments of forces *F* and *P* is located at the intersection of these two concurrent vectors, point O.

However, Eq. (2-92) shows that, if point A is located anywhere along a line with $y_A = \dfrac{9}{15}x_A = \dfrac{3}{5}x_A$, $M_A = 0$. See from Eq. (2-91) that this is the equation of the line of action of resultant vector *R*, so if point A is located anywhere along this line, moment $M_A = 0$. This is confirmed for a specific selection of point A in the next part.

Part (e): Find perpendicular distances from point A to the lines of action of both *F* and *P*; then use scalar operations to confirm the magnitude M_A for a specific point A.

1, 2. *Conceptualize, Categorize*: If you place point A along the line of action of resultant vector *R* (Fig. 2-38), the moment magnitude M_A will equal zero. Assume that $x_A = 10$ m and $y_A = 6$ m. From Eqs. (2-72), unit vectors along *F* and *P* are

$$n_F = \frac{F}{F} = 0.6i + 0.8j \quad \text{and} \quad n_P = \frac{P}{P} = 0.923i + 0.385j \qquad \text{(2-93)}$$

3. *Analyze*: Using Eqs. (2-62) through (2-64), first find components of position vector r_{AO} along (or parallel to) vectors *P* and *F* using dot products, then use vector subtraction to find the components of r_{AO} which are perpendicular to *P* and *F* (i.e., $r_{AO\perp P}$ and $r_{AO\perp F}$). The magnitudes of these perpendicular components are the perpendicular distances d_P and d_F in Fig. 2-38. The vector operations and results are

$$d_P = \left| r_{AO\perp P} \right| = \left| r_{AO} - r_{AO//P} \right| = \left| r_{AO} - (r_{AO} \cdot n_P)n_P \right| \text{ and}$$

$$d_F = \left| r_{AO} - (r_{AO} \cdot n_F)n_F \right| \qquad \text{(2-94)}$$

where $r_{AO} = (-10)i + (-6)j$ m for the selection of point A along the line of action of resultant *R* (Fig. 2-38). Evaluating the expressions in Eq. (2-94) gives $d_P = 1.692$ m and $d_F = 4.4$ m.

4. *Finalize*: The following scalar calculation of M_A shows that $M_A = 0$ as expected since the resultant *R* has a line of action through the point about which you are summing moments:

$$M_A = \left| M_A \right| = |P|d_P - |F|d_F = (13\,\text{kN})(1.692\,\text{m}) - (5\,\text{kN})(4.4\,\text{m}) = 0 \quad \text{(2-95)}$$

As a final note, observe that the magnitude of the cross product has a geometric interpretation: It is the positive area of the parallelogram in Fig. 2-38 having *P* and *F* as its sides. From Eq. (2-88), this area is 33 kN^2.

FIGURE 2-38

Perpendicular distances from A to vectors *P* and *F*

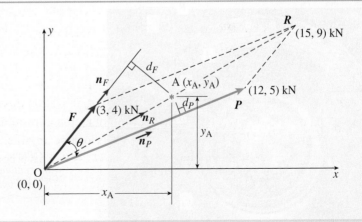

Varignon's Theorem

Varignon's theorem states that moment M_O about point O in Fig. 2-39 is equal to the sum of the moments of the components of the force about the same point. Cartesian components of force (F_x, F_y) are commonly used in statics, so the calculation of moment M_O can be simplified by using rectangular components of both force and distance to find M_O in many two-dimensional (and some three-dimensional) problems. Moment M_O is a vector in the z-direction, but its magnitude is

$$M_O = |M_O| = F_x d_y + F_y d_x \qquad \text{(2-96)}$$

In Fig. 2-39, perpendicular distances are $d_x = d\cos(\theta)$, $d_y = d\sin(\theta)$.

If Cartesian components of force are not used (Fig. 2-40), Varignon's theorem in vector format is required:

$$M_O = r_{OA} \times R = r_{OA} \times (F_1 + F_2) = r_{OA} \times F_1 + r_{OA} \times F_2 \qquad \text{(2-97)}$$

Forces F_1 and F_2 are the nonrectangular components of resultant R (where $R = F_1 + F_2$), and position vector r_{OA} goes from point O to any point along the line of action of R. Equation (2-97) illustrates the *distributive law* for cross products [see also Eq. (2-90)]. The magnitude of moment M_O is computed using scalar operations as

$$M_O = |M_O| = R d_R = F_1 d_1 + F_2 d_2 \qquad \text{(2-98)}$$

The challenge of computing perpendicular distances d_R, d_1, and d_2 remains, but this is accomplished using vector operations as described in Eq. (2-94). Both scalar and vector solution approaches are used in Example 2-5 to find moments about a point.

Varignon's theorem is now extended to three dimensions for use with either rectangular or nonrectangular components of forces or systems of concurrent forces. For the concurrent forces F_1, F_2, and F_3 applied at point A in Fig. 2-41a, the sum of the moments about an axis through some point O is

$$M_O = r_{OA} \times F_1 + r_{OA} \times F_2 + r_{OA} \times F_3 = r_{OA} \times (F_1 + F_2 + F_3) \qquad \text{(2-99)}$$

which is equivalent to

$$M_O = r_{OA} \times R \qquad \text{(2-100)}$$

FIGURE 2-39

Moment M_O of force F about point O using Varignon's theorem

$$M_O = Fd = F_x d_y + F_y d_x$$
$$d = \sqrt{d_x^2 + d_y^2}$$

Jaggat Rashidi/Shutterstock.com

FIGURE 2-40

Moment M_O of force R with nonrectangular components using Varignon's Theorem

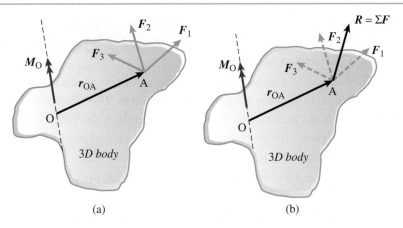

FIGURE 2-41

(a,b) Varignon's theorem in three dimensions

(a)   (b)

Vector **R** is the resultant or sum of the three forces (Fig. 2-41b). Equation (2-99) uses the distributive law for cross products. Equations (2-99) and (2-100) state that the moment of a system of concurrent forces about an axis through some point O can be computed using either the individual moments of each force or the moment of the resultant of the system of concurrent forces. Computation of moments about an axis through a point for both two- and three-dimensional bodies is illustrated in the following examples.

Example 2-5

In Example 1-3, you evaluated the action of force vectors F_1 and F_2 applied to an automobile parking brake at point P (see Fig. 2-42). The force magnitudes are $F_1 = 13$ lb and $F_2 = 15$ lb. This example extends Example 1-3 to find the moment of forces F_1 and F_2 about point O using both scalar and vector solution methods.

FIGURE 2-42

(a, b) Forces applied to automobile parking brake

(a)   (b)

Solution:

Use the four-step problem-solving approach to find the moment of forces about point O using both scalar and vector solution methods.

FIGURE 2-43

Cartesian components of F_1, F_2, and R

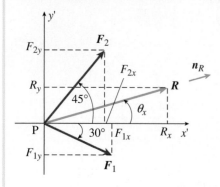

1. *Conceptualize*: The resultant of forces F_1 and F_2 (renamed R here) was computed in Example 1-3 (see Fig. 2-43), as were their Cartesian components.

2. *Categorize*: Use both scalar and vector (cross product) operations to find the moment at base point O. The separate solutions provide an opportunity to confirm the validity of Varignon's theorem. Begin by summarizing relevant results from Example 1-3.

The x and y components of vectors F_1 and F_2 are

$$F_1 = F_{1x}i + F_{1y}j = (11.26)i + (-6.5)j \text{ lb} = \begin{pmatrix} 11.26 \\ -6.5 \end{pmatrix} \text{ lb} \tag{2-101}$$

$$F_2 = F_{2x}i + F_{2y}j = (10.61)i + (10.61)j \text{ lb} = \begin{pmatrix} 10.61 \\ 10.61 \end{pmatrix} \text{ lb}$$

The *sum* of the two vectors is the resultant R whose magnitude is $R = 22.25$ lb:

$$R = F_1 + F_2 = \left(F_{1x} + F_{2x}\right)i + \left(F_{1y} + F_{2y}\right)j$$

$$= (21.86)i + (4.11)j \text{ lb} = \begin{pmatrix} 21.86 \\ 4.11 \end{pmatrix} \text{ lb} \tag{2-102}$$

where $R_x = 21.86$ lb and $R_y = 4.11$ lb. The direction of resultant R is given by direction cosine $\theta_x = \cos^{-1}\left(\dfrac{R_x}{R}\right) = 10.64°$ measured CCW from the x axis. A unit vector along resultant R is

$$n_R = \frac{R}{R} = \frac{(21.86)i + (4.11)j \text{ lb}}{22.25 \text{ lb}} = (0.983)i + (0.185)j = \begin{pmatrix} 0.983 \\ 0.185 \end{pmatrix} \tag{2-103}$$

A position vector from base point O to force application point P is

$$r_{OP} = (-10)i + (12)j \text{ in.} = \begin{pmatrix} -10 \\ 12 \end{pmatrix} \text{in.} \quad \text{and} \quad r_{OP} = 15.62 \text{ in.} \tag{2-104}$$

A unit vector n_{OP} along line OP is

$$n_{OP} = \frac{r_{OP}}{r_{OP}} = \left(\frac{-10}{15.62}\right)i + \left(\frac{12}{15.62}\right)j = (-0.64)i + (0.768)j = \begin{pmatrix} -0.64 \\ 0.768 \end{pmatrix} \tag{2-105}$$

3. *Analyze*: The moment M_O can be computed using a variety of different approaches; in all cases, moment M_O is clockwise (CW) and so is negative by the right-hand rule (RHR). Some possible approaches are presented in parts (a) to (f) below.

Part (a): Varignon's theorem.

$R_x = F_{1x} + F_{2x}$ and $R_y = F_{1y} + F_{2y}$: $M_O = (R_x)(12 \text{ in.}) + (R_y)(10 \text{ in.})$ (see Fig. 2-44). Using scalar calculations, the magnitude of moment M_O is

$$M_O = (21.86 \text{ lb})(12 \text{ in.}) + (4.11 \text{ lb})(10 \text{ in.}) = 303 \text{ lb-in.} \tag{2-106}$$

FIGURE 2-44

M_O using Varignon's theorem

FIGURE 2-45

M_O using component R_x only

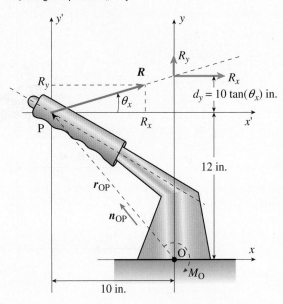

Part (b): Vector $M_O = r_{OP} \times R$, then $M_O = |M_O|$ (see Figs. 2-42 and 2-43). Recall that $i \times j = k$, $j \times i = -k$, and $i \times i = j \times j = 0$. Using the cross product to find M_O proceed as

$$M_O = r_{OP} \times R = \left[(-10)i + (12)j\right] \text{ in.} \times \left[(21.86)i + (4.11)j\right] \text{ lb}$$
$$= (-41.1 - 262.3)k \text{ lb-in.} = -303.4k \text{ lb-in.} \tag{2-107}$$

and

$$M_O = 303.4 \text{ lb-in.}$$

Part (c): Slide R so that R_y is aligned with the y axis; find d_y then $M_O = (R_x)(12 \text{ in.} + d_y)$ (see Fig. 2-45). Component R_y has a line of action through point O, so it does not contribute to moment M_O.

$$M_O = R_x(12 \text{ in.} + d_y) = (21.86 \text{ lb})\left[12 \text{ in.} + (10 \text{ in.})\tan(\theta_x)\right]$$
$$= (21.86 \text{ lb})(12 \text{ in.} + 1.878 \text{ in.}) = 303 \text{ lb-in} \tag{2-108}$$

Part (d): Slide R so that R_x is aligned with the x axis; find d_x then $M_O = (R_y)(10 \text{ in.} + d_x)$ (see Fig. 2-46) as

$$M_O = (R_y)(10 \text{ in.} + d_x) = (4.11 \text{ lb})\left[10 \text{ in.} + (12 \text{ in.})/\tan(\theta_x)\right]$$
$$= (4.11 \text{ lb})(10 \text{ in.} + 63.8 \text{ in.}) = 303 \text{ lb-in} \tag{2-109}$$

Part (e): Use the dot product to find the projection of r_{OP} on unit vector n_R, then find $d_R = |r_{OP\perp}|$, which is the perpendicular distance from point O to the line of action of R [see Eqs. (2-62) through (2-64)]; then magnitude $M_O = (R)(d_R)$ (see Fig. 2-47).

FIGURE 2-46

M_O using component R_y only

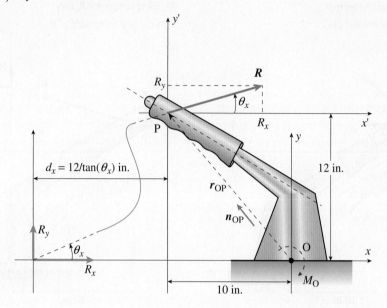

$$r_{OP//} = r_{OP} \cdot n_R = ((-10)i + (12)j) \text{ in.} \cdot ((0.983)i + (0.185)j)$$
$$= -7.613 \text{ in.}$$

so

$$d_R = r_{OP\perp} = \sqrt{r_{OP}^2 - r_{OP//}^2} = \sqrt{(15.62 \text{ in.})^2 - (-7.613 \text{ in.})^2}$$
$$= \sqrt{186.04} \text{ in.} = 13.64 \text{ in.}$$

and

$$M_O = (R)(d_R) = (22.25 \text{ lb})(13.64 \text{ in.}) = 303 \text{ lb-in.} \qquad \text{(2-110)}$$

In summary, a vector expression for use in computing d_R is

$$d_R = |r_{OP\perp}| = |r_{OP} - r_{OP//}| = |r_{OP} - (r_{OP} \cdot n_R)n_R| \qquad \text{(2-111)}$$

Alternatively, find angle $\gamma = \tan^{-1}\left(\dfrac{12}{10}\right) = 50.194°$; then use the law of sines

(Fig. 2-47) to show that $d_R = \dfrac{r_{OP}}{\sin(90°)} \sin(\gamma + \theta_x) = 13.64$ in.

Part (f): Find perpendicular component of R, i.e., $R_\perp$ [see Eqs. (2-62) through (2-64)]; then $M_O = (R_\perp)(r_{OP})$ (see Fig. 2-48).

$$R_{//} = (R \cdot n_{OP})n_{OP} = (6.94)i + (-8.33)j \text{ lb}$$
$$R_\perp = R - R_{//} = (14.92)i + (12.44)j \text{ lb} \qquad \text{(2-112)}$$

Note that $R = |R| = 22.25$ lb, $R_{//} = |R_{//}| = 10.84$ lb, and $R_\perp = |R_\perp| = 19.43$ lb.

Using the Pythagorean theorem, confirm that $\sqrt{10.84^2 + 19.43^2}$ lb $= 22.25$ lb.

Finally, using scalar quantities, moment M_O is once again:

$$M_O = (R_\perp)(r_{OP}) = (19.43 \text{ lb})(15.62 \text{ in.}) = 303 \text{ lb-in.} \qquad \text{(2-113)}$$

FIGURE 2-47

M_O using perpendicular distance d_R

FIGURE 2-48

M_O using perpendicular component of R

4. *Finalize*: The six different approaches used in parts (a) to (f) to find M_O all produce the same result. For this 2D problem with well-defined geometry, it is likely that the solution approach presented in part (a) [using Cartesian components of resultant R and Varignon's theorem] will be the preferred method. The resulting moment vector M_O is perpendicular to the x-y plane and is in the $(-z)$ direction, so it can be expressed as $M_O = -(M_O)k$. Recall that the RHR is used to determine the sign of the vector result.

Moments About an Axis

At this point, you would like to know the moment M_β of force F about an axis β through point O (see Fig. 2-49). First find a unit vector n_β along the β axis, then find the component of moment $M_O = r_{OA} \times F$ along the β axis using a dot product:

$$M_\beta = \left(M_O \cdot n_\beta\right)n_\beta = \left(r_{OA} \times F \cdot n_\beta\right)n_\beta \qquad \textbf{(2-114)}$$

The scalar expression $r_{OA} \times F \cdot n_\beta$ is the magnitude of the component of M_O along the β axis. This expression is known as the *triple scalar product* and is evaluated using the following determinant:

$$M_\beta = \left|M_\beta\right| = r_{OA} \times F \cdot n_\beta = \left|\left|\begin{array}{ccc} r_{OAx} & r_{OAy} & r_{OAz} \\ F_x & F_y & F_z \\ l_\beta & m_\beta & n_\beta \end{array}\right|\right| \qquad \textbf{(2-115)}$$

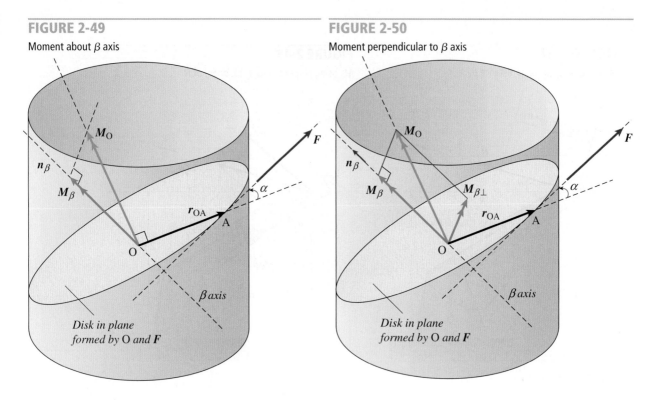

FIGURE 2-49
Moment about β axis

FIGURE 2-50
Moment perpendicular to β axis

Entries in the third row $\left(l_\beta, \, m_\beta, \, n_\beta \right)$ are the direction cosines of unit vector $\boldsymbol{n}_\beta$. [Moment component $\boldsymbol{M}_\beta$ can be labeled $\boldsymbol{M}_{O//}$ if the notation used in Eqs. (2-62) through (2-64) is repeated here.]

The component of moment $\boldsymbol{M}_O$ perpendicular to the β axis (Fig. 2-50) is computed as

$$\boldsymbol{M}_{\beta\perp} = \boldsymbol{M}_O - \boldsymbol{M}_\beta = \boldsymbol{r}_{OA} \times \boldsymbol{F} - \left(\boldsymbol{r}_{OA} \times \boldsymbol{F} \cdot \boldsymbol{n}_\beta \right) \boldsymbol{n}_\beta \qquad \textbf{(2-116)}$$

Example 2-6

A hand winch is used to hoist and trim the sails on a sailboat (Fig. 2-51). A force with magnitude $F = 70$ N is applied at A on the winch handle and at the instant shown in Fig. 2-52, has a line of action from A to B. Find moment $\boldsymbol{M}_O$ of force $\boldsymbol{F}$ at point O at the base of the winch and also moment $\boldsymbol{M}_D$ at the end of the winch arm using both scalar and vector solution methods. Find the components of $\boldsymbol{M}_D$ about and perpendicular to the axis of the winch arm DC.

Solution:

Use the four-step problem-solving approach to find moments $\boldsymbol{M}_O$ and $\boldsymbol{M}_D$ and the components of $\boldsymbol{M}_D$ about and perpendicular to the axis of the winch arm DC.

FIGURE 2-51

Hand winch on sailboat

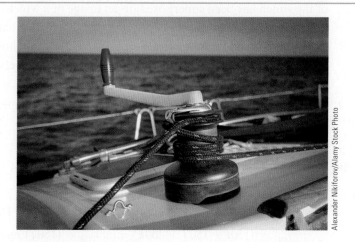

Alexander Nikiforov/Alamy Stock Photo

FIGURE 2-52

Force **F** applied to winch handle at A

1. *Conceptualize*: A 3D Cartesian coordinate system with origin at O at the base of the winch has been superposed on an image of a winch. The winch shaft OD is vertical and is aligned with the y axis. The winch handle CA is also vertical and is parallel to the y axis. Winch arm CD lies in a horizontal plane parallel to the x-z plane and is assumed to be a straight line between C and D with a projection OC' on the x-z plane.

2. *Categorize*: Use both scalar and vector (cross product) operations to find the moments M_O and M_D.

3. *Analyze*: Proceed with the following steps.

(a) **Find key position and unit vectors:**

Position vectors:

$$r_{OA} = (150i + 290j + 200k) \text{ mm}, \quad r_{OB} = (180j + 160k) \text{ mm}$$
$$r_{OC} = (150i + 150j + 200k) \text{ mm}, \quad r_{OD} = (150j) \text{ mm}$$
$$r_{AB} = r_{OB} - r_{OA} = (-150i - 110j - 40k) \text{ mm} \tag{2-117}$$
$$r_{DC} = r_{OC} - r_{OD} = (150i + 200k) \text{ mm}$$
$$r_{DA} = r_{OA} - r_{OD} = (150i + 140j + 200k) \text{ mm}$$

Unit vectors:

$$n_{OA} = r_{OA}/r_{OA} = (0.392i + 0.757j + 0.522k)$$
$$n_{AB} = r_{AB}/r_{AB} = (-0.788i - 0.578j - 0.21k) \tag{2-118}$$
$$n_{DC} = r_{DC}/r_{DC} = (0.6i + 0.8k)$$

(b) **Find force vector F:**

$$F = Fn_{AB} = (70 \text{ N})n_{AB} = (-55.19i - 40.47j - 14.72k) \text{ N} \tag{2-119}$$

(c) **Find moments M_O and M_D using cross product expressions:**

Moment M_O (see Fig. 2-53):

$$M_O = r_{OA} \times F = (150i + 290j + 200k) \text{ mm} \times (-55.19i - 40.47j - 14.72k) \text{ N}$$

$$= \begin{vmatrix} i & j & k \\ r_{OAx} & r_{OAy} & r_{OAz} \\ F_x & F_y & F_z \end{vmatrix} = \begin{vmatrix} i & j & k \\ 0.15 & 0.29 & 0.20 \\ -55.19 & -40.47 & -14.72 \end{vmatrix} \text{ N} \cdot \text{m} \tag{2-120}$$

$$M_O = (3.83i - 8.83j + 9.93k) \text{ N} \cdot \text{m}, \quad M_O = 13.83 \text{ N} \cdot \text{m} \quad \Leftarrow$$

Moment M_D:

$$M_D = r_{DA} \times F = (150i + 140j + 200k) \text{ mm} \times (-55.19i - 40.47j - 14.72k) \text{ N}$$

$$= \begin{vmatrix} i & j & k \\ 0.15 & 0.14 & 0.20 \\ -55.19 & -40.47 & -14.72 \end{vmatrix} \text{ N} \cdot \text{m}$$

$$M_D = (6.03i - 8.83j + 1.66k) \text{ N} \cdot \text{m}, \quad M_D = 10.82 \text{ N} \cdot \text{m} \quad \Leftarrow \tag{2-121}$$

FIGURE 2-53

Force vector **F** at A and
moment **M**₀ at O

d. **Confirm the magnitude of M_O using scalar operations and *Varignon's Theorem*:** The moment arm distances along the Cartesian coordinate axes from point O to point A are $d_x = 150$ mm, $d_y = 290$ mm, and $d_z = 200$ m (Fig. 2-53). Write separate expressions for the components of M_O about each axis (with signs determined by using the right-hand rule) then use the Pythagorean theorem to confirm magnitude M_O in Eq. (2-120):

$$M_{Ox} = -F_y d_z + F_z d_y = 3.826 \text{ N·m}$$
$$M_{Oy} = -F_z d_x + F_x d_z = -8.83 \text{ N·m}$$
$$M_{Oz} = -F_x d_y + F_y d_x = 9.934 \text{ N·m} \qquad \textbf{(2-122)}$$
$$\text{and } M_O = \sqrt{M_{Ox}^2 + M_{Oy}^2 + M_{Oz}^2} = 13.83 \text{ N·m}$$

Similar expressions for the Cartesian components of moment M_D confirm the magnitude given in Eq. (2-121).

$$M_{Dx} = -F_y d_z + F_z(d_y - 0.15\ \text{m}) = 6.034\ \text{N} \cdot \text{m}$$
$$M_{Dy} = -F_z d_x + F_x d_z = -8.83\ \text{N} \cdot \text{m}$$
$$M_{Dz} = -F_x(d_y - 0.15\ \text{m}) + F_y d_x = 1.656\ \text{N} \cdot \text{m} \qquad \textbf{(2-123)}$$
$$M_D = \sqrt{M_{Dx}^2 + M_{Dy}^2 + M_{Dz}^2} = 10.82\ \text{N} \cdot \text{m}$$

e. **You can also use the dot product to find angle θ, then the law of sines to find the perpendicular distance d_F from O to the line of action of force F (Fig. 2-54):**
It follows that $M_O = (F)(d_F)$ (a similar procedure can be used to find M_D):

$$\theta = \cos^{-1}(n_{AB} \cdot n_{OA}) = 31.066°$$

then

$$d_F = r_{OA}\left(\frac{\sin\theta}{\sin 90°}\right) = 197.6\ \text{mm}$$

and

$$M_O = (F)(d_F) = (70\ \text{N})(0.1976\ \text{m}) = 13.83\ \text{N} \cdot \text{m} \qquad \textbf{(2-124)}$$

f. **Last, find the components of M_D about (parallel to) and perpendicular to the axis of the winch arm DC:**

FIGURE 2-54

Distance d_F from point O to line AB

First, use the *triple scalar product* to find the component of M_D parallel to the axis DC of the arm of the winch:

$$M_{D//} = (M_D \cdot n_{DC})n_{DC} = (r_{DA} \times F \cdot n_{DC})n_{DC} = (4.945 \text{ N} \cdot \text{m})n_{DC}$$

$$M_{D//} = (2.967i + 3.956k) \text{ N} \cdot \text{m}, \; M_{D//} = |M_{D//}| = 4.945 \text{ N} \cdot \text{m} \qquad \text{(2-125)}$$

This is the *torsional*, or twisting, moment (or torque) that acts on arm DC. The component of M_D that is perpendicular to arm DC is referred to as the *bending moment*:

$$M_{D\perp} = M_D - M_{D//} = (3.067i - 8.83j - 2.3k) \text{ N} \cdot \text{m} \qquad \text{(2-126)}$$

Equation (2-126) reveals that the arm DC is experiencing *biaxial bending*.

4. *Finalize*: Three-dimensional moment calculations usually require vector algebra unless the geometry is quite simple, so scalar calculations can be used. Varignon's theorem in part (d) provided an economical solution approach for moment magnitudes M_O and M_D. Only one position of the winch was studied under one loading condition, but a number of different positions and applied loadings on the winch must be evaluated to determine the worst case that controls the overall design. Static analyses of the type presented in this example are an important starting point in such engineering analyses.

FIGURE 2-55

Components of M_D parallel to and perpendicular to axis of winch arm DC

FIGURE 2-56

Couple

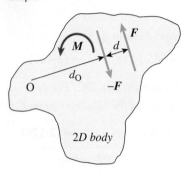

2D body

Couple and Force-Couple Systems

A couple is formed from two force vectors F that are equal in magnitude, opposite in sense, and acting along parallel lines of action on a body (Fig. 2-56). The forces are a distance d apart and are positioned at distance d_O from some point O. The sum of the forces in every direction is zero, but the pair of forces exerts a counterclockwise turning effect or moment M on the body about the moment axis through point O. The magnitude of the couple is computed as

$$M = F(d_O + d) - Fd_O = Fd \qquad \text{(2-127)}$$

The moment of the two forces about point O depends only on the perpendicular distance d between the forces and not on the distance d_O from point O. As a result, the CCW couple magnitude in Eq. (2-127) is M and not M_O and is the same for any moment axis through any point on the body.

FIGURE 2-57

Couple applied to lug wrench

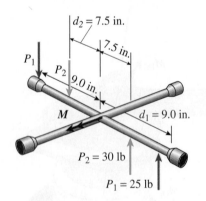

A practical example of a couple is shown in Fig. 2-57. Forces P are applied to a lug wrench to loosen a wheel bolt, and the couple magnitude is $M = 450$ lb-in. The motorist can apply the couple to the wrench as $(P_1)(2d_1) = (25\text{ lb})(18\text{ in.})$ or as $(P_2)(2d_2) = (30\text{ lb})(15\text{ in.})$. The two couples are said to be *equivalent* because *any* choice of force times distance (two possible choices of position of forces along one arm of the wrench are shown in Fig. 2-57) that results in a moment of 450 lb-in. has the same turning effect on the wheel bolt.

For some problems (especially three-dimensional problems), it is more convenient to use the vector cross product to find couple M. Consider the two-dimensional body in Fig. 2-58a (or the three-dimensional body in Fig. 2-58b) with parallel forces F, equal in magnitude and opposite in sense, applied along lines of action through points A and B. The moment about some point O is

$$M = r_{OA} \times F + r_{OB} \times (-F) = r_{OA} \times F + (r_{OA} - r_{BA}) \times (-F)$$

or

$$M = r_{BA} \times F \qquad \text{(2-128)}$$

The moment M depends only on the relative position of the two forces and not on the location of point O about which moments are taken. The couple is the same about all points on the body, so moment vector M is a *free* vector whose sign is established using the right-hand rule. The magnitude of moment M is $M = Fd$ where d is again the perpendicular distance between the lines of action of forces F (Fig. 2-58b). Couples can be combined using the rules for addition and subtraction of vectors.

Force-Couple System When considering the effect of force F acting on a body, you are often interested in the effect of that force at some other point on the body. You can use the concept of a couple to replace the original force acting at some point A (Fig. 2-59a) with an equivalent force and moment at another point of interest B (Fig. 2-59b). In transferring force F from point A to point B, you must also transfer the turning effect of the force (M_B). The process depicted in Fig. 2-59b allows you to do this correctly; *the two forces added at B* $(+F, -F)$ *have no resultant force and no moment*. The same transfer process is illustrated in 3D in Fig. 2-60 with original force F at point A being replaced with the force-couple system F and M_B at point B.

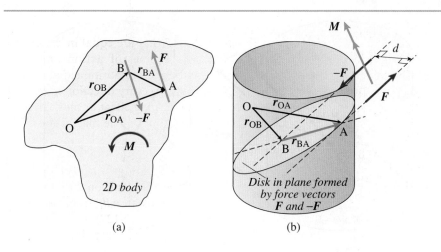

FIGURE 2-58

(a,b) Couple using vector notation

(a)   (b)

Note that the process shown in Figs. 2-59 and 2-60 also works in reverse. A force F and moment M_B acting at B can be moved by some distance d to a parallel line of action and still retain the turning effect of the force with respect to point B.

An example of a force-couple system is shown in Fig. 2-61. Downward force P applied to the handle of a socket wrench at A is replaced at B with

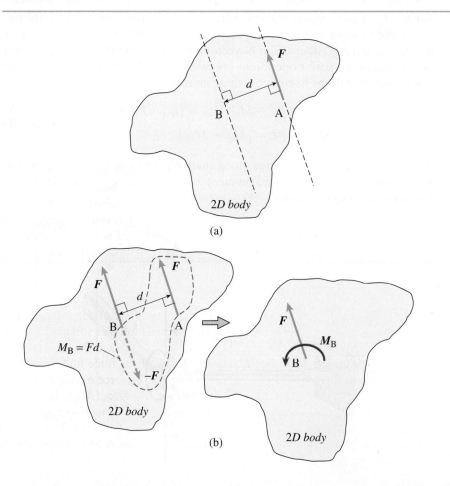

(a)

(b)

FIGURE 2-60

Force-couple system in 3D

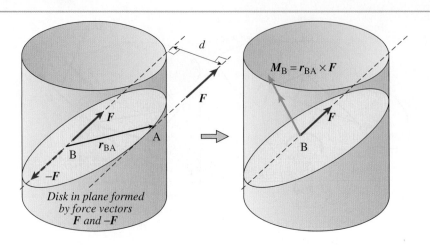

parallel force P and couple moment (or torque) T_B at B. You may want to know the equivalent force-couple system at C where the turning effect is applied to a bolt head. Replace force P at point A with downward force P at point C and couple moment M_C (Figs. 2-62 and 2-63). Couple vector M_C lies in the plane of the socket wrench (plane ABC) and acts normal to position vector r_{CA}.

The magnitude of the couple is $M_C = (P)(|r_{CA}|) = P\sqrt{d^2 + L^2}$, and moments along ($M_{C//}$) and perpendicular to ($M_{C\perp}$) shaft BC (see Fig. 2-63) are the *torsional* and *bending* moments, respectively, in shaft BC. These moments may be computed using either scalar or vector operations, but the simple geometry here suggests that scalar operations are easier to do. The recommended scalar and vector operations to compute these moment components are

$$M_{C//} = Pd = M_C \cdot n_{BC} = (r_{CA} \times P) \cdot n_{BC}$$
$$M_{C\perp} = PL = |M_C - M_{C//}|$$

(2-129)

where vector n_{BC} is a unit vector along shaft BC. Here, moment Pd is the *torsional* moment in shaft BC and moment PL is the *bending* moment that acts normal to the axis of the shaft.

FIGURE 2-61

Force-couple system at B on socket wrench

FIGURE 2-62

Force-couple system at C

FIGURE 2-63

Top view

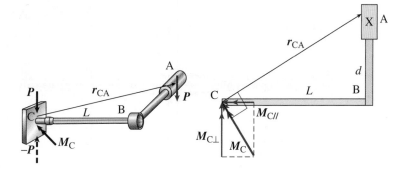

Example 2-7

A mountain bike rider going uphill (Fig. 2-64) applies two forces F by pulling on each of the two handlebar extenders DE (Fig. 2-65). Because of symmetry of the handlebars and the loading about the centerline at A (Fig. 2-65), consider the right half of the handlebar assembly only (assume the bars are fixed at the fork at A). Find the equivalent force-couple system at A. Assume that $F = 15$ lb, $d = 4$ in., $L_1 = 2$ in., $L_2 = 1.2$ in., and $L_3 = 8.5$ in.

Solution:

Use the four-step problem-solving approach to find the equivalent force-couple system at A.

FIGURE 2-64

Mountain bike rider going uphill

FIGURE 2-65

Mountain bike handlebars

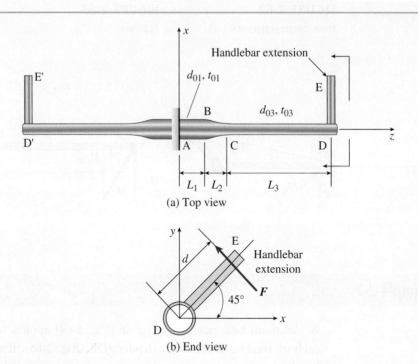

(a) Top view

(b) End view

1. *Conceptualize*: A 3D Cartesian coordinate system with origin at A has been superposed on the handlebar drawings in Figs. 2-65 and 2-66. The extenders lie in a vertical plane parallel to the x-y plane and the axis of the handle bar ABCD is the z axis. Force F is applied normal to extender DE (Fig. 2-65b) and also lies in the plane containing extender DE. A simple sketch shows a perspective view of ADE (Fig. 2-66).

2. *Categorize*: Use both scalar and vector (cross product) operations to find equivalent force-couple system at A (F, M_A). A scalar approach alone would be sufficient because of the simple geometry. A modified version of Fig. 2-66

FIGURE 2-66

Sketch showing bar ADE in a perspective view

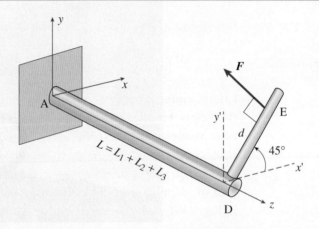

FIGURE 2-67

Cartesian components of
force-couple system F_A and M_A

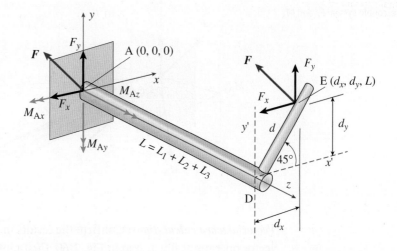

shows each of the components of couple M_A in the actual direction that each
component acts in accordance with the right-hand rule (Fig. 2-67).

3. *Analyze*: First find force-couple system F_A and M_A using scalar operations,
then confirm the results using vector operations.

Scalar calculations: As illustrated in Figs. 2-59 and 2-60, first move force F to A
so $F_A = F$. The components of F are

$$F_x = F_y = F \sin(45°) = (15 \text{ lb})(0.707) = 10.61 \text{ lb} \qquad \text{◂━ (2-130)}$$

Next use Varignon's theorem to find the Cartesian components of couple M_A
from the components of force F:

$$M_{Ax} = F_y L = \left(\frac{15}{\sqrt{2}} \text{ lb} \right)(2 + 1.2 + 8.5) \text{ in.}$$

$$= (10.61 \text{ lb})(11.7 \text{ in.}) = 124.1 \text{ lb-in.}$$

$$M_{Ay} = F_x L = M_{Ax} = 124.1 \text{ lb-in.} \qquad \text{◂━ (2-131)}$$

$$M_{Az} = F_x d_y + F_y d_x = 2 \left(\frac{15}{\sqrt{2}} \text{ lb} \right) \left(\frac{4}{\sqrt{2}} \text{ in.} \right) = 60 \text{ lb-in.}$$

Couple moment components M_{Ax} and M_{Ay} lie in the x-y plane in the directions shown in Fig. 2-67 and are negative in accordance with the right-hand
rule. Moments M_{Ax} and M_{Ay} are the *bending moments* that are used to compute stresses and make design calculations for handlebar AD. Component
M_{Az} acts in the +z direction and is the *torsional moment* for shaft AD. Magnitude M_{Az} could have been computed as $[(F)(d)]$ (like that of the socket wrench
in Fig. 2-61).

FIGURE 2-68

Force-couple system F_A and M_A

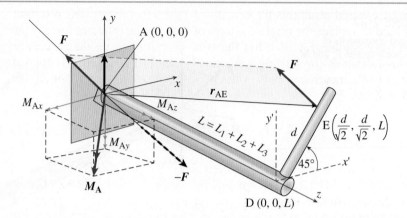

Vector algebra calculations: Confirm the results in (2-130) and (2-131) using the vector operations illustrated in Fig. 2-60. First move force F to A so $F_A = F$.

$$F_A = F = F_x(-i) + F_y j = (-10.61\text{ lb})i + (10.61\text{ lb})j \quad \text{(2-132)}$$

Next, couple moment M_A is computed as

$$
\begin{aligned}
M_A &= r_{AE} \times F = (d_x i + d_y j + Lk) \times (-F_x i + F_y j) \\
&= -F_y Li - F_x Lj + (F_x d_y + F_y d_x)k \\
&= (-10.61\text{ lb})(11.7\text{ in.})i + (-10.61\text{ lb})(11.7\text{ in.})j + (2)(10.61\text{ lb})\left(\frac{4}{\sqrt{2}}\text{ in.}\right)k \quad \text{(2-133)} \\
&= (-124.1\text{ lb-in.})i + (-124.1\text{ lb-in.})j + (60\text{ lb-in.})k
\end{aligned}
$$

Couple moment M_A is normal to the plane formed by the two force vectors F (see shaded plane in Fig. 2-68). The x and y rectangular components of M_A are the bending moments, and the z component is the torsional moment acting on shaft AD. The results in Eqs. (2-133) confirm the scalar calculation results in Eqs. (2-131) and also show the proper signs in accordance with the right-hand rule.

4. *Finalize:* Three-dimensional moment calculations usually require vector algebra unless the geometry is quite simple, as in this example, so that scalar calculations can be used. Varignon's theorem in Eq. (2-131) provided an economical solution approach for finding moment magnitude M_A. Moments (and forces) computed at a variety of different points along the bicycle handlebar are essential information for the engineer charged with its *design*. Note that the handlebar has a larger cross section over segment AB, then it tapers to a reduced cross section over transition region BC. This suggests that the bending stresses in region AB are larger than those in segment CD. You will also need to look at a variety of different loading conditions (not just the one loading considered in this example) to determine the worst case that controls the overall design of the handlebar and extenders. Static analyses of the type presented in this example are an important starting point in such engineering analyses.

Resultant of a System of Forces

Section 2.1 covered resultants for systems of *concurrent* forces. The resultant force is used to replace the original system of forces. The resultant force is the simplest force combination that has the same overall effect on the particle or rigid body as that of the starting system of forces. The resultant R is applied at the point of concurrency and is obtained by simple vector addition of the individual forces, F, as

$$R = F_1 + F_2 + F_3 = \sum_{i=1}^{3} F_i \qquad (2\text{-}134)$$

Both a *polygon law graphical approach* (see Figs. 2-13 and 2-14) and a *vector summation approach* [Eq. (2-134)] are used to find the resultant. When a Cartesian coordinate system is used, the magnitudes of the x, y, and z components of the resultant are equal to the sums of the corresponding components of the individual forces:

$$R_x = \Sigma F_x,\ R_y = \Sigma F_y,\ R_z = \Sigma F_z$$
$$R = \sqrt{R_x^2 + R_y^2 + R_z^2} = \sqrt{(\Sigma F_x)^2 + (\Sigma F_y)^2 + (\Sigma F_z)^2} \qquad (2\text{-}135)$$

For both two-dimensional and three-dimensional concurrent force systems, the lines of action of all forces meet at a point, so there are no moments about the point of concurrency. Since there are no moments, the effect of the system of concurrent forces on either a particle or a rigid body is the same.

If you move away from the point of concurrency, however, the moment about any other point is not equal to zero [unless the resultant is zero or the point of interest lies on the line of action of the resultant—see Eq. (2-95) and Fig. 2-38]. You can use Varignon's theorem (see Figs. 2-40 and 2-41) to find the moment about some arbitrary point O using either the resultant R or the individual components [such as F_1, F_2, F_3 in Eq. (2-134) or R_x, R_y, R_z in Eq. (2-135)]. Example 2-5 showed this for an automobile parking brake for which moment M_O was computed first using the components and then the resultant to arrive at the same moment M_O.

Now moving beyond concurrent forces to find resultants for general systems of forces, the objective remains the same: *To find the simplest force and moment combination that has the same overall effect on the rigid body as that of the system of forces and couples that you started with.*

Nonconcurrent Forces

Consider the force system F_1, F_2, and F_3 applied to a planar rigid body in Fig. 2-69a. You want to find the simplest resultant R and its line of action through a convenient reference point, such as point O. Move the forces to point O one at a time, as illustrated for F_3 in Fig. 2-69b, by creating an equivalent force-couple combination at O for each force. The resulting forces and couples at point O for all three forces are shown in Fig. 2-69c. In Fig. 2-69d, the force-couples are combined using the following equation:

$$R = \Sigma F, M_O = \Sigma M = \Sigma(Fd) = Rd \qquad (2\text{-}136\text{a,b})$$

A force polygon (see Figs. 2-13 and 2-14) is used in Fig. 2-69d to find the resultant R and its line of action through point O. In a final step, move resultant R

a distance d away from point O (but on a parallel line of action) to find the simplest expression of the overall effect of the original system of forces on the rigid body (Fig. 2-69e). The distance d is determined from Eq. (2-136b) as $d = M_O/R$. Equation (2-136b) represents an extension of Varignon's theorem to nonconcurrent force systems and is known as the *principle of moments*. This principle states that the moment of a system of forces about some point O is equal to the moment of the resultant force R about that same point and is found by taking the sum of the moments about O of the individual forces. Note that the resultant force could be zero, but it does not follow that the resultant couple is zero.

The process of creating a simple resultant for a system of coplanar forces, shown in Fig. 2-69, also applies to three-dimensional systems of forces except that vector operations are usually required. This process is represented in 3D

FIGURE 2-69

Process used to find simplest resultant for noncurrent coplanar force system

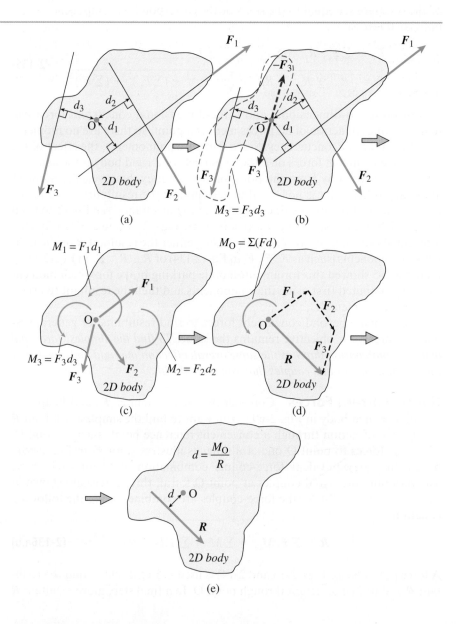

in Fig. 2-70, and the resultant force R and moment M at some arbitrary point O are computed as

$$R = \sum_{i=1}^{3} F_i = F_1 + F_2 + F_3$$

$$M = \Sigma(r \times F) = \sum_{i=1}^{3} M_i = M_1 + M_2 + M_3$$

(2-137)

where $M_i = r_{Oi} \times F_i$.

The magnitude and direction of resultant moment M depend on the choice of point O, but recall that M is a free vector and can be moved to any parallel position. Resultant force R has the same magnitude and direction regardless of the choice of point O.

Parallel Forces   The magnitude of resultant force R of a system of parallel forces is simply the algebraic sum of the magnitudes of the individual forces. The resultant moment M_O is obtained by summing the moments of the individual parallel forces [Eq. (2-137)] using a position vector from O to each force. To find the *simplest resultant*, move resultant vector R a distance d from point O, where d is computed using Eq. (2-136b), so that R alone represents the entire force-couple system. This procedure is illustrated in Example 2-8.

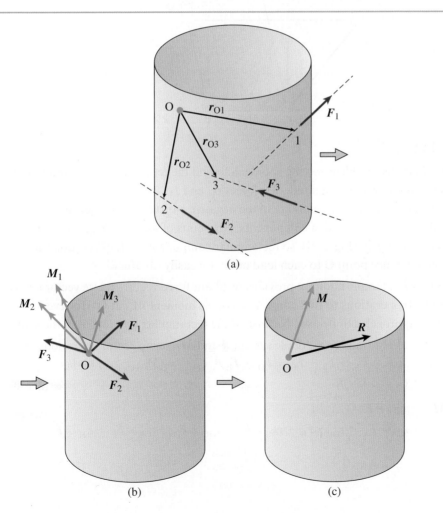

(a)

(b)   (c)

FIGURE 2-70

Process used to find simplest resultant for noncurrent 3D force system

Example 2-8

A foundation slab for a machine has equal loads of 5 kN each at A, D, and E and loads of 7.5 kN each at B, C, and F (Fig. 2-71). Let $L = 3$ m.

(a) Find the resultant R and the x and y coordinates of the point through which the resultant acts if R alone represents the entire force-couple system.

(b) Find revised magnitudes for forces F_D and F_F so that resultant R passes through point G (other forces at A, B, C, and E are unchanged).

FIGURE 2-71

Foundation slab and loading

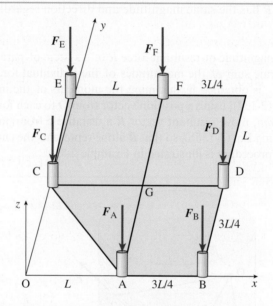

Solution:

Use the four-step problem-solving approach to find the resultant R, the x and y coordinates, and the revised magnitudes for forces F_D and F_F.

1. *Conceptualize*: All six forces are in the negative z-direction, so they are easy to define using Cartesian components. All points of load application lie on a rectangular grid (Fig. 2-71), so rectangular components (i, j) of position vectors from reference point O to each load point are easily obtained.

2. *Categorize*: Simple addition provides resultant force vector R. Use vector (cross product) operations to find equivalent couple moment M_O at O. However, a scalar approach can be used because of the rectangular grid layout of the slab.

3. *Analyze*: First find force-couple system R and M_O using vector operations:

$$R = \sum F = F_A + F_B + F_C + F_D + F_E + F_F = -[(3)(5 + 7.5)\ \text{kN}]k$$
$$= (-37.5\ \text{kN})k \qquad \text{(2-138)}$$

$$M_O = \sum(r \times F)$$
$$= r_{OA} \times F_A + r_{OB} \times F_B + r_{OC} \times F_C + r_{OD} \times F_D + r_{OE} \times F_E + r_{OF} \times F_F \qquad \text{(2-139)}$$

The position vectors in Eq. (2-139) are

$$r_{OA} = Li, \; r_{OB} = \frac{7}{4}Li, \; r_{OC} = \frac{3}{4}Lj, \; r_{OD} = \frac{7}{4}Li + \frac{3}{4}Lj,$$

$$r_{OE} = \frac{7}{4}Lj, \; r_{OF} = Li + \frac{7}{4}Lj \tag{2-140}$$

Part (a): Evaluating Eq. (2-139) gives $M_O = (-93.75)i + (103.125)\,j$ kN·m. Now use the principle of moments:

$$M_O = r \times R = (xi + yj) \times (-37.5 \text{ kN})k \tag{2-141}$$

then equate these two expression for M_O and solve for coordinates x and y. The coordinates through which resultant R must pass are $x = 2.75$ m and $y = 2.5$ m (Fig. 2-72) (and coordinate z may take on any value) if R alone is to represent the entire force-couple system. Also, distance d_R from point O to the point at which R pierces the slab is

$$d_R = \sqrt{x^2 + y^2} = \frac{|M_O|}{|R|} = 3.717 \text{ m}$$

Part (b): Repeat the procedure described above. However, position vector r in Eq. (2-141) is now known:

$$r_{OG} = Li + \frac{3}{4}Lj = (3)i + (2.25)j \text{ m} \tag{2-142}$$

but force vectors F_D and F_F are unknown, so Eq. (2-138) now becomes

$$\begin{aligned} R = \sum F &= F_A + F_B + F_C + F_D + F_E + F_F \\ &= -[(2)(5 + 7.5) \text{ kN}]k + (-F_D)k + (-F_F)k \\ &= (-25 \text{ kN} - F_D - F_F)k \end{aligned} \tag{2-143}$$

FIGURE 2-72

Resultant R for a parallel force system

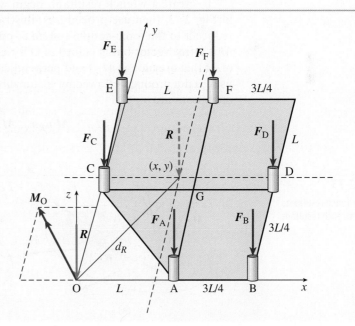

Eq. (2-141) is altered as follows for part (b):

$$M_O = r_{OG} \times R = (3i + 2.25j) \text{ m} \times (-25 \text{ kN} - F_D - F_F)k$$
$$= (-2.25(F_D + F_F) - 56.25)i + (3(F_D + F_F) + 75)j \text{ kN·m}$$

(2-144)

And Eq. (2-139) becomes for part (b):

$$M_O = \Sigma(r \times F) = (-2.25F_D - 5.25F_F - 43.125)i \text{ kN·m}$$
$$+ (5.25F_D + 3F_F + 54.375)j \text{ kN·m}$$

(2-145)

Equating the expressions for M_O in Eqs. (2-144) and (2-145), then solving for force magnitudes F_D and F_F gives: $F_D = 9.17$ kN, $F_F = 4.38$ kN. So if resultant R must pass through point G (Fig. 2-72), force magnitudes at D and F must be changed from 5 kN and 7.5 kN to 9.17 kN and 4.38 kN, respectively.

4. *Finalize*: Vector calculations provide an efficient and organized procedure for reducing a system of parallel forces to a simple force-couple system at point O or a single force positioned at some other point on the slab [i.e., a point having coordinates x and y in part (a), and at point G in part (b)]. Repeat the solutions for parts (a) and (b) using scalar calculations to confirm the results and to gain additional insight into the solution procedure used in this example.

Wrench Resultant   For some three-dimensional force systems, it is possible to reduce the force system to a *wrench resultant* consisting of resultant force R and a moment M along the line of action of R. If R and M point in the same direction (Fig. 2-73a), it is a *positive* wrench resultant; otherwise, it is a *negative* wrench (Fig. 2-73b). A positive wrench is sometimes referred to as a "screwdriver" because the combined effect of R and M mimics the action of tightening a screw: pushing inward while turning the screwdriver clockwise.

To create a wrench resultant, begin with the general force system shown in Fig. 2-73, then use procedures illustrated in Eq. (2-137) and Fig. 2-73 to reduce it to the force-couple system at point O shown in Fig. 2-74a. Moment M is a free vector but is placed at O for convenience. Next, find components of M that are along ($M_{//}$) and perpendicular to ($M_\perp$) the line of action of R, using a dot product followed by vector subtraction [see Eq. (2-129)]. The vector operations are

$$M_{//} = M \cdot n_R n_R, \quad M_\perp = M - M_{//}$$

(2-146)

FIGURE 2-73

(a) Positive wrench resultant;
(b) negative wrench resultant

(a)   (b)

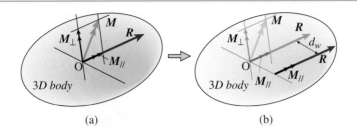

where n_R is a unit vector along R. Finally, to create the simplest possible resultant, move the wrench resultant a distance d_w to a parallel line of action (Fig. 2-74b) thereby capturing the turning action of component $M_\perp$. Distance d_w is computed as

$$d_W = \frac{|M_\perp|}{|R|} \qquad \textbf{(2-147)}$$

The line of action of the wrench is the same as that of resultant R, and the wrench lies in a plane normal to that containing R and M. By creating the wrench resultant, you have satisfied the stated objective for this section: *To find the simplest force and moment combination that has the same overall effect on the rigid body as that of the system of forces and couples that you started with.* However, as a practical matter, the wrench has limited use because you are usually more interested in finding resultants at other points on the body (such as locations of supports or the center of mass of the body). The following examples provide some additional insight into the creation and location of wrench resultants.

Example 2-9

A standard socket wrench was used to illustrate the concept of force-couple systems (see Figs. 2-61 and 2-62). In many cases, bolts are in hard-to-reach locations, so a socket wrench that can pivot at the head (Fig. 2-75) combined with an extension bar (component BC in Fig. 2-76) must be used. Consider here the rotated position of the pivot socket wrench and extender shown in Fig. 2-76. The mechanic has applied downward force P with his right hand to the handle at A to tighten a bolt at C. To keep the extender BC aligned with the z-axis, he also applies a balancing upward force P and inward thrust force F at B. Forces P are applied normal to plane ABC. Assume that force magnitude $P = 18$ lb and dimension $L = 15$ in.

FIGURE 2-75

Pivot socket wrench

FIGURE 2-76

Pivot socket wrench
with bar extension

(a) Find the resultant force-couple system at C.

(b) Determine a wrench resultant and the coordinates of the point at which the
line of action of the wrench passes through the *x-y* plane.

Solution:

Use the four-step problem-solving approach to find the resultant force-couple system
at C and the wrench resultant and coordinates of the point at which the line of action
of the wrench passes through the *x-y* plane.

1. *Conceptualize*: Start by sketching projection views of the socket wrench and extension in each of the Cartesian planes to be sure that correct dimensions and vector
orientations are used in all calculations. The three views are shown in Fig. 2-77.

FIGURE 2-77

Projection views

2. *Categorize*: Begin by finding the equivalent force-couple system at C. The wrench resultant can be found using Eqs. (2-146) and (2-147). You will need position vectors from C to the two load points (A and B) and a unit vector normal to plane ABC (vector $\boldsymbol{n}_P$ in Fig. 2-77b) to define the lines of action of the two $\boldsymbol{P}$ forces.

3. *Analyze*: First find position vectors $\boldsymbol{r}_{CA}$, $\boldsymbol{r}_{CB}$, and $\boldsymbol{r}_{BA}$:

$$\boldsymbol{r}_{CA} = x_A \boldsymbol{i} + y_A \boldsymbol{j} + z_A \boldsymbol{k} = 9\boldsymbol{i} + 2.598\boldsymbol{j} + 18.5\boldsymbol{k} \text{ in.}$$

$$\boldsymbol{r}_{CB} = L\boldsymbol{k} = 15\boldsymbol{k} \text{ in.} \tag{2-148}$$

$$\boldsymbol{r}_{BA} = (x_A - x_B)\boldsymbol{i} + (y_A - y_B)\boldsymbol{j} + (z_A - z_B)\boldsymbol{k} = 9\boldsymbol{i} + 2.598\boldsymbol{j} + 3.5\boldsymbol{k} \text{ in.}$$

Wrench arm AB has length $d = 2L/3 = 10$ in., so the y coordinate of point A is

$$y_A = \sqrt{d^2 - (0.9d)^2 - (0.35d)^2} = 0.25981d = 2.598 \text{ in.}$$

Unit vector $\boldsymbol{n}_P$ is normal to plane ABC, so use a cross product of position vectors $\boldsymbol{r}_{CA}$ and $\boldsymbol{r}_{BA}$ to find $\boldsymbol{n}_P$:

$$\boldsymbol{n}_P = \frac{\boldsymbol{r}_{CA} \times \boldsymbol{r}_{BA}}{|\boldsymbol{r}_{CA} \times \boldsymbol{r}_{BA}|} = (-0.27735)\boldsymbol{i} + (0.96077)\boldsymbol{j} \tag{2-149}$$

The force vectors at A and B are written as

$$\boldsymbol{F}_A = -P\boldsymbol{n}_P = (4.992)\boldsymbol{i} + (-17.294)\boldsymbol{j} \text{ lb}$$

$$\boldsymbol{F}_B = P\boldsymbol{n}_P + \left(-\frac{3}{5}P\right)\boldsymbol{k} = (-4.992)\boldsymbol{i} + (17.294)\boldsymbol{j} + (-10.8)\boldsymbol{k} \text{ lb} \tag{2-150}$$

Part (a): Using vector operations, resultant force-couple system R and M_C at point C are

$$\boldsymbol{R} = \boldsymbol{F}_A + \boldsymbol{F}_B = (-10.8)\boldsymbol{k} \text{ lb} \qquad \longleftarrow \tag{2-151}$$

$$\boldsymbol{M}_C = \boldsymbol{r}_{CA} \times \boldsymbol{F}_A + \boldsymbol{r}_{CB} \times \boldsymbol{F}_B = (60.5)\boldsymbol{i} + (17.5)\boldsymbol{j} + (-168.6)\boldsymbol{k} \text{ lb-in.}$$

The original force system and the resultant force-couple system at C are shown in Fig. 2-78.

Part (b): Resultant R has a z-direction component only and is therefore normal to the x-y plane.

As a result, the *wrench resultant* has a line of action normal to the x-y plane and is found using Eqs. (2-146) and (2-147).

 i. The moment component of the wrench resultant has the same line of action as $\boldsymbol{R}$ as expressed by the unit vector: $\boldsymbol{n}_R = -\boldsymbol{k}$, so

$$\boldsymbol{M}_{//} = \boldsymbol{M}_C \cdot \boldsymbol{n}_R \boldsymbol{n}_R = (-168.6)\boldsymbol{k} \text{ lb-in.} \qquad \longleftarrow \tag{2-152}$$

 ii. The moment component perpendicular to the wrench resultant is computed by vector subtraction as

$$\boldsymbol{M}_{\perp} = \boldsymbol{M}_C - \boldsymbol{M}_{//} = (60.5)\boldsymbol{i} + (17.5)\boldsymbol{j} \text{ lb-in.} \tag{2-153}$$

 iii. The perpendicular distance from point C to the line of action of the wrench resultant is computed using Eq. (2-147):

$$d_W = \frac{|\boldsymbol{M}_{\perp}|}{|\boldsymbol{R}|} = \frac{63 \text{ lb-in.}}{10.8 \text{ lb}} = 5.83 \text{ in.} \tag{2-154}$$

FIGURE 2-78

Resultant force-couple
system at point C

iv. Finally, the coordinates of the point at which line of action of the wrench
     passes through the x-y plane (see Fig. 2-79) are determined as

$$(x\mathbf{i} + y\mathbf{j}) \times \mathbf{R} = \mathbf{M}_\perp$$

or   **(2-155)**

$$(-10.8y)\mathbf{i} + (10.8x)\mathbf{j} \text{ lb} = (60.5)\mathbf{i} + (17.5)\mathbf{j} \text{ lb-in.}$$

Equating like components in Eq. (2-155) gives $x = 1.618$ in. and $y = -5.605$ in.
Distance d_w in Fig. 2-79 is $\sqrt{(1.618 \text{ in.})^2 + (-5.605 \text{ in.})^2} = 5.83$ in.   ◄─

FIGURE 2-79

Line of action of positive
wrench resultant

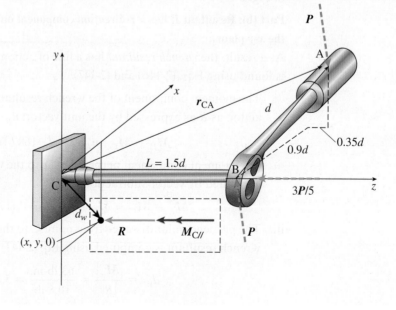

4. *Finalize*: Vector calculations provide an efficient and organized procedure for reducing a system of forces at A and B to a simple force-couple system at point C. The results apply only for this load system and for this particular orientation of the pivot socket wrench. Reducing the force-couple system further to obtain a positive wrench resultant and its line of action at some distance d_w from point C is an interesting exercise, but as a practical matter, the resultants at C are likely to be of greater interest to the mechanic who is attempting to tighten the bolt located in an inaccessible location.

Example 2-10

A hand winch used to hoist and trim the sails on a sailboat (Fig. 2-80) was the subject of analysis in Example 2-6. A force with magnitude $F = 70$ N is applied at A on the winch handle and has a line of action from A to B. Now, downward force $S = 50$ N is added at D.

FIGURE 2-80

Forces *F* and *S* applied to hand winch on sailboat

(a) Find the resultant force-couple system at O.

(b) Determine a wrench resultant and the coordinates of the point at which the line of action of the wrench passes through the y-z plane.

Solution:

Use the four-step problem-solving approach to find the resultant force-couple system and the wrench resultant and point coordinates.

1. *Conceptualize*: Use the same 3D Cartesian coordinate system with origin at O at the base of the winch as that used in Example 2-6 (see Fig. 2-80). The winch shaft OD is vertical and is aligned with the y axis. Force S is applied downward at point D and has a line of action from D to O.

2. *Categorize*: Moment M_O was computed in Example 2-6 for the case of force F applied at A. The added force S acts through base point O, so it does not add any additional moment to M_O. Force S must be considered when computing the resultant force R, however.

3. *Analyze*: Find the resultant force as

$$R = F + S = (-55.19i - 40.47j - 14.72k) \text{ N} + (-50)j \text{ N}$$
$$R = (-55.19i - 90.47j - 14.72k) \text{ N} \quad \text{(2-156)}$$

Moment M_O is known [see Eq. (2-120) in Example 2-6]. Rewriting this equation and adding the moment from force S confirms that force S makes no contribution to M_O (since $j \times j = 0$ in the second cross product in the following equation):

$$M_O = r_{OA} \times F + r_{OD} \times S$$
$$= (150i + 290j + 200k) \text{ mm} \times (-55.19i - 40.47j - 14.72k) \text{ N}$$
$$+ (150j) \text{ mm} \times (-50j) \text{ N} \quad \text{(2-157)}$$
$$M_O = (3.83i - 8.83j + 9.93k) \text{ N·m}, \ M_O = 13.83 \text{ N·m}$$

Part (a): The resultant force-couple system at point O is R and M_O, as shown in Fig. 2-81.

Part (b): Find the wrench resultant for this force-couple system.

Moment M_D and its components along ($M_{D//}$) and perpendicular to ($M_{D\perp}$) arm DC were computed in Example 2-6 [see Eq. (2-121)]. Using the same approach here gives components of moment M_O along ($M_{O//}$) and perpendicular to ($M_{O\perp}$) to resultant force R. A unit vector along R is

$$n_R = \frac{R}{|R|} = (-0.516)i + (-0.846)j + (-0.138)k \quad \text{(2-158)}$$

Moment $M_{O//}$ along resultant force R (shown in Fig. 2-81) is

$$M_{O//} = (M_O \cdot n_R)n_R = (-2.128)i + (-3.489)j + (-0.568)k \text{ N·m} \quad \text{(2-159)}$$

FIGURE 2-81

Resultant force-couple
system at point O

Moment component $M_{O//}$ is directed along resultant force R, so this is a positive wrench resultant. Component $M_{O\perp}$ perpendicular to resultant force R from vector subtraction is

$$M_{O\perp} = M_O - M_{O//} = (5.955)\,i + (-5.341)\,j + (10.501)\,k \text{ N·m} \qquad \textbf{(2-160)}$$

From Eq. (2-147), the perpendicular distance from point O to the line of action of the wrench resultant is

$$d_W = \frac{|M_{O\perp}|}{|R|} = \frac{13.201 \text{ N·m}}{107 \text{ N}} = 123.4 \text{ mm} \qquad \textbf{(2-161)}$$

Finally, the coordinates of the point at which line of action of the wrench passes through the y-z plane (see Fig. 2-82) are

$$(y\,j + z\,k) \times R = M_{O\perp}$$

or $\qquad\qquad\qquad\qquad\qquad\qquad\qquad\qquad\qquad\qquad\qquad\qquad$ **(2-162)**

$$(-55.187z)\,j + (55.187y)\,k \text{ N} = (-5.341)\,j + (10.501)\,k \text{ N·m}$$

Equating like components in Eq. (2-162) gives $y = 190.3$ mm and $z = 96.8$ mm.

4. *Finalize*: The conclusions here are much the same as those stated at the end of Example 2-9. Finding the wrench resultant is an interesting exercise, but the resulting force and couple are not all that useful for an engineer charged with designing the various components of the winch (handle, arm, and main shaft). The y-direction components of $\boldsymbol{R}$ and $\boldsymbol{M}_O$, for example, will be used in the design of the main shaft of the winch. However, other positions of the winch and other loadings will have to be considered to find the worst case that governs the overall design.

FIGURE 2-82

Line of action of positive wrench resultant

In Chapter 2, you studied the action of **forces** and **moments** on a particle or rigid body. The major concepts discussed in Chapter 2 are given here.

1. Two-dimensional forces and force systems usually employ **scalar** operations, but three-dimensional force systems often require **vector algebra** to determine their **components** in directions of interest.

2. The **principle of transmissibility** is used to slide force vectors to the point at which their lines of action cross.

3. A system of **concurrent forces** can be reduced to a **resultant** force at the point of concurrency.

4. Separate forces (such as **F** or **T** in the figures below) can be combined into a **resultant** force **R**. The process can be applied in reverse by finding the **components** of a force in stated directions or along specified axes. The **triangle law** and the **parallelogram** law are used to find the **resultant** of two force vectors; or an extension of the triangle law, the **polygon law**, is used if three or more vectors are involved as illustrated below.

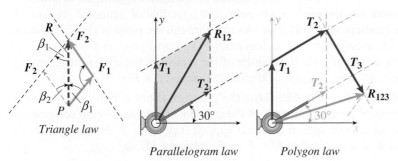

Triangle law

Parallelogram law *Polygon law*

5. Both general and rectangular Cartesian coordinate systems are used when finding **components** of a force **R** along directions of interest. Rectangular components R_x, R_y, and R_z are shown in the figure.

6. The **law of sines** and the **law of cosines** are very useful in carrying out the calculations required to find vector components and resultants for both rectangular and nonrectangular force systems.

7. In three dimensions, it is convenient to use a **unit vector** (n_R) comprised of the **direction cosines** [$\cos(\theta_x)$, $\cos(\theta_y)$, $\cos(\theta_z)$—see figure] of the original vector to define its line of action within a Cartesian reference frame; force vector **R** is then expressed as the product of its magnitude [$\|R\| = R$] times the unit vector: $R = Rn_R$.

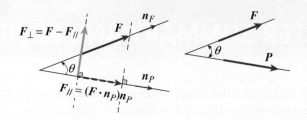

8. The **dot product** (or scalar product) is used to find the components of a force F along ($F_{//}$) the line of action of another force P (defined by unit vector n_P in the figure). Vector subtraction is then used to find the perpendicular component ($F_\perp$). The dot product also can be used to find the angle θ between vectors F and P as $\theta = \cos^{-1}\left[\dfrac{F \cdot P}{FP}\right]$.

9. A **moment** about a point is defined as the product of the magnitude of a force and the perpendicular distance from the point to the line of action of the force. A moment vector is perpendicular to the plane defined by the point and the force, and the sign of the moment vector is determined using the **right-hand rule**.

10. The moment M of a force F is expressed using vector algebra as a **cross product**: $M = r \times F$. Vector r is a position vector from the point about which the moment is computed to any point along the line of action of the force F. The moment vector M about an axis through the point is perpendicular to the plane defined by the vectors r and F. The cross product can be computed using a determinant. If the order of the vectors in the cross product expression is reversed, the sign of the result changes: $r \times F = -(F \times r)$.

11. **Varignon's theorem** states that the moment about a point (M_O) can be computed by using the sum of the moments of the individual components of a concurrent system of forces applied at point A. If the resultant of the system of forces is $R = F_1 + F_2 + F_3$, moment M_O is

$$M_O = r_{OA} \times R = r_{OA} \times \left(F_1 + F_2 + F_3\right) = r_{OA} \times F_1 + r_{OA} \times F_2 + r_{OA} \times F_3$$

12. Use a **triple scalar product** to find the scalar component of a moment vector M_β *along* a line of interest (defined by unit vector n_β in the equations below); vector subtraction is used to find the vector component $M_{\beta\perp}$ *perpendicular* to that line. The vector operations are

$$M_\beta = \left|M_\beta\right| = r_{OA} \times F \cdot n_\beta$$
$$M_{\beta\perp} = M_O - M_\beta = r_{OA} \times F - \left(r_{OA} \times F \cdot n_\beta\right)n_\beta$$

13. The moment of a **couple** is a free vector representing the turning effect of a pair of forces equal in magnitude but opposite in sense, acting along parallel lines of action. Use a **force-couple system** to replace a force acting at some point A with an equivalent force and moment at another point of interest B (see figure). The process works in reverse, so a force-couple can be simplified to single resultant force by moving the force an appropriate distance away from the point of interest.

14. The **simplest resultant** for the cases of noncurrent and parallel force systems is often of interest. A **wrench resultant** is the combination of a resultant force R and resultant moment M having the same line of action. Wrench resultants are positive if both force and moment vectors have the same sign (as in the figure), and negative if not.

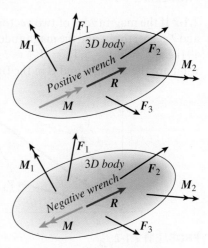

PROBLEMS

2.1 Forces: 2D, 3D

Introductory Problems

2.1-1 (a) If vector $V_1 = 4i + j$ and vector $V_2 = 2i + \alpha j$, and vectors V_1 and V_2 are *orthogonal*, what is α?

(b) What is α if the angle φ between V_1 and V_2 is 30 degrees?

(c) If vector magnitudes are $V_1 = 4$ and $V_2 = 8$ and the magnitude of the sum of V_1 and V_2 is $V_s = 12$, what is angle φ?

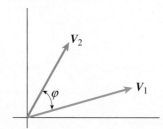

PROBLEM 2.1-1

2.1-2 If the magnitudes of two vectors V_a and V_b are 5 and 7, respectively, and the magnitude of their sum V_s is 10 (i.e., $V_a = |V_a| = 5$, $V_b = |V_b| = 7$, $V_s = |V_s| = 10$), what are the angles θ and β in the figure?

PROBLEM 2.1-2

2.1-3 A 500-lb force F is applied at O.

(a) Find the components of F along the x-y axes.

(b) Find the components of F along the α-β axes.

(c) Find the projections of F onto the α-β axes.

PROBLEM 2.1-3

2.1-4 Force F has magnitude of 2.5 kN and acts at point A (see figure).

(a) Find the components of force F parallel to the x-y axes.

(b) Find the components of F parallel to the n-t axes (the n-t axes are rotated 60° clockwise from the x-y axes).

PROBLEM 2.1-4

2.1-5 Forces F_1 to F_5 act on structural members that meet at joint A (see figure). It is known that $F_1 = 100$ lb, $F_3 = 20$ lb, and $F_5 = 150$ lb. Find the magnitudes of forces F_2 and F_4 if the magnitude of the resultant of all five forces is known to be zero.

PROBLEM 2.1-5

2.1-6 Tension forces F_1, F_2, and F_4 and compression force F_3 act on truss members that meet at joint A.

(a) Find the *resultant* R_{12} of forces F_1 and F_2 and the angle it makes with the x axis.

(b) Repeat part (a) for forces F_3 and F_4.

(c) Find the resultant R of all four forces.

PROBLEM 2.1-6

2.1-7 Cable forces of magnitude $T = 80$ lb act on a pin-supported pulley of radius $r = 4.0$ in. (see figure). Find the magnitude and direction of the resultant force R acting at point O.

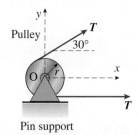

PROBLEM 2.1-7

2.1-8 A cylinder is supported on an inclined plane by contact forces N_1 and N_2 (see figure). The force magnitudes are known to be $N_1 = 200$ N and $N_2 = 347$ N.

(a) Find the *resultant* R of forces N_1 and N_2 and the angle it makes with the x axis.

(b) Find the components of resultant R along the n-t axes (see figure).

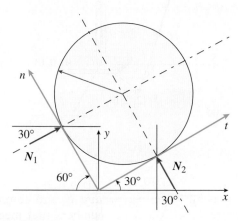

PROBLEM 2.1-8

2.1-9 Three cables are attached to an eye bolt (see figure). The forces in the cables are known to be $T_1 = 800$ lb, $T_2 = 600$ lb, and $T_3 = 1200$ lb. Find the resultant R of the three cable forces and the angle it makes with the x axis.

PROBLEM 2.1-9

Representative Problems

2.1-10 Two cables are part of the support system for an entrance canopy on a building. Cable tension forces are $T_1 = 440$ N and $T_2 = 560$ N. Both cables lie in the x-y plane (see figure). Find the resultant R of the two cable forces in terms of unit vectors i and j and the angle it makes with the x axis.

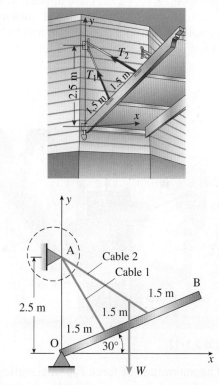

PROBLEM 2.1-10

2.1-11 Continuous cable ABC supports an inclined rectangular panel with dimensions $a = 15$ ft and $b = 25$ ft. The tension force in the cable is $T = 5000$ lb. Angles θ and α are known to be $\theta = 94.4°$ and $\alpha = 54.9°$. If resultant R of cable tension forces T lies along the y axis, find R in terms of unit vectors i and j. Also find the angles β_1 and β_2 that the cable forces make with the vertical axis.

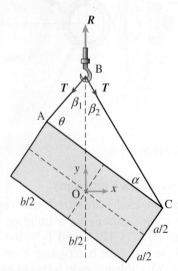

PROBLEM 2.1-11

2.1-12 Forces $F_1 = F_3 = 250$ N are applied at angle $\theta = 30°$, and $F_2 = 180$ N is applied downward (includes both hands) to a leg exercise machine (see figure). Find resultant force vector R for these three forces.

PROBLEM 2.1-12

2.1-13 The magnitude of force F_{EB} in elevating cylinder EB of a mobile crane is 4500 lb. Assume that $\alpha = 80°$ and $\beta = 60°$.

(a) Write force F_{EB} at E in terms of unit vectors $i, j,$ and k.

(b) Find the projection F_{CA} of F_{EB} onto line CA.

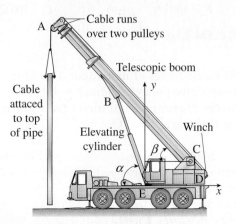

PROBLEM 2.1-13

2.1-14 Bracket ADO is acted on by forces F and T (see figure). Find resultant force vector R at point O for each of the *three different load cases* shown in Figs. a to c.

Assume $F = 8$ N, $T = 9$ N, and that joint coordinates are in meters.

(a)

(b)

(c)

PROBLEM 2.1-14

2.1-15 Multiple cables anchor a cargo ship to the dock. When taut, the tension in one cable is $T = 2800$ lb. The cable is attached to the ship at point A whose coordinates are (5 ft, 20 ft, 30 ft) relative to point O. Write cable force T as a vector. Neglect cable sag.

Rrrainbow/Shutterstock.com

PROBLEM 2.1-15

2.1-16 A radio tower is held in a vertical position by four cables. The tension forces in three of the four cables are given in the figure. Find the tension force T in the 4th cable if the resultant R must act along the vertical axis of the tower.

PROBLEM 2.1-16

2.1-17 Multiple cables are used to support an entrance canopy. The magnitude of cable force $T_{CE} = 375$ lb. Write cable force T_{CE} as a vector. Find the projection T_{CA} of T_{CE} onto line CA. Assume that dimensions are $a = 20$ ft, $b = 8$ ft, and $c = 6$ ft.

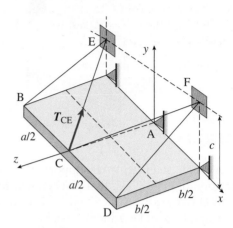

PROBLEM 2.1-17

2.1-18 Four cables are used to support an entrance canopy. The magnitude of cable forces at C are $T_{CE} = T_{CF} = 1700$ N. Write cable forces T_{CE} and T_{CF} as vectors. Find the resultant R of these two vectors.

Assume that dimensions are $a = 10$ m, $b = 4$ m, and $c = 3$ m.

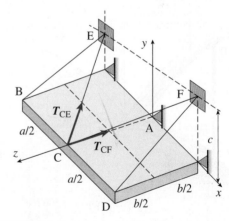

PROBLEM 2.1-18

2.1-19 Four cables are used to support an entrance canopy. The magnitudes of two of the cable forces are known: $T_{EC} = 375$ lb and $T_{EB} = 300$ lb.

(a) Find the *magnitude* of resultant R of forces T_{EC} and T_{EB}.

(b) Find the *projection* of vector R onto lines EB and EC.

Assume that dimensions are $a = 20$ ft, $b = 8$ ft, and $c = 6$ ft.

103

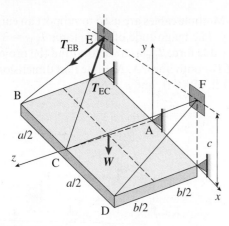

PROBLEM 2.1-19

2.1-20 A single steel strut AB supports the vehicle engine hood of mass 20 kg that pivots about hinges at C and D (see figure parts a and b). The strut lies in a vertical plane and is attached to a bolt at A. The force in the strut is $F_S = 154$ N.

(a) Write force F_S at A in terms of unit vectors $i, j,$ and k.

(b) Find the projection F_{AD} of F_S onto line AD and the angle θ between F_S and AD.

(a)

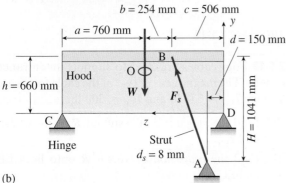

(b)

PROBLEM 2.1-20

2.1-21 Force F with magnitude of 350 lb is directed from point O to A on a right circular cylinder of radius $r = 2.5$ ft and height $h = 7.5$ ft.

(a) Write force F at O in terms of unit vectors $i, j,$ and k.

(b) Find the projection F_{BC} of F onto line BC and the angle θ between F and BC.

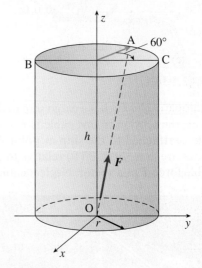

PROBLEM 2.1-21

2.1-22 Three cables are used to lift a concrete slab. The tension force in cable B is directed from point B to point C and has known magnitude $T_B = 3700$ N. Find a unit vector along cable B. Also find the x-y-z components of force T_B which acts along line BC. The coordinates (meters) of points A to D are given in the figure.

PROBLEM 2.1-22

2.1-23 A ceiling access panel is supported in the open position by cable AB. The tension force in the cable is $T_{AB} = 15$ lb.

(a) Write force T_{AB} at A in terms of unit vectors $i, j,$ and k.

(b) Find the projection of force T_{AB} onto line CD.

(c) Find the projection of force T_{AB} onto line CB.

PROBLEM 2.1-23

2.1-24 Force P has a magnitude of 3 kN and a line of action from C to E; force Q has magnitude of 4 kN and acts from F to E (see figure). Coordinates of points A through E are given in meters.

(a) Express forces P and Q as vectors in terms of unit vectors i, j, and k.

(b) Find the resultant R of forces P and Q and the angle that R makes with the $+y$ axis.

(c) Find the components of P along and perpendicular to line CA.

(d) Repeat part (c) for Q and line BA.

(e) What is the angle between P and Q?

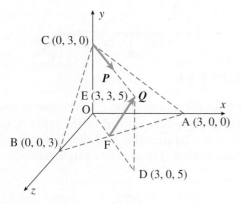

PROBLEM 2.1-24

2.1-25 Three cables support an L-shaped panel (see figure). The tension force in cable 2 is known to be $T_2 = 5200$ lb. Find a unit vector along cable 2 and the x-y-z components of force T_2. Also, find the magnitude of the projection of T_2 along line BD.

PROBLEM 2.1-25

2.1-26 Force F has a magnitude of 25 kN and a line of action from origin O to point A. Coordinates of point A are given in meters.

(a) Express force F as a vector in terms of unit vectors i, j, and k.

(b) Find a unit vector n along vector F.

(c) Find the components of F in the x'-y'-z' coordinate system which is rotated 30° clockwise about the z axis from the x-y-z system.

PROBLEM 2.1-26

2.1-27 Force P has a magnitude of 10 kips and a line of action from H to B (see figure).

(a) Find the x-y-z components of P and express P as a vector.

(b) Find the components of P parallel to and perpendicular to line HC.

(c) Find the projection P_{DG} of P onto line DG and the angle θ between P and DG

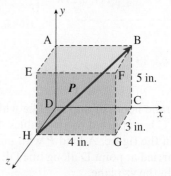

PROBLEM 2.1-27

2.1-28 Force P is applied at point A and has x-y-z components of $(30, 25, -75)$ kN. The coordinates of points A, B, and C are given in the figure.

(a) Express force P as a vector in terms of unit vectors $i, j,$ and k.

(b) Find the projection of P along line BC.

(c) What is the angle between P and BC?

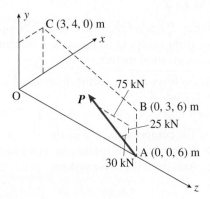

PROBLEM 2.1-28

2.1-29 Four forces with magnitudes $T_{DB} = T_{DC}$ $= 10$ kips and $T_{DE} = T_{DF} = 15$ kips are applied at point D on a framework (see figure). Coordinates of points A through G are given in feet in the figure.

(a) Express each force as a vector in terms of unit vectors $i, j,$ and k.

(b) Find the resultant R of the four force vectors.

PROBLEM 2.1-29

2.1-30 A crane boom of mass 450 kg with its center of mass at C is stabilized by two cables AQ and BQ as shown in the figure. A load P with magnitude of 20 kN is carried at point D along line DE. The crane boom lies in the yz plane.

(a) Express boom weight W and applied force P as vectors in terms of unit vectors $i, j,$ and k.

(b) Find the resultant R of the two force vectors.

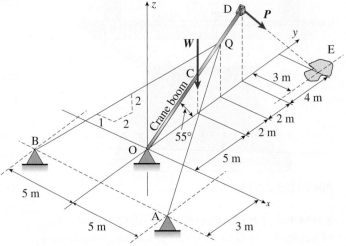

PROBLEM 2.1-30

Moments and couples: 2D, 3D

Introductory Problems

2.2-1 Forces $P = 10$ kips and moment $M = 25$ in-kip are applied to a cantilever beam (see figure). Find the resultant force and couple at point A. Also find the x intercept of the resultant force R if treated as a single vector.

Let $L = 6$ ft and $\theta = 45°$.

PROBLEM 2.2-1

2.2-2 Find the resultant force and couple at point A for the force system applied to the truss structure shown in the figure. Also find the y intercept of the resultant force R if treated as a single vector.

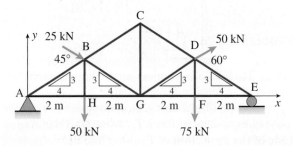

PROBLEM 2.2-2

2.2-3 Three forces and a couple are applied to beam ABCDE (see figure). Let $L = 12$ ft and $\theta = 60°$. Find the resultant force and couple at point A.

PROBLEM 2.2-3

2.2-4 Forces F are applied to the vertices of a hexagon (see figure). Find the resultant force and couple at O.

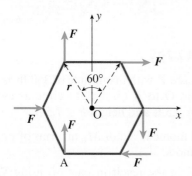

PROBLEM 2.2-4

2.2-5 The force F_{EB} in elevating cylinder EB of a mobile crane has a magnitude of 2500 lb. Assume that $\alpha = 80°$, $\beta = 45°$, and that dimension BC = 18 ft.

(a) Find moment vector M_C in terms of unit vectors i, j, and k.

(b) Confirm the result in part (a) using Varignon's theorem.

PROBLEM 2.2-5

2.2-6 Forces F, each with magnitude of 110 kN, are applied at points A and B (see figure). Replace the two forces by a moment at O. The coordinates of points A and B are given in meters.

PROBLEM 2.2-6

2.2-7 Force F has a magnitude of 10 kips and a line of action from origin O to point A. Coordinates of points A, B, and C are given in inches.

(a) Find the moment of force F about point B.

(b) Repeat part (a) for point C.

(c) Find moments of F about and perpendicular to line CB.

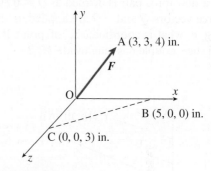

PROBLEM 2.2-7

Representative Problems

2.2-8 Force P is applied at point A and has x-y-z components of $(30, 25, -75)$ kN. The coordinates of points A, B, and C are given in the figure.

(a) Find moment vector M_B in terms of unit vectors i, j, and k.

(b) Confirm the result in part (a) using Varignon's theorem.

PROBLEM 2.2-8

(c) Find the moment of force **P** about and perpendicular to line BC.

(d) Find moment vector M_O then find the moment about and perpendicular to line OB.

2.2-9 Force **P** is applied at point A and has x-y-z components of $(6, 12, -18)$ kips (Fig. a).

(a) Find the coordinates of point D where the line of action of **P** pierces the x-y plane.

(b) If force $-P$ is added at O to form a *couple* (Fig. b), show that the moment vectors M_O, M_A, M_D, and M_C have the same magnitude.

(c) Find the perpendicular distance d between the forces **P** at O and A.

(d) Confirm moment magnitude M_O in part (b) using Varignon's theorem.

(e) If a new force pair is defined as $Q = 0.80\ P$ and force vectors **Q** and $-Q$ are applied as shown in Fig. c, what is coordinate z_Q of point B so that this new couple has magnitude M_O?

(a)

(b)

(c)

PROBLEM 2.2-9

2.2-10 Force **P** is applied at A, force $-P$ is applied at O, and force **F** is applied at C (see figure). The magnitude of **F** is equal to that of **P**. Find the resultant force and couple at O, then find a *wrench* resultant and the coordinates of the point at which it pierces the y-z plane.

PROBLEM 2.2-10

2.2-11 Force **F** with magnitude of 350 lb is directed from point O to A on a right circular cylinder of radius $r = 2.5$ ft and height $h = 7.5$ ft.

(a) Find moment vector M_B in terms of unit vectors i, j, and k.

(b) Confirm the result in part (a) using Varignon's theorem.

(c) Find the moment of force **F** about and perpendicular to line BD.

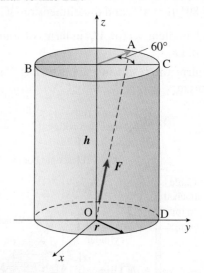

PROBLEM 2.2-11

2.2-12 Force **F** with magnitude of 520 N acts from C to A (see figure).

(a) Find moment vector M_B in terms of unit vectors i, j, and k.

(b) Confirm the result in part (a) using Varignon's theorem.

(c) Find the moment of force **F** about and perpendicular to line BD.

(d) Find moment vector **M**$_O$ then find the moment about and perpendicular to line OD.

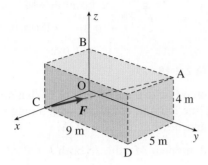

PROBLEM 2.2-12

2.2-13 A ceiling access panel is supported in the open position by cable AB. The tension force T_{AB} in the cable has a magnitude of 15 lb.

(a) Find the moment of force T_{AB} about point O in terms of unit vectors **i**, **j**, and **k**.

(b) Confirm the result in part (a) using Varignon's theorem.

(c) Find the moment of force T_{AB} about and perpendicular to line CD.

PROBLEM 2.2-13

2.2-14 Bracket ADO lies in the x-z plane. Force $F_{DO} = 2.5$ N is applied downward at D, force $T_{AE} = 4$ N is applied at A toward E and moment $M_y = 5$ N·m is applied on member OD (see figure). Coordinates for each joint are given in meters.

(a) Find moment vector **M**$_O$ in terms of unit vectors **i**, **j**, and **k**.

(b) Repeat part (a) using Varignon's theorem.

PROBLEM 2.2-14

2.2-15 A 40-lb force is applied downward at C on the end of the handle BC of a winch (see figure). Segment OA is aligned with the y axis; arm AB lies in a plane which is parallel to the x-z plane; and handle BC is parallel to the y axis.

(a) Find the resultant force and moment vectors at point O.

(b) Find the moment about line AB, which is the 5.0 in.-long segment of the handle.

PROBLEM 2.2-15

2.2-16 Bracket ADO lies in the x-z plane. Force $F_{DO} = 2.5$ N is applied downward at D, force $T_{AB} = 4$ N is applied at A toward B, and moment $M_y = 5$ N·m is applied on member OD (see figure). Coordinates for each joint are given in meters.

(a) Find moment M_O in terms of unit vectors **i**, **j**, and **k**.

(b) Repeat part (a) using Varignon's theorem.

(c) Find moments about and perpendicular to line OC.

(d) What is the shortest distance d from point O to the line of action of force T_{AB}?

PROBLEM 2.2-16

2.2-17 Find the moment vectors M_A, M_D, and M_O for force-couple system F (see figure). Assume that $F = 15$ lb and that joint coordinates shown in the figure are in feet.

PROBLEM 2.2-17

2.2-18 Force F with a magnitude of 500 N acts from B to A (see figure). Coordinates of all points are given in meters.

(a) Find moment of F about point C.

(b) Find moment of F about line CD using a cross product.

(c) Find the shortest distance d between component $F_\perp$ (perpendicular to line CD) and line CD.

PROBLEM 2.2-18

2.2-19 Force F with a magnitude of 20 lb is applied at D along line DB and couple M with a magnitude of 50 in-lb acts normal to plane BCA.

(a) Find moment M_A due to both F and M.

(b) Find the moment about line AC.

(c) What is the perpendicular distance d (inches) from force F to line AC?

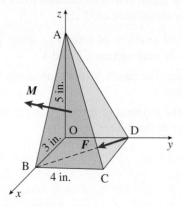

PROBLEM 2.2-19

2.2-20 Couple M_1 acts normal to plane BCA, couple M_2 acts normal to plane CDA, and couple M_3 is normal to plane BODC (see figure). Find a single couple that is statically equivalent to the three couples. Each couple moment has magnitude M_0.

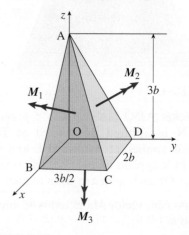

PROBLEM 2.2-20

2.2-21 Forces $F_1 = 12$ lb and $F_2 = 18$ lb are applied at G and couple $C = 14$ lb-ft acts normal to plane OGB (see figure). Find:

(a) Moment of force F_1 about E.

(b) Moment of force F_2 about E.

(c) Moment of couple C at E.

(d) Moment of resultant of F_1, F_2, and C about line EG.

(e) Moment of resultant of F_1, F_2, and C about line EB.

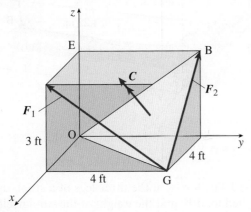

PROBLEM 2.2-21

2.2-22 Forces are applied in the y-direction to members of frame ABCDEF (see figure). Each force acts at the center of the bar segment to which it is applied. All turns of ABCDEF are 90° angles. The length of each bar segment is given in meters.

(a) Find the resultant for this parallel force system.

(b) Find the moment at B.

(c) Find the simplest resultant for this force system and its line of action.

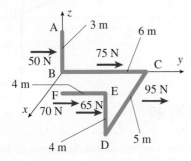

PROBLEM 2.2-22

2.2-23 Forces are applied at B on a steel door handle. Find the resultant force and moment at A.

(a) Use Varignon's theorem.

(b) Use a vector approach.

(c) Find the coordinates of the point at which a *wrench resultant* pierces the y-z plane.

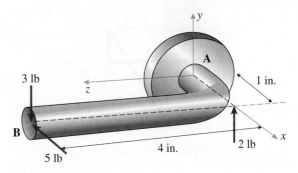

PROBLEM 2.2-23

2.2-24 Find the statically equivalent *wrench resultant* (screwdriver) for the force system shown. Also find the coordinates of the point at which the wrench pierces the x-y plane.

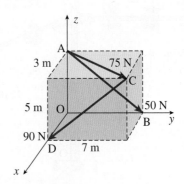

PROBLEM 2.2-24

2.2-25 Find the resultant force and moment at O for the force system shown in the figure.

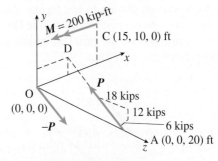

PROBLEM 2.2-25

2.2-26 Find the resultant force-couple system at A, then find the statically equivalent *wrench resultant* (screwdriver) for the force system shown. Also find the coordinates of the point at which the wrench pierces the y-z plane. All turns of ABCDE are 90° angles.

PROBLEM 2.2-26

2.2-27 A traffic signal pole at an intersection supports five signals each weighing 175 lb (see figure). Angle $\beta = 120°$. Find the resultant force and moment at O.

PROBLEM 2.2-27

2.2-28 Find the resultant force-couple system at D for the force system shown in the figure. Also, find moments about and perpendicular to line DF. Lastly, find the coordinates of the point at which a *wrench resultant* pierces the *x-y* plane.

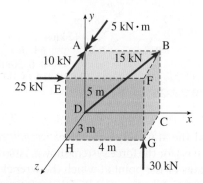

PROBLEM 2.2-28

112

2.2-29 Force F with a magnitude of 25 lb is applied to the handle at B and force N with a magnitude of 15 lb is applied to the head of a socket wrench at A to tighten a bolt at O.

(a) Find the resultant force and moment at point O.

(b) Find the coordinates of the point at which a *wrench resultant* pierces the *x-y* plane.

PROBLEM 2.2-29

2.2-30 The forces in the three legs of a stool, due to applied load W and the weight of the stool itself, are known to have magnitudes of $F_1 = F_2 = 150$ N and $F_3 = 75$ N. Find the coordinates of the point at which the resultant force passes through the *x-y* plane.

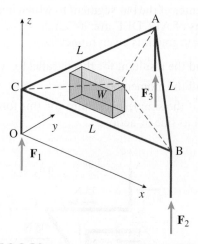

PROBLEM 2.2-30

2.2-31 Forces and moments are applied to bracket OABC (see figure). Member AB is parallel to the *y* axis, and member BC is parallel to the *x* axis. Find the resultant force and moment at O.

(a) Use Varignon's theorem.

(b) Use a vector algebra approach.

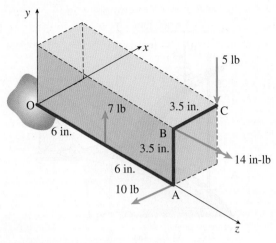

PROBLEM 2.2-31

2.2-32 Forces and moments act on a 3 m × 3 m square rigid plate (see figure). Find the equivalent force-couple system at point O at the center of the plate.

PROBLEM 2.2-32

2.2-33 Space frame OABC is acted on at B by couple M_x with magnitude of 100 in-kip and at C by the force F with magnitude of 25 kips due to cable CD.

(a) Replace the couple and force by a force-couple system at O.

(b) Find the moment about and perpendicular to member OA.

PROBLEM 2.2-33

2.2-34 Force P has a magnitude of 3 kN and a line of action from C to E; force Q has magnitude of 4 kN and acts from F to E (see figure). Coordinates of points A through E are given in meters.

(a) Find the equivalent force-couple system at point at O.

(b) Find the moment about and perpendicular to line AD.

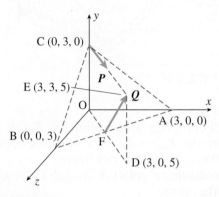

PROBLEM 2.2-34

2.2-35 Force P has a magnitude of 10 kips and a line of action from H to B (see figure).

(a) Find the equivalent force-couple system at point D.

(b) Find the moment about and perpendicular to line DG.

(c) Find the perpendicular distance from point A to the line of action of force P.

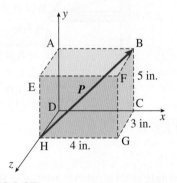

PROBLEM 2.2-35

2.2-36 Four cables support an entrance canopy. The magnitude of cable force T_{CE} is 50 kN.

(a) Find the equivalent force-couple system at point A for cable force T_{CE}.

(b) Find the moment of force T_{CE} about and perpendicular to line AD.

Assume that $a = 10$ m, $b = 4$ m, and $c = 3$ m.

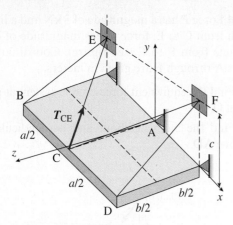

PROBLEM 2.2-36

2.2-37 Four forces with magnitudes $T_{DB} = T_{DC} = 10$ kips, and $T_{DE} = T_{DF} = 15$ kips are applied at point D on a framework (see figure).

Coordinates of points A through G are given in feet in the figure.

(a) Find the equivalent force-couple system at point at A.

(b) Find the moment about vertical member AG.

PROBLEM 2.2-37

2.2-38 A single steel strut AB supports the vehicle engine hood of mass 20 kg which pivots about hinges at C and D (see figure parts a and b). The strut lies in a vertical plane and is attached to a bolt at A. The magnitude of the force in the strut is $F_S = 154$ N.

(a) Find the equivalent force-couple system at point at A for forces W and F_S.

(b) Find the moment of these two forces about hinge line CD.

(a)

(b)

PROBLEM 2.2-38

2.2-39 A crane boom is stabilized by two cables AQ and BQ as shown in the figure. A load P with magnitude of 4500 lb is carried at point D along line DE. The crane boom lies in the y-z plane. Find the equivalent force-couple system at point at O for boom weight W with magnitude of 975 lb and applied force P.

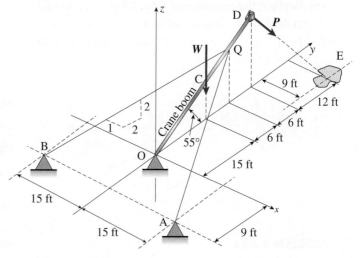

114

Equilibrium of Components and Rigid Bodies

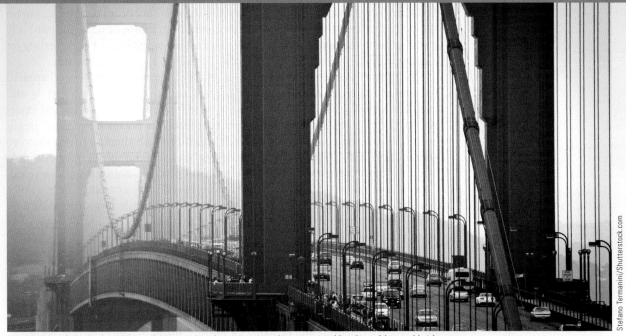

Stefano Termanini/Shutterstock.com

The weight of the bridge deck, vehicles and pedestrians must be supported by the cables and bridge towers.

CHAPTER OBJECTIVES

- Present the *equations of static equilibrium* for two-dimensional (2D) and then for three-dimensional (3D) bodies, expressed in both vector algebra and scalar algebra.

- Describe different *support conditions* that define possible restraint conditions for a structure or component of interest.

- Show how to draw a proper *free-body diagram* of a structure or component, which is an essential first step in its static equilibrium analysis.

- Discuss *stability* and static *determinacy* of both 2D and 3D structures.

- Examine *friction forces* between two bodies or components in contact; consider dry friction forces in the equilibrium analysis of a variety of structures and machines.

Chapter 3 Outline:

3.1 Introduction

The most important topic in statics is evaluation of *equilibrium* of structures and machines under the action of systems of forces. If equilibrium is to be achieved, a body must be supported or restrained in such a way that it is stable and at rest, and therefore the *resultant* of all forces and the *resultant* of all couples acting on the body each must sum to zero. A restrained body cannot undergo rigid-body motion due to the application of *static* forces. One exception is a body resting on a rough surface whose motion is opposed by static and kinetic friction.

Start by constructing a *free-body diagram* of the body to display all applied (or active) and reactive forces. To do so, first *isolate* the body under study and then apply the *equations of equilibrium* to the free-body diagram to find external reaction forces and couples or internal forces at points of interest. Begin by writing the basic static equilibrium equations and then applying them to the solution of a variety of structures using both scalar and vector operations. Statics is most often concerned with bodies at rest, so assume that both velocity and acceleration of the body are zero. You can also consider the case of a body in motion—but having zero acceleration—with the motion opposed by kinetic friction forces.

3.2 Free-Body Diagrams

A static equilibrium analysis is impossible without a correct free-body diagram. The free-body diagram represents the plan for carrying out an analysis to solve for unknown reactions and possibly for internal forces as well. The free-body diagram is drawn during the *conceptualization stage* of the problem-solving activity prior to writing equilibrium equations or selecting a mathematical approach. This section first examines the equilibrium equations needed for either 2D or 3D solutions in vector or scalar algebra format. Then seven different support or restraint conditions for use in constructing a stable model of the component are presented. Finally, the importance of free-body diagram construction is illustrated through a number of practical examples.

Equilibrium Equations

The resultant force R and resultant couple M of *all* forces acting on either a rigid or deformable body *in equilibrium* are both zero. The resulting equilibrium equations are expressed in *vector form* as

$$R = \Sigma F = 0 \tag{3-1}$$

$$M = \Sigma M = \Sigma (r \times F) = 0 \tag{3-2}$$

where F is one of a number of vectors of forces acting on the body and r is a position vector from the point where moments are taken to a point along the line of application of force F. It is often convenient to write the equilibrium equations using *scalar algebra* within a rectangular Cartesian coordinate system. In two dimensions (x, y), the equilibrium equations can be written as

$$\Sigma F_x = 0 \quad \Sigma F_y = 0 \quad \Sigma M_z = 0 \tag{3-3}$$

where the z axis is normal to the plane of the body through any convenient point on or off the body. Equation (3-3) can be used for two-dimensional or planar problems, but in three dimensions (x, y, z), three force and three moment equations are required:

$$\Sigma F_x = 0 \quad \Sigma F_y = 0 \quad \Sigma F_z = 0 \tag{3-4}$$

$$\Sigma M_x = 0 \quad \Sigma M_y = 0 \quad \Sigma M_z = 0 \tag{3-5}$$

If the number of unknowns is equal to the number of independent equilibrium equations, Eqs. (3-4) and (3-5) are sufficient to solve for all unknown reactions or internal forces in the body, and the problem is referred to as *statically determinate*, provided that the body is stable. If the body or structure is constrained by additional or *redundant* supports, it is *statically indeterminate*, and a solution is not possible using the laws of static equilibrium alone. The equilibrium equations are the conditions for equilibrium. If the body is statically determinate and rigid, they are necessary and sufficient to establish equilibrium of the body.

Reactive Forces and Support Conditions

Proper restraint of the body or structure is essential if the equilibrium equations are to be satisfied. A *sufficient number* and *appropriate arrangement* of supports must be present to insure a stable model and prevent rigid-body motion under the action of static forces. A reaction force at a support is represented by a single arrow with a slash drawn through it, while a moment restraint at a support is shown as a double-headed or curved arrow with a slash. Reactions usually result from the action of applied (or active) forces shown as arrows without a slash. Representative reaction forces are shown in Table 3-1.

Review the action of forces on mechanical systems so you can correctly represent them on free-body diagrams. Different support conditions may be assumed, depending on whether the problem is 2D or 3D. Table 3-1 lists some of the most commonly used idealizations for 2D and 3D supports.

The reaction and constraint forces described above are essential elements of a free-body diagram. Include all forces acting on the free-body diagram to obtain a correct static equilibrium solution for all reactions and internal forces.

Importance of Free-Body Diagrams

A free-body diagram is an *essential first step* in a static analysis of a rigid or deformable body. All forces and couples acting on the body, or component part of the body, must be displayed to obtain a correct equilibrium solution. This includes applied forces, reactions, and any contact forces if a body is broken into separate components and analyzed. Some examples of the importance of sketching a correct free-body diagram follow in this chapter.

A *statics sign convention* is used in the solution for unknown forces. Forces acting in the positive directions of the coordinate axes are positive, and counterclockwise moment vectors are positive. While the direction of reaction or contact forces may be obvious in many problems, it is good practice to use a statics sign convention so that the sign of the final result reveals the correct sense of all unknowns.

Free-body diagrams are critical in solving statics problems, so spend some time studying how to convert mechanical systems into free-body diagrams. It is important to carefully review this first step in the *four-step problem-solving approach* to calculating unknown forces and not rush to solution of the resulting equations. *An incorrect free-body diagram can lead to erroneous equilibrium equations and results.*

Table 3-1	Type of Support or Connection	Simplified Sketch of Support or Connection	Display of Restraint Forces and Moments, or Connection Forces
Modeling reaction forces and support conditions in 2D or 3D static analysis			

1. Roller Support: A single reaction force R is developed and is normal to the rolling surface; force R opposes motion into or away from the rolling surface. The rolling surface may be horizontal, vertical, or inclined at some angle θ. If friction is present, then include a force F opposing the movement of the support and tangential to the rolling surface. In 3D, the roller moves in the x-z plane and reaction R_y is normal to that plane.

Bridge with roller support (see 1.1, 1.2)

Bridge with *rocker* support (see 1.3)

(1.1)   (1.2)   (1.3)

Horizontal *roller* support [(1.1), (1.2)]; or alternate representation as *rocker* support [(1.3)] Both downward and uplift motions are restrained.

Vertical roller restraints

Rotated or inclined roller support

3D roller support

(a) **Two-dimensional roller support** (friction force $F = 0$ for smooth rolling surface)

(b) **Three-dimensional roller support** (friction force $F = 0$ for smooth rolling surface; reaction R_y acts normal to plane x-z on which roller translates)

2. Pin Support: A single resultant force, usually shown using two rectangular components R_x and R_y in 2D but three components in 3D, resists motion in any direction normal to the pin. The pin support cannot resist moment, and the pin is free to rotate about the z axis. In 3D, the pin becomes a ball-and-socket joint or support.

Two-dimensional pin

(a) **Two-dimensional pin support**

Ball-and-socket joint model
(camera mount, 3D)

Hip prosthesis for hip
replacement

Ball-and-socket joint model

(b) **Three-dimensional pin
support**

3. Sliding Support: A support that translates without rotation is a sliding support. Examples are a collar sliding along a sleeve or a flange moving within a slot. Reactions in 2D are a force R_x normal to the sleeve and a moment M_z representing resistance to rotation relative to the sleeve. In 3D, the sliding support translates on frictionless plane y-z and reaction moment components M_y and M_z prevent rotation relative to that plane.

Sliding support for column light
stand ·

Frictionless sleeve
on vertical shaft

Two-dimensional sliding
support

Friction F opposes
motion in $+y$ direction in
2D along sliding surface;
F is zero if smooth
surface is assumed.

In 3D, add restraint
moment M_x to prevent
rotation about x axis.

(a) **Two-dimensional sliding
support** (support translates
on frictionless path along
$+y$ or $-y$ direction)

(b) **Three-dimensional sliding
support** (support translates
on frictionless y-z plane)

(Continued)

Type of Support or Connection	Simplified Sketch of Support or Connection	Display of Restraint Forces and Moments, or Connection Forces

4. Fixed Support: No translation or rotation occurs between member and support in a fixed support. This requires three reaction components in 2D: force components R_x and R_y and moment M_z. In 3D, three force-reaction components and three moment-reaction components are required.

Steel bollard anchored in concrete

Column bolted to footing

Fixed support at base of sign post

(a) **Two-dimensional fixed support**

Horizontal member

Vertical member

(b) **Three-dimensional fixed support**

5. Elastic or Spring Support: In 2D, there may be a longitudinal or normal translational spring or a combination of both. For linear springs, the support reaction at the base of the spring is the product of the spring constant k times the displacement δ in the direction of the spring axis. If joint A translates in $+x$ (δ_x) and $+y$ (δ_y) directions, reaction forces R_x and R_y are created in $-x$ and $-y$ directions, respectively, at the supports of linear translational springs. Alternatively, the support may be pinned for translation but have moment spring k_r for rotation. If joint A rotates about the $+z$ axis (θ_z), reaction moment M_z is created in the $-z$ direction at the base of the rotational spring. In 3D, a fully elastic support consists of three translational springs (k_x, k_y, k_z) and three rotational springs (k_{rx}, k_{ry}, k_{rz}), and an arbitrary joint displacement results in three reaction forces and three reaction moments. In the limit, as each spring constant value approaches infinity, the elastic support becomes a fully fixed support like that shown in Section 4b above.

Translational spring support for heavy equipment

In **3D**, add spring in $+z$ direction k_z with reaction force $R_z = -k_z \delta_z$.

In **3D**, add rotational *flexural* spring about $+y$ direction k_{ry} with reaction moment $M_y = -k_{ry}\theta_y$ and add rotational *torsional* spring about $+x$

(a) **Translational spring (k) in 2D**

$R_x = -k_x \delta_x$

Rotational spring in a clothespin

direction with reaction
moment $M_x = -k_{rx}\theta_x$.
   Torsion springs are
found in window shades
and as part of the lift
mechanism in power garage
door-opening systems.

(b) **Rotational spring (k_r) in 2D**

$$M_z = -k_r\,\theta_z$$

6. Wheel on Rail Support: This support is a particular form of the 3D roller support
(see Section 1b above). Now general movement in the *x-z* plane is constrained by
normal force R_y and lateral force R_x, both acting normal to the rail or slot on which
the wheel travels. If friction is considered, friction force F is added along the rail in the
direction opposing the wheel translation.

Wheel rolls on rail or in
slot along *z* axis; friction
force opposing motion
is neglected; R_x is lateral
constraint force, R_y is
normal force.

Cross section through guide rail

7. Thrust-Bearing Support: A thrust bearing constrains translational motion along
the shaft axis while allowing rotary motion to occur about that axis. Support reaction
forces and moment components act in all directions except for reaction moment
$M_x = 0$ about the thrust axis (in the absence of friction). A special case is the journal
bearing for which axial thrust restraint component $R_x = 0$.

Thrust bearing has support
reaction force (R_x, R_y, R_z)
and reaction moment
components (M_y, M_z)—no
moment M_x about the
thrust or rotation (*x*) axis.

Journal bearing has no
axial thrust reaction force
$(R_x = 0)$ in addition to
$(M_x = 0)$.

Pillow block bearing

Free-Body Diagram Examples

The following examples demonstrate how to convert a physical structure or mechanical system into a simplified model for analysis using the laws of static equilibrium. Begin by selecting the structure of interest and then *isolate* it from any surrounding or attached bodies. Show reactions at points of support, and then add forces at points where the structure comes into contact with other bodies. You also need to add forces due to gravity and magnetic attraction between bodies (body forces). Applied (or active) and reactive forces may be concentrated or distributed; for now, treat gravitational forces as concentrated forces acting at the center of gravity of the body.

Plane Truss   The plane truss in Fig. 3-1 is supported at A and B and is subjected to vehicle loads ($2P$, P) at joints along the bottom chord of the truss. External load P is a known load value (such as the weight of a small car); the weight of a truck might be twice this value, as shown in the figure. For simplicity, the weight of each of the truss members is assumed to be small compared to load value P, so it is not considered here. Also, truss dimensions are omitted but are needed if you use force and moment equilibrium equations to find reactive force unknowns at A and B. The *pin* support at A provides resistance to both horizontal and vertical translation [see Table 3-1(2)], so concentrated reactive force components A_x and A_y are shown at joint A in the free-body diagram (Fig. 3-1b). The support at B is a *roller* [see Table 3-1(1)] capable of resisting vertical force only, so only force component B_y is shown in Fig. 3-1b. [Alternatively, support B could be a *rocker* support that differs physically from a roller while having the same behavior; see Table 3-1(1)]. The overall truss structure is treated as a single rigid body in constructing the free-body diagram in Fig. 3-1b. For now, only reactions at A and B are of interest, so details pertaining to the individual truss members and joints can be ignored; hence, the individual members are concealed by shading and only the boundary of the structure is shown in the figure.

Cable-Supported Beam   Beam AB in Fig. 3-2 is supported by cable BC. The beam is bolted to the wall at A, so no moment restraint is provided, and support A may be treated as a pin support with reactive force components A_x and A_y. You are only interested in forces acting on the beam, so isolate the beam by cutting through cable BC (see dashed outline isolation box), thereby exposing its internal force T. Since the line of action from B to C is known, use the *principle of transmissibility* to slide force T to point B. Components of the cable

---

FIGURE 3-1

(a) Plane truss structure and
(b) free-body diagram

(a)   (b)

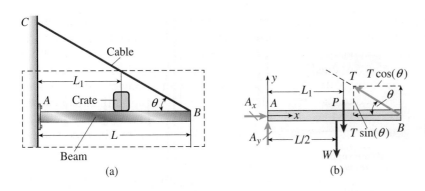

force can be expressed in terms of angle θ and applied at B on the free-body diagram on the right. The uniformly distributed weight of the beam is represented as a concentrated force W at the beam center of gravity, while the load due to the crate is applied as force P at distance $x = L_1$ to the right of A.

Airplane   The airplane in Fig. 3-3a is being towed to the hanger. A static equilibrium analysis reveals forces acting on the landing gear and also the required towing force at the front support at B. A two-dimensional free-body diagram can be used if wing loads and landing gear reactions are assumed to be equal in magnitude for the right and left sides of the airplane. The 2D free-body diagram displays:

- Towing force F applied to the front landing gear.
- Weight W_1 for the main fuselage (including passengers and cargo).
- Weight W_2 for the structure plus fuel of both wings.
- Weight W_3 for the two jet engines.

Reaction force A_x is the combined rear gear drag due to rolling friction. Force A_y is the combined reaction force, which is equal to the sum of forces at each of the rear landing gear, and force B_y is that at the front landing gear. Of course, the exact location of each of the forces is needed before a static equilibrium analysis can be performed.

Automobile Lift Frame   Figure 3-4 shows a free-body diagram of a small automobile lift. The reaction forces (A_x, A_y) and reaction moment (M_A) at the base of the lift frame are needed for design of both the framework and the foundation. Assume that the centers of gravity of both the frame and the car

FIGURE 3-4

(a) Automobile supported on lift
and (b) two-dimensional free-body
diagram of lift frame

(a)   (b)

lie in the same x-y plane so that a 2D free-body diagram can be used. Applied loadings on the frame are

- Weight W_1 of the horizontal frame member
- Weight W_2 of the vertical frame member
- Weight C of the car.

Weights W are applied at the centers of gravity of each of the frame parts and half of the total car weight C is applied at each ramp support on the free-body diagram in Fig. 3-4b.

Container Crane   The container crane in Fig. 3-5 loads and unloads containerized cargo from a ship. The overall crane structure moves along rails that run parallel to the dock and is supported at four points (two along rail A and two along rail B). Once in position to unload a ship, the crane is anchored to create pin supports at A and B. If all loads act in the x-y plane that contains the center of gravity of the crane, a two-dimensional free-body diagram can be used in a static equilibrium analysis to find reaction forces at pin supports A and B. Applied loads include:

- Weight W_1 of the rear ballast.
- Weight W_2 of the machine shed.
- Overall weight W_3 of the crane frame members.
- Weight P of the trolley (operator cabin, cables, and spreader) plus the container cargo.

FIGURE 3-5

(a) Container crane and
(b) two-dimensional free-body
diagram of container crane

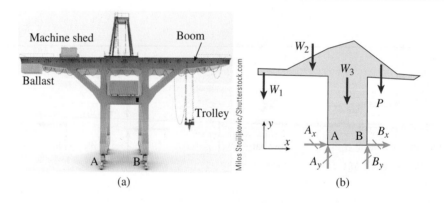

(a)   (b)

To lift different pieces of cargo, the crane operator moves the trolley along the crane boom, so the exact location of force P must be clearly specified before an equilibrium analysis can be performed. To examine forces between portions of the structure, create separate free-body diagrams for those isolated portions and then carry out an equilibrium analysis for each portion.

Mobile Construction Crane A small mobile crane is used to lift a section of drill pipe of weight P (Fig. 3-6a). A cable attached to the top end of the pipe runs over two pulleys at A and is anchored to a winch on the crane boom near D. Isolate the telescopic boom $ABCD$ to do a static equilibrium analysis to find the forces acting at B and C along the boom (see dashed box in Fig. 3-6b). Sketch a free-body diagram (Fig. 3-6c) for use in a static equilibrium analysis. To isolate boom $ABCD$, cut through the cable just above the suspended pipe near A and apply downward cable force P along the line of action of the lift cable. Also cut through the elevating cylinder between B and E or simply detach the cylinder from the boom at B and replace it with lifting force F_{EB}. Finally, remove the pin at C that attaches the boom to the truck. The cable winch is attached to the boom at D, so there is no need to cut the cable along the top side of the boom. The weight W of the boom is applied at its center of gravity.

FIGURE 3-6

(a) Mobile crane, (b) mobile crane lifting drill pipe, and (c) two-dimensional free-body diagram of crane boom

Reaction force components C_x and C_y represent the pin reaction at C. Finally, force F_{EB} in elevating cylinder EB is applied at B to hold the crane arm in the position defined by angles α and β.

Traffic Signal Pole   Next, construct free-body diagrams for both 2D and 3D models of traffic signal poles at a road intersection (Fig. 3-7). The framework is a vertical cantilever, so only base reactions at A must be included as reaction unknowns (these reactions are used to design both the pole and the base foundation for the pole assembly). In 2D, forces W_1 and W_2 represent the weights of the signal arm and pole, respectively, and each force P is the weight of one signal structure. For a 2D cantilever, there are only two force reactions (A_x, A_y) and one moment reaction (M_A). For a 3D signal structure (Figs. 3-7c to e), apply forces W and P on each arm, and additional weight W_3 for the pole. There are three reaction force components and three reaction moment components at A (Fig. 3-7e).

Cabinet Door   Consider a cabinet door supported by hinges at A and B (Fig. 3-8a). The hinges are considered small in relation to overall door dimensions, so each hinge supports only reaction forces at support points A and B. The door is closed against a magnetic switch at D. The weight of the door (W) is applied at its center of gravity. Closing force P is applied to the door at the location of the door latch at E. Each hinge provides reaction force components in x-y-z directions, and the door stop reaction is force D_z. Ignoring friction, the hinge provides no restraining moment about the y axis, so components M_{Ay} and M_{By} are both zero and are not shown on the free-body diagram.

FIGURE 3-7

(a) Traffic signal pole assembly (2D), (b) two-dimensional free-body diagram of signal pole, (c) photo of signal pole assembly (3D), (d) traffic signal pole assembly (3D), and (e) three-dimensional free-body diagram of signal structure

(a)   (b)   (c)

(d)   (e)

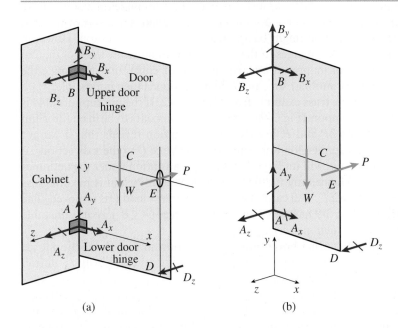

FIGURE 3-8

(a) Cabinet door supported on hinges (3D) and (b) free-body diagram of cabinet door

(a)   (b)

In summary, proper construction of the free-body diagram is an important first step in a static analysis of a structure or mechanical system leading to an equilibrium solution for reactive forces that support a body. Additional free-body diagrams are needed to find contact forces between and internal forces within components that make up a body. Development of the free-body diagram occurs during the *conceptualization stage* of a well-conceived problem-solving strategy and is an indispensable step on the path to a correct solution. Errors made in development of the free-body diagram will lead to erroneous results. Problems at the end of this chapter give you practice in sketching correct free-body diagrams—both for the structure as a whole or for portions of a structure. In addition, detailed examples at the end of Section 3.3 begin with the construction of free-body diagrams and illustrate their use in carrying out the static equilibrium solution using the *four-step problem-solving approach*.

3.3 Equilibrium in 2D and 3D

Before looking at detailed examples of structures or components in equilibrium, you must ensure that the structure is *properly restrained* so that an equilibrium solution is possible. In addition, the structure must not have more *constraints* (or supports) than available independent equations of equilibrium; otherwise it is classified as *statically indeterminate*. A discussion of constraints, mobility, and static determinacy is presented first, followed by examples where support reactions are found for a variety of statically determinate structures.

Constraints and Mobility

If equilibrium Eqs. (3-1) through (3-5) are satisfied, the body is in a state of static equilibrium. There must be a *sufficient number* and *appropriate arrangement* of supports or constraints to prevent rigid-body motion without resistance under any potential loading applied to that body. All of the possible constraints on a 2D or 3D body are summarized in Table 3-1, and many were used in the free-body diagram examples presented previously. In all of these examples, the bodies were stable and in static equilibrium—not only for the applied forces shown but also for any other loadings that might be applied to the body in question. However, it is rather easy to make changes to these structures to illustrate the fact that an insufficient number or inappropriate arrangement of constraints can make the structure unstable (or mobile); an equilibrium solution is then impossible.

Consider the **plane truss** example from Section 3.2. If the pin support at A is changed to a roller support (Fig. 3-9a), the structure appears to be in equilibrium under vertical loads $2P$ and P, but closer examination reveals that it cannot resist an additional applied loading, such as some force Q in the x direction, so it is unstable. The truss will undergo rigid-body translation under the action of horizontal forces. Hence, any further consider of equilibrium is meaningless because the equation $\Sigma F_x = 0$ cannot be satisfied unless load Q is removed. The truss structure in Fig. 3-9 has only two reaction forces (A_y, B_y), so it is unstable because of an *insufficient number of constraints*.

Even if you add another horizontal roller support at C with reaction force C_y in Fig. 3-10a, the truss remains unstable, making an equilibrium solution impossible. There are three reactions (A_y, B_y, C_y) and three equations of equilibrium [Eq. (3-3)], but now there is an *inappropriate arrangement* of constraints (three horizontal roller supports), and $\Sigma F_x = 0$ cannot be satisfied. The structure is mobile; it cannot resist an applied loading in the x direction, such as force Q (or any inclined loading with a component in the x direction).

Another example of an *inappropriate arrangement* of constraints is shown in Fig. 3-11a. The pin support has been restored to point A, but the roller support at B has been removed and the roller at C has been rotated so that the line of action of reaction force R_C passes through point A. There are now three constraints (A_x, A_y, and R_C in Fig. 3-11b), so equilibrium equations $\Sigma F_x = 0$ and $\Sigma F_y = 0$ are written using scalar algebra and the Cartesian components of R_C as

$$\Sigma F_x = 0 \text{ so } A_x + R_C \cos\theta + Q = 0 \quad \text{or} \quad A_x + R_C \cos\theta = -Q$$
$$\Sigma F_y = 0 \text{ so } A_y + R_C \sin\theta - 2P - P = 0 \quad \text{or} \quad A_y + R_C \sin\theta = 3P$$

However, the line of action of reaction C passes through point A and makes it impossible to satisfy the third equilibrium equation $\Sigma M_A = 0$:

$$\Sigma M_A = 0 \text{ so } (-R_C \cos\theta)(H) + (R_C \sin\theta)(2L) - QH - 2PL - P(2L) = 0$$
$$\text{or} \quad R_C(-H \cos\theta + 2L \sin\theta) = QH + 4PL$$

However, the moment arm $(-H \cos\theta + 2L \sin\theta)$ is 0, so you are left with the meaningless result that $QH + 4PL = 0$, which confirms that the structure

FIGURE 3-9

(a) Unstable plane truss with two constraints and (b) free-body diagram of unstable plane truss

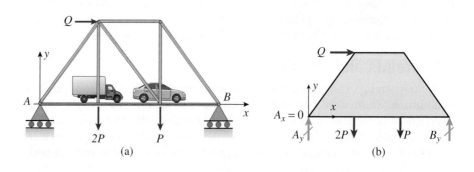

FIGURE 3-10

(a) Unstable plane truss with three constraints, and (b) free-body diagram of unstable plane truss

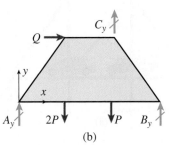

is unstable. Applied loads Q, $2P$, and P produce clockwise moments about A, but reaction force R_C has a resisting moment of zero about A.

Even the addition of extra but inappropriate constraints at B and D on the plane truss (Fig. 3-11c) does not fix the problem because all reaction forces still have a line of action through point A. Finally, if the pin at A is changed to a roller support and roller supports are in place at B and C as shown in Fig. 3-11d, all reaction forces now pass through a point D, and you are not able to satisfy the equation $\Sigma M_D = 0$, so the structure is still unstable because of an inappropriate arrangement of supports.

So you see that *external stability* depends on the number and/or arrangement of constraints. It is also important to consider *internal stability* to make sure there are a sufficient number and appropriate arrangement of *members* to prevent collapse of the truss under load. For example, if the center diagonal member is removed in the pin-jointed truss in Fig. 3-12a, the structure will collapse because of an insufficient number of members (Fig. 3-12b).

FIGURE 3-11

(a) *Inappropriate arrangement* of constraints, (b) free-body diagram of unstable plane truss, (c) unstable plane truss (moment at A), and (d) unstable plane truss (moment at D)

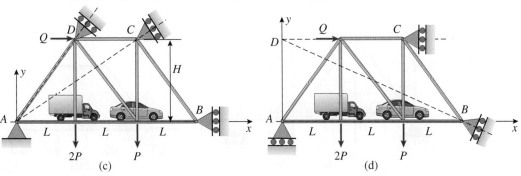

FIGURE 3-12

(a) Stable plane truss structure and
(b) unstable due to insufficient
number of members

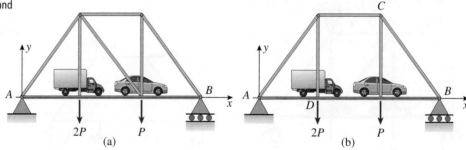

Independent Equations of Equilibrium
For two-dimensional struc-
tures, three *independent* equations are required to establish static equilibrium
and to solve for all unknown reaction forces or moments (in three dimensions,
six independent equations are needed—see Example 3-5). In many cases, the
obvious choice is $\Sigma F_x = 0$, $\Sigma F_y = 0$, and $\Sigma M_z = 0$. For the plane truss example in Fig. 3-13a, use the free-body diagram in Fig. 3-13b to obtain three equilibrium equations and a solution for reactions at A and B as

$$\Sigma F_x = 0 \quad \text{so} \quad A_x = 0 \tag{3-6a}$$

$$\Sigma M_A = 0 \quad \text{so} \quad B_y(3L) - 2P(L) - P(2L) = 0 \quad \text{or} \quad B_y = \frac{4}{3}P \tag{3-6b}$$

$$\Sigma F_y = 0 \quad \text{so} \quad A_y + B_y - 2P - P = 0 \quad \text{or} \quad A_y = 3P - B_y = \frac{5}{3}P \tag{3-6c}$$

A *statics sign convention* was used in writing these equations: Forces in $+x$
and $+y$ directions are assumed positive, and a counterclockwise moment about
the $+z$ axis is positive. If a negative answer is obtained, the reaction force or
moment is opposite in direction to that shown on the free-body diagram.

An alternate approach is to use *independent moment equilibrium equations*
$\Sigma M_A = 0$, $\Sigma M_B = 0$, and $\Sigma M_C = 0$ to arrive at the same solution for reactions A_x, A_y, and B_y as in Eq. (3-6):

- $\Sigma M_A = 0$ leads to the solution for reaction B_y
- $\Sigma M_B = 0$ results in the solution for reaction A_y
- $\Sigma M_C = 0$ confirms that $A_x = 0$

FIGURE 3-13

(a) Plane truss structure and
(b) free-body diagram

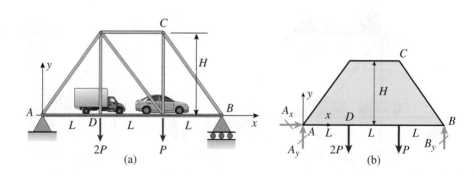

Other valid approaches for finding reactions at A and B use the following equations.

1. $\Sigma M_A = 0$, $\Sigma M_B = 0$, and $\Sigma F_x = 0$
2. $\Sigma F_x = 0$, $\Sigma F_y = 0$, and $\Sigma M_C = 0$
3. $\Sigma F_x = 0$, $\Sigma F_y = 0$, and $\Sigma M_D = 0$
4. $\Sigma F_y = 0$, $\Sigma M_C = 0$, and $\Sigma M_D = 0$

Use of $\Sigma M_A = 0$, $\Sigma M_B = 0$, and $\Sigma M_D = 0$ will not lead to a solution because $\Sigma M_D = 0$ provides duplicate information that confirms the relative values of reactions A_y and B_y but does not lead to a solution for A_x. In general, three moment equations written for any three points along a straight line, such as line ADB in Fig. 3-13b, are not independent, and a full solution for all unknown reactions cannot be obtained using these equations.

Static Determinacy

Three *independent* equations are required for two-dimensional structures to establish static equilibrium, and the solution of these equations gives all unknown reaction forces or moments. In three dimensions, six independent equations are needed to obtain a full solution for all reaction unknowns. Once you have established that the structure is stable, you can compare the number of unknowns to the number of independent equations of equilibrium.

Plane Truss   The constraints are *external* to the plane truss in Fig. 3-13a, and this is emphasized in the free-body diagram sketch in Fig. 3-13b where the internal members are concealed by shading. If the number of unknowns is equal to the number of available equations, the problem is described as *externally statically determinate*. For this *statically determinate* plane truss, the three equilibrium equations and their solution presented in Eq. (3-6) illustrate the approach to be used to find all three reaction forces at supports A and B.

If roller support B in the plane truss is changed to a pin support (Fig. 3-14a), there are now four reaction unknowns, as shown in the updated free-body diagram in Fig. 3-14b. Equation (3-6a) must be revised to state that $\Sigma F_x = 0$, so $A_x + B_x = 0$ and now $A_x = -B_x$. However, a unique solution for either A_x or B_x cannot be found—only that they are equal in magnitude and opposite in sign. With three independent equations of equilibrium and four unknown reactions, the structure is now *one degree statically indeterminate externally*.

Space Truss   The space truss structure in Fig. 3-15a supports a small awning (Fig. 3-15b) over the entrance to a coffee shop. Individual truss members are assumed to be hinge-connected to each other, so in response to any applied loads such as forces P in Fig. 3-15a, each member is assumed to develop an internal axial force whose line of action lies along the axis of the member itself. The structure is a *simple* space truss made up of two tetrahedra ($ABCG$ and $DEFG$). The

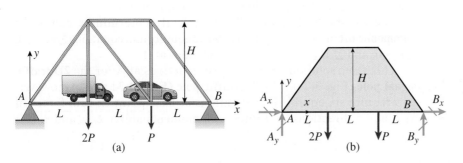

FIGURE 3-14

(a) Statically indeterminate plane truss structure and (b) free-body diagram

FIGURE 3-15

(a) Space truss structure made up using two tetrahedra, (b) awning supported by truss structure, and (c) free-body diagram of space truss structure

(a)

(b)

(c)

truss is pin-supported at joints C and F, and there are additional reaction forces at joints A and D (Fig. 3-15a). The space truss appears to be stable. There are *nine* external reaction forces, so solution for the reaction forces using six independent static equilibrium equations is not possible using the free-body diagram for the structure as a whole (Fig. 3-15c). The truss is *three-degrees statically indeterminate externally*.

FIGURE 3-2 (Repeated)

(b) Free-body diagram of cable-supported beam

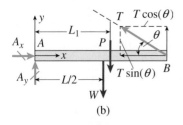

(b)

FIGURE 3-3 (Repeated)

(b) Free-body diagram of airplane

(b)

Additional Examples   The **cable-supported beam** in Fig. 3-2 appears to be stable under the applied loading. The beam model has a sufficient number and appropriate arrangement of constraints, so it is statically determinate. An equilibrium solution for reaction force components A_x and A_y, as well as cable force T, is possible using the free-body diagram in Fig. 3-2b. Suitable equilibrium equations are $\Sigma F_x = 0$, $\Sigma F_y = 0$, and $\Sigma M_A = 0$.

The same is true for the **airplane** example in Fig. 3-3. Reaction forces A_x, A_y, and B_y can be computed using the free-body diagram in Fig. 3-3b to write the three independent equations of static equilibrium.

The free-body diagram of the statically determinate cantilever **automobile lift** in Fig. 3-4b is used to write equilibrium equations $\Sigma F_x = 0$, $\Sigma F_y = 0$, and $\Sigma M_A = 0$, leading to reactions at A.

This same solution approach gives reactions at A for the **2D traffic signal pole** (Fig. 3-7b). The **3D traffic pole** (Figs. 3-7d and e), however, will require six equilibrium equations at base point A: sum forces in x, y, and z directions at A, then sum moments about x, y, and z axes

FIGURE 3-4 (Repeated)

(b) Free-body diagram of automobile lift

(b)

FIGURE 3-5 (Repeated)

(b) Free-body diagram of container crane

(b)

FIGURE 3-6 (Repeated)

(c) Free-body diagram of crane boom

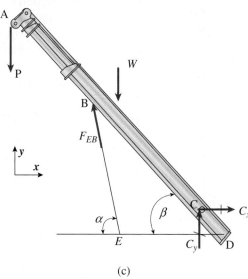

(c)

at A. Both two and three-dimensional signal poles are statically determinate structures.

The free-body diagram of the telescopic boom (Fig. 3-6c) on the **mobile crane structure** is also statically determinate. Solution for hydraulic strut force F_{EB} and pin force components C_x and C_y involves summing moments about C to find F_{EB} followed by summation of forces in x and y directions, respectively, to find C_x and C_y.

Consider next the two-dimensional free-body diagram of the **container crane** (Fig. 3-5b repeated here). The crane moves on rails at A and B, which are parallel to the z axis. The crane is moved into position along the rails and then is pinned at A and B to load or unload the container ship. There are *four* reaction forces (A_x, A_y, B_x, and B_y) but only *three* independent equilibrium equations (such as $\Sigma F_x = 0$, $\Sigma F_y = 0$, and $\Sigma M_A = 0$), so the crane structure model is *one-degree statically indeterminate externally*. There is one extra reaction, or *redundant*, beyond the minimum required for equilibrium (like the plane truss free-body diagram in Fig. 3-14b). If either A_x or B_x is removed, the structure is still in equilibrium under applied forces W_1, W_2, W_3, and P. As with the plane truss, it is not possible to determine the magnitudes of reactions A_x and B_x using statics alone. You can only find that $\Sigma F_x = 0$, so $A_x + B_x = 0$. Hence A_x and B_x are equal in magnitude but are of opposite sign for the loading shown in Fig. 3-5b.

As a final example, consider again the 3D model of the **cabinet door** supported on two hinges (Fig. 3-8b). There are *seven* reaction forces but only *six* independent equations of equilibrium, so the structure is *one degree statically indeterminate externally*. In particular, you cannot use statics alone to determine how weight W is distributed between supports A and B; you can only say, based on $\Sigma F_y = 0$, that $A_y + B_y = W$. A reasonable assumption might be that $A_y = B_y = \dfrac{W}{2}$ (or perhaps that the entire weight of the door is supported by either the upper or lower hinge alone) and the static equilibrium analysis for

FIGURE 3-7 (Repeated)

(b) free-body diagram of signal pole

(b)

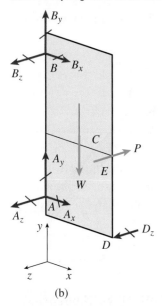

(b)

the remaining reaction forces can then proceed. A possible next step would be a summation of moments about the y axis leading to a direct solution for reaction force D_z. Note that the structure is unstable without reaction force D_z because it is not possible to satisfy the equation $\sum M_y = 0$.

For the most part, statics problems in this text are concerned with statically determinate structures having the minimum number of constraints required for static equilibrium. In this case, all reactions can be found using the available independent equations of static equilibrium.

Equilibrium in 2D: Examples

The structures studied in these examples are stable and statically determinate. Assume that the structures in this section, even if made up of three-dimensional bodies, may be represented by two-dimensional models lying in the x-y plane because both the structure itself and the applied loadings are symmetric with respect to the x-y plane. The *four-step problem-solving approach* is used for all examples; construction of a proper *free-body diagram* (or diagrams) is an important first step. The objective is to find all reactions required for static equilibrium of the structure. A scalar algebra approach can be used for most 2D problems. Scalar and vector algebra solution approaches are compared in several of the examples.

Example 3-1

The plane truss structure in Fig. 3-16a was analyzed previously to find support reactions at joints A and B using the overall FBD in Fig. 3-13b; reaction forces are given in Eqs. (3-6).

Now consider the equilibrium of forces *at joint C* (see Fig. 3-16b) using a separate free-body diagram of joint C alone. Assume that the analysis of joint C reveals that the resulting internal forces in members 1 and 3 have magnitudes (in terms of applied force variable P) of $F_1 = 5P/3$ and $F_3 = 4P/3$, respectively. Use the given forces F_1 and F_3 to find the force F_2 in diagonal member 2 and the force F_4 in horizontal member 4. Neglect the weight of the truss.

FIGURE 3-16

(a) Forces acting at joints D and C on plane truss, (b) forces in members at joint C in plane truss, and (c) photo of actual truss joint

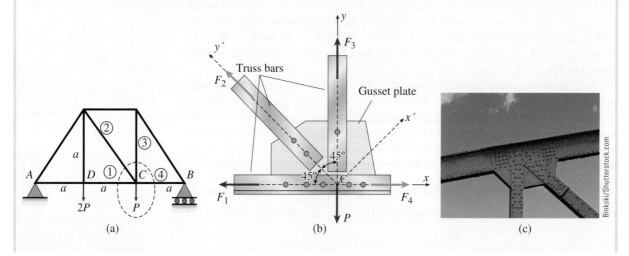

Solution:

Use the four-step problem-solving approach to find the force F_2 in diagonal member 2 and the force F_4 in horizontal member 4.

FIGURE 3-17

Five concurrent forces

1. *Conceptualize* [*hypothesize, sketch*]: Figure 3-16b is a *free-body diagram* of joint C and must be in equilibrium under the action of the five forces. Simplify the diagram by showing the member forces as five *concurrent* forces as shown in Fig. 3-17.

2. *Categorize* [*simplify, classify*]: The five forces are concurrent, so no moment equation is needed in a statement of the equilibrium of joint C. Equilibrium of joint C requires that the forces in x and y directions must sum to zero. A solution approach based on scalar algebra may be easier to carry out rather than one based on vector algebra.

3. *Analyze* [*evaluate; select relevant equations, carry out mathematical solution*]: Before proceeding with a solution, observe the following.

(a) A summation of forces in the y direction will eliminate unknown force F_4, resulting in one equation with one unknown (F_2).

(b) In the same way, using alternate rotated coordinate system x'-y', a summation of forces in the x' direction will eliminate unknown force F_2 and a simple direct solution for force F_4 can be carried out.

Solution approach (a) uses coordinate axes x and y. You should confirm that approach (b) leads to the same solution for member forces F_2 and F_4.

Solution using scalar algebra equations in x-y coordinate system: Summing forces in the x-y coordinate system provides two equations as

$$\Sigma F_x = 0 \qquad -F_1 - F_2 \cos(45°) + F_4 = 0 \qquad \text{(a)}$$

or
$$-F_2 \cos(45°) + F_4 = F_1 = \frac{5}{3}P \qquad \text{(b)}$$

$$\Sigma F_y = 0 \qquad -P + F_2 \sin(45°) + F_3 = 0 \qquad \text{(c)}$$

or
$$F_2 \sin(45°) = P - F_3 = P - \frac{4}{3}P = -\frac{1}{3}P \qquad \text{(d)}$$

From Eqs. (c) and (d), force F_2 is

$$F_2 = -\frac{\sqrt{2}}{3}P = -0.471P \qquad \text{(e)}$$

Now substitute F_2 into Eq. (b) to find

$$F_4 = \frac{5}{3}P + \left(-\frac{\sqrt{2}}{3}P\right)\left(\frac{1}{\sqrt{2}}\right) = \frac{4}{3}P = 1.333P \qquad \text{(f)}$$

Force F_2 is negative, which means that the force vector is actually directed *toward* joint C—not away from it as initially assumed, while force F_4 is positive and is shown correctly in Fig. 3-17. Based on this, you can say that member 2 is a *compression member* while member 4 is a *tension member* in the plane truss.

Repeat solution using vector algebra: a vector expression of force equilibrium, with i and j as the unit vectors along the coordinate axis directions, can be written as

$$\Sigma F = 0 \qquad -F_1 i - (F_2 \cos(45°))i + (F_2 \sin(45°))j + F_3 j + F_4 i - Pj = 0 \quad \text{(g)}$$

$$\text{or} \qquad \left(-\frac{5}{3}P\right)i - (F_2 \cos(45°))i + (F_2 \sin(45°))j + \left(\frac{4}{3}P\right)j + F_4 i - Pj = 0 \quad \text{(h)}$$

Separate Eq. (h) into two scalar equations containing terms with coefficient i and terms with coefficient j, respectively, to obtain Eqs. (b) and (d), then carry out the solution as described previously.

4. Finalize [*conclude; examine answer—Does it make sense? Are units correct? How does it compare to similar problem solutions?*]: In Chapter 2, you found the resultant R of a concurrent system of forces and saw that the system of forces and resultant form a closed polygon [see Eq. (2-21) and Figs. 2-13 and 2-14]. If the system of forces is in equilibrium, the resultant force is zero. In this example, application of the *polygon law* (with all forces drawn with some care to appropriate scale) is shown in Fig. 3-18, with known forces P, F_1, and F_3 drawn first (head-to-tail starting at joint C), then unknown forces F_2 and F_4 (whose lines of action are known) added to close the polygon. Measurement of lines 2 and 4 in Fig. 3-18 (using line P for a reference scale) will confirm the mathematical solutions for F_2 and F_4 presented earlier.

FIGURE 3-18

Force polygon to find forces F_2 and F_4

Example 3-2

An eye bolt is attached to a wall and supports three cables with tension forces T_1, T_2, and T_3 (see Fig. 3-19). The magnitude of the force in cable 1 is known to be $T_1 = 3600$ N, and the magnitude of reaction force Q at the center of the eye bolt is known to be $Q = 8500$ N. Find the force magnitudes T_2 and T_3.

FIGURE 3-19

Cable forces on eye bolt

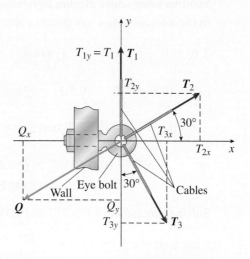

Solution:

Use the four-step problem-solving approach to find cable force magnitudes T_2 and T_3.

1. *Conceptualize*: Recall that this structure was studied in Chapter 2 as part of a discussion of *components of a force* [see Figs. 2-11 through 2-14 and Eqs. (2-18) through (2-21)]. A free-body diagram of the eye bolt–cable structure must be in equilibrium under the action of the four forces. You can simplify the diagram by showing the reaction and cable forces as four *concurrent* forces, as seen in Fig. 3-20.

2. *Categorize*: No moment equation is needed in a statement of the equilibrium of the bolt joint because the four forces are concurrent. Equilibrium requires that the forces in x and y directions must sum to zero. Solution approaches based on both scalar and vector algebra are carried out in step 3.

   Note that unknown force vectors T_2 and T_3 are *orthogonal*, that is $T_2 \cdot T_3 = 0$.

FIGURE 3-20

Concurrent forces at center of eye bolt

3. *Analyze*: As in Example 3-1, observe that, by using alternate rotated coordinate system x'-y', a summation of forces in the x' direction will eliminate unknown force vector T_3 and a simple direct solution for cable force T_2 can be obtained. Next, a summation of forces in the y' direction will eliminate T_2, and force T_3 can be found directly. This more efficient solution approach is a consequence of the orthogonality of vectors T_2 and T_3 and avoids the solution of simultaneous equations presented here.

   The scalar algebra and vector algebra solution approaches presented use the Cartesian x-y coordinate axes shown in Fig. 3-20.

Solution using scalar algebra equations in *x-y* coordinate system: Summing forces in the *x-y* coordinate system provides two equations:

$$\Sigma F_x = 0 \quad T_2 \cos(30°) + T_3 \sin(30°) - Q\cos(40°) = 0 \qquad \text{(a)}$$

or $\quad (0.86603)T_2 + (0.5)T_3 = Q\cos(40°) = (8500 \text{ N})(0.76604) = 6511 \text{ N}$ **(b)**

$$\Sigma F_y = 0 \quad T_2 \sin(30°) - T_3 \cos(30°) - Q\sin(40°) + T_1 = 0 \qquad \text{(c)}$$

or $\quad (0.5)T_2 - (0.86603)T_3 = Q\sin(40°) - T_1$ **(d)**
$$= (8500 \text{ N})(0.64279) - 3600 \text{ N} = 1864 \text{ N}$$

Solving simultaneous Eqs. (b) and (d) gives cable force magnitudes T_2 and T_3:

$$T_2 = 6571 \text{ N}, \quad T_3 = 1642 \text{ N} \qquad \leftarrow \text{(e)}$$

Repeat solution using vector algebra: A vector expression of force equilibrium with i and j as the unit vectors along the coordinate axis directions is

$$\Sigma F = 0 \quad T_1 j + (T_2 \cos(30°))i + (T_2 \sin(30°))j + (T_3 \sin(30°))i$$
$$- (T_3 \cos(30°))j - (Q\cos(40°))i - (Q\sin(40°))j = 0 \qquad \text{(f)}$$

If Eq. (f) is separated into two scalar equations containing terms with coefficient i and terms with coefficient j, respectively, you obtain Eqs. (a) and (c). Solution of these two equations will produce the cable force results shown in Eq. (e).

4. *Finalize:* For this two-dimensional concurrent force system, only two equations of force equilibrium are available for the solution. No moment equation is available. However, with only two unknown forces, the two force equilibrium equations are sufficient for a solution of this statically determinate problem, so you can solve for the unknowns T_2 and T_3.

Finally, confirm the numerical solution using a graphical display of the *force polygon* for this structure. The qualitative sketch shown in Fig. 3-21 can be redrawn to scale starting with known force vectors Q and T_1 linked head-to-tail. Next, add unknown force vectors T_2 and T_3: first T_2 from the head of vector T_1, then vector T_3 from the head of T_2 (the line of action of each is known). The head of vector T_3 must meet the tail of vector Q for the polygon to close, indicating overall force equilibrium. Measurement of the length of vectors T_2 and T_3 will confirm the mathematical results listed in Eq. (e).

FIGURE 3-21

Force polygon

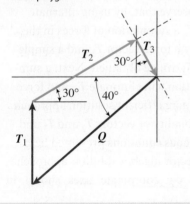

Example 3-3

In Examples 1-3 and 2-5, you studied the action of force vectors F_1 and F_2 applied to an automobile parking brake at point P (see Fig. 3-22a). The force magnitudes are $F_1 = 13$ lb and $F_2 = 15$ lb. Now extend Examples 1-3 and 2-5 to find the reaction force R_O and moment M_O at the base (Fig. 3-22b). Neglect the weight of the hand brake.

FIGURE 3-22

(a) Forces applied to automobile parking brake and (b) Cartesian components of applied forces and reactions at base

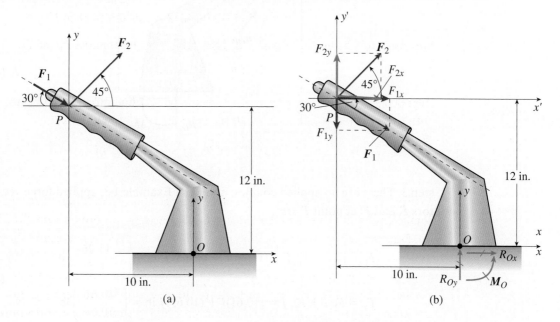

(a)   (b)

Solution:

Use the four-step problem-solving approach to find the reaction force R_O and moment M_O at the base of the hand brake.

1. *Conceptualize*: Begin by drawing a free-body diagram of the hand brake structure showing reaction force components R_{Ox} and R_{Oy} at O, as well as reaction moment M_O, since the base is fixed against both translation in x and y directions and rotation about z. Cartesian components of applied force vectors F_1 and F_2 from Example 1-3 are shown in Fig. 3-22b. The free-body diagram of the hand brake is given in Fig. 3-23.

2. *Categorize*: Equilibrium in two dimensions means that the forces in x and y directions sum to zero, and the moment of all forces and couples about an arbitrary axis parallel to z are also zero. Cartesian axes x and y are shown in Fig. 3-23, and you will use moments about the z axis (through point O). Solution approaches based on both scalar and vector algebra are carried out in

FIGURE 3-23

Free-body diagram of hand brake
structure

step 3. There are no applied couples; and from Example 1-3, applied force vectors F_1 and F_2 at point P are

$$F_1 = F_{1x}i = F_{1y}j = (11.26)i + (-6.5)j \text{ lb} = \begin{pmatrix} 11.26 \\ -6.5 \end{pmatrix} \text{lb}$$

$$F_2 = F_{2x}i + F_{2y}j = (10.61)i + (10.61)j \text{ lb} = \begin{pmatrix} 10.61 \\ 10.61 \end{pmatrix} \text{lb}$$

(a)

A position vector from base point O to force application point P is

$$r_{OP} = (-10)i + (12)j \text{ in.} = \begin{pmatrix} -10 \\ 12 \end{pmatrix} \text{ in.} \quad \text{and} \quad r_{OP} = 15.62 \text{ in.} \qquad \textbf{(b)}$$

3. *Analyze*: The reaction force (R_O) and moment (M_O) at point O are found using the two-dimensional force and moment equilibrium equations.

Scalar algebra equations: First sum forces in x and y directions to find components of reaction force R_O:

$$\Sigma F_x = 0 \quad R_{Ox} + F_{1x} + F_{2x} = 0$$

so $\qquad R_{Ox} = -(F_{1x} + F_{2x}) = -(11.26 + 10.61) \text{ lb} = -21.87 \text{ lb}$

$$\Sigma F_y = 0 \quad R_{Oy} + F_{1y} + F_{2y} = 0$$

(c)

so $\qquad R_{Oy} = -(F_{1y} + F_{2y}) = -(-6.5 + 10.61) \text{ lb} = -4.11 \text{ lb}$

The magnitude of reaction force R_O is

$$R_O = \sqrt{(-21.87)^2 + (-4.11)^2} \text{ lb} = \sqrt{495.2} \text{ lb} = 22.2 \text{ lb} \qquad \text{(d)}$$

Using Varignon's theorem (see Example 2-5), the magnitude of reaction moment M_O is

$$\sum M_z = 0 \qquad M_O = (21.87 \text{ lb})(12 \text{ in.}) + (4.11 \text{ lb})(10 \text{ in.}) = 303 \text{ lb-in.} \qquad \text{(e)}$$

Repeat using vector algebra:

Force equilibrium requires $R = \sum F = 0$ where $R = R_O + \sum F$, so reaction force $R_O = -\sum F$. Hence, the reaction force vector at base point O is $R_O = -[(21.87)i + (4.11)j]$ lb, which is in agreement with the scalar algebra solution. The magnitude of the reaction force is $R_O = \sqrt{R_{Ox}^2 + R_{Oy}^2} = 22.2$ lb, which is in agreement with Eq. (d).

Moment equilibrium requires $M = \sum M = 0$, so $M_O + r_{OP} \times (F_1 + F_2) = 0$. Solving for the reaction moment at the base gives

$$M_O = -r_{OP} \times (F_1 + F_2) = -[(-10)i + (12)j] \text{ in.} \times [(21.87)i + (4.11)j] \text{ lb}$$
$$= -(-41.1 - 262.3)k \text{ lb-in.} = 303.4k \text{ lb-in.}$$

and

$$M_O = 303 \text{ lb-in.}$$

(f)

The magnitude of the reaction moment M_O agrees with the result shown in Eq. (e).

4. *Finalize*: Two different approaches were used to find base reactions R_O and M_O, and the results are in agreement. For this 2D problem with well-defined geometry, it is likely that the scalar algebra solution approach will be the preferred method. Note that moment vector M_O is perpendicular to the x-y plane and is in the (+z) direction, so it can be expressed as $M_O = (M_O)k$. Recall that the right-hand rule is used to determine the sign of the vector result.

Example 3-4

A continuous cable with tension force T is attached to a horizontal frame member at B and C to support a sign structure (see Fig. 3-24). The cable passes over a small frictionless pulley at D. (This structure was analyzed in Example 2-1 to find the resultant cable force at D). Find the tension T in the cable and the forces in the pins at A and D. Assume that the weight of the sign, frame member ABC, and cable BDC are included in applied load W.

FIGURE 3-24

Sign structure supported by cable
BDC

Solution:

Use the four-step problem-solving approach to find the tension T in the cable and the forces in the pins at A and D involves the following steps.

1. *Conceptualize*: Begin by drawing a free-body diagram of the sign structure showing pin reactions at A and D and structure weight W split into two equal forces $W/2$ at support points E and F. Pin reaction A is shown in terms of rectangular components A_x and A_y; the free-body diagram of the sign structure is shown in Fig. 3-25.

FIGURE 3-25

Free-body diagram of entire sign
structure

2. *Categorize*: Overall static equilibrium requires that the forces in x and y directions sum to zero and that the moment about the z axis through any chosen point be zero. Neglect the size of the pulley and assume that cable forces at B and C are directed at point D (see Example 2-1 for a discussion of the effect of

pulley size). Writing a moment equation about D is a convenient way to eliminate the reaction force at D and the reaction force component A_y from consideration. This more efficient solution approach results in one equation and one unknown: reaction force component A_x.

For this 2D problem, a solution approach based on scalar algebra is easier to carry out than one based on vector algebra.

3. *Analyze*: The solution steps are given here.

    i.   Sum moments about point D to **find reaction component A_x** (counterclockwise moments are positive):

$$\Sigma M_D = 0 \qquad A_x(2 \text{ m}) - \frac{W}{2}(0.75 \text{ m} + 2.75 \text{ m}) = 0$$

$$A_x = \frac{6800 \text{ N}}{2}\left(\frac{3.5 \text{ m}}{2 \text{ m}}\right) = 5950 \text{ N} \qquad \text{(a)}$$

    ii.   Sum forces in the x direction **find reaction component D_x**:

$$\Sigma F_x = 0 \qquad A_x + D_x = 0 \qquad \text{or} \qquad D_x = -A_x = -5950 \text{ N} \qquad \text{(b)}$$

The minus sign means that D_x acts in the negative x direction.

FIGURE 3-26

Free-body diagram of joint D

    iii.   Sum forces in the x direction at joint D using the free-body diagram of joint D (Fig. 3-26) to **find the force T in the continuous cable BDC**. First compute angles α_B and α_C (see Fig. 3-25):

$$\alpha_B = \tan^{-1}\left(\frac{2}{1.5}\right) = 53.13° \qquad \alpha_C = \tan^{-1}\left(\frac{2}{3}\right) = 33.69° \qquad \text{(c)}$$

Applying equilibrium equation $\Sigma F_x = 0$ at joint D in Fig. 3-26 gives

$$D_x + T(\cos(\alpha_B) + \cos(\alpha_C)) = 0$$

$$\text{or} \quad T = \frac{-D_x}{\cos(\alpha_B) + \cos(\alpha_C)} = \frac{-(-5950 \text{ N})}{\cos(\alpha_B) + \cos(\alpha_C)} = 4155 \text{ N} \qquad \text{(d)}$$

    iv.   Now, to **find the vertical reaction D_y** at joint D where $T_B = T_C = T$ (because the cable is one continuous cable), apply equilibrium equation $\Sigma F_y = 0$ to obtain:

$$D_y = T(\sin(\alpha_B) + \sin(\alpha_C)) = 5629 \text{ N} \qquad \text{(e)}$$

The plus sign means that D_y acts in the positive y direction.

    v.   Using components D_x and D_y, **solve for the resultant force in the pin at D**:

$$D_{res} = \sqrt{D_x^2 + D_y^2} = \sqrt{(-5950 \text{ N})^2 + (5629 \text{ N})^2} = 8190 \text{ N} \qquad \text{(f)}$$

vi. Sum forces in the y direction for the entire free-body diagram (Fig. 3-25) to **find reaction component** A_y:

$$\Sigma F_y = 0 \qquad A_y + D_y = W$$

or
$$A_y = -D_y + W = -5629 \text{ N} + 6800 \text{ N} = 1171 \text{ N} \impliedby \text{(g)}$$

vii. Finally, the **resultant force in the pin at** A is

$$A_{res} = \sqrt{A_x^2 + A_y^2} = \sqrt{(5950 \text{ N})^2 + (1171 \text{ N})^2} = 6064 \text{ N} \qquad \text{(h)}$$

4. *Finalize*: As a final check, confirm the equilibrium solution using a free-body diagram of pipe ABC alone (see Fig. 3-27, where $T_B = T_C = T$).

Here, create the free-body diagram by passing a horizontal plane through the entire structure cutting through the cable and exposing cable force T. The following equations expressing force equilibrium, as well as moment equilibrium about A, each sum to zero as required:

$$\Sigma F_x = 0 \quad A_x - T(\cos(\alpha_B) + \cos(\alpha_C)) = 0$$
$$\Sigma F_y = 0 \quad A_y - W + T(\sin(\alpha_B) + \sin(\alpha_C)) = 0 \qquad \text{(i)}$$

$$\Sigma M_A = 0 \quad T\sin(\alpha_B)(1.5 \text{ m}) + T\sin(\alpha_C)(3 \text{ m}) - \frac{W}{2}(0.75 \text{ m} + 2.75 \text{ m}) = 0$$

FIGURE 3-27

Free-body diagram of lower part of structure

Equilibrium in 3D: Examples

The 3D structures studied in the following examples are stable and statically determinate. Use the *four-step problem-solving approach* for all examples. As illustrated in the 2D examples, construction of a proper *free-body diagram* is an essential first step. The objective is to find specified forces required for static equilibrium of the structure or an isolated portion of the structure. A vector algebra approach is often more convenient for many 3D problems, but a scalar algebra approach is presented as well for comparison in several of these examples.

Example 3-5

A cargo ship is tied down to marine bollards (Fig. 3-28) at a number of points along its length, while its cargo is unloaded by a container-handling crane (Fig. 3-29). Each bollard is fastened to the wharf using anchor bolts (Figs. 3-30 and 3-31). Three cables with known tension force magnitudes $F_1 = 25$ kips, $F_2 = 19$ kips and $F_3 = 20$ kips are attached to one bollard at a point A with coordinates (0, 1.5 ft, 0) in the x-y-z coordinate system shown in Fig. 3-30. Force F_1 is directed from point A to an attachment point on the ship having coordinates (10 ft, 30 ft, 0); force F_2 is directed to a point with coordinates (21 ft, 28 ft, 6.5 ft); and force F_3 is directed to a point with coordinates (26 ft, 30 ft, 16 ft). Find the reaction force magnitudes R_x, R_y, and R_z and also reaction moment magnitudes M_x and M_z at the base (point O) of the bollard (Figs. 3-30 and 3-31).

FIGURE 3-28
Marine bollard

FIGURE 3-29
Cargo ship secured to wharf structure

csp_le_cyclope/Fotosearch

FIGURE 3-30
Bollard anchored in wharf

FIGURE 3-31
Plan view of bollard (*Sec. a–a*)

Solution:

Use the four-step problem-solving approach to find the reaction force magnitudes R_x, R_y, and R_z at point O at the base of the bollard (Figs. 3-30 and 3-31) and also reaction moment magnitudes M_x and M_z.

FIGURE 3-32

Free-body diagram of bollard with cable forces F_1, F_2, and F_3 applied at point A

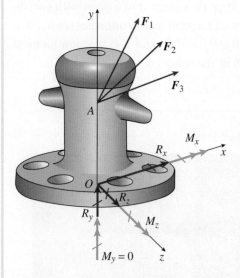

1. *Conceptualize:* A free-body diagram of the bollard is shown in Fig. 3-32. The bollard is in equilibrium under the action of all applied and reactive forces and moments. Reaction moment component M_y is zero because all three applied force vectors act through the y axis.

2. *Categorize:* Replace the three cable forces F_1, F_2, and F_3 by their resultant Q (see Fig. 3-33) to simplify the solution. Use the Cartesian components of Q (Q_x, Q_y, and Q_z) in force and moment equilibrium equations to find reactions at the base of the bollard. Once components of Q are known, a solution approach based on scalar algebra may be easier to carry out than one based on vector algebra.

3. *Analyze:* Begin by finding vector expressions for cable forces F_1, F_2, and F_3 where n_{A1}, n_{A2} and n_{A3} are unit vectors along each force vector:

$$F_1 = F_1 n_{A1} = 25 \text{ kips} \left(\frac{10i + (30 - 1.5)j}{\sqrt{10^2 + 28.5^2}} \right)$$
$$= 8.28i + 23.59j \text{ kips}$$

$$F_2 = F_2 n_{A2} = 19 \text{ kips} \left(\frac{21i + (28 - 1.5)j + 6.5k}{\sqrt{21^2 + 26.5^2 + 6.5^2}} \right)$$
$$= 11.59i + 14.62j + 3.59k \text{ kips}$$

$$F_3 = F_3 n_{A3} = 20 \text{ kips} \left(\frac{26i + (30 - 1.5)j + 16k}{\sqrt{26^2 + 28.5^2 + 16^2}} \right)$$
$$= 12.45i + 13.65j + 7.66k \text{ kips} \tag{a}$$

The resultant of cable forces F_1, F_2 and F_3 is

$$Q = F_1 + F_2 + F_3 = 32.32i + 51.86j + 11.25k \text{ kips} \tag{b}$$

So resultant force components are $Q_x = 32.32$ kips, $Q_y = 51.86$ kips, and $Q_z = 11.25$ kips. Now proceed with solution approaches based on either scalar algebra or vector algebra.

Scalar algebra equations: First sum forces in x and y directions to find components of reaction force R_O:

$$\Sigma F_x = 0 \qquad R_x + Q_x = 0 \quad \text{so} \quad R_x = -Q_x = -32.32 \text{ kips}$$
$$\Sigma F_y = 0 \qquad R_y + Q_y = 0 \quad \text{so} \quad R_y = -Q_y = -51.86 \text{ kips} \blacktriangleleft \textbf{(c)}$$
$$\Sigma F_z = 0 \qquad R_z + Q_z = 0 \quad \text{so} \quad R_z = -Q_z = -11.25 \text{ kips}$$

FIGURE 3-33

Free-body diagram of bollard with resultant Q applied at point A

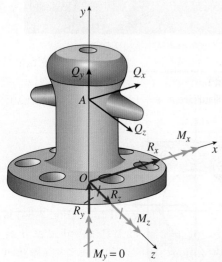

The magnitude of reaction force R_O is

$$R_O = \sqrt{(-32.32)^2 + (-51.86)^2 + (-11.25)^2} \text{ kips} = \sqrt{3861} \text{ kips} = 62.1 \text{ kips}$$

Using Varignon's theorem (see Example 2-5), the magnitudes of reaction moment components are

$$\Sigma M_x = 0 \qquad M_x + Q_z (1.5 \text{ ft}) = 0 \quad \text{so} \quad M_x = -(11.25 \text{ kips})(1.5 \text{ ft})$$
$$= -16.9 \text{ kip-ft}$$

$$\Sigma M_y = 0 \qquad M_y = 0 \qquad\qquad\qquad\qquad\qquad\qquad \text{(d)}$$

$$\Sigma M_z = 0 \qquad M_z - Q_x (1.5 \text{ ft}) = 0 \quad \text{so} \quad M_z = (32.32 \text{ kips})(1.5 \text{ ft})$$
$$= 48.5 \text{ kip-ft}$$

And the magnitude of reaction moment M_O is

$$M_O = \sqrt{(-16.9)^2 + (48.5)^2} \text{ kip-ft} = 51.3 \text{ kip-ft}$$

Repeat using vector algebra: Reaction force R_O is expressed simply as $R_O = -\Sigma F = -Q$ since cable forces F_1, F_2, and F_3 are the only applied forces. The results are presented in Eq. (c).

Moment equilibrium requires that $M = \Sigma M = 0$, so $M_O + r_{OA} \times (F_1 + F_2 + F_3) = 0$. Solving for the reaction moment at the base gives

$$M_O = -r_{OA} \times (F_1 + F_2 + F_3) = -r_{OA} \times Q$$
$$= -(1.5 \text{ ft})\,j \times (32.32i + 51.86j + 11.25k \text{ kips})$$

$$= -\begin{vmatrix} i & j & k \\ r_{OAx} & r_{OAy} & r_{OAz} \\ Q_x & Q_y & Q_z \end{vmatrix} = \begin{vmatrix} i & j & k \\ 0 & -1.5 & 0 \\ 32.32 & 51.86 & 11.25 \end{vmatrix} \text{ kip-ft} \qquad \text{(e)}$$

$$M_O = M_x i + M_y j + M_z k = (-16.9i + 48.5k)\,\text{kip-ft}$$

with magnitude $M_O = 51.3$ kip-ft

4. *Finalize:* The three cable forces are *concurrent* at point A, so they can be replaced by a single *resultant* force at that location. Use of the single resultant force simplifies subsequent force and moment equilibrium equations. Scalar and vector algebra results are in agreement for the reaction force and moment components at the base of the bollard.

Example 3-6

A small portion of a roof truss supporting an airport terminal roof (Fig. 3-34) is being tested to check its strength and stiffness. For the test, load P is applied at A by tensioning a cable anchored at Q (Fig. 3-35). At the conclusion of the test, it is known that member 2 has compressive axial force $F_2 = -4$ kN. Find (a) cable force P and forces F_1 and F_3 in members 1 and 3, and (b) support reaction forces at joints B, C, and O. Neglect the weight of the truss and cable. The coordinates of all joints are given in meters in Fig. 3-35.

FIGURE 3-34

Roof truss in airport terminal structure

FIGURE 3-35

Free-body diagram of space truss

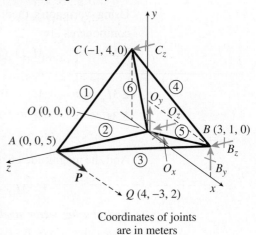

Coordinates of joints
are in meters

Solution:

Use the four-step problem-solving approach to find cable force P and the forces F_1 and F_3 in members 1 and 3 and find all of the reaction forces.

1. *Conceptualize*: First sketch several free-body diagrams. The first is of joint A (Fig. 3-36a) showing all forces at joint A where only one (F_2) is known. Joint A must be in equilibrium under the action of the four *concurrent* forces: F_1, F_2, F_3, and P. The second free-body diagram is of the entire truss (Fig. 3-36b).

FIGURE 3-36

Free-body diagrams: (a) for joint A;
(b) for entire space truss

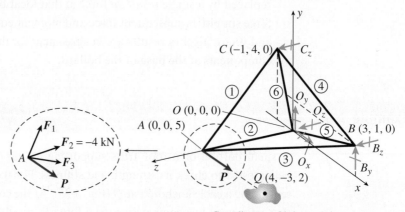

Coordinates of joints are in meters

(a)   (b)

2. *Categorize*: In Fig. 3-36a, the three member forces (F_1, F_2, F_3) and applied force P are concurrent, so no moment equation is needed in a statement of the equilibrium of joint A. Equilibrium requires that the force components in x, y, and z directions must sum to zero: $\Sigma F = 0$.

In Fig. 3-36b, note that there are six reaction force unknowns. There are six static equilibrium equations, so the structure is statically determinate, and a solution for all reactions can be obtained.

3. *Analyze*:

Part (a): Cable force P and member forces F_1 and F_3.

The components of applied force P and member forces F_1, F_2, and F_3 are found using vector operations and the free-body diagram in Fig. 3-36a, as

$$P = P n_{AQ} = P \frac{r_{AQ}}{|r_{AQ}|} = P(0.686i - 0.514j - 0.514k)$$

$$F_1 = F_1 n_{AC} = F_1 \frac{r_{AC}}{|r_{AC}|} = F_1(-0.154i + 0.617j - 0.772k) \tag{a}$$

$$F_2 = (-4.0 \text{ kN}) n_{AO} = (-4.0 \text{ kN})(-k) = (-4.0 \text{ kN})k$$

$$F_3 = F_3 n_{AB} = F_3 \frac{r_{AB}}{|r_{AB}|} = F_3(0.507i + 0.169j - 0.845k)$$

where, for example, r_{AQ} is a position vector along force vector P from point A to point Q and n_{AQ} is the corresponding unit vector. Member force vectors F_1, F_2, and F_3 are also written in terms of their unit vectors.

Force equilibrium at joint A requires that $\Sigma F = 0$, so

$$P + F_1 + F_2 + F_3 = 0 \tag{b}$$

Substituting Eq. (a) into Eq. (b) gives

$$P(0.686i - 0.514j - 0.514k) + F_1(-0.154i + 0.617j - 0.772k)$$
$$+ (4.0 \text{ kN})k + F_3(0.507i + 0.169j - 0.845k) = 0 \tag{c}$$

Equating each of the coefficients of the i, j, and k unit vectors in Eq. (c) to zero gives three simultaneous force equilibrium equations:

$$-0.154F_1 + 0.507F_3 + 0.686P = 0$$
$$0.617F_1 + 0.169F_3 - 0.514P = 0 \tag{d}$$
$$-0.772F_1 - 0.845F_3 - 0.514P + 4.0 \text{ kN} = 0$$

A matrix solution can be used:

$$\begin{bmatrix} F_1 \\ F_3 \\ P \end{bmatrix} = \begin{bmatrix} -0.154 & 0.507 & 0.686 \\ 0.617 & 0.169 & -0.514 \\ -0.772 & -0.845 & -0.514 \end{bmatrix}^{-1} \begin{bmatrix} 0 \\ 0 \\ -4.0 \end{bmatrix} = \begin{bmatrix} 8.65 \\ -7.91 \\ 7.79 \end{bmatrix} \text{ kN} \tag{e}$$

So if member 2 has a measured compression force of 4.0 kN, the required magnitudes of member forces F_1 and F_3 and applied force P to satisfy static equilibrium at joint A are

$$F_1 = 8.65 \text{ kN}, \quad F_3 = -7.91 \text{ kN}, \quad \text{and} \quad P = 7.79 \text{ kN} \qquad \blacktriangleleft \text{ (f)}$$

A statics sign convention was used, so the minus signs on forces F_2 and F_3 indicate that the assumed directions of the force vectors $\boldsymbol{F_2}$ and $\boldsymbol{F_3}$ in Fig. 3-36 are incorrect. Instead, both members are in *compression*, not tension, under the action of applied force $\boldsymbol{P}$.

Part (b): Reaction forces at joints B, C, and O.

Next use the overall structure free-body diagram (Fig. 3-36b) to find support reaction forces at joints B, C, and O. From Eq. (f), $P = 7.79$ kN. While it is possible—and perhaps more efficient—to solve for unknown reaction forces in a targeted manner (such as $\Sigma M_z = 0$ gives $B_y = 0$, then $\Sigma F_y = 0$ gives $O_y = 4$ kN), a systematic approach using six independent static equilibrium equations is used to find reactions:

$$\Sigma F_x = 0 \quad O_x + P_x = 0 \quad \text{so} \quad O_x = -P_x = -5.34 \text{ kN}$$
$$\Sigma F_y = 0 \quad B_y + O_y + P_y = 0 \quad \text{or} \quad B_y + O_y = -P_y = 4.0 \text{ kN}$$
$$\Sigma F_z = 0 \quad B_z + C_z + O_z + P_z = 0 \quad \text{or} \quad B_z + C_z + O_z = -P_z = 4.0 \text{ kN}$$

$$\Sigma M_x = 0 \quad B_z(1 \text{ m}) + C_z(4 \text{ m}) - P_y(5 \text{ m}) = 0 \quad \text{or} \quad B_z + 4C_z = 5P_y \text{ kN} \cdot \text{m}$$
$$\Sigma M_y = 0 \quad -B_z(3 \text{ m}) + C_z(1 \text{ m}) + P_x(5 \text{ m}) = 0 \quad \text{or} \quad -3B_z + C_z = -5P_x \qquad \text{(g)}$$
$$\Sigma M_z = 0 \quad B_y(3 \text{ m}) = 0 \quad \text{so} \quad B_y = 0$$

With O_x and B_y known, simultaneous solution of Eqs. (g) gives four reaction forces: $B_z = 6.67$ kN, $C_z = -6.67$ kN, $O_y = 4.0$ kN, and $O_z = 4.0$ kN. $\blacktriangleleft$

4. *Finalize*: Force $\boldsymbol{P}$ and member forces F_1 and F_3 were computed in part (a). The magnitude of the resultant force at joint A is zero, so joint A is in equilibrium and Eq. (b) is satisfied. In part (b), solution of simultaneous equilibrium equations led to reaction forces at joints B, C, and O. Other targeted solution strategies for finding reactions are possible. For example, (a) sum moments about line BC to find reaction O_z; (b) then sum moments about line OB to find reaction C_z; and (c) using C_z and O_z, sum forces in the z direction to find B_z. Once all reactions have been found, you can apply $\Sigma F = 0$ at joints B, C, and O in turn to find member forces F_4, F_5, and F_6 to complete the analysis of this structure.

Example 3-7

Two separate cables AC and BC support a sign structure of weight $W = 1575$ lb attached to a building (see Fig. 3-37) (see Example 2-2 for more discussion of this structure). The sign is also supported by a pin (ball-and-socket) support at O and a lateral restraint in the z-direction at D. Find the tension in each cable and the reaction forces at O and D.

FIGURE 3-37

Sign structure supported by cables AC and BC

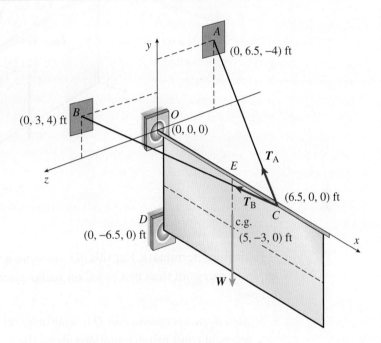

Solution:

Use the four-step problem-solving approach to find cable forces T_A and T_B and reaction forces O at pin support O and force D_z at lateral restraint D.

1. *Conceptualize*: Start by drawing a free-body diagram of the sign structure showing pin reactions at O and D, cable forces T_A and T_B, and sign weight W. If all forces are shown in terms of their x, y, and z components, the free-body diagram of the sign structure is that shown in Fig. 3-38. Reaction force components and cable forces are shown in their assumed positive directions in accordance with a statics sign convention.

2. *Categorize*: Use three force equilibrium equations ($\Sigma F = 0$) and three moment equilibrium equations ($\Sigma M = 0$) to find reaction forces at O and D and cable forces T_A and T_B. With six unknowns and six equations, the structure

FIGURE 3-38

Free-body diagram

is statically determinate. For this 3D problem, a vector algebra solution is easier to carry out than one based on scalar algebra. The solution steps are as follows.

i.   *Sum moments about point O* to eliminate reaction forces at O, and write moment equilibrium equations about the x, y, and z axes using the remaining forces. All forces except reaction D_z pass through the x axis, so summing moments about the x axis reveals that D_z is zero for this loading.

ii. Write moment equations about the y and z axes through point O to *solve for the components of cable forces T_A and T_B* because there are two equations and two unknowns.

iii. Finally, use force equilibrium equations in the x, y, and z directions to *solve for pin reaction components O_x, O_y, and O_z*.

Another solution strategy, which involves *summing moments about a line*, is used in step 3 to confirm cable and reaction force results obtained from this approach.

3. *Analyze*: Follow the solution approach described in step 2.

i.   *The moment about point O* is

$$M_O = r_{OA} \times T_A + r_{OB} \times T_B + r_{OE} \times W + r_{OE} \times D \qquad \text{(a)}$$

where r_{OA}, r_{OB}, r_{OE}, and r_{OD} are position vectors from point O to the lines of actions of force vectors T_A, T_B, W, and D, respectively. Cable force vectors meet at C, so M_O also can be written as

$$M_O = r_{OC} \times (T_A + T_B) + r_{OE} \times W + r_{OD} \times D \qquad \text{(b)}$$

Position vectors r_{OC}, r_{OE}, and r_{OD} are

$$r_{OC} = 6.5i \text{ ft}, \quad r_{OE} = 5i \text{ ft}, \quad r_{OD} = -6.5j \text{ ft} \qquad \text{(c)}$$

Force vectors in Eq. (b) are written in terms of unit vectors as

$$T_A = T_A n_{CA} = T_A \frac{-6.5i + 6.5j - 4k}{\sqrt{(-6.5)^2 + 6.5^2 + (-4)^2}} = T_A(-0.648i + 0.648j - 0.399k)$$

$$T_B = T_B n_{CB} = T_B \frac{-6.5i + 3j + 4k}{\sqrt{(-6.5)^2 + 3^2 + (4)^2}} = T_B(-0.793i + 0.366j + 0.488k) \qquad \text{(d)}$$

$$W = -1575j \text{ lb}, \quad D = D_z k \text{ lb}$$

Substituting Eqs. (c) and (d) into Eq. (b) gives for moment M_O:

$$M_O = (-6.5D_z)i + (2.5935T_A - 3.1705T_B)j \\ + (4.2145T_A + 2.3779T_B - 7875 \text{ lb})k \qquad \text{(e)}$$

ii.   Equating the coefficients of i, j, and k to zero and solving gives $D_z = 0$, $T_A = 1278$ lb, and $T_B = 1046$ lb. From Eq. (d), the cable forces are

$$T_A = (-829i + 829j - 510k) \text{ lb}$$
$$T_B = (-829i + 383j + 510k) \text{ lb} \qquad \Longleftarrow \text{(f)}$$

iii.   Finally, force equilibrium requires that all forces shown in Fig. 3-38 sum to zero, so the *reaction force at O* is

$$O = O_x i + O_y j + O_z k = -[T_A + T_B + W + D] = 1658i + 363j \text{ lb} \Longleftarrow \text{(g)}$$

Reaction force components $O_z = D_z = 0$ as expected since there is no force (such as wind) applied normal to the surface of the sign in this problem. So the resultant force on the pin at O is

$$\sqrt{O_x^2 + O_y^2} = \sqrt{1658^2 + 363^2} \text{ lb} = 1697 \text{ lb} \qquad \text{(h)}$$

Alternate Solution Approach: Begin with the expression for the moment M_O about point O given in Eq. (a) and then use a dot product to find the moment about line OA. Only sign weight W and cable force T_B remain in the moment equation so you can directly solve for cable force magnitude T_B.

A unit vector along line OA is $n_{OA} = 0.852j - 0.524k$, so the moment about line OA is

$$M_{OA} = M_O \cdot \frac{r_{OA}}{|r_{OA}|} = M_O \cdot n_{OA} = -3.9464T_B + 2.6205W \qquad \text{(i)}$$

Moment M_{OA} is equal to zero for equilibriumm, so the solution for cable force magnitude T_B gives $T_B = 0.664W = 0.664(1575\ \text{lb}) = 1046\ \text{lb}$.

Next use a dot product to find the moment about line OB, then solve for the magnitude of cable force T_A. A unit vector along line OB is $n_{OB} = 0.6j + 0.8k$ so the moment about line OB is

$$M_{OB} = M_O \cdot n_{OB} = 4.928T_A - 4W \tag{j}$$

Moment equilibrium requires that $M_{OB} = 0$, so the magnitude of the force in cable AC is $T_A = 0.812W = 0.812(1575\ \text{lb}) = 1278\ \text{lb}$.

Finally, force equilibrium equations in the x, y, and z directions lead to the solutions for pin reaction components $O_x, O_y,$ and O_z and lateral restraint force D_z listed previously. Note also that the moment about line AB can be used to solve directly for reaction force component O_x.

4. *Finalize*: Results of the alternate solution approach are in agreement with the solution for cable forces T_A and T_B in Eq. (f). Using a moment about a line is often an effective way to isolate unknowns leading to a simpler and more direct solution.

Example 3-8

Framework $ABCD$ (Fig. 3-39) moves along rails 1 and 2—each parallel to the z axis—to carry loads applied at C. In the position shown (Fig. 3-39), the frame carries force P of 2.5 kN applied at C and directed at point F. The wheel at A is locked to form a pin support, and a cable is attached from B to O to provide additional restraint against lateral (z-direction) movement. Find the tension in the cable T and reaction components at A and D.

FIGURE 3-39

Framework *ABCD* on two rails

Solution:

Use the four-step problem-solving approach to find cable and reaction forces as outlined here.

1. *Conceptualize*: A free-body diagram of the framework shows applied force P and cable force T (Fig. 3-40). The free-body diagram of the entire system displays all applied and reactive forces. The wheel reaction force at A has three components because the wheel has been restrained to prevent z-direction translation. The wheel at D cannot move in the x or y directions (normal to the rail, Fig. 3-41). The overall framework is in equilibrium under the action of the applied force P, cable force T, and reaction forces at A and D.

FIGURE 3-40

Free-body diagram of framework
ABCD

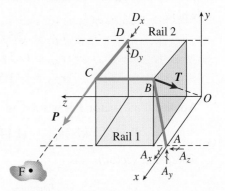

FIGURE 3-41

Cross section at rail 2: (a) photo;
(b) reaction force components at D

(a)

(b)

2. *Categorize*: Equilibrium requires that the forces in x, y, and z directions sum to zero, and moments about x, y, and z axes through an arbitrary point or about any axis are also zero. There are six unknown forces (cable force T, two reaction components at D, and three reaction components at A). The addition of restraint

force A_z prevents unrestrained motion along the rails and the structure is stable under applied load P. With six equations of equilibrium and six unknowns, the structure is statically determinate. A solution approach based in part on vector algebra is easier to carry out than one based on scalar algebra alone.

3. *Analyze*: A key step in the solution is to find cable force T resulting from application of force $P = 2.5$ kN at C. After T is found, reaction force components at A and D are obtained using scalar algebra. By summing moments about line AD, unknown reaction force components at A and D can be eliminated, leading to a straightforward solution for cable force T (one equation, one unknown). First write an expression for moment M_A (you could select M_D instead):

$$M_A = r_{AC} \times P + r_{AB} \times T + r_{AD} \times D \tag{a}$$

where

$$r_{AC} = (2.5j + 2.5k)\text{ m}$$
$$r_{AB} = (2.5j + 0.5k)\text{ m}$$
$$r_{AD} = (-1.5i + 2.5j + 2.5k)\text{ m}$$

Applied force vector P is written as

$$P = (2.5 \text{ kN})n_{CF} = (1.14374i - 1.90623j + 1.14374k)\text{ kN}$$

where unit vector n_{CF} is

$$n_{CF} = \frac{r_{CF}}{|r_{CF}|} = \frac{(1.5i - 2.5j + 1.5k)}{\sqrt{(1.5)^2 + (-2.5)^2 + (1.5)^2}}$$

The unknown cable force T is

$$T = Tn_{BO} = T(-0.50709i - 0.84515j - 0.16903k)$$

and unit vector n_{BO} is

$$n_{BO} = \frac{r_{BO}}{|r_{BO}|} = \frac{-(1.5i + 2.5j + 0.5k)}{\sqrt{(1.5)^2 + (2.5)^2 + (0.5)^2}}$$

Reaction force $D = D_x i + D_y j$ acts through line AD, so it is eliminated from Eq. (a) when a dot product is used to find the moment about line AD:

$$M_{AD} = M_A \cdot n_{AD} = (0.66018T - 2.978)\text{ kN} \cdot \text{m} \tag{b}$$

with $n_{AD} = \dfrac{r_{AD}}{|r_{AD}|} = \dfrac{(-1.5i + 2.5j + 2.5k)}{\sqrt{(-1.5)^2 + (2.5)^2 + (2.5)^2}} = -0.391i + 0.651j + 0.651k.$

Solving Eq. (b) for cable force magnitude T gives

$$T = 4.51 \text{ kN} \tag{c}$$

The components of T are

$$T = Tn_{BO} = (-2.29i - 3.81j - 0.762k) \text{ kN} \qquad \text{(d)}$$

With T known, the following sequence of steps leads to the reaction force components at A and D.

  i. Sum moments about the x axis:

$$-2.5D_y - 2.5P_y + 2.5P_z = 0 \qquad D_y = -P_y + P_z = 3.05 \text{ kN} \quad \text{⬅ (e)}$$

  ii. Sum moments about a vertical axis through A:

$$2.5D_x + 2.5P_x + 0.5T_x = 0 \qquad D_x = -P_x - \frac{T_x}{5} = -0.686 \text{ kN} \quad \text{⬅ (f)}$$

  iii. Sum forces in the x direction:

$$A_x + D_x + T_x + P_x = 0 \qquad A_x = -(D_x + T_x + P_x) = 1.83 \text{ kN} \quad \text{⬅ (g)}$$

  iv. Sum forces in the y direction:

$$A_y + D_y + T_y + P_y = 0 \qquad A_y = -(D_y + T_y + P_y) = 2.67 \text{ kN} \quad \text{⬅ (h)}$$

  v. Sum forces in the z direction:

$$A_z + T_z + P_z = 0 \qquad A_z = -(T_z + P_z) = -0.381 \text{ kN} \quad \text{⬅ (i)}$$

4. *Finalize*: Use of a moment about a line often provides the opportunity to eliminate many of the unknown forces in a problem because those forces have a line of action through that line. This leads to a direct and simpler solution, which avoids solution of simultaneous equations. Here, cable force T was found using one equilibrium equation with one unknown [Eq. (b)]. You should also look for opportunities to sum moments about the coordinate axes themselves, or about lines that are parallel to one of the coordinate axes [see Eq. (f)], to solve for unknown reactions. Finally, recall that, when using a statics sign convention, a minus sign means that the force component acts in the negative direction of the associated coordinate axis.

3.4 Dry Friction

In many cases, you can assume the forces acting between bodies in contact are normal to the contacting surfaces. If the surfaces are rough rather than smooth, however, a tangential component of contact force known as a *friction force* must also be considered. The friction force acts to oppose any motion of one body along the surface of another. Friction is undesirable in bodies such as gears and engines where lubrication is used to minimize its effects. Friction can serve a useful—even essential—purpose as in automotive brakes and drive belts. Side effects of friction include energy dissipation in the form of heat and also wear of the contacting surfaces. *Tribology*, a branch of mechanical engineering and materials science, is the study of interacting surfaces in relative motion and includes the study and application of the principles of friction, lubrication, and wear.

Types of Friction

Our focus in statics is on *dry friction*, which acts to oppose relative motion between solid surfaces in contact. Other types of friction, such as *fluid* and *internal* friction, are also important but are not considered in this text. Fluid friction, considered in the study of fluid mechanics, occurs between layers in a viscous fluid and depends on both the velocity gradient and the viscosity of the fluid. Internal friction occurs when a solid body undergoes plastic deformation during cyclic loading. Shearing deformation of inelastic bodies leads to considerable amounts of internal friction and is studied in detail in materials science.

Dry friction can be subdivided into *static friction*, which acts between two bodies at rest, and *kinetic friction*, which acts as relative motion occurs between the two bodies. In both cases, the nature of the interaction between the two surfaces is characterized by *coefficients* of static and kinetic friction. Larger coefficients reflect strong interaction, and smaller coefficients indicate lesser interaction, resulting in a smaller friction force. Coefficients vary with the *types of materials* in contact and the *condition* of the contacting surfaces. Coefficient values for a variety of materials are given in Table 3-2.

Dry (Coulomb) Friction

Consider a block of weight W resting on a horizontal plane (Fig. 3-42a). Begin by applying a gradually increasing force P in the $+x$ (or $+t$) direction parallel to the surface. A free-body diagram of the block (Fig. 3-42b) shows all forces acting on the block: applied force P, gravity force W in the $-y$ (or $-n$) direction, normal reaction force N, and friction force f in the $-x$ (or $-t$) direction. The resultant of forces f and N is R, which represents the action of the surface on the bottom of the block.

Stage 1—Static Friction
As force P is increased, the friction force increases in proportion from zero to some value f_{max} (Fig. 3-42c). During this first stage, the block is at rest, and *static equilibrium* requires that $\Sigma F_x = 0$ (*or* $\Sigma F_t = 0$), so the friction force is equal but opposite to the applied force, and $f = P$. This is the static friction stage and, since the block is at rest, you may use static equilibrium to find resisting force f (or f_s) for any level of applied force P. In stage 1, the friction force f_s is

$$f = f_s < f_{max} \tag{3-7}$$

Transition Point—Impending Motion
When the block is on the verge of slipping, the static frictional resistance has reached a maximum value for the two surfaces in contact. Friction force f_{max} is proportional to the normal force N and may be expressed as

$$f_{max} = \mu_s N \tag{3-8}$$

where the constant of proportionality is μ_s, which is the *coefficient of static friction*. Static equilibrium still applies since motion has not yet begun, but friction force f_s only equals $\mu_s N$ when motion is impending and force f_s has reached its maximum value. Coefficient μ_s only applies to a *pair of surfaces*, which could be the same or different materials in contact.

(a) Block on horizontal surface   (b) Free-body diagram of block

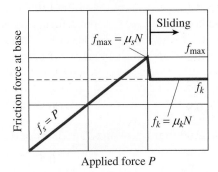

(c) Friction force f vs. applied force P

Stage 2—Kinetic Friction

Just after f reaches f_{max}, the block slips abruptly and moves in the direction of applied force P. The friction force drops slightly (Fig. 3-42c) in this second stage as the block moves with increasing velocity. Friction force f still acts in opposition to applied force P but is no longer equal to P; *static equilibrium no longer applies*. In stage 2, the friction force f_k is

$$f = f_k = \mu_k N \tag{3-9}$$

where the constant of proportionality is μ_k, which is the *coefficient of kinetic* friction.

Coefficients of Static (μ_s) and Kinetic (μ_k) Friction

The friction coefficients for a pair of surfaces can be determined using simple experiments and are dependent on the condition of the surfaces. Friction coefficient values are subject to uncertainty, and friction forces are usually considered an unreliable source of resistance to applied loads in engineering analysis of structures. Hence, the values recorded in Table 3-2 should be used as representative but approximate values for the surfaces listed in the table.

Friction Angles; Angle of Repose

Now consider the block from Fig. 3-42 to be at rest on a ramp with coefficient of static friction μ_s and inclination angle θ acted on by gravity alone (Fig. 3-43a). If you increase the slope of the ramp, you can find the angle θ where the block begins to slide down the plane. This angle, θ_{max}, is known as the *angle of repose*. From the free-body diagram in Fig. 3-42b:

$$\Sigma F_t = 0 \quad f - W \sin\theta = 0 \quad f = W \sin\theta$$
$$\Sigma F_n = 0 \quad N - W \cos\theta = 0 \quad N = W \cos\theta \tag{3-10a, b}$$

Use coordinates t and n (rather than x and y), which are tangential and normal to the ramp, respectively, to avoid resolving forces f and N into components. Solving for W in Eq. (3-10b) gives $W = N/\cos\theta$, and substituting this expression into Eq. (3-10a) gives $f = N \tan\theta$. With the block on the verge of

FIGURE 3-43

Friction force for a block on an inclined surface

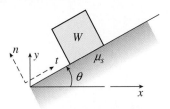

(a) Block on inclined surface

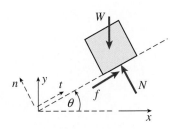

(b) Free-body diagram of block

Table 3-2	Contacting Materials	Static Coefficient, μ_s	Kinetic Coefficient, μ_k
Approximate values of coefficients of friction*	Aluminum on aluminum	1.1	1.0
	Aluminum on steel	0.6	0.5
	Brass on brass	0.9	0.8
	Brass on steel	0.5	0.4
	Cast iron on cast iron	1.1	1.0
	Cast iron on copper	1.1	1.0
	Copper on copper	1.2	1.1
	Copper on steel	0.5	0.4
	Glass on glass	0.9	0.8
	Metal on ice	0.04	0.02
	Metal on stone	0.5	0.4
	Rope on pulley	0.2	0.15
	Rubber on concrete	0.8	0.7
	Rubber on ice	0.1	0.05
	Rubber on metal	0.6	0.4
	Steel on steel	0.7	0.5
	Steel on concrete	0.5	0.4
	Teflon on steel	0.04	0.04
	Wood on concrete	0.6	0.5
	Wood on metal	0.4	0.3

*To be used solely for solving problems in this book. Surface cleanliness, finish, pressure, and lubrication can lead to significant changes in these values.

sliding, f becomes f_{max} and angle θ is now θ_{max}. From Eq. (3-8), it follows that $f_{max} = \mu_s N = N \tan\theta_{max}$, so

$$\mu_s = \tan\theta_{max} \qquad \text{or} \qquad \theta_{max} = \tan^{-1}\mu_s \qquad \textbf{(3-11)}$$

By conducting a simple experiment—measuring the angle at which a block of *unknown* weight W begins to slide down a plane—you can find the *coefficient of static friction* μ_s between the two surfaces in contact. See Examples 3-9 and 3-11 for a more detailed examination of a block or group of blocks resting on an inclined plane and acted on by gravity and an applied force P.

Sliding versus Overturning An important application of friction is that associated with possible sliding versus overturning of mechanical and electrical cabinets subject to lateral forces, such as those caused by earthquake ground motion. Many mechanical and electrical cabinets are supported on leveling mounts or rails, so friction and normal reaction forces occur along the cabinet boundary or at specific points near the four corners of the cabinet. Therefore, sliding or overturning of the 3D cabinet must be evaluated for either horizontal axis (Fig. 3-44). The following discussion is intended to provide an introduction to the topic of sliding versus overturning. A more detailed examination of a cabinet resting on a horizontal plane and subjected to either sliding or overturning is presented in Example 3-10.

Consider the 2D model of a cabinet of mass m and dimensions $B \times H$ shown in Fig. 3-45a; assume that $H = 3B$. Because of the distribution of internal components, the center of gravity (point C in Fig. 3-45) is not located at the geometric center of the cabinet. In accordance with seismic code procedures, equivalent lateral earthquake force F is applied at the center of gravity at height $2H/3$ above the base and can act in either the $+x$ or $-x$ direction. Frictional forces are assumed to act over the entire base of the cabinet. You want to find the conditions where the cabinet will *slide* or *overturn* (either to the left or right) due to increasing levels of force F.

FIGURE 3-44

Unanchored switchgear cabinets are susceptible to sliding or overturning due to lateral forces

DarkWeapon/Shutterstock.com

Force F in $+x$ Direction   The equations of static equilibrium for lateral load F acting in the $+x$ direction, based on the free-body diagram in Fig. 3-45b, are

$$\Sigma F_x = 0 \quad F - f = 0 \quad \text{or} \quad F = f$$
$$\Sigma F_y = 0 \quad N - mg = 0 \quad \text{or} \quad N = mg \qquad \text{(3-12a, b, c)}$$
$$\Sigma M_E = 0 \quad -F\left(\frac{2H}{3}\right) + mgd_R = 0 \quad \text{or} \quad d_R = \frac{2FH}{3mg}$$

The resultant of friction and normal forces acting on the base is assumed to be located at point E, which is a distance d_R in the positive x direction from point C.

FIGURE 3-45

(a) Cabinet with lateral load F, (b) free-body diagram: force F in $+x$ direction, and (c) free-body diagram: force F in $-x$ direction

(a)

(b)

(c)

If the cabinet is *on the verge of sliding*, $f = \mu_s N = \mu_s mg = F$. Substituting for F and H in Eq. (3-12c) leads to the following:

$$d_R = \frac{2FH}{3mg} = \frac{2(\mu_s mg)(3B)}{3mg} = 2\mu_s B \qquad (3\text{-}13)$$

Distance d_R must be less that $2B/3$ *if sliding is to occur first*, so for the case of applied force F in the $+x$ direction,

$$d_R = 2\mu_s B < \frac{2}{3}B \qquad \text{or} \qquad \mu_s < \frac{1}{3} \qquad (3\text{-}14)$$

As an alternative, you could carry out an analysis that assumes that overturning occurs before sliding. For overturning to the right (clockwise rotation), the cabinet must rotate about point D. Summing moments about D gives

$$\Sigma M_D = 0 \quad F\left(\frac{2}{3}H\right) = mg\left(\frac{2}{3}B\right) \quad \text{or} \quad f\left(\frac{2}{3}(3B)\right) = mg\left(\frac{2}{3}B\right) \quad \text{so} \quad f = \frac{mg}{3} \quad (3\text{-}15)$$

Overturning is assumed, so sliding must not occur before overturning. This means that *for overturning to occur first* for force F in the $+x$ direction:

$$f = \frac{mg}{3} < \mu_s N = \mu_s mg \qquad \text{so} \qquad \frac{1}{3} < \mu_s \qquad (3\text{-}16)$$

So Eqs. (3-14) and (3-16) provide the conditions needed for either sliding [Eq. (3-14)] or overturning [Eq. (3-16)] of the cabinet under applied force F applied in the $+x$ direction. Of course, sliding and overturning occur at the same time if $\mu_s = 1/3$.

Force F in $-x$ Direction   The cabinet may be more sensitive to overturning when force F is applied in the $-x$ direction because the cabinet center of gravity is shifted to the left in this example. To develop the static equilibrium equations, you find that Eqs. (3-12a and b) are unchanged. Sum moments about E in Fig. 3-18c to get

$$\Sigma M_E = 0 \qquad F\left(\frac{2H}{3}\right) - mgd_L = 0 \quad \text{or} \quad d_L = \frac{2FH}{3mg} \qquad (3\text{-}17)$$

Now, distance d_L must be less than $B/3$ *if sliding is to occur first*, so for the case of applied force F in the $-x$ direction,

$$d_L = 2\mu_s B < \frac{1}{3}B \qquad \text{or} \qquad \mu_s < \frac{1}{6} \qquad (3\text{-}18)$$

For overturning to the left (counterclockwise rotation), the cabinet must rotate about point A. Summing moments about point A gives

$$\Sigma M_A = 0 \quad F\left(\frac{2}{3}H\right) = mg\left(\frac{1}{3}B\right) \quad \text{or} \quad f\left(\frac{2}{3}(3B)\right) = mg\left(\frac{1}{3}B\right) \quad \text{so} \quad f = \frac{mg}{6} \quad (3\text{-}19)$$

So *for overturning to occur first* for the case of force F applied in the $-x$ direction,

$$f = \frac{mg}{6} < \mu_s N = \mu_s mg \qquad \text{so} \qquad \frac{1}{6} < \mu_s \qquad (3\text{-}20)$$

If force F is applied in the $-x$ direction, Eqs. (3-18) and (3-20) provide the conditions needed for either sliding [Eq. (3-18)] or overturning [Eq. (3-20)].

Of course, sliding and overturning occur at the same time if $\mu_s = 1/6$. This reduced value of μ_s for force F applied in the $-x$ direction suggests that counterclockwise overturning is likely as expected.

Many electrical switchgear cabinets are tall and slender ($H = 3B$ or $4B$) and not bolted to the floor, so they are susceptible to damage due to sliding and overturning in earthquakes. A steel cabinet resting on a concrete floor is likely to have μ_s values near 0.4 or 0.5 (see Table 3-2). Damage to unanchored cabinets (Fig. 3-46) can lead to loss of power in buildings, resulting in severe social and economic disruption in the aftermath of even moderate earthquakes.

FIGURE 3-46

Overturned electrical control cabinets

Rotational Motion and Slip   The discussion of friction so far considered static and kinetic frictional forces associated with *sliding* and possible *overturning* of blocks on either horizontal or inclined surfaces. A third example of friction is impending *rotational motion* of balls, disks, wheels, or cylinders subject to applied *moments*. Automobiles depend on friction not only for braking but also for *traction* on paved surfaces. If the road is steep and covered with ice, frictional forces are severely reduced (compare friction coefficients for "*Rubber on concrete*" and "*Rubber on ice*" in Table 3-2).

Consider a disk of mass m resting on an incline at B having slope θ and also supported by a wall at A (Fig. 3-47a). The t axis is taken along the incline and axis n is aligned with the wall. Clockwise moment M is applied to the disk and static coefficient of friction μ_s applies at both contact points. Find the conditions under which the disk (a) slips in place while maintaining contact at both A and B or (b) loses contact at A and starts to roll up the incline.

Static equilibrium of the disk can be assessed using the free-body diagram in Fig. 3-47b. Summing forces in t and n directions and summing moments about the center of gravity C results in the following equilibrium equations:

$$\Sigma F_t = 0 \qquad N_A + f_B - mg\sin\theta = 0$$
$$\Sigma F_n = 0 \qquad -f_A + N_B - mg\cos\theta = 0 \qquad \textbf{(3-21a, b, c)}$$
$$\Sigma M_C = 0 \qquad (f_A + f_B)r - M = 0$$

FIGURE 3-47

Friction forces for wheel on an inclined surface

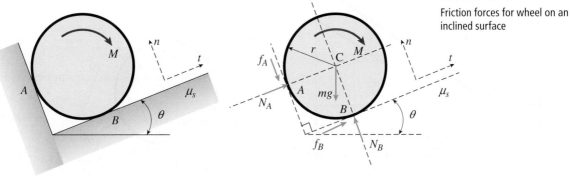

(a) Disk on incline supported by wall   (b) Free-body diagram of disk

If the disk is on the verge of slipping at both A and B, with no loss of contact at A, then

$$f_A \to f_{A,\text{max}} = \mu_s N_A \qquad \text{and} \qquad f_B \to f_{B,\text{max}} = \mu_s N_B \qquad \textbf{(3-22)}$$

Substituting Eq. (3-22) into Eqs. (3-21) gives

$$\Sigma F_t = 0 \qquad N_A + \mu_s N_B = mg\sin\theta$$
$$\Sigma F_n = 0 \qquad -\mu_s N_A + N_B = mg\cos\theta \qquad \textbf{(3-23a, b, c)}$$
$$\Sigma M_C = 0 \qquad \mu_s(N_A + N_B)r = M$$

Solving Eqs. (3-23a and b) (two equations with two unknowns, N_A and N_B) results in the following expressions:

$$N_A = \frac{mg}{1+\mu_s^2}(\sin\theta - \mu_s\cos\theta) \quad \text{and} \quad N_B = \frac{mg}{1+\mu_s^2}(\mu_s\sin\theta + \cos\theta) \textbf{(3-24a, b)}$$

If $N_A > 0$, the disk is still in contact at A and will **slip in place** if: (1) From Eq. (3-24a), the expression $(\sin\theta - \mu_s\cos\theta)$ is greater than zero, which means that the slope angle θ must be larger than the angle of repose, $\tan^{-1}(\mu_s)$; and (2) Applied moment $M > \mu_s(N_A + N_B)r$.

If Eq. (3-24a) is plotted, normal force N_A versus incline angle θ over the range $0 \leq \theta \leq 60°$ and for several coefficients of static friction, say $\mu_s = 0.25$ and 0.5, normal force N_A is positive (acting in the $+t$ direction in Fig. 3-47b) only for $\theta > \tan^{-1}(\mu_s)$ for each value of μ_s (Fig. 3-48). Both friction force f_A and normal force N_A are zero for values of θ from zero to $\theta = \tan^{-1}(0.25) = 14.04°$ and $\theta = \tan^{-1}(0.5) = 26.56°$. Of course, negative values of force N_A are meaningless because the disk simply rests against the wall at point A.

If $f_A = 0$ and $N_A = 0$, the disk has lost contact with the wall at A and is about to **roll up the incline**. The equilibrium equations for this case can be obtained from the free-body diagram in Fig. 3-49 or simply by setting both f_A and N_A equal to zero in Eq. (3-21). The disk will roll up the incline provided that (1) incline angle $\theta < \theta_{\text{max}} = \tan^{-1}(\mu_s)$ [from Eq. (3-24a)]; and (2) from summing moments about B after contact is lost at A (Fig. 3-49), applied moment $M > (mg\sin\theta)(r)$.

See Example 3-13 for a more detailed analysis of rotational motion and friction, illustrating how friction force f varies with applied moment M, for a gear-driven sheet feeder system.

FIGURE 3-48

Normal force at A as a function of incline angle θ and coefficient of static friction μ_s

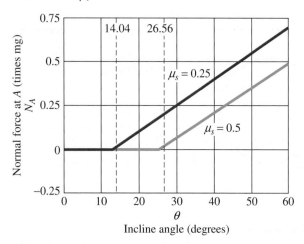

FIGURE 3-49

Free-body diagram of disk for $N_A = 0$, $f_A = 0$

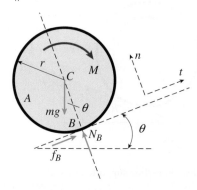

Example 3-9

A block of weight $W = 250$ lb rests on an inclined plane with slope $\theta = 22.5°$. Force P is applied to the block in the $(+)\,t$ direction, as shown in Fig. 3-50a. The coefficient of static friction between the block and plane is $\mu_s = 0.3$. Solve the following.

(a) Find the maximum and minimum values of force P if the block is just on the verge of sliding.

(b) Repeat part (a) if the line of action of force P is at an angle $\alpha = 40°$ from the line of movement of the block (see Fig. 3-50b).

(c) If $P = 0$ and an undeformed linear spring, with spring constant $k = 80$ lb/in, is attached to the block at rest, how far (distance d_{max}, in.) can the block be displaced *down* the slope and still remain at rest in its new position (see Fig. 3-50c)? Neglect the weight of the spring.

FIGURE 3-50

(a) Force P restrains block on inclined plane, (b) inclined force P restrains block, and (c) linear spring restrains block on inclined plane

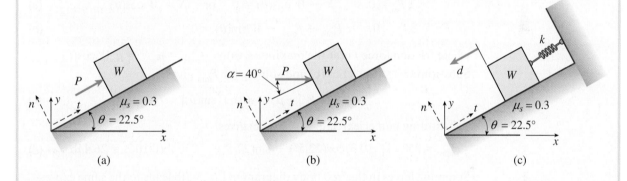

(a)   (b)   (c)

Solution:

Part (a): Use the four-step problem-solving approach to find the maximum and minimum values of force P.

1. *Conceptualize*: The block cannot be held in place by static friction alone because the coefficient of static friction $\mu_s = 0.3$ is less than $\tan(\theta) = 0.414$. The angle of repose $\phi = \tan^{-1}(\mu_s) = 16.7°$ is less than slope angle θ so an additional force P is required to prevent sliding. Figs. 3-51a and b show the free-body diagrams of the block for the two extreme cases: (a) a minimum force P together with static friction force f prevent the block from sliding *down* the plane and (b) a maximum force P opposes both gravity and friction as the block is about to slide *up* the slope. In both cases, the friction force is at a maximum value ($f_{max} = \mu_s N$) because the block is on the verge of sliding.

FIGURE 3-51

(a) Free-body diagram for minimum force and P and (b) free-body diagram for maximum force P

(a)   (b)

2. *Categorize*: For static equilibrium, it is most convenient to sum forces in the rotated n-t coordinate system that is normal to (n-direction) and along (tangent to, t-direction) the plane to avoid resolving P, f, and N forces into x-y components. Note that using the x-y coordinate system, however, will lead to the same results.

3. *Analyze*: Enforce equilibrium in the n and t coordinate directions. The resulting equations based on the free-body diagram in Fig. 3-51a are

$$\Sigma F_n = 0 \qquad N - W\cos(\theta) = 0 \quad \text{or} \quad N = W\cos(\theta) \tag{a}$$

$$\Sigma F_t = 0 \qquad P_{min} + f_{max} - W\sin(\theta) = 0 \tag{b}$$

where friction force f is at the maximum value, $f_{max} = \mu_s N = \mu_s W\cos(\theta)$. Substituting f_{max} into Eq. (b) and solving for P_{min} gives

$$P_{min} = W[-\mu_s \cos(\theta) + \sin(\theta)] \tag{c}$$

Substituting numerical data into Eq. (c) gives

$$P_{min} = 250 \text{ lb } [-(0.3)\cos(22.5°) + \sin(22.5°)] = 250 \text{ lb}(0.106) = 26.4 \text{ lb} \quad \leftarrow \tag{d}$$

Summing forces in the free-body diagram in Fig. 3-51b leads to the same expression for N in Eq. (a), but in Eq. (b), the sign of f_{max} is now negative, so

$$\Sigma F_t = 0 \qquad P_{max} - f_{max} - W\sin(\theta) = 0 \tag{e}$$

The solution expression for P_{max} is

$$P_{max} = W[\mu_s \cos(\theta) + \sin(\theta)] \tag{f}$$

and the numerical result for P_{max} is

$$P_{max} = 250 \text{ lb } [(0.3)\cos(22.5°) + \sin(22.5°)] = 250 \text{ lb } (0.66) = 165 \text{ lb} \quad \leftarrow \tag{g}$$

4. *Finalize*: For the case in which force P acts in the positive t direction, the values of P_{min} and P_{max} vary as follows:
   i. If the coefficient of static friction is increased from 0.3 to tan (22.5°) = 0.414 in Eq. (d), then P_{min} goes to zero. Force P_{min} is zero because static friction alone is sufficient to hold the block in place without sliding down the plane. However, P_{max} is larger due to the increased friction force.

ii. Alternatively, if $\mu_s = 0.3$ but angle θ is reduced from 22.5° to $\phi = \tan^{-1}(0.3) = 16.7°$, $P_{min} = 0$ and P_{max} is smaller than the value in Eq. (g). Once again, force $P_{min} = 0$ because static friction alone is sufficient to hold the block in place without sliding down the plane.

iii. For $P < P_{min}$ and $P > P_{max}$, the block will slide down and up, respectively, the inclined plane. Recall that Eqs. (b) and (e) still apply as long as the block moves with zero acceleration.

The full range of possibilities is seen if you plot friction force f versus applied force P (from zero to beyond P_{max}) for the case of $W = 250$ lb, $\mu_s = 0.3$, $\mu_k = 0.2$ (kinetic friction), and $\theta = 22.5°$. Friction force f is plotted above the P-axis (Fig. 3-52) when it opposes applied force P for the case of motion up the plane shown in Fig. 3-51b. Once applied force P exceeds P_{max} or drops below P_{min}, the block begins to slide on the plane, and the friction force resistance drops to a *constant value of kinetic friction force*

$$f_k = \mu_k (W \cos(\theta)) = 46.2 \text{ lb}.$$

FIGURE 3-52

Plot of friction force f vs. applied force P

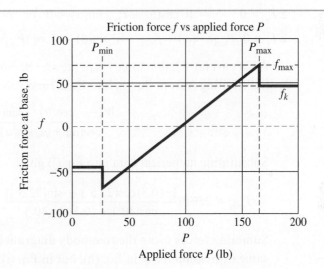

Friction force f vs applied force P

Part (b): Use the four-step problem-solving approach to find the maximum and minimum values of required force P if the line of action of force P is at an angle $\alpha = 40°$ from the line of movement of the block (see Fig. 3-50b).

1. *Conceptualize*: Figures 3-53a and b show the *revised free-body diagrams* of the block for the two extreme cases: a *minimum* force P together with static friction force f prevent the block from sliding *down* the plane and a *maximum* force P acts against both gravity and friction as the block is about to slide *up* the slope. As in part (a), the friction force is at a maximum value ($f_{max} = \mu_s N$) because the block is on the verge of sliding.

(a)   (b)

2. *Categorize*: Only a component of *P* acts along the *t*-direction, so a larger value of *P* is required to either hold the block from sliding down the plane (Fig. 3-53a) or put it on the verge of moving up the plane (Fig. 3-53b). The component of force *P* in the *n* direction increases the value of normal force *N* so the maximum static friction force f_{max} is larger than that in part (a).

3. *Analyze*: The equilibrium equations resulting from the free-body diagram in Fig. 3-53a are

$$\Sigma F_n = 0 \quad N - W\cos(\theta) - P_{min}\sin(\alpha) = 0 \quad \text{or} \quad N = W\cos(\theta) + P_{min}\sin(\alpha) \quad \textbf{(h)}$$

$$\Sigma F_t = 0 \quad P_{min}\cos(\alpha) + f_{max} - W\sin(\theta) = 0 \quad \text{or} \quad P_{min}\cos(\theta) + \mu_s N - W\sin(\theta) = 0 \textbf{ (i)}$$

As in part (a), friction force *f* is at the maximum value $f_{max} = \mu_s N$. Substituting *N* from Eq. (h) into Eq. (i) and solving for P_{min} gives

$$P_{min} = \frac{W[-\mu_s \cos(\theta) + \sin(\theta)]}{\cos(\alpha) + \mu_s \sin(\alpha)} \quad \textbf{(j)}$$

Substituting numerical data into Eq. (j) gives

$$P_{min} = 250 \text{ lb}\frac{[-(0.3)\cos(22.5°) + \sin(22.5°)]}{\cos(40°) + (0.3)\sin(40°)} = 250 \text{ lb}(0.11) = 27.5 \text{ lb} \quad \textbf{(k)}$$

Summing forces using the free-body diagram in Fig. 3-53b leads to the same expression for *N* in Eq. (h), but in Eq. (i), the sign of f_{max} is now negative, so

$$\Sigma F_t = 0 \qquad P_{max}\cos(\alpha) - f_{max} - W\sin(\theta) = 0 \qquad \textbf{(l)}$$

The solution expression for P_{max} is

$$P_{max} = W\frac{[\mu_s \cos(\theta) + \sin(\theta)]}{\cos(\alpha) - \mu_s \sin(\alpha)} \qquad \textbf{(m)}$$

and the numerical result for P_{max} is

$$P_{max} = 250 \text{ lb}\frac{[(0.3)\cos(22.5°) + \sin(22.5°)]}{\cos(40°) - (0.3)\sin(40°)} = 250 \text{ lb }(1.151) = 288 \text{ lb} \quad \textbf{(n)}$$

4. *Finalize*: If angle $\alpha = 0$, Eq. (j) reduces to Eq. (c) and Eq. (m) becomes Eq. (f) as expected. The value of P_{min} increases slightly, but a 75% increase in P_{max} is required if the load P is applied at downward angle α.

   As in part (a), you can plot the variation in friction force f versus applied force P (Fig. 3-54) for a full range of values of P (from zero to beyond P_{max}). Now the downward inclination of force P means that P contributes to normal force N [see Eq. (h)]. As a result, for values of force P greater than P_{max}, normal force N increases as well and thus kinetic friction force f_k is no longer constant. This is also true for the kinetic friction forces for $P < P_{min}$.

FIGURE 3-54

Plot of friction force f versus applied force P inclined at angle α

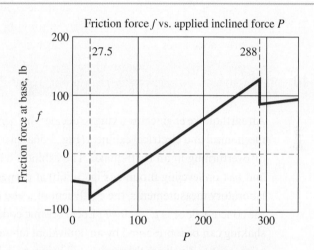

Friction force f vs. applied inclined force P

Applied force P (lb) at angle alpha

Part (c): Finally, find how far *down* the slope (distance d_{max}, in.) the block can be displaced and still remain at rest in its new position if restrained by a linear spring (Fig. 3-50c).

FIGURE 3-55

Free-body diagram for maximum force P

1. *Conceptualize*: Figure 3-55 is the *revised* free-body diagram of the block showing a maximum *spring* force F_{max}, which opposes both gravity and friction as the displaced block is about to slide *up* the slope. Force F represents the action of the spring *on* the block. As in parts (a) and (b), the friction force is at a maximum value ($f_{max} = \mu_s N$) because the block is on the verge of sliding.

2. *Categorize*: The free-body diagram in Fig. 3-55 is the same as that in Fig. 3-51b, so the equilibrium analysis in part (a) applies here. It follows that $F_{max} = P_{max} = 165$ lb [Eq. (g)].

3. *Analyze*: Spring force F is linearly related to spring displacement d by spring constant k so $F = kd$. Solving for spring elongation d_{max} gives

$$d_{max} = \frac{F_{max}}{k} = \frac{165 \text{ lb}}{80 \text{ lb/in}} = 2.06 \text{ in} \qquad \blacktriangleleft \text{(p)}$$

4. *Finalize*: The linear spring is a deformable component defined by its spring constant k (or *stiffness coefficient*). The spring applies a force to the block in proportion to its displacement. Comparison of the free-body diagrams in Figs. 3-51b and 3-55 reveals the similarity of the systems in parts (a) and (c).

One final thought: if you move the block *up* the incline to compress the spring, what is d_{max}?

Example 3-10

An earthquake engineer is trying to decide whether or not to use anchor bolts to tie down mechanical and electrical cabinets (Fig. 3-56a) in a building to prevent them from sliding or overturning in an earthquake. The cabinets have weight W, width B, and height H and rest on leveling mounts (Fig. 3-56b) at corners A and D (see Fig. 3-56). Based on laboratory measurements, the coefficient of static friction between the steel mounts and the concrete floor is $\mu_s = 0.47$. Local seismic codes specify that the earthquake ground shaking can be represented by an equivalent lateral force F applied at the cabinet center of mass at C. Evaluate the static equilibrium of cabinets 1 and 2. Will they slide or overturn due to earthquake force F applied in the $+x$ direction at C?

FIGURE 3-56

(a) Lateral force F applied to cabinets 1 and 2, and (b) leveling mount

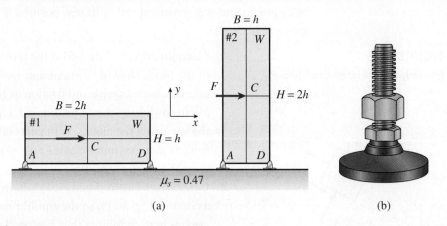

(a)   (b)

Solution:

Use the four-step problem-solving approach to find the lateral force F required for the sliding or overturning of cabinets 1 and 2.

FIGURE 3-57

(a) Free-body diagram for cabinet 1,
and (b) Free-body diagram for cabinet 2

(a)

(b)

1. *Conceptualize*: Start by drawing the free-body diagrams for cabinets 1 and 2 (see Figs. 3-57a and b). The mass of each cabinet is assumed to be uniformly distributed, so gravity force W is applied at the geometric center of each cabinet. Friction and normal forces act only at the support points at A and D where the steel leveling mounts rest on the concrete floor. In both cases, if the cabinet is on the verge of sliding, the friction forces f_A and f_D will be at their maximum values ($f_{max} = \mu_s N$).

2. *Categorize*: For static equilibrium, sum forces in the x and y directions then sum moments about right support D. For convenience, write expressions in terms of cabinet dimensions B and H, then substitute values of B and H for either cabinet 1 ($B = 2h$, $H = h$) or cabinet 2 ($B = h$, $H = 2h$). First, find the force required to cause the *sliding* of cabinets 1 (F_{s1}) and 2 (F_{s2}), then the forces required to cause the *overturning* of each cabinet (F_{o1}, F_{o2}). If $F_s < F_o$, the cabinet will slide rather than overturn. Because of symmetry, only the case of earthquake force F applied in the $+x$ direction need be considered.

3. *Analyze*:

   Sliding of cabinets 1 and 2: The static equilibrium equations for the free body diagrams in Fig. 3-57 are

$$\Sigma F_x = 0 \qquad F_s - f_A - f_D = 0 \qquad \text{(a)}$$
$$\Sigma F_y = 0 \qquad N_A + N_D - W = 0 \qquad \text{(b)}$$
$$\Sigma M_D = 0 \qquad -F_s \frac{H}{2} - N_A B + W \frac{B}{2} = 0 \qquad \text{(c)}$$

Each cabinet is assumed to be on the verge of sliding, so friction force f is at the maximum value $f_{max} = \mu_s N$. Substituting f_{max} for f_A and f_D into Eq. (a) gives

$$F_s - \mu_s(N_A + N_D) = 0 \qquad \text{(d)}$$

or

$$F_s = \mu_s(N_A + N_D) \qquad \text{(e)}$$

From Eq. (b), $N_A + N_D = W$, so the force required to cause sliding of either cabinet 1 or 2 is the same:

$$F_s = \mu_s W \qquad \text{(f)}$$

Only Eqs. (a) and (b) were used to arrive at this result (two equations and two unknowns, N_A and N_D), so what purpose does Eq. (c) serve here? Use Eq. (c) to confirm that reaction force N_A is positive (in the $+y$ direction in

Figs. 3-57a and b), otherwise overturning—not sliding as assumed—has occurred. Solving for N_A in Eq. (c) and substituting F_s from Eq. (f) gives

$$N_A = \frac{1}{2B}[WB - F_sH] = \frac{1}{2B}[WB - (\mu_sW)H] = \frac{W}{2}\left[1 - \mu_s\frac{H}{B}\right] \tag{g}$$

If $\left[1 - \mu_s\dfrac{H}{B}\right] > 0$, then reaction N_A is positive and the expression for sliding

force F_s in Eq. (f) is valid. Substituting $\mu_s = 0.47$ for cabinets 1 and 2 gives

$$\textbf{Cabinet 1:} \qquad \left[1 - \mu_s\frac{H}{B}\right] = \left[1 - 0.47\frac{h}{2h}\right] = 0.765$$

$$\textbf{Cabinet 2:} \qquad \left[1 - \mu_s\frac{H}{B}\right] = \left[1 - 0.47\frac{2h}{h}\right] = 0.060 \tag{h}$$

So reaction N_A is positive and the force required to cause *sliding* of cabinets 1 and 2 is

$$F_{s1} = F_{s2} = \mu_sW = 0.47W \tag{i}$$

Overturning of cabinets 1 and 2: If the cabinets are about to overturn, each rotates about support D for the case of force F applied in the $+x$ direction, and support A uplifts. As a result, $f_A = N_A = 0$. Replace force F_s with force F_o in Eq. (c) to find:

$$-F_o\frac{H}{2} + W\frac{B}{2} = 0 \quad \text{or} \quad F_o = \frac{B}{H}W \tag{j}$$

Substituting the dimensions of each cabinet into Eq. (j) gives

$$\textbf{Cabinet 1:} \qquad F_{o1} = \frac{2h}{h}W = 2W$$

$$\textbf{Cabinet 2:} \qquad F_{o2} = \frac{h}{2h}W = \frac{W}{2} \tag{k}$$

Comparing F_s and F_o from Eqs. (i) and (k), *both cabinets 1 and 2 will slide before they overturn.*

4. *Finalize*: By equating the expressions for F_s and F_o in Eqs. (f) and (j), cabinet height H must *exceed* B/μ_s for overturning to control. For cabinet 1 ($B = 2h$, $H = h$), height $h < 2h/0.47 = 4.26h$ and for cabinet 2 ($B = h$, $H = 2h$), height $2h < h/0.47 = 2.13h$; so in both cases, sliding controls. If cabinet 2 dimensions are altered slightly to $B = h$, $H = 2.5h$, then $2.5h > h/0.47 = 2.13h$ and overturning will control for cabinet 2.

Example 3-11

FIGURE 3-58

One possible arrangement of three packages on access ramp

A package delivery carrier temporarily unloads three packages so she can locate a missing package. She wants to place packages A, B, and C on the truck access ramp in any order but with the requirement that *none of the packages, alone or in groups, slides down the ramp*. The packages are placed in contact at rest. Package weights are $W_A = 35$ lb, $W_B = 55$ lb, and $W_C = 50$ lb, and each has a different coefficient of static friction with the ramp (see Fig. 3-58). The ramp incline angle $\theta = 13.5°$. Find the permissible arrangement(s) for the three packages on the ramp.

Solution:

Use the four-step problem-solving approach to find permissible arrangement(s) for the three packages on the ramp.

1. *Conceptualize*: There are six different ways to arrange the three packages on the ramp: *A-B-C* (see Fig. 3-58); *B-C-A*; *C-A-B*; *B-A-C*; *A-C-B*; and *C-B-A* (the number of possible combinations equals $3! = 3 \cdot 2 \cdot 1 = 6$). The packages may slide *alone, acting in pairs*, or *all three packages may act together*. A free-body diagram of any one of the three packages acting alone is shown in Fig. 3-59a ($i = A$, B, or C), and a diagram for all three packages acting together is shown in Fig. 3-59b. Any package i if *acting alone* will slide if its angle of repose $\phi_i = \tan^{-1}(\mu_i)$ is less than the incline angle θ; *acting alone* means that it is not restrained by a package (or packages) located down the slope from its position. This means that package i acting alone will slide if the force acting down the incline, $W_i \sin(\theta)$, exceeds the maximum possible resisting friction force for that package $f_i = \mu_i N_i$. Observe that only package B will slide if acting alone because only friction angle $\phi_B = \tan^{-1}(0.20) = 11.3°$ is less than incline angle $\theta = 13.5°$.

FIGURE 3-59

(a) Free-body diagram for one package and (b) free-body diagram for all three packages

(a)

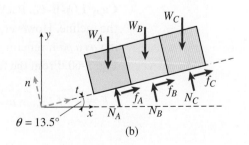

(b)

2. *Categorize*: The six possible arrangements of packages on the ramp are labeled as Cases 1–6 as given in Table 3-3.

Table 3-3	Case Number	Package Order (bottom to top)	Behavior Description
Package arrangement cases	1	A-B-C	OK if A + B do not slide down the incline
	2	B-C-A	**Not permissible**; lowest package B will slide alone down incline
	3	C-A-B	OK if C + A do not slide down the incline
	4	B-A-C	**Not permissible**; lowest package B will slide alone down incline
	5	A-C-B	OK if A + C do not slide down the incline and can restrain B
	6	C-B-A	OK if C + B do not slide down the incline

3. *Analyze*:

Sliding of all three packages: To evaluate all three packages acting together in some arrangement, compare the sum of the maximum possible resisting friction force R_{max} to the sum of the gravity forces G acting down the incline:

$$R_{max} = \sum_{i=1}^{3} f_i = [\mu_A W_A + \mu_B W_B + \mu_C W_C]\cos(13.5°)$$

$$= [0.30(35\text{ lb}) + 0.20(55\text{ lb}) + 0.25(50\text{ lb})]\cos(13.5°) = 33.1\text{ lb}$$

(a)

$$G = \sum_{i=1}^{3} W_i \sin\theta = (35\text{ lb} + 55\text{ lb} + 50\text{ lb})\sin(13.5°) = 32.7\text{ lb}$$

(b)

Since $R_{max} > G$, an equilibrium solution is possible. Of course for static equilibrium, $R = G = 32.7$ lb. If R_{max} is less than or equal to G, there is no solution to this problem.

Now examine Cases 1, 3, 5, and 6 in Table 3-3.

Case 1 (A-B-C): Packages A and C will remain at rest if each is placed alone on the incline. However, package B alone will slide. Will packages A and B, acting *as a pair*, remain at rest? The free-body diagrams for Case 1 are shown in Fig. 3-60. From the free-body diagram of package C alone:

$$\sum F_t = 0 \quad f_{C,max} = \mu_C N_C = 0.25(50\text{ lb})\cos(13.5°) = 12.16\text{ lb} > W_C\sin(13.5°)$$
$$= 11.67\text{ lb}$$

(c)

From Eq. (c), the force from C acting on B, F_{CB} in Fig. 3-60, is equal to zero because package C does not require assistance from packages A and B below.

FIGURE 3-60

Free-body diagrams for A and B acting together and C acting alone

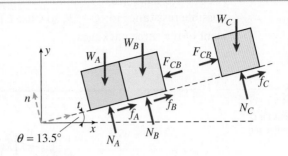

As the topmost package, C will remain at rest. Now evaluate A and B acting as a pair. Use the free-body diagram of packages A and B in Fig. 3-60 to compare gravity and resistance forces:

$$R_{A+B,\text{max}} = [\mu_A W_A + \mu_B W_B]\cos(13.5°) = [0.30(35\text{ lb}) + 0.20(55\text{ lb})]\cos(13.5°)$$
$$= 20.9\text{ lb} \tag{d}$$

$$G_{A+B} = [W_A + W_B]\sin\theta = (35\text{ lb} + 55\text{ lb})\sin(13.5°) = 21.0\text{ lb}$$

From Eq. (d), the gravity load down the ramp (G) is greater than the maximum possible resistance for $A + B$, so Case 1 (A-B-C) is not a permissible arrangement of the three packages.

Cases 3 (C-A-B) and 5 (A-C-B): Packages A and C will remain at rest if each is placed alone on the incline so $A + C$ and $C + A$ are also stable pairs. From Eqs. (a) and (b), note that A and C acting as a pair provide sufficient resistance to prevent B at the top from sliding.

Case 6 (C-B-A): Finally, will packages C and B acting together remain at rest? The free-body diagrams for this case are shown in Fig. 3-61. From the free-body diagram of package A alone:

$$\Sigma F_t = 0 \quad f_{A,\text{max}} = \mu_A N_A = 0.30(35\text{ lb})\cos(13.5°)$$
$$= 10.21\text{ lb} > W_A\sin(13.5°) = 8.17\text{ lb} \tag{e}$$

From Eq. (e), the force from A acting on B—F_{AB} in Fig. 3-60—is equal to zero because package A does not require assistance from packages C and B below. As the topmost package, A will remain at rest. Now evaluate C and B as a pair. From the free-body diagram of packages C and B in Fig. 3-61:

$$R_{C+B,\text{max}} = [\mu_C W_C + \mu_B W_B]\cos(13.5°)$$
$$= [0.25(50\text{ lb}) + 0.20(55\text{ lb})]\cos(13.5°) = 22.8\text{ lb} \tag{f}$$

$$G_{C+B} = [W_C + W_B]\sin\theta = (50\text{ lb} + 55\text{ lb})\sin(13.5°) = 24.5\text{ lb}$$

From Eq. (f), the gravity load down the ramp (G) is greater than the maximum possible resistance for $C + B$, so Case 6 (C-B-A) is not a permissible arrangement of the three packages.

FIGURE 3-61

Free-body diagrams for C and B acting together and A acting alone

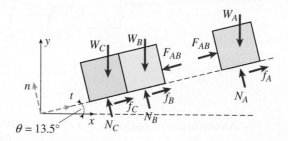

Table 3-4	Case Number	Package Order (bottom to top)	Behavior Description
Final assessment of package arrangement cases	1	A-B-C	**Not permissible**; $A + B$ *will slide down the incline* [Eq. (d)]
	2	B-C-A	**Not permissible**; lowest package B will slide alone down incline
	3	C-A-B	*OK based on* Eqs. (a) *and* (b); *B is at top and is restrained by* $C + A$
	4	B-A-C	**Not permissible**; lowest package B will slide alone down incline
	5	A-C-B	*OK based on* Eqs. (a) *and* (b); *B is at top and is restrained by* $A + C$
	6	C-B-A	**Not permissible**; $C + B$ *will slide down the incline* [Eq. (f)]

4. *Finalize*: As shown in Table 3-4, Cases 3 and 5 are the only permissible arrangement of the three packages that will result in no sliding of any packages down the ramp. (It is unlikely that the package carrier, in her rush to make deliveries on time, will have time to make the assessment outlined above. Trial and error may lead to a quicker solution in the field!)

Example 3-12

FIGURE 3-62

Pole being lowered by cable

A cable and pulley system at D is used to lower a 230-kg pole (ACB) shown in Fig. 3-62. The cable has tension force T and is attached to the pole at C. The pulley at D is assumed to be frictionless. The length L of the pole is 6.0 m. The pole was initially in a vertical position ($\theta = 0°$) and pivots about a contact point at A on a rough surface as it is lowered. Find the coefficient of static friction μ_s at A if the pole slips at ($\theta = 30°$).

Solution:

Use the four-step problem-solving approach to find the coefficient of static friction.

1. *Conceptualize*: Begin with a sketch of the free-body diagram of the pole (Fig. 3-63). Forces acting on the pole are tension force T in the cable at attachment point C; weight W of the pole at the center of gravity of the pole; normal force N_A at the base of the pole; and friction force f_A opposing slippage of the pole to the right at A.

FIGURE 3-63

Free-body diagram of pole

FIGURE 3-64

Angle β

2. *Categorize*: For static equilibrium, sum forces and moments in the x-y coordinate system that is aligned with normal and friction forces at A. When rotated to a position at which angle $\theta = 30°$, the pole is on the verge of slipping at A, so friction force f_A becomes $f_{A,\max} = \mu_s N_A$.

3. *Analyze*: Use a statics sign convention. Equilibrium of the pole just prior to slipping at *A* requires:

$$\sum F_x = 0 \quad T\cos(\beta) - f_{A,\max} = 0 \qquad \text{so } T = \frac{f_{A,\max}}{\cos(\beta)} \tag{a}$$

$$\sum F_y = 0 \quad -T\sin(\beta) - W + N_A = 0 \text{ so } W = N_A - T\sin(\beta) \tag{b}$$

Substitute the expression for *T* in Eq. (a) into Eq. (b) to obtain for *W*:

$$W = N_A - f_{A,\max}\frac{\sin(\beta)}{\cos(\beta)} = N_A - f_{A,\max}\tan(\beta) = N_A(1 - \mu_s\tan(\beta)) \tag{c}$$

Finally, sum moments about point C to get

$$\sum M_C = 0 \quad N_A(5\text{ m})\sin(\theta) - f_{A,\max}(5\text{ m})\cos(\theta) - W(2\text{ m})\sin(\theta) = 0 \tag{d}$$

Substituting the expression for *W* from Eq. (c) into Eq. (d) and using $f_{A,\max} = \mu_s N_A$ results in the following simplified form of Eq. (d):

$$\mu_s[-(5\text{ m})\cos(\theta) + (2\text{ m})\sin(\theta)\tan(\beta)] + (3\text{ m})\sin(\theta) = 0 \tag{e}$$

Finally, solving for coefficient μ_s results in the following expression for friction coefficient μ_s:

$$\mu_s = \frac{(3)\sin(\theta)}{(5)\cos(\theta) - (2)\sin(\theta)\tan(\beta)} \tag{f}$$

Solution for angle β: Find an expression for $\tan(\beta)$ in Eq. (f) where β is the angle between the cable and a horizontal axis. From Fig. 3-64,

$$\tan(\beta) = \frac{(5)\cos(\theta)}{4 + (5)\sin(\theta)} \tag{g}$$

Solve for angle β for the case of $\theta = 30°$:

$$\beta = \tan^{-1}\left[\frac{(5)\cos(30°)}{(4) + (5)\sin(30°)}\right] = 33.67° \tag{h}$$

Solution for coefficient of static friction: Use Eq. (f) to find that the required coefficient of static friction for the case of the pole slipping at *A* when angle $\theta = 30°$ and $\beta = 33.67°$:

$$\mu_s = \frac{(3)\sin(30°)}{(5)\cos(30°) - (2)\sin(30°)\tan(33.67°)} = \frac{1.5}{3.664} = 0.409 \quad \longleftarrow \text{(i)}$$

4. *Finalize*: If the specified angle at which the pole slips (θ) is increased, the required coefficient of static friction also increases. For example, if $\theta = 40°$, $\beta = 27.97°$ [Eq. (h)] and the required coefficient of static friction is $\mu_s = 0.613$ from Eq. (i).

Equilibrium solution continued: If desired, you can continue the equilibrium solution to find other items of interest. For example, from Eqs. (a) and (b), the cable force T for the case of $\theta = 30°$, $\mu_s = 0.409$, $\beta = 33.67°$, and $W = 230 \text{ kg}(9.81 \text{ m/s}^2) = 2.256 \text{ kN}$ is

$$T = \frac{W\mu_s}{\cos(\beta) - \mu_s \sin(\beta)} = \frac{2.256 \text{ kN}(0.409)}{\cos(33.67°) - (0.409)\sin(33.67°)} = 1.524 \text{ kN} \qquad \textbf{(j)}$$

Also from Eqs. (a) and (b), the normal force and maximum friction force at A are

$$N_A = \frac{T}{\mu_s}\cos(\beta) = \frac{1.524 \text{ kN}}{0.409}\cos(33.67°) = 3.10 \text{ kN} \qquad \textbf{(k)}$$

$$f_{A,\text{max}} = T\cos(\beta) \quad \text{or} \quad f_{A,\text{max}} = \mu_s N_A = 0.409(3.10 \text{ kN}) = 1.268 \text{ kN} \qquad \textbf{(l)}$$

General solution: In the solution leading to Eqs. (f) and (g), substitution of numerical data was delayed for as long as possible to develop a general relationship between the coefficient of static friction μ_s and pole angle θ. Substitute Eq. (g) into Eq. (f) to obtain the following functional relationship between μ_s and θ:

$$\mu_s(\theta) = \frac{(3)\sin(\theta)}{(5)\cos(\theta) - (2)\sin(\theta)\left[\dfrac{(5)\cos(\theta)}{4 + (5)\sin(\theta)}\right]} \qquad \textbf{(m)}$$

$$= \frac{12\sin(\theta) - 15\cos(\theta)^2 + 15}{5\cos(\theta)[4 + 3\sin(\theta)]}$$

Now use computer software to plot Eq. (m) to see a range of solutions for the required coefficient of static friction for different angles of pole inclination, say for $0 \le \theta \le 60°$ (see Fig. 3-65). This plot reflects the nonlinear relationship between the required coefficient of static friction μ_s and pole inclination angle θ. Similar plots can be found for cable force T as a function of inclination angle θ.

FIGURE 3-65

Plot of coefficient of static friction versus angle θ

Pole inclination angle, degrees

Example 3-13

Counterclockwise moment M is applied to a drive gear of radius r_1, which applies only a tangential force at A to rotate a drum of weight $W = 75$ lb. The tangential force is applied to a larger gear of radius r_2 attached to the end of the drum. The drum is also supported by a guide at B that rolls without friction. The drum rests on a plywood sheet at D and rotates to move the sheet through a machine from right to left along a series of small frictionless rollers below the sheet. However, the sheet has jammed in the feeder, but the drive gear continues to apply moment M (Fig. 3-66).

Find the friction force f_D at D if applied moment M is 40 lb-in. Repeat if M is 60 lb-in. What is the maximum possible friction force $f_{D,\text{max}}$ that can act at D to free the jam and force the sheet through the feeder? Assume that static and kinetic friction coefficients between drum and sheet are $\mu_s = 0.4$ and $\mu_k = 0.3$, respectively. Also assume that points A and D on the drum are at radial distance r_2 from the center of drum rotation at point C.

FIGURE 3-66

Moment M applied to gear at A to apply friction force at D

Solution:

Use the four-step problem-solving approach to find the friction force f_D at D for various applied moments M.

1. *Conceptualize*: First construct the *free-body diagrams* of the gear and drum (Fig. 3-67). The free-body diagram of the drive gear shows that moment M is statically equivalent to tangential force F_A applied at distance r_1. To construct the drum free-body diagram, extract the drum from the overall assemblage and apply forces *on the drum* due to its interaction with other components such as the gear drive at A, the guide at B, and the jammed plywood sheet at D.

FIGURE 3-67

Free-body diagrams: (a) drive gear; (b) drum with drive force at A, guide force at B, and friction force at D

Normal and friction forces at D, N_D, and f_D, represent the action of the restrained plywood sheet on the drum and are aligned with the x and y axes. Tangential force F_A at A is the drive force on the drum resulting from applied moment M from the drive gear. The force at B represents the action of the guide on the drum, so F_B is normal to the drum and has a line of action through drum center of gravity at C. Assume there are no friction forces at B. The drum weight W acts at the center of gravity of the drum.

2. *Categorize*: For static equilibrium of the drum, it is most convenient to sum forces in the x-y coordinate system since three of the five forces acting on the drum in Fig. 3-67 are aligned with either x or y.

3. *Analyze*: Applied moment M is statically equivalent to tangential gear force F_A applied to the drum at point A:

$$F_A = \frac{M}{r_1} \qquad \text{(a)}$$

Summing moments about the center of gravity of the drum, point C, reveals

$$\sum M_C = 0 \quad -F_A r_2 + f_D r_2 = 0 \qquad \text{(b)}$$

Substitute Eq. (a) into Eq. (b) to obtain

$$f_D = F_A = \frac{M}{r_1} \qquad \text{(c)}$$

However, for this result to be valid, friction force f_D must not exceed the maximum possible static friction force $f_{D,\max} = \mu_s N_D$. Find normal force N_D in order to make this assessment. Force equilibrium in the x direction leads to

$$\sum F_x = 0 \quad f_D + F_A \cos\alpha - F_B \cos\theta = 0 \qquad \text{(d)}$$

Substituting $F_A = f_D$ into Eq. (d) and solving for F_B gives

$$F_B = f_D \left[\frac{1 + \cos\alpha}{\cos\theta} \right] \qquad \text{(e)}$$

Force equilibrium in the y direction leads to

$$\sum F_y = 0 \quad N_D + F_A \sin\alpha - F_B \sin\theta - W = 0 \qquad \text{(f)}$$

$$\text{or} \qquad N_D + f_D \sin\alpha - f_D (1 + \cos\alpha)\tan\theta - W = 0 \qquad \text{(g)}$$

If the drum is on the verge of slipping at D, friction force f_D equals $f_{D,max} = \mu_s N_D$. Solve for normal force N_D in Eq. (g) to obtain:

$$N_D = \frac{W}{1 + \mu_s[\sin\alpha - (1 + \cos\alpha)\tan\theta]} \tag{h}$$

However, if slipping has already occurred, static moment equilibrium in Eq. (b) is no longer valid, so forces F_A and f_D are no longer equal. Friction force f_D is then reduced to the level of kinetic friction force $f_{Dk} = \mu_k N_D$, while tangential force F_A continues to increase with increasing values of applied moment M [see Eq. (a)] and Eq. (c) no longer holds. Upon slipping, the drum is now rotating, so $\sum M_C \neq 0$ (unless the angular velocity is constant). Force equilibrium is still valid in the x and y directions, so $\sum F_x = 0$ and $\sum F_y = 0$. Force equilibrium can be used to find an appropriate expression for normal force N_D different from that in Eq. (h). Using Eq. (d), force F_B is

$$F_B = \frac{\dfrac{M}{r_1}\cos\alpha + \mu_k N_D}{\cos\theta} \tag{i}$$

Substituting Eq. (i) into Eq. (f) results in the expression for normal force N_D, which is valid when the drum is slipping at D:

$$N_{Dk} = \frac{W + \dfrac{M}{r_1}[\cos\alpha\tan\theta - \sin\alpha]}{1 - \mu_k\tan\theta} \tag{j}$$

Now there are two expressions for normal force N_D: N_{Ds} if static friction controls and N_{Dk} if kinetic friction controls. It follows that

$$f_{D,max} = \mu_s N_{Ds} = \frac{\mu_s W}{1 + \mu_s[\sin\alpha - (1 + \cos\alpha)\tan\theta]} \tag{k}$$

$$f_D = \frac{M}{r_1} \quad \text{if } f_D < f_{D,max} \tag{l}$$

$$f_{Dk} = \mu_k N_{Dk} = \frac{\mu_k W + \dfrac{\mu_k M}{r_1}[\cos\alpha\tan\theta - \sin\alpha]}{1 - \mu_k\tan\theta} \quad \text{if } f_D > f_{D,max} \tag{m}$$

Now consider the two possible numerical values of applied moment M. Substitute numerical values for the variables in Eqs. (k) through (m) and find the friction force at D for each case.

Case 1—Applied moment $M = 40$ lb-in.: From Eq. (c), the friction force at D is

$$f_D = \frac{M}{r_1} = \frac{40 \text{ lb-in.}}{1.5 \text{ in.}} = 26.7 \text{ lb} \quad \longleftarrow \tag{n}$$

From Eq. (k), the maximum static friction force that can be developed at D is

$$f_{D,max} = \frac{0.40(75 \text{ lb})}{1 + 0.40[\sin 30° - (1 + \cos 30°)\tan 22.5°]} = 33.7 \text{ lb} \qquad \text{(o)}$$

Since $f_D = 26.7$ lb $< f_{D,max} = 33.7$ lb, if applied moment $M = 40$ lb-in., slip has not yet occurred at D and this friction force is valid.

Case 2—Applied moment $M = 60$ lb - in.: From Eq. (c), the friction force at D is

$$f_D = \frac{M}{r_1} = \frac{60 \text{ lb-in.}}{1.5 \text{ in.}} = 40.0 \text{ lb} \qquad \text{(p)}$$

However, this force exceeds the maximum possible static friction force of 33.7 lb [Eq. (o)], so slip has occurred under applied moment $M = 60$ lb-in. As a result, the friction force is now limited to f_{Dk} in Eq. (m), which can be computed as

$$f_{Dk} = \frac{0.3(75 \text{ lb}) + \dfrac{0.3(60 \text{ lb-in.})}{1.5 \text{ in.}}[\cos 30°\tan 22.5° - \sin 30°]}{1 - 0.3\tan 22.5°} = 23.8 \text{ lb} \qquad \text{(q)}$$

4. *Finalize*: From Eq. (o), the maximum moment that can be applied to the drive gear at A before slip of the drum occurs at D is

$$M_{max} = f_{D,max}r_1 = 33.67 \text{ lb}(1.5 \text{ in.}) = 50.5 \text{ lb-in.} \qquad \text{(r)}$$

Values of applied moment M below this value do not result in slip at D while values of M greater than M_{max} will cause the drum to slip on the plywood sheet at D as illustrated in Case 2. Use computer software to plot Eqs. (k) through (m) to show how friction force f_D varies with applied moment M over a range of possible values, $0 \leq M \leq 90$ lb-in. The result is shown in Fig. 3-68. Note that friction force f_{Dk} is not constant in the region beyond M_{max} where kinetic friction controls because N_{Dk} [Eq. (j)], and hence f_{Dk} [Eq. (m)] both vary with applied moment M.

FIGURE 3-68

Variation of friction force at D with applied moment M

Friction force at D vs. applied moment M

Friction force at base of drum (D), lb

Moment M (lb-in.)

Application of Dry Friction to Structures and Machines

Now look at a number of applications of dry friction in various structures and machines such as wedges, screws, and flat and V-belts.

Wedge   A wedge is a simple machine made up of two inclined planes separated by an angle β (Fig. 3-69a). Wedges are usually used for such tasks as holding doors against movement (Fig. 3-69a), aligning two objects when used as shims in construction (Fig. 3-69b), or forcing two objects apart as in splitting of wood (Fig. 3-69c). Use of wedges results in large normal and friction forces but very little relative motion between the objects. Static equilibrium analysis requires that separate free-body diagrams be drawn for each wedge and object to obtain a solution for the associated forces when both are on the verge of movement.

Consider the use of a wedge with angle β and of negligible mass used to move a concrete block of mass m_1, as shown in Fig. 3-70. The wedge is inserted in the space between an inclined wall and the left face of the block. Force P is applied downward on the wedge to move the block in the $+x$ direction along a horizontal surface. The coefficients of static friction are μ_1 between block and horizontal surface and μ_w for both wedge faces. Find *the minimum value of force P* to initiate movement of the block.

A critical first step is to *disassemble the wedge-block system* and then draw *free-body diagrams* of each component, showing all applied and interacting forces in their proper directions. The resulting free-body diagrams of wedge and block are shown in Fig. 3-71a and b, respectively; study these carefully before proceeding to the equilibrium solution. Both wedge and block are on the verge of sliding, so all friction forces are at the level of $f_{max} = \mu N$ and are

---

(a) Wedge used to restrain door; (b) wedge used to align flooring; (c) wedge used to split wood

(a)   (b)   (c)

FIGURE 3-70

Wedge used to move block with mass m_1

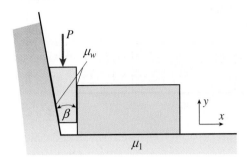

FIGURE 3-71

Free-body diagrams of wedge and block

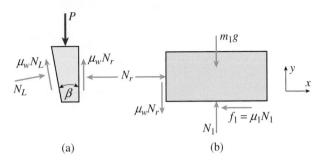

directed to oppose the motion of block or wedge. Normal force N_1 acts on the base of the block (Fig. 3-71b), and normal forces N_r and N_L act on the right and left faces of the wedge (Fig. 3-71a).

Static equilibrium of the block requires that all forces sum to zero in both x and y directions. Two equations, $\sum F_x = 0$ and $\sum F_y = 0$, allow you to solve for the two unknowns, N_1 and N_r. The equilibrium equations for the block are

$$\sum F_x = 0 \quad N_r - \mu_1 N_1 = 0$$
$$\sum F_y = 0 \quad N_1 - \mu_w N_r = m_1 g \tag{3-25a,b}$$

Solution of Eq. 3-25 gives

$$N_r = \frac{m_1 g \mu_1}{1 - \mu_1 \mu_w} \quad \text{and} \quad N_1 = \frac{N_r}{\mu_1} = \frac{m_1 g}{1 - \mu_1 \mu_w} \quad \text{with} \quad f_1 = \mu_1 N_1 = N_r \tag{3-26}$$

With normal force N_r now known, summing forces in x and y directions for the wedge provides a solution for the remaining two unknowns, N_L and P. The equilibrium equations for the wedge are

$$\sum F_x = 0 \quad N_L \cos \beta - \mu_w N_L \sin \beta = N_r$$
$$\sum F_y = 0 \quad N_L \sin \beta + \mu_w N_L \cos \beta - P = -\mu_w N_r \tag{3-27}$$

The general solution expressions for N_L and P from Eq. (3-27) are lengthy so numerical values for all variables are now inserted to look at a specific solution. If $m_1 = 100$ kg, $g = 9.807$ m/s^2, $\beta = 5°$, $\mu_1 = 0.7$ and $\mu_w = 0.5$, Eqs. (3-26) lead to $N_r = f_1 = 1056$ N and $N_1 = 1509$ N. Eqs. (3-27) are easily solved to find normal force N_L and *required wedge force P*: $N_L = 1109$ N and $P = 1177$ N.

Wedge Between Two Blocks

What happens if you replace the wall at left in Fig. 3-70 with a second block of mass $m_2 = 1.5 \, m_1$ (Fig. 3-72)? Will application of wedge force P result in movement of both blocks or only of block 1? The free-body diagrams in Fig. 3-71 are now expanded to include block 2 (Fig. 3-73). The previous solution is still valid provided that block 2 is *not* on the verge of movement and therefore acts just like the wall in Fig. 3-70. This means that friction force f_2 on block 2 must be less than $f_{2,max} = \mu_2 N_2$. From

FIGURE 3-72

Wedge used to separate two blocks

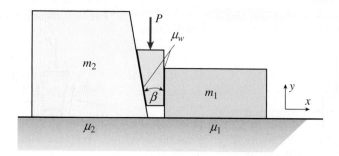

FIGURE 3-73

Free-body diagrams of wedge
and two blocks

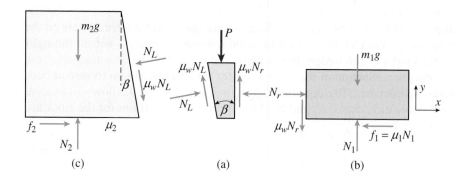

overall equilibrium of the wedge and two blocks (Fig. 3-74), $f_2 = f_1 = 1056$ N since $\sum F_x = 0$. Normal force N_2 is obtained using $\sum F_y = 0$ for block 2 (Fig. 3-73c):

$$N_2 = m_2 g + N_L \sin \beta + \mu_w N_L \cos \beta$$

$$= 150 \text{ kg} \left(9.807 \ \frac{\text{m}}{\text{s}^2} \right) + 1109 \text{ N}(0.087) + 0.5(1109 \text{ N})(0.996) = 2120 \text{ N} \quad \textbf{(3-28)}$$

The conclusion is that block 2 is not about to slide (unlike block 1) because

$$f_2 = 1056 \text{ N} < \mu_2 N_2 \text{ provided that } \mu_2 > \frac{f_2}{N_2} = 0.498 \qquad \textbf{(3-29)}$$

FIGURE 3-74

Overall free-body diagram of wedge
and two blocks

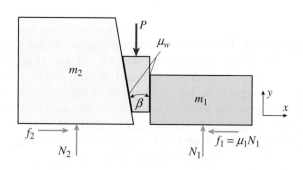

Any value of static friction coefficient μ_2 between block 2 and the horizontal surface that exceeds 0.498 (such as $\mu_2 = \mu_1 = 0.7$) will insure that block 2 will not move under the action of applied wedge force P.

Wedge Between Two Identical Blocks

You can modify the two-block example to illustrate the notion of *symmetry-anti-symmetry analysis*, which is an important concept in mechanics. The basic idea is that symmetric structures under symmetric (or anti-symmetric) loading can be analyzed using only one half of the overall model. Consider the modified two-block structure plus wedge shown in Fig. 3-75 where a symmetric wedge is now used (with angle 2β) and the two blocks have the same mass ($m_1 = m_2$). Wedge force P is applied on the plane of symmetry. An analysis of the full structure in Fig. 3-75 shows that each block is on the verge of sliding simultaneously and that, for $m_1 = m_2 = 100$ kg, $g = 9.807$ m/s^2, $2\beta = 10°$, $\mu_1 = \mu_2 = 0.7$, and $\mu_w = 0.5$, the solution for all forces leads to $N_1 = N_2 = 1721$ N, $f_2 = -f_1 = 1204$ N, $N_L = -N_r = 1264$ N, and $P = 1480$ N. The key point here is that you can obtain the same results from *analysis of only one half of the overall structure* (see right half of structure marked with a dash outline in Fig. 3-75). To create this symmetric model, split the wedge into two parts, divide load P into two equal parts, and add a reaction force N_{sym} on the plane of symmetry to prevent asymmetric motion of the symmetric model. Carry out and compare the analyses of the full and symmetric models to confirm these results.

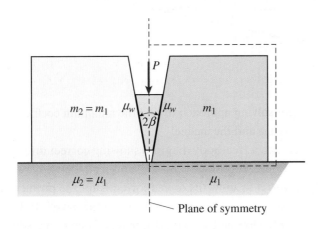

FIGURE 3-75

Symmetric two block structure with symmetric load

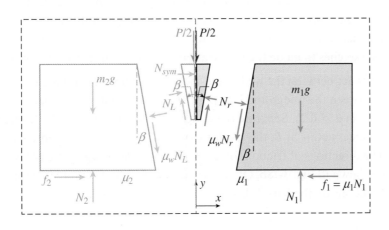

FIGURE 3-76

Symmetric model for two-block structure with symmetric load

Example 3-14 illustrates the use of a wedge to prevent rotational motion involving possible sliding of a barrel on an inclined surface.

Example 3-14

A dock worker is rolling a 60 kg barrel up an incline with slope $\theta = 15°$ (Fig. 3-77). Part way up the incline, the worker inserts a 22° wedge to hold the barrel in place temporarily. The coefficient of static friction μ between barrel and incline, and also between wedge and barrel, is 0.2. What is the minimum coefficient of static friction μ_w between the wedge and the incline if the barrel is to remain stationary in this temporary position? Neglect the weight of the wedge.

FIGURE 3-77

Barrel on incline supported by wedge

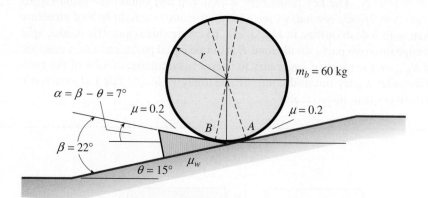

$$\alpha = \beta - \theta = 7°$$
$$\mu = 0.2$$
$$\beta = 22°$$
$$\theta = 15°$$
$$\mu_w$$
$$m_b = 60 \text{ kg}$$
$$\mu = 0.2$$

Solution:

Use the four-step problem-solving approach to find the minimum coefficient of static friction μ_w between the wedge and the incline.

1. *Conceptualize*: Start with a free-body diagram showing correct directions of all friction and normal forces. Envision the forces acting on the wedge *as it is being inserted beneath the barrel* to show the forces correctly for the problem at hand: *these forces act to squeeze the wedge out* from beneath the barrel. In Fig. 3-78a, friction forces f act to the left on the wedge to oppose force F to insert the wedge. However, friction forces f act in the opposite direction, to the right (Fig. 3-78b), as the wedge resists the tendency of the barrel to force the wedge down the incline. Note that Figs. 3-78a and b are not free-body diagrams but rather initial sketches to guide the preparation of a correct free-body diagram.

   Friction forces act between the wedge and incline (with unknown static coefficient μ_w); between the wedge and barrel at B ($\mu = 0.2$); and between barrel and incline at A ($\mu = 0.2$). Separate the wedge and barrel to expose the friction and normal forces acting at B; the resulting free-body diagrams are shown in Fig. 3-79. The directions of friction forces f_B and f_w *on the wedge* (Fig. 3-79b) match those shown in Fig. 3-78b. Of course, normal and friction forces act in opposite directions on the barrel (Fig. 3-79a).

FIGURE 3-78

(a) Friction forces oppose wedge insertion force *F*; (b) friction forces oppose wedge being pushed out

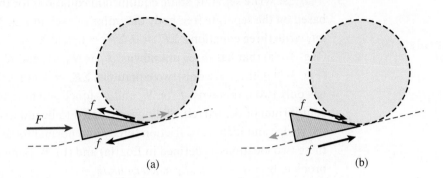

(a)   (b)

2. *Categorize*: For static equilibrium of wedge and barrel, sum forces in the *x-y* coordinate system since angles α and θ are easily identified in Fig. 3-79 for both components.

For objects like these with multiple contact surfaces (barrel-incline, barrel-wedge, and wedge-incline), it is not clear at the start where sliding might occur. All three contact points could be on the verge of sliding, but this is unlikely. *Assume* here that sliding is imminent at *A* and also at the wedge-incline contact surface, but not between wedge and barrel at *B*. If so, the following expressions apply:

$$f_A = \mu N_A \tag{a}$$

$$f_w = \mu_w N_w \tag{b}$$

$$f_B < \mu N_B \tag{c}$$

After all forces have been computed, confirm that Eq. (c) is correct to certify the solution.

FIGURE 3-79

Free-body diagrams of (a) barrel and (b) wedge

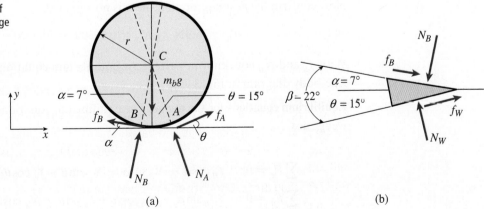

(a)   (b)

3. *Analyze*: Write separate static equilibrium equations for the barrel and wedge based on the separate free-body diagrams of each in Fig. 3-79. At the outset, write three equations: $\Sigma F_x = 0$, $\Sigma F_y = 0$, and $\Sigma M_C = 0$ for the **barrel** (Fig. 3-79a) that has three unknowns: f_A or N_A, f_B, and N_B. For the **wedge** (Fig. 3-79b), there are only two equations: $\Sigma F_x = 0$ and $\Sigma F_y = 0$, but the wedge has only two unknowns: f_w or N_w and μ_w (once you have found f_B and N_B from equilibrium of the barrel). Sliding is assumed to be imminent at A and between the wedge and incline, so friction force f and normal force N are related at these two locations as defined in Eqs. (a) and (b). Without this assumption, the problem becomes *statically indeterminate*.

Equilibrium equations for the barrel: Using the free-body diagram in Fig. 3-79a, the three equilibrium equations for the barrel are

$$\sum F_x = 0 \qquad -f_B \cos\alpha + f_A \cos\theta + N_B \sin\alpha - N_A \sin\theta = 0 \tag{d}$$

$$\sum F_y = 0 \qquad f_B \sin\alpha + f_A \sin\theta + N_B \cos\alpha + N_A \cos\theta = m_b g \tag{e}$$

$$\sum M_C = 0 \qquad f_A r - f_B r = 0 \text{ or } f_A = f \tag{f}$$

Based on Eqs. (c) and (f), it follows that:

$$f_A = f_B = \mu N_A \tag{g}$$

Angles α and θ are given in Fig. 3-79 as 7° and 15°, respectively. Substituting Eq. (g) into Eqs. (d) and (e), inserting $m_b = 60$ kg and $g = 9.807$ m/s², and then solving for N_A and N_B gives

$$N_A = 184.3 \text{ N} \quad N_B = 399 \text{ N} \tag{h}$$

From Eq. (g), friction forces at A and B are

$$f_B = f_A = 0.2(184.3 \text{ N}) = 36.9 \text{ N} \tag{i}$$

Now confirm the assumption that there is no slip at B:

$$f_B = 36.9 \text{N} < \mu N_B = 0.2(399) = 79.8 \text{ N} \tag{j}$$

With f_B and N_B now known, write static equilibrium equations for the wedge free-body diagram in Fig. 3-79b.

Equilibrium equations for the wedge: The equilibrium equations for the wedge are

$$\sum F_x = 0 \qquad f_B \cos\alpha - N_B \sin\alpha - N_w \sin\theta + f_w \cos\theta = 0$$
$$\sum F_y = 0 \qquad -f_B \sin\alpha - N_B \cos\alpha + f_w \sin\theta + N_w \cos\theta = 0 \tag{k}$$

In Eq. (k), f_B and N_B are known and after substituting the expression $f_w = \mu_w N_w$, there are two equations and two unknowns (N_w and μ_w):

$$\sum F_x = 0 \quad -N_w \sin\theta + \mu_w N_w \cos\theta = 12.04 \text{ N}$$

$$\sum F_y = 0 \quad \mu_w N_w \sin\theta + N_w \cos\theta = 400.52 \text{ N}$$

(l)

Solve Eqs. (l) to find that

$$N_w = 384 \text{ N} \quad \text{and} \quad \mu_w = 0.3 \quad \text{so} \quad f_w = \mu_w N_w = 0.3(384 \text{ N}) = 115.2 \text{ N} \hspace{10pt} \textbf{(m)}$$

4. *Finalize:* The minimum coefficient of static friction μ_w between the wedge and the incline is 0.3 for the barrel to remain at rest on the incline. This should not be difficult to achieve. From Table 3-2, a wood wedge on a concrete incline has a static coefficient of friction considerably larger than $\mu_w = 0.3$ if the two surfaces are in good condition.

   Now that all forces are known, *check the overall equilibrium* of the barrel-wedge system using the overall free-body diagram shown in Fig. 3-80. The equilibrium equations are

$$\sum F_x = 0 \quad -N_w \sin\theta + f_w \cos\theta + f_A \cos\theta - N_A \sin\theta = 0$$

or
$$-384 \text{ N}(0.259) + 115.2 \text{ N}(0.966) + 36.9 \text{ N}(0.966) -$$
$$184.3 \text{ N}(0.259) \approx 0$$

(n)

$$\sum F_y = 0 \quad N_w \cos\theta + f_w \sin\theta + N_A \cos\theta + f_A \sin\theta = m_b g = 588.4 \text{ N}$$

or
$$384 \text{ N}(0.966) + 115.2 \text{ N}(0.259) + 184.3 \text{ N}(0.966) +$$
$$36.9 \text{ N}(0.259) = 588.4 \text{ N}$$

Equations (n) show that the overall equilibrium of the wedge-barrel system in Fig. 3-80 is satisfied.

FIGURE 3-80

Overall free-body diagram of barrel-wedge system

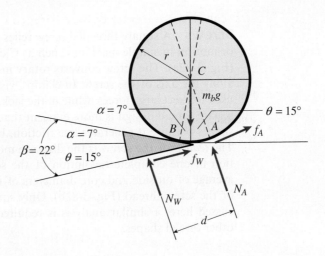

Next, use the overall free-body diagram in Fig. 3-80 to find the distance d from point A to the location of the reaction R_w beneath the wedge where R_w is the *resultant* of forces f_w and N_w. The dimensions of the wedge are not provided so you cannot write a moment equation to locate force N_w as part of the equilibrium equations for the wedge [see Eqs. (k)]. Instead, sum moments about point A in Fig. 3-80, to get

$$\sum M_A = 0 \ (m_b g \sin\theta)r = N_w d \ \text{ so } \ d = \frac{m_b g \sin\theta}{N_w}r = \frac{588.4 \text{ N}(0.259)}{384 \text{ N}} = 0.397r \quad \text{(o)}$$

Finally, if static coefficients of friction are the same on all interacting surfaces (so $\mu_w = \mu = 0.2$), how does the wedge-barrel assemblage perform? Now $\mu_w = 0.2 < 0.3$, where $\mu_w = 0.3$ is the required value from Eq. (m), so a sliding failure is likely if μ_w is reduced. One way to prevent this is to add a force F to the wedge (Fig. 3-81) and use the modified equilibrium equations for the wedge to solve for required force F. Equilibrium of the barrel has not changed, so the solutions for f_B and N_B in Eqs. (h) and (i) are still valid. However, Eqs. (k) and (l) must be changed to specify that $\mu_w = 0.2$ and to add the x-direction force F. The modified equilibrium equations for the wedge become

FIGURE 3-81

Free-body diagram of wedge with added force F

$$\sum F_x = 0 \ \ -N_w \sin\theta + (0.2)N_w \cos\theta + F = 12.04 \text{ N}$$
$$\sum F_y = 0 \ \ (0.2)N_w \sin\theta + N_w \cos\theta = 400.52 \text{ N} \quad \text{(p)}$$

Solution of Eqs. (p) provides new values for N_w, for $f_w = \mu N_w$, and also for required force F. The updated solution is

$$N_w = 394 \text{ N}, \ f_w = \mu N_w = 78.7 \text{ N} \ \text{ and } \ F = 37.9 \text{ N} \quad \text{(q)}$$

Repeat of Eqs. (n), with F added to the first equation and using the new values of f_w and N_w in Eq. (q), shows that overall equilibrium of the wedge-barrel system is satisfied.

Screws   A square threaded screw relies on friction to transmit force, power, or motion in simple machines such as C clamps, bench vises, and screw jacks (Fig. 3-82a). The screw converts rotary motion to linear translational motion along the axis of the screw. In clamps, vises, or testing machines, the screw is used to exert large forces, while in the jack, the screw is used to lift a weight. A square thread design is more efficient than a triangular or other screw thread shapes because it generates less friction, but it is more expensive to machine. The *lead L* of the screw is the distance moved by the screw in the axial direction in one complete revolution of the screw. The *mean diameter d_m* is the average of outside and core diameters of the screw. Angle α is the helix angle of the screw thread (Fig. 3-82b). Only square single-threaded screws are discussed here; a similar analysis is required for multiple-threaded screws and other thread shapes.

FIGURE 3-82

(a) Screw jack under force F and moment M; (b) square-threaded screw detail

Examine the force R in Fig. 3-83, which is applied by the jack base (Fig. 3-82a) on a small part of the square screw thread, to see that it acts upward to oppose downward applied force F but is inclined to the vertical because of thread angle α and also because of friction on the thread. If the screw is on the verge of sliding due to applied moment M, force R is inclined to a normal to the thread by the friction angle ϕ where $\phi = \tan(\mu)$ and μ is the coefficient of static friction. Sum all forces R acting on all threads that are in contact with the screw thread to obtain

$$\sum F_{\text{vertical}} = 0 \quad F = (\sum R)[\cos(\phi + \alpha)] \quad \textbf{(3-30)}$$

Next, sum moments about the screw axis to find that moment M applied to raise the jack and oppose downward force F is

$$\sum M_{\text{vertical}} = 0 \quad M = (\sum R)\left[\frac{d_m}{2}\sin(\phi + \alpha)\right] \quad \textbf{(3-31)}$$

Solve for $\sum R$ in Eq. (3-30) then substitute the resulting expression into Eq. (3-31) to relate applied moment M and force F in

$$M = \frac{F}{[\cos(\phi + \alpha)]}\left[\frac{d_m}{2}\sin(\phi + \alpha)\right] = \frac{Fd_m}{2}\tan(\phi + \alpha) \quad \textbf{(3-32)}$$

A *simpler equivalent model* can be used to represent the *entire screw* in contact with the base if you unroll one wrap of the thread and apply force F and resisting force R to create the force diagram in Fig. 3-84a. This force diagram represents the action of *raising* the jack screw against force F so friction

FIGURE 3-83

Representative force R applied on screw thread

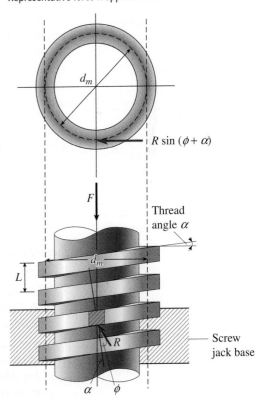

force μN acts to oppose the upward motion. Think of this as the action of a single force P_r to *raise* the load F, pushing the thread up the incline that has thread helix angle $\alpha = \tan^{-1}(L/\pi d_m)$. Thus, applied moment $M = P_r d_m/2$. The static equilibrium equations for raising the load F from Fig. 3-84a are

$$\sum F_x = 0 \quad P_r - \mu N \cos\alpha - N \sin\alpha = 0$$

$$\sum F_y = 0 \quad N \cos\alpha - \mu N \sin\alpha - F = 0$$

(3-33)

Solving Eqs. (3-33) gives the following expressions for normal force N and equivalent force P_r used to *raise* load F:

$$N = \frac{F}{\cos\alpha - \mu\sin\alpha}$$

$$P_r = F\left[\frac{\sin\alpha + \mu\cos\alpha}{\cos\alpha - \mu\sin\alpha}\right] = F\left[\frac{\mu + \tan\alpha}{1 - \mu\tan\alpha}\right] = F\left[\frac{\pi\mu d_m + L}{\pi d_m - L\mu}\right]$$

(3-34a, b)

FIGURE 3-84

Force diagrams for (a) raising and (b) lowering the applied load F

(a)

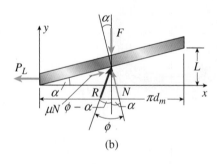

(b)

Substituting $\mu = \tan(\phi)$ in Eq. (3-34b) and using the expression for P_r in the moment expression for *raising the load*, $M = M_r = P_r d_m/2$ results in the same expression as in Eq. (3-32):

$$M_r = \frac{P_r d_m}{2} = \frac{F d_m}{2}\left[\frac{\tan\phi + \tan\alpha}{1 - \tan\phi\tan\alpha}\right] = \frac{F d_m}{2}\tan(\phi + \alpha) \quad \text{(3-35)}$$

Repeat the process to develop a *simpler equivalent model* to represent the *entire screw* for the action of *lowering* the jack screw against force F by starting with equilibrium equations for Fig. 3-84b, noting that friction force μN now changes direction and acts to oppose the lowering of the screw. First, solve for N and P_L and then simplify the resulting equations to obtain expressions for equivalent force P_L and the required moment M_L for *lowering the load*:

$$P_L = F\left[\frac{\mu\cos\alpha - \sin\alpha}{\cos\alpha + \mu\sin\alpha}\right] = F\left[\frac{\mu - \tan\alpha}{1 + \mu\tan\alpha}\right] = F\left[\frac{\pi\mu d_m - L}{\pi d_m + L\mu}\right]$$

(3-36)

$$M_L = \frac{P_L d_m}{2} = \frac{F d_m}{2}\left[\frac{\tan\phi - \tan\alpha}{1 + \tan\phi\tan\alpha}\right] = \frac{F d_m}{2}\tan(\phi - \alpha)$$

Conclusions for square-threaded screws are determined from the previous equations.

- The moment (or torque) M_r required to overcome friction in the square threads and *raise* the load is given by Eq. (3-35); the moment vector acts upward as shown in Fig. 3-82b.

- The moment required to lower the load M_L is given by Eq. (3-36b); the moment vector acts downward provided that $\tan\phi = \mu > \tan\alpha = L/\pi d_m$ in Eq. (3-36a); moment M_L for lowering the screw is now positive. The screw will not unwind by itself and is said to be ***self-locking***.

- If lead L is large or friction coefficient μ is small, the screw will unwind on its own and lower the load. If helix angle α is larger than friction angle ϕ, moment M_L changes sign (acts upward) and is needed to prevent unwinding of the screw.

Example 3-15 investigates screw friction in two different designs of a jack stand used to manually raise cars for tire changes and other automobile maintenance.

Example 3-15

A motorist is shopping for a replacement car jack for her car. She finds both single arm (Fig. 3-85a) and scissor-type (Fig. 3-85b) jacks at the auto parts store. The screw of each jack has a single square thread with mean diameter of 3/8 in., lead of 1/8 in., and coefficient of friction of 0.2 for the threads. For each jack, the screw is anchored with a pivot bearing at C and acts through a screw collar at D. Find the required force applied to the handle at G to both raise (Q_r) and lower (Q_L) a 1250-lb load W at E. Neglect the weight of the jacks in your calculations.

FIGURE 3-85

Manual car jacks: (a) single arm and (b) scissor-type

(a)   (b)

Solution:

Use the four-step problem-solving approach to find the required force applied to the handle of each jack to both raise (Q_r) and lower (Q_L) a 1250-lb load W.

FIGURE 3-86

Free-body diagrams: (a) single
jack arm *BCE* and (b) scissor
jack joints *E* and *D*

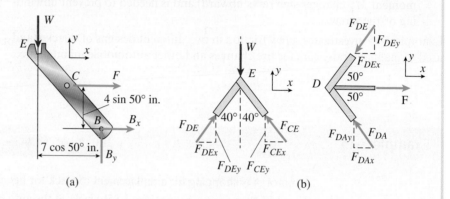

(a)

(b)

1. *Conceptualize*: First sketch free-body diagrams of each jack, or portions
   thereof, that show the force *F* in the screw caused by load *W*. Then find
   moment *M*, and finally find the required handle force *Q*. Figure 3-86a shows
   the forces acting on arm *BCE* for the single jack, and Fig. 3-86b contains free-
   body diagrams of joints *E* and *D* for the scissor jack. Apply static equilibrium
   to each of these free-body diagrams to find the force *F* in each screw.

2. *Categorize*: For static equilibrium of **single jack** arm *BCE* (Fig. 3-86a), sum
   moments about *B* in the *x-y* coordinate system to find force *F* in one step. For the
   scissor jack, two steps are required: (a) First use the free-body diagram of joint *E* to
   find scissor arm forces F_{DE} and F_{CE}; (b) Then use force F_{DE} and the free-body diagram of joint *D* (Fig. 3-86b) to find forces F_{DA} and screw force *F*.

3. *Analyze*: The mean diameter of each screw is $d_m = 3/8$ in. and the
   lead is $L = 1/8$ in., from which you can compute the helix angle
   $\alpha = \tan^{-1}(L/\pi d_m) = 6.06°$. The friction angle $\phi = \tan^{-1}(0.2) = 11.31°$. Since
   $\phi > \alpha$, each jack is *self-locking* and will not unwind under applied load *W*.

   Single arm jack: From the free-body diagram in Fig. 3-86a, summation of
   moments about *B* gives

   $$\sum M_B = 0 \quad -F(4\sin 50° \text{ in.}) + W(7\cos 50° \text{ in.}) = 0$$

   $$\text{so} \quad F = \frac{1250 \text{ lb}(7\cos 50° \text{ in.})}{4\sin 50° \text{ in.}} = 1836 \text{ lb} \tag{a}$$

   Next find moments required to *raise* and *lower* the jack from Eqs. 3-35 and 3-36
   as

   $$M_r = \frac{Fd_m}{2}\tan(\phi + \alpha) = 1836 \text{ lb}\left[\frac{3/8 \text{ in.}}{2}\right]\tan(17.37°) = 107.6 \text{ in-lb} \tag{b}$$

   $$M_L = \frac{Fd_m}{2}\tan(\phi - \alpha) = 1836 \text{ lb}\left[\frac{3/8 \text{ in.}}{2}\right]\tan(5.25°) = 31.6 \text{ in-lb} \tag{c}$$

Finally, the required handle forces to raise and lower the *single arm* jack are

$$Q_r = \frac{M_r}{6 \text{ in.}} = 17.94 \text{ lb} \quad \text{and} \quad Q_L = \frac{M_L}{6 \text{ in.}} = 5.27 \text{ lb} \qquad \text{(d)}$$

Scissor jack: From the free-body diagram of *joint E* in Fig. 3-86b, summation of forces gives

$$\Sigma F_x = 0 \quad F_{DE} \cos 40° - F_{CE} \cos 40° = 0 \quad \text{so} \quad F_{DE} = F_{CE}$$

$$\Sigma F_y = 0 \quad F_{DE} \sin 40° + F_{CE} \sin 40° = W \quad \text{so} \quad F_{DE} = \frac{W}{2 \sin 40°} = \frac{1250 \text{ lb}}{1.286} = 972.3 \text{ lb} \qquad \text{(e)}$$

Next, using the free-body diagram of *joint D* in Fig. 3-86b, sum forces as

$$\sum F_x = 0 \quad -F_{DE} \cos 50° - F_{DA} \cos 50° + F = 0 \qquad \text{(f)}$$

$$\sum F_y = 0 \quad -F_{DE} \sin 50° + F_{DA} \sin 50° = 0 \quad \text{so} \quad F_{DA} = F_{DE} = 972.3 \text{ lb} \qquad \text{(g)}$$

Now find force *F* in the screw from Eq. (f):

$$F = 2F_{DE} \cos 50° = 2(972.3 \text{ lb})(0.643) = 1250 \text{ lb} \qquad \text{(h)}$$

With the scissor jack, force *F* has been reduced from 1836 lb [for the single arm jack, Eq. (a)] to 1250 lb. From Eqs. (b) and (c), the moments required to raise and lower the scissor jack under load *W* are

$$M_r = \frac{1250}{1836} 107.6 \text{ in-lb} = 73.3 \text{ in-lb}$$

$$M_L = \frac{1250}{1836} 31.6 \text{ in-lb} = 21.5 \text{ in-lb} \qquad \text{(i)}$$

Finally, the required handle forces to raise and lower the *scissor jack* are

$$Q_r = \frac{M_r}{6 \text{ in.}} = 12.22 \text{ lb} \quad \text{and} \quad Q_L = \frac{M_L}{6 \text{ in.}} = 3.58 \text{ lb} \qquad \text{(j)}$$

4. *Finalize*: All forces and moments associated with the scissor jack are 68% of the corresponding values for the single arm jack [from Eqs. (a) and (h), 1250/1836 = 0.68], so the scissor jack will be the preferred tool if the required applied force is a key factor in the motorist's purchasing decision.

   Also, since $\phi > \alpha$, each jack is *self-locking* and will not unwind under applied load *W*. Both jacks have eight threads per inch. Set $\phi = \alpha$ to find that a minimum of 4.24 threads per inch ensures that the screw is self-locking for the case of static coefficient of friction $\mu = 0.2$ and $d_m = 3/8$ in.

Flexible Belts   Flexible belts are used in a wide variety of engineering applications that rely on friction between the belt and rotating pulley or drum for proper function. For example, flat belts (or bands) are used in band brakes, strap-type oil filter wrenches, conveyors, and clothes dryers. V-belts are commonly used to mechanically link two or more rotating shafts to deliver engine power to drive components such as the alternator, power-steering pump, and air-conditioning compressor in automobiles. The principles of belt friction also apply to ropes wrapped around sailboat winches or around bollards on docks where ships are anchored and to equipment used by mountain climbers.

The focus here is on flat belts (Fig. 3-87a) and V belts (Fig. 3-87b) wrapped around pulleys, but the resulting formulas apply to all the applications listed earlier. Flat belts are usually made of leather or fabric, but many are now made of rubber or polymers. V belts are made using rubber or polymer and may have fibers embedded in the rubber or polymer for strength and reinforcement. Flat belts are more efficient than V belts because they generate less heat, but V belts have two surfaces in contact with the pulley, so larger belt friction forces can be developed.

The goal is to both find the relationship between the tension forces on the two sides of the belt *as it is about to slip* on a restrained pulley as well as on a pulley that is rotating at constant speed. The bending stiffness and shear stiffness of the belt are assumed to be small, so they can be ignored. If the belt is moving at constant speed, any associated inertia effects are assumed small and not considered.

Consider the flat belt wrapped around a circular pulley of radius r that has been restrained from translation by force R and from counterclockwise rotation by moment M (Fig. 3-88). Belt tension T_2 is larger than tension T_1 because sum of moments about the center of the pulley requires for equilibrium that $T_2 = T_1 + M/r$. The belt is in contact with the pulley over angle $\theta = 0$ to $\theta = \beta$.

The relationship between belt tension forces T_1 and T_2 is established by evaluating equilibrium of a differential segment of the belt of length $rd\theta$ (Fig. 3-89). Belt tension force varies from T to $T + dT$ as you move counterclockwise over the segment of length $rd\theta$. Differential force dN acts normal to the element, and belt friction force μdN opposes imminent counterclockwise slippage along the pulley. Summing forces in the t-direction to establish equilibrium leads to

$$\sum F_t = 0 \quad -(T + dT)\cos\frac{d\theta}{2} + \mu dN + T\cos\frac{d\theta}{2} = 0 \tag{3-37}$$

Lyroky/Alamy Stock Photo

FIGURE 3-87

Flexible belts on pulleys: (a) flat belt; (b) V belt

(a)

(b)

© Barry Goodno

FIGURE 3-88

Flat belt on restrained pulley

FIGURE 3-89

Differential segment of a flat belt on restrained pulley

If products of differentials are neglected and $\cos(d\theta/2) \to 1$ in the limit, Eq. (3-37) reduces to

$$dT = \mu\, dN \qquad \textbf{(3-38)}$$

Next sum forces to find conditions for equilibrium in the n-direction (Fig. 3-89) as

$$\sum F_n = 0 \quad -(T + dT)\sin\frac{d\theta}{2} + dN - T\sin\frac{d\theta}{2} = 0 \qquad \textbf{(3-39)}$$

Simplify Eq. (3-39), again ignoring products of differentials and using $\sin(d\theta/2) \to d\theta/2$ in the limit to obtain

$$T\, d\theta = dN \qquad \textbf{(3-40)}$$

Finally, from Eqs. (3-38) and (3-40),

$$\frac{dT}{T} = \mu\, d\theta \qquad \textbf{(3-41)}$$

Integrate Eq. (3-41) between tension force limits T_1 and T_2 and between contact angle limits 0 and β (Fig. 3-89) to obtain *for the flat belt*:

$$\int_{T_1}^{T_2} \frac{dT}{T} = \int_0^\beta \mu\, d\theta \quad \text{or} \quad \ln\frac{T_2}{T_1} = \mu\beta \ \text{ or } \ T_2 = T_1 e^{\mu\beta} \qquad \textbf{(3-42)}$$

where $\ln(T_2/T_1)$ is the natural logarithm of the tension force ratio. Recall that force $T_2 > T_1$ and that contact angle β must be in radians. If a strap or rope is wrapped n times around a drum or cylinder, contact angle $\beta = 2\pi n$. There is no dependence on pulley radius r (Fig. 3-88) in Eq. (3-42), so this relationship between T_2 and T_1 holds just as well for a noncircular contact surface with total contact angle β.

V-Belt Friction   The equilibrium equations above are modified slightly for the case of a V belt having a trapezoidal cross section and running in a grooved pulley (Fig. 3-90). The V angle (2α in Fig. 3-90a) typically varies between

FIGURE 3-90

V belt: (a) cross section showing belt in grooved pulley; (b) tension forces on restrained pulley

(a)

(b)

30 to 38 degrees. The V belt develops friction forces on its two sloped sides rather than along the bottom of the belt (Fig. 3-91). As a result, Eqs. (3-37) and (3-38) must be modified as follows for equilibrium in the *t*-direction:

$$\sum F_t = 0 \quad -(T + dT)\cos\frac{d\theta}{2} + 2\mu dN + T\cos\frac{d\theta}{2} = 0$$

(3-43a, b)

or

$$dT = 2\mu dN$$

Similarly, Eqs. (3-39) and (3-40) are modified as follows for equilibrium in the *n*-direction:

$$\sum F_n = 0 \quad -(T + dT)\sin\frac{d\theta}{2} + 2dN\sin\alpha - T\sin\frac{d\theta}{2} = 0$$

(3-44a, b)

or

$$Td\theta = 2dN\sin\alpha$$

FIGURE 3-91

V belt: (a) differential segment of a V belt on restrained pulley; (b) friction forces on sides of V belt

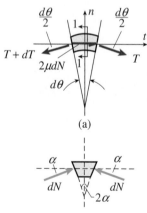

(a)

Section 1–1

(b)

Combine Eqs. (3-43b) and (3-44b) to get the following modification of Eq. (3-41):

$$\frac{dT}{T} = \frac{\mu}{\sin\alpha}d\theta$$

(3-45)

Integrate Eq. (3-45) between tension force limits T_1 and T_2 and between contact angle limits 0 and β (Fig. 3-90b) to obtain *for the V belt*:

$$\int_{T_1}^{T_2}\frac{dT}{T} = \int_0^\beta \frac{\mu}{\sin\alpha}d\theta \quad \text{or} \quad \ln\frac{T_2}{T_1} = \frac{\mu}{\sin\alpha}\beta \quad \text{or} \quad T_2 = T_1 e^{\frac{\mu\beta}{\sin\alpha}}$$

(3-46)

Comparing Eq. (3-42) for the flat belt with Eq. (3-46) for the V belt, you see that the expressions differ only in the exponent term where the sine of one half of the V angle appears in the denominator. However, use of a *V belt* with the half V angle of 19° results in a more than three-fold increase in friction coefficient μ compared to a *flat belt* with the same coefficient of friction (i.e., 1/sin (19°) = 3.07). Alternatively, this friction increase from use of a V belt is equivalent to wrapping a flat belt (or strap or rope) more than three times around a cylinder [exponent is now $\mu k\beta$, with $k = 3.07$, rather than $\mu\beta$ in Eq. (3-42)].

Example 3-16 illustrates the proper application of the belt friction formulas and examines the effect of both belt and surface friction in a simple example involving equilibrium of two blocks.

Example 3-16

FIGURE 3-92

Two blocks connected by a flexible strap

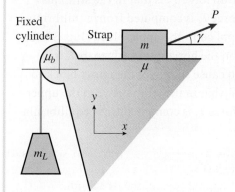

A block of mass $m = 15$ kg rests on a horizontal surface and is connected to a load of mass $m_L = 20$ kg by a flexible strap that passes over a fixed cylinder. Load P is applied to the right on the upper block at inclination angle $\gamma = 20°$ (Fig. 3-92). The coefficient of static friction between the upper block and the horizontal surface is $\mu = 0.3$ and between the strap and the cylinder is $\mu_b = 0.2$. Find the minimum and maximum values of load P for equilibrium of the system (neither block moves). *Repeat if the flat strap is replaced by one of trapezoidal cross section with angle $2\alpha = 38°$. Neglect the bending stiffness, shear stiffness, and weight of the straps in your calculations. Assume that the upper block is not susceptible to overturning.

Solution:

Use the four-step problem-solving approach to find the range of values of force P when the block does not slide on the horizontal surface.

1. *Conceptualize*: The two cases of interest that define the range of values of load P are shown in Fig. 3-93. Force P_{min} is that required to just prevent the upper block from sliding to the left due to the gravity load from the lower block (Fig. 3-93a), while force P_{max} is the force applied to the upper block just before it begins to slide to the right. These figures show which strap force, on either side of the cylinder, is the larger tension T_2 and which is lower tension T_1 in the belt friction formula [Eq. (3-42), $T_2 = T_1 e^{\mu\beta}$]. Using Fig. 3-93 for guidance, draw the free-body diagrams for the two cases of interest (Fig. 3-94).

FIGURE 3-93

Tension forces in strap when loads P_{min} and P_{max} are applied to upper block

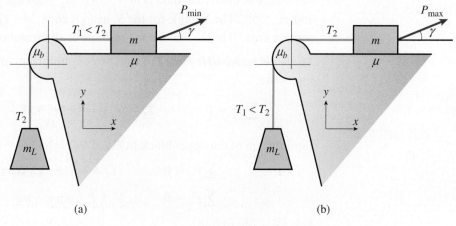

(a)   (b)

2. *Categorize*: When force P_{min} is applied to the upper block (Fig. 3-94a), force P when combined with friction on the horizontal surface along with belt friction acts to prevent the upper block from sliding to the left due to weight of the lower block. As a result, the larger strap force T_2 is that in the strap just above the lower block, and the smaller force T_1 is computed from equilibrium of the upper block. When force P_{max} is applied to the upper block (Fig. 3-94b), applied force P must overcome friction on the horizontal surface, belt friction, and the weight of the lower block to cause sliding of the upper block to the right. Now, the larger strap force T_2 is that in the strap between the upper block and the cylinder, and the smaller force T_1 is computed from equilibrium of the lower block.

FIGURE 3-94

Free-body diagrams: (a) load P_{min} applied to upper block and (b) P_{max} applied to upper block

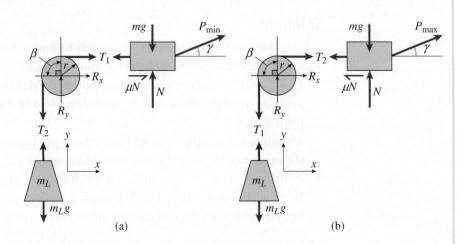

(a)   (b)

---

3. *Analyze*: The numerical data is $m = 15$ kg, $m_L = 20$ kg, $\gamma = 20°$, $\mu = 0.3$, and $\mu_b = 0.2$. The V angle for the V strap is $2\alpha = 38°$. The strap–cylinder contact angle β is 90°, but use $\pi/2$ radians in all formulas.

Minimum permissible force P: From force equilibrium of the lower block in Fig. 3-94a:

$$\sum F_y = 0 \qquad T_2 = m_L g = 20 \text{ kg}\left(9.81 \frac{\text{m}}{\text{s}^2}\right) = 196.2 \text{ N} \qquad \textbf{(a)}$$

Equilibrium of the upper block in Fig. 3-94a requires

$$\sum F_x = 0 \qquad -T_1 + \mu N + P_{min} \cos\gamma = 0$$

$$\sum F_y = 0 \qquad N + P_{min} \sin\gamma - mg = 0 \qquad \textbf{(b)}$$

Solve Eqs. (b) to find

$$T_1 = \mu\, mg + P_{min}(\cos\gamma - \mu\sin\gamma) \qquad \textbf{(c)}$$

Substitute Eq. (c) into the belt friction formula for the *flexible flat strap* [Eq. (3-42)], then solve for $P_{\min}$:

$$\frac{T_2}{T_1} = e^{\mu\beta} \quad \text{so} \quad P_{\min} = \frac{m_L g e^{\frac{-\pi\mu_b}{2}} - \mu\, mg}{\cos\gamma - \mu\sin\gamma} \tag{d}$$

Substitution of numerical data into Eq. (d) gives $P_{\min}$ for the case of a *flat strap*:

$$P_{\min} = \frac{20\text{ kg }(9.81\text{ m/s}^2)e^{-\pi(0.2)/2} - 0.3(15\text{ kg})(9.81\text{ m/s}^2)}{\cos 20° - 0.3(\sin 20°)} = 118.4\text{ N} \quad \longleftarrow \text{(e)}$$

This analysis also applies to the flexible strap of *trapezoidal cross section*, but you need to change the belt friction formula in Eq. (d) to that for a V belt [Eq. (3-46)]. The modified forms of Eqs. (d) and (e) yield

$$\frac{T_2}{T_1} = e^{\frac{\mu\beta}{\sin\alpha}} \quad \text{so} \quad P_{\min} = \frac{m_L g e^{\frac{-\pi\mu_b}{2\sin\alpha}} - \mu\, mg}{\cos\gamma - \mu\sin\gamma} = 36.6\text{ N} \tag{f}$$

Now $\sin(\alpha) = \sin(19°) = 0.326$ appears in the denominator of the exponential term, and the result is a substantial reduction in the required minimum force: $P_{\min}$ drops to 36.6 *N* or 31% of the value for $P_{\min}$ in Eq. (e) for a flat strap. The V strap provides considerable additional friction to prevent leftward sliding of the upper block.

Maximum permissible force P: Force equilibrium of the lower block in Fig. 3-94b gives $T_1 = m_L g = 196.2$ N. Equilibrium of the upper block results in

$$\sum F_x = 0 \qquad -T_2 - \mu N + P_{\max}\cos\gamma = 0$$
$$\sum F_y = 0 \qquad N + P_{\max}\sin\gamma - mg = 0 \tag{g}$$

Solution of Eqs. (g) gives

$$T_2 = -\mu\, mg + P_{\max}(\cos\gamma + \mu\sin\gamma) \tag{h}$$

Now substitute Eq. (h) into the belt friction formula for the *flexible flat strap* and solve for $P_{\max}$:

$$\frac{T_2}{T_1} = e^{\mu\beta} \quad \text{so} \quad P_{\max} = \frac{m_L g e^{\frac{\pi\mu_b}{2}} + \mu mg}{\cos\gamma + \mu\sin\gamma} \tag{i}$$

Substitution of numerical data into Eq. (i) results in $P_{\max}$ for the case of a *flat strap*:

$$P_{\max} = \frac{20\text{ kg }(9.81\text{ m/s}^2)e^{\pi(0.2)/2} + 0.3(15\text{ kg})(9.81\text{ m/s}^2)}{\cos 20° + 0.3(\sin 20°)} = 300\text{ N} \quad \longleftarrow \text{(j)}$$

For the flexible strap of *trapezoidal cross section*, the modified formulas and result for P_{max} are

$$\frac{T_2}{T_1} = e^{\frac{\mu\beta}{\sin\alpha}} \qquad \text{so} \qquad P_{\text{max}} = \frac{m_L g e^{\frac{\pi\mu_b}{2\sin\alpha}} + \mu\, mg}{\cos\gamma + \mu\sin\gamma} = 536 \text{ N} \qquad \longleftarrow \textbf{(k)}$$

The force P_{max} required to slide the upper block to the right is much larger than that required to prevent leftward movement, because the rightward force must overcome gravity force $m_L g$, belt friction, and friction between the upper block and horizontal surface. V strap friction is much larger than that for a flat strap. As a result, P_{max} is 1.8 times larger for a V strap than for a flat strap.

4. *Finalize*: The radius r of the cylinder does not enter into any calculations, so no numerical value is required for variable r. Only the coefficient of friction μ_b between the strap and the cylinder and the angle of contact β (in radians, not degrees) are needed in the belt friction formulas.

CHAPTER SUMMARY AND REVIEW

Chapter 3 covered the *equilibrium* of structures and components. A summary of the major concepts discussed in Chapter 3 is given here.

1. The **static equilibrium equations** are expressed in *vector form* as

$$\mathbf{R} = \Sigma \mathbf{F} = 0 \quad \text{and} \quad \mathbf{M} = \Sigma \mathbf{M} = \Sigma(\mathbf{r} \times \mathbf{F}) = 0$$

where $\mathbf{R}$ is the resultant force and $\mathbf{M}$ is the resultant couple acting on either a rigid or deformable body. In two dimensions (x, y), the equilibrium equations usually are written in *scalar form* as

$$\Sigma F_x = 0 \qquad \Sigma F_y = 0 \qquad \Sigma M_z = 0$$

where the z axis is normal to the plane of the body through any convenient point on or off the body. In three dimensions (x, y, z), if scalar format is used, three force and three moment equations are required:

$$\Sigma F_x = 0 \quad \Sigma F_y = 0 \quad \Sigma F_z = 0 \quad \text{and} \quad \Sigma M_x = 0 \quad \Sigma M_y = 0 \quad \Sigma M_z = 0$$

A simpler and more efficient solution can be obtained if moments are summed about a point or about a line through which multiple unknowns pass so they can be eliminated from the equation. If the body is statically determinate and rigid, the equilibrium equations are necessary and are sufficient to establish equilibrium of the body.

2. Seven different categories of restraints are presented and illustrated in Table 3-1 as a guide in the development of simple models of actual structures or components acted on by applied forces and couples. **Roller, pin, sliding, fixed, elastic, rail**, and **thrust bearing** supports are defined, in both 2D and 3D, and sketches show the reactions associated with each type (see summary of 2D supports shown in the figures here). The appropriate reactions are used as part of a **free-body diagram** of the isolated structure or component whose equilibrium is of interest.

Roller   Pin

Sliding   Fixed   Elastic-translational   Elastic-rotational

Wheel on rail   Thrust bearing

3. Isolating the body of interest and sketching the associated **free-body diagram** is the **essential first step** in a static equilibrium analysis. All problem solutions are presented here using the *four-step problem-solving approach*, and formulation of the free-body diagram is carried out in the initial *conceptualization stage*. All applied (or active) and reactive forces and couples must be shown on the properly drawn free-body diagram. A variety of 2D and 3D examples show how a structure, or portion thereof, must be isolated and then represented by an appropriate free-body diagram.

4. An equilibrium solution only can be obtained if the body is **stable** and **statically determinate**. For stability, there must be a *sufficient number* and *appropriate arrangement of supports*, so rigid body motion under any possible applied loading is prevented. Once stability is satisfied, static determinacy can be investigated. The number of unknown reactions must be equal to the number of **independent equations of equilibrium** if a solution is to be found using statics alone. Plane and space truss structure examples demonstrate the process of assessing stability and determinacy of simple structure models. Similar beam examples are shown: (1) *beam (a)* is stable and statically determinate because there are *three* independent equations of equilibrium (such as $\Sigma F_x = 0$, $\Sigma F_y = 0$, $\Sigma M_A = 0$, or $\Sigma F_x = 0$, $\Sigma M_A = 0$, $\Sigma M_B = 0$) and *three* unknown support reactions (A_x, A_y, B_y); (2) *beam (b)* is unstable because rigid-body motion in the x direction is not restrained; and (3) *beam (c)* is statically indeterminate because there are *three* independent equations of equilibrium but *four* unknown support reactions (A_x, A_y, M_{Az}, B_y).

(a) Stable, statically determinate

(b) Unstable

(c) Stable, statically indeterminate

5. The 2D and 3D **example problems** illustrate a complete static equilibrium solution procedure using the *four-step problem-solving approach*. Truss, beam, frame, and cable-supported structures were among those considered. Both scalar algebra and vector algebra solution strategies were demonstrated. **Solution efficiency, possible pitfalls, and interpretation of results** were discussed in the finalization stage for each example problem.

6. Finally, **dry (Coulomb) friction** and its **application to structures and machines** is discussed and illustrated in a variety of example problems. Two **ranges of behavior** are possible and are shown in the figure for the case of a block of weight W acted on by applied load P and resting on a horizontal surface. The first stage is **static friction**, where the friction force f is less than the maximum possible value, $f < f_{max}$. Summing forces parallel to the surface leads to a linear relationship between static friction force f_s and load P until the maximum friction force $f_{max} = \mu_s N$ is reached, at which point the body is on the verge of sliding. When load P exceeds f_{max}, **kinetic friction** controls as the block slides along the surface, and the force f_k opposing motion between the body and surface depends on the normal force N and the coefficient of kinetic friction μ_k. The coefficient of kinetic friction μ_k is usually less than the coefficient of static friction μ_s; Table 3-2 provides values for coefficients μ_s and μ_k for a variety of different materials in contact. If the block rests on a plane with inclination θ, it will slide down the plane if $\theta > \tan^{-1}(\mu_s)$; this friction angle is known as the *angle of repose*. For blocks that are tall and slender, **overturning** under lateral force P may occur before sliding begins. Once sliding begins, the equation of static equilibrium in the direction of motion *does not apply* unless the acceleration of the body is equal to zero. A number of example problems illustrated the static equilibrium of simple structures for different cases of static and kinetic friction. Application of the principles of dry friction to **disks** undergoing **rotational motion** as well as **wedges, screws**, and **flexible belts** demonstrated the importance of this topic in the study of equilibrium in practical engineering problems.

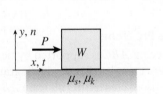

(a) Block on horizontal surface

(b) Free-body diagram of block

Friction force f vs applied force P

(c) Friction force f vs. applied force P

PROBLEMS

3.2 Free-Body Diagrams

Introductory Problems

3.2-1 Uniform bar AB of weight W is supported by springs of stiffness k at A and B. Draw a complete free-body diagram of the bar.

PROBLEM 3.2-1

3.2-2 Uniform bar AB of weight W is pin supported at A and rests on a smooth inclined surface at B. Draw a complete free-body diagram of the bar.

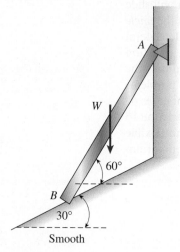

PROBLEM 3.2-2

3.2-3 Bar AB is fixed at A and supported by continuous cable BCD at B. The cable runs over a frictionless pulley at C. Draw an overall free-body diagram of the entire structure. Also draw a FBD of bar AB alone.

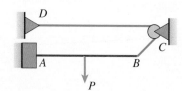

PROBLEM 3.2-3

3.2-4 Blocks A and C, each of weight W, rest on smooth surfaces and are connected by a cable that runs over a frictionless pulley at B. Block A is restrained by a horizontal spring with stiffness k. Draw an overall free-body diagram of the entire system. Also sketch a FBD of each block alone.

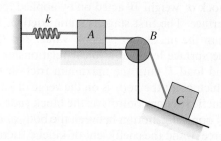

PROBLEM 3.2-4

3.2-5 Two circular disks (radii r_A, r_B and weights W_A, W_B) rest on a circular plane of radius R (see figure). Draw an overall free-body diagram of the entire system. Also sketch a FBD of each disk alone.

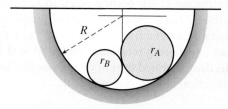

PROBLEM 3.2-5

Representative Problems

3.2-6 Cantilever beam ABC is supported by tie rod BD (see figure). Draw a complete free-body diagram of the entire structure. Also sketch a FBD of beam ABC alone.

PROBLEM 3.2-6

3.2-7 Bar ACB is supported by a strut DC and carries downward load P at B. Draw a complete free-body diagram of the bar and strut structure. Also sketch a FBD of bar ACD alone and another FBD of strut DC.

3.2-8 A cross section of a retaining wall and shoring support struts is shown in the figure along with a simplified model to be used for analysis. Draw a complete free-body diagram of the wall and strut support. Also sketch free-body diagrams of each component (wall, strut) alone.

3.2-9 A pickup truck tailgate of weight W_T supports a crate of weight W_C (see figure). The tailgate pivots about a pin at B and is restrained by a cable attached at A. Draw a complete free-body diagram of the tailgate.

3.2-10 A cable and pulley system at D is used to bring a 230-kg pole (ACB) to a vertical position as shown in the figure. The cable has tensile force T and is attached at C. Draw a complete free-body diagram of the pole.

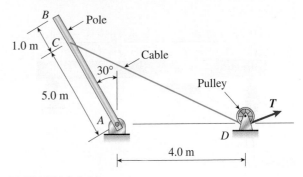

3.2-11 Two gondolas on a ski lift are supported at B and C on a cable that runs from a tower at A to a tower at D (see figure). Draw a complete free-body diagram of each gondola showing the gondola weight and also the cable forces supporting it.

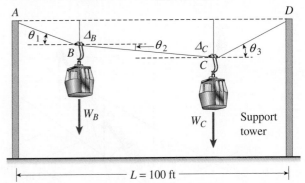

3.2-12 Beam ABC rests on beam DBE at B. Neglect beam weights and also friction at supports D and E and at B. Draw a complete free-body diagram of the entire structure. Also draw FBD's of beam ABC and of beam DBE.

3.2-13 Rigid bar $ABCD$ is spring supported at A and D and has a pin support at B. Draw a complete free-body diagram of bar $ABCD$.

PROBLEM 3.2-13

3.2-14 A crane boom of mass 450 kg with its center of gravity at C is stabilized by two cables AQ and BQ as shown in the figure. A load $P = 20$ kN is applied at point D. The crane boom lies in the y–z plane. Draw a complete free-body diagram of crane boom OD.

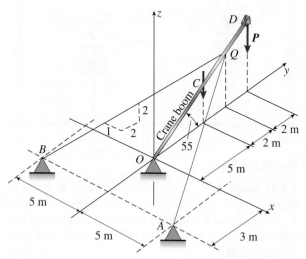

PROBLEM 3.2-14

3.3 Equilibrium in 2D and 3D

Introductory Problems

3.3-1 Three cables are attached to an eye bolt (see figure). The forces in the cables are known to be: $T_1 = 800$ lb, $T_2 = 600$ lb and $T_3 = 1200$ lb. Find the reaction force components at the eye bolt (assume the eye bolt is a pin support).

210

PROBLEM 3.3-1

3.3-2 Cable forces of magnitude $T = 360 N$ act on a pin-supported pulley of radius $r = 90$ mm (see figure). Find the reaction force components at pin support O.

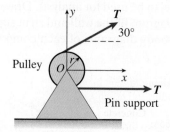

PROBLEM 3.3-2

3.3-3 A cylinder of weight $W = 90$ lb is supported at two points, 1 and 2, as shown. Neglect friction at the two points of contact. Find normal forces N_1 and N_2.

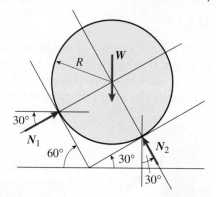

PROBLEM 3.3-3

3.3-4 Two cylinders are in contact along an inclined plane. The upper cylinder rests against the lower cylinder, and both are supported by a wall where the contact force is N_1. The contact force between the upper cylinder and the inclined plane is N_3 and that for the lower cylinder is N_2 (see figure). Assume that all friction forces are negligible. Find an expression for the contact force N_1 in terms of load variable W. Also find an expression for the contact force F between the cylinders.

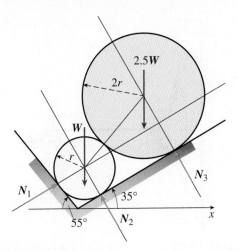

PROBLEM 3.3-4

3.3-5 Repeat Prob. 3.3-4 for the two cylinders shown in the figure.

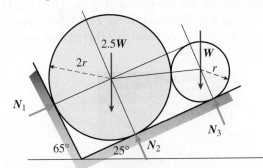

PROBLEM 3.3-5

3.3-6 Repeat Prob. 3.3-4 for the two cylinders shown in the figure.

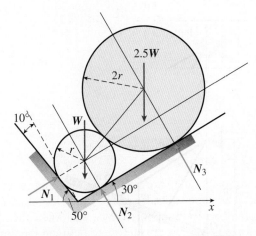

PROBLEM 3.3-6

3.3-7 Repeat Prob. 3.3-4 for the two cylinders shown in the figure.

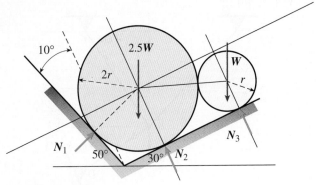

PROBLEM 3.3-7

3.3-8 A weight W (50 N) is supported by continuous cable ABC as shown (see figure). The cable is of length L (3.5 m) and is attached to supports at A and C. The cable tension T is the same in the left (length L_1) and right (length L_2) cable segments because it passes through a smooth ring attached to the weight at B. Horizontal distance $d = 2.4$ m. Find cable tension T and *also* the cable segment lengths L_1 and L_2.

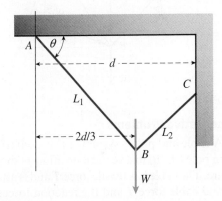

PROBLEM 3.3-8

3.3-9 A known weight W is supported at B by two separate cables, AB and BC. Later cable BD with known tensile force $T_{BD} = W/2$ is added to prevent oscillations. Find expressions for cable forces T_{AB} and T_{BC} in terms of weight W.

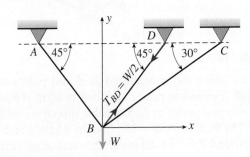

PROBLEM 3.3-9

3.3-10 Repeat Prob. 3.3-9 for the cable-supported weight W shown in the figure. Force $T_{BD} = W/3$.

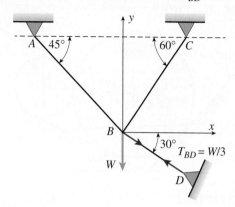

PROBLEM 3.3-10

3.3-11 Repeat Prob. 3.3-9 for the cable-supported weight W shown in the figure. Force $T_{BD} = W/5$.

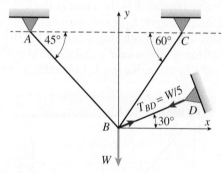

PROBLEM 3.3-11

Representative Problems

3.3-12 A cable and pulley system at D is used to bring a 230-kg pole (ACB) to a vertical position as shown in the figure. The cable has tensile force T and is attached at C. Find cable force T and the reaction forces at A and D for the pole position shown in the figure.

PROBLEM 3.3-12

3.3-13 Cable OBC is anchored at O and passes over a small frictionless pulley attached to strut AB at B and a second pulley at C. The cable supports a weight $W = 150$ lb. If angle $\alpha = 35°$, find the length L and the tension force F in strut AB required for static equilibrium.

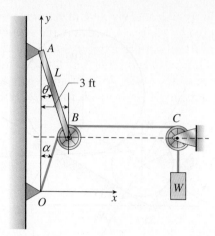

PROBLEM 3.3-13

3.3-14 A continuous cable runs over a small frictionless pulley at A to support one segment of an entrance canopy for a building (Fig. a). Use the 2D sketch in Fig. b to find cable force T and pin-support reactions at O and A. Assume that this part of canopy has weight $W = 3000$ N.

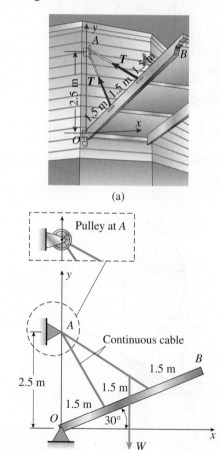

PROBLEM 3.3-14

3.3-15 Uniform slender bar ABC of mass $m = 0.35$ slug and length $L = 12$ ft is pinned at A and supported by a cable at B and a roller support at C. The cable runs over small frictionless pulleys at B and D and supports a mass $M = 0.7$ slug (see figure). Let $\theta = 30°$. Cable segment DC is vertical. Find reactions at A and C.

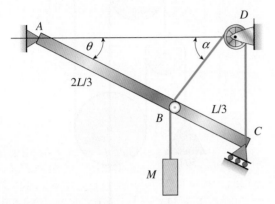

PROBLEM 3.3-15

3.3-16 Uniform slender bar ABC of mass $m = 5$ kg and length $L = 3$ m is pinned at A and has an inclined roller support at F. A cable supported at D passes over frictionless pulleys at B and E. The cable is attached to bar ABC at C and supports a mass $M = 10$ kg (see figure). Let $\theta = 30°$. Find reaction force components at A and F.

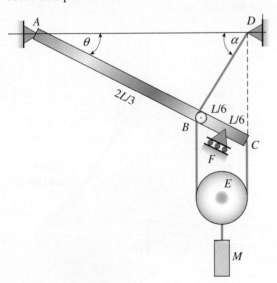

PROBLEM 3.3-16

3.3-17 Bracket $ABCD$ is fixed at A. Two loads P are applied at D. Find expressions for reaction forces and moment at A in terms of load P and dimension variable b.

PROBLEM 3.3-17

3.3-18 A radio tower is held in a vertical position by four cables. The tension forces in three of the four cables are given in the figure. Find the tension force T in the fourth cable and the reaction force components at pin supports A, O, and B. Assume that weight of the tower is 20 kN and includes the weight of the cables.

PROBLEM 3.3-18

213

3.3-19 A mobile crane lifts a 7500-lb load (see figure). The crane weighs 32 kips with center of gravity at C and is supported by stabilizers at A and B. Find the reaction forces at A and B.

PROBLEM 3.3-19

3.3-20 Find the force T at pulley C required for equilibrium of the system shown. Let $W = 200$ N.

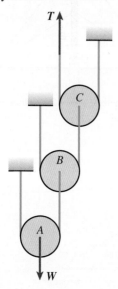

PROBLEM 3.3-20

3.3-21 Find the force F required to hold the 50-lb weight in equilibrium.

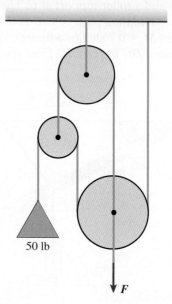

PROBLEM 3.3-21

3.3-22 A steel plate of dimensions $2.5 \times 1.5 \times 0.08$ m and weighing 23.1 kN is hoisted by two separate steel cables with lengths $L_1 = 3.2$ m and $L_2 = 3.9$ m that are each attached to the plate by a clevis and pin (see figure). The measured orientation angles are $\theta = 80°$ and $\alpha = 25°$. For these conditions, determine the cable forces T_1 and T_2.

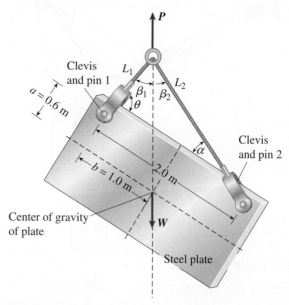

PROBLEM 3.3-22

3.3-23 Segments AB and BC of beam ABC are pin-connected a small distance to the right of joint B (see figure). Axial loads act at A and at the mid-span of AB. A concentrated moment is applied at joint B. Find reactions at supports A, B, and C.

PROBLEM 3.3-23

3.3-24 Segments AB and BCD of beam $ABCD$ are pin-connected at $x = 4$ m. The beam is supported by a sliding support at A and roller supports at C and D (see figure). A 160 N load acts on BC. A concentrated moment is applied at joint D. Find reactions at supports A, C, and D.

PROBLEM 3.3-24

3.3-25 A plane truss structure is subjected to forces $P_1 = 50$ lb at D and inclined load $P_2 = 100$ lb at A. Find reactions at C and F. Assume that $L = 3$ ft.

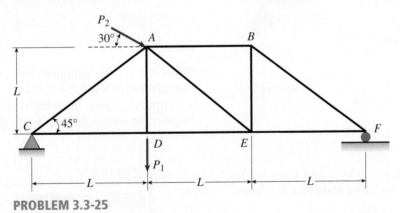

PROBLEM 3.3-25

3.3-26 Find reactions at A and E for the roof truss shown in the figure. Let $P = 25$ kN and $L = 2$ m.

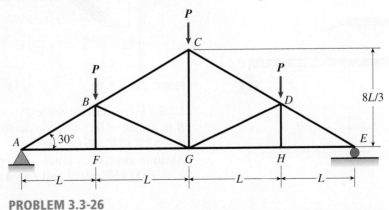

PROBLEM 3.3-26

215

3.3-27 A plane truss structure is loaded by cable force $P = 10$ kips. The cable is anchored at H and runs over a small frictionless pulley at D. Find support reactions at A and E. Assume that dimension $L = 10$ ft.

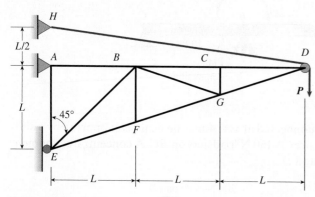

PROBLEM 3.3-27

3.3-28 Find reaction forces at A and C (in terms of load variable P) for the frame shown in the figure.

PROBLEM 3.3-28

3.3-29 Find reaction forces at A and C (in terms of load variable P) for the frame shown in the figure.

PROBLEM 3.3-29

3.3-30 The traffic signal shown in the photo is modeled as a cantilever framework. The model is subjected to signal weights $P = 200$ N at joints A, B, C, D, and E and also a road light $P_1 = 60$ N at F. Find support reaction components at support H. Let $L = 2$ m.

PROBLEM 3.3-30

3.3-31 Bar ACB is supported by a strut DC and carries downward load $P = 350$ lb at B. Find reactions at supports A and D. What is the pin force at C? Neglect the weight of bars ACB and DC.

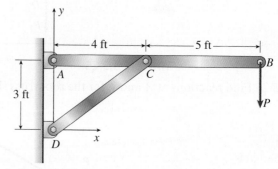

PROBLEM 3.3-31

3.3-32 When lateral force $P = 65$ kN is applied at B to frame ABC, the elastic support at A rotates $\theta = 1.25°$ CW. Find the reactions at A and C. Assume that rotational spring constant at A is $k_r = 7000$ kN·m/rad and $L = 4$ m.

PROBLEM 3.3-32

3.3-33 A plane frame is restrained at joints A and C, as shown in the figure. Members AB and BC are pin-connected at B. A lateral load of 540 lb acts on AB. A concentrated moment is applied at joint C. Find reactions at supports A and C.

PROBLEM 3.3-33

3.3-34 A plane frame is restrained at joints A and D, as shown in the figure. Members AB and BCD are pin-connected at B. A lateral load of 160 N acts on CD as shown. An inclined concentrated force of 200 N is

3.3-36 A plane frame with a pin support at A and roller supports at C and E has a cable attached at E, which runs over frictionless pulleys at D and B (see figure). The cable force is known to be 400 N. There is a pin connection just to the left of joint C. Find reactions at supports A, C, and E.

applied at the mid-span of BC. Find reactions at supports A and D.

PROBLEM 3.3-34

3.3-35 A wall on a building is braced to prevent collapse after being damaged in an earthquake (see photo). Members AB and BC in the A-frame are pin-connected at A. Lateral force variable P is estimated to be 5 kips. Find the reactions at B and C due to the lateral loading from the damaged wall. Let $L = 45$ ft.

Maciej Bledowski/Shutterstock.com

PROBLEM 3.3-35

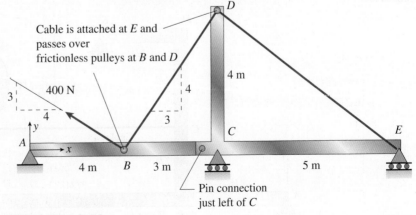

PROBLEM 3.3-36

217

3.3-37 A 200-lb trap door (*AB*) is supported by a strut (*BC*) that is pin-connected to the door at *B* (see figure). Find reactions at supports *A* and *C*.

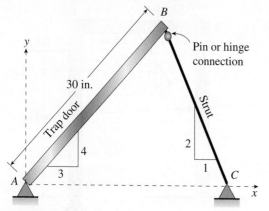

PROBLEM 3.3-37

3.3-38 L-shaped frame *BED* carries load $P = 210$ kN from cable *AED*, which runs over frictionless pulleys at *E* and *D*. Find reactions at supports *A*, *B*, and *C*. Let $L = 2$ m.

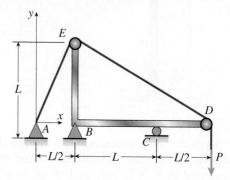

PROBLEM 3.3-38

3.3-39 A plane frame with a roller support at *A* and a pin support at *E* has a cable attached at *C*. The cable runs over a frictionless pulley at *F* (see figure). The cable force is known to be 500 lb. Find reactions at supports *A* and *E*.

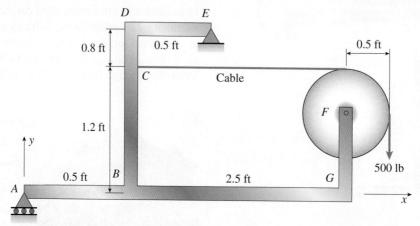

PROBLEM 3.3-39

3.3-40 A plane frame with a pin support at *A*, and roller supports at *E* and *G* has a cable attached at *C*. The cable runs over a frictionless pulley at *F* (see figure). Member *BCDE* is pin-connected to member *AGF* at *B*. Find pin force components B_x and B_y (in terms of upward cable force variable *T*) for the pin connection at *B*. Also find reaction G_y.

PROBLEM 3.3-40

3.3-41 Repeat Prob. 3.3-40 for the cable loading T shown in the figure.

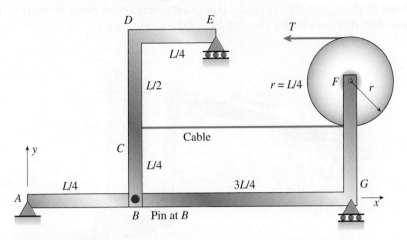

PROBLEM 3.3-41

3.3-42 Repeat Prob. 3.3-40 for the cable loading T shown in the figure.

3.3-43 Forces A_x and B_z are applied to a hand drill (see figure) that is drilling a hole at O. If the drill bit is locked at O, what are the reaction forces and moments at point O?

PROBLEM 3.3-42

PROBLEM 3.3-43

3.3-44 A hand pump is used to extract fluid from a cylinder. In the position shown in the figure, force $P = 90$ N is applied at O normal to the pump arm. If pump arm OAB pivots about point A, what is the pivot force at A and the vertical force F at B? (Start by drawing a FBD of the pump arm.)

PROBLEM 3.3-44

3.3-45 Bracket $ABCD$ is fixed at A. Load components P_x and P_z are applied at D. Find reaction forces and moments at A. Assume that dimension variable $b = 5$ in.

PROBLEM 3.3-45

3.3-46 A rectangular slab with c.g. at O is supported by three cables (A, B, and D). Find the force in all three cables. The slab weight $W = 8500$ N.

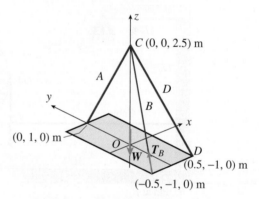

PROBLEM 3.3-46

3.3-47 A 9-in. thick, L-shaped concrete slab (see figure) is supported by three cables. Find the force in all three cables. The slab weight $W = 12{,}150$ lb.

PROBLEM 3.3-47

3.3-48 Bracket *ADO* is fixed at *O* and is acted on by forces F and T (see figures). Find support reaction force and moment components at point *O* for each of the *three different load cases* shown in Figs. a though c.

Assume that $F = 8$ N and $T = 9$ N and that joint coordinates are in meters.

(a)

(b)   (c)

PROBLEM 3.3-48

3.3-49 Bracket *ABCD* is fixed at *A*. Moment components M_x and M_z are applied at *D*. Find reactions at *A*. Assume that dimension variable $b = 5$ in.

PROBLEM 3.3-49

3.3-50 Space frame *ABCD* is clamped at *A*, except it is free to translate in the *x* direction. There is also a roller support at *D* that is normal to line *CDE*. A load $Q_z = 75$ N acts on *AB* in the positive *z* direction. Forces $P_x = 60$ N and $P_z = 45$ N are applied at joint *C*, and a concentrated moment $M_y = 120$ N·m acts at the mid-span of member BC. Find reactions at supports *A* and *D*.

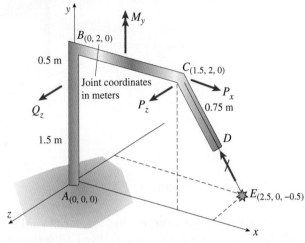

PROBLEM 3.3-50

3.3-51 Space frame *ABC* is clamped at *A*, except it is free to rotate at *A* about the *x* and *y* axes. Cables *DC* and *EC* support the frame at *C*. Force $P_y = 50$ lb is applied at the mid-span of *AB*, and a concentrated moment $M_x = 20$ in.-lb acts at joint *B*. (a) Find reactions at support *A*. (b) Find cable tension forces.

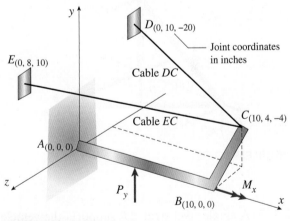

PROBLEM 3.3-51

3.3-52 A soccer goal is subjected to gravity loads (in the $-z$ direction, $w = 73$ N/m for DG, BG, and BC; $w = 29$ N/m for all other members; see figure) and a force $F = 200$ N applied eccentrically at the midheight of member DG. Find reactions at supports C, D, and H.

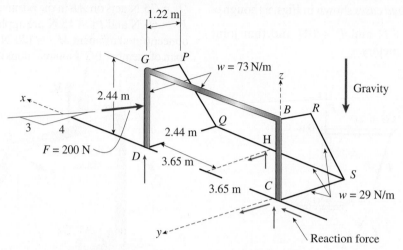

PROBLEM 3.3-52

3.3-53 A space truss structure has force restraints at joints O, A, B, and C (see figure). Find all restraint forces due to loads P at A and $2P$ at C. Assume that $P = 200$ lb and $L = 12$ ft.

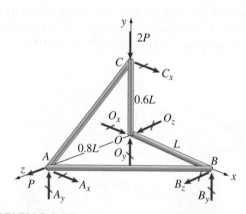

PROBLEM 3.3-53

3.3-54 A single steel strut AB supports the vehicle engine hood of mass 20 kg that pivots about hinges at C and D (see figure parts a and b). The strut lies in a vertical plane and is attached to a bolt at A. Find the force F_S in the strut and the pin support reaction components at A.

(a)

(b)

PROBLEM 3.3-54

222

3.3-55 Bracket $ABCD$ is fixed at A. Force P of magnitude 150 lb is applied at D and directed along line DH. Find reactions at A. Assume that dimension variable $b = 5$ in.

PROBLEM 3.3-55

3.3-56 The rear hatch of a van ($BDCF$ in figure part a) is supported by two hinges at B_1 and B_2 and by two struts attached to the van body with pins at A_1 and A_2. If a closing force $P = 50$ N is applied at G, what is the force F in each strut? Hatch weight $W_h = M_h g = (45 \text{ kg})(g)$ is concentrated at C. (Use the free-body diagram of one half of the hatch in the figure part b.)

(a)

(b)

PROBLEM 3.3-56

3.3-57 Four struts [BE, CE, CF, DF] are used to support a canopy that weighs 1500 lb. The canopy is also attached to a wall at A and G, and reaction force components at A and G are shown in the figure.

(a) Comment on the stability and static determinacy of the structure.

(b) If the struts are replaced by *one continuous cable BECFD* that runs over smooth pulleys at E, C, and F, what is the stability and determinacy of this system?

(c) For the cable supported canopy in part (b), find reactions at A and G and cable force T due the canopy weight alone.

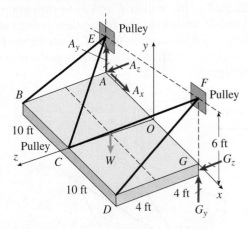

PROBLEM 3.3-57

223

3.3-58 A 5 m × 8 m canopy of weight $W = 6$ kN is supported by a *continuous* cable *EBCDF*. The cable is attached to the wall at *E*, *C*, and *F*; to the canopy at *B* and *D*; and runs over frictionless pulleys at *B*, *C*, and *D*.

(a) Find the tension force *T* in the cable.

(b) Find reaction force components at wall supports at *A* and *G* (see figure).

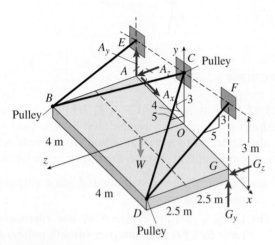

PROBLEM 3.3-58

3.3-59 An 8 ft. × 20 ft. canopy of weight $W = 1700$ lb is supported by a *continuous* cable *EBCDF*. The cable is attached to the wall at *E* and *F*; to the canopy at *B*, *C* and *D*; and runs over frictionless pulleys at *C*, *E* and *F*. Find the tension force *T* in the cable. Find reaction force components at wall supports at *A* and *G* (see figure).

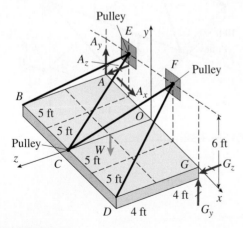

PROBLEM 3.3-59

3.3-60 Rigid bar *ABCD* is supported by linear springs at *A* and *D* and by a pin at *B* (Fig. a). The measured rotation of the bar about *B* is 2.5°.

(a) What is the applied force *P*?

(b) Repeat (a) if a rotational spring $k_r = 6.9$ kN · m/rad is added at *B*.

PROBLEM 3.3-60

3.3-61 Uniform slender bar *ABC* of mass $m = 0.35$ slugs and length $L = 12$ ft is pinned at A and then attached to an undeformed spring at *C* (Fig. a). *Find* the spring deformation δ (in.) and the reaction forces at *A* and *C* due to the weight of bar ABC. *Next*, a cable is attached to the bar at *B* (Fig. b). The cable runs over a small frictionless pulley at *D* and supports a mass *M*. Find mass *M* so that the spring force is 2 lb or less. Let θ = 30° and $k = 3$ lb/in.

(a)

PROBLEM 3.3-61

(b)

PROBLEM 3.3-61 (*continued*)

3.3-62 Uniform slender bar *ABC* of mass $m = 5$ kg and length $L = 3$ m is pinned at *A* and then attached to an undeformed rotational spring at *C* (Fig. a). *Find* the spring deformation φ (radians), the reaction forces at *A*, and moment reaction at *C* due to the weight of bar *ABC*. Next, a cable is attached to the bar at *B* (Fig. b). The cable runs over a small frictionless pulley at *D* and supports a mass *M*. Find mass *M* so that the spring moment at *C* is 50 N · m or less. Let $\theta = 30°$ and $k_r = 9750$ N · m/radian.

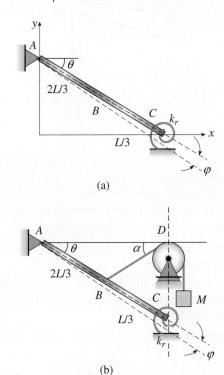

(a)

(b)

PROBLEM 3.3-62

3.4 Dry Friction

Introductory Problems

3.4-1 Bar *AB* of weight *W* is supported by a smooth roller at *B* and rests on a horizontal surface at *A*, which has coefficient of static friction μ_s. If inclination angle $\beta = 50°$, what is the minimum value of μ_s if equilibrium is to be maintained?

PROBLEM 3.4-1

3.4-2 Bar *AB* of weight *W* is supported by a smooth roller at *B* and rests on a horizontal surface at *A*, which has coefficient of static friction $\mu_s = 0.30$. What is the minimum value of inclination angle β if equilibrium is to be maintained?

PROBLEM 3.4-2

3.4-3 A prismatic bar of weight *W* and length *L* is supported at *A* and *B* between two walls a distance *s* apart (see figure). The bar is inclined at an angle of 40°. The coefficient of static friction $\mu_s = 0.30$ at *A* and *B*. What is the maximum value of ratio *L/s* if equilibrium is to be maintained?

PROBLEM 3.4-3

3.4-4 A rigid bar ABC of weight W is supported by a frictionless roller at A with a radius $r_A = 0.5$ m. Find the coefficient of friction for which the bar is on the verge of motion. Let $W = 50$ kN, $L = 3$ m.

PROBLEM 3.4-4

3.4-5 A block of weight $W = 250$ lb rests on a rough surface. An undeformed spring with stiffness constant k is attached to the block.

(a) In Fig. a, if $k = 50$ lb/in and $\mu = 0.3$, what is the maximum displacement that can be applied to the block without slip?

(b) In Fig. b, if $\theta = 30°$ and $k = 50$ lb/in, what is the minimum coefficient of friction μ required for equilibrium if the block is displaced by 0.2 in.?

(c) In Fig. c, if $\theta = 30°$, $\mu = 0.4$ and $k = 50$ lb/in, what is the minimum weight of the block required for equilibrium if the block is displaced by 0.25 in.?

PROBLEM 3.4-5

3.4-6 A uniform bar of mass $m = 10$ kg and length $L = 1.5$ m is restrained at the top by a vertical roller support and a rotational spring with spring constant $k_r = 12$ N · m/radian. The bar rests on a rough curved surface with radius L.

(a) If $\mu = 0.3$, what is the maximum angular displacement θ that can be applied to the bar without slip?

(b) Repeat part (a) if $k_r = 0$.

PROBLEM 3.4-6

3.4-7 Force P is applied to Block 2 parallel to an inclined smooth surface to hold the two blocks in equilibrium (see figure). The blocks are connected by a cable that runs over a frictionless pulley at A.

(a) Find the required range of values of force P to maintain equilibrium.

(b) Revise your solution if friction between Block 2 and the inclined surface ($\mu_s = 0.2$) is now considered.

Let $W_1 = 10$ kips, $W_2 = 40$ kips, $\theta_2 = 60°$, and $\theta_1 = 30°$.

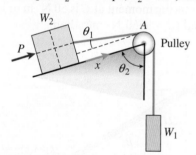

PROBLEM 3.4-7

3.4-8 Blocks 1 and 2 rest on rough surfaces with coefficients of friction μ_1 and μ_2, respectively. The blocks are connected by a cable that runs over a smooth pulley and are in equilibrium as shown in the figure.

(a) If Block 2 is on the verge of sliding *down* the plane, what is the minimum weight W_1 required for equilibrium?

(b) If Block 2 is on the verge of sliding *up* the plane, what is the minimum weight W_1 required for equilibrium?

Assume that $W_2 = 50$ N, $\theta_1 = 30°$, $\theta_2 = 40°$, $\mu_1 = 0.3$, and $\mu_2 = 0.25$.

PROBLEM 3.4-8

3.4-9 Uniform bar AB of weight = 100 lb and length $L = 6$ ft is supported at A and B at angle $\theta = 55°$. What is the minimum value of μ_s at both A and B if equilibrium is to be maintained?

PROBLEM 3.4-9

3.4-10 Uniform bar AB of mass $m = 50$ kg and length $L = 2$ m is supported at A and B at some angle θ. Find the minimum angle θ for which equilibrium is possible for the following two cases:

(a) The coefficient of friction at both A and B is 0.3.
(b) The coefficient of friction at both A and B is 0.25.

PROBLEM 3.4-10

Representative Problems

3.4-11 Uniform bar ABC rests on a rough surface at A and is supported at B on a smooth cylindrical surface with radius b. Find the minimum coefficient of friction at A so that the bar will not slip.

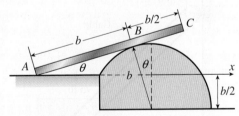

PROBLEM 3.4-11

3.4-12 Uniform rigid bar ABC of weight W is supported on a rough surface at C by a single continuous cable. Find the coefficient of static friction if the bar is observed to slip for each of the following positions of the cable:

(a) Cable attached at A only (Position 1)
(b) Cable attached at B (at $L_2/2$ along bar) only (Position 2).

Assume that $\theta = 35°$, $L_1 = 4$ m, $L_2 = 1.5$ m, and $W = 445$ N.

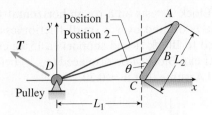

PROBLEM 3.4-12

3.4-13 Block A rests on a rough horizontal surface. A cable is draped over a small frictionless pulley attached to Block A and supports weight B. Find the minimum required coefficient of friction to prevent sliding of Block A in Fig. a. Repeat for the configuration shown in Fig. b.

PROBLEM 3.4-13

227

3.4-14 Block *A* rests on a rough horizontal surface. A cable is draped over a small frictionless pulley attached to Block *A* and supports a 15 kg cylinder. How far can Block *A* be displaced (x_{max}) before slip occurs? Assume that $\mu = 0.30$.

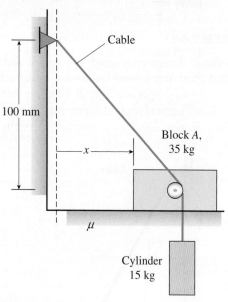

PROBLEM 3.4-14

3.4-15 Block *A* rests on a rough horizontal surface. A cable is draped over a small frictionless pulley attached to Block *A* and supports a 20 lb cylinder. How far can Block *A* be displaced (x_{max}) before slip occurs? Assume that $\mu = 0.30$.

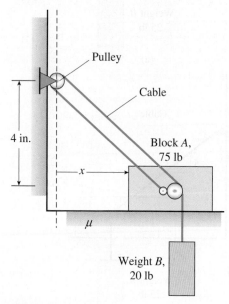

PROBLEM 3.4-15

3.4-16 Block *A* of mass 40 kg rests on a rough horizontal surface. A cable, supporting a weight *W*, is draped over two frictionless pulleys one of which is attached to Block *A* (see figure). Find the minimum required coefficient of friction to prevent sliding of Block *A*. Let $W = 200$ N.

PROBLEM 3.4-16

3.4-17 Block *A* rests on a rough horizontal surface. A cable is draped over a frictionless pulley attached to Block *A* and supports weight *B*. Find the minimum required coefficient of friction to prevent sliding of Block *A*.

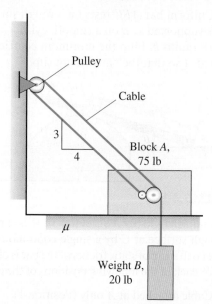

PROBLEM 3.4-17

228

3.4-18 A rigid L-frame of weight $W = 15$ N is supported by a roller at C and rests on a fixed semicircular support at A. Horizontal force P is applied 1 m above joint C to hold the frame in equilibrium. If the coefficient friction at A between the semicircular surface and the frame is 0.5, find the range of force P needed to keep the system in equilibrium.

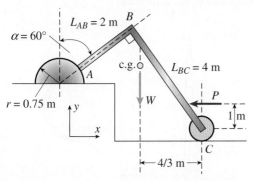

PROBLEM 3.4-18

3.4-19 Three blocks are connected by the cable-pulley system shown in the figure. Find the minimum required coefficient of friction (μ) to maintain equilibrium. Consider:

(a) Case 1: No slip between Block 2 and Block 3.

(b) Case 2: Slip occurs between Block 2 and Block 3.

Numerical data: $\mu_s = 0.75$, $W_1 = 4$ kips, $W_2 = 0.5$ kips, $W_3 = 2$ kips, $\theta_1 = 45°$, and $\theta_2 = 30°$.

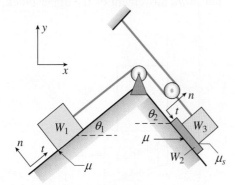

PROBLEM 3.4-19

3.4-20 A block of weight $W = 30$ N rests on a rough surface ($\mu = 0.33$) and is connected to an unstretched vertical spring of length $L = 0.75$ m with spring constant $k = 250$ N/m.

(a) What is the maximum horizontal displacement x that can be applied before the block slips?

(b) If the block is displaced $x = 0.32$ m, what is the minimum required coefficient of friction to prevent slip?

(c) If the block is displaced $x = 0.32$ m and $\mu = 0.33$, what is the minimum weight of the block required to prevent slip?

PROBLEM 3.4-20

3.4-21 A block of weight $W = 7.5$ lb rests on a rough inclined surface ($\theta = 15°$) and is connected to an unstretched vertical spring of length $L = 29$ in. with spring constant $k = 17$ lb/ft. Find the minimum required coefficient of friction:

(a) If the block is displaced 10 in. *down* the plane (Fig. a).

(b) If the block is displaced 10 in. *up* the plane (Fig. a).

PROBLEM 3.4-21

229

3.4-22 A uniform, massless bar is used to hold a rigid cylinder (weight = W) against a wall as shown in Fig. a. Force P is applied at the top of the bar to maintain equilibrium.

(a) If force P is gradually increased, where will slip occur first: between the wall and the cylinder or between the cylinder and the bar? What is P (in terms of W) when first slip occurs?

(b) Repeat part (a) if P is now applied at $0.75L$ from A (Fig. b).

Assume that $\mu_2 = 0.45$ and $\mu_1 = 0.2$.

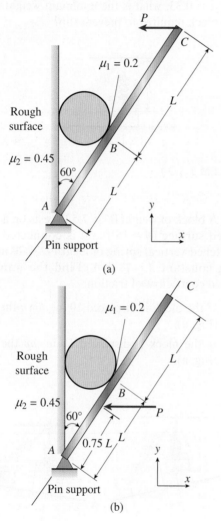

(a)

(b)

PROBLEM 3.4-22

3.4-23 A workman plans to use a wedge ($\beta = 5°$) to make a small adjustment to center the cover on a concrete storage tank. The tank has outer diameter $d_2 = 8$ ft, inner diameter $d_1 = 6.5$ ft, and the cover weighs 780 lb. Find the minimum force P required to move the cover. Assume that the coefficient of static friction is 0.25 for both wedge surfaces and is 0.4 between cover and tank.

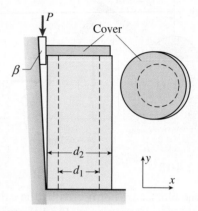

PROBLEM 3.4-23

3.4-24 If the coefficient of static friction is 0.3 for all surfaces:

(a) Find horizontal force H required to *raise* the upper block.

(b) Find *minimum* horizontal force H to prevent *downward* movement of the upper block.

PROBLEM 3.4-24

3.4-25 A three-piece wedge jack is used to level machinery (see figure). Wedges 2 and 3 rest in base 1. Clockwise moment M is applied to the adjusting bolt to move wedge 2 to the right, raising wedge 3. Find the moment M needed to *raise* load $F = 3000$ lb. Assume that the coefficient of static friction between separate pieces of the wedge jack is $\mu_w = 0.35$. The square threaded adjustment bolt has mean diameter $d_m = 1.0$ in. and pitch $L = 0.08$ in. The coefficient of friction in the square threads is $\mu_s = 0.25$.

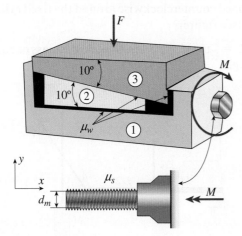

PROBLEM 3.4-25

3.4-26 Uniform slender bar ABC of mass $m = 5$ kg and length $L = 3$ m is pinned at A and supported by a cable attached at B. The cable runs over a frictionless pulley at E, but the small pulley at D has seized and will not rotate; the coefficient of friction between the cable and the seized pulley at D is 0.2. The cable supports a cylinder of mass M (see figure). *Find the mass M required to initiate counterclockwise (CCW) rotation of bar ABC.* Let $\alpha = 30°$ and $\theta = 25°$.

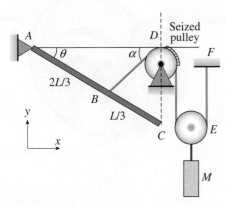

PROBLEM 3.4-26

3.4-27 Blocks A and B, each of weight W, rest on surfaces with friction coefficient μ between block and surface. The two blocks are connected by a cable that runs over a fixed cylinder at C. The coefficient of friction between cable and fixed cylinder is μ_c. Find the minimum force P on Block A (see figure) needed to initiate motion of the blocks if: (a) $\mu_C = 0$; (b) $\mu_C = 0.3$. Let $W = 100$ lb and $\mu = 0.25$.

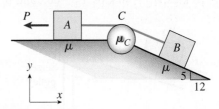

PROBLEM 3.4-27

3.4-28 Blocks 1 and 2 rest on rough surfaces with coefficients of friction μ_1 and μ_2, respectively. The blocks are connected by a cable that runs over a fixed cylinder and are in equilibrium as shown in the figure. The coefficient of friction between cable and fixed cylinder is μ_1.

(a) If Block 2 is on the verge of sliding *down* the plane, what is the minimum weight W_1 required for equilibrium?

(b) If Block 2 is on the verge of sliding *up* the plane, what is the minimum weight W_1 required for equilibrium?

Assume that $W_2 = 50$ N., $\theta_1 = 30°$, $\theta_2 = 40°$, $\mu_1 = 0.3$, and $\mu_2 = 0.25$.

PROBLEM 3.4-28

3.4-29 Force P is applied to Block 2 parallel to an inclined rough surface ($\mu = 0.2$) to hold the two blocks in equilibrium (see figure). The blocks are connected by a cable that runs over a pulley at A. However, the pulley at A has seized and will not rotate; the coefficient of friction between the cable and the seized pulley at A is 0.1. What is the minimum required force P to prevent sliding of Block 2?

Let $W_1 = 10$ kips, $W_2 = 40$ kips, $\theta_2 = 60°$, and $\theta_1 = 30°$.

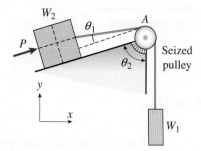

PROBLEM 3.4-29

3.4-30 A dockhand must apply force P to a rope wrapped around a mooring bollard to bring a boat to dock. The boat exerts a force of 2 kN. The friction coefficient between rope and bollard is 0.30. Find the minimum required force P.

PROBLEM 3.4-30

3.4-31 A cable is attached to bar AB at B, runs over a fixed cylinder at C and supports mass M. The coefficient of friction between the cable and fixed cylinder is μ. Load P is applied at the mid-span of AB. Find the minimum force P required to prevent counter-clockwise rotation of bar AB if (a) $\mu = 0$ and (b) $\mu = 0.3$

Let $L = 15$ ft and $M = 1.75$ slug. Assume that the mass of bar AB is 0.25 slug.

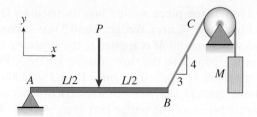

PROBLEM 3.4-31

3.4-32 Repeat Problem 3.4-31 for the case of the cable wrapped counterclockwise around the fixed cylinder at C (see figure).

Let $L = 5$ m, $M = 25$ kg. Assume the mass of bar AB is 5 kg.

PROBLEM 3.4-32

3.4-33 A C clamp is used to hold a brake caliper in place in an automotive disk brake system (see figure). The compression force on the caliper is 150 lb. Find force F, applied normal to the C clamp arm at distance d, required to: (a) tighten the clamp; (b) loosen the clamp. Assume that the C clamp thread has mean diameter of 1/2 in., has four square threads per inch, and a static coefficient of friction in the threads of 0.20.

PROBLEM 3.4-33

http://www.atthetipwebs.com/technologyinstructions/brakes_front_s70.htm

PROBLEM 3.4-33 (*continued*)

3.4-34 A fully loaded clothes washing machine must be adjusted to reduce vibration during the spin cycle. The downward load on one leg is estimated to be $F_y = 1300$ N. Find the force F_z that must be applied to the wrench to raise or lower one corner of the washer. Assume that the leveling screw has a single square thread with mean diameter of 6.5 mm, lead of 2.25 mm, and coefficient of friction of 0.2 for the threads. Wrench force F_z is applied at $d = 70$ mm.

PROBLEM 3.4-34

3.4-35 Two blocks of equal weight $W = 295$ lb are to be moved slightly closer together using a double-acting single-threaded turnbuckle (see figures). The static coefficient of friction between block and ground is $\mu = 0.30$ while that in the turnbuckle threads is $\mu_t = 0.20$. The pitch of the threads is 0.08 in. What is the mean diameter d_m of the turnbuckle threads if the blocks are on the verge of slipping under moment $M_t = 20$ lb-in.?

PROBLEM 3.4-35

3.4-36 A steel bar clamp is being used to hold a metal veneer on a wood beam while the epoxy cures. Assume that the clamp thread has mean diameter of 10 mm, pitch of 2 mm, and a static coefficient of friction in the threads of 0.25. What is the clamping force if the applied moment to tighten the clamp is 400 N·mm? What moment is needed to loosen the clamp?

PROBLEM 3.4-36

Structural Applications

Ashwin/Shutterstock.com

Engineering mechanics principles are used in the analysis and design of structures such as bridges which are integral components of a transportation network. This 280-meter, DNA spiral-looking bridge is a pedestrian bridge in Singapore linking Marina Centre with Marina South.

CHAPTER OBJECTIVES

- Learn about the *stability* and *static determinacy* of *plane* and *space trusses* that are made up of *two-force members*.

- Study two methods for analysis of statically determinate plane and space trusses: the *method of joints* and the *method of sections*; illustrate both methods using example problems and the problem-solving approach.

- Learn about the *stability* and *static determinacy* of *plane* and *space frames* and *machines* that are made up of *multi-force members*.

- Solve for *support* reaction forces, *pin-connection* forces and *internal* axial and shear forces and moments using both 2D and 3D examples of frames and machines using the problem-solving approach.

Chapter Outline

4.1 Introduction

In this chapter, you can now apply the laws of equilibrium to analysis of various kinds of structures such as trusses, frames, and machines. In Chapter 3, you viewed the structure as a whole and then drew free-body diagrams so that you could find *external* reaction forces and moments at the supports. Now, you will disassemble the structures at the joints—those locations where two or more members are connected together to make up the structure—to find interaction forces between individual members. By disassembling the individual elements or members of a structure, the forces of action–reaction between the members that are internal to the structure as a whole are exposed. The forces of action–reaction between interacting structure members stem from Newton's third law, which states: *Action and reaction forces act between bodies in contact, and these forces are equal in magnitude, opposite in direction, and collinear.*

You will study the equilibrium of plane and space trusses, plane and space frames, and also machines. Prior to attempting an equilibrium analysis of any structure, you will evaluate its stability using the procedures discussed in Chapter 3. If the structure is found to be stable, you will then investigate its static determinacy. If statically determinate, you can use an appropriate free-body diagram of the structure to carry out an equilibrium analysis leading to support reactions, forces at connecting points between interacting parts of the structure, and even forces in the members themselves.

Section 4.2 presents the equilibrium analysis of plane trusses using two different approaches. The analysis of space trusses follows in Section 4.3. Finally, frames and machines are considered in Section 4.4. Each structure type and analysis approach is illustrated using example solutions.

4.2 Plane Trusses

Trusses are lightweight, stiff, and efficient load-carrying systems that are commonly used for towers, construction cranes, bridges, long span roof structures, and other applications. Two-dimensional trusses that lie in one plane are known as plane trusses and consist of an assemblage of members connected at pinned joints. Figure 4-1a shows examples of plane trusses being used side-by-side to carry the deck and rails for a railroad bridge, and Fig. 4-1b shows parallel plane trusses as part of a retractable roof for a sports arena.

FIGURE 4-1

Plane trusses used in (a) a railroad bridge and (b) a retractable roof for a sports arena

(a)   (b)

FIGURE 4-2

(a, b) A plane truss is an assemblage of pin-connected members that form triangles

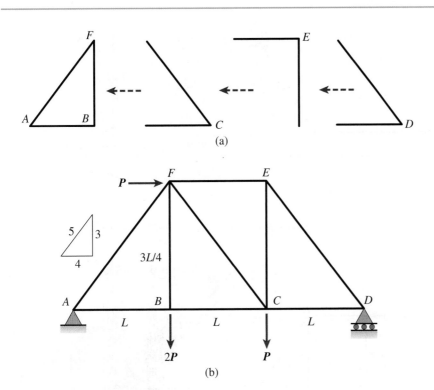

(a)

(b)

The basic unit of construction of a plane truss is a triangle composed of three (usually) straight members (Fig. 4-2a). Adding additional members in pairs enlarges the truss, allowing it to be used to span a river, as in a bridge or a sports arena roof. Concentrated loads are assumed to be applied at the joints only, and pin and roller supports are used in a proper arrangement to provide stability and restraint (Fig. 4-2b). The distributed weight W of an individual truss member is usually neglected because (normally) it is small compared to the applied loads such as P in Fig. 4-2b. If not, weight W can be applied equally to each end of the member as statically equivalent concentrated joint loads $W/2$ at each end joint if the member is uniform in size along its length.

Individual members or bars of the truss are often referred to as *two-force members* in that they transmit applied loads to the supports by collinear tension (T) or compression (C) forces acting along the axis of the member (Fig. 4-3a). The member force is constant along the entire length of the truss member and is equal in magnitude but opposite in direction to the force that acts on the pin used to join members. For example, members AB, BC, and BF are seen to be in tension in Fig. 4-3b; the member force acting on the pin at joint B is in the opposite direction to that acting on the end of the member.

Pinned joints, along with the assumption of joint loads only, ensure that there is only axial force with no transfer of moment from one member to the next. In reality, gusset plates (like that shown in Fig. 4-4) are commonly used to join two or more members at a truss joint. However, the line of action of each truss member is set to meet at a common point, as in Fig. 4-4b, thereby avoiding misalignment of member axes that causes a moment in the joint. Photos of common gusset-plate connections in trusses are shown in Fig. 4-5.

FIGURE 4-3

(a) Truss members (or bars) are two-force members in axial tension (T) or compression (C); (b) equal but opposite forces act on the member end and joint

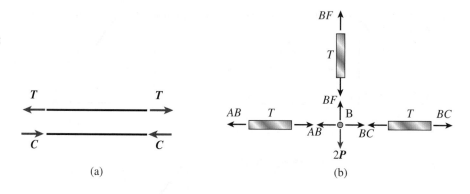

(a)   (b)

FIGURE 4-4

Plane truss joint with member lines of action meeting at common point C

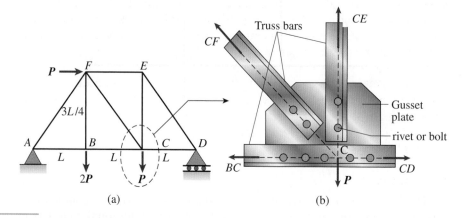

(a)   (b)

FIGURE 4-5

(a) Joint with steel gusset plate in wood truss; (b) steel truss gusset plate

(a)

(b)

There are two basic approaches for finding member forces in stable and statically determinate trusses: the *method of joints* and the *method of sections*. Each method will be applied to plane truss *ADFE* in Figs. 4-2b and 4-4a. Additional examples follow to further illustrate each solution method in detail.

Method of Joints

Using free-body diagrams of pins at each joint in a plane truss, write force equilibrium equations at each joint to get a set of simultaneous equations with the set of forces in each member as the basic unknowns. However, a more efficient solution strategy involves moving from one joint to the next, eliminating unknowns as you go. First, find the reaction forces at supports using a free-body diagram of the entire structure (Fig. 4-6). For example, solution of equations $\Sigma F_x = 0$, $\Sigma F_y = 0$, and $\Sigma M_A = 0$ gives the reaction components A_x, A_y, and D_y in terms of load variable P. The shading in Fig. 4-6 emphasizes that the internal arrangement of members need not be considered to establish the external equilibrium of the overall structure and to solve for support reaction forces.

With reaction forces known, you can now draw a free-body diagram of the pin at each joint and apply equilibrium equations $\Sigma F_x = 0$ and $\Sigma F_y = 0$ at each joint to find member forces. Free-body diagrams of the pins at all joints are shown in the composite of Fig. 4-7a, where the free-body diagram of pin *A* is extracted and displayed in Fig. 4-7b.

FIGURE 4-6

Overall free-body diagram of truss and solution for reactions at *A* and *D*

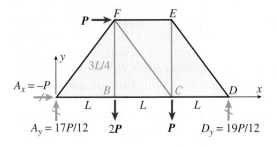

FIGURE 4-7

(a) Free-body diagrams of pins at all joints; (b) free-body diagram of pin at joint *A*

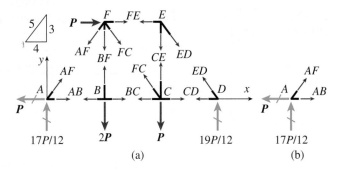

Members are named from the joints at each end, and each member is initially assumed to be in tension (hence the member force arrow acts away from the joint along the axis of the member), so the free-body diagram of joint *A* includes members *AF* and *AB*.

Using the free-body diagram of joint *A* and the 3-4-5 triangle geometry of diagonal members in the truss, write the two force equilibrium equations for joint *A*. First sum forces in the *y* direction to solve for member force *AF* using one equation with one unknown to avoid solving two simultaneous equations. The force equilibrium equations for the pin at joint *A* and solution for member forces *AF* and *AB* are

$$\Sigma F_y = 0 \text{ at joint } A \quad \frac{17}{12}P + \frac{3}{5}AF = 0 \quad \text{so} \quad AF = -\frac{5}{3}\left(\frac{17}{12}P\right)$$

$$= -\frac{85}{36}P = -2.35P \text{ (C)}$$

(4-1)

$$\Sigma F_x = 0 \quad -P + \frac{4}{5}AF + AB = 0 \quad \text{so} \quad AB = P - \frac{4}{5}\left(-\frac{85}{36}P\right)$$

$$= \left(\frac{26}{9}\right)P = 2.89P \text{ (T)}$$

The minus sign for member force *AF* means that the assumed direction of force *AF* on pin *A* in Fig. 4-7b is incorrect. The force vector is in fact in the opposite direction, and member *AF* is in compression (C). Force *AB* is positive, however, so vector *AB* is correctly drawn and member *AB* is in tension (T) as initially assumed.

The method of joints analysis at joint *B* is easier because there are no inclined members incident upon joint *B*; members *AB* and *BC* are in the *x* direction, and member *BF* is in the *y* direction. In addition, there are two unknown member forces at joint *B* (*BC*, *BF*) and two available equations, so forces *BC* and *BF* can be found. [Had you moved from joint *A* to joint *F* next, you would find three unknowns—forces *BF*, *FC*, and *FE* all acting on pin *F*—but with only two equations, $\Sigma F_x = 0$ and $\Sigma F_y = 0$. So you could not find these member forces using the equilibrium of joint *F* alone]. Using the free-body diagram of the pin at joint *B* in Fig. 4-7a, application of force equilibrium leads to member forces *BC* and *BF* as

$$\Sigma F_x = 0 \text{ at joint } B \quad -AB + BC = 0 \quad \text{so} \quad BC = AB = 2.89P \text{ (T)} \quad \textbf{(4-2)}$$

$$\Sigma F_y = 0 \quad BF - 2P = 0 \quad \text{so} \quad BF = 2P \text{ (T)}$$

FIGURE 4-8

Reaction and member forces for plane truss example

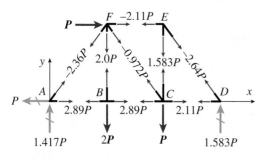

Next, with AF and BF known, you can move to joint F and use $\Sigma F_y = 0$ to find force FC and then $\Sigma F_x = 0$ to find force FE. Using $\Sigma F_x = 0$ at joint E gives force ED followed by $\Sigma F_y = 0$ to find force CE. At joint D, $\Sigma F_x = 0$ gives force CD, and $\Sigma F_y = 0$ at D confirms that the vertical force component in ED is equal to reaction D_y. Finally, arriving at joint C, confirm that x and y direction forces sum to zero in each direction, using the previously computed forces in members BC, CD, FC, and CE. The final results for the plane truss—both reaction and member forces—are summarized in Fig. 4-8.

This joint-by-joint solution strategy allows you to solve for (at most) two unknown member forces at each joint. An alternative approach is to write $2j$ force equilibrium equations (j = number of joints in the truss) for the truss as a whole and to then solve these equations simultaneously.

Stability and Determinacy Issues

Section 3.3 considered issues of stability and determinacy for the plane truss structure analyzed previously by the method of joints. Stability requires that a structure has a *sufficient number and appropriate arrangement of constraints*, so it can maintain its static equilibrium position under any system of applied loads. The addition of one more constraint, say reaction force component D_x at joint D (so there are pin supports at both A and D), will make this structure *externally statically indeterminate* to the first degree. The number of external reactions (r) cannot exceed 3 if the plane truss is to be externally determinate. If $r > 3$, application of the laws of static equilibrium alone ($\Sigma \boldsymbol{F} = 0$, $\Sigma \boldsymbol{M} = 0$) does not lead to a solution for reaction and member forces. However, if $r < 3$, the structure is unstable and may collapse under a particular loading condition (see Section 3.3). Note that the plane truss analyzed here is both stable and determinate with $m + r = 9 + 3 = 12$ and $2j = 2(6) = 12$, so a complete solution for all reaction and member forces is possible using the method of joints.

You must also consider *internal* stability and determinacy of these truss structures. In the method of joints, there are $2j$ equilibrium equations where j = number of joints in the plane truss. If the number of member (m) and reaction force (r) unknowns exceeds the number of available equilibrium equations ($2j$) for the plane truss, the truss is either externally or internally indeterminate. So the first step in a solution for reaction and member forces in a plane truss is the assessment of its stability (or possible mobility), as discussed in Section 3.3. If the truss is stable, you can consider its static determinacy (both internal and external). Some possible determinacy cases for stable plane trusses are

- $m + r < 2j$ more equations than member and reaction force unknowns; no solution
- $m + r = 2j$ statically determinate
- $m + r > 2j$ statically indeterminate (either *externally* or *internally*); $r > 3$ means *externally* indeterminate
- $m + 3 > 2j$ *internally* indeterminate

Additional Considerations

Several situations often occur in analysis of plane trusses and should be recognized when using the method of joints.

Zero-Force Members   Careful inspection of a truss may reveal that some members carry no force for the loading under consideration. For example, if there is no load at joint B (Fig. 4-9a), $\Sigma F_y = 0$ at B reveals that member BF is a zero-force member (see Eq. 4-2b but delete load $2P$). Another instance is shown in Fig. 4-9b, where the roller support is now at C and a horizontal load is applied at D. Apply $\Sigma F_y = 0$ at D and see that member ED is a zero-force member because its vertical component of force must be zero. It follows that the horizontal component of force in ED is also zero. Last, $\Sigma F_y = 0$ at E shows that member CE is a zero-force member.

Collinear Members If three members meet at joint J as shown in Fig. 4-10a, and members IJ and JK are collinear on axis x', the equilibrium of forces along rotated axis y' ($\Sigma F_{y'} = 0$), reveals that force component JL_y must be zero, so member JL is a zero-force member (unless some load having a component along axis y' is applied at joint J). It follows from ($\Sigma F_{x'} = 0$) that member forces IJ and JK are equal. In general, for collinear members meeting at a common joint J (with no load at that joint, Fig. 4-10b), $\Sigma F_{y'} = 0$ means that $JL = JM$, and $\Sigma F_{y''} = 0$ means that $IJ = JK$.

Cables in Cross Bracing In some instances, cables are used for the diagonal members of a plane (or space) truss. Figure 4-11 shows two different possible arrangements for use of cables for diagonal members in a plane truss: cable pair 1 (BE/FC) and cable pair 2 (AE/FD). In cable pair 1, diagonal member FC in the original truss (Fig. 4-8 repeated) has been replaced by a cable, and cable member BE has been added to the truss. In cable pair 2, diagonal member FC in the original truss has been replaced by a cable from F to D, and cable member AE has been added to the truss. Cables can

FIGURE 4-9

(a, b) Zero-force members in a plane truss

(a)

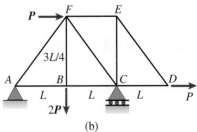

(b)

FIGURE 4-10

(a, b) Force equilibrium for collinear members at joint J

FIGURE 4-11

(a, b) Cables are sometimes used for diagonals in trusses

Cable pair 1
(FC in compression)

(a)

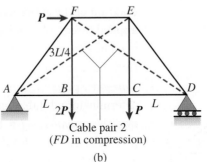

Cable pair 2
(FD in compression)

(b)

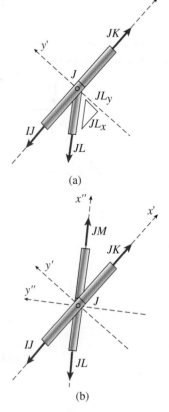

(a)

(b)

FIGURE 4-8 (Repeated)

Reaction and member forces for plane truss example

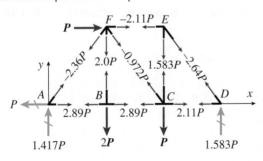

support only tension forces, so they become zero-force members in a truss if the analysis reveals that they must carry a compression force for an applied loading. Note that addition of *both* cables in the pair makes the truss statically indeterminate *internally* because $m + 3 = 13 > 2j = 12$, and an equilibrium solution is not possible using statics. However, since cable *FC* in pair 1 and *FD* in pair 2 are in compression, they can be treated as zero-force members. [Note that *BE* in pair 1 and *AE* in pair 2 are each in tension for the loading shown in Fig. 4-11]. As a result, you can delete cable members *FC* and *FD* and proceed with an equilibrium analysis of each truss using only the remaining members. However, if a different loading system leads to *tension* in *both* cables in each pair, an equilibrium solution is not possible $(m + 3 = 13 > 2j)$ based on statics alone; and if *both* cables in each pair are in *compression* $(m + 3 = 11 < 2j)$, the trusses are unstable.

Example 4-1

FIGURE 4-12

Plane truss model

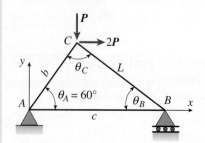

A plane truss is pin supported at *A* and has a roller support at *B* (Fig. 4-12). Joint loads $2P$ and $-P$ are applied at joint *C*. Find support reactions at joints *A* and *B*, then solve for forces in members *AB*, *AC*, and *BC*. Use the following numerical data: $P = 35$ kips, $L = 10$ ft, $\theta_A = 60°$, and $b = 7.1$ ft.

Solution:

Use the four-step problem-solving approach and the method of joints to find reaction forces at *A* and *B*, then find forces in the three members.

FIGURE 4-13

Free-body diagram (FBD) of truss model

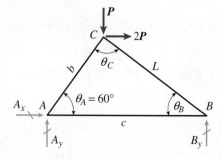

1. *Conceptualize* [*hypothesize, sketch*]: First sketch a free-body diagram of the entire truss model (Fig. 4-13). Only known applied forces at *C* and unknown reaction forces at *A* and *B* need be shown and then used in an equilibrium analysis to find the reactions.

2. *Categorize* [*simplify, classify*]: The truss is externally statically determinate and stable (*unknowns: $m + r = 3 + 3 = 6$ and available equations: $2j = 6$*). First find the three reaction components A_x, A_y, and B_y. Then sketch free-body diagrams of each joint to find the forces in the three members. Use a *statics sign convention* to find external reactions and a *deformation sign convention* to solve for member forces.

3. *Analyze* [*evaluate; select relevant equations, carry out mathematical solution*]:

   Geometry: Find angles θ_B and θ_C and length *c* of member *AB*. Using the law of sines:

   $$\theta_B = \sin^{-1}\left[\frac{b}{L}\sin(\theta_A)\right] = 37.943° \quad \text{so} \quad \theta_C = 180° - (\theta_A + \theta_B) = 82.057°$$

   and $\quad c = L\left[\frac{\sin(\theta_C)}{\sin(\theta_A)}\right] = 11.436$ ft $\quad$ or $\quad c = b\cos(\theta_A) + L\cos(\theta_B) = 11.436$ ft

Alternatively, the law of cosines gives length c:

$$c = \sqrt{b^2 + L^2 - 2bL\cos(\theta_C)} = 11.436 \text{ ft}$$

Support reactions: Equilibrium of the overall truss model in Fig. 4-13 is expressed in the following three equations. Solving Eq. (4-3) gives the support reactions.

$$
\begin{array}{lll}
\Sigma F_x = 0 & A_x + 2P = 0 & A_x = -2P = -70 \text{ kips} \\
\Sigma M_A = 0 & B_y(c) - P(b\cos(\theta_A)) - 2P(b\sin(\theta_A)) = 0 & B_y = 48.5 \text{ kips} \\
\Sigma F_y = 0 & A_y + B_y = P & A_y = -13.5 \text{ kips}
\end{array}
\qquad \textbf{(4-3)}
$$

FIGURE 4-14

Free-body diagram (FBD) of pin at each truss joint

Based on the statics sign convention, reaction component A_x acts in the $-x$ direction and component A_y acts in the $-y$ direction.

Member forces: In the method of joints, draw a free-body diagram of the pin at each joint (Fig. 4-14). Using a deformation sign convention, assume that each member is in tension (so the member force arrows act away from the two joints where each member is connected). The forces are concurrent at each joint, so apply force equilibrium at each location to find the unknown member forces. Force equilibrium at joint A gives

$$\Sigma F_y = 0 \text{ at joint } A \quad F_{AC} = 13.5 \text{ kips}/\sin(60°) = 15.59 \text{ kips}$$
$$\Sigma F_x = 0 \text{ at joint } A \quad F_{AB} = -F_{AC}\cos(60°) + 70 \text{ kips} = 62.2 \text{ kips}$$

Moving to joint B and enforcing force equilibrium in either the x or y direction gives the force in member BC:

$$\Sigma F_x = 0 \text{ at joint } B \qquad F_{BC}\cos(37.943°) + F_{AB} = 0 \text{ so } F_{BC} = -78.9 \text{ kips}$$
$$\text{or } \Sigma F_y = 0 \text{ at joint } B \quad F_{BC}\sin(37.943°) + 48.5 \text{ kips} = 0$$

As a final step, confirm the computed member forces in AC and BC by checking equilibrium in both x and y directions at joint C:

$$\Sigma F_x = 0 \text{ at joint } C \qquad -F_{AC}\cos(60°) + F_{BC}\cos(37.943°) + 2P = 0$$
$$\text{and } \Sigma F_y = 0 \text{ at joint } C \qquad -F_{AC}\sin(60°) - F_{BC}\sin(37.943°) - P = 0$$

4. *Finalize* [*conclude; examine answer—Does it make sense? Are units correct? How does it compare to similar problem solutions?*]: After finding external support reactions, the method of joints usually involves working your way around the structure and enforcing force equilibrium at each joint. Ideally, you can solve a series of equations, one at a time, each equation containing only one unknown

force so you do not have to solve simultaneous equations containing multiple unknowns.

It may be faster in this problem to start at pin C, find forces F_{AC} and F_{BC} before going to pin B to find forces in F_{AB} and B_y. Lastly, go to pin A to find A_x and A_y. There are $2j = 6$ equilibrium equations for the truss as a whole. Applying $\Sigma F_x = 0$ and $\Sigma F_y = 0$ at each joint in succession followed by a computer solution of these equations would be a direct path to a solution for the three reaction forces and three member forces.

Example 4-2

An orthopedic surgeon (who obtained an engineering degree before entering medical school) is attempting to diagnose and treat persistent foot injuries in a patient who is an elite marathon runner. The physician develops a plane truss model to aid in the evaluation of the flow of forces in the runner's foot (Fig. 4-15). It is known from studies of foot strike injuries that the impact force on the foot can be three or more times the runner's weight (W). Find the member forces in the plane truss model due to force components B_x and B_y at the top of the foot.

FIGURE 4-15

Plane truss model of a human foot

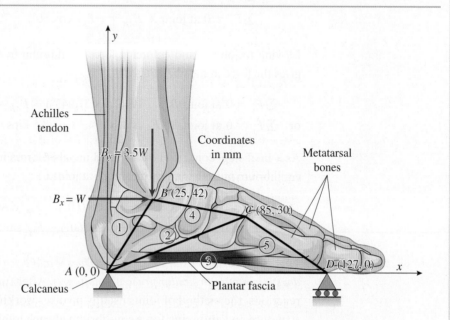

FIGURE 4-16

Free-body diagram (FBD) of truss model of foot

Solution:

Use the four-step problem-solving approach and the method of joints to find reaction forces at A and D, then find forces in members 1 to 5.

1. *Conceptualize*: First sketch a free-body diagram of the entire truss model (Fig. 4-16). Only known applied forces at B and unknown reaction forces at A and D need be shown and then used in an equilibrium analysis to find the reactions.

2. *Categorize*: Overall equilibrium requires that the force components in x and y directions and the moment about the z axis must sum to zero: $\Sigma F = 0$ and $\Sigma M = 0$. The truss is statically determinate (*unknowns*: $m + r = 5 + 3 = 8$ and *available equations*: $2j = 8$). Solve for the three reaction components A_x, A_y, and D_y. Then sketch free-body diagrams of each joint to find the forces in the five members. Use a *statics sign convention* to compute external reactions and a *deformation sign convention* to solve for member forces.

3. *Analyze*:

Support reactions: The overall equilibrium of the truss model in Fig. 4-16 is expressed in the following equations whose solution gives the support reactions. Use of a statics sign convention reveals that the sign of force A_x is negative, so it is actually directed to the left in Fig. 4-16.

$$\Sigma F_x = 0 \quad A_x + B_x = 0 \qquad\qquad A_x = -B_x = -W$$

$$\Sigma F_y = 0 \quad A_y + D_y - B_y = 0 \qquad A_y + D_y = B_y = 3.5W \quad \textbf{(4-4a,b,c)}$$

$$\Sigma M_A = 0 \quad D_y(127 \text{ mm}) - B_x(42 \text{ mm}) - B_y(25 \text{ mm}) = 0 \quad D_y = 1.02W$$

Substitution of D_y from Eq. (4-4c) into (4-4b) gives

$$A_y = 3.5W - D_y = 3.5W - 1.02W = 2.48W$$

Member forces: Draw a free-body diagram of the pin at each joint (Fig. 4-17). Assume that every member is in tension (which is positive in a deformation sign convention), so the member force arrows act away from the two joints where the member is connected. The forces are concurrent at each joint, so apply force equilibrium at each location to find the unknown member forces. Using the joint coordinates given in the figure, compute angles needed to find force components, for example at joint A (Fig. 4-17): $\tan^{-1}\left(\dfrac{25}{42}\right) = 30.76°$. One possible sequence of calculations based on the method of joints is shown next. Start at joint D to obtain one equation in one unknown. (Find the vertical component of member force F_5 in terms of reaction D_y and then use F_5 to find F_3.) Force equilibrium at joint D requires

$$\sum F_y = 0 \text{ at joint } D \quad F_5 = -1.02W / \sin(35.54°) = -1.755W$$

$$\sum F_x = 0 \text{ at joint } D \quad F_3 = -F_5\cos(35.54°) = -(-1.755W)(0.814) = 1.428W$$

Force F_3 is the tension force in the plantar fascia (Fig. 4-15).

FIGURE 4-17

Free-body diagram of each truss joint

Next move to joint C. Knowing force F_5, forces F_2 and F_4 can computed using two simultaneous equations:

$$\sum F_x = 0 \text{ at joint } C \quad F_2\sin(70.56°) + F_4\cos(11.31°) = F_5\sin(54.46°) = -1.428W$$

$$\sum F_y = 0 \text{ at joint } C \quad -F_2\cos(70.56°) + F_4\sin(11.31°) = F_5\cos(54.46°) = -1.02W$$

Solving these two equations gives $F_2 = 1.409W$ and $F_4 = -2.81W$. Lastly, summing horizontal force components at joint B gives force F_1:

$$-F_1\sin(30.76°) + F_4\cos(11.31°) + W = 0 \quad so \quad F_1 = -3.43W$$

Members 1, 4, and 5 are in compression (negative in the deformation sign convention), while members 2 and 3 are in tension for the applied loading at B.

4. *Finalize:* The truss analysis should provide new and useful information to aid the physician in an assessment of forces developed in the runner's foot. Knowledge of these forces, along with results of treadmill tests in an exercise science laboratory, will lead to appropriate treatment for the injured runner.

Examine a truss problem carefully before rushing into a solution. It may be possible to isolate one equation and one unknown and then work your way from joint to joint around the structure to solve for more unknown member forces rather than setting up and having to solve simultaneous equations. In the solution, external reactions at A and D were found and then used in a solution for internal member forces starting with F_5 using $\sum F_y = 0$. Next, force F_3

was computed using $\Sigma F_x = 0$ at D since force F_5 was known before moving to joint C and then joint B. Lastly, applying $\Sigma F_x = 0$ and $\Sigma F_y = 0$ at joint A confirms that support reactions at A and member forces F_1, F_2, and F_3 sum to zero in x and y directions, respectively.

Of course, there are $2j = 8$ equilibrium equations obtained by applying $\Sigma F_x = 0$ and $\Sigma F_y = 0$ at each joint in succession, but then a computer solution of these equations would most likely be required to find the three reaction forces and five member forces. The eight force equilibrium equations are as follows (first at joint A, then joint B, and so on, where $\boldsymbol{n}_{AB}$ is a unit vector along member AB), and their solution yields the same reaction and member forces as listed previously:

$$\Sigma \boldsymbol{F} = 0 \text{ at joint } A \quad F_1\boldsymbol{n}_{AB} + F_2\boldsymbol{n}_{AC} + F_3\boldsymbol{n}_{AD} + \begin{pmatrix} A_x \\ A_y \end{pmatrix} = 0$$

$$\Sigma \boldsymbol{F} = 0 \text{ at joint } B \quad F_1(-\boldsymbol{n}_{AB}) + F_4\boldsymbol{n}_{BC} + \begin{pmatrix} W \\ -3.5\ W \end{pmatrix} = 0$$

$$\Sigma \boldsymbol{F} = 0 \text{ at joint } C \quad F_2(-\boldsymbol{n}_{AC}) + F_4(-\boldsymbol{n}_{BC}) + F_5\boldsymbol{n}_{CD} = 0$$

$$\Sigma \boldsymbol{F} = 0 \text{ at joint } D \quad F_5(-\boldsymbol{n}_{CD}) + F_3(-\boldsymbol{n}_{AD}) + \begin{pmatrix} 0 \\ D_y \end{pmatrix} = 0$$

Expanding the vector equilibrium equation at joint A to scalar form gives, for example:

$$\Sigma F_x = 0 \text{ at joint } A \quad 0.512F_1 + 0.943F_2 + F_3 = -A_x$$
$$\Sigma F_y = 0 \text{ at joint } A \quad 0.859F_1 + 0.333F_2 = -A_y$$

where $\sin(30.76°) = 0.512$ and $\cos(19.44°) = 0.943$. Expanding the force equilibrium equations at joints B, C, and D will provide similar equations for forces in x and y directions at these joints.

In Example 4-4, this plane truss is reanalyzed using the *method of sections* to find selected member forces of interest. The method of sections approach is different than the method of joints, but the computed member forces are the same.

Method of Sections

The method of sections offers an alternative approach to the method of joints and is often used when forces in only a few members in the truss are of interest. In the method of sections, portions of a plane truss are cut through to create separate free-body diagrams and thereby expose the internal forces in selected members. The cut must pass completely through the truss and display the internal force in the member or members of interest. Two free-body diagrams are produced by making the cut through the structure, and either free-body diagram

(a)

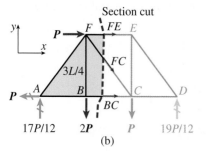

(b)

(left–right, upper–lower) can be used in the subsequent equilibrium analysis. *If the whole structure is in equilibrium, then any section also must be in equilibrium.* Internal forces in members in these free-body diagrams are shown with arrows pointing away from cut members to show the assumed tension force in a cut member. If the computed force turns out to be negative, the member is actually in compression.

Several section cuts and the resulting free-body diagrams are shown in Figs. 4-18 and 4-19. Note that the free-body diagrams in Figs. 4-18b and 4-19b are the *left-hand* and *right-hand* free-body diagrams that result from a section cut through the middle panel of the truss, while Figs. 4-18a and 4-19a are free-body diagrams obtained by separate section cuts at either end of the truss. Each shaded area shown is treated as a single rigid body in equilibrium; the shading emphasizes that uncut members internal to the body are not involved in the equilibrium analysis of the free-body diagram. Section cuts must pass through the members and not through any joints.

While the method of joints utilizes only the force equilibrium equations (in scalar form, $\Sigma F_x = 0$, $\Sigma F_y = 0$ at each of j joints), the method of sections allows you to use *both* force and moment equilibrium for the free-body diagram created by the section cut through the structure. For example, the sum of moments about joint F in Fig. 4-18a allows you to solve for member force AB as follows (since you earlier solved for reaction forces using the structure free-body diagram—see Fig. 4-6):

$$\Sigma M_F = 0 \quad AB\left(\frac{3L}{4}\right) = P\left(\frac{3L}{4}\right) + \frac{17P}{12}(L) = \left(\frac{26}{12}\right)PL$$

$$\text{so} \quad AB = \left(\frac{26}{9}\right)P = 2.89P\,(T) \tag{4-5}$$

This tension force in member AB is in agreement with Eq. (4-1b) and results shown in Fig. 4-8. Note that you summed moments about joint F (through which force AF acts) that does not actually lie on the free-body diagram. Next compute moments about joint C for the free-body diagram in Fig. 4-18b to find force FE. Forces FC and BC act through C, so they are eliminated from the moment equation:

$$\Sigma M_C = 0 \quad FE\left(\frac{3L}{4}\right) = -P\left(\frac{3L}{4}\right) + 2P(L) - \left(\frac{17}{12}\right)P(2L) = -\frac{19}{12}PL$$

$$\text{so} \quad FE = -\left(\frac{19}{9}\right)P = -2.11P\,(C) \tag{4-6}$$

Member force FE is negative, which indicates that the initial assumption of tension is incorrect: FE is in compression (in agreement with Fig. 4-8). As a final

(a)

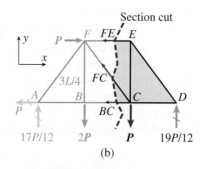

(b)

illustration of the method of sections, consider the right-hand free-body diagram (shaded area) in Fig. 4-19a. Summing forces in the vertical (y) direction will give member force CE as:

$$\Sigma F_y = 0 \quad -CE + \frac{19}{12}P = 0 \quad \text{or} \quad CE = \frac{19}{12}P = 1.583P \text{ (T)} \qquad \textbf{(4-7)}$$

The equilibrium analyses show that the use of either force or moment equilibrium (and scalar calculations) can be used to find selected forces of interest. As expected, member force results in Eqs. (4-5) through (4-7) are in agreement with results in Fig. 4-8 obtained using the method of joints.

In many cases, a combination of the methods of joints and sections is the most effective way to perform an equilibrium analysis of a truss structure, especially for space truss structures.

Example 4-3

The plane truss in Fig. 4-20 has seven joints and 11 members. Find the member forces in members 6, 10, and 11.

FIGURE 4-20

Plane truss model

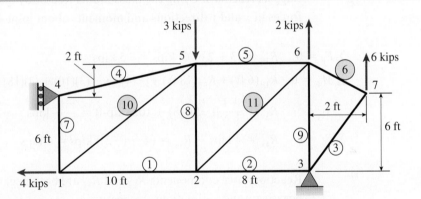

Solution:

Use the four-step problem-solving approach and the method of sections to find reaction forces at joints 3 and 4 and then find forces in members 6, 10, and 11.

1. *Conceptualize*: First sketch a free-body diagram of the entire truss model (Fig. 4-21). Shading is used in the figure to emphasize that only known applied forces at joints 1, 5, 6, and 7 and unknown reaction forces at joints 3 and 4 are needed in an equilibrium analysis to find the external reactions.

2. *Categorize*: Overall equilibrium requires that the force components in the x and y directions and the moment about the z axis must sum to zero. The truss is statically determinate (*unknowns*: $m + r = 11 + 3 = 14$ and *available equations*: $2j = 14$). Note that it is not necessary to solve for reactions in all cases if you are interested in only a few selected member forces. As before, use a

FIGURE 4-21

Free-body diagram of truss model

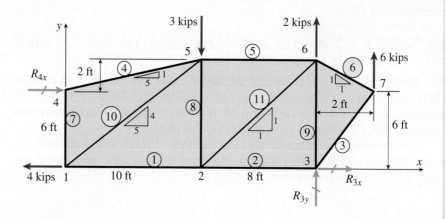

statics sign convention when computing external reactions and a deformation sign convention when solving for member forces.

3. *Analyze*:

Support reactions: Using the truss model free-body diagram in Fig. 4-21, sum forces in x and y directions and moments about joint 4:

$$\Sigma F_y = 0 \quad R_{3y} = (3 - 2 - 6) \text{ kips} = -5 \text{ kips}$$

$$\Sigma M_4 = 0 \quad R_{3x}(6 \text{ ft}) + R_{3y}(18 \text{ ft}) + [-(4)(6) - (3)(10) + (2)(18) + (6)(20)] \text{ kip-ft} = 0$$

$$\text{or} \quad R_{3x} = \frac{-1}{6 \text{ ft}}[(-5)(18) + 102] \text{ kip-ft} = \frac{-12}{6} \text{ kips} = -2 \text{ kips} \tag{4-8}$$

$$\Sigma F_x = 0 \quad R_{4x} = 4 \text{ kips} - R_{3x} = (4 - (-2)) \text{ kips} = 6 \text{ kips}$$

Reaction force components R_{3x} and R_{3y} are both negative, so they act in the negative x and y directions, respectively.

Force in member 6: Make a section cut through members 3 and 6, and use the free-body diagram to the right of the section cut (Fig. 4-22). (Another possible cut is a diagonal cut through members 2, 9, and 6—not shown). Summing moments about joint 3, on the line of action of member F_3, provides one equation in one unknown [note that support reactions from Eq. (4-8) are not needed in this moment equilibrium equation]; x and y components of F_6 are assumed to be located just left of joint 7:

$$\Sigma M_3 = 0 \quad F_{6x}(6 \text{ ft}) + F_{6y}(2 \text{ ft}) + (6 \text{ kips})(2 \text{ ft}) = 0$$

$$\text{where} \quad F_{6x} = F_{6y} = F_6 / \sqrt{2} \quad \text{so} \quad F_6 = \frac{-12}{(6 + 2)}\sqrt{2} = -2.12 \text{ kips} \tag{4-9}$$

FIGURE 4-22

Section cut and resulting free-body diagram for finding F_6

Alternatively, a vector algebra solution for F_6 is expressed as

$$\sum M_3 = 0 \quad [2\text{ ft }\boldsymbol{i} + 6\text{ ft }\boldsymbol{j}] \times [\frac{F_6}{\sqrt{2}}(-\boldsymbol{i} + \boldsymbol{j}) + 6\text{ kips }\boldsymbol{j}] = [4\sqrt{2}F_6 + 12]\boldsymbol{k}\text{ kip-ft} = 0$$

so $F_6 = -\dfrac{3}{\sqrt{2}}$ kips $= -2.12$ kips (C)

Member 6 is in compression (negative in the deformation sign convention), so force F_6 acts *toward* joint 7 in Fig. 4-22.

Force in member 10: Make a vertical section cut through members 1, 10, and 4 and use the free-body diagram to the left of the section cut (Fig. 4-23).

If the force in member 4 is known, sum moments about joint 2 to find F_{10}. Using the method of joints at joint 4:

$$\sum F_x = 0 \quad \text{at joint 4} \quad R_{4x} + F_4(5/\sqrt{26}) = 0$$

so $\qquad\qquad F_4 = -\dfrac{\sqrt{26}}{5}(6\text{ kips}) = -6.119$ kips (C)

FIGURE 4-23

Section cut and resulting free-body diagram for finding F_{10}

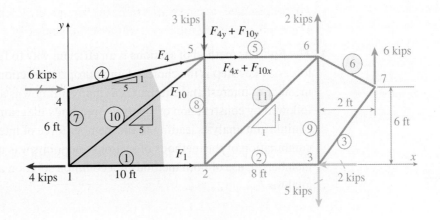

Now slide force vectors F_4 and F_{10} along their lines of action to joint 5 so that their y components have a line of action through joint 2 (see joint 5 in Fig. 4-23); now only their x components are needed in the moment equation. Summing moments about joint 2 in the free-body diagram in Fig. 4-23 gives

$$\Sigma M_2 = 0 \quad -R_{4x}(6 \text{ ft}) - F_{4x}(8 \text{ ft}) - F_{10x}(8 \text{ ft}) = 0$$

$$F_{10x} = \frac{1}{8 \text{ ft}}[-R_{4x}(6 \text{ ft}) - F_{4x}(8 \text{ ft})] = \frac{-1}{8}[6(6) + (-6)(8)] \text{ kips} = \frac{3}{2} \text{ kips} \quad \textbf{(4-10)}$$

so $\quad F_{10} = \frac{\sqrt{41}}{5}\left(\frac{3}{2}\right) = 1.921 \text{ kips (T)}$

Force in member 11: Last, make a vertical section cut through members 2, 11, and 5 and use the free-body diagram to the right of the section cut (Fig. 4-24). Summing forces in the y direction gives the force in member 11:

$$\Sigma F_y = 0 \quad F_{11y} = R_{3y} + (2 + 6) \text{ kips} = (-5 + 2 + 6) \text{ kips} = 3 \text{ kips}$$

$$\text{and} \quad F_{11} = \sqrt{2}F_{11y} = 3\sqrt{2} \text{ kips} = 4.24 \text{ kips (T)} \quad \textbf{(4-11)}$$

FIGURE 4-24

Section cut and resulting free-body diagram for finding F_{11}

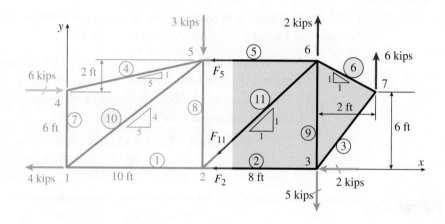

4. *Finalize*: The method of sections is an efficient way to find selected member forces. A key step is the choice of an appropriate section cut that isolates the member of interest and eliminates as many unknowns as possible. This is followed by construction of a correct free-body diagram for use in the static equilibrium analysis leading to the member force of interest. In this problem, a combination of the methods of sections and joints was used to find the force in member 10; use of both methods is a common solution approach in plane truss analysis.

Example 4-4

FIGURE 4-25

Plane truss model of human foot: external reaction forces at *A* and *D*

Reanalyze the plane truss model of the human foot from Example 4-2. Horizontal load $B_x = W$ and downward load $B_y = 3.5W$ at joint *B* represent the impact force at the top of the foot during running, where *W* is the weight of the runner. Use external reaction forces at *A* and *D* obtained in Example 4-2 (such as $A_x = -W$, $A_y = 2.48W$, $D_y = 1.02W$).

Solution:

Use the four-step problem-solving approach and the method of sections to find forces in members 1 to 5.

1. *Conceptualize*: In the method of sections, make a series of cuts through the truss to expose internal forces in cut members and create free-body diagrams for use in equilibrium analysis. The section cuts divide the structure and create both left and right free-body diagrams or upper and lower free-body diagrams as needed. Then apply force and/or moment equilibrium equations, $\Sigma F = 0$ and $\Sigma M = 0$, to find the member forces. For example, a cut through members 2, 3, and 4 and a free-body diagram of the truss to the left of the cut (see Fig. 4-26) leads to the forces in these cut members.

2. *Categorize*: In the method of joints, only force equilibrium, $\Sigma F = 0$ at each joint, is used to find forces in members connected to that joint. Now in the method of sections, moment equilibrium is also used to find member forces. For example, use $\Sigma M_A = 0$ in Fig. 4-26 to isolate and then find member force F_4 (since forces F_2 and F_3 have a line of action through joint *A*). Next, using the same left-hand FBD, use $\Sigma M_C = 0$ to find member force F_3. A series of carefully selected and drawn section cuts often can be used to find all of the member forces of interest in the truss.

3. *Analyze*: The method of sections is often used to find a subset of the forces in the truss. However, to illustrate the method here, sketch section cuts and perform equilibrium analyses to find all of the member forces for the truss in Fig. 4-25.

FIGURE 4-26

Section cut to create left-hand FBD

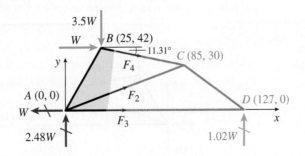

In Example 4-3, mostly scalar equilibrium equations were used. For comparison, most calculations here use vector algebra.

Section cut through members 2-3-4 (Figs. 4-26, 4-27, and 4-28): The left-hand free-body diagram in Fig. 4-26 is used to find force F_4:

$$\Sigma M_A = 0 \quad r_{AB} \times (F_4 n_{BC} + B) = 0$$

Solving gives

$$F_4 = -2.81W$$

where $r_{AB} = 25i + 42j$ mm, $n_{BC} = (0.981)i + (0.196)(-j)$, and $B = (W)i + (3.5W)(-j)$.

A solution for member force F_4 in scalar form uses the x and y components of F_4, so $F_{4x} = F_4 \cos(11.31°)$ *and* $F_{4y} = F_4 \sin(11.31°)$ and applied force components B_x and B_y. The moment equation about joint A becomes

$$\Sigma M_A = 0 \quad [F_4 \cos(11.31°) + W](42 \text{ mm}) + [F_4 \sin(11.31°) + 3.5W](25 \text{ mm}) = 0$$

Solution of this equation again gives $F_4 = -2.81W$.

Next, use the free-body diagram in Fig. 4-26 to find member force F_3:

$$\Sigma M_C = 0 \quad r_{CA} \times (F_3 n_{AD} + A) + r_{CB} \times B = 0$$

Solving gives

$$F_3 = 1.428W$$

where $r_{CA} = 85(-i) + 30(-j)$ mm, $n_{AD} = i$, $r_{CB} = 60(-i) + 12j$ mm, and $A = (W)(-i) + (2.48W)j$.

By extending lines AD and BC to their intersection at point O (Fig. 4-27), a solution for member force F_2 is obtained:

$$\Sigma M_O = 0 \quad r_{OC} \times (F_2 n_{AC}) + r_{OB} \times B + r_{OA} \times A = 0$$

Solving gives

$$F_2 = 1.409W$$

where $r_{OC} = 150(-i) + 30j$ mm, $n_{AC} = (0.943)i + (0.333)j$, $r_{OB} = 210(-i) + 42j$ mm, and $r_{OA} = 235(-i)$ mm.

Now also find member forces F_2, F_3, and F_4 using a right-hand free-body diagram (Fig. 4-28). Sum moments about A to find F_4; then sum moments about C to find F_3; and finally sum moments about D or O to find force F_2.

FIGURE 4-27

Modified *left-hand* FBD with section cut through members 2–3–4

FIGURE 4-28

Modified *right-hand* free-body diagram with section cut through members 2–3–4

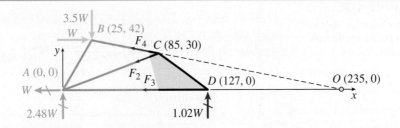

Section cut through members 3 and 5 (Fig. 4-29): Summing moments about joint C gives confirmation of the solution above for force F_3:

$$\sum M_C = 0 \quad r_{CD} \times D + r_{CA} \times (F_3 n_{DA}) = 0$$

Solving gives

$$F_3 = 1.428W$$

where $r_{CD} = 42i + 30(-j)$ mm, $D = 1.02Wj$, $r_{CA} = 85(-i) + 30(-j)$ mm, and $n_{DA} = -i$.

Next, summing moments about joint A in Fig. 4-29 results in a solution for member force F_5:

$$\sum M_A = 0 \quad r_{AC} \times (F_5 n_{DC}) + r_{AD} \times D = 0$$

Solving gives

$$F_5 = -1.754W$$

where $r_{AC} = -r_{CA}$, $n_{DC} = 0.814(-i) + 0.581j$, $r_{AD} = 127i$ mm, and $D = 1.02Wj$.

FIGURE 4-29

Right-hand free-body diagram with section cut through members 3 and 5

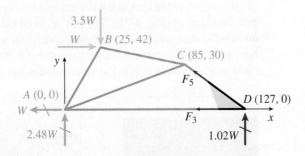

Section cut through members 1 and 4 (Fig. 4-30): Lastly, use the *upper* free-body diagram around joint B in Fig. 4-30 and sum moments about C to eliminate force F_4 and solve for force F_1:

$$\sum M_C = 0 \quad r_{CB} \times [B + F_1 n_{BA}] = 0$$

Solving gives

$$F_1 = -3.43W$$

where $r_{CB} = 60(-i) + 12j$ mm and $n_{BA} = 0.511(-i) + 0.859(-j)$.

FIGURE 4-30

Upper free-body diagram with section cut through members 1 and 4

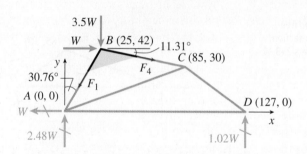

4. *Finalize*: The method of sections approach is quite different from the method of joints used in Example 2-2, but the computed member forces are the same for this loading condition. The focus in this example was on the use of *moment equilibrium equations* leading to the isolation of one unknown in each equation and a direct calculation of the member force of interest. Of course *force equilibrium* can also be used for each of the free-body diagrams in Figs. 4-26 through 4-30 to find member forces. For example in Fig. 4-30, summing forces in the x direction leads to a solution for member force F_1 (assuming that force $F_4 = -2.81W$ is already known) as

$$\Sigma F_x = 0 \quad -F_1 \sin(30.76°) + F_4 \cos(11.31°) + W = 0$$

so
$$F_1 = \frac{(-2.81W(0.981) + W)}{\sin(30.76°)} = -3.43W$$

4.3 Space Trusses

A space truss is the three-dimensional version of a plane truss. Space trusses are often used for communication towers and for roofs covering sports arenas and airport concourses (Fig. 4-31). The pin between members in a plane truss becomes a *ball-and-socket joint* in a space truss, consistent with the assumption that all members are two-force members and no moment is transferred between interconnected members. All loads applied to space trusses are assumed to act at the joints only.

FIGURE 4-31

Space trusses used for (a) communication tower and (b) airport concourse

(a)   (b)

In a *simple* space truss, the basic unit of construction is a *tetrahedron* rather than a *triangle*. Four joints (O, $A–C$) and six members make up the tetrahedron, and the addition of one joint (D) and three members extends the simple space truss (Fig. 4-32a). With joint E and three more members, the space truss is enlarged further (Fig. 4-32b).

Concentrated force components are applied at joints in the x, y, and z directions, and no moments are transferred or supported at the joints. A space truss has m two-force members, and the r support reactions provide restraints against joint translation in the x, y, and z directions to create a stable structure. There are j joints, so $3j$ force equilibrium equations are available in a method of joints solution for the $m + r$ unknown forces in a space truss as a whole.

Stability and Determinacy Issues for Space Trusses

As for plane trusses, the first step is to assess the stability of the space truss. If the space truss is stable, you can then consider its static determinacy (both internal and external). Some possible determinacy cases for stable space trusses are given here.

- $m + r < 3j$ more equations than unknowns; no solution
- $m + r = 3j$ statically determinate
- $m + r > 3j$ statically indeterminate (either *externally* or *internally*); $r > 6$ means *externally* indeterminate
- $m + 6 > 3j$ *internally* indeterminate

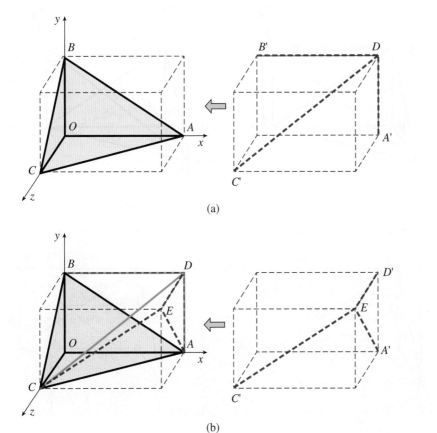

FIGURE 4-32

(a) Simple space truss structure; (b) enlarged simple space truss structure

(a)

(b)

Space Truss Analysis Using the Method of Joints   The simple space truss composed of two tetrahedra in Fig. 4-33a is used to show how to solve for all unknown forces in a space truss by the method of joints. The assumed loading consists of force vector $P = 3i + 7j + 4k$ kN at joint D (Fig. 4-33a). The truss has nine members (m) and six support restraints ($r = 6$, three restraints at O, two at C and one at B) for a total of 15 unknown forces. There are five joints (j) so there are 15 equilibrium equations ($3j$) available for a solution using the method of joints. Overall, the truss is stable and statically determinate [$m + r = 3j$].

Start at joint D where load P is applied and where there are only three unknown member forces (AD, BD, and CD). [Joint A (Fig. 4-33d) has more than three member force unknowns and joints O, B, and C have both reaction and member force unknowns, so solution of equilibrium equations at any of these joints alone will not lead to immediate member or reaction force results.] Three equilibrium equations in the x, y, and z directions are obtained from the free-body diagram of joint D in Fig. 4-33c:

$$\Sigma F_z = 0 \text{ at joint } D \qquad CD = \frac{-\sqrt{50}}{3}(4 \text{ kN}) = -9.43 \text{ kN}$$

$$\Sigma F_x = 0 \text{ at joint } D \qquad BD = \frac{-5}{\sqrt{50}}CD + 3 \text{ kN} = 9.67 \text{ kN} \qquad \textbf{(4-12)}$$

$$\Sigma F_y = 0 \text{ at joint } D \qquad AD = -\frac{4}{\sqrt{50}}CD + 7 \text{ kN} = 12.33 \text{ kN}$$

FIGURE 4-33

(a) Simple space truss; (b) overall free-body diagram; (c) FBD of joint D; (d) free-body diagram of joint A

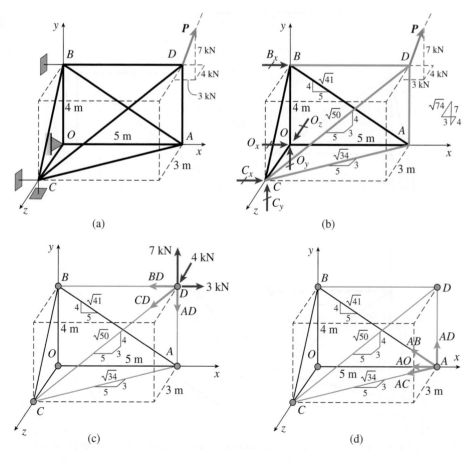

As with plane trusses, each member is initially assumed to be in tension (arrow pointing away from pin in Fig. 4-33c), so a positive member force means tension and a negative force indicates compression. Next, use member force results from Eq. 4-12 to find forces in members at joint A:

$$\Sigma F_y = 0 \text{ at joint } A \qquad AB = \frac{\sqrt{41}}{4}(-AD) = -19.74 \text{ kN}$$

$$\Sigma F_z = 0 \text{ at joint } A \qquad AC = 0 \qquad \text{(4-13)}$$

$$\Sigma F_x = 0 \text{ at joint } A \qquad AO = \frac{-5}{\sqrt{34}}AC - \frac{5}{\sqrt{41}}AB = 15.42 \text{ kN}$$

Moving to joint B and using results from Eqs. 4-12 and 4-13 leads to member forces BC and BO as well as reaction force B_x:

$$\Sigma F_z = 0 \text{ at joint } B \qquad BC = 0$$

$$\Sigma F_y = 0 \text{ at joint } B \qquad BO = \frac{-4}{5}BC - \frac{4}{\sqrt{41}}AB = 12.33 \text{ kN} \qquad \text{(4-14)}$$

$$\Sigma F_x = 0 \text{ at joint } B \qquad B_x = -BD - \frac{5}{\sqrt{41}}AB = 5.75 \text{ kN}$$

Summation of forces at joint C gives reaction force components C_x and C_y and member force CO:

$$\Sigma F_x = 0 \text{ at joint } C \qquad C_x = \frac{-5}{\sqrt{34}}AC - \frac{5}{\sqrt{50}}CD = 6.67 \text{ kN}$$

$$\Sigma F_y = 0 \text{ at joint } C \qquad C_y = \frac{-4}{5}BC - \frac{4}{\sqrt{50}}CD = 5.33 \text{ kN} \qquad \text{(4-15)}$$

$$\Sigma F_z = 0 \text{ at joint } C \qquad CO = \frac{-3}{5}BC - \frac{3}{\sqrt{50}}CD - \frac{3}{\sqrt{34}}AC = 4.0 \text{ kN}$$

Finally, reaction force components at O are computed from member forces at that joint:

$$\Sigma F_x = 0 \text{ at joint } O \qquad O_x = -AO = -15.42 \text{ kN}$$

$$\Sigma F_y = 0 \text{ at joint } O \qquad O_y = -BO = -12.33 \text{ kN} \qquad \text{(4-16)}$$

$$\Sigma F_z = 0 \text{ at joint } O \qquad O_z = -CO = -4.0 \text{ kN}$$

As a final check, summing forces for the overall free-body diagram of the truss in Fig. 4-33b shows that overall equilibrium is satisfied:

$$\Sigma F_x = 0 \qquad O_x + B_x + C_x + P_x = (-15.42 + 5.75 + 6.67 + 3.0) \text{ kN} = 0$$

$$\Sigma F_y = 0 \qquad O_y + C_y + P_y = (-12.33 + 5.33 + 7.0) \text{ kN} = 0 \qquad \text{(4-17)}$$

$$\Sigma F_z = 0 \qquad O_z + P_z = (-4.0 + 4.0) \text{ kN} = 0$$

In many cases, a systematic equilibrium analysis like that used for this simple space truss, moving from joint to joint and finding unknown forces in each step, will not be possible. Alternatively, as the first step in the solution, you can use overall equilibrium of the free-body diagram in Fig. 4-33b to find reaction forces at joints O, B, and C. With reaction forces known, find member forces AO, BO, and CO from the reaction force components instead of the reverse procedure used in Eq. (4-16).

Vector operations and a combination of the methods of joints and sections are often needed to solve for member forces in a space truss. If only a few member forces are of interest, the method of sections is well suited to the task because it may be possible to find a moment axis that eliminates all but the member force of interest. This is illustrated next in the partial reanalysis of the simple space truss in Fig. 4-33.

Space Truss Analysis Using the Method of Sections

Many different section cuts can be made through the space truss of Fig. 4-33a to form free-body diagrams to expose member forces throughout the truss. Recall that a section cut must pass completely through the structure, thereby exposing forces in members of interest. An inclined section cut through the truss just below joint D reveals the member forces AD, BD, and CD, as shown in Fig. 4-34a. Cut members are initially assumed to be in tension, so the member force arrow points away from the section cut.

Summing moments about the x axis in this free-body diagram (Fig. 4-34a) leads to a solution for member force CD:

$$\sum M_x = 0 \quad \left[\frac{3}{\sqrt{50}} CD\right](4\ m) + (4\ kN)(4\ m) = 0$$

$$\text{so}\ \ CD = \left[\frac{-\sqrt{50}}{3}\right](4\ kN) = -9.43\ kN \tag{4-18}$$

Force AD passes through the x axis and force BD is parallel to the x axis, so neither is included in the moment expression in Eq. (4-18). Next, summing forces in x and y directions using the free-body diagram in Fig. 4-34a gives member force $AD = 12.33\ kN$ and $BD = 9.67\ kN$, which are in agreement with forces in Eq. (4-12) from the method of joints. [Note that force BD also could be found by summing moments about line AC, leading to an equation with BD as the only unknown, but this is a more labor-intensive approach.]

A second section cut obtained by passing a vertical plane through the truss between joints B and D leads to the right-hand free-body diagram shown in Fig. 4-34b. If member force AO is of interest, start with an expression for moment M_B:

$$M_B = r_{BD} \times [P + CDn_{DC}] + r_{BA} \times [AO(-i) + ACn_{AC}] \tag{4-19}$$

where n_{DC} and n_{AC} are unit vectors along DC and AC, respectively. Member forces AB and BD are not included in Eq. (4-19) because they have a line of

---

FIGURE 4-34

(a, b) Sample section cuts and free-body diagrams for space truss structure

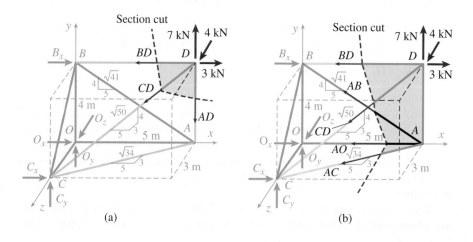

(a)   (b)

action through point B. Now find moment M_{BC}, which is the moment about line BC, and set M_{BC} equal to zero as required for equilibrium. All member forces in the free-body diagram in Fig. 4-34b pass through line BC except force AO so moment equilibrium about line BC leads to

$$M_{BC} = M_B \cdot n_{BC} = 0 \text{ so } [r_{BD} \times P + r_{BO} \times AO(-i)] \cdot n_{BC} = 0 \qquad \textbf{(4-20)}$$

Vectors in Eq. (4-20) are $n_{BC} = \dfrac{4}{5}(-j) + \dfrac{3}{5}k$, $r_{BD} = 5 \text{ m}(-i)$,

$P = 3i + 7j + 4k$ kN, and $r_{BO} = 4 \text{ m}(-j)$. Substituting these vectors into Eq. (4-20) gives

$$[20(-j) + (35 - 4AO)k \text{ kN}]\left[\frac{4}{5}(-j) + \frac{3}{5}k\right] = 0 \text{ or } \left[37 - \frac{12}{5}AO\right]\text{kN} = 0$$

Solution for member force AO leads to $AO = 5(37)/12 \text{ kN} = 15.42 \text{ kN}$, as in Eq. (4-13).

Other section cuts are possible in the space truss, but the two cuts used in Fig. 4-34 illustrate the basic procedures to be used in the method of sections. More detailed analysis of several space truss structures using the methods of joints and sections are presented in the following examples.

Example 4-5

FIGURE 4-35

Space truss model

The space truss in Fig. 4-35 was created from the plane truss in Example 4-1 by moving joint A along the z axis, while holding B on the x axis, and constraining C to lie on the y axis. At the same time, member lengths (L, b, and c) from the plane truss model are also used here. Joint loads $2P$ and $-P$ are applied at joint C in the $+x$ and $-y$ directions, respectively. Support A is a 3D pin support (ball-and-socket), while joint B is restrained in the y and z directions by rigid links (reaction forces B_y, B_z), and joint C is restrained by a link in the z direction (reaction force C_z). Find support reactions at joints A, B, and C. Then solve for forces in members AB, AC, and BC. Use the following numerical data: $P = 35$ kips, $L = 10$ ft, $b = 7.1$ ft, and $c = 11.436$ ft.

Solution:

Use the four-step problem-solving approach and the method of joints to find reaction forces at A, B, and C. Then find forces in the three members.

1. *Conceptualize*: First sketch a free-body diagram of the entire truss model (Fig. 4-36). Only known applied forces at C and unknown reaction forces at A, B, and C need be shown and used in an equilibrium analysis to find the reactions.

FIGURE 4-36

Free-body diagram of truss model

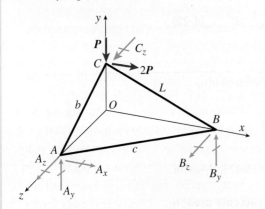

2. *Categorize*: The space truss is externally statically determinate (*unknowns*: $m + r = 3 + 6 = 9$ and *knowns*: $3j = 9$), so the six reaction components can be found using Fig. 4-36. Once all six reactions are known, sketch free-body diagrams of the pin at each joint to find the forces in the three members. Each member lies in a global plane (x–y, x–z, and y–z), so it is likely that this planar structure that lies in a rotated plane can be analyzed using the method of joints. Use a *statics sign convention* when computing external reactions and a *deformation sign convention* when solving for member forces.

3. *Analyze*: Find projections of each member along the coordinate axes as

$$OB = \sqrt{\frac{L^2 - B^2 + c^2}{2}} = 9.4968 \text{ ft} \quad OC = \sqrt{\frac{L^2 + b^2 - c^2}{2}} = 3.1323 \text{ ft}$$

$$OA = \sqrt{\frac{-L^2 + b^2 + c^2}{2}} = 6.3717 \text{ ft}$$

Support reactions: The equilibrium of the overall truss model in Fig. 4-36 is expressed in equilibrium equations whose solution gives the support reactions. First, sum moments about a line through joint A, parallel to the y axis, to isolate and then solve for reaction B_z:

$$B_z(OB) + 2P(OA) = 0 \quad \text{so} \quad B_z = -2P\left(\frac{OA}{OB}\right) = -47 \text{ kips} \quad \text{◄}$$

The minus sign means that reaction B_z acts in the $-z$ direction. Next, sum moments about the z axis to find reaction B_y, followed by summing forces in the y direction to get reaction A_y:

$$B_y = 2P\left(\frac{OC}{OB}\right) = 23.1 \text{ kips} \quad \text{so} \quad A_y = P - B_y = 11.91 \text{ kips} \quad \text{◄}$$

Summing moments about the x axis gives reaction C_z:

$$C_z = A_y\left(\frac{OA}{OC}\right) = 24.2 \text{ kips} \quad \text{◄}$$

Reactions A_x and A_z are found by summing forces in the x and z directions for the overall truss:

$$A_x = -2P = -70 \text{ kips} \quad \text{and} \quad A_z = -C_z - B_z = 22.7 \text{ kips} \quad \text{◄}$$

FIGURE 4-37

Free-body diagram of pin at each truss joint

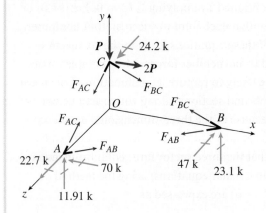

Based on the statics sign convention, reaction component A_x acts in the $-x$ direction, and component A_z acts in the $+z$ direction. All support reactions are now known, so you can find member forces proceeding joint by joint as discussed next.

Member forces: In the method of joints, begin by drawing free-body diagrams of the pin at each joint (Fig. 4-37). Recall that dimensions $AC = b$, $BC = L$, and $AB = c$. Use a deformation sign convention, assuming initially that each member is in tension (so the member force arrows act away from the two joints where each member is connected). The forces are concurrent at each joint, so apply force equilibrium at each location to find the unknown member forces.

First sum forces in the x direction at joint A:

$$\frac{OB}{AB}F_{AB} = 70 \text{ kips} \quad \text{so} \quad F_{AB} = \left(\frac{AB}{OB}\right)70 \text{ kips} = 84.3 \text{ kips} \quad \leftarrow$$

Next sum forces in the y direction at joint A:

$$\frac{OC}{AC}F_{AC} = -11.91 \text{ kips} \quad \text{so} \quad F_{AC} = \left(\frac{AC}{OC}\right)(-11.91) \text{ kips} = -27 \text{ kips} \quad \leftarrow$$

Finally, sum forces in the y direction at joint B:

$$\frac{OC}{AC}F_{BC} = -23.1 \text{ kips} \quad \text{so} \quad F_{BC} = \left(\frac{BC}{OC}\right)(-23.1) \text{ kips} = -73.7 \text{ kips} \quad \leftarrow$$

Method of sections: In the method of joints, you move from joint to joint to find member forces in all members incident upon each joint, which is a straightforward task for this small space truss. In general, a combination of both methods—joints and sections—is needed to find any and all member forces of interest. For example, to find member force F_{BC}, use the lower free-body diagram in Fig. 4-38 containing joints A and B, cutting through members AC and BC. Summing moments about joint A confirms that member force $F_{BC} = -73.7$ kips:

$$\Sigma M_A = r_{AB} \times \left[\begin{pmatrix} 0 \\ 23.1 \\ -47 \end{pmatrix} + F_{BC}\frac{r_{BC}}{|r_{BC}|} \right] = 0$$

$$\text{where } r_{AB} = \begin{pmatrix} OB \\ 0 \\ -OA \end{pmatrix} = (OB)\mathbf{i} + (OA)(-\mathbf{k}) \text{ and } r_{BC} = \begin{pmatrix} -OB \\ OC \\ 0 \end{pmatrix}$$

Also, summing moments about joint B in Fig. 4-38 confirms that $F_{AC} = -27$ kips, and so on.

FIGURE 4-38

Lower free-body diagram

4. *Finalize*: Of course there are $3j = 9$ equilibrium equations for the simple space truss considered here, and these are obtained by applying $\Sigma F_x = 0$, $\Sigma F_y = 0$, and $\Sigma F_z = 0$ at each joint in succession. A computer solution of these equations leads to the six reaction forces and three member forces. More complex space trusses are likely to require a combination of the methods of joints and sections, along with use of vector algebra because of the three-dimensional geometry involved.

Note that the three vector force equilibrium equations (or nine scalar equations) using the method of joints ($\Sigma F = 0$) are expressed as

$$F_{AB}\boldsymbol{n}_{AB} + F_{AC}\boldsymbol{n}_{AC} + \boldsymbol{A} = 0$$
$$F_{AB}(-\boldsymbol{n}_{AB}) + F_{BC}\boldsymbol{n}_{BC} + \boldsymbol{B} = 0$$
$$F_{AC}(-\boldsymbol{n}_{AC}) + F_{BC}(-\boldsymbol{n}_{BC}) + \boldsymbol{C} = 0$$

Vector $\boldsymbol{n}_{AB}$ is a unit vector along member AB (and so on for $\boldsymbol{n}_{AC}$ and $\boldsymbol{n}_{BC}$), and vectors $\boldsymbol{A}$, $\boldsymbol{B}$, and $\boldsymbol{C}$ contain the reaction force components at joints A, B, and C. Solution of these equations will provide the member and reaction forces displayed in Fig. 4-37.

Example 4-6

The space truss in Fig. 4-39 is a model of the end section of a tower crane (Fig. 4.40). Tetrahedra 4–2–6–7, 4–2–3–7, and 4–5–6–7 make up the bulk of the simple space truss. The truss model has ball-and-socket support joints 1, 2, and 3. Load $\boldsymbol{P}_5$ is applied at joint 5 and load $\boldsymbol{P}_7$ is applied at joint 7; the force components in x-y-z directions in kilonewtons (kN) are given in the figure for each load vector. Coordinates of all joints (meters) are shown next to joint names in Fig. 4-39. Find forces in selected members of the truss.

Solution:

Both the method of joints and the method of sections are used to find the forces in a variety of different members of the truss. Use the four-step problem-solving approach as in the following steps.

1. *Conceptualize*: In the method of sections, make careful cuts through the truss at appropriate locations to expose the internal forces in members of interest. This can be challenging for space truss structures. For example, a diagonal plane through members 4, 10, and 11 and just below joint 5 (Fig. 4-41) exposes the forces in these three members. The applied force $\boldsymbol{P}_5$ at joint 5 is part of

FIGURE 4-39

Space truss model

FIGURE 4-40

Tower crane

this section cut and must be retained, but the force P_7 at joint 7 is eliminated. Other section cuts leading to free-body diagrams of various joints are needed to develop appropriate equilibrium equations whose solution leads to member forces of interest.

2. *Categorize*: The truss is externally statically determinate (*unknowns*: $m + r = 12 + 9 = 21$, *available equations*: $3j = 21$), so you can find the nine reaction components if desired. Instead use the methods of joints and sections to find only selected member forces. A vector algebra solution is most effective for this 3D structure. Use a *deformation sign convention* when solving for member forces (tension is positive).

3. *Analyze*: The section cuts and associated equilibrium analyses presented here offer one possible solution approach for this space truss. Begin with the upper free-body diagram near joint 5 shown in Fig. 4-41 that is obtained by passing a diagonal cutting plane through members 4, 10, and 11 and retaining the free-body diagram above the plane.

Members 4, 10, and 11: Plane 5–6–7 [passing through joints 5, 6, and 7] is normal to line 4–5 in Fig. 4-41, so take advantage of this to write a scalar force equilibrium equation using the method of joints to find the magnitude of force F_4:

$$\Sigma F_x = 0 \qquad F_4 = P_{5x} = 2 \text{ kN}$$

Forces F_{10} and F_{11} pass through line 6–7, so summing moments about line 6–7 in Fig. 4-41 will also lead to a (more laborious) solution for F_4:

$$\Sigma M_6 = 0 \qquad M_6 = r_{65} \times [F_4 + F_{10} + F_{11} + P_5] = 0$$
$$\text{with} \qquad P_5 = 2i + 3(-j) + 2(-k) \text{ kN}$$
$$\text{then} \qquad M_6 \cdot n_{67} = 0 \quad \text{with} \quad n_{67} = -k$$

Next, sum moments about line 6–2 [passing through joints 6 and 2] to find force F_{11}, eliminating both force F_{10}, which passes through line 6–2, and force F_4, which is parallel to line 6–2:

$$M_6 \cdot n_{62} = M_6 \cdot (-i) = 1.342F_{11} + 5.25 = 0$$
Solving gives
$$F_{11} = -3.91 \text{ kN}$$

Alternatively, use scalar algebra operations to find force F_{11} using y and z components of F_{11} and force P_5 (see Fig. 4-41) in the moment equilibrium equation about line 6–2:

$$\Sigma M_{62} = 0 \quad 1.5(F_{11z} + P_{5z}) + 0.75(F_{11y} + P_{5y})$$
$$= 1.5\left(\frac{F_{11}}{\sqrt{5}} + 2\right) + 0.75\left(\frac{2F_{11}}{\sqrt{5}} + 3\right) = 0$$

so
$$F_{11} = -\frac{7}{4}\sqrt{5} \text{ kN} = -3.91 \text{ kN}$$

Finally, with forces F_4 and F_{11} known, use force equilibrium at joint 5 to find force F_{10}:

$$\sum F_5 = 0 \quad F_4 + F_{10}n_{56} + F_{11} + P_5 = 0$$

Solving gives $F_{10} = 0.559$ kN, where $n_{56} = 0.894(-j) + 0.447k$.

The equivalent scalar force equilibrium equation in the y direction at joint 5 is

$$\sum F_{5y} = 0 \quad \frac{2}{\sqrt{5}}F_{10} + \frac{2}{\sqrt{5}}F_{11} + P_{5y} = \frac{2}{\sqrt{5}}F_{10} + \frac{2}{\sqrt{5}}\left(-\frac{7}{4}\sqrt{5}\text{ kN}\right) + 3\text{ kN} = 0$$

Solving gives $F_{10} = \dfrac{\sqrt{5}}{4}$ kN $= 0.559$ kN.

Members 12, 5, and 6: A diagonal section cut through members 5, 6, 10, and 12 is shown in Fig. 4-42. Sum moments about line 4–5 so that all forces in the free-body diagram are eliminated except force F_{12}, which must be zero:

$$\sum M_5 = 0 \quad M_5 = r_{56} \times [F_5 + F_6 + F_{10} + F_{12}] = 0$$

Then $\quad M_5 \cdot n_{45} = 0 \quad$ with $\quad n_{45} = -i, \quad$ so $\quad F_{12} = 0.$

FIGURE 4-42

Section cut 5–6–10–12
around joint 6

(Continues)

Applying force equilibrium to the free-body diagram in Fig. 4-42 will give member forces F_5 and F_6:

$$\Sigma F_6 = 0 \qquad F_5 n_{64} + F_6 n_{62} + F_{10} n_{65} + F_{12} = 0$$

Solving gives

$$F_5 = -0.612 \text{ kN}, \ F_6 = 0.25 \text{ kN}$$

where $n_{64} = 0.408(-i) + 0.816j + 0.409(-k)$, $n_{62} = -i$, and $n_{65} = -n_{56}$.

Alternatively, scalar operations result in the same force magnitude values for F_5 and F_6:

$$\Sigma F_{6y} = 0 \quad \frac{2}{\sqrt{6}} F_5 + \frac{2}{\sqrt{5}} F_{10} = 0 \ \text{ so } \ F_5 = -\frac{\sqrt{6}}{4} \text{ kN} = -0.612 \text{ kN}$$

$$\Sigma F_{6x} = 0 \quad \frac{1}{\sqrt{6}} F_5 + F_6 = 0 \ \text{ so } \ F_6 = -\frac{1}{\sqrt{6}} \left[-\frac{\sqrt{6}}{4} \text{ kN} \right] = \frac{1}{4} \text{ kN} = 0.25 \text{ kN}$$

Member 1: Find member force F_1 from joint force equilibrium at joint 1, but recall that there is no solution for reactions so use the method of sections here instead. Passing a diagonal plane through member 1 provides the free-body diagram shown in Fig. 4-43. Sum moments about line 2–3, which eliminates reaction forces at joints 2 and 3 because these forces pass through line 2–3. The moment about line 2–3 gives the magnitude of force F_1:

FIGURE 4-43

Section cut 1 through member 1 near joint 1

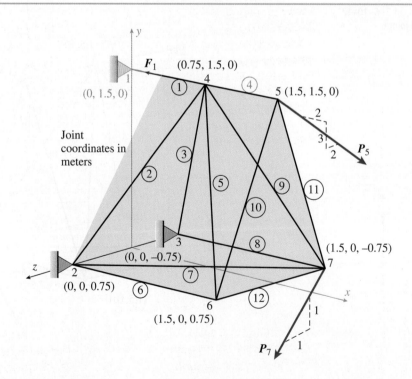

$$\sum M_2 = 0 \quad M_2 = r_{24} \times F_1(-i) + r_{25} \times P_5 + r_{27} \times P_7 = 0$$

with $r_{24} = 0.75i + 1.5j + 0.75(-k)$ m, $r_{25} = 1.5i + 1.5j + 0.75(-k)$ m, $r_{27} = 1.5i + 1.5(-k)$ m, and $n_{23} = -k$.

The moment about line 2–3 is computed by using a dot product:

$$M_2 \cdot n_{23} = M_2 \cdot (-k) = 0$$

Solving gives $F_1 = 6.0$ kN.

In scalar form, the moment about line 2–3 and the solution for F_1 is

$$\sum M_{23} = 0 \quad 1.5F_1 - 1.5(P_{5x} + P_{5y} + P_{7y}) = 0$$

so $\quad F_1 = P_{5x} + P_{5y} + P_{7y} = (2 + 3 + 1)$ kN $= 6$ kN

Member 3: A final application of the method of sections is illustrated by the section cut around joint 4 in Fig. 4-44, which exposes member forces 1–2–3–4–5–9. Sum moments about line 2–7 to find member force magnitude F_3. Forces F_2 and F_9 are unknown but pass through line 2–7, so they do not contribute to moment M_2. Forces F_1, F_4, and F_5 were computed previously so the moment about line 2–7 gives force F_3 as

$$\sum M_2 = 0 \quad M_2 = r_{24} \times [F_1 + F_3 n_{34} + F_4 + F_5] = 0$$

where $r_{24} = 0.75i + 1.5j + 0.75(-k)$ m and $n_{34} = 0.408i + 0.816j + 0.408k$.

FIGURE 4-44

Section cut 1–2–3–4–5–9
around joint 4

Joint
coordinates in
meters

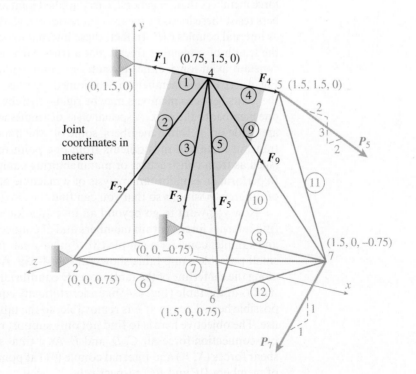

The moment about line 2–7 is computed as

$$M_2 \cdot n_{27} = M_2 \cdot [0.707i + 0.707(-k)] = 0$$

Solving gives $F_3 = -5.51$ kN.

4. *Finalize*: There are $3j = 21$ equilibrium equations for the space truss, which are obtained by applying $\sum F_x = 0$, $\sum F_y = 0$, and $\sum F_z = 0$ at each joint in succession. A computer solution of these equations is required to find the nine reaction forces and twelve member forces. The seven vector force equilibrium equations (or 21 scalar equations) are obtained using the method of joints ($\sum F = 0$). For example, the vector equation at joint 6 (one of the seven joint equilibrium equations) is

$$F_5 n_{64} + F_6 n_{62} + F_{10} n_{65} + F_{12} n_{67} = 0$$

Vector n_{64} is a unit vector along member 5 (and so on for n_{62}, n_{65}, and n_{67}). Computer solution of these equations provides *all* member and reaction forces for the space truss.

4.4 Frames and Machines

Unlike the truss structures discussed previously, frames are made up of multi-force members that, in general, carry applied joint and member loads. The members resist these loads by means of *internal* axial (N) and shear (V) forces as well as internal couples (M). In fact, these internal forces and couples can vary over the length of a member (recall that a truss bar is a two-force member and has constant axial force, either tension or compression, over its entire length—see Fig. 4-3a). Frame members may be joined together using pinned connections, or the joints between members may be rigid—thereby transmitting moments from one member to the next. A special class of frames is referred to as *machines* that are made up of frame members, some of which move to different positions to transmit force and/or moments from one point to another during some task (such as from construction or manufacturing equipment). It may be necessary to perform an equilibrium analysis of a machine or its components in a variety of different positions so that you can find the maximum internal forces.

Now you want to go beyond an investigation of external equilibrium and find internal forces within members and at connections between members. As an example, consider a simple two-dimensional frame structure, such as the small portable camp table shown in Fig. 4-45a. A point load of $2P$ is applied along the centerline of the table, so a 2D equilibrium analysis of the front frame *ABDG* of the table (Fig. 4-45b) under statically equivalent load P at point G is possible here. The pin at E is removable, so the table can be folded when not in use. The objective here is to find not only support reactions at A and B but also pin-connection forces at C, D, and E. As a final step, find internal axial and shear forces (N, V) and internal couple (M) at points J and H at the midpoints of members DE and BC, respectively.

FIGURE 4-45

(a) Portable camp table;
(b) 2D model of table

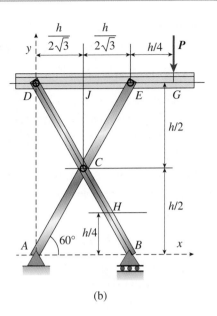

(a)

(b)

FIGURE 4-46

Free-body diagram of camp table

There are three reaction forces and three equations of equilibrium, so the 2D model of the table is *statically determinate externally.* Reaction force components at A and B are determined using the overall free-body diagram shown in Fig. 4-46. Shading is used in Fig. 4-46 to emphasize that internal details of the model, such as the pins at C, D, and E, are not relevant to the following expressions of overall equilibrium and solution for reaction forces:

$$\Sigma F_x = 0 \qquad A_x = 0$$

$$\Sigma M_A = 0 \qquad B_y\left(\frac{h}{\sqrt{3}}\right) - P\left(\frac{h}{\sqrt{3}} + \frac{h}{4}\right) = 0$$

(4-21)

so $\quad B_y = \dfrac{P}{4}(\sqrt{3} + 4) = 1.433P\ [\uparrow]$

$$\Sigma F_y = 0 \qquad A_y + B_y = P \text{ so } A_y = -0.433P\ [\downarrow]$$

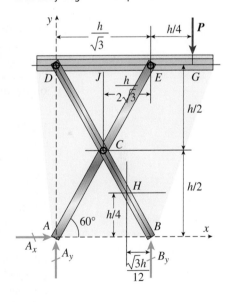

Reaction B_y is positive, so it is directed upward in the $+y$ direction as shown in Fig. 4-46 [arrows following the reaction forces in Eqs. (4-21) indicate the direction of the force component]. However, component A_y is negative, so it acts in the $-y$ direction; the table will overturn clockwise due to load P at G if not tied down at pin support A.

Pin-Connection Forces at C, D, and E   Next disassemble the frame model by removing pins at C, D, and E to create separate free-body diagrams of each member (Fig. 4-47). Note that Newton's third law (action–reaction) is carefully followed: forces at C, D, and E are shown in their assumed directions but are in opposite directions on each segment model.

FIGURE 4-47

Separate free-body diagrams of each table member

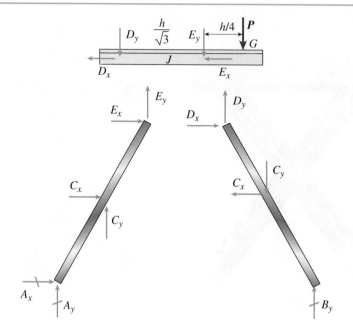

Starting with **element DEG**, enforce equilibrium to find pin force components D_y and E_y:

$$\sum M_D = 0 \;\; E_y\left(\frac{h}{\sqrt{3}}\right) = -P\left(\frac{h}{\sqrt{3}} + \frac{h}{4}\right) \;\; \text{so} \;\; E_y = -P\left(\frac{\sqrt{3} + 4}{4}\right) = -1.433P \,[\uparrow]$$

(4-22)

$$\sum F_y = 0 \;\; D_y = -E_y - P = 0.433P \,[\downarrow]$$

Arrows following the pin-connection forces indicate the direction of the force component *on that element* (which is opposite to the force component direction on the adjacent element). Now from **element BCD**:

$$\sum M_C = 0 \;\; D_x\left(\frac{h}{2}\right) = (B_y - D_y)\left(\frac{h}{2\sqrt{3}}\right) \;\; \text{so} \;\; D_x = \frac{P}{\sqrt{3}} = 0.577P \,[\rightarrow]$$

$$\sum F_x = 0 \;\; C_x = D_x = 0.577P \,[\leftarrow]$$

(4-23)

$$\sum F_y = 0 \;\; C_y = B_y + D_y = 1.866P \,[\downarrow]$$

Finally, returning to **element DEG**, force component E_x is

$$\sum F_x = 0 \;\; D_x + E_x = 0 \;\; \text{so} \;\; E_x = -D_x = 0.577P \,[\rightarrow]$$

(4-24)

All reaction and pin-connection force components are shown in Fig. 4-48 in their correct directions. Equilibrium of forces in the x and y directions is confirmed on each element free-body diagram.

Internal Forces at Point J   Now that the reaction and pin force components are known, next find internal forces (N, V) and internal moment (M) at point J on element DEG. Cutting through element DEG midway between joints D and E exposes these internal forces (N_J, V_J) and internal moment (M_J) as shown in their assumed directions in Fig. 4-49. Note that force components and the moment at J are equal in magnitude but opposite in direction on either side of the cut through member DEG. Using a left-hand free-body diagram (i.e., left of the section cut through DEG), find the following results.

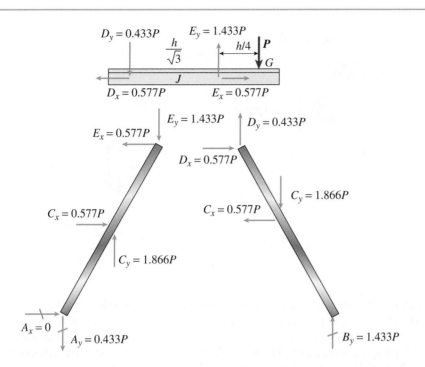

FIGURE 4-48

Final reaction and pin connection
force components

$$\sum F_x = 0 \quad -0.577P + N_J = 0 \quad \text{so} \quad N_J = 0.577P \, (\text{T})$$

$$\sum F_y = 0 \quad -0.433P - V_J = 0 \quad \text{so} \quad V_J = -0.433P \, [\uparrow]$$ **(4-25)**

$$\sum M_{\text{at pt. } J} = 0 \quad M_J + 0.433P\left(\frac{h}{2\sqrt{3}}\right) = 0 \quad \text{so} \quad M_J = -0.125Ph \, [\text{CW}]$$

Of course, the same expressions for N_J, V_J, and M_J in Eq. (4-25) can be
obtained using the free-body diagram of member segment JG instead of DJ.
The negative signs for V_J and M_J indicate that the assumed directions shown
in Fig. 4-49b are incorrect and are in fact opposite to that shown.

Internal Forces at Point H   Next find the internal forces and moment
at point H that is midway between joints B and C (see Fig. 4-45b). Using the
lower free-body diagram BH (Fig. 4-50b), enforce equilibrium to find

$$\sum F_y = 0 \quad F_H = -B_y = -1.433P \, [\downarrow]$$

so $\qquad N_H = F_H \sin 60° = -1.241P \, (\text{C})$

and $\qquad V_H = F_H \cos 60° = -0.717P$ **(4-26)**

$$\sum M_{\text{at pt. } H} = 0 \quad M_H = -B_y\left(\frac{\sqrt{3}h}{12}\right) = -0.207Ph \, [\text{CW}]$$

All three internal force quantities in Eq. (4-26) are negative, so the directions of
N_H, V_H, and M_H are opposite to those shown in Fig. 4-50b. Axial force N_H for
example is negative, so all of member BH is in compression as expected.

Stability and Determinacy of Frames   In examining the truss structures, unknown forces were compared to available equilibrium equations and

(a)

(b)

FIGURE 4-50

(a) Internal forces and moment at point *H*; (b) lower free-body diagram *BH*

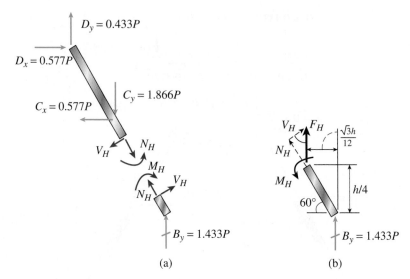

(a)   (b)

cases were listed where an equilibrium solution is possible for stable trusses. These cases are dependent upon the relationship between unknown member and reaction forces ($m + r$) and available equilibrium equations ($2j$ for a plane truss, $3j$ for a space truss). For 2D frames, there are $3m$ internal force unknowns (N, V, and M) and r reaction force unknowns. There are $3j$ available equilibrium equations ($\Sigma F_x = 0$, $\Sigma F_y = 0$, and $\Sigma M_z = 0$) plus c additional moment equations ($\Sigma M_{pin} = 0$) at each of the pin connections (c = number of pinned connections in the frame) once the frame members are disassembled by removing pin attachments. In summary the available combinations for stable 2D frames are

- $3m + r < 3j + c$ more equations than unknowns; no solution
- $3m + r = 3j + c$ statically determinate
- $3m + r > 3j + c$ statically indeterminate (either *externally* or *internally*); $r > 3$ means *externally* indeterminate
- $3m + 3 > 3j + c$ *internally* indeterminate

For the camping table frame analyzed previously, $m = 6$ (AC, BC, CD, CE, DE, and EG), $r = 3$ (A_x, A_y, and B_y), $j = 6$ (joints A, B, C, D, E, and G), and $c = 3$ (pins at C, D, and E). Note that points J and H are not joints but rather locations on the frame where internal forces are of interest. As a result, $3m + r = 3j + c = 21$, and the frame is statically determinate (both externally and internally). For space frames, $3m$ and $3j$ become $6m$ and $6j$, respectively, and pinned connections can become ball-and-socket joints.

The solution approaches presented are now applied to several frame and machine structures in the following examples.

Example 4-7

The plane frame in Fig. 4-51 will be part of a new machine in a manufacturing facility. A cable attached at *C* runs over a pulley at *F* and carries a 500-lb downward load. Two alternate designs also are being considered (Figs. 4-52 and 4-53). In Design 2, support *A* is now a pin, and segment *BCDE* is pin-connected at *B*. In Design 3, support *A* is a roller, *BCDE* is pin-connected at *B*, and brace *HJK* is pin-connected at *H* and *K*. For

FIGURE 4-51

Frame Design 1

FIGURE 4-52

Frame Design 2

FIGURE 4-53

Frame Design 3

each design, find the reaction forces at *A* and *E* and all force components at pin attachments between members. Neglect the weights of all members.

Solution:

Use the four-step problem-solving approach to find the solution.

1. *Conceptualize*: Begin by sketching overall free-body diagrams that display the external loads and external support reactions only, ignoring any details of attachments between members or segments of the frames (such as the pins at *B*, *H*, and *K* in Design 3). It is apparent from the three free-body diagrams (Figs. 4-54a, b, and c) that Designs 1 and 3 will have the same support reactions at joints *A* and *E*. Design 2, however, has an additional reaction component A_x at *A* (Fig. 4-54b) and will require removal of the pin at *B* to create two separate free-body diagrams: one of lower segment *ABGF* and the other of upper segment *BCDE*.

FIGURE 4-54

(a, b, c) Overall free-body diagrams of Designs 1–3

2. *Categorize*: Frame Designs 1 and 3 are *externally* statically determinate. For Design 1, there are 18 *unknowns* and 18 *available equations* of equilibrium [$3m + r = 15 + 3 = 18$ and $3j = 18$ where m = number of members, r = number of restraints, and j = number of joints]. For Design 3, $3m + r = 27 + 3 = 30$ and $3j + c = 27 + 3 = 30$ where c = number of pin connections. So you can find the three reaction components for both designs without removing any pin connections between members. However, although Design 2 is statically determinate overall [i.e., $3m + r = 15 + 4 = 19$ and $3j + c = 18 + 1 = 19$], it is *externally* statically *indeterminate*. There are four reaction components but only three equilibrium equations, so you must remove the pin at B to find another equilibrium equation. Here, the moment between frame segments *ABGF* and *BCDE* is zero, so use $\sum M_B = 0$ for the upper and lower free-body diagrams to obtain one more equation.

3. *Analyze*: The analyses for the reaction forces at A and E and all force components at pin attachments between members are presented next. Each set of analyses is based on the equilibrium of the segment shown in free-body diagrams.

External reactions—Designs 1 and 3: Using the free-body diagram for either Design 1 or 3 (Fig. 4-54a or c), find external reaction forces at A and C as

$$\Sigma F_x = 0 \quad E_x = 0$$

$$\Sigma M_A = 0 \quad E_y(1.0 \text{ ft}) - 500 \text{ lb}(3.5 \text{ ft}) = 0 \text{ so } E_y = 1750 \text{ lb}$$

$$\Sigma F_y = 0 \quad A_y + E_y - 500 \text{ lb} = 0 \text{ so } A_y = -1750 \text{ lb} + 500 \text{ lb} = -1250 \text{ lb}$$

Reaction component A_y is negative, so it is directed downward at joint A in accordance with the statics sign convention.

Pin-connection force components—Design 3: Using the reaction force components, now dissect the Design 3 frame to find forces at pin connections at B, F, H, and K (Fig. 4-55). Removing the four pins and separating the resulting structure segments gives the four free-body diagrams in Fig. 4-55. Note that pin-connection forces at B, H, and K are in opposite directions on adjoining free-body diagrams in accordance with Newton's third law (equal in magnitude, opposite in direction, and collinear). Using the reaction components at A and E and assuming a statics sign convention, solve for pin-connection forces as

FBD Pulley: $\Sigma F_x = 0 \quad F_x = 500 \text{ lb and } \Sigma F_y = 0 \quad F_y = 500 \text{ lb}$

FBD $ABKF$: $\Sigma M_B = 0 \quad K_y(1.5) = A_y(0.5) - 500\,(0.7) + 500(2.5) \text{ or } K_y = 183.3 \text{ lb}$

$\qquad\qquad\quad \Sigma F_y = 0 \quad B_y = -A_y - K_y + 500 \quad \text{or } B_y = 1567 \text{ lb}$

FBD BHE: $\Sigma F_y = 0 \quad H_y = E_y - B_y = 183.3 \text{ lb}$

FBD HK: $\Sigma M_K = 0 \quad H_x(1.0) + H_y(1.5) = 0 \text{ so } H_x = -275 \text{ lb}$

$\qquad\qquad \Sigma F_x = 0 \quad K_x = H_x = -275 \text{ lb}$

FBD $ABKF$: $\Sigma F_x = 0 \quad B_x = -K_x + 500\,(0.7) \text{ so } B_x = 775 \text{ lb}$

where forces are in pounds and distances are in feet.

FIGURE 4-55

Separate free-body diagrams for elements of frame Design 3

External reactions and pin-connection forces—Design 2: This frame is statically indeterminate externally so first disassemble it to create three free-body diagrams (Fig. 4-56) by removing the pins at B and F. The pulley force components at F are the same as those discussed earlier, but now use $\Sigma M_B = 0$ for free-body diagrams BE and ABF to find one additional relationship among the unknown forces. The equilibrium analysis to find internal connection forces at pins proceeds as follows.

FIGURE 4-56

Separate free-body diagrams for Frame Design 2

FBD ABF:	$\Sigma M_B = 0$	$A_y(0.5) = 500(0.7) - 500\,(2.5)$ so $A_y = -1800$ lb
	$\Sigma F_y = 0$	$B_y = -A_y + 500$ so $B_y = 2300$ lb
FBD BE:	$\Sigma F_y = 0$	$E_y = B_y = 2300$ lb
	$\Sigma M_B = 0$	$E_x(2) + 500(1.2) - E_y(0.5) = 0$ so $E_x = 275$ lb
	$\Sigma F_x = 0$	$B_x = 500 + E_x = 775$ lb
FBD ABF:	$\Sigma F_x = 0$	$A_x = -B_x + 500 = -275$ lb

where forces are in pounds and distances are in feet.

4. *Finalize*: Carefully drawn free-body diagrams are an important first step in a correct equilibrium analysis of each of the frames. If the structure is externally statically determinate, begin with an overall free-body diagram to find external reaction forces and moments, and then proceed to separate free-body diagrams for structure segments if connection forces between individual segments or internal member forces are of interest. If there are more unknown reactions than equations of equilibrium, you must dissect the structure into its component parts (as in Design 2) to find additional equilibrium equations leading to a solution.

As a check on the reaction components obtained using the separate free-body diagrams in Fig. 4-56, note that equilibrium of forces and moments for the overall free-body diagram of the frame Design 2 in Fig. 4-54b is satisfied.

Example 4-8

Pole *ABCD* is supported laterally by struts *CE* and *BF* and by a pre-tensioned cable from *C* to a ground anchorage point at *G* (Fig. 4-57). Supports at *A*, *E*, and *F* and the strut-to-pole connections at *B* and *C* are all ball-and-socket joints. The cable pre-tension force T_{CG} is 10 kN. A force of 7 kN is applied in the (−*x*) direction at mid-span of strut *FB*, and a force of 5 kN is applied to the top of pole at *D* in the (−*z*) direction. Find the reaction force components at supports *A*, *E*, and *F*, and find all force components at pin attachments at *B* and *C*. Neglect the weights of all members.

FIGURE 4-57

Pole supported by struts and cable

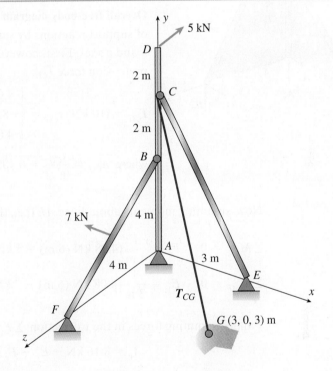

Solution:

Use the four-step problem-solving approach to find the equilibrium solution.

1. *Conceptualize*: The overall free-body diagram of the pole structure is shown in Fig. 4-58. Only applied forces (including the cable pre-tension force T_{CG}) and the reaction force components at *A*, *E*, and *F* are shown; pin-connection forces at *B* and *C* are internal forces and will not be displayed until the structure is broken into separate free-body diagrams. Either a vector algebra approach or scalar equations employing the *x*-*y*-*z* force components can be used for this 3D structure.

FIGURE 4-58

Overall free-body diagram of pole structure

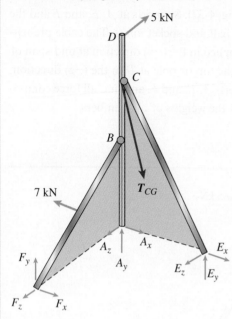

2. *Categorize*: Strut members *EC* and *FB* lie in global coordinate planes *x-y* and *y-z*, respectively, so distances between reaction force components at *E* and *F* and pin-connection force components at *B* and *C* will be easy to obtain. This suggests that scalar moment equations can be written, thereby eliminating the need for vector algebra for the most part.

3. *Analyze*: There are a variety of different paths to a solution for reaction forces at *A*, *E*, and *F* and pin-connection forces at *B* and *C*. One such step-by-step solution approach is presented here.

Overall free-body diagram (Fig. 4-58): Find *y* components of support reactions by summing moments about the *x* and *z* axes. First, however, find the components of cable pretension force T_{CG}:

$$T_{CG} = (10 \text{ kN})n_{CG} = \begin{Bmatrix} 4.08 \\ -8.16 \\ 4.08 \end{Bmatrix} \text{kN}$$

where $n_{CG} = \dfrac{r_{CG}}{|r_{CG}|} = 0.408i - 0.816j + 0.408k$.

Now, summing moments about lines *AE* (i.e., the *x* axis) and *AF* (*z* axis) gives

$$\sum M_{AE} = 0 \quad F_y = \frac{1}{4 \text{ m}}(4.08 \text{ kN} (6 \text{ m}) - 5 \text{ kN} (8 \text{ m})) = -3.88 \text{ kN}$$

$$\sum M_{AF} = 0 \quad E_y = \frac{1}{3 \text{ m}}(4.08 \text{ kN} (6 \text{ m}) - 7 \text{ kN} (2 \text{ m})) = 3.49 \text{ kN}$$

Finally, summing forces in the *y* direction: $\sum F_y = 0$, so

$$A_y = 8.16 \text{ kN} - E_y - F_y = 8.55 \text{ kN}$$

(A more complicated alternative requiring vector algebra is to sum moments about diagonal line *EF* to obtain force A_y.)

Next, disassemble the structure, removing the pins at *B* and *C*, to evaluate the equilibrium of struts *CE* and *BF*. Write moment equations based on the free-body diagram for each strut (Fig. 4-59) and note that moments are zero at each end of each strut. Pin-connection force components at *B* and *C* are shown in opposite directions on pole and strut, which is consistent with the action–reaction principle at these ball-and-socket joints.

Free-body diagram of strut *CE* (Fig. 4-59): Reaction force component E_y was computed previously, so now find the three pin-connection force components at *C* and components E_x and E_z by enforcing equilibrium on strut *CE*.

FIGURE 4-59

Disassembled pole structure: free-body diagrams of pole and struts

$$\sum M_{Ey} = 0 \;\; (\text{or} \; \sum M_{Ex} = 0) \quad C_z = 0$$

$$\sum F_z = 0 \qquad\qquad\qquad\qquad E_z = 0$$

$$\sum M_{Ez} = 0 \qquad\qquad\qquad\qquad C_x = \frac{-3\text{ m}}{6\text{ m}}\, C_y = 1.747\text{ kN}$$

$$\sum F_x = 0 \qquad\qquad\qquad\qquad E_x = -C_x = -1.747\text{ kN}$$

In these equations, $\sum M_{Ey}$ refers to moments taken about a line through joint E, which is parallel to the y axis (and so on for $\sum M_{Ez}$).

Free-body diagram of strut BF (Fig. 4-59): Equilibrium of strut BF yields five more force components at B and F:

$$\sum M_{Fz} = 0 \;\; (\text{or} \sum M_{Fy} = 0) \quad B_x = \frac{1}{4\text{ m}}(7\text{ kN}(2\text{ m})) = 3.5\text{ kN}$$

$$\sum F_x = 0 \qquad\qquad\qquad\qquad F_x = 7\text{ kN} - B_x = 3.5\text{ kN}$$

$$\sum F_y = 0 \qquad\qquad\qquad\qquad B_y = -F_y = 3.88\text{ kN}$$

$$\sum M_{Fx} = 0 \qquad\qquad\qquad\qquad B_z = \frac{1}{4\text{ m}}(-B_y(4\text{ m})) = -3.88\text{ kN}$$

$$\sum F_z = 0 \qquad\qquad\qquad\qquad F_z = -B_z = 3.88\text{ kN}$$

Overall free-body diagram (Fig. 4-58): Last, use either the pole free-body diagram in Fig. 4-59 or the overall free-body diagram of the entire structure in

Fig. 4-58 to find remaining reaction force unknowns at A. Using Fig. 4-58, the components A_x and A_z are

$$\Sigma F_x = 0 \quad A_x = 7 \text{ kN} - E_x - F_x - 4.08 \text{ kN} = 1.167 \text{ kN}$$
$$\Sigma F_z = 0 \quad A_z = 5 \text{ kN} - 4.08 \text{ kN} - E_z - F_z = -2.96 \text{ kN}$$

A statics sign convention is used here, so a negative force value means that the direction is opposite to that assumed and shown in Figs. 4-58 and 4-59.

4. *Finalize*: The free-body diagrams in Figs. 4-58 and 4-59 were essential to carrying out a correct equilibrium analysis of the overall structure and each of the supporting struts. As a final check on results, use the free-body diagram of pole *ABCD* in Fig. 4-59 to check that equilibrium is satisfied for the pole itself. It is important to use the force component vectors shown in their assumed directions in Fig. 4-59 in these equilibrium equations, using numerical values along with the associated sign (plus or minus) listed. The following equations show that the pole itself is in equilibrium as required:

$$\Sigma F_x = 0 \quad A_x - B_x - C_x + 4.08 \text{ kN} = (1.167 - 3.5 - 1.747 + 4.08) \text{ kN} = 0$$
$$\Sigma F_y = 0 \quad A_y - B_y - C_y - 8.16 \text{ kN} = (8.55 - 3.88 + 3.49 - 8.16) \text{ kN} = 0$$
$$\Sigma F_z = 0 \quad A_z - B_z - C_z + 4.08 \text{ kN} - 5 \text{ kN} = (-2.96 + 3.88 + 4.08 - 5) \text{ kN} = 0$$
$$\Sigma M_x = 0 \quad -B_z(4 \text{ m}) - C_z(6 \text{ m}) + 4.08 \text{ kN} (6 \text{ m}) - 5 \text{ kN} (8 \text{ m})$$
$$= [3.88(4) + 4.08(6) - 5(8)] \text{ kN} \cdot \text{m} = 0$$
$$\Sigma M_z = 0 \quad B_x(4 \text{ m}) + C_x(6 \text{ m}) - 4.08 \text{ kN} (6 \text{ m})$$
$$= [3.5(4) + 1.747(6) - 4.08(6)] \text{ kN} \cdot \text{m} = 0$$
$$\Sigma M_y = 0 \quad \text{by inspection}$$

With all reaction and pin-connection forces now known, you can find internal forces and couples (N, V, M) at selected locations on members of interest. This can be a formidable task for 3D frames, however, because there are three internal force components (N_x, V_y, V_z) and three internal couple components (M_x, M_y, M_z) to be found at each section cut.

Example 4-9

In the position shown in Fig. 4-60, a mobile crane supports a load P of 15 kips. Crane boom *CDE* weighs 2.5 kips and is pinned to continuous member *ACB*. The crane boom position is controlled by two hydraulic rams (*BD*). Find the force in each ram and also the force components in the pin at *C*. Use the following dimensions: $L_1 = 10$ ft, $L_2 = 7$ ft, $L_3 = 3.5$ ft, $H_1 = H_2 = 7$ ft, $H_3 = 14$ ft, and $H_4 = 3.5$ ft.

FIGURE 4-60

(a) Mobile crane;
(b) plane frame model

(a)   (b)

Solution:

Use the four-step problem-solving approach to analyze the crane boom.

1. *Conceptualize*: First sketch the crane segment *A–E* and label dimensions and
    required angles (Fig. 4-60b). Next remove the pins at *C* and *D*, detach boom
    CDE from base frame *ACB*, and isolate the crane boom for analysis. Begin by
    sketching a free-body diagram showing all forces on *CDE* that must be included
    in the analysis: crane boom weight W_{CE}, crane load P, pin force components C_x
    and C_y, and hydraulic ram force F_{BD} are shown in their assumed positive direc-
    tions in the free-body diagram in Fig. 4-61.

FIGURE 4-61

Free-body diagram of crane
boom *CDE*

2. *Categorize*: Summation of forces and moments leads to the unknown forces of interest. First, find angles α_{CE} and α_{BD} and dimension CDE_x in Fig. 4-61.

3. *Analyze*: The angles α_{CE} and α_{BD} and the projection of CDE along the x axis are

$$\alpha_{CE} = \tan^{-1}\left[\frac{H_2}{L_2 - L_3}\right] = \tan^{-1}\left[\frac{7 \text{ ft}}{7 \text{ ft} - 3.5 \text{ ft}}\right] = \tan^{-1}(2) = 63.44°$$

$$\alpha_{BD} = \tan^{-1}\left[\frac{H_1 + H_2}{L_3}\right] = \tan^{-1}\left[\frac{14 \text{ ft}}{3.5 \text{ ft}}\right] = \tan^{-1}(4) = 75.96°$$

$$CDE_x = \frac{H_2 + H_3}{\tan(\alpha_{CE})} = \frac{(7 + 14)\text{ ft}}{2} = 10.5 \text{ ft}$$

Force in hydraulic jack *BD*: Summing moments about *C* gives hydraulic ram force magnitude F_{BD}:

$$\Sigma M_C = 0 \ \ [F_{BD}\sin(\alpha_{BD})](L_2 - L_3) + [F_{BD}\cos(\alpha_{BD})](H_2) - (P)(CDE_x) - W_{CE}\left[\frac{CDE_x}{2}\right] = 0$$

$$F_{BD} = \frac{(P)(CDE_x) + W_{CE}\left[\dfrac{CDE_x}{2}\right]}{(L_2 - L_3)\sin(\alpha_{BD}) + H_2\cos(\alpha_{BD})} = \frac{(15 \text{ kips})(10.5 \text{ ft}) + (2.5 \text{ kips})(5.25 \text{ ft})}{(3.5 \text{ ft})(0.97) + (7 \text{ ft})(0.243)}$$
$$= 33.5 \text{ kips} \quad \leftarrow$$

Force F_{DB} is the combined force in both rams and is positive, so it is directed upward and to the left at joint *D* as assumed in Fig. 4-61 (in accordance with the statics sign convention).

Pin *C* connection force components: With hydraulic ram force F_{BD} now known, summation of forces in x and y directions for the free-body diagram in Fig. 4-61 gives pin force components C_x and C_y:

$$\Sigma F_x = 0 \ \ C_x = F_{BD}\cos(\alpha_{BD}) = (33.5 \text{ kips})(0.243) = 8.125 \text{ kips}$$
$$\Sigma F_y = 0 \ \ C_y = P + W_{CE} - F_{BD}\sin(\alpha_{BD}) = 15 \text{ kips} + 2.5 \text{ kips} - (33.5 \text{ kips})(0.97)$$
$$= -15.0 \text{ kips} \quad \leftarrow$$

The minus sign for component C_y means that force C_y is actually directed downward in Fig. 4-61. The resultant force acting on the pin at *C* is

$$C_{res} = \sqrt{C_x^2 + C_y^2} = \sqrt{(8.125)^2 + (-15.0)^2} \text{ kips} = 17.06 \text{ kips}$$

4. *Finalize*: Summing moments about another point such as point *D* on the free-body diagram of *CDE* (Fig. 4-61) provides a quick check on the solution for pin force components C_x and C_y:

$$\Sigma M_D = 0 \ \ C_x H_2 - C_y(L_2 - L_3) - (P)[CDE_x - (L_2 - L_3)] - W_{CE}\left[\frac{CDE_x}{2} - (L_2 - L_3)\right] = 0$$

Inserting numerical values from step 3 shows that this equation is satisfied. Next, use the force F_{BD} in the hydraulic rams to compute the pressure p_{BD} against the piston inside each of the two hydraulic cylinders to make sure that it is within acceptable limits. For example, if each piston has a diameter d of 5 in., pressure p_{BD} is

$$p_{BD} = \frac{F_{BD}/2}{A_p} = \frac{F_{BD}/2}{\pi d^2/4} = \frac{16.75 \text{ kips}}{19.64 \text{ in}^2} = 853 \text{ psi}$$

Further analysis of sub-frame ACB is not possible at this time because it is *statically indeterminate*. There are four unknown reaction components (two each at supports A and B) but only three equations of statics.

Example 4-10

An excavator lifts a load of soil and rock to place it in a dump truck (see Fig. 4-62a). The weight of the main boom is $W_1 = 8$ kN; the dipper weight is $W_2 = 5$ kN; and the bucket and soil weight is $W_3 = 12$ kN. Each weight is applied at the corresponding component center of gravity (see Fig. 4-62b). Find:

(a) The force in each of the two hydraulic cylinders BC (Fig. 4-62a).

(b) The pressure in cylinder DE if the piston diameter is 95 mm.

(c) The pressure in cylinder GH if the piston diameter is 70 mm.

FIGURE 4-62

(a) Excavator; (b) free-body diagram 1 of boom, dipper, and bucket

(a)   (b)

Solution:

Use the four-step problem-solving approach to analyze the excavator.

1. *Conceptualize*:

   Free-body diagram 1—Remove the pins at A and B to detach the boom-dipper-bucket from the tractor and isolate it for analysis. The resulting free-body diagram is shown in Fig. 4-62b and includes pin force components A_x and A_y at A as well as the two hydraulic cylinder forces F_{BC}. The hydraulic cylinders are two force members, so forces F_{BC} are directed along the axis of each ram. The weight of each component is positioned at the center of gravity of that component.

   Free-body diagram 2—To find the force in hydraulic cylinder DE (F_{DE}), detach the dipper-bucket from the boom. Remove the pins at E and F to reveal force F_{DE} and pin-force components F_x and F_y at F (Fig. 4-63a), all acting on the dipper-bucket assemblage. Dipper weight W_2 and bucket-soil weight W_3 must also be included.

   Free-body diagram 3—Last, remove pins K and L to detach the bucket from the dipper (Fig. 4-63b). This exposes link force F_{KH} and pin-force components L_x and L_y. Once link force F_{KH} is known, force equilibrium at joint H will lead to hydraulic cylinder force F_{GH}.

FIGURE 4-63

(a) Free-body diagram 2 of dipper and bucket; (b) free-body diagram 3 of dipper with force from link KH

(a)   (b)

2. *Categorize*: *The key initial step in the solution is creation of the free-body diagrams.* Summation of forces and moments in each will lead to the unknown forces in hydraulic cylinders *BC*, *DE*, and *GH*. All forces acting on key components must be included in the free-body diagrams in Figs. 4-62 and 4-63.

3. *Analyze*:

Force in hydraulic cylinder BC: Sum moments about joint *A* in FBD 1 to find the magnitude of hydraulic cylinder force F_{BC} in each ram:

$$\Sigma M_A = 0 \quad 2F_{BC}\left[\sin(77.1°)(0.3\text{ m}) + \cos(77.1°)(0.4\text{ m})\right]$$

$$= W_1(1.4\text{ m}) + W_2(3.1\text{ m}) + W_3(2.6\text{ m})$$

$$F_{BC} = \frac{1}{2}\left[\frac{8\text{ kN }(1.4\text{ m}) + 5\text{ kN }(3.1\text{ m}) + 12\text{ kN }(2.6\text{ m})}{\sin(77.1°)(0.3\text{ m}) + \cos(77.1°)(0.4\text{ m})}\right] = 75.8\text{ kN}$$

Pressure in hydraulic cylinder DE: Now use summation of moments about joint *F* in FBD 2 to find F_{DE}:

$$\Sigma M_F = 0 \quad F_{DE}\left[\cos(24.5°)(0.6\text{ m}) + \sin(24.5°)(0.4\text{ m})\right] = -W_2(0.4\text{ m}) + W_3(0.4\text{ m})$$

$$F_{DE} = \left[\frac{-5\text{ kN }(0.4\text{ m}) + 12\text{ kN }(0.4\text{ m})}{\cos(24.5°)(0.6\text{ m}) - \sin(24.5°)(0.4\text{ m})}\right] = 7.37\text{ kN}$$

The hydraulic pressure in cylinder *DE* is $p_{DE} = \dfrac{F_{DE}}{A_{DE}}$ where

$$A_{DE} = \frac{\pi}{4}(95\text{ mm})^2 = 7.008(10^{-3})\text{ m}^2$$

So cylinder pressure is

$$p_{DE} = \frac{F_{DE}}{A_{DE}} = \frac{7.37\text{ kN}}{7.008(10^{-3})\text{ m}^2} = 1.04\text{ MPa}$$

Summation of forces in FBD 2 gives pin *F* force components F_x and F_y, which are needed in FBD 3:

$$\Sigma F_x = 0 \quad F_x = -F_{DE}\cos(24.5°) = -6.703\text{ kN}$$

$$\Sigma F_y = 0 \quad F_y = -F_{DE}\sin(24.5°) + W_2 + W_3 = 13.945\text{ kN}$$

Pressure in hydraulic cylinder GH: First find link force F_{KH} by summing moments about joint *L* in FBD 3 [Fig. 4-63b]:

$$\Sigma M_L = 0 \quad F_{KH} = \frac{F_{DE}\cos(24.5°)(3.5\text{ m}) + F_x(2.9\text{ m}) + F_y(0.4\text{ m})}{\cos(36°)(0.7\text{ m}) - \sin(36°)(0.1\text{ m})} = 18.915\text{ kN}$$

Next remove pins *G*, *J*, and *K* to isolate two-force link members *KH*, *JH*, and cylinder *GH* (see Fig. 4-64). Summation of forces in *x* and *y* directions at joint *H* (Fig. 4-64b) results in two equations:

$$\Sigma F_x = 0 \quad F_{JH}\cos(23°) + F_{GH}\sin(9.9°) = F_{KH}\sin(36°)$$

$$\Sigma F_y = 0 \quad F_{JH}\sin(23°) + F_{GH}\cos(9.9°) = -F_{KH}\cos(36°)$$

FIGURE 4-64

(a) Detail at bucket;
(b) forces on pin at *H*

(a)   (b)

Solving these equations to find link force F_{JH} and cylinder force F_{KH} gives $F_{JH} = 16.2$ kN and $F_{GH} = -22.0$ kN. Cylinder *GH* is in compression, and the pressure in 70 mm-diameter ram *GH* is

$$p_{GH} = \frac{|F_{GH}|}{A_{GH}} = \frac{22.0 \text{ kN}}{3.848(10^{-3}) \text{ m}^2} = 5.7 \text{ MPa}$$

4. *Finalize*: Solutions for the forces and pressures in the hydraulic cylinders of this excavator could not have been accomplished without properly drawn free-body diagrams.

   Force $\boldsymbol{F}_{BC}$ acts on cylinder *BC* as shown in Fig. 4-62b, so cylinder *BC* is in compression. Force $\boldsymbol{F}_{DE}$ acts on pin *E* as shown in Fig. 4-63, so cylinder *DE* is also in compression. Last, the solution for cylinder force F_{GH} leads to a negative result, so $\boldsymbol{F}_{GH}$ acts opposite to the direction shown in Fig. 4-64b and is in compression for the loading considered in this problem.

In Chapter 4, you applied **principles of equilibrium** to analyze **plane** and **space trusses** and **frames** and **machines**. All structures considered for analysis must be both **stable** and **statically determinate**. A carefully drawn **free-body diagram** of the structure under study is an important first step in carrying out a solution for support reactions and member forces in trusses and reactions, pin-connection forces, and internal forces and moments in frames. Newton's third law regarding **action–reaction forces** must be strictly followed so that pin-connection forces in frames, for example, are shown as collinear and equal but opposite forces on adjacent free-body diagrams of disassembled members of a frame.

Here are the major concepts covered in this chapter.

1. The basic unit of construction of a **plane truss** is the **triangle**, while that of the **space truss** is a **tetrahedron**. Truss members are referred to as **two-force members** in that they only carry axial forces that are constant over the length of the member and are either tensile or compressive. **Pinned connections** are used in plane trusses, and **ball-and-socket connections** are used in space trusses, which is consistent with the assumption that no moment is transferred between interconnected members in the model of a truss. All loads applied to trusses are assumed to act at the joints only.

2. If m = number of **members**, r = number of support **reactions**, and j = number of **joints** in a truss, a **plane truss** is *statically determinate* if the number of unknown forces $(m + r)$ is equal to the number of available equations of equilibrium $(2j)$. For a **space truss**, the requirement for determinacy is $m + r = 3j$ because there are three force equilibrium equations per joint. In both cases, the truss must first be **stable**; otherwise any discussion of determinacy is meaningless.

3. In the **method of joints** for truss analysis, use free-body diagrams of all joints to formulate a set of simultaneous force equilibrium equations ($\Sigma F_x = 0$, $\Sigma F_y = 0$, $\Sigma F_z = 0$) with the forces in each member as the basic unknowns. Alternatively, you can use the **method of sections** to find forces in selective members. Here, a section cut is made through the entire structure and passes through the members of interest to expose their axial forces in the resulting free-body diagram. Usually a moment equilibrium equation is then written about some joint in order to eliminate as many unknowns as possible and isolate the member force of interest.

4. In many cases, **scalar equations** can be used in the analysis of plane trusses. For space trusses, **vector operations** are often required along with a combined solution approach, involving both the methods of joints and sections.

5. Careful inspection of a truss before rushing into a solution may reveal that some members are zero-force members that carry no axial force for the loading under consideration. **Zero-force members** often occur at joints where two collinear members are pin-connected to a member from another direction and there is no applied load at that joint. Also, if **cables** are used in cross bracing, one of the cable braces may be in compression and can be treated as a zero-force member in the analysis for that loading.

6. **Frame** and **machine** structures are composed of **multi-force members** that, unlike trusses, can carry both joint and member loads and can transfer both forces and moments to adjacent members (unless a pin attachment between selected members is used). A frame or machine structure must be disassembled to break it into separate free-body diagrams thereby exposing the **action–reaction forces** acting on pins where two or more **multi-force** members are joined.

7. Multi-force members in frames and machines carry applied joint and member loads and develop *internal* axial (N) and transverse shear (V) forces as well as internal moments (M) as they resist these loads. The internal forces and moments may vary along the length of the member. The member under consideration must be cut to expose the **internal forces and moment** (N, V, and M).

8. In a frame, if m = number of **members**, r = number of support **reactions**, j = number of **joints**, and c = number of **pinned connections**, a **plane frame** is *statically determinate* if the number of unknown forces ($3m + r$) is equal to the number of available equations of equilibrium ($3j + c$). If $r > 3$, the plane frame is *externally indeterminate*, but an equilibrium solution still may be found if additional equations can be obtained by disassembly of the structure to expose zero internal moments at pin attachments. As with trusses, **stability** of the structure is a required precondition to consideration of its static determinacy.

PROBLEMS

4.2 Plane Trusses

Introductory Problems

4.2-1 Find support reactions at joints A and B; then solve for forces in members AB, AC, and BC for the plane truss shown in the figure. Assume that $P = 35$ kips, $L = 10$ ft, $\theta_A = 60°$, and $b = 7.1$ ft.

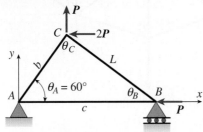

PROBLEM 4.2-1

4.2-2 Find support reactions at E and F; then find all bar forces in the plane truss shown in the figure.

PROBLEM 4.2-2

4.2-3 Find support reactions at F and H; then find all bar forces in the plane truss shown in the figure. Let $P = 10$ kips and $L = 8$ ft.

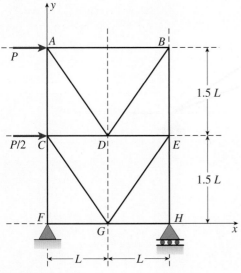

PROBLEM 4.2-3

4.2-4 Find reaction forces at A and D; then find member forces AB, BC, BD, and DC using the method of joints.

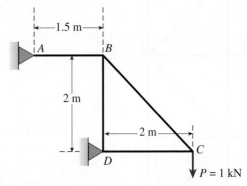

PROBLEM 4.2-4

4.2-5 Find all member forces in the small truss in the figure using the method of joints. Let $P = 300$ lb.

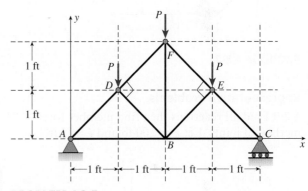

PROBLEM 4.2-5

4.2-6 Find all member forces in the plane truss in the figure using the method of joints.

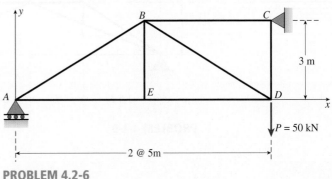

PROBLEM 4.2-6

291

4.2-7 Find support reactions at *A* and *J*; then find member forces *BC*, *FE*, *FG*, and *IJ* using the method of joints.

4.2-8 Find support reactions at *A* and *E*; then find all member forces using the method of joints.

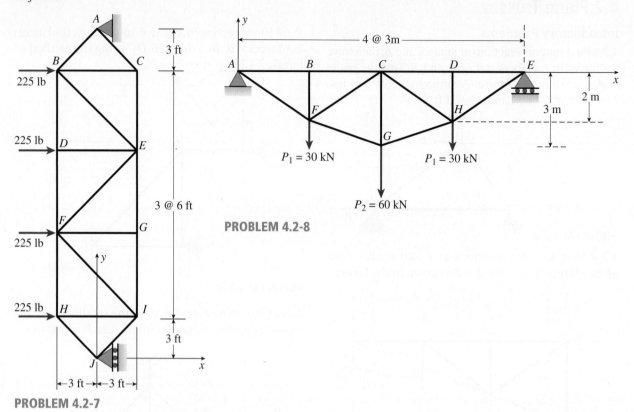

PROBLEM 4.2-8

PROBLEM 4.2-7

Representative Problems

4.2-9 Find all reactions then find member forces *GH* and *CD* in the plane truss shown in the figure. Use the method of joints. Let *P* = 200 lb and *L* = 3 ft.

PROBLEM 4.2-9

4.2-10 Find support reactions at joints 3 and 4; then find member forces in members 6, 10, and 11 of the plane truss in the figure.

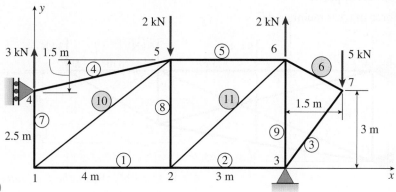

PROBLEM 4.2-10

4.2-11 A plane truss has a pin support at A and a roller support at E.

(a) Find reactions at the supports.

(b) Find the axial force in truss member FE.

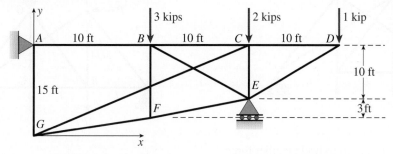

PROBLEM 4.2-11

4.2-12 A plane truss has a pin support at F and a roller support at E.

(a) Find reactions at the supports.

(b) Find the axial force in truss member FE.

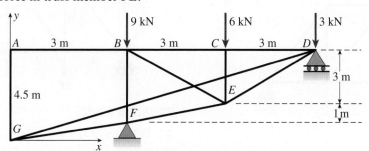

PROBLEM 4.2-12

4.2-13 A plane truss has a pin support at A and a roller support at G.

(a) Find reactions at the supports.

(b) Find the axial force in truss member AE.

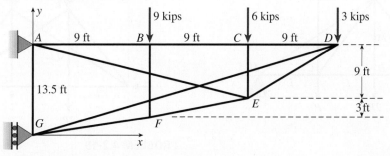

PROBLEM 4.2-13

293

4.2-14 A plane truss has a pin support at A and a roller support at G.

(a) Find reactions at the supports.

(b) Find the axial force in truss member AE.

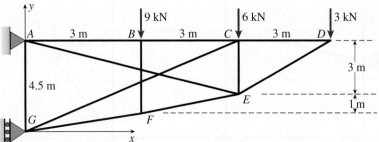

PROBLEM 4.2-14

4.2-15 Use the *method of sections* to find member forces in bars AG and FD of the plane truss shown in the figure.

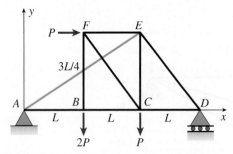

PROBLEM 4.2-15

4.2-16 Use the *method of sections* to find member forces in bars AE and FC of the plane truss shown in the figure.

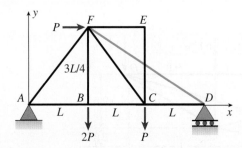

PROBLEM 4.2-16

4.2-17 Use the *method of sections* to find member forces in bars FC and FD of the plane truss shown in the figure.

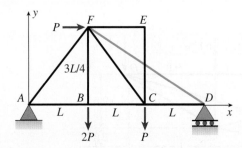

PROBLEM 4.2-17

4.2-18 Use only the *method of joints* to find the force in member GB of the plane truss in Fig. a. Repeat for the plane truss in Fig. b.

(a)

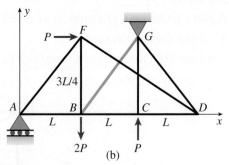

(b)

PROBLEM 4.2-18

4.2-19 Use only the *method of sections* to find the force in member FD of the plane truss in Fig. a. Repeat for member GA in the plane truss in Fig. b.

(a)

PROBLEM 4.2-19

PROBLEM 4.2-19 (*Continued*)

4.2-20 Find reaction forces at *A* and *G* in terms of load variable *P*; then find member forces *BG* and *FG*.

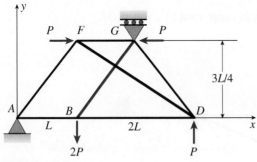

PROBLEM 4.2-20

4.2-21 Find reaction forces at *A* and *G* in terms of load variable *P*; then find member forces *AG* and *FD*.

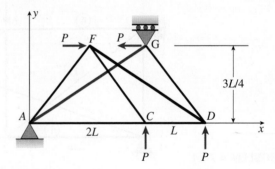

PROBLEM 4.2-21

4.2-22 Find reactions at support joints 3 and 5 for the plane truss shown in the figure. Also, find axial forces in truss members 11 and 13.

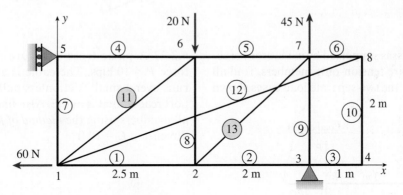

PROBLEM 4.2-22

4.2-23 Find all support reactions for the plane truss shown in the figure. Also, find axial forces in truss members 11 and 13.

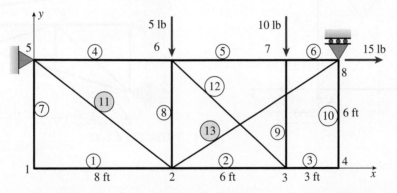

PROBLEM 4.2-23

295

4.2-24 Find all support reactions for the plane truss shown in the figure. Also, find axial forces in truss members 11 and 13.

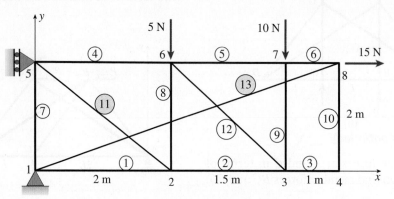

PROBLEM 4.2-24

4.2-25 Find all reactions and member forces for the plane truss supported by cable *CDF*.

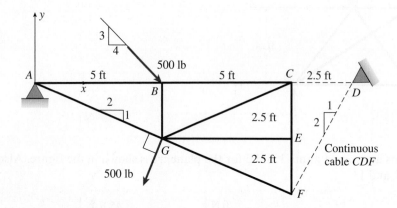

PROBLEM 4.2-25

4.2-26 A plane truss is supported by cable cross bracing. If the cables are tension-only members, find all member forces for the two separate load cases shown in the figure.

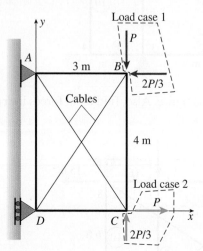

PROBLEM 4.2-26

4.2-27 A plane truss structure is loaded by cable force *P* = 10 kips. The cable is anchored at *H* and runs over a small frictionless pulley at *D*. Find support reactions at *A* and *E*. Also find member forces in all members using the *method of joints*. Assume that dimension *L* = 10 ft.

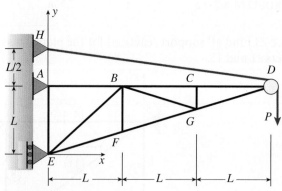

PROBLEM 4.2-27

296

4.2-28 A cable anchored at D runs over a pulley at A. The cable carries a 1.2 kN force and is directed parallel to member AG. Find bar forces GC and CF.

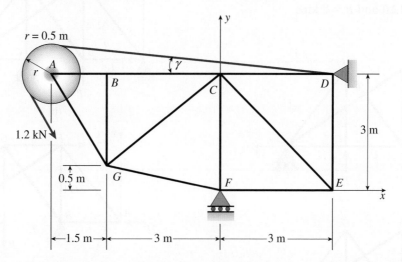

PROBLEM 4.2-28

4.2-29 A cable is anchored at A and runs over pulleys at D and K, each with radius of 6 in. Find forces in members CD, DH, and HI. Assume that $L = 9$ ft and $P = 500$ lb.

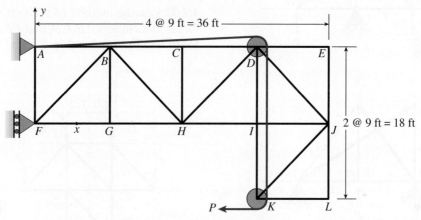

PROBLEM 4.2-29

4.2-30 A crane structure is supported at joints S and T. The crane carries a load $W = 50$ kN by means of a cable that is anchored at S and runs over small pulleys at K, C, A, B, and J (see figure). Find support reactions at S and T and also forces in members FG, NO, and LM.

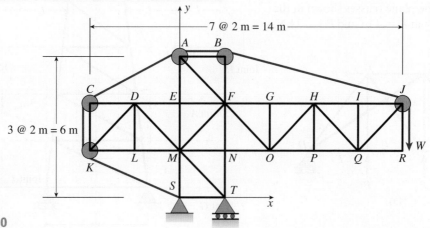

PROBLEM 4.2-30

4.2-31 Find the forces in members *FD* and *GB* for each of the plane trusses shown in the figures. Assume that $L = 12$ ft and $P = 8$ kips.

(a)

(b)

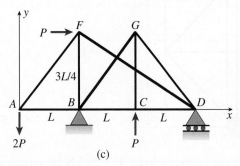

(c)

PROBLEM 4.2-31

4.2-32 Find the forces in members *FD* and *GB* for each of the plane trusses shown in the figures. Assume that $L = 4$ m and $P = 15$ kN.

(a)

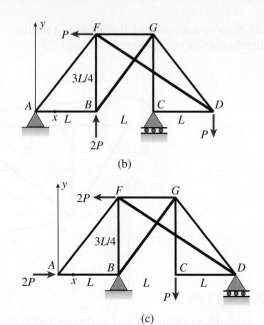

(b)

(c)

PROBLEM 4.2-33

4.2-33 Use the method of sections to find members forces in members *AB*, *CD*, and *EF* of a transmission tower (see figure). Let $L = 9$ ft and $P = 3$ kips.

PROBLEM 4.2-33

4.2-34 Find reaction forces at A and C. Also find the force in member BF using (a) method of joints only; and (b) repeat part (a) using only the method of sections.

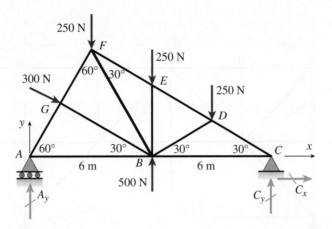

PROBLEM 4.2-34

4.2-35 A cantilever truss has a pin support at A and roller support at B. Find reaction forces at A and B; then find the force in member IJ using the method of joints and the force in EF using the method of sections. Let $P = 10$ kips.

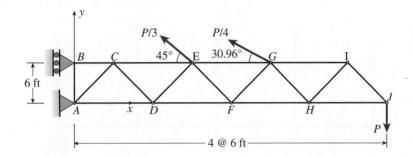

PROBLEM 4.2-35

4.2-36 Point load R is applied on member CE that rests on truss joints C and E (see figure). Find member forces FG, AB, and AG using the method of sections.

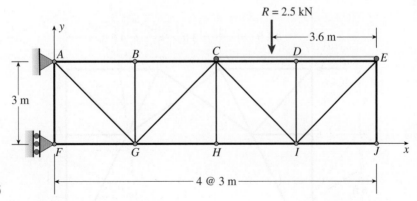

PROBLEM 4.2-36

4.2-37 A truss system is loaded by horizontal loads (*P*) and a vertical load (*V*) as shown in the figure. Find member forces *BC*, *GC*, and *GH* by the method of sections. Let *P* = 3 kips and *V* = 5 kips.

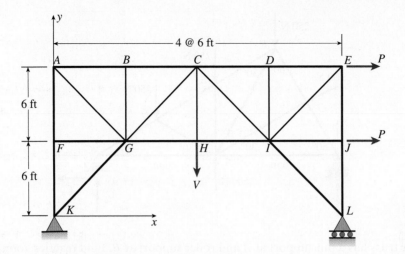

PROBLEM 4.2-37

4.2-38 A truss bridge carries downward loads at *A* through *E* (see figure). If *P* = 7.5 kN, determine the member forces *BC*, *HI*, *IJ*, and *JK* using method of sections.

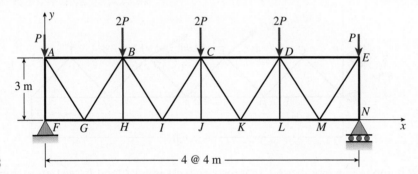

PROBLEM 4.2-38

4.2-39 Use the method of sections to find forces in members *EI*, *BF*, *AF*, and *CF*. Let *P* = 1.25 kips.

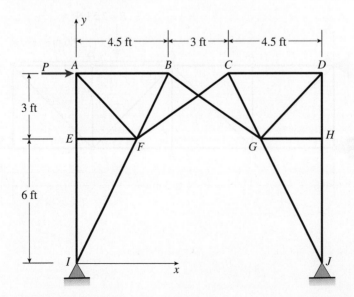

PROBLEM 4.2-39

4.2-40 Use the method of sections to find forces in members BC, HI, and CH of the inclined truss bridge shown in the figure. Let $P = 5$ kN.

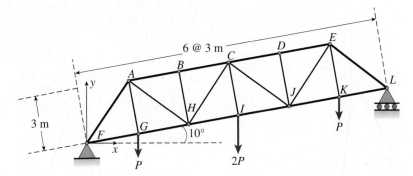

PROBLEM 4.2-40

4.3 Space Trusses

Introductory Problems

4.3-1 For the space truss in the figure, find support reactions at joints A, B, and C, then solve for forces in members AB, AC, and BC. Assume that $P = 35$ kips.

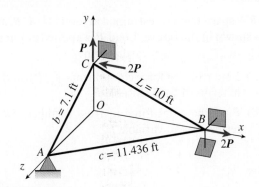

PROBLEM 4.3-1

4.3-2 For the space truss in the figure, find support reactions at joints A, B, and C, then solve for forces in members AB, AC, and BC. Assume that $P = 1.0$ kN.

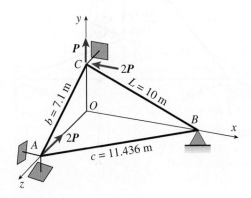

PROBLEM 4.3-2

4.3-3 A space truss has 3D pin supports at joints O, B, and C. Load P is applied at joint A and acts toward point Q. Coordinates of all joints are given in ft (see figure).

(a) Find reaction force components B_x, B_z, and O_z.
(b) Find the axial force in truss member AC.

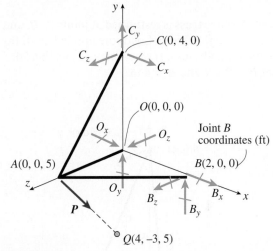

PROBLEM 4.3-3

4.3-4 Find all support reactions and member forces for the space truss shown in the figure. Let $P = 2.5$ kN and $L = 1.5$ m.

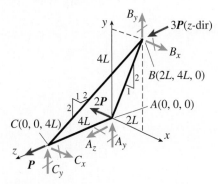

PROBLEM 4.3-4

4.3-5 Find all support reactions and member forces for the space truss shown in the figure. Let $P = 2.5$ kips and $L = 3$ ft.

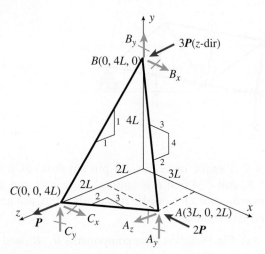

PROBLEM 4.3-5

4.3-6 A space truss is restrained at joints A, B, and C as shown in the figure. Loads are applied at all three joints. Find all support reactions and member forces. Let $P = 9$ kN and $L = 4$ m.

PROBLEM 4.3-6

4.3-7 A space truss is restrained at joints A, B, and C as shown in the figure. Loads are applied at all three joints. Find all support reactions and member forces. Let $P = 6$ kips and $L = 3$ ft.

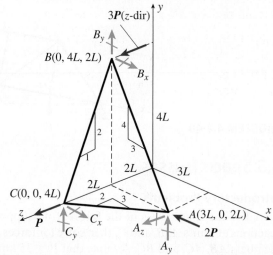

PROBLEM 4.3-7

4.3-8 A space truss is restrained at joints O, A, B, and C as shown in the figure. Load P is applied at joint A and load $2P$ acts downward at joint C.

(a) Find reaction force components A_x, B_y, and B_z in terms of load variable P.

(b) Find the axial force in truss member AB in terms of load variable P.

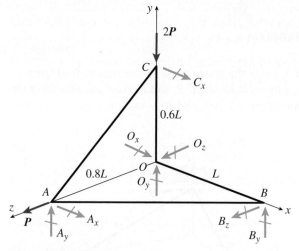

PROBLEM 4.3-8

4.3-9 A space truss is restrained at joints A, B, and C as shown in the figure. Load $2P$ is applied at in the $-x$-direction at joint A, load $3P$ acts in the $+z$-direction at joint B and load $\boldsymbol{P}$ is applied in the $+z$-direction at joint C. Coordinates of all joints are given in terms of dimension variable L (see figure).

 (a) Find reaction force components A_y and A_z in terms of load variable P.

 (b) Find the axial force in truss member AB in terms of load variable P.

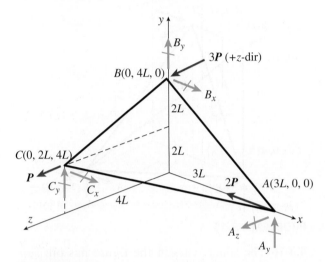

PROBLEM 4.3-9

4.3-10 A space truss is restrained at joints A, B, and C as shown in the figure. Load $\boldsymbol{P}$ acts in the $+z$ direction at joint B and in the $-z$ direction at joint C. Coordinates of all joints are given in terms of dimension variable L (see figure). Let $P = 5$ kN and $L = 2$ m.

 (a) Find reaction force components A_z and B_x.

 (b) Find the axial force in truss member AB.

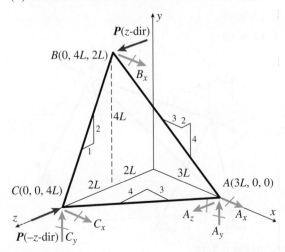

PROBLEM 4.3-10

Representative Problems

4.3-11 Find member forces BC and BD in the space truss shown in the figure. Also find reactions at joints A and F.

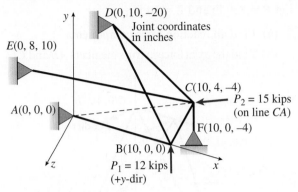

PROBLEM 4.3-11

4.3-12 A space truss is restrained at joints A, B, and C as shown in the figure. Load $\boldsymbol{P}$ acts in the $+z$ direction at joint B, in the $-z$ direction at joint C, and in the $+x$ direction at joint O. Coordinates of all joints are given in terms of dimension variable L (see figure). Let $P = 5$ kN and $L = 2.5$ m.

 (a) Find all reaction force components.

 (b) Find the axial force in truss members AB and CO.

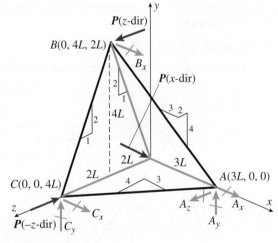

PROBLEM 4.3-12

303

4.3-13 A space truss is restrained at joints A, B, and C as shown in the figure. Load $\boldsymbol{P}$ acts in the $+z$ direction at joint B, in the $-z$ direction at joint C, and in the $+x$ direction at joint O. Coordinates of all joints are given in terms of dimension variable L (see figure). Let $P = 850$ lb and $L = 10$ ft.

(a) Find all reaction force components.

(b) Find the axial force in truss members AB and CO.

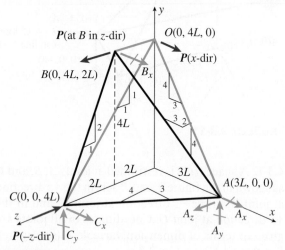

PROBLEM 4.3-13

4.3-14 A space truss is restrained at joints A, B, and C as shown in the figure. Load $\boldsymbol{P}$ acts in the $+z$ direction at joint B, in the $-z$ direction at joint C, and in the $+x$ direction at joint O. Coordinates of all joints are given in terms of dimension variable L (see figure). Let $P = 7$ kN and $L = 3$ m.

(a) Find all reaction force components.

(b) Find the axial force in truss members AB and CO.

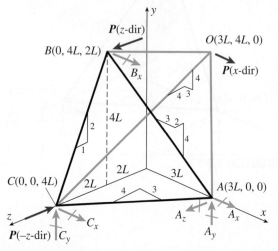

PROBLEM 4.3-14

4.3-15 A space truss is restrained at joints A, B, and C as shown in the fig. Load $\boldsymbol{P}$ acts in the $+z$ direction at joint B, in the $-z$ direction at joint C, and in the $+x$ direction at joint O. Coordinates of all joints are given in terms of dimension variable L (see figure). Let $P = 500$ lb and $L = 4$ ft.

(a) Find all reaction force components.

(b) Find the axial force in truss members AB and CO.

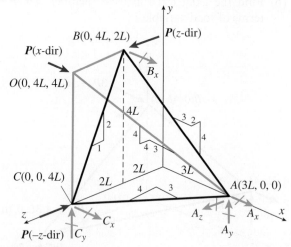

PROBLEM 4.3-15

4.3-16 The space truss in the figure has pin supports at joints C, D, E, and F. Loads $P_y = 50$ N and $P_z = 10$ N are applied as shown at both joint A and joint B. Find all member forces.

The coordinates of all joints (in meters) are

A (1.5, 0, 3)

B (1.5, 6, 3)

C (0, 0, 0)

D (0, 6, 0)

E (3, 0, 0)

F (3, 6, 0)

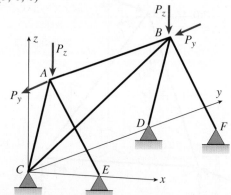

PROBLEM 4.3-16

304

4.3-17 Find all support reactions and member forces for the space truss shown in the figure. Let $P_x = P_y = P_z = 7$ kips and $L = 3$ ft.

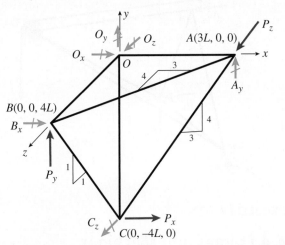

PROBLEM 4.3-17

4.3-18 A space truss is supported at A, B, C, and D (see figure). Load P is applied in the x direction at joint F. Find all support reactions and member forces. Assume that $P = 15$ kN and $L = 2.5$ m.

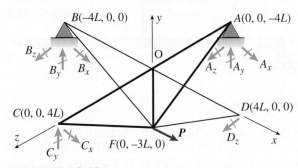

PROBLEM 4.3-18

4.3-19 A space truss is supported at A, B, C, D, and F (see figure). Load P is applied in the y direction at B and in the x direction at D. Find all support reactions and member forces. Assume that $P = 12$ kips and $L = 3$ ft.

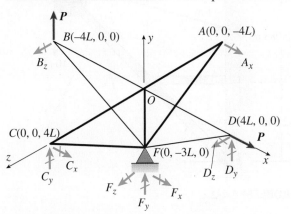

PROBLEM 4.3-19

4.3-20 A space truss is supported at O, A, D, and C (see figure). Loads P are applied in the y direction at joints B and C and in the x direction at joint D. Find all support reactions and member forces. Assume that $P = 10$ kN and $L = 1.5$ m.

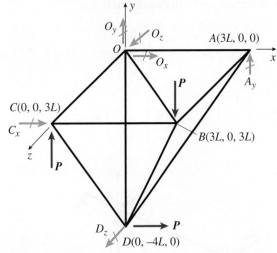

PROBLEM 4.3-20

4.3-21 A space truss is supported at joints A, B, D, E, and F as shown in the figure. Find the six support reactions, then find member forces CB, DA, and DF using the method of joints. The load at C has components

$$P_C = \begin{pmatrix} -300 \\ -600 \\ 500 \end{pmatrix} \text{ lb. Coordinates of joints are:}$$

$$D = \begin{pmatrix} 0 \\ 0 \\ 0 \end{pmatrix} \text{ft} \quad A = \begin{pmatrix} 0 \\ 9 \\ 0 \end{pmatrix} \text{ft} \quad B = \begin{pmatrix} 12 \\ 9 \\ 0 \end{pmatrix} \text{ft}$$

$$C = \begin{pmatrix} 9 \\ 9 \\ 4 \end{pmatrix} \text{ft} \quad E = \begin{pmatrix} 12 \\ 0 \\ 0 \end{pmatrix} \text{ft} \quad F = \begin{pmatrix} 9 \\ 0 \\ 4 \end{pmatrix} \text{ft}$$

PROBLEM 4.3-21

305

4.3-22 A space truss is supported at joints A, B, C, and D as shown in the figure. Find the six support reactions, then find member forces BE and AE using the method of joints. Let $P_B = 100$ N in the $-z$ direction and $P_E = 50$ N in the $-y$ direction coordinates of joints are

$$A = \begin{pmatrix} 0 \\ 3 \\ 3 \end{pmatrix} \text{m} \quad B = \begin{pmatrix} 0 \\ 3 \\ 0 \end{pmatrix} \text{m} \quad C = \begin{pmatrix} 0 \\ 0 \\ 3 \end{pmatrix} \text{m}$$

$$D = \begin{pmatrix} 0 \\ 0 \\ 0 \end{pmatrix} \text{m} \quad E = \begin{pmatrix} 3 \\ 0 \\ 1.5 \end{pmatrix} \text{m}$$

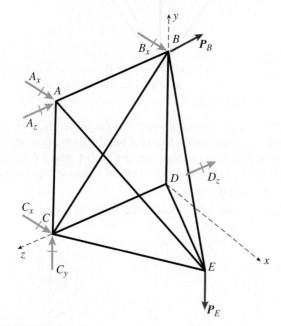

PROBLEM 4.3-22

4.3-23 A cable supporting weight $W = 1000$ lb passes over a frictionless pulley at D on a space truss and is restrained at point E. Find member forces DA, DB, and DC using the method of joints. The coordinates of the joints are

$$A = \begin{pmatrix} 0 \\ 6 \\ 4.5 \end{pmatrix} \text{ft} \quad C = \begin{pmatrix} 0 \\ 0 \\ 9 \end{pmatrix} \text{ft} \quad B = \begin{pmatrix} 0 \\ 0 \\ 0 \end{pmatrix} \text{ft}$$

$$D = \begin{pmatrix} 9 \\ 6 \\ 4.5 \end{pmatrix} \text{ft} \quad E = \begin{pmatrix} 0 \\ 6.5 \\ 4.5 \end{pmatrix} \text{ft}$$

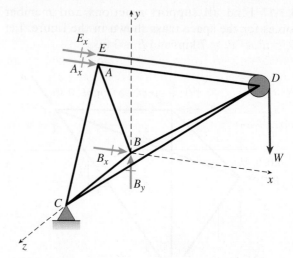

PROBLEM 4.3-23

4.4 Frames and Machines

Introductory Problems

4.4-1 Find the reaction forces at A and C for the beam shown below. Start by drawing a *FBD* of beam *ABCD*.

PROBLEM 4.4-1

4.4-2 A frame made up of members AC and BD is pin connected at C. A cable is anchored at E and passes over a frictionless pulley at D. Find reaction force components at A and B and pin connection forces at C due to an downward cable force of 10 kN.

PROBLEM 4.4-2

4.4-3 Frame *ABCDE* has a pin support at *A* and roller support at *D*. A cable with known tension force *P* is attached at *B*, runs over a frictionless pulley at *E* and has a horizontal line of action to the right of *E* (see figure). *Find* reaction forces at supports *A* and *D*. *Repeat* if the cable force to the right of *E* now has a line of action from *E* to *D*. Start by drawing a FBD of the frame.

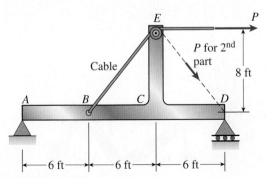

PROBLEM 4.4-3

4.4-4 Frame *ABCD* has a cable with force *P* anchored at *C*. The cable passes over a pulley at *D* and force *P* acts in the –*x* direction. There is a moment release just left of *B*. *Neglect the self-weight of the beam and cable.* Assume that *P* = 450 N, and *L* = 0.25 m. Find reactions at *A* and *C*.

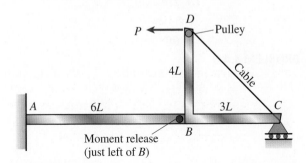

PROBLEM 4.4-4

4.4-5 A 150-lb rigid bar *AB*, with frictionless rollers at each end, is held in the position shown in the figure by a continuous cable *CAD*. The cable is pinned at *C* and *D* and runs over a pulley at *A*. Find reactions at supports *A* and *B*. Also find the force in the cable. Start by drawing a FBD of the bar.

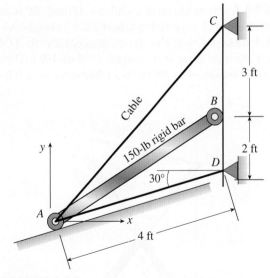

PROBLEM 4.4-5

4.4-6 Find reaction forces at supports *B*, *C*, and *D* for the frame shown in the figure. All turns of *ABCDE* are 90° angles. The support at *D* is a ball and socket. Support *B* is restrained against translation in the *y* and *z* directions only and support *C* is restrained in the *z* direction only. Each *y* direction force acts at the center of the bar segment to which it is applied.

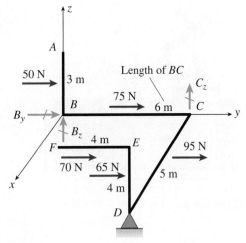

PROBLEM 4.4-6

Representative Problems

4.4-7 A frame made up of members AE and BC is pin connected at D. A cable is anchored at C, passes over a frictionless pulley at E and then runs parallel to ADE. Find reaction force components at A and B and pin connection forces at D due to a cable force of 500 lb.

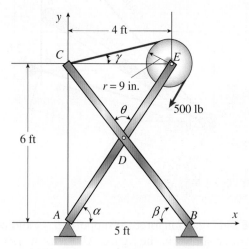

PROBLEM 4.4-7

4.4-8 A small truck crane is used to lift a 150 kg mass (see figure). For the position shown in the figure, find the force in hydraulic cylinder BD and the resultant force acting on the pin at C.

PROBLEM 4.4-8

4.4-9 Space frame ABC is clamped at A *but* it is free to rotate at A about the x axis (Fig. a). The frame is also supported at C by a spring that is normal to plane ABC. The spring and frame member BC lie in a plane that is parallel to the yz plane (Fig. b). Forces $P = 3$ kips are applied downward at mid-span of both AB and BC. Find reactions at support A *and* the force in the spring. Let $L = 12$ ft.

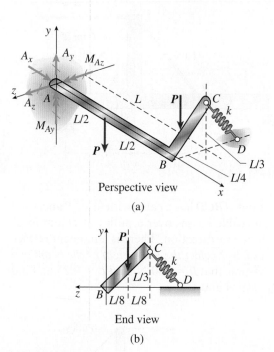

PROBLEM 4.4-9

4.4-10 A small bucket crane (also known as a cherry picker, see Fig. a) is used to lift a man and materials to make repairs to a building façade. Find the force in hydraulic cylinder *BE* (Fig. b). The cylinder is pin-connected to the boom at *E* and to pin-connected links *BA* and *BC*. Assume that the mass of the bucket and contents is $m_1 = 190$ kg, and the combined mass of the boom, cylinder, and links is $m_2 = 950$ kg.

(a)   (b)

PROBLEM 4.4-10

4.4-11 The boom on a mobile crane supports a 500 lb pipe at *A* (figure part a). Find the force F_{EB} in the elevating cylinder and the pin support reaction components at *C*. What is the fluid pressure inside the 4-in. diameter hydraulic cylinder *EB*? Assume that boom angle $\beta = 60°$, cylinder angle $\alpha = 80°$, and boom weight $W = 3600$ lb (see figure part b).

(a)   (b)

PROBLEM 4.4-11

4.4-12 An excavator lifts a section of concrete pipe of mass 1200 kg (see Fig. a). The weight of the main boom (ACF) is $W_1 = 8$ kN; the dipper (EFL) weight is $W_2 = 5$ kN; and the bucket weight is $W_3 = 0.5$ kN (see Fig. b).

(a) Find the force in each of the two hydraulic cylinders BC; in cylinder DE; and in cylinder GH.

(b) What is the pressure in cylinder GH if the piston diameter is 95 mm? (See Fig. c for dipper-bucket dimensions.)

(c)

(a)

(b)

PROBLEM 4.4-12

4.4-13 Two hydraulic rams are used to raise a truck bed as it rotates about D (one on each side of the bed; ram AB is shown and is pinned at both A and B). The truck bed and contents weigh $W = 75$ kips with center of gravity at C (see figure).

(a) Find the force in each ram for a 30° rotation of the bed.

(b) Find the pressure (psi) against the piston in each cylinder if the piston diameter is 5 in.

Assume that the truck load does not move during the rotation.

PROBLEM 4.4-13

CHAPTER

5

Centroids, Center of Mass, and Moments of Inertia

Steel members come in a wide variety of shapes; the properties of the cross section are needed for analysis and design.

Chapter Objectives

- Use *first moments* to locate the *centroids of areas, lines, and volumes for basic shapes.*
- Find centroids of *assemblages* of areas, lines and volumes referred to as *composite* shapes.
- Expand the concept of centroids to find the *center of mass* and *center of gravity* for various bodies.
- Use the two *theorems of Pappus* to find *surface areas and volumes* for bodies created by revolving a line or plane area, respectively, about a nonintersecting axis.
- Find *moments of inertia* of plane areas (also known as second moments of area), including *polar moments of inertia* and *products of inertia* of plane areas.

- Study the *parallel-axis* (or transfer) *theorem* that is used to transfer moments of inertia between different parallel axes for a plane area.
- Find the rotated axis position where the moments of inertia with respect to x and y axes in the cross section have maximum and minimum values known as *principal moments of inertia.*
- Use a graphical approach known as *Mohr's circle* to calculate moments of inertia for any rotated position of a plane area.

Chapter Outline

311

5.1 Introduction

This chapter begins with definitions and formulas pertaining to *centroids* of plane areas, lines, and volumes. The term *centroid* refers to a geometrical property of a body. The focus here is on centroids of plane *areas*. Once centroids of plane areas are understood, you can easily reduce an area to a single curve (or *line*) or expand the area to a *volume* and modify the formulas accordingly.

The primary interest in this text is the behavior of actual physical bodies that are acted upon by *static* forces. One such force is the self-weight of the body, which depends on the gravitational field of the earth. If the gravitational field is uniform, parallel gravitational forces act on all infinitesimal elements of the body. The distributed gravitational forces can be replaced by a statically equivalent concentrated force, W, at the *center of gravity* of the body. If the body of interest is subjected to dynamic forces, you must find its *center of mass*: the point where acceleration is computed as the applied force divided by the body's mass.[1] If the gravity field is uniform and parallel, the center of gravity and center of mass coincide. No gravitational forces act on a body removed from the earth's gravitational field, so the notion of a center of gravity for that body has no meaning; however, you can still find the body's center of mass. The centroid and center of mass of a body are the same if the density of the body is uniform throughout the body.

The latter part of the chapter considers the *moments of inertia* of plane areas. Here centroids and centers of gravity and mass are found using *first moments*, while moments of inertia are *second moments* of area. The terminology used here may appear puzzling. For instance, the term "moment of inertia" is clearly a misnomer when referring to properties of an area, since in this instance, no mass is involved. Even the word "area" is used inappropriately. The phrase "plane area" means "plane surface." Strictly speaking, area is a measure of the *size* of a surface and is not the same thing as the surface itself. In spite of its deficiencies, the terminology used in this book is so entrenched in the engineering literature that it rarely causes confusion. Later you will see that bending of beams and twisting of shafts are related to the moment of inertia of the beam or shaft cross section. For these and other applications, determination of moments of inertia of plane areas is an important topic in mechanics.

5.2 Centroids of Areas, Lines, and Volumes

Centroid of a Plane Area

The position of the centroid of a plane area is an important geometric property. To obtain formulas for locating centroid, refer to Fig. 5-1, which shows a plane area of irregular shape with its centroid at point C. The x-y coordinate system is oriented arbitrarily with its origin at any point O. The **area** of the geometric figure is defined by the integral:

$$A = \int dA \tag{5-1}$$

where dA is a differential element of area having coordinates x and y (Fig. 5-1) and A is the total area of the figure.

---

[1]Dynamic forces are the subject of a follow-on course in mechanics.

The **first moments** of the area with respect to the x and y axes are defined, respectively, as

$$Q_x = \int y\, dA \qquad Q_y = \int x\, dA \qquad \text{(5-2a,b)}$$

Thus, the sums of products of the differential areas and their distances from the reference axes (usually the coordinate axes) are the first moments of the areas. First moments may be positive or negative depending on the relative position of the areas with respect to the reference axes. Also, first moments have units of length raised to the third power; for instance, in³ or mm³.

The coordinates $\bar{x}$ and $\bar{y}$ of the **centroid** C (Fig. 5-1) are equal to the first moments divided by the area:

$$\bar{x} = \frac{Q_y}{A} = \frac{\int x\, dA}{\int dA} \qquad \bar{y} = \frac{Q_y}{A} = \frac{\int y\, dA}{\int dA} \qquad \text{(5-3a,b)}$$

If the boundaries of the area are defined by simple mathematical expressions, you can evaluate the integrals appearing in Eqs. (5-3a and b) in closed form and thereby obtain formulas for $\bar{x}$ and $\bar{y}$. The formulas listed in Appendix B were obtained in this manner. In general, the coordinates $\bar{x}$ and $\bar{y}$ may be positive or negative, depending upon the position of the centroid with respect to the reference axes.

If an area is **symmetric about an axis**, the centroid must lie on that axis because the first moment about an axis of symmetry equals zero. For example, the centroid of the single symmetric area shown in Fig. 5-2 must lie on the x axis, which is the axis of symmetry. Therefore, only one coordinate must be calculated in order to locate the centroid C.

If an area has **two axes of symmetry**, as illustrated in Fig. 5-3, the position of the centroid can be determined by inspection because it lies at the intersection of the axes of symmetry.

An area of the type shown in Fig. 5-4 is **symmetric about a point**. It has no axes of symmetry, but there is a point (called the **center of symmetry**) such that every line drawn through that point contacts the area in a symmetrical manner. The centroid of such an area coincides with the center of symmetry, and therefore, the centroid can be located by inspection.

If an area has **irregular boundaries** not defined by simple mathematical expressions, locate the centroid by numerically evaluating the integrals in Eqs. (5-3a and b). The simplest procedure is to divide the geometric figure into small finite

FIGURE 5-1

Plane area of arbitrary shape with centroid C

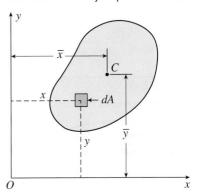

FIGURE 5-2

Area with one axis of symmetry

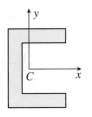

FIGURE 5-3

Area with two axes of symmetry

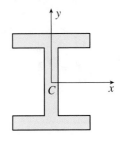

FIGURE 5-4

Area that is symmetric about a point

The centroid of wide-flange steel sections lies at the intersection of the axes of symmetry

(a)

(b)

elements and replace the integrations with summations. If the area of the ith element is denoted by ΔA_i, the expressions for the summations are

$$A = \sum_{i=1}^{n} \Delta A_i \quad Q_x = \sum_{i=1}^{n} \bar{y}_i \Delta A_i \quad Q_y = \sum_{i=1}^{n} \bar{x}_i \Delta A_i \qquad \text{(5-4a,b,c)}$$

where n is the total number of elements, $\bar{y}_i$ is the y coordinate of the centroid of the ith element, and $\bar{x}_i$ is the x coordinate of the centroid of the ith element. Replace the integrals in Eqs. (5-3a and b) by the corresponding summations to obtain the following formulas for the coordinates of the centroid:

$$\bar{x} = \frac{Q_y}{A} = \frac{\displaystyle\sum_{i=1}^{n} \bar{x}_i \Delta A_i}{\displaystyle\sum_{i=1}^{n} \Delta A_i} \qquad \bar{y} = \frac{Q_x}{A} = \frac{\displaystyle\sum_{i=1}^{n} \bar{y}_i \Delta A_i}{\displaystyle\sum_{i=1}^{n} \Delta A_i} \qquad \text{(5-5a,b)}$$

The accuracy of the calculations for $\bar{x}$ and $\bar{y}$ depends upon how closely the selected elements fit the actual area. If they fit exactly, the results are exact. Many computer programs for locating centroids use a numerical scheme similar to the one expressed by Eqs. (5-5a and b).

Centroid of a Curved Surface

If the area of interest is a curved surface (Fig. 5-5) instead of a plane area (Fig. 5-1), you must add a third centroidal coordinate to those listed in Eqs. (5-3). Assume the surface has constant thickness. The coordinates of the centroid of the surface are $(\bar{x}, \bar{y}, \bar{z})$ where the third coordinate is

$$\bar{z} = \frac{\int z\,dA}{\int dA} \qquad \text{(5-6)}$$

In general, the centroid C does not lie on the curved surface.

Centroid of a Line

If the curved surface is reduced to a uniform curved line (Fig. 5-6), replace the elemental area dA in Eqs. (5-3) and (5-6) by elemental *length* ds, where the length of the line is $L = \int ds$, and centroidal coordinates for the line are found using

$$\bar{x} = \frac{\int x\,ds}{L} \qquad \bar{y} = \frac{\int y\,ds}{L} \qquad \bar{z} = \frac{\int z\,ds}{L} \qquad \text{(5-7)}$$

In general, the centroid C does not lie on the line itself. If the line is in one of the coordinate planes, only two of the three coordinates in Eq. (5-7) are needed.

Centroid of a Volume

If the curved surface is now expanded to a homogeneous volume (Fig. 5-7), the element of volume is dV and the total volume is $V = \int dV$. The centroidal coordinates of the volume are computed as

$$\bar{x} = \frac{\int x\,dV}{V} \qquad \bar{y} = \frac{\int y\,dV}{V} \qquad \bar{z} = \frac{\int z\,dV}{V} \qquad \text{(5-8)}$$

FIGURE 5-5

Centroid of a curved surface

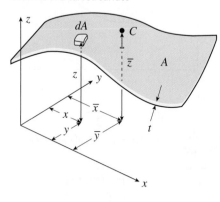

FIGURE 5-6

Centroid of a line

FIGURE 5-7

Centroid of a volume

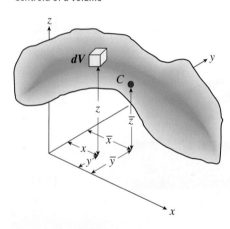

Examples 5-1, 5-2, and 5-3 present the analyses required to find the centroids of a line, an area, and a volume using the formulas presented in this section.

Example 5-1

Find the centroid of the parabolic line segment shown in Fig. 5-8. Assume that $b = 2$ and $h = 1$.

FIGURE 5-8

Centroid of a parabolic line segment

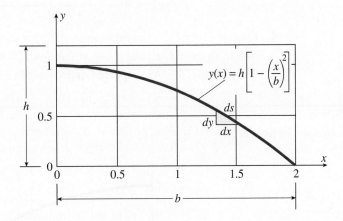

Solution:

Use the four-step problem-solving approach and first moments to find the centroid of the line segment that lies in the x-y plane.

1. *Conceptualize* [*hypothesize, sketch*]: Use first moments to find coordinates x_L and y_L of centroid C. The centroid does not lie on the line itself.

2. *Categorize* [*simplify, classify*]: If ds is an elemental length of the line segment, total length L is easily obtained as $L = \int ds$, where from Fig. 5-8,

$$ds = \sqrt{dx^2 + dy^2} = \sqrt{1 + \left(\frac{dy}{dx}\right)^2}\, dx \qquad \textbf{(5-9)}$$

First moments of elemental length ds about x and y axes in turn give coordinates x_L and y_L of centroid C as

$$x_L = \frac{\int x\,ds}{L} \qquad y_L = \frac{\int y\,ds}{L} \qquad \textbf{(5-10)}$$

3. *Analyze* [*evaluate; select relevant equations, carry out mathematical solution*]:

The length of the parabolic line segment defined by $y(x) = h\left[1 - \left(\frac{x}{b}\right)^2\right]$ is

$$L = \int ds = \int_0^b \sqrt{1 + \left(\frac{dy}{dx}\right)^2}\, dx = \int_0^b \sqrt{1 + \left(-2\frac{hx}{b^2}\right)^2}\, dx$$

$$= \int_0^2 \sqrt{1 + \left(-\frac{x}{2}\right)^2}\, dx = 2.296 \qquad \textbf{(5-11)}$$

using numerical values $b = 2$ and $h = 1$. Finally, find the centroid of the line segment from Eqs. (5-10):

$$x_L = \frac{\int_0^2 x\sqrt{1 + \left(-\frac{x}{2}\right)^2}\,dx}{L} = 1.062$$

$$y_L = \frac{\int_0^2 \left[1 - \left(\frac{x}{2}\right)^2\right]\sqrt{1 + \left(-\frac{x}{2}\right)^2}\,dx}{L} = 0.634 \qquad \blacktriangleleft\ \textbf{(5-12)}$$

4. **Finalize** [*conclude; examine answer—Does it make sense? Are units correct? How does it compare to similar problem solutions?*]: The location of the centroid is shown in Fig. 5-9. As suspected, it does not lie on the line itself.

FIGURE 5-9

Centroid of a parabolic line segment at coordinates (x_L, y_L)

You can confirm, using the method presented in this example, that if line segment $y(x)$ is instead straight line $y(x) = h\left[1 - \dfrac{x}{b}\right]$, the centroid has coordinates $(b/2, h/2)$ and lies on the line itself (Fig. 5-10).

FIGURE 5-10

Centroid of a straight line segment at coordinates (x_L, y_L)

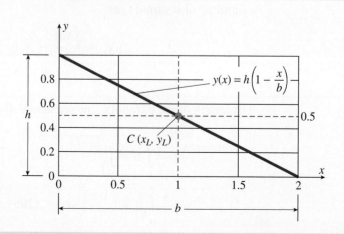

Example 5-2

FIGURE 5-11

(a) Centroid of area bounded by parabola *AB*

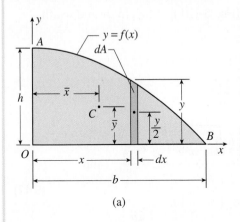

(a)

Area *OAB* is bounded by the x axis, the y axis, and a parabolic curve having its vertex at A [Fig. 5-11(a)]. The equation of the curve is $y = f(x) = h\left(1 - \left[\dfrac{x}{b}\right]^2\right)$. Find the centroid C of this area. Assume that $b = 2$ and $h = 1$.

Solution:

Use the four-step problem-solving approach and first moments of area to find the centroid of the plane area that lies in the x-y plane.

1. *Conceptualize:* Use first moments of area to find coordinates $\bar{x}$ and $\bar{y}$ of centroid C. Either a thin vertical strip $dA = ydx$ [Fig. 5-11(a)] or a thin horizontal strip ($dA = xdy$, not shown) may be selected as an element of area.

2. *Categorize:* The area of the vertical differential element is

$$dA = ydx = h\left(1 - \left[\frac{x}{b}\right]^2\right)dx \tag{5-13}$$

The area bounded by parabola *AB* is

$$A = \int dA = \int_0^b h\left(1 - \left[\frac{x}{b}\right]^2\right)dx = \frac{2bh}{3} = \frac{2(2)(1)}{3} = \frac{4}{3} \tag{5-14}$$

First moments of elemental area dA about the x and y axes in turn lead to a solution for coordinates $\bar{x}$ and $\bar{y}$ of centroid C.

3. *Analyze:* The first moment of an element of area dA (see Eq. 5-13) with respect to an axis is obtained by multiplying the area of the element by the distance from its centroid to the axis. Since the x and y coordinates of the centroid of the element shown in Fig. 5-11(a) are x and $y/2$, respectively, the first moments of the element with respect to the x and y axes are

$$Q_x = \int \frac{y}{2}dA = \int_0^b \frac{h^2}{2}\left(1 - \left[\frac{x}{b}\right]^2\right)^2 dx = \frac{4bh^2}{15}$$

$$Q_y = \int x\, dA = \int_0^b hx\left(1 - \left[\frac{x}{b}\right]^2\right)dx = \frac{b^2h}{4} \tag{5-15}$$

Now determine the coordinates of the centroid C:

$$\bar{x} = \frac{Q_y}{A} = \frac{3b}{8} = \frac{3(2)}{8} = \frac{3}{4} \quad \text{and} \quad \bar{y} = \frac{Q_x}{A} = \frac{2h}{5} = \frac{2(1)}{5} = \frac{2}{5} \tag{5-16}$$

FIGURE 5-11 (continued)

(b) centroid of area using horizontal strip

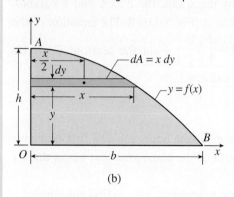

(b)

4. *Finalize:* These results agree with the formulas listed in Appendix B, Case 17.

Note that the centroid C of the area bounded by parabola AB also may be located by taking the element of area dA as a *horizontal strip* [Fig. 5-11(b)] of height dy and width $x = b\sqrt{1 - \dfrac{y}{h}}$, so

$$A = \int_0^h b\sqrt{1 - \frac{y}{h}}\, dy = \frac{2bh}{3}$$

$$\bar{x} = \frac{\displaystyle\int_0^h \frac{x}{2}\, x\, dy}{A} = \frac{\displaystyle\int_0^h \frac{1}{2}\left[b\sqrt{1 - \frac{y}{h}}\right]^2 dy}{A} = \frac{3b}{8} \quad \text{and}$$

$$\bar{y} = \frac{\displaystyle\int_0^h yx\, dy}{A} = \frac{\displaystyle\int_0^h y\left[b\sqrt{1 - \frac{y}{h}}\right] dy}{A} = \frac{2h}{5}$$

(5-17)

Example 5-3

The area bounded by parabola AB from Example 5-2 creates a volume referred to here as a paraboloid (Fig. 5-12) if revolved about the y axis. The equation of the curve that defines the top surface and lies in the x-y plane is $y = f(x) = h\left(1 - \left[\dfrac{x}{b}\right]^2\right)$. Find the centroid C of the paraboloid. Assume that $b = 2$ and $h = 1$.

Solution:

Use the four-step problem-solving approach and first moments of volume to find the centroid of the volume of the paraboloid.

1. *Conceptualize:* The equation of the curve that defines the top surface and lies in the y-z plane is $y = f(z) = h\left(1 - \left[\dfrac{z}{b}\right]^2\right)$, and the base of the paraboloid in the x-z plane is a circle $x^2 + z^2 = b^2$. Use first moments of volume to find coordinate $\bar{y}$ of centroid C. Note at the outset that $\bar{x} = \bar{z} = 0$ due to symmetry.

2. *Categorize:* Either a thin horizontal disk with radius x and thickness dy [$dV = (\pi x^2)dy$ at distance y above the x-z plane, Fig. 5-12a] or a thin vertical cylindrical shell with radius x and thickness dx [$dV = y(2\pi x)dx$ of height y and centroid at $y/2$, Fig. 5-12b] may be selected as an element of volume. Using the horizontal disk, the volume of the paraboloid is

$$V = \int_0^h \pi x^2\, dy = \int_0^h \pi \left[b\sqrt{1 - \frac{y}{h}}\right]^2 dy = \frac{\pi b^2 h}{2} = \frac{\pi (2)^2 (1)}{2} = 2\pi \quad \text{(5-18)}$$

FIGURE 5-12

(a) Element of volume of paraboloid is a horizontal disk; (b) element of volume of paraboloid is a vertical cylindrical shell

(a)

(b)

The first moment of elemental volume dV about the x or z axis leads to a solution for coordinate $\bar{y}$ of centroid C.

3. *Analyze:* The first moment of an element of volume dV with respect to either the x or z axis is obtained by multiplying the volume of the element by the distance from its centroid to the axis. The first moments of the elemental volume (horizontal disk in Fig. 5-12a) with respect to the x and z axes are

$$Q_x = \int y\,dV = \int_0^h y(\pi x^2)\,dy = \int_0^h y\left(\pi\left[b\sqrt{1-\frac{y}{h}}\right]^2\right)dy = \frac{\pi b^2 h^2}{6}$$

$$Q_z = \int y\,dV = \int_0^h y(\pi z^2)\,dy = \int_0^h y\left(\pi\left[b\sqrt{1-\frac{z}{h}}\right]^2\right)dy = \frac{\pi b^2 h^2}{6}$$

(5-19)

Now determine the $\bar{y}$ coordinate of the centroid C using either Q_x or Q_z:

$$\bar{y} = \frac{Q_x}{V} = \frac{h}{3} = \frac{1}{3} \qquad \qquad \text{(5-20)}$$

4. *Finalize:* As noted here, the centroid C of the paraboloid also may be located by taking the element of volume dV as a *thin vertical cylindrical shell* with radius x and thickness dx $[dV = y(2\pi x)dx$ of height y and centroid at $y/2$, Fig. 5-12b]. Now the volume is computed as

$$V = \int_0^b y(2\pi x)\,dx = \int_0^b 2\pi xh\left(1 - \left[\frac{x}{b}\right]^2\right)dx = \frac{\pi b^2 h}{2} = \frac{\pi (2)^2 (1)}{2} = 2\pi \quad \text{(5-21)}$$

and centroidal distance $\bar{y}$ is confirmed as

$$\bar{y} = \frac{Q_x}{V} = \frac{\displaystyle\int_0^b \frac{2\pi x}{2}\left[h\left[1 - \left(\frac{x}{b}\right)^2\right]\right]^2 dx}{V} = \frac{h}{3}$$

Note: The theorems of Pappus (Section 5.5) provide a much simpler path to the solution for the volume and centroid location for this paraboloid (see Example 5-6).

5.3 Centroids of Composite Bodies

Composite Areas

In engineering work, you rarely need to locate centroids by integration because the centroids of common geometric figures are already known and tabulated. However, you frequently need to locate the centroids of areas composed of several parts, each part having a familiar geometric shape, such as a rectangle or a circle. Examples of such **composite areas** are the cross sections of beams and columns, which usually consist of rectangular elements (for instance, see Figures 5-2, 5-3, and 5-4).

The **areas and first moments** of composite areas may be calculated by summing the corresponding properties of the component parts. Assume that a composite area is divided into a total of n parts, and denote the area of the ith part as A_i. Then you can obtain the area and first moments by the following summations:

$$A = \sum_{i=1}^{n} A_i \qquad Q_x = \sum_{i=1}^{n} \bar{y}_i A_i \qquad Q_y = \sum_{i=1}^{n} \bar{x}_i A_i \qquad \text{(5-22a,b,c)}$$

where $\bar{x}_i$ and $\bar{y}_i$ are the coordinates of the centroid of the ith part. The coordinates of each part of a composite body are always measured with respect to the reference coordinate system for the overall area as a whole.

The **coordinates of the centroid** of the composite area are

$$\bar{x} = \frac{Q_y}{A} = \frac{\sum\limits_{i=1}^{n} \bar{x}_i A_i}{\sum\limits_{i=1}^{n} A_i} \qquad \bar{y} = \frac{Q_x}{A} = \frac{\sum\limits_{i=1}^{n} \bar{y}_i A_i}{\sum\limits_{i=1}^{n} A_i} \qquad \textbf{(5-23a,b)}$$

Since the composite area is represented exactly by the n parts, the preceding equations give exact results for the coordinates of the centroid.

To illustrate the use of Eqs. (5-23a and b), consider the L-shaped area (or angle section) shown in Fig. 5-13a. This area has side dimensions b and c and thickness t. The area can be divided into two rectangles of areas A_1 and A_2 with centroids C_1 and C_2, respectively (Fig. 5-13b). The areas and centroidal coordinates of these two parts are

$$A_1 = bt \qquad \bar{x}_1 = \frac{t}{2} \qquad \bar{y}_1 = \frac{b}{2}$$

$$A_2 = (c - t)t \qquad \bar{x}_2 = \frac{c + t}{2} \qquad \bar{y}_2 = \frac{t}{2}$$

Therefore, the area and first moments of the composite area [from Eqs. (5-22a, b, and c)] are

$$A = A_1 + A_2 = t(b + c - t)$$

$$Q_x = \bar{y}_1 A_1 + \bar{y}_2 A_2 = \frac{t}{2}(b^2 + ct - t^2)$$

$$Q_y = \bar{x}_1 A_1 + \bar{x}_2 A_2 = \frac{t}{2}(bt + c^2 - t^2)$$

Finally, compute the coordinates $\bar{x}$ and $\bar{y}$ of the centroid C of the composite area (Fig. 5-13b) from Eqs. (5-23a and b):

$$\bar{x} = \frac{Q_y}{A} = \frac{bt + c^2 - t^2}{2(b + c - t)} \qquad \bar{y} = \frac{Q_x}{A} = \frac{b^2 + ct - t^2}{2(b + c - t)} \qquad \textbf{(5-24a,b)}$$

A similar procedure can be used for more complex areas, as illustrated in Examples 5-4 and 5-5.

Note 1: When a composite area is divided into only two parts, the centroid C of the entire area lies on the line joining the centroids C_1 and C_2 of the two parts (as shown in Fig. 5-13b for the L-shaped area).

Note 2: When using the formulas for composite areas [see Eqs. (5-22) and (5-23)], you can handle the *absence* of an area by subtraction. This procedure is useful when there are cutouts or holes in the figure.

For instance, consider the area shown in Fig. 5-14a. You can analyze this figure as a composite area by subtracting the properties of the inner rectangle *efgh* from the corresponding properties of the outer rectangle *abcd*. (From another viewpoint, think of the outer rectangle as a "positive area" and the inner rectangle as a "negative area.")

Similarly, if an area has a hole (Fig. 5-14b), subtract the properties of the area of the hole from those of the outer rectangle. (Again, the same effect is achieved if you treat the outer rectangle as a "positive area" and the hole as a "negative area.")

Last, the centroid of the L-shaped area in Fig. 5-13 might be computed by using "positive area" bc, and "negative area" $(b-t)(c-t)$.

FIGURE 5-13

Centroid of a composite area consisting of two parts

(a)

(b)

FIGURE 5-14

(a,b) Composite areas with a cutout and a hole;
(c) cutouts in beams must be considered in centroid and moment of inertia calculations

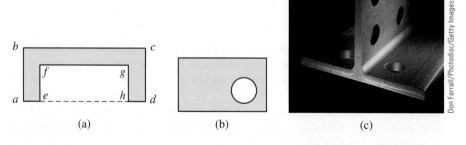

(a) (b) (c)

Don Farrall/Photodisc/Getty Images

Composite Lines and Volumes

Equations (5-22) and (5-23) pertaining to areas can be altered for application to composite assemblages of lines and volumes as follows. For *lines*,

$$L = \sum_{i=1}^{n} L_i \qquad Q_x = \sum_{i=1}^{n} \bar{y}_i L_i \qquad Q_y = \sum_{i=1}^{n} \bar{x}_i L_i \qquad \text{(5-25)}$$

where $\bar{x}_i$ and $\bar{y}_i$ are the coordinates of the centroid of the ith line segment and Q_x and Q_y are the first moments of the assemblage of lines about x and y axes, respectively. The coordinates of the centroid of a *two-dimensional* composite assemblage of lines can then be found as

$$\bar{x} = \frac{Q_y}{L} = \frac{\sum_{i=1}^{n} \bar{x}_i L_i}{\sum_{i=1}^{n} L_i} \qquad \bar{y} = \frac{Q_x}{L} = \frac{\sum_{i=1}^{n} \bar{y}_i L_i}{\sum_{i=1}^{n} L_i} \qquad \text{(5-26)}$$

Finally, the coordinates of an ***assemblage of volumes*** may be found in *three dimensions* in a similar manner using

$$\bar{x} = \frac{\sum_{i=1}^{n} \bar{x}_i V_i}{\sum_{i=1}^{n} V_i} \qquad \bar{y} = \frac{\sum_{i=1}^{n} \bar{y}_i V_i}{\sum_{i=1}^{n} V_i} \qquad \bar{z} = \frac{\sum_{i=1}^{n} \bar{z}_i V_i}{\sum_{i=1}^{n} V_i} \qquad \text{(5-27)}$$

where $\bar{x}_i$, $\bar{y}_i$ and $\bar{z}_i$, are the coordinates of the centroid of the ith volume component. The calculation of the centroid location for an assemblage of lines, and also for an assemblage of volumes, is illustrated at the end of Example 5-4.

Example 5-4

The area bounded by parabola AB from Example 5-2 is now reduced in area by two cutouts: a right triangle ($b \times h/6$) and a circle of radius $b/6$ centered at coordinates $x = b/3$, $y = h/2$ (Fig. 5-15). Find the centroid C of the composite area. Assume that $b = 2$ and $h = 1$.

FIGURE 5-15

Centroid of a composite area: area bounded by parabola AB with two cutouts

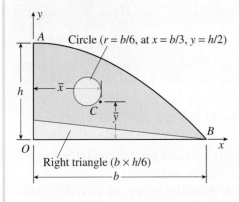

Solution:

Use the four-step problem-solving approach and first moments of area to find the centroid of the composite plane area.

1. *Conceptualize:* Use known formulas for areas and first moments for each of the component shapes that make up the composite figure. Treat the cutout areas as negative areas and subtract them from the overall area bounded by parabola AB.

2. *Categorize:* Begin by collecting areas and centroidal coordinates for each of the component areas shown in Fig. 5-15. From Example 5-2 (or Case 17, Appendix B), the area bounded by parabola AB is

$$A_p = \frac{2bh}{3} \qquad \bar{x}_p = \frac{3b}{8} \qquad \bar{y}_p = \frac{2h}{5} \qquad (5\text{-}28)$$

Areas and centroidal coordinates for the triangle cutout are listed in Case 6, and for the circle cutout in Case 9, of Appendix B:

$$A_t = -\frac{1}{2}(b)\left(\frac{h}{6}\right) = -\frac{bh}{12} \qquad \bar{x}_t = \frac{b}{3} \qquad \bar{y}_t = \frac{1}{3}\left(\frac{h}{6}\right) = \frac{h}{18}$$

$$A_c = -\pi\left(\frac{b}{6}\right)^2 = -\frac{\pi b^2}{36} \qquad \bar{x}_c = \frac{b}{3} \qquad \bar{y}_c = \frac{h}{2} \qquad (5\text{-}29)$$

3. *Analyze:* The composite area is obtained by summing component areas with cutouts listed as negative areas as noted:

$$A = A_p + A_t + A_c = \frac{8bh}{12} - \frac{bh}{12} - \frac{\pi b^2}{36} = \frac{7bh}{12} - \frac{\pi b^2}{36} = 0.818 \quad (5\text{-}30)$$

The first moments of the component areas with respect to the x and y axes are

$$Q_x = y_p A_p + y_t A_t + y_c A_c = \frac{283}{1080}bh^2 - \frac{\pi}{72}b^2 h = 0.350$$

$$Q_y = x_p A_p + x_t A_t + x_c A_c = \frac{2}{9}b^2 h - \frac{\pi}{108}b^3 = 0.656 \qquad (5\text{-}31)$$

Now insert numerical dimensions $b = 2$ and $h = 1$ and find the $\bar{x}$ and $\bar{y}$ coordinates of the centroid C using Q_x and Q_y as

$$\bar{x} = \frac{Q_y}{A} = \frac{b(\pi b - 24h)}{3(\pi b - 21h)} = \frac{0.656}{0.818} = 0.803$$

$$\bar{y} = \frac{Q_x}{A} = h\frac{(283h - 15\pi b)}{630h - 30\pi b} = \frac{0.350}{0.818} = 0.428 \qquad \text{←} \quad (5\text{-}32)$$

As an alternative to this approach based on use of formulas for each part, use numerical values for dimensions and organize the calculations in the following tabular format. The final computed values of centroid coordinates $\bar{x}$ and $\bar{y}$ are as in Eq. (5-32).

Part	Area A	$\bar{x}$	$\bar{y}$	$\bar{x}A$	$\bar{y}A$
Parabola	1.333	0.750	0.400	1.000	0.533
Triangle	-0.167	0.667	0.056	-0.111	-0.009
Circle	-0.349	0.667	0.500	-0.233	-0.175
Sum	$A = 0.818$			$Q_y = 0.656$	$Q_x = 0.350$

FIGURE 5-16

Centroid of a composite line assemblage:
parabola plus a circle

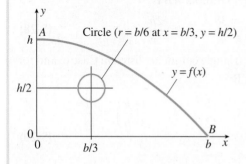

4. *Finalize:* Use superposition of the tabulated results for known component areas from Appendix B to find the centroid of a composite area and summation rather than integration to obtain the desired results.

The same summation methodology can be used if the centroids of **assemblages of lines or volumes** are of interest. For example, if Fig. 5-15 is replaced by a **composite line** *assemblage* (Fig. 5-16) consisting of parabola AB and a circle of radius $b/6$, use the first moment of lines instead of areas to locate the centroid of the assemblage. From Example 5-1, the length and centroidal coordinates of the parabola (for $b = 2$, $h = 1$) are $L_1 = 2.296$, $x_L = 1.062$, and $y_L = 0.634$. The circle has circumference $L_c = 2\pi r = 2\pi(b/6) = 2.094$ and centroid at its center ($x_c = 2/3$, $y_c = 1/2$). The centroid of the assemblage of the parabola and circle is computed as

$$\bar{x}_L = \frac{L_1 x_L + L_c x_c}{L_1 + L_c} = \frac{2.296(1.062) + 2.094(0.667)}{2.296 + 2.094} = \frac{3.835}{4.39} = 0.873$$

$$\bar{y}_L = \frac{L_1 y_L + L_c y_c}{L_1 + L_c} = \frac{2.296(0.634) + 2.094(0.5)}{2.296 + 2.094} = \frac{2.503}{4.39} = 0.57$$

(5-33)

If the composite line assemblage in Fig. 5-16 is extended to create a **composite volume** consisting of a paraboloid with a toroidal void inside (Fig. 5-17), you can locate the composite centroid. From Example 5-3, the properties of the paraboloid are $V_p = \dfrac{\pi b^2 h}{2}$, $\bar{x}_p = \bar{z}_p = 0$, and $\bar{y}_p = \dfrac{h}{3}$. If the toroidal void has radii $R = b/3$ and $r = b/6$, its properties are $V_t = (2\pi R)(\pi r^2) = 2(\pi r)^2 R = \dfrac{\pi^2 b^3}{54}$, $\bar{x}_t = \bar{z}_t = 0$, and $\bar{y}_t = \dfrac{h}{2}$. Because of symmetry, $\bar{x}_V = \bar{z}_V = 0$. First moments about x or z give the y coordinate of the centroid as (again assuming that $b = 2$, $h = 1$):

$$\bar{y}_V = \frac{V_p y_p - V_t y_t}{V_p - V_t} = \frac{h\,(\pi b - 18h)}{2\,(\pi b - 27h)} = \frac{1}{2}\frac{(2\pi - 18(1))}{(2\pi - 27(1))}$$

$$= \frac{-11.72}{-41.43} = 0.283$$

(5-34)

The composite volume centroid location has $\bar{y}$ coordinate location below that of the paraboloid alone, as expected.

FIGURE 5-17

(a) Centroid of a composite volume: paraboloid minus a torus; (b) centroid of a composite volume: paraboloid minus a torus

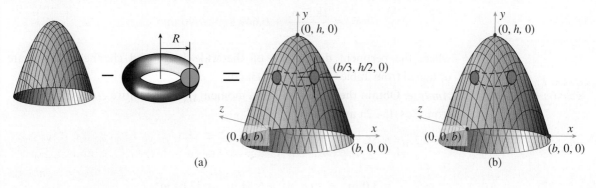

(a)   (b)

Example 5-5

FIGURE 5-18

Centroid of a composite area

The cross section of a steel beam is constructed of a W 18 × 71 wide-flange section with a 6 in. × 1/2 in. cover plate welded to the top flange and a C 10 × 30 channel section welded to the bottom flange (Fig. 5-18).

Locate the centroid C of the cross-sectional area.

Solution:

Use the four-step problem-solving approach to find the centroid of the composite area.

1. *Conceptualize:* Denote the areas of the cover plate, the wide-flange section, and the channel section as areas A_1, A_2, and A_3, respectively. The centroids of these three areas are labeled C_1, C_2, and C_3, respectively, in Fig. 5-18. Note that the composite area has an axis of symmetry; therefore all centroids lie on that axis. The three partial areas are

$$A_1 = (6 \text{ in.})(0.5 \text{ in.}) = 3.0 \text{ in}^2 \quad A_2 = 20.8 \text{ in}^2 \quad A_3 = 8.81 \text{ in}^2$$

where areas A_2 and A_3 are obtained from Tables C-1(a) and C-3(a) of Appendix C.

2. *Categorize:* Place the origin of the x and y axes at the centroid C_2 of the wide-flange section. Then the distances from the x axis to the centroids of the three areas are

$$\bar{y}_1 = \frac{18.5 \text{ in.}}{2} + \frac{0.5 \text{ in.}}{2} = 9.50 \text{ in.} \qquad \bar{y}_2 = 0$$

$$\bar{y}_3 = \frac{18.5 \text{ in.}}{2} + 0.649 \text{ in.} = 9.90 \text{ in.}$$

where the pertinent dimensions of the wide-flange and channel sections are obtained from Tables C-1(a) and C-3(a).

3. *Analyze:* Obtain the area A and first moment Q_x of the entire cross section from Eqs. (5-22a and b) as

$$A = \sum_{i=1}^{n} A_i = A_1 + A_2 + A_3$$

$$= 3.0 \text{ in}^2 + 20.8 \text{ in}^2 + 8.81 \text{ in}^2 = 32.61 \text{ in}^2$$

$$Q_x = \sum_{i=1}^{n} \bar{y}_i A_i = \bar{y}_1 A_1 + \bar{y}_2 A_2 + \bar{y}_3 A_3$$

$$= (9.50 \text{ in.})(3.0 \text{ in}^2) + 0 - (9.90 \text{ in.})(8.81 \text{ in}^2) = -58.71 \text{ in}^3$$

Now obtain the coordinate $\bar{y}$ to the centroid C of the composite area from Eq. (5-23b):

$$\bar{y} = \frac{Q_x}{A} = \frac{-58.71 \text{ in}^3}{32.61 \text{ in}^2} = -1.80 \text{ in.}$$

Since $\bar{y}$ is positive in the positive direction of the y axis, the minus sign means that the centroid C of the composite area is located below the x axis, as shown in Fig. 5-18. Thus, the distance $\bar{c}$ between the x axis and the centroid C is

$$\bar{c} = 1.80 \text{ in.} \qquad \longleftarrow$$

It may be more convenient to organize the calculations in the following tabular format that results in the same value of centroid coordinate $\bar{y}$ and distance $\bar{c}$ as listed previously.

Part	Area A (in^2)	$\bar{y}$ (in.)	$\bar{y}A$ (in^3)
1	3.0	9.50	28.5
2	20.8	0	0
3	8.81	−9.90	−87.21
Sum	$A = 32.61$		$Q_x = -58.71$

> 4. *Finalize:* Note that the position of the reference axis (the x axis) is arbitrary; however, this example places it through the centroid of the wide-flange section because it simplifies the calculations.

5.4 Center of Mass and Center of Gravity

It is a straightforward task to develop integral expressions that define the centers of mass and gravity from similar expressions for centroids presented in Sections 5.2 and 5.3. The center of mass is of most interest for bodies subject to dynamic forces because the acceleration of the *center of mass* of a body is equal to the dynamic force divided by the mass of the body (see Newton's second law in Section 1.2). In the case of static forces on a body due to gravity, the distributed gravitational forces acting throughout the body can be replaced by a single force equal to the total weight of the body and acting at its *center of gravity*. The modified expressions for each case are presented here.

Center of Mass

Begin by defining the density ρ of a body as the mass per unit volume. As a result, an element of mass is $dm = \rho\,dV$ for an element of volume dV. If the density varies throughout the body, use first moments to find the center of mass of the body of mass $m = \int \rho\,dV$ as

$$\bar{x} = \frac{\int x\rho\,dV}{\int \rho\,dV} \qquad \bar{y} = \frac{\int y\rho\,dV}{\int \rho\,dV} \qquad \bar{z} = \frac{\int z\rho\,dV}{\int \rho\,dV} \qquad \textbf{(5-35)}$$

If the density is constant, variable ρ cancels from numerator and denominator in each of Eqs. (5-35), and Eqs. (5-35) reduce to Eqs. (5-8); hence the center of mass of the body and the centroid of the volume of the body are seen to be the same point.

Composite Body   In the case of a *composite body*, sum first moments for each part where mass m_i and mass center coordinates $(\bar{x}_i, \bar{y}_i, \bar{z}_i)$ are known for the ith part of the n parts, which make up the body. The integrals in Eqs. (5-35) are replaced by summations:

$$\bar{x} = \frac{\sum\limits_{i}^{n} m_i \bar{x}_i}{\sum\limits_{i}^{n} m_i} \qquad \bar{y} = \frac{\sum\limits_{i}^{n} m_i \bar{y}_i}{\sum\limits_{i}^{n} m_i} \qquad \bar{z} = \frac{\sum\limits_{i}^{n} m_i \bar{z}_i}{\sum\limits_{i}^{n} m_i} \qquad \textbf{(5-36)}$$

Example 5-5 Continued: Center of Mass   If you remove the top steel cover plate and the bottom steel C shape on the W 18×71 beam in Example 5-5 and add a segment of concrete slab, you can create the composite beam-slab assemblage shown in Fig. 5-19. This is an example of *composite construction*, where a number of steel beams act together with a concrete slab to support the load on a floor slab in a building or on the road surface of a bridge. Each beam is then paired with an effective width of slab, and its cross-sectional properties are used to find stresses in each component. The objective here is to find the *center of mass* of the composite assemblage. The mass densities of steel and concrete are listed in Table F-1 of Appendix F; for steel, $\rho_s = 15.2$ slugs/ft^3, and for concrete, $\rho_c = 4.7$ slugs/ft^3.

For a one-foot-long segment of the composite section, the masses of steel and concrete segments are 2.196 and 15.667 slugs, respectively. Noting that $\bar{x} = 0$ due to symmetry about the y axis, to find the *center of mass* just find $\bar{y}$, which here is measured from the centroid C of the steel beam (Fig. 5-19). Distance $\bar{y}$ is found using the summation form of Eq. (5-36b) as

$$\bar{y} = \frac{\sum\limits_{i=1}^{2} m_i \bar{y}_i}{\sum\limits_{i=1}^{2} m_i} = \frac{(m_s)(0) + (m_c)(9.25 \text{ in.} + 4 \text{ in.})}{m_s + m_c} \tag{5-37}$$

$$= \frac{(15.667 \text{ slugs})(13.25 \text{ in.})}{(2.196 + 15.667) \text{ slugs}} = \frac{207.59}{17.86} \text{ in.} = 11.62 \text{ in.}$$

Had you used cross-sectional areas (A_i) instead of masses (m_i) in Eq. (5-37), distance $\bar{y}$ would have been computed *incorrectly* as 12.70 in.

Center of Gravity

The gravitational force $W = mg = \int g\,dm$ is the *weight* of the body and acts at the center of gravity of the body. Constant g is used to calculate the force due to gravity. A parallel system of gravitational forces $dW = g\,dm$ acts on each element of the body. Using first moments to find the center of gravity results in the following expressions:

$$\bar{x} = \frac{\int x\,dW}{W} \qquad \bar{y} = \frac{\int y\,dW}{W} \qquad \bar{z} = \frac{\int z\,dW}{W} \tag{5-38}$$

FIGURE 5-19

Center of mass of composite beam-slab

If the gravitational field is both uniform and parallel, constant g cancels in numerator and denominator in Eqs. (5-38). In this case, the center of mass and the center of gravity coincide and are defined as

$$\bar{x} = \frac{\int x\,dm}{m} \qquad \bar{y} = \frac{\int y\,dm}{m} \qquad \bar{z} = \frac{\int z\,dm}{m} \qquad \textbf{(5-39)}$$

If a body is not subject to a gravitational field, the notion of center of gravity is irrelevant, and you should refer instead to its center of mass.

Composite Body In the case of a *composite body*, sum first moments for each part where weight W_i and center of gravity coordinates $(\bar{x}_i, \bar{y}_i, \bar{z}_i)$ are known for the ith part of the n parts that make up the body. Equations (5-39) are replaced by the following summations:

$$\bar{x} = \frac{\displaystyle\sum_i^n W_i \bar{x}_i}{\displaystyle\sum_i^n W_i} \qquad \bar{y} = \frac{\displaystyle\sum_i^n W_i \bar{y}_i}{\displaystyle\sum_i^n W_i} \qquad \bar{z} = \frac{\displaystyle\sum_i^n W_i \bar{z}_i}{\displaystyle\sum_i^n W_i} \qquad \textbf{(5-40)}$$

5.5 Theorems of Pappus

Pappus was Greek mathematician who lived in the third century in Alexandria, and his name is associated with a number of theorems in geometry. Two of his theorems utilize first moments of lines and areas, respectively, to simplify the calculation of surfaces and volumes created by revolving that line or area about a nonintersecting axis. In the ***first theorem***, you compute the surface area A of a surface of revolution (Fig. 5-20) by multiplying the arc length $L = \int ds$ of the curve by the circumferential distance traveled by the centroid of the curve, $2\pi\bar{y}$, to find the surface area A:

$$A = 2\pi \int y\,ds = 2\pi \bar{y} L \qquad \textbf{(5-41)}$$

Avoid integration by taking advantage of the following equation of first moments for computing the centroid $\bar{y}$ of a curve:

$$\bar{y} L = \int y\,ds \qquad \textbf{(5-42)}$$

For example, consider a cylinder generated by revolving a line $y = r$ about the x axis (Fig. 5-21). The centroid of the line is shown in the figure at $\bar{y} = r$. From Eq. (5-41) for a line of length b, the surface area A is computed without integration as

$$A = 2\pi \bar{y} L = 2\pi(r)(b) = 2\pi rb \qquad \textbf{(5-43)}$$

In the ***second theorem***, compute the volume obtained by revolving a *plane area* about a nonintersecting axis (Fig. 5-22). The volume V of the solid of revolution is equal to the product of the area $A = \int dA$, and the distance traveled by the centroid $\bar{y}$ of the area during the revolution:

$$V = 2\pi \int y\,dA = 2\pi \bar{y} A \qquad \textbf{(5-44)}$$

FIGURE 5-20

Line segment used to create a surface of revolution

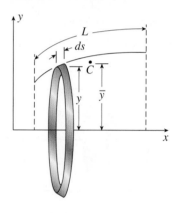

FIGURE 5-21

(a) Elemental area; (b) surface area of a cylinder of revolution

(a)

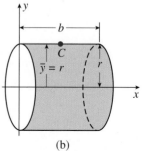

(b)

FIGURE 5-22

Plane area used to create a volume
of revolution

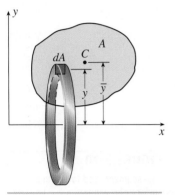

Again, avoid integration by using the following relation between first moments of an area and the elemental areas of which it is composed:

$$\bar{y}A = \int y\,dA \tag{5-45}$$

As an example, if you revolve a rectangle, $b \times r$, about the x axis (Fig. 5-23), you obtain the volume V of the cylinder from Eq. (5-44) as

$$V = 2\pi\bar{y}A = 2\pi\left(\frac{r}{2}\right)(br) = \pi b r^2 \tag{5-46}$$

If the revolution of a line through angle θ (in radians) is less than 2π, Eq. (5-41) for surface area becomes

$$A = \theta\bar{y}L \tag{5-47}$$

Correspondingly, for revolution of a plane *area* through arbitrary angle θ, modify Eq. (5-44) so that the volume V of the solid of revolution is now equal to

$$V = \theta\bar{y}A \tag{5-48}$$

For the previous example, a cylinder was created by revolving line $y = r$ of length b about the x axis, so if the angle of revolution is now $\theta = \pi/4$, the surface area is (Fig. 5-24)

$$A = \theta\bar{y}b = \frac{\pi r b}{4} \tag{5-49}$$

Revolving the plane area $b \times r$ about the x axis through angle $\theta = \pi/4$ results in volume of revolution (Fig. 5-25):

$$V = \theta\bar{y}A = \left(\frac{\pi}{4}\right)(r)br = \frac{\pi b r^2}{4} \tag{5-50}$$

FIGURE 5-23

Volume of a cylinder of revolution

Finally, note that the theorems of Pappus can be used to find the centroids of plane curves and plane areas if the surface area and volume of the associated bodies of revolution are known in advance. From Eq. (5-41), if line length L and surface area A are known, find centroid location $\bar{y}$ as

$$\bar{y} = \frac{A}{2\pi L} \tag{5-51}$$

FIGURE 5-24

Partial surface area

FIGURE 5-25

Partial volume

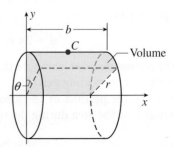

Likewise, if plane area A and volume V are known in Eq. (5-44), the centroid of the plane area is located at

$$\bar{y} = \frac{V}{2\pi A} \qquad \textbf{(5-52)}$$

Example 5-6

FIGURE 5-26

Volume and surface area of a paraboloid

Use the theorems of Pappus to find the volume V and surface area A_s of the paraboloid of Example 5-3. The paraboloid was created by revolving the parabolic segment OAB in Example 5-2 about the y axis (Fig. 5-26).

Solution:

Use the four-step problem-solving approach to obtain the volume and surface area of the paraboloid.

1. *Conceptualize:* Revolving the plane curve described by the equation $y = f(x) = h\left(1 - \left[\dfrac{x}{b}\right]^2\right)$ about the y axis creates the paraboloid shown in Fig. 5-26. Its *surface area* is computed using the first theorem of Pappus, and the *volume* generated by revolving the area under the curve is obtained using the second theorem of Pappus.

2. *Categorize:* From Example 5-1, the x coordinate of the centroid of the parabolic line is $x_L = \dfrac{\int x\,ds}{L} = \dfrac{\int x\,ds}{\int ds}$. From the first theorem of Pappus, the surface area is $A_s = 2\pi x_L L$. Next using integration in Example 5-2, the area under the curve is $A = \int y\,dx$. Then in Example 5-3, the volume V and centroid $\bar{y}$ of the *paraboloid* were obtained using an integral approach. Now use the simple method of Pappus to confirm the computed volume in Example 5-3.

3. *Analyze:*

   Surface Area: From Example 5-1, $L = 2.296$ and $x_L = 1.062$ for the case of dimension variables $h = 1, b = 2$. The surface area is easily computed as

   $$A_s = 2\pi x_L L = 2\pi(1.062)(2.296) = 15.32 \qquad \longleftarrow \textbf{(5-53)}$$

   Volume: From Example 5-2, *the area under the curve is* $A = \dfrac{2bh}{3} = \dfrac{4}{3}$ and $\bar{x} = \dfrac{3b}{8} = \dfrac{3}{4}$ for the case of dimension variables $h = 1, b = 2$. The volume of the paraboloid is easily computed as

   $$V = 2\pi\bar{x}A = 2\pi\left(\frac{2bh}{3}\right)\left(\frac{3b}{8}\right) = \frac{\pi b^2 h}{2} = \frac{\pi(2)^2(1)}{2} = 2\pi \qquad \longleftarrow \textbf{(5-54)}$$

4. *Finalize:* The volume of the paraboloid from the second theorem of Pappus (Eq. 5-54) is in agreement with that found in Example 5-3 using an integral approach.

5.6 Moments of Inertia of Plane Areas and Composite Areas

Having mastered the application of *first moments* for location of centroids (as well as centers of gravity and mass) of areas, lines, and volumes, you are now ready to consider the use of *second moments* of area to find moments of inertia of plane areas. Begin by defining different moments of inertia of a plane area with respect to centroidal or other parallel axes of the cross section. Use integral expressions to find formulas for moments of inertia of basic shapes, such as rectangles and circles. Then assemblages of basic shapes often can be used to represent composite areas. Finally, rotate the axes of the cross section to find the principal axes orientation where the moment of inertia properties are either a maximum or a minimum.

Rectangular Moments of Inertia

The **moments of inertia** of a plane area (Fig. 5-27) with respect to the x and y axes, respectively, are defined by the integrals

$$I_x = \int y^2 dA \qquad I_y = \int x^2 dA \qquad \text{(5-55a,b)}$$

FIGURE 5-27

Plane area of arbitrary shape

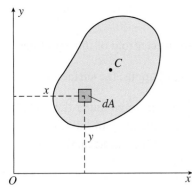

where x and y are the coordinates of the differential element of area dA. Because the element dA is multiplied by the square of the distance from the reference axis, moments of inertia are also called **second moments of area**. Also, moments of inertia of areas (unlike first moments) are always positive quantities.

To illustrate how moments of inertia are obtained by integration, consider a rectangle having width b and height h (Fig. 5-28). The x and y axes have their origin at the centroid C. For convenience, use a differential element of area dA in the form of a thin horizontal strip of width b and height dy (therefore, $dA = bdy$). Since all parts of the elemental strip are the same distance from the x axis, express the moment of inertia I_x with respect to the x axis as

$$I_x = \int y^2 dA = \int_{-h/2}^{h/2} y^2 b\, dy = \frac{bh^3}{12} \qquad \text{(5-56)}$$

FIGURE 5-28

Moments of inertia of a rectangle

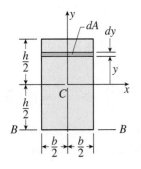

In a similar manner, use an element of area in the form of a vertical strip with area $dA = hdx$ and obtain the moment of inertia with respect to the y axis:

$$I_y = \int x^2 dA = \int_{-b/2}^{b/2} x^2 h\, dx = \frac{hb^3}{12} \qquad \text{(5-57)}$$

If a different set of axes is selected, the moments of inertia have different values. For instance, consider axis BB at the base of the rectangle (Fig. 5-28). If this axis is selected as the reference, you must define y as the coordinate distance from that axis to the element of area dA. Then the calculations for the moment of inertia become

$$I_{BB} = \int y^2 dA = \int_0^h y^2 b\, dy = \frac{bh^3}{3} \qquad \text{(5-58)}$$

Note that the moment of inertia with respect to axis *BB* is larger than the moment of inertia with respect to the centroidal *x* axis. In general, the moment of inertia increases as the reference axis is moved parallel to itself farther from the centroid.

Radius of Gyration

A distance known as the **radius of gyration** is occasionally encountered in mechanics. Radius of gyration of a plane area is defined as the square root of the moment of inertia of the area divided by the area itself; thus,

$$r_x = \sqrt{\frac{I_x}{A}} \qquad r_y = \sqrt{\frac{I_y}{A}} \qquad \textbf{(5-59a,b)}$$

where r_x and r_y denote the radii of gyration with respect to the *x* and *y* axes, respectively. Since the moment of inertia has units of length to the fourth power and the area has units of length to the second power, the radius of gyration has units of length.

   Although the radius of gyration of an area does not have an obvious physical meaning, consider it to be the distance (from the reference axis) where the entire area could be concentrated and still have the same moment of inertia as the original area.

Example 5-7

FIGURE 5-29

Moments of inertia of area bounded by parabola *AB*

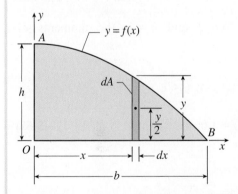

Determine the moments of inertia I_x and I_y for the area bounded by parabola *AB* shown in Fig. 5-29. The equation of the parabolic boundary is

$$y = f(x) = h\left(1 - \frac{x^2}{b^2}\right) \qquad \textbf{(5-60)}$$

Also find the radii of gyration about each axis. Assume that $b = 2$ and $h = 1$. [This same area was considered previously in Example 5-2.]

Solution:

Use the four-step problem-solving approach to find the moments of inertia by integration.

1. *Conceptualize:* To determine the moments of inertia by integration, use Eqs. (5-55a and b). The differential element of area dA is selected as a vertical strip of width dx and height y, as shown in Fig. 5-29. The area of this element is

$$dA = y\,dx = h\left(1 - \frac{x^2}{b^2}\right)dx \qquad \textbf{(5-61)}$$

As in Example 5-2, the area of the entire area bounded by parabola *AB* is

$$A = \int dA = \int_0^b h\left(1 - \left[\frac{x}{b}\right]^2\right)dx = \frac{2bh}{3} = \frac{2(2)(1)}{3} = \frac{4}{3} \qquad \textbf{(5-62)}$$

2. *Categorize:* Use second moments of area to find moments of inertia about both x and y axes.

3. *Analyze:*

Moments of inertia: Since every point in this element is at the same distance from the y axis, the moment of inertia of the element with respect to the y axis is $x^2 dA$. Therefore, the moment of inertia of the entire area with respect to the y axis is obtained as

$$I_y = \int x^2 \, dA = \int_0^b x^2 h \left(1 - \frac{x^2}{b^2} \right) dx = \frac{2hb^3}{15} \qquad \text{(5-63)}$$

To obtain the moment of inertia with respect to the x axis, note that the differential element of area dA has a moment of inertia dI_x with respect to the x axis equal to

$$dI_x = \frac{1}{3}(dx)y^3 = \frac{y^3}{3} dx$$

as obtained from Eq. (5-58). Hence, the moment of inertia of the entire area with respect to the x axis is

$$I_x = \int_0^b \frac{y^3}{3} dx = \int_0^b \frac{h^3}{3} \left(1 - \frac{x^2}{b^2} \right)^3 dx = \frac{16bh^3}{105} \qquad \text{(5-64)}$$

If you now insert numerical values $b = 2$ and $h = 1$, the moments of inertia are

$$I_x = \frac{16}{105}bh^3 = \frac{16}{105}(2)(1)^3 = 0.305$$

$$I_y = \frac{2}{15}b^3 h = \frac{2}{15}(2)^3(1) = 1.067 \qquad \text{(5-65)}$$

Radii of gyration: From Eqs. (5-59),

$$r_x = \sqrt{\frac{I_x}{A}} = 2\sqrt{\frac{2}{35}}h = 0.478 \qquad r_y = \sqrt{\frac{I_y}{A}} = \frac{b}{\sqrt{5}} = 0.894 \qquad \text{(5-66)}$$

4. *Finalize:* These same results for I_x and I_y can be obtained by using the element in the form of a horizontal strip of area $dA = x\,dy$ or by using a rectangular element of area $dA = dx\,dy$ and performing a double integration. Also, note that the preceding formulas for I_x and I_y agree with those given in Case 17 of Appendix B.

If the horizontal strip option is used $\left(\text{with } x = b\sqrt{1 - \dfrac{y}{h}} \right)$, integration for moments of inertia proceeds as

$$I_x = \int_0^h y^2 \left[b\sqrt{1 - \frac{y}{h}} \right] dy = \frac{16}{105} bh^3 \quad \text{and}$$

$$I_y = \int_0^h \frac{\left[b\sqrt{1 - \dfrac{y}{h}} \right]^3}{3} \, dy = \frac{2}{15} b^3 h \tag{5-67}$$

Transfer of Axes (Parallel-Axis Theorem)

In this section, a very useful theorem pertaining to moments of inertia of plane areas is derived. Known as the **parallel-axis theorem**, it gives the relationship between the moment of inertia with respect to a centroidal axis and the moment of inertia with respect to any parallel axis.

To derive the theorem, consider an area of arbitrary shape with centroid C (Fig. 5-30). Also consider two sets of coordinate axes: (1) the $x_c y_c$ axes with origin at the centroid and (2) a set of parallel x-y axes with origin at any point O. The distances between the two sets of parallel axes are denoted d_1 and d_2, which are shown in Fig. 5-30 as positive but may be positive or negative. Also, identify an element of area dA having coordinates x and y with respect to the centroidal axes.

From the definition of moment of inertia, write the following equation for the moment of inertia I_x with respect to the x axis:

$$I_x = \int (y + d_1)^2 dA = \int y^2 dA + 2d_1 \int y \, dA + d_1^2 \int dA \tag{5-68}$$

The first integral on the right-hand side is the moment of inertia I_{xc} with respect to the x_c axis. The second integral is the first moment of the area with respect to the x_c axis (this integral equals zero because the x_c axis passes through the centroid). The third integral is the area A itself. Therefore, the preceding equation reduces to

$$I_x = I_{xc} + Ad_1^2 \tag{5-69a}$$

Proceed in the same manner for the moment of inertia with respect to the y axis to obtain

$$I_y = I_{yc} + Ad_2^2 \tag{5-69b}$$

Equations (5-69a and b) represent the **parallel-axis theorem for moments of inertia:**

The moment of inertia of an area with respect to any axis in its plane is equal to the moment of inertia with respect to a parallel centroidal axis plus the product of the area and the square of the distance between the two axes.

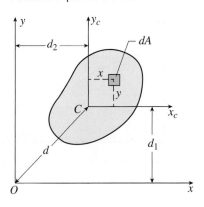

FIGURE 5-30

Derivation of parallel-axis theorem

FIGURE 5-28 (Repeated)

Moments of inertia of a rectangle

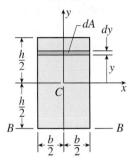

To illustrate the use of this theorem, consider again the rectangle shown in Fig. 5-28. Knowing that the moment of inertia about the x axis, which is through the centroid, is equal to $bh^3/12$ [see Eq. (5-56)], you can determine the moment of inertia I_{BB} about the base of the rectangle by using the parallel-axis theorem:

$$I_{BB} = I_x + Ad^2 = \frac{bh^3}{12} + bh\left(\frac{h}{2}\right)^2 = \frac{bh^3}{3}$$

This result agrees with the moment of inertia obtained previously by integration [see Eq. (5-58)].

From the parallel-axis theorem, the moment of inertia increases as the axis is moved parallel to itself farther from the centroid. Therefore, the moment of inertia about a centroidal axis is the *least moment of inertia* of an area (for a given direction of the axis).

When using the parallel-axis theorem, it is essential to remember that one of the two parallel axes *must* be a centroidal axis. If it is necessary to find the moment of inertia I_2 about a noncentroidal axis 2–2 (Fig. 5-31) when the moment of inertia I_1 about another noncentroidal (and parallel) axis 1–1 is known, apply the parallel-axis theorem twice. First, find the centroidal moment of inertia I_{xc} from the known moment of inertia I_1:

FIGURE 5-31

Plane area with two parallel noncentroidal axes (axes 1–1 and 2–2)

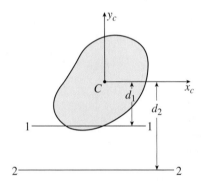

$$I_{xc} = I_1 - Ad_1^2 \tag{5-70}$$

Then find the moment of inertia I_2 from the centroidal moment of inertia:

$$I_2 = I_{xc} + Ad_2^2 = I_1 + A(d_2^2 - d_1^2) \tag{5-71}$$

This equation shows again that the moment of inertia increases with increasing distance from the centroid of the area.

Example 5-8

FIGURE 5-32

Parallel-axis (transfer) theorem for area bounded by parabola AB

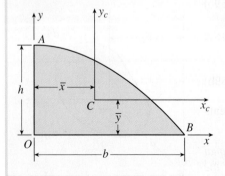

The area bounded by parabola AB shown in Fig. 5-32 has base b and height h. Using the parallel-axis theorem, determine the moments of inertia I_{xc} and I_{yc} with respect to the centroidal axes x_c and y_c.

Solution:

Use the parallel-axis (or transfer) theorem and the four-step problem-solving approach to find the centroidal moments of inertia of the area bounded by parabola AB.

1. *Conceptualize:* In Example 5-2, you found the centroid, and in Example 5-7, you computed the moments of inertia of the area bounded by parabola AB in Fig. 5-32 with respect to the x and y axes using integration. In many cases, centroidal moments of inertia of a plane area are of more interest.

2. *Categorize:* Use the parallel-axis theorem (rather than integration) to find the centroidal moments of inertia because the area A, the centroidal coordinates $\bar{x}$ and $\bar{y}$, and the moments of inertia I_x and I_y with respect to the x and y axes are already known. These quantities were obtained earlier in Examples 5-2 and 5-7. They are repeated here:

$$A = \frac{2bh}{3} \quad \bar{x} = \frac{3b}{8} \quad \bar{y} = \frac{2h}{5} \quad I_x = \frac{16bh^3}{105} \quad I_y = \frac{2hb^3}{15}$$

3. *Analyze:* To obtain the moment of inertia with respect to the x_c axis, use Eq. (5-70) and write the parallel-axis theorem as

$$I_{xc} = I_x - A\bar{y}^2 = \frac{16bh^3}{105} - \frac{2bh}{3}\left(\frac{2h}{5}\right)^2 = \frac{8bh^3}{175} \quad \text{(5-72)}$$

In a similar manner, obtain the moment of inertia with respect to the y_c axis:

$$I_{yc} = I_y - A\bar{x}^2 = \frac{2hb^3}{15} - \frac{2bh}{3}\left(\frac{3b}{8}\right)^2 = \frac{19hb^3}{480} \quad \text{(5-73)}$$

4. *Finalize:* The parallel-axis theorem also applies to radii of gyration. This is easily confirmed using the previous expressions. The radii of gyration for each axis are

$$r_x = \sqrt{\frac{I_x}{A}} = \sqrt{\frac{8h^2}{35}} \qquad r_y = \sqrt{\frac{I_y}{A}} = \sqrt{\frac{b^2}{5}}$$

$$r_{xc} = \sqrt{\frac{I_{xc}}{A}} = \sqrt{\frac{12h^2}{175}} \qquad r_{yc} = \sqrt{\frac{I_{yc}}{A}} = \sqrt{\frac{19b^2}{320}}$$

Dividing through by $A = 2bh/3$ in the transfer theorem expression in Eq. (5-72) gives

$$\frac{I_x}{A} = r_x^2 = \frac{I_{xc}}{A} + \frac{A}{A}\bar{y}^2 = r_{xc}^2 + \bar{y}^2 \quad \text{so} \quad r_x^2 = \frac{12}{175}h^2 + \left[\frac{2h}{5}\right]^2 = \frac{8}{35}h^2$$

Similarly, from Eq. (5-73),

$$r_y^2 = r_{yc}^2 + \bar{x}^2 = \frac{19b^2}{320} + \left[\frac{3b}{8}\right]^2 = \frac{b^2}{5}$$

Composite Areas

The moment of inertia of a **composite area** with respect to any particular axis its the sum of the moments of inertia of its parts with respect to that same axis. An example is the hollow box section shown in Fig. 5-33a, where the x and y axes are axes of symmetry through the centroid C. The moment of inertia I_x with respect to the x axis is equal to the algebraic sum of the moments of inertia of the outer and inner rectangles. (Think of the inner rectangle as a "negative area" and the outer rectangle as a "positive area.") Therefore,

$$I_x = \frac{bh^3}{12} - \frac{b_1 h_1^3}{12} \tag{5-74}$$

This same formula applies to the channel section shown in Fig. 5-33b, where the cutout is considered to be a "negative area."

For the hollow box section, use a similar technique to obtain the moment of inertia I_y with respect to the vertical axis. However, in the case of the channel section, the centroidal y axis for the composite shape is not in the same location for the positive and negative rectangular areas, so the determination of the moment of inertia I_y requires the use of the parallel-axis theorem discussed earlier.

Formulas for moments of inertia are listed in Appendix B. For shapes not shown, the moments of inertia usually can be obtained by using the listed formulas in conjunction with the parallel-axis theorem. If an area is of such irregular shape that its moments of inertia cannot be obtained in this manner, then use numerical methods. Divide the area into small elements of area ΔA_i, multiply each such area by the square of its distance from the reference axis, and then sum the products.

FIGURE 5-33

Composite areas

(a)   (b)

Example 5-9

Determine the moment of inertia I_c with respect to the horizontal axis $C\text{–}C$ through the centroid C of the beam cross section shown in Fig. 5-34. (The position of the centroid C was determined previously in Example 5-5.

Solution:

Use the parallel-axis (or transfer) theorem and the four-step problem-solving approach to find the moment of inertia of the reinforced W shape in Fig. 5-34 about axis $C\text{–}C$.

1. *Conceptualize:* In Example 5-5, you found the centroid of this composite shape. Now use tabulated properties for the three component areas to find I_c about horizontal axis $C\text{–}C$ through the centroid in Fig. 5-34.

FIGURE 5-34

Parallel-axis (transfer) theorem for a composite W shape

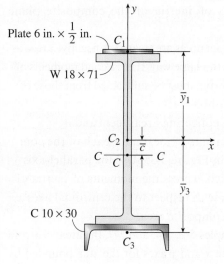

2. *Categorize:* Determine the moment of inertia I_c with respect to axis C–C by applying the parallel-axis theorem to each individual part of the composite area. The area divides naturally into three parts: (1) the cover plate, (2) the wide-flange section, and (3) the channel section. The following areas and centroidal distances were obtained previously in Example 5-5:

$$A_1 = 3.0 \text{ in}^2 \qquad A_2 = 20.8 \text{ in}^2 \qquad A_3 = 8.81 \text{ in}^2$$
$$\bar{y}_1 = 9.485 \text{ in.} \qquad \bar{y}_2 = 0 \qquad \bar{y}_3 = 9.884 \text{ in.} \qquad \bar{c} = 1.80 \text{ in.}$$

3. *Analyze:* The moments of inertia I_2 and I_3 are obtained from Tables C-1(a) and C-3(a), respectively, of Appendix C.

   Now use the parallel-axis theorem to calculate the moments of inertia about axis C–C for each of the three parts of the composite area:

$$(I_c)_1 = I_1 + A_1(\bar{y}_1 + \bar{c})^2 = 0.063 \text{ in}^4 + (3.0 \text{ in}^2)(11.28 \text{ in.})^2 = 382 \text{ in}^4$$
$$(I_c)_2 = I_2 + A_2\bar{c}^2 = 1170 \text{ in}^4 + (20.8 \text{ in}^2)(1.80 \text{ in.})^2 = 1240 \text{ in}^4$$
$$(I_c)_3 = I_3 + A_3(\bar{y}_3 - \bar{c})^2 = 3.94 \text{ in}^4 + (8.81 \text{ in}^2)(8.084 \text{ in.})^2 = 580 \text{ in}^4$$

The sum of these individual moments of inertia gives the moment of inertia of the entire cross-sectional area about its centroidal axis C–C:

$$I_c = (I_c)_1 + (I_c)_2 + (I_c)_3 = 2200 \text{ in}^4$$

4. *Finalize:* This example shows how to calculate moments of inertia of composite areas by using the parallel-axis theorem.

   Note: Axis C–C is the neutral axis for bending of this beam; therefore, the moment of inertia I_c must be determined in order to calculate the stresses and deflections of this beam.

Example 5-10

Find the moments of inertia I_x and I_y with respect to centroidal axes x_C and y_C, respectively, for the area bounded by parabola AB with two cutouts from Example 5-4. The two cutouts are a right triangle ($b \times h/6$) and a circle of radius $b/6$ centered at coordinates $x = b/3$, $y = h/2$ (Fig. 5-35). Assume that $b = 2$ and $h = 1$.

FIGURE 5-35

Moments of inertia of a composite area: area bounded by parabola AB with two cutouts

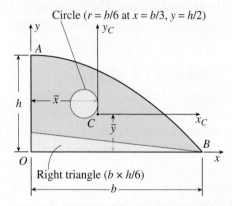

Solution:

Use tabulated solutions for centroid locations and moments of inertia and the four-step problem-solving approach to find the centroidal moments of inertia of the composite plane area.

1. *Conceptualize:* The cutout areas are treated as negative areas as in Example 5-4 and, in the same way, their second moments of area, or moments of inertia, must be subtracted from those of the area bounded by parabola AB.

2. *Categorize:* Begin by collecting areas, centroidal coordinates, and moments of inertia for each of the component areas shown in Figure 5-35. Use the parallel-axis (or transfer) theorem to express the moments of inertia of each component area with respect to the centroidal axes x_C and y_C of the entire composite area.

Area bounded by parabola AB: From Examples 5-2, 5-4, and 5-7, the area, centroid location, and moments of inertia about the x and y axes for the area bounded by parabola AB are

$$A_p = \frac{2bh}{3} \quad \overline{x}_p = \frac{3b}{8} \quad \overline{y}_p = \frac{2h}{5} \quad I_{px} = \frac{16}{105}bh^3 \quad I_{py} = \frac{2}{15}b^3h \quad \textbf{(5-75)}$$

Treat this area as positive. The parallel-axis theorem gives the moments of inertia of the area bounded by parabola AB about its own centroidal axes (see Example 5-8):

$$I_{pxC} = I_{px} - A_p\overline{y}_p^2 = \frac{8}{175}bh^3 \qquad I_{pyC} = I_{py} - A_p\overline{x}_p^2 = \frac{19}{480}b^3h \quad \textbf{(5-76)}$$

Right triangle cutout: The triangle cutout is treated as a negative area. From Case 6, Appendix B, areas, centroidal coordinates, and moments of inertia for the triangle cutout are

$$A_t = -\frac{1}{2}(b)\left(\frac{h}{6}\right) = -\frac{bh}{12} \quad \overline{x}_t = \frac{b}{3} \quad \overline{y}_t = \frac{1}{3}\left(\frac{h}{6}\right) = \frac{h}{18}$$

$$I_{txC} = \frac{-b\left[\dfrac{h}{6}\right]^3}{36} = \frac{-bh^3}{7776} \qquad I_{tyC} = \frac{-b^3\left[\dfrac{h}{6}\right]}{36} = \frac{-b^3h}{216} \qquad \textbf{(5-77)}$$

Circle cutout: The circle cutout is also treated as a negative area. From Case 9 of Appendix B, areas, centroidal coordinates, and moments of inertia for the circle cutout are

$$A_c = -\pi\left[\frac{b}{6}\right]^2 = -\frac{\pi b^2}{36} \qquad \overline{x}_c = \frac{b}{3} \qquad \overline{y}_c = \frac{h}{2}$$

$$I_{cxC} = I_{cyC} = \frac{-\pi\left[\dfrac{b}{6}\right]^4}{4} = -\frac{\pi b^4}{5184}$$

(5-78)

3. *Analyze:* The moments of inertia with respect to the overall centroidal axes x_C and y_C (Fig. 5-35) for the entire composite area are obtained by (a) transferring individual moments of inertia from the centroid of each component area to axes x_C and y_C and (b) then summing component inertias with cutout properties treated as negative quantities. Recall from Example 5-4 (using numerical dimensions $b = 2$, $h = 1$) that the coordinates of the centroid of the composite area (labeled as point C in Fig. 5-35) are $\overline{x} = 0.803$, $\overline{y} = 0.428$. The numerical results for each component area are obtained by substituting b and h values into Eqs. (5-75) to (5-78) to obtain:

Area bounded by parabola: $\qquad I_{pxC} + A_p(\overline{y} - y_p)^2 = 0.092$

$$I_{pyC} + A_p(\overline{x} - x_p)^2 = 0.32$$

Triangle cutout: $\qquad I_{txC} + A_p(\overline{y} - y_t)^2 = -0.023$

$$I_{tyC} + A_p(\overline{x} - x_t)^2 = -0.04$$

Circle cutout : $\qquad I_{cxC} + A_p(\overline{y} - y_c)^2 = -0.012$

(5-79)

$$I_{cyC} + A_p(\overline{x} - x_c)^2 = -0.016$$

Summation of these results $\left[\text{such as } I_x = \displaystyle\sum_{i=p,t,c} I_{ixC} + \sum_{i=p,t,c} A_i(\overline{y} - y_i)^2\right]$ provides the numerical values for centroidal moments of inertia as

$$I_x = 0.058 \qquad I_y = 0.264 \qquad \text{(5-80)}$$

If dimensions b and h have units, the units of I_x and I_y in Eq. (5-80) would be in^4 in USCS or mm^4 in SI.

As an alternative to the previous approach based on use of formulas for each part, use numerical values for dimensions ($b = 2$, $h = 1$) and organize the calculations in Eqs. (5-76) to (5-79) in the following tabular format. The final computed values of centroid coordinates I_x and I_y are found by adding table columns 5 and 7 and columns 6 and 8 to obtain the same results as in Eq. (5-80):

$$I_x = 0.08148 + (-0.02388) = 0.05760 \approx 0.058$$
$$I_y = 0.26993 + (-0.00585) = 0.26408 \approx 0.264$$

Part i	Area A	x_i	y_i	I_{ixC}	I_{iyC}	$A_i(\bar{y} - y_i)^2$ $\bar{y} = 0.42752$	$A_i(\bar{x} - x_i)^2$ $\bar{x} = 0.80257$
(1)	(2)	(3)	(4)	(5)	(6)	(7)	(8)
Parabola	1.3333	0.75000	0.40000	0.09143	0.31667	0.00101	0.00368
Triangle	−0.16667	0.66667	0.05556	−0.00026	−0.03704	−0.02306	−0.00308
Circle	−0.34907	0.66667	0.50000	−0.00970	−0.00970	−0.00183	−0.00645
Sum	**0.81760**			**0.08148**	**0.26993**	**−0.02388**	**−0.00585**

4. *Finalize:* Superposition of tabulated results for known component areas from Appendix B was used to find the moments of inertia of a composite area; summation rather than integration was used to obtain the desired results. This same summation approach can be used if the centroids of assemblages of lines, volumes, or masses are of interest.

You can easily compute the moments of inertia of the area bounded by parabola AB about its own centroidal axes using Eqs. 5-76 to find

$$I_{pxC} = \frac{8}{175}bh^3 = \frac{8}{175}(2)(1)^3 = 0.091$$

$$I_{pyC} = \frac{19}{480}b^3h = \frac{19}{480}(2)^3(1) = 0.317$$

(5-81)

Both values are larger than corresponding properties in Eq. (5-80), as expected, because negative contributions due to the triangle and circle cutouts have been removed.

Note that this summation approach does not apply to the calculation of radii of gyration for a composite shape, however. The final total area and moment of inertia must be used in the expression $r = \sqrt{\dfrac{I}{A}}$ to find the radius of gyration about a given axis.

Polar Moment of Inertia

The moments of inertia discussed in the preceding sections are defined with respect to axes lying in the plane of the area itself, such as the x and y axes in Fig. 5-36. Now consider an axis *perpendicular* to the plane of the area and intersecting the plane at the origin O. The moment of inertia with respect to this perpendicular axis is called the **polar moment of inertia** and is denoted by the symbol I_P.

The polar moment of inertia with respect to an axis through O perpendicular to the plane of the figure is defined by the integral:

$$I_P = \int \rho^2 dA$$

(5-82)

where ρ is the distance from point O to the differential element of area dA (Fig. 5-36). This integral is similar in form to those for moments of inertia I_x and I_y [see Eqs. (5-55a and b)].

Because as $\rho^2 = x^2 + y^2$, where x and y are the rectangular coordinates of the element dA, you obtain the following expression for I_P:

$$I_P = \int \rho^2 dA = \int (x^2 + y^2) dA = \int x^2 dA + \int y^2 dA$$

Thus, you obtain the important relationship

$$I_P = I_x + I_y \qquad \text{(5-83)}$$

This equation shows that the polar moment of inertia with respect to an axis perpendicular to the plane of the figure at any point O is equal to the sum of the moments of inertia with respect to *any* two perpendicular axes x and y passing through that same point and lying in the plane of the figure.

For convenience, I_P is usually referred to simply as the polar moment of inertia with respect to point O, without mentioning that the axis is perpendicular to the plane of the figure. Also, to distinguish them from **polar** moments of inertia, refer to I_x and I_y as **rectangular** moments of inertia.

Polar moments of inertia with respect to various points in the plane of an area are related by the **parallel-axis theorem for polar moments of inertia**. Derive this theorem by referring again to Fig. 5-30. Denote the polar moments of inertia with respect to the origin O and the centroid C by $(I_P)_O$ and $(I_P)_C$, respectively. Then, using Eq. (5-83), write

$$(I_P)_O = I_x + I_y \qquad (I_P)_C = I_{xc} + I_{yc} \qquad \text{(5-84)}$$

Now refer to the parallel-axis theorems derived earlier for rectangular moments of inertia [see Eqs. (5-69a and b)]. Add those two equations to get

$$I_x + I_y = I_{xc} + I_{yc} + A(d_1^2 + d_2^2)$$

Substitute from Eqs. (5-84), and also note that $d^2 = d_1^2 + d_2^2$ (Fig. 5-30), to obtain

$$(I_P)_O = (I_P)_C + Ad^2 \qquad \text{(5-85)}$$

This equation represents the **parallel-axis theorem** for polar moments of inertia:

The polar moment of inertia of an area with respect to any point O in its plane is equal to the polar moment of inertia with respect to the centroid C plus the product of the area and the square of the distance between points O and C.

To illustrate the determination of polar moments of inertia and the use of the parallel-axis theorem, consider a circle of radius r (Fig. 5-37). Take a differential element of area dA in the form of a thin ring of radius ρ and thickness $d\rho$ (thus, $dA = 2\pi\rho\, d\rho$). Since every point in the element is at the same distance ρ from the center of the circle, the polar moment of inertia of the entire circle with respect to the center is

$$(I_P)_C = \int \rho^2 dA = \int_0^r 2\pi\rho^3 d\rho = \frac{\pi r^4}{2} \qquad \text{(5-86)}$$

This result is listed in Case 9 of Appendix B.

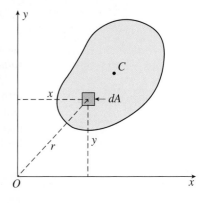

FIGURE 5-36

Plane area of arbitrary shape

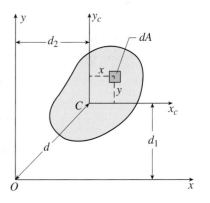

FIGURE 5-30 (Repeated)

Derivation of parallel-axis theorem

FIGURE 5-37

Polar moment of inertia of a circle

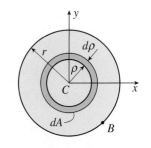

The polar moment of inertia of the circle with respect to any point B on its circumference (Fig. 5-37) can be obtained from the parallel-axis theorem:

$$(I_P)_B = (I_P)_C + Ad^2 = \frac{\pi r^4}{2} + \pi r^2 (r^2) = \frac{3\pi r^4}{2} \qquad (5\text{-}87)$$

As an incidental matter, note that the polar moment of inertia has its *smallest value* when the reference point is the centroid of the area.

A circle is a special case in which the polar moment of inertia can be determined by integration. However, most of the shapes encountered in engineering work do not lend themselves to this technique. Instead, polar moments of inertia are usually obtained by summing the rectangular moments of inertia for two perpendicular axes (Eq. 5-83).

Product of Inertia

The product of inertia of a plane area is defined with respect to a set of perpendicular axes lying in the plane of the area. Thus, referring to the area shown in Fig. 5-38, the **product of inertia** is defined with respect to the x and y axes as

$$I_{xy} = \int xy \, dA \qquad (5\text{-}88)$$

From this definition, you see that each differential element of area dA is multiplied by the product of its coordinates. As a consequence, products of inertia may be positive, negative, or zero, depending upon the position of the x-y axes with respect to the area.

If the area lies entirely in the first quadrant of the axes (as in Fig. 5-38), the product of inertia is positive because every element dA has positive coordinates x and y. If the area lies entirely in the second quadrant, the product of inertia is negative because every element has a positive y coordinate and a negative x coordinate. Similarly, areas entirely within the third and fourth quadrants have positive and negative products of inertia, respectively. When the area is located in more than one quadrant, the sign of the product of inertia depends upon the distribution of the area within the quadrants.

A special case arises when one of the axes is an **axis of symmetry** of the area. For instance, consider the area shown in Fig. 5-39, which is symmetric about the y axis. For every element dA having coordinates x and y, there exists an equal and symmetrically located element dA having the same y coordinate but an x coordinate of opposite sign. Therefore, the products $xy \, dA$ cancel each other, and the integral

FIGURE 5-38

Plane area of arbitrary shape

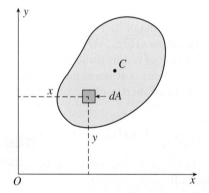

FIGURE 5-39

The product of inertia equals zero when one axis is an axis of symmetry

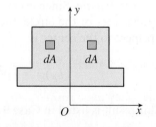

in Eq. (5-88) vanishes. Thus, *the product of inertia of an area is zero with respect to any pair of axes where at least one axis is an axis of symmetry of the area.*

As examples of the preceding rule, the product of inertia I_{xy} equals zero for the areas shown in Figs. 5-28, 5-33, and 5-39. In contrast, the product of inertia I_{xy} has a positive nonzero value for the L-shaped area shown in Figure 5-13. (These observations are valid for products of inertia with respect to the particular *x-y* axes shown in the figures. If the axes are shifted to another position, the product of inertia may change.)

Products of inertia of an area with respect to parallel sets of axes are related by a **parallel-axis theorem** that is analogous to the corresponding theorems for rectangular moments of inertia and polar moments of inertia. To obtain this theorem, consider the area shown in Fig. 5-40, which has centroid C and centroidal $x_c y_c$ axes. The product of inertia I_{xy} with respect to any other set of axes parallel to the $x_c y_c$ axes is

FIGURE 5-40

Plane area of arbitrary shape

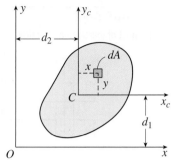

$$I_{xy} = \int (x + d_2)(y + d_1)\,dA$$
$$= \int xy\,dA + d_1 \int x\,dA + d_2 \int y\,dA + d_1 d_2 \int dA$$

where d_1 and d_2 are the coordinates of the centroid C with respect to the *x-y* axes (thus, d_1 and d_2 may have positive or negative values).

The first integral in the last expression is the product of inertia $I_{x_c y_c}$ with respect to the centroidal axes; the second and third integrals equal zero because they are the first moments of the area with respect to the centroidal axes; and the last integral is the area A. Therefore, the preceding equation reduces to

$$I_{xy} = I_{x_c y_c} + A d_1 d_2 \qquad \textbf{(5-89)}$$

This equation represents the **parallel-axis theorem for product of inertia**:

The product of inertia of an area with respect to any pair of axes in its plane is equal to the product of inertia with respect to parallel centroidal axes plus the product of the area and the coordinates of the centroid with respect to the pair of axes.

To demonstrate the use of this parallel-axis theorem, find the product of inertia of a rectangle with respect to *x-y* axes having their origin at point O at the lower left-hand corner of the rectangle (Fig. 5-41). The product of inertia with respect to the centroidal $x_c y_c$ axes is zero because of symmetry. Also, the coordinates of the centroid with respect to the *x-y* axes are

FIGURE 5-41

Parallel-axis theorem for products of inertia

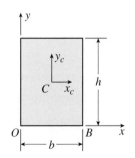

$$d_1 = \frac{h}{2} \qquad d_2 = \frac{b}{2}$$

Substituting into Eq. (5-89), you obtain

$$I_{xy} = I_{x_c y_c} + A d_1 d_2 = 0 + bh\left(\frac{h}{2}\right)\left(\frac{b}{2}\right) = \frac{b^2 h^2}{4} \qquad \textbf{(5-90)}$$

This product of inertia is positive because the entire area lies in the first quadrant. If the *x-y* axes are translated horizontally so that the origin moves to point B at the lower right-hand corner of the rectangle (Fig. 5-41), the entire area lies in the second quadrant and the product of inertia becomes $-b^2 h^2/4$ because now d_2 is negative $(-b/2)$.

The following example also illustrates the use of the parallel-axis theorem for products of inertia.

Example 5-11

FIGURE 5-42

Product of inertia of a Z section

Determine the product of inertia I_{xy} of the Z section shown in Fig. 5-42. The section has width b, height h, and constant thickness t.

SOLUTION:

Use the parallel-axis (or transfer) theorem and the four-step problem-solving approach to find the product of inertia of the Z shape in Fig. 5-42 for the centroidal x-y axes shown in the figure.

1. *Conceptualize:* To obtain the product of inertia with respect to the x-y axes through the centroid, divide the area into three parts and use the parallel-axis theorem. The three parts are labeled A_1 to A_3 in Fig. 5-42.

2. *Categorize:* The parts are (a) a rectangle of width $(b - t)$ and thickness t in the upper flange, (b) a similar rectangle in the lower flange, and (c) a web rectangle with height h and thickness t.

3. *Analyze:* The product of inertia of the web rectangle with respect to the x-y axes is zero (from symmetry). The product of inertia $(I_{xy})_1$ of the upper-flange rectangle (with respect to the x-y axes) is determined by using the parallel-axis theorem:

$$(I_{xy})_1 = I_{xcyc} + Ad_1d_2 \qquad \text{(5-91)}$$

where I_{xcyc} is the product of inertia of the rectangle with respect to its own centroid, A is the area of the rectangle, d_1 is the y coordinate of the centroid of the rectangle, and d_2 is the x coordinate of the centroid of the rectangle. Thus,

$$I_{xcyc} = 0 \qquad A = (b - t)(t) \qquad d_1 = \frac{h}{2} - \frac{t}{2} \qquad d_2 = \frac{b}{2}$$

Substitute into Eq. (5-91) to obtain the product of inertia of the rectangle in the upper flange:

$$(I_{xy})_1 = I_{xcyc} + Ad_1d_2 = 0 + (b - t)(t)\left(\frac{h}{2} - \frac{t}{2}\right)\left(\frac{b}{2}\right) = \frac{bt}{4}(h - t)(b - t)$$

The product of inertia of the rectangle in the lower flange is the same. Therefore, the product of inertia of the entire Z section is twice $(I_{xy})_1$, or

$$I_{xy} = \frac{bt}{2}(h - t)(b - t) \qquad \text{(5-92)}$$

4. *Finalize:* Note that this product of inertia is positive because the flanges lie in the first and third quadrants.

5.7 Rotation of Axes for Moments of Inertia

FIGURE 5-43

Rotation of axes

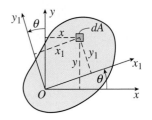

The moments of inertia and product of inertia of a plane area depend upon the position of the origin and the orientation of the reference axes. For a given origin, the moments and product of inertia vary as the axes are rotated about that origin. The manner in which they vary, and the magnitudes of the maximum and minimum values, are discussed in this and the following section.

Consider the plane area shown in Fig. 5-43, and assume that the x-y axes are a pair of arbitrarily located reference axes. The moments and products of inertia with respect to those axes are

$$I_x = \int y^2 dA \qquad I_y = \int x^2 dA \qquad I_{xy} = \int xy \, dA \qquad \text{(5-93a,b,c)}$$

where x and y are the coordinates of a differential element of area dA.

The x_1-y_1 axes have the same origin as the x-y axes but are rotated through a counterclockwise angle θ with respect to those axes. The moments and product of inertia with respect to the x_1-y_1 axes are denoted I_{x1}, I_{y1}, and I_{x1y1}, respectively. To obtain these quantities, use the coordinates of the element of area dA with respect to the x_1-y_1 axes. These coordinates may be expressed in terms of the x-y coordinates and the angle θ by geometry, as

$$x_1 = x \cos\theta + y \sin\theta \qquad y_1 = y \cos\theta - x \sin\theta \qquad \text{(5-94a,b)}$$

Then the moment of inertia with respect to the x_1 axis is

$$I_{x1} = \int y_1^2 dA = \int (y \cos\theta - x \sin\theta)^2 dA$$

$$= \cos^2\theta \int y^2 dA + \sin^2\theta \int x^2 dA - 2\sin\theta\cos\theta \int xy \, dA$$

or, by using Eqs. (5-93a, b, and c)

$$I_{x1} = I_x \cos^2\theta + I_y \sin^2\theta - 2I_{xy} \sin\theta\cos\theta \qquad \text{(5-95)}$$

Now use the following trigonometric identities:

$$\cos^2\theta = \frac{1}{2}(1 + \cos 2\theta) \qquad \sin^2\theta = \frac{1}{2}(1 - \cos 2\theta)$$

$$2\sin\theta\cos\theta = \sin 2\theta$$

Then Eq. (5-95) becomes

$$I_{x1} = \frac{I_x + I_y}{2} + \frac{I_x - I_y}{2}\cos 2\theta - I_{xy} \sin 2\theta \qquad \text{(5-96)}$$

In a similar manner, obtain the product of inertia with respect to the $x_1 y_1$ axes:

$$I_{x1y1} = \int x_1 y_1 \, dA = \int (x\cos\theta + y\sin\theta)(y\cos\theta - x\sin\theta) \, dA$$

$$= (I_x - I_y)\sin\theta\cos\theta + I_{xy}(\cos^2\theta - \sin^2\theta) \qquad \text{(5-97)}$$

Again use the trigonometric identities to obtain

$$I_{x1y1} = \frac{I_x - I_y}{2}\sin 2\theta + I_{xy} \cos 2\theta \qquad \text{(5-98)}$$

Equations (5-96) and (5-98) give the moment of inertia I_{x1} and the product of inertia I_{x1y1} with respect to the rotated axes in terms of the moments and product of inertia for the original axes. These equations are called the **transformation equations for moments and products of inertia**.

The moment of inertia I_{y1} may be obtained by the same procedure used for finding I_{x1} and I_{x1y1}. However, a simpler procedure is to replace θ with $\theta + 90°$ in Eq. (5-96). The result is

$$I_{y1} = \frac{I_x + I_y}{2} - \frac{I_x - I_y}{2}\cos 2\theta + I_{xy}\sin 2\theta \qquad \text{(5-99)}$$

This equation shows how the moment of inertia I_{y1} varies as the axes are rotated about the origin.

A useful equation related to moments of inertia is obtained by taking the sum of I_{x1} and I_{y1} [Eqs. (5-96) and (5-99)]. The result is

$$I_{x1} + I_{y1} = I_x + I_y \qquad \text{(5-100)}$$

This equation shows that the sum of the moments of inertia with respect to a pair of axes remains constant as the axes are rotated about the origin. This sum is the polar moment of inertia of the area with respect to the origin.

5.8 Principal Axes and Principal Moments of Inertia

The transformation equations for moments and products of inertia [Eqs. (5-96), (5-98), and (5-99)] show how the moments and products of inertia vary as the angle of rotation θ varies. The maximum and minimum values of the moment of inertia are of special interest. These values are known as the **principal moments of inertia**, and the corresponding axes are known as **principal axes**.

Principal Axes

To find the values of the angle θ that make the moment of inertia I_{x1} a maximum or a minimum, take the derivative with respect to θ of the expression on the right-hand side of Eq. (5-96) and set it equal to zero:

$$(I_x - I_y)\sin 2\theta + 2I_{xy}\cos 2\theta = 0 \qquad \text{(5-101)}$$

Solving for θ from this equation gives

$$\tan 2\theta_p = -\frac{2I_{xy}}{I_x - I_y} \qquad \text{(5-102)}$$

where θ_p denotes the angle defining a principal axis. This same result is obtained by taking the derivative of I_{y1} [Eq. (5-99)].

Equation (5-102) yields two values of the angle $2\theta_p$ in the range from 0 to 360°; these values differ by 180°. The corresponding values of θ_p differ by 90° and define the two perpendicular principal axes. One of these axes corresponds to the maximum moment of inertia and the other corresponds to the minimum moment of inertia.

Now examine the variation in the product of inertia I_{x1y1} as θ changes [see Eq. (5-98)]. If $\theta = 0$, $I_{x1y1} = I_{xy}$, as expected. If $\theta = 90°$, $I_{x1y1} = -I_{xy}$. Thus, during a 90° rotation the product of inertia changes sign, meaning for an intermediate orientation of the axes, the product of inertia must equal zero. To determine this orientation, set I_{x1y1} [Eq. (5-98)] equal to zero:

$$(I_x - I_y)\sin 2\theta + 2I_{xy}\cos 2\theta = 0$$

This equation is the same as Eq. (5-101), which defines the angle θ_p to the principal axes. Therefore, *the product of inertia is zero for the principal axes*.

Earlier the product of inertia of an area was shown to be zero with respect to a pair of axes if at least one of the axes is an axis of symmetry. It follows that if an area has an axis of symmetry, that axis and any axis perpendicular to it constitute a set of principal axes.

The preceding observations may be summarized as

1. Principal axes through an origin O are a pair of orthogonal axes for which the moments of inertia are a maximum and a minimum.
2. The orientation of the principal axes is given by the angle θ_p obtained from Eq. (5-102).
3. The product of inertia is zero for principal axes.
4. An axis of symmetry is always a principal axis.

Principal Points

Now consider a pair of principal axes with origin at a given point O. If there exists a *different* pair of principal axes through that same point, then *every* pair of axes through that point is a set of principal axes. Furthermore, the moment of inertia must be constant as the angle θ is varied.

The preceding conclusions follow from the nature of the transformation equation for I_{x1} [Eq. (5-96)]. Because this equation contains trigonometric functions of the angle 2θ, there is one maximum value and one minimum value of I_{x1} as 2θ varies through a range of 360° (or as θ varies through a range of 180°). If a second maximum exists, then the only possibility is that I_{x1} remains constant, which means that every pair of axes is a set of principal axes and all moments of inertia are the same.

A point so located that every axis through the point is a principal axis, and hence the moments of inertia are the same for all axes through the point, is called a **principal point**.

As an illustration, consider the rectangle of width $2b$ and height b shown in Fig. 5-44. The x-y axes, with origin at point O, are principal axes of the rectangle because the y axis is an axis of symmetry. The x'-y' axes, with the same origin, are also principal axes because the product of inertia $I_{x'y'}$ equals zero (because the triangles are symmetrically located with respect to the x' and y' axes). It follows that every pair of axes through O is a set of principal axes, and every moment of inertia is the same (and equal to $2b^4/3$). Therefore, point O is a principal point for the rectangle. (A second principal point is located where the y axis intersects the upper side of the rectangle.)

A useful corollary of the concepts described in the preceding four paragraphs applies to axes through the centroid of an area. Consider an area having *two different pairs* of centroidal axes such that at least one axis in each pair is an axis of symmetry. In other words, there exist two different axes of symmetry that are not perpendicular to each other. Then it follows that the centroid is a principal point.

Two examples, a square and an equilateral triangle, are shown in Figure 5-45. In each case, the x-y axes are principal centroidal axes because

FIGURE 5-44

Rectangle where every axis (in the plane of the area) through point O is a principal axis

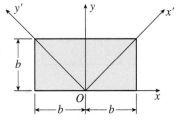

FIGURE 5-45

Examples of areas where every centroidal axis is a principal axis and the centroid C is a principal point

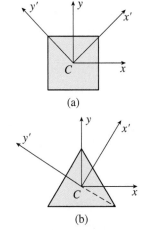

(a)

(b)

their origin is at the centroid C and at least one of the two axes is an axis of symmetry. In addition, a second pair of centroidal axes (the x'-y' axes) has at least one axis of symmetry. It follows that both the x-y and x'-y' axes are principal axes. Therefore, every axis through the centroid C is a principal axis, and every such axis has the same moment of inertia.

If an area has *three different axes of symmetry*—even if two of them are perpendicular—the conditions described in the preceding paragraph are automatically fulfilled. Therefore, if an area has three or more axes of symmetry, the centroid is a principal point and every axis through the centroid is a principal axis and has the same moment of inertia. These conditions are fulfilled for a circle, for all regular polygons (equilateral triangle, square, regular pentagon, regular hexagon, and so on), and for many other symmetric shapes.

In general, every plane area has two principal points. These points lie equidistant from the centroid on the principal centroidal axis having the larger principal moment of inertia. A special case occurs when the two principal centroidal moments of inertia are equal; then the two principal points merge at the centroid, which becomes the sole principal point.

FIGURE 5-46

Geometric representation of Eq. (5-102)

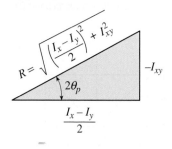

Principal Moments of Inertia

Now determine the principal moments of inertia, assuming that I_x, I_y, and I_{xy} are known. One method is to determine the two values of θ_p (differing by 90°) from Eq. (5-102) and then substitute these values into Eq. (5-96) for I_{x1}. The resulting two values are the principal moments of inertia denoted by I_1 and I_2. The advantage of this method is that you know which of the two principal angles θ_p corresponds to each principal moment of inertia.

It is also possible to obtain general formulas for the principal moments of inertia. Note from Eq. (5-102) and Fig. 5-46 [which is a geometric representation of Eq. (5-102)] that

$$\cos 2\theta_p = \frac{I_x - I_y}{2R} \quad \sin 2\theta_p = \frac{-I_{xy}}{R} \qquad \text{(5-103a,b)}$$

where

$$R = \sqrt{\left(\frac{I_x - I_y}{2}\right)^2 + I_{xy}^2} \qquad \text{(5-104)}$$

is the hypotenuse of the triangle. When evaluating R, always take the positive square root.

Now substitute the expressions for $\cos 2\theta_p$ and $\sin 2\theta_p$ [from Eqs. (5-103a and b)] into Eq. (5-96) for I_{x1} and obtain the algebraically larger of the two principal moments of inertia, denoted by the symbol I_1:

$$I_1 = \frac{I_x + I_y}{2} + \sqrt{\left(\frac{I_x - I_y}{2}\right)^2 + I_{xy}^2} \qquad \text{(5-105a)}$$

The smaller principal moment of inertia, denoted as I_2, may be obtained from

$$I_1 + I_2 = I_x + I_y$$

[see Eq. (5-100)]. Substitute the expression for I_1 into this equation and solve for I_2 to get

$$I_2 = \frac{I_x + I_y}{2} - \sqrt{\left(\frac{I_x - I_y}{2}\right)^2 + I_{xy}^2} \qquad \text{(5-105b)}$$

Equations (5-105a and b) provide a convenient way to calculate the principal moments of inertia.

The next example illustrates the method for locating the principal axes and determining the principal moments of inertia.

Mohr's Circle   The transformation equations for moments and products of inertia [Eqs. (5-96) and (5-98)] can be displayed in graphical form as a plot known as Mohr's circle. The plot is a useful way of showing the relationships between inertias I_{x1}, I_{y1}, and I_{x1y1} for various rotated positions of the x_1y_1 axes in Fig. 5-43.

First rearrange Eq. (5-96) to the form:

$$I_{x1} - \left(\frac{I_x + I_y}{2}\right) = \frac{I_x - I_y}{2}\cos 2\theta + I_{xy}\sin 2\theta \qquad \text{(5-106)}$$

Square both sides of Eq. (5-106) and also Eq. (5-98) and add the two results to eliminate parameter 2θ and obtain

$$\left(I_{x1} - \frac{I_x + I_y}{2}\right)^2 + I_{x1y1}^2 = \left(\frac{I_x - I_y}{2}\right)^2 + I_{xy}^2 \qquad \text{(5-107)}$$

To simplify Eq. (5-107), introduce the notation:

$$I_{ave} = \frac{I_x + I_y}{2} \qquad R = \sqrt{\left(\frac{I_x - I_y}{2}\right)^2 + I_{xy}^2} \qquad \text{(5-108)}$$

Substituting the expressions in Eqs. (5-108) into Eq. (5-107) gives the following equation of a circle:

$$\left(I_{x1} - I_{ave}\right)^2 + I_{x1y1}^2 = R^2 \qquad \text{(5-109)}$$

Plot the moment of inertia (I) on the horizontal axis and the product of inertia (I_{xy}) on the vertical axis (*positive downward*). The center of the circle has coordinates (I_{ave}, 0) and the radius is R (Fig. 5-46).

To construct Mohr's circle for moments and products of inertia, follow these steps:

1. Label point A at coordinates (I_x, $-I_{xy}$) [i.e., $\theta = 0°$ in Fig. 5-46] and point B at coordinates (I_y, I_{xy}) [i.e., $\theta = 90°$ in Fig. 5-46]; points A and B represent inertias for the body with respect to the x-y axes in Fig. 5-43.

2. Draw a line from A to B; this is a diameter of the circle and passes through the center at C [coordinates are (I_{ave}, 0)].

3. Use point C as the center and draw a circle through points A and B; the circle has radius R.

4. Clockwise (CW) or counterclockwise (CCW) movement on the circle through some angle 2θ (e.g., from A to A' and B to B' in Fig. 5-46) corresponds to rotation of the axes by angle θ in Fig. 5-43. CW movement on the circle corresponds to CW rotation of the axes, and so on for CCW rotation.

5. The most important use of Mohr's circle is to identify and display the orientation of the principal axes for the body; the principal axes are the x_1–y_1 axes in Fig. 5-43 where the product of inertia is zero and moments of inertia I_{x1} and I_{y1} reach maximum and minimum values, as discussed in the previous section and as illustrated in Example 5-12.

FIGURE 5-46

Construction of Mohr's circle for moments and products of inertia

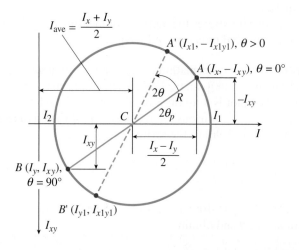

Example 5-12

FIGURE 5-47

Principal axes and principal moments of inertia for a Z section

Determine the orientations of the principal centroidal axes and the magnitudes of the principal centroidal moments of inertia for the cross-sectional area of the Z section shown in Fig. 5-47. Use the following numerical data: height $h = 200$ mm, width $b = 90$ mm, and constant thickness $t = 15$ mm.

Solution:

Use the transformation equations for moments and products of inertia and the four-step problem-solving approach to find the principal centroidal moments of inertia of the Z shape in Fig. 5-47. Confirm the solution using Mohr's circle for moments and products of inertia.

1. *Conceptualize:* Use the x–y axes (Fig. 5-47) as the reference axes through the centroid C. The moments and product of inertia with respect to these axes can be obtained by dividing the area into three rectangles and using the parallel-axis theorems, as illustrated in Example 5-11.

2. *Categorize:* The calculations for centroidal moments of inertia I_x, I_y, and I_{xy} lead to the following numerical results:

$$I_x = 29.29 \times 10^6 \text{ mm}^4 \qquad I_y = 5.667 \times 10^6 \text{ mm}^4$$

$$I_{xy} = -9.366 \times 10^6 \text{ mm}^4$$

Substitute these values into the equation for the angle θ_p (Eq. 5-102) to get

$$\tan 2\theta_p = -\frac{2I_{xy}}{I_x - I_y} = 0.7930 \qquad 2\theta_p = 34.4° \text{ and } 218.4°$$

Thus, the two values of θ_p are

$$\theta_p = 19.2° \text{ and } 109.2°$$

The principal axes are shown in Fig. 5-47 as the x_1–y_1 axes.

3. *Analyze*:

   Transformation equations: Use these values of θ_p in the transformation equation for I_{x1} [Eq. (5-96)] to find $I_{x1} = 32.6 \times 10^6 \text{ mm}^4$ and $2.4 \times 10^6 \text{ mm}^4$, respectively. These same values are obtained by substituting into Eqs. (5-105a and b). Thus, the principal moments of inertia and the angles to the corresponding principal axes are

$$I_1 = 32.6 \times 10^6 \text{ mm}^4 \qquad \theta_{p1} = 19.2°$$
$$I_2 = 2.4 \times 10^6 \text{ mm}^4 \qquad \theta_{p2} = 109.2°$$

   Mohr's circle: Alternatively, create Mohr's circle by plotting coordinates $(I_x, -I_{xy})$ to locate point A and coordinates (I_y, I_{xy}) to locate point B on Mohr's circle with moments of inertia I along the horizontal axis and product of inertia I_{xy} along the vertical axis (*positive downward*), as shown in Fig. 5-48 (multiply distance labels such as 32.6 by 10^6 mm^4 in Fig. 5-48). A counterclockwise (CCW) rotation through angle $2\theta_p$ on Mohr's circle corresponds to a CCW rotation of the x axis into principal centroidal axis x_1 through angle θ_p in Fig. 5-47. The maximum and minimum values of inertia I along the horizontal axis in Fig. 5-48 identify principal moments of inertia I_1 and I_2, respectively.

FIGURE 5-48

Mohr's circle for moments and products of inertia

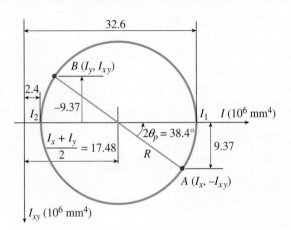

4. *Finalize:* Principal centroidal moments of inertia I and I, and principal rotation angle θ_p obtained from Mohr's circle are in agreement with results obtained using the transformation equations.

The center of Mohr's circle lies along the horizontal I axis at coordinates $[I_{\text{ave}}, 0]$ where

$$I_{\text{ave}} = \frac{I_x + I_y}{2} = 17.48(10^6 \text{ mm}^4)$$

The radius R of the circle is

$$R = \sqrt{\left[\frac{I_x - I_y}{2}\right]^2 + I_{xy}^2} = 15.07(10^6 \text{ mm}^4)$$

Note that, as expected, the *polar moment of inertia* of the Z section is invariant with respect to the set of centroidal axes that are used:

$$I_1 + I_2 = (32.6 + 2.4)(10^6 \text{ mm}^4) = (35.0)(10^6 \text{ mm}^4)$$
$$I_x + I_y = (29.3 + 5.7)(10^6 \text{ mm}^4)$$

CHAPTER SUMMARY AND REVIEW

In Chapter 5, **centroids of areas**, **lines**, **and volumes** are found using integral expressions based on first moments of areas, lengths, or volumes for a variety of different shapes. The resulting formulas are tabulated in **Appendix B** for plane areas such as rectangles, circles, and triangles. Assemblages of basic shapes with known centroid locations are then used to find centroid locations of **composite areas** to avoid evaluation of integral expressions. **Centers of mass and gravity** are defined. The two **theorems of Pappus** are used to compute **surface areas and volumes** for bodies created by revolving a line or plane area, respectively, about a nonintersecting axis.

Second moments of area, known as **moments of inertia**, are defined for selected x and y axes in a plane area and solutions developed for a variety of different shapes. The **parallel-axis** (or transfer) **theorem** is used to transfer moments of inertia between different parallel axes for a plane area. Moments of inertia for any rotated position of the coordinate axes for a plane area are computed from transformation equations. **Principal moments of inertia** are defined for that rotated position of the coordinate axes that gives maximum and minimum inertia values. The transformation equations can be represented in a graphical display known as **Mohr's circle** to facilitate the calculation of moments of inertia for any rotated position of the axes for a plane area of interest.

Some of the major concepts presented in Chapter 5 are given here.

1. The coordinates $(\bar{x}, \bar{y})$ of the **centroid C of a plane area** $A = \int dA$ are found by taking first moments of area about each coordinate axis as

$$\bar{x} = \frac{\int x\, dA}{\int dA} \qquad \bar{y} = \frac{\int y\, dA}{\int dA}$$

2. If the plane area has one or more **axes of symmetry**, the centroid C must lie on that axis because the first moment of an area is zero with respect to an axis of symmetry.

Area with one axis of symmetry

Area with two axes of symmetry

3. If the plane area is replaced by a **three-dimensional surface**, three centroidal coordinates are required; coordinates $\bar{x}$ and $\bar{y}$ are as stated in concept (1), and the third coordinate is

$$\bar{z} = \frac{\int z\, dA}{\int dA}$$

355

4. If the plane area is reduced to a **line** (or curve) of length $L = \int ds$, its centroid usually does not lie on the line but is found using first moments:

$$\bar{x} = \frac{\int x\,ds}{L} \qquad \bar{y} = \frac{\int y\,ds}{L} \qquad \bar{z} = \frac{\int z\,ds}{L}$$

5. If the curved surface is expanded to a **volume**, $V = \int dV$, its centroidal coordinates are

$$\bar{x} = \frac{\int x\,dV}{V} \qquad \bar{y} = \frac{\int y\,dV}{V} \qquad \bar{z} = \frac{\int z\,dV}{V}$$

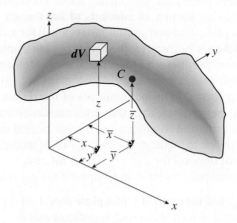

6. A composite area may be represented as an assemblage of basic shapes such as rectangles and triangles whose areas and centroidal coordinates are known. The centroid of a **composite area** is then found by summation rather than integration:

$$\bar{x} = \frac{Q_y}{A} = \frac{\sum\limits_{i=1}^{n} \bar{x}_i A_i}{\sum\limits_{i=1}^{n} A_i} \qquad \bar{y} = \frac{Q_x}{A} = \frac{\sum\limits_{i=1}^{n} \bar{y}_i A_i}{\sum\limits_{i=1}^{n} A_i}$$

7. If a body has mass density ρ defined as mass per unit volume, it then has mass $m = \int \rho\,dV$, and its **center of mass** is computed using first moments of mass as

$$\bar{x} = \frac{\int x\rho\,dV}{\int \rho\,dV} \qquad \bar{y} = \frac{\int y\rho\,dV}{\int \rho\,dV} \qquad \bar{z} = \frac{\int z\rho\,dV}{\int \rho\,dV}$$

If the density is constant throughout the body, ρ may be canceled in numerator and denominator in the previous expressions, and the center of mass and the volume centroid coincide.

8. In a gravitational field, a body of weight $W = \int dW = \int g\,dm$ has its **center of gravity** at coordinates

$$\bar{x} = \frac{\int x\,dW}{W} \qquad \bar{y} = \frac{\int y\,dW}{W} \qquad \bar{z} = \frac{\int z\,dW}{W}$$

9. The **first theorem of Pappus** states that the surface area A of a body created by revolving a *line* of length $L = \int ds$ and having centroid $\bar{y}$ about the x axis is computed without need for integration as

$$A = 2\pi \int y\,ds = 2\pi \bar{y} L$$

Moreover, the **second theorem of Pappus** states that the volume obtained by revolving a *plane area* A about a nonintersecting axis is

$$V = 2\pi \int y\,dA = 2\pi \bar{y} A.$$

10. Rectangular **moments of inertia** (also known as *second moments of area*) are defined for a plane area as

$$I_x = \int y^2 dA \qquad I_y = \int x^2 dA$$

The associated radii of gyration are computed as the square root of the moment of inertia divided by the plane area A:

$$r_x = \sqrt{\frac{I_x}{A}} \qquad r_y = \sqrt{\frac{I_y}{A}}$$

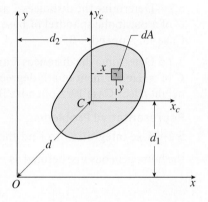

11. The **parallel axis theorem** is used to transfer the moment of inertia with respect to a centroidal axis (I_{xc} and I_{yc}) to parallel x-y axes as

$$I_x = I_{xc} + A d_1^2 \qquad I_y = I_{yc} + A d_2^2$$

12. Moments of inertia of **composite areas** are found by superposition of those for basic shapes (such as the shapes shown in Appendix B). Holes or cutouts in the composite area can be accounted for by using negative properties.

13. The **polar moment of inertia** for a plane area is the moment of inertia with respect to axis that is perpendicular to the plane of the figure and is defined as $I_p = \int r^2 dA$ where $r^2 = x^2 + y^2$. It follows that $I_p = I_x + I_y$.

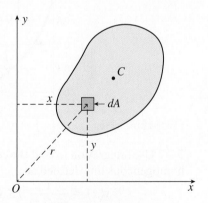

14. The **product of inertia** of a plane area is defined as $I_{xy} = \int xy\,dA$. The parallel-axis theorem applies for both polar moment of inertia and product of inertia.

15. **Rotation of axes** is used to find moments and products of inertia with respect to any set of axes x_1-y_1 that are rotated through angle θ from the x-y system. The transformation equations are

$$I_{x1} = \frac{I_x + I_y}{2} + \frac{I_x - I_y}{2}\cos 2\theta - I_{xy}\sin 2\theta$$

$$I_{y1} = \frac{I_x + I_y}{2} - \frac{I_x - I_y}{2}\cos 2\theta + I_{xy}\sin 2\theta$$

$$I_{x1y1} = \frac{I_x - I_y}{2}\sin 2\theta + I_{xy}\cos 2\theta$$

16. For some angle $\theta_p = \frac{1}{2}\tan^{-1}\left[\dfrac{-2I_{xy}}{I_x - I_y}\right]$, moments of inertia I_{x1} and I_{y1} have maximum and minimum values and the product of inertia I_{x1y1} is zero; these are known as **principal axes** and the moments of inertia are referred to as the **principal moments of inertia** I_1 and I_2. Principal rotation angle θ_p and principal inertias I_1 and I_2 are easily identified on **Mohr's circle**.

357

Problems

5.2 Centroids of Areas, Lines, and Volumes

The problems for Section 5.2 are to be solved by integration.

Introductory Problems

5.2-1 Determine the distances $\bar{x}$ and $\bar{y}$ to the centroid C of a right triangle having base b and altitude h (see Case 6, Appendix B).

5.2-2 Determine the distance $\bar{y}$ to the centroid C of a trapezoid having bases a and b and altitude h (see Case 8, Appendix B).

5.2-3 Determine the distance $\bar{y}$ to the centroid C of a semicircle of radius r (see Case 10, Appendix B).

5.2-4 Determine the distances $\bar{x}$ and $\bar{y}$ to the centroid C of a parabolic spandrel of base b and height h (see Case 18, Appendix B).

5.2-5 Determine the distances $\bar{x}$ and $\bar{y}$ to the centroid C of a semisegment of nth degree having base b and height h (see Case 19, Appendix B).

Representative Problems

5.2-6 Use integration to find the centroid $(\bar{x}, \bar{y})$ of the homogeneous line defined as $y(x) = \dfrac{1}{x^2}$ $1 \leq x \leq 3$.

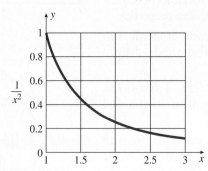

PROBLEM 5.2-6

5.2-7 Use integration to find the centroid $(\bar{x}, \bar{y})$ of the homogeneous line defined as

$$y(x) = \frac{1}{\sin(x^2)} \quad 1 \leq x \leq 1.5$$

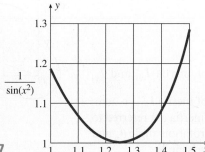

PROBLEM 5.2-7

358

5.2-8 Use integration to find the centroid $(\bar{x}, \bar{y})$ of the homogeneous line defined as

$$y(x) = \frac{\sin(x)}{x} \quad 0 \leq x \leq 5$$

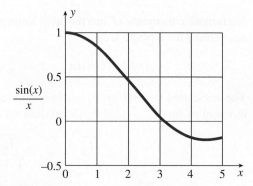

PROBLEM 5.2-8

5.2-9 Use integration to find the centroid $(\bar{x}, \bar{y})$ of the homogeneous line defined as $\dfrac{x^2}{a^2} + \dfrac{y^2}{b^2} = 1$. Let $a = 2, b = 1$.

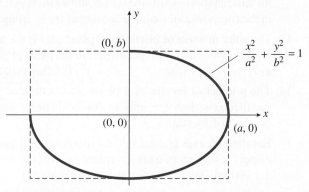

PROBLEM 5.2-9

5.2-10 Use integration to find the centroid $(\bar{x}, \bar{y})$ of the homogeneous parabolic line AOB. The coordinates of points A and B are given in meters.

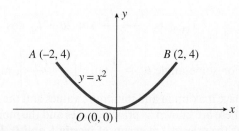

PROBLEM 5.2-10

5.2-11 Find the centroidal coordinates $(\bar{x}, \bar{y}, \bar{z})$ of a solid right circular cone (see figure). Let $r = 6$ in. and $h = 8$ in.

PROBLEM 5.2-11

5.2-12 Find the centroidal coordinates $(\bar{x}, \bar{y}, \bar{z})$ of a semi-circular solid cylinder (see figure). Let $r = 30$ mm and $h = 100$ mm.

PROBLEM 5.2-12

5.2-13 Find the centroidal coordinates $(\bar{x}, \bar{y}, \bar{z})$ of a semi-circular cylindrical solid wedge (see figure). Let $R = 6$ in. and $h = 8$ in.

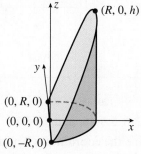

PROBLEM 5.2-13

5.2-14 Find the centroidal coordinates $(\bar{x}, \bar{y}, \bar{z})$ of a solid nonprismatic bar with rectangular cross section (see figure). Coordinates of joints A through H are given in meters.

$A\,(-2, 0, 2)$
$B\,(-2, 0, -2)$
$C\,(2, 0, -2)$
$D\,(2, 0, 2)$
$E\,(-1, 4, 1)$
$F\,(-1, 4, -1)$
$G\,(1, 4, -1)$
$H\,(1, 4, 1)$

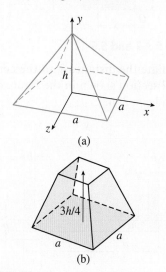

PROBLEM 5.2-14

5.2-15 Find the centroidal coordinates $(\bar{x}, \bar{y}, \bar{z})$ of a solid homogeneous trihedral $ABCD$.

(a) Let $a = b = c = 3$.

(b) Repeat for $a = 4$, $b = 5$, and $c = 6$.

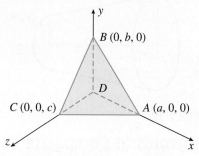

PROBLEM 5.2-15

5.2-16 Find the centroidal coordinates $(\bar{x}, \bar{y}, \bar{z})$ of homogeneous solid $ABCD$. Let $a = b = c = 3$.

$$y = b - \left[\left(\frac{x}{a}\right)^2 + \left(\frac{z}{c}\right)^2 \right]$$

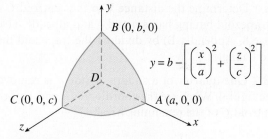

PROBLEM 5.2-16

5.2-17 (a) Find the centroidal coordinates $(\bar{x}, \bar{y}, \bar{z})$ of a solid square pyramid (see Fig. a). Assume that a = 5 in. and h = 4 in.

(b) Repeat if the pyramid is truncated at height $3h/4$ (see Fig. b).

(a)

(b)

PROBLEM 5.2-17

5.2-18 Find expressions for the centroidal coordinates $(\bar{x}, \bar{y}, \bar{z})$ of a homogeneous tapered circular solid cylinder (see figure). Assume that $r_a = h/3$.

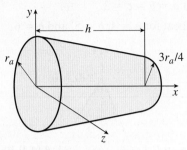

PROBLEM 5.2-18

5.3 Centroids of Composite Bodies

The problems for Section 5.3 are to be solved by using the formulas for composite areas.

Introductory Problems

5.3-1 Determine the distance $\bar{y}$ to the centroid C of a trapezoid having bases a and b and altitude h (see Case 8, Appendix B) by dividing the trapezoid into two triangles.

5.3-2 One quarter of a square of side a is removed (see figure). What are the coordinates $\bar{x}$ and $\bar{y}$ of the centroid C of the remaining area?

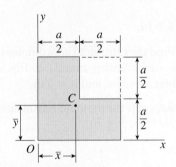

PROBLEMS 5.3-2 and 5.6-12

5.3-3 Calculate the distance $\bar{y}$ to the centroid C of the channel section shown in the figure if $a = 6$ in., $b = 1$ in., and $c = 2$ in.

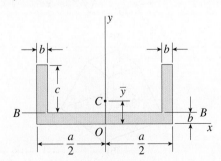

PROBLEMS 5.3-3, 5.3-4, and 5.6-15

5.3-4 What must be the relationship between the dimensions a, b, and c of the channel section shown in the figure in order that the centroid C will lie on line BB?

Representative Problems

5.3-5 The cross section of a beam constructed of a W 24 × 162 wide-flange section with an 8 in. × 3/4 in. cover plate welded to the top flange is shown in the figure.

Determine the distance $\bar{y}$ from the base of the beam to the centroid C of the cross-sectional area.

PROBLEMS 5.3-5 and 5.6-17

5.3-6 Determine the distance $\bar{y}$ to the centroid C of the composite area shown in the figure.

PROBLEMS 5.3-6, 5.6-16, and 5.6-34

5.3-7 Determine the coordinates $\bar{x}$ and $\bar{y}$ of the centroid C of the L-shaped area shown in the figure.

PROBLEMS 5.3-7, 5.6-7, 5.6-19, and 5.6-35

5.3-8 Determine the coordinates $\bar{x}$ and $\bar{y}$ of the centroid C of the area shown in the figure.

PROBLEM 5.3-8

5.3-9 Find the coordinates of the centroid $(\bar{x}, \bar{y})$ of the assemblage of lines shown in the figure.

PROBLEM 5.3-9

5.3-10 Find the coordinates of the centroid $(\bar{x}, \bar{y})$ of the T-shaped arrangement of lines shown in the figure.

PROBLEM 5.3-10

5.3-11 If the centroid $(0, \bar{y})$ of the Y-shaped arrangement of lines is 4 in. from the base, find required dimension h (see figure).

PROBLEM 5.3-11

5.3-12 Find the coordinates of the centroid $(\bar{x}, \bar{y}, \bar{z})$ of the set of perpendicular lines shown in the figure.

PROBLEM 5.3-12

5.3-13 Find the coordinates of the centroid $(\bar{x}, \bar{y}, \bar{z})$ of the set of lines shown in the figure. The coordinates of all points are given in inches.

PROBLEM 5.3-13

5.3-14 (a) Find the coordinates of the centroid $(\bar{x}, \bar{y})$ of the semicircular *area* with a rectangular cutout shown in the figure.

(b) Repeat if the body consists only of the perimeter *lines*.

PROBLEM 5.3-14

5.3-15 (a) Find the coordinates of the centroid $(\bar{x}, \bar{y})$ of the composite *area* shown in the figure.

(b) Repeat if the body consists only of the perimeter *lines*.

Assume that $b = 2r$.

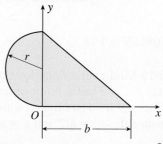

PROBLEM 5.3-15

5.3-16 Find the coordinates of the centroid $(\bar{x}, \bar{y}, \bar{z})$ of a homemade soccer goal (see figure). Assume that all members are of uniform size.

PROBLEM 5.3-16

5.3-17 Find the coordinates of the centroid $(\bar{x}, \bar{y}, \bar{z})$ of the homogeneous solid body shown in the figure. Assume that $L = 8$ in., $h = 2$ in., $t = 4$ in., $a = 0.5$ in., $b = 2$ in., and $c = 2$ in.

PROBLEM 5.3-17

5.3-18 Find expressions for the centroidal coordinates $(\bar{x}, \bar{y}, \bar{z})$ of a homogeneous circular solid cylinder (see figure) with a cylindrical cutout of length $h/2$. Assume that $r_a = h/3$ and $r_b = h/4$.

PROBLEM 5.3-18

5.3-19 Find expressions for the centroidal coordinates $(\bar{x}, \bar{y}, \bar{z})$ of a homogeneous tapered circular cylindrical solid (see figure) with a cylindrical cutout. Assume that $r_a = h/2$ and $r_b = h/4$.

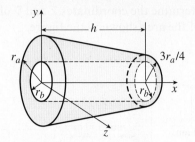

PROBLEM 5.3-19

5.3-20 Compare the centroid locations of the two bodies shown in the figures. In Fig. a, a solid cone of height h is attached to a solid cylinder of height t and radius r. In Fig. b, a cone of height h is attached to a solid hemisphere of radius r. Let $h = 75$ mm, $t = 15$ mm, and $r = 20$ mm.

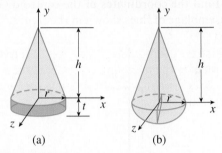

(a)　　　　(b)

PROBLEM 5.3-20

5.3-21 Find the coordinates of the centroid $(\bar{x}, \bar{y}, \bar{z})$ of a homogeneous cube (see figure) with a conical cutout. Assume that $a = 2r = 50$ mm.

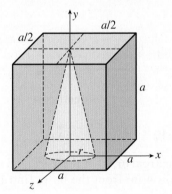

PROBLEM 5.3-21

5.4 Center of Mass and Center of Gravity

Introductory Problems

5.4-1 Find the coordinates of the *center of mass* $(\bar{x}, \bar{y})$ of the assemblage of uniform bars shown in the figure. Assume that top and bottom bars have mass per unit length 2ρ while that of the vertical bar is ρ.

PROBLEM 5.4-1

5.4-2 A straight circular steel bar of length 0.45 m and diameter of 15 mm has mass per unit length $\rho = \rho_0\left(\frac{x}{2}\right)$. If $\rho_0 = 7850$ kg/m^3, locate the *center of mass* of the bar.

PROBLEM 5.4-2

5.4-3 Find the coordinates of the *center of mass* $(\overline{x}, \overline{y}, \overline{z})$ of the set of perpendicular uniform bars shown in the figure. Bars in the *x-z* plane have mass per unit length $\rho/2$ while that of the vertical bar is ρ.

PROBLEM 5.4-3

5.4-4 Find the coordinates of the *center of mass* $(\overline{x}, \overline{y})$ of the T-shaped arrangement of bars shown in the figure. Assume that the horizontal bar has mass per unit length 2ρ while that of the vertical bar is ρ.

PROBLEM 5.4-1

Representative Problems

5.4-5 Find the coordinates of the *center of mass* $(\overline{x}, \overline{y}, \overline{z})$ of the set of uniform bars shown in the figure. The coordinates of all points are given in inches. Assume that bars AB and AC have mass per unit length 2ρ while that of all remaining bars is ρ.

PROBLEM 5.4-5

5.4-6 Find the coordinates of the *center of mass* $(\overline{x}, \overline{y})$ of the semicircular *area* with a rectangular cutout shown in the figure. The mass per unit area of area 1 is 20 kg/m^2 and that of area 2 is 30 kg/m^2.

PROBLEM 5.4-6

5.4-7 Find the coordinates of the *center of mass* $(\overline{x}, \overline{y})$ of the composite *area* shown in the figure. The semicircular area has mass per unit area of 2ρ while that of the triangular area is 1.5ρ. Assume that $b = 2r$.

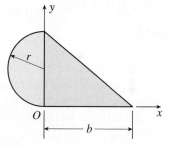

PROBLEM 5.4-7

5.4-8 Find the *center of gravity* of the semicircular plate segment shown in the figure. Segment 1 weighs 500 N/m^2 and segment 2 weighs 1200 N/m^2. Let $b = 60$ mm.

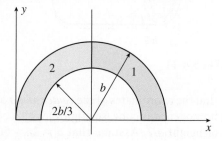

PROBLEM 5.4-8

363

5.4-9 Find the *center of gravity* of the uniform parallelogram plate shown in the figure. Triangle 1 weighs 6 lb/ft^2, and triangle 2 weighs 12 lb/ft^2. Let $b = 6$ in. and $h = 9$ in.

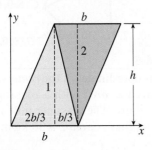

PROBLEM 5.4-9

5.4-10 Find the *center of gravity* of the assembly shown in the figure. The coordinates of all points are given in meters. Triangular plate ABC weighs 50 N. Bars BD and CF each weigh 15 N, and bars DE and FG each weigh 10 N.

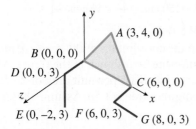

PROBLEM 5.4-10

5.4-11 Find the *center of mass* $(\bar{x}, \bar{y}, \bar{z})$ of a homogeneous circular solid cylinder (see figure) with a cylindrical insert of length $h/2$. Assume that $r_a = h/3$ and $r_b = h/4$. The outer cylinder has mass density ρ while that of the insert is 2ρ.

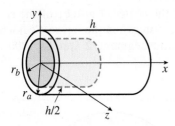

PROBLEM 5.4-11

5.4-12 Find the coordinates of the *center of mass* $(\bar{x}, \bar{y}, \bar{z})$ of a homogeneous cube (see figure) with a conical cavity of height $a/2$. Assume that $a = 2r = 50$ mm. Let $\rho =$ mass per unit volume.

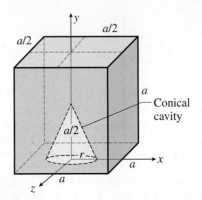

PROBLEM 5.4-12

5.5 Theorems of Pappus

Introductory Problems

5.5-1 A circular arc is rotated 360° about the y-axis to form a portion of a spherical shell. Find the surface area of the shell.

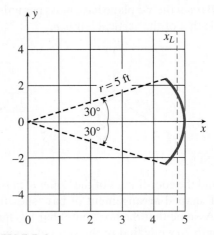

PROBLEM 5.5-1

5.5-2 If the parabolic curve OA is rotated 360° about the x axis, find the surface area of the resulting body.

5.5-3 Repeat Problem 5.5-2, but now rotate the parabolic curve through 80° about the y axis.

PROBLEMS 5.5-2 AND 5.5-3

364

5.5-4 If an elliptical curve is rotated 360° about the y axis, find the surface area of the resulting body.
Let $a = 2$ and $b = 1$.

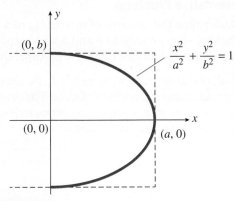

PROBLEM 5.5-4

5.5-5 If the parabolic curve BA is rotated 360° about the x axis, find the surface area of the resulting body.

PROBLEM 5.5-5

Representative Problems

5.5-6 Find the surface area of the truncated cone formed by revolving line AB 360° about the y axis.

PROBLEM 5.5-6

5.5-7 Find the volume of the body formed by revolving the L-shaped figure through 360° about the y axis.

PROBLEM 5.5-7

5.5-8 Find the surface area and volume of the body formed by revolving the isosceles triangular area shown in the figure through 360° about the y axis.

PROBLEM 5.5-8

5.5-9 Find the volume of a retaining wall formed by revolving the trapezoidal area shown in the figure through 60° about the y axis.

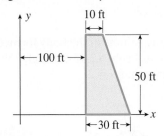

PROBLEM 5.5-9

5.5-10 Find the volume of the hollow tube wall formed by revolving the ring shaped area shown in the figure through 180° about the y axis.

PROBLEM 5.5-10

5.5-11 A small arch bridge is formed by revolving an isosceles trapezoidal area through 90° about the z axis (see figure). Find the volume of the bridge.

PROBLEM 5.5-11

365

5.5-12 The shaded area in the figure is composed of an isosceles triangle with a circular cutout of radius 4 mm. The center of the circular cutout is coincident with the centroid of the triangle. Find the volume of the body formed by revolving the shaded area through 360°: (a) about the y axis and (b) about the x axis.

PROBLEM 5.5-12

5.5-13 Find the volume of the body formed by revolving the body in Prob. 5.3-15 through 360° about the y axis. Assume that $b = 2r$.

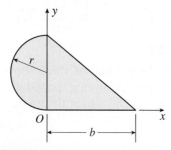

PROBLEM 5.5-13

5.6 Moments of Inertia for Plane and Composite Areas

Problems 5.6-1 through 5.6-4 are to be solved by integration.

Introductory Problems

5.6-1 Determine the moment of inertia I_x of a triangle of base b and altitude h with respect to its base (see Case 4, Appendix B).

5.6-2 Determine the moment of inertia I_{BB} of a trapezoid having bases a and b and altitude h with respect to its base (see Case 8, Appendix B).

5.6-3 Determine the moment of inertia I_x of a parabolic spandrel of base b and height h with respect to its base (see Case 18, Appendix B).

5.6-4 Determine the moment of inertia I_x of a circle of radius r with respect to a diameter (see Case 9, Appendix B).

Problems 5.6-5 through 5.6-9 are to be solved by considering the area to be a composite area.

Representative Problems

5.6-5 Determine the moment of inertia I_{BB} of a rectangle having sides of lengths b and h with respect to a diagonal of the rectangle (see Case 2, Appendix B).

5.6-6 Calculate the moment of inertia I_x for the composite circular area shown in the figure. The origin of the axes is at the center of the concentric circles, and the three diameters are 20, 40, and 60 mm.

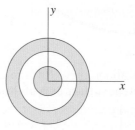

PROBLEM 5.6-6

5.6-7 Calculate the moments of inertia I_x and I_y with respect to the x and y axes for the L-shaped area shown in the figure for Prob. 5.3-7.

5.6-8 A semicircular area of radius 150 mm has a rectangular cutout of dimensions 50 mm × 100 mm (see figure).

Calculate the moments of inertia I_x and I_y with respect to the x and y axes. Also, calculate the corresponding radii of gyration r_x and r_y.

PROBLEM 5.6-8

5.6-9 Calculate the moments of inertia I_1 and I_2 of a W 16 × 100 wide-flange section using the cross-sectional dimensions given in Table C-1, Appendix C. (Disregard the cross-sectional areas of the fillets.) Also, calculate the corresponding radii of gyration r_1 and r_2, respectively.

5.6-10 Calculate the moments of inertia I_x and I_y with respect to the x and y axes for the semicircular segment shown in the figure ($b = 60$ mm). Also, calculate the corresponding radii of gyration r_x and r_y.

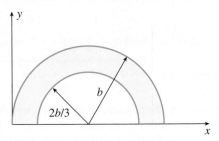

PROBLEMS 5.6-10, 5.6-20, and 5.6-28

5.6-11 Calculate the moments of inertia I_x and I_y with respect to the x and y axes for the parallelogram shown in the figure ($b = 6$ in., $h = 9$ in.). Also, calculate the corresponding radii of gyration r_x and r_y.

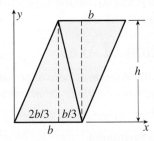

PROBLEMS 5.6-11, 5.6-21, and 5.6-27

Parallel-Axis Theorem

Introductory Problems

5.6-12 Determine the moment of inertia I_c with respect to an axis through the centroid C and parallel to the x axis for the geometric figure described in Prob. 5.3-2.

5.6-13 Calculate the moment of inertia I_b of a W 12 × 50 wide-flange section with respect to its base. (Use data from Table C-1, Appendix C.)

5.6-14 The moment of inertia with respect to axis 1–1 of the scalene triangle shown in the figure is 90×10^3 mm^4. Calculate its moment of inertia I_2 with respect to axis 2–2.

5.6-15 For the channel section described in Prob. 5.3-3, calculate the moment of inertia I_x, with respect to an axis through the centroid C and parallel to the x axis.

PROBLEM 5.6-14

Representative Problems

5.6-16 Calculate the moment of inertia I_{xc} with respect to an axis through the centroid C and parallel to the x axis for the composite area shown in the figure for Prob. 5.3-6.

5.6-17 For the beam cross section described in Prob. 5.3-5, calculate the centroidal moments of inertia I_{xc} and I_{yc} with respect to axes through the centroid C such that the x_c axis is parallel to the x axis and the y_c axis coincides with the y axis.

5.6-18 The wide-flange beam section shown in the figure has a total height of 250 mm and a constant thickness of 15 mm.

Determine the flange width b if it is required that the centroidal moments of inertia I_x and I_y be in the ratio 3 to 1, respectively.

PROBLEM 5.6-18

5.6-19 Calculate the centroidal moments of inertia I_{xc} and I_{yc} with respect to axes through the centroid C and parallel to the x and y axes, respectively, for the L-shaped area shown in the figure for Prob. 5.3-7.

5.6-20 Determine the moment of inertia I_{xc} with respect to an axis through the centroid C and parallel to the x axis for the geometric figure described in Prob. 5.6-10. Note that $b = 60$ mm.

5.6-21 Determine the moment of inertia I_{xc} with respect to an axis through the centroid C and parallel to the x axis for the geometric figure described in Prob. 5.6-11. Note that $b = 6$ in. and $h = 9$ in.

367

Polar Moments of Inertia

Introductory Problems

5.6-22 Determine the polar moment of inertia I_p of an isosceles triangle of base b and altitude h with respect to its apex (see Case 5, Appendix B).

5.6-23 Determine the polar moment of inertia I_p for a W 8 × 21 wide-flange section with respect to one of its outermost corners.

5.6-24 Determine the polar moment of inertia $(I_p)_C$ with respect to the centroid C for a circular sector (see Case 13, Appendix B).

Representative Problems

5.6-25 Obtain a formula for the polar moment of inertia I_p with respect to the midpoint of the hypotenuse for a right triangle of base b and height h (see Case 6, Appendix B).

5.6-26 Determine the polar moment of inertia $(I_p)_C$ with respect to the centroid C for a quarter-circular spandrel (see Case 12, Appendix B).

5.6-27 Determine the polar moment of inertia $(I_P)_C$ with respect to an axis through the centroid C for the geometric figure described in Prob. 5.6-11. Note that $b = 6$ in. and $h = 9$ in.

5.6-28 Determine the polar moment of inertia $(I_P)_C$ with respect to an axis through the centroid C for the geometric figure described in Prob. 5.6-10. Note that $b = 60$ mm.

Products of Inertia
Introductory Problems

5.6-29 Using integration, determine the product of inertia I_{xy} for the parabolic semisegment shown in Fig. 5-11 (see also Case 17 in Appendix B).

5.6-30 Using integration, determine the product of inertia I_{xy} for the quarter-circular spandrel shown in Case 12, Appendix B.

Representative Problems

5.6-31 Find the relationship between the radius r and the distance b for the composite area shown in the figure in order that the product of inertia I_{xy} will be zero.

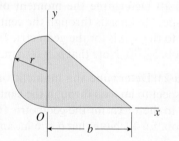

PROBLEM 5.6-31

368

5.6-32 Obtain a formula for the product of inertia I_{xy} of the symmetrical L-shaped area shown in the figure.

PROBLEM 5.6-32

5.6-33 Calculate the product of inertia I_{12} with respect to the centroidal axes 1–1 and 2–2 for an L 6 in. × 6 in. × 1 in. angle section (see Table C-4, Appendix C). (Disregard the cross-sectional areas of the fillet and rounded corners.)

5.6-34 Calculate the product of inertia I_{xy} for the composite area shown in Prob. 5.3-6.

5.6-35 Determine the product of inertia I_{xcyc} with respect to centroidal axes x_c and y_c parallel to the x and y axes, respectively, for the L-shaped area shown in Prob. 5.3-7.

5.7 Rotation of Axes

The problems for Section 5.7 are to be solved by using the transformation equations for moments and products of inertia.

Introductory Problems

5.7-1 Determine the moments of inertia I_{x1} and I_{y1} and the product of inertia I_{x1y1} for a square with sides b, as shown in the figure. (Note that the $x_1 y_1$ axes are centroidal axes rotated through an angle θ with respect to the x-y axes.)

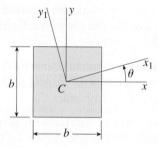

PROBLEM 5.7-1

5.7-2 Determine the moments and product of inertia with respect to the x_1–y_1 axes for the rectangle shown in the figure. (Note that the x_1 axis is a diagonal of the rectangle.)

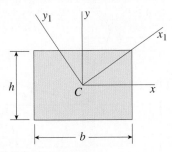

PROBLEM 5.7-2

5.7-3 Calculate the moment of inertia I_d for a W 12 × 50 wide-flange section with respect to a diagonal passing through the centroid and two outside corners of the flanges. (Use the dimensions and properties given in Table C-1.)

Representative Problems

5.7-4 Calculate the moments of inertia I_{x1} and I_{y1} and the product of inertia I_{x1y1} with respect to the axes for x_1–y_1 the L-shaped area shown in the figure if $a = 150$ mm, $b = 100$ mm, $t = 15$ mm, and $\theta = 30°$.

PROBLEM 5.7-4

5.7-5 Calculate the moments of inertia I_{x1} and I_{y1} and the product of inertia I_{x1y1} with respect to the x_1–y_1 axes for the Z section shown in the figure if $b = 3$ in., $h = 4$ in., $t = 0.5$ in., and $\theta = 60°$.

5.7-6 Solve the preceding problem if $b = 80$ mm, $h = 120$ mm, $t = 12$ mm, and $\theta = 30°$.

PROBLEMS 5.7-5 and 5.7-6

5.8 Principal Axes and Principal Moments of Inertia

Introductory Problems

5.8-1 An ellipse with major axis of length $2a$ and minor axis of length $2b$ is shown in the figure.

(a) Determine the distance c from the centroid C of the ellipse to the principal points P on the minor axis (y axis).

(b) For what ratio a/b do the principal points lie on the circumference of the ellipse?

(c) For what ratios do they lie inside the ellipse?

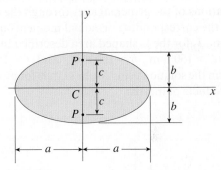

PROBLEM 5.8-1

5.8-2 Demonstrate that the two points P_1 and P_2, located as shown in the figure, are the principal points of the isosceles right triangle.

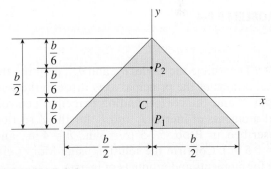

PROBLEM 5.8-2

5.8-3 Determine the angles θ_{p1} and θ_{p2}, defining the orientations of the principal axes through the origin O for the right triangle shown in the figure if $b = 6$ in. and $h = 8$ in. Also, calculate the corresponding principal moments of inertia I_1 and I_2. Confirm the solution using Mohr's circle for moments and products of inertia.

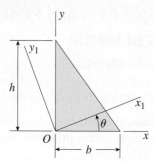

PROBLEM 5.8-3

Representative Problems

5.8-4 Determine the angles θ_{p1} and θ_{p2}, defining the orientations of the principal axes through the origin O and the corresponding principal moments of inertia I_1 and I_2 for the L-shaped area described in Prob. 5.7-4 ($a = 150$ mm, $b = 100$ mm, and $t = 15$ mm) Confirm the solution using Mohr's circle for moments and products of inertia.

PROBLEM 5.8-4

5.8-5 Determine the angles θ_{p1} and θ_{p2}, defining the orientations of the principal axes through the centroid C and the corresponding principal centroidal moments of inertia I_1 and I_2 for the Z section described in Prob. 5.7-5 ($b = 3$ in., $h = 4$ in., and $t = 0.5$ in.). Confirm the solution using Mohr's circle for moments and products of inertia.

5.8-6 Solve the preceding problem for the Z section described in Prob. 5.7-6 if $b = 80$ mm, $h = 120$ mm, and $t = 12$ mm.

PROBLEM 5.8-5 and 5.8-6

5.8-7 Determine the angles θ_{p1} and θ_{p2} defining the orientations of the principal axes through the centroid C for the right triangle shown in the figure if $h = 2b$. Also, determine the corresponding principal centroidal moments of inertia I_1 and I_2. Confirm the solution using Mohr's circle for moments and products of inertia.

PROBLEM 5.8-7

5.8-8 Determine the angles θ_{p1} and θ_{p2}, defining the orientations of the principal centroidal axes and the corresponding principal moments of inertia I_1 and I_2 for the L-shaped area shown in the figure if $a = 80$ mm, $b = 150$ mm, and $t = 16$ mm. Confirm the solution using Mohr's circle for moments and products of inertia.

5.8-9 Solve the preceding problem if $a = 3$ in., $b = 6$ in., and $t = 5/8$ in.

PROBLEMS 5.8-8 and 5.8-9

Internal Effects in Bars, Shafts, Beams, and Frames

This heavy mining drill machine is made up of a variety of structural elements that are subjected to axial, torsional, and bending effects.

CHAPTER OBJECTIVES

- Study *internal forces and moments* in bars, shafts, beams, and frames.
- Derive *differential and integral relationships* between internal *axial force $N(x)$* and *distributed axial load $q(x)$* for bars acted on by axial forces.
- Derive *differential and integral relationships* between internal *torsional moment $T(x)$* and distributed *applied torsional loading $t(x)$* for shafts with applied torsional moments.
- Define different *types of beams*, various types of loads that act on beams, and different *support conditions* for beams.

- Derive *differential and integral relationships* between *distributed transverse load $q(x)$*, internal shear force $V(x)$, and bending moment $M(x)$ for *beams*.
- Draw *axial-force diagrams* (*AFD*) for bars, *torsional moment diagrams* (*TMD*) for shafts, and *shear* (*V*) and *bending-moment* (*M*) *diagrams* for beams.
- Use diagrams to find maximum axial force, torsional moment, and shear and moment in bars, shafts and beams, respectively, which are important in design.

Chapter Outline

371

6.1 Introduction

Once you have established the overall static equilibrium of a structure, having computed reactions at the supports (see Chapter 3), you can make section cuts normal to the axis of each member to find internal forces and moments at any point. Internal axial forces are needed in the calculation of axial deformation of **bars**; internal torsional moments are needed in the calculation of torsional deformations of **shafts**; and internal shear forces and bending moments are needed in the calculation of shearing and bending deformations of **beams**. These internal effects are also needed to compute stresses and strains at locations along the length of these elements. Maximum values of internal forces and moments govern the design of these components.

6.2 Bars with Axial Loads

Structural components subjected only to *tension* or *compression* are known as **axially loaded members**. Solid bars with straight longitudinal axes, such as truss members, are the most common type, although cables and coil springs also carry axial loads. Other examples of axially loaded bars are connecting rods in engines, spokes in bicycle wheels, columns in buildings, and struts in aircraft engine mounts.

Internal Forces in Bars

To begin, consider the prismatic cantilever bar of length L subjected to applied force P_0 at the free end (Fig. 6-1a). Load P_0 is applied (at $x = L$) along the longitudinal centroidal axis of the bar. Reaction force R at the fixed support (at $x = 0$) is easily found using $\Sigma F_x = 0$, so $R + P_0 = 0$ or $R = -P_0$. From the right-hand free body (RHFB, Fig. 6-1b), the **internal axial force** $N(x) = P_0$. In fact, axial force N is a *tensile force* and is *constant* over the entire length of the bar.

Distributed Axial Load　If the axial load on the bar is a distributed load, such as the axial centroidal **uniform load** of intensity q_0 in Fig. 6-1c, multiply load intensity q_0 by the distance along the bar over which it acts. Hence reaction force R at the support is found using $\Sigma F_x = 0$, so $R + q_0L = 0$ or $R = -q_0L$. A common alternative way to show uniform axial load q_0 is presented in Fig. 6-1d, where a rectangle of constant height q_0 is used and the distributed load is shown above the bar for clarity. It is understood that the load is actually applied along the longitudinal centroidal axis. Now the total axial load on the bar is seen to be the area under the distributed load or $F_q = \int_0^L q_0 dx = q_0L$ and once again, enforcing overall static equilibrium, gives $R = -F_q = -q_0L$. The **internal axial force** $N(x)$ at some position x along the bar can be found from the RHFB in either Fig. 6-1c or d as

$$N(x) = q_0(L - x) \tag{6-1}$$

Now, axial force N is a *tensile force* that varies linearly with x over the entire length of the bar; at $x = 0$, N is equal to the reaction force R while at $x = L$, $N = 0$.

FIGURE 6-1

(a) Cantilever bar with applied axial force P_0 at $x = L$; (b) cantilever bar with constant internal axial force $N(x)$; (c) cantilever bar with uniform axial load of intensity q_0; and (d) alternative way to show uniform axial load of intensity q_0 on cantilever bar

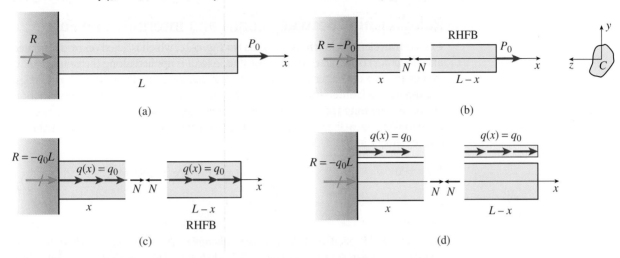

Last, consider the **linearly varying** axial centroidal load $q(x)$ shown in Fig. 6-2a. The load has intensity of zero at $x = 0$ (that is, $q(0) = 0$) and has peak intensity q_0 at $x = L$ (so $q(L) = q_0$), and the intensity as a function of x is $q(x) = q_0\left(\dfrac{x}{L}\right)$. The total axial load on the bar is $F_q = \int_0^L q(x)dx = \dfrac{1}{2}q_0 L$, and F_q is the area under the load triangle. Static equilibrium requires that $\Sigma F_x = 0$ or $R + F_q = 0$, so $R = -\dfrac{1}{2}q_0 L$.

From $\Sigma F_x = 0$ using the left-hand free-body diagram in Fig. 6-2b, **internal axial force** $N(x)$ is equal to

$$N(x) = \frac{1}{2}q_0 L - \frac{1}{2}q(x)x = \frac{1}{2}q_0 L\left(1 - \frac{x^2}{L^2}\right) \qquad \textbf{(6-2)}$$

You can plot Eq. (6-2) to show the variation of internal force $N(x)$ over the entire length of the bar.

Sign Convention In accordance with Newton's third law (action–reaction), internal forces and moments are equal in magnitude but opposite in direction on the two faces of each cut section in Figs. 6-1 and 6-2. If internal axial force vector N acts away from the cut section, the segment upon which it acts is in *tension* (which is usually regarded as positive); if N acts toward the cut face, the segment is in *compression* (usually regarded as negative). For this reason, the *deformation sign convention* to be used for internal axial force N is as follows:

> *An **axial force (N) vector** acting away from the face of a cut section in a bar is positive; a force vector pointing toward the cut section is negative.*

FIGURE 6-2

(a) Cantilever bar with linearly varying axial load with peak intensity q_0, and (b) left-hand free-body diagram

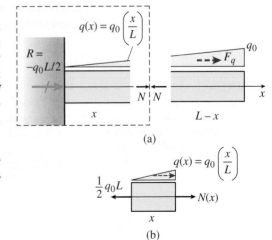

If the computed result for internal force N turns out to be negative, an incorrect assumption for direction was made, and the direction of N must be reversed on the free-body diagram of the bar.

Relationships Between Loads and Internal Axial Force

Consider an element taken from the bar in Fig. 6-1c, which is acted on by distributed axial load $q(x)$ (see Fig. 6-3). The distributed load is applied along the centroidal (x) axis of the bar, and internal axial forces $N(x)$ are developed within the bar to resist the applied load; differential force dN accounts for any changes in internal axial force as you move along the bar. The internal forces are shown in the assumed positive directions in Fig. 6-3b in accordance with a deformation sign convention. Equilibrium of the element in Fig. 6-3b requires that

$$\Sigma F_x = 0 \qquad\qquad -N + q(x)dx + (N + dN) = 0$$

$$\text{or} \qquad\qquad \frac{dN}{dx} = -q \qquad\qquad \textbf{(6-3)}$$

Equation (6-3) specifies that *the rate of change of the axial force within the bar is equal to the negative of the intensity of the distributed load at that same point along the bar*. If there is no applied distributed load over some region of the bar (that is, $q = 0$ as for the bar in Fig. 6-1b), then $\dfrac{dN}{dx} = 0$ and the internal axial force N is constant in that region. In the same way, if the axial load is *uniformly distributed* over a segment of the bar (Fig. 6-1c), dN/dx is constant and internal axial force N varies linearly with position x in that segment of the bar [see Eq. (6-1)]. And if the distributed axial load varies *linearly* (Fig. 6-2), the internal axial force $N(x)$ is a quadratic function of x as shown in Eq. (6-2).

To obtain a second relationship between load q and axial force N, multiply both sides of Eq. (6-3) by dx and integrate between any two points A and B along the axis of the bar:

$$\int_A^B dN = N_B - N_A = -\int_A^B q(x)dx \qquad\qquad \textbf{(6-4)}$$

Expressing Eq. (6-4) in words:

The change in internal axial force between any two points along a bar is equal to the negative of the area under the axial load curve between those same two points.

FIGURE 6-3

(a, b) Element of a bar used in deriving relationships between loads and internal axial force (distributed axial load q and internal axial force N are shown in their positive directions)

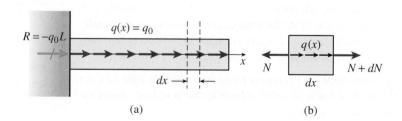

(a)   (b)

Confirm this for the bar with linearly varying axial loading in Fig. 6-2. If point A is taken at $x = 0$ and point B is at $x = L$,

$$N_B - N_A = -\frac{1}{2}q_0 L = -F_q \qquad \text{(6-5)}$$

where F_q is the area under the triangular load curve.

Axial-Force Diagram (AFD)

A graphical display of the variation in internal axial force over the length of a bar helps a designer to quickly identify critical regions of the bar, such as the location of maximum internal axial force $N_{\max}$. This display is called the *axial-force diagram* (AFD). Guidelines for construction of axial-force diagrams are based on the differential and integral relationships between $q(x)$ and $N(x)$ in Eqs. (6-3) and (6-4). These guidelines are given here.

- The slope at any point on the AFD is equal to the $(-)$ ordinate on the distributed axial-load diagram (that is, $q(x)$) at that same point.
- The change in internal axial force between any two points along a bar is equal to the negative of the area under the axial load diagram between those same two points.
- At the point(s) where the axial-load diagram crosses the reference axis, the value of the internal axial force on the axial-force diagram is a local maximum or minimum.
- At points of application of concentrated loads, there is a discontinuity, or change in the ordinate, on the AFD.

Using these guidelines, you can quickly sketch the AFD for the bar with linearly varying axial force in Fig. 6-2 (repeated in Fig. 6-4a and 6-4b). The cantilever bar with a linearly varying distributed axial load is again shown in Fig. 6-4a along with the overall FBD of the bar in Fig. 6-4b. The resulting AFD is shown in Fig. 6-4c. In Fig. 6-4c, observe the following.

1. The ordinate on the AFD at $x = 0$ is $+q_0 L / 2$ where the $(+)$ sign indicates tension in the bar in accordance with the deformation sign convention.
2. The slopes of tangents to the AFD at $x = 0$ and $x = L$ are the corresponding $(-)$ ordinates on the axial-load diagram at those same points.
3. Maximum axial force $(N_{\max})$ occurs at $x = 0$ where the axial-load diagram ordinate is zero, so the AFD slope is equal to zero.
4. The change in ordinate on the AFD between $x = 0$ and $x = L$ is equal to the $(-)$ area under the axial-load diagram $(-F_q)$.

Several examples that analyze bars acted upon by axial loads of various kinds are presented in Examples 6-1 and 6-2.

FIGURE 6-4

(a) Cantilever bar with linearly varying axial load; (b) axial-load diagram; and (c) axial-force diagram

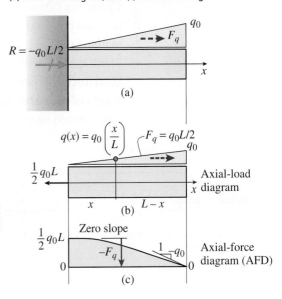

Example 6-1

A prismatic cantilever bar is loaded by centroidal axial forces along its length (loads at B, C, and D; see Fig. 6-5). Find the support reaction R at A; then find internal axial forces $N(x)$ in segments AB, BC, and CD. Plot the axial-force diagram. Neglect the self-weight of the bar.

FIGURE 6-5

Prismatic bar subjected to
concentrated centroidal axial
forces

Solution:

Use the four-step problem-solving approach to find internal axial forces $N(x)$; then plot the axial-force diagram.

1. *Conceptualize* [*hypothesize, sketch*]: Start with a free-body diagram of the bar as a whole to find the support reaction at A. Separate free-body diagram sketches, each with a section cut through one bar segment, reveal the internal axial force in that segment.

   The overall FBD of the bar showing applied loads and reaction force R at A is shown in Fig. 6-6. Separate FBDs showing internal forces N in each segment are given in Fig. 6-7.

2. *Categorize* [*simplify, classify*]: The bar is subjected to axial forces that act along the centroidal axis of the bar so only internal axial force $N(x)$ is present at any section cut along the bar; no shear forces or moments are developed in the bar. There is no distributed axial load of intensity q acting on this bar, so internal axial force N is constant within each segment. The lateral force due to the self-weight of the bar is assumed to be negligible compared to the applied loads P so is ignored here.

FIGURE 6-6

Free-body diagram of entire bar

FIGURE 6-7

(a, b, c) Free-body diagrams
of each bar segment

3. *Analyze* [*evaluate; select relevant equations, carry out mathematical solution*]:

Solution for external reaction force R: The cantilever bar is stable and statically determinate. Reaction R is easily determined in a static equilibrium analysis by summing forces in the x direction in Fig. 6-6 as

$$\sum F_x = 0 \qquad\qquad R + 3P - 2P + P = 0 \quad \text{so} \quad R = -2P \;(\leftarrow) \qquad \text{◂ (6-6)}$$

Based on the *statics sign convention* used in Eq. (6-6), the negative result indicates that reaction force R acts in the $-x$ direction.

Solution for internal axial forces N **in each bar segment:** Either free body—the left hand (LHFB) or the right hand (RHFB)—in Figs. 6-7 can be used to find the internal axial force in each segment in succession. Start with **segment** AB and use either the LHFB or RHFB to find

$$\sum F_x = 0 \qquad\qquad N_{AB} = -R = 2P \,(\text{LHFB}, 0 \le x \le L/4) \text{ or}$$
$$N_{AB} = 3P - 2P + P = 2P \,(\text{RHFB}) \qquad \text{◂ (6-7)}$$

A positive result for N_{AB} means that the internal axial force is tensile, as assumed in Fig. (6-7a) using a *deformation sign convention*. Either FBD can be used; the result is the same.

Next, equilibrium of horizontal forces in **segments** BC **and** CD leads to internal forces N_{BC} and N_{CD} as

$$\sum F_x = 0 \qquad N_{BC} = -R - 3P = -P \,(\text{LHFB}, L/4 \le x \le L/2) \quad \text{or}$$
$$N_{BC} = -2P + P = -P \,(\text{RHFB})$$
$$\sum F_x = 0 \qquad N_{CD} = -R - 3P + 2P = P \,(\text{LHFB}, L/2 \le x \le L) \quad \text{or} \qquad \text{◂ (6-8)}$$
$$N_{CD} = P \,(\text{RHFB})$$

FIGURE 6-8

Overall FBD and axial-force diagram (AFD) for entire bar

From the sign of these results, segment BC is in *compression* while segment CD is in *tension*, which is in accordance with the *deformation sign convention*.

Axial-force diagram (AFD): As expected, internal axial forces are constant within each segment of the bar, so the AFD can be plotted as three straight line segments (Fig. 6-8). Plot tension above the reference axis and compression below the axis.

4. *Finalize* [*conclude; examine answer—Does it make sense? Are units correct? How does it compare to similar problem solutions?*]: The AFD allows you to quickly identify the maximum tensile internal force in segment AB and the maximum compressive force in segment BC. Maximum force N_{max} and its location are needed to compute maximum stresses and strains, as well as axial deformations, in the bar.

With distributed load intensity $q = 0$ for the entire bar, note that the slope of each segment of the AFD is zero and there is no change in magnitude of $N(x)$ within each bar segment. This confirms the previous discussion about the differential relationship, $\dfrac{dN}{dx} = -q$ [Eq. (6-3)], which can be interpreted as *the slope on the AFD is equal to the (−) ordinate on the distributed axial-load diagram (q(x))*. In addition, *there is no area under the axial-load diagram, so there is no change in ordinate from start to end of each segment of the AFD.*

This is in agreement with Eq. (6-4), which states that $N_B - N_A = -\displaystyle\int_A^B q\,dx$.

Example 6-2

A section of steel drill pipe hangs from an oil drilling rig just before the start of drilling operations (Fig. 6-9). The pipe has length $L = 10$ m, diameter $d = 120$ mm, and wall thickness $t = 12$ mm. Find the weight W of the pipe, support reaction R at the support point at the top of the pipe, and internal axial forces $N(x)$ due to gravity. Plot the axial-force diagram.

FIGURE 6-9

(a, b) Prismatic pipe hanging from drill rig

(b)

Jim Blecha

(a)

Solution:

FIGURE 6-10

Free-body diagram of entire pipe

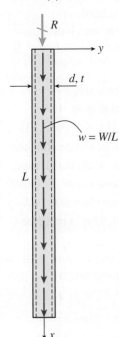

Use the four-step problem-solving approach to find pipe weight W, support reaction R, and internal axial forces $N(x)$. Plot the axial-force diagram (AFD).

1. *Conceptualize*: The drill pipe is prismatic and has a constant cross-sectional area A. The total pipe weight is W, so the weight per unit length is a uniformly distributed axial load of constant intensity $w = W/L$. The x axis is the longitudinal centroidal axis of the pipe and runs from the support at the top to the free end of the pipe at the bottom. The top support is shown as clamped, but this represents axial restraint only in this problem. Start with a free-body diagram of the pipe as a whole (Fig. 6-10) to find the support reaction R at the top. A section cut through the pipe at distance x from the top and $L - x$ from the bottom breaks the pipe structure into upper and lower free-body diagrams (Fig. 6-11) and reveals the internal axial force $N(x)$.

2. *Categorize*: The drill pipe is subjected to axial forces only; there are no lateral forces or moments applied to the pipe at the start of drilling operations. Only internal axial force $N(x)$ is present at any section cut along the pipe; no shear forces or moments are developed in the pipe.

3. *Analyze*:

   Pipe weight W and support reaction R: The pipe has circular cross section, so the cross sectional area of the pipe is computed as

   $$A = \frac{\pi}{4}\left(d^2 - (d - 2t)^2\right) = \frac{\pi}{4}\left((120)^2 - (120 - 24)^2\right) \text{ mm}^2 = 4072 \text{ mm}^2 \quad \textbf{(6-9)}$$

FIGURE 6-11

Upper and lower free-body diagrams

The weight density of steel is $\gamma = 77.0$ kN/m^3 (Appendix Table F-1). So the pipe weight per unit length is $w = \gamma A = 313.5$ N/m, and the total pipe weight is $W = wL = (313.5 \text{ N/m})(10 \text{ m}) = 3135$ N. The hanging pipe is statically determinate (if only centroidal axial forces are applied). Reaction R is easily determined by summing forces in the x direction in Fig. 6-10 as

$$\Sigma F_x = 0 \qquad R + wL = 0 \quad \text{so} \quad R = -wL = -W = -3135 \text{ N} (\uparrow) \quad \text{(6-10)}$$

Based on the *statics sign convention* used in Eq. (6-10), the negative result means that reaction force R acts *upward* in the $-x$ direction.

Internal axial forces $N(x)$ in hanging pipe: Either the upper FBD or the lower FBD in Fig. 6-11 can be used to find the internal axial force at any position x along the pipe. Starting with the upper free-body diagram, to find

$$\Sigma F_x = 0 \quad R + (w)(x) + N(x) = 0 \text{ (UFBD)} \quad \text{or} \quad N(x) = w(L - x) \quad \text{(6-11)}$$

Equilibrium of the lower free-body diagram in Fig. 6-11 yields the same result:

$$\Sigma F_x = 0 \quad -N(x) + w(L - x) = 0 \text{ (LFBD)} \quad \text{or} \quad N(x) = w(L - x) \quad \text{(6-12)}$$

The internal axial force is a linear function of x and is tensile, as assumed in Fig. 6-11a using a *deformation sign convention*. Substituting numerical values in Eq. (6-11) or (6-12) shows that maximum tensile force is $N_{\max} = 3135$ N at the top $(x = 0)$ near the support, decreasing linearly to zero at the free end at the bottom of the pipe $(x = L)$.

FIGURE 6-12

Free-body diagram and axial-force diagram (AFD) for entire pipe

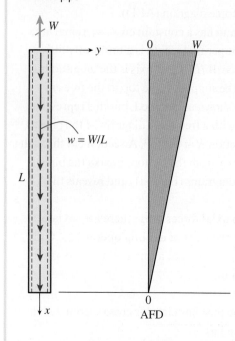

Axial-force diagram (AFD): Internal axial force varies linearly from top to bottom as shown in the AFD (Fig. 6-12). Plot tension to the right of the reference axis (in the $+y$ direction).

4. *Finalize*: From the axial-force diagram, the maximum tensile internal force is at the top support. As expected, $N_{\max}$ is equal to the pipe weight W.

Here, the distributed load intensity is constant, $q = w$, over the entire pipe. The differential relationship is $dN/dx = -q$ [see Eq. (6-3)], so the constant slope of the AFD in the x-y coordinate system is $-(W/L)$. This confirms that *the slope on the AFD is equal to the (−) ordinate on the distributed axial load diagram (w(x))*. In addition, the area under the axial load diagram is $(-W)(L)$, which is equal to the change in ordinate from start to end of the pipe on the AFD. This is in agreement with Eq. (6-4), which states

$$N_{\text{bottom}} - N_{\text{top}} = -W = -\int_0^L w\, dx = -wL$$

6.3 Shafts with Torsional Moments

Components acted on by *twisting* or *torsional moments* are referred to as **shafts in torsion**. These moments or torques cause twisting rotation about the longitudinal axis of the shaft. Some examples of such members are drive shafts and axles in automobiles, propeller shafts in ships, steering rods in vehicles of all kinds, and drill bits in hand and power drills.

Internal Torsional Moments in Shafts

Consider the cantilever shaft in Fig. 6-13a with applied torsional moment T_0 acting in the $+x$ direction at the centroid C of the cross section at the free end. Summation of moments about the x axis (that is, $\Sigma M_x = 0$ so $M + T_0 = 0$) leads to a solution for reaction moment $M = -T_0$. Further equilibrium analysis using the RHFB in Fig. 6-13b shows that **internal torsional moment** $T(x) = T_0$, so $T(x)$ is constant over the entire length of the shaft. Note that internal torsional moment vectors in Fig. 6-13 point *away from* the face of the cut section on both the left and right faces of the cut.

Distributed Torsional Moment Loading   If the longitudinal centroidal moment on the shaft is a distributed load, such as the centroidal **uniform moment** of constant intensity t_0 in Fig. 6-13c, multiply moment intensity t_0 by the distance along the shaft over which it acts to obtain the total applied moment on that segment of the shaft. If distributed moment $t(x)$ acts over the entire length of the shaft, reaction moment M at the support is found using $\Sigma M_x = 0$, so $M + t_0 L = 0$ or $M = -t_0 L$. A common alternative way to show uniformly distributed moment t_0 is presented in Fig. 6-13d, where a rectangle of constant height t_0 is used and the distributed moment is shown above the shaft for clarity with the

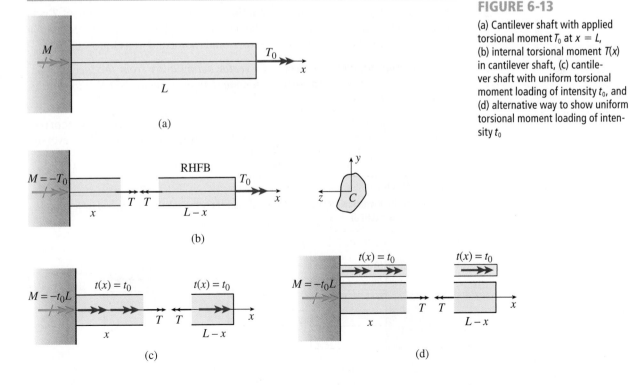

FIGURE 6-13

(a) Cantilever shaft with applied torsional moment T_0 at $x = L$, (b) internal torsional moment $T(x)$ in cantilever shaft, (c) cantilever shaft with uniform torsional moment loading of intensity t_0, and (d) alternative way to show uniform torsional moment loading of intensity t_0

understanding that the load is actually applied along the longitudinal centroidal axis. Now the total applied moment on the shaft is seen to be the area under the distributed load or $M_t = \int_0^L t_0 dx = t_0 L$—once again enforcing overall static equilibrium gives $M = -M_t = -t_0 L$. The **internal moment** $T(x)$ can be found from the right-hand free-body diagram (RHFB) in either Fig. 6-13c or d as

$$T(x) = t_0(L - x) \qquad (6\text{-}13)$$

Now, internal moment T varies linearly with x over the entire length of the shaft; at $x = 0$, T is equal to the reaction moment M, while at $x = L$, $T = 0$.

Finally, consider the **linearly varying** torsional moment loading $t(x)$ shown in Fig. 6-14a. The load has peak intensity t_0 at $x = 0$ (that is, $t(0) = t_0$), has zero intensity at $x = L$ (so $t(L) = 0$), and the intensity as a function of x is $t(x) = t_0\left(1 - \dfrac{x}{L}\right)$. The total torsional moment applied to the shaft is $M_t = \int_0^L t(x)dx = \dfrac{1}{2}t_0 L$, and M_t is the area under the load triangle. Static equilibrium requires that $\sum M_x = 0$ or $M + M_t = 0$, so $M = -\dfrac{1}{2}t_0 L$.

From $\sum M_x = 0$, using the left-hand free-body diagram (LHFB) in Fig. 6-14b, the **internal torsional moment** $T(x)$ is equal to

$$T(x) = \frac{1}{2}t_0 L - \frac{x}{2}(t_0 + t(x)) = \frac{t_0}{2L}(L - x)^2 \qquad (6\text{-}14)$$

You can plot Eq. (6-14) to show the variation of internal moment $T(x)$ over the length of the shaft.

Sign Convention

In accordance with Newton's third law (action–reaction), internal moments are equal in magnitude but opposite in direction on the two faces of each cut section in Figs. 6-13 and Fig. 6-14. If internal torsional moment vector T acts away from the cut section, this is taken as positive; if moment T acts toward the cut face, this is assumed to be negative. For this reason, the *deformation sign convention* to be used for internal torsional moment T is as follows:

> An **internal torsional moment (T) vector** acting away from the face of a cut section in a shaft is positive; a torsional moment vector pointing toward the cut section is negative.

If the computed result for moment T turns out to be negative, an incorrect assumption for direction was made, and the direction of T must be reversed on the free-body diagram of the shaft.

FIGURE 6-14

(a) Cantilever shaft with linearly varying torsional moment with peak intensity t_0, and (b) left-hand free-body diagram (LHFB)

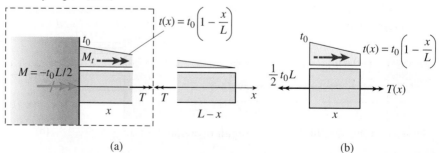

(a)   (b)

Relationships Between Loads and Internal Torsional Moment

Twisting of straight shafts due to applied torsional moments is an important problem in engineering mechanics. First consider an element taken from a shaft acted on by applied distributed torsional moment $t(x)$ (Fig. 6-15). Internal torques T are developed within the shaft and vary along the shaft; differential moment dT accounts for changes in internal torsional moment along the shaft. Internal torques on each face of the element are assumed to act away from the face in accordance with a deformation sign convention.

Equilibrium of the element in Fig. 6-15b requires that

$$\Sigma M_x = 0 \qquad -T + t(x)dx + (T + dT) = 0$$

or

$$\frac{dT}{dx} = -t$$

(6-15)

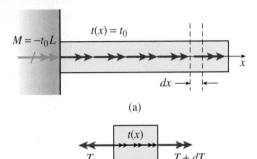

FIGURE 6-15

(a, b) Element of a shaft used in deriving relationships between loads and internal torsional moment (distributed torsional load t and internal torsional moment T are shown in their positive directions)

Equation (6-15) specifies that *the rate of change of the torsional moment within the shaft is equal to the negative of the intensity of the distributed torsional load at that same point along the shaft.* This differential relationship between applied torque t and internal torque T is mathematically the same as that between axial load q and internal axial force N presented in Eq. (6-3). So if there is no applied distributed moment over some region of the shaft (that is, $t(x) = 0$ as in Fig. 6-13b), then $dT/dx = 0$, and the internal torsional moment is constant in that region. In the same way, if the torsional moment load is *uniformly distributed* over a segment of the shaft (see Figs. 6-13c and 6-15), then dT/dx is constant and internal torsional moment T varies linearly in that segment of the bar. And for the case of *linearly* distributed torsional moment loading (Fig. 6-14), the quadratic expression for internal torsional moment is given in Eq. (6-14). Take the derivative of $T(x)$ in Eq. (6-14) to get

$$\frac{dT}{dx} = \frac{d}{dx}\left(\frac{t_0}{2L}(L-x)^2\right) = -t_0\left(1 - \frac{x}{L}\right) = -t(x)$$

(6-16)

This result agrees with Eq. (6-15).

A second relationship between distributed torsional load of intensity t and internal torsional moment T is obtained by multiplying both sides of Eq. (6-15) by dx and then integrating between any two points A and B along the axis of the shaft:

$$\int_A^B dT = T_B - T_A = -\int_A^B t(x)dx$$

(6-17)

This is the same expression as that in Eq. (6-4) but is now applied to internal torsional moments T rather than internal axial forces N. Expressing Eq. (6-17) in words:

> *The change in internal torsional moment T between any two points along a shaft is equal to the negative of the area under the torsional load curve between those same two points.*

For the shaft with linearly varying torsional loading in Fig. 6-14, if point A is taken at $x = 0$ and point B is at $x = L$, then

$$T_B - T_A = -\frac{1}{2}t_0 L = -M_t$$

(6-18)

where M_t is the area under the triangular torsional loading curve.

Torsional Moment Diagram (TMD)

A graphical display of the variation in internal torsional moment over the length of a shaft is constructed in the same manner as that described for the AFD in Section 6.2. This display of $T(x)$, referred to here as the *torsional moment diagram* (TMD), helps a designer to quickly identify the location of maximum internal torsional moment T_{max}. Guidelines for construction of torsional moment diagrams are based on the differential and integral relationships between $t(x)$ and $T(x)$ in Eqs. (6-15) and (6-17). A slight modification of the guidelines given previously for constructing an axial-force diagram (AFD) leads to those needed to plot a TMD; they are as follows.

- The slope at any point on the TMD is equal to the ($-$) ordinate on the distributed torsional moment loading diagram (that is, $t(x)$) at that same point.
- The change in internal torsional moment between any two points along a shaft is equal to the negative of the area under the torsional load diagram between those same two points.
- At the point(s) where the torsional load diagram crosses the reference axis, the value of the internal torsional moment on the torsional moment diagram is a local maximum or minimum.
- At points of application of concentrated torsional moments, there is a discontinuity, or change in the ordinate, in the TMD.

Using these guidelines, you can sketch the TMD for the shaft with linearly varying torsional moment in Fig. 6-14. The cantilever shaft with linearly varying distributed torsional load is again shown in Fig. 6-16a along with the overall FBD of the shaft in Fig. 6-16b. The resulting TMD is shown in Fig. 6-16c. Some key features of the TMD are as follows.

FIGURE 6-16

(a,b,c) Torsional load and moment diagrams for shaft with linearly varying torsional moment loading

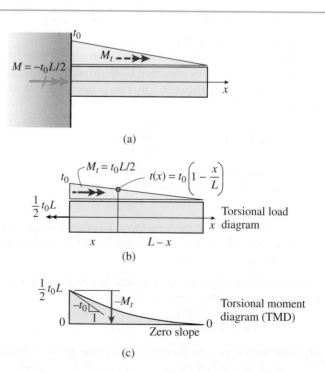

1. The ordinate on the TMD at $x = 0$ is $+t_0L/2$, where the (+) sign indicates that the internal torsional moment vector points away from the cut section in accordance with the deformation sign convention.
2. The slopes of tangents to the TMD at $x = 0$ and $x = L$ are the corresponding (−) ordinates on the torsional load diagram at those same points.
3. Maximum torsional moment (T_{max}) occurs at $x = 0$.
4. The change in ordinate on the TMD between $x = 0$ and $x = L$ is equal to the (−) area under the torsional load diagram ($-M_t$).

Examples 6-3 and 6-4 provide further discussion of analysis of shafts to determine internal torsional moments and their associated torsional moment diagrams.

Example 6-3

A stepped circular shaft is fixed at A and has three gears that transmit the torques shown in Fig. 6-17. Find the reaction torque M_{Ax} at A; then find the internal torsional moments in segments AB, BC, and CD. Use properly drawn free-body diagrams in your solution. Plot the torsional moment diagram (TMD).

Solution:

Use the four-step problem-solving approach to find internal torsional moments $T(x)$; then plot the torsional moment diagram (TMD).

1. *Conceptualize*: The cantilever shaft structure is statically determinate. The solution for the reaction moment at A (M_{Ax}) must begin with a proper drawing of the FBD of the overall structure (Fig. 6-18). The FBD shows all applied and reactive torques. Separate FBDs showing internal torques T in each segment are obtained by cutting the shaft in regions AB, BC, and CD in succession and are given in Fig. 6-19a through c. Each cut produces a left-hand and a right-hand free-body diagram.

FIGURE 6-17

Stepped circular shaft subjected to concentrated torques

FIGURE 6-18

Free-body diagram of shaft

2. *Categorize*: The shaft is subjected to applied torques that act along the centroidal axis of the shaft, so only the internal torsional moment $T(x)$ is present at any section cut along the shaft. There is no distributed torque of intensity t acting on this shaft, so the internal torsional moment T is constant within each segment.

3. *Analyze*:

   Solution for external reaction moment M_{Ax}: Sum the moments about the x axis to find the reaction moment M_{Ax}. This structure is statically determinate because there is one available equation from statics ($\sum M_x = 0$) and one reaction unknown (M_{Ax}). A statics sign convention is used (such as the right-hand rule or CCW is positive).

$$M_{Ax} - 17{,}000 \text{ lb-in.} + 9000 \text{ lb-in.} + 5000 \text{ lb-in.} = 0$$
$$M_{Ax} = -(-17000 \text{ lb-in.} + 9000 \text{ lb-in.} + 5000 \text{ lb-in.})$$
$$= 3000 \text{ lb-in.}$$

   The computed result for M_{Ax} is positive, so the reaction moment vector is in the positive x direction as assumed.

Solution for internal torsional moments T in each shaft segment:
Start with **segment AB** and use either FBD in Fig. 6-19a to find

Left FBD:

$$T_{AB} = -M_{Ax} = -3000 \text{ lb-in.}$$

Right FBD:

$$T_{AB} = -17,000 \text{ lb-in.} + 9000 \text{ lb-in.}$$
$$+ 5000 \text{ lb-in.} = -3000 \text{ lb-in.} \quad \twoheadleftarrow$$

Next consider **segment BC**. Summing moments about the x axis in Fig. 6-19b gives

Left FBD:

$$T_{BC} = -M_{Ax} + 17,000 \text{ lb-in.}$$
$$= 14,000 \text{ lb-in.}$$

Right FBD:

$$T_{BC} = 9000 \text{ lb-in.} + 5000 \text{ lb-in.}$$
$$= 14,000 \text{ lb-in.} \quad \twoheadleftarrow$$

Finally, moment equilibrium about the x axis leads to a solution for the internal torsional moment in **segment CD**:

Left FBD:

$$T_{CD} = -M_{Ax} + 17,000 \text{ lb-in.}$$
$$- 9000 \text{ lb-in.} = 5000 \text{ lb-in.}$$

Right FBD:

$$T_{CD} = 5000 \text{ lb-in.} \quad \twoheadleftarrow$$

In each segment, the internal torsional moments computed using either the left or right FBDs are the same.

Torsional moment diagram: As expected, internal torsional moments are constant within each segment of the shaft, so the TMD can be plotted as three straight line segments (Fig. 6-20). Positive torsional moments (i.e., moment vector points away from face of cut section) are plotted above the reference axis and negative moments below the axis.

4. *Finalize*: The TMD shows that segment BC has the maximum internal torsional moment (and positive moment) and segment AB has the maximum negative torsional moment.

FIGURE 6-20

Overall FBD and torsional moment diagram (TMD) for entire shaft

$M_{Ax} = 3000$ lb-in.   17,000 lb-in.   9000 lb-in.   5000 lb-in.

A B C D

$T_{BC} = 14,000$ lb-in.

$T_{CD} = 5,000$ lb-in.

0   TMD $\twoheadleftarrow$

$T_{AB} = -3,000$ lb-in.

With distributed torque intensity $t = 0$ for the entire shaft, the slope of each segment of the TMD is zero, and there is no change in magnitude of $T(x)$ within each shaft segment. This confirms the previous discussion about the differential relationship $dT/dx = -t$ [Eq. (6-15)]. This relationship can be stated as given here.

> *First, the slope on the TMD is equal to the $(-)$ ordinate on the distributed torsional moment load diagram $(t(x))$. Second, there is no area under the torsional load diagram, so there is no change in ordinate from start to end of each segment of the TMD.*

This is in agreement with Eq. (6-17), which states that $T_B - T_A = -\int\limits_{A}^{B} t(x)\, dx$.

The differential and integral expressions for bars with axial loads (Section 6.2) and shafts acted on by torsional moments (Section 6.3) have the same mathematical form, so the axial-force (AFD) and torsional moment (TMD) diagrams are the same except for the applied loading.

Example 6-4

FIGURE 6-21

(a) Paddle mixer shaft subjected to concentrated and distributed torques and (b) paddle mixer shaft segment *BC*, which is subjected to distributed torque of intensity t_0

(a)

(b)

A paddle mixer is being used to mix oil and powder in an industrial plant. The two sections (AB, BC) of the mixer shaft are shown in Fig. 6-21. The shaft is attached to the mixer motor at A, and a hand stabilizer applies an additional concentrated torque $2T_0$ at $x = 2L/5$. The paddle is immersed in the powder over length BC (see Fig. 6-21b) and uniformly distributed torque intensity $t_0 = 3T_0/L$ is applied over this segment of the shaft. Find expressions for internal torques $T(x)$ over the length of the mixer shaft. Plot the torsional moment diagram (TMD).

Solution:

Use the four-step problem-solving approach to find internal torsional moments $T(x)$ then plot the torsional moment diagram (TMD).

1. *Conceptualize*: Find the reaction moment at A (R_A) using the FBD of the overall structure shown in Fig. 6-22. The FBD shows the applied torques and the reactive torque at A. Obtain separate FBD's showing internal torques $T_i(x)$ in each segment by cutting the shaft in the three separate regions labeled 1 through 3 in Fig. 6-22. The left-hand FBD's for each region are shown in Figs. 6-23a through c. Each cut produces a left-hand and a right-hand free-body diagram, but only the left-hand FBD is shown for convenience.

FIGURE 6-22

Free-body diagram of overall shaft

2. *Categorize*: The shaft is subjected to applied and resisting torques that act along the centroidal axis of the shaft. Expressions for internal torsional moments $T_i(x)$ must be found at each section cut along the shaft (Fig. 6-23). The internal torsional moment T is constant within segments 1 and 2. However, a function is needed in section 3 to account for the variation in internal torque caused by the distributed torque of intensity t_0 acting on BC.

3. *Analyze*:

Solution for external reaction moment R_A: Sum torsional moments about the x axis of the structure to find

$$\sum M_x = 0 \qquad R_A + 2T_0 - t_0\left(\frac{2L}{5}\right) = 0$$

so $\qquad R_A = -2T_0 + \left(\frac{3T_0}{L}\right)\left(\frac{2L}{5}\right) = \frac{-4T_0}{5}$ **(a)**

Reaction R_A is negative, which means that the reactive torsional moment vector is in the $(-x)$ direction based on a *statics sign convention*.

FIGURE 6-23

(a, b, c) Left-hand free-body diagrams for each shaft segment

(a)

(b)

(c)

Solution for internal torsional moments T in each shaft segment: From the FBD of segment 1 (Fig. 6-23a), the internal torsional moment is constant and is equal to reactive torque R_A. Torque $T_1(x)$ is positive because the torsional moment vector points away from the cut section of the pipe in accordance with the *deformation sign convention*:

$$T_1(x) = \frac{4}{5}T_0 \qquad 0 \le x \le \frac{2}{5}L \qquad \text{(b)}$$

Next, a FBD of segment 2 of the pipe structure (Fig. 6-23b) gives

$$T_2(x) = \frac{4}{5}T_0 - 2T_0 = \frac{-6}{5}T_0 \qquad \frac{2}{5}L \le x \le \frac{3}{5}L \qquad \text{(c)}$$

where $T_2(x)$ is also constant and the minus sign means that $T_2(x)$ actually points in the negative x direction.

Finally, the FBD of segment 3 of the pipe structure (Fig. 6-23c) provides the following expression for internal torsional moment $T_3(x)$:

$$T_3(x) = \frac{4}{5}T_0 - 2T_0 + t_0\left(x - \frac{3}{5}L\right) = 3T_0\left(\frac{x}{L} - 1\right) \qquad \frac{3}{5}L \le x \le L \qquad \text{(d)}$$

Evaluating Eq. (d) at B and C gives the torsional moment at B as

$$T_3\left(\frac{3}{5}L\right) = 3T_0\left(\frac{3}{5} - 1\right) = \frac{-6}{5}T_0$$

and at C,

$$T_3(L) = 3T_0(1 - 1) = 0$$

Now plot Eqs. (b), (c), and (d) to get a *torsional moment diagram* (TMD) (Fig. 6-24), which displays the variation of internal torsional moment over the length of the pipe structure ($x = 0$ to $x = L$).

FIGURE 6-24

Overall FBD and torsional moment diagram (TMD) for entire shaft

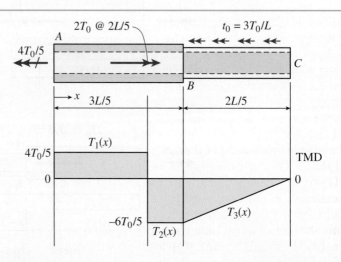

Torsional moment diagram (TMD): Internal torsional moments $T_i(x)$ are constant in sections 1 and 2 and linearly varying in x in section 3, so the TMD can be plotted as three straight line segments (Fig. 6-24). Positive torsional moments (where the moment vector points away from the face of the cut section) are plotted above the reference axis and negative moments below the axis.

4. *Finalize*: From the TMD, the maximum internal torsional moment T_{max} is in section 2 $\left[\dfrac{2L}{5} \leq x \leq \dfrac{3L}{5}\right]$.

The distributed torque intensity $t = 0$ for the first region of the shaft $\left[0 \leq x \leq \dfrac{2L}{5}\right]$, so the slope of the TMD is zero and there is no change in magnitude of $T(x)$ in this region since the slope of the TMD is given by the differential relationship $\dfrac{dT}{dx} = -t$ [see Eq. (6-15)]. For $\left[\dfrac{2L}{5} \leq x \leq \dfrac{3L}{5}\right]$, torque intensity $t = 0$ and the slope is again zero. However, in segment BC, the torque intensity is $t(x) = -t_0 = -\left(-\dfrac{3T_0}{L}\right) = \dfrac{3T_0}{L}$, so the slope is constant and positive as shown in the TMD plot (Fig. 6-24). Finally, on segment BC, the *area under the torsional load diagram* is from Eq. (6-17),

$$T_C - T_B = -\int_B^C t(x)\, dx = -\int_B^C [-t_0]\, dx = -\left[\dfrac{-3T_0}{L}\right]\left[\dfrac{2L}{5}\right] = \dfrac{6T_0}{5}, \text{ which is equal to}$$

the change in ordinate on the TMD from B to C. The TMD returns to zero at $x = L$ since there is no applied torque at C.

6.4 Beams and Frames with Transverse Loads and Applied Moments

Beams are members that *bend* when subjected to transverse loads along their lengths, that is, forces or moments having their vectors perpendicular to the axis of the beam. The loads may be *concentrated* or *distributed* and may also include applied bending moments at specific points along the beam. Examples of beams are the members of a floor system in a building or bridge, the wing of an airplane, and frame members in a stadium that support the seating. The focus here is the determination of both internal forces and moments in these structural components that are essential to an assessment of their behavior under applied loadings and also to their design. In general, beams are more complicated load-carrying elements than the bars with axial loads and shafts with torsional moments discussed previously, so beams are studied in more depth than bars and shafts.

The beams shown in Fig. 6-25 are classified as *planar structures* because they lie in a single plane. If all loads act in that same plane, and if all deflections (shown by the dashed lines) occur in that plane, that plane is referred to as the **plane of bending**. This

FIGURE 6-25

Examples of beams subjected to lateral loads

chapter discusses shear forces and bending moments in beams and how these quantities are related to each other and to the loads.

Finding the shear forces and bending moments is an essential step in the *design* of any beam. You usually need to know not only the maximum values of these quantities but also the manner in which they vary along the axis. Once the shear forces and bending moments are known, you can find the stresses, strains, and deflections in beams of known size, or you can design (that is, find the required size of) new beams.

Types of Beams, Supports, and Loads

Beams are usually described by the manner in which they are supported. For instance, a beam with a pin support at one end and a roller support at the other (Fig. 6-26a) is called a **simply supported beam** or a **simple beam**. As you saw in Section 3.2 (see Table 3-1), the essential feature of a **pin support** is that it prevents translation at the end of a beam but does not prevent rotation. Thus, end A of the beam of Fig. 6-26a cannot move horizontally or vertically, but the axis of the beam can rotate in the plane of the figure. Consequently, a pin support is capable of developing a force reaction with both horizontal and vertical components (H_A and R_A), but it cannot develop a moment reaction.

At end B of the beam (Fig. 6-26a), the **roller support** prevents translation in the vertical direction but not in the horizontal direction; hence, this support can resist a vertical force (R_B) but not a horizontal force. Of course, the axis of the beam is free to rotate at B just as it is at A. The vertical reactions at roller supports and pin supports may act *either* upward or downward, and the horizontal reaction at a pin support may act either to the left or to the right. Recall that, in the figures, reactions are indicated by slashes across the arrows in order to distinguish them from loads, as explained in Section 3.2.

The beam shown in Fig. 6-26b, which is fixed at one end and free at the other, is called a **cantilever beam**. At the **fixed support** (or *clamped support*), the beam can neither translate nor rotate, whereas at the free end, it may do both. Consequently, both force and moment reactions may exist at the fixed support.

FIGURE 6-26

Types of beams: (a) simple beam, (b) cantilever beam, and (c) beam with an overhang

The third example in the figure is a **beam with an overhang** (Fig. 6-26c). This beam is simply supported at points A and B (that is, it has a pin support at A and a roller support at B) but it also projects beyond the support at B. The overhanging segment BC is similar to a cantilever beam except that the beam axis may rotate at point B. Reaction force component H_A is zero for the loadings shown in Fig. 6-26c so is not included in the figure.

When drawing sketches of beams, identify the supports using **conventional symbols**, such as those discussed in Section 3.2 and shown in Fig. 6-26. These symbols indicate the manner in which the beam is restrained; therefore, they also indicate the nature of the reactive forces and moments. However, *the symbols do not represent the actual physical construction*. For instance, consider the examples shown in Fig. 6-27. Figure 6-27a shows a wide-flange beam supported on a concrete wall and held down by anchor bolts that pass through slotted holes in the lower flange of the beam. This connection restrains the beam against vertical movement (either upward or downward) but does not prevent horizontal movement. Also, any restraint against rotation of the longitudinal axis of the beam is small and ordinarily may be disregarded. Consequently, this type of support is usually represented by a roller, as shown in Figure 6-27b.

The second example is a beam-to-column connection (Fig. 6-27c) where the beam is attached to the column flange by bolted angles. This type of support is usually assumed to restrain the beam against horizontal and vertical movement but not against rotation (restraint against rotation is slight because both the angles and the column can bend). Thus, this connection is usually represented as a pin support for the beam (Fig. 6-27d).

The last example (Fig. 6-27e/f) is a metal pole welded to a base plate that is anchored to a concrete pier embedded deep in the ground. Since the base of the pole is fully restrained against both translation and rotation, it is represented as a fixed support (Fig. 6-27f).

FIGURE 6-27

Beam supported on a wall: (a) actual construction and (b) representation as a roller support. Beam-to-column connection: (c) actual construction and (d) representation as a pin support. Pole anchored to a concrete pier: (e) actual construction and (f) representation as a fixed support

Joe Gough/Shutterstock.com

Beam-to-column connection with one beam attached to column flange and other attached to column web

The task of representing a real structure by an **idealized model**, as illustrated by the beams shown in Fig. 6-26, is an important aspect of engineering work. The model should be simple enough to facilitate mathematical analysis and yet complex enough to represent the actual behavior of the structure with reasonable accuracy. Of course, every model is an approximation to nature. For instance, the actual supports of a beam are never perfectly rigid, and so there is always a small amount of translation at a pin support and a small amount of rotation at a fixed support. Also, supports are never entirely free of friction, and so there is always a small amount of restraint against translation at a roller support. In most circumstances, especially for statically determinate beams, these deviations from the idealized conditions have little effect on the action of the beam and can be disregarded safely.

Types of Loads

Assume in this discussion that the loads act in the plane of the figure, which means that all forces must have their vectors in the plane of the figure and all couples must have their moment vectors perpendicular to the plane of the figure. Furthermore, the beam itself must be symmetric about that plane, which means that every cross section of the beam must have a vertical axis of symmetry. Under these conditions, the beam deflects only in the *plane of bending* (the plane of the figure).

Several types of loads that act on beams are illustrated in Fig. 6-26. When a load is applied over a very small area, it may be idealized as a **concentrated load**, which is a single force. Examples are the loads P_1, P_2, P_3, and P_4 in the figure. Another kind of load is a **couple**, illustrated by the moment M_1 acting on the overhanging beam (Fig. 6-26c).

When a load is spread along the axis of a beam, it is represented as a **distributed load**, such as the load q in part (a) of the figure. Like bars with axial loads (Section 6.2) and shafts with torsional moments (Section 6.3), *distributed loads* are measured by their **intensity**, which is expressed in units of force per unit distance (for example, newtons per meter or pounds per foot). A **uniformly distributed load**, or **uniform load**, has constant intensity q per unit distance (Fig. 6-26a). A varying load has an intensity that changes with distance along the axis; for instance, the **linearly varying load** of Fig. 6-26b has an intensity that varies linearly from q_1 to q_2. In most cases for beams, distributed loads are assumed to act either upward or downward and transverse to the member.

Distributed Loads

Replace a distributed load with its resultant concentrated load when carrying out a static analysis to find reactions at supports. The resultant concentrated load is *statically equivalent* to the distributed load and acts at the *centroid* of the area representing the distributed load. For example, if the distributed load has constant intensity q and acts over some length L, replace it with resultant F_q acting at $L/2$ as shown in Fig. 6-28a. In a similar manner, a distributed load with *linearly varying intensity* $q(x)$ over distance L, with an intensity of q_0 at the start and zero at the end (Fig. 6-28b), may be replaced by a statically equivalent concentrated load $F_{q0} = \dfrac{1}{2}q_0 L$ at the centroid of the triangular shape where $x = L/3$.

If the distributed load area is a trapezoid with a load intensity that varies linearly between values of q_1 and q_2 over distance L (see Fig. 6-28c), the statically equivalent concentrated load is $F_q = \dfrac{L}{2}(q_1 + q_2)$ and F_q acts at centroid location $\bar{x} = \dfrac{L}{3}\left(\dfrac{q_1 + 2q_2}{q_1 + q_2}\right)$. Here use results of the solutions for *centroids of*

FIGURE 6-28

Statically equivalent resultants
of distributed loads

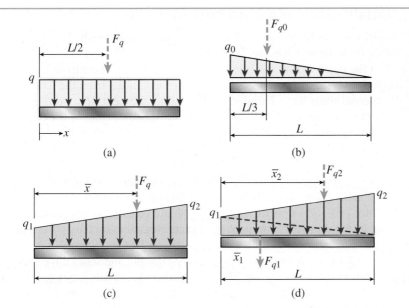

(a)   (b)

(c)   (d)

plane areas presented in Sections 5.2 and 5.3 and formulas summarized in Appendix B (see Case 8 for the trapezoid). You can also use the notion of composite areas (Section 5.3) to break the trapezoid into *two triangles* (see Fig. 6-28d). Use first moments of the two triangular load areas to confirm the above for $\bar{x}$ as

$$\bar{x} = \frac{\bar{x}_1 F_{q1} + \bar{x}_2 F_{q2}}{F_{q1} + F_{q2}} = \frac{\left(\dfrac{L}{3}\right)\left(\dfrac{1}{2}q_1 L\right) + \left(\dfrac{2L}{3}\right)\left(\dfrac{1}{2}q_2 L\right)}{\dfrac{1}{2}q_1 L + \dfrac{1}{2}q_2 L} = \frac{L}{3}\left(\frac{q_1 + 2q_2}{q_1 + q_2}\right) \quad \textbf{(6-19)}$$

If the distributed load cannot be reduced to a composite of plane areas with known centroid locations, use the integral expressions developed in Section 5.2 to find the location of the centroid of the distributed load. Equating first moments of the load area and resultant gives

$$\bar{x} = \frac{\int xq(x)dx}{F_q} = \frac{\int xq(x)dx}{\int q(x)dx} \quad \textbf{(6-20)}$$

For example, if the distributed load intensity is $q(x) = \sqrt{\dfrac{x}{L}}q_0$ (see Fig. 6-29), $F_q = \displaystyle\int_0^L q(x)dx = \frac{2}{3}q_0 L$ and the resultant load F_q should be placed at the centroid $\bar{x}$, which is determined as

$$\bar{x} = \frac{\displaystyle\int_0^L x\sqrt{\dfrac{x}{L}}q_0 dx}{\dfrac{2}{3}q_0 L} = \frac{3}{5}L \quad \textbf{(6-21)}$$

FIGURE 6-29

Statically equivalent resultant
of distributed load $q(x)$

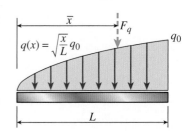

Internal Forces and Moments in Beams

Consider the cantilever beam in Fig. 6-30a. Beams resist applied loads by a combination of *shear* and *bending*. In the beam in Fig. 6-30, load P_0 is applied transverse to the beam axis at the free end ($x = L$) and acts at centroid C in the ($-y$) direction. At a section cut through the beam a distance x from the fixed support, both internal shear force V and internal bending moment M are present and act together to resist the applied load P_0. From static equilibrium of the entire beam, find reaction force R_1 as follows: $\Sigma F_y = 0$, so $R_1 - P_0 = 0$ or $R_1 = P_0$. Summation of moments about the z axis at the left end of the beam (at $x = 0$) gives reaction moment R_2: $\Sigma M_z = 0$, so $R_2 - P_0 L = 0$ or $R_2 = P_0 L$. (Note the use of a *statics sign convention* to solve for reaction force and moment as in Chapters 3 and 4). Finally, using the RHFB (Fig. 6-30b) and assuming that positive internal shear force ($+V$) acts upward on the right-hand face of the cut section and that positive internal bending moment ($+M$) acts clockwise on that same face, solve for shear and moment as follows: $\Sigma F_y = 0$, so $V(x) = P_0$; and $\Sigma M_z = 0$, so $M(x) = -P_0(L - x)$. For this introductory example, the shear force is constant at all locations along the beam. However, internal moment is not constant over the length of the beam. In general, it is necessary to plot functions $V(x)$ and $M(x)$ as separate **diagrams** to see how each varies with longitudinal coordinate x and to find the location along the beam where each reaches a maximum value. Before doing this, however, the *sign convention* to be used to compute and then display plots of the internal shear force V and bending moment M must be defined.

Sign Convention   Now consider the sign conventions for shear forces and bending moments. It is customary to assume that shear forces and bending moments are positive when they act in the directions shown in Fig. 6-30b. Note that on the *right-hand face* [which is part of the right-hand free-body diagram (RHFB)] the shear force tends to rotate the beam segment clockwise, and the bending moment tends to compress the upper part of the beam and elongate the lower part. The action of shear force V and moment M against the *left-hand part* of the

---

FIGURE 6-30

(a) Cantilever beam with applied transverse load P_0 at $x = L$ and (b) cantilever beam with internal shear force $V(x)$ and internal moment $M(x)$

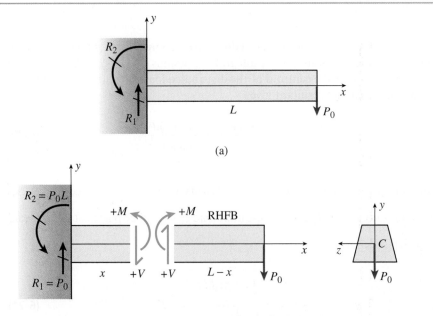

(a)

(b)

beam is also shown in Fig. 6-30b. The directions of both quantities are now reversed—the shear force acts downward, and the bending moment acts counterclockwise. However, the shear force still tends to rotate the material clockwise, and the bending moment still tends to compress the upper part of the beam and elongate the lower part.

Therefore, the algebraic sign of an internal shear force or bending moment is determined by how it deforms the material on which it acts, rather than by its direction in space. In the case of a beam, follow these sign conventions.

> *A positive shear force acts clockwise against the material* (Fig. 6-30b) *and a negative shear force acts counterclockwise against the material. Also, a positive bending moment compresses the upper part of the beam* (Fig. 6-30b), *and a negative bending moment compresses the lower part.*

To make these conventions clear, both positive and negative shear forces and bending moments are shown in Fig. 6-31. The forces and moments are shown acting on an element of a beam cut out between two cross sections that are a small distance apart.

The *deformations* of an element caused by both positive and negative shear forces and bending moments are sketched in Fig. 6-32. A positive shear force tends to deform the element by causing the right-hand face to move downward with respect to the left-hand face, and, as already mentioned, a positive bending moment compresses the upper part of a beam and elongates the lower part.

Sign conventions for internal shear forces and bending moments are called **deformation sign conventions** because they are based upon how the material is deformed. For instance, you previously used a deformation sign convention in dealing with axial forces in a bar. An axial force producing elongation (or tension) in a bar is positive and an axial force producing shortening (or compression) is negative. Thus, the sign of an internal axial force depends upon how it deforms the material, not upon its direction in space.

By contrast, when writing equations of equilibrium, use a **static sign convention** where forces are positive or negative according to their directions along the coordinate axes. For instance, when summing forces in the y direction, forces acting in the positive direction of the y axis are taken as positive and forces acting in the negative direction are taken as negative.

Relationships Between Loads, Shear Forces, and Bending Moments

Some important relationships between loads, shear forces, and bending moments in beams are needed when investigating the shear forces and bending moments throughout the entire length of a beam. They are especially helpful when constructing shear-force and bending-moment diagrams. Follow the same procedures here used in Sections 6.2 and 6.3 to find differential and integral relationships between axial loads and internal axial forces in bars as well as between torsional moment loadings and internal torsional moments in shafts.

As a means of obtaining the relationships, consider an element of a beam cut out between two cross sections that are distance dx apart (Fig. 6-33). The load acting on the top surface of the element may be a distributed load, a concentrated load, or a couple, as shown in Figs. 6-33a, b, and c, respectively. Follow the **sign conventions** for these loads.

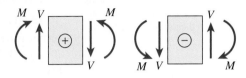

FIGURE 6-31

Sign conventions for shear force *V* and bending moment *M*

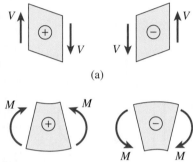

FIGURE 6-32

Deformations (highly exaggerated) of a beam element caused by (a) shear forces, and (b) bending moments

(a)

(b)

FIGURE 6-33

Element of a beam used in deriving the relationships between loads, shear forces, and bending moments (All loads, shears, and moments are shown in their positive directions)

(a)

(b)

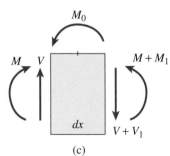

(c)

Distributed loads and concentrated loads are positive when they act downward on the beam and negative when they act upward. A couple acting as a load on a beam is positive when it is counterclockwise and negative when it is clockwise.

If other sign conventions are used, changes may occur in the sign of the terms appearing in the equations derived in this section.

The shear forces and bending moments acting on the sides of the element are shown in their positive directions in Fig. 6-32. In general, the shear forces and bending moments vary along the axis of the beam. Therefore, their values on the right-hand face of the element may be different from their values on the left-hand face.

In the case of a distributed load (Fig. 6-33a), the increments in V and M are infinitesimal, so they are denoted by dV and dM, respectively. The corresponding shear and moment on the right-hand face are $V + dV$ and $M + dM$.

In the case of a concentrated load (Fig. 6-33b) or a couple (Fig. 6-33c), the increments may be finite, so they are denoted V_1 and M_1. The corresponding shear and moment on the right-hand face are $V + V_1$ and $M + M_1$.

For each type of loading, write two equations of equilibrium for the element—one equation for equilibrium of forces in the vertical direction and one for equilibrium of moments. The first of these equations gives the relationship between the load and the shear force, and the second gives the relationship between the shear force and the bending moment.

Distributed Loads   The first type of loading is a distributed load of intensity q, as shown in Fig. 6-33a. First consider its relationship to the shear force and second its relationship to the bending moment.

Shear Force. Equilibrium of forces in the vertical direction (upward forces are positive) gives

$$\Sigma F_{\text{vert}} = 0 \qquad\qquad V - q\,dx - (V + dV) = 0$$

or

$$\frac{dV}{dx} = -q \qquad\qquad\text{(6-22)}$$

From this equation, the rate of change of the shear force at any point on the axis of the beam is equal to the negative of the intensity of the distributed load at that same point. (*Note:* If the sign convention for the distributed load is reversed so that q is positive upward instead of downward, the minus sign is omitted in the preceding equation.) Note that the relationship between shear force V and distributed load intensity q in Eq. (6-22) is similar to that for axial load intensity q and internal axial force N in Eq. (6-3), and for applied torsional moment intensity t and internal torsional moment T in Eq. (6-15).

Some useful relations are immediately obvious from Eq. (6-22). For instance, if there is no distributed load on a segment of the beam (that is, if $q = 0$), $dV/dx = 0$ and the shear force is constant in that part of the beam. Also, if the distributed load is uniform along part of the beam ($q = $ constant), dV/dx is also constant and the shear force varies linearly in that part of the beam.

As a demonstration of Eq. (6-22), consider the cantilever beam with a linearly varying load (see Fig. 6-34). The load on the beam is

$$q = \frac{q_0 x}{L}$$

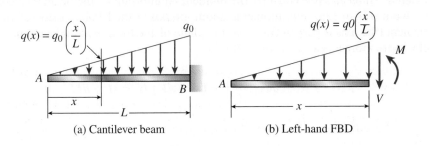

(a) Cantilever beam   (b) Left-hand FBD

which is positive because it acts downward. Also, the shear force is obtained from Fig. 6-34b as

$$V(x) = -\frac{1}{2}q_0\left[\frac{x}{L}\right]x = -\frac{q_0 x^2}{2L}$$

Taking the derivative dV/dx gives

$$\frac{dV}{dx} = \frac{d}{dx}\left(-\frac{q_0 x^2}{2L}\right) = -\frac{q_0 x}{L} = -q$$

which agrees with Eq. (6-22).

A useful relationship pertaining to the shear forces at two different cross sections of a beam can be obtained by integrating Eq. (6-22) along the axis of the beam. To do so, multiply both sides of Eq. (6-22) by dx and then integrate between any two points A and B on the axis of the beam; thus,

$$\int_A^B dV = -\int_A^B q\,dx \qquad\qquad \textbf{(6-23)}$$

by assuming that as x increases it moves from point A to point B. The left-hand side of this equation equals the difference $(V_B - V_A)$ of the shear forces at B and A. The integral on the right-hand side represents the area of the loading diagram between A and B, which in turn is equal to the magnitude of the resultant of the distributed load acting between points A and B. Thus, from Eq. (6-23),

$$V_B - V_A = -\int_A^B q\,dx \qquad\qquad \textbf{(6-24)}$$
$$= -(\text{area of the loading diagram between } A \text{ and } B)$$

In other words, the change in shear force between two points along the axis of the beam is equal to the negative of the total downward load between those points. The area of the loading diagram may be positive (if q acts downward) or negative (if q acts upward). Note that the relationship between change in shear force V and area under the distributed load diagram in Eq. (6-24) is similar to that for change in internal axial force N and area under the axial load diagram for *bars* in Eq. (6-4); and between change in internal torsional moment T and applied torsional moment intensity t for *shafts* in Eq. (6-17).

Because Eq. (6-22) was derived for an element of the beam subjected *only* to a distributed load (or to no load), Eq. (6-22) cannot be used at a point where a concentrated load is applied (because the *intensity* of load is not defined for a concentrated load). For the same reason, Eq. (6-24) cannot be used if a concentrated load P acts on the beam between points A and B.

Bending Moment. Now consider the moment equilibrium of the beam element shown in Fig. 6-33a. Sum moments about an axis at the left-hand side of the element (the axis is perpendicular to the plane of the figure) and take counter-clockwise moments as positive to obtain

$$\Sigma M = 0 \qquad\qquad -M - qdx\left(\frac{dx}{2}\right) - (V + dV)dx + M + dM = 0$$

Discard products of differentials (because they are negligible compared to the other terms), to obtain the following relationship:

$$\frac{dM}{dx} = V \tag{6-25}$$

This equation shows that the rate of change of the bending moment at any point on the axis of a beam is equal to the shear force at that same point. For instance, if the shear force is zero in a region of the beam, the bending moment is constant in that same region.

Equation (6-25) applies only in regions where distributed loads (or no loads) act on the beam. At a point where a concentrated load acts, a sudden change (or discontinuity) in the shear force occurs, and the derivative dM/dx is undefined at that point.

Again using the free-body diagram of the cantilever beam (see Fig. 6-34b), note that the bending moment is

$$M(x) = -\frac{1}{2}q(x)\left(\frac{x}{3}\right) = -\frac{q_0 x^3}{6L}$$

Therefore, the derivative dM/dx is

$$\frac{dM}{dx} = \frac{d}{dx}\left(-\frac{q_0 x^3}{6L}\right) = -\frac{q_0 x^2}{2L}$$

which is equal to the shear force in the beam.

Integrating Eq. (6-25) between two points A and B on the beam axis gives

$$\int_A^B dM = \int_A^B V dx \tag{6-26}$$

The integral on the left-hand side of this equation is equal to the difference $(M_B - M_A)$ of the bending moments at points B and A. The integral on the right-hand side represents the area below the shear-force diagram between A and B. Therefore, Eq. (6-26) can be expressed in the following manner:

$$M_B - M_A = \int_A^B V dx \tag{6-27}$$
$$= \text{(area of the shear-force diagram between } A \text{ and } B)$$

This equation is valid even when concentrated loads act on the beam between points A and B. However, it is not valid if a couple acts between A and B. A couple produces a sudden change in the bending moment, and the left-hand side of Eq. (6-26) cannot be integrated across such a discontinuity.

Concentrated Loads Now consider a concentrated load P acting on the beam element (Fig. 6-33b). From equilibrium of forces in the vertical direction,

$$V - P - (V + V_1) = 0 \quad \text{or} \quad V_1 = -P \tag{6-28}$$

This result means that an abrupt change in the shear force occurs at any point where a concentrated load acts. Passing from left to right through the point of load application, the shear force decreases by an amount equal to the magnitude of the downward load P.

From the equilibrium of moments about the left-hand face of the element (Fig. 6-33b),

$$-M - P\left(\frac{dx}{2}\right) - (V + V_1)dx + M + M_1 = 0$$

or

$$M_1 = P\left(\frac{dx}{2}\right) + Vdx + V_1dx \tag{6-29}$$

Since the length dx of the element is infinitesimally small, the increment M_1 in the bending moment is also infinitesimally small. *Thus, the bending moment does not change when passing through the point of application of a concentrated load.*

Even though the bending moment M does not change at a concentrated load, its rate of change dM/dx undergoes an abrupt change. At the left-hand side of the element (Fig. 6-33b), the rate of change of the bending moment [see Eq. (6-25)] is $dM/dx = V$. At the right-hand side, the rate of change is $dM/dx = V + V_1 = V - P$. *Therefore, at the point of application of a concentrated load P, the rate of change dM/dx of the bending moment changes abruptly by an amount equal to P.*

Loads in the Form of Couples The last case to be considered is a load in the form of a couple M_0 (Fig. 6-33c). From equilibrium of the element in the vertical direction, $V_1 = 0$, which shows that the *shear force does not change at the point of application of a couple.*

Equilibrium of moments about the left-hand side of the element gives

$$-M + M_0 - (V + V_1)dx + M + M_1 = 0$$

Disregarding terms that contain differentials (because they are negligible compared to the finite terms) gives

$$M_1 = -M_0 \tag{6-30}$$

This equation shows that the bending moment decreases by M_0 when moving from left to right through the point of load application. *Thus, the bending moment changes abruptly at the point of application of a couple.*

Equations (6-22) through (6-30) are useful when making a complete investigation of the shear forces and bending moments in a beam, as discussed in the next section.

Shear-Force and Bending-Moment Diagrams You gained some insight into calculation and graphical display of internal axial forces N in bars subject to axial centroidal loading in Section 6.2 and of internal torsional moments T in shafts under applied torsional moments in Section 6.3. Here the concepts used to plot axial-force diagrams (AFD) in bars and torsional-moment diagrams (TMD) for shafts are extended to display of internal shear and moment in beams.

When designing a beam, it is important to know how the shear forces and bending moments vary throughout the length of the beam. Of special importance are the maximum and minimum values of these quantities. Information of this kind is usually provided by graphs where the shear force and bending moment are plotted as ordinates and the distance x along the axis of the beam is plotted as the abscissa. Such graphs are called **shear-force and bending-moment diagrams**.

To provide a clear understanding of these diagrams, it is necessary to review how they are constructed and interpreted for three basic loading conditions—a single concentrated load, a uniform load, and several concentrated loads. In addition, Examples 6-5 to 6-7 at the end of this section provide detailed illustration of the techniques for handling various kinds of loads, including the case of a couple acting as a load on a beam.

Concentrated Load   Begin with a simple beam AB supporting a concentrated load P (Fig. 6-35a). The load P acts at distance a from the left-hand support and distance b from the right-hand support. Consider the entire beam as a free body to determine the reactions of the beam from equilibrium; the results are

$$R_A = \frac{Pb}{L} \qquad R_B = \frac{Pa}{L}$$ **(6-31a,b)**

Now cut through the beam at a cross section to the left of the load P and at distance x from the support at A. Then draw a free-body diagram of the left-hand part of the beam (Fig. 6-35b). From the equations of equilibrium for this free body, obtain the shear force V and bending moment M at distance x from the support:

$$V = R_A = \frac{Pb}{L} \qquad M = R_A x = \frac{Pbx}{L} \qquad (0 < x < a)$$ **(6-32a,b)**

These expressions are valid only for the part of the beam to the left of load P.

FIGURE 6-35

Shear-force (V) and bending-moment (M) diagrams for a simple beam with a concentrated load

Next, cut through the beam to the right of the load P (that is, in the region $a < x < L$) and again draw a free-body diagram of the left-hand part of the beam (Fig. 6-35c). From the equations of equilibrium for this free body, obtain the following expressions for the shear force and bending moment:

$$V = R_A - P = \frac{Pb}{L} - P = -\frac{Pa}{L} \quad (a < x < L) \tag{6-33a}$$

and

$$M = R_A x - P(x - a) = \frac{Pbx}{L} - P(x - a)$$
$$= \frac{Pa}{L}(L - x) \qquad\qquad (a < x < L) \tag{6-33b}$$

Note that these equations are valid only for the right-hand part of the beam.

The equations for the shear forces and bending moments [Eqs. (6-32) and (6-33)] are plotted below the sketches of the beam. Figure 6-35d is the *shear-force diagram* and Fig. 6-35e is the *bending-moment diagram*.

From Fig. 6-35d, the shear force at end A of the beam ($x = 0$) is equal to the reaction R_A. Then it remains constant to the point of application of load P. At that point, the shear force decreases abruptly by an amount equal to load P. In the right-hand part of the beam, the shear force is again constant but equal numerically to the reaction at B.

As shown in Fig. 6-35e, the bending moment in the left-hand part of the beam increases linearly from zero at the support to Pab/L at the concentrated load ($x = a$). In the right-hand part, the bending moment is again a linear function of x, varying from Pab/L at $x = a$ to zero at the support ($x = L$). Thus, the maximum bending moment is

$$M_{max} = \frac{Pab}{L} \tag{6-34}$$

and it occurs under the concentrated load.

When deriving the expressions for the shear force and bending moment to the right of load P [Eqs. (6-33a and b)], consider the equilibrium of the left-hand part of the beam (Fig. 6-35c). This free body is acted upon by the forces R_A and P in addition to V and M. It is slightly simpler in this particular example to consider the right-hand portion of the beam as a free body because then only one force (R_B) appears in the equilibrium equations (in addition to V and M). Of course, the final results are unchanged.

Certain characteristics of the shear-force and bending-moment diagrams (Figs. 6-35d and e) now may be seen. Note first that the slope dV/dx of the shear-force diagram is zero in the region $0 < x < a$ and $a < x < L$, which is in accord with the equation $dV/dx = -q$ [Eq. (6-22)]. Also, in these same regions, the slope dM/dx of the bending-moment diagram is equal to V [Eq. (6-25)]. To the left of load P, the slope of the moment diagram is positive and equal to Pb/L; to the right, it is negative and equal to $-Pa/L$. Thus, at the point of application of load P, there is an abrupt change in the shear-force diagram (equal to the magnitude of load P) and a corresponding change in the slope of the bending-moment diagram.

Now consider the *area* of the shear-force diagram. In moving from $x = 0$ to $x = a$, the area of the shear-force diagram is $(Pb/L)a$, or Pab/L. This quantity represents the increase in bending moment between these same two points [see Eq. (6-27)]. From $x = a$ to $x = L$, the area of the shear-force diagram is $-Pab/L$, which means that in this region the bending moment decreases by that amount. Consequently, the bending moment is zero at end B of the beam, as expected.

If the bending moments at both ends of a beam are zero, as is usually the case with a simple beam, the area of the shear-force diagram between the ends of the beam must be zero—provided no couples act on the beam.

As mentioned previously, the maximum and minimum values of the shear forces and bending moments are needed when designing beams. For a simple beam with a single concentrated load, the maximum shear force occurs over the region of the beam of length a or b—whichever is smaller—and the maximum bending moment occurs under the load itself.

Uniform Load   A simple beam with a uniformly distributed load of constant intensity q is shown in Fig. 6-36a. Because the beam and its loading are symmetric, each of the reactions (R_A and R_B) is equal to $qL/2$. Therefore, the shear force and bending moment at distance x from the left-hand end are

$$V = R_A - qx = \frac{qL}{2} - qx \tag{6-35a}$$

and

$$M = R_A x - qx\left(\frac{x}{2}\right) = \frac{qLx}{2} - \frac{qx^2}{2} \tag{6-35b}$$

These equations, which are valid throughout the length of the beam, are plotted as shear-force and bending-moment diagrams in Figs. 6-36c and d, respectively.

The shear-force diagram consists of an inclined straight line having ordinates at $x = 0$ and $x = L$ that are equal numerically to the reactions. The slope of the line is $-q$, as expected from Eq. (6-22). The bending-moment diagram is a parabolic curve that is symmetric about the midpoint of the beam. At each cross section, the slope of the bending-moment diagram is equal to the shear force [see Eq. (6-25)]:

$$\frac{dM}{dx} = \frac{d}{dx}\left(\frac{qLx}{2} - \frac{qx^2}{2}\right) = \frac{qL}{2} - qx = V$$

FIGURE 6-36

Shear-force (V) and bending-moment (M) diagrams for a simple beam with a uniform load

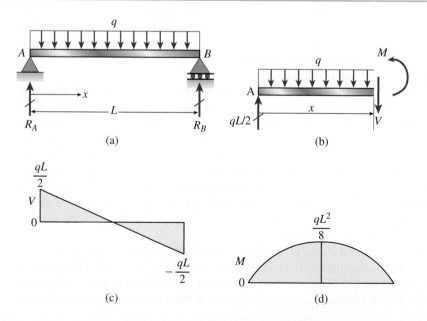

The maximum value of the bending moment occurs at the midpoint of the beam where both dM/dx and the shear force V are equal to zero. Therefore, substitute $x = L/2$ into Eq. (6-35b) to obtain

$$M_{max} = \frac{qL^2}{8} \qquad \textbf{(6-36)}$$

as shown on the bending-moment diagram.

The diagram of load intensity (Fig. 6-36a) has area qL, and according to Eq. (6-24), the shear force V must decrease by this amount when moving along the beam from A to B. This is indeed the case, because the shear force decreases from $qL/2$ to $-qL/2$.

The area of the shear-force diagram between $x = 0$ and $x = L/2$ is $qL^2/8$, and this area represents the increase in the bending moment between those same two points (Eq. 6-27). In a similar manner, the bending moment decreases by $qL^2/8$ in the region from $x = L/2$ to $x = L$.

Several Concentrated Loads If several concentrated loads act on a simple beam (Fig. 6-37a), expressions for the shear forces and bending moments may be determined for each segment of the beam between the points of load application. Again use free-body diagrams of the left-hand part of the beam and measure the distance x from end A to obtain the following equations for the first segment of the beam:

$$V = R_A \qquad M = R_A x \qquad (0 < x < a_1) \qquad \textbf{(6-37a,b)}$$

For the second segment,

$$V = R_A - P_1 \qquad M = R_A x - P_1(x - a_1) \quad (a_1 < x < a_2) \qquad \textbf{(6-38a,b)}$$

For the third segment of the beam, it is advantageous to consider the right-hand part of the beam rather than the left because fewer loads act on the corresponding free body. Hence,

$$V = -R_B + P_3 \qquad \textbf{(6-39a)}$$

$$M = R_B(L - x) - P_3(L - b_3 - x) \qquad (a_2 < x < a_3) \qquad \textbf{(6-39b)}$$

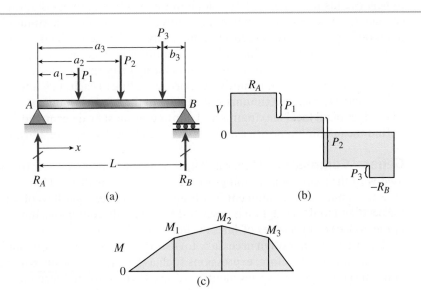

FIGURE 6-37

Shear-force (V) and bending-moment (M) diagrams for a simple beam with several concentrated loads

Finally, for the fourth segment of the beam,

$$V = -R_B \qquad M = R_B(L - x) \qquad (a_3 < x < L) \qquad \textbf{(6-40a,b)}$$

Equations (6-37) through (6-40) can be used to construct the shear-force and bending-moment diagrams (Figs. 6-37b and c).

From the shear-force diagram, note that the shear force is constant in each segment of the beam and changes abruptly at every load point with the amount of each change being equal to the load. Also, the bending moment in each segment is a linear function of x; therefore, the corresponding part of the bending-moment diagram is an inclined straight line. To assist in drawing these lines, find the bending moments under the concentrated loads by substituting $x = a_1$, $x = a_2$, and $x = a_3$ into Eqs. (6-37b), (6-38b), and (6-39b), respectively. In this manner, obtain the following bending moments:

$$M_1 = R_A a_1 \qquad M_2 = R_A a_2 - P_1(a_2 - a_1) \qquad M_3 = R_B b_3 \qquad \textbf{(6-41a,b,c)}$$

Knowing these values, you can readily construct the bending-moment diagram by connecting the points with straight lines.

At each discontinuity in the shear force, there is a corresponding change in the slope dM/dx of the bending-moment diagram. Also, the change in bending moment between two load points equals the area of the shear-force diagram between those same two points [see Eq. (6-27)]. For example, the change in bending moment between loads P_1 and P_2 is $M_2 - M_1$. Substitute from Eqs. (6-41a and b) to get

$$M_2 - M_1 = (R_A - P_1)(a_2 - a_1)$$

which is the area of the rectangular shear-force diagram between $x = a_1$ and $x = a_2$.

The maximum bending moment in a beam having only concentrated loads *must* occur under one of the loads or at a reaction. To show this, recall that the slope of the bending-moment diagram is equal to the shear force. Therefore, whenever the bending moment has a maximum or minimum value, the derivative dM/dx (and hence the shear force) must change sign. However, in a beam with only concentrated loads, the shear force can change sign only under a load.

If, when moving along the x axis, the shear force changes from positive to negative (as in Fig. 6-37b), the slope in the bending-moment diagram also changes from positive to negative. Therefore, there must be a maximum bending moment at this cross section. Conversely, a change in shear force from a negative to a positive value indicates a minimum bending moment. Theoretically, the shear-force diagram can intersect the horizontal axis at several points, although this is quite unlikely. Corresponding to each such intersection point, there is a local maximum or minimum in the bending-moment diagram. The values of all local maximums and minimums must be determined in order to find the maximum positive and negative bending moments in a beam.

General Comments   The terms "maximum" and "minimum" are frequently used with their common meanings of "largest" and "smallest." Consequently, "the maximum bending moment in a beam" is used here regardless of whether the bending-moment diagram is described by a smooth, continuous function (as in Fig. 6-36d) or by a series of lines (as in Fig. 6-37c).

Furthermore, it is often necessary to distinguish between positive and negative quantities. Therefore, expressions such as "maximum positive moment" and "maximum negative moment" are used. In both of these cases, the expression refers to the numerically largest quantity; that is, the term "maximum negative moment" really means "numerically largest negative moment." Analogous comments apply to other beam quantities, such a shear forces and deflections.

The maximum positive and negative bending moments in a beam may occur at the following places: (1) a cross section where a concentrated load is applied and the shear force changes sign (see Figs. 6-35 and 6-37); (2) a cross section where the shear force equals zero (see Fig. 6-36); (3) a point of support where a vertical reaction is present; and (4) a cross section where a couple is applied. The preceding discussions and the following examples illustrate all of these possibilities.

When several loads act on a beam, the shear-force and bending-moment diagrams can be obtained by superposition (or summation) of the diagrams obtained for each of the loads acting separately. For instance, the shear-force diagram of Fig. 6-37b is actually the sum of three separate diagrams, each of the type shown in Fig. 6-35d for a single concentrated load. An analogous comment can be made for the bending-moment diagram of Fig. 6-37c. Superposition of shear-force and bending-moment diagrams is permissible because shear forces and bending moments in statically determinate beams are linear functions of the applied loads.

Computer programs are readily available for drawing shear-force and bending-moment diagrams (see discussion at end of Example 6-5). After you have developed an understanding of the nature of the diagrams by constructing them manually, you should feel secure in using computer programs to plot the diagrams and obtain numerical results. For convenient reference, the differential relationships used in drawing shear-force and bending-moment diagrams are summarized in the Summary and Review following Example 6-7.

Example 6-5

A simple beam with an overhang is supported at points A and B (Fig. 6-38). A uniform load of intensity $q = 400$ lb/ft acts throughout the length of the beam, and a concentrated load $P = 2400$ lb acts at a point 9 ft from the left-hand support. The span length is 24 ft, and the length of the overhang is 6 ft.

Calculate the shear force V and bending moment M at cross section D located 15 ft from the left-hand support, then plot the shear and moment diagrams.

FIGURE 6-38

Beam with an overhang and uniform and concentrated loads

Solution:

Find external reactions at supports, then internal shear force and bending moment at point D. Plot shear force and bending-moment diagrams. Use the four-step problem-solving approach to find the solution.

1. *Conceptualize*: Find the reaction forces R_A and R_B using the FBD of the overall structure shown in Fig. 6-39. The horizontal reaction component at the pin at A is zero because only vertical loading is applied to the beam. Internal shear force V_D and bending moment M_D at point D (Fig. 6-40) can be obtained by cutting the beam at D. Either the left-hand or right-hand free-body diagram in Fig. 6-40 may be used to find V_D and M_D.

2. *Categorize*: The free-body diagrams in Fig. 6-40 show internal shear force V_D and bending moment M_D in their assumed positive directions based on a deformation sign convention. Shear and moment functions $V(x)$ and $M(x)$ are discontinuous due to concentrated load P and reaction force R_B, so three separate cuts have to be made to find $V(x)$ and $M(x)$ over the entire length of the beam (Fig. 6-41). Start by finding reaction forces R_A and R_B for this statically determinate beam.

3. *Analyze*:

   Solution for external reactions: Sum forces in the y direction and moments about point A and use a *statics sign convention* to find

   $$\Sigma F_y = 0 \quad R_A + R_B = P + qL = 2400 \text{ lb} + (400 \text{ lb/ft})(30 \text{ ft}) = 14{,}400 \text{ lb} \quad \textbf{(6-42a,b)}$$
   $$\Sigma M_A = 0 \quad R_B(24 \text{ ft}) = (2400 \text{ lb})(9 \text{ ft}) + (400 \text{ lb/ft})(30 \text{ ft})(15 \text{ ft}) = 201{,}600 \text{ lb-ft}$$

FIGURE 6-39

Free-body diagram of beam

FIGURE 6-40

Left- and right-hand free-body diagrams from section cut at D

In Eq. (6-42b), the distributed load has been replaced by its resultant acting at $x = (24 + 6)/2$ ft $= 15$ ft. Solve Eq. (6-42b) to obtain $R_B = 8400$ lb, then find $R_A = 6000$ lb using Eq. (6-42a). These reaction forces are shown in Fig. 6-40.

Solution for internal force V and moment M at point D: Use the left-hand FBD in Fig. 6-40 to find that

$$\Sigma F_y = 0 \quad V_D = 6000 \text{ lb} - 2400 \text{ lb} - (400 \text{ lb/ft})(9 + 6)\text{ft} = -2400 \text{ lb} \; (\uparrow)$$

$$\Sigma M_D = 0$$

$$M_D = 6000 \text{ lb}(15 \text{ ft}) - 2400 \text{ lb}(6 \text{ ft}) - (400 \text{ lb/ft})(15 \text{ ft})(15 \text{ ft}/2)$$

$$= 30{,}600 \text{ lb-ft (CCW)}$$

(6-43)

Solution for internal force $V(x)$ and moment $M(x)$: One option to plot shear and moment diagrams is to find the functions $V(x)$ and $M(x)$ over the entire length of the beam and then use computer software to create the V and M diagram plots. Cut the beam at locations 1 through 3 (Fig. 6-41); then use either the left-hand or right-hand free-body diagrams as

Shear $V(x)$:

$$\text{Section 1 (LHFB)} \quad V(x) = R_A - qx \qquad 0 \le x \le 9 \text{ ft}$$
$$\text{Section 2 (LHFB)} \quad V(x) = R_A - qx - P \qquad 9 \text{ ft} < x < 24 \text{ ft} \quad \textbf{(6-44)}$$
$$\text{Section 3 (RHFB)} \quad V(x) = q(30 \text{ ft} - x) \qquad x \ge 24 \text{ ft}$$

In Section 3, it is easier to find $V(x)$ if the RHFB is used rather than the LHFB. If the LHFB had been used, the expression for $V(x)$ would include R_A and P but would simplify to that shown in Eq. (6-44c) after numerical values were substituted. Next find the bending-moment functions:

Moment $M(x)$:

$$\text{Section 1 (LHFB)} \quad M(x) = R_A x - q\frac{x^2}{2} = 6000x - 200x^2 \text{ (lb-ft)} \qquad 0 \le x \le 9 \text{ ft}$$

$$\text{Section 2 (LHFB)} \quad M(x) = R_A x - q\frac{x^2}{2} - P(x - 9 \text{ ft}) = 3600x - 200x^2 + 21{,}600 \text{ (lb-ft)} \quad 9 \text{ ft} \le x \le 24 \text{ ft} \; \textbf{(6-45)}$$

$$\text{Section 3 (RHFB)} \quad M(x) = \frac{-q}{2}(30 \text{ ft} - x)^2 = -200(30 - x)^2 \text{ (lb-ft)} \qquad x \ge 24 \text{ ft}$$

The maximum value of the shear force occurs to the right of support A and to the left of support B, and the maximum value of the moment occurs at the point at which $V = 0$. The point of maximum moment can be located by setting the expression for $V(x)$ [Eq. (6-44a)] equal to zero, then solving for $x_m = 9$ ft. Solving for $M(x_m)$ [Eq. (6-45a)] provides the expression for $M_{max} = 37{,}800$ lb-ft.

Shear and moment diagrams: The preferred alternative approach for constructing the shear and moment diagrams relies on graphical interpretation of the differential and integral relationships between load, shear, and moment. The resulting diagrams are presented in Fig. 6-42. Based on this graphical approach, observe from these diagrams the following results.

a. The shear force V a small distance to the right of A is $+R_A$. As you move further to the right of A, the shear force decreases by load intensity q (lb/ft). At distance $x = 9$ ft, shear V has dropped from 6000 lb to [6000 lb $-$ 9 ft (400 lb/ft)] = 2400 lb.

b. The slope of the tangent to the shear curve is equal to the negative ordinate on the load curve since $dV/dx = -q(x)$. Load $q(x)$ is constant, so the shear decreases linearly.

c. Just to the right of the point of application of load P, the shear is 2400 lb minus P, or zero. This is the location ($x_m = 9$ ft) of the maximum positive moment, M_{max}.

d. The change in moment from support A (where the moment is zero) to the point of maximum moment at $x_m = 9$ ft is equal to the area under the shear diagram between these two points. This trapezoidal area under the shear diagram (labeled A_1 in Fig. 6-42a) is

$$M_{max} = \frac{1}{2}[6000 \text{ lb} + 2400 \text{ lb}](x_m) = 37{,}800 \text{ lb-ft}.$$

e. The integral relationships between load and shear and between shear and moment mean that, since the load curve $q(x)$ has constant magnitude, the shear diagram $V(x)$ is a linear plot and moment $M(x)$ diagram is a quadratic plot. [This is shown in Eqs. (6-44) and (6-45)].

f. The shear at D is $(-6 \text{ ft})(q) = -2400$ lb less than that at location $x = x_m$. The moment at D is M_{max} minus the triangular area under the shear curve from $x = x_m$ to $x = 15$ ft, or

$$M_{max} - \frac{1}{2}(2400 \text{ lb})(6 \text{ ft}) = [37{,}800 - \frac{1}{2}(2400)(6)] \text{ lb-ft} = 30{,}600 \text{ lb-ft}.$$

By subtracting the entire triangular area under the shear curve from $x = x_m$ to $x = 24$ ft, find $M_B = M_{max} - \frac{1}{2}(6000 \text{ lb})(15 \text{ ft}) =$

[37,800 $-$ 45,000] lb-ft $= -7200$ lb-ft.

g. As you move past B on the shear diagram, the shear increases from -6000 lb to $+2400$ lb, which is an increase of $+8400$ lb due to reaction force R_B.

h. The moment is zero just to the left of B at the *inflection point*. Equate the two areas labeled A_1 on the shear diagram (the trapezoidal area A_1 is computed in part (d) and can be equated to the triangular area labeled $-A_1$) to find that

FIGURE 6-42

(a,b) Shear and moment diagrams

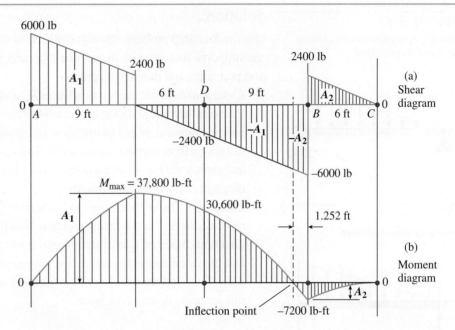

the inflection point is 1.252 ft to the left of B. The area under the shear curve from B to C (labeled A_2) is equal to 7200 lb-ft; adding this to moment M_B takes the moment diagram back to zero at C.

4. *Finalize*: Graphical interpretation of the differential and integral relationships between load, shear, and moment was used to construct the diagrams shown in Fig. 6-42. The functions $V(x)$ and $M(x)$ [Eqs. (6-44) and (6-45)] can be plotted using widely available computer software (such as Maple, Mathcad, Mathematica, and MATLAB) to confirm these plots.

Note that the moment diagram is plotted on the compression side of the beam. Hence, most of the top of the beam from A to B (from point A to the inflection point) is in compression, while the top of the beam from the inflection point to C is in tension. This is important information to consider in the design of beams made of certain materials.

Example 6-6

A simply supported beam (Fig. 6-43) of length L is subjected to a distributed load of linearly varying intensity $q(x) = \dfrac{x}{L}q_0$. Find reaction forces at supports A and B; then find expressions for shear force $V(x)$ and bending moment $M(x)$ at a distance x from point A on the beam. Plot the shear and moment diagrams.

FIGURE 6-43

Simply supported beam with a linearly varying distributed load

FIGURE 6-44

Free-body diagram of beam

FIGURE 6-45

Left-hand free-body diagram

Solution:

Use the four-step problem-solving approach to find external reactions at the supports, find internal shear force $V(x)$ and bending moment $M(x)$, and plot shear and moment diagrams.

1. *Conceptualize*: Find the reaction forces R_A and R_B using the FBD of the overall structure shown in Fig. 6-44. Internal shear force V and bending moment M can be displayed by cutting the beam at some distance x from support A (Fig. 6-45). This cut produces left-hand and right-hand free-body diagrams, but only the left-hand free-body diagram is shown for convenience.

2. *Categorize*: The left-hand free-body diagram (Fig. 6-45) shows internal shear force V and bending moment M in their assumed positive directions based on a deformation sign convention. Summing forces and then moments at the cut section provides functions $V(x)$ and $M(x)$. First find reaction force R_A, which is a straightforward task for this statically determinate beam.

3. *Analyze*:

Solution for external reactions: Sum forces in the y direction and moments about point A and use a *statics sign convention* to find

$$\Sigma F_y = 0 \quad R_A + R_B - \frac{1}{2}q_0 L = 0$$

$$\Sigma M_A = 0 \quad R_B L - \frac{1}{2}(q_0 L)\left(\frac{2}{3}L\right) = 0$$

$$R_B = \frac{1}{3}q_0 L$$

Substituting R_B in the first equation gives

$$R_A = -R_B + \frac{1}{2}q_0 L = \frac{1}{6}q_0 L$$

The reaction at A carries one-third of the applied load and at B carries two-thirds of the load. Note that the x-direction reaction at pin A is zero by inspection because no horizontal load or load component is applied.

Solution for internal force $V(x)$ and moment $M(x)$: Sum forces and moments for the FBD in Fig. 6-45 to get the following expressions that show the variation in V and M over the length of the beam:

$$\Sigma F_y = 0 \quad V(x) = R_A - \frac{1}{2}\left(\frac{x}{L}q_0\right)(x)$$

$$V(x) = \frac{q_0 L}{6} - \frac{1}{2}\left(\frac{x}{L}q_0\right)(x) \qquad \textbf{(6-46)}$$

$$V(x) = \frac{q_0(L^2 - 3x^2)}{6L}$$

$$\sum M = 0 \quad M(x) = R_A x - \frac{1}{2}\left(\frac{x}{L}q_0\right)(x)\left(\frac{x}{3}\right) \tag{6-47}$$

$$= \left(\frac{q_0 L}{6}\right)x - \frac{1}{2}\left(\frac{x}{L}q_0\right)(x)\left(\frac{x}{3}\right) = \frac{q_0 x(L^2 - x^2)}{6L} \quad \leftarrow$$

Now, when $x = 0$, $V(0) = R_A$ and $M(0) = 0$ at joint A. When $x = L$, $V(L) = -R_B$ and once again $M(L) = 0$, because no moment is applied at the roller support at B. The numerically largest shear is at B, where R_B is twice the value of R_A:

$$V_{\text{max}} = \frac{-q_0 L}{3} \tag{6-48}$$

It is not readily apparent where along the beam the maximum moment occurs. However, if you differentiate the expression for $M(x)$, equate it to zero, and solve for x, you will find the point (x_m) where a local maxima or minima occurs in the function $M(x)$. Solving for x_m and then substituting x_m into the moment expression gives

$$\frac{d}{dx}(M(x)) = \frac{d}{dx}\left[\frac{q_0 x(L^2 - x^2)}{6L}\right] = \frac{q_0(L^2 - 3x^2)}{6L} = 0$$

This results in

$$x_m = \frac{L}{\sqrt{3}}$$

and so

$$M_{\text{max}} = M(x_m) = \frac{\sqrt{3}}{27}q_0 L^2 \tag{6-49}$$

The expression that results from $d/dx(M(x))$ is the same as that for $V(x)$ in Eq. (6-46). This is as expected based on the differential relationship between shear and moment (i.e., $dM/dx = V$) derived previously [see Eq.(6-25)].

Shear and moment diagrams: The shear-force and bending-moment diagrams (Fig. 6-46) can be obtained by plotting the expressions above for $V(x)$ [Eq. (6-46)] and $M(x)$ [Eq. (6-47)]. The slope of the shear-force diagram at any point x along the beam is equal $-q(x)$, and the slope of the bending-moment diagram at any point x is equal to $V(x)$. The maximum value of the shear force occurs at support B where $x = L$ [see Eq. (6-48)], and the maximum value of the moment occurs at the point where $V = 0$. The point of maximum moment can be located by setting the expression for $V(x)$ [Eq. (6-46)] equal to zero, then solving for x_m as before. Solving for $M(x_m)$ [Eq. (6-47)] provides the expression for M_{max} shown in Eq. (6-49).

The preferred alternative approach for constructing the shear and moment diagrams is based on graphical interpretation of the differential and integral

FIGURE 6-46

(a,b,c) Overall free-body diagram and shear and moment diagrams

(a)

(b)

(c)

relationships between load, shear, and moment. Using this graphical approach, observe that:

- The shear force V at distance x from the support at A is obtained by subtracting the triangular area under the distributed load diagram from the reaction force R_A. The change in shear from A to B is equal to the negative value of the area under the entire distributed-load diagram $(-A_q)$, as shown in Fig. 6-46b [also see Eq. (6-24)].

- The slope of the tangent to the shear curve is equal to the negative ordinate on the load curve since $dV/dx = -q(x)$. Hence, the slope is zero at A and $(-q_0)$ at B (Fig. 6-46b).

- The slope of the tangent to the moment curve is equal to the ordinate on the shear diagram since $dM/dx = V$. Hence, the slope is R_A at A, $(-R_B)$ at B, and zero at the location of maximum moment (Fig. 6-46c).

- The change in moment from support A to the point of maximum moment at $x_m = \dfrac{L}{\sqrt{3}}$ is equal to the area under the shear diagram (A_V) between these two points [see Eq. (6-27)]. The area under the shear diagram is $(-A_V)$ from $x = x_m$ to $x = L$ so the moment diagram returns to zero at B.

- The integral relationships between load and shear and between shear and moment mean that, since the load curve $q(x)$ varies linearly, the shear diagram $V(x)$ is quadratic and moment $M(x)$ diagram is cubic.

4. *Finalize*: Graphical interpretation of the differential and integral relationships between shear and moment offers an efficient way to draw shear and moment diagrams for beams. The resulting diagrams are the same as those obtained by plotting functional expressions for $V(x)$ and $M(x)$ using computer software, but the graphical approach offers much more insight into the problem. You saw earlier (Sections 6.2 and 6.3) that this same graphical approach works well for constructing axial-force diagrams (AFD) in bars with axial loads and for sketching torsional-moment diagrams (TMD) for shafts in torsion.

Example 6-7

The plane frame (Fig. 6-47) is loaded by both concentrated and distributed loads. Moment M_A is applied at pin-supported joint A, and load F_B is applied at joint B. A uniform load with intensity q_1 acts on member BC, and a linearly distributed load with peak intensity q_0 is applied downward on member ED. Find the support reactions at joints A and D; then solve for internal forces at the top of member BC. Plot the axial-force, shear, and moment diagrams for all members. Use the numerical properties given.

FIGURE 6-47

Plane frame with both concentrated and distributed loads

Numerical data (lb, ft):

$a = 10$ ft $b = 6$ ft $c = 18$ ft $d = 8$ ft

$M_A = 280$ lb-ft $F_B = 45$ lb $q_0 = 5$ lb/ft $q_1 = 10$ lb/ft

Solution:

Use the four-step problem-solving approach to solve for external reactions at supports and internal forces and moments in all members. Plot the axial-force, shear-force, and bending-moment diagrams.

1. *Conceptualize*: Find the reaction forces at supports A and D using the FBD of the overall structure shown in Fig. 6-48. Distributed loads on members ED and BC can be replaced by their static equivalents (F_{q0} and F_{q1}), each acting at the centroid of its distributed load. The statically equivalent forces are equal in magnitude to the area under the distributed load curves. Internal axial force N_C, shear force V_C, and bending moment M_C just below joint C are obtained by cutting the frame as shown in Fig. 6-49. Either the upper or lower free-body diagram in Fig. 6-49 may be used to find internal forces and moment at that location.

2. *Categorize*: The overall frame is statically determinate. First solve for reaction forces at A and D using the overall FBD in Fig. 6-48; then use either FBD in Fig. 6-49 to find internal forces and moment below joint C. Treat members ABC and EDC as separate beams, each having its own longitudinal centroidal

FIGURE 6-48

Free-body diagram of frame

FIGURE 6-49

Upper and lower free-body diagrams from section cut just below joint C

axis x_m (see Fig. 6-49); then find axial-force, shear-force, and moment functions $[N(x), V(x),$ and $M(x)]$ and the associated diagrams for each member.

3. *Analyze*:

Statically equivalent forces and force components: Distributed forces are replaced by their static equivalents (F_{q0} and F_{q1}). Load F_{q0} acts downward at $c/3 = 6$ ft to the right of E and load F_{q1} acts at the center of BC. The static equivalents and the components of the inclined concentrated force at B are computed as

$$F_{q0} = \frac{1}{2}q_0 c = \frac{1}{2}(5 \text{ lb/ft})(18 \text{ ft}) = 45 \text{ lb} \quad F_{q1} = q_1 b = (10 \text{ lb/ft})(6 \text{ ft}) = 60 \text{ lb} \quad \textbf{(6-50)}$$

$$F_{Bx} = \frac{4}{5}F_B = \frac{4}{5}(45 \text{ lb}) = 36 \text{ lb} \qquad F_{By} = \frac{3}{5}F_B = \frac{3}{5}(45 \text{ lb}) = 27 \text{ lb}$$

External reactions: Use a *statics sign convention* when solving for reaction forces:

$$\Sigma F_x = 0 \quad A_x = -F_{q1} + F_{Bx} = -24 \text{ lb}$$

$$\Sigma M_A = 0 \quad D_y = \frac{1}{d}\left[F_{q0}\left(d + \frac{2c}{3}\right) - F_{q1}\left(a + \frac{b}{2}\right) + F_{Bx}(a) - M_A\right] = 25 \text{ lb} \quad \Longleftarrow \text{(6-51)}$$

$$\Sigma F_y = 0 \quad A_y = F_{q0} - D_y + F_{By} = 47 \text{ lb}$$

Internal forces N and V and moment M just below point C: Use a *deformation sign convention* to find internal quantities N_C, V_C, and M_C that are shown in their assumed positive directions in Fig. 6-49. Equilibrium of the upper free-body diagram in Fig. 6-49 gives

$$N_C = D_y - F_{q0} = -20 \text{ lb} (\uparrow)$$

$$V_C = 0 \quad \Longleftarrow \text{(6-52)}$$

$$M_C = F_{q0}\left(d + \frac{2c}{3}\right) - D_y(d) = 700 \text{ lb-ft (CW)}$$

Moment M_C is clockwise (CW) as assumed in Fig. 6-49, so the result is positive.

Axial-force, shear-force, and moment functions [$N(x)$, $V(x)$, and $M(x)$]: Treat each member as a beam and use the centroidal axis x_m for each member to write axial-force, shear-force, and moment functions over the entire length of the member.

Beam EDC (Fig. 6-49, upper FBD): Split the triangular load on ED into triangular load segments (see dashed line in Fig. 6-50) to make it easier to write expressions for $V(x_m)$ and $M(x_m)$ for the two parts of EDC as

$$N(x_m) = 0 \qquad\qquad 0 \le x_m \le c+d$$

$$V(x_m) = -\frac{1}{2}(q_0 + q_x)(x_m) \quad \text{with} \quad q_x = \left[\frac{c - x_m}{c}\right]q_0 \quad 0 \le x_m \le c$$

$$= -\frac{1}{2}(q_0)(c) + D_y \qquad\qquad c \le x_m \le c + d \qquad \text{(6-53)}$$

$$M(x_m) = -\frac{1}{2}(q_0 x_m)\left(\frac{2x_m}{3}\right) - \frac{1}{2}(q_x)\left(\frac{x_m^2}{3}\right) \qquad 0 \le x_m \le c$$

$$= -\frac{1}{2}(q_0 c)\left(x_m - \frac{c}{3}\right) + D_y(x_m - c) \qquad c \le x_m \le c + d$$

FIGURE 6-50

Triangular load on *ED*

Beam ABC (Fig. 6-49, lower FBD): Column ABC can be treated as a beam for the purpose of writing the axial-force, shear-force, and moment expressions for the two segments, AB and BC. The origin of member axis x_m is at A (Fig. 6-49); member ABC may be more conveniently viewed if rotated clockwise to a horizontal position. Functions $N(x_m)$, $V(x_m)$, and $M(x_m)$ for the two parts of ABC are

$$N(x_m) = -A_y \quad 0 \leq x_m \leq a, \quad N(x_m) = -A_y + F_{By} \quad a \leq x_m \leq a+b$$

$$V(x_m) = -A_x \quad 0 \leq x_m \leq a, \quad V(x_m) = -A_x + F_{Bx} - q_1(x_m - a) \quad a \leq x_m \leq a+b$$

$$M(x_m) = M_A - A_x x_m \qquad\qquad\qquad\qquad\qquad 0 \leq x_m \leq a \qquad \textbf{(6-54)}$$

$$= M_A - A_x x_m + F_{Bx}(x_m - a) - \frac{q_1}{2}(x_m - a)^2 \qquad\qquad a \leq x_m \leq a+b$$

As a check on the functions in Eq. (6-54), note that $N(a + b) = N_C$, $V(a + b) = V_C$, and $M(a + b) = M_C$, where internal forces and moment at C are given in Eq. (6-52).

Axial-force, shear-force, and bending-moment diagrams: An alternative approach for constructing the axial-force, shear-force, and moment diagrams is that based on the graphical rules. These rules are based on the differential and integral relationships between load, shear, and moment. The resulting diagrams are presented in Fig. 6-51. Based on this graphical approach, observe from these diagrams that:

(a) The axial force is constant and compressive in the two column segments AB and BC (see AFD, Fig. 6-51a). In AB, $N(x_m) = -A_y$, and in BC, $N(x_m) = N_C$ [from Eq. (6-52)].

(b) The slope of the tangent to the shear curve at E (Fig. 6-51b) is equal to $-q_0$, since $dV/dx = -q(x)$; at D, the load ordinate $q(c) = 0$ is zero, so the slope is zero on shear curve. The area under the load curve from E to D is F_{q0}, so the change in shear from E to D is $-F_{q0}$. The change in shear at D is $+D_y = 25$ lb, and the shear remains constant from D to C. The shear just left of joint C is equal to N_C (Fig. 6-49).

(c) The area under the shear curve from E to D (A_{VED}), is equal to the change in moment from E to D (Figs. 6-51b and c). The change in moment from D to C equals the negative rectangular area under the shear curve from D to C [that is $(-20 \text{ lb})(d) = -160 \text{ lb-ft}$].

(d) The shear from A to B is constant because the slope is equal to the distributed load magnitude, but $q = 0$ in this segment. At B, the shear increases by the magnitude of load component $F_{Bx} = 36$ lb [Eq. (6-50)] and then varies linearly with the slope of $(-q_1)$ to a value of $V_C = 0$ below joint C. The decrease in shear from B to C equals $-F_{q1}$, which is the area under the load curve from B to C.

(a) Axial-force diagram, $N(x)$ (lb)

(a)

(b) Shear-force diagram, $V(x)$ (lb)

(b)

(c) Moment diagram, $M(x)$ (lb-ft)

(c)

(e) The change in moment from support A (where moment M_A is applied) to point B is equal to the positive area under the shear curve from A to B, so $A_{VAB} = (24 \text{ lb})(10 \text{ ft}) = 240$ lb-ft. The moment from B to C increases further by the triangular area under the shear curve, which is equal to $\frac{1}{2}(60 \text{ lb})(b) = 180$ lb-ft.

(f) From the diagrams in Fig. 6-51a: (i) maximum axial force occurs on segment AB; (ii) maximum shear force is just above joint B; and (iii) maximum moment can be found on either side of joint C.

4. *Finalize*: Axial-force, shear-force, and moment diagrams for frames are frequently plotted on top of the frame, as shown in Fig. 6-51. There is no axial force in beam EDC, but column ABC is in compression over its entire height. The moment diagram shows that the left side of column AB and the entire underside of beam EDC are in compression.

FIGURE 6-52

Internal forces and moments on either side of joint C

The connection between beam EDC and column ABC will be an important design feature of this frame if it is to fulfill its intended function. Use the N, V, and M diagrams and a FBD of joint C to get a better idea of the manner in which forces flow into and through this connection. Then use the forces to find stresses in the connection components. From the FBD of the connection at C (Fig. 6-52), observe that force and moment equilibrium are satisfied for this joint. Shear to the left of $C(V_{CD})$ equilibrates axial force in $BC(N_{CB})$, while shear below joint $C(V_{CB})$ equals axial force (N_{CD}) at the end of member DC.

Chapter 6 reviewed the analysis of statically determinate bars, shafts, beams, and simple frames to find support reactions and internal forces and moments (N, T, V, and M), then plotted **axial-force**, **torsional-moment**, **shear**, and **bending-moment diagrams** to show the variation of these quantities throughout the structure. A variety of different support conditions, and both concentrated and distributed loadings, were considered in assembling models of structures for analysis. A **statics sign convention** was used when solving for external forces and moments and a **deformation sign convention** when finding internal quantities at selected points along a member.

Some of the major concepts presented in Chapter 6 are given here.

1. If the structure is **statically determinate**, the laws of statics alone are sufficient to solve for all values of support reaction forces and moments, as well as the magnitude of the internal axial force (N) in bars, internal torsional moment (T) in shafts, and shear force (V) and bending moment (M) at any location in a beam or frame member.

2. Positive values of internal quantities N, T, V, and M are defined using a **deformation sign convention** as shown in the following figures.

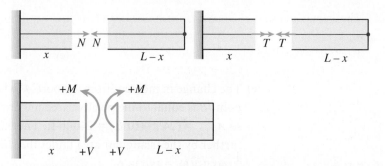

3. Graphical displays or **diagrams** showing the variation of N, T, V, and M over a structure are useful in design because they readily show the location of maximum values of N, T, V, and M needed in **design**.

4. **Differential** and **integral relationships** between internal **axial force** $N(x)$ and **distributed axial load** $q(x)$ are derived by taking an elemental slice from the bar (see figure). Similar relationships are obtained for internal **torsional moment** $T(x)$ and distributed **torsional moment loading** $t(x)$ using an elemental slice from a shaft in torsion (see figure). Assuming that A and B are two locations along the bar or shaft, those relationships are as follows.

For a bar with distributed axial centroidal loading:

$$\frac{dN}{dx} = -q \text{ and } N_B - N_A = -\int_A^B q(x)dx$$

For a shaft with distributed torsional moment:

$$\frac{dT}{dx} = -t \text{ and } T_B - T_A = -\int_A^B t(x)dx$$

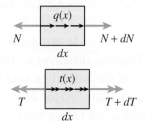

5. Using the differential and integral relationships above, the guidelines for **drawing axial-force diagrams (AFD)** for straight bars with axial loading may be summarized as follows.

- The slope at any point on the AFD is equal to the ($-$) ordinate on the distributed axial-load diagram (that is, $q(x)$) at that same point.
- The change in internal axial force N between any two points along a bar is equal to the negative of the area under the load diagram between those same two points.
- At the point(s) where the axial-load diagram crosses the reference axis, the value of the internal axial force N on the axial-force diagram is a local maximum or minimum.
- At points of application of concentrated loads, there is a change in ordinate (or discontinuity) on the AFD.

6. Using the differential and integral relationships in part 4, the guidelines for **drawing torsional-moment diagrams (TMD)** for straight shafts with torsional moments may be summarized as follows.

- The slope at any point on the TMD is equal to the ($-$) ordinate on the distributed torsional moment loading diagram (that is, $t(x)$) at that same point.
- The change in internal torsional moment T between any two points along a shaft is equal to the negative of the area under the load diagram between those same two points.
- At the point(s) where the torsional load diagram crosses the reference axis, the value of the internal torsional moment T on the torsional moment diagram is a local maximum or minimum.
- At points of application of concentrated torsional moments, there is a change in ordinate (or discontinuity) on the TMD.

7. A **distributed load** q on a beam is replaced by its resultant concentrated load F_q when carrying out a static analysis to find reactions at supports. The resultant concentrated load is *statically equivalent* to the distributed load and acts at the *centroid* of the area representing the distributed load. The location $\overline{x}$ of the centroid of the distributed load is obtained by equating first moments of the load area and resultant as

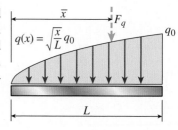

$$\overline{x} = \frac{\int xq(x)dx}{F_q} = \frac{\int xq(x)dx}{\int q(x)dx}$$

8. The **differential** and **integral relationships** between distributed load $q(x)$, internal shear force $V(x)$, and bending moment $M(x)$ for a **beam** may be derived using the elemental slice in the figure. Those relationships are

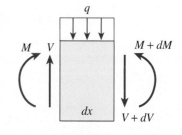

$$\frac{dV}{dx} = -q \text{ and } V_B - V_A = -\int_A^B q(x)dx; \text{ also } \frac{dM}{dx} = V \text{ and } M_B - M_A = \int_A^B V(x)dx$$

9. The guidelines for **drawing shear** and **bending-moment diagrams** for beams and frames are based on the differential and integral relationships in part 8 and may be summarized as follows.

- The ordinate on the distributed load curve (q) is equal to the negative of the slope dV/dx on the shear diagram.

- The difference in shear values between any two points A and B on the shear diagram is equal to the (−) area under the distributed load curve between those same two points.

- The ordinate on the shear diagram (V) is equal to the slope dM/dx on the bending-moment diagram.

- The difference in values between any two points on the moment diagram is equal to the area under the shear diagram between those same two points.

- At those points where the shear curve crosses the reference axis ($V = 0$), the value of the moment on the bending-moment diagram is a local maximum or minimum.

- At points of application of concentrated loads or bending moments, there are changes in ordinates (or discontinuities) on the shear and/or moment diagrams.

10. With a deformation sign convention, positive moment causes compression at the top and tension at the bottom of the beam. As a result, the bending-moment diagram is plotted on the **compression side** of a structural member or portion of a member.

11. The point at where the bending-moment diagram crosses the reference axis ($M = 0$) is called an **inflection point**.

12. **Axial-force**, **torsional-moment**, and **shear** and **bending-moment** diagrams help to identify values and locations of maximum axial force, torsional moment and shear and moment so that **maximum stresses and strains** can be found in members.

13. **Axial-force**, **torsional-moment**, and **bending-moment diagrams** are used to create **plots** that show the variation of axial displacements [that is, an axial-displacement diagram (ADD)], torsional-displacements (TDD), and deflections of beams (equation of the elastic curve).

PROBLEMS

6.2 Bars with Axial Loads

Introductory Problems

6.2-1 Aluminum bar AD (see figure) is loaded by forces $P_1 = 1700$ lb, $P_2 = 1200$ lb, and $P_3 = 1300$ lb. The lengths of the segments of the bar are $a = 60$ in., $b = 24$ in., and $c = 36$ in. Find internal axial forces in each segment, and then plot the AFD.

PROBLEM 6.2-1

6.2-2 A two-story building has steel columns AB in the first floor and BC in the second floor, as shown in the figure. The roof load P_1 equals 400 kN, and the second-floor load P_2 equals 720 kN. Each column has length $L = 3.75$ m. Find internal axial forces in each column, and then plot the AFD. Ignore the self-weight of the columns.

PROBLEM 6.2-2

6.2-3 Find internal axial forces in each segment of the nonprismatic bar shown in the figure, and then plot the AFD.

PROBLEM 6.2-3

Representative Problems

6.2-4 A wood pile, driven into the earth, supports a load P entirely by friction along its sides (see Fig. a). The friction force f per unit length of the pile is assumed to be uniformly distributed over the surface of the pile.

(a) Find internal axial force $N(y)$ in the pile in terms of load P (Fig. a), and then plot the AFD.

(b) Repeat if skin friction f varies linearly with depth (see Fig. b).

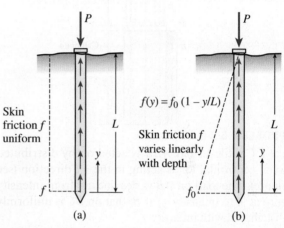

PROBLEM 6.2-4

6.2-5 A long, slender bar in the shape of a right circular cone with length L and base diameter d hangs vertically under its own weight (see figure). The weight of the cone is W. Derive a formula for the internal axial force $N(x)$ in the bar due to its own weight, and then plot the AFD.

PROBLEM 6.2-5

423

6.2-6 A hollow circular pipe hangs from a support at A (see figure) and has applied load $P = 20$ kN at B (which is uniformly distributed around a cap plate at the top of the lower pipe). In addition, uniform load $q = 2$ kN/m is applied upward along the centroidal axis of pipe segment 2. Pipe lengths are $L_1 = 2$ m and $L_2 = 3$ m. Find the internal axial force $N(x)$ in each bar segment, and then plot the AFD. Neglect the self-weight of the pipes. Assume that cap plate thickness is small compared to L_1 and L_2.

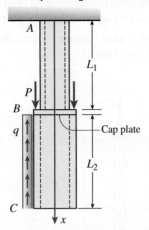

PROBLEM 6.2-6

6.2-7 Cantilever bar ABC is acted on by distributed axial centroidal loads acting in the $+x$ direction (see figure). The load on AB varies linearly from intensity zero at A to intensity q_0 at B; that on BC is uniformly distributed with intensity q_0.

(a) Find the internal axial force $N(x)$ in each bar segment, and then plot the AFD.

(b) Repeat part (a) if the load on BC is now applied in the $(-x)$ direction.

PROBLEM 6.2-7

6.2-8 Bar ABC of length $L = 9$ m has a pin support at A and a roller support at C. A linearly varying axial centroidal load with peak intensity $q_0 = 1.5$ kN/m acts to the right on BC. Find the internal axial force $N(x)$ in each bar segment, and then plot the AFD.

PROBLEM 6.2-8

424

6.2-9 Cantilever bar ABC is acted on by linearly varying axial centroidal loads shown in the figure. The load on AB acts in the $+x$ direction while that on BC acts in the $(-x)$ direction. Assume that peak intensity $q_0 = 250$ lb/ft and $L = 12$ ft. Find the internal axial force $N(x)$ in each bar segment, and then plot the AFD.

PROBLEM 6.2-9

6.2-10 Bar AB is compressed by axial centroidal load $q(x)$ (see figure). Assume that $L = 8$ m and peak intensity $q_0 = 4$ kN/m. Find the internal axial force $N(x)$ in the bar, and then plot the AFD.

PROBLEM 6.2-10

6.3 Shafts with Torsional Moments

Introductory Problems

6.3-1 Aluminum shaft AD (see figure) is loaded by torques $T_1 = 1700$ lb-in., $T_2 = 1200$ lb-in., and $T_3 = 1300$ lb-in. The lengths of the segments of the shaft are $a = 60$ in., $b = 24$ in., and $c = 36$ in. Find internal torsional moments $T(x)$ in each segment, and then plot the TMD.

PROBLEM 6.3-1

6.3-2 Torques $T_a = 100$ N·m and $T_b = 150$ N·m are applied to shaft ACB as shown. Let $L = 10$ m.

(a) Find the reaction torque at A, and then plot the TMD.

(b) (Repeat part (a) if torsional moment vector T_b is directed to the left instead of to the right.

PROBLEM 6.3-2

6.3-3 Stepped shaft $ABCD$ is subjected to the torques shown in the figure. Find the reaction torque at A, and then plot the TMD.

12.5 k-in. 9.8 k-in. 9.2 k-in.

3.5 in. 2.75 in. 2.5 in.

—25 in.——25 in.——25 in.—

PROBLEM 6.3-3

Representative Problems

6.3-4 A prismatic bar AB of length L and solid circular cross section is loaded by a distributed torque of constant intensity t per unit distance (see figure). Find the reaction torque at A, and then plot the TMD.

PROBLEM 6.3-4

6.3-5 Gear shaft $ABCDE$ is subjected to the torques shown in the figure. Find the internal torque in each segment, and then plot the TMD. Assume that the spacing between gears is constant.

$T_1 =$ 1000 lb-in. $T_2 =$ 500 lb-in. $T_3 =$ 800 lb-in. $T_4 =$ 500 lb-in. $T_5 =$ 800 lb-in.

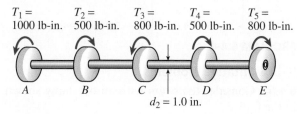

$d_2 = 1.0$ in.

PROBLEM 6.3-5

6.3-6 A circular shaft AB of length L is fixed against rotation at the ends and loaded by a distributed torque $t(x)$ that varies linearly in intensity from zero at end A to t_0 at end B (see figure). The reaction torque at B is known to be $t_0 L/3$. Find the reaction torque at A, and then plot the TMD.

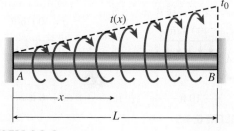

$t(x)$ t_0

PROBLEM 6.3-6

6.3-7 The nonprismatic steel pipe of constant thickness t and variable diameter d has applied torque $T_0/2$ at joint 2 and applied torque $3T_0/2$ at joint 3. Find the reaction torque at joint 1, and then plot the TMD.

$2d$ t d t d $3T_0/2$

$T_0/2$

$\dfrac{L}{2}$ $\dfrac{L}{2}$

PROBLEM 6.3-7

6.3-8 A hollow steel shaft ACB is held against rotation at ends A and B (see figure). Horizontal forces $P = 3$ kN are applied at the ends of a vertical arm that is welded to the shaft at point C. The reaction torque at B is known to be -720 N·m. Find the reaction torque at A, and then plot the TMD.

200 mm

200 mm

P C B

P

600 mm

400 mm

PROBLEM 6.3-8

6.3-9 Cantilever shaft ABC is acted on by linearly varying centroidal torsional moments as shown in the figure. The load on AB acts in the $+x$ direction while that on BC acts in the $(-x)$ direction. Assume that peak intensity $t_0 = 150$ lb-ft/ft and $L = 12$ ft. Find the internal torsional moment $T(x)$ in each bar segment, and then plot the TMD.

t_0

A B C x

$2L/3$ $L/3$

PROBLEM 6.3-9

425

6.3-10 The cantilever shaft shown in the figure is subjected to distributed torsional moment $t(x) = t_0 \left[1 - \left(\dfrac{x}{L} \right)^2 \right]$. Find the reaction torque at A, and then plot the TMD.

PROBLEM 6.3-10

6.4 Beams and Frames with Transverse Loads and Applied Moments

Shear Forces and Bending Moments

Introductory Problems

6.4-1 Calculate the shear force V and bending moment M at a cross section just to the right of the 800-lb load acting on the simple beam AB shown in the figure.

PROBLEM 6.4-1

6.4-2 Determine the shear force V and bending moment M just right of the 6-kN load on the simple beam AB shown in the figure.

PROBLEM 6.4-2

6.4-3 Determine the shear force V and bending moment M at the midpoint of the beam with overhangs (see figure). Note that one load acts downward and the other upward, and clockwise moments Pb are applied at each support. Repeat if moments Pb are moved to the ends of the beam (Fig. b).

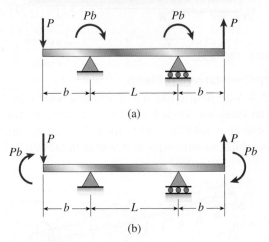

PROBLEM 6.4-3

6.4-4 Calculate the shear force V and bending moment M at a cross section located just right of the 4-kN load on the cantilever beam AB shown in the figure.

PROBLEM 6.4-4

Representative Problems

6.4-5 Consider the beam with an overhang shown in the figure.

(a) Determine the shear force V and bending moment M at a cross section located 18 ft from the left-hand end A.

(b) Find the required magnitude of load intensity q acting on the right half of member BC that will result in a zero shear force on the cross section 18 ft from A.

PROBLEM 6.4-5

6.4-6 The beam ABC shown in the figure is simply supported at A and B and has an overhang from B to C. The loads consist of a horizontal force $P_1 = 4.0$ kN acting at the end of a vertical arm and a vertical force $P_2 = 8.0$ kN acting at the end of the overhang.

(a) Determine the shear force V and bending moment M at a cross section located 3.0 m from the left-hand support. *Note:* Disregard the widths of the beam and vertical arm and use centerline dimensions when making calculations.

(b) Find the value of load P_2 that results in $V = 0$ at a cross section located 2.0 m from the left-hand support.

(c) If $P_2 = 8.0$ kN, find the value of load P_1 that results in $M = 0$ at a cross section located 2.0 m from the left-hand support.

PROBLEM 6.4-6

6.4-7 The beam $ABCD$ shown in the figure has overhangs at each end and carries a uniform load of intensity q (Fig. a). For what ratio b/L will the bending moment at the midpoint of the beam be zero?

Repeat for a triangular load with peak intensity q_0 at $L/2$ (Fig. b).

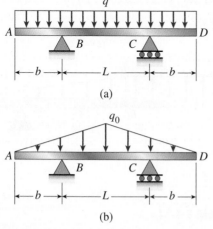

PROBLEM 6.4-7

6.4-8 At a full draw, an archer applies a pull of 130 N to the bowstring of the bow shown in the figure. Determine the bending moment at the midpoint of the bow.

PROBLEM 6.4-8

6.4-9 A curved bar ABC is subjected to loads in the form of two equal and opposite forces P, as shown in the figure. The axis of the bar forms a semicircle of radius r.

Determine the axial force N, shear force V, and bending moment M acting at a cross section defined by the angle θ.

PROBLEM 6.4-9

6.4-10 Under cruising conditions, the distributed load acting on the wing of a small airplane has the idealized variation shown in the figure.

Calculate the shear force V and bending moment M at 4 m from the tip of the wing.

(a)
Wings of a small airplane have distributed uplift loads

(b)

PROBLEM 6.4-10

6.4-11 A beam *ABCD* with a vertical arm *CE* is supported as a simple beam at *A* and *D* (see Fig. a). A cable passes over a small pulley that is attached to the arm at *E*. One end of the cable is attached to the beam at point *B*.

(a) What is the force *P* in the cable if the bending moment in the beam just to the left of point *C* is equal numerically to 640 lb-ft? *Note:* Disregard the widths of the beam and vertical arm and use centerline dimensions when making calculations.

(b) Repeat part (a) if a roller support is added at *C* and a shear release is inserted just left of *C* (see Fig. b).

(a)

(b)

PROBLEM 6.4-11

6.4-12 A simply supported beam *AB* supports a trapezoidally distributed load (see figure). The intensity of the load varies linearly from 50 kN/m at support *A* to 25 kN/m at support *B*.

Calculate the shear force *V* and bending moment *M* at the midpoint of the beam.

PROBLEM 6.4-12

6.4-13 Beam *ABCD* represents a reinforced-concrete foundation beam that supports a uniform load of intensity $q_1 = 3500$ lb/ft (see figure). Assume that the soil pressure on the underside of the beam is uniformly distributed with intensity q_2.

(a) Find the shear force V_B and bending moment M_B at point *B*.

(b) Find the shear force V_m and bending moment M_m at the midpoint of the beam.

PROBLEM 6.4-13

6.4-14 Find shear (*V*) and moment (*M*) at $x = 3L/4$ for the beam shown in Fig. a. Let $M_A = 24$ kN·m, $P = 48$ kN, $L = 6$ m, and $q_0 = 8$ kN/m. Repeat for the beam in Fig. b. (First, solve for the reaction moment at fixed support *A*.)

(a)

(b)

PROBLEM 6.4-14

428

6.4-15 Find expressions for shear force V and moment M at mid-span of beam AB in terms of peak load intensity q_0 and beam length variables a and L. Let $a = 5L/6$.

PROBLEM 6.4-15

6.4-16 Find expressions for shear force V and moment M at $x = 2L/3$ of beam (a) in terms of peak load intensity q_0 and beam length variable L. Repeat for beam (b) but at $x = L/2$.

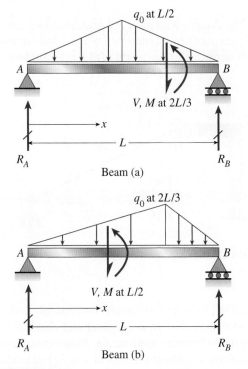

Beam (a)

Beam (b)

PROBLEM 6.4-16

6.4-17 Find expressions for shear force V and moment M at $x = 2L/3$ of beam (a) in terms of peak load intensity q_0 and beam length variable L. Repeat for beam (b).

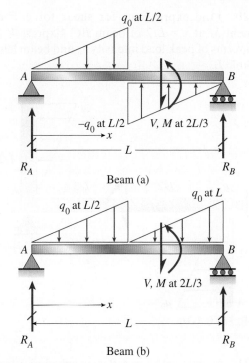

Beam (a)

Beam (b)

PROBLEM 6.4-17

6.4-18 Find expressions for shear force V and moment M at $x = x_0$ of beam AB in terms of peak load intensity q_0 and beam length variable L. Let $x_0 = L/2$.

PROBLEM 6.4-18

6.4-19 Find expressions for shear force V and moment M at $x = x_0$ of beam AB in terms of peak load intensity q_0 and beam length variable L. Let $x_0 = 2L/3$.

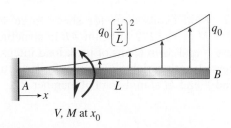

PROBLEM 6.4-19

6.4-20 Find expressions for shear force V and moment M at $x = L/2$ of beam BC. Express V and M in terms of peak load intensity q_0 and beam length variable L.

PROBLEM 6.4-20

6.4-21 A cable with force P is attached to a frame at A and runs over a frictionless pulley at D. Find expressions for shear force V and moment M at $x = L/2$ of beam BC.

PROBLEM 6.4-21

6.4-22 Find expressions for shear force V and moment M at $x = L/2$ of beam AB in structure (a). Express V and M in terms of peak load intensity q_0 and beam length variable L. Repeat for structure (b) but find V and M at mid-span of member BC.

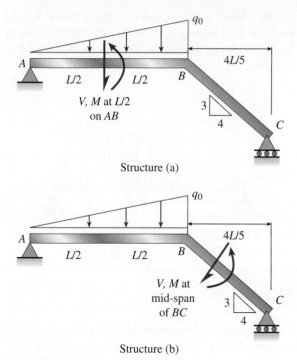

Structure (a)

Structure (b)

PROBLEM 6.4-22

6.4-23 A cable with force P is attached to a frame at D and runs over a frictionless pulley at B. Find expressions for shear force V and moment M at $x = L/3$ of beam AB.

PROBLEM 6.4-23

6.4-24 Frame $ABCD$ carries two concentrated loads ($2P$ at C and P at D, see figure) and also a linearly varying distributed load on AB. Find expressions for shear force V and moment M at $x = L/3$ of beam AB in terms of peak load intensity q_0, force P, and beam length variable L. Let $q_0 = P/L$.

PROBLEM 6.4-24

6.4-25 Frame *ABC* has a moment release just left of joint *B*. Find axial force *N*, shear force *V*, and moment *M* at the top of column *AB*. Write variables *N*, *V*, and *M* in terms of variables *P* and *L*.

PROBLEM 6.4-25

6.4-26 The simply supported beam *ABCD* is loaded by a weight $W = 27$ kN through the arrangement shown in Fig. a. The cable passes over a small frictionless pulley at *B* and is attached at *E* to the end of the vertical arm.

(a) Calculate the axial force *N*, shear force *V*, and bending moment *M* at section *C*, which is just to the left of the vertical arm. *Note:* Disregard the widths of the beam and vertical arm and use centerline dimensions when making calculations.

(b) Repeat part (a) if a roller support is added at *C* and a moment release is inserted just left of *C* (see Fig. b).

(a)

(b)

PROBLEM 6.4-26

Shear-Force and Bending-Moment Diagrams

Draw the shear-force and bending-moment diagrams approximately to scale and label all critical ordinates, including the maximum and minimum values.

Introductory Problems

6.4-27 Draw the shear-force and bending-moment diagrams for a simple beam *AB* supporting two equal concentrated loads *P* (see figure). Repeat if the left-hand load is upward and the right-hand load is downward.

PROBLEM 6.4-27

6.4-28 A simple beam AB is subjected to a counterclockwise couple of moment M_0 acting at distance a from the left-hand support (see figure). Draw the shear-force and bending-moment diagrams for this beam.

Also draw the shear-force and bending-moment diagrams if a second moment M_0 is added at distance a from support B.

PROBLEM 6.4-28

6.4-29 Draw the shear-force and bending-moment diagrams for a cantilever beam AB carrying a uniform load of intensity q over one-half of its length (see figure).

PROBLEM 6.4-29

6.4-30 The cantilever beam AB shown in the figure is subjected to a concentrated load P at the midpoint and a counterclockwise couple of moment $M_1 = PL/4$ at the free end.

Draw the shear-force and bending-moment diagrams for this beam.

PROBLEM 6.4-30

6.4-31 Cantilever beam AB carries an upward uniform load of intensity q_1 from $x = 0$ to $L/2$ (see Fig. a) and a downward uniform load of intensity q from $x = L/2$ to L.

432

(a) Find q_1 in terms of q if the resulting moment at A is zero. Draw V and M diagrams for the case of both q and q_1 as applied loadings.

(b) Repeat part (a) for the case of an upward triangularly distributed load with peak intensity q_0 (see Fig. b). For part (b), find q_0 instead of q_1.

(a)

(b)

PROBLEM 6.4-31

6.4-32 The simple beam AB shown in the figure is subjected to a concentrated load P and a clockwise couple $M_1 = PL/3$ acting at the third points.

Draw the shear-force and bending-moment diagrams for this beam.

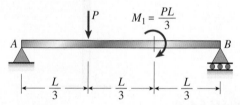

PROBLEM 6.4-32

6.4-33 A simple beam AB subjected to couples M_1 and $3M_1$ acting at the third points is shown in the figure.

Draw the shear-force and bending-moment diagrams for this beam.

PROBLEM 6.4-33

6.4-34 A simply supported beam ABC is loaded by a vertical load P acting at the end of a bracket BDE (see figure).

(a) Draw the shear-force and bending-moment diagrams for beam ABC.

(b) Now assume that load P at E is directed to the right. The vertical dimension BD is $L/5$. Draw axial-force, shear-force, and bending-moment diagrams for ABC.

PROBLEM 6.4-34

6.4-35 A simply supported beam ABC is loaded at the end of a bracket BDE (see figure). Draw axial-force, shear-force, and bending-moment diagrams for ABC.

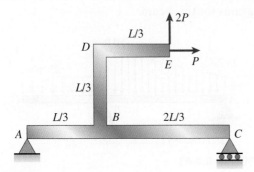

PROBLEM 6.4-35

Representative Problems

6.4-36 A beam ABC is simply supported at A and B and has an overhang BC (see figure). The beam is loaded by two forces P and a clockwise couple of moment Pa at D that act through the arrangement shown.

(a) Draw the shear-force and bending-moment diagrams for beam ABC.

(b) If moment Pa at D is replaced by moment M, find an expression for M in terms of variables P and a so that the reaction at B goes to zero. Plot the associated shear-force and bending-moment diagrams for beam ABC.

PROBLEM 6.4-36

6.4-37 Beam $ABCD$ is simply supported at B and C and has overhangs at each end (see Fig. a). The span length is L and each overhang has length $L/3$. A uniform load of intensity q acts along the entire length of the beam.

(a) Draw the shear-force and bending-moment diagrams for this beam.

(b) Repeat part (a) if the uniform load is replaced with a triangularly distributed load with peak intensity $q_0 = q$ at mid-span (see Fig. b).

(a)

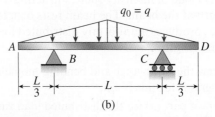

(b)

PROBLEM 6.4-37

6.4-38 Draw the shear-force and bending-moment diagrams for a cantilever beam AB acted upon by two different load cases.

(a) A distributed load with linear variation and maximum intensity q_0 (see Fig. a).

(b) A distributed load with parabolic variation and maximum intensity q_0 (see Fig. b).

(a)

(b)

PROBLEM 6.4-38

6.4-39 The simple beam AB supports a triangular load of maximum intensity $q_0 = 10$ lb/in. acting over one-half of the span and a concentrated load $P = 80$ lb acting at midspan (see figure). Draw the shear-force and bending-moment diagrams for this beam.

$q_0 = 10$ lb/in.

$P = 80$ lb

A B

$\dfrac{L}{2} = 40$ in. $\dfrac{L}{2} = 40$ in.

PROBLEM 6.4-39

6.4-40 The beam AB shown in the figure supports a uniform load of intensity 3000 N/m acting over half the length of the beam. The beam rests on a foundation that produces a uniformly distributed load over the entire length.

(a) Draw the shear-force and bending-moment diagrams for this beam.

(b) Repeat part (a) for the distributed load variation shown in Fig. b.

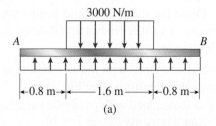

3000 N/m

A B

−0.8 m−   1.6 m   −0.8 m−

(a)

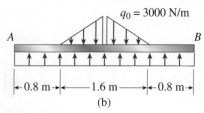

$q_0 = 3000$ N/m

A B

−0.8 m−   1.6 m   −0.8 m−

(b)

PROBLEM 6.4-40

6.4-41 A cantilever beam AB supports a couple and a concentrated load, as shown in the figure. Draw the shear-force and bending-moment diagrams for this beam.

200 lb

400 lb-ft

A B

5 ft   5 ft

PROBLEM 6.4-41

6.4-42 The cantilever beam AB shown in the figure is subjected to a triangular load acting over one-half of its length and a concentrated load acting at the free end.

Draw the shear-force and bending-moment diagrams for this beam.

2.0 kN/m

2.5 kN

A B

2 m   2 m

PROBLEM 6.4-42

6.4-43 Beam ABC has simple supports at A and B, an overhang BC and the distributed loading shown in the figure.

Draw the shear-force and bending-moment diagrams for this beam.

50 lb/in.

25 lb/in.   25 lb/in.

A B C

72 in.   48 in.

PROBLEM 6.4-43

6.4-44 Beam ABC with an overhang at one end supports a partial uniform load of intensity 12 kN/m and a concentrated moment of magnitude 4 kN·m at C (see figure).

Draw the shear-force and bending-moment diagrams for this beam.

12 kN/m

4 kN · m

A B C

1.6 m   1.6 m   1.6 m

PROBLEM 6.4-44

6.4-45 Consider the two beams shown in the figures. Which beam has the larger maximum moment?

First, find support reactions; then plot axial-force (N), shear (V), and moment (M) diagrams for both beams. Label all critical N, V, and M values and also the *distance* to points where N, V, and/or M are zero.

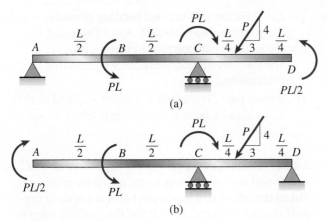

(a)

(b)

PROBLEM 6.4-45

6.4-46 The three beams in the figure have the same loading. However, one has a *moment release* just to the left of *C*, the second has a *shear release* just to the right of *C*, and the third has an *axial release* just to the left of *C*. Which beam has the largest maximum moment?

First, find support reactions; then plot axial-force (*N*), shear (*V*), and moment (*M*) diagrams for all three beams. *Label* all critical *N, V*, and *M* values and also the *distance* to points where *N, V*, and/or *M* are zero.

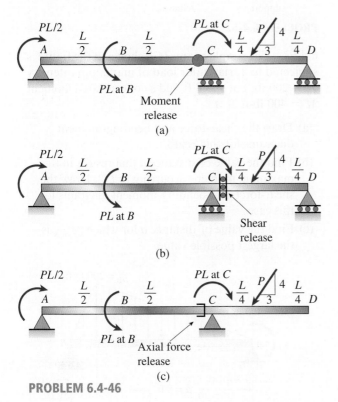

PROBLEM 6.4-46

6.4-47 The beam *ABC* shown in the figure is simply supported at *A* and *B* and has an overhang from *B* to *C*.

Draw the shear-force and bending-moment diagrams for beam *ABC*. *Note:* Disregard the widths of the beam and vertical arm and use centerline dimensions when making calculations.

PROBLEM 6.4-47

6.4-48 A simple beam *AB* is loaded by two segments of uniform load and two horizontal and vertical forces acting at the ends of a vertical arm (see figure).

Draw the shear-force and bending-moment diagrams for this beam.

PROBLEM 6.4-48

6.4-49 Two beams (see figure) are loaded the same and have the same support conditions. However, the location of internal *axial, shear*, and *moment releases* is different for each beam (see figures). Which beam has the larger maximum moment?

First, find support reactions; then plot axial-force (*N*), shear (*V*), and moment (*M*) diagrams for both beams. *Label* all critical *N, V*, and *M* values and also the *distance* to points where *N, V*, and/or *M* are zero.

PROBLEM 6.4-49

435

6.4-50 The beam *ABCD* shown in the figure has overhangs that extend in both directions for a distance of 4.2 m from the supports at *B* and *C*, which are 1.2 m apart.

Draw the shear-force and bending-moment diagrams for this overhanging beam.

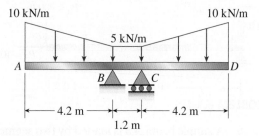

PROBLEM 6.4-50

6.4-51 A beam *ABCD* with a vertical arm *CE* is supported as a simple beam at *A* and *D* (see figure). A cable passes over a small pulley that is attached to the arm at *E*. One end of the cable is attached to the beam at point *B*. The tensile force in the cable is 1800 lb.

(a)

(b)

PROBLEM 6.4-51

(a) Draw the shear-force and bending-moment diagrams for beam *ABCD*. *Note:* Disregard the widths of the beam and vertical arm and use centerline dimensions when making calculations.

(b) Repeat part (a) if a roller support is added at *C* and a shear release is inserted just left of *C* (see Fig. b).

6.4-52 Beams *ABC* and *CD* are supported at *A*, *C*, and *D* and are joined by a hinge (or *moment release*) just to the left of *C*. The support at *A* is a sliding support (hence reaction $A_y = 0$ for the loading shown below). Find all support reactions; then plot shear (*V*) and moment (*M*) diagrams. *Label* all critical *V* and *M* values and also the *distance* to points where either *V* and/or *M* are zero.

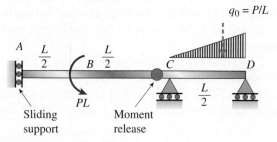

PROBLEM 6.4-52

6.4-53 The simple beam *ACB* shown in the figure is subjected to a triangular load of maximum intensity $q_0 = 200$ lb/ft at $a = 8$ ft and a concentrated moment $M = 400$ lb-ft at *A*.

(a) Draw the shear-force and bending-moment diagrams for this beam.

(b) Find the value of distance *a* that results in the maximum moment occurring at *L/2*. Draw the shear-force and bending-moment diagrams for this case.

(c) Find the value of distance *a* for which M_{max} is the largest possible value.

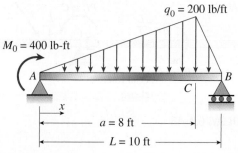

PROBLEM 6.4-53

436

6.4-54 A beam with simple supports is subjected to a trapezoidally distributed load (see figure). The intensity of the load varies from 1.0 kN/m at support A to 2.5 kN/m at support B.

(a) Draw the shear-force and bending-moment diagrams for this beam. Assume that M_0 at B is zero.
(b) Find the required moment M_0 at B so that the maximum moment in the beam does not exceed 1.0 kN $\cdot$ m.

PROBLEM 6.4-54

6.4-55 A beam of length L is designed to support a uniform load of intensity q (see figure). If the supports of the beam are placed at the ends, creating a simple beam, the maximum bending moment in the beam is $qL^2/8$. However, if the supports of the beam are moved symmetrically toward the middle of the beam (as shown), the maximum bending moment is reduced.

(a) Determine the distance a between the supports so that the maximum bending moment in the beam has the smallest possible numerical value. Draw the shear-force and bending-moment diagrams for this condition.
(b) Repeat part (a) if the uniform load is replaced with a triangularly distributed load with peak intensity $q_0 = q$ at mid-span (see Fig. b).

PROBLEM 6.4-55

6.4-56 The compound beam $ABCDE$ shown in the figure consists of two beams (AD and DE) joined by a hinged connection (or moment release) at D. The moment release can transmit a shear force but not a bending moment. Draw the shear-force and bending-moment diagrams for this compound beam.

PROBLEM 6.4-56

6.4-57 Draw the shear-force and bending-moment diagrams for beam AB with a sliding support at A and an elastic support with spring constant k at B acted upon by two different load cases:

(a) A distributed load with linear variation and maximum intensity q_0 (see Fig. a).
(b) A distributed load with parabolic variation with maximum intensity q_0 (see Fig. b).

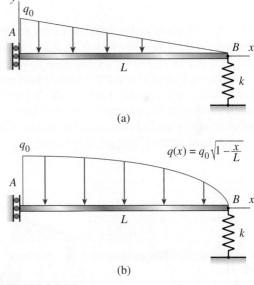

PROBLEM 6.4-57

6.4-58 The shear-force diagram for a simple beam is shown in the figure.

Determine the loading on the beam and draw the bending-moment diagram, assuming that no couples act as loads on the beam.

PROBLEM 6.4-58

6.4-59 The shear-force diagram for a beam is shown in the figure. Assuming that no couples act as loads on the beam, determine the forces acting on the beam and draw the bending-moment diagram.

PROBLEM 6.4-59

6.4-60 A compound beam (see figure) has an internal *moment release* just to the left of B and a *shear release* just to the right of C. Reactions have been computed at A, C, and D and are shown in the figure.

First, confirm the reaction expressions using statics; then plot shear (V) and moment (M) diagrams. *Label* all critical V and M values and also the *distance* to points where either V and/or M are zero.

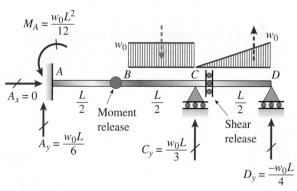

PROBLEM 6.4-60

6.4-61 A compound beam (see figure) has an *shear release* just to the left of C and a *moment release* just to the right of C. A plot of the moment diagram is provided below the beam for applied load P at B and triangular distributed loads $w(x)$ on segments BC and CD.

First, solve for reactions using statics; then plot axial-force (N) and shear force (V) diagrams. Confirm that the moment diagram is that shown below. *Label* all critical N, V, and M values and also the *distance* to points where N, V, and/or M are zero.

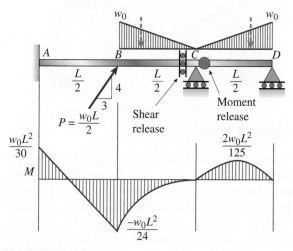

PROBLEM 6.4-61

6.4-62 A simple beam AB supports two connected wheel loads $3P$ and $2P$ that are a distance d apart (see figure). The wheels may be placed at any distance x from the left-hand support of the beam. (Assume $P = 12$ kN, $d = 2$ m, and $L = 15$ m.)

(a) Determine the distance x that will produce the maximum shear force in the beam, and also determine the maximum shear force V_{max}.
(b) Determine the distance x that will produce the maximum bending moment in the beam, and also draw the corresponding bending-moment diagram.

PROBLEM 6.4-62

6.4-63 The inclined beam represents a ladder with the following applied loads: the weight (W) of the house painter and the distributed weight (w) of the ladder itself.

(a) Find support reactions at A and B; then plot axial-force (N), shear (V), and moment (M) diagrams. *Label* all critical N, V, and M values and also the *distance* to points where any critical ordinates are zero. Plot N, V, and M diagrams normal to the inclined ladder.

(b) Repeat part (a) for the case of the ladder suspended from a pin at B and traveling on a roller support perpendicular to the floor at A.

PROBLEM 6.4-63

6.4-64 Beam ABC is supported by a tie rod CD as shown. Two configurations are possible: pin support at A and downward triangular load on AB or pin at B and upward load on AB. Which has the larger maximum moment?

First, find all support reactions; then plot axial-force (N), shear (V), and moment (M) diagrams for ABC only and *label* all critical N, V, and M values. Label the *distance* to points where any critical ordinates are zero.

PROBLEM 6.4-64

439

6.4-65 A plane frame (see figure) consists of column AB and beam BC that carries a triangular distributed load (see Fig. a). Support A is fixed, and there is a roller support at C. Beam BC has a shear release just right of joint B.

(a) Find the support reactions at A and C; then plot axial-force (N), shear-force (V), and bending-moment (M) diagrams for both members. *Label* all critical N, V, and M values and also the *distance* to points where any critical ordinates are zero.

(b) Repeat part (a) if a parabolic lateral load acting to the right is now added on column AB (Fig. b).

(a)

PROBLEM 6.4-65

6.4-66 The plane frame shown in the figure is part of an elevated freeway system. Supports at A and D are fixed, but there are *moment releases* at the base of both columns (AB and DE) as well as in column BC and at the end of beam BE.

Find all support reactions; then plot axial-force (N), shear (V), and moment (M) diagrams for all beam and column members. *Label* all critical N, V, and M values and also the *distance* to points where any critical ordinates are zero.

PROBLEM 6.4-66

Tension, Compression, and Shear

© Can Stock Photo Inc./ronyzmbow

Check stresses in all components and also in connections to be sure that the structure meets all strength requirements with an appropriate factor of safety.

CHAPTER OBJECTIVES

- Define *mechanics of materials*, which examines the stresses, strains, and displacements in structures made of various materials acted on by a variety of different loads.

- Study normal stress (σ) and normal strain (ε) in materials used for structural applications.

- Identify key properties of various materials, such as the modulus of elasticity (E) and yield (σ_y) and ultimate (σ_u) stresses, from plots of stress (σ) versus strain (ε).

- Plot shear stress (τ) versus shear strain (γ) and identify the shearing modulus of elasticity (G).

- Study *Hooke's law* for normal stress and strain ($\sigma = E\varepsilon$) and also for shear stress and strain ($\tau = G\gamma$).

- Investigate changes in lateral dimensions and volume of a bar, which depend upon Poisson's ratio (ν) for the material of the bar.

- Study normal, shear, and bearing stresses in simple bolted connections between members.

- Use factors of safety to establish *allowable values* of stresses.

- Introduce basic concepts of *design*: the iterative process by which the appropriate size of structural members is determined to meet a variety of both strength and stiffness requirements.

Chapter Outline

441

7.1 Introduction to Mechanics of Materials

Mechanics of materials is a branch of applied mechanics that deals with the behavior of solid bodies subjected to various types of loading. Other names for this field of study are *strength of materials* and *mechanics of deformable bodies*. The solid bodies considered in this book include bars with axial loads, shafts in torsion, beams in bending, and columns in compression.

The principal objective of mechanics of materials is to determine the stresses, strains, and displacements in structures and their components due to the loads acting on them. An understanding of mechanical behavior is essential for the safe design of all types of structures, whether airplanes and antennas, buildings and bridges, machines and motors, or ships and spacecraft. That is why mechanics of materials is a basic subject in so many engineering fields. Most problems in mechanics of materials begin with an examination of the external and internal forces acting on a stable deformable body. First the loads acting on the body are defined, along with its support conditions, then reaction forces at supports and internal forces in its members or elements are determined using the basic laws of static equilibrium (provided that the body is statically determinate).

In mechanics of materials you study the stresses and strains inside real bodies, that is, bodies of finite dimensions that deform under loads. To determine the stresses and strains, use the physical properties of the materials as well as numerous theoretical laws and concepts. Mechanics of materials provides additional essential information, based on the deformations of the body, to solve statically indeterminate problems (not possible using the laws of static equilibrium alone).

Theoretical analyses and experimental results have equally important roles in mechanics of materials. Theories are used to derive formulas and equations for predicting mechanical behavior but these expressions cannot be used in practical design unless the physical properties of the materials are known. Such properties are available only after careful experiments have been carried out in the laboratory. Furthermore, not all practical problems are amenable to theoretical analysis alone, and in such cases physical testing is a necessity.

The historical development of mechanics of materials is a fascinating blend of both theory and experiment—theory has pointed the way to useful results in some instances, and experiment has done so in others. Such famous persons as Leonardo da Vinci (1452–1519) and Galileo Galilei (1564–1642) performed experiments to determine the strength of wires, bars, and beams, although they did not develop adequate theories (by today's standards) to explain their test results. By contrast, the famous mathematician Leonhard Euler (1707–1783) developed the mathematical theory of columns and calculated the critical load of a column in 1744, long before any experimental evidence existed to show the significance of his results. Without appropriate tests to back up his theories, Euler's results remained unused for over a hundred years, although today they are the basis for the design and analysis of most columns.

7.2 Normal Stress and Strain

Now that reactive and internal forces are known from statics, it is time to examine internal actions more closely for a deformable body of interest. The most fundamental concepts in mechanics of materials are **stress** and **strain**. These concepts can be illustrated in their most elementary form by considering a prismatic bar

FIGURE 7-1

Structural members subjected to axial loads (the tow bar is in tension and the landing gear strut is in compression)

Landing gear strut

Tow bar

subjected to axial forces. A **prismatic bar** is a straight structural member having the same cross section throughout its length, and an **axial force** is a load directed along the axis of the member, resulting in either tension or compression in the bar. Examples are shown in Fig. 7-1, where the tow bar is a prismatic member in tension and the landing gear strut is a member in compression. Other examples include the members of a bridge truss, connecting rods in automobile engines, spokes of bicycle wheels, columns in buildings, and wing struts in small airplanes.

For discussion purposes, consider the tow bar of Fig. 7-1 and isolate a segment of it as a free body (Fig. 7-2a) (see Fig. 3-3 for a FBD of the entire airplane). When drawing this free-body diagram, disregard the weight of the bar itself and assume that the only active forces are the axial forces P at the ends. Next, consider two views of the bar: the first showing the same bar *before* the loads are applied (Fig. 7-2b) and the second showing it *after* the loads are applied (Fig. 7-2c). The original length of the bar is denoted by the letter L, and the increase in length due to the loads is denoted by the Greek letter δ (delta).

The internal actions in the bar are exposed by making an imaginary cut through the bar at section mn (Fig. 7-2c). Because this section is taken perpendicular to the longitudinal axis of the bar, it is called a **cross section**.

Now isolate the part of the bar to the left of cross section mn as a free body (Fig. 7-2d). At the right-hand end of this free body (section mn), you can see the action of the removed part of the bar (that is, the part to the right of section mn) upon the part that remains. This action consists of continuously distributed *stresses* acting over the entire cross section, and the axial force P acting at the cross section is the *resultant* of those stresses. (The resultant force is shown with a dashed line in Fig. 7-2d.)

Stress has units of force per unit area and is denoted by the Greek letter σ (sigma). In general, the stresses σ acting on a plane surface may be uniform throughout the area or may vary in intensity from one point to another. Assume that the stresses acting on cross section mn (Fig. 7-2d) are *uniformly distributed* over the area. Then the resultant of those stresses must be equal to the magnitude of the stress times the cross-sectional area A of the bar, that is, $P = \sigma A$. Therefore, you can obtain the following expression for the magnitude of the stresses:

$$\sigma = \frac{P}{A} \tag{7-1}$$

This equation gives the intensity of uniform stress in an axially loaded, prismatic bar of arbitrary cross-sectional shape.

FIGURE 7-2

Prismatic bar in tension: (a) free-body diagram of a segment of the bar, (b) segment of the bar before loading, (c) segment of the bar after loading, and (d) normal stresses in the bar

$P \longleftarrow \qquad \longrightarrow P$

(a)

$\longleftarrow L \longrightarrow$

(b)

$\qquad\qquad m$

$P \longleftarrow \qquad \longrightarrow P$

$\qquad\qquad n$

$\longleftarrow L + \delta \longrightarrow$

(c)

$\qquad\qquad m$

$P \longleftarrow \qquad \dashrightarrow P$

$d \qquad n$

$\sigma = \dfrac{P}{A}$

(d)

When the bar is stretched by the forces P, the stresses are **tensile stresses**; if the forces are reversed in direction, causing the bar to be compressed, they are **compressive stresses**. **Normal stresses** stresses act in a direction perpendicular to the cut surface. Normal stresses may be either tensile or compressive. **Shear stresses** discussed in Section 7.6 act parallel to the surface.

It is customary to define tensile stresses as positive and compressive stresses as negative. Because the normal stress σ is obtained by dividing the axial force by the cross-sectional area, it has units of force per unit of area. Stress is customarily expressed in pounds per square inch (psi) or kips per square inch (ksi).[1] For instance, suppose that the bar of Fig. 7-2 has a diameter d of 2.0 inches, and the load P has a magnitude of 6 kips. Then the stress in the bar is

$$\sigma = \frac{P}{A} = \frac{P}{\pi d^2/4} = \frac{6\text{k}}{\pi(2.0 \text{ in.})^2/4} = 1.91 \text{ ksi (or 1910 psi)}$$

In this example the stress is tensile, or positive.

When SI units are used, force is expressed in newtons (N) and area in square meters (m^2). Consequently, stress has units of newtons per square meter (N/m^2), which is equal to a pascal (Pa). However, the pascal is such a small unit of stress that it is necessary to work with large multiples, usually the megapascal (MPa).

A pascal is so small that it takes almost 7000 pascals to make 1 psi.[2] The stress in the bar described in the preceding example (1.91 ksi) converts to 13.2 MPa, which is 13.2×10^6 pascals. Although it is not recommended in SI, you will sometimes find stress given in newtons per square millimeter (N/mm^2), which is a unit equal to the megapascal (MPa).

Limitations

The equation $\sigma = P/A$ is valid only if the stress is uniformly distributed over the cross section of the bar. This condition is realized if the axial force P acts through the centroid of the cross-sectional area, as demonstrated later in this section. However, in this book (as in common practice), it is understood that axial forces are applied at the centroids of the cross sections unless specifically stated otherwise.

The uniform stress condition pictured in Fig. 7-2d exists throughout the length of the bar except near the ends. The stress distribution at the end of a bar depends upon how the load P is transmitted to the bar. If the load is distributed uniformly over the end, the stress pattern at the end is the same as everywhere else. However, it is more likely that the load is transmitted through a pin or a bolt, producing high localized stresses called *stress concentrations*.

One possibility is illustrated by the eyebar shown in Fig. 7-3. In this instance, the loads P are transmitted to the bar by pins that pass through the holes (or eyes) at the ends of the bar. Thus, the forces shown in the figure are actually the resultants of bearing pressures between the pins and the eyebar, and the stress distribution around the holes is quite complex. However, as you move

FIGURE 7-3

Steel eyebar subjected to tensile loads P

---

[1]One kip, or kilopound, equals 1000 lb.

[2]Conversion factors between USCS units and SI units are listed in Section 1.4.

away from the ends and toward the middle of the bar, the stress distribution gradually approaches the uniform distribution pictured in Fig. 7-2d.

As a practical rule, the formula $\sigma = P/A$ may be used with good accuracy at any point within a prismatic bar that is at least as far away from the stress concentration as the largest lateral dimension of the bar. In other words, the stress distribution in the steel eyebar of Fig. 7-3 is uniform at distances b or greater from the enlarged ends where b is the width of the bar, and the stress distribution in the prismatic bar of Fig. 7-2 is uniform at distances d or greater from the ends where d is the diameter of the bar (Fig. 7-2d). Of course, even when the stress is *not* uniformly distributed, the equation $\sigma = P/A$ still may be useful because it gives the *average* normal stress on the cross section.

Normal Strain

As already observed, a straight bar changes in length when loaded axially, becoming longer when in tension and shorter when in compression. For instance, consider again the prismatic bar of Fig. 7-2. The elongation δ of this bar (Fig. 7-2c) is the cumulative result of the stretching of all elements of the material throughout the volume of the bar. Assume that the material is the same everywhere in the bar. Then half of the bar (length $L/2$) has an elongation equal to $\delta/2$, and one-fourth of the bar has an elongation equal to $\delta/4$.

In general, the elongation of a segment is equal to its length divided by the total length L and multiplied by the total elongation δ. Therefore, a unit length of the bar has an elongation equal to $1/L \times \delta$. This quantity is called the *elongation per unit length*, or **strain**, and is denoted by the Greek letter ϵ (epsilon). Strain is given by

$$\varepsilon = \frac{\delta}{L} \tag{7-2}$$

If the bar is in tension, the strain is called a **tensile strain**, representing an elongation or stretching of the material. If the bar is in compression, the strain is a **compressive strain** and the bar shortens. Tensile strain is usually taken as positive and compressive strain as negative. The strain ϵ is called a **normal strain** because it is associated with normal stresses.

Because normal strain is the ratio of two lengths, it is a **dimensionless quantity**, that is, it has no units. Therefore, strain is expressed simply as a number that is independent of any system of units. Numerical values of strain are usually very small, because bars made of structural materials undergo only small changes in length when loaded.

As an example, consider a steel bar having length L equal to 2.0 m. When heavily loaded in tension, this bar might elongate by 1.4 mm, which means that the strain is

$$\varepsilon = \frac{\delta}{L} = \frac{1.4 \text{ mm}}{2.0 \text{ m}} = 0.0007 = 700 \times 10^{-6}$$

In practice, the original units of δ and L are sometimes attached to the strain itself, and then the strain is recorded in forms such as mm/m, μm/m, and in./in. For instance, the strain ϵ in the preceding example could be given as 700 μm/m or 700×10^{-6} in./in. Strain is sometimes expressed as a percent, especially when the strains are large. (In the preceding example, the strain is 0.07%.)

Uniaxial Stress and Strain

The definitions of normal stress and normal strain are based upon purely static and geometric considerations, which means that Eqs. (7-1) and (7-2) can be used for loads of any magnitude and for any material. The principal requirement is that the deformation of the bar be uniform throughout its volume, which in turn requires that the bar be prismatic, the loads act through the centroids of the cross sections, and the material be **homogeneous** (that is, the same throughout all parts of the bar). The resulting state of stress and strain is called **uniaxial stress and strain** (although lateral strain exists, as discussed later in Section 7.5).

Line of Action of the Axial Forces for a Uniform Stress Distribution

Throughout the preceding discussion of stress and strain in a prismatic bar, the normal stress σ was assumed to be distributed uniformly over the cross section. Note that this condition is met if the line of action of the axial forces is through the centroid of the cross-sectional area.

(a)

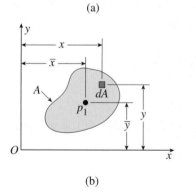

(b)

Consider a prismatic bar of arbitrary cross-sectional shape subjected to axial forces P that produce uniformly distributed stresses σ (Fig. 7-4a). Let p_1 represent the point in the cross section where the line of action of the forces intersects the cross section (Fig. 7-4b). Construct a set of x-y axes in the plane of the cross section and denote the coordinates of point p_1 by $\bar{x}$ and $\bar{y}$. To determine these coordinates, observe that the moments M_x and M_y of the force P about the x and y axes, respectively, must be equal to the corresponding moments of the uniformly distributed stresses.

The moments of the force P are

$$M_x = P\bar{y} \qquad M_y = -P\bar{x} \qquad \textbf{(7-3a,b)}$$

in which a moment is considered positive when its vector (using the right-hand rule) acts in the positive direction of the corresponding axis.

The moments of the distributed stresses are obtained by integrating over the cross-sectional area A. The differential force acting on an element of area dA (Fig. 7-4b) is equal to $\sigma\,dA$. The moments of this elemental force about the x and y axes are $\sigma y\,dA$ and $-\sigma x\,dA$, respectively, in which x and y denote the coordinates of the element dA. The total moments are obtained by integrating over the cross-sectional area:

$$M_x = \int \sigma y\,dA \qquad M_y = -\int \sigma x\,dA \qquad \textbf{(7-3c,d)}$$

These expressions give the moments produced by the stresses σ.

Next, equate the moments M_x and M_y obtained from the force P [Eqs. (7-3a and b)] to the moments obtained from the distributed stresses [Eqs. (7-3c and d)]:

$$P\bar{y} = \int \sigma y\,dA \qquad P\bar{x} = -\int \sigma x\,dA$$

Because the stresses σ are uniformly distributed, they are constant over the cross-sectional area A and can be placed outside the integral signs. Also, σ is equal to P/A. Therefore, you can obtain the following formulas for the coordinates of point p_1:

$$\bar{y} = \frac{\int y\,dA}{A} \quad \bar{x} = \frac{\int x\,dA}{A} \qquad \textbf{(7-4a,b)}$$

These equations are the same as the equations defining the coordinates of the centroid of an area (see Section 5.2). Therefore, the important conclusion here is:

> *In order to have uniform tension or compression in a prismatic bar, the axial force must act through the centroid of the cross-sectional area.*

Always assume that these conditions are met unless it is specifically stated otherwise.

The following examples illustrate the calculation of stresses and strains in prismatic bars. In the first example, disregard the weight of the bar and in the second, include it. (It is customary when solving textbook problems to omit the weight of the structure unless specifically instructed to include it.)

Example 7-1

An elastic spring rests on a base plate that is on top of rigid tube B (see Fig. 7-5). The spring is enclosed by rigid tube A but is longer than tube A by an amount s. Force P is then applied to a cap plate to compress the spring. Both tubes have outer diameter d_O, but the inner diameters are d_A and d_B for tubes A and B, respectively. Assume that spring stiffness $k = 24$ kips/in., $d_O = 3$ in., $d_A = 2.5$ in., $d_B = 2.25$ in., and $s = 0.125$ in.

(a) If applied load $P = 2500$ lb, what are the axial normal stresses in tubes A and B?

(b) Repeat part (a) if $P = 5000$ lb.

(c) What is P if the normal stress in tube A is 800 psi? What is the associated stress in tube B?

Solution:

Use the four-step problem-solving approach.

1. *Conceptualize* [*hypothesize, sketch*]: The two possible states of the assemblage are shown in the free-body diagrams in Fig. 7-6. In Fig. 7-6a, an upper-section cut through both the spring and tube A creates an upper free-body diagram that reveals a spring force $(k)(x)$ for the case of downward cap displacement x

FIGURE 7-5

FIGURE 7-5

Elastic spring inside rigid tubes

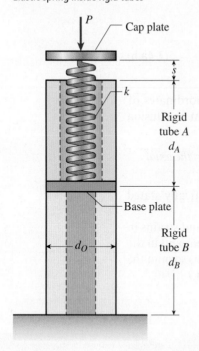

FIGURE 7-6

(a, b) Free-body diagrams ($x < s$, $x = s$)

that is less than gap width s. In Fig. 7-6b, cap displacement x is equal to gap width s, so the spring force now equals $(k)(s)$. Figure 7-6 also shows lower free-body diagrams for both cases in which a section cut through tube B shows that the internal compressive force in tube B is equal to applied load P. (Internal forces in tubes A and B are shown as two arrows, one at each tube wall, indicating that N_A and N_B are actually uniformly distributed forces acting on the circular cross section of each tube.)

2. *Categorize* [*simplify, classify*]: The force P required to close gap s is $(k)(s)$. This is also the maximum force that can be developed in the spring. If applied force P is too small to close the gap s, force P will be transferred to the base plate and into rigid tube B; tube A will be unaffected by the load. However, if force P is large enough to compress the spring to close the gap s, the spring and tube A will share the load P applied to the cap plate and together will transfer it to tube B through the base plate. In summary, the free-body diagrams in Fig. 7-6 show that, if the spring is compressed by load P an amount x, the compressive internal forces in the spring and the two tubes are

$$N_s = P = kx, \ N_A = 0, \ N_B = P \qquad \text{for} \quad x < s$$
$$N_s = ks, \ N_A = P - ks, \ N_B = N_s + N_A = P \quad \text{for} \quad x = s$$

3. *Analyze* [*evaluate; select relevant equations, carry out mathematical solution*]:

Force P required to close gap s: The gap closes when force P is equal to ks

$$P = ks = \left(24 \frac{\text{kips}}{\text{in.}} \right)(0.125 \text{ in.}) = 3000 \text{ lb}$$

Tube stresses for applied load $P = 2500$ lb: The cap will displace downward a distance $x = P/k = 0.104$ in. ($< s$), so tube internal forces are $N_A = 0$ and $N_B = P$. Tube cross-sectional areas are

$$A_A = \frac{\pi}{4}\left(d_O^2 - d_A^2\right) = \frac{\pi}{4}\left(3^2 - 2.5^2\right)\text{in}^2 = 2.16 \text{ in}^2$$

$$A_B = \frac{\pi}{4}\left(d_O^2 - d_B^2\right) = \frac{\pi}{4}\left(3^2 - 2.25^2\right)\text{in}^2 = 3.093 \text{ in}^2$$

The resulting axial normal compressive stresses in tubes A and B are

$$\sigma_A = \frac{N_A}{A_A} = 0 \quad \sigma_B = \frac{N_B}{A_B} = 808 \text{ psi} \qquad \leftarrow$$

Tube stresses for applied load $P = 5000$ lb: Cap downward displacement is now $x = P/k = 0.208$ in. ($> s$), so tube internal forces are $N_A = P - (k)(s) = (5000 - 3000) \text{ lb} = 2000 \text{ lb}$ and $N_B = P$. The normal stresses in tubes A and B are now:

$$\sigma_A = \frac{2000 \text{ lb}}{2.16 \text{ in}^2} = 926 \text{ psi} \quad \sigma_B = \frac{5000 \text{ lb}}{3.093 \text{ in}^2} = 1617 \text{ psi} \qquad \leftarrow$$

Applied load P if stress in tube A is 800 psi: Force P must exceed $(k)(s) = 3000 \text{ lb}$ for the gap to close, leading to a force in tube A and a normal stress of $\sigma_A = 800$ psi. The normal compressive force in tube A is $N_A = (\sigma_A)(A_A) = 1728 \text{ lb}$. It follows that applied force P is now $P = N_A + ks = 1728 \text{ lb} + 3000 \text{ lb} = 4728 \text{ lb}$. Internal force $N_B = P$, so the normal compressive stress in tube B is now $\sigma_B = \dfrac{4728 \text{ lb}}{3.093 \text{ in}^2} = 1529$ psi. $\leftarrow$

4. *Finalize* [*conclude; examine answer—Does it make sense? Are units correct? How does it compare to similar problem solutions?*]: If tube A is elastic instead of rigid as assumed here, tube A can be modeled as another spring that is parallel to the spring it encloses. Now a more advanced analysis procedure will be needed to find tube force N_A for the case of $P > (k)(s)$. Force N_A is no longer equal to $P - (k)(s)$, and downward displacement x can be larger than s.

Example 7-2

An antenna and receiver are suspended on a steel wire from a helicopter to measure the effects of wind turbines on a local radar installation (see Fig. 7-7). Obtain a *formula for the maximum stress* in the wire, taking into account the weight of the wire itself. Calculate the *maximum stress in the wire* in MPa using the following numerical properties: $L_1 = 6$ m, $L_2 = 5$ m, $d = 9.5$ mm; antenna weight is $W_1 = 380$ N; receiver weight is $W_2 = 700$ N. Note that the weight density γ of steel is 77.0 kN/m^3 (from Table F-1 in Appendix F).

FIGURE 7-7

(a, b) Instruments suspended on wire from helicopter

William Floyd Holdman, Jr./Getty Images

(a)   (b)

FIGURE 7-8

Free-body diagram

Solution:

Use the four-step problem-solving approach.

1. *Conceptualize*: A free-body diagram of the suspended instrument package is shown in Fig. 7-8. The antenna (W_1) and receiver (W_2) weights are concentrated forces at specified locations along the wire; the weight of the wire is a uniformly distributed axial force expressed as $w(x) = \gamma A$, where A is the cross-sectional area of the wire. Cutting the wire at some point x leads to upper and lower free-body diagrams (Fig. 7-9); either can be used to find the internal axial force $N(x)$ at the location of the cut section. The internal axial force in the wire is a maximum at the point at which it is attached to the helicopter ($x = 0$).

2. *Categorize*: Start by solving for the reaction force R at the top of the wire and then cut the wire a short distance below the support to find $N_{\max}$. The wire is prismatic, so simply divide $N_{\max}$ by cross-sectional area A to find the maximum axial normal stress $\sigma_{\max}$.

3. *Analyze:*

Reaction force R: Use the free-body diagram in Fig. 7-8 to obtain

$$R = -\left[W_1 + W_2 + w(L_1 + L_2)\right] = -\left[W_1 + W_2 + \gamma A(L_1 + L_2)\right]$$

The minus sign indicates that reaction force R acts in the $(-x)$ direction, or upward in Figs. 7-8 and 7-9.

Internal axial forces $N(x)$ in hanging wire: The internal axial force in the wire varies over the length of the wire. Cutting through the wire in upper and lower segments (the lower segment is cut in Fig. 7-9) gives

$$N(x) = W_1 + W_2 + w(L_1 + L_2 - x) \quad 0 \le x \le L_1$$
$$N(x) = W_2 + w(L_1 + L_2 - x) \qquad L_1 \le x \le L_1 + L_2$$

Internal force $N(x)$ is shown as a pair of forces acting away from the cut section in accordance with a deformation sign convention in which the wire is initially assumed to be in tension and that tension is positive. The maximum force in the wire is at $x = 0$: $N_{max} = N(0) = W_1 + W_2 + w(L_1 + L_2)$.

Formula for maximum stress in the wire: The cross-sectional area A of the wire is constant, so dividing N_{max} by A leads to a formula for maximum stress in the wire:

$$\sigma_{max} = \frac{N_{max}}{A} = \frac{W_1 + W_2 + w(L_1 + L_2)}{A} = \frac{W_1 + W_2}{A} + \gamma(L_1 + L_2) \quad \Longleftarrow$$

Numerical calculations: The cross-sectional area of the wire is

$$A = \frac{\pi}{4}d^2 = \frac{\pi}{4}(9.5 \text{ mm})^2 = 70.88 \text{ mm}^2$$

Solving for maximum normal stress gives

$$\sigma_{max} = \frac{W_1 + W_2}{A} + \gamma(L_1 + L_2) = \frac{380 \text{ N} + 700 \text{ N}}{70.88 \text{ mm}^2} + 77.0\frac{\text{kN}}{\text{m}^3}(6 \text{ m} + 5 \text{ m}) = 16.08 \text{ MPa} \quad \Longleftarrow$$

4. *Finalize:* If the weight of the wire is ignored, the maximum normal stress is reduced to 15.24 MPa, which is a decrease of more than 5%. Although the stresses are low here, eliminating the self-weight of the wire from the stress calculation is not recommended.

FIGURE 7-9

Axial force $N(x)$ in wire

7.3 Mechanical Properties of Materials

The design of machines and structures so that they function properly requires an understanding of the **mechanical behavior** of the materials being used. Ordinarily, the only way to determine how materials behave when they are subjected to loads is to perform experiments in the laboratory. The usual procedure is to place small specimens of the material in testing machines, apply the loads, and then measure the resulting deformations (such as changes in length and changes in diameter).

Tensile-test machine with automatic data-processing system

A typical **tensile-test machine** is shown in Fig. 7-10. The test specimen is installed between the two large grips of the testing machine and then loaded in tension. Measuring devices record the load and deformations, and the automatic control and data-processing systems tabulate and graph the results.

A more detailed view of a **tensile-test specimen** is shown in Fig. 7-11. The ends of the specimen are enlarged where they fit in the grips that failure will not occur near the grips themselves. A failure at the ends would not produce the desired information about the material because the stress distribution near the grips is not uniform, as explained in Section 7.2. In a properly designed specimen, failure occurs in the prismatic portion of the specimen where the stress distribution is uniform and the bar is subjected only to pure tension. This situation is shown in Fig. 7-11 and Fig. 7-14, where the steel specimen is about to fracture under load. The device in front, which is attached by two arms to the specimen, is an **extensometer** that measures the elongation during loading.

In order that test results stay comparable, the dimensions of test specimens and the methods of applying loads must be standardized. One of the major standards organizations in the United States is the American Society for Testing and Materials (ASTM), which is a technical society that publishes specifications and standards for materials and testing. Other standardizing organizations include the American Standards Association (ASA) and the National Institute of Standards and Technology (NIST). Similar organizations exist in other countries.

The ASTM standard tension specimen has a diameter of 0.505 in. and a **gage length** of 2.0 in. between the gage marks, which are the points where the extensometer arms are attached to the specimen (see Fig. 7-11). As the specimen is pulled, the axial load is measured and recorded, either automatically or by reading from a dial. The elongation over the gage length is measured simultaneously, either by mechanical gages of the kind shown in Fig. 7-11 or by electrical-resistance strain gages.

In a **static test**, the load is applied slowly and the precise *rate* of loading is not of interest because it does not affect the behavior of the specimen. However, in a **dynamic test**, the load is applied rapidly and sometimes in a cyclical manner. Since the nature of a dynamic load affects the properties of the materials, the rate of loading must be measured.

Compression tests of metals are customarily made on small specimens in the shape of cubes or circular cylinders. For instance, cubes may be 2.0 in. on a side, and cylinders may have diameters of 1 in. and lengths from 1 to 12 in. Both the load applied by the machine and the shortening of the specimen may be measured. The shortening should be measured over a gage length that is less than the total length of the specimen in order to eliminate end effects.

Concrete is tested in compression on important construction projects to ensure that the required strength has been obtained. One type of concrete test specimen is 6 in. in diameter, 12 in. in length, and 28 days old (the age of concrete is important because concrete gains strength as it cures). Similar but somewhat smaller specimens are used when performing compression tests of rock (see Fig. 7-12 on the next page).

Stress-Strain Diagrams

Test results generally depend upon the dimensions of the specimen being tested. Since it is unlikely that you will design a structure having parts that are the same size as the test specimens, you need to express the test results in a form that can

Typical tensile-test specimen with extensometer attached

be applied to members of any size. A simple way to achieve this objective is to convert the test results to stresses and strains.

The axial stress σ in a test specimen is calculated by dividing the axial load P by the cross-sectional area A (Eq. 7-1). When the initial area of the specimen is used in the calculation, the stress is called the **nominal stress** (other names are *conventional stress* and *engineering stress*). A more exact value of the axial stress, called the **true stress**, can be calculated by using the actual area of the bar at the cross section where failure occurs. Since the actual area in a tension test is always less than the initial area (see Fig. 7-14), the true stress is larger than the nominal stress.

The average axial strain ε in the test specimen is found by dividing the measured elongation δ between the gage marks by the gage length L. If the initial gage length is used in the calculation (for instance, 2.0 in.), the **nominal strain** is obtained. Since the distance between the gage marks increases as the tensile load is applied, you can calculate the **true strain** (or *natural strain*) at any value of the load by using the actual distance between the gage marks. In tension, true strain is always smaller than nominal strain. However, for most engineering purposes, nominal stress and nominal strain are adequate.

After performing a tension or compression test and determining the stress and strain at various magnitudes of the load, you can plot a diagram of stress versus strain. Such a **stress-strain diagram** is a characteristic of the particular material being tested and conveys important information about the mechanical properties and type of behavior.

Stresses and Strains for Structural Steel   **Structural steel**, also known as *mild steel* or *low-carbon steel*, is one of the most widely used metals found in buildings, bridges, cranes, ships, towers, vehicles, and many other types of construction. A stress-strain diagram for a typical structural steel in tension is shown in Fig. 7-13. Strains are plotted on the horizontal axis and stresses on the vertical axis.

The diagram begins with a straight line from the origin O to point A, showing that the relationship between stress and strain in this initial region is not only *linear* but also *proportional*.[3] Beyond point A, the proportionality between stress

FIGURE 7-12

Rock sample being tested in compression to obtain compressive strength, elastic modulus, and Poisson's ratio

Courtesy of MTS Systems Corporation

FIGURE 7-13

Stress-strain diagram for a typical structure steel in tension (not to scale)

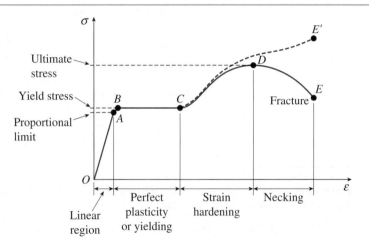

FIGURE 7-14

Necking of a mild-steel bar in
tension (© Barry Goodno)

Load

Region
of
necking

Region
of
impending
fracture

© Barry Goodno

Load

and strain no longer exists; hence the stress at *A* is called the **proportional limit**. For low-carbon steels, this limit is in the range 30 to 50 ksi (210 to 350 MPa), but high-strength steels (with higher carbon content plus other alloys) can have proportional limits of more than 80 ksi (550 MPa). The slope of the straight line from *O* to *A* is called the **modulus of elasticity**. Because the slope has units of stress divided by strain, modulus of elasticity has the same units as stress. (See Section 7.4.)

With an increase in stress beyond the proportional limit, the strain begins to increase more rapidly for each increment in stress. Consequently, the stress-strain curve has a smaller and smaller slope, until, at point *B*, the curve becomes horizontal (see Fig. 7-13). Beginning at this point, considerable elongation of the test specimen occurs with no noticeable increase in the tensile force (from *B* to *C*). This phenomenon is known as **yielding** of the material, and point *B* is called the **yield point**. The corresponding stress is known as the **yield stress** of the steel.

In the region from *B* to *C* (see Fig. 7-13), the material becomes **perfectly plastic**, which means that it deforms without an increase in the applied load. The elongation of a mild-steel specimen in the perfectly plastic region is typically 10 to 15 times the elongation that occurs in the linear region (between the onset of loading and the proportional limit). The presence of very large strains in the plastic region (and beyond) is the reason for not plotting this diagram to scale.

After undergoing the large strains that occur during yielding in the region *BC*, the steel begins to **strain harden**. During strain hardening, the material undergoes changes in its crystalline structure, resulting in increased resistance of the material to further deformation. Elongation of the test specimen in this region requires an increase in the tensile load; therefore, the stress-strain diagram has a positive slope from *C* to *D*. The load eventually reaches its maximum value, and the corresponding stress (at point *D*) is called the **ultimate stress**. Further stretching of the bar is actually accompanied by a reduction in the load, and fracture finally occurs at a point such as *E* in Fig. 7-13.

The yield stress and ultimate stress of a material are also called the **yield strength** and **ultimate strength**, respectively. **Strength** is a general term that refers to the capacity of a structure to resist loads. For instance, the yield strength of a beam is the magnitude of the load required to cause yielding in the beam, and the ultimate strength of a truss is the maximum load it can support, that is, the failure load. However, when conducting a tension test of a particular material, define load-carrying capacity by the stresses in the specimen rather than by the total loads acting on the specimen. As a result, the strength of a material is usually stated as a stress.

When a test specimen is stretched, **lateral contraction** occurs. The resulting decrease in cross-sectional area is too small to have a noticeable effect on the calculated values of the stresses up to about point *C* in Fig. 7-13, but beyond that point the reduction in area begins to alter the shape of the curve. In the vicinity of the ultimate stress, the reduction in area of the bar becomes clearly visible and a pronounced **necking** of the bar occurs (see Fig. 7-14).

If the actual cross-sectional area at the narrow part of the neck is used to calculate the stress, the **true stress-strain curve** (the dashed line *CE'* in Fig. 7-13) is obtained. The total load the bar can carry does indeed diminish

---

[3]Two variables are said to be *proportional* if their ratio remains constant. Therefore, a proportional relationship may be represented by a straight line through the origin. Although a proportional relationship is linear, the converse is not necessarily true, because a relationship represented by a straight line that does *not* pass through the origin is linear but not proportional. "Directly proportional" is synonymous with "proportional".

after the ultimate stress is reached (as shown by curve *DE*), but this reduction is due to the decrease in area of the bar and not to a loss in strength of the material itself. In reality, the material withstands an increase in true stress up to failure (point *E'*). Because most structures are expected to function at stresses below the proportional limit, the **conventional stress-strain curve** *OABCDE*, which is based upon the original cross-sectional area of the specimen and is easy to determine, provides satisfactory information for use in engineering design.

The diagram of Fig. 7-13 shows the general characteristics of the stress-strain curve for mild steel, but its proportions are not realistic because the strain that occurs from *B* to *C* may be more than ten times the strain occurring from *O* to *A*. Furthermore, the strains from *C* to *E* are many times greater than those from *B* to *C*. The correct relationships are portrayed in Fig. 7-15, which shows a stress-strain diagram for mild steel drawn to scale. In this figure, the strains from the zero point to point *A* are so small in comparison to the strains from point *A* to point *E* that they cannot be seen, and the initial part of the diagram appears to be a vertical line.

Ductility   The presence of a clearly defined yield point followed by large plastic strains is an important characteristic of structural steel that is sometimes utilized in practical design. Metals such as structural steel that undergo large *permanent* strains before failure are classified as **ductile**. Ductility is the property that enables a bar of steel to be bent into a circular arc or drawn into a wire without breaking. A desirable feature of ductile materials is that visible distortions occur if the loads become too large, thus providing an opportunity to take remedial action before an actual fracture occurs. Also, materials exhibiting ductile behavior are capable of absorbing large amounts of strain energy prior to fracture.

Structural steel is an alloy of iron containing about 0.2% carbon; therefore, it is classified as a low-carbon steel. With increasing carbon content, steel becomes less ductile but stronger (higher yield stress and higher ultimate stress). The physical properties of steel are also affected by heat treatment, the presence of other metals, and manufacturing processes such as rolling. Other materials that behave in a ductile manner (under certain conditions) include aluminum, copper, magnesium, lead, molybdenum, nickel, brass, bronze, monel metal, nylon, and teflon.

Aluminum Alloys   Although they have considerable ductility, **aluminum alloys** typically do not have a clearly definable yield point, as shown by the stress-strain diagram of Fig. 7-16. However, they do have an initial linear region with a recognizable proportional limit. Alloys produced for structural purposes have proportional limits in the range 10 to 60 ksi (70 to 410 MPa) and ultimate stresses in the range 20 to 80 ksi (140 to 550 MPa).

When a material such as aluminum does not have an obvious yield point yet undergoes large strains after the proportional limit is exceeded, an *arbitrary* yield stress may be determined by the **offset method**. A straight line is drawn on the stress-strain diagram parallel to the initial linear part of the curve (Fig. 7-17) but offset by some standard strain, such as 0.002 (or 0.2%). The intersection of the offset line and the stress-strain curve (point *A* in the figure) defines the yield stress. Because this stress is determined by an arbitrary rule and is not an inherent physical property of the material, it should be distinguished from a true yield stress by referring to it

FIGURE 7-15

Stress-strain diagram for a typical structural steel in tension (drawn to scale)

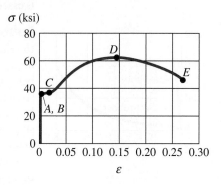

FIGURE 7-16

Typical stress-strain diagram for an aluminum alloy

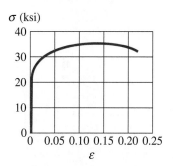

FIGURE 7-17

Arbitrary yields stress determined by the offset method

FIGURE 7-18

Stress-strain curves for two kinds of rubber in tension

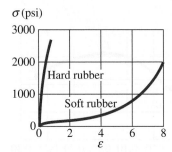

FIGURE 7-18

Stress-strain curves for two kinds of rubber in tension

as the **offset yield stress**. For a material such as aluminum, the offset yield stress is slightly above the proportional limit. In the case of structural steel, with its abrupt transition from the linear region to the region of plastic stretching, the offset stress is essentially the same as both the yield stress and the proportional limit.

Rubber   Rubber maintains a linear relationship between stress and strain up to relatively large strains (as compared to metals). The strain at the proportional limit may be as high as 0.1 or 0.2 (10 or 20%). Beyond the proportional limit, the behavior depends upon the type of rubber (Fig. 7-18). Some kinds of soft rubber stretch enormously without failure, reaching lengths several times their original lengths. The material eventually offers increasing resistance to the load, and the stress-strain curve turns markedly upward. You can easily sense this characteristic behavior by stretching a rubber band with your hands. (Note that although rubber exhibits very large strains, it is not a ductile material because the strains are not permanent. It is, of course, an elastic material; see Section 7.4.)

Ductility and Elongation   The ductility of a material in tension can be characterized by its elongation and by the reduction in area at the cross section where fracture occurs. The **percent elongation** is defined as

$$\text{Percent elongation} = \frac{L_1 - L_0}{L_0}(100) \tag{7-5}$$

in which L_0 is the original gage length and L_1 is the distance between the gage marks at fracture. Because the elongation is not uniform over the length of the specimen but is concentrated in the region of necking, the percent elongation depends upon the gage length. Therefore, when stating the percent elongation, the gage length should always be given. For a 2-in. gage length, steel may have an elongation in the range from 3 to 40%, depending upon composition; in the case of structural steel, values of 20 or 30% are common. The elongation of aluminum alloys varies from 1 to 45%, depending upon composition and treatment.

The **percent reduction in area** measures the amount of necking that occurs and is defined as

$$\text{Percent reduction} = \frac{A_0 - A_1}{A_0}(100) \tag{7-6}$$

where A_0 is the original cross-sectional area and A_1 is the final area at the fracture section. For ductile steels, the reduction is about 50%.

FIGURE 7-19

Typical stress-strain diagram for a brittle material showing the proportional limit (point *A*) and fracture stress (point *B*)

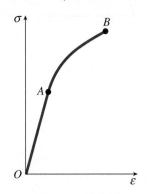

Brittle Materials   Materials that fail in tension at relatively low values of strain are classified as **brittle**. Examples are concrete, stone, cast iron, glass, ceramics, and a variety of metallic alloys. Brittle materials fail with only little elongation after the proportional limit (the stress at point *A* in Fig. 7-19) is exceeded. Furthermore, the reduction in area is insignificant, and so the nominal fracture stress (point *B*) is the same as the true ultimate stress. High-carbon steels have very high yield stresses—over 100 ksi (700 MPa) in some cases—but they behave in a brittle manner and fracture occurs at an elongation of only a few percent.

Ordinary glass is a nearly ideal brittle material because it exhibits almost no ductility. The stress-strain curve for glass in tension is essentially a straight line, with failure occurring before any yielding takes place. The ultimate stress is about 10,000 psi (70 MPa) for certain kinds of plate glass, but great variations exist, depending upon the type of glass, the size of the specimen, and the presence of microscopic defects. **Glass fibers** can develop enormous strengths, and ultimate stresses over 1,000,000 psi (7 GPa) have been attained.

Plastics   Many types of **plastics** are used for structural purposes because of their light weight, resistance to corrosion, and good electrical insulation properties. Their mechanical properties vary tremendously, with some plastics being brittle and others ductile. When designing with plastics, it is important to realize that their properties are greatly affected by both temperature changes and the passage of time. For instance, the ultimate tensile stress of some plastics is cut in half merely by raising the temperature from 50° to 120°F. Also, a loaded plastic may stretch gradually over time until it is no longer serviceable. For example, a bar of polyvinyl chloride subjected to a tensile load that initially produces a strain of 0.005 may have that strain doubled after one week, even though the load remains constant. (This phenomenon, known as *creep*, is discussed in the next section.)

Ultimate tensile stresses for plastics are generally in the range 2 to 50 ksi (14 to 350 MPa) and weight densities vary from 50 to 90 lb/ft^3 (8 to 14 kN/m^3). One type of nylon has an ultimate stress of 12 ksi (80 MPa) and weighs only 70 lb/ft^3 (11 kN/m^3), which is only 12% heavier than water. Because of its light weight, the strength-to-weight ratio for nylon is about the same as for structural steel (see Prob. 7.3-4).

Composites   A **filament-reinforced material** consists of a base material (or *matrix*) in which high-strength filaments, fibers, or whiskers are embedded. The resulting composite material has much greater strength than the base material. As an example, the use of glass fibers can more than double the strength of a plastic matrix. Composites are widely used in aircraft, boats, rockets, and space vehicles where high strength and light weight are needed.

Compression

Stress-strain curves for materials in compression differ from those in tension. Ductile metals such as steel, aluminum, and copper have proportional limits in compression very close to those in tension, and the initial regions of their compressive and tensile stress-strain diagrams are about the same. However, after yielding begins, the behavior is quite different. In a tension test, the specimen is stretched, necking may occur, and fracture ultimately takes place. When the material is compressed, it bulges outward on the sides and becomes barrel shaped, because friction between the specimen and the end plates prevents lateral expansion. With increasing load, the specimen is flattened out and offers greatly increased resistance to further shortening (which means that the stress-strain curve becomes very steep). These characteristics are illustrated in Fig. 7-20, which shows a compressive stress-strain diagram for copper. Since the actual cross-sectional area of a specimen tested in compression is larger than the initial area, the true stress in a compression test is smaller than the nominal stress.

Brittle materials loaded in compression typically have an initial linear region followed by a region in which the shortening increases at a slightly higher rate than does the load. The stress-strain curves for compression and tension often have similar shapes, but the ultimate stresses in compression are much higher than those in tension. Also, unlike ductile materials, which flatten out when compressed, brittle materials actually break at the maximum load.

Tables of Mechanical Properties

Properties of materials are listed in the tables of Appendix F at the back of the book. The data in the tables are typical of the materials and are suitable for solving problems in this book. However, properties of materials and stress-strain curves vary greatly—even for the same material—because of different manufacturing processes, chemical composition, internal defects, temperature, and many other factors.

FIGURE 7-20

Stress-strain diagrams for copper in compression

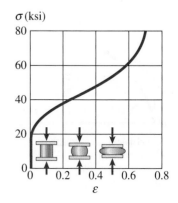

For these reasons, data obtained from Appendix F (or other tables of a similar nature) should not be used for specific engineering or design purposes. Instead, the manufacturers or materials suppliers should be consulted for information about a particular product.

7.4 Elasticity, Plasticity, and Creep

Stress-strain diagrams portray the behavior of engineering materials when the materials are loaded in tension or compression, as described in the preceding section. Now consider what happens when the load is removed and the material is *unloaded*.

Assume, for instance, that you apply a load to a tensile specimen so that the stress and strain go from the origin O to point A on the stress-strain curve of Fig. 7-21a. Suppose that when the load is removed, the material follows exactly the same curve back to the origin O. This property of a material, by which it returns to its original dimensions during unloading, is called **elasticity**, and the material itself is said to be *elastic. Note that the stress-strain curve from O to A need not be linear in order for the material to be elastic.*

Now suppose that you load this same material to a higher level, so that point B is reached on the stress-strain curve (Fig. 7-21b). When unloading occurs from point B, the material follows line BC on the diagram. This unloading line is parallel to the initial portion of the loading curve; that is, line BC is parallel to a tangent to the stress-strain curve at the origin. When point C is reached, the load has been entirely removed, but a **residual strain**, or *permanent strain*, which is represented by line OC, remains in the material. As a consequence, the bar being tested is longer than it was before loading. This residual elongation of the bar is called the **permanent set**. Of the total strain OD developed during loading from O to B, the strain CD has been recovered elastically and the strain OC remains as a permanent strain. Thus, during unloading the bar returns partially to its original shape, and the material is said to be **partially elastic**.

Between points A and B on the stress-strain curve (Fig. 7-21b), there must be a point before which the material is elastic and beyond which the material is partially elastic. To find this point, load the material to some selected value of stress and then remove the load. If there is no permanent set (that is, if the elongation of the bar returns to zero), the material is fully elastic up to the selected value of the stress.

The process of loading and unloading can be repeated for successively higher values of stress. Eventually, a stress will be reached such that not all the strain is recovered during unloading. By this procedure, it is possible to determine the stress at the upper limit of the elastic region, for instance, the stress at point E in Figs. 7-21a and b. The stress at this point is known as the **elastic limit** of the material.

Many materials, including most metals, have linear regions at the beginning of their stress-strain curves (for example, see Figs. 7-13 and 7-16). The stress at the upper limit of this linear region is the proportional limit. The elastic limit is usually the same as, or slightly above, the proportional limit. Hence, for many materials the two limits are assigned the same numerical value. In the case of mild steel, the yield stress is also very close to the proportional limit, so for practical purposes, the yield stress, the elastic limit, and the proportional limit are assumed to be equal. Of course, this situation does not hold for all materials. Rubber is an outstanding example of a material that is elastic far beyond the proportional limit.

The characteristic of a material that undergoes inelastic strains beyond the strain at the elastic limit is known as **plasticity**. Thus, on the stress-strain curve

FIGURE 7-21

Stress-strain diagrams illustrating (a) elastic behavior and (b) partially elastic behavior

(a)

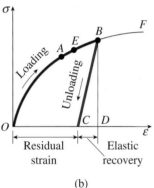

(b)

of Fig. 7-21a, an elastic region is followed by a plastic region. When large deformations occur in a ductile material loaded into the plastic region, the material is said to undergo **plastic flow**.

Reloading of a Material

If the material remains within the elastic range, it can be loaded, unloaded, and loaded again without significantly changing its behavior. However, when loaded into the plastic range, the internal structure of the material is altered and its properties change. For instance, you have already observed that a permanent strain exists in the specimen after unloading from the plastic region (Fig. 7-21b). Now suppose that the material is reloaded after such an unloading (Fig. 7-22). The new loading begins at point C on the diagram and continues upward to point B, which is the point where unloading began during the first loading cycle. The material then follows the original stress-strain curve toward point F. Thus, for the second loading, imagine that there is a new stress-strain diagram with its origin at point C.

During the second loading, the material behaves in a linearly elastic manner from C to B, with the slope of line CB being the same as the slope of the tangent to the original loading curve at the origin O. The proportional limit is now at point B, which is at a higher stress than the original elastic limit (point E). Thus, by stretching a material such as steel or aluminum into the inelastic or plastic range, the *properties of the material are changed*—the linearly elastic region is increased, the proportional limit is raised, and the elastic limit is raised. However, the ductility is reduced because in the "new material" the amount of yielding beyond the elastic limit (from B to F) is less than in the original material (from E to F).[4]

Creep

The stress-strain diagrams described previously were obtained from tension tests involving static loading and unloading of the specimens, and the passage of time did not enter the discussion. However, when loaded for long periods of time, some materials develop additional strains and are said to **creep**.

Creep can manifest itself in a variety of ways. For instance, suppose that a vertical bar (Fig. 7-23a) is loaded slowly by a force P, producing an elongation equal to δ_0. Assume that the loading and corresponding elongation take place during a time interval of duration t_0 (Fig. 7-23b). Subsequent to time t_0, the load remains constant. However, due to creep, the bar may gradually lengthen, as shown in Fig. 7-23b, even though the load does not change. This behavior occurs with many materials, although sometimes the change is too small to be of concern.

As another manifestation of creep, consider a wire that is stretched between two immovable supports so that it has an initial tensile stress σ_0 (Fig. 7-24). Again, denote the time during which the wire is initially stretched as t_0. With the elapse of time, the stress in the wire gradually diminishes, eventually reaching a constant value, even though the supports at the ends of the wire do not move. This process is called **relaxation** of the material.

Creep is usually more important at high temperatures than at ordinary temperatures, therefore it should always be considered in the design of engines,

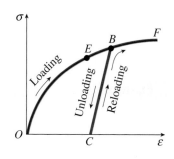

FIGURE 7-22

Reloading of a material and raising of the elastic and proportional limits

FIGURE 7-23

Creep in a bar under constant load

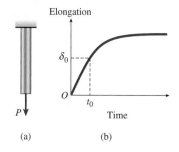

FIGURE 7-24

Relaxation of stress in a wire under constant strain

---

[4]The study of material behavior under various environmental and loading conditions is an important branch of applied mechanics. For more detailed engineering information about materials, consult a textbook devoted solely to this subject.

furnaces, and other structures that operate at elevated temperatures for long periods of time. However, materials such as steel, concrete, and wood will creep slightly—even at atmospheric temperatures. For example, creep of concrete over long periods of time can create undulations in bridge decks because of sagging between the supports. (One remedy is to construct the deck with an upward **camber**, which is an initial displacement above the horizontal, so that when creep occurs, the spans lower to the level position.)

Example 7-3

A machine component slides along a horizontal bar at A and moves in a vertical slot B. The component is represented as a rigid bar AB (length $L = 5$ ft, weight $W = 985$ lb) with roller supports at A and B (neglect friction). When not in use, the machine component is supported by a single wire (diameter $d = 1/8$ in.) with one end attached at A and the other end supported at C (see Fig. 7-25). The wire is made of a copper alloy; the stress-strain relationship for the wire is

$$\sigma(\varepsilon) = \frac{17,500\varepsilon}{1+240\varepsilon} \qquad 0 \le \varepsilon \le 0.03 \quad (\sigma \text{ in ksi})$$

(a) Plot a stress-strain diagram for the material; What is the modulus of elasticity E (ksi)? What is the 0.2% offset yield stress (ksi)?

(b) Find the tensile force T (lb) in the wire.

(c) Find the normal axial strain ϵ and elongation δ (in.) of the wire.

(d) Find the permanent set of the wire if all forces are removed.

FIGURE 7-25

Example 7-3: (a) Rigid bar supported by copper alloy wire and (b) close up view at A

(a)   (b)

Solution:

Use the four-step problem-solving approach to find the modulus of elasticity, yield stress, tensile force, normal strain and elongation, and the permanent set of copper alloy wire AC.

1. *Conceptualize:* The copper alloy has considerable ductility but will have a stress-strain curve without a well-defined yield point. Define the yield point

using an *offset method* as illustrated in Fig. 7-17. Find the residual strain and then the *permanent set* of the wire, as shown in Fig. 7-21b.

2. *Categorize*: The given analytical expression for the stress-strain curve $\sigma(\varepsilon)$ is based on measured laboratory data for the copper alloy used to manufacture this wire. Hence, the analytical expression is an approximation of the actual behavior of this material and was formulated based on test data. Analytical representations of actual stress-strain curves are often used in computer programs to model and analyze structures of different materials under applied loads of various kinds.

3. *Analyze*:

   Part (a): Plot a stress-strain diagram for the material. What is the modulus of elasticity E (ksi)? What is the 0.2% offset yield stress (ksi)?

   Plot the function $\sigma(\varepsilon)$ for strain values between 0 and 0.03 (Fig. 7-26). The stress at strain $\varepsilon = 0.03$ is 64 ksi.

   $$\sigma(\varepsilon) = \frac{17{,}500\varepsilon}{1 + 240\varepsilon} \qquad \varepsilon = 0, 0.001, \ldots, 0.03$$

   $$\sigma(0) = 0 \qquad \sigma(0.03) = 64 \text{ ksi}$$

FIGURE 7-26

Stress-strain curve for copper alloy wire in Example 7-3

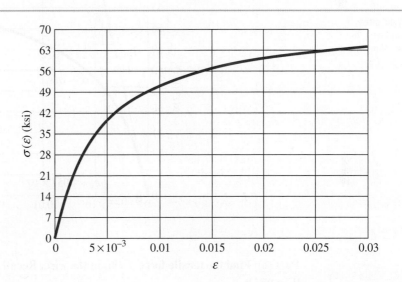

The slope of the tangent to the stress-strain curve at strain $\varepsilon = 0$ is the modulus of elasticity E (see Fig. 7-27). Take the derivative of $\sigma(\varepsilon)$ to get the slope of the tangent to the $\sigma(\varepsilon)$ curve, and evaluate the derivative at strain $\varepsilon = 0$ to find E:

$$E(\varepsilon) = \frac{d}{d\varepsilon}\sigma(\varepsilon) = \frac{17{,}500}{(240\varepsilon + 1)^2}$$

$$E = E(0) \qquad E = 17{,}500 \text{ ksi}$$

Next, find an expression for the yield strain ε_y, which is the point where the 0.2% offset line crosses the stress-strain curve (see Fig. 7-27). Substitute the expression ε_y into the $\sigma(\varepsilon)$ expression and then solve for yield $\sigma(\varepsilon_y) = \sigma_y$:

$$\varepsilon_y = 0.002 + \frac{\sigma_y}{E} \quad \text{and} \quad \sigma(\varepsilon_y) = \sigma_y \quad \text{or} \quad \sigma_y = \frac{17{,}500\varepsilon_y}{1+240\varepsilon_y}$$

Rearranging the equation in terms of σ_y gives

$$\sigma_y^2 + \left(\frac{37E}{6000} - \frac{875}{12}\right)\sigma_y - \frac{7E}{48} = 0$$

Solving this quadratic equation for the 0.2% offset yield, stress σ_y gives $\sigma_y = 36$ ksi.

The yield strain is computed as

$$\varepsilon_y = 0.002 + \frac{\sigma_y}{E(\text{ksi})} = 4.057 \times 10^{-3}$$

FIGURE 7-27

Modulus of elasticity E, 0.2% offset line, and yield stress σ_y and strain ε_y for copper alloy wire in Example 7-3

Part (b): Find the tensile force T (lb) in the wire. Recall that bar weight $W = 985$ lb.

Find the angle between the x-axis and cable attachment position at C:

$$\alpha_C = \arctan\left(\frac{1.5}{4}\right) = 20.556°$$

Sum the moments about A to obtain one equation and one unknown. The reaction B_x acts to the left:

$$B_x = \frac{-W(2\text{ ft})}{3\text{ ft}} = -656.667 \text{ lb}$$

Next, sum the forces in the x direction to find the cable force T_C:

$$T_C = \frac{-B_x}{\cos(\alpha_C)} \qquad T_C = 701 \text{ lb}$$

Part (c): Find the normal axial strain ε and elongation δ (in.) of the wire.

Compute the normal stress; then find the associated strain from stress-strain plot (or from the $\sigma(\varepsilon)$ equation). The wire elongation is strain times wire length.

The wire diameter, cross-sectional area, and length are

$$d = \frac{1}{8} \text{ in.} \quad A = \frac{\pi}{4} d^2 = 0.0123 \text{ in}^2$$

$$L_C = \sqrt{(4 \text{ ft})^2 + (1.5 \text{ ft})^2} = 4.272 \text{ ft}$$

Now compute the stress and strain in the wire and the elongation of the wire as

$$\sigma_C = \frac{T_C}{A} = 57.1 \text{ ksi}$$

Note that the stress in the wire exceeds the 0.2% offset yield stress of 36 ksi. The corresponding normal strain is found from the $\sigma(\varepsilon)$ plot or by rearranging the $\sigma(\varepsilon)$ equation to give

$$\varepsilon(\sigma) = \frac{\sigma}{17{,}500 - 240\sigma}$$

Then,

$$\varepsilon(\sigma_C) = \varepsilon_C, \quad \text{or} \quad \varepsilon_C = \frac{\sigma_C}{17{,}500 \text{ ksi} - 240\sigma_C} = 0.015$$

Finally, the wire elongation is

$$\delta_C = \varepsilon_C L_C = 0.774 \text{ in.}$$

Part (d): Find the permanent set of the wire if all forces are removed.

If the load is removed from the wire, the stress in the wire will return to zero following the unloading line in Fig. 7-28 (see also Fig. 7-21b). The elastic recovery strain is

$$\varepsilon_{er} = \frac{\sigma_C}{E} = 3.266 \times 10^{-3}$$

The residual strain is the difference between the total strain (ε_C) and the elastic recovery strain (ε_{er})

$$\varepsilon_{res} = \varepsilon_C - \varepsilon_{er} = 0.012$$

Finally, the permanent set of the wire is the product of the residual strain and the length of the wire:

$$\varepsilon_{res} L_C = 0.607 \text{ in.}$$

FIGURE 7-28

Residual strain (ε_{res}) and elastic recovery strain (ε_{er}) for copper alloy wire in Example 7-3

4. *Finalize*: This example presents an analytical model of the stress-strain relationship for a copper alloy. The computed values of modulus of elasticity E and yield stress σ_y are consistent with values listed in Appendix F. The tensile force, normal strain and elongation, and permanent set are computed for the wire when stressed beyond the apparent yield point of the material.

7.5 Linear Elasticity, Hooke's Law, and Poisson's Ratio

Many structural materials, including most metals, wood, plastics, and ceramics, behave both elastically and linearly when first loaded. Consequently, their stress-strain curves begin with a straight line passing through the origin. An example is the stress-strain curve for structural steel (Fig. 7-13), where the region from the origin O to the proportional limit (point A) is both linear and elastic. Other examples are the regions below *both* the proportional limits and the elastic limits on the diagrams for aluminum (Fig. 7-16), brittle materials (Fig. 7-19), and copper (Fig. 7-20).

When a material behaves elastically and also exhibits a linear relationship between stress and strain, it is said to be **linearly elastic**. By designing structures and machines to function in this region, engineers avoid permanent deformations due to yielding.

Hooke's Law

The linear relationship between stress and strain for a bar in simple tension or compression is expressed by the equation

$$\sigma = E\varepsilon \tag{7-7}$$

where σ is the axial stress, ϵ is the axial strain, and E is a constant of proportionality known as the **modulus of elasticity** for the material. The modulus of elasticity is the slope of the stress-strain diagram in the linearly elastic region, as mentioned previously in Section 7.3. Since strain is dimensionless, the units of E are the same as the units of stress. Typical units of E are psi or ksi in USCS units and pascals (or multiples thereof) in SI units.

The equation $\sigma = E\varepsilon$ is commonly known as **Hooke's law**, named for the famous English scientist Robert Hooke (1635–1703). Hooke was the first person to investigate scientifically the elastic properties of materials, and he tested such diverse materials as metal, wood, stone, bone, and sinew. He measured the stretching of long wires supporting weights and observed that the elongations "always bear the same proportions one to the other that the weights do that made them." Thus, Hooke established the linear relationship between the applied loads and the resulting elongations.

Equation (7-7) is actually a very limited version of Hooke's law because it relates only to the longitudinal stresses and strains developed in simple tension or compression of a bar (*uniaxial stress*). To deal with more complicated states of stress, such as those found in most structures and machines, more extensive equations of Hooke's law are needed (see Sections 11.5 and 11.6).

The modulus of elasticity has relatively large values for materials that are very stiff, such as structural metals. Steel has a modulus of approximately 30,000 ksi (210 GPa); for aluminum, values around 10,600 ksi (73 GPa) are typical. More flexible materials have a lower modulus—values for plastics range from 100 to 2000 ksi (0.7 to 14 GPa). Some representative values of E are listed in Table F-2, Appendix F. For most materials, the value of E in compression is nearly the same as in tension.

Modulus of elasticity is often called **Young's modulus**, after another English scientist, Thomas Young (1773–1829). In connection with an investigation of tension and compression of prismatic bars, Young introduced the idea of a "modulus of the elasticity." However, his modulus was not the same as the one in use today because it involved properties of the bar as well as of the material.

Poisson's Ratio

When a prismatic bar is loaded in tension, the axial elongation is accompanied by **lateral contraction** (that is, contraction normal to the direction of the applied load). This change in shape is pictured in Fig. 7-29, where part (a) shows the bar before loading and part (b) shows it after loading. In part (b), the dashed lines represent the shape of the bar prior to loading.

Lateral contraction is easily seen by stretching a rubber band, but in metals, the changes in lateral dimensions (in the linearly elastic region) are usually too small to be visible. However, they can be detected with sensitive measuring devices.

The **lateral strain** ε' at any point in a bar is proportional to the axial strain ϵ at that same point if the material is linearly elastic. The ratio of these strains is a property of the material known as **Poisson's ratio**. This dimensionless ratio, usually denoted by the Greek letter ν (nu), can be expressed by

$$\nu = -\frac{\text{lateral strain}}{\text{axial strain}} = -\frac{\varepsilon'}{\varepsilon} \qquad \textbf{(7-8)}$$

The minus sign is inserted in the equation to compensate for the fact that the lateral and axial strains normally have opposite signs. For instance, the

FIGURE 7-29

Axial elongation and lateral contraction of a prismatic bar in tension: (a) bar before loading and (b) bar after loading (The deformations of the bar are highly exaggerated)

(a)

(b)

axial strain in a bar in tension is positive, and the lateral strain is negative (because the width of the bar decreases). The opposite is true for compression, with the bar becoming shorter (negative axial strain) and wider (positive lateral strain). Therefore, for ordinary materials, Poisson's ratio has a positive value. Some materials such as low-density open-cell polymer foams, however, can have a negative Poisson's ratio so, in the linear elastic range, Poisson's ratio lies between -1 and $+0.5$.

When Poisson's ratio for a material is known, you can obtain the lateral strain from the axial strain as

$$\varepsilon' = -\nu\varepsilon \tag{7-9}$$

When using Eqs. (7-8) and (7-9), always keep in mind that they apply only to a bar in uniaxial stress, that is, a bar for which the only stress is the normal stress σ in the axial direction.

Poisson's ratio is named for the famous French mathematician Siméon Denis Poisson (1781–1840), who attempted to calculate this ratio by a molecular theory of materials. For isotropic materials, Poisson found $v = 1/4$. More recent calculations based upon better models of atomic structure give $v = 1/3$. Both of these values are close to actual measured values, which are in the range 0.25 to 0.35 for most metals and many other materials. Materials with an extremely low value of Poisson's ratio include cork, for which v is practically zero, and concrete, for which ν is about 0.1 or 0.2. A theoretical upper limit for Poisson's ratio is 0.5, as explained later in Section 11.5. Rubber comes close to this limiting value. Note however, that Poisson's ratio may be as low as –1.0 for materials such as low-density open-cell polymer foams. Hence, in the elastic range, Poisson's ratio varies between -1.0 and $+0.5$.

A table of Poisson's ratios for various materials in the linearly elastic range is given in Appendix F (see Table F-2). For most purposes, Poisson's ratio is assumed to be the same in both tension and compression.

When the strains in a material become large, Poisson's ratio changes. For instance, in the case of structural steel, the ratio becomes almost 0.5 when plastic yielding occurs. Thus, Poisson's ratio remains constant only in the linearly elastic range. When the material behavior is nonlinear, the ratio of lateral strain to axial strain is often called the *contraction ratio*. Of course, in the special case of linearly elastic behavior, the contraction ratio is the same as Poisson's ratio.

Limitations

For a particular material, Poisson's ratio remains constant throughout the linearly elastic range, as explained previously. Therefore, at any given point in the prismatic bar of Fig. 7-29, the lateral strain remains proportional to the axial strain as the load increases or decreases. However, for a given value of the load (which means that the axial strain is constant throughout the bar), additional conditions must be met if the lateral strains are to be the same throughout the entire bar.

First, the material must be **homogeneous**, that is, it must have the same composition (and hence the same elastic properties) at every point. However, having a homogeneous material does not mean that the elastic properties at a particular point are the same in all *directions*. For instance, the modulus of elasticity could be different in the axial and lateral directions, as in the case of a wood pole. Therefore, a second condition for uniformity in the lateral strains is that the elastic properties must be the same in all directions *perpendicular* to the longitudinal axis. When the preceding conditions are met, as is often the case with metals, the lateral strains in a prismatic bar subjected to uniform tension will be the same at every point in the bar and the same in all lateral directions.

Materials having the same properties in all directions (whether axial, lateral, or any other direction) are said to be **isotropic**. If the properties differ in various directions, the material is **anisotropic**.

In this book, all examples and problems are solved with the assumption that the material is linearly elastic, homogeneous, and isotropic—unless a specific statement is made to the contrary.

Example 7-4

A hollow plastic circular pipe (length L_p, inner and outer diameters d_1 and d_2, respectively; see Fig. 7-30) is inserted as a liner inside a cast iron pipe (length L_c, inner and outer diameters d_3 and d_4, respectively).

(a) Derive a formula for the required initial length L_p of the plastic pipe so that, when it is compressed by some force P, the final length of both pipes is the same and at the same time the final outer diameter of the plastic pipe is equal to the inner diameter of the cast iron pipe.

(b) Using the numerical data given, find the initial length L_p (m) and final thickness t_p (mm) for the plastic pipe.

(c) What is the required compressive force P (N)? What are the final normal stresses (MPa) in both pipes?

(d) Compare the initial and final volumes (mm^3) for the plastic pipe.

Numerical data and pipe cross-section properties are

$$L_c = 0.25 \text{ m} \quad E_c = 170 \text{ GPa} \quad E_p = 2.1 \text{ GPa} \quad v_c = 0.3 \quad v_p = 0.4$$

$$d_1 = 109.8 \text{ mm} \quad d_2 = 110 \text{ mm} \quad d_3 = 110.2 \text{ mm}$$

$$d_4 = 115 \text{ mm} \quad t_p = \frac{d_2 - d_1}{2} = 0.1 \text{ mm}$$

FIGURE 7-30

Example 7-4: Plastic pipe compressed inside cast iron pipe

Force P compresses plastic pipe

d_2

d_1

d_2

Inner plastic pipe liner

L_p

L_c

d_3

d_4

Outer cast iron pipe

Solution:

Use the four-step problem-solving approach to find the dimensions and force required to fit a plastic liner into a cast iron pipe.

1. *Conceptualize*: Application of a compressive force P results in compressive normal strains and extensional lateral strains in the plastic pipe, while the cast iron pipe is stress-free. The initial length of the plastic pipe (L_p) is greater than that of the cast iron pipe (L_c). With full application of force P, the lengths are made equal.

   The initial cross-sectional areas of the plastic and cast iron pipes are

   $$A_p = \frac{\pi}{4}(d_2^2 - d_1^2) = 34.526 \text{ mm}^2 \quad A_c = \frac{\pi}{4}(d_4^2 - d_3^2) = 848.984 \text{ mm}^2$$

2. *Categorize*: The two requirements are (i) compression of the plastic pipe must close the gap ($d_3 - d_2$) between the plastic pipe and the inner surface of the cast iron pipe and (ii) the final lengths of the two pipes are the same. The first requirement depends on lateral strain and the second on normal strain. Each requirement leads to an expression for shortening of the plastic pipe. Equating the two expressions (i.e., enforcing *compatibility of displacements*) leads to a solution for the required length of the plastic pipe.

3. *Analyze*:

   Part (a): Derive a formula for the required initial length L_p of the plastic pipe.

   The lateral strain resulting from compression of the plastic pipe must close the gap ($d_3 - d_2$) between the plastic pipe and the inner surface of the cast iron pipe. The required *extensional* lateral strain is positive (here, $\varepsilon_{\text{lat}} = \varepsilon'$):

   $$\varepsilon_{\text{lat}} = \frac{d_3 - d_2}{d_2} = 1.818 \times 10^{-3}$$

   The accompanying *compressive* normal strain in the plastic pipe is obtained using Eq. (7-9), which requires Poisson's ratio for the plastic pipe and also the required lateral strain:

   $$\varepsilon_p = \frac{-\varepsilon_{\text{lat}}}{\nu_p} \text{ or } \varepsilon_p = \frac{-1}{\nu_p}\left(\frac{d_3 - d_2}{d_2}\right) = -4.545 \times 10^{-3}$$

   Use the compressive normal strain ε_p to compute the *shortening* δ_{p1} of the plastic pipe as

   $$\delta_{p1} = \varepsilon_p L_p$$

   The required *shortening* of the plastic pipe (so that it has the same final length as that of the cast iron pipe) is

   $$\delta_{p2} = -(L_p - L_c)$$

Equating δ_{p1} and δ_{p2} leads to a formula for the required initial length L_p of the plastic pipe:

$$L_p = \frac{L_c}{1 + \varepsilon_p} \quad \text{or} \quad L_p = \frac{L_c}{1 - \dfrac{d_3 - d_2}{v_p d_2}}$$

Part (b): Now substitute the numerical data to find the initial length L_p, change in thickness Δt_p, and final thickness t_{pf} for the plastic pipe.

As expected, L_p is greater than the length of the cast iron pipe, $L_c = 0.25$ m, and the thickness of the compressed plastic pipe increases by Δt_p:

$$L_p = \frac{L_c}{1 - \left(\dfrac{d_3 - d_2}{v_p d_2}\right)} = 0.25114 \text{ m}$$

$$\Delta t_p = \varepsilon_{\text{lat}} t_p = 1.818 \times 10^{-4} \text{ mm} \quad \text{so} \quad t_{pf} = t_p + \Delta t_p = 0.10018 \text{ mm}$$

Part (c): Next find the required compressive force P and the final normal stresses in both pipes.

A check on the normal compressive stress in the plastic pipe, computed using Hooke's law [Eq. (7-7)], shows that it is well below the ultimate stress for selected plastics (see Table F-3, Appendix F); this is also the final normal stress in the plastic pipe:

$$\sigma_p = E_p \varepsilon_p = -9.55 \text{ MPa}$$

The required downward force to compress the plastic pipe is

$$P_{\text{reqd}} = \sigma_p A_p = -330 \text{ N}$$

Both the initial and final stresses in the cast iron pipe are zero because no force is applied to the cast iron pipe.

Part (d): Lastly, compare the initial and final volumes of the plastic pipe.

The initial cross-sectional area of the plastic pipe is

$$A_p = 34.526 \text{ mm}^2$$

The final cross-sectional area of the plastic pipe is

$$A_{pf} = \frac{\pi}{4} [d_3^2 - (d_3 - 2t_{pf})^2] = 34.652 \text{ mm}^2$$

The initial volume of the plastic pipe is

$$V_{\text{pinit}} = L_p A_p = 8671 \text{ mm}^3$$

and the final volume of the plastic pipe is

$$V_{p\text{final}} = L_c A_{pf} \text{ or } V_{p\text{final}} = 8663 \text{ mm}^3$$

4. *Finalize*: The ratio of final to initial volume reveals little change in the volume of the plastic pipe:

$$\frac{V_{p\text{final}}}{V_{p\text{init}}} = 0.99908$$

The numerical results obtained in this example illustrate that the dimensional changes in structural materials under normal loading conditions are extremely small. In spite of their smallness, changes in dimensions can be important in certain kinds of analysis (such as the analysis of statically indeterminate structures) and in the experimental determination of stresses and strains.

7.6 Shear Stress and Strain

Diagonal bracing for an elevated walkway showing a clevis and a pin in double shear

The preceding sections discussed the effects of normal stresses produced by axial loads acting on straight bars. These stresses are called "normal stresses" because they act in directions *perpendicular* to the surface of the material. Now consider another kind of stress, called a **shear stress**, that acts *tangential* to the surface of the material.

As an illustration of the action of shear stresses, consider the bolted connection shown in Fig. 7-31a. This connection consists of a flat bar A, a clevis C, and a bolt B that pass through holes in the bar and clevis. Under the action of the tensile loads P, the bar and clevis press against the bolt in **bearing**, and contact stresses, called **bearing stresses**, are developed. In addition, the bar and clevis tend to shear the bolt, that is, cut through it, and this tendency is resisted by shear stresses in the bolt. As an example, see the bracing for an elevated pedestrian walkway shown in the photograph.

To show more clearly the actions of the bearing and shear stresses, look at this type of connection in a schematic side view (Fig. 7-31b). With this view in mind, draw a free-body diagram of the bolt (Fig. 7-31c). The bearing stresses exerted by the clevis against the bolt appear on the left-hand side of the free-body diagram and are labeled 1 and 3. The stresses from the bar appear on the right-hand side and are labeled 2. The actual distribution of the bearing stresses is difficult to determine, so it is customary to assume that the stresses are uniformly distributed. Based upon the assumption of uniform distribution, calculate an **average bearing** stress σ_b by dividing the total bearing force F_b by the bearing area A_b:

$$\sigma_b = \frac{F_b}{A_b} \tag{7-10}$$

FIGURE 7-31

Bolted connection in which the bolt is loaded in double shear

(a)

(b)   (c)   (d)   (e)

The **bearing area** is defined as the projected area of the curved bearing surface. For instance, consider the bearing stresses labeled 1. The projected area A_b on which they act is a rectangle having a height equal to the thickness of the clevis and a width equal to the diameter of the bolt. Also, the bearing force F_b represented by the stresses labeled 1 is equal to $P/2$. The same area and the same force apply to the stresses labeled 3.

Now look at the bearing stresses between the flat bar and the bolt (the stresses labeled 2). For these stresses, the bearing area A_b is a rectangle with height equal to the thickness of the flat bar and width equal to the bolt diameter. The corresponding bearing force F_b is equal to the load P.

The free-body diagram of Fig. 7-31c shows that there is a tendency to shear the bolt along cross sections mn and pq. From a free-body diagram of the portion $mnpq$ of the bolt (see Fig. 7-31d), note that shear forces V act over the cut surfaces of the bolt. There are two planes of shear (mn and pq), and the bolt is said to be in **double shear**. In double shear, each of the shear forces is equal to one-half of the total load transmitted by the bolt, that is, $V = P/2$.

The shear forces V are the resultants of the shear stresses distributed over the cross-sectional area of the bolt. For instance, the shear stresses acting on cross section mn are shown in Fig. 7-31e. These stresses act parallel to the cut surface. The exact distribution of the stresses is not known, but they are highest near the center and become zero at certain locations on the edges. As indicated in Fig. 7-31e, shear stresses are customarily denoted by the Greek letter τ (tau).

A bolted connection in single shear is shown in Fig. 7-32a, where the axial force P in the metal bar is transmitted to the flange of the steel column through a bolt. A cross-sectional view of the column (Fig. 7-32b) shows the connection in more detail. Also, a sketch of the bolt (Fig. 7-32c) shows the assumed distribution of the bearing stresses acting on the bolt. The actual distribution of these bearing stresses is much more complex than shown in the figure. Furthermore, bearing stresses are also developed against the inside surfaces of the bolt head and nut. Thus, Fig. 7-32c is *not* a free-body diagram—only the idealized bearing stresses acting on the shank of the bolt are shown in the figure.

FIGURE 7-32

Bolted connection in which the bolt is loaded in single shear

(a)   (b)   (c)   (d)

Cutting through the bolt at section *mn* reveals the diagram shown in Fig. 7-32d. This diagram includes the shear force *V* (equal to the load *P*) acting on the cross section of the bolt. This shear force is the resultant of the shear stresses that act over the cross-sectional area of the bolt.

The deformation of a bolt loaded almost to fracture in single shear is shown in Fig. 7-33 (compare with Fig. 7-32c).

The preceding discussion of bolted connections disregarded **friction** (produced by tightening of the bolts) between the connecting elements. The presence of friction means that part of the load is carried by friction forces, thereby reducing the loads on the bolts. Since friction forces are unreliable and difficult to estimate, it is common practice to err on the conservative side and omit them from the calculations.

The **average shear stress** on the cross section of a bolt is obtained by dividing the total shear force *V* by the area *A* of the cross section on which it acts, as

$$\tau_{\text{aver}} = \frac{V}{A}$$

(7-11)

FIGURE 7-33

Failure of a bolt in single shear

Bearing stress acting on bolt

Bearing stress acting on bolt

© Barry Goodno

In the example of Fig. 7-32, which shows a bolt in *single shear*, the shear force V is equal to the load P, and the area A is the cross-sectional area of the bolt. However, in the example of Fig. 7-31, where the bolt is in *double shear*, the shear force V equals $P/2$.

Equation (7-11) shows that shear stresses, like normal stresses, represent intensity of force, or force per unit of area. Thus, the **units of** shear stress are the same as those for normal stress, namely, psi or ksi in USCS units and pascals or multiples thereof in SI units.

The loading arrangements shown in Figs. 7-31 and 7-32 are examples of **direct shear** (or *simple shear*) in which the shear stresses are created by the direct action of the forces trying to cut through the material. Direct shear arises in the design of bolts, pins, rivets, keys, welds, and glued joints.

Equality of Shear Stresses on Perpendicular Planes

FIGURE 7-34

Small element of material subjected to shear stresses

To obtain a more complete picture of the action of shear stresses, consider a small element of material in the form of a rectangular parallelepiped having sides of lengths a, b, and c in the x, y, and z directions, respectively (Fig. 7-34).[5] The front and rear faces of this element are free of stress.

Now assume that a shear stress τ_1 is distributed uniformly over the right-hand face, which has area bc. In order for the element to be in equilibrium in the y direction, the total shear force $\tau_1 bc$ acting on the right-hand face must be balanced by an equal but oppositely directed shear force on the left-hand face. Since the areas of these two faces are equal, it follows that the shear stresses on the two faces must be equal.

The forces $\tau_1 bc$ acting on the left- and right-hand side faces (Fig. 7-34) form a couple having a moment about the z axis of magnitude $\tau_1 bc$, acting counterclockwise in the figure.[6] Equilibrium of the element requires that this moment be balanced by an equal and opposite moment resulting from shear stresses acting on the top and bottom faces of the element. If the stresses on the top and bottom faces are labeled as τ_2, the corresponding horizontal shear forces equal $\tau_2 ac$. These forces form a clockwise couple of moment $\tau_2 abc$. From moment equilibrium of the element about the z axis, $\tau_1 abc$ equals $\tau_2 abc$, or

$$\tau_1 = \tau_2 \qquad\qquad \textbf{(7-12)}$$

Therefore, the magnitudes of the four shear stresses acting on the element are equal, as shown in Fig. 7-35a.

The following are observations regarding shear stresses acting on a rectangular element:

1. Shear stresses on opposite (and parallel) faces of an element are equal in magnitude and opposite in direction.

2. Shear stresses on adjacent (and perpendicular) faces of an element are equal in magnitude and have directions such that both stresses point toward, or both point away from, the line of intersection of the faces.

---

[5]A **parallelepiped** is a prism whose bases are parallelograms; thus, a parallelepiped has six faces, each of which is a parallelogram. Opposite faces are parallel and identical parallelograms. A **rectangular parallelepiped** has all faces in the form of rectangles.

[6]A **couple** consists of two parallel forces that are equal in magnitude and opposite in direction.

(a)

(b)

These observations apply to an element subjected only to shear stresses (no normal stresses), as pictured in Figs. 7-34 and 7-35. This state of stress is called **pure shear**.

For most purposes, the preceding conclusions remain valid even when normal stresses act on the faces of the element. The reason is that the normal stresses on opposite faces of a small element usually are equal in magnitude and opposite in direction.

Shear Strain

Shear stresses acting on an element of material (Fig. 7-35a) are accompanied by *shear strains*. As an aid in visualizing these strains, note that the shear stresses have no tendency to elongate or shorten the element in the x, y, and z directions—in other words, the lengths of the sides of the element do not change. Instead, the shear stresses produce a change in the *shape* of the element (Fig. 7-35b). The original element, which is a rectangular parallelepiped, is deformed into an oblique parallelepiped, and the front and rear faces become rhomboids.[7]

Because of this deformation, the angles between the side faces change. For instance, the angles at points q and s, which were $\pi/2$ before deformation, are reduced by a small angle γ to $\pi/2 - \gamma$ (Fig. 7-35b). At the same time, the angles at points p and r are increased to $\pi/2 + \gamma$. The angle γ is a measure of the **distortion**, or change in shape, of the element and is called the shear strain. Because **shear strain** is an angle, it is usually measured in degrees or radians.

Sign Conventions for Shear Stresses and Strains

As an aid in establishing sign conventions for shear stresses and strains, you need a scheme for identifying the various faces of a stress element (Fig. 7-35a). Henceforth, the faces oriented toward the positive directions of the axes are referred to as the positive faces of the element. In other words, a positive face has its outward normal directed in the positive direction of a coordinate axis. The opposite faces are negative faces. Thus, in Fig. 7-35a, the right-hand, top, and front faces are the positive x, y, and z faces, respectively, and the opposite faces are the negative x, y, and z faces.

Using the terminology described in the preceding paragraph, the sign convention for shear stresses is as follows:

A shear stress acting on a positive face of an element is positive if it acts in the positive direction of one of the coordinate axes and negative if it acts in the negative direction of an axis. A shear stress acting on a negative face of an element is positive if it acts in the negative direction of an axis and negative if it acts in a positive direction.

---

[7]An **oblique angle** can be either acute or obtuse, but it is *not* a right angle. A **rhomboid** is a parallelogram with oblique angles and adjacent sides *not* equal. (A *rhombus* is a parallelogram with oblique angles and all four sides equal, sometimes called a *diamond-shaped figure*.)

Thus, all shear stresses shown in Fig. 7-35a are positive.

The sign convention for shear strains is as follows:

Shear strain in an element is positive when the angle between two positive faces (or two negative faces) is reduced. The strain is negative when the angle between two positive (or two negative) faces is increased.

Thus, the strains shown in Fig. 7-35b are positive, and the positive shear stresses are accompanied by positive shear strains.

Hooke's Law in Shear

The properties of a material in shear can be determined experimentally from direct-shear tests or from torsion tests. The latter tests are performed by twisting hollow, circular tubes, thereby producing a state of pure shear. The results of these tests are used to plot **shear stress-strain diagrams** (that is, diagrams of shear stress τ versus shear strain γ). These diagrams are similar in shape to tension-test diagrams (σ versus ε) for the same materials, although they differ in magnitudes.

From shear stress-strain diagrams, you can obtain material properties such as the proportional limit, modulus of elasticity, yield stress, and ultimate stress. Numerical values of these properties in shear are usually about half as large as those in tension. For instance, the yield stress for structural steel in shear is 0.5 to 0.6 times the yield stress in tension.

For many materials, the initial part of the shear stress-strain diagram is a straight line through the origin, just as it is in tension. For this linearly elastic region, the shear stress and shear strain are proportional, resulting in the following equation for **Hooke's law in shear**:

$$\tau = G\gamma \tag{7-13}$$

where G is the **shear modulus of elasticity** (also called the *modulus of rigidity*).

The shear modulus G has the same units as the tension modulus E, namely, psi or ksi in USCS units and pascals (or multiples thereof) in SI units. For mild steel, typical values of G are 11,000 ksi or 75 GPa; for aluminum alloys, typical values are 4000 ksi or 28 GPa. Additional values are listed in Table F-2, Appendix F.

The moduli of elasticity in tension and shear are related by

$$G = \frac{E}{2(1 + \nu)} \tag{7-14}$$

in which ν is Poisson's ratio. This relationship shows that E, G, and ν are not independent elastic properties of the material. Because the value of Poisson's ratio for ordinary materials is between zero and one-half, from Eq. (7-14) G must be from one-third to one-half of E.

The following examples illustrate some typical analyses involving the effects of shear. Example 7-5 is concerned with shear stresses in a plate, Example 7-6 involves finding shear stresses and shear strains in an elastomeric bearing pad subjected to a horizontal shear force, and Example 7-7 deals with normal and shear stresses in a bolted bracket.

Example 7-5

A punch for making holes in steel plates is shown in Fig. 7-36a. Assume that a punch having diameter $d = 3/4$ in. is used to punch a hole in an 3/10 in. plate, as shown in the cross-sectional view (Fig. 7-36b).

If a force $P = 24$ kips is required to create the hole, what is the average shear stress in the plate and the average compressive stress in the punch?

FIGURE 7-36

Example 7-5: Punching a hole in a steel plate

(a)   (b)

Solution:

Use the four-step problem-solving approach to find average shear stress in the plate and average compressive stress in the punch.

1. *Conceptualize*: Assume that the shaft of the punch is in compression over its entire length due to applied load P. Force P acts downward on the plate and is applied as a uniformly distributed force along a circle of diameter d as the punch passes through the plate.

2. *Categorize*: The average shear stress in the plate is obtained by dividing the force P by the shear area of the plate. The shear area is the cylindrical area of the plate that is exposed when the punch passes through the plate. The compressive stress of interest is the one acting on a circular cross section through the lower segment of the punch (Fig. 7-36).

3. *Analyze*: The shear area A_s is equal to the circumference of the hole times the thickness of the plate, or

$$A_s = \pi d t = \pi(3/4 \text{ in.})(3/10 \text{ in.}) = 0.707 \text{ in}^2$$

where d is the diameter of the punch and t is the thickness of the plate. Therefore, the average shear stress in the plate is

$$\tau_{aver} = \frac{P}{A_s} = 24 \text{ kips}/0.707 \text{ in}^2 = 34 \text{ ksi}$$

The average compressive stress in the punch is

$$\sigma_c = \frac{P}{A_{punch}} = \frac{P}{\pi d^2 / 4} = 24 \text{ kips}/\pi (0.75 \text{ in.})^2 / 4 = 54.3 \text{ ksi}$$

in which A_{punch} is the cross-sectional area of the lower segment of the punch.

4. *Finalize*: The normal and shear stress distributions are not uniform due to stress concentration effects; hence, the calculations result in "average" stresses. In addition, this analysis is highly idealized because impact effects that occur when a punch is rammed through a plate are not part of this analysis.

Example 7-6

A bearing pad of the kind used to support machines and bridge girders consists (see photos) of a linearly elastic material (usually an elastomer, such as rubber) capped by a steel plate (Fig. 7-37a). Assume that the thickness of the elastomer is h, the dimensions of the plate are $a \times b$, and the pad is subjected to a horizontal shear force V.

Obtain formulas for the average shear stress τ_{aver} in the elastomer and the horizontal displacement d of the plate (Fig. 7-37b).

FIGURE 7-37

Example 7-6:
Bearing pad in shear

(a)   (b)

Solution:

Use the four-step problem-solving approach to find average shear stress in the elastomer and horizontal displacement d of the plate.

1. *Conceptualize*: The distortion of the bearing pad under shear force V is assumed to be linear through the thickness h, as shown in Fig. 7-37b.

2. *Categorize*: Assume that the shear stresses in the elastomer are uniformly distributed throughout its entire volume and that the shear strain γ is small.

3. *Analyze*: The shear stress on any horizontal plane through the elastomer equals the shear force V divided by the area ab of the plane (Fig. 7-37a):

$$\tau_{\text{aver}} = \frac{V}{ab} \qquad \qquad \text{(7-15)}$$

The corresponding shear strain [from Hooke's law in shear; Eq. (7-13)] is

$$\gamma = \frac{\tau_{\text{aver}}}{G_e} = \frac{V}{abG_e} \qquad \qquad \text{(7-16)}$$

where G_e is the shear modulus of the elastomeric material. Finally, the horizontal displacement d is equal to $h \tan \gamma$ (from Fig. 7-37b):

$$d = h \tan \gamma = h \tan\left(\frac{V}{abG_e}\right) \qquad \qquad \text{(7-17)}$$

In most practical situations, the shear strain γ is a small angle, and in such cases, replace $\tan\gamma$ with γ and obtain

$$d = h\gamma = \frac{hV}{abG_e} \qquad \qquad \text{(7-18)}$$

For example, if $V = 0.8$ kN, $a = 75$ mm, $b = 60$ mm, $h = 20$ mm, and $G_e = 1.25$ MPa, Eq. (7-17) results in $d = 2.86$ mm, while Eq. (7-18) gives $d = 2.84$ mm.

4. *Finalize*: Equations (7-17) and (7-18) give approximate results for the horizontal displacement of the plate because they are based upon the assumption that the shear stress and strain are constant throughout the volume of the elastomeric material. In reality, the shear stress is zero at the edges of the material (because there are no shear stresses on the free vertical faces); therefore, the deformation of the material is more complex than pictured in Fig. 7-37b. However, if the length *a* of the plate is large compared with the thickness *h* of the elastomer, the preceding results are satisfactory for design purposes.

Example 7-7

A basketball player hangs on the rim after dunking the ball. The player applies a downward force at point A with an estimated magnitude of $P = 400$ lb (Fig. 7-38a). Later, the player dunks again and hangs on the rim with two hands: one at A and one at B (see Fig. 7-38b, c). The rim and support bracket are bolted to the backboard using four bolts with washers (Fig. 7-38d). Find and compare the stresses in the bolted connection at bolt location 2 for the one-handed and two-handed load cases. Assume that the backboard is a fixed support and that bolt and washer diameters are $d_b = 3/8$ in. and $d_w = 5/8$ in., respectively. The support bracket has thickness $t = 3/16$ in.

FIGURE 7-38

(a) Load case 1—downward force on rim at A; (b) load case 2—forces applied at A and B; (c) top view of rim; (d) support bracket and bolt detail

(a)
One-handed

(b)
Two-handed

(c)
Top view

(d)
Support bracket

Solution:

Use the four-step problem-solving approach.

1. *Conceptualize:* Find reaction forces and moments at support point O and then distribute forces and moments to each bolt location. The rim and bracket act

as a cantilever beam. From the free-body diagrams in Fig. 7-39, the reactions for each load case are as follows.

Reactions—Load case 1: Load P is applied in the $(-y)$ direction at A. Sum forces and moments to find:

$$\Sigma F_y = 0 \quad R_y = P = 400 \text{ lb}$$

$$\Sigma M_z = 0 \quad M_{z1} = P(5 + 18) \text{ in.} = (400)(23) \text{ lb-in.} = 9200 \text{ lb-in.}$$ **(a)**

Reactions—Load case 2: Loads $P/2$ are applied at both A and B. Reactions at O are

$$\Sigma F_y = 0 \quad R_y = P = 400 \text{ lb}$$

$$\Sigma M_z = 0 \quad M_{z2} = \frac{P}{2}(5 + 9) \text{ in.} + \frac{P}{2}(23 \text{ in.}) = [200(14) + 200(23)] \text{ lb-in.} = 7400 \text{ lb-in.} \quad \textbf{(b)}$$

$$\Sigma M_x = 0 \quad T_x = -\frac{P}{2}(9 \text{ in.}) = -200(9) \text{ lb-in.} = -1800 \text{ lb-in.}$$

FIGURE 7-39

(a) Support reactions at O for load case 1; (b) support reactions at O for load case 2

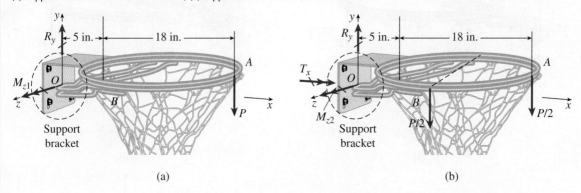

(a)   (b)

Forces at bolt 2—Load case 1: Use the negatives of the reactions at O to find normal and shear forces acting on the bolts. From reaction R_y, downward shear force $P/4 = 100$ lb acts at each of the four bolt locations (Fig. 7-40a). Replace moment M_{z1} [Eq. (a)] with two force couples, each equal to $(N_1)(h)$ (see Fig. 7-40a) so normal force N_1 acts in the $(+x)$ direction at bolt 2 and is computed as

$$N_1 = \frac{M_{z1}}{2h} = \frac{23P}{2(3 \text{ in.})} = \frac{9200 \text{ lb-in.}}{6 \text{ in.}} = 1533 \text{ lb}$$ **(c)**

Forces at bolt 2—Load case 2: Reaction R_y is the same in load cases 1 and 2, so downward shear force $P/4 = 100$ lb acts at bolt 2 (Fig. 7-40b). Replace moment M_{z2} [Eq. (b)] with two force couples (Fig. 7-40b), so the tension force on bolt 2 is

$$N_2 = \frac{M_{z2}}{2h} = \frac{7400 \text{ lb-in.}}{6 \text{ in.}} = 1233 \text{ lb}$$ **(d)**

FIGURE 7-40

(a) Bolt 2 forces for load case 1; (b) bolt 2 forces for load case 2

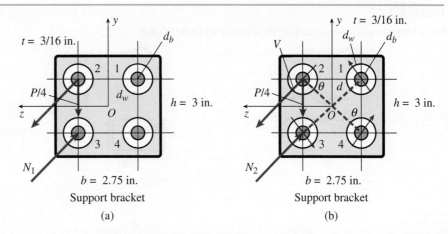

(a)

(b)

Load case 2 also creates a torsional reaction moment T_x [see Eq. (b)] that can be replaced by two counterclockwise force couples each equal to $(V)(d)$ (see Fig. 7-40b) where $d = \sqrt{b^2 + h^2} = 4.0697$ in. Compute the additional in-plane shear force on bolt 2 as

$$V = \frac{T_x}{2d} = \frac{1800 \text{ lb-in.}}{2(4.0697 \text{ in.})} = 221.15 \text{ lb} \qquad \textbf{(e)}$$

The line of action of force V is shown in Fig. 7-40(b) at angle $\theta = \tan^{-1}\left(\frac{b}{h}\right) = 42.51°$. The total in-plane shear force on bolt 2 is the resultant R of forces V and $P/4$ computed as

$$R = \sqrt{(V\cos\theta)^2 + \left(\frac{P}{4} + V\sin\theta\right)^2} = \sqrt{(163.02)^2 + (100 + 149.43)^2} \text{ lb} = 298 \text{ lb} \qquad \textbf{(f)}$$

FIGURE 7-41

Resultant R using parallelogram law

Resultant R also can be found using the parallelogram law, as shown in Fig. 7-41 with $\beta = 14.32°$ and $\gamma = 132.51°$.

2. *Categorize*: Use the forces acting on bolt 2 in simple formulas to compute average stresses in the bolt and on the washer and support bracket at bolt location 2. The five connection stresses of interest are (i) normal stress in bolt (Fig. 7-42a); (ii) shear stress in bolt (Fig. 7-42b); (iii) bearing stress on shank of bolt (Fig. 7-42c); (iv) bearing stress on washer (Fig. 7-42d); and (v) shear stress through bracket on periphery of washer (Fig. 7-42e).

3. *Analyze*: The five connection stresses at bolt location 2 are listed in Table 7-1 for load cases 1 and 2. Numerical dimensions for the bolt, washer, and bracket are $d_b = 3/8$ in., $d_w = 5/8$ in., and $t = 3/16$ in., respectively. Areas needed in stress calculations are

Cross-sectional area of bolt:

$$A_b = \frac{\pi}{4}d_b^2 = 0.1104 \text{ in}^2$$

FIGURE 7-42

Five average stresses in the bolt and on the washer and support bracket

(a)   (b)   (c)

(d)   (e)

Surface area of washer:

$$A_w = \frac{\pi}{4}\left(d_w^2 - d_b^2\right) = 0.1963 \text{ in}^2$$

Cylindrical area through bracket on periphery of washer:

$$A_p = \pi d_w t = 0.3682 \text{ in}^2$$

4. *Finalize*: Shear and bearing stresses on bolt 2 (items b and c in Table 7-1) are increased three-fold for load case 2 when torsional moment T_x [see Eq. (b)] is applied to the bracket. The other three bolt stresses differ by about 25% for the two load cases. The average stresses illustrated in Fig. 7-42 and listed in Table 7-1 are only approximations to the true state of stress at one of the four bolt locations on the support bracket. The true maximum stresses are likely to be higher for a variety of reasons, such as localized stress concentrations, pretensioning of bolts, and impact aspects of the loading. The stress levels computed here are low. If stress values are higher, a more detailed and sophisticated analysis using computer models employing the finite element method may be required.

Table 7-1	**Connection Stress**	**Load Case 1**	**Load Case 2**
Five connection stresses at bolt location 2	a. Normal stress in bolt	$\dfrac{N_1}{A_b} = 13{,}880$ psi	$\dfrac{N_2}{A_b} = 11{,}164$ psi
	b. Shear stress in bolt	$\dfrac{P}{4A_b} = 905$ psi	$\dfrac{R}{A_b} = 2698$ psi
	c. Bearing stress on shank of bolt	$\dfrac{P}{4d_b t} = 1422$ psi	$\dfrac{R}{d_b t} = 4238$ psi
	d. Bearing stress on washer	$\dfrac{N_1}{A_w} = 7808$ psi	$\dfrac{N_2}{A_w} = 6280$ psi
	e. Shear stress through bracket on periphery of washer	$\dfrac{N_1}{A_p} = 4164$ psi	$\dfrac{N_2}{A_p} = 3349$ psi

7.7 Allowable Stresses and Allowable Loads

Engineers design a seemingly endless variety of objects to serve the basic needs of society. These needs include housing, agriculture, transportation, communication, and many other aspects of modern life. Factors to be considered in design include functionality, strength, appearance, economics, and environmental effects. However, when studying mechanics of materials, our principal design interest is **strength**, that is, *the capacity of the object to support or transmit loads*. Objects that must sustain loads include buildings, machines, containers, trucks, aircraft, ships, and the like. For simplicity, all such objects are referred to as **structures**; thus, a *structure is any object that must support or transmit loads*.

Factors of Safety

Strength is the ability of a structure to resist loads. *The actual strength of a structure must exceed the required strength.* The ratio of the actual strength to the required strength is called the **factor of safety** n:

$$\text{Factor of safety } n = \frac{\text{Actual strength}}{\text{Required strength}} \tag{7-19}$$

Of course, the factor of safety must be greater than 1.0 if failure is to be avoided. Depending upon the circumstances, factors of safety from slightly above 1.0 to as much as 10 are used.

The incorporation of factors of safety into design is not a simple matter, because both strength and failure have many different meanings. Strength may be measured by the load-carrying capacity of a structure, or it may be measured by the stress in the material. Failure may mean the fracture and complete collapse of a structure, or it may mean that the deformations have become so large that the structure can no longer perform its intended functions. The latter kind of failure may occur at loads much smaller than those that cause actual collapse.

The determination of a factor of safety must also take into account such matters as the following: probability of accidental overloading of the structure by loads that exceed the design loads; types of loads (static or dynamic); whether the loads are applied once or are repeated; how accurately the loads are known; possibilities for fatigue failure; inaccuracies in construction; variability in the quality of workmanship; variations in properties of materials; deterioration due to corrosion or other environmental effects; accuracy of the methods of analysis; whether failure is gradual (ample warning) or sudden (no warning); consequences of failure (minor damage or major catastrophe); and other such considerations. If the factor of safety is too low, the likelihood of failure will be high and the structure will be unacceptable; if the factor is too large, the structure will be wasteful of materials and perhaps unsuitable for its function (for instance, it may be too heavy).

Because of these complexities and uncertainties, factors of safety must be determined on a probabilistic basis. They usually are established by groups of experienced engineers who write the codes and specifications used by other designers, and sometimes they are even enacted into law. The provisions of codes and specifications are intended to provide reasonable levels of safety without unreasonable costs.

In aircraft design, it is customary to speak of the **margin of safety** rather than the factor of safety. The margin of safety is defined as the factor of safety minus one:

$$\text{Margin of safety} = n - 1 \tag{7-20}$$

Margin of safety is often expressed as a percent, in which case the value given above is multiplied by 100. Thus, a structure having an actual strength that is 1.75 times the required strength has a factor of safety of 1.75 and a margin of safety of 0.75 (or 75%). When the margin of safety is reduced to zero or less, the structure (presumably) will fail.

Allowable Stresses

Factors of safety are defined and implemented in various ways. For many structures, it is important that the material remain within the linearly elastic range in order to avoid permanent deformations when the loads are removed. Under these conditions, the factor of safety is established with respect to yielding of the structure. Yielding begins when the yield stress is reached at *any* point within the structure. Therefore, by applying a factor of safety with respect to the yield stress (or yield strength), you obtain an **allowable stress** (or *working stress*) that must not be exceeded anywhere in the structure. Thus,

$$\text{Allowable stress} = \frac{\text{Yield strength}}{\text{Factor of safety}} \tag{7-21}$$

or for tension and shear, respectively,

$$\sigma_{\text{allow}} = \frac{\sigma_Y}{n_1} \quad \text{and} \quad \tau_{\text{allow}} = \frac{\tau_Y}{n_2} \tag{7-22a,b}$$

in which σ_Y and τ_Y are the yield stresses and n_1 and n_2 are the corresponding factors of safety. In building design, a typical factor of safety with respect to yielding in tension is 1.67; thus, a mild steel having a yield stress of 36 ksi has an allowable stress of 21.6 ksi.

Sometimes the factor of safety is applied to the **ultimate stress** instead of the yield stress. This method is suitable for brittle materials, such as concrete and some plastics, and for materials without a clearly defined yield stress, such as wood and high-strength steels. In these cases, the allowable stresses in tension and shear are

$$\sigma_{\text{allow}} = \frac{\sigma_U}{n_3} \quad \text{and} \quad \tau_{\text{allow}} = \frac{\tau_U}{n_4} \tag{7-23}$$

in which σ_U and τ_U are the ultimate stresses (or ultimate strengths). Factors of safety with respect to the ultimate strength of a material are usually larger than those based upon yield strength. In the case of mild steel, a factor of safety of 1.67 with respect to yielding corresponds to a factor of approximately 2.8 with respect to the ultimate strength.

Allowable Loads

After the allowable stress has been established for a particular material and structure, the **allowable load** on that structure can be determined. The relationship between the allowable load and the allowable stress depends upon the type of structure. This chapter is concerned only with the most elementary kinds of structures, namely, bars in tension or compression and pins (or bolts) in direct shear and bearing.

In these kinds of structures, the stresses are uniformly distributed (or at least *assumed* to be uniformly distributed) over an area. For instance, in the case of a bar in tension, the stress is uniformly distributed over the cross-sectional area provided the resultant axial force acts through the centroid of the cross section. The same is true of a bar in compression provided the bar is not subject to buckling. In the case of a pin subjected to shear, consider only the average shear stress on the cross section, which is equivalent to assuming that the shear stress is uniformly distributed. Similarly, consider only an average value of the bearing stress acting on the projected area of the pin.

Therefore, in all four of the preceding cases, the **allowable load** (also called the *permissible load* or the *safe load*) is equal to the allowable stress times the area over which it acts:

$$\text{Allowable load} = (\text{Allowable stress})(\text{Area}) \tag{7-24}$$

For bars in direct *tension* and *compression* (no buckling), this equation becomes

$$P_{\text{allow}} = \sigma_{\text{allow}} A \tag{7-25}$$

in which σ_{allow} is the permissible normal stress and A is the cross-sectional area of the bar. If the bar has a hole through it, the *net area* is normally used when the bar is in tension. The **net area** is the gross cross-sectional area minus the area removed by the hole. For compression, the gross area may be used if the hole is filled by a bolt or pin that can transmit the compressive stresses.

For pins in *direct shear*, Eq. (7-24) becomes

$$P_{\text{allow}} = \tau_{\text{allow}} A \tag{7-26}$$

in which τ_{allow} is the permissible shear stress and A is the area over which the shear stresses act. If the pin is in single shear, the area is the cross-sectional area of the pin; in double shear, it is twice the cross-sectional area.

Finally, the permissible load based upon *bearing* is

$$P_{allow} = \sigma_b A_b \tag{7-27}$$

in which σ_b is the allowable bearing stress and A_b is the projected area of the pin or other surface over which the bearing stresses act.

The following example illustrates how allowable loads are determined when the allowable stresses for the material are known.

Example 7-8

A steel bar serving as a vertical hanger to support heavy machinery in a factory is attached to a support by the bolted connection shown in Fig. 7-43. Two clip angles (thickness $t_c = 9.5$ mm) are fastened to an upper support by bolts 1 and 2 each with a diameter of 12 mm; each bolt has a washer with a diameter of $d_w = 28$ mm. The main part of the hanger is attached to the clip angles by a single bolt (bolt 3 in Fig. 7-43a) with a diameter of $d = 25$ mm. The hanger has a rectangular cross section with a width of $b_1 = 38$ mm and thickness of $t = 13$ mm, but at the bolted connection, the hanger is enlarged to a width of $b_2 = 75$ mm. Determine the allowable value of the tensile load P in the hanger based upon the following considerations.

(a) The allowable tensile stress in the main part of the hanger is 110 MPa.

(b) The allowable tensile stress in the hanger at its cross section through the bolt 3 hole is 75 MPa. (The permissible stress at this section is lower because of the stress concentrations around the hole.)

(c) The allowable bearing stress between the hanger and the shank of bolt 3 is 180 MPa.

(d) The allowable shear stress in bolt 3 is 45 MPa.

(e) The allowable normal stress in bolts 1 and 2 is 160 MPa.

(f) The allowable bearing stress between the washer and the clip angle at either bolt 1 or 2 is 65 MPa.

(g) The allowable shear stress through the clip angle at bolts 1 and 2 is 35 MPa.

Solution:

Use the four-step problem-solving approach to find the allowable value of the tensile load P in the hanger based upon a variety of different allowable stresses in the different connection components.

1. *Conceptualize*: Start by sketching a series of free-body diagrams to find the forces acting on each connection component. Express the force on each component in terms of an allowable stress times the associated area upon which it acts. This force is the allowable value of applied load P for that stress condition. Each of the seven stress states [(a)–(g) in the problem statement] and the associated applied load are illustrated in this example's figures; each is adjacent to the corresponding calculations in step (3).

2. *Categorize*: Compute seven different values of the allowable load P, each based on an allowable stress and a corresponding area. The minimum value of load P will control.

   Numerical data for the hanger connection design shown in Fig. 7-43 are as follows.

   Connection component dimensions:

   $$t_c = 9.5 \text{ mm} \quad t = 13 \text{ mm} \quad b_1 = 38 \text{ mm} \quad b_2 = 75 \text{ mm}$$
   $$d_1 = 12 \text{ mm} \quad d = 25 \text{ mm} \quad d_w = 28 \text{ mm}$$

   Allowable stresses:

   $$\sigma_a = 110 \text{ MPa} \quad \sigma_{a3} = 75 \text{ MPa} \quad \sigma_{ba3} = 180 \text{ MPa} \quad \tau_{a3} = 45 \text{ MPa}$$
   $$\tau_{a1} = 35 \text{ MPa} \quad \sigma_{a1} = 160 \text{ MPa} \quad \sigma_{ba1} = 65 \text{ MPa}$$

3. *Analyze*:

   Part (a): Find the allowable load based upon the stress in the main part of the hanger (Fig. 7-43c). This is equal to the allowable stress in tension (110 MPa) times the cross-sectional area of the hanger [Eq. (7-25)]:

   $$P_a = \sigma_a b_1 t = (110 \text{ MPa})(38 \text{ mm} \times 13 \text{ mm}) = 54.3 \text{ kN}$$

   A load greater than this value will overstress the main part of the hanger (that is, the actual stress will exceed the allowable stress), thereby reducing the factor of safety.

FIGURE 7-43

Example 7-8: Vertical hanger subjected to a tensile load P: (a) front view of bolted connection and (b) side view of connection

(a)   (b)

FIGURE 7-43
(Continued)

Example 7-8: Vertical hanger subjected to a tensile load P: (c) and (d) tensile stresses on hanger; (e) bearing stress on bolt

$P_{\max(a)}$
(c)

$P_{\max(b)}$
(d)

$P_{\max(c)}$
(e)

Part (b): Find the allowable load based upon the allowable tensile stress (75 MPa) in the hanger at its cross section through the bolt 3 hole.

At the cross section of the hanger through the bolt hole (Fig. 7-43d), make a similar calculation but with a different allowable stress and a different area. The net cross-sectional area (that is, the area that remains after the hole is drilled through the bar) is equal to the net width times the thickness. The net width is equal to the gross width b_2 minus the diameter d of the hole. Thus, the equation for the allowable load P_b at this section is

$$P_b = \sigma_{a3}(b_2 - d)t = (75 \text{ MPa})(75 \text{ mm} - 25 \text{ mm})(13 \text{ mm}) = 48.8 \text{ kN}$$

Part (c): Now find the allowable load based upon the allowable bearing stress (180 MPa) between the hanger and the shank of bolt 3.

The allowable load based upon bearing between the hanger and the bolt (Fig. 7-43e) is equal to the allowable bearing stress times the bearing area. The bearing area is the projection of the actual contact area, which is equal to the bolt diameter times the thickness of the hanger. Therefore, the allowable load [Eq. (7-27)] is

$$P_c = \sigma_{ba3}dt = 58.5 \text{ kN} = (180 \text{ MPa})(25 \text{ mm})(13 \text{ mm}) = 58.5 \text{ kN}$$

Part (d): Determine the allowable load based upon the allowable shear stress (45 MPa) in bolt 3.

The allowable load P_d based upon shear in the bolt (Fig. 7-43f) is equal to the allowable shear stress times the shear area [Eq. (7-26)]. The shear area is twice the area of the bolt because the bolt is in double shear; thus,

$$P_d = 2\tau_{a3}\left(\frac{\pi}{4}d^2\right) = 2(45 \text{ MPa})\left[\frac{\pi}{4}(25 \text{ mm})^2\right] = 44.2 \text{ kN}$$

Part (e): Find the allowable load based upon the allowable normal stress (160 MPa) in bolts 1 and 2.

The allowable normal stress in bolts 1 and 2 is 160 MPa. Each bolt carries one half of the applied load P (see Fig. 7-43g). The allowable total load P_e is the product of the allowable normal stress in the bolt and the sum of the cross-sectional areas of bolts 1 and 2:

$$P_e = \sigma_{a1}(2)\left(\frac{\pi}{4}d_1^2\right) = (160 \text{ MPa})(2)\left[\frac{\pi}{4}(12 \text{ mm})^2\right] = 36.2 \text{ kN}$$

Part (f): Now find the allowable load based upon the allowable bearing stress (65 MPa) between the washer and the clip angle at either bolt 1 or 2.

The allowable bearing stress between the washer and the clip angle at either bolt 1 or 2 is 65 MPa. Each bolt (1 or 2) carries one-half of the applied load P (see Fig. 7-43h). The bearing area here is the ring-shaped circular area of the washer (the washer is assumed to fit snugly against the bolt). The allowable total load P_f is the allowable bearing stress on the washer times twice the area of the washer:

$$P_f = \sigma_{ba1}(2)\left[\frac{\pi}{4}(d_w^2 - d_1^2)\right] = (65 \text{ MPa})(2)\left\{\frac{\pi}{4}[(28 \text{ mm})^2 - (12 \text{ mm})^2]\right\}$$
$$= 65.3 \text{ kN}$$

Part (g): Finally, determine the allowable load based upon the allowable shear stress (35 MPa) through the clip angle at bolts 1 and 2.

The allowable shear stress through the clip angle at bolts 1 and 2 is 35 MPa. Each bolt (1 or 2) carries one half of the applied load P (see Fig. 7-43i). The shear area at each bolt is equal to the circumference of the hole ($\pi \times d_w$) times the thickness of the clip angle (t_c).

The allowable total load P_g is the the allowable shear stress times twice the shear area:

$$P_g = \tau_{a1}(2)(\pi d_w t_c) = (35 \text{ MPa})(2)(\pi \times 28 \text{ mm} \times 9.5 \text{ mm}) = 58.5 \text{ kN}$$

4. *Finalize*: All seven conditions were used to find the allowable tensile loads in the hanger. Comparing the seven preceding results shows that the smallest value of the load is $P_{\text{allow}} = 36.2$ kN. This load is based upon normal stress in ← bolts 1 and 2 [see part (e)] and is the allowable tensile load for the hanger.

A more refined analysis that includes the weight of the entire hanger assembly can be carried out as shown in Example 7-2. As in Example 7-7, note that these computed stresses are average values only and do not include localized effects such as stress concentrations around bolt holes.

7.8 Design for Axial Loads and Direct Shear

The preceding section discussed the determination of allowable loads for simple structures, and earlier sections showed how to find the stresses, strains, and deformations of bars. The determination of such quantities is known as **analysis**. In the context of mechanics of materials, analysis consists of determining the *response* of a structure to loads, temperature changes, and other physical actions. The response of a structure means the stresses, strains, and deformations produced by the loads.

Response also refers to the load-carrying capacity of a structure; for instance, the allowable load on a structure is a form of response.

A structure is said to be *known* (or *given*) when there is a complete physical description of the structure, that is, all of its *properties* are known. The properties of a structure include the types of members and how they are arranged, the dimensions of all members, the types of supports and where they are located, the materials used, and the properties of the materials. Thus, when analyzing a structure, *the properties are given, and the response is to be determined.*

The inverse process is called **design**. When designing a structure, *you must determine the properties of the structure in order that the structure will support the loads and perform its intended functions.* For instance, a common design problem in engineering is to determine the size of a member to support given loads. Designing a structure is usually a much lengthier and more difficult process than analyzing it—indeed, analyzing a structure, often more than once, is typically part of the design process.

This section covers design in its most elementary form by calculating the required sizes of simple tension and compression members as well as pins and bolts loaded in shear. In these cases, the design process is quite straightforward. Knowing the loads to be transmitted and the allowable stresses in the materials, you can calculate the required areas of members from the following general relationship [compare with Eq. (7-24)]:

$$\text{Required area} = \frac{\text{Load to be transmitted}}{\text{Allowable stress}} \qquad \textbf{(7-28)}$$

Apply this equation to any structure where the stresses are uniformly distributed over the area. (The use of this equation for finding the size of a cable in tension and the size of a pin in shear is illustrated in Example 7-9.)

In addition to **strength** considerations, as shown by Eq. (7-28), the design of a structure is likely to involve **stiffness** and **stability**. Stiffness refers to the ability of the structure to resist changes in shape (for instance, to resist stretching, bending, or twisting), and stability refers to the ability of the structure to resist buckling under compressive stresses. Limitations on stiffness are sometimes necessary to prevent excessive deformations, such as large deflections of a beam that might interfere with its performance. Buckling is the principal consideration in the design of columns, which are slender compression members (Chapter 14).

Another part of the design process is **optimization**, which is the task of designing the best structure to meet a particular goal, such as minimum weight. For instance, there are many structures that can support a given load, but in some circumstances the best structure is the lightest one. Of course, a goal such as minimum weight usually must be balanced against more general considerations, including the aesthetic, economic, environmental, political, and technical aspects of the particular design project.

When analyzing or designing a structure, the forces that act on it are **loads** and **reactions**. Loads are *active forces* that are applied to the structure by some external cause, such as gravity, water pressure, wind, and earthquake ground motion. Reactions are *passive forces* that are induced at the supports of the structure—their magnitudes and directions are determined by the nature of the structure itself. Thus, reactions must be calculated as part of the analysis, whereas loads are known in advance.

Example 7-9

Continuous cable *ADB* runs over a small frictionless pulley at *D* to support beam *OABC*, which is part of an entrance canopy for a building (see Fig. 7-44). Load $P = 1000$ lb is applied at the end of the canopy at *C*. Assume that the canopy segment has weight $W = 1700$ lb.

(a) Find cable force *T* and pin support reactions at *O* and *D*.

(b) Find the required cross-sectional area of cable *ADB* if the allowable normal stress is 18 ksi.

(c) Determine the required diameter of the pins at *O*, *A*, *B*, and *D* if the allowable stress in shear is 12 ksi.

(*Note*: The pins at *O*, *A*, *B*, and *D* are in double shear. Also, consider only load *P* and the weight *W* of the canopy; disregard the weight of cable *ADB*.)

FIGURE 7-44

(a) Inclined canopy at entrance to building; (b) two-dimensional model of one beam and supporting cable

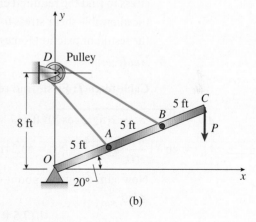

(a)   (b)

Solution:

Use the four-step problem-solving approach.

1. *Conceptualize*: Begin with a free-body diagram of beam *OABC* (Fig. 7-45a). Also sketch free-body diagrams of the entire structure (Fig. 7-45b) and of joint *D* alone (Fig. 7-45c). Show cable force *T* and all applied and reaction force components.

FIGURE 7-45

Free-body diagrams of
(a) beam *OABC*; (b) beam and
cable structure; (c) joint *D*

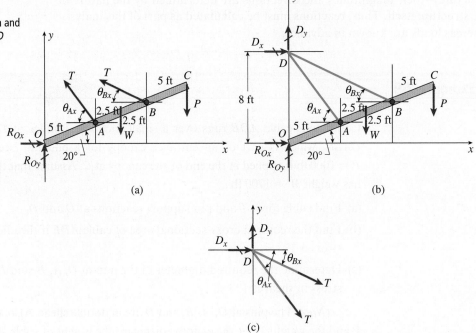

(a)

(b)

(c)

2. *Categorize*: First, use the free-body diagram of beam *OABC* (Fig. 7-45a) to find cable force *T* and reaction force components at *O*. Then use Fig. 7-45b or Fig. 7-45c to find reaction forces at *D*. Use cable force *T* and the allowable normal stress to find the required cross-sectional area of the cable. Also use force *T* and the allowable shear stress to find the required diameter of the pins at *A* and *B*. Use the resultant reaction forces at *O* and *D* to find pin diameters at these locations.

3. *Analyze*:

Cable force *T*: First, find required distances and angles in Fig. 7-45a:

$$AD = \sqrt{5^2 + 8^2 - 2(5)(8)\cos 70°} \text{ ft} = 7.851 \text{ ft} \quad BD = \sqrt{10^2 + 8^2 - 2(10)(8)\cos 70°} \text{ ft} = 10.454 \text{ ft}$$

$$\theta_{Ax} = \sin^{-1}\left(\frac{8 \text{ ft}}{AD}\sin 70°\right) - 20° = 53.241° \qquad \theta_{Bx} = \sin^{-1}\left(\frac{8 \text{ ft}}{BD}\sin 70°\right) - 20° = 25.983°$$

Now sum moments about *O* in Fig. 7-45a to find tension *T* in continuous cable *ADB*:

$$T = \frac{W(7.5 \text{ ft})(\cos 20°) + P(15 \text{ ft})(\cos 20°)}{d_1 + d_2} = 2177 \text{ lb} \qquad \text{◀ (a)}$$

where

$$d_1 = \cos\theta_{Ax}(8 \text{ ft} - AD\sin\theta_{Ax}) + \sin\theta_{Ax}(5 \text{ ft}\cos 20°) = 4.788 \text{ ft}$$
$$d_2 = \cos\theta_{Bx}(8 \text{ ft} - BD\sin\theta_{Bx}) + \sin\theta_{Bx}(10 \text{ ft}\cos 20°) = 7.191 \text{ ft}$$

Reaction force at *O*: Sum forces in Fig. 7-45a to find reaction force components at *O*:

$$\Sigma F_x = 0 \quad R_{Ox} = T(\cos\theta_{Ax} + \cos\theta_{Bx}) = 3260 \text{ lb}$$

$$\Sigma F_y = 0 \quad R_{Oy} = -T(\sin\theta_{Ax} + \sin\theta_{Bx}) + W + P = 2 \text{ lb}$$

The resultant reaction force at *O* is $R_{Ores} = \sqrt{R_{Ox}^2 + R_{Oy}^2} = 3260$ lb   **(b)**

Reaction force at *D*: Sum forces in Fig. 7-45c to find reaction force components at *D*:

$$\Sigma F_x = 0 \quad D_x = -T(\cos\theta_{Ax} + \cos\theta_{Bx}) = -3260 \text{ lb}$$

$$\Sigma F_y = 0 \quad D_y = T(\sin\theta_{Ax} + \sin\theta_{Bx}) = 2698 \text{ lb}$$

The resultant reaction force at *D* is

$$D_{\text{res}} = \sqrt{D_x^2 + D_y^2} = 4231 \text{ lb} \qquad \textbf{(c)}$$

Cross-sectional area of cable *ADB*: Use the allowable normal stress of 18 ksi and cable force $T = 2177$ lb [Eq. (a)] to find the required cross-sectional area of the cable:

$$A_{\text{cable}} = \frac{2177 \text{ lb}}{18 \text{ ksi}} = 0.121 \text{ in}^2$$

Required diameter of the pins at *O*, *A*, *B*, and *D*: All pins are in double shear. The allowable shear stress is $\tau_{\text{allow}} = 12$ ksi. Required diameters of each pin are computed as

Pins A, B: $\quad A_{\text{reqd}} = \dfrac{T}{2\tau_{\text{allow}}} = \dfrac{2177 \text{ lb}}{2(12 \text{ ksi})} = 0.091 \text{ in}^2 \quad$ so $d = \sqrt{\dfrac{4}{\pi}(0.091 \text{ in}^2)} = 0.340$ in.

Pin O: $\quad A_{\text{reqd}} = \dfrac{R_{Ores}}{2\tau_{\text{allow}}} = \dfrac{3260 \text{ lb}}{2(12 \text{ ksi})} = 0.136 \text{ in}^2 \quad$ so $d = \sqrt{\dfrac{4}{\pi}(0.136 \text{ in}^2)} = 0.416$ in.

Pin D: $\quad A_{\text{reqd}} = \dfrac{D_{\text{res}}}{2\tau_{\text{allow}}} = \dfrac{4231 \text{ lb}}{2(12 \text{ ksi})} = 0.176 \text{ in}^2 \quad$ so $d = \sqrt{\dfrac{4}{\pi}(0.176 \text{ in}^2)} = 0.474$ in.

4. *Finalize*: In practice, other loads besides the weight of the canopy would have to be considered before making a final decision about the sizes of the cables and pins. Loads that could be important include wind loads, earthquake loads, and the weights of objects that might have to be supported temporarily by the structure. In addition, if cables *AD* and *BD* are *separate cables* (instead of one continuous cable *ADB*), the forces in the two cables are not equal in magnitude. The structure is now *statically indeterminate*, and the cable forces and the reactions at *O* and *D* cannot be determined using the equations of static equilibrium alone. Problems of this type are discussed in Chapter 8, Section 8.4 (see Example 8-7).

Chapter 7 covered mechanical properties of construction materials. Normal stresses and strains in bars loaded by centroidal axial loads were computed. Shear stresses and strains (as well as bearing stresses) in pin connections used to assemble simple structures, such as trusses, were evaluated. Allowable levels of stress were calculated from appropriate factors of safety and used to set allowable loads that could be applied to the structure.

Some of the major concepts presented in this chapter are given here.

1. The principal objective of mechanics of materials is to determine the **stresses**, **strains**, and **displacements** in structures and their components due to the loads acting on them. These components include bars with axial loads, shafts in torsion, beams in bending, and columns in compression.

2. Prismatic bars subjected to tensile or compressive loads acting through the centroid of their cross section (to avoid bending) experience **normal stress** (σ) and **strain** (ε):

$$\sigma = \frac{P}{A} \qquad \varepsilon = \frac{\delta}{L}$$

and either extension or contraction proportional to their lengths. These stresses and strains are **uniform** except near points of load application where high localized stresses, or **stress concentrations**, occur.

3. The **mechanical behavior** of various materials was displayed in a stress-strain diagram. **Ductile** materials such as mild steel have an initial linear relationship between normal stress and strain up to the **proportional limit** and are **linearly elastic** with stress and strain related by **Hooke's law**:

$$\sigma = E\varepsilon$$

They also have a well-defined yield point. Other ductile materials, such as aluminum alloys, typically do not have a clearly definable yield point, so an arbitrary yield stress is determined using the **offset method**.

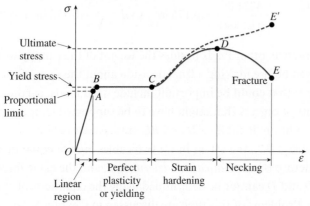

Mild steel

4. Materials that fail in tension at relatively low values of strain (such as concrete, stone, cast iron, glass ceramics, and a variety of metallic alloys) are classified as **brittle**. Brittle materials fail with only little elongation after the proportional limit.

5. If a material remains within the elastic range, it can be loaded, unloaded, and loaded again without significantly changing its behavior. However, when loaded into the plastic range, the internal structure of the material is altered and its properties change. Loading and unloading behavior of materials depends on their **elasticity** and **plasticity** properties such as the **elastic limit** and the possibility of **permanent set** (residual strain). Sustained loading over time may lead to **creep** and **relaxation**.

6. Axial elongation of bars loaded in tension is accompanied by lateral contraction; the ratio of lateral strain to normal strain is known as **Poisson's ratio** (v):

$$v = -\frac{\text{lateral strain}}{\text{axial strain}} = -\frac{\varepsilon'}{\varepsilon}$$

Poisson's ratio remains constant throughout the linearly elastic range, provided the material is homogeneous and isotropic. Most of the examples and problems in the text are solved with the assumption that the material is linearly elastic, homogeneous, and **isotropic**.

Lateral contraction

7. **Normal** stresses (σ) act perpendicular to the surface of the material, and **shear stresses** (τ) act tangential to the surface. In bolted connections between plates, the bolts are subjected to either single or double shear (τ_{aver}) where the average shear stress is

$$\tau_{\text{aver}} = \frac{V}{A}$$

Average **bearing** stresses (σ_b) act on the rectangular projected area (A_b) of the actual curved contact surface between a bolt and plate:

$$\sigma_b = \frac{F_b}{A_b}$$

Brittle material

Partially elastic behavior

$P_{\max(a)}$

Normal stresses

$P_{\max(c)}$

Bearing stresses on a bolt passing through a bar

$P_{\max(d)}$

Shear stresses on a bolt passing through a bar

8. An element of material acted on by only shear stresses and strains is in a state of stress referred to as **pure shear**. Shear strain (γ) is a measure of the distortion or change in shape of the element in pure shear. Hooke's law in shear relates shear stress (τ) to shear strain by the shearing modulus of elasticity G:

$$\tau = G\gamma$$

Moduli E and G are not independent elastic properties of the material. Compute modulus G from E using Poisson's ratio:

$$G = \frac{E}{2(1+\nu)}$$

9. **Strength** is the capacity of a structure or component to support or transmit loads. **Factors of safety** relate actual to required strength of structural members and account for a variety of uncertainties, such as variations in material properties, uncertain magnitudes or distributions of loadings, and probability of accidental overload. Because of these uncertainties, factors of safety ($n_1, n_2, n_3,$ and n_4) must be determined using probabilistic methods.

10. Yield or ultimate-level stresses are divided by factors of safety to produce allowable values for use in design. For **ductile** materials,

$$\sigma_{\text{allow}} = \frac{\sigma_Y}{n_1} \qquad \tau_{\text{allow}} = \frac{\tau_Y}{n_2}$$

while for **brittle** materials,

$$\sigma_{\text{allow}} = \frac{\sigma_U}{n_3} \qquad \tau_{\text{allow}} = \frac{\tau_U}{n_4}$$

A typical value of n_1 and n_2 is 1.67, while n_3 and n_4 might be 2.8.

For a pin-connected member in axial tension, the **allowable load** depends on the allowable stress times the appropriate area (such as the net cross-sectional area for bars acted on by centroidal tensile loads, cross-sectional area of pin for pins in shear, and projected area for bolts in bearing). If the bar is in compression, the net cross-sectional area need not be used, but buckling may be an important consideration.

11. **Design** is the iterative process by which the appropriate size of structural members is determined to meet a variety of both **strength and stiffness requirements**. Incorporation of factors of safety into design is not a simple matter because both strength and failure have many different meanings.

PROBLEMS

7.2 Normal Stress and Strain

Introductory Problems

7.2-1 A hollow circular post ABC (see figure) supports a load $P_1 = 1700$ lb acting at the top. A second load P_2 is uniformly distributed around the cap plate at B. The diameters and thicknesses of the upper and lower parts of the post are $d_{AB} = 1.25$ in., $t_{AB} = 0.5$ in., $d_{BC} = 2.25$ in., and $t_{BC} = 0.375$ in., respectively.

(a) Calculate the normal stress σ_{AB} in the upper part of the post.

(b) If you want the lower part of the post to have the same compressive stress as the upper part, what should be the required magnitude of load P_2?

(c) If P_1 remains at 1700 lb and P_2 is set at 2260 lb, what new thickness of BC will result in the same compressive stress in both parts?

PROBLEM 7.2-1

7.2-2 A circular nylon pipe supports a downward load $P_A = 10$ kN, which is uniformly distributed around a cap plate at the top of the lower pipe. A second load $P_B = 20$ kN is applied upward at the top. The inner and outer diameters of the upper and lower parts of the pipe are $d_1 = 50$ mm, $d_2 = 60$ mm, $d_3 = 55$ mm, and $d_4 = 65$ mm, respectively. The bottom pipe has length 400 mm and the upper pipe has length 300 mm.

(a) Calculate the axial normal stress in each pipe segment.

(b) Calculate the strain in each pipe segment if the upward displacement at B on the upper pipe is 3.29 mm and the upward displacement at A on the bottom part is 1.25 mm.

PROBLEM 7.2-2

7.2-3 A circular tube AB is fixed at one end and free at the other end. The tube is subjected to axial force at joint B. If the outer diameter of the tube is 3 in. and the thickness is ¾ in., calculate the maximum normal stress in the tube.

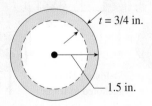

PROBLEM 7.2-3

497

7.2-4 A force P of 70 N is applied by a rider to the front hand brake of a bicycle (P is the resultant of an evenly distributed pressure). As the hand brake pivots at A, a tension T develops in the 460-mm long brake cable ($A_e = 1.075$ mm^2), which elongates by $\delta = 0.214$ mm. Find the normal stress δ and strain ϵ in the brake cable.

Brake cable, $L = 460$ mm   Hand brake pivot A

37.5 mm A

T

P (Resultant of distributed pressure)

50 mm

100 mm

Uniform hand brake pressure

PROBLEM 7.2-4

Representative Problems

7.2-5 A bicycle rider wants to compare the effectiveness of cantilever hand brakes (see figure part a) versus V brakes (figure part b).

(a) Calculate the braking force R_B at the wheel rims for each of the bicycle brake systems shown. Assume that all forces act in the plane of the figure and that cable tension $T = 45$ lb. Also, what is the average compressive normal stress σ_c on the brake pad ($A = 0.625$ in^2)?

T

D

$T_{DC} = T_{DE}$

$45°$

T_{DE}

4 in.

T_{DC_v}

E

T_{DE}

$90°$

T_{DC}

T_{DCh}

C

5 in.

2 in.

F

B

1 in.

R_B

1 in.

H_A

G

A

Pivot points anchored to frame

V_A

(a) Cantilever brakes

(b) For each braking system, what is the stress in the brake cable if the effective cross-sectional area is 0.00167 in^2?

Hint: Because of symmmetry, use only the right half of each figure in your analysis.

T

4 in.

C

D

4.25 in.

E

B

1 in.

F

Pivot points anchored to frame

A

(b) V brakes

PROBLEM 7.2-5

7.2-6 A circular aluminum tube with a lenth of $L = 420$ mm is loaded in compression by forces P (see figure). The hollow segment of length $L/3$ has outside and inside diameters of 60 mm and 35 mm, respectively. The solid segment of length $2L/3$ has a diameter of 60 mm. A strain gage is placed on the outside of the hollow segment of the bar to measure normal strains in the longitudinal direction.

(a) If the measured strain in the hollow segment is $\varepsilon_h = 470 \times 10^{-6}$, what is the strain ε_s in the solid part? *Hint:* The strain in the solid segment is equal to that in the hollow segment multiplied by the ratio of the area of the hollow to that of the solid segment.

(b) What is the overall shortening δ of the bar?

(c) If the compressive stress in the bar cannot exceed 48 MPa, what is the maximum permissible value of load P?

PROBLEM 7.2-6

7.2-7 The cross section of a concrete corner column that is loaded uniformly in compression is shown in the figure. A circular pipe chase cut-out of 10 in. in diameter runs the height of the column (see figure).

(a) Determine the average compression stress σ_c in the concrete if the load is equal to 3500 kips.

(b) Determine the coordinates x_c and y_c of the point where the resultant load must act in order to produce uniform normal stress in the column.

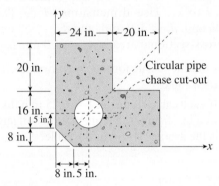

PROBLEM 7.2-7

7.2-8 A car weighing 130 kN when fully loaded is pulled slowly up a steep inclined track by a steel cable (see figure). The cable has an effective cross-sectional area of 490 mm², and the angle α of the incline is 30°.

(a) Calculate the tensile stress σ_t in the cable.

(b) If the allowable stress in the cable is 150 MPa, what is the maximum acceptable angle of the incline for a fully loaded car?

PROBLEM 7.2-8

7.2-9 Two steel wires support a moveable overhead camera weighing $W = 28$ lb (see figure part a) used for close-up viewing of field action at sporting events. At some instant, wire 1 is at an angle $\alpha = 22°$ to the horizontal and wire 2 is at angle $\beta = 40°$. Wires 1 and 2 have diameters of 30 and 35 mils, respectively. (Wire diameters are often expressed in mils; one mil equals 0.001 in.)

(a) Determine the tensile stresses σ_1 and σ_2 in the two wires.

(b) If the stresses in wires 1 and 2 must be the same, what is the required diameter of wire 1?

(c) To stabilize the camera for windy outdoor conditions, a third wire is added (see figure part b). Assume the three wires meet at a common point coordinates (0, 0, 0) above the camera at the instant shown in figure part b. Wire 1 is attached to a support at coordinates (75 ft, 48 ft, 70 ft). Wire 2 is supported at (−70 ft, 55 ft, 80 ft). Wire 3 is supported at (−10 ft, −85 ft, 75 ft). Assume that all three wires have a diameter of 30 mils. Find the tensile stresses in all three wires.

Plan view of camera suspension system

PROBLEM 7.2-9

7.2-10 A long retaining wall is braced by wood shores set at an angle of 30° and supported by concrete thrust blocks, as shown in the first part of the figure. The shores are evenly spaced at 3 m apart.

For analysis purposes, the wall and shores are idealized as shown in the second part of the figure. Note that the base of the wall and both ends of the shores are assumed to be pinned. The pressure of the soil against the wall is assumed to be triangularly distributed, and the resultant force acting on a 3-meter length of the walls is $F = 190$ kN.

If each shore has a 150 mm × 150 mm square cross section, what is the compressive stress σ_c in the shores?

PROBLEM 7.2-11

PROBLEM 7.2-10

7.2-11 A pickup truck tailgate supports a crate where $W_C = 150$ lb, as shown in the figure. The tailgate weighs $W_T = 60$ lb and is supported by two cables (only one is shown in the figure). Each cable has an effective cross-sectional area $A_e = 0.017$ in.²

(a) Find the tensile force T and normal stress σ in each cable.

(b) If each cable elongates $\delta = 0.01$ in. due to the weight of both the crate and the tailgate, what is the average strain in the cable?

7.2-12 Solve the preceding problem if the mass of the tailgate is $M_T = 27$ kg and that of the crate is $M_C = 68$ kg. Use dimensions $H = 305$ mm, $L = 406$ mm, $d_C = 460$ mm, and $d_T = 350$ mm. The cable cross-sectional area is $A_e = 11.0$ mm².

(a) Find the tensile force T and normal stress σ in each cable.

(b) If each cable elongates $\delta = 0.25$ mm due to the weight of both the crate and the tailgate, what is the average strain in the cable?

PROBLEM 7.2-12

7.2-13 An L-shaped reinforced concrete slab 12 ft × 12 ft, with a 6 ft × 6 ft cut-out and thickness $t = 9.0$ in, is lifted by three cables attached at O, B, and D, as shown in the figure. The cables are are combined at point Q, which is 7.0 ft above the top of the slab and directly above the center of mass at C. Each cable has an effective cross-sectional area of $A_e = 0.12$ in^2.

(a) Find the tensile force $T_i (i = 1, 2, 3)$ in each cable due to the weight W of the concrete slab (ignore weight of cables).

(b) Find the average stress σ_i in each cable. (See Table F-1 in Appendix F for the weight density of reinforced concrete.)

(c) Add cable AQ so that OQA is one continuous cable, with each segment having force T_1, which is connected to cables BQ and DQ at point Q. Repeat parts (a) and (b). *Hint*: There are now three forced equilibrium equations and one *constraint equation*, $T_1 = T_4$.

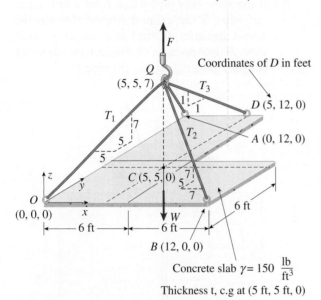

Coordinates of D in feet

Concrete slab $\gamma = 150 \dfrac{\text{lb}}{\text{ft}^3}$

Thickness t, c.g at (5 ft, 5 ft, 0)

PROBLEM 7.2-13

7.2-14 A crane boom of mass 450 kg with its center of gravity at C is stabilized by two cables AQ and BQ ($A_e = 304$ mm^2 for each cable) as shown in the figure. A load $P = 20$ kN is supported at point D. The crane boom lies in the y-z plane.

(a) Find the tension forces in each cable: T_{AQ} and T_{BQ} (kN). Neglect the mass of the cables, but include the mass of the boom in addition to load P.

(b) Find the average stress (σ) in each cable.

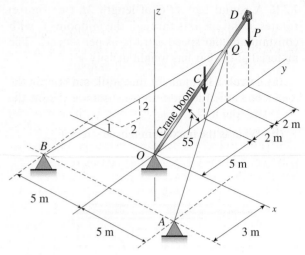

PROBLEM 7.2-14

7.2-15 Two gondolas on a ski lift are locked in the position shown in the figure while repairs are being made elsewhere. The distance between support towers is $L = 100$ ft. The length of each cable segment under gondolas weighing $W_B = 450$ lb and $W_C = 650$ lb are $D_{AB} = 12$ ft, $D_{BC} = 70$ ft, and $D_{CD} = 20$ ft. The cable sag at B is $\Delta_B = 3.9$ ft and that at C is $\Delta_C = 7.1$ ft. The effective cross-sectional area of the cables is $A_e = 0.12$ in^2.

(a) Find the tension force in each segment; neglect the mass of the cable.

(b) Find the average stress (σ) in each cable segment.

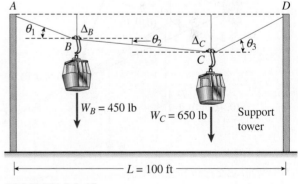

PROBLEM 7.2-15

7.2-16 A round bar ABC of length $2L$ (see figure) rotates about an axis through the midpoint C with constant angular speed ω (radians per second). The material of the bar has weight density γ.

(a) Derive a formula for the tensile stress σ_x in the bar as a function of the distance x from the midpoint C.

(b) What is the maximum tensile stress σ_{max}?

PROBLEM 7.2-16

7.2-17 Two separate cables AC and BC support a sign structure of weight $W = 1575$ lb attached to a building. The sign is also supported by a pin support at O and a lateral restraint in the z-direction at D.

(a) Find the tension in each cable. Neglect the mass of the cables.

(b) Find the average stress in each cable if the area of each cable is $A_e = 0.471$ in^2.

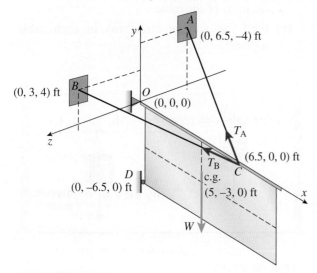

PROBLEM 7.2-17

7.3 Mechanical Properties of Materials

Introductory Problems

7.3-1 Imagine that a long steel wire hangs vertically from a high-altitude balloon.

(a) What is the greatest length (feet) it can have without yielding if the steel yields at 40 ksi?

(b) If the same wire hangs from a ship at sea, what is the greatest length? (Obtain the weight densities of steel and sea water from Table F-1, Appendix F.)

7.3-2 A steel riser pipe hangs from a drill rig located offshore in deep water (see figure).

(a) What is the greatest length (meters) it can have without breaking if the pipe is suspended in the air and the ultimate strength (or breaking strength) is 550 MPa?

(b) If the same riser pipe hangs from a drill rig at sea, what is the greatest length? (Obtain the weight densities of steel and sea water from Table F-1, Appendix F. Neglect the effect of buoyant foam casings on the pipe.)

PROBLEM 7.3-2

7.3-3 Three different materials, designated A, B, and C, are tested in tension using test specimens having diameters of 0.505 in. and gage lengths of 2.0 in. (see figure). At failure, the distances between the gage marks are found to be 2.13, 2.48, and 2.78 in., respectively. Also, at the failure cross sections, the diameters are found to be 0.484, 0.398, and 0.253 in., respectively.

Determine the percent elongation and percent reduction in area of each specimen. Using your own judgment, classify each material as brittle or ductile.

PROBLEM 7.3-3

Representative Problems

7.3-4 The *strength-to-weight ratio* of a structural material is defined as its load-carrying capacity divided by its weight. For materials in tension, use a characteristic tensile stress obtained from a stress-strain curve as a measure of strength. For instance, either the yield stress or the ultimate stress could be used, depending upon the particular application. Thus, the strength-to-weight ratio $R_{S/W}$ for a material in tension is defined as

$$R_{S/W} = \frac{\sigma}{\gamma}$$

in which σ is the characteristic stress and γ is the weight density. Note that the ratio has units of length.

Using the ultimate stress σ_U as the strength parameter, calculate the strength-to-weight ratio (in units of meters) for each of the following materials: aluminum alloy 6067-T6, Douglas fir (in bending), nylon, structural steel ASTM-A572, and a titanium alloy. Obtain the material properties from Tables F-1 and F-3 of Appendix F. When a range of values is given in a table, use the average value.

7.3-5 A symmetrical framework consisting of three pin-connected bars is loaded by a force P (see figure). The angle between the inclined bars and the horizontal is $\alpha = 52°$. The axial strain in the middle bar is measured as 0.036.

Determine the tensile stress in the outer bars if they are constructed of a copper alloy having the following stress-strain relationship:

$$\sigma = \frac{18,000\varepsilon}{1 + 300\varepsilon} \quad 0 \le \varepsilon \le 0.03 \quad (\sigma = \text{ksi})$$

PROBLEM 7.3-5

7.3-6 A specimen of a methacrylate plastic is tested in tension at room temperature (see figure), producing the stress-strain data listed in the accompanying table.

Plot the stress-strain curve and determine the proportional limit, modulus of elasticity (which is the slope of the initial part of the stress-strain curve), and the yield stress at 0.2% offset. Is the material ductile or brittle?

PROBLEM 7.3-6

STRESS-STRAIN DATA FOR PROB. 7.3-6

Stress (MPa)	Strain
8.0	0.0032
17.5	0.0073
25.6	0.0111
31.1	0.0129
39.8	0.0163
44.0	0.0184
48.2	0.0209
53.9	0.0260
58.1	0.0331
60.2	0.0384
62.0	0.0429
62.1	Fracture

7.3-7 The data shown in the accompanying table are from a tensile test of high-strength steel. The test specimen has a diameter of 0.505 in. and a gage length of 2.00 in. (see figure for Prob. 7.3-3). At fracture, the elongation between the gage marks is 0.12 in. and the minimum diameter is 0.42 in.

Plot the conventional stress-strain curve for the steel and determine the proportional limit, modulus of elasticity (the slope of the initial part of the stress-strain curve), yield stress at 0.1% offset, ultimate stress, percent elongation in 2.00 in., and percent reduction in area.

TENSILE-TEST DATA FOR PROB. 7.3-7

Load (lb)	Elongation (in.)
1000	0.0002
2000	0.0006
6000	0.0019
10,000	0.0033
12,000	0.0039
12,900	0.0043
13,400	0.0047
13,600	0.0054
13,800	0.0063
14,000	0.0090
14,400	0.0102
15,200	0.0130
16,800	0.0230
18,400	0.0336
20,000	0.0507
22,400	0.1108
22,600	Fracture

7.4 Elasticity, Plasticity, and Creep

Introductory Problems

7.4-1 A bar made of structural steel having the stress-strain diagram shown in the figure has a length of 60 in. The yield stress of the steel is 50 ksi, and the slope of the initial linear part of the stress-strain curve is 29,000 ksi.

(a) The bar is loaded axially until it elongates 0.2 in. and then the load is removed. How does the final length of the bar compare with its original length?

(b) If the bar has a circular cross section with a diameter $d = 1.5$ in. and is loaded by tensile forces $P = 80$ kips, what is the stress in the bar? What is the permanent set of the bar?

PROBLEM 7.4-1

7.4-2 A bar of length 2.0 m is made of a structural steel having the stress-strain diagram shown in the figure. The yield stress of the steel is 250 MPa, and the slope of the initial linear part of the stress-strain curve (modulus of elasticity) is 200 GPa. The bar is loaded axially until it elongates 6.5 mm, and then the load is removed.

How does the final length of the bar compare with its original length? *Hint:* Use the concepts illustrated in Fig. 7-21b.

PROBLEM 7.4-2

7.4-3 A bar made of structural steel having the stress-strain diagram shown in the figure has a length of 48 in. The yield stress of the steel is 42 ksi, and the slope of the initial linear part of the stress-strain curve (modulus of elasticity) is 30×10^3 ksi. The bar is loaded axially until it elongates 0.20 in., and then the load is removed.

How does the final length of the bar compare with its original length? *Hint:* Use the concepts illustrated in Fig. 7-21b.

σ(ksi)

PROBLEM 7.4-3

Representative Problems

7.4-4 A circular bar of magnesium alloy is 750 mm long. The stress-strain diagram for the material is shown in the figure. The bar is loaded in tension to an elongation of 6.0 mm, and then the load is removed.

(a) What is the permanent set of the bar?

(b) If the bar is reloaded, what is the proportional limit?

Hint: Use the concepts illustrated in Figs. 7-21b and 7-22.

PROBLEM 7.4-4

7.4-5 An aluminum bar has length $L = 6$ ft and diameter $d = 1.375$ in. The stress-strain curve for the aluminum is shown in Fig. 7-16. The initial straight-line part of the curve has a slope (modulus of elasticity) of 10.6×10^6 psi. The bar is loaded by tensile forces $P = 44.6$ k and then unloaded.

(a) What is the permanent set of the bar?

(b) If the bar is reloaded, what is the proportional limit?

Hint: Use the concepts illustrated in Figs. 7-21b and 7-22.

7.4-6 A continuous cable (diameter 6 mm) with tension force T is attached to a horizontal frame member at B and C to support a sign structure. The cable passes over a small frictionless pulley at D. The wire is made of a copper alloy, and the stress-strain relationship for the wire is

$$\sigma(\varepsilon) = \frac{124,000\varepsilon}{1 + 300\varepsilon} \quad 0 \leq \varepsilon \leq 0.03 \quad (\sigma \text{ in MPa})$$

(a) Find the axial normal strain in the cable and its elongation due to the load $W = 6.8$ kN.

(b) If the forces are removed, what is the permanent set of the cable?

Hint: Start with constructing the stress-strain diagram and determine the modulus of elasticity, E, and the 0.2% offset yield stress.

PROBLEM 7.4-6

7.4-7 A wire of length $L = 4$ ft and diameter $d = 0.125$ in. is stretched by tensile forces $P = 600$ lb. The wire is made of a copper alloy having a stress-strain relationship that may be described mathematically by

$$\sigma = \frac{18,000\varepsilon}{1 + 300\varepsilon} \quad 0 \leq \varepsilon \leq 0.03 \quad (\sigma = \text{ksi})$$

in which ϵ is nondimensional and σ has units of kips per square inch (ksi).

(a) Construct a stress-strain diagram for the material.

(b) Determine the elongation of the wire due to the forces P.

(c) If the forces are removed, what is the permanent set of the bar?

(d) If the forces are applied again, what is the proportional limit?

505

7.5 Linear Elasticity, Hooke's Law, and Poisson's Ratio

When solving the problems for Section 7.5, assume that the material behaves linearly elastically.

Introductory Problems

7.5-1 A high-strength steel bar used in a large crane has a diameter $d = 2.00$ in. (see figure). The steel has a modulus of elasticity $E = 29 \times 10^6$ psi and Poisson's ratio is $\nu = 0.29$. Because of clearance requirements, the diameter of the bar is limited to 2.001 in. when it is compressed by axial forces.

What is the largest compressive load P_{max} that is permitted?

PROBLEM 7.5-1

7.5-2 A round bar of 10 mm diameter is made of aluminum alloy 7075-T6 (see figure). When the bar is stretched by axial forces P, its diameter decreases by 0.016 mm.

Find the magnitude of the load P. Obtain the material properties from Appendix F.

PROBLEM 7.5-2

7.5-3 A polyethylene bar with a diameter $d_1 = 4.0$ in. is placed inside a steel tube with an inner diameter $d_2 = 4.01$ in. (see figure). The polyethylene bar is then compressed by an axial force P.

At what value of the force P will the space between the polyethylene bar and the steel tube be closed? For polyethylene, assume $E = 200$ ksi and $\nu = 0.4$.

PROBLEM 7.5-3

7.5-4 A square plastic bar (length L_p, side dimension $s_p = 193$ mm) is inserted inside a hollow, square, cast iron tube (length $L_c = 400$ mm, side $s_c = 200$ mm, and thickness $t_c = 3$ mm).

(a) What is the required initial length L_p of the plastic bar so that, when it is compressed by some force P, the final length of bar and tube are equal to length L_c and, at the same time, the gap between plastic bar and cast iron tube is closed?

(b) Compare initial and final volumes for the plastic bar.

Assume that $E_c = 170$ GPa, $E_p = 2.1$ GPa, $\nu_c = 0.3$, and $\nu_p = 0.4$.

PROBLEM 7.5-4

7.5-5 A polyethylene bar having rectangular cross section with a width 7.35 in. and depth 7 in. is placed inside a hollow steel square section with side dimension of 8 in. The polyethylene bar is then compressed by an axial force P. At what value of the force P will the gap between the polyethylene bar and the steel tube be closed for the first time on one side? What is the remaining gap between the polyethylene bar and the steel tube on the other side? For polyethylene, assume $E = 200$ ksi and $\nu = 0.4$.

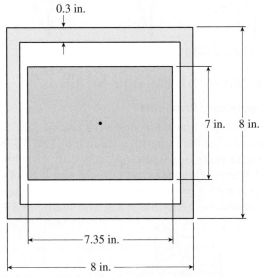

PROBLEM 7.5-5

Representative Problems

7.5-6 A circular aluminum tube of length $L = 600$ mm is loaded in compression by forces P (see figure). The outside and inside diameters are $d_2 = 75$ mm and $d_1 = 63$ mm, respectively. A strain gage is placed on the outside of the tube to measure normal strains in the longitudinal direction. Assume that $E = 73$ GPa and Poisson's ratio is $\nu = 0.33$.

(a) If the compressive stress in the tube is 57 MPa, what is the load P?

(b) If the measured strain is $\varepsilon = 781 \times 10^{-6}$, what is the shortening δ of the tube? What is the percent change in its cross-sectional area? What is the volume change of the tube?

(c) If the tube has a constant outer diameter of $d_2 = 75$ mm along its entire length L but now has increased *inner* diameter d_3 with a nor-

mal stress of 70 MPa over the middle third (see figure part b) while the rest of the tube remains at normal stress of 57 MPa, what is the diameter d_3?

(a)

(b)

PROBLEM 7.5-6

7.5-7 A bar of monel metal with a length $L = 9$ in. and a diameter $d = 0.225$ in. is loaded axially by a tensile force P (see figure). If the bar elongates by 0.0195 in., what is the decrease in diameter d? What is the magnitude of the load P? Use the data in Table F-2, Appendix F.

PROBLEM 7.5-7

7.5-8 A tensile test is performed on a brass specimen 10 mm in diameter using a gage length of 50 mm (see figure). When the tensile load P reaches a value of 20 kN, the distance between the gage marks has increased by 0.122 mm.

(a) What is the modulus of elasticity E of the brass?

(b) If the diameter decreases by 0.00830 mm, what is Poisson's ratio?

PROBLEM 7.5-8

7.5-9 A hollow, brass circular pipe ABC (see figure) supports a load $P_1 = 26.5$ kips acting at the top. A second load $P_2 = 22.0$ kips is uniformly distributed around the cap plate at B. The diameters and thicknesses of the upper and lower parts of the pipe are $d_{AB} = 1.25$ in., $t_{AB} = 0.5$ in., $d_{BC} = 2.25$ in., and $t_{BC} = 0.375$ in., respectively. The modulus of elasticity is 14,000 ksi. When both loads are fully applied, the wall thickness of pipe segment BC increases by 200×10^{-6} in.

(a) Find the increase in the inner diameter of pipe segment BC.

(b) Find Poisson's ratio for the brass.

(c) Find the increase in the wall thickness of pipe segment AB and the increase in the inner diameter of segment AB.

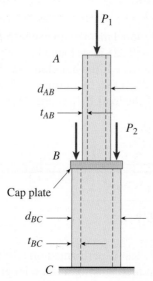

PROBLEM 7.5-9

7.5-10 Three round, copper alloy bars having the same length L but different shapes are shown in the figure. The first bar has a diameter d over its entire length, the second has a diameter d over one-fifth of its length, and the third has a diameter d over one-fifteenth of its length. Elsewhere, the second and third bars have a diameter $2d$. All three bars are subjected to the same axial load P.

Use the following numerical data: $P = 1400$ kN, $L = 5$ m, $d = 80$ mm, $E = 110$ GPa, and $\nu = 0.33$.

(a) Find the change in length of each bar.

(b) Find the change in volume of each bar.

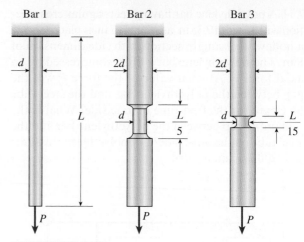

PROBLEM 7.5-10

7.6 Shear Stress and Strain

Introductory Problems

7.6-1 An angle bracket having a thickness $t = 0.75$ in. is attached to the flange of a column by two 5/8-inch diameter bolts (see figure). A uniformly distributed load from a floor joist acts on the top face of the bracket with a pressure $p = 275$ psi. The top face of the bracket has a length $L = 8$ in. and width $b = 3.0$ in.

Angle bracket

Distributed pressure on angle bracket

Floor slab

Floor joist

Angle bracket

PROBLEM 7.6-1

Determine the average bearing pressure σ_b between the angle bracket and the bolts and the average shear stress τ_{aver} in the bolts. Disregard friction between the bracket and the column.

7.6-2 Truss members supporting a roof are connected to a 26-mm-thick gusset plate by a 22-mm diameter pin, as shown in the figure and photo. The two end plates on the truss members are each 14 mm thick.

(a) If the load $P = 80$ kN, what is the largest bearing stress acting on the pin?

(b) If the ultimate shear stress for the pin is 190 MPa, what force P_{ult} is required to cause the pin to fail in shear?

Disregard friction between the plates.

Truss members supporting a roof

PROBLEM 7.6-2

7.6-3 The upper deck of a football stadium is supported by braces, each of which transfers a load $P = 160$ kips to the base of a column (see figure part a). A cap plate at the bottom of the brace distributes the load P to four flange plates ($t_f = 1$ in.) through a pin ($d_p = 2$ in.) to two gusset plates ($t_g = 1.5$ in.) (see figure parts b and c).

Determine the following quantities.

(a) The average shear stress τ_{aver} in the pin.

(b) The average bearing stress between the flange plates and the pin (σ_{bf}), and also between the gusset plates and the pin (σ_{bg}).

Disregard friction between the plates.

(a) Stadium brace

PROBLEM 7.6-3

(b) Detail at bottom of brace

(c) Section through bottom of brace

7.6-4 The inclined ladder AB supports a house painter (85 kg) at C and the weight ($q = 40$ N/m) of the ladder itself. Each ladder rail ($t_r = 4$ mm) is supported by a shoe ($t_s = 5$ mm) that is attached to the ladder rail by a bolt of diameter $d_p = 8$ mm.

(a) Find support reactions at A and B.

(b) Find the resultant force in the shoe bolt at A.

(c) Find maximum average shear τ and bearing (σ_b) stresses in the shoe bolt at A.

$H = 7.5$ m

Shoe bolt at A

$q = 40$ N/m

$a = 1.8$ m

$b = 0.7$ m

Assume no slip at A

Typical rung

Ladder rail ($t_r = 4$ mm)

Shoe bolt ($d_p = 8$ mm)

Ladder shoe ($t_s = 5$ mm)

Section at base

PROBLEM 7.6-4

7.6-5 The force in the brake cable of the V-brake system shown in the figure is $T = 45$ lb. The pivot pin at A has a diameter $d_p = 0.25$ in. and length $L_p = 5/8$ in.

Use the dimensions shown in the figure. Neglect the weight of the brake system.

(a) Find the average shear stress τ_{aver} in the pivot pin where it is anchored to the bicycle frame at B.

(b) Find the average bearing stress $\sigma_{b,aver}$ in the pivot pin over segment AB.

T

Lower end of front brake cable

D

Brake pads

3.25 in.

C

1.0 in.

B

A

Pivot pins anchored to frame (d_p)

L_p

PROBLEM 7.6-5

7.6-6 A steel plate of dimensions $2.5 \times 1.5 \times 0.08$ m and weighing 23.1 kN is hoisted by steel cables with lengths $L_1 = 3.2$ m and $L_2 = 3.9$ m that are each attached to the plate by a clevis and pin (see figure). The pins through the clevises are 18 mm in diameter and are located 2.0 m apart. The orientation angles are measured to be $\theta = 94.4°$ and $\alpha = 54.9°$.

For these conditions, first determine the cable forces T_1 and T_2, then find the average shear stress τ_{aver} in both pin 1 and pin 2, and then the average bearing stress σ_b between the steel plate and each pin. Ignore the weight of the cables.

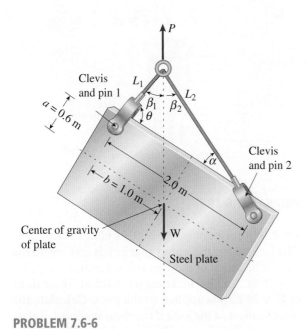

Center of gravity of plate

PROBLEM 7.6-6

7.6-7 A special-purpose eye bolt with a shank diameter $d = 0.50$ in. passes through a hole in a steel plate of thickness $t_p = 0.75$ in. (see figure) and is secured by a nut with thickness $t = 0.25$ in. The hexagonal nut bears directly against the steel plate. The radius of the circumscribed circle for the hexagon is $r = 0.40$ in., so each side of the hexagon has a length 0.40 in. The tensile forces in three cables attached to the eye bolt are $T_1 = 800$ lb, $T_2 = 550$ lb, and $T_3 = 1241$ lb.

(a) Find the resultant force acting on the eye bolt.

(b) Determine the average bearing stress σ_b between the hexagonal nut on the eye bolt and the plate.

(c) Determine the average shear stress τ_{aver} in the nut and also in the steel plate.

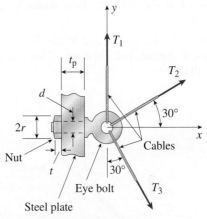

PROBLEM 7.6-7

Representative Problems

7.6-8 An elastomeric bearing pad consisting of two steel plates bonded to a chloroprene elastomer (an artificial rubber) is subjected to a shear force V during a static loading test (see figure). The pad has dimensions $a = 125$ mm and $b = 240$ mm, and the elastomer has a thickness $t = 50$ mm. When the force V equals 12 kN, the top plate is found to have displaced laterally 8.0 mm with respect to the bottom plate.

What is the shear modulus of elasticity G of the chloroprene?

PROBLEM 7.6-8

7.6-9 A joint between two concrete slabs A and B is filled with a flexible epoxy that bonds securely to the concrete (see figure). The height of the joint is $h = 4.0$ in., its length is $L = 40$ in., and its thickness is $t = 0.5$ in. Under the action of shear forces V, the slabs displace vertically through the distance $d = 0.002$ in. relative to each other.

(a) What is the average shear strain γ_{aver} in the epoxy?

(b) What is the magnitude of the forces V if the shear modulus of elasticity G for the epoxy is 140 ksi?

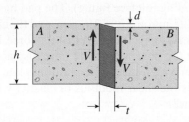

PROBLEM 7.6-9

7.6-10 A steel punch consists of two shafts: upper shaft and lower shaft. Assume that the upper shaft has a diameter $d_1 = 24$ mm and the bottom shaft has a diameter $d_2 = 16$ mm. The punch is used to insert a hole in a 4 mm plate, as shown in the figure. If a force $P = 70$ kN is required to create the hole, what is the average shear stress in the plate and the average compressive stress in the upper and lower shaft of the punch?

PROBLEM 7.6-10

7.6-11 A joint between two glass plates A and B is filled with a flexible epoxy that bonds securely to the glass. The height of the joint is $h = 0.5$ in., its length is $L = 30$ in., and its thickness is $t = 0.5$ in. Shear force of $V = 25$ kips is applied to the joint. Calculate the displacement of the joint if the shear modulus of elasticity G of the epoxy is 100 ksi. Calculate the average shear strain in the epoxy.

PROBLEM 7.6-11

7.6-12 A punch for making a slotted hole in ID cards is shown in the figure part a. Assume that the hole produced by the punch can be described as a rectangle (12 mm × 3 mm) with two half circles (r = 1.5 mm) on the left and right sides. If P = 10 N and the thickness of the ID card is 1 mm, what is the average shear stress in the card?

(a)

(b)

PROBLEM 7.6-12

7.6-13 A steel riser pipe hangs from a drill rig located offshore in deep water (see figure). Separate segments are joined using bolted flange plages (see figure part b and photo). Assume that there are six bolts at each pipe segment connection. Assume that the total length of the riser pipe is $L = 5000$ ft; outer and inner diameters are $d_2 = 16$ in. and $d_1 = 15$ in.; flange plate thickness $t_f = 1.75$ in.; and bolt and washer diameters are $d_b = 1.125$ in., and $d_w = 1.875$ in., respectively.

(a) If the entire length of the riser pipe is suspended in air, find the average normal stress σ in each bolt, the average bearing stress σ_b beneath each washer, and the average shear stress τ through the flange plate at each bolt location for the topmost bolted connection.

(b) If the same riser pipe hangs from a drill rig at sea, what are the normal, bearing, and shear stresses in the connection? Obtain the weight densities of steel and sea water from Table F-1, Appendix F. Neglect the effect of buoyant foam casings on the riser pipe.

Flange plate (t_f), typical bolt (d_b), and washer (d_w)

Riser pipe (d_2, d_1, L)

(a)

Flange plate on riser pipe

Flange plate on riser pipe– plan view ($n = 6$ bolts shown)

(b)

(c)

PROBLEM 7.6-13

7.6-14 A flexible connection consisting of rubber pads (thickness $t = 9$ mm) bonded to steel plates is shown in the figure. The pads are 160 mm long and 80 mm wide.

(a) Find the average shear strain γ_{aver} in the rubber if the force $P = 16$ kN and the shear modulus for the rubber is $G = 1250$ kPa.

(b) Find the relative horizontal displacement δ between the interior plate and the outer plates.

160 mm

Rubber pad

Rubber pad

$\frac{P}{2}$

$\frac{P}{2}$

P

80 mm

$t = 9$ mm

$t = 9$ mm

Section X–X

PROBLEM 7.6-14

513

7.6-15 A hitch-mounted bicycle rack is designed to carry up to four 30-lb bikes mounted on and strapped to two arms *GH* (see bike loads in the figure part a). The rack is attached to the vehicle at *A* and is assumed to be like a cantilever beam *ABCDGH* (figure part b). The weight of fixed segment *AB* is $W_1 = 10$ lb, centered 9 in. from *A* (see figure part b) and the rest of the rack weighs $W_2 = 40$ lb, centered 19 in. from *A*. Segment *ABCDG* is a steel tube of 2×2 in. with a thickness $t = 1/8$ in. Segment *BCDGH* pivots about a bolt at *B* with a diameter $d_B = 0.25$ in. to allow access to the rear of the vehicle without removing the hitch rack.

When in use, the rack is secured in an upright position by a pin at *C* (diameter of pin $d_p = 5/16$ in.) (see photo and figure part c). The overturning effect of the bikes on the rack is resisted by a force couple $F \cdot h$ at *BC*.

(a) Find the support reactions at *A* for the fully loaded rack.

(b) Find forces in the bolt at *B* and the pin at *C*.

(c) Find average shear stresses τ_{aver} in both the bolt at *B* and the pin at *C*.

(d) Find average bearing stresses σ_b in the bolt at *B* and the pin at *C*.

(a)

(b)

PROBLEM 7.6-15

(c) Section *a–a*

7.6-16 The clamp shown in the figure supports a load hanging from the lower flange of a steel beam. The clamp consists of two arms (*A* and *B*) joined by a pin at *C*. The pin has a diameter *d* = 12 mm. Because arm *B* straddles arm *A*, the pin is in double shear.

Line 1 in the figure defines the line of action of the resultant horizontal force *H* acting between the lower flange of the beam and arm *B*. The vertical distance from this line to the pin is *h* = 250 mm. Line 2 defines the line of action of the resultant vertical force *V* acting between the flange and arm *B*. The horizontal distance from this line to the centerline of the beam is *c* = 100 mm. The force conditions between arm *A* and the lower flange are symmetrical with those given for arm *B*.

Determine the average shear stress in the pin at *C* when the load *P* = 18 kN.

7.6-17 A shock mount constructed as shown in the figure is used to support a delicate instrument. The mount consists of an outer steel tube with inside diameter *b*, a central steel bar of diameter *d* that supports the load *P*, and a hollow rubber cylinder (height *h*) bonded to the tube and bar.

(a) Obtain a formula for the shear stress τ in the rubber at a radial distance *r* from the center of the shock mount.

(b) Obtain a formula for the downward displacement δ of the central bar due to the load *P*, assuming that *G* is the shear modulus of elasticity of the rubber and that the steel tube and bar are rigid.

PROBLEM 7.6-17

PROBLEM 7.6-16

7.6-18 A removable sign post on a hurricane evacuation route (see figure part a) consists of an upper pole with a slotted base plate bolted to a short post anchored in the ground. The lower post is capped with a separate conventional base plate having four holes of diameter d_b. The upper base plate has slots at locations 1 to 4 and is bolted to the lower base plate at these four points (see figure part b). Each of the four bolts has a diameter of d_b and a washer with a diameter of d_w. The bolts are arranged in a rectangular pattern ($b \times h$). Consider only wind force W_y applied in the y direction at the center of pressure (C.P.) of the sign structure at height $z = L$ above the base. Neglect the weight of the sign and post and the friction between the upper and lower base plates. Assume that the lower base plate and short anchored post are rigid.

(a) Find the average shear stress τ (MPa) at bolts 1 and 4 (see figure part c) due to the wind force W_y.

(b) Find the average bearing stress σ_b (MPa) between the bolt and the upper base plate (thickness t) at bolts 1 and 4.

(c) Find the average bearing stress σ_b (MPa) between the upper base plate and washer at bolt 4 due to the wind force W_y (assume the initial bolt pretension is zero).

Numerical data

$H = 150$ mm	$b = 96$ mm
$h = 108$ mm	$t = 14$ mm
$d_b = 12$ mm	$d_w = 22$ mm
$L = 2.75$ m	$W_y = 667$ N

(a)

(b)

Bolt and washer
(d_b, d_w)

Square base plate
$(H \times H)$ of
thickness t

Slot in upper base plate
$(d_b = \text{slot width})$

Plan view
of upper
base plate

(c)

(d) Find the average shear stress τ (MPa) through the upper base plate at bolt 4 due to the wind force W_y.

(e) Find an expression for the normal stress σ in bolt 3 due to the wind force W_y.

(See problem 7.7-17 for additional discussion of wind on a sign and the resulting forces acting on a conventional base plate.)

7.6-19 A spray nozzle for a garden hose requires a force $F = 5$ lb to open the spring-loaded spray chamber AB. The nozzle hand grip pivots about a pin through a flange at O. Each of the two flanges has a thickness $t = 1/16$ in., and the pin has a diameter $d_p = 1/8$ in. (see figure part a). The spray nozzle is attached to the garden hose with a quick release fitting at B (see figure part b). Three brass balls (diameter $d_b = 3/16$ in.) hold the spray head in place under a water pressure force $f_p = 30$ lb at C (see figure part c). Use dimensions given in figure part a.

(a) Find the force in the pin at O due to applied force F.

(b) Find average shear stress τ_{aver} and bearing stress σ_b in the pin at O.

(c) Find the average shear stress τ_{aver} in the brass retaining balls at C due to water pressure force f_p.

PROBLEM 7.6-19

7.6-20 A single steel strut AB with a diameter $d_s = 8$ mm supports the vehicle engine hood of a mass 20 kg that pivots about hinges at C and D (see figure parts a and b). The strut is bent into a loop at its end and then attached to a bolt at A with a diameter $d_b = 10$ mm. Strut AB lies in a vertical plane.

(a) Find the strut force F_s and average normal stress σ in the strut.

(b) Find the average shear stress τ_{aver} in the bolt at A.

(c) Find the average bearing stress σ_b on the bolt at A.

(a)

(b)

PROBLEM 7.6-20

7.6-21 The top portion of a pole saw used to trim small branches from trees is shown in the figure part a. The cutting blade BCD (see figure parts a and c) applies a force P at point D. Ignore the effect of the weak return spring attached to the cutting blade below B. Use properties and dimensions given in the figure.

(a) Find the force P on the cutting blade at D if the tension force in the rope is $T = 25$ lb (see free-body diagram in figure part b).

(b) Find force in the pin at C.

(c) Find average shear stress τ_{aver} and bearing stress σ_b in the support pin at C (see section a–a through cutting blade in figure part c).

(a) Top part of pole saw

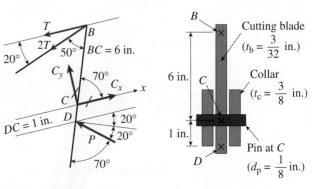

(b) Free-body diagram

(c) Section a–a

PROBLEM 7.6-21

7.6-22 A cargo ship is tied down to marine bollards (see figure part a) at a number of points along its length while its cargo is unloaded by a container handling crane. Each bollard is fastened to the wharf using anchor bolts. Three cables having known tension force magnitudes $F_1 = 110$ kN, $F_2 = 85$ kN, and $F_3 = 90$ kN are secured to one bollard at a point A with coordinates (0, 0.45 m, 0) in the x-y-z coordinate system shown in the figure part b. Each cable force is directed at an attachment point on the ship. Force F_1 is directed from point A to a point on the ship having coordinates (3 m, 9 m, 0); force F_2 is directed at a point with coordinates (6.5 m, 8.5 m, 2 m); and force F_3 is directed at a point with coordinates (8 m, 9 m, 5 m). The diameter of each anchor bolts is $d_b = 24$ mm.

(a) Find the reaction forces and reaction moments at the base of the bollard.

(b) Calculate the average shear stress in the anchor bolts (in the x-z plane). Assume each bolt carries an equal share of the total force.

(a)

(b)

(c)

PROBLEM 7.6-22

7.6-23 A basketball player hangs on the rim after a dunk. He applies equal forces $P_1 = P_2 = 110$ lb at both A and B (see joint coordinates in the figure). Forces P_1 and P_2 act parallel to the y-z plane.

(a) Find the *reactions* at the support bracket (assume that the bracket-rim assembly is a cantilever beam).

(b) Find connection shear stresses at bolt 2.

PROBLEM 7.6-23 Support bracket

7.6-24 A bicycle chain consists of a series of small links, where each are 12 mm long between the centers of the pins (see figure). You might wish to examine a bicycle chain and observe its construction. Note particularly the pins, which have a diameter of 2.5 mm.

To solve this problem, make two measurements on a bicycle (see figure): (1) the length L of the crank arm from main axle to pedal axle and (2) the radius R of the sprocket (the toothed wheel, sometimes called the chainring).

(a) Using your measured dimensions, calculate the tensile force T in the chain due to a force $F = 800$ N applied to one of the pedals.

(b) Calculate the average shear stress τ_{aver} in the pins.

PROBLEM 7.6-24

7.7 Allowable Stresses and Allowable Loads

Introductory Problems

7.7-1 A bar of solid circular cross section is loaded in tension by forces P (see figure). The bar has a length $L = 16.0$ in. and diameter $d = 0.50$ in. The material is a magnesium alloy having a modulus of elasticity $E = 6.4 \times 10^6$ psi. The allowable stress in tension is $\sigma_{\text{allow}} = 17,000$ psi, and the elongation of the bar must not exceed 0.04 in.

What is the allowable value of the forces P?

PROBLEM 7.7-1

7.7-2 A torque T_0 is transmitted between two flanged shafts by means of ten 20-mm bolts (see figure and photo). The diameter of the bolt circle is $d = 250$ mm.

If the allowable shear stress in the bolts is 85 MPa, what is the maximum permissible torque? (Disregard friction between the flanges.)

Drive shaft coupling on a ship propulsion motor
PROBLEM 7.7-2

7.7-3 A tie-down on the deck of a sailboat consists of a bent bar bolted at both ends, as shown in the figure. The diameter d_B of the bar is 1/4 in., the diameter d_W

of the washers is 7/8 in., and the thickness t of the fiberglass deck is 3/8 in.

If the allowable shear stress in the fiberglass is 300 psi, and the allowable bearing pressure between the washer and the fiberglass is 550 psi, what is the allowable load P_{allow} on the tie-down?

PROBLEM 7.7-3

7.7-4 Two steel tubes are joined at B by four pins ($d_p = 11$ mm), as shown in the cross section a–a in the figure. The outer diameters of the tubes are $d_{AB} = 41$ mm and $d_{BC} = 28$ mm. The wall thickness are $t_{AB} = 6.5$ mm and $t_{BC} = 7.5$ mm. The yield stress in tension for the steel is $\sigma_Y = 200$ MPa and the ultimate stress in *tension* is $\sigma_U = 340$ MPa. The corresponding yield and ultimate values in *shear* for the

Section a–a

PROBLEM 7.7-4

pin are 80 MPa and 140 MPa, respectively. Finally, the yield and ultimate values in *bearing* between the pins and the tubes are 260 MPa and 450 MPa, respectively. Assume that the factors of safety with respect to yield stress and ultimate stress are 3.5 and 4.5, respectively.

(a) Calculate the allowable tensile force P_{allow} considering tension in the tubes.

(b) Recompute P_{allow} for shear in the pins.

(c) Finally, recompute P_{allow} for bearing between the pin and the tubes. Which is the controlling value of P?

7.7-5 A steel pad supporting heavy machinery rests on four short, hollow, cast iron piers (see figure). The ultimate strength of the cast iron in compression is 50 ksi. The outer diameter of the piers is $d = 4.5$ in., and the wall thickness is $t = 0.40$ in.

Using a factor of safety of 3.5 with respect to the ultimate strength, determine the total load P that can be supported by the pad.

PROBLEMS 7.7-5 and 7.7-6

7.7-6 A steel pad supporting heavy machinery rests on four short, hollow, cast iron piers (see figure). The ultimate strength of the cast iron in compression is 400 MPa. The total load P that may be supported by the pad is 900 kN. Using a factor of safety 3.0 with respect to ultimate strength, determine the outer diameter of the pier if the thickness is of the cross section is 12 mm.

7.7-7 A steel riser pipe hangs from a drill rig. Individual segments of equal length $L = 50$ ft are joined together using bolted flange plates (see figure part b). There are six bolts at each pipe segment connection. The outer and inner pipe diameters are $d_2 = 14$ in. and $d_1 = 13$ in.; flange plate thickness $t_f = 1.5$ in.; and bolt and washer diameters are $d_b = 1.125$ in. and $d_w = 1.875$ in. Find the number n of permissible segments of pipe based on following allowable stresses.

(a) The allowable tensile stress in the pipe is 50 ksi.

(b) The allowable tensile stress in a bolt is 120 ksi.

Find number of segments n for two cases: pipe hanging in air and pipe hanging in seawater.

522

Riser pipe (d_2, d_1, L)

(a)

Flange plate on riser pipe plan view ($n = 6$ bolts shown)

PROBLEM 7.7-7   (b)

Representative Problems

7.7-8 The rear hatch of a van ($BDCG$ in figure part a) is supported by two hinges at B_1 and B_2 and by two struts A_1B_1 and A_2B_2 (diameter $d_s = 10$ mm), as shown in figure part b. The struts are supported at A_1 and A_2 by pins, each with a diameter $d_p = 9$ mm and passing through an eyelet of thickness $t = 8$ mm at the end of the strut (figure part b). A closing force $P = 50$ N is applied at G, and the mass of the hatch $M_h = 43$ kg is concentrated at C.

(a) What is the force F in each strut? (Use the free-body diagram of one half of the hatch in the figure part c.)

(b) What is the maximum permissible force in the strut, F_{allow}, if the allowable stresses are compressive stress in the strut, 70 MPa; shear stress in the pin, 45 MPa; and bearing stress between the pin and the end of the strut, 110 MPa?

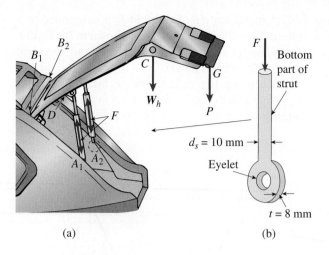

(a)

Bottom
part of
strut

F

$d_s = 10$ mm

Eyelet

$t = 8$ mm

(b)

127 mm

505 mm

505 mm

75 mm

B_x

D

710 mm

C

$\dfrac{M_h}{2} \times g$

G

$\dfrac{P}{2}$

B_y

10°

460 mm

A ← Pin support

F

PROBLEM 7.7-8

(c)

7.7-9 A lifeboat hangs from two ship's davits, as shown in the figure. A pin of diameter $d = 0.80$ in. passes through each davit and supports two pulleys, one on each side of the davit.

Cables attached to the lifeboat pass over the pulleys and wind around winches that raise and lower the lifeboat. The lower parts of the cables are vertical and the

upper parts make an angle $\alpha = 15°$ with the horizontal. The allowable tensile force in each cable is 1800 lb, and the allowable shear stress in the pins is 4000 psi.

If the lifeboat weighs 1500 lb, what is the maximum weight that can be carried in the lifeboat?

7.7-10 A cable and pulley system in the figure part a supports a cage of a mass 300 kg at B. Assume that this includes the mass of the cables as well. The thickness of each of the three steel pulleys is $t = 40$ mm. The pin diameters are $d_{pA} = 25$ mm, $d_{pB} = 30$ mm, and $d_{pC} = 22$ mm (see figure part a and part b).

(a) Find expressions for the resultant forces acting on the pulleys at A, B, and C in terms of cable tension T.

(b) What is the maximum weight W that can be added to the cage at B based on the following allowable stresses? Shear stress in the pins is 50 MPa; bearing stress between the pin and the pulley is 110 MPa.

a

$d_{pA} = 25$ mm

C

L_1

A

Cable

$d_{pC} = 22$ mm

a

L_2

$d_{pB} = 30$ mm

B

Cage

W

(a)

PROBLEM 7.7-9

Cable

Pulley

t

Pin

d_p

Support bracket

Section a–a: pulley support detail at A and C

Cable

d_{pB}

t_B

Cage at B

Section a–a: pulley support detail at B

(b)

PROBLEM 7.7-10

523

7.7-11 A ship's spar is attached at the base of a mast by a pin connection (see figure). The spar is a steel tube of outer diameter $d_2 = 3.5$ in. and inner diameter $d_1 = 2.8$ in. The steel pin has a diameter $d = 1$ in., and the two plates connecting the spar to the pin have a thickness $t = 0.5$ in. The allowable stresses are compressive stress in the spar, 10 ksi; shear stress in the pin, 6.5 ksi; and bearing stress between the pin and the connecting plates, 16 ksi.

Determine the allowable compressive force P_{allow} in the spar.

PROBLEM 7.7-11

7.7-12 What is the maximum possible value of the clamping force C in the jaws of the pliers shown in the figure if the ultimate shear stress in the 5-mm diameter pin is 340 MPa?

What is the maximum permissible value of the applied load P to maintain a factor of safety of 3.0 with respect to failure of the pin?

PROBLEM 7.7-12

7.7-13 A metal bar AB of a weight W is suspended by a system of steel wires arranged as shown in the figure. The diameter of the wires is 5/64 in., and the yield stress of the steel is 65 ksi.

Determine the maximum permissible weight W_{max} for a factor of safety of 1.9 with respect to yielding.

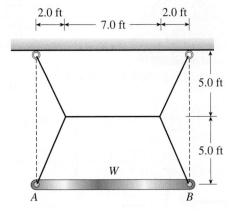

PROBLEM 7.7-13

7.7-14 A plane truss is subjected to loads $2P$ and P at joints B and C, respectively, as shown in the figure part a. The truss bars are made of two L 102 × 76 × 6.4 steel angles (see Table C-5(b): cross-sectional area of the two angles, $A = 2180$ mm^2, and figure part b) having an ultimate stress in tension equal to 390 MPa. The angles are connected to a 12-mm-thick gusset plate at C (figure part c) with 16-mm diameter rivets; assume each rivet transfers an equal share of the member force to the gusset plate. The ultimate stresses in shear and bearing for the rivet steel are 190 MPa and 550 MPa, respectively.

Determine the allowable load P_{allow} if a safety factor of 2.5 is desired with respect to the ultimate load that can be carried. Consider tension in the bars, shear in the rivets, bearing between the rivets and the bars, and also bearing between the rivets and the gusset plate. Disregard friction between the plates and the weight of the truss itself.

(a)

Truss bars

Gusset plate

Rivet

(c)

(b) Section a–a

PROBLEM 7.7-14

7.7-15 A solid bar of circular cross section (diameter d) has a hole of diameter $d/5$ drilled laterally through the center of the bar (see figure). The allowable average tensile stress on the net cross section of the bar is σ_{allow}.

(a) Obtain a formula for the allowable load P_{allow} that the bar can carry in tension.

(b) Calculate the value of P_{allow} if the bar is made of brass with a diameter $d = 1.75$ in. and $\sigma_{allow} = 12$ ksi.

Hint: Use the formulas of Case 15, Appendix B.

PROBLEM 7.7-15

7.7-16 A solid steel bar of a diameter $d_1 = 60$ mm has a hole of a diameter $d_2 = 32$ mm drilled through it (see figure). A steel pin of a diameter d_2 passes through the hole and is attached to supports.

Determine the maximum permissible tensile load P_{allow} in the bar if the yield stress for shear in the pin is $\tau_Y = 120$ MPa, the yield stress for tension in the bar is $\sigma_Y = 250$ MPa, and a factor of safety of 2.0 with respect to yielding is required. *Hint*: Use the formulas of Case 15, Appendix B.

PROBLEM 7.7-16

7.7-17 A sign of weight W is supported at its base by four bolts anchored in a concrete footing. Wind pressure p acts normal to the surface of the sign; the resultant of the uniform wind pressure is force F at the center of pressure (C.P.). The wind force is assumed to create equal shear forces $F/4$ in the y direction at each bolt (see figure parts a and c). The overturning effect of the wind force also causes an uplift force R at bolts A and C and a downward force $(-R)$ at bolts B and D (see figure part b). The resulting effects of the wind and the associated ultimate stresses for each stress condition are normal stress in each bolt ($\sigma_u = 60$ ksi); shear through the base plate ($\tau_u = 17$ ksi); horizontal shear and bearing on each bolt ($\tau_{hu} = 25$ ksi and $\sigma_{bu} = 75$ ksi); and bearing on the bottom washer at B (or D) ($\sigma_{bw} = 50$ ksi).

Find the maximum wind pressure p_{max} (psf) that can be carried by the bolted support system for the sign if a safety factor of 2.5 is desired with respect to the ultimate wind load that can be carried.

Use the following numerical data: bolt $d_b = \frac{3}{4}$ in.; washer $d_w = 1.5$ in.; base plate $t_{bp} = 1$ in.; base plate dimensions $h = 14$ in. and $b = 12$ in.; $W = 500$ lb; $H = 17$ ft; sign dimensions $L_v = 10$ ft. $\times L_h = 12$ ft.; pipe column diameter $d = 6$ in.; and pipe column thickness $t = 3/8$ in.

7.7-18 The piston in an engine is attached to a connecting rod AB, which in turn is connected to a crank arm BC (see figure). The piston slides without friction in a cylinder and is subjected to a force P (assumed to be constant) while moving to the right in the figure. The connecting rod, with diameter d and length L, is attached at both ends by pins. The crank arm rotates about the axle at C with the pin at B moving in a circle of radius R. The axle at C, which is supported by bearings, exerts a resisting moment M against the crank arm.

(a) Obtain a formula for the maximum permissible force P_{allow} based upon an allowable compressive stress σ_c in the connecting rod.

(b) Calculate the force P_{allow} for the following data: $\sigma_c = 160$ MPa, $d = 9.00$ mm, and $R = 0.28L$.

(c)

PROBLEM 7.7-17

PROBLEM 7.7-18

7.8 Design for Axial Loads and Direct Shear

Introductory Problems

7.8-1 An aluminum tube is required to transmit an axial tensile force $P = 33$ k (see figure part a). The thickness of the wall of the tube is 0.25 in.

(a) What is the minimum required outer diameter d_{min} if the allowable tensile stress is 12,000 psi?

(b) Repeat part (a) if the tube has a hole of a diameter $d/10$ at mid-length (see figure parts b and c).

(a)

(b)

(c)

PROBLEM 7.8-1

7.8-2 A copper alloy pipe with a yield stress $\sigma_Y = 290$ MPa is to carry an axial tensile load $P = 1500$ kN (see figure part a). Use a factor of safety of 1.8 against yielding.

(a) If the thickness t of the pipe is one-eighth of its outer diameter, what is the minimum required outer diameter d_{min}?

(b) Repeat part (a) if the tube has a hole of diameter $d/10$ drilled through the entire tube, as shown in the figure part b.

(a)

(b)

PROBLEM 7.8-2

7.8-3 A horizontal beam AB with cross-sectional dimensions $(b = 0.75$ in.$) \times (h = 8.0$ in.$)$ is supported by an inclined strut CD and carries a load $P = 2700$ lb at joint B (see figure part a). The strut, which consists of two bars each of thickness $5b/8$, is connected to the beam by a bolt passing through the three bars meeting at joint C (see figure part b).

(a) If the allowable shear stress in the bolt is 13,000 psi, what is the minimum required diameter d_{min} of the bolt at C?

(b) If the allowable bearing stress in the bolt is 19,000 psi, what is the minimum required diameter d_{min} of the bolt at C?

(a)

(b)

PROBLEM 7.8-3

527

Representative Problems

7.8-4 Lateral bracing for an elevated pedestrian walkway is shown in the figure part a. The thickness of the clevis plate $t_c = 16$ mm and the thickness of the gusset plate $t_g = 20$ mm (see figure part b). The maximum force in the diagonal bracing is expected to be $F = 190$ kN.

If the allowable shear stress in the pin is 90 MPa and the allowable bearing stress between the pin and both the clevis and gusset plates is 150 MPa, what is the minimum required diameter d_{min} of the pin?

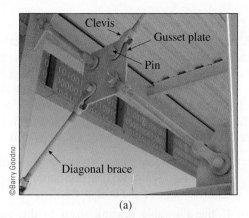

Clevis
Gusset plate
Pin
Diagonal brace
©Barry Goodno

(a)

Gusset plate
t_c
t_g
d_{min}
Clevis

(b)

PROBLEM 7.8-4

7.8-5 A plane truss has joint loads P, $2P$, and $3P$ at joints D, C, and B, respectively (see figure) where load variable $P = 5200$ lb. All members have two end plates (see figure for Prob. 7.6-2) that are pin-connected to gusset plates. Each end plate has a thickness $t_p = 0.625$ in., and all gusset plates have a thickness $t_g = 1.125$ in. If the allowable shear stress in each pin is 12,000 psi and the allowable bearing stress in each pin is 18,000 psi, what is the minimum

required diameter d_{min} of the pins used at either end of member BE?

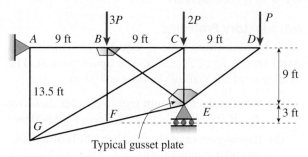

Typical gusset plate

PROBLEM 7.8-5

7.8-6 Cable DB supports canopy beam $OABC$ as shown in the figure. Find the required cross-sectional area of cable BD if the allowable normal stress is 125 MPa. Determine the required diameter of the pins at O, B, and D if the allowable stress in shear is 80 MPa. Assume that canopy beam weight is $W = 8$ kN. *Note*: The pins at O, A, B, and D are in double shear. Consider only the weight of the canopy; disregard the weight of cable DB.

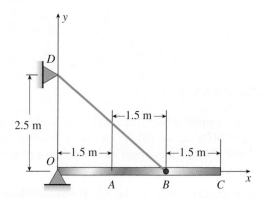

PROBLEM 7.8-6

7.8-7 Continuous cable ADB runs over a small friction-less pulley at D to support beam $OABC$ that is part of an entrance canopy for a building (see figure). Assume that the canopy segment has a weight $W = 1700$ lb.

(a) Find the required cross-sectional area of cable ADB if the allowable stress is 18 ksi.

(b) Determine the required diameter of the pins at O, A, B, and D if the allowable stress in shear is 12 ksi.

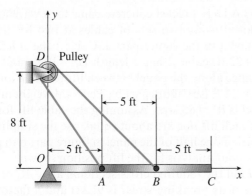

PROBLEM 7.8-7

7.8-8 A suspender on a suspension bridge consists of a cable that passes over the main cable (see figure) and supports the bridge deck, which is far below. The suspender is held in position by a metal tie that is prevented from sliding downward by clamps around the suspender cable.

Let P represent the load in each part of the suspender cable, and let θ represent the angle of the suspender cable just above the tie. Let σ_{allow} represent the allowable tensile stress in the metal tie.

(a) Obtain a formula for the minimum required cross-sectional area of the tie.

(b) Calculate the minimum area if $P = 130$ kN, $\theta = 75°$, and $\sigma_{allow} = 80$ MPa.

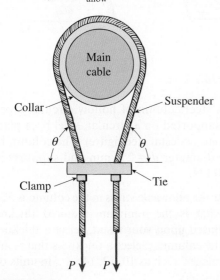

PROBLEM 7.8-8

7.8-9 A square steel tube of a length $L = 20$ ft and width $b_2 = 10.0$ in. is hoisted by a crane (see figure). The tube hangs from a pin of diameter d that is held by the cables at points A and B. The cross section is a hollow square with an inner dimension $b_1 = 8.5$ in.

and outer dimension $b_2 = 10.0$ in. The allowable shear stress in the pin is 8,700 psi, and the allowable bearing stress between the pin and the tube is 13,000 psi.

Determine the minimum diameter of the pin in order to support the weight of the tube. *Note:* Disregard the rounded corners of the tube when calculating its weight.

PROBLEM 7.8-9

7.8-10 A cable and pulley system at D is used to bring a 230-kg pole (ACB) to a vertical position, as shown in the figure part a. The cable has tensile force T and is attached at C. The length L of the pole is 6.0 m, the outer diameter is $d = 140$ mm, and the wall thickness is $t = 12$ mm. The pole pivots about a pin at A in figure part b. The allowable shear stress in the pin is 60 MPa and the allowable bearing stress is 90 MPa.

Find the minimum diameter of the pin at A in order to support the weight of the pole in the position shown in the figure part a.

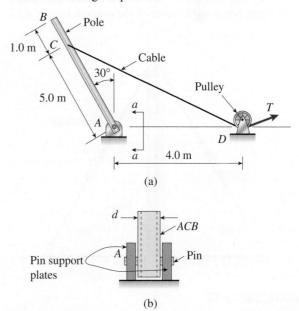

PROBLEM 7.8-10

529

7.8-11 A pressurized circular cylinder has a sealed cover plate fastened with steel bolts (see figure). The pressure p of the gas in the cylinder is 290 psi, the inside diameter D of the cylinder is 10.0 in., and the diameter d_B of the bolts is 0.50 in.

If the allowable tensile stress in the bolts is 10,000 psi, find the number n of bolts needed to fasten the cover.

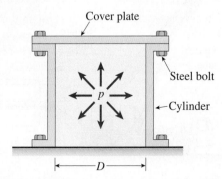

Cover plate

Steel bolt

p

Cylinder

D

PROBLEM 7.8-11

7.8-12 A tubular post of outer diameter d_2 is guyed by two cables fitted with turnbuckles (see figure). The cables are tightened by rotating the turnbuckles, producing tension in the cables and compression in the post. Both cables are tightened to a tensile force of 110 kN. The angle between the cables and the ground is 60°, and the allowable compressive stress in the post is $\sigma_c = 35$ MPa.

If the wall thickness of the post is 15 mm, what is the minimum permissible value of the outer diameter d_2?

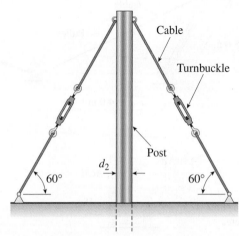

Cable

Turnbuckle

Post

d_2

60°   60°

PROBLEM 7.8-12

7.8-13 A large precast concrete panel for a warehouse is raised using two sets of cables at two lift lines, as shown in the figure part a. Cable 1 has a length $L_1 = 22$ ft, cable 2 has a length $L_2 = 10$ ft, and the distance along the panel between lift points B and D is $d = 14$ ft (see figure part b). The total weight of the panel is $W = 85$ kips. Assuming the cable lift forces F at each lift line are about equal, use the simplified model of one half of the panel in figure part b to perform your analysis for the lift position shown.

Find the required cross-sectional area AC of the cable if its breaking stress is 91 ksi and a factor of safety of 4 with respect to failure is desired.

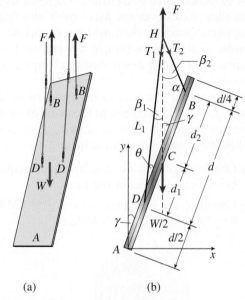

(a)   (b)

PROBLEM 7.8-13

7.8-14 A steel column of hollow circular cross section is supported on a circular, steel base plate and a concrete pedestal (see figure). The column has an outside diameter $d = 250$ mm and supports a load $P = 750$ kN.

(a) If the allowable stress in the column is 55 MPa, what is the minimum required thickness t? Based upon your result, select a thickness for the column. (Select a thickness that is an even integer, such as 10, 12, 14, . . ., in units of millimeters.)

(b) If the allowable bearing stress on the concrete pedestal is 11.5 MPa, what is the minimum required diameter D of the base plate if it is designed for the allowable load P_{allow} that the column with the selected thickness can support?

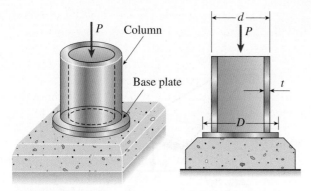

PROBLEM 7.8-14

7.8-15 An elevated jogging track is supported at intervals by a wood beam AB ($L = 7.5$ ft) that is pinned at A and supported by steel rod BC and a steel washer at B. Both the rod ($d_{BC} = 3/16$ in.) and the washer ($d_B = 1.0$ in.) were designed using a rod tension force of $T_{BC} = 425$ lb. The rod was sized using a factor of safety of 3 against reaching the ultimate stress $\sigma_u = 60$ ksi. An allowable bearing stress $\sigma_{ba} = 565$ psi was used to size the washer at B.

A small platform HF is suspended below a section of the elevated track to support some mechanical and electrical equipment. The equipment load is uniform load $q = 50$ lb/ft and concentrated load $W_E = 175$ lb at mid-span of beam HF. The plan is to drill a hole through beam AB at D and install the same rod (d_{BC}) and washer (d_B) at both D and F to support beam HF.

(a) Use σ_u and σ_{ba} to check the proposed design for rod DF and washer d_F; are they acceptable?

(b) Re-check the normal tensile stress in rod BC and bearing stress at B; if either is inadequate under the additional load from platform HF, redesign them to meet the original design criteria.

7.8-16 A flat bar of a width $b = 60$ mm and thickness $t = 10$ mm is loaded in tension by a force P (see figure). The bar is attached to a support by a pin of a diameter d that passes through a hole of the same size in the bar. The allowable tensile stress on the net cross section of the bar is $\sigma_T = 140$ MPa, the allowable shear stress in the pin is $\tau_S = 80$ MPa, and the allowable bearing stress between the pin and the bar is $\sigma_B = 200$ MPa.

(a) Determine the pin diameter d_m for which the load P is a maximum.

(b) Determine the corresponding value P_{max} of the load.

PROBLEM 7.8-15

PROBLEM 7.8-16

7.8-17 Continuous cable *ADB* runs over a small frictionless pulley at *D* to support beam *OABC*, which is part of an entrance canopy for a building (see figure). The canopy segment has a weight $W = 1700$ lb that acts as a concentrated load in the middle of segment *AB*.

(a) What is the maximum permissible value of load *P* at *C* if the allowable force in the cable is 4200 lb?

(b) If $P = 2300$ lb, what is the required diameter of pins *A*, *B*, and *D*? Assume that the pins are in *double shear* and the allowable shear stress in the pins is 10 ksi.

PROBLEM 7.8-18

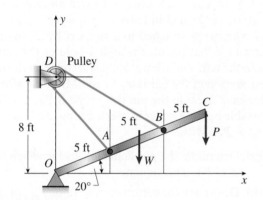

PROBLEM 7.8-17

7.8-19 Two bars *AC* and *BC* of the same material support a vertical load *P* (see figure). The length *L* of the horizontal bar is fixed, but the angle θ can be varied by moving support *A* vertically and changing the length of bar *AC* to correspond with the new position of support *A*. The allowable stresses in the bars are the same in tension and compression.

When the angle θ is reduced, bar *AC* becomes shorter, but the cross-sectional areas of both bars increase because the axial forces are larger. The opposite effects occur if the angle θ is increased. Thus, the weight of the structure (which is proportional to the volume) depends upon the angle θ.

Determine the angle θ so that the structure has minimum weight without exceeding the allowable stresses in the bars. *Note*: The weights of the bars are very small compared to the force *P* and may be disregarded.

7.8-18 Continuous cable *ADB* runs over a small frictionless pulley at *D* to support beam *OABC*, which is part of an entrance canopy for a building (see figure). A downward distributed load with peak intensity $q_0 = 5$ kN/m at *O* acts on the beam (see figure). Assume that canopy weight $W = 8$ kN and that the cable cross-sectional area is 100 mm^2.

What is the required diameter of pins *A*, *B*, and *D* if the pins are in *double shear* and the allowable shear stress is 80 MPa? Note that dimensions $OA = AB = BC = 1.5$ m.

PROBLEM 7.8-19

Axially Loaded Members

Joe Raedle/Reportage/Getty Images

An oil drilling rig is comprised of axially loaded members that must be designed for a variety of loading conditions, including self-weight, impact, and temperature effects.

CHAPTER OBJECTIVES

- Study changes in lengths of axially loaded members based on a *force-displacement relation*.

- Find support reactions in *statically indeterminate bars* acted on by concentrated and distributed axial forces.

- Find changes in lengths of bars due to temperature, misfit, and prestrain effects.

- Find both normal and shear stresses on inclined sections at points of interest on axially loaded bars.

Chapter Outline

8.1 Introduction

Structural components subjected only to tension or compression are known as **axially loaded members**. Solid bars with straight longitudinal axes are the most common type, although cables and coil springs also carry axial loads. Examples of axially loaded bars include truss members, connecting rods in engines, spokes in bicycle wheels, columns in buildings, and struts in aircraft engine mounts. Chapter 7 covered the stress-strain behavior of such members and also how to obtain equations for the stresses acting on cross sections ($\sigma = P/A$) and the strains in longitudinal directions ($\varepsilon = \delta/L$).

8.2 Changes in Lengths of Axially Loaded Members

FIGURE 8-1

Spring subjected to an axial
load P

When determining the changes in lengths of axially loaded members, it is convenient to begin with a **coil spring** (Fig. 8-1). Springs of this type are used in large numbers in many kinds of machines and devices—for instance, there are dozens of them in every automobile.

When a load is applied along the axis of a spring, as shown in Fig. 8-1, the spring gets longer or shorter depending upon the direction of the load. If the load acts away from the spring, the spring elongates, and the spring is loaded in *tension*. If the load acts toward the spring, the spring shortens, and it is in *compression*. However, it should not be inferred from this terminology that the individual coils of a spring are subjected to direct tensile or compressive stresses; rather, the coils act primarily in direct shear and torsion (or twisting). Nevertheless, the overall stretching or shortening of a spring is analogous to the behavior of a bar in tension or compression, and so the same terminology is used.

Springs

The elongation of a spring is pictured in Fig. 8-2, where the upper part of the figure shows a spring in its **natural length** L (also called its *unstressed length, relaxed length,* or *free length*), and the lower part of the figure shows the effects of applying a tensile load. Under the action of the force P, the spring lengthens by an amount δ and its final length becomes $L + \delta$. If the material of the spring is **linearly elastic**, the load and elongation will be proportional:

FIGURE 8-2

Elongation of an axially loaded
spring

$$P = k\delta \qquad \delta = fP \qquad \text{(8-1a,b)}$$

in which k and f are constants of proportionality.

The constant k is called the **stiffness** of the spring and is defined as the force required to produce a unit elongation, that is, $k = P/\delta$. Similarly, the constant f is known as the **flexibility** and is defined as the elongation produced by a load of unit value, that is, $f = \delta/P$. Although a spring in tension is used for this discussion, it should be obvious that Eqs. (8-1a and b) also apply to springs in compression.

From the preceding discussion, you can see that the stiffness and flexibility of a spring are the reciprocal of each other:

$$k = \frac{1}{f} \qquad f = \frac{1}{k} \qquad \text{(8-2a,b)}$$

The flexibility of a spring easily can be determined by measuring the elongation produced by a known load, and then the stiffness can be calculated from Eq. (8-2a). Other terms for the stiffness and flexibility of a spring are the **spring constant** and **compliance**, respectively.

The spring properties given by Eqs. (8-1) and (8-2) can be used in the analysis and design of various mechanical devices involving springs, as illustrated later in Example 8-1.

Prismatic Bars

Axially loaded bars elongate under tensile loads and shorten under compressive loads, just as springs do. To analyze this behavior, let us consider the prismatic bar shown in Fig. 8-3. **A prismatic bar** is a structural member having a straight longitudinal axis and constant cross section throughout its length. Although circular bars are used in these illustrations, structural members may have a variety of cross-sectional shapes, such as those shown in Fig. 8-4.

The **elongation** δ of a prismatic bar subjected to a tensile load P is shown in Fig. 8-5. If the load acts through the centroid of the end cross section, the uniform normal stress at cross sections away from the ends is given by the formula $\sigma = P/A$, where A is the cross-sectional area. Furthermore, if the bar is made of a homogeneous material, the axial strain is $\varepsilon = \delta/L$, where δ is the elongation and L is the length of the bar.

Assume that the material is **linearly elastic**, which means that it follows Hooke's law. Then the longitudinal stress and strain are related by the equation $\sigma = E\varepsilon$, where E is the modulus of elasticity. Combining these basic relationships results in the following equation for the elongation of the bar:

$$\delta = \frac{PL}{EA} \tag{8-3}$$

This equation shows that the elongation is directly proportional to the load P and the length L and inversely proportional to the modulus of elasticity E and the cross-sectional area A. The product EA is known as the **axial rigidity** of the bar.

Although Eq. (8-3) was derived for a member in tension, it applies equally well to a member in compression, in which case δ represents the shortening of the bar. Usually you know by inspection whether a member gets longer or shorter; however, there are occasions when a **sign convention** is needed (for

FIGURE 8-3

Prismatic bar of circular cross section

FIGURE 8-4

Typical cross sections of structural members

Solid cross sections

Hollow or tubular cross sections

Thin-walled open cross sections

FIGURE 8-5

Elongation of a prismatic bar in tension

instance, when analyzing a statically indeterminate bar). When that happens, elongation is usually taken as positive and shortening as negative. This is known as a *deformation sign convention*.

The change in length of a bar is normally very small in comparison to its length, especially when the material is a structural metal, such as steel or aluminum. As an example, consider an aluminum strut that is 75.0 in. long and subjected to a moderate compressive stress of 7000 psi. If the modulus of elasticity is 10,500 ksi, the shortening of the strut (from Eq. (8-3) with P/A replaced by σ) is $\delta = 0.050$ in. Consequently, the ratio of the change in length to the original length is 0.05/75, or 1/1500, and the final length is 0.999 times the original length. Under ordinary conditions similar to these, you can use the original length of a bar (instead of the final length) in calculations.

The stiffness and flexibility of a prismatic bar are defined in the same way as for a spring. The stiffness is the force required to produce a unit elongation, or P/δ, and the flexibility is the elongation due to a unit load, or δ/P. Thus, from Eq. (8-3) the **stiffness** and **flexibility** of a prismatic bar are, respectively,

$$k = \frac{EA}{L} \qquad f = \frac{L}{EA} \qquad \textbf{(8-4a,b)}$$

Stiffnesses and flexibilities of structural members, including those given by Eqs. (8-4a and b), have a special role in the analysis of large structures by computer-oriented methods.

Steel cables on a pulley

Cables

Cables are used to transmit large tensile forces, for example, when lifting and pulling heavy objects, raising elevators, guying towers, and supporting suspension bridges. Unlike springs and prismatic bars, cables cannot resist compression. Furthermore, they have little resistance to bending and therefore may be curved as well as straight. Nevertheless, a cable is considered to be an axially loaded member because it is subjected only to tensile forces. Because the tensile forces in a cable are directed along the axis, the forces may vary in both direction and magnitude, depending upon the configuration of the cable.

Cables are constructed from a large number of wires wound in some particular manner. While many arrangements are available depending upon how the cable will be used, a common type of cable, shown in Fig. 8-6, is formed by six *strands* wound helically around a central strand. Each strand is in turn constructed of many wires, also wound helically. For this reason, cables are often referred to as **wire rope**.

The cross-sectional area of a cable is equal to the total cross-sectional area of the individual wires, called the **effective area** or **metallic area**. This area is less than the area of a circle having the same diameter as the cable because there are spaces between the individual wires. For example, the actual cross-sectional area (effective area) of a particular 1.0-in. diameter cable is only 0.471 in^2, whereas the area of a 1.0-in. diameter circle is 0.785 in^2.

Under the same tensile load, the elongation of a cable is greater than the elongation of a solid bar of the same material and same metallic cross-sectional area, because the wires in a cable "tighten up" in the same manner as the fibers in a rope. Thus, the modulus of elasticity (called the **effective modulus**) of a cable is less than the modulus of the material of which it is made. The effective modulus of steel cables is about 20,000 ksi (140 GPa), whereas the steel itself has a modulus of about 30,000 ksi (210 GPa).

When determining the **elongation** of a cable from Eq. (8-3), the effective modulus should be used for E and the effective area should be used for A.

In practice, the cross-sectional dimensions and other properties of cables are obtained from the manufacturers. However, for use in solving problems in this book (and definitely *not* for use in engineering applications), Table 8-1 lists the properties of a particular type of cable. Note that the last column contains the *ultimate load*, which is the load that would cause the cable to break. The *allowable load* is obtained from the ultimate load by applying a safety factor that may range from 3 to 10, depending upon how the cable is to be used. The individual wires in a cable are usually made of high-strength steel, and the calculated tensile stress at the breaking load can be as high as 200,000 psi (1400 MPa).

The following examples illustrate techniques for analyzing simple devices containing springs and bars. The solutions require the use of free-body diagrams, equations of equilibrium, and equations for changes in length. The problems at the end of the chapter provide many additional examples.

FIGURE 8-6

Typical arrangement of strands and wires in a steel cable

Tom Grundy/Shutterstock.com

Nominal Diameter		Approximate Weight		Effective Area		Ultimate Load		**Table 8-1**
in.	mm	lb/ft	N/m	in^2	mm^2	lb	kN	Properties of steel cables[1]
0.50	12	0.42	6.1	0.119	76.7	23,100	102	
0.75	20	0.95	13.9	0.268	173	51,900	231	
1.00	25	1.67	24.4	0.471	304	91,300	406	
1.25	32	2.64	38.5	0.745	481	144,000	641	
1.50	38	3.83	55.9	1.08	697	209,000	930	
1.75	44	5.24	76.4	1.47	948	285,000	1260	
2.00	50	6.84	99.8	1.92	1230	372,000	1650	

[1]To be used solely for solving problems in this book.

Example 8-1

A machine component is modeled as two rigid bars connected to each other by a spring at CD and supported by linearly elastic springs at B and E (see Fig. 8-7). Before load P at C is applied, the bars are parallel and the springs are without stress. Ignore the weight of the bars.

(a) Derive formulas for downward displacement δ_C at C and rotation θ of bar DEF when load P is applied at C. (Assume that the bars rotate through very small angles under the action of the load P.)

(b) Find the forces in both translational springs and the moment in the rotational spring at E.

FIGURE 8-7

Two rigid bars supported by and connected by elastic springs

Solution:

Use the four-step problem-solving approach to find the rotations of both bars, the forces in translational springs, and the moment in the rotational spring.

1. *Conceptualize* [*hypothesize, sketch*]: Bar ABC is expected to displace downward under load P, and bar DEF is expected to rotate counterclockwise, as shown by the dashed lines in Fig. 8-8. As a result, spring forces F_B and F_{CD} and moment M_E in the rotational spring will act as shown in Fig. 8-8. Now sketch a free-body diagram of each of the bars by replacing the translational spring support at B with force $F_B = k\delta_B$ and the rotational spring support at E with moment $M_E = k_R\theta$. The spring connecting the two bars at C and D applies equal but opposite forces $F_{CD} = k(\delta_C - \delta_D)$ at points C and D, which depend on the relative displacement $(\delta_C - \delta_D)$. Force P is

FIGURE 8-8

Free-body diagrams of upper and lower bars joined by a spring connecting points C and D

applied downward at joint C, so translation δ_C is larger than δ_D. A negative result means that the associated force or moment acts opposite to that shown in Fig. 8-8.

2. *Categorize* [*simplify, classify*]: The bars are rigid and the displacements and rotations are small, so it follows that $\delta_B = \delta_C/2$ and $\theta = \delta_D/L$.

3. *Analyze* [*evaluate; select relevant equations, carry out mathematical solution*]:

Equilibrium of bar *ABC*: Sum moments about point A in the upper free-body diagram in Fig. 8-8:

$$\Sigma M_A = 0 \qquad F_B \frac{L}{2} + F_{CD}L - PL = 0 \quad \text{or} \quad \frac{k}{2}\delta_B + k(\delta_C - \delta_D) = P$$

Substitute $\delta_B = \delta_C/2$ and simplify to obtain

$$\frac{5k}{4}\delta_C - k\delta_D = P \tag{a}$$

Equilibrium of bar *DEF*: Next sum moments about point F in the lower free-body diagram in Fig. 8-8:

$$\Sigma M_F = 0 \qquad F_{CD}L - M_E = 0 \quad \text{or} \quad k(\delta_C - \delta_D)L = k_R\theta$$

Substitute $\theta = \dfrac{\delta_D}{L}$ and $k_R = \dfrac{2}{5}kL^2$ and simplify to get

$$\delta_C = \frac{7}{5}\delta_D \tag{b}$$

Finally, substitute Eq. (b) into Eq. (a) and solve for displacements δ_C, δ_D, and rotation θ as

$$\delta_D = \frac{4}{3}\frac{P}{k}, \ \delta_C = \frac{28}{15}\frac{P}{k}, \text{ and } \theta = \frac{4}{3}\frac{P}{kL} \qquad \longleftarrow \text{(c)}$$

Forces in both translational springs and the moment in the rotational spring at *E*: Using the displacements in Eqs. (b) and (c), the forces in both translational springs and the moment in the rotational spring support at E are computed as

$$F_B = k\delta_B = \frac{14}{15}P, \ F_{CD} = k(\delta_C - \delta_D) = \frac{8}{15}P, \text{ and } M_E = k_R\theta = \frac{8}{15}PL \qquad \longleftarrow$$

4. *Finalize* [*conclude; examine answer—Does it make sense? Are units correct? How does it compare to similar problem solutions?*]: The spring force and moment results are all positive, so the assumed directions of force and moment vectors in Fig. 8-8 are correct. Both translational springs are in compression, and the rotational spring at E acts to oppose counterclockwise rotation of bar *DEF*, as expected.

Example 8-2

Two frames are constructed using a rigid beam (ABC) and an elastic column (BD, see Fig. 8-9). Frame 1 has a sliding support at A and load P applied at C. Frame 2 has a pin support at A and moment $M = PL/3$ applied at C. For both frames, column BD has cross-sectional area $A = 8500$ mm^2 and height $L = 725$ mm. Assume that the columns are made of steel with $E = 205$ GPa.

(a) For each frame, find the allowable value of load variable P if the displacement δ_C of point C is limited to 0.75 mm.

(b) If load variable $P = 1200$ kN, what is the required cross-sectional area of each column so that $\delta_C = 0.75$ mm?

FIGURE 8-9

Two frames: rigid beam supported by an elastic column

Solution:

Use the four-step problem-solving approach to find the allowable load P and required cross-sectional areas.

1. *Conceptualize*: The free-body diagrams of rigid beam ABC in Frames 1 and 2 are shown in Fig. 8-10. Use these diagrams to find the force F_{BD} in column BD for each frame.

2. *Categorize*: Rigid beam ABC in Frame 1 cannot rotate because of the sliding support at A, so it will translate downward as shown in the displacement diagram in Fig. 8-11a. As a result, the downward displacement δ_B applied to

FIGURE 8-10

Free-body diagrams of rigid beam ABC in (a) Frame 1 and (b) Frame 2

FIGURE 8-11

Displacement diagrams
for rigid beam *ABC* in
(a) Frame 1 and (b) Frame 2

column *BD* is equal to δ_C. In Frame 2, rigid beam *ABC* must rotate about
the pin support at *A* as shown in the displacement diagram in Fig. 8-11b.
From similar triangles, the resulting downward displacement δ_B at the top of
column *BD* is equal to $\delta_C/2$. All displacements in Fig. 8-11 are assumed to be
small.

3. *Analyze:*

Allowable load variable *P* for Frame 1: Sum forces in the *y* direction in
Fig. 8-10a to find that $F_{BD1} = P$. Use the *force-displacement relation* and
$\delta_B = \delta_C$ to find an expression for allowable load P_{a1} on Frame 1 that satisfies
the requirement that downward displacement at *C* cannot exceed δ_C. Substitute
numerical values to obtain

$$P_{a1} = \frac{EA}{L}\delta_C = \frac{(205 \text{ GPa})(8500 \text{ mm}^2)}{725 \text{ mm}}(0.75 \text{ mm}) = 1803 \text{ kN}$$

Allowable load variable *P* for Frame 2: Sum moments about joint *A* in Fig. 8-10b
to find that $F_{BD2} = 2P/3$. Use the *force-displacement relation* and $\delta_B = \delta_C/2$
to find an expression for allowable load P_{a2} on Frame 2 and then substitute
numerical values:

$$\delta_B = \frac{F_{BD2}L}{EA} = \frac{(2P/3)L}{EA} = \frac{\delta_C}{2}$$

so

$$P_{a2} = \frac{3}{4}\frac{EA}{L}\delta_C = \frac{3}{4}\frac{(205 \text{ GPa})(8500 \text{ mm}^2)}{725 \text{ mm}}(0.75 \text{ mm}) = 1352 \text{ kN}$$

It follows that the allowable value of applied moment *M* at joint *C* is
$P_{a2}L/3 = 327 \text{ kN·m}$.

Required cross-sectional area of column *BD*: Now find the required cross-
sectional area of column *BD* in each frame based on the limiting value of
displacement δ_C but for a specified load variable *P* = 1200 kN. From the
force-displacement relation,

$$A_{BD1} = \frac{F_{BD1}L}{E\delta_B} = \frac{(1200 \text{ kN})(725 \text{ mm})}{(205 \text{ GPa})(0.75 \text{ mm})} = 5659 \text{ mm}^2$$

$$A_{BD2} = \frac{F_{BD2}L}{E\delta_B} = \frac{\left(\frac{2}{3}1200 \text{ kN}\right)(725 \text{ mm})}{(205 \text{ GPa})\left(\dfrac{0.75 \text{ mm}}{2}\right)} = 7545 \text{ mm}^2$$

4. *Finalize*: The allowable load values P_{a1} and P_{a2} computed here are based on a displacement limitation at C: $\delta_C \leq 0.75$ mm. Now also check the axial normal stresses σ in column BD in each frame to be sure that each is in the elastic range when these allowable load values are applied. The computed stresses in column BD in Frames 1 and 2 are

$$\sigma_{BD1} = \frac{F_{BD1}}{A} = \frac{1803 \text{ kN}}{8500 \text{ mm}^2} = 212 \text{ MPa} \quad \text{and} \quad \sigma_{BD2} = \frac{F_{BD2}}{A} = \frac{901 \text{ kN}}{8500 \text{ mm}^2} = 106 \text{ MPa}$$

Both stress values are acceptable, since the yield stress for steel is 250 MPa (see Table F-3 in Appendix F). Note also that both columns are in compression, so the possibility of *buckling instability* will have to be considered in future discussions.

8.3 Changes in Lengths Under Nonuniform Conditions

When a prismatic bar of linearly elastic material is loaded only at the ends, you can obtain its change in length from the equation $\delta = PL/EA$, as described in the preceding section. In this section, you will see how this same equation can be used in more general situations.

Bars with Intermediate Axial Loads

Suppose, for instance, that a prismatic bar is loaded by one or more axial loads acting at intermediate points along the axis (Fig. 8-12a). You can determine the change in length of this bar algebraically by adding the elongations and shortenings of the individual segments. The procedure is as follows.

1. Identify the segments AB, BC, and CD of the bar as 1, 2, and 3, respectively.

2. Determine the internal axial forces N_1, N_2, and N_3 in segments 1, 2, and 3, respectively, from the free-body diagrams of Fig. 8-12b, c, and d. Note that the internal axial forces are denoted by the letter N to distinguish them from the external loads P. By summing forces in the vertical direction, you obtain the following expressions for the axial forces:

$$N_1 = -P_B + P_C + P_D \qquad N_2 = P_C + P_D \qquad N_3 = P_D$$

The deformation sign convention was used in writing these equations (internal axial forces are positive when in tension and negative when in compression). You can also sketch the *axial-force diagram* (AFD) for this bar, which displays the variation of internal axial force $N(x)$ over the length of the bar (Fig. 8-12e).

3. Determine the changes in the lengths of the segments from Eq. (8-3):

$$\delta_1 = \frac{N_1 L_1}{EA} \qquad \delta_2 = \frac{N_2 L_2}{EA} \qquad \delta_3 = \frac{N_3 L_3}{EA}$$

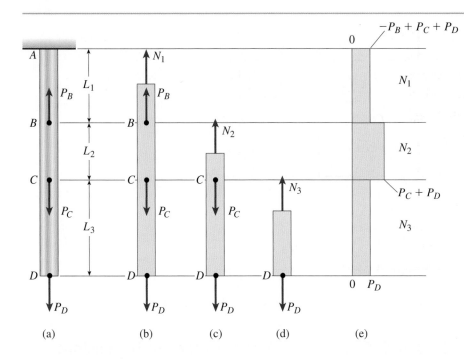

FIGURE 8-12
(a) Bar with external loads acting at intermediate points; (b), (c), and (d) free-body diagrams showing the internal axial forces N_1, N_2 and N_3; (e) axial-force diagram

in which L_1, L_2, and L_3 are the lengths of the segments and EA is the axial rigidity of the bar.

4. Add δ_1, δ_2, and δ_3 to obtain δ, which is the change in length of the entire bar:

$$\delta = \sum_{i=1}^{3} \delta_i = \delta_1 + \delta_2 + \delta_3$$

The changes in lengths must be added algebraically, with elongations being positive and shortenings negative. A plot of displacements at the ends of each bar segment is referred to as the *axial-displacement diagram* (ADD) for the structure.

Bars Consisting of Prismatic Segments

This same general approach can be used when the bar consists of several prismatic segments, each having different axial forces, different dimensions, and different materials (Fig. 8-13a). The change in length is obtained from

$$\delta = \sum_{i=1}^{n} \frac{N_i L_i}{E_i A_i} \tag{8-5}$$

in which the subscript i is a numbering index for the various segments of the bar and n is the total number of segments. Note especially that N_i is not an external load but is the internal axial force in segment i. A sketch of the AFD is given in Fig. 8-13b, and the ADD is shown in Fig. 8-13c. The displacement at A, δ_A, is the sum of segment shortenings $\delta_{AB} + \delta_{BC}$ and that at B, δ_B, is equal to the shortening of segment BC (δ_{BC}). Support C is restrained, so $\delta_C = 0$. The internal forces in each segment are constant within each segment, but internal deformations vary linearly within each segment (Fig. 8-13c).

FIGURE 8-13

(a) Bar consisting of prismatic segments having different axial forces, different dimensions, and different materials; (b) AFD; (c) ADD

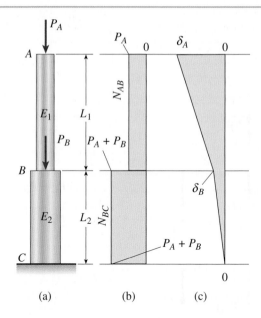

(a)   (b)   (c)

Bars with Continuously Varying Loads or Dimensions

Sometimes the axial force N and the cross-sectional area A vary continuously along the axis of a bar, as illustrated by the tapered bar of Fig. 8-14a. This bar not only has a continuously varying cross-sectional area but also a continuously varying axial force. In this illustration, the load consists of two parts: a single force P_B acting at end B of the bar and distributed forces $p(x)$ acting along the axis. (A distributed force has units of force per unit distance, such as pounds per inch or newtons per meter.) A distributed axial load may be produced by such factors as centrifugal forces, friction forces, or the weight of a bar hanging in a vertical position.

Under these conditions, Eq. (8-5) is no longer useful to obtain the change in length. Instead, determine the change in length of a differential element of the bar and then integrate over the length of the bar.

Select a differential element at a distance x from the left-hand end of the bar (Fig. 8-14c). The internal axial force $N(x)$ acting at this cross section (Fig. 8-14b) may be determined from equilibrium using either segment AC or segment CB as a free body. In general, this force is a function of x. Also, knowing the dimensions of the bar, the cross-sectional area $A(x)$ is expressed as a function of x.

FIGURE 8-14

Bar with varying cross-sectional area and varying axial force

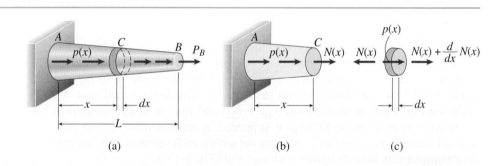

(a)   (b)   (c)

The elongation $d\delta$ of the differential element (Fig. 8-14c) may be obtained from $\delta = PL/EA$ by substituting $N(x)$ for P, dx for L, and $A(x)$ for A, as

$$d\delta = \frac{N(x)dx}{EA(x)} \tag{8-6}$$

The elongation of the entire bar is obtained by integrating over the length:

$$\delta = \int_0^L d\delta = \int_0^L \frac{N(x)dx}{EA(x)} \tag{8-7}$$

If the expressions for $N(x)$ and $A(x)$ are not too complicated, the integral can be evaluated analytically and a formula for δ or $\delta(x)$ can be obtained, as illustrated later in Example 8-4. However, if formal integration is either difficult or impossible, a numerical method for evaluating the integral should be used.

Limitations

Equations (8-5) through (8-7) apply only to bars made of linearly elastic materials, as shown by the presence of the modulus of elasticity E in the formulas. Also, the formula $\delta = PL/EA$ was derived using the assumption that the stress distribution is uniform over every cross section (because it is based on the formula $\sigma = P/A$). This assumption is valid for prismatic bars but not for tapered bars; therefore, Eq. (8-7) gives satisfactory results for a tapered bar only if angle between the sides of the bar is small.

As an illustration, if the angle between the sides of a bar is 20°, the stress calculated from the expression $\sigma = P/A$ (at an arbitrarily selected cross section) is 3% less than the exact stress for that same cross section (calculated by more advanced methods). For smaller angles, the error is even less. Consequently, Eq. (8-7) is satisfactory if the angle of taper is small. If the taper is large, more accurate methods of analysis are needed.

Axial-Displacement Diagram (ADD)

Equilibrium was used to develop a graphical display of the variation in internal axial force $N(x)$ over the length of a bar so critical regions of the bar, such as the location of maximum internal axial force N_{max} (which may also be the location of maximum axial normal stress), could be identified. This display is called the *axial-force diagram* (AFD). Rules for construction of axial-force diagrams are based on the differential and integral relationships between $q(x)$ and $N(x)$. Now use the *force–displacement relation* in Eq. (8-7) to create a display of the variation of axial displacement $\delta(x)$ over the length of the bar, referred to here as an *axial-displacement diagram* (ADD).

To construct the ADD, first note in Eq. 8-7 that internal axial force $N(x)$ is part of the integrand in the expression for axial displacement $\delta(x)$. It follows that the ordinate $N(x)$ on the axial-force diagram is *proportional to* the slope on the axial displacement diagram; the full expression for slope must include the term $EA(x)$ in the denominator in Eq. 8-7. Second, the integral expression indicates that the change in axial displacement between any two points A and B is proportional to the area under the AFD between those same two points:

$$\delta_B - \delta_A = \int_A^B \frac{N(x)}{EA(x)}dx$$

Use these observations to develop the following rules or guidelines for constructing the ADD from the AFD:

- The slope at any point on the ADD is equal to the ordinate on the AFD at that same point divided by the axial rigidity of the bar EA at that same location.
- The change in axial displacement between any two points along a bar is equal to the area under the axial-force diagram between those same two points divided by the axial rigidity of the bar EA over that same interval.

If the bar is piecewise prismatic (so that EA is constant in each bar segment) and the variation of $N(x)$ is constant or linear in x, the ADD can be constructed using simple geometry as illustrated below. Otherwise, analytical or numerical integration of Eq. (8-7) will be required to develop the ADD.

Axial-force diagrams were plotted in Figs. 8-12e and 8-13b for several bars acted on by axial forces. Now consider the equilibrium of the bar in Figs. 8-15a and b but use the AFD in Fig. 8-15c to construct the ADD in Fig. 8-15d. The ADD in Fig. 8-15d is based on the force–displacement relation in Eq. 8-7 but take advantage of the graphical interpretations expressed in the two rules listed above. First, note that support displacement $\delta_A = 0$. Next, the ordinate on the AFD between A and B is constant ($+2P$) and the area under the AFD between A and B is $+PL/2EA$, so the ADD varies linearly with slope $+2P/EA$ and the change in ordinate over this segment is $\delta_B - \delta_A = \delta_B = +PL/2EA$. The change in ordinate from B to C is $-PL/4EA$, so $\delta_C = +PL/4EA$. Finally, the change in ordinate over segment CD is $+PL/2EA$, so $\delta_D = \delta_{max} = +3PL/4EA$. Use a statics sign convention in developing the ADD: Axial displacement $\delta(x)$ is considered positive (and plotted above the reference axis on the ADD) if the axial displacement vector points in the $+x$ direction. Examples 8-3 and 8-4 provide further application of these concepts.

FIGURE 8-15

(a) Prismatic bar subjected to intermediate axial centroidal loads; (b, c, d) overall FBD, axial-force diagram (AFD), and axial-displacement diagram (ADD)

Example 8-3

A section of steel drill pipe hangs from an oil drilling rig just before the start of drilling operations (Fig. 8-16). The pipe has a length L, diameter d, and wall thickness t. Find expressions for the weight W of the pipe, support reaction R at the support point at the top of the pipe, and internal axial force $N(x)$ and displacement $\delta(x)$ at any point x along the pipe due to gravity. Plot both the axial-force diagram (AFD) and the axial-displacement diagram (ADD).

FIGURE 8-16

Prismatic pipe hanging from drill rig

(a)   (b)

Solution:

Use the four-step problem-solving approach.

FIGURE 8-17

Free-body diagram of entire pipe

1. *Conceptualize:* The drill pipe is prismatic and the total pipe weight is W, so the weight per unit length is a uniformly distributed axial load of constant intensity $w = W/L$. The x axis is the longitudinal centroidal axis of the pipe and runs from the support at the top to the free end of the pipe at the bottom. The top support is shown as a fixed support, but this represents axial restraint only. Start with a free-body diagram of the pipe as a whole (Fig. 8-17) to find the support reaction R at the top. A section cut through the pipe at distance x from the top and $L - x$ from the bottom breaks the pipe structure into upper and lower free-body diagrams (Fig. 8-18) and shows the internal axial force $N(x)$.

2. *Categorize:* The drill pipe is subjected to axial forces only at the start of drilling operations. Only internal axial force $N(x)$ is present at any section cut along the pipe; no shear forces or moments are developed in the pipe at this stage. The pipe will increase in length due to uniformly distributed gravitational force $w(x)$.

FIGURE 8-18

Upper and lower free-body diagrams

3. *Analyze*:

Pipe weight W and support reaction R: The cross-sectional area of the circular pipe is

$$A = \frac{\pi}{4}\left(d^2 - (d-2t)^2\right) = \pi t(d-t)$$

The weight density of steel is γ (see Appendix Table F-1 for numerical values of weight density for various materials), so the pipe weight per unit length is $w = \gamma A$ and the total pipe weight is $W = wL$. The hanging pipe is statically determinate, and only axial centroidal forces are considered here. Summing forces in the x direction in Fig. 8-17 gives reaction R:

$$\Sigma F_x = 0 \qquad\qquad R + wL = 0 \quad \text{so} \quad R = -wL = -W$$

Based on a statics sign convention, the negative result indicates that reaction force R acts *upward* in the $-x$ direction.

Internal axial forces $N(x)$ in hanging pipe: Either the upper or lower free-body diagram in Fig. 8-18 can be used to find the internal axial force at any position x along the pipe. Using the upper free-body diagram (UFBD) gives

$$\Sigma F_x = 0 \qquad R + (w)(x) + N(x) = 0 \text{ (UFBD)} \quad \text{or} \quad N(x) = w(L-x)$$

Equilibrium of the lower free-body diagram (LFBD) in Fig. 8-18 yields the same result:

$$\Sigma F_x = 0 \qquad -N(x) + (w)(L-x) = 0 \text{ (LFBD)} \quad \text{or} \quad N(x) = w(L-x)$$

The internal axial force is a linear function of x and is tensile as assumed in Fig. 8-18 based on a deformation sign convention. Substituting values of x into the equation for $N(x)$ shows that maximum tensile force is $N(0) = N_{\max} = W$ at the top support ($x = 0$), decreasing linearly to $N(L) = 0$ at the free end at the bottom of the pipe ($x = L$).

Axial-force diagram (AFD): A plot of internal axial force $N(x)$ is shown in Fig. 8-19 and is referred to as an axial-force diagram. The entire pipe is in tension, so $N(x)$ is plotted to the right of the reference axis (in the $+y$ direction). The plot has constant slope equal to $\dfrac{dN}{dx} = -w$.

Axial-displacement diagram (ADD): Use the force–displacement relation [Eq. (8-7)] (with zeta as a dummy variable for integration) to find an expression for displacement $\delta(x)$ at any location x along the pipe:

$$\delta(x) = \int_0^x \frac{N(\zeta)d\zeta}{EA} = \int_0^x \frac{w(L-\zeta)d\zeta}{EA} = \frac{wL^2}{2EA}\left[\frac{2x}{L} - \left(\frac{x}{L}\right)^2\right]$$

Axial force $N(x)$ is a linear function of x, and displacement $\delta(x)$ is a quadratic function equal to zero at the top support and a maximum at $x = L$:

$$\delta(0) = 0, \quad \delta_{\max} = \delta(L) = \frac{wL^2}{2EA} = \frac{WL}{2EA}$$

FIGURE 8-19

(a) Axial-force diagram (AFD) and (b) axial-displacement diagram (ADD) for entire pipe

(a) AFD, $N(x)$ (b) ADD, $\delta(x)$

The maximum displacement at the lower end of the pipe is one-half of that which would occur if the entire pipe weight W had been applied as a concentrated force at $x = L$. A display of displacement $\delta(x)$ over the height of the pipe is referred to as an axial-displacement diagram, as plotted in Fig. 8-19b.

4. *Finalize*: From the axial-force diagram (Fig. 8-19a), the maximum tensile internal force N_{max} is at the top support. As expected, N_{max} is equal to the pipe weight W. The pipe is prismatic, so maximum stress and strain also occur at the top support.

The distributed load intensity w is constant over the entire pipe. The differential relationship between load intensity w and internal axial force N is $\dfrac{dN}{dx} = -w$, as shown in Fig. 8-14(c) [after replacing load intensity $p(x)$ with $w(x)$ here]. Note that the constant slope of the AFD in the x-y coordinate system is $-w = -(W/L)$, which confirms that *the slope on the AFD is equal to the $(-)$ ordinate on the distributed axial-load diagram as $w(x)$*. In addition, *the area under the axial-load diagram is $-wL$, which is equal to the change in internal force ordinate from start to end of the pipe on the AFD*. This is expressed as

$$N_{bottom} - N_{top} = -W = -\int_0^L w\,dx = -wL.$$

Similar geometric relationships between internal axial force $N(x)$ and displacement $\delta(x)$ are apparent if the AFD and ADD are compared in Fig. 8-19. Note in Eq. (8-7) that internal axial force $N(x)$ is part of the integrand in the expression for axial displacement $\delta(x)$, so it follows that the ordinate $N(x)$ on the axial-force diagram is proportional to the slope on the axial-displacement diagram [the full expression for slope must include the term $EA(x)$ in the denominator in Eq. (8-7)]. In addition, the integral expression means that the change in axial displacement between any two points is proportional to the *area* under the AFD between those same two points. If top $(x = 0)$ and bottom $(x = L)$ of pipe are selected as the two locations, the entire area under the AFD is $\frac{1}{2}\,WL$ and

$$\delta_{bottom} - \delta_{top} = \delta_{max} - 0 = \frac{WL}{2EA}.$$

Example 8-4

FIGURE 8-20

T-frame structure

A T-frame structure is composed of prismatic beam ABC and nonprismatic column DBF; these are joined at B by a frictionless pin connection. The beam has a pin support at A, and the column is pin supported at F (see Fig. 8-20). Beam ABC and column segment DB have a cross-sectional area A; column segment BF has area $2A$. The modulus of elasticity E is the same for both members. Load $2P$ is applied downward at C, and load P acts at D. Consider also the self-weight of both members: The weight density of the material is γ. Find expressions for the downward displacements of column DBF at $D\,(\delta_D)$ and also at $B\,(\delta_B)$. Plot axial-force (AFD) and axial-displacement (ADD) diagrams for column DBF.

Solution:

Use the four-step problem-solving approach to find displacements at joints B and D for this T-frame structure.

1. *Conceptualize*: Remove the pin at B and split the frame into two separate free-body diagrams: one for beam ABC and the other for column DBF (Fig. 8-21). Show reaction force components at A and F; equal but opposite pin force components at B; applied loads $2P$ and P at joints C and D, respectively; uniformly distributed self-weight transverse load γA on beam ABC; and uniformly distributed self-weight axial loads γA on DB and $2\gamma A$ on BF. Use forces in column segments DB and BF in the *force-displacement relations* to find the axial deformations of DB and BF and then joint displacements δ_B and δ_D.

FIGURE 8-21

Free-body diagrams of (a) beam ABC
and (b) column DBF

(a)

(b)

2. *Categorize*: Displacement δ_B at B results from the compression of column segment BF, while displacement δ_D at D is the sum of δ_B and the compression of column segment DB. The separate contributions to displacements from the loads shown in Fig. 8-21b are summarized in Table 8-2. Recall from Example 8-3 that the displacement due to a uniformly distributed axial load w is one-half of that due to a concentrated load or

$$\frac{wL^2}{2EA} = \frac{(\gamma A)L^2}{2EA}$$

Table 8-2	(1) Load Component in Figure 8-21b	(2) Column Segment *DB*	(3) Column Segment *BF*
Contributions of loads in Fig. 8-21b to downward displacement at *D*	Force P at D	$\dfrac{P(L/2)}{EA}$	$\dfrac{P(L/2)}{2EA}$
	Uniform axial load γA on DB	$\dfrac{\gamma A(L/2)^2}{2EA}$	$\dfrac{\gamma A(L/2)(L/2)}{2EA}$
	Force B_y at B	none	$\dfrac{B_y(L/2)}{2EA}$
	Uniform axial load $2\gamma A$ on BF	none	$\dfrac{2\gamma A(L/2)^2}{2(2EA)}$

3. *Analyze*:

Statics—pin force component B_y: Sum moments about joint A in Fig. 8-21a to find pin force B_y:

$$\Sigma M_A = 0 \qquad B_y = \frac{1}{(2L/3)}\left(\gamma AL\left(\frac{L}{2}\right) + 2PL\right) = \frac{3}{4}\gamma AL + 3P \qquad \textbf{(a)}$$

Summing forces in the y direction in Fig. 8-21b gives reaction force component F_1 at the base of column DBF:

$$F_y = P + B_y + (\gamma A + 2\gamma A)\frac{L}{2} = 4P + \frac{9}{4}\gamma AL \qquad \textbf{(b)}$$

Downward displacement at B: After substituting the expression for force B_y from Eq. (a), sum the elements in column 3 of Table 8-2 to find displacement δ_B:

$$\delta_B = \frac{7}{16}\frac{\gamma L^2}{E} + \frac{PL}{EA} \qquad \text{◄ (c)}$$

Downward displacement at D: To find displacement δ_D, sum the elements in column 2 of Table 8-2 and then add δ_B from Eq. (c):

$$\delta_D = \delta_B + \frac{\gamma L^2}{8E} + \frac{PL}{2EA} = \frac{9}{16}\frac{\gamma L^2}{E} + \frac{3}{2}\frac{PL}{EA} \qquad \text{◄ (d)}$$

Now, to simplify axial force and displacement plots, assume that γAL is equal to applied load P. Substitute this expression for γA into Eqs. (a) to (d) to obtain the simplified expressions for forces F_y and B_y and for displacements at B and D:

$$B_y = \frac{15}{4}P \text{ and } F_y = \frac{25}{4}P; \text{ also } \delta_B = \frac{23}{16}\frac{PL}{EA} \text{ and } \delta_D = \frac{33}{16}\frac{PL}{EA}$$

Axial-force diagram (AFD): Start with the free body diagram (Fig. 8-22a) then plot the axial-force diagram (Fig. 8-22b) that shows the variation of internal axial force $N(y)$ over the height of column DBF. Longitudinal centroidal axis y is measured

FIGURE 8-22

(a) Free-body diagram of *DBF*;
(b) internal axial forces *N*;
and (c) axial displacements δ

$F_y = 25P/4$

(a)

AFD, *N(y)*

(b)

ADD, δ*(y)*

(c)

upward from the base at joint *F*. Using a deformation sign convention, negative $N(y)$ indicates compressive internal force. At the base, internal force $N(0) = -F_y$. At the top of column *DBF*, $N(L) = -P$. Just below *B*, the compressive axial force increases by B_y. Between *D* and *B* and between *D* and *F*, $N(y)$ varies linearly, so the AFD consists of straight lines connecting known values of axial force *N* at $y = 0$, $L/2$, and *L*.

Axial-displacement diagram (ADD): The axial-displacement diagram (Fig. 8-22c) is a plot of axial displacement δ(y) over the height of column *DBF*. Note that $\delta(0) = 0$, $\delta(L/2) = \delta_B$, and $\delta(L) = \delta_D$. Internal force $N(y)$ varies linearly over the height of column *DBF*, so downward displacement δ(y) varies quadratically in segments *DB* and *BF*.

4. *Finalize:* Plots of axial force and axial displacement identify locations of maximum force and displacement in the structure. For column *DBF*, the maximum compressive force is at the base and the maximum downward displacement is at the top as expected. Free-body diagrams like those in Fig. 8-21 are essential to the solution for internal forces $N(y)$. Internal forces are then used in the force–displacement relations to find the axial displacement distribution δ(y) over the height of the column.

The diagrams in Fig. 8-22 were drawn by inserting lines between points of known force (AFD) or displacement (ADD). A more rigorous approach is to cut column *DBF* into upper and lower free-body diagrams (see Fig. 8-23) and then to write two functions for internal force $N(y)$ for segments *BF* and *DB*:

$$N(y) = -\frac{25}{4}P + \frac{2P}{L}y \ \text{ if } \ y \le \frac{L}{2} \qquad N(y) = -P - \frac{P}{L}(L - y) \ \text{ if } \ \frac{L}{2} \le y \le L$$

Internal forces *N(y)* are negative because column *DBF* is in compression over its entire height. The axial displacement distribution δ(y) over the height of

FIGURE 8-23

Upper and lower free-body diagrams of column *DBF*

the column is obtained by inserting forces $N(y)$ into the force–displacement relations to obtain the following two functions:

$$\delta(y) = \int_0^y \frac{\left(-\dfrac{25}{4}P + \dfrac{2P}{L}\zeta\right)}{2EA}\, d\zeta = \frac{P}{EA}\left(\frac{y^2}{2L} - \frac{25}{8}y\right) \text{ if } y \leq \frac{L}{2}$$

$$\delta(y) = -\frac{23}{16}\frac{PL}{EA} - \int_{L/2}^y \frac{P + \left(\dfrac{P}{L}\right)(L-\zeta)}{EA}\, d\zeta = \frac{P}{EA}\left(\frac{y^2}{2L} - 2y - \frac{9L}{16}\right) \quad \frac{L}{2} \leq y \leq L$$

Displacements $\delta(y)$ are downward and so are negative. Plots of functions $N(y)$ and $\delta(y)$ are shown in Figs. 8-22b and c, respectively.

Example 8-5

A flat bar has a length L, constant thickness t, and a rectangular cross section whose width varies linearly between b_2 at the fixed support to b_1 at the free end (see Fig. 8-24a). Assume that the taper of the bar is small. The bar has a modulus of elasticity E.

(a) Find an expression for the elongation δ_C of the bar due to axial centroidal load P_C at the free end. What is the elongation δ_B at B due to P_C alone?

(b) Calculate the elongations δ_B and δ_C if $P_C = 62$ kips, $L = 5$ ft, $t = \frac{3}{4}$ in., $b_1 = 4$ in., $b_2 = 6$ in., and $E = 10{,}400$ ksi.

FIGURE 8-24

Flat bar with rectangular cross section: (a) axial load P_C only; (b) axial loads P_B and P_C

(a)   (b)

(c) What load P_B (kips) is required so that $\delta_C = 0$ when both loads are applied as shown in Fig. 8-24b.

Solution:

Use the four-step problem-solving approach to find changes in length of this tapered bar.

1. *Conceptualize*: In Fig. 8-24a, the bar is in tension over its entire length, and the internal axial force $N(x)$ is equal to applied force P_C (see free-body diagram in Fig. 8-25a). When load P_B is added in Fig. 8-24b, the internal tensile force $N(x)$ is

$$N(x) = P_C - P_B \quad \text{for} \quad 0 \leq x \leq L/2$$
$$N(x) = P_C \quad \quad \text{for} \quad L/2 \leq x \leq L$$

Several free-body diagram sketches (Fig. 8-25b and 8-25c) are used as guides to obtain these equations. In all cases, internal force $N(x)$ is initially assumed to be tensile in accordance with a deformation sign convention, so arrowheads for $N(x)$ are directed away from each face of the cut section.

2. *Categorize*: Bar *ABC* has constant axial force over each bar segment but continuously varying cross-sectional dimensions, so an integral form of the *force–displacement relation* [see Eq. (8-7)] must be used to find displacements at *B* and *C*. The cross-sectional area varies linearly with x and depends on bar widths b_1 and b_2 and thickness t as $A(x) = t\left[b_2 - \dfrac{x}{L}(b_2 - b_1)\right]$.

FIGURE 8-25

Internal axial tension force $N(x)$: (a) Free-body diagram for case of applied axial load P_C only; (b) two section cuts in *AB* and *BC* for case of both axial loads P_B and P_C

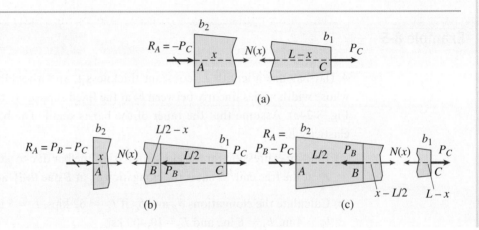

(a)

(b)   (c)

3. *Analyze*:

Part (a): Bar elongations at B and C due to load P_C only.

Apply load P_C at $x = L$ and use Eq. (8-7) to find the elongation of nonprismatic bar *ABC*:

$$\delta_C = \int_0^L \frac{P_C}{EA(x)}dx = \int_0^L \frac{P_C}{Et\left[b_2 - \frac{x}{L}(b_2 - b_1)\right]}dx = \frac{P_C L}{Et(b_2 - b_1)}\ln\left[\frac{b_2}{b_1}\right] \quad\leftarrow$$

This same formula can also be used to find displacement δ_B at *B* due to P_C alone if *L* is replaced with $L/2$ and b_1 is replaced with the bar width at $x = L/2 : b_B = (b_1 + b_2)/2$. The displacement at *B* is

$$\delta_B = \int_0^{L/2} \frac{P_C}{Et\left[b_2 - \frac{x}{L}(b_2 - b_B)\right]}dx = \frac{P_C L/2}{Et(b_2 - b_B)}\ln\left[\frac{b_2}{b_B}\right] = \frac{P_C L}{Et(b_2 - b_1)}\ln\left[\frac{2b_2}{b_1 + b_2}\right] \quad\leftarrow$$

Part (b): Numerical calculations, bar elongations at B and C due to load P_C only.
Bar elongations at *B* and *C* are obtained by substituting the numerical properties into the previous equations for δ_B and δ_C, where $P_C = 62$ kips, $L = 5$ ft, $t = \frac{3}{4}$ in., $b_1 = 4$ in., $b_2 = 6$ in., and $E = 10{,}400$ ksi:

$$\delta_B = \frac{62 \text{ kips } (5 \text{ ft})(12 \text{ in/ft})}{10{,}400 \text{ ksi } (0.75 \text{ in.})2 \text{ in.}}\ln\left[\frac{12}{10}\right] = 0.0435 \text{ in.} \quad\leftarrow$$

$$\delta_C = \frac{62 \text{ kips } (5 \text{ ft})(12 \text{ in/ft})}{10{,}400 \text{ ksi } (0.75 \text{ in.})2 \text{ in.}}\ln\left[\frac{6}{4}\right] = 0.0967 \text{ in.} \quad\leftarrow$$

Part (c): Required load P_B so net displacement at joint C is zero due to loads P_B and P_C. If load P_B *alone* is applied in the negative *x* direction, joints *B* and *C* will displace the same amount and are computed using the expression for δ_B given here, except that P_C is replaced with $(-P_B)$.

Compatibility requires that displacements at *C* sum to zero when both P_B and P_C are applied, which leads to the following solution for the required load P_B:

$$P_B = \frac{Et(b_2 - b_B)}{\frac{L}{2}\ln\left[\frac{b_2}{b_B}\right]}\delta_C = \frac{10{,}400 \text{ ksi } (0.75 \text{ in.})(6 \text{ in.} - 5 \text{ in.})}{2.5 \text{ ft } (12 \text{ in/ft})\ln\left[\frac{6 \text{ in.}}{5 \text{ in.}}\right]}(0.0967 \text{ in.}) = 137.9 \text{ kips} \quad\leftarrow$$

4. *Finalize*:

Note 1: Computed average stresses at various locations *x* along the bar are $\sigma(x) = N(x)/A(x)$. For the combined loading in part (c), stresses are well below yield for this 7075-T6 aluminum alloy (see Table F-3: $\sigma_y = 70$ ksi):

$$\sigma_{nearA} = \frac{(-137.9 + 62)\text{ kips}}{(6\text{ in.})(0.75\text{ in.})} = -16.86\text{ ksi},\quad \sigma_{nearB} = \frac{62\text{ kips}}{(5\text{ in.})(0.75\text{ in.})} = 16.53\text{ ksi},$$

$$\sigma_{nearC} = \frac{62\text{ kips}}{(4\text{ in.})(0.75\text{ in.})} = 20.67\text{ ksi}$$

Note 2: In the limit, as width b_2 approaches width b_1 to produce a prismatic bar with constant cross-sectional area $A = (b_1)(t)$ (and constant thickness t), the force–displacement relation reduces to $\delta = PL/EA$ [see Eq. (8-3)]. To show this, express the above equation for δ_C in terms of $\xi = b_2/b_1$. Then, use L'Hospital's Rule with $f(\xi) = \ln(\xi)$ and $g(\xi) = b_1(\xi - 1)$ to find the displacement δ_C as $b_2 \rightarrow b_1$:

$$\lim_{x\to 1}\frac{f(\xi)}{g(\xi)} = \lim_{x\to 1}\frac{\ln(\xi)}{b_1(\xi-1)} = \lim_{x\to 1}\frac{f'(\xi)}{g'(\xi)} = \frac{1}{b_1}\quad \text{so}\quad \delta_C = \frac{P_C L}{E(b_1 t)} = \frac{P_C L}{EA}$$

8.4 Statically Indeterminate Structures

The springs, bars, and cables covered in the preceding sections have one important feature in common—their reactions and internal forces can be determined solely from free-body diagrams and equations of equilibrium. Structures of this type are classified as **statically determinate**. Note especially that the forces in a statically determinate structure can be found without knowing the properties of the materials. Consider, for instance, the bar AB shown in Fig. 8-26. The calculations for the internal axial forces in both parts of the bar, as well as for the reaction R at the base, are independent of the material of which the bar is made.

Most structures are more complex than the bar of Fig. 8-26, and their reactions and internal forces cannot be found by statics alone. This situation is illustrated in Fig. 8-27, which shows a bar AB fixed at *both* ends. There are now two vertical reactions (R_A and R_B) but only one useful equation of equilibrium—the equation for summing forces in the vertical direction. Since this equation contains two unknowns, it is not sufficient for finding the reactions. Structures of this kind are classified as **statically indeterminate**. To analyze such structures, you must supplement the equilibrium equations with additional equations pertaining to the displacements of the structure.

To see how a statically indeterminate structure is analyzed, consider the example of Fig. 8-28. The prismatic bar AB is attached to rigid supports at both ends and is axially loaded by a force P at an intermediate point C. As already discussed, the reactions R_A and R_B. cannot be found by statics alone, because only one **equation of equilibrium** is available:

$$\Sigma F_{\text{vert}} = 0 \qquad\qquad R_A - P + R_B = 0 \qquad\qquad \text{(8-8)}$$

An additional equation is needed in order to solve for the two unknown reactions.

The additional equation is based upon the observation that a bar with both ends fixed does not change in length. If you separate the bar from its supports

FIGURE 8-26

Statically determinate bar

(Fig. 8-28), you obtain a bar that is free at both ends and loaded by the three forces; R_A, R_B, and P. These forces cause the bar to change in length by an amount δ_{AB}, which must be equal to zero:

$$\delta_{AB} = 0 \tag{8-9}$$

This equation, called an **equation of compatibility**, expresses the fact that the change in length of the bar must be compatible with the conditions at the supports.

In order to solve Eqs. (8-8) and (8-9), express the compatibility equation in terms of the unknown forces R_A and R_B. The relationships between the forces acting on a bar and its changes in length are known as **force–displacement relations**. These relations have various forms depending upon the properties of the material. If the material is linearly elastic, the equation $\delta = PL/EA$ can be used to obtain the force–displacement relations.

Assume that the bar of Fig. 8-28 has a cross-sectional area A and is made of a material with a modulus E. Then the changes in lengths of the upper and lower segments of the bar are, respectively,

$$\delta_{AC} = \frac{R_A a}{EA} \qquad \delta_{CB} = -\frac{R_B b}{EA} \tag{8-10a,b}$$

where the minus sign indicates a shortening of the bar. Equations (8-10a and b) are the force–displacement relations.

Now solve simultaneously the three sets of equations (the equation of *equilibrium*, the equation of *compatibility*, and the *force–displacement* relations). This illustration begins by combining the force–displacement relations with the equation of compatibility:

$$\delta_{AB} = \delta_{AC} + \delta_{CB} = \frac{R_A a}{EA} - \frac{R_B b}{EA} = 0 \tag{8-11}$$

Note that this equation contains the two reactions as unknowns.

FIGURE 8-27

Statically indeterminate bar

FIGURE 8-28

Analysis of a statically indeterminate bar: (a) statically indeterminate bar; (b) free-body diagram; (c) axial-force diagram; (d) axial-displacement diagram

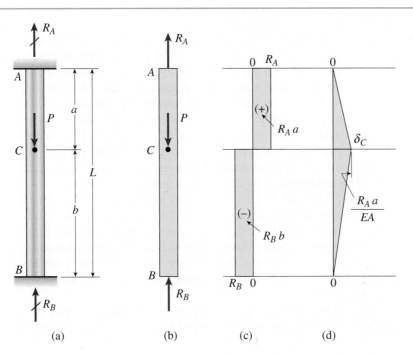

(a)   (b)   (c)   (d)

The next step is to solve simultaneously the equation of equilibrium (Eq. 8-8) and the preceding equation [Eq. (8-11)]. The results are

$$R_A = \frac{Pb}{L} \qquad R_B = \frac{Pa}{L} \qquad \text{(8-12a,b)}$$

With the reactions known, all other force and displacement quantities can be determined. Suppose, for instance, that you wish to find the downward displacement δ_C of point C. This displacement is equal to the elongation of segment AC:

$$\delta_C = \delta_{AC} = \frac{R_A a}{EA} = \frac{Pab}{LEA} \qquad \text{(8-13)}$$

Also, you can find the stresses in the two segments of the bar directly from the internal axial forces (e.g., $\sigma_{AC} = R_A/A = Pb/AL$). A plot of the distribution of internal axial forces N (the AFD) is shown in Fig. 8-28c. The variation of axial displacements δ over the height of the bar (the ADD) is given in Fig. 8-28d. The bar is fully restrained at A and B, so displacements $\delta_A = \delta_B = 0$ as shown on the ADD.

General Comments

From the preceding discussion, the analysis of a statically indeterminate structure involves setting up and solving equations of *equilibrium*, equations of *compatibility*, and *force–displacement* relations. The equilibrium equations relate the loads acting on the structure to the unknown forces (which may be reactions or internal forces), and the compatibility equations express conditions on the displacements of the structure. The force–displacement relations are expressions that use the dimensions and properties of the structural members to relate the forces and displacements of those members. In the case of axially loaded bars that behave in a linearly elastic manner, the relations are based upon the equation $\delta = PL/EA$. Finally, all three sets of equations may be solved simultaneously for the unknown forces and displacements.

In the engineering literature, various terms are used for the conditions expressed by the equilibrium, compatibility, and force–displacement equations. The equilibrium equations are also known as *static* or *kinetic* equations; the compatibility equations are sometimes called *geometric* equations, *kinematic* equations, or equations of *consistent deformations*; and the force–displacement relations are often referred to as *constitutive relations* (because they deal with the *constitution*, or physical properties, of the materials).

For the relatively simple structures discussed in this chapter, the preceding method of analysis is adequate. However, more formalized approaches are needed for complicated structures. Two commonly used methods, the *flexibility method* (also called the *force method*) and the *stiffness method* (also called the *displacement method*), are described in detail in textbooks on structural analysis. Even though these methods are normally used for large and complex structures requiring the solution of hundreds and sometimes thousands of simultaneous equations, they still are based upon the concepts described previously, that is, equilibrium equations, compatibility equations, and force–displacement relations.

The following three examples illustrate the methodology for analyzing statically indeterminate structures consisting of axially loaded members.

Example 8-6

Find expressions for all support reaction forces in the plane frame with load $2P$ applied at C shown in Fig. 8-29. Assume that member AC is a flat prismatic bar of length L, width b, and thickness t. Beam ABC is pinned to column DBF at mid-height (point B). Column DBF has constant thickness t and tapers linearly from width $b_1 = b$ at D to width $b_2 = 5b/4$ at F. Consider load $2P$ at C only; neglect the weights of all members. The modulus of elasticity E is the same for both members.

FIGURE 8-29

T-frame structure

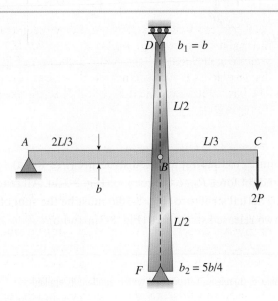

Solutions:

Use the four-step problem-solving approach to find support reactions for this T-frame structure.

1. *Conceptualize*: The frame is one-degree statically indeterminate: There are five reaction force components (two each at joints A and F, one at D; see Fig. 8-30a) but only three overall equilibrium equations plus one additional equation due to the pin at B. Remove the pin at B and split the frame into two separate free-body diagrams: One of beam ABC and the other of column DBF (Fig. 8-30b). Show equal but opposite pin force components at B on each separate free-body diagram.

2. *Categorize*: Select reaction R_D as the redundant force and use superposition to develop one additional equation based on the *compatibility* of displacements at joint D.

   Superposition of released structures: Redundant force R_I is removed to create the released, or statically determinate, structure. Apply the actual load, $2P$

FIGURE 8-30

(a) Overall free-body diagram; (b) free-body diagrams of beam *ABC* and column *DBF*

(a)   (b)

at joint *C*, to the first released structure (Fig. 8-31a) and apply the unknown redundant force R_D to the second (Fig. 8-31b). All reactions and displacements in the actual structure (Fig. 8-30a) must be the sum of corresponding items in the two released structures (Fig. 8-31a and b).

FIGURE 8-31

(a) Released structure 1 under actual load; (b) released structure 2 with redundant R_D applied

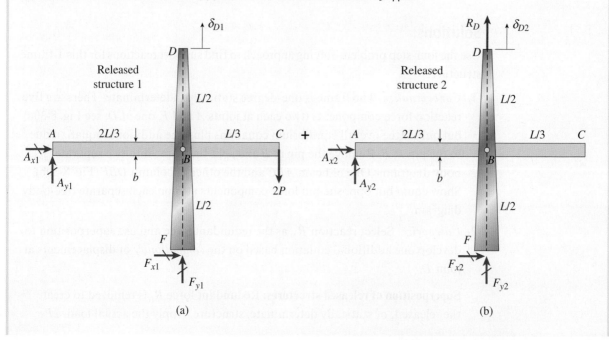

(a)   (b)

Compatibility equation: The vertical displacement at joint D in the actual structure (Fig. 8-30a) is zero. Superposition of displacements in Fig. 8-31a and b at joint D gives the compatibility equation:

$$\delta_D = 0 = \delta_{D1} + \delta_{D2} \tag{a}$$

Applying the force–displacement relation to the released structure in Fig. 8-31a will lead to an expression for δ_{D1} in terms of applied load $2P$ at joint C. Similarly, solution for displacement δ_{D2} in Fig. 8-31b will give an expression in terms of unknown reaction R_D. Substitution of both expressions into Eq. (a) will give reaction R_D. The final step is the application of static *equilibrium* to find all remaining reactions.

3. *Analyze:*

Find displacement δ_{D1} in released structure 1: Remove the pin at B and draw separate free-body diagrams for beam ABC and column DBF (Fig. 8-32a). Sum moments about joint A in beam ABC to find that $B_{y1} = 3P$. Apply B_{y1} to column DBF to see that the internal force in segment DB is zero and segment BF has a compressive internal force of $3P$. The width of the tapered bar is $b_B = 9b/8$ at B and $b_2 = 5b/4$ at F. The taper of bar DBF is small, so use the displacement expressions derived in Example 8-5 to find that the downward displacement at D in released structure 1 is

$$\delta_{D1} = \frac{(-3P)\left(\dfrac{L}{2}\right)}{Et(b_2 - b_B)} \ln\left(\frac{b_2}{b_B}\right) = -\frac{PL}{Ebt}(12)\ln\left(\frac{10}{9}\right) = -1.2643\frac{PL}{Ebt} \tag{b}$$

FIGURE 8-32

Beam and column free-body diagrams for (a) released structure 1 and (b) released structure 2

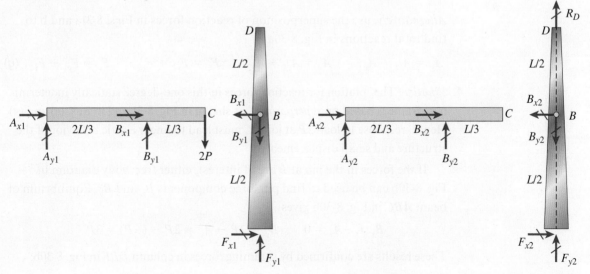

(a)   (b)

Find displacement δ_{D2} in released structure 2: Summing moments about joint A in Fig. 8-32b gives $B_{y2} = 0$. Next sum forces in the y direction in the free-body diagram of column DBF to find that $F_{y2} = -R_D$. Column DBF has a constant internal force equal to applied load R_D, and the width of bar DBF is $b_1 = b$ at D. Thus, the upward displacement at D in released structure 2 is

$$\delta_{D2} = \frac{(R_D L)}{Et(b_2 - b_1)} \ln\left(\frac{b_2}{b_1}\right) = \frac{R_D L}{Ebt}(4)\ln\left(\frac{5}{4}\right) = 0.8926\frac{R_D L}{Ebt} \qquad \textbf{(c)}$$

Solve compatibility equation to find redundant R_D: Substitute expressions for δ_{D1} and δ_{D2} into Eq. (a) and solve for redundant reaction force R_D:

$$R_D = P\frac{1.2643}{0.8926} = 1.416P \qquad \text{←} \quad \textbf{(d)}$$

Equilibrium to find remaining reaction forces: With redundant R_D now known, reaction force components at joints A and F are found using the free-body diagrams in Fig. 8-30. From beam ABC in Fig. 8-30b,

$$\Sigma M_B = 0 \qquad A_y = \frac{-1}{\left(\dfrac{2L}{3}\right)}\left(2P\left(\frac{L}{3}\right)\right) = -P \qquad \text{←} \quad \textbf{(e)}$$

Now use the overall free-body diagram in Fig. 8-30 to find all remaining reactions:

$$\Sigma M_F = 0 \qquad A_x = \frac{-1}{\left(\dfrac{L}{2}\right)}\left(A_y\left(\frac{2L}{3}\right) + 2P\left(\frac{L}{3}\right)\right) = 0$$

$$\Sigma F_x = 0 \qquad F_x = -A_x = 0 \qquad \text{←} \quad \textbf{(f)}$$
$$\Sigma F_y = 0 \qquad F_y = 2P - A_y - R_D = 1.584P$$

Alternatively, use the superposition of reaction forces in Figs. 8-31a and b to find total reactions in Fig. 8-30a as

$$A_x = A_{x1} + A_{x2} \qquad A_y = A_{y1} + A_{y2} \qquad F_x = F_{x1} + F_{x2} \qquad F_y = F_{y1} + F_{y2} \quad \textbf{(g)}$$

4. *Finalize:* The solution for reaction forces in this one-degree statically indeterminate frame is based on *superposition*, as shown in Fig. 8-31 and in Eqs. (a) and (g). Hence, response to load $2P$ at joint C must lead to linear elastic behavior of the structure and small displacements.

If the forces in the pin at B are of interest, either free-body diagram in Fig. 8-30b can be used to find pin force components B_x and B_y. Equilibrium of beam ABC in Fig. 8-30b gives

$$B_x = -A_x = 0 \qquad B_y = 2P - A_y = 2P - (-P) = 3P$$

These results are confirmed by summing forces in column DBF in Fig. 8-30b:

$$B_x = -F_x = 0 \qquad B_y = R_D + F_y = 1.416P + 1.584P = 3P$$

Example 8-7

Rigid beam *OABC* is part of the support structure for a building entrance canopy (Fig. 8-33a). Beam *OABC* is pinned at end *O* and supported by two cables (*AD* and *BD*) at points *A* and *B* (Fig. 8-33b). The combined canopy and cable weight of $W = 1700$ lb is applied as a concentrated load between *A* and *B* and, vertical load *P* acts at joint *C*. The beam has a length of 15 ft. Both cables have a nominal diameter of 0.50 in., an effective cross-sectional area of 0.119 in^2 (see Table 8-1), and an effective modulus of 20,000 ksi. Find the maximum load *P* that can act at *C* if the maximum permissible force in each cable is 4.6 kips.

FIGURE 8-33

(a) Inclined canopy at entrance to building; (b) two-dimensional model of one beam and two supporting cables

(a)   (b)

Solution:

Use the four-step problem-solving approach.

1. *Conceptualize*: Start with a free-body diagram of the entire structure (Fig. 8-34a), and then pass a horizontal plane through the cables to create upper and lower free-body diagrams. The lower free-body diagram in Fig. 8-34b will be of most use here because the forces T_1 and T_2 in cables *AD* and *BD*, respectively, are exposed.

2. *Categorize*: This structure was studied in Example 7-9, but in that example, a continuous cable running over a pulley at *D* supported the canopy in place of the two separate cables *AD* and *BD* used here. Two unknown cable forces T_1 and T_2 replace the single unknown cable force *T* in Example 7-9, so beam OABC in Fig. 8-34b is *one degree statically indeterminate*: There are four unknowns (R_{Ox}, R_{Oy}, T_1, and T_2) but only three independent *equations of equilibrium*. A *compatibility equation* is needed to find another relationship between the two unknown cable forces.

FIGURE 8-34

Free-body diagrams of (a) beam and cable structure and (b) beam *OABC*

             (a)   (b)

3. *Analyze*:

Equation of equilibrium: Sum moments about point *O* in Fig. 8-34b to find an equilibrium equation containing cable forces T_1 and T_2:

$$\Sigma M_O = 0 \qquad T_1(x_A \sin\theta_{Ax} + y_A \cos\theta_{Ax}) + T_2(x_B \sin\theta_{Bx} + y_B \cos\theta_{Bx})$$
$$= Wx_W + Px_C \tag{a}$$

where numerical values of variables in Eq. (a) are computed as shown in Table 8-3.

Table 8-3	Moment arm distances from point *O*	$x_A = (60 \text{ in.})\cos(20°) = 56.382 \text{ in.}$
Numerical values of variables for use in Eq. (a)		$y_A = (60 \text{ in.})\sin(20°) = 20.521 \text{ in.}$
		$x_B = 112.763 \text{ in.}, \quad y_B = 41.042 \text{ in.}$
		$x_C = 169.145 \text{ in.}, \quad x_W = 84.572 \text{ in.}$
	Initial cable lengths using law of cosines	$L_1 = \sqrt{5^2 + 8^2 - 2(5)(8)\cos 70°} \text{ ft} = 7.851 \text{ ft}$
		$L_2 = \sqrt{10^2 + 8^2 - 2(10)(8)\cos 70°} \text{ ft} = 10.454 \text{ ft}$
	Angles between cables and beam using law of sines	$\theta_A = \sin^{-1}\left(\dfrac{8 \text{ ft}}{L_1}\sin 70°\right) = 73.241°$
		$\theta_B = \sin^{-1}\left(\dfrac{8 \text{ ft}}{L_2}\sin 70°\right) = 45.983°$
		$\theta_{Ax} = \theta_A - 20° = 53.241°, \quad \theta_{Bx} = \theta_B - 20° = 25.983°$

Inserting numerical values from Table 8-3 into moment equilibrium Eq. (a) using inch and pound units gives

$$57.452T_1 + 86.296T_2 = 143{,}773 \text{ lb-in} + 169.14P \qquad \textbf{(b)}$$

Equation of compatibility: Rigid beam $OABC$ rotates clockwise about point O due to loads W and P. The displaced position of the beam is straight line $OA'B'C'$ in Fig. 8-35 with the understanding that all displacements are very small. Displacements Δ_A and Δ_B of points A and B, respectively, are normal to line $OABC$. Use of similar triangles OAA' and OBB' gives the compatibility equation:

$$\Delta_B = 2\Delta_A \qquad \textbf{(c)}$$

Force-displacement relations: Displacements in Fig. 8-35 are very small but are shown sufficiently large so that geometric relationships between displacements Δ_A and Δ_B of points A and B and cable elongations δ_1 for cable AD and δ_2 for cable BD can be written as

$$\Delta_A = \frac{\delta_1}{\sin\theta_A} \text{ and } \Delta_B = \frac{\delta_2}{\sin\theta_B}$$

Substitute these expressions for Δ_A and Δ_t into Eq. (c) to find:

$$\delta_2 = 2\left(\frac{\sin\theta_B}{\sin\theta_A}\right)\delta_1$$

The cables behave in a linearly elastic manner, so their elongations are written in terms of their flexibilities f in the *force–displacement relations*:

$$\delta_1 = \left(\frac{L_1}{EA}\right)T_1 = f_1 T_1 \quad \text{and} \quad \delta_2 = \left(\frac{L_2}{EA}\right)T_2 = f_2 T_2$$

It follows that

$$f_2 T_2 = 2\left(\frac{\sin\theta_B}{\sin\theta_A}\right)f_1 T_1 \quad \text{or} \quad T_2 = 2\left(\frac{\sin\theta_B}{\sin\theta_A}\right)\left(\frac{f_1}{f_2}\right)T_1$$

FIGURE 8-35

Displacement diagram

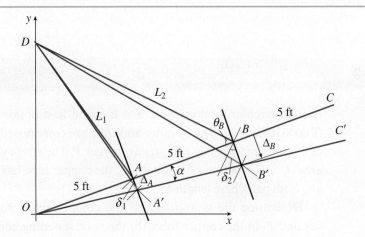

Inserting numerical values gives

$$T_2 = 2\left(\frac{\sin 45.983°}{\sin 73.241°}\right)\left(\frac{3.958(10^{-5})}{5.271(10^{-5})}\right)T_1 = 1.1281\,T_1 \qquad \textbf{(d)}$$

This relationship between cable forces T_1 and T_2 is a result of inserting the force–displacement relations into the compatibility equation [Eq. (c)].

Solution of equations: A constraint here is that neither cable force T_1 nor T_2 can exceed 4600 lb. From Eq. (d), force T_2 is larger than T_1 and will reach the 4600 lb limit first as force P at point C is increased in magnitude. Substitution of Eq. (d) into the equilibrium equation [Eq. (b)] and substitution of $T_2 = 4600$ lb leads to a solution for P_{max}:

$$P_{max} = \frac{1}{169.14}\left(57.452\left(\frac{T_2}{1.1281}\right) + 86.296\,T_2 - 143,773\right)\text{lb} = 2882\text{ lb} \qquad \textbf{(e)}$$

4. *Finalize*: When $P_{max} = 2882$ lb at point C, cable BD reaches the limit force value of $T_2 = 4600$ lb, and cable AD has tension force $T_1 = \dfrac{4600\text{ lb}}{1.1281} = 4077$ lb.

Cable flexibilities f are known, so use the force–displacement relations to find cable elongations of $\delta_1 = f_1 T_1 = 0.1614$ in. and $\delta_2 = f_2 T_2 = 0.2425$ in. Next, downward deflections at points A and B (normal to beam $OABC$, see Fig. 8-35)

are $\Delta_A = \dfrac{\delta_1}{\sin\theta_A} = 0.1686$ in. and $\Delta_B = \dfrac{\delta_2}{\sin\theta_B} = 0.3371$ in. So deflections in

x and y directions at point B are, for example, $\Delta_{Bx} = \Delta_B \sin(20°) = 0.1153$ in., $\Delta_{By} = -\Delta_B \cos(20°) = -0.3168$ in.

Using deflection Δ_B, the clockwise rotation of beam $OABC$ is computed as $\alpha = \Delta_B/120$ in. $= 2.81(10^{-3})$ radians $= 0.161°$, which confirms the assumption of small displacements. Finally, reactions at supports O and D (see Fig. 8-34a) can be computed using statics now that cable forces T_1 and T_2 are known.

Example 8-8

A solid circular steel cylinder S is encased in a hollow circular copper tube C (Fig. 8-36a and b). The cylinder and tube are compressed between the rigid plates of a testing machine by compressive forces P. The steel cylinder has cross-sectional area A_s and modulus of elasticity E_s, the copper tube has area A_c and modulus E_c, and both parts have length L.

Determine the quantities: (a) the compressive forces P_s in the steel cylinder and P_c in the copper tube; (b) the corresponding compressive stresses σ_s and σ_c; and (c) the shortening δ of the assembly.

FIGURE 8-36

Example 8-8: Analysis of
a statically indeterminate
structure

(a)

(b)

(c)

(d)

Solution:

The four-step problem-solving approach is used in the solution; only essential steps are shown in each part.

Part (a): Compressive forces in the steel cylinder and copper tube.

1. *Conceptualize*: Begin by removing the upper plate of the assembly in order to expose the compressive forces P_s and P_c acting on the steel cylinder and copper tube, respectively (Fig. 8-36c). The force P_s is the resultant of the uniformly distributed stresses acting over the cross section of the steel cylinder, and the force P_c is the resultant of the stresses acting over the cross section of the copper tube.

2. *Categorize*:

   Equation of equilibrium: A free-body diagram of the upper plate is shown in Fig. 8-36d. This plate is subjected to the force P and to the unknown compressive forces P_s and P_c; thus, the equation of equilibrium is

$$\Sigma F_{\text{vert}} = 0 \qquad P_s + P_c - P = 0 \qquad \textbf{(a)}$$

$$P_s + P_c - P = 0$$

This equation, which is the only nontrivial equilibrium equation available, contains two unknowns. Therefore, the structure is statically indeterminate.

Equation of compatibility: Because the end plates are rigid, the steel cylinder and copper tube must shorten by the same amount. Denoting the shortening of the steel and copper parts by δ_s and δ_c, respectively, results in the equation of compatibility:

$$\delta_s = \delta_c \tag{b}$$

Force–displacement relations: The changes in lengths of the cylinder and tube can be obtained from the general equation $\delta = PL/EA$. Therefore, in this example the force–displacement relations are

$$\delta_s = \frac{P_s L}{E_s A_s} \qquad \delta_c = \frac{P_c L}{E_c A_c} \tag{c,d}$$

3. *Analyze:* Now solve simultaneously the three sets of equations. First, substitute the force–displacement relations in the equation of compatibility [Eq. (b)], which gives

$$\frac{P_s L}{E_s A_s} = \frac{P_c L}{E_c A_c} \tag{e}$$

This equation expresses the compatibility condition in terms of the unknown forces.

Next, solve simultaneously the equation of equilibrium [Eq. (a)] and the preceding equation of compatibility [Eq. (e)] and obtain the axial forces in the steel cylinder and copper tube:

$$P_s = P\left(\frac{E_s A_s}{E_s A_s + E_c A_c}\right) \qquad P_c = P\left(\frac{E_c A_c}{E_s A_s + E_c A_c}\right) \tag{8-14a,b}$$

4. *Finalize:* These equations show that the compressive forces in the steel and copper parts are directly proportional to their respective axial rigidities and inversely proportional to the sum of their rigidities.

Part (b): Compressive stresses in the steel cylinder and copper tube.

3. *Analyze:* Knowing the axial forces, now obtain the compressive stresses in the two materials:

$$\sigma_s = \frac{P_s}{A_s} = \frac{P E_s}{E_s A_s + E_c A_c} \qquad \sigma_c = \frac{P_c}{A_c} = \frac{P E_c}{E_s A_s + E_c A_c} \tag{8-15a,b}$$

4. *Finalize:* Note that the ratio σ_s/σ_c of the stresses is equal to the ratio E_s/E_c of the moduli of elasticity, showing that in general the "stiffer" material always has the larger stress.

Part (c): Shortening of the assembly.

2. *Categorize*: The shortening δ of the entire assembly can be obtained from either Eq. (c) or (d). Substituting the forces [from Eqs. (8-14a and b)] gives

$$\delta = \frac{P_s L}{E_s A_s} = \frac{P_c L}{E_c A_c} = \frac{PL}{E_s A_s + E_c A_c} \quad \text{(8-16)}$$

4. *Finalize*: This result shows that the shortening of the assembly is equal to the total load divided by the sum of the stiffness of the two parts [recall from Eq. (8-4a) that the stiffness of an axially loaded bar is $k = EA/L$].

Alternative solution of the equations: Instead of substituting the force–displacement relations [Eqs. (c) and (d)] into the equation of compatibility, those relations can be written in the form

$$P_s = \frac{E_s A_s}{L}\delta_s \qquad P_c = \frac{E_c A_c}{L}\delta_c \qquad \text{(f,g)}$$

and substituted into the equation of equilibrium [Eq. (a)];

$$\frac{E_s A_s}{L}\delta_s + \frac{E_c A_c}{L}\delta_c = P \qquad \text{(h)}$$

This equation expresses the equilibrium condition in terms of the unknown displacements. Then solve simultaneously the equation of compatibility [Eq. (b)] and the preceding equation, thus obtaining the displacements:

$$\delta_s = \delta_c = \frac{PL}{E_s A_s + E_c A_c} \qquad \text{(i)}$$

which agrees with Eq. (8-16). Finally, substitute expression (i) into Eqs. (f) and (g) and obtain the compressive forces P_s and P_c [(see Eqs. (8-14a and b)].

Note: The alternative method of solving the equations is a simplified version of the **stiffness** (or **displacement**) **method of analysis**, and the first method of solving the equations is a simplified version of the **flexibility** (or **force**) **method**. The names of these two methods arise from the fact that Eq. (h) has displacements as unknowns and stiffnesses as coefficients [see Eq. (8-4a)], whereas Eq. (e) has forces as unknowns and flexibilities as coefficients [see Eq. (8-4b)].

8.5 Thermal Effects, Misfits, and Prestrains

External loads are not the only sources of stresses and strains in a structure. Other sources include *thermal effects* arising from temperature changes, *misfits* resulting from imperfections in construction, and *prestrains* that are produced by initial deformations. Still other causes are settlements (or movements) of supports, inertial loads resulting from accelerating motion, and natural phenomenon such as earthquakes.

Thermal effects, misfits, and prestrains are commonly found in both mechanical and structural systems and are described in this section. As a general rule, they are much more important in the design of statically indeterminate structures than in statically determinate ones.

Thermal Effects

Changes in temperature produce expansion or contraction of structural materials, resulting in **thermal strains** and **thermal stresses**. A simple illustration of thermal expansion is shown in Fig. 8-37, where the block of material is unrestrained and therefore free to expand. When the block is heated, every element of the material undergoes thermal strains in all directions, and consequently, the dimensions of the block increase. If corner A is taken as a fixed reference point and side AB maintains its original alignment, the block will have the shape shown by the dashed lines.

For most structural materials, thermal strain ε_T is proportional to the temperature change ΔT; that is,

$$\varepsilon_T = \alpha(\Delta T) \tag{8-17}$$

in which α is a property of the material called the **coefficient of thermal expansion**. Since strain is a dimensionless quantity, the coefficient of thermal expansion has units equal to the reciprocal of temperature change. In SI units, the dimensions of α can be expressed as either 1/K (the reciprocal of kelvins) or 1/°C (the reciprocal of degrees Celsius). The value of α is the same in both cases because a *change* in temperature is numerically the same in both kelvins and degrees Celsius. In USCS units, the dimensions of α are 1/°F (the reciprocal of degrees Fahrenheit). Typical values of α are listed in Table F-4 of Appendix F.

When a **sign convention** is needed for thermal strains, assume that expansion is positive and contraction is negative.

To demonstrate the relative importance of thermal strains, compare thermal strains with load-induced strains in the following manner. Suppose you have an axially loaded bar with longitudinal strains given by the equation $\varepsilon = \sigma/E$, where σ is the stress and E is the modulus of elasticity. Then suppose you have an identical bar subjected to a temperature change ΔT, which means that the bar has thermal strains given by Eq. (8-17). Equating the two strains gives

$$\sigma = E\alpha(\Delta T)$$

Use this equation to calculate the axial stress σ that produces the same strain as does the temperature change ΔT. For instance, consider a stainless steel bar with $E = 30 \times 10^6$ psi and $\alpha = 9.6 \times 10^{-6}$ /°F. A quick calculation from the preceding equation for σ shows that a change in temperature of 100°F produces the same strain as a stress of 29,000 psi. This stress is in the range of typical allowable stresses for stainless steel. Thus, a relatively modest change in temperature produces strains of the same magnitude as the strains caused by ordinary loads, which shows that temperature effects can be important in engineering design.

Ordinary structural materials expand when heated and contract when cooled; therefore, an increase in temperature produces a positive thermal strain. Thermal strains usually are reversible in the sense that the member returns to its original shape when its temperature returns to the original value. However, a few special metallic alloys have recently been developed that do not behave in the customary manner. Instead, over certain temperature ranges, their dimensions decrease when heated and increase when cooled.

Water is also an unusual material from a thermal standpoint—it expands when heated at temperatures above 4°C and also expands when cooled below 4°C. Thus, water has its maximum density at 4°C.

Now return to the block of material shown in Fig. 8-37. Assume that the material is homogeneous and isotropic and that the temperature increase ΔT is uniform throughout the block. Calculate the increase in *any* dimension of the block by multiplying the original dimension by the thermal strain. For instance, if one of the dimensions is L, then that dimension will increase by the amount

$$\delta_T = \varepsilon_T L = \alpha(\Delta T)L \qquad \textbf{(8-18)}$$

Equation (8-18) is a **temperature–displacement relation**, analogous to the force–displacement relations described in the preceding section. It can be used to calculate changes in lengths of structural members subjected to uniform temperature changes, such as the elongation δ_T of the prismatic bar shown in Fig. 8-38. (The transverse dimensions of the bar also change, but these changes are not shown in the figure since they usually have no effect on the axial forces being transmitted by the bar.)

The preceding discussions of thermal strains assumed that the structure had no restraints and was able to expand or contract freely. These conditions exist when an object rests on a frictionless surface or hangs in open space. In such cases, no stresses are produced by a uniform temperature change throughout the object, although nonuniform temperature changes may produce internal stresses. However, many structures have supports that prevent free expansion and contraction, in which case **thermal stresses** will develop even when the temperature change is uniform throughout the structure.

To illustrate some of these ideas about thermal effects, consider the two-bar truss ABC of Fig. 8-39 and assume that the temperature of bar AB is changed by ΔT_1 and the temperature of bar BC is changed by ΔT_2. Because the truss is statically determinate, both bars are free to lengthen or shorten, resulting in a displacement of joint B. However, there are no stresses in either bar and no reactions at the supports. This conclusion applies generally to **statically determinate structures**; that is, uniform temperature changes in the members produce thermal strains (and the corresponding changes in lengths) without producing any corresponding stresses.

A **statically indeterminate structure** may or may not develop temperature stresses, depending upon the character of the structure and the nature of the temperature changes. To illustrate some of the possibilities, consider the statically indeterminate truss shown in Fig. 8-40. Because the supports of this structure permit joint D to move horizontally, no stresses are developed when the *entire* truss is heated uniformly. All members increase in length in proportion to their original lengths, and the truss becomes slightly larger in size.

However, if some bars are heated and others are not, thermal stresses will develop because the statically indeterminate arrangement of the bars prevents free expansion. To visualize this condition, imagine that just one bar is heated. As this bar becomes longer, it meets resistance from the other bars; therefore, stresses develop in all members.

FIGURE 8-38

Increase in length of a prismatic bar due to a uniform increase in temperature [Eq. (8-18)]

FIGURE 8-39

Statically determinate truss with a uniform temperature change in each member

Forces can develop in statically indeterminate trusses due to temperature and prestrain

The analysis of a statically indeterminate structure with temperature changes is based upon the concepts discussed in the preceding section, namely equilibrium equations, compatibility equations, and force–displacement relations. The principal difference is that temperature–displacement relations [Eq. (8-18)] are used in addition to force–displacement relations (such as $\delta = PL/EA$) when performing the analysis. The following two examples illustrate the procedures in detail.

Example 8-9

A prismatic bar AB of length L is made of linearly elastic material and is held between immovable supports (Fig. 8-41a). The bar has a modulus of elasticity E and a coefficient of thermal expansion α.

(a) If the temperature of the bar is raised uniformly by an amount ΔT, derive a formula for the thermal stress σ_T developed in the bar.

(b) Modify the formula in part (a) if the rigid support at B is replaced by an elastic support having a spring constant k (Fig. 8-41b); assume that only bar AB is subject to the uniform temperature increase ΔT.

FIGURE 8-41

Example 8-9: (a) Statically indeterminate
bar with uniform temperature
increase ΔT, (b) statically
indeterminate bar with elastic
support and uniform temperature
increase ΔT, and (c) statically
indeterminate bar with elastic
support and nonuniform
temperature increase $\Delta T(x)$

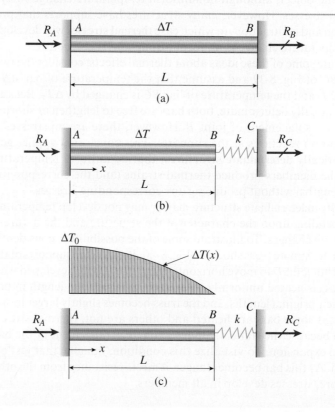

(c) Repeat part (b), but now assume that the bar is heated nonuniformly such that the temperature increase at distance x from A is given by $\Delta T(x) = \Delta T_0\left(1 - x^2/L^2\right)$ (see Fig. 8-41c).

Solution:

The four-step problem-solving approach is used in the solution; only essential steps are shown in each part.

Part (a): Determine thermal stress in the bar fixed at A and B subjected to uniform temperature increase ΔT.

1. *Conceptualize*: Because the temperature increases, the bar tends to elongate but is restrained by the rigid supports at A and B. Therefore, reactions R_A and R_B are developed at the supports, and the bar is subjected to uniform compressive stresses.

2. *Categorize*:

   Equation of Equilibrium: The only nontrivial equation of static equilibrium is that reactions R_A and R_B must sum to zero. So there is one equation but two unknowns, which is a *one-degree statically indeterminate problem*:

$$\Sigma F_x = 0 \qquad\qquad R_A + R_B = 0 \qquad\qquad \textbf{(a)}$$

   Select reaction R_B as the *redundant* and use the *superposition* of two statically determinate "released" structures (Fig. 8-41d) to develop an additional equation: an equation of *compatibility*. The first released structure is subjected to the temperature increase ΔT and hence elongates by amount δ_T. The second elongates δ_B under redundant R_B, which is applied as a load. Use a statics sign convention so that forces and displacements in the x direction are assumed to be positive.

   Equation of compatibility: The equation of compatibility expresses the fact that the net change in length of the bar is zero because supports A and B are fully restrained:

$$\delta_T + \delta_B = 0 \qquad\qquad \textbf{(b)}$$

   Temperature–displacement and force–displacement relations: The increase in length of the bar due to temperature is [Eq. (8-18)]

$$\delta_T = \alpha(\Delta T)L \qquad\qquad \textbf{(c)}$$

FIGURE 8-41 Continued

Example 8-9: (d) Statically determinate bars with support B removed (i.e., *released* structures)

(d)

$\delta_T + \delta_B = 0$

$\delta_T = \alpha L \Delta T$

$\alpha L \Delta T + R_B f_{AB} = 0$

where α is the coefficient of thermal expansion of the material. The increase in bar length due to unknown applied force R_B is obtained from the force–displacement relation:

$$\delta_B = R_B \left(\frac{L}{EA}\right) = R_B f_{AB} \qquad \text{(d)}$$

in which E is the modulus of elasticity, A is the bar cross-sectional area, and f_{AB} is the flexibility of the bar.

3. *Analyze*: Substituting Eqs. (c) and (d) into the compatibility equation Eq. (b) and solving for redundant R_B gives

$$R_B = \frac{-\alpha(\Delta T)L}{f_{AB}} = -EA\alpha(\Delta T) \qquad \text{(e)}$$

and the equilibrium equation Eq. (a) gives

$$R_A = -R_B = EA\alpha(\Delta T) \qquad \text{(f)}$$

Based on a statics sign convention, R_B is in the negative x direction, while R_A is in the positive x direction. As a final step, compute the comprehensive stress in the bar (assuming that ΔT is positive and therefore an increase in temperature) to be

$$\sigma_T = \frac{R_A}{A} = E\alpha(\Delta T) \qquad \text{(g)}$$

4. *Finalize*:

Note 1: In this example, the reactions are independent of the length of the bar, and the stress is independent of both the length and the cross-sectional area [see Eqs. (f) and (g)]. Thus, again, the usefulness of a symbolic solution is demonstrated because these important features of the bar's behavior might not be noticed in a purely numerical solution.

Note 2: When determining the thermal elongation of the bar [Eq. (c)], the material was assumed to be homogeneous and the increase in temperature uniform throughout the volume of the bar. Also, when determining the increase in length due to the reactive force [Eq. (d)], linearly elastic behavior of the material was assumed. These limitations always should be kept in mind when writing equations, such as Eqs. (c) and (d).

Note 3: The bar in this example has zero longitudinal displacements, not only at the ends but also at every cross section. Thus, there are *no axial strains in this bar*, and the special situation exists of *longitudinal stresses without longitudinal strains*. Of course, there are transverse strains in the bar from both the temperature change and the axial compression.

Part (b): Determine thermal stress in the bar fixed at A with elastic support at B and subjected to uniform temperature change ΔT.

2. *Categorize*: The structure in Fig. 8-41b is one-degree statically indeterminate, so select reaction R_C as the redundant and once again use the superposition of two released structures to solve the problem.

3. *Analyze*: First, *static equilibrium* of the original indeterminate structure requires that

$$R_A + R_C = 0 \tag{h}$$

while *compatibility of displacements* at joint C for the two released structures is expressed as

$$\delta_T + \delta_C = 0 \tag{i}$$

In the first released structure, apply uniform temperature change ΔT to bar AB only, so

$$\delta_T = \alpha(\Delta T)L \tag{c, repeated}$$

Note that the spring displaces in the positive x direction but is not deformed by the temperature change. Next, redundant R_C is applied to the end of the spring in the second released structure, resulting in displacement in the positive x direction. Both bar AB and the spring are subject to force R_C, so the total displacement at C is the sum of the elongations of bar and spring:

$$\delta_C = R_C\left(\frac{L}{EA}\right) + \frac{R_C}{k} = R_C(f_{AB} + f) \tag{j}$$

where $f = 1/k$ is the flexibility of the spring. Substituting the temperature–displacement equation [Eq. (c)] and force–displacement equation [Eq. (j)] into the compatibility equation [Eq. (i)], then solving for redundant R_C gives

$$R_C = \frac{-\alpha(\Delta T)L}{f_{AB} + f} = \frac{-\alpha(\Delta T)L}{\dfrac{L}{EA} + \dfrac{1}{k}} \quad \text{or} \quad R_C = -\left[\frac{EA\alpha(\Delta T)}{1 + \dfrac{EA}{kL}}\right] \tag{k}$$

Then equilibrium [Eq. (h)] leads to

$$R_A = -R_C = \frac{EA\alpha(\Delta T)}{1 + \dfrac{EA}{kL}} \tag{l}$$

Recall that, based on a statics sign convention, reaction force R_A is in the positive x direction, while reaction force R_C is in the negative x direction. Finally, the compressive stress in the bar is

$$\sigma_T = \frac{R_A}{A} = \frac{E\alpha(\Delta T)}{1 + \dfrac{EA}{kL}} \tag{m}$$

4. *Finalize*: Note that *if the spring stiffness k goes to infinity*, Eq. (l) becomes Eq. (f) and Eq. (m) becomes Eq. (g). In effect, use of an infinitely stiff spring moves the rigid support from *C* back to *B*.

Part (c): Determine thermal stress in the bar fixed at *A* with elastic support at *B* and subjected to *nonuniform* temperature change.

2. *Categorize*: The structure in Fig. 8-41c is one-degree statically indeterminate. So, once again, select reaction R_C as the redundant and, as in parts (a) and (b), use superposition of two released structures to solve the one-degree statically indeterminate problem (Fig. 8-41e, f).

FIGURE 8-41 Continued

Example 8-9: (e) Statically determinate bar with support *C* removed (i.e., *released* structure) under nonuniform temperature increase; (f) statically determinate bar with redundant R_C applied

3. *Analyze*: The equation of *static equilibrium* for the overall structure is Eq. (h), and the equation of *compatibility* is Eq. (i). First, solve for displacement δ_T in the released structure (Fig. 8-41e) as

$$\delta_T = \int_0^L \alpha[\Delta T(x)]dx = \int_0^L \alpha\left\{\Delta T_0\left[1 - \left(\frac{x}{L}\right)^2\right]\right\}dx = \frac{2}{3}\alpha(\Delta T_0)L \quad \textbf{(n)}$$

and δ_C for the second released structure (Fig. 8-41f) is the same as Eq. (j), giving

$$\delta_C = R_C(f_{AB} + f) \quad \textbf{(j, repeated)}$$

Substituting the temperature–displacement equation [Eq. (n)] and the force–displacement equation [Eq. (j)] into the compatibility equation [Eq. (i)] gives

$$R_C = \frac{\dfrac{-2}{3}\alpha(\Delta T_0)L}{f_{AB} + f} = \frac{-2\alpha(\Delta T_0)L}{3\left(\dfrac{L}{EA} + \dfrac{1}{k}\right)} \quad \text{or} \quad R_C = -\left(\frac{2}{3}\right)\left[\frac{EA\alpha(\Delta T_0)}{1 + \dfrac{EA}{kL}}\right] \quad \textbf{(o)}$$

From the static equilibrium equation [Eq. (h)],

$$R_A = -R_C = \left(\frac{2}{3}\right)\left[\frac{EA\alpha(\Delta T_0)}{1 + \dfrac{EA}{kL}}\right] \quad \textbf{(p)}$$

Finally, the compressive stress in the bar under nonuniform temperature change $\Delta T(x) = \Delta T_0(1 - (x/L)^2)$ is

$$\sigma_T = \frac{R_A}{A} = \left(\frac{2}{3}\right)\left[\frac{E\alpha(\Delta T_0)}{1 + \dfrac{EA}{kL}}\right] \quad \longleftarrow \text{(q)}$$

4. *Finalize*: Note once again that use of an infinitely stiff spring eliminates the EA/kL term from Eq. (q) and provides the solution for a prismatic bar fixed at A and B with nonuniform temperature change as $\Delta T(x) = \Delta T_0(1 - (x/L)^2)$.

Example 8-10

A sleeve in the form of a circular tube of length L is placed around a bolt and fitted between washers at each end (Fig. 8-42a). The nut is then turned until it is just snug. The sleeve and bolt are made of different materials and have different cross-sectional areas. (Assume that the coefficient of thermal expansion α_S of the sleeve is greater than the coefficient α_B of the bolt.)

(a) If the temperature of the entire assembly is raised by an amount ΔT, what stresses σ_S and σ_B are developed in the sleeve and bolt, respectively?

(b) What is the increase δ in the length L of the sleeve and bolt?

FIGURE 8-42

Example 8-10: Sleeve and bolt assembly with uniform temperature increase ΔT

Solution:

The four-step problem-solving approach is used in the solution; only essential steps are shown in each part.

Part (a): Stresses in sleeve and bolt.

1. *Conceptualize*: Because the sleeve and bolt are of different materials, they will elongate by different amounts when heated and allowed to expand freely. However, when they are held together by the assembly, free expansion cannot occur, and thermal stresses are developed in both materials. To find these stresses, use the same concepts as in any statically indeterminate analysis—equilibrium equations, compatibility equations, and displacement relations. However, you cannot formulate these equations until you disassemble the structure.

2. *Categorize*: A simple way to cut the structure is to remove the head of the bolt, thereby allowing the sleeve and bolt to expand freely under the temperature change ΔT (Fig. 8-42b). The resulting elongations of the sleeve and bolt are denoted δ_1 and δ_2, respectively, and the corresponding *temperature–displacement relations* are

$$\delta_1 = \alpha_S(\Delta T)L \qquad \delta_2 = \alpha_B(\Delta T)L \tag{a,b}$$

Since α_S is greater than δ_B, the elongation δ_1 is greater than δ_2, as shown in Fig. 8-42b.

The axial forces in the sleeve and bolt must be such that they shorten the sleeve and stretch the bolt until the final lengths of the sleeve and bolt are the same. These forces are shown in Fig. 8-42c, where P_S denotes the compressive force in the sleeve and P_B denotes the tensile force in the bolt. The corresponding shortening δ_3 of the sleeve and elongation δ_4 of the bolt are

$$\delta_3 = \frac{P_S L}{E_S A_S} \qquad \delta_4 = \frac{P_B L}{E_B A_B} \tag{c,d}$$

in which $E_S A_S$ and $E_B A_B$ are the respective axial rigidities. Equations (c) and (d) are the *load–displacement relations*.

3. *Analyze*: Now write an *equation of compatibility* expressing the fact that the final elongation δ is the same for both the sleeve and bolt. The elongation of the sleeve is $\delta_1 - \delta_3$ and of the bolt is $\delta_2 + \delta_4$; therefore,

$$\delta = \delta_1 - \delta_3 = \delta_2 + \delta_4 \tag{e}$$

Substituting the temperature–displacement and load–displacement relations [Eqs. (a) to (d)] into this equation gives

$$\delta = \alpha_S(\Delta T)L - \frac{P_S L}{E_S A_S} = \alpha_B(\Delta T)L + \frac{P_B L}{E_B A_B} \tag{f}$$

Rearranging Eq. (f) gives

$$\frac{P_S L}{E_S A_S} + \frac{P_B L}{E_B A_B} = \alpha_S(\Delta T)L - \alpha_B(\Delta T)L \tag{g}$$

which is a modified form of the compatibility equation. Note that it contains the forces P_S and P_B as unknowns.

An *equation of equilibrium* is obtained from Fig. 8-42c, which is a free-body diagram of the part of the assembly remaining after the head of the bolt is removed. Summing forces in the horizontal direction gives

$$P_S = P_B \tag{h}$$

which expresses the obvious fact that the compressive force in the sleeve is equal to the tensile force in the bolt.

Now solve simultaneously Eqs. (g) and (h) and obtain the axial forces in the sleeve and bolt:

$$P_S = P_B = \frac{(\alpha_S - \alpha_B)(\Delta T)E_S A_S E_B A_B}{E_S A_S + E_B A_B} \tag{8-19}$$

4. *Finalize*: This equation was derived assuming that the temperature increased and that the coefficient α_S was greater than the coefficient α_B. Under these conditions, P_S is the compressive force in the sleeve and P_B is the tensile force in the bolt.

The results will be quite different if the temperature increases, but the coefficient α_S is less than the coefficient α_B. Under these conditions, a gap will open between the bolt head and the sleeve, and there will be no stresses in either part of the assembly.

Stresses in the sleeve and bolt: Expressions for the stresses σ_S and σ_B in the sleeve and bolt, respectively, are obtained by dividing the corresponding forces by the appropriate areas:

$$\sigma_S = \frac{P_S}{A_S} = \frac{(\alpha_S - \alpha_B)(\Delta T)E_S E_B A_B}{E_S A_S + E_B A_B} \quad \longleftarrow (8\text{-}20a)$$

$$\sigma_B = \frac{P_B}{A_B} = \frac{(\alpha_S - \alpha_B)(\Delta T)E_S A_S E_B}{E_S A_S + E_B A_B} \quad \longleftarrow (8\text{-}20b)$$

Under the assumed conditions, the stress σ_S in the sleeve is compressive and the stress σ_B in the bolt is tensile. It is interesting to note that these stresses are independent of the length of the assembly and their magnitudes are inversely proportional to their respective areas (that is, $\sigma_S/\sigma_B = A_B/A_S$).

Part (b): Increase in length of the sleeve and bolt.

2, 3. *Categorize, Analyze*: The elongation δ of the assembly can be found by substituting either P_S or P_B from Eq. (8-19) into Eq. (f), yielding

$$\delta = \frac{(\alpha_S E_S A_S + \alpha_B E_B A_B)(\Delta T)L}{E_S A_S + E_B A_B} \quad \longleftarrow (8\text{-}21)$$

With the preceding formulas available, you can readily calculate the forces, stresses, and displacements of the assembly for any given set of numerical data.

4. *Finalize*: As a partial check on the results, note that Eqs. (8-19), (8-20), and (8-21) reduce to known values in simplified cases. For instance, suppose that the bolt is rigid and therefore unaffected by temperature changes. You can represent this situation by setting $\alpha_B = 0$ and letting E_B become infinitely large, thereby creating an assembly in which the sleeve is held between rigid supports. Substituting these values into Eqs. (8-19), (8-20), and (8-21) gives

$$P_S = E_S A_S \alpha_S(\Delta T) \qquad \sigma_S = E_S \alpha_S(\Delta T) \qquad \delta = 0$$

These results agree with those of Example 8-9 for a bar held between rigid supports.

As a second special case, suppose that the sleeve and bolt are made of the same material. Then both parts will expand freely and will lengthen the same amount when the temperature changes. No forces or stresses will be developed. To see if the derived equations predict this behavior, substitute $\alpha_S = \alpha_B = \alpha$ into Eqs. (8-19), (8-20), and (8-21) and obtain

$$P_S = P_B = 0 \qquad \sigma_S = \sigma_B = 0 \qquad \delta = \alpha(\Delta T)L$$

which are the expected results.

Misfits and Prestrains

FIGURE 8-43

Statically determinate structure with a small misfit

(a)

(b)

Suppose that a member of a structure is manufactured with its length slightly different from its prescribed length. Then the member will not fit into the structure in its intended manner, and the geometry of the structure will be different from what was planned. Situations of this kind are called **misfits**. Sometimes misfits are intentionally created in order to introduce strains into the structure at the time it is built. Because these strains exist before any loads are applied to the structure, they are called **prestrains**. Accompanying the prestrains are prestresses, and the structure is said to be **prestressed**. Common examples of prestressing are spokes in bicycle wheels (which would collapse if not prestressed), the pretensioned faces of tennis racquets, shrink-fitted machine parts, and prestressed concrete beams.

If a structure is **statically determinate**, small misfits in one or more members will not produce strains or stresses, although there will be departures from the theoretical configuration of the structure. To illustrate this statement, consider a simple structure consisting of a horizontal beam AB supported by a vertical bar CD (Fig. 8-43a). If bar CD has exactly the correct length L, the beam will be horizontal at the time the structure is built. However, if the bar is slightly longer than intended, the beam will make a small angle with the horizontal. Nevertheless, there will be no strains or stresses in either the bar or the beam attributable to the incorrect length of the bar. Furthermore, if a load P acts at the end of the beam (Fig. 8-43b), the stresses in the structure due to that load will be unaffected by the incorrect length of bar CD.

In general, if a structure is statically determinate, the presence of small misfits will produce small changes in geometry but no strains or stresses. Thus, the effects of a misfit are similar to those of a temperature change.

The situation is quite different if the structure is **statically indeterminate** because then the structure is not free to adjust to misfits (just as it is not free to adjust to certain kinds of temperature changes). To show this, consider a beam supported by two vertical bars (Fig. 8-44a). If both bars have exactly the correct length L, the structure can be assembled with no strains or stresses and the beam will be horizontal.

Suppose, however, that bar CD is slightly longer than the prescribed length. Then, in order to assemble the structure, bar CD must be compressed by external forces (or bar EF stretched by external forces), the bars must be fitted into place, and then the external forces must be released. As a result, the beam will deform and rotate, bar CD will be in compression, and bar EF will be in tension. In other words, prestrains will exist in all members and the structure will be prestressed, even though no external loads are acting. If a load P is now added (Fig. 8-44b), additional strains and stresses will be produced.

The analysis of a statically indeterminate structure with misfits and prestrains proceeds in the same general manner as described previously for loads and temperature changes. The basic ingredients of the analysis are equations of equilibrium, equations of compatibility, force-displacement relations, and (if appropriate) temperature–displacement relations. The methodology is illustrated in Example 8-12.

FIGURE 8-44

Statically indeterminate structure with a small misfit

(a)

(b)

Bolts and Turnbuckles

Prestressing a structure requires that one or more parts of the structure be stretched or compressed from their theoretical lengths. A simple way to produce a change in length is to tighten a bolt or a turnbuckle. In the case of a **bolt** (Fig. 8-45), each turn of the nut will cause the nut to travel along the bolt a distance equal to the spacing p of the threads (called the *pitch* of the threads). Thus, the distance δ traveled by the nut is

FIGURE 8-45

The *pitch* of the threads is the distance from one thread to the next

$$\delta = np \qquad \textbf{(8-22)}$$

in which n is the number of revolutions of the nut (not necessarily an integer). Depending upon how the structure is arranged, turning the nut can stretch or compress a member.

In the case of a **double-acting turnbuckle** (Fig. 8-46), there are two end screws. Because a right-hand thread is used at one end and a left-hand thread at the other, the device either lengthens or shortens when the buckle is rotated. Each full turn of the buckle causes it to travel a distance p along each screw, where again p is the pitch of the threads. Therefore, if the turnbuckle is tightened by one turn, the screws are drawn closer together by a distance $2p$, and the effect is to shorten the device by $2p$. For n turns, the distance is

$$\delta = 2np \qquad \textbf{(8-23)}$$

Turnbuckles are often inserted in cables and then tightened, thus creating initial tension in the cables, as illustrated in the following example.

FIGURE 8-46

Double-acting turnbuckle (Each full turn of the turnbuckle shortens or lengthens the cable by $2p$, where p is the pitch of the screw threads)

Example 8-11

The mechanical assembly shown in Fig. 8-47a consists of a copper tube, a rigid end plate, and two steel cables with turnbuckles. The slack is removed from the cables by rotating the turnbuckles until the assembly is snug but with no initial stresses. (Further tightening of the turnbuckles will produce a prestressed condition where the cables are in tension and the tube is in compression.)

(a) Determine the forces in the tube and cables (Fig. 8-47a) when the turnbuckles are tightened by n turns.

(b) Determine the shortening of the tube.

FIGURE 8-47

Example 8-11: Statically indeterminate assembly with a copper tube in compression and two steel cables in tension

Solution:

The four-step problem-solving approach is used in the solution; only essential steps are shown in each part.

Part (a): Forces in the cables and tube.

1. *Conceptualize*: Begin the analysis by removing the plate at the right-hand end of the assembly so that the tube and cables are free to change in length (Fig. 8-47b). Rotating the turnbuckles through n turns will shorten the cables by a distance

$$\delta_1 = 2np \tag{a}$$

as shown in Fig. 8-47b.

The tensile forces in the cables and the compressive force in the tube must be such that they elongate the cables and shorten the tube until their final lengths are the same. These forces are shown in Fig. 8-47c, where P_s denotes the tensile force in one of the steel cables and P_c denotes the compressive force in the copper tube. The elongation of a cable due to the force P_s is

$$\delta_2 = \frac{P_s L}{E_s A_s} \qquad \text{(b)}$$

in which $E_s A_s$ is the axial rigidity and L is the length of a cable. Also, the compressive force P_c in the copper tube causes it to shorten by

$$\delta_3 = \frac{P_c L}{E_c A_c} \qquad \text{(c)}$$

in which $E_c A_c$ is the axial rigidity of the tube. Equations (b) and (c) are the *load–displacement relations*.

2. *Categorize*: The final shortening of one of the cables is equal to the shortening δ_1 caused by rotating the turnbuckle minus the elongation δ_2 caused by the force P_s. This final shortening of the cable must equal the shortening δ_3 of the tube:

$$\delta_1 - \delta_2 = \delta_3 \qquad \text{(d)}$$

which is the *equation of compatibility*.

Substituting the turnbuckle relation [Eq. (a)] and the load–displacement relations [Eqs. (b) and (c)] into the preceding equation yields

$$2np - \frac{P_s L}{E_s A_s} = \frac{P_c L}{E_c A_c} \qquad \text{(e)}$$

or

$$\frac{P_s L}{E_s A_s} + \frac{P_c L}{E_c A_c} = 2np \qquad \text{(f)}$$

which is a modified form of the compatibility equation. Note that it contains P_s and P_c as unknowns.

Using Fig. 8-47c, which is a free-body diagram of the assembly with the end plate removed, write the following equation of equilibrium:

$$2P_s = P_c \qquad \text{(g)}$$

3. *Analyze*: **Forces in the cables and tube:** Now solve simultaneously Eqs. (f) and (g) and obtain the axial forces in the steel cables and copper tube, respectively:

$$P_s = \frac{2npE_c A_c E_s A_s}{L(E_c A_c + 2E_s A_s)} \qquad P_c = \frac{4npE_c A_c E_s A_s}{L(E_c A_c + 2E_s A_s)} \qquad \text{⟵ (8-24a,b)}$$

Recall that the forces P_s are tensile forces and the force P_c is compressive. If desired, the stresses σ_s and σ_c in the steel and copper now can be obtained by dividing the forces P_s and P_c by the cross-sectional areas A_s and A_c, respectively.

Part (b): Shortening of the tube.

2, 3. *Categorize, Analyze*: The decrease in length of the tube is the quantity δ_3 [see (Fig. 8-47) and Eq. (c)]:

$$\delta_3 = \frac{P_c L}{E_c A_c} = \frac{4npE_s A_s}{E_c A_c + 2E_s A_s} \qquad \Longleftarrow \text{(8-25)}$$

4. *Finalize*: With the preceding formulas available, you can readily calculate the forces, stresses, and displacements of the assembly for any given set of numerical data.

Example 8-12

FIGURE 8-48

Elastic spring inside
aluminum tubes

FIGURE 8-48

Elastic spring inside
aluminum tubes

P

Cap plate

s

k

Elastic
tube *A*
d_A, L_A

Base plate

Elastic
tube *B*
d_B, L_B

d_O

An elastic spring is enclosed by tube *A* but is longer than tube *A* by an amount *s*. The spring rests on a rigid base plate, and the base plate rests on top of tube *B* (see Fig. 8-48). Force *P* is applied to a rigid cap plate to compress the spring and the tubes. Tubes *A* and *B* are made of an aluminum alloy and have an outer diameter d_O and inner diameters d_A and d_B, respectively. If applied load $P = 140$ kN, find the axial normal compressive stresses in tubes *A* and *B* and plot the axial force and displacement diagrams for the tube structure. Assume that $E = 72$ GPa, spring stiffness $k = 200$ MN/m, $d_O = 76$ mm, $d_A = 62$ mm, $d_B = 58$ mm, and $s = 0.5$ mm. The lengths of tubes *A* and *B* are $L_A = 0.35$ m and $L_B = 0.4$ m.

Solution:

Use the four-step problem-solving approach.

1. *Conceptualize*: The elastic spring does not fit in tube *A*, so force *P* acts to compress the spring to close the gap; the force in tube *A* is zero until the gap is closed. In Fig. 8-49a, an upper section cut through the spring and also through tube *A* creates an upper free-body diagram that shows the spring force $N_s = (k)(x) = P$ for downward cap displacement *x*. The internal force N_A in tube *A* is zero until the cap plate rests on tube *A* when the gap is closed. In Fig. 8-49b, force *P* now compresses both the spring and tube *A* when cap displacement *x* exceeds gap *s*. Figure 8-49 also gives the lower free-body diagram with a section cut through tube *B* and with the internal compressive force N_B in tube *B* equal to the applied load *P*. (Internal forces in tubes *A* and *B* are shown as two arrows, one at each tube wall, indicating that N_A and N_B

FIGURE 8-49

(a, b) Free-body diagrams
($x \leq s$, $x > s$)

(b)

(a)

are actually uniformly distributed forces acting on the circular cross section of each tube.)

2. *Categorize*: If force P is large enough to close gap s, the structure becomes statically indeterminate; part of force P is carried by the spring and part by tube A, as shown in Fig. 8-49b. *Equilibrium* and *compatibility* equations are needed to determine the force in the spring N_s and internal force N_A in tube A. The force P required to close gap s is $(k)(s)$. After that, the spring and tube A deform together in parallel.

Equilibrium: Summing axial forces in the x direction in the lower free-body diagram in Fig. 8-49a gives the compressive force in tube B as $N_B = P$. Summing forces in the upper free-body diagram in Fig. 8-49b leads to

$$N_A + N_s = P \tag{a}$$

This one equation has two unknowns, so the structure is *one degree statically indeterminate*.

Compatibility: Once the gap is closed by compression of the spring by force P, tube A and the spring begin to deform together. The compatibility equation is

$$\frac{N_A}{k_A} = \frac{N_s}{k} - s \tag{b}$$

where the axial stiffness of tube A is $k_A = \dfrac{EA_A}{L_A}$ with A_A as the cross-sectional area of tube A.

3. *Analyze*:

Solve equilibrium and compatibility equations: Solve for N_A in Eq. (b) to obtain $N_A = \dfrac{k_A}{k} N_s - k_A s$. Substitute this expression into Eq. (a) to find that the spring force is $N_s = \dfrac{k}{k + k_A}(P + k_A s)$. Then $N_A = P - N_s$ or $N_A = \dfrac{k_A}{k + k_A}(P - ks)$. Finally, since $N_s = kx$, equilibrium Eq. (a) also can be written as $kx + k_A(x - s) = P$.

Stresses in tubes A and B: The axial normal *compressive* stresses in tubes A and B can be computed now that internal forces N_A and N_B are known. Tube cross-sectional areas are

$$A_A = \frac{\pi}{4}\left(d_O^2 - d_A^2\right) = 1517\,\text{mm}^2 \quad A_B = \frac{\pi}{4}\left(d_O^2 - d_B^2\right) = 1894\,\text{mm}^2$$

If $E = 72$ GPa and $L_A = 0.35$ m, tube A has axial stiffness:

$$k_A = \frac{EA_A}{L_A} = \frac{(72\,\text{GPa})(1517\,\text{mm}^2)}{350\,\text{mm}} = 312\,\frac{\text{MN}}{\text{m}} \quad \text{and}$$

$$N_A = \frac{k_A}{k + k_A}(P - ks) = \frac{312}{200 + 312}\left(140\,\text{kN} - 200\,\frac{\text{MN}}{\text{m}}(0.5\,\text{mm})\right) = 24.4\,\text{kN}$$

The resulting stresses in tubes A and B are

$$\sigma_A = \frac{N_A}{A_A} = \frac{24.4\,\text{kN}}{1517\,\text{mm}^2} = 16.1\,\text{MPa} \quad \text{and} \quad \sigma_B = \frac{N_B}{A_B} = \frac{140\,\text{kN}}{1894\,\text{mm}^2} = 73.9\,\text{MPa} \;\;\leftarrow$$

Axial-force and displacement diagrams: From equilibrium, the force in the spring is $N_s = P - N_A = 140\,\text{kN} - 24.4\,\text{kN} = 115.6\,\text{kN}$, so the total compression of the spring is $x = \dfrac{N_s}{k} = \dfrac{115.6\,\text{kN}}{200\,\text{MN/m}} = 0.578\,\text{mm}$. This confirms that force P is large enough to close gap s and compress tube A an amount equal to $x - s = 0.078$ mm.

The self-weight of the tubes is small, so it is not included in this analysis. Hence, forces N_A and N_B are taken as constant over the height of each tube, as shown in the axial-force diagram (AFD) in Fig. 8-50a. The AFD is composed of constant force segments, so the ADD (Fig. 8-50b) varies linearly over the height of the tube structure. The change in axial displacement between the base and the top of tube B is equal to the area under the AFD over distance L_B divided by EA_B: $\dfrac{(140\,\text{kN})L_B}{EA_B} = 0.411$ mm. Add to this the compression $(x - s)$ of tube A to find the total downward displacement 0.489 mm at the top.

FIGURE 8-50

(a, b) Axial-force and displacement diagrams for tubes *A* and *B*

4. *Finalize*: Once the gap *s* is closed, tube *A* and spring *k* become *springs in parallel*, and each carries part of force *P* in proportion to its axial stiffness (k or k_A) in relation to the total axial stiffness of the spring and tube *A* together ($k + k_A$). Here, as the gap closes, the spring alone carries $(k)(s) = 100$ kN; after that the remaining force $(P - ks)$ is distributed to tube *A* and the spring based on the ratios of axial stiffness: $\dfrac{k_A}{k + k_A}$ for tube *A* and $\dfrac{k}{k + k_A}$ for the elastic spring.

In the end, the spring carries $N_s = 100$ kN + 15.6 kN = 115.6 kN (83% of applied load *P*) and tube *A* carries $N_A = 24.4$ kN (17% of applied load *P*).

Recall that Example 7-1 used a model of this structure in which tubes *A* and *B* are assumed to be *rigid*. The assumption of rigid tubes leads to a simpler analysis procedure and may be appropriate if spring stiffness *k* is much smaller that the axial stiffness of tube *A*; this is not true in this example, in which $k/k_A = 0.64$. In Example 7-1, the elastic spring is almost 50 times more flexible than that used here.

If tube *A* is assumed to be *rigid* in this example, the spring force is $N_s = (k)(s) = 100$ kN, which is 13.5% lower than computed previously. At the same time, the force in tube *A* is $N_A = P - N_s = (140 - 100)$ kN = 40 kN, which is approximately 64% larger than that obtained in the statically indeterminate analysis, leading to $N_A = 24.4$ kN. The assumption of rigid tubes and elastic spring in this example leads to erroneous results.

8.6 Stresses on Inclined Sections

In the previous discussions of tension and compression in axially loaded members, the only stresses considered were the normal stresses acting on cross sections. These stresses are pictured in Fig. 8-51, which shows a bar AB subjected to axial loads P.

Cutting the bar at an intermediate cross section by a plane mn (perpendicular to the x axis) reveals the normal stresses shown in Fig. 8-51b. The normal stresses acting over the cut section may be calculated from the formula $\sigma_x = P/A$ provided that the stress distribution is uniform over the entire cross-sectional area A. As explained in Chapter 7, this condition exists if the bar is prismatic, the material is homogeneous, the axial force P acts at the centroid of the cross-sectional area, and the cross section is away from any localized stress concentrations. Of course, there are no shear stresses acting on the cut section because it is perpendicular to the longitudinal axis of the bar.

For convenience, show the stresses in a two-dimensional view of the bar (Fig. 8-51c) rather than the more complex three-dimensional view (Fig. 8-51b). However, when working with two-dimensional figures, do not forget that the bar has a thickness perpendicular to the plane of the figure. This third dimension must be considered when making derivations and calculations.

Stress Elements

The most useful way of representing the stresses in the bar of Fig. 8-51 is to isolate a small element of material, such as the element labeled C in Fig. 8-51c, and then show the stresses acting on all faces of this element. An element of this

FIGURE 8-51

Prismatic bar in tension showing the stresses acting on cross section mn: (a) bar with axial forces P, (b) three-dimensional view of the cut bar showing the normal stresses, and (c) two-dimensional view

(a)

(b)

(c)

kind is called a **stress element**. The stress element at point C is a small rectangular block (it doesn't matter whether it is a cube or a rectangular parallelepiped) with its right-hand face lying in cross section mn.

The dimensions of a stress element are assumed to be infinitesimally small, but for clarity, draw the element to a large scale, as in Fig. 8-52a. In this case, the edges of the element are parallel to the x, y, and z axes, and the only stresses are the normal stresses σ_x acting on the x faces (recall that the x faces have their normals parallel to the x axis). Because it is more convenient, draw a two-dimensional view of the element (Fig. 8-52b) instead of a three-dimensional view.

Stresses on Inclined Sections

The stress element of Fig. 8-52 provides only a limited view of the stresses in an axially loaded bar. To obtain a more complete picture, you need to investigate the stresses acting on **inclined sections**, such as the section cut by the inclined plane pq in Fig. 8-53a. Because the stresses are the same throughout the entire bar, the stresses acting over the inclined section must be uniformly distributed, as pictured in the free-body diagrams of Fig. 8-53b (three-dimensional view) and Fig. 8-53c (two-dimensional view). From the equilibrium of the free body, the resultant of the stresses must be a horizontal force P. (The resultant is drawn with a dashed line in Figs. 8-53b and c.)

As a preliminary matter, the **orientation** of the inclined section pq must be specified. A standard method is to specify the angle θ between the x axis and the normal n to the section (see Fig. 8-54a). Thus, the angle θ for the inclined section shown in the figure is approximately 30°. By contrast, cross section mn (Fig. 8-51a) has an angle θ equal to zero (because the normal to the section is the x axis). For additional examples, consider the stress element of Fig. 8-52. The angle θ for the right-hand face is 0, for the top face is 90° (a longitudinal section of the bar), for the left-hand face is 180°, and for the bottom face is 270° (or −90°).

Now return to the task of finding the stresses acting on section pq (Fig. 8-54b). As already mentioned, the resultant of these stresses is a force P acting in the x direction. This resultant may be resolved into two components, a normal force N that is perpendicular to the inclined plane pq and a shear force V that is tangential to it. These force components are

$$N = P\cos\theta \qquad V = P\sin\theta \qquad\qquad \textbf{(8-26a,b)}$$

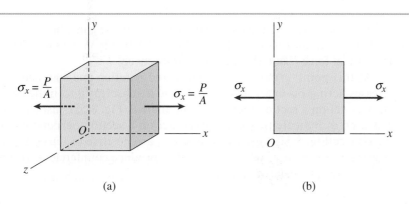

FIGURE 8-52

Stress element at point C of the axially loaded bar shown in Fig. 8-51c: (a) three-dimensional view of the element and (b) two-dimensional view of the element

(a)

(b)

FIGURE 8-53

Prismatic bar in tension showing the stresses acting on an inclined section pq: (a) bar with axial forces P, (b) three-dimensional view of the cut bar showing the stresses, and (c) two-dimensional view

(a)

(b)

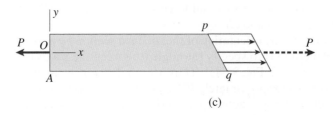

(c)

Associated with the forces N and V are normal and shear stresses that are uniformly distributed over the inclined section (Figs. 8-54c and d). The normal stress is equal to the normal force N divided by the area of the section, and the shear stress is equal to the shear force V divided by the area of the section. Thus, the stresses are

$$\sigma = \frac{N}{A_1} \qquad \tau = \frac{V}{A_1} \tag{8-27a,b}$$

in which A_1 is the area of the inclined section, as

$$A_1 = \frac{A}{\cos\theta} \tag{8-28}$$

As usual, A represents the cross-sectional area of the bar. The stresses σ and τ act in the directions shown in Figs. 8-54c and d, that is, in the same directions as the normal force N and shear force V, respectively.

At this point, establish a standardized **notation and sign convention** for stresses acting on inclined sections. Use a subscript θ to indicate that the stresses act on a section inclined at an angle θ (Fig. 8-55), just as a subscript x is used to indicate that the stresses act on a section perpendicular to the x axis (see Fig. 8-51). Normal stresses σ_θ are positive in tension, and shear stresses τ_θ are positive when they tend to produce counterclockwise rotation of the material, as shown in Fig. 8-55.

FIGURE 8-54

Prismatic bar in tension showing the stresses acting on an inclined section pq

(a)

(b)

(c)

(d)

For a bar in tension, the normal force N produces positive normal stresses σ_θ (see Figs. 8-54c), and the shear force V produces negative shear stresses τ_θ (see Fig. 8-54d). These stresses are given by [see Eqs. (8-26), (8-27), and (8-28)]

$$\sigma_\theta = \frac{N}{A_1} = \frac{P}{A}\cos^2\theta \qquad \tau_\theta = -\frac{V}{A_1} = -\frac{P}{A}\sin\theta\cos\theta$$

Introducing the notation $\sigma_x = P/A$, in which σ_x is the normal stress on a cross section, and also using the trigonometric relations

$$\cos^2\theta = \frac{1}{2}(1 + \cos 2\theta) \qquad \sin\theta\cos\theta = \frac{1}{2}(\sin 2\theta)$$

results in the following expressions for the **normal and shear stresses**:

$$\sigma_\theta = \sigma_x \cos^2 \theta = \frac{\sigma_x}{2}(1 + \cos 2\theta) \tag{8-29a}$$

$$\tau_\theta = -\sigma_x \sin\theta \cos\theta = -\frac{\sigma_x}{2}(\sin 2\theta) \tag{8-29b}$$

These equations give the stresses acting on an inclined section oriented at an angle θ to the x axis (Fig. 8-55).

It is important to recognize that Eqs. (8-29a and b) were derived only from statics, and therefore they are independent of the material. Thus, these equations are valid for any material, whether it behaves linearly or nonlinearly, elastically or inelastically.

Maximum Normal and Shear Stresses

The manner in which the stresses vary as the inclined section is cut at various angles is shown in Fig. 8-56. The horizontal axis gives the angle θ as it varies from $-90°$ to $+90°$, and the vertical axis gives the stresses σ_θ and τ_θ. Note that a positive angle θ is measured counterclockwise from the x axis (Fig. 8-55) and a negative angle is measured clockwise.

As shown on the graph, the normal stress σ_θ equals σ_x when $\theta = 0$. Then, as θ increases or decreases, the normal stress diminishes until at $\theta = \pm90°$ it becomes zero because there are no normal stresses on sections cut parallel to the longitudinal axis. The **maximum normal stress** occurs at $\theta = 0$ and is

$$\sigma_{max} = \sigma_x \tag{8-30}$$

Also, note that when $\theta = \pm45°$, the normal stress is one-half the maximum value.

The shear stress τ_θ is zero on cross sections of the bar ($\theta = 0$) as well as on longitudinal sections ($\theta = \pm90°$). Between these extremes, the stress varies as shown on the graph, reaching the largest positive value when $\theta = -45°$ and the largest negative value when $\theta = +45°$. These **maximum shear stresses** have the same magnitude:

$$\tau_{max} = \frac{\sigma_x}{2} \tag{8-31}$$

but they tend to rotate the element in opposite directions.

The maximum stresses in a **bar in tension** are shown in Fig. 8-57. Two stress elements are selected—element A is oriented at $\theta = 0°$ and element B is oriented at $\theta = 45°$. Element A has the maximum normal stresses [Eq. (8-30)] and element B has the maximum shear stresses [Eq. (8-31)]. In the case of element A (Fig. 8-57b), the only stresses are the maximum normal stresses (no shear stresses exist on any of the faces).

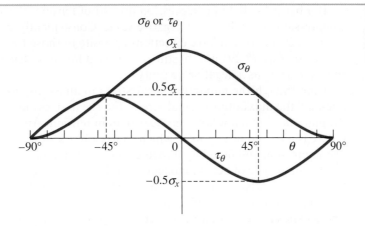

FIGURE 8-56

Graph of normal stress σ_θ and shear stress τ_θ versus angle θ of the inclined section [see Fig. 8-55 and Eqs. (8-29a and b)]

In the case of element B (Fig. 8-57c), both normal and shear stresses act on all faces (except, of course, the front and rear faces of the element). Consider, for instance, the face at 45° (the upper right-hand face). On this face, the normal and shear stresses [from Eqs. (8-29a and b)] are $\sigma_x/2$ and $-\sigma_x/2$, respectively. Hence, the normal stress is tension (positive) and the shear stress acts clockwise (negative) against the element. The stresses on the remaining faces are obtained in a similar manner by substituting $\theta = 135°, -45°$, and $-135°$ into Eqs. (8-29a and b).

Thus, in this special case of an element oriented at $\theta = 45°$, the normal stresses on all four faces are the same (equal to $\sigma_x/2$), and all four shear stresses have the maximum magnitude (equal to $\sigma_x/2$). Also, note that the shear stresses acting on perpendicular planes are equal in magnitude and have directions either toward, or away from, the line of intersection of the planes, as discussed in detail in Section 7.6.

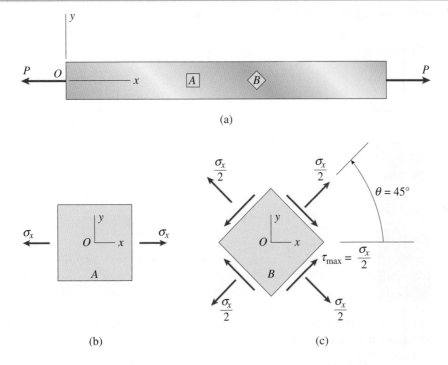

FIGURE 8-57

Normal and shear stresses acting on stress elements oriented at $\theta = 0°$ and $\theta = 45°$ for a bar in tension

(a)

(b)   (c)

Load

Load

Jim Gere

If a bar is loaded in compression instead of tension, the stress σ_x will be compression and will have a negative value. Consequently, all stresses acting on stress elements will have directions opposite to those for a bar in tension. Of course, Eqs. (8-29a and b) still can be used for the calculations simply by substituting σ_x as a negative quantity.

Even though the maximum shear stress in an axially loaded bar is only one-half the maximum normal stress, the shear stress may cause failure if the material is much weaker in shear than in tension. An example of a shear failure is pictured in Fig. 8-58, which shows a block of wood that was loaded in compression and failed by shearing along a 45° plane.

A similar type of behavior occurs in mild steel loaded in tension. During a tensile test of a flat bar of low-carbon steel with polished surfaces, visible *slip bands* appear on the sides of the bar at approximately 45° to the axis (Fig. 8-59). These bands indicate that the material is failing in shear along the planes on which the shear stress is maximum. Such bands were first observed by G. Piobert in 1842 and W. Lüders in 1860, and today they are called either *Lüders' bands* or *Piobert's bands*. They begin to appear when the yield stress is reached in the bar (point *B* in Fig. 7-13 of Section 7.3).

Uniaxial Stress

The state of stress described throughout this section is called **uniaxial stress** for the obvious reason that the bar is subjected to simple tension or compression in just one direction. The most important orientations of stress elements for uniaxial stress are $\theta = 0$ and $\theta = 45°$ (Figs. 8-57b and c); the former has the maximum normal stress and the latter has the maximum shear stress. If sections are cut through the bar at other angles, the stresses acting on the faces of the corresponding stress elements can be determined from Eqs. (8-29a and b), as illustrated in Examples 8-13 and 8-14.

Uniaxial stress is a special case of a more general stress state known as *plane stress*, which is described in detail in Chapter 11.

Load

Load

Jim Gere

Example 8-13

A compression bar having a square cross section of width b must support a load $P = 8000$ lb (Fig. 8-60a). The bar is constructed from two pieces of material that are connected by a glued joint (known as a *scarf joint*) along plane pq, which is at an angle $\alpha = 40°$ to the vertical. The material is a structural plastic for which the allowable stresses in compression and shear are 1100 psi and 600 psi, respectively. Also, the allowable stresses in the glued joint are 750 psi in compression and 500 psi in shear.

Determine the minimum width b of the bar.

Solution:

Use the four-step problem-solving approach in the solution.

1. *Conceptualize*: For convenience, rotate a segment of the bar to a horizontal position (Fig. 8-60b) that matches the figures used in deriving the equations for the stresses on an inclined section (see Figs. 8-54 and 8-55). With the bar in this position, observe that the normal n to the plane of the glued joint (plane pq) makes an angle $\beta = 90° - \alpha$, or $50°$, with the axis of the bar. Since the angle θ is defined as positive when counterclockwise (Fig. 8-55), angle $\theta = -50°$ for the glued joint.

   The cross-sectional area of the bar is related to the load P and the stress σ_x acting on the cross sections by

$$A = \frac{P}{\sigma_x} \tag{a}$$

2. *Categorize*: To find the required area, determine the value of σ_x corresponding to each of the four allowable stresses. Then the smallest value of σ_x will determine the required area. The values of σ_x are obtained by rearranging Eqs. (8-29a and b) as

$$\sigma_x = \frac{\sigma_\theta}{\cos^2 \theta} \qquad \sigma_x = -\frac{\tau_\theta}{\sin\theta\cos\theta} \tag{8-32a,b}$$

   Now apply these equations to the glued joint and to the plastic.

3. *Analyze*:

Part (a): Values of σ_x based upon the allowable stresses in the glued joint.

For compression in the glued joint, $\sigma_\theta = -750$ psi and $\theta = -50°$. Substitute into Eq. (8-32a) to get

$$\sigma_x = \frac{-750 \text{ psi}}{(\cos (-50°))^2} = -1815 \text{ psi} \tag{b}$$

For shear in the glued joint, the allowable stress is 500 psi. However, it is not immediately evident whether τ_θ is $+500$ psi or -500 psi. One approach is to substitute both $+500$ psi and -500 psi into Eq. (8-32b) and then select the value of σ_x that is negative. The other value of σ_x will be positive (tension) and does not apply to this bar. Another approach is to inspect the bar itself (Fig. 8-60b) and observe from the directions of the loads that the shear stress will act

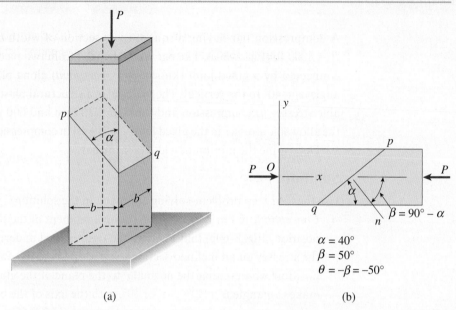

$\alpha = 40°$

$\beta = 50°$

$\theta = -\beta = -50°$

(a)   (b)

clockwise against plane pq, which means that the shear stress is negative. Therefore, substitute $\tau_\theta = -500$ psi and $\theta = -50°$ into Eq. (8-32b) and obtain

$$\sigma_x = -\frac{-500 \text{ psi}}{(\sin{(-50°)})(\cos{(-50°)})} = -1015 \text{ psi} \qquad \textbf{(c)}$$

Part (b): Values of σ_x based upon the allowable stresses in the plastic.

The maximum compressive stress in the plastic occurs on a cross section. Therefore, since the allowable stress in compression is 1100 psi, you know immediately that

$$\sigma_x = -1100 \text{ psi} \qquad \textbf{(d)}$$

The maximum shear stress occurs on a plane at 45° and is numerically equal to $\sigma_x/2$ [see Eq. (8-31)]. Since the allowable stress in shear is 600 psi, it follows that

$$\sigma_x = -1200 \text{ psi} \qquad \textbf{(e)}$$

This same result can be obtained from Eq. (8-32b) by substituting $\tau_\theta = 600$ psi and $\theta = 45°$.

Part (c): Minimum width of the bar.

Compare the four values of σ_x [Eqs. (b), (c), (d), and (e)] and note that the smallest is $\sigma_x = -1015$ psi. Therefore, this value governs the design. Substitute into Eq. (a) and use only numerical values to obtain the required area:

$$A = \frac{8000 \text{ lb}}{1015 \text{ psi}} = 7.88 \text{ in}^2$$

4. *Finalize*: Since the bar has a square cross section ($A = b^2$), the minimum width is

$$b_{min} = \sqrt{A} = \sqrt{7.88 \text{ in}^2} = 2.81 \text{ in.}$$

Any width larger than b_{min} will ensure that the allowable stresses are not exceeded.

Example 8-14

A prismatic brass bar with a length of $L = 0.5$ m and a cross-sectional area $A = 1200$ mm^2 is compressed by an axial load $P = 90$ kN (Fig. 8-61a).

(a) Determine the complete state of stress acting on an inclined section pq cut through the bar at an angle $\theta = 25°$, and show the stresses on a properly oriented stress element.

(b) If the bar is now fixed between supports A and B (Fig. 8-61b) and then subjected to a temperature increase of $\Delta T = 33°C$, the compressive stress on plane rs is known to be 65 MPa. Find the shear stress τ_θ on plane rs. What is angle θ? (Assume the modulus of elasticity is $E = 110$ GPa and coefficient of thermal expansion is $\alpha = 20 \times 10^{-6}/°C$.)

(c) If the allowable normal stress is ± 82 MPa and the allowable shear stress is ± 40 MPa, find the maximum permissible temperature *increase* (ΔT) in the bar if allowable stress values in the bar are not to be exceeded.

FIGURE 8-61

Example 8-14: (a) Stresses on inclined section pq through bar and (b) stresses on inclined section rs through bar

(a)

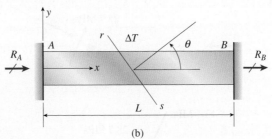

(b)

Solution:

The four-step problem-solving approach is used in the solution. Some steps are combined and only essential steps are shown in each part.

Part (a): Determine the complete state of stress on stress element aligned with inclined section pq.

1, 2. *Conceptualize, Categorize*: To find the state of stress on inclined section pq, start by finding compressive normal stress σ_x due to applied load P:

$$\sigma_x = \frac{-P}{A} = \frac{-90 \text{ kN}}{1200 \text{ mm}^2} = -75 \text{ MPa}$$

3. *Analyze*: Next find normal and shear stresses from Eqs. (8-29a and b) with $\theta = 25°$ as

$$\sigma_\theta = \sigma_x \cos(\theta)^2 = (-75 \text{ MPa}) \cos(25°)^2 = -61.6 \text{ MPa}$$
$$\tau_\theta = -\sigma_x \sin(\theta) \cos(\theta) = -(-75 \text{ MPa}) \sin(25°) \cos(25°)$$
$$= 28.7 \text{ MPa}$$

These stresses are shown acting on the inclined section *pq* in Fig. 8-61c. Stress element **face ab** (Fig. 8-61d) is aligned with section *pq*. Note that the normal stress σ_θ is *negative* (compressive), and the shear stress τ_θ is *positive* (counterclockwise). Now use Eqs. (8-29a and b) to find normal and shear stresses on the remaining three faces of the stress element (see Fig. 8-61d).

The normal and shear stresses on **face cb** are computed using angle $\theta + 90° = 115°$ in Eqs. (8-29a and b):

$$\sigma_{cb} = \sigma_x \cos(115°)^2 = (-75 \text{ MPa}) \cos(115°)^2 = -13.4 \text{ MPa}$$
$$\tau_{cb} = -\sigma_x \sin(115°) \cos(115°) = -(-75 \text{ MPa})[\sin(115°) \cos(115°)]$$
$$= -28.7 \text{ MPa}$$

4. *Finalize*: The stresses on the opposite **face cd** are the same as those on **face ab**, which can be verified by substituting $\theta = 25° + 180° = 205°$ into Eqs. (8-29a and b). For **face ad**, substitute $\theta = 25° - 90° = -65°$ into Eqs. (8-29a and b). The complete state of stress is shown in Fig. (8-61d).

FIGURE 8-61 Continued

Example 8-14: (c) Stresses on element at inclined section *pq* through bar and (d) complete state of stress on element at inclined section *pq* through bar

(c)

(d)

FIGURE 8-61 Continued

Example 8-14: (e) Normal and
shear stresses on element at
inclined section *rs* through bar

(e)

Part (b): Determine the normal and shear stresses due to temperature increase on the stress element aligned with inclined section *rs*.

2. *Categorize*: From Example 8-9, reactions R_A and R_B (Fig. 8-61b) due to temperature increase $\Delta T = 33°$ are

$$R_A = -R_B = EA\alpha(\Delta T) \tag{a}$$

and the resulting axial compressive thermal stress is

$$\sigma_T = \frac{R_A}{A} = E\alpha(\Delta T) \tag{b}$$

So

$$\sigma_x = -(110 \text{ GPa})[20 \times 10^{-6}/°C)](33°C) = -72.6 \text{ MPa}$$

3. *Analyze*: Since the compressive stress on plane *rs* is known to be 65 MPa, angle θ for inclined plane *rs* from Eq. (8-29a) is

$$\theta_{rs} = \cos^{-1}\left(\sqrt{\frac{\sigma_\theta}{\sigma_x}}\right) = \cos^{-1}\left(\sqrt{\frac{-65 \text{ MPa}}{-72.6 \text{ MPa}}}\right) = 18.878°$$

and from Eq. (8-29b), shear stress τ_θ on inclined plane *rs* is

$$\tau_\theta = -\sigma_x(\sin(\theta_{rs})\cos(\theta_{rs})) = -(-72.6 \text{ MPa})\sin(18.878°) \cos(18.878°)$$
$$= 22.2 \text{ MPa} \qquad \longleftarrow$$

Part (c): Determine the maximum permissible temperature *increase* (ΔT) in the bar based on allowable stress values.

2, 3. *Categorize, Analyze*: The maximum normal stress σ_{max} occurs on a stress element inclined at $\theta = 0$ [Eq. (8-30)], so $\sigma_{max} = \sigma_x$. Equate *thermal stress* from Eq. (b) to allowable normal stress $\sigma_a = 82$ MPa to find the value of ΔT_{max} based on allowable normal stress:

$$\Delta T_{max1} = \frac{\sigma_a}{E\alpha} = \frac{82 \text{ MPa}}{(110 \text{ GPa})[20 \times 10^{-6}/°C]} = 37.3°C \tag{c}$$

From Eq. (8-31), maximum shear stress τ_{max} occurs at a section inclination of 45° where $\tau_{max} = \sigma_x/2$. Using the given allowable shear stress value, $\tau_a = 40$ MPa, and the relationship between maximum normal and shear stresses in Eq. (8-31), compute a second value for ΔT_{max} as

$$\Delta T_{max2} = \frac{2\tau_a}{E\alpha} = \frac{2(40 \text{ MPa})}{(110 \text{ GPa})[20 \times 10^{-6}/°C]} = 36.4°C \qquad \longleftarrow$$

4. *Finalize*: The lower temperature increase value, based on not exceeding allowable shear stress τ_a, controls. You could have anticipated this because $\tau_{allow} < \sigma_{allow}/2$.

Chapter 8 discussed the behavior of axially loaded bars acted on by distributed loads, such as self-weight, and also temperature changes and prestrains. Force–displacement relations were derived for use in computing changes in lengths of bars under both uniform and nonuniform conditions. Equilibrium and compatibility equations were developed for statically indeterminate structures. Equations were developed for normal and shear stresses on inclined sections. These are the major concepts presented in this chapter.

1. The elongation or shortening (δ) of prismatic bars subjected to tensile or compressive centroidal loads is proportional to both the load (P) and the length (L) of the bar and inversely proportional to the axial rigidity (EA) of the bar; this relationship is called a **force–displacement relation**.

$$\delta = \frac{PL}{EA}$$

2. Cables are **tension-only elements**, and an effective modulus of elasticity (E_e) and effective cross-sectional area (A_e) should be used to account for the tightening effect that occurs when cables are placed under load.

3. The axial rigidity per unit length of a bar is referred to as its **stiffness** (k), and the inverse relationship is the **flexibility** (f) of the bar.

$$\delta = Pf = \frac{P}{k} \qquad f = \frac{L}{EA} = \frac{1}{k}$$

4. The summation of the displacements of the individual segments of a nonprismatic bar equals the elongation or shortening of the entire bar (δ).

$$\delta = \sum_{i=1}^{n} \frac{N_i L_i}{E_i A_i}$$

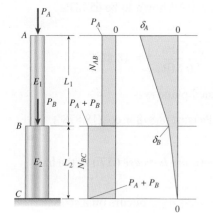

A graphical display of the internal forces N_i is called an **axial-force diagram** (AFD). The AFD is used along with the *force–displacement relation* to create the displaced shape of the bar referred to as the **axial-displacement diagram** (ADD). These diagrams are useful visual aids that help to identify locations of maximum force and stress, displacement, and strain.

5. **Free-body diagrams** are used to find the axial force (N_i) in each segment i; if axial forces and/or cross-sectional areas vary continuously, an integral expression is required.

$$\delta = \int_0^L d\delta = \int_0^L \frac{N(x)dx}{EA(x)}$$

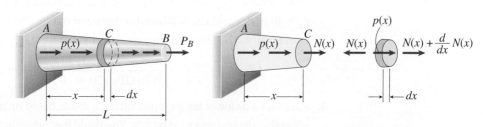

6. If the bar structure is **statically indeterminate**, additional equations beyond those available from statics are required to solve for unknown forces. **Compatibility equations** are used to relate bar displacements to support conditions and thereby generate additional relationships among the unknowns. It is convenient to use a **superposition** of "released" (or statically determinate) structures to represent the actual statically indeterminate bar structure.

7. **Thermal effects** result in displacements that are proportional to the temperature change (ΔT) and the length (L) of the bar, but not stresses in statically determinate structures. The coefficient of thermal expansion (α) of the material also is required to compute axial strains (ε_T) and axial displacements (δ_T) due to thermal effects.

$$\varepsilon_T = \alpha(\Delta T) \qquad \delta_T = \varepsilon_T L = \alpha(\Delta T)L$$

8. **Misfits** and **prestrains** induce axial forces only in statically indeterminate bars.

9. **Maximum normal** (σ_{max}) and **shear stresses** (τ_{max}) are obtained by considering an inclined stress element for a bar loaded by axial forces. The maximum normal stress occurs along the axis of the bar, but the maximum shear stress occurs at an inclination of 45° to the axis of the bar, and the maximum shear stress is one-half of the maximum normal stress.

$$\sigma_{max} = \sigma_x \qquad \tau_{max} = \frac{\sigma_x}{2}$$

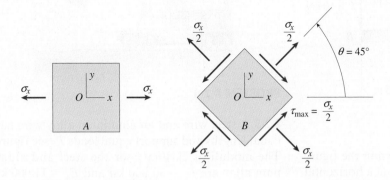

PROBLEMS

8.2 Changes in Lengths of Axially Loaded Members

Introductory Problems

8.2-1 A 10-ft rigid bar AB is supported with a vertical translational spring at A and a pin at B. The bar is subjected to a linearly varying distributed load with maximum intensity q_0. Calculate the vertical deformation of the spring if the spring constant is 4 kips/in.

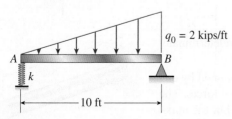

PROBLEM 8.2-1

8.2-2 Rigid bar ABC is supported with a pin at A and an elastic steel rod at C. The elastic rod has a diameter of 25 mm and modulus of elasticity $E = 200$ GPa. The bar is subjected to a uniform load q on span AC and a point load at B. Calculate the change in length of the elastic rod. What is the vertical displacement at point B?

PROBLEM 8.2-2

8.2-3 The L-shaped arm $ABCD$ shown in the figure lies in a vertical plane and pivots about a horizontal pin at A. The arm has a constant cross-sectional area and total weight W. A vertical spring of stiffness k supports the arm at point B.

(a) Obtain a formula for the elongation of the spring due to the weight of the arm.

(b) Repeat part (a) if the pin support at A is moved to D.

PROBLEM 8.2-3

8.2-4 A steel cable with a nominal diameter of 25 mm (see Table 8-1) is used in a construction yard to lift a bridge section weighing 38 kN, as shown in the figure. The cable has an effective modulus of elasticity $E = 140$ GPa.

(a) If the cable is 14 m long, how much will it stretch when the load is picked up?

(b) If the cable is rated for a maximum load of 70 kN, what is the factor of safety with respect to failure of the cable?

PROBLEM 8.2-4

8.2-5 A steel wire and an aluminum alloy wire have equal lengths and support equal loads P (see figure). The moduli of elasticity for the steel and aluminum alloy are $E_s = 30,000$ ksi and $E_a = 11,000$ ksi, respectively.

(a) If the wires have the same diameters, what is the ratio of the elongation of the aluminum alloy wire to the elongation of the steel wire?

(b) If the wires stretch the same amount, what is the ratio of the diameter of the aluminum alloy wire to the diameter of the steel wire?

602

(c) If the wires have the same diameters and same load P, what is the ratio of the initial length of the aluminum alloy wire to that of the steel wire if the aluminum alloy wire stretches 1.5 times that of the steel wire?

(d) If the wires have the same diameters, same initial length, and same load P, what is the material of the upper wire if it elongates 1.7 times that of the steel wire?

Aluminum alloy wire

Steel wire

P

P

PROBLEM 8.2-5

8.2-6 By what distance h does the cage shown in the figure move downward when the weight W is placed inside it? (See the figure.)

Consider only the effects of the stretching of the cable, which has axial rigidity $EA = 10,700$ kN. The pulley at A has a diameter $d_A = 300$ mm and the pulley at B has a diameter $d_B = 150$ mm. Also, the distance $L_1 = 4.6$ m, the distance $L_2 = 10.5$ m, and the weight $W = 22$ kN. *Note:* When calculating the length of the cable, include the parts of the cable that go around the pulleys at A and B.

8.2-7 Rigid bar ACB is supported by an elastic circular strut DC having an outer diameter of 15 in. and inner diameter of 14.4 in. The strut is made of steel with a modulus elasticity of $E = 29,000$ ksi. Point load $P = 5$ kips is applied at B. Calculate the change in length of the circular strut DC. What is the vertical displacement of the rigid bar at point B?

PROBLEM 8.2-7

8.2-8 A device consists of a horizontal beam ABC supported by two vertical bars BD and CE. Bar CE is pinned at both ends but bar BD is fixed to the foundation at its lower end. The distance from A to B is 600 mm and from B to C is 350 mm. Bars BD and CE have lengths of 350 mm and 450 mm, respectively, and their cross-sectional area is 720 mm^2. The bars are made of steel having a modulus of elasticity $E = 200$ GPa. If load P is 20 kN, calculate the displacement at point A.

PROBLEM 8.2-6

PROBLEM 8.2-8

Representative Problems

8.2-9 A safety valve on the top of a tank containing steam under pressure p has a discharge hole of diameter d (see figure). The valve is designed to release the steam when the pressure reaches the value p_{max}.

If the natural length of the spring is L and its stiffness is k, what should be the dimension h of the valve? (Express your result as a formula for h.)

PROBLEM 8.2-9

8.2-10 The device shown in the figure consists of a prismatic rigid pointer ABC supported by a uniform translational spring of stiffness $k = 950$ N/m. The spring is positioned at distance $b = 165$ mm from the pinned end A of the pointer. The device is adjusted so that, when there is no load P, the pointer reads zero on the angular scale.

(a) If the load $P = 11$ N, at what distance x should the load be placed so that the pointer will read $\theta = 2.5°$ on the scale (see figure part a)?

(b) Repeat part (a) if a rotational spring $k_r = kb^2$ is added at A (see figure part b).

(c) Let $x = 7b/8$. What is P_{max} (N) if θ cannot exceed 2°? Include spring k_r in your analysis.

(d) Now, if the weight of the pointer ABC is known to be $W_p = 3$ N and the weight of the spring is $W_s = 2.75$ N, what initial angular position (i.e., θ in degrees) of the pointer will result in a zero reading on the angular scale once the pointer is released from rest? Assume $P = k_r = 0$.

(e) If the pointer is rotated to a vertical position (see figure part c), find the required load P applied at mid-height of the pointer that will

result in a pointer reading of $\theta = 2.5°$ on the scale. Consider the weight of the pointer W_p in your analysis.

PROBLEM 8.2-10

8.2-11 A small lab scale has a rigid L-shaped frame ABC consisting of a horizontal arm AB (length $b = 10$ in.) and a vertical arm BC (length $c = 7$ in.) pivoted at point B. The pivot is attached to the outer frame BCD that stands on a laboratory bench.

The position of the pointer at C is controlled by a spring, $k = 5$ lb/in., that is attached to a threaded rod. The pitch of the threads is $p = 1/16$ in. Under application of load W, 12 revolutions of the nut are required to bring the pointer back to the mark. Calculate the weight W.

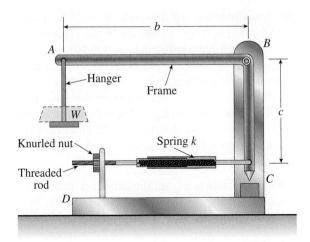

PROBLEM 8.2-11

8.2-12 A small lab scale has a rigid L-shaped frame ABC consisting of a horizontal arm AB (length $b = 30$ cm) and a vertical arm BC (length $c = 20$ cm) pivoted at point B. The pivot is attached to the outer frame BCD that stands on a laboratory bench. The

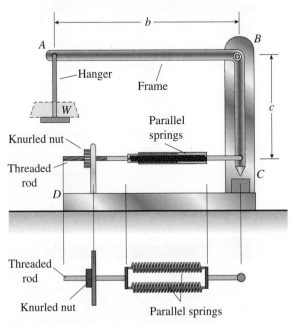

Top view of the springs

PROBLEM 8.2-12

position of the pointer at C is controlled by two parallel springs, each having a spring constant $k = 3650$ N/m, that are attached to a threaded rod. The pitch of the threads is $p = 1.5$ mm. If the weight is 65 N, how many revolutions of the nut are required to bring the pointer back to the mark?

8.2-13 Two rigid bars are connected to each other by two linearly elastic springs. Before loads are applied, the lengths of the springs are such that the bars are parallel and the springs are without stress.

(a) Derive a formula for the displacement δ_4 at point 4 when the load P is applied at joint 3 and moment PL is applied at joint 1, as shown in the figure part a. (Assume that the bars rotate through very small angles under the action of load P.)

(b) Repeat part (a) if a rotational spring, $k_r = kL^2$, is now added at joint 6. What is the ratio of the deflection δ_4 in the figure part a to that in the figure part b ?

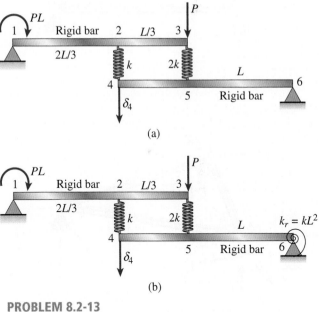

PROBLEM 8.2-13

605

8.2-14 The three-bar truss ABC shown in the figure part a has a span $L = 3$ m and is constructed of steel pipes having a cross-sectional area $A = 3900$ mm² and modulus of elasticity $E = 200$ GPa. Identical loads P act both vertically and horizontally at joint C, as shown.

(a) If $P = 475$ kN, what is the horizontal displacement of joint B?

(b) What is the maximum permissible load value P_{max} if the displacement of joint B is limited to 1.5 mm?

(c) Repeat parts (a) and (b) if the plane truss is replaced by a space truss (see figure part b).

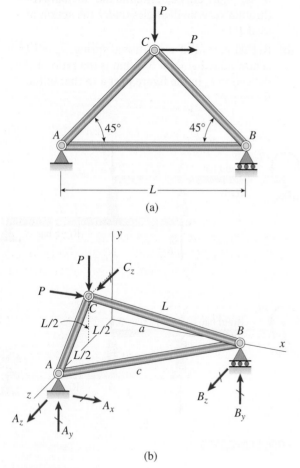

(a)

(b)

PROBLEM 8.2-14

8.2-15 An aluminum wire having a diameter $d = 1/10$ in. and length $L = 12$ ft is subjected to a tensile load P (see figure). The aluminum has a modulus of elasticity $E = 10,600$ ksi.

If the maximum permissible elongation of the wire is 1/8 in. and the allowable stress in tension is 10 ksi, what is the allowable load P_{max}?

PROBLEM 8.2-15

8.2-16 A uniform bar AB of weight $W = 25$ N is supported by two springs, as shown in the figure. The spring on the left has a stiffness $k_1 = 300$ N/m and natural length $L_1 = 250$ mm. The corresponding quantities for the spring on the right are $k_2 = 400$ N/m and $L_2 = 200$ mm. The distance between the springs is $L = 350$ mm, and the spring on the right is suspended from a support that is a distance $h = 80$ mm below the point of support for the spring on the left. Neglect the weight of the springs.

(a) At what distance x from the left-hand spring (figure part a) should a load $P = 18$ N be placed in order to bring the bar to a horizontal position?

(a)

(b)

PROBLEM 8.2-16

(b) If P is now removed, what new value of k_1 is required so that the bar (figure part a) will hang in a horizontal position under weight W?

(c) If P is removed and $k_1 = 300$ N/m, what distance b should spring k_1 be moved to the right so that the bar (figure part a) will hang in a horizontal position under weight W?

(d) If the spring on the left is now replaced by two springs in series ($k_1 = 300$ N/m, k_3) with overall natural length $L_1 = 250$ mm (see figure part b), what value of k_3 is required so that the bar will hang in a horizontal position under weight W?

8.2-17 A hollow, circular, cast-iron pipe ($E_c = 12,000$ ksi) supports a brass rod ($E_b = 14,000$ ksi) and weight $W = 2$ kips, as shown. The outside diameter of the pipe is $d_c = 6$ in.

(a) If the allowable compressive stress in the pipe is 5000 psi and the allowable shortening of the pipe is 0.02 in., what is the minimum required wall thickness $t_{c,\text{min}}$? (Include the weights of the rod and steel cap in your calculations.)

(b) What is the elongation of the brass rod δ_r due to both load W and its own weight?

(c) What is the minimum required clearance h?

8.2-18 The horizontal rigid beam $ABCD$ is supported by vertical bars BE and CF and is loaded by vertical forces $P_1 = 400$ kN and $P_2 = 360$ kN acting at points A and D, respectively (see figure). Bars BE and CF are made of steel ($E = 200$ GPa) and have cross-sectional areas $A_{BE} = 11,100$ mm² and $A_{CF} = 9280$ mm². The distances between various points on the bars are shown in the figure.

Determine the vertical displacements δ_A and δ_D of points A and D, respectively.

PROBLEM 8.2-18

8.2-19 Two pipe columns (AB, FC) are pin-connected to a *rigid* beam (BCD), as shown in the figure. Each pipe column has a modulus of E, but heights (L_1 or L_2) and outer diameters (d_1 or d_2) are different for each

PROBLEM 8.2-17

PROBLEM 8.2-19

column. Assume the inner diameter of each column is 3/4 of outer diameter. Uniformly distributed downward load $q = 2P/L$ is applied over a distance of $3L/4$ along BC, and concentrated load $P/4$ is applied downward at D.

(a) Derive a formula for the displacement δ_D at point D in terms of P and column flexibilities f_1 and f_2.

(b) If $d_1 = (9/8)d_2$, find the L_1/L_2 ratio so that beam BCD displaces downward to a horizontal position under the load system in part (a).

(c) If $L_1 = 2L_2$, find the d_1/d_2 ratio so that beam BCD displaces downward to a horizontal position under the load system in part (a).

(d) If $d_1 = (9/8)d_2$ and $L_1/L_2 = 1.5$, at what horizontal distance x from B should load $P/4$ be placed so that beam BCD displaces downward to a horizontal position under the load system in part (a)?

8.2-20 A framework ABC consists of two rigid bars AB and BC, each having a length b (see the first part of the figure part a). The bars have pin connections at A, B, and C and are joined by a spring of stiffness k. The spring is attached at the midpoints of the bars. The framework has a pin support at A and a roller support at C, and the bars are at an angle α to the horizontal.

When a vertical load P is applied at joint B (see the second part of the figure part a) the roller support C moves to the right, the spring is stretched, and the angle of the bars decreases from α to the angle θ.

(a) Determine the angle θ and the increase δ in the distance between points A and C. Also find reactions at A and C. (Use the following data: $b = 200$ mm, $k = 3.2$ kN/m, $\alpha = 45°$, and $P = 50$ N.)

(b) Repeat part (a) if a translational spring $k_1 = k/2$ is added at C and a rotational spring $k_r = kb^2/2$ is added at A (see figure part b).

8.2-21 Solve the preceding problem for the following data: $b = 8.0$ in., $k = 16$ lb/in., $\alpha = 45°$, and $P = 10$ lb.

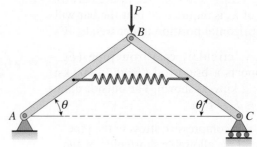

(a) Initial and displaced structures

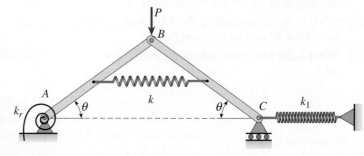

(b) Displaced structure

PROBLEMS 8.2-20 and 8.2-21

8.3 Changes in Lengths Under Nonuniform Conditions

Introductory Problems

8.3-1 The length of the end segments of the bar (see figure) is 20 in. and the length of the prismatic middle segment is 50 in. Also, the diameters at cross sections A, B, C, and D are 0.5, 1.0, 1.0, and 0.5 in., respectively, and the modulus of elasticity is 18,000 ksi.

(a) Calculate the elongation of a copper bar of solid circular cross section with tapered ends when it is stretched by axial loads of magnitude 3.0 kips (see figure).

(b) If the total elongation of the bar cannot exceed 0.025 in., what are the required diameters at B and C? Assume that diameters at A and D remain at 0.5 in.

3.0 kips

A B C D

20 in.

50 in.

20 in.   3.0 kips

PROBLEM 8.3-1

8.3-2 A long, rectangular copper bar under a tensile load P hangs from a pin that is supported by two steel posts (see figure). The copper bar has a length of 2.0 m, a cross-sectional area of 4800 mm^2, and a modulus of elasticity $E_c = 120$ GPa. Each steel post has a height of 0.5 m, a cross-sectional area of 4500 mm^2, and a modulus of elasticity $E_s = 200$ GPa.

(a) Determine the downward displacement δ of the lower end of the copper bar due to a load $P = 180$ kN.

(b) What is the maximum permissible load P_{max} if the displacement δ is limited to 1.0 mm?

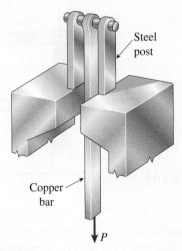

Steel post

Copper bar

P

PROBLEM 8.3-2

8.3-3 An aluminum bar AD (see figure) has a cross-sectional area of 0.40 in^2 and is loaded by forces $P_1 = 1700$ lb, $P_2 = 1200$ lb, and $P_3 = 1300$ lb. The lengths of the segments of the bar are $a = 60$ in., $b = 24$ in., and $c = 36$ in.

(a) Assuming that the modulus of elasticity is $E = 10.4 \times 10^6$ psi, calculate the change in length of the bar. Does the bar elongate or shorten?

(b) By what amount P should the load P_3 be increased so that the bar does not change in length when the three loads are applied?

(c) If P_3 remains at 1300 lb, what revised cross-sectional area for segment AB will result in no change of length when all three loads are applied?

P_1 P_2

A B C D P_3

a b c

PROBLEM 8.3-3

8.3-4 A vertical bar consists of three prismatic segments A_1, A_2, and A_3 with cross-sectional areas of 6000 mm^2, 5000 mm^2, and 4000 mm^2, respectively. The bar is made of steel with $E = 200$ GPa. Calculate the displacements at points B, D, and E. Ignore the weight of the bar.

A

$P_B = 50$ N   500 mm

B

250 mm

C

$P_C = 250$ N   250 mm

D

500 mm

E

$P_E = 350$ N

PROBLEMS 8.3-4 and 8.3-6

8.3-5 A vertical bar is loaded with axial loads at points B, C, and D, as shown in the figure. The bar is made of steel with a modulus of elasticity $E = 29,000$ ksi. The bar has a cross-sectional area of 8.24 in^2. Calculate the displacements at points B, C, and D. Ignore the weight of the bar.

PROBLEMS 8.3-5 and 8.3-7

8.3-6 Repeat Problem 8.3-4, but now include the weight of the bar. See Table F-1 in Appendix F for the weight density of steel.

8.3-7 Repeat Problem 8.3-5, but now include the weight of the bar. See Table F-1 in Appendix F for the weight density of steel.

Representative Problems

8.3-8 A rectangular bar of length L has a slot in the middle half of its length (see figure). The bar has width b, thickness t, and modulus of elasticity E. The slot has width $b/4$.

(a) Obtain a formula for the elongation δ of the bar due to the axial loads P.

(b) Calculate the elongation of the bar if the material is high-strength steel, the axial stress in the middle region is 160 MPa, the length is 750 mm, and the modulus of elasticity is 210 GPa.

(c) If the total elongation of the bar is limited to $\delta_{max} = 0.475$ mm, what is the maximum length of the slotted region? Assume that the axial stress in the middle region remains at 160 MPa.

PROBLEMS 8.3-8 and 8.3-9

8.3-9 Solve the preceding problem if the axial stress in the middle region is 24,000 psi, the length is 30 in., and the modulus of elasticity is 30×10^6 psi. In part (c), assume that $\delta_{max} = 0.02$ in.

8.3-10 A two-story building has steel columns AB in the first floor and BC in the second floor, as shown in the figure. The roof load P_1 equals 400 kN, and the second-floor load P_2 equals 720 kN. Each column has a length $L = 3.75$ m. The cross-sectional areas of the first- and second-floor columns are 11,000 mm^2 and 3900 mm^2, respectively.

(a) Assuming that $E = 206$ GPa, determine the total shortening δ_{AC} of the two columns due to the combined action of the loads P_1 and P_2.

(b) How much additional load P_0 can be placed at the top of the column (point C) if the total shortening δ_{AC} is not to exceed 4.0 mm?

PROBLEM 8.3-10

8.3-11 A steel bar is 8.0 ft long and has a circular cross section of diameter $d_1 = 0.75$ in. over one-half of its length and diameter $d_2 = 0.5$ in. over the other half (see figure on following page part a). The modulus of elasticity is $E = 30 \times 10^6$ psi.

(a) How much will the bar elongate under a tensile load $P = 5000$ lb?

(b) If the same volume of material is made into a bar of constant diameter d and length 8.0 ft, what will be the elongation under the same load P?

(c) If the uniform axial centroidal load $q = 1250$ lb/ft is applied to the left over segment 1 (see figure part b), find the ratio of the total elongation of the bar to that in parts (a) and (b).

(a)

(b)

PROBLEM 8.3-11

8.3-12 A bar ABC of length L consists of two parts of equal lengths but different diameters. Segment AB has diameter $d_1 = 100$ mm, and segment BC has diameter $d_2 = 60$ mm. Both segments have a length $L/2 = 0.6$ m. A longitudinal hole of diameter d is drilled through segment AB for one-half of its length (distance $L/4 = 0.3$ m). The bar is made of plastic having a modulus of elasticity $E = 4.0$ GPa. Compressive loads $P = 110$ kN act at the ends of the bar.

(a) If the shortening of the bar is limited to 8.0 mm, what is the maximum allowable diameter d_{max} of the hole? (See figure part a.)

(b) Now, if d_{max} is instead set at $d_2/2$, at what distance b from end C should load P be applied to limit the bar shortening to 8.0 mm? (See figure part b.)

(c) Finally, if loads P are applied at the ends and $d_{max} = d_2/2$, what is the permissible length x of the hole if shortening is to be limited to 8.0 mm? (See figure part c.)

8.3-13 A wood pile, driven into the earth, supports a load P entirely by friction along its sides (see figure part a). The friction force f per unit length of the pile is assumed to be uniformly distributed over the surface of the pile. The pile has a length L, cross-sectional area A, and modulus of elasticity E.

(a) Derive a formula for the shortening δ of the pile in terms of P, L, E, and A.

(b) Draw a diagram showing how the compressive stress σ_c varies throughout the length of the pile.

(c) Repeat parts (a) and (b) if skin friction f varies linearly with depth (see figure part b).

(a)

(b)

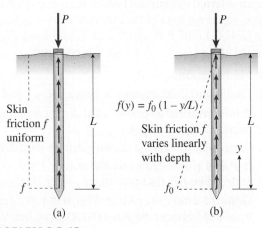

(a)   (b)

PROBLEM 8.3-13

PROBLEM 8.3-12

8.3-14 Consider the copper tubes joined in the figure using a "sweated" joint. Use the properties and dimensions given.

(a) Find the total elongation of segment 2-3-4 (δ_{2-4}) for an applied tensile force of $P = 5$ kN. Use $E_c = 120$ GPa.

(b) If the yield strength in shear of the tin-lead solder is $\tau_y = 30$ MPa and the tensile yield strength of the copper is $\sigma_y = 200$ MPa, what is the maximum load P_{max} that can be applied to the joint if the desired factor of safety in shear is $FS_\tau = 2$ and in tension is $FS_\sigma = 1.7$?

(c) Find the value of L_2 at which tube and solder capacities are equal.

Sweated joint

Solder joints

Segment number

L_2 L_3 L_4

$d_0 = 18.9$ mm
$t = 1.25$ mm

$d_0 = 22.2$ mm
$t = 1.65$ mm

$L_3 = 40$ mm
$L_2 = L_4 = 18$ mm

Tin-lead solder in space between copper tubes; assume thickness of solder equals zero

© Barry Goodno

PROBLEM 8.3-14

8.3-15 The nonprismatic cantilever circular bar shown has an internal cylindrical hole of diameter $d/2$ from 0 to x, so the net area of the cross section for segment 1 is $(3/4)A$. Load P is applied at x, and load $P/2$ is applied at $x = L$. Assume that E is constant.

(a) Find reaction force R_1.

(b) Find internal axial forces N_i in segments 1 and 2.

(c) Find x required to obtain axial displacement at joint 3 of $\delta_3 = PL/EA$.

(d) In part (c), what is the displacement at joint 2, δ_2?

(e) If P acts at $x = 2L/3$ and $P/2$ at joint 3 is replaced by βP, find β so that $\delta_3 = PL/EA$.

(f) Draw the *axial-force* (AFD: $N(x)$, $0 \le x \le L$) and *axial-displacement diagrams* (ADD: $\delta(x)$, $0 \le x \le L$) using results from parts (b) through (d).

PROBLEM 8.3-15

8.3-16 A prismatic bar AB of length L, cross-sectional area A, modulus of elasticity E, and weight W hangs vertically under its own weight (see figure).

(a) Derive a formula for the downward displacement δ_C of point C, located at distance h from the lower end of the bar.

(b) What is the elongation δ_B of the entire bar?

(c) What is the ratio β of the elongation of the upper half of the bar to the elongation of the lower half of the bar?

(d) If bar AB is a riser pipe hanging from a drill rig at sea, what is the total elongation of the pipe? Let $L = 1500$ m, $A = 0.0157$ m², and $E = 210$ GPa. See Appendix F for weight densities of steel and sea water. (See Probs. 7.3-2 and 7.6-13 for additional figures.)

PROBLEM 8.3-16

8.3-17 A flat bar of rectangular cross section, length L, and constant thickness t is subjected to tension by forces P (see figure). The width of the bar varies linearly from b_1 at the smaller end to b_2 at the larger end. Assume that the angle of taper is small.

(a) Derive the following formula for the elongation of the bar:

$$\delta = \frac{PL}{Et(b_2 - b_1)} \ln \frac{b_2}{b_1}$$

PROBLEM 8.3-17

(b) Calculate the elongation, assuming $L = 5$ ft, $t = 1.0$ in., $P = 25$ kips, $b_1 = 4.0$ in., $b_2 = 6.0$ in., and $E = 30 \times 10^6$ psi.

8.3-18 A flat brass bar has length L, constant thickness t, and a rectangular cross section whose width varies linearly between b_2 at the fixed support to b_1 at the free end (see figure). Assume that the taper of the bar is small. The bar has modulus of elasticity E. Calculate the displacements δ_B and δ_C if $P = 200$ kN, $L = 2$ m, $t = 20$ mm, $b_1 = 100$ mm, $b_2 = 115$ mm, and $E = 96$ GPa.

PROBLEM 8.3-18

8.3-19 Repeat Problem 8.3-18, but assume that the bar is made of copper alloy. Calculate the displacements δ_B and δ_C if $P = 50$ kips, $L = 5$ ft, $t = 3/8$ in., $b_1 = 2.75$ in., $b_2 = 3$ in., and $E = 16,000$ ksi.

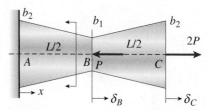

PROBLEM 8.3-19

8.3-20 Repeat Problem 8.3-18, but assume that the bar is made of aluminum alloy. If $P_2 = 200$ kN, what is P_1 so that displacement $\delta_C = 0$? What is displacement δ_B? Assume that $L = 2$ m, $t = 20$ mm, $b_1 = 100$ mm, $b_2 = 115$ mm, and $E = 72$ GPa.

PROBLEM 8.3-20

8.3-21 A slightly tapered bar AB of solid circular cross section and length L is supported at end B and subjected to a tensile load P at the free end A. The diameters of the bar at ends A and B are d_A and d_B, respectively. Determine the length of the bar if the elongation of the bar due to the load $P = 45$ kips is 0.02 in. Assume that E = 10,400 ksi.

PROBLEM 8.3-21

8.3-22 A circular aluminum alloy bar of length $L = 1.8$ m has a slot in the middle half of its length (see figure). The bar has a radius $r = 36$ mm and modulus of elasticity $E = 72$ GPa. The slot has a height $2a = r/4$. Calculate the elongation of the bar due to forces P applied at the ends if the axial stress in the middle region is known to be 180 MPa.

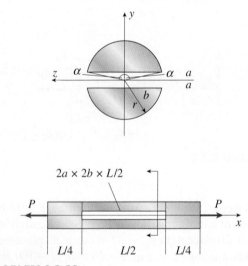

PROBLEM 8.3-22

8.3-23 A long, slender bar in the shape of a right circular cone with length L and base diameter d hangs vertically under the action of its own weight (see figure). The weight of the cone is W and the modulus of elasticity of the material is E.

Derive a formula for the increase δ in the length of the bar due to its own weight. (Assume that the angle of taper of the cone is small.)

PROBLEM 8.3-23

8.3-24 A post AB supporting equipment in a laboratory is tapered uniformly throughout its height H (see figure). The cross sections of the post are square, with dimensions $b \times b$ at the top and $1.5b \times 1.5b$ at the base.

Derive a formula for the shortening δ of the post due to the compressive load P acting at the top. (Assume that the angle of taper is small and disregard the weight of the post itself.)

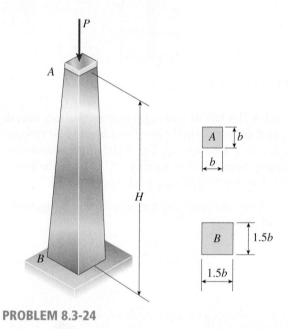

PROBLEM 8.3-24

8.3-25 The main cables of a suspension bridge (see figure part a) follow a curve that is nearly parabolic because the primary load on the cables is the weight of the bridge deck, which is uniform in intensity along the horizontal. Therefore, represent the central region AOB of one of the main cables (see part b of the figure) as a parabolic cable supported at points A and B and carrying a uniform load of intensity q

along the horizontal. The span of the cable is L, the sag is h, the axial rigidity is EA, and the origin of coordinates is at midspan.

(a) Derive the following formula for the elongation of cable AOB shown in part b of the figure:

$$\delta = \frac{qL^3}{8hEA}\left(1 + \frac{16h^2}{3L^2}\right)$$

(b) Calculate the elongation δ of the central span of one of the main cables of the Golden Gate Bridge for which the dimensions and properties are $L = 4200$ ft, $h = 470$ ft, $q = 12,700$ lb/ft, and $E = 28,800,000$ psi. The cable consists of 27,572 parallel wires of diameter 0.196 in.

Hint: Determine the tensile force T at any point in the cable from a free-body diagram of part of the cable; then determine the elongation of an element of the cable of length ds; finally, integrate along the curve of the cable to obtain an equation for the elongation δ.

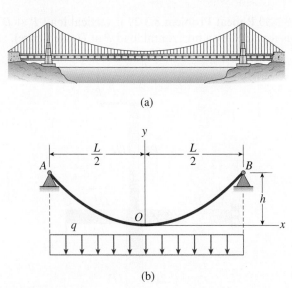

(a)

(b)

PROBLEM 8.3-25

8.3-26 A uniformly tapered tube AB of circular cross section and length L is shown in the figure. The average diameters at the ends are d_A and $d_B = 2d_A$. Assume E is constant. Find the elongation δ of the tube when it is subjected to loads P acting at the ends. Use the following numerical data: $d_A = 35$ mm, $L = 300$ mm, $E = 2.1$ GPa, and $P = 25$ kN. Consider the following cases.

(a) A hole of *constant* diameter d_A is drilled from B toward A to form a hollow section of length $x = L/2$.

(b) A hole of *variable* diameter $d(x)$ is drilled from B toward A to form a hollow section of length $x = L/2$ and constant thickness $t = d_A/20$.

(a)

(b)

PROBLEM 8.3-26

8.3-27 A vertical steel bar ABC is pin-supported at its upper end and loaded by a force P_1 at its lower end. A horizontal beam BDE is pinned to the vertical

PROBLEM 8.3-27

615

bar at joint B and supported at point D. Load P_2 and moment M are applied at end E. Calculate the vertical displacement δ_C at point C if the loads are $P_1 = 2.5$ kip, $P_2 = 1$ kip, and $M = 25$ kip-in. The modulus of elasticity is $E = 29,000$ ksi and cross-sectional areas are $A_1 = 0.25$ in² and $A_2 = 0.15$ in². Ignore the weight of the bar.

8.3-28 A T-frame structure is composed of a prismatic beam ABC and a nonprismatic column DBF. The beam and the column have a pin support at A and D, respectively. Both members are connected with a pin at B. The lengths and properties of the members are shown in the figure. Find the vertical displacement of the column at points F and B. Plot axial-force (AFD) and axial-displacement (ADD) diagrams for column DBF.

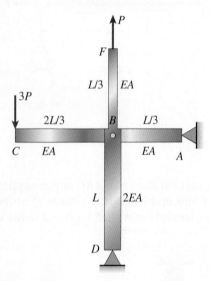

PROBLEM 8.3-28

8.3-29 A T-frame structure is composed of prismatic beam ABC and nonprismatic column DBF that are joined at B by a frictionless pin connection. The beam has a sliding support at A and the column is pin-supported at F (see figure). Beam ABC and column segment DB have cross-sectional area A; column segment BF has area $2A$. The modulus of elasticity E is the same for both members. Load $2P$ is applied downward at C, and load P acts at D. Find expressions for the downward displacements of column DBF at D (δ_D) and also at B (δ_B). Plot axial-force (AFD) and axial-displacement (ADD) diagrams for column DBF.

PROBLEM 8.3-29

8.3-30 Repeat Problem 8.3-29 if vertical load P at D is replaced by a horizontal load P at D (see figure).

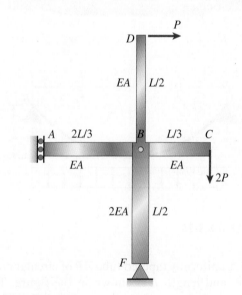

PROBLEM 8.3-30

8.3-31 A bar ABC revolves in a horizontal plane about a vertical axis at the midpoint C (see figure). The bar, which has a length $2L$ and cross-sectional area A, revolves at constant angular speed ω. Each half of the bar (AC and BC) has a weight W_1 and supports a weight W_2 at its end.

Derive the following formula for the elongation of one-half of the bar (that is, the elongation of either AC or BC):

$$\delta = \frac{L^2 \omega^2}{3gEA}(W_1 + 3W_2)$$

in which E is the modulus of elasticity of the material of the bar and g is the acceleration of gravity.

PROBLEM 8.3-31

8.4 Statically Indeterminate Structures

Introductory Problems

8.4-1 The assembly shown in the figure consists of a brass core (diameter $d_1 = 0.25$ in.) surrounded by a steel shell (inner diameter $d_2 = 0.28$ in., outer diameter $d_3 = 0.35$ in.). A load P compresses the core and shell that both have a length $L = 4.0$ in. The moduli of elasticity of the brass and steel are $E_b = 15 \times 10^6$ psi and $E_s = 30 \times 10^6$ psi, respectively.

(a) What load P will compress the assembly by 0.003 in.?

(b) If the allowable stress in the steel is 22 ksi and the allowable stress in the brass is 16 ksi, what is the allowable compressive load P_{allow}? (*Suggestion:* Use the equations derived in Example 8-8.)

PROBLEM 8.4-1

8.4-2 A cylindrical assembly consisting of a brass core and an aluminum collar is compressed by a load P (see figure). The length of the aluminum collar and brass core is 350 mm, the diameter of the core is 25 mm, and the outside diameter of the collar is 40 mm. Also, the moduli of elasticity of the aluminum and brass are 72 GPa and 100 GPa, respectively.

(a) If the length of the assembly decreases by 0.1% when the load P is applied, what is the magnitude of the load?

(b) What is the maximum permissible load P_{max} if the allowable stresses in the aluminum and brass are 80 MPa and 120 MPa, respectively?

PROBLEM 8.4-2

8.4-3 A steel bar with a uniform cross section is fixed at both ends. A load $P = 2.5$ kips is applied at point C. The bar has a cross-sectional area of 8 in.^{2}. Calculate the reactions at joints A and B and the displacement at joint C. Assume that the modulus of elasticity $E = 29,000$ ksi.

PROBLEM 8.4-3

8.4-4 A horizontal rigid bar ABC is pinned at end A and supported by two cables at points B and C. A vertical load $P = 10$ kN acts at end C of the bar. The two cables are made of steel with a modulus elasticity $E = 200$ GPa and have the same cross-sectional area. Calculate the minimum cross-sectional area of each cable if the yield stress of the cable is 400 MPa and the factor of safety is 2.0. Consider load P only; ignore the weight of bar ABC and the cables.

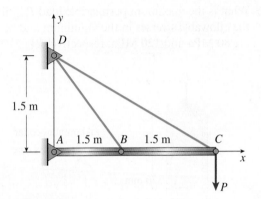

PROBLEM 8.4-4

8.4-5 A solid circular steel cylinder S is encased in a hollow circular aluminum tube A. The cylinder and tube are compressed between the rigid plates of a testing machine which applies forces P. Calculate the allowable value of the compressive force if the yield stresses of steel and aluminum are $\sigma_{yS} = 50$ ksi and $\sigma_{yA} = 60$ ksi, respectively. Assume that $A_S = 12$ in^2, $A_A = 6$ in^2, $L = 20$ in., $E_S = 29,000$ ksi, and $E_A = 10,600$ ksi

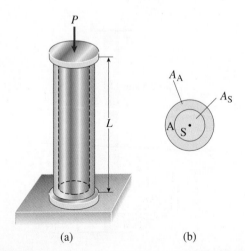

    (a)   (b)

PROBLEM 8.4-5

8.4-6 Three prismatic bars, two of material A and one of material B, transmit a tensile load P (see figure). The two outer bars (material A) are identical. The cross-sectional area of the middle bar (material B) is 50% larger than the cross-sectional area of one of the outer bars. Also, the modulus of elasticity of material A is twice that of material B.

(a) What fraction of the load P is transmitted by the middle bar?

(b) What is the ratio of the stress in the middle bar to the stress in the outer bars?

(c) What is the ratio of the strain in the middle bar to the strain in the outer bars?

PROBLEM 8.4-6

Representative Problems

8.4-7 A circular bar ACB of a diameter d having a cylindrical hole of length x and diameter $d/2$ from A to C is held between rigid supports at A and B.

(a)

(b)

PROBLEM 8.4-7

A load P acts at $L/2$ from ends A and B. Assume E is constant.

(a) Obtain formulas for the reactions R_A and R_B at supports A and B, respectively, due to the load P (see figure part a).

(b) Obtain a formula for the displacement δ at the point of load application (see figure part a).

(c) For what value of x is $R_B = (6/5)R_A$? (See figure part a.)

(d) Repeat part (a) if the bar is now rotated to a vertical position, load P is removed, and the bar is hanging under its own weight (assume mass density $= \rho$). (See figure part b.) Assume that $x = L/2$.

8.4-8 Bar ABC is fixed at both ends (see figure) and has load P applied at B. Find reactions at A and C and displacement δ_B if $P = 200$ kN, $L = 2$ m, $t = 20$ mm, $b_1 = 100$ mm, $b_2 = 115$ mm, and $E = 96$ GPa.

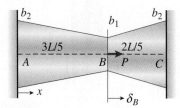

PROBLEM 8.4-8

8.4-9 Repeat Problem 8.4-8, but assume that the bar is made of aluminum alloy and that BC is prismatic. Assume that $P = 20$ kips, $L = 3$ ft, $t = 1/4$ in., $b_1 = 2$ in., $b_2 = 2.5$ in., and $E = 10,400$ ksi.

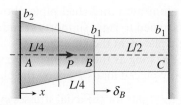

PROBLEM 8.4-9

8.4-10 A plastic rod AB of length $L = 0.5$ m has a diameter $d_1 = 30$ mm (see figure). A plastic sleeve CD of length $c = 0.3$ m and outer diameter $d_2 = 45$ mm is securely bonded to the rod so that no slippage can occur between the rod and the sleeve. The rod is made of an acrylic with a modulus of elasticity

$E_1 = 3.1$ GPa, and the sleeve is made of a polyamide with $E_2 = 2.5$ GPa.

(a) Calculate the elongation δ of the rod when it is pulled by axial forces $P = 12$ kN.

(b) If the sleeve is extended for the full length of the rod, what is the elongation?

(c) If the sleeve is removed, what is the elongation?

PROBLEM 8.4-10

8.4-11 Three steel cables jointly support a load of 12 kips (see figure). The diameter of the middle cable is 3/4 in. and the diameter of each outer cable is 1/2 in. The tensions in the cables are adjusted so that each cable carries one-third of the load (i.e., 4 kips). Later, the load is increased by 9 kips to a total load of 21 kips.

(a) What percent of the total load is now carried by the middle cable?

(b) What are the stresses σ_M and σ_O in the middle and outer cables, respectively? *Note:* See Table 8-1 in Section 8.2 for properties of cables.

PROBLEM 8.4-11

619

8.4-12 The fixed-end bar $ABCD$ consists of three prismatic segments, as shown in the figure. The end segments have a cross-sectional area $A_1 = 840$ mm^2 and length $L_1 = 200$ mm. The middle segment has a cross-sectional area $A_2 = 1260$ mm^2 and length $L_2 = 250$ mm. Loads P_B and P_C are equal to 25.5 kN and 17.0 kN, respectively.

(a) Determine the reactions R_A and R_D at the fixed supports.

(b) Determine the compressive axial force F_{BC} in the middle segment of the bar.

PROBLEM 8.4-12

8.4-13 A tube structure is acted on by loads at B and D, as shown in the figure. The tubes are joined using two flange plates at C that are bolted together using six 0.5-in. diameter bolts.

(a) Derive formulas for the reactions R_A and R_E at the ends of the bar.

(b) Determine the axial displacements δ_B, δ_C, and δ_D at points B, C, and D, respectively.

(c) Draw an axial-displacement diagram (ADD) in which the abscissa is the distance x from support A to any point on the bar and the ordinate is the horizontal displacement δ at that point.

(d) Find the maximum value of the load variable P if allowable normal stress in the bolts is 14 ksi.

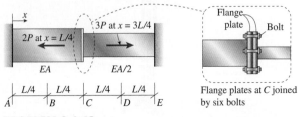

PROBLEM 8.4-13

8.4-14 A hollow circular pipe (see figure) supports a load P that is uniformly distributed around a cap plate at the top of the lower pipe. The inner and outer diameters of the upper and lower parts of the

pipe are $d_1 = 50$ mm, $d_2 = 60$ mm, $d_3 = 57$ mm, and $d_4 = 64$ mm, respectively. Pipe lengths are $L_1 = 2$ m and $L_2 = 3$ m. Neglect the self-weight of the pipes. Assume that cap plate thickness is small compared to L_1 and L_2. Let $E = 110$ MPa.

(a) If the tensile stress in the upper part is $\sigma_1 = 10.5$ MPa, what is load P? Also, what are reactions R_1 at the upper support and R_2 at the lower support? What is the stress σ_2 (MPa) in the lower part?

(b) Find displacement δ (mm) at the cap plate. Plot the axial-force diagram (AFD) $[N(x)]$ and axial-displacement diagram (ADD) $[\delta(x)]$.

(c) Add the uniformly distributed load q along the centroidal axis of pipe segment 2. Find q (kN/m) so that $R_2 = 0$. Assume that load P from part (a) is also applied.

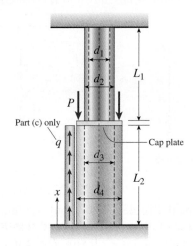

PROBLEM 8.4-14

8.4-15 The aluminum and steel pipes shown in the figure are fastened to rigid supports at ends A and B and to a rigid plate C at their junction. The aluminum pipe is twice as long as the steel pipe. Two equal and symmetrically placed loads P act on the plate at C.

(a) Obtain formulas for the axial stresses σ_a and σ_s in the aluminum and steel pipes, respectively.

(b) Calculate the stresses for the following data:
$P = 12$ kips
$A_a = 8.92$ in^2, cross-sectional area of aluminum pipe
$A_s = 1.03$ in^2, cross-sectional area of steel pipe
$E_a = 10 \times 10^6$ psi, modulus of elasticity of aluminum
$E_s = 29 \times 10^6$ psi, modulus of elasticity of steel

PROBLEM 8.4-15

8.4-16 A rigid bar of weight $W = 800$ N hangs from three equally spaced vertical wires (length $L = 150$ mm, spacing $a = 50$ mm): two of steel and one of aluminum. The wires also support a load P acting on the bar. The diameter of the steel wires is $d_s = 2$ mm, and the diameter of the aluminum wire is $d_a = 4$ mm. Assume $E_s = 210$ GPa and $E_a = 70$ GPa.

(a) What load P_{allow} can be supported *at the midpoint of the bar* $(x = a)$ if the allowable stress in

the steel wires is 220 MPa and in the aluminum wire is 80 MPa? (See figure part a.)

(b) What is P_{allow} if the load is positioned at $x = a/2$? (See figure part a.)

(c) Repeat part (b) if the second and third wires are *switched* as shown in the figure part b.

8.4-17 A *bimetallic* bar (or composite bar) of square cross section with dimensions $2b \times 2b$ is constructed of two different metals having moduli of elasticity E_1 and E_2 (see figure). The two parts of the bar have the same cross-sectional dimensions. The bar is compressed by forces P acting through rigid end plates. The line of action of the loads has an eccentricity e of such magnitude that each part of the bar is stressed uniformly in compression.

(a) Determine the axial forces P_1 and P_2 in the two parts of the bar.

(b) Determine the eccentricity e of the loads.

(c) Determine the ratio σ_1/σ_2 of the stresses in the two parts of the bar.

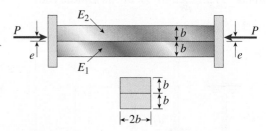

PROBLEM 8.4-17

8.4-18 Three-bar truss ABC (see figure) is constructed of steel pipes having a cross-sectional area $A = 3500$ mm^2 and a modulus of elasticity $E = 210$ GPa. Member BC is of length $L = 2.5$ m, and the angle between members AC and AB is known to be 60°. Member AC length is $b = 0.71$ L. Loads

(a)

(b)

PROBLEM 8.4-16

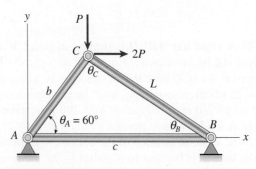

PROBLEM 8.4-18

621

$P = 185$ kN and $2P = 370$ kN act vertically and horizontally at joint C, as shown. Joints A and B are pinned supports. (Use the law of sines and law of cosines to find missing dimensions and angles in the figure.)

(a) Find the support reactions at joints A and B. Use horizontal reaction B_x as the redundant.

(b) What is the maximum permissible value of load variable P if the allowable normal stress in each truss member is 150 MPa?

8.4-19 A horizontal rigid bar of weight $W = 7200$ lb is supported by three slender circular rods that are equally spaced (see figure). The two outer rods are made of aluminum ($E_1 = 10 \times 10^6$ psi) with diameter $d_1 = 0.4$ in. and length $L_1 = 40$ in. The inner rod is magnesium ($E_2 = 6.5 \times 10^6$ psi) with diameter d_2 and length L_2. The allowable stresses in the aluminum and magnesium are 24,000 psi and 13,000 psi, respectively.

If it is desired to have all three rods loaded to their maximum allowable values, what should be the diameter d_2 and length L_2 of the middle rod?

PROBLEM 8.4-20

8.4-21 A rigid bar AB of a length $L = 66$ in. is hinged to a support at A and supported by two vertical wires attached at points C and D (see figure). Both wires have the same cross-sectional area ($A = 0.0272$ in.^{2}) and are made of the same material (modulus $E = 30 \times 10^6$ psi). The wire at C has a length $h = 18$ in. and the wire at D has a length twice that amount. The horizontal distances are $c = 20$ in. and $d = 50$ in.

(a) Determine the tensile stresses σ_C and σ_D in the wires due to the load $P = 340$ lb acting at end B of the bar.

(b) Find the downward displacement δ_B at end B of the bar.

W = weight of rigid bar

PROBLEM 8.4-19

PROBLEM 8.4-21

8.4-20 A rigid bar $ABCD$ is pinned at point B and supported by springs at A and D (see figure). The springs at A and D have stiffnesses $k_1 = 10$ kN/m and $k_2 = 25$ kN/m, respectively, and the dimensions a, b, and c are 250 mm, 500 mm, and 200 mm, respectively. A load P acts at point C.

If the angle of rotation of the bar due to the action of the load P is limited to 3°, what is the maximum permissible load P_{max}?

8.4-22 Find expressions for all support reaction forces in the plane frame with load $2P$ applied at C, as shown in the figure. Joint A is a sliding support, joint D is pinned, and joint F is a roller support. Assume that member AC is a flat prismatic bar of length L, width b, and thickness t. Beam ABC is pinned to column DBF at mid-height (point B). Column DBF has constant thickness t and width b for DB but width $2b$ for BF. Consider load $2P$ at C only; neglect the weights of all members. The modulus of elasticity E is the same for both members. Select reaction R_F as the *redundant*.

A
2L/3
EA

EA
B L/3 C

EA

2P

2EA L/2

EA L/2

D

F

PROBLEM 8.4-22

8.4-23 A trimetallic bar is uniformly compressed by an axial force $P = 9$ kips applied through a rigid end plate (see figure). The bar consists of a circular steel core surrounded by brass and copper tubes. The steel core has a diameter of 1.25 in., the brass tube has an outer diameter of 1.75 in., and the copper tube has an outer diameter of 2.25 in. The corresponding moduli of elasticity are $E_s = 30,000$ ksi, $E_b = 16,000$ ksi, and $E_c = 18,000$ ksi.

Calculate the compressive stresses σ_s, σ_b, and σ_c in the steel, brass, and copper, respectively, due to the force P.

PROBLEM 8.4-23

8.4-24 Find expressions for all support reaction forces in the plane frame with load $3P$ applied at C as shown in the figure. Joints A and D are pin supported, and there is a roller support at joint F. The lengths and the properties of the members are shown in the figure. Neglect the weights of all members. Select R_F as the redundant.

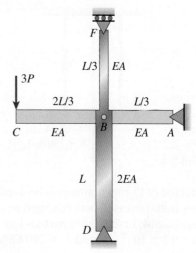

PROBLEM 8.4-24

8.5 Thermal Effects, Misfits, and Prestrains

Thermal Effects

Introductory Problems

8.5-1 The rails of a railroad track are welded together at their ends (to form continuous rails and thus eliminate the clacking sound of the wheels) when the temperature is 60°F.

What compressive stress σ is produced in the rails when they are heated by the sun to 120°F if the coefficient of thermal expansion $\alpha = 6.5 \times 10^{-6}$/°F and the modulus of elasticity $E = 30 \times 10^6$ psi?

8.5-2 An aluminum pipe has a length of 60 m at a temperature of 10°C. An adjacent steel pipe at the same temperature is 5 mm longer than the aluminum pipe.

At what temperature (degrees Celsius) will the aluminum pipe be 15 mm longer than the steel pipe? (Assume that the coefficients of thermal expansion of aluminum and steel are $\alpha_a = 23 \times 10^{-6}$/°C and as $\alpha_S = 12 \times 10^{-6}$/°C, respectively.)

8.5-3 A rigid bar of weight $W = 750$ lb hangs from three equally spaced wires: two of steel and one of aluminum (see figure). The diameter of the wires is 1/8 in. Before they were loaded, all three wires had the same length.

What temperature increase ΔT in all three wires will result in the entire load being carried by the steel wires? (Assume $E_S = 30 \times 10^6$ psi, $\alpha_S = 6.5 \times 10^{-6} / °F$, and $\alpha_a = 12 \times 10^{-6}/°F$.)

$W = 750$ lb

PROBLEM 8.5-3

8.5-4 A steel rod of 15-mm diameter is held snugly (but without any initial stresses) between rigid walls by the arrangement shown in the figure part a. (For the steel rod, use $\alpha = 12 \times 10^{-6}/°C$ and $E = 200$ GPa.)

(a) Calculate the temperature drop ΔT (degrees Celsius) at which the average shear stress in the 18-mm diameter bolt becomes 45 MPa. Also, what is the normal stress in the rod?

(b) What are the average bearing stresses in the bolt and clevis at A and between the washer ($d_w = 20$ mm) and wall ($t = 18$ mm) at B?

12-mm diameter bolt

ΔT

Washer, $d_w = 20$ mm

B

A

Clevis, $t = 10$ mm

15 mm

18 mm

(a)

12-mm diameter bolt

ΔT

Bolt and washer (d_b, d_w)

A

Clevis, $t = 10$ mm

15 mm

Mounting plate (t)

(b)

PROBLEM 8.5-4

624

(c) If the connection to the wall at B is changed to an end plate with two bolts (see figure part b), what is the required diameter d_b of each bolt if the temperature drop is $\Delta T = 38°C$ and the allowable bolt stress is 90 MPa?

Representative Problems

8.5-5 A bar AB of length L is held between rigid supports and heated nonuniformly in such a manner that the temperature increase ΔT at a distance x from end A is given by the expression $\Delta T = \Delta T_B x^3/L^3$, where ΔT_B is the increase in temperature at end B of the bar (see figure part a).

(a) Derive a formula for the compressive stress σ_c in the bar. (Assume that the material has modulus of elasticity E and coefficient of thermal expansion α).

(b) Now modify the formula in part (a) if the rigid support at A is replaced by an elastic support at A having a spring constant k (see figure part b). Assume that only bar AB is subject to the temperature increase.

(a)

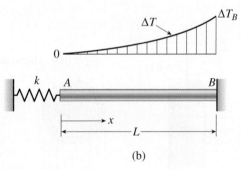

(b)

PROBLEM 8.5-5

8.5-6 A beam is constructed using two angle sections (L102 × 76 × 6.4) arranged back to back, as shown in the figure. The beam is fixed at joint A and attached to an elastic support having a spring constant $k = 1750$ kN/m at joint B. Assume only the beam is subjected to temperature increase $\Delta T = 45°C$. Calculate the thermal stress developed in the beam and the displacement at point B. Assume that $\alpha = 12 \times 10^{-6}/°C$. Let $E = 205$ GPa.

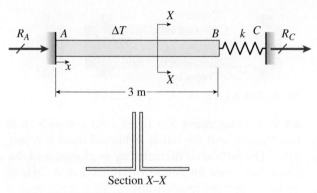

Section X–X

PROBLEM 8.5-6

Also, the modulus of elasticity E is 6.0 GPa, and the coefficient of thermal expansion α is $100 \times 10^{-6}/°C$. The bar is subjected to a uniform temperature increase of 30°C.

(a) Calculate the following quantities: (1) the compressive force N in the bar; (2) the maximum compressive stress σ_c; and (3) the displacement δ_C of point C.

(b) Repeat part (a) if the rigid support at A is replaced by an elastic support having spring constant $k = 50$ MN/m (see figure part b; assume that only the bar ACB is subject to the temperature increase).

8.5-7 A W 8×28 beam of a length 10 ft is held between immoveable supports. The beam has a modulus of elasticity $E = 29{,}000$ ksi and coefficient of thermal expansion $\alpha = 6.5 \times 10^{-6}/°F$. If the temperature of the beam is raised uniformly by an amount $\Delta T = 20°F$, calculate the thermal stress σ_T in the beam.

8.5-9 A flat aluminum alloy bar is fixed at both ends. Segment AB has a slight taper. If the temperature of the bar is raised uniformly by an amount $\Delta T = 20°F$, find reactions at A and C. What is the displacement at B? Assume that $L = 3$ ft, $t = 1/4$ in., $b_1 = 2$ in., $b_2 = 2.5$ in., $E = 10{,}400$ ksi, and the coefficient of thermal expansion $\alpha = 13 \times 10^{-6}/°F$.

PROBLEM 8.5-7

8.5-8 A plastic bar ACB having two different solid circular cross sections is held between rigid supports, as shown in the figure. The diameters in the left- and right-hand parts are 50 mm and 75 mm, respectively. The corresponding lengths are 225 mm and 300 mm.

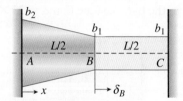

PROBLEM 8.5-9

8.5-10 Repeat Problem 8.5-9 for the flat bar shown in the figure but assume that $\Delta T = 30°C$, and that $\alpha = 19 \times 10^{-6}/°C$, $L = 2$ m, $t = 20$ mm, $b_1 = 100$ mm, $b_2 = 115$ mm, and $E = 96$ GPa.

(a)

(b)

PROBLEM 8.5-8

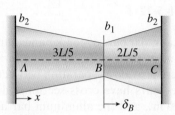

PROBLEM 8.5-10

625

8.5-11 A circular steel rod AB (diameter $d_1 = 1.0$ in., length $L_1 = 3.0$ ft) has a bronze sleeve (outer diameter $d_2 = 1.25$ in., length $L_2 = 1.0$ ft) shrunk onto it so that the two parts are securely bonded (see figure).

Calculate the total elongation δ of the steel bar due to a temperature rise $\Delta T = 500°F$. (Material properties are as follows: for steel, $E_S = 30 \times 10^6$ psi and $\alpha_S = 6.5 \times 10^{-6}/°F$; for bronze, $E_b = 15 \times 10^6$ psi and $\alpha_b = 11 \times 10^{-6}/°F$.)

PROBLEM 8.5-11

8.5-12 A circular, aluminum alloy bar of a length $L = 1.8$ m has a slot in the middle half of its length (see figure). The bar has a radius $r = 36$ mm and modulus of elasticity $E = 72$ GPa. The slot has a height $2a = r/4$. If the temperature of the beam is raised uniformly by an amount $\Delta T = 15°C$, calculate the thermal stress σ_T developed in the bar. Assume that $\alpha = 23 \times 10^{-6}/°C$.

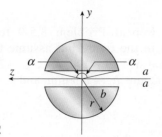

PROBLEM 8.5-12

8.5-13 Rectangular bars of copper and aluminum are held by pins at their ends, as shown in the figure. Thin spacers provide a separation between the bars. The copper bars have cross-sectional dimensions 0.5 in. $\times 2.0$ in., and the aluminum bar has dimensions 1.0 in. $\times 2.0$ in.

Determine the shear stress in the 7/16-in. diameter pins if the temperature is raised by 100°F. (For copper, $E_c = 18,000$ ksi and $\alpha_c = 9.5 \times 10^{-6}/°F$; for aluminum, $E_a = 10,000$ ksi and $\alpha_a = 13 \times 10^{-6}/°F$.) *Suggestion:* Use the results of Example 8-10.

PROBLEM 8.5-13

8.5-14 A brass sleeve S is fitted over a steel bolt B (see figure), and the nut is tightened until it is just snug. The bolt has a diameter $d_B = 25$ mm, and the sleeve has inside and outside diameters $d_1 = 26$ mm and $d_2 = 36$ mm, respectively.

Calculate the temperature rise ΔT that is required to produce a compressive stress of 25 MPa in the sleeve. (Use material properties as follows: for the sleeve, $\alpha_S = 21 \times 10^{-6}/°C$ and $E_S = 100$ GPa; for the bolt, $\alpha_B = 10 \times 10^{-6}/°C$ and $E_B = 200$ GPa.)

PROBLEM 8.5-14

8.5-15 A rigid triangular frame is pivoted at C and held by two identical horizontal wires at points A and B (see figure). Each wire has an axial rigidity $EA = 120$ kips and coefficient of thermal expansion $\alpha = 12.5 \times 10^{-6}/°F$.

(a) If a vertical load $P = 500$ lb acts at point D, what are the tensile forces T_A and T_B in the wires at A and B, respectively?

(b) If both wires have their temperatures raised by 180°F while the load P is acting, what are the forces T_A and T_B?

(c) What further increase in temperature will cause the wire at B to become slack?

PROBLEM 8.5-15

8.5-16 A rigid bar $ABCD$ is pinned at end A and supported by two cables at points B and C (see figure). The cable at B has a nominal diameter $d_B = 12$ mm and the cable at C has a nominal diameter $d_C = 20$ mm. A load P acts at end D of the bar.

What is the allowable load P if the temperature rises by 60°C and each cable is required to have a factor of safety of at least 5 against its ultimate load?

Note: The cables have an effective modulus of elasticity $E = 140$ GPa and coefficient of thermal expansion $\alpha = 12 \times 10^{-6}/°C$. Other properties of the cables can be found in Table 8-1, Section 8.2.

PROBLEM 8.5-16

Misfits and Prestrains

Introductory Problems

8.5-17 A copper bar AB with a length 25 in. and diameter 2 in. is placed in position at room temperature with a gap of 0.008 in. between end A and a rigid restraint (see figure). The bar is supported at end B by an elastic spring with a spring constant $k = 1.2 \times 10^6$ lb/in.

(a) Calculate the axial compressive stress σ_c in the bar if the temperature *of the bar only* rises 50°F. (For copper, use $\alpha = 9.6 \times 10^{-6}/°F$ and $E = 16 \times 10^6$ psi.)

(b) What is the force in the spring? (Neglect gravity effects.)

(c) Repeat part (a) if $k \to \infty$.

PROBLEM 8.5-17

8.5-18 A steel wire AB is stretched between rigid supports (see figure). The initial prestress in the wire is 42 MPa when the temperature is 20°C.

(a) What is the stress σ in the wire when the temperature drops to 0°C?

(b) At what temperature T will the stress in the wire become zero? (Assume $\alpha = 14 \times 10^{-6}/°C$ and $E = 200$ GPa.)

Steel wire

PROBLEM 8.5-18

8.5-19 The mechanical assembly shown in the figure consists of an aluminum tube, a rigid end plate, and two steel cables. The slack is removed from the cables by rotating the turnbuckles until the assembly is snug but with no initial stresses. Afterward, the turnbuckles are tightened by 1.5 turns. Calculate the forces in the tube and the cables and determine the shortening of the tube. $A_s = 0.85$ in.^{2} for each cable, $A_A = 4.5$ in.^{2}, $L = 20$ in., $E_s = 29,000$ ksi, $E_A = 10,600$ ksi, and $p = 1/16$ in.

Aluminum tube   Steel cable   Turnbuckle   Rigid plate

PROBLEM 8.5-19

8.5-20 A bar AB having a length L and axial rigidity EA is fixed at end A (see figure). At the other end, a small gap of dimension s exists between the end of the bar and a rigid surface. A load P acts on the bar at point C, which is two-thirds of the length from the fixed end.

If the support reactions produced by load P are to be equal in magnitude, what should be the size s of the gap?

PROBLEM 8.5-20

8.5-21 Pipe 2 has been inserted snugly into Pipe 1, but the holes for a connecting pin do not line up; there is a gap s. The user decides to apply *either* force P_1 to Pipe 1 *or* force P_2 to Pipe 2, whichever is smaller. Determine the following using the numerical properties in the box.

(a) If only P_1 is applied, find P_1 (kips) required to close gap s; if a pin is then inserted and P_1 removed, what are reaction forces R_A and R_B for this load case?

(b) If only P_2 is applied, find P_2 (kips) required to close gap s; if a pin is inserted and P_2 removed, what are reaction forces R_A and R_B for this load case?

(c) What is the maximum *shear* stress in the pipes, for the loads in parts (a) and (b)?

(d) If a temperature increase ΔT is to be applied to the entire structure to close gap s (*instead of applying forces P_1 and P_2*), find the ΔT required to close the gap. If a pin is inserted after the gap has closed, what are reaction forces R_A and R_B for this case?

(e) Finally, if the structure (with pin inserted) then cools to the *original* ambient temperature, what are reaction forces R_A and R_B?

Numerical properties:
$E_1 = 30{,}000$ ksi, $E_2 = 14{,}000$ ksi
$\alpha_1 = 6.5 \times 10^{-6}/°F$, $\alpha_2 = 11 \times 10^{-6}/°F$
Gap $s = 0.05$ in.
$L_1 = 56$ in., $d_1 = 6$ in., $t_1 = 0.5$ in., $A_1 = 8.64$ in^2
$L_2 = 36$ in., $d_2 = 5$ in., $t_2 = 0.25$ in., $A_2 = 3.73$ in^2

PROBLEM 8.5-21

8.5-22 A nonprismatic bar ABC made up of segments AB (length L_1, cross-sectional area A_1) and BC (length L_2, cross-sectional area A_2) is fixed at end A and free at end C (see figure). The modulus of elasticity of the bar is E. A small gap of dimension s exists between the end of the bar and an elastic spring of length L_3 and spring constant k_3. If bar ABC only (*not the spring*) is subjected to temperature increase ΔT, determine the following.

(a) Write an expression for reaction forces R_A and R_D if the elongation of ABC exceeds gap length s.

(b) Find expressions for the displacements of points B and C if the elongation of ABC exceeds gap length s.

PROBLEM 8.5-22

8.5-23 Wires B and C are attached to a support at the left-hand end and to a pin-supported rigid bar at the right-hand end (see figure). Each wire has cross-sectional area $A = 0.03$ in^2 and modulus of elasticity $E = 30 \times 10^6$ psi. When the bar is in a vertical position, the length of each wire is $L = 80$ in. However, before being attached to the bar, the length of wire B was 79.98 in. and wire C was 79.95 in.

Find the tensile forces T_B and T_C in the wires under the action of a force $P = 700$ lb acting at the upper end of the bar.

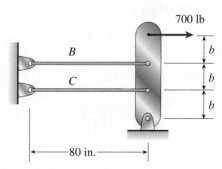

PROBLEM 8.5-23

8.5-24 A rigid steel plate is supported by three posts of high-strength concrete each having an effective cross-sectional area $A = 40,000 \text{ mm}^2$ and length $L = 2 \text{ m}$ (see figure). Before the load P is applied, the middle post is shorter than the others by an amount $s = 1.0 \text{ mm}$.

Determine the maximum allowable load P_{allow} if the allowable compressive stress in the concrete is $\sigma_{\text{allow}} = 20 \text{ MPa}$. (Use $E = 30 \text{ GPa}$ for concrete.)

PROBLEM 8.5-24

Representative Problems

8.5-25 A capped cast-iron pipe is compressed by a brass rod, as shown. The nut is turned until it is just snug, then an additional quarter turn is added to pre-compress the cast-iron pipe. The pitch of the threads of the bolt is $p = 52$ mils (a mil is one-thousandth of an inch). Use the numerical properties provided.

(a) What stresses σ_p and σ_r will be produced in the cast-iron pipe and brass rod, respectively, by the additional quarter turn of the nut?

(b) Find the bearing stress σ_b beneath the washer and the shear stress τ_c in the steel cap.

8.5-26 A plastic cylinder is held snugly between a rigid plate and a foundation by two steel bolts (see figure).

Determine the compressive stress σ_p in the plastic when the nuts on the steel bolts are tightened by one complete turn.

Data for the assembly are as follows: length $L = 200 \text{ mm}$, pitch of the bolt threads $p = 1.0 \text{ mm}$, modulus of elasticity for steel $E_s = 200 \text{ GPa}$, modulus of elasticity for the plastic $E_P = 7.5 \text{ GPa}$, cross-sectional area of one bolt $A_s = 36.0 \text{ mm}^2$, and cross-sectional area of the plastic cylinder $A_P = 960 \text{ mm}^2$.

Nut and washer
$\left(d_w = \dfrac{3}{4} \text{ in.}\right)$

Steel cap
$(t_c = 1 \text{ in.})$

Cast iron pipe
$(d_o = 6 \text{ in.},$
$d_i = 5.625 \text{ in.})$

$L_{ci} = 4 \text{ ft}$

Brass rod
$\left(d_r = \dfrac{1}{2} \text{ in.}\right)$

Modulus of elasticity, E:
Steel (30,000 ksi)
Brass (14,000 ksi)
Cast iron (12,000 ksi)

PROBLEM 8.5-25

Steel bolt

L

PROBLEMS 8.5-26 and 8.5-27

8.5-27 Solve the preceding problem if the data for the assembly are as follows: length $L = 10 \text{ in.}$, pitch of the bolt threads $p = 0.058 \text{ in.}$, modulus of elasticity for steel $E_s = 30 \times 10^6 \text{ psi}$, modulus of elasticity for the plastic $E_P = 500 \text{ ksi}$, cross-sectional area of one bolt $A_s = 0.06 \text{ in}^2$, and cross-sectional area of the plastic cylinder $A_P = 1.5 \text{ in}^2$.

8.5-28 Consider the sleeve made from two copper tubes joined by tin-lead solder over distance s. The sleeve has brass caps at both ends that are held in place by a steel bolt and washer with the nut turned just snug at the outset. Then, two "loadings" are applied: $n = 1/2$ turn applied to the nut; at the same time, the internal temperature is raised by $\Delta T = 30°C$.

(a) Find the forces in the sleeve and bolt, P_S and P_B, due to both the prestress in the bolt and the temperature increase. For copper, use $E_c = 120$ GPa and $\alpha_c = 17 \times 10^{-6}/°C$; for steel, use $E_s = 200$ GPa and $\alpha_s = 12 \times 10^{-6}/°C$. The pitch of the bolt threads is $p = 1.0$ mm. Assume $s = 26$ mm and bolt diameter $d_B = 5$ mm.

(b) Find the required length of the solder joint, s, if shear stress in the sweated joint cannot exceed the allowable shear stress $\tau_{aj} = 18.5$ MPa.

(c) What is the final elongation of the entire assemblage due to both temperature change ΔT and the initial prestress in the bolt?

8.5-29 A polyethylene tube (length L) has a cap that when installed compresses a spring (with undeformed length $L_1 > L$) by an amount $\delta = (L_1 - L)$. Ignore deformations of the cap and base. Use the force at the base of the spring as the redundant. Use numerical properties given in the boxes.

(a) What is the resulting force in the spring, F_k?
(b) What is the resulting force in the tube, F_t?
(c) What is the final length of the tube, L_f?
(d) What temperature change ΔT inside the tube will result in zero force in the spring?

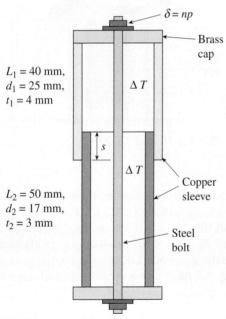

$L_1 = 40$ mm,
$d_1 = 25$ mm,
$t_1 = 4$ mm

ΔT

$\delta = np$

Brass cap

ΔT

Copper sleeve

$L_2 = 50$ mm,
$d_2 = 17$ mm,
$t_2 = 3$ mm

Steel bolt

PROBLEM 8.5-28

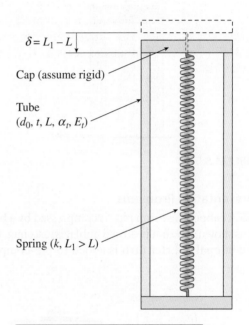

$\delta = L_1 - L$

Cap (assume rigid)

Tube
$(d_0, t, L, \alpha_t, E_t)$

Spring $(k, L_1 > L)$

| Modulus of elasticity: |
| Polyethylene tube ($E_t = 100$ ksi) |

| Coefficients of thermal expansion: |
| $\alpha_t = 80 \times 10^{-6}/°F$, $\alpha_k = 6.5 \times 10^{-6}/°F$ |

| Properties and dimensions: |
| $d_0 = 6$ in. $t = \dfrac{1}{8}$ in. |
| $L_1 = 12.125$ in. $> L = 12$ in. $k = 1.5 \dfrac{\text{kips}}{\text{in.}}$ |

PROBLEM 8.5-29

8.5-30 Prestressed concrete beams are sometimes manufactured in the following manner. High-strength steel wires are stretched by a jacking mechanism that applies a force Q, as represented schematically in part a of the figure. Concrete is then poured around the wires to form a beam, as shown in the figure part b.

After the concrete sets properly, the jacks are released, and the force Q is removed (see part c of the figure). Thus, the beam is left in a prestressed condition with the wires in tension and the concrete in compression.

Assume that the prestressing force Q produces in the steel wires an initial stress $\sigma_0 = 620$ MPa. If the moduli of elasticity of the steel and concrete are in the ratio 12:1 and the cross-sectional areas are in the ratio 1:50, what are the final stresses σ_s and σ_c in the two materials?

(a)

(b)

(c)

PROBLEM 8.5-30

8.5-31 A polyethylene tube (length L) has a cap that is held in place by a spring (with undeformed length $L_1 < L$). After installing the cap, the spring is post-tensioned by turning an adjustment screw by an amount δ. Ignore deformations of the cap and base. Use the force at the base of the spring as the redundant. Use numerical properties in the boxes below the figure.

(a) What is the resulting force in the spring, F_k?
(b) What is the resulting force in the tube, F_t?
(c) What is the final length of the tube, L_f?
(d) What temperature change ΔT inside the tube will result in zero force in the spring?

Cap (assume rigid)

Tube $(d_0, t, L, \alpha_t, E_t)$

Spring $(k, L_1 < L)$

$\delta = L - L_1$

Adjustment screw

Modulus of elasticity:
Polyethylene tube ($E_t = 100$ ksi)

Coefficients of thermal expansion:
$\alpha_t = 80 \times 10^{-6}/°F$, $\alpha_k = 6.5 \times 10^{-6}/°F$

Properties and dimensions:
$d_0 = 6$ in. $t = \dfrac{1}{8}$ in.
$L = 12$ in. $L_1 = 11.875$ in. $k = 1.5\dfrac{\text{kips}}{\text{in.}}$

PROBLEM 8.5-31

8.6 Stresses on Inclined Sections

Introductory Problems

8.6-1 A steel bar of rectangular cross section (1.5 in. × 2.0 in.) carries a tensile load P (see figure). The allowable stresses in tension and shear are 14,500 psi and 7,100 psi, respectively. Determine the maximum permissible load P_{max}.

2.0 in.

P P

1.5 in.

PROBLEM 8.6-1

8.6-2 A circular steel rod of diameter d is subjected to a tensile force $P = 3.5$ kN (see figure). The allowable stresses in tension and shear are 118 MPa and 48 MPa, respectively. What is the minimum permissible diameter d_{min} of the rod?

PROBLEM 8.6-2

8.6-3 A standard brick (dimensions 8 in. × 4 in. × 2.5 in.) is compressed lengthwise by a force P, as shown in the figure. If the ultimate shear stress for brick is 1200 psi and the ultimate compressive stress is 3600 psi, what force P_{max} is required to break the brick?

PROBLEM 8.6-3

8.6-4 A brass wire of diameter $d = 2.42$ mm is stretched tightly between rigid supports so that the tensile force is $T = 98$ N (see figure). The coefficient of thermal expansion for the wire is $19.5 \times 10^{-6}/°C$, and the modulus of elasticity is $E = 110$ GPa.

(a) What is the maximum permissible temperature drop ΔT if the allowable shear stress in the wire is 60 MPa?

(b) At what temperature change does the wire go slack?

PROBLEMS 8.6-4 and 8.6-5

8.6-5 A brass wire of diameter $d = 1/16$ in. is stretched between rigid supports with an initial tension T of 37 lb (see figure). Assume that the coefficient of thermal expansion is $10.6 \times 10^{-6}/°F$ and the modulus of elasticity is 15×10^6 psi.

(a) If the temperature is lowered by 60°F, what is the maximum shear stress τ_{max} in the wire?

(b) If the allowable shear stress is 10,000 psi, what is the maximum permissible temperature drop?

(c) At what temperature change ΔT does the wire go slack?

8.6-6 A steel bar with a diameter $d = 12$ mm is subjected to a tensile load $P = 9.5$ kN (see figure).

(a) What is the maximum normal stress σ_{max} in the bar?

(b) What is the maximum shear stress τ_{max}?

(c) Draw a stress element oriented at 45° to the axis of the bar and show all stresses acting on the faces of this element.

(d) Repeat part (c) for a stress element oriented at 22.5° to the axis of the bar.

PROBLEM 8.6-6

8.6-7 During a tension test of a mild-steel specimen (see figure), the extensometer shows an elongation of 0.00120 in. with a gage length of 2 in. Assume that the steel is stressed below the proportional limit and that the modulus of elasticity $E = 30 \times 10^6$ psi.

(a) What is the maximum normal stress σ_{max} in the specimen?

(b) What is the maximum shear stress τ_{max}?

(c) Draw a stress element oriented at an angle of 45° to the axis of the bar, and show all stresses acting on the faces of this element.

PROBLEM 8.6-7

Representative Problems

8.6-8 A copper bar with a rectangular cross section is held without stress between rigid supports (see figure). Subsequently, the temperature of the bar is raised 50°C.

(a) Determine the stresses on all faces of the elements A and B, and show these stresses on sketches of the elements. (Assume $\alpha = 17.5 \times 10^{-6}/°C$ and $E = 120$ GPa.)

(b) If the shear stress at B is known to be 48 MPa at some inclination θ, find angle θ and show the stresses on a sketch of a properly oriented element.

PROBLEM 8.6-8

8.6-9 A prismatic bar with a length $L = 3$ ft and cross-sectional area $A = 8\,\text{in}^2$ is compressed by an axial centroidal load $P = 10$ kips. Determine the complete state of stress acting on an inclined section pq that is cut through the bar at an angle $\theta = 35°$, and show the stresses on a properly oriented stress element.

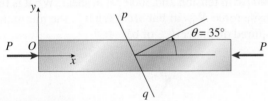

PROBLEM 8.6-9

8.6-10 A prismatic bar with a length $L = 1$ m and cross-sectional area $A = 1200\,\text{mm}^2$ is supported at the ends. The bar is then subjected to a temperature increase of $\Delta T = 25°C$. Calculate the complete state of stress acting on an inclined section rs that is cut through the bar at an angle $\theta = 45°$. Use $E = 200$ GPa and the coefficient of thermal expansion $\alpha = 12 \times 10^{-6}/°C$.

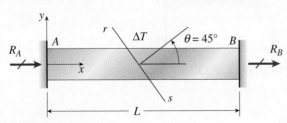

PROBLEM 8.6-10

8.6-11 The plane truss in the figure is assembled from steel C 10 × 20 shapes [see Table C-3(a) in Appendix C]. Assume that $L = 10$ ft and $b = 0.71L$.

(a) If load variable $P = 49$ kips, what is the maximum shear stress τ_{max} in each truss member?

(b) What is the maximum permissible value of load variable P if the allowable normal stress is 14 ksi and the allowable shear stress is 7.5 ksi?

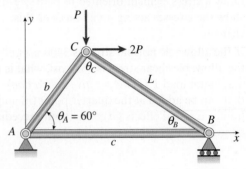

PROBLEM 8.6-11

8.6-12 A plastic bar of diameter $d = 32$ mm is compressed in a testing device by a force $P = 190$ N that is applied as shown in the figure.

(a) Determine the normal and shear stresses acting on all faces of stress elements oriented at (1) an angle $\theta = 0°$, (2) an angle $\theta = 22.5°$, and (3) an angle $\theta = 45°$. In each case, show the stresses on a sketch of a properly oriented element. What are σ_{max} and τ_{max}?

(b) Find σ_{max} and τ_{max} in the plastic bar if a re-centering spring of stiffness k is inserted into the testing device, as shown in the figure. The spring stiffness is 1/6 of the axial stiffness of the plastic bar.

PROBLEM 8.6-12

633

8.6-13 A plastic bar of rectangular cross section ($b = 1.5$ in. and $h = 3$ in.) fits snugly between rigid supports at room temperature (68°F) but with no initial stress (see figure). When the temperature of the bar is raised to 160°F, the compressive stress on an inclined plane pq at mid-span becomes 1700 psi.

(a) What is the shear stress on plane pq? (Assume $\alpha = 60 \times 10^{-6}$/°F and $E = 450 \times 10^3$ psi.)

(b) Draw a stress element oriented to plane pq and show the stresses acting on all faces of this element.

(c) If the allowable normal stress is 3400 psi and the allowable shear stress is 1650 psi, what is the maximum load P (*in the positive x direction*), that can be added at the quarter point (in addition to thermal effects given) without exceeding allowable stress values in the bar?

Load P for part (c) only

PROBLEM 8.6-13

8.6-14 A copper bar of rectangular cross section ($b = 18$ mm and $h = 40$ mm) is held snugly (but without any initial stress) between rigid supports (see figure). The allowable stresses on the inclined plane pq at midspan, for which $\theta = 55°$, are specified as 60 MPa in compression and 30 MPa in shear.

(a) What is the maximum permissible temperature rise ΔT if the allowable stresses on plane pq are not to be exceeded? (Assume $\alpha = 17 \times 10^{-6}$/°C and $E = 120$ GPa.)

(b) If the temperature increases by the maximum permissible amount, what are the stresses on plane pq?

(c) If the temperature rises $\Delta T = 28°C$, how far to the right of end A (distance βL, which is expressed as a fraction of length L) can load $P = 15$ kN be applied without exceeding allowable stress values in the bar? Assume that $\sigma_a = 75$ MPa and $\tau_a = 35$ MPa.

Load for part (c) only

PROBLEM 8.6-14

8.6-15 A circular brass bar with a diameter d is member AC in truss ABC that has load $P = 5000$ lb applied at joint C. Bar AC is composed of two segments brazed together on a plane pq, making an angle $\alpha = 36°$ with the axis of the bar (see figure). The allowable stresses in the brass are 13,500 psi in tension and 6500 psi in shear. On the brazed joint, the allowable stresses are 6000 psi in tension and 3000 psi in shear. What is the tensile force N_{AC} in bar AC? What is the minimum required diameter d_{min} of bar AC?

PROBLEM 8.6-15

8.6-16 Two boards are joined by gluing along a scarf joint, as shown in the figure. For purposes of cutting and gluing, the angle α between the plane of the joint and the faces of the boards must be between $10°$ and $40°$. Under a tensile load P, the normal stress in the boards is 4.9 MPa.

(a) What are the normal and shear stresses acting on the glued joint if $\alpha = 20°$?

(b) If the allowable shear stress on the joint is 2.25 MPa, what is the largest permissible value of the angle α?

(c) For what angle α will the shear stress on the glued joint be numerically equal to twice the normal stress on the joint?

PROBLEM 8.6-16

8.6-17 Acting on the sides of a stress element cut from a bar in uniaxial stress are tensile stresses of 10,000 psi and 5000 psi, as shown in the figure.

(a) Determine the angle θ and the shear stress τ_θ, and show all stresses on a sketch of the element.

(b) Determine the maximum normal stress σ_{max} and the maximum shear stress τ_{max} in the material.

PROBLEM 8.6-17

8.6-18 A prismatic bar is subjected to an axial force that produces a tensile stress $\sigma_\theta = 65$ MPa and a shear stress $\tau_\theta = 23$ MPa on a certain inclined plane (see figure). Determine the stresses acting on all faces of a stress element oriented at $\theta = 30°$, and show the stresses on a sketch of the element.

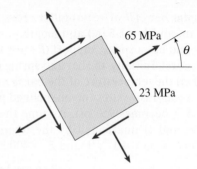

PROBLEM 8.6-18

8.6-19 The normal stress on plane pq of a prismatic bar in tension (see figure) is found to be 7500 psi. On plane rs, which makes an angle $\beta = 30°$ with plane pq, the stress is found to be 2500 psi.

Determine the maximum normal stress σ_{max} and maximum shear stress τ_{max} in the bar.

PROBLEM 8.6-19

8.6-20 A tension member is to be constructed of two pieces of plastic glued along plane pq (see figure). For purposes of cutting and gluing, the angle θ must be between $25°$ and $45°$. The allowable stresses on the glued joint in tension and shear are 5.0 MPa and 3.0 MPa, respectively

(a) Determine the angle θ so that the bar will carry the largest load P. (Assume that the strength of the glued joint controls the design.)

(b) Determine the maximum allowable load P_{max} if the cross-sectional area of the bar is 225 mm^2.

PROBLEM 8.6-20

8.6-21 Plastic bar AB of rectangular cross section ($b = 0.75$ in. and $h = 1.5$ in.) and length $L = 2$ ft is fixed at A and has a spring support ($k = 18$ kips/in.) at C (see figure). Initially, the bar and spring have no stress. When the temperature of the bar is *raised* by 100°F, the *compressive* stress on an inclined plane pq at $L_\theta = 1.5$ ft becomes 950 psi. Assume the spring is massless and is unaffected by the temperature change. Let $\alpha = 55 \times 10^{-6}/°F$ and $E = 400$ ksi.

(a) What is the shear stress τ_θ on plane pq? What is angle θ?

(b) Draw a stress element oriented to plane pq, and show the stresses acting on all faces of this element.

(c) If the allowable normal stress is ±1000 psi and the allowable shear stress is ±560 psi, what is the maximum permissible value of spring constant k if the allowable stress values in the bar are not to be exceeded?

(d) What is the maximum permissible length L of the bar if the allowable stress values in the bar are not be exceeded? (Assume $k = 18$ kips/in.)

(e) What is the maximum permissible temperature increase (ΔT) in the bar if the allowable stress values in the bar are not to be exceeded? (Assume $L = 2$ ft and $k = 18$ kips/in.)

8.6-22 A compression bar having a square cross section with sides $b = 50$ mm is subjected to load P. The bar is constructed from two pieces of wood that are connected by a glued joint along plane pq that is inclined at angle $\alpha = 35°$. The allowable stress in the wood in compression is 11.5 MPa and in shear is 4.5 MPa. Also, the allowable stress in the glued joint in compression is 3.5 MPa and in shear is 1.25 MPa. Determine the maximum load P that can be applied to the bar.

PROBLEM 8.6-21

PROBLEM 8.6-22

Torsion

Four sensors are screwed into the thigh bone (yellow) to measure stresses and test the torsional stability of an artificial hip. The prosthesis is mostly inside the thigh bone with the protruding ball part held by the clamp at the top of the test machine.

CHAPTER OBJECTIVES

- Study twisting of circular bars and hollow shafts acted on by torsional moments.

- Develop the *torsion formula*, which relates the variation of shear stress with increasing radial distance in the cross section.

- Define the *torque–displacement relation*, which shows that the angle of twist is proportional to the product of the internal torsional moment and the torsional flexibility of a circular bar.

- Find both normal and shear stresses at points of interest on inclined sections of circular shafts.

- Define the relationship between modulus of elasticity E and shearing modulus G.

- Study *power transmission* by circular shafts to find a relation among power P, torque T, and rotational speed n.

- Find support reaction moments in *statically indeterminate shafts* acted on by torsional moments.

Chapter Objectives

9.1 Introduction

Chapters 7 and 8 discussed the behavior of the simplest type of structural member—namely, a straight bar subjected to axial loads. Now a slightly more complex type of behavior known as **torsion** is considered. Torsion refers to the twisting of a straight bar when it is loaded by moments (or torques) that tend to produce rotation about the longitudinal axis of the bar. For instance, when you turn a screwdriver (Fig. 9-1a), your hand applies a torque T to the handle (Fig. 9-1b) and twists the shank of the screwdriver. Other examples of bars in torsion are drive shafts in automobiles, axles, propeller shafts, steering rods, and drill bits.

(a)

(b)

An idealized case of torsional loading is pictured in Fig. 9-2a, which shows a straight bar supported at one end and loaded by two pairs of equal and opposite forces. The first pair consists of the forces P_1 acting near the midpoint of the bar and the second pair consists of the forces P_2 acting at the end. Each pair of forces forms a **couple** that tends to twist the bar about its longitudinal axis. As you know from statics, the **moment of a couple** is equal to the product of one of the forces and the perpendicular distance between the lines of action of the forces; thus, the first couple has a moment $T_1 = P_1 d_1$ and the second has a moment $T_2 = P_2 d_2$.

Typical USCS **units** for moment are the pound-foot (lb-ft) and the pound-inch (lb-in.). The SI unit for moment is the newton meter (N·m).

The moment of a couple may be represented by a **vector** in the form of a double-headed arrow (Fig. 9-2b). The arrow is perpendicular to the plane containing the couple, and in this case, both arrows are parallel to the axis of the bar. The direction (or *sense*) of the moment is indicated by the *right-hand rule* for moment vectors—namely, using your right hand, let your fingers curl in the direction of the moment, and then your thumb will point in the direction of the vector.

An alternative representation of a moment is a curved arrow acting in the direction of rotation (Fig. 9-2c). Both the curved arrow and vector representations are in common use and are used in this book. The choice depends upon convenience and personal preference.

Moments that produce the twisting of a bar, such as the moments T_1 and T_2 in Fig. 9-2, are called **torques** or **twisting moments**. Cylindrical members that are subjected to torques and transmit power through rotation are called **shafts**; for instance, the drive shaft of an automobile or the propeller shaft of a ship. Most shafts have circular cross sections, either solid or tubular.

This chapter begins by developing formulas for the deformations and stresses in circular bars subjected to torsion. Then analysis of the state of stress known as *pure shear* and the relationship between the moduli of elasticity E and G in tension and shear, respectively, are presented. Next, rotating shafts are analyzed to determine the power they transmit. Finally, several additional topics related to torsion are covered, namely, statically indeterminate members, and noncircular prismatic shafts.

$$T_1 = P_1 d_1 \qquad T_2 = P_2 d_2$$
(a)

(b)

(c)

9.2 Torsional Deformations of a Circular Bar

First consider a prismatic bar of circular cross section twisted by torques T acting at the ends (Fig. 9-3a). Since every cross section of the bar is identical, and since every cross section is subjected to the same internal torque T, the bar is in **pure torsion**. From considerations of symmetry, it can be proved that cross sections of the bar do not change in shape as they rotate about the longitudinal axis. In other words, all cross sections remain plane and circular, and all radii remain straight. Furthermore, if the angle of rotation between one end of the bar and the other is small, neither the length of the bar nor its radius will change.

To aid in visualizing the deformation of the bar, imagine that the left-hand end of the bar (Fig. 9-3a) is fixed in position. Then, under the action of the torque T, the right-hand end will rotate (with respect to the left-hand end) through a small angle ϕ, known as the **angle of twist** (or *angle of rotation*). Because of this rotation, a straight longitudinal line pq on the surface of the bar will become a helical curve pq', where q' is the position of point q after the end cross section has rotated through the angle ϕ (Fig. 9-3b).

The angle of twist changes along the axis of the bar and, at intermediate cross sections, will have a value $\phi(x)$ that is between zero at the left-hand end and ϕ at the right-hand end. If every cross section of the bar has the same radius and is subjected to the same torque (pure torsion), the angle $\phi(x)$ will vary linearly between the ends.

Shear Strains at the Outer Surface

Now consider an element of the bar between two cross sections distance dx apart (see Fig. 9-4a). This element is shown enlarged in Fig. 9-4b. On its outer surface, a small element $abcd$ is identified with sides ab and cd that initially are parallel to the longitudinal axis. During twisting of the bar, the right-hand cross section rotates with respect to the left-hand cross section through a small angle of twist $d\phi$, so that points b and c move to b' and c', respectively. The lengths of the sides of the element, which is now element $ab'c'd$, do not change during this small rotation.

However, the angles at the corners of the element (Fig. 9-4b) are no longer equal to 90°. The element is therefore in a state of **pure shear**, which means that the element is subjected to shear strains but no normal strains (see Fig. 7-35 of Section 7.6). The magnitude of the shear strain at the outer surface of the bar, denoted γ_{max}, is equal to the decrease in the angle at point a, that is, the decrease in angle bad. From Fig. 9-4b, the decrease in this angle is

$$\gamma_{max} = \frac{bb'}{ab} \tag{9-1}$$

FIGURE 9-3

Deformations of a circular bar in pure torsion

(a)

(b)

(a)

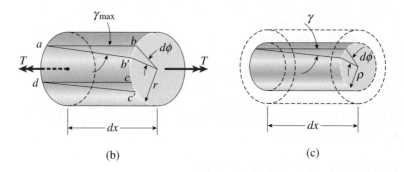

(b)   (c)

where $\gamma_{\max}$ is measured in radians, bb' is the distance through which point b moves, and ab is the length of the element (equal to dx). With r denoting the radius of the bar, express the distance bb' as $rd\phi$, where $d\phi$ also is measured in radians. Thus, the preceding equation becomes

$$\gamma_{\max} = \frac{rd\phi}{dx} \qquad (9\text{-}2)$$

This equation relates the shear strain at the outer surface of the bar to the angle of twist.

The quantity $d\phi/dx$ is the rate of change of the angle of twist ϕ with respect to the distance x measured along the axis of the bar. Denote $d\phi/dx$ by the symbol θ and refer to it as the **rate of twist**, or the **angle of twist per unit length**:

$$\theta = \frac{d\phi}{dx} \qquad (9\text{-}3)$$

With this notation, now write the equation for the shear strain at the outer surface [Eq. (9-2)] as

$$\gamma_{\max} = \frac{rd\phi}{dx} = r\theta \qquad (9\text{-}4)$$

For convenience, a bar in pure torsion was discussed when deriving Eqs. (9-3) and (9-4). However, both equations are valid in more general cases of torsion, such as when the rate of twist θ is not constant but varies with the distance x along the axis of the bar.

In the special case of pure torsion, the rate of twist is equal to the total angle of twist ϕ divided by the length L, that is, $\theta = \phi/L$. Therefore, *for pure torsion only*, the result is

$$\gamma_{\max} = r\theta = \frac{r\phi}{L} \qquad (9\text{-}5)$$

This equation can be obtained directly from the geometry of Fig. 9-3a by noting that γ_{max} is the angle between lines pq and pq', that is, γ_{max} is the angle qpq'. Therefore, $\gamma_{max}L$ is equal to the distance qq' at the end of the bar. But since the distance qq' also equals $r\phi$ (Fig. 9-3b), the relation $r\phi = \gamma_{max}L$ is produced, which agrees with Eq. (9-5).

Shear Strains within the Bar

The shear strains within the interior of the bar can be found by the same method used to find the shear strain γ_{max} at the surface. Because radii in the cross sections of a bar remain straight and undistorted during twisting, the preceding discussion for an element $abcd$ at the outer surface (Fig. 9-4b) will also hold for a similar element situated on the surface of an interior cylinder of radius ρ (Fig. 9-4c). Thus, interior elements are also in pure shear with the corresponding shear strains given by the equation [compare with Eq. (9-4)]:

$$\gamma = \rho\theta = \frac{\rho}{r}\gamma_{max} \tag{9-6}$$

This equation shows that the shear strains in a circular bar vary linearly with the radial distance ρ from the center with the strain being zero at the center and reaching a maximum value γ_{max} at the outer surface.

Circular Tubes

A review of the preceding discussions will show that the equations for the shear strains [Eqs. (9-2) to (9-4)] apply to **circular tubes** (Fig. 9-5) as well as to solid circular bars. Figure 9-5 shows the linear variation in shear strain between the maximum strain at the outer surface and the minimum strain at the interior surface. The equations for these strains are

$$\gamma_{max} = \frac{r_2\phi}{L} \quad \gamma_{min} = \frac{r_1}{r_2}\gamma_{max} = \frac{r_1\phi}{L} \tag{9-7a,b}$$

in which r_1 and r_2 are the inner and outer radii, respectively, of the tube.

All of the preceding equations for the strains in a circular bar are based upon geometric concepts and do not involve the material properties. Therefore, the equations are valid for any material, whether it behaves elastically or inelastically, linearly or nonlinearly. However, the equations are limited to bars having small angles of twist and small strains.

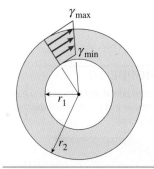

FIGURE 9-5

Shear strains in a circular tube

9.3 Circular Bars of Linearly Elastic Materials

Now that the shear strains in a circular bar in torsion have been investigated (see Figs. 9-3 to 9-5), it is time to determine the directions and magnitudes of the corresponding shear stresses. The directions of the stresses can be determined by inspection, as illustrated in Fig. 9-6a. Observe that the torque T tends to rotate the right-hand end of the bar counterclockwise when viewed from the right. Therefore, the shear stresses τ acting on a stress element located on the surface of the bar will have the directions shown in the figure.

For clarity, the stress element shown in Fig. 9-6a is enlarged in Fig. 9-6b, where both the shear strain and the shear stresses are shown. As explained previously in Section 8.6, stress elements are customarily drawn in two

FIGURE 9-6

Shear stresses in a circular bar in torsion

(a)

(b)   (c)

dimensions, as in Fig. 9-6b. But always remember that stress elements are actually three-dimensional objects with a thickness perpendicular to the plane of the figure.

The magnitudes of the shear stresses can be determined from the strains by using the stress-strain relation for the material of the bar. If the material is linearly elastic, use **Hooke's law in shear** [Eq. (7-13)]:

$$\tau = G\gamma \tag{9-8}$$

in which G is the shear modulus of elasticity and γ is the shear strain in radians. Combine this equation with the equations for the shear strains [Eqs. (9-2) and (9-4)] to get

$$\tau_{max} = Gr\theta \qquad \tau = G\rho\theta = \frac{\rho}{r}\tau_{max} \tag{9-9a,b}$$

in which τ_{max} is the shear stress at the outer surface of the bar (radius r), τ is the shear stress at an interior point (radius ρ), and θ is the rate of twist. (In these equations, θ has units of radians per unit of length.)

Equations (9-9a and b) show that the shear stresses vary linearly with the distance from the center of the bar, as illustrated by the triangular stress diagram in Fig. 9-6c. This linear variation of stress is a consequence of Hooke's law. If the stress-strain relation is nonlinear, the stresses will vary nonlinearly, and other methods of analysis will be needed.

The shear stresses acting on a cross-sectional plane are accompanied by shear stresses of the same magnitude acting on longitudinal planes (Fig. 9-7). This is because equal shear stresses always exist on mutually perpendicular planes, as explained in Section 7.6. If the material of the bar is weaker in shear on longitudinal planes than on cross-sectional planes, as is typical of wood when the grain runs parallel to the axis of the bar, the first cracks due to torsion will appear on the surface in the longitudinal direction.

The state of pure shear at the surface of a bar (Fig. 9-6b) is equivalent to equal tensile and compressive stresses acting on an element oriented at an angle of 45°, as explained later in Section 9.5. Therefore, a rectangular element with sides at 45° to the axis of the shaft will be subjected to tensile and compressive stresses, as shown in Fig. 9-8. If a torsion bar is made of a material that is weaker in tension than in shear, failure will occur in tension along a helix inclined at 45° to the axis, as you can demonstrate by twisting a piece of classroom chalk.

The Torsion Formula

The next step in this analysis is to determine the relationship between the shear stresses and the torque T. Once this is accomplished, you will be able to calculate the stresses and strains in a bar due to any set of applied torques.

FIGURE 9-9

Determination of the resultant of the shear stresses acting on a cross section

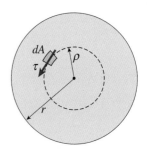

The distribution of the shear stresses acting on a cross section is pictured in Figs. 9-6c and 9-7. Because these stresses act continuously around the cross section, they have a resultant in the form of a moment—a moment equal to the torque T acting on the bar. To determine this resultant, consider an element of area dA located at radial distance ρ from the axis of the bar (Fig. 9-9). The shear force acting on this element is equal to $\tau\, dA$, where τ is the shear stress at radius ρ. The moment of this force about the axis of the bar is equal to the force times its distance from the center, or $\tau\rho dA$. Substitute for the shear stress τ from Eq. (9-9b) to express this elemental moment as

$$dM = \tau\rho dA = \frac{\tau_{max}}{r}\rho^2 dA$$

The resultant moment (equal to the torque T) is the summation over the entire cross-sectional area of all such elemental moments:

$$T = \int_A dM = \frac{\tau_{max}}{r}\int_A \rho^2 dA = \frac{\tau_{max}}{r}I_p \tag{9-10}$$

in which

$$I_p = \int_A \rho^2 dA \tag{9-11}$$

is the **polar moment of inertia** of the circular cross section.

For a **circle** of radius r and diameter d, the polar moment of inertia is

$$I_p = \frac{\pi r^4}{2} = \frac{\pi d^4}{32} \tag{9-12}$$

as given in Appendix B, Case 9. Note that moments of inertia have units of length to the fourth power.

An expression for the maximum shear stress is obtained by rearranging Eq. (9-10) as

$$\tau_{max} = \frac{Tr}{I_p} \tag{9-13}$$

This equation, known as the **torsion formula**, shows that the maximum shear stress is proportional to the applied torque T and inversely proportional to the polar moment of inertia I_p.

Typical **units** used with the torsion formula are as follows. In SI, the torque T is usually expressed in newton meters ($N \cdot m$), the radius r in meters (m), the polar moment of inertia I_p in meters to the fourth power (m^4), and the shear stress τ

in pascals (Pa). If USCS units are used, T is often expressed in pound-feet (lb-ft) or pound-inches (lb-in.), r in inches (in.), I_p in inches to the fourth power (in^4), and τ in pounds per square inch (psi).

Substitute $r = d/2$ and $I_p = \pi d^4/32$ into the torsion formula to get the following equation for the maximum stress:

$$\tau_{max} = \frac{16T}{\pi d^3} \tag{9-14}$$

This equation applies only to bars of *solid circular cross section*, whereas the torsion formula itself [Eq. (9-13)] applies to both solid bars and circular tubes, as explained later. Equation (9-14) shows that the shear stress is inversely proportional to the cube of the diameter. Thus, if the diameter is doubled, the stress is reduced by a factor of eight.

The shear stress at distance ρ from the center of the bar is

$$\tau = \frac{\rho}{r}\tau_{max} = \frac{T\rho}{I_p} \tag{9-15}$$

which is obtained by combining Eq. (9-9b) with the torsion formula [Eq. (9-13)]. Equation (9-15) is a *generalized torsion formula*, and once again, the shear stresses vary linearly with the radial distance from the center of the bar.

Angle of Twist

The angle of twist of a bar of linearly elastic material now can be related to the applied torque T. Combine Eq. (9-9a) with the torsion formula to get

$$\theta = \frac{T}{GI_p} \tag{9-16}$$

in which θ has units of radians per unit of length. This equation shows that the rate of twist θ is directly proportional to the torque T and inversely proportional to the product GI_p, which is known as the **torsional rigidity** of the bar.

For a bar in **pure torsion**, the total angle of twist ϕ, which is equal to the rate of twist times the length of the bar (that is, $\phi = \theta L$), is

$$\phi = \frac{TL}{GI_p} \tag{9-17}$$

where ϕ is measured in radians. The use of the preceding equations in both analysis and design is illustrated later in Examples 9-1 and 9-2.

The quantity GI_p/L, called the **torsional stiffness** of the bar, is the torque required to produce a unit angle of rotation. The **torsional flexibility** is the reciprocal of the stiffness, or L/GI_p, and is defined as the angle of rotation produced by a unit torque. Thus, the expressions for torsional stiffness and flexibility are

$$k_T = \frac{GI_p}{L} \quad f_T = \frac{L}{GI_p} \tag{9-18a,b}$$

These quantities are analogous to the axial stiffness $k = EA/L$ and axial flexibility $f = L/EA$ of a bar in tension or compression [compare with Eqs. (8-4a and b)]. Stiffnesses and flexibilities have important roles in structural analysis.

The equation for the angle of twist [Eq. (9-17)] provides a convenient way to determine the shear modulus of elasticity G for a material. Conduct a torsion test on a circular bar and measure the angle of twist ϕ produced by a known torque T. Then the value of G can be calculated from Eq. (9-17).

Circular Tubes

Circular tubes are more efficient than solid bars in resisting torsional loads. The shear stresses in a solid circular bar are maximum at the outer boundary of the cross section and zero at the center. Therefore, most of the material in a solid shaft is stressed significantly below the maximum shear stress. Furthermore, the stresses near the center of the cross section have a smaller moment arm ρ for use in determining the torque [see Fig. 9-9 and Eq. (9-10)].

By contrast, in a typical hollow tube, most of the material is near the outer boundary of the cross section where both the shear stresses and the moment arms are highest (Fig. 9-10). Thus, if weight reduction and savings of material are important, it is advisable to use a circular tube. For instance, large drive shafts, propeller shafts, and generator shafts usually have hollow circular cross sections.

The analysis of the torsion of a circular tube is almost identical to that for a solid bar. The same basic expressions for the shear stresses may be used [for instance, Eqs. (9-9a and b)]. Of course, the radial distance ρ is limited to the range r_1 to r_2, where r_1 is the inner radius and r_2 is the outer radius of the bar (Fig. 9-10).

The relationship between the torque T and the maximum stress is given by Eq. (9-10), but the limits on the integral for the polar moment of inertia [Eq. (9-11)] are $\rho = r_1$ and $\rho = r_2$. Therefore, the polar moment of inertia of the cross-sectional area of a tube is

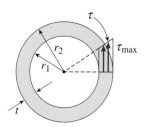

$$I_p = \frac{\pi}{2}(r_2^4 - r_1^4) = \frac{\pi}{32}(d_2^4 - d_1^4) \qquad \text{(9-19)}$$

The preceding expressions also can be written in the forms:

$$I_p = \frac{\pi r t}{2}(4r^2 + t^2) = \frac{\pi d t}{4}(d^2 + t^2) \qquad \text{(9-20)}$$

where r is the *average radius* of the tube equal to $(r_1 + r_2)/2$; d is the *average diameter* equal to $(d_1 + d_2)/2$; and t is the *wall thickness* (Fig. 9-10) equal to $r_2 - r_1$. Of course, Eqs. (9-19) and (9-20) give the same results, but sometimes the latter is more convenient.

If the tube is relatively thin so that the wall thickness t is small compared to the average radius r, disregard the terms t^2 in Eq. (9-20). This simplification leads to the following *approximate formulas* for the polar moment of inertia:

$$I_p \approx 2\pi r^3 t = \frac{\pi d^3 t}{4} \qquad \text{(9-21)}$$

These expressions are given in Case 22 of Appendix B.

Reminders: In Eqs. (9-20) and (9-21), the quantities r and d are the average radius and diameter, not the maximums. Also, Eqs. (9-19) and (9-20) are exact; Eq. (9-21) is approximate.

The torsion formula [Eq. (9-13)] may be used for a circular tube of linearly elastic material provided I_p is evaluated according to Eq. (9-19), Eq. (9-20), or, if appropriate, Eq. (9-21). The same comment applies to the general equation for shear stress [Eq. (9-15)], the equations for rate of twist and angle of twist [Eqs. (9-16) and Eq. (9-17)], and the equations for stiffness and flexibility [Eqs. (9-18a and b)].

The shear stress distribution in a tube is pictured in Fig. 9-10. From the figure, note that the average stress in a thin tube is nearly as great as the maximum stress. This means that a hollow bar is more efficient in the use of material than is a solid bar, as explained previously and as demonstrated later in Examples 9-2 and 9-3.

When designing a circular tube to transmit a torque, be sure that the thickness t is large enough to prevent wrinkling or buckling of the wall of the tube. For instance, a maximum value of the radius-to-thickness ratio, such as $(r_2/t)_{max} = 12$, may be specified. Other design considerations include environmental and durability factors, which also may impose requirements for minimum wall thickness. These topics are discussed in courses and textbooks on mechanical design.

Limitations

The equations derived in this section are limited to bars of circular cross section (either solid or hollow) that behave in a linearly elastic manner. In other words, the loads must be such that the stresses do not exceed the proportional limit of the material. Furthermore, the equations for stresses are valid only in parts of the bars away from stress concentrations (such as holes and other abrupt changes in shape) and away from cross sections where loads are applied.

Finally, it is important to emphasize that the equations for the torsion of circular bars and tubes cannot be used for bars of other shapes. Noncircular bars, such as rectangular bars and bars having I-shaped cross sections, behave quite differently than do circular bars. For instance, their cross sections do *not* remain plane, and their maximum stresses are *not* located at the farthest distances from the midpoints of the cross sections. Thus, these bars require more advanced methods of analysis, such as those presented in books on theory of elasticity and advanced mechanics of materials. (A brief overview of torsion of noncircular prismatic shafts is presented in Section 9.9.)

Example 9-1

A solid steel bar of circular cross section (Fig. 9-11) has a diameter $d = 1.5$ in., length $L = 54$ in., and shear modulus of elasticity $G = 11.5 \times 10^6$ psi. The bar is subjected to torques T acting at the ends.

(a) If the torques have a magnitude $T = 250$ lb-ft, what is the maximum shear stress in the bar? What is the angle of twist between the ends?

(b) If the allowable shear stress is 6000 psi and the allowable angle of twist is 2.5°, what is the maximum permissible torque?

FIGURE 9-11

Example 9-1: Bar
in pure torsion

$d = 1.5$ in.

T T

$L = 54$ in.

Solution:

Use the four-step problem-solving approach. Combine steps as needed for an efficient solution.

Part (a): Maximum shear stress and angle of twist.

1, 2. *Conceptualize* [*hypothesize, sketch*], *Categorize* [*simplify, classify*]: Because the bar has a solid circular cross section, compute the maximum shear stress from Eq. (9-14) as

$$\tau_{max} = \frac{16T}{\pi d^3} = \frac{16(250 \text{ lb-ft})(12 \text{ in/ft})}{\pi(1.5 \text{ in.})^3} = 4530 \text{ psi}$$

In a similar manner, the angle of twist is obtained from Eq. (9-17) with the polar moment of inertia given by Eq. (9-12).

3. *Analyze* [*evaluate; select relevant equations, carry out mathematical solution*]:

$$I_p = \frac{\pi d^4}{32} = \frac{\pi(1.5 \text{ in.})^4}{32} = 0.4970 \text{ in}^4$$

$$\phi = \frac{TL}{GI_p} = \frac{(250 \text{ lb-ft})(12 \text{ in/ft})(54 \text{ in.})}{(11.5 \times 10^6 \text{ psi})(0.4970 \text{ in}^4)} = 0.02834 \text{ rad} = 1.62°$$

4. *Finalize* [*conclude; examine answer—Does it make sense? Are units correct? How does it compare to similar problem solutions?*]: The formulas used here apply to bars of circular cross section only. The twist distortion of the bar is very small as expected.

Part (b): Maximum permissible torque.

1, 2. *Conceptualize, Categorize*: The maximum permissible torque is determined either by the allowable shear stress or by the allowable angle of twist.

3. *Analyze*: Beginning with the shear stress, rearrange Eq. (9-14) and calculate as

$$T_1 = \frac{\pi d^3 \tau_{allow}}{16} = \frac{\pi}{16}(1.5 \text{ in.})^3(6000 \text{ psi}) = 3980 \text{ lb-in.} = 331 \text{ lb-ft}$$

Any torque larger than this value will result in a shear stress that exceeds the allowable stress of 6000 psi.

Using a rearranged Eq. (9-17), now calculate the torque based upon the angle of twist:

$$T_2 = \frac{GI_p\phi_{\text{allow}}}{L} = \frac{(11.5 \times 10^6 \text{ psi})(0.4970 \text{ in}^4)(2.5°)(\pi \text{ rad}/180°)}{54 \text{ in.}}$$

$$= 4618 \text{ lb-in.} = 385 \text{ lb-ft}$$

4. *Finalize*: Any torque larger than T_2 will exceed the allowable angle of twist. The maximum permissible torque is the smaller of T_1 and T_2:

$$T_{\text{max}} = 331 \text{ lb-ft}$$

In this example, the allowable shear stress provides the limiting condition.

Example 9-2

A steel shaft is to be manufactured either as a solid circular bar or as a circular tube (Fig. 9-12). The shaft is required to transmit a torque of 1200 N·m without exceeding an allowable shear stress of 40 MPa nor an allowable rate of twist of 0.75°/m. (The shear modulus of elasticity of the steel is 78 GPa.)

(a) Determine the required diameter d_0 of the solid shaft.

(b) Determine the required outer diameter d_2 of the hollow shaft if the thickness t of the shaft is specified as one-tenth of the outer diameter.

(c) Determine the ratio of diameters (that is, the ratio d_2/d_0) and the ratio of weights of the hollow and solid shafts.

Solution:

The four-step problem-solving approach is used in the solution; only essential steps are shown in each part.

Part (a): Solid shaft.

1,2. *Conceptualize, Categorize*: The required diameter d_0 is determined either from the allowable shear stress or from the allowable rate of twist.

FIGURE 9-12

Example 9-2: Torsion
of a steel shaft

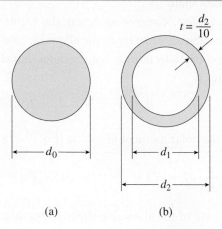

(a)   (b)

3. *Analyze*: In the case of the allowable shear stress, rearrange Eq. (9-14) and obtain

$$d_0^3 = \frac{16T}{\pi\tau_{allow}} = \frac{16(1200 \text{ N} \cdot \text{m})}{\pi(40 \text{ MPa})} = 152.8 \times 10^{-6} \text{ m}^3$$

so the required diameter is

$$d_0 = 0.0535 \text{ m} = 53.5 \text{ mm}$$

In the case of the allowable rate of twist, start by finding the required polar moment of inertia [see Eq. (9-16)]:

$$I_p = \frac{T}{G\theta_{allow}} = \frac{1200 \text{ N} \cdot \text{m}}{(78 \text{ GPa})(0.75°/\text{m})(\pi \text{ rad}/180°)} = 1175 \times 10^{-9} \text{ m}^4$$

Since the polar moment of inertia is equal to $\pi d^4/32$, the required diameter is

$$d_0^4 = \frac{32I_p}{\pi} = \frac{32(1175 \times 10^{-9} \text{ m}^4)}{\pi} = 11.97 \times 10^{-6} \text{ m}^4$$

or

$$d_0 = 0.0588 \text{ m} = 58.8 \text{ mm}$$

4. *Finalize*: Comparing the two values of d_0, note that the rate of twist governs the design and the required diameter of the solid shaft is

$$d_0 = 58.8 \text{ mm}$$

In a practical design, you would select a diameter slightly larger than the calculated value of d_0; for instance, 60 mm.

Part (b): Hollow shaft.

1,2. *Conceptualize, Categorize*: Again, the required diameter is based upon either the allowable shear stress or the allowable rate of twist.

3. *Analyze*: Begin by noting that the outer diameter of the bar is d_2 and the inner diameter is

$$d_1 = d_2 - 2t = d_2 - 2(0.1d_2) = 0.8d_2$$

Thus, the polar moment of inertia [Eq. (9-19)] is

$$I_p = \frac{\pi}{32}(d_2^4 - d_1^4) = \frac{\pi}{32}\left[d_2^4 - (0.8d_2)^4\right] = \frac{\pi}{32}(0.5904d_2^4) = 0.05796d_2^4$$

In the case of the allowable shear stress, use the torsion formula [Eq. (9-13)] as

$$\tau_{allow} = \frac{Tr}{I_p} = \frac{T(d_2/2)}{0.05796d_2^4} = \frac{T}{0.1159d_2^3}$$

Rearrange to get

$$d_2^3 = \frac{T}{0.1159\tau_{allow}} = \frac{1200\ \text{N} \cdot \text{m}}{0.1159(40\ \text{MPa})} = 258.8 \times 10^{-6}\ \text{m}^3$$

Solving for d_2 gives

$$d_2 = 0.0637\ \text{m} = 63.7\ \text{mm}$$

which is the required outer diameter based upon the shear stress.

In the case of the allowable rate of twist, use Eq. (9-16) with θ replaced by θ_{allow} and I_p replaced by the previously obtained expression; thus,

$$\theta_{allow} = \frac{T}{G(0.05796d_2^4)}$$

from which

$$d_2^4 = \frac{T}{0.05796G\theta_{allow}}$$

$$= \frac{1200\ \text{N} \cdot \text{m}}{0.05796(78\ \text{GPa})(0.75°/\text{m})(\pi\ \text{rad}/180°)} = 20.28 \times 10^{-6}\ \text{m}^4$$

Solving for d_2 gives

$$d_2 = 0.0671\ \text{m} = 67.1\ \text{mm}$$

which is the required diameter based upon the rate of twist.

4. *Finalize*: Compare the two values of d_2 and note that the rate of twist governs the design. The required outer diameter of the hollow shaft is

$$d_2 = 67.1 \text{ mm}$$

The inner diameter d_1 is equal to $0.8d_2$, or 53.7 mm. (As practical values, you might select $d_2 = 70$ mm and $d_1 = 0.8d_2 = 56$ mm.)

Part (c): Ratios of diameters and weights.

1, 2. *Conceptualize, Categorize*: The ratio of the outer diameter of the hollow shaft to the diameter of the solid shaft (using the calculated values) is

$$\frac{d_2}{d_0} = \frac{67.1 \text{ mm}}{58.8 \text{ mm}} = 1.14$$

3. *Analyze*: Since the weights of the shafts are proportional to their cross-sectional areas, express the ratio of the weight of the hollow shaft to the weight of the solid shaft as

$$\frac{W_H}{W_S} = \frac{A_H}{A_S} = \frac{\pi(d_2^2 - d_1^2)/4}{\pi d_0^2/4} = \frac{d_2^2 - d_1^2}{d_0^2}$$

$$= \frac{(67.1 \text{ mm})^2 - (53.7 \text{ mm})^2}{(58.8 \text{ mm})^2} = 0.47$$

4. *Finalize*: These results show that the hollow shaft uses only 47% as much material as does the solid shaft, while its outer diameter is only 14% larger.

Note: This example illustrates how to determine the required sizes of both solid bars and circular tubes when allowable stresses and allowable rates of twist are known. It also illustrates the fact that circular tubes are more efficient in the use of materials than are solid circular bars.

Example 9-3

A hollow shaft and a solid shaft constructed of the same material have the same length and the same outer radius R (Fig. 9-13). The inner radius of the hollow shaft is $0.6R$.

(a) Assuming that both shafts are subjected to the same torque, compare their shear stresses, angles of twist, and weights.

(b) Determine the strength-to-weight ratios for both shafts.

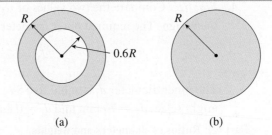

(a)   (b)

Solution:

Use the four-step problem-solving approach. Combine steps as needed for an efficient solution.

Part (a): Comparison of shear stresses.

2, 3. *Categorize, Analyze*: The maximum shear stresses given by the torsion formula [Eq. (9-13)] are proportional to $1/I_p$ inasmuch as the torques and radii are the same. For the hollow shaft,

$$I_p = \frac{\pi R^4}{2} - \frac{\pi(0.6R)^4}{2} = 0.4352\pi R^4$$

and for the solid shaft,

$$I_p = \frac{\pi R^4}{2} = 0.5\pi R^4$$

Therefore, the ratio β_1 of the maximum shear stress in the hollow shaft to that in the solid shaft is

$$\beta_1 = \frac{\tau_H}{\tau_S} = \frac{0.5\pi R^4}{0.4352\pi R^4} = 1.15$$

where the subscripts H and S refer to the hollow shaft and the solid shaft, respectively.

Comparison of angles of twist.

2, 3. *Categorize, Analyze*: The angles of twist [Eq. (9-17)] are also proportional to $1/I_p$ because the torque T, length L, and shearing modulus G are the same for both shafts. Therefore, the ratio of angles is the same as for the shear stresses:

$$\beta_2 = \frac{\phi_H}{\phi_S} = \frac{0.5\pi R^4}{0.4352\pi R^4} = 1.15$$

Comparison of weights.

2,3. *Categorize, Analyze*: The weights of the shafts are proportional to their cross-sectional areas; consequently, the weight of the solid shaft is proportional to πR^2 and the weight of the hollow shaft is proportional to

$$\pi R^2 - \pi (0.6R)^2 = 0.64\pi R^2$$

Therefore, the ratio of the weight of the hollow shaft to the weight of the solid shaft is

$$\beta_3 = \frac{W_H}{W_S} = \frac{0.64\pi R^2}{\pi R^2} = 0.64 \qquad \longleftarrow$$

4. *Finalize*: From the preceding ratios, note again the inherent advantage of hollow shafts. In this example, the hollow shaft has 15% greater stress and 15% greater angle of rotation than the solid shaft but 36% less weight.

Part (b): Strength-to-weight ratios.

2. *Categorize*: The relative efficiency of a structure is sometimes measured by its *strength-to-weight ratio*, which is defined for a bar in torsion as the allowable torque divided by the weight.

3. *Analyze*: The allowable torque for the hollow shaft of Fig. 9-13a (from the torsion formula) is

$$T_H = \frac{\tau_{max} I_p}{R} = \frac{\tau_{max}(0.4352\pi R^4)}{R} = 0.4352\pi R^3 \tau_{max}$$

and for the solid shaft is

$$T_S = \frac{\tau_{max} I_p}{R} = \frac{\tau_{max}(0.5\pi R^4)}{R} = 0.5\pi R^3 \tau_{max}$$

The weights of the shafts are equal to the cross-sectional areas times the length L times the weight density γ of the material:

$$W_H = 0.64\pi R^2 L\gamma \qquad W_S = \pi R^2 L\gamma$$

Thus, the strength-to-weight ratios S_H and S_S for the hollow and solid bars, respectively, are

$$S_H = \frac{T_H}{W_H} = 0.68\frac{\tau_{max} R}{\gamma L} \qquad S_S = \frac{T_S}{W_S} = 0.5\frac{\tau_{max} R}{\gamma L} \qquad \longleftarrow$$

4. *Finalize*: In this example, the strength-to-weight ratio of the hollow shaft is 36% greater than the strength-to-weight ratio for the solid shaft, demonstrating once again the relative efficiency of hollow shafts. For a thinner shaft, the percentage will increase; for a thicker shaft, it will decrease.

9.4 Nonuniform Torsion

As explained in Section 9.2, *pure torsion* refers to torsion of a prismatic bar subjected to torques acting only at the ends. **Nonuniform torsion** differs from pure torsion in that the bar need not be prismatic, and the applied torques may act anywhere along the axis of the bar. Bars in nonuniform torsion can be analyzed by applying the formulas of pure torsion to finite segments of the bar and then by adding the results or by applying the formulas to differential elements of the bar and then integrating.

Consider three cases of nonuniform torsion. Other cases can be handled by techniques similar to those described here.

Case 1. *Bar consisting of prismatic segments with constant torque throughout each segment* (Fig. 9-14). The bar shown in part (a) of the figure has two different diameters and is loaded by torques acting at points A, B, C, and D. Consequently, divide the bar into segments in such a way that each segment is prismatic and subjected to a constant torque. In this example, there are three such segments, AB, BC, and CD. Each segment is in pure torsion; therefore, all of the formulas derived in the preceding section may be applied to each part separately.

The first step in the analysis is to determine the magnitude and direction of the internal torque in each segment. Usually, the torques can be determined by inspection, but if necessary, they can be found by cutting sections through the bar, drawing free-body diagrams, and solving equations of equilibrium. This process is illustrated in Figs. 9-14b, c, and d. The first cut is made anywhere in segment CD, thereby exposing the internal torque T_{CD}. From the free-body diagram (Fig. 9-14b), T_{CD} is equal to $-T_1 - T_2 + T_3$. From the next diagram, T_{BC} equals to $-T_1 - T_2$, and from the last, T_{AB} equals $-T_1$. Thus,

$$T_{CD} = -T_1 - T_2 + T_3 \qquad T_{BC} = -T_1 - T_2 \qquad T_{AB} = -T_1 \qquad \textbf{(9-22a,b,c)}$$

Each of these torques is constant throughout the length of its segment.

To find the shear stresses in each segment, only the magnitudes of these internal torques are needed, since the directions of the stresses are not of interest. However, when finding the angle of twist for the entire bar, the direction of twist in each segment must be known in order to combine the angles of twist correctly. Therefore, establish a *sign convention* for the internal torques. A convenient rule in many cases is

An internal torque is positive when its vector points away from the cut section and negative when its vector points toward the section.

Thus, all of the internal torques shown in Figs. 9-14b, c, and d are pictured in their positive directions. If the calculated torque [from (Eq. 9-22a, b, or c)] turns out to have a positive sign, it means that the torque acts in the assumed direction; if the torque has a negative sign, it acts in the opposite direction.

The maximum shear stress in each segment of the bar is readily obtained from the torsion formula [Eq. (9-13)] using the appropriate cross-sectional dimensions and internal torque. For instance, the maximum stress in segment BC (Fig. 9-14) is found using the diameter of that segment and the torque T_{BC} calculated from Eq. (9-22b). The maximum stress in the entire bar is the largest stress from among the stresses calculated for each of the three segments.

FIGURE 9-14

Bar in nonuniform torsion (Case 1)

(a)

(b)

(c)

(d)

$$T_{CD} + T_2 + T_1 - T_3 = 0$$

$$T_{CD} = -T_1 - T_2 + T_3$$

$$\gamma = \frac{Tr}{I_p}$$

The angle of twist for each segment is found from Eq. (9-17), again using the appropriate dimensions and torque. The total angle of twist of one end of the bar with respect to the other is then obtained by algebraic summation, as

$$\phi = \phi_1 + \phi_2 + \dots + \phi_n \qquad \textbf{(9-23)}$$

where ϕ_1 is the angle of twist for segment 1, ϕ_2 is the angle for segment 2, and so on, while n is the total number of segments. Since each angle of twist is found from Eq. (9-17), the general formula used to find the total angle of twist is

$$\phi = \sum_{i=1}^{n} \phi_i = \sum_{i=1}^{n} \frac{T_i L_i}{G_i (I_p)_i} \qquad \textbf{(9-24)}$$

where the subscript i is a numbering index for the various segments. For segment i of the bar, T_i is the internal torque (found from equilibrium, as illustrated in Fig. 9-14), L_i is the length, G_i is the shear modulus, and $(I_p)_i$ is the polar moment of inertia. Some of the torques (and the corresponding angles of twist) may be positive, and some may be negative. By summing *algebraically* the angles of twist for all segments, you obtain the total angle of twist ϕ between the ends of the bar. This process is illustrated later in Example 9-4.

Case 2. *Bar with continuously varying cross sections and constant torque* (Fig. 9-15). When the torque is constant, the maximum shear stress in a solid bar always occurs at the cross section having the smallest diameter, as shown by Eq. (9-14). Furthermore, this observation usually holds for tubular bars. If this is the case, you only need to investigate the smallest cross section in order to calculate the maximum shear stress. Otherwise, it may be necessary to evaluate the stresses at more than one location in order to obtain the maximum.

To find the angle of twist, consider an element of length dx at distance x from one end of the bar (Fig. 9-15). The differential angle of rotation $d\phi$ for this element is

$$d\phi = \frac{T dx}{G I_p(x)} \qquad \textbf{(9-25)}$$

where $I_p(x)$ is the polar moment of inertia of the cross section at distance x from the end. The angle of twist for the entire bar is the summation of the differential angles of rotation:

$$\phi = \int_0^L d\phi = \int_0^L \frac{T dx}{G I_p(x)} \qquad \textbf{(9-26)}$$

If the expression for the polar moment of inertia $I_p(x)$ is not too complex, this integral can be evaluated analytically. In other cases, it must be evaluated numerically.

Case 3. *Bar with continuously varying cross sections and continuously varying torque* (Fig. 9-16). The bar shown in part (a) of the figure is subjected to a *distributed torque* of intensity t per unit distance along the axis of the bar. As a result, the internal torque $T(x)$ varies continuously along the axis (Fig. 9-16b). The internal torque can be evaluated with the aid of a free-body diagram and an equation of equilibrium. As in Case 2, the polar moment of inertia $I_p(x)$ can be evaluated from the cross-sectional dimensions of the bar.

FIGURE 9-15

Bar in nonuniform torsion (Case 2)

FIGURE 9-16

Bar in nonuniform torsion (Case 3)

(a)

(b)

Knowing both the torque and polar moment of inertia as functions of x, use the torsion formula to determine how the shear stress varies along the axis of the bar. The cross section of maximum shear stress then can be identified, and the maximum shear stress can be determined.

The angle of twist for the bar of Fig. 9-16a can be found in the same manner as described for Case 2. The only difference is that the torque, like the polar moment of inertia, also varies along the axis. Consequently, the equation for the angle of twist becomes

$$\phi = \int_0^L d\phi = \int_0^L \frac{T(x)dx}{GI_p(x)} \tag{9-27}$$

This integral can be evaluated analytically in some cases, but usually it must be evaluated numerically.

Limitations

The analyses described in this section are valid for bars made of linearly elastic materials with circular cross sections (either solid or hollow). Also, the stresses determined from the torsion formula are valid in regions of the bar *away* from stress concentrations, which are highly localized stresses that occur wherever the diameter changes abruptly and wherever concentrated torques are applied. However, stress concentrations have relatively little effect on the angle of twist; therefore, the equations for ϕ are generally valid.

Finally, keep in mind that the torsion formula and the formulas for angles of twist were derived for prismatic bars with circular cross sections (see Section 9.9 for a brief discussion of noncircular bars in torsion). You can safely apply them to bars with varying cross sections only when the changes in diameter are small and gradual. As a rule of thumb, the formulas given here are satisfactory as long as the angle of taper (the angle between the sides of the bar) is less than 10°.

Torsional-Displacement Diagram (TDD)

A graphical display of the variation in internal torsional moment $T(x)$ over the length of a shaft is presented here, so critical regions of the shaft (such as the location of maximum internal torsional moment T_{max} that also may be the location of maximum shear stress) can be identified. This display is referred to as the *torsional-moment diagram* (TMD). Rules for construction of torsional-moment diagrams are based on the differential and integral relationships between the applied torque of intensity $t(x)$ and the internal torsional moment $T(x)$. Now the *torque–displacement relation* in Eq. (9-27) is used to create a display of the variation of torsional displacement (or twist) $\phi(x)$ over the length of the shaft, referred to here as a *torsional-displacement diagram* (TDD).

To construct the TDD, first note in Eq. (9-27) that the internal torsional moment $T(x)$ is part of the integrand in the expression for torsional displacement $\phi(x)$. It follows that the slope on the torsional-displacement diagram is *proportional to* the ordinate on the torsional-moment diagram; the full expression for slope must include the torsional rigidity term $GI_p(x)$ in the denominator in Eq. (9-27). Second, the integral expression indicates that the change in torsional displacement between any two points is proportional to the area under the TMD between those same two points:

$$\phi_B - \phi_A = \int_A^B \frac{T(x)}{GI_p(x)}dx$$

These observations lead to the following rules or guidelines to be used in developing the TDD from the TMD:

- The slope at any point on the TDD is equal to the ordinate on the TMD at the same point divided by the torsional rigidity of the shaft (GI_p) at the same location.
- The change in torsional displacement between any two points along a shaft is equal to the area under the torsional-moment diagram between those same two points divided by the torsional rigidity of the shaft (GI_p) over that same interval.

If the shaft is piecewise prismatic (so that GI_p is constant in each shaft segment) and the internal torsional moment $T(x)$ is constant, the TDD can be constructed using simple geometry. For instance, if the internal torsional moment is constant within each segment, the twist will vary linearly within that segment. Otherwise, analytical or numerical integration of Eq. (9-27) will be required to develop the TDD.

Construction of the TMD is illustrated using a cantilever shaft subjected to triangularly distributed torque $t(x)$ (see Figs. 9-17a and b). The torsional loading has linear variation, so the internal torque $T(x)$ has a quadratic variation as shown on the TMD in Fig. 9-17c. It follows that the twist displacement $\phi(x)$ has a cubic variation on the TDD (Fig. 9-17d). Comparison of the TMD and TDD plots in Figs. 9-17c and d leads to the following observations: (a) the slope of the tangent to the TDD curve at $x = 0$ is equal to the TMD ordinate $t_0 L/2$ divided by the torsional rigidity GI_p; (b) the TMD ordinate is zero at $x = L$, so the TDD has zero slope at that location, and (c) the area A_m

FIGURE 9-17

(a) Prismatic shaft subject to triangularly distributed torque; (b, c, d) overall FBD, torsional-moment diagram (TMD), and torsional-displacement diagram (TDD) under the entire TMD is proportional to the change in twist from $x = 0$ to $x = L$, so A_m/GI_p is equal to the maximum twist at $x = L$. Use a *statics sign convention* in developing the TDD: twist displacement $\phi(x)$ is considered positive (and plotted above the reference axis on the TDD) if the twist displacement vector points in the $+x$ direction. Examples 9-4 and 9-5 provide further application of these concepts.

Example 9-4

FIGURE 9-18

Example 9-4: Steel shaft in torsion

Oleksandr Kostiuchenko/Shutterstock.com

A solid steel shaft *ABCDE* (Fig. 9-18) having a diameter $d = 30$ mm turns freely in bearings at points *A* and *E*. The shaft is driven by a gear at *C*, which applies a torque $T_2 = 450$ N·m in the direction shown in the figure. Gears at *B* and *D* are driven by the shaft and have resisting torques $T_1 = 275$ N·m and $T_3 = 175$ N·m, respectively, acting in the opposite direction to the torque T_2. Segments *BC* and *CD* have lengths $L_{BC} = 500$ mm and $L_{CD} = 400$ mm, respectively, and the shear modulus is $G = 80$ GPa.

Determine the maximum shear stress in each part of the shaft and the angle of twist between gears *B* and *D*.

Solution:

Use the four-step problem-solving approach to find the maximum shear stress and angle of twist between *B* and *D*.

1. *Conceptualize:* Each segment of the bar is prismatic and subjected to a constant torque (Case 1). Therefore, the analysis procedure is to determine the torques acting in the segments, followed by calculation of the shear stresses and angles of twist in each segment. Cutting the shaft in segments *BC* and *CD* gives the free-body diagrams (FBDs) shown in Fig. 9-19.

FIGURE 9-19

Free-body diagrams for Example 9-4

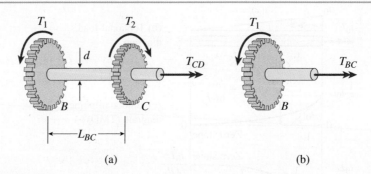

(a)   (b)

2. *Categorize*: Sum moments about the shaft axis in each FBD to find the internal torques. Internal torques T_{BC} and T_{CD} are shown in their assumed positive directions in Fig. 9-19.

Torques acting in the segments: The torques in the end segments (AB and DE) are zero, since any friction in the bearings at the supports is assumed to be too small to consider. Therefore, the end segments have no stresses and no angles of twist.

The torque T_{CD} in segment CD is found by cutting a section through the segment and constructing a free-body diagram, as in Fig. 9-19a. The torque is assumed to be positive; therefore, its vector points away from the cut section. From equilibrium of the free body, torque T_{CD} is

$$T_{CD} = T_2 - T_1 = 450 \text{ N} \cdot \text{m} - 275 \text{ N} \cdot \text{m} = 175 \text{ N} \cdot \text{m}$$

The positive sign in the result means that T_{CD} acts in the assumed positive direction.

The torque in segment BC is found in a similar manner, using the free-body diagram of Fig. 9-19b:

$$T_{BC} = -T_1 = -275 \text{ N} \cdot \text{m}$$

Note that this torque has an negative sign, which means that its direction is opposite to the direction shown in the figure.

With all internal torques known, plot the TMD as shown in Fig. 9-20a.

FIGURE 9-20

(a) TMD and (b) TDD for steel shaft in torsion

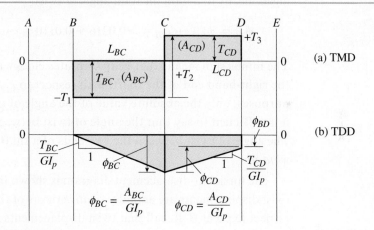

(a) TMD

(b) TDD

$$\phi_{BC} = \frac{A_{BC}}{GI_p} \qquad \phi_{CD} = \frac{A_{CD}}{GI_p}$$

3. *Analyze*:

Shear stresses: The maximum shear stresses in segments BC and CD are found from the modified form of the torsion formula [Eq. (9-14)]; thus,

$$\tau_{BC} = \frac{16T_{BC}}{\pi d^3} = \frac{16(275 \text{ N} \cdot \text{m})}{\pi (30 \text{ mm})^3} = 51.9 \text{ MPa} \qquad \curvearrowleft$$

$$\tau_{CD} = \frac{16T_{CD}}{\pi d^3} = \frac{16(175 \text{ N} \cdot \text{m})}{\pi (30 \text{ mm})^3} = 33.0 \text{ MPa} \qquad \curvearrowleft$$

Since the direction of shear stresses are not of the interest of this example, only absolute values of the torques are used in the preceding calculations.

Angles of twist: The angle of twist ϕ_{BD} between gears B and D is the algebraic sum of the angles of twist for the intervening segments of the bar, as given by Eq. (9-23); thus,

$$\phi_{BD} = \phi_{BC} + \phi_{CD}$$

When calculating the individual angles of twist, use the polar moment of inertia of the cross section, which is computed as

$$I_p = \frac{\pi d^4}{32} = \frac{\pi (30 \text{ mm})^4}{32} = 79{,}520 \text{ mm}^4$$

Now find the angles of twist, as

$$\phi_{BC} = \frac{T_{BC} L_{BC}}{G I_p} = \frac{(-275 \text{ N} \cdot \text{m})(500 \text{ mm})}{(80 \text{ GPa})(79{,}520 \text{ mm}^4)} = -0.0216 \text{ rad}$$

and

$$\phi_{CD} = \frac{T_{CD} L_{CD}}{G I_p} = \frac{(175 \text{ N} \cdot \text{m})(400 \text{ mm})}{(80 \text{ GPa})(79{,}520 \text{ mm}^4)} = 0.0110 \text{ rad}$$

Note that in this example the angles of twist have opposite directions. Add algebraically to obtain the total angle of twist:

$$\phi_{BD} = \phi_{BC} + \phi_{CD} = -0.0216 + 0.0110 = -0.0106 \text{ rad} = -0.61° \quad \leftarrow$$

The minus sign means that gear D rotates clockwise (when viewed from the right-hand end of the shaft) with respect to gear B. However, for most purposes, only the absolute value of the angle of twist is needed; therefore, it is sufficient to say that the angle of twist between gears B and D is 0.61°. The angle of twist between the two ends of a shaft is sometimes called the *wind-up*.

The torsional-displacement diagram is shown in Fig. 9-20b. Here the twist displacements are actually *relative* twists of the shaft segments with respect to point B. Recall that twist displacements are positive if the displacement vector points in the positive direction of the shaft axis. Study Figs. 9-20a and b to confirm the geometric relationships labeled on the figures.

4. *Finalize*: The procedures illustrated in this example can be used for shafts having segments of different diameters or of different materials, as long as the dimensions and properties remain constant within each segment.

Only the effects of torsion are considered in this example and in the problems at the end of the chapter. Bending effects are considered later, beginning with Chapter 10.

Example 9-5

Two sections (*AB, BC*) of steel drill pipe, joined by bolted flange plates at *B*, are being tested to assess the adequacy of both the pipe and the bolted connection (see Fig. 9-21). In the test, the pipe structure is fixed at *A*, a concentrated torque $2T_0$ is applied at $x = 2L/5$, and a uniformly distributed torque with intensity $t_0 = 3T_0/L$ is applied on pipe *BC*.

(a) Find the maximum shear stress τ_{max} in the pipes and its location. Assume that load variable $T_0 = 2000$ kip-in. Let $G = 11,800$ ksi, and assume that both pipes have the same inner diameter $d = 10$ in. Pipe *AB* has a thickness of $t_{AB} = 3/4$ in., while pipe *BC* has a thickness of $t_{BC} = 5/8$ in.

(b) Find the expressions for twist rotation $\phi(x)$ over the length of the pipe structure. If the maximum allowable twist of the pipe structure is $\phi_{allow} = 0.5°$, find the maximum permissible value of *load variable* T_0 (kip-in.). Let $L = 10$ ft. Also, plot the torsional-moment diagram (TMD) and the torsional-displacement diagram (TDD).

(c) Use maximum T_0 from part (b) to find the number of $d_b = 7/8$-in. diameter bolts at radius $r = 15$ in. required in the flange plate connection at *B*. Assume that the allowable shear stress for the bolts is $\tau_a = 28$ ksi.

Solution:

Use the four-step problem-solving approach to find maximum shear stress, maximum twist, and the number of connection bolts at *B*.

1. *Conceptualize*: Make section cuts along the shaft to find internal torques $T(x)$; then plot the torsional-moment diagram (TMD) as shown in Fig. 9-21c. From the TMD, the maximum internal torque is $6T_0/5$ in the central region of the structure. It appears that the maximum shear stress may be in the thinner pipe segment *BC* just right of the connection at *B*.

2. *Categorize*: Use the torsion formula [Eq. (9-13)] to find shear stresses on the surface of each segment. Use the torque–displacement relation [Eq. (9-17)] to find the twist rotations over the length of the pipe that are displayed as the TDD. Finally, examine shear stresses in the bolted connection to find the required number of bolts.

FIGURE 9-21

Example 9-5: (a, b) Two pipes
in nonuniform torsion;
(c) torsional-moment diagram
(TMD) for entire shaft

(a) Nonprismatic pipe

(b) Section at B

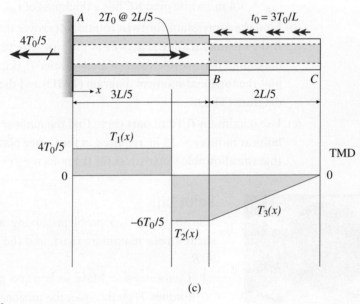

(c)

3. *Analyze:*

Part (a): Maximum shear stress in pipe τ_{max}.

Use the torsion formula [Eq. (9-13)] to compute the shear stress in the pipe. The maximum shear stress is on the surface of the pipe. The polar moment of inertia of each pipe is computed as

$$l_{pAB} = \frac{\pi}{32}\Big[(d + 2t_{AB})^4 - (d)^4\Big]$$

$$= \frac{\pi}{32}\left[\left[10 \text{ in.} + 2\left(\frac{3}{4}\text{ in.}\right)\right]^4 - (10 \text{ in.})^4\right] = 735.335 \text{ in}^4$$

and

$$I_{pBC} = \frac{\pi}{32}\Big[(d + 2t_{BC})^4 - (d)^4\Big]$$

$$= \frac{\pi}{32}\left[\left[10 \text{ in.} + 2\left(\frac{5}{8}\text{ in.}\right)\right]^4 - (10 \text{ in.})^4\right] = 590.822 \text{ in}^4$$

The shear modulus G is constant, so the torsional rigidity of AB is 1.245 times that of BC. From the TMD (Fig. 9-21c), the maximum torsional moments in both AB and BC (each equal to $6T_0/5$) are near joint B. Applying the torsion formula to pipes AB and BC near B gives

$$\tau_{\max AB} = \frac{\left(\dfrac{6}{5}T_0\right)\left(\dfrac{d + 2t_{AB}}{2}\right)}{I_{pAB}}$$

$$= \frac{\left(\dfrac{6}{5}2000 \text{ kip-in.}\right)\left[\dfrac{10 \text{ in.} + 2\left(\dfrac{3}{4}\text{ in.}\right)}{2}\right]}{735.335 \text{ in}^4} = 18.8 \text{ ksi}$$

$$\tau_{\max BC} = \frac{\left(\dfrac{6}{5}T_0\right)\left(\dfrac{d + 2t_{BC}}{2}\right)}{I_{pBC}}$$

$$= \frac{\left(\dfrac{6}{5}2000 \text{ kip-in.}\right)\left[\dfrac{10 \text{ in.} + 2\left(\dfrac{5}{8}\text{ in.}\right)}{2}\right]}{590.822 \text{ in}^4} = 22.8 \text{ ksi} \quad \longleftarrow$$

So the maximum shear stress in the pipe is *just to the right of* the flange plate connection at joint B. "Just to the right of" means an appropriate distance away from the connection to avoid any stress-concentration effects at the point of attachment of the two pipes in accordance with St. Venant's principle.

Part (b): Twist rotations $\phi(x)$.

Next, use the *torque–displacement relation* from Eqs. (9-24) through (9-27) to find the variation of twist rotation ϕ over the length of the pipe structure. Support A is fixed, so $\phi_A = \phi(0) = 0$. The internal torque from $x = 0$ to $x = 2L/5$ (segment 1) is constant, so use Eq. (9-24) to find twist rotation $\phi_1(x)$, which varies linearly from $x = 0$ to $x = 2L/5$:

$$\phi_1(x) = \frac{T_1(x)(x)}{GI_{pAB}} = \frac{\left(\dfrac{4T_0}{5}\right)(x)}{GI_{pAB}} = \frac{4T_0 x}{5GI_{pAB}} \qquad 0 \le x \le \frac{2L}{5} \qquad \textbf{(a)}$$

Evaluate Eq. (a) at $x = 2L/5$ to find the twist rotation at the point of application of torque $2T_0$ as

$$\phi_1\left(\frac{2L}{5}\right) = \frac{T_1\left(\dfrac{2L}{5}\right)\left(\dfrac{2L}{5}\right)}{GI_{pAB}} = \frac{\left(\dfrac{4T_0}{5}\right)\left(\dfrac{2L}{5}\right)}{GI_{pAB}} = \frac{8T_0 L}{25GI_{pAB}} = \frac{0.32T_0 L}{GI_{pAB}} \qquad \textbf{(b)}$$

Next, find an expression for the variation of twist angle $\phi_2(x)$ from $x = 2L/5$ to $x = 3L/5$ (point B). As with $\phi_1(x)$, twist $\phi_2(x)$ varies linearly over segment 2, because torque $T_2(x)$ is constant (Fig. 9-21c). Use Eq. (9-24) to get

$$\phi_2(x) = \phi_1\left(\frac{2L}{5}\right) + \frac{T_2(x)\left(x - \frac{2L}{5}\right)}{GI_{pAB}} = \frac{8T_0L}{25GI_{pAB}} + \frac{\left(\frac{-6}{5}T_0\right)\left(x - \frac{2L}{5}\right)}{GI_{pAB}} \quad \text{(c)}$$

$$= \frac{2T_0(2L - 3x)}{5GI_{pAB}} \quad \frac{2L}{5} \leq x \leq \frac{3L}{5}$$

Finally, develop an expression for twist over segment 3 (or pipe BC). The internal torsional moment now has a linear variation (Fig. 9-21c), so an integral form of the torque–displacement relation [Eq. (9-27)] is required. Insert an expression for $T_3(x)$ and add the torsional displacement at B to get a formula for the variation of twist in BC (note that variable ζ is a dummy variable of integration):

$$\phi_3(x) = \phi_2\left(\frac{3L}{5}\right) + \int_{\frac{3L}{5}}^{x} \frac{\left[3T_0\left(\frac{\zeta}{L} - 1\right)\right]}{GI_{pBC}} \, d\zeta$$

$$= \frac{2T_0\left[2L - 3\left(\frac{3L}{5}\right)\right]}{5GI_{pAB}} + \int_{\frac{3L}{5}}^{x} \frac{\left[3T_0\left(\frac{\zeta}{L} - 1\right)\right]}{GI_{pBC}} \, d\zeta$$

Torque $T_3(x)$ has a linear variation, so evaluating the integral yields a quadratic expression for variation of twist in BC:

$$\phi_3(x) = \frac{2T_0L}{25GI_{pAB}} + \frac{3T_0(21L^2 - 50Lx + 25x^2)}{50GI_{pBC}L} \quad \frac{3L}{5} \leq x \leq L \quad \text{(d)}$$

Substitute $x = 3L/5$ to find the twist at B:

$$\phi_3\left(\frac{3L}{5}\right) = \frac{2T_0L}{25GI_{pAB}}$$

At $x = L$, the twist at C is

$$\phi_3(L) = \frac{2T_0L}{25GI_{pAB}} - \frac{6T_0L}{25GI_{pBC}} = -0.219\frac{T_0L}{GI_{pAB}}$$

Now assume that $I_{pAB} = 1.245I_{pBC}$ (based on the numerical properties here), and plot the variation of twist over the length of the pipe structure (Fig. 9-22), noting that ϕ_{max} occurs at $x = 2L/5$ [see Eq. (b)].

FIGURE 9-22

Example 9-5: Torsional-displacement diagram (TDD)

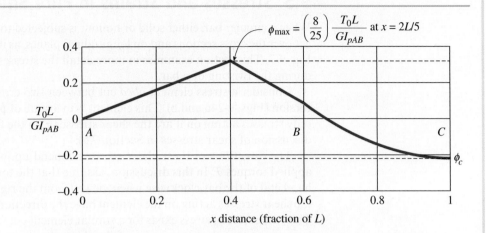

$$\phi_{\max} = \left(\frac{8}{25}\right)\frac{T_0 L}{GI_{pAB}} \text{ at } x = 2L/5$$

x distance (fraction of L)

Finally, restrict $\phi_{\max}$ to the allowable value of 0.5° and solve for the maximum permissible value of *load variable* T_0 (kip-in.) using the numerical properties given previously:

$$T_{0\,\max} = \frac{GI_{pAB}}{0.32L}(\phi_{\text{allow}}) = \frac{(11{,}800 \text{ ksi})(735.335 \text{ in}^4)}{0.32\left[10 \text{ ft}\left(12\dfrac{\text{in.}}{\text{ft}}\right)\right]}(0.5°) \tag{e}$$

$$= 1972 \text{ kip-in.}$$

Part (c): Number of bolts required in flange plate.

Now use $T_{0,\max}$ from Eq. (e) to find the required number of $d_b = 7/8$-in. diameter bolts at radius $r = 15$ in. in the flange plate connection at B. The allowable shear stress in the bolts is $\tau_a = 28$ ksi. Assume that each bolt carries an equal share of the torque at B, so each of n bolts carries shear force F_b at a distance r from the centroid of the cross section (Fig. 9-23).

The maximum shear force F_b per bolt is τ_a times the bolt cross-sectional area A_b, and the total torque at B is $6T_{0,\max}/5$ (see TMD in Fig. 9-21c), so

FIGURE 9-23

Example 9-5: Flange plate bolts at B

$$nF_b r = \frac{6}{5}T_{0\max} \text{ or } n = \frac{\dfrac{6}{5}T_{0\max}}{\tau_a A_b r} = \frac{\dfrac{6}{5}(1972 \text{ kip-in.})}{(28 \text{ ksi})\left[\dfrac{\pi}{4}\left(\dfrac{7}{8}\text{ in.}\right)^2\right](15 \text{ in.})} = 9.37$$

Use ten 7/8-in. diameter bolts at a radius of 15 in. in the flange plate connection at B.

4. *Finalize*: Confirm the TDD in Fig. 9-22 using the TMD in Fig. 9-21c along with the rules presented in Section 9.4. Values and locations of maximum torsional moment and maximum twist are readily visible on these diagrams and are essential information for use in an engineering evaluation of this pipe structure.

9.5 Stresses and Strains in Pure Shear

When a circular bar, either solid or hollow, is subjected to torsion, shear stresses act over the cross sections and on longitudinal planes, as illustrated previously in Fig. 9-7. This section examines in more detail the stresses and strains produced during the twisting of a bar.

Consider a stress element *abcd* cut between two cross sections of a bar in torsion (Figs. 9-24a and b). This element is in a state of **pure shear**, because the only stresses acting on it are the shear stresses τ on the four side faces (see the discussion of shear stresses in Section 7.6).

The directions of these shear stresses depend upon the directions of the applied torques T. In this discussion, assume that the torques rotate the right-hand end of the bar clockwise when viewed from the right (Fig. 9-24a); hence, the shear stresses acting on the element have the directions shown in the figure. This same state of stress exists for a similar element cut from the interior of the bar, except that the magnitudes of the shear stresses are smaller because the radial distance to the element is smaller.

The directions of the torques shown in Fig. 9-24a are intentionally chosen so that the resulting shear stresses (Fig. 9-24b) are positive according to the sign convention for shear stresses described previously in Section 7.6. This **sign convention** is repeated here:

> *A shear stress acting on a positive face of an element is positive if it acts in the positive direction of one of the coordinate axes and negative if it acts in the negative direction of an axis. Conversely, a shear stress acting on a negative face of an element is positive if it acts in the negative direction of one of the coordinate axes and negative if it acts in the positive direction of an axis.*

Apply this sign convention to the shear stresses acting on the stress element of Fig. 9-24b, and see that all four shear stresses are positive. For instance, the stress on the right-hand face (which is a positive face because the x axis is directed to the right) acts in the positive direction of the y axis; therefore, it is a positive shear stress. Also, the stress on the left-hand face (which is a negative face) acts in the negative direction of the y axis; therefore, it is a positive shear stress. Analogous comments apply to the remaining stresses.

Stresses on Inclined Planes

Now determine the stresses acting on *inclined planes* cut through the stress element in pure shear. Follow the same approach as the one used in Section 8.6 for investigating the stresses in uniaxial stress.

A two-dimensional view of the stress element is shown in Fig. 9-25a. As explained previously in Section 8.6, a two-dimensional view is usually drawn for convenience, but always be aware that the element has a third dimension (thickness) perpendicular to the plane of the figure.

Now cut from the element a wedge-shaped (or "triangular") stress element having one face oriented at an angle θ to the x axis (Fig. 9-25b). Normal stresses σ_θ and shear stresses τ_θ act on this inclined face and are shown in their

FIGURE 9-24

Stresses acting on a stress element cut from a bar in torsion (pure shear)

(a)

(b)

positive directions in the figure. The **sign convention** for stresses σ_θ and τ_θ was described previously in Section 8.6 and is repeated here:

Normal stresses σ_θ are positive in tension, and shear stresses τ_θ are positive when they tend to produce counterclockwise rotation of the material.

(Note that this sign convention for the shear stress τ_θ acting on an inclined plane is different from the sign convention for ordinary shear stresses τ that act on the sides of rectangular elements oriented to a set of *x-y* axes.)

The horizontal and vertical faces of the triangular element (Fig. 9-25b) have positive shear stresses τ acting on them, and the front and rear faces of the element are free of stress. Therefore, all stresses acting on the element are visible in this figure.

The stresses σ_θ and τ_θ now may be determined from the equilibrium of the triangular element. The *forces* acting on its three side faces can be obtained by multiplying the stresses by the areas over which they act. For instance, the force on the left-hand face is equal to τA_0, where A_0 is the area of the vertical face. This force acts in the negative *y* direction and is shown in the *free-body diagram* of Fig. 9-25c. Because the thickness of the element in the *z* direction is constant, the area of the bottom face is $A_0 \tan \theta$ and the area of the inclined face is $A_0 \sec \theta$. Multiplying the stresses acting on these faces by the corresponding areas enables us to obtain the remaining forces and thereby complete the free-body diagram (Fig. 9-25c).

Now write two equations of equilibrium for the triangular element: one in the direction of σ_θ and the other in the direction of τ_θ. When writing these equations, the forces acting on the left-hand and bottom faces must be resolved into components in the directions of σ_θ and τ_θ. Thus, the first equation, obtained by summing forces in the direction of σ_θ, is

$$\sigma_\theta A_0 \sec \theta = \tau A_0 \sin \theta + \tau A_0 \tan \theta \cos \theta$$

or

$$\sigma_\theta = 2\tau \sin \theta \cos \theta \qquad \text{(9-28a)}$$

The second equation is obtained by summing forces in the direction of τ_θ:

$$\tau_\theta A_0 \sec \theta = \tau A_0 \cos \theta - \tau A_0 \tan \theta \sin \theta$$

or

$$\tau_\theta = \tau(\cos^2 \theta - \sin^2 \theta) \qquad \text{(9-28b)}$$

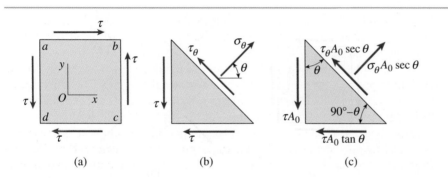

(a)   (b)   (c)

FIGURE 9-25

Analysis of stresses on inclined planes: (a) element in pure shear, (b) stresses acting on a triangular stress element, and (c) forces acting on the triangular stress element (free-body diagram)

These equations can be expressed in simpler forms by introducing the following trigonometric identities (see Appendix A):

$$\sin 2\theta = 2 \sin \theta \cos \theta \qquad \cos 2\theta = \cos^2 \theta - \sin^2 \theta$$

Then the equations for σ_θ and τ_θ become

$$\sigma_\theta = \tau \sin 2\theta \qquad \tau_\theta = \tau \cos 2\theta \qquad \text{(9-29a,b)}$$

Equations (9-29a and b) give the normal and shear stresses acting on any inclined plane in terms of the shear stresses τ acting on the x and y planes (Fig. 9-25a) and the angle θ defining the orientation of the inclined plane (Fig. 9-25b).

The manner in which the stresses σ_θ and τ_θ vary with the orientation of the inclined plane is shown by the graph in Fig. 9-26, which is a plot of Eqs. (9-29a and b). For $\theta = 0$, which is the right-hand face of the stress element in Fig. 9-25a, the graph gives $\sigma_\theta = 0$ and $\tau_\theta = \tau$. This latter result is expected, because the shear stress τ acts counterclockwise against the element and therefore produces a positive shear stress τ_θ.

For the top face of the element ($\theta = 90°$), $\sigma_\theta = 0$ and $\tau_\theta = -\tau$. The minus sign for τ_θ means that it acts clockwise against the element, that is, to the right on face ab (Fig. 9-25a), which is consistent with the direction of the shear stress τ. Note that the numerically largest shear stresses occur on the planes for which $\theta = 0$ and $90°$, as well as on the opposite faces ($\theta = 180°$ and $270°$).

In the graph, the normal stress σ_θ reaches a maximum value at $\theta = 45°$. At that angle, the stress is positive (tension) and equal numerically to the shear stress τ. Similarly, σ_θ has its minimum value (which is compressive) at $\theta = -45°$. At both of these 45° angles, the shear stress τ_θ is equal to zero. These conditions are pictured in Fig. 9-27, which shows stress elements oriented at $\theta = 0$ and $\theta = 45°$. The element at 45° is acted upon by equal tensile and compressive stresses in perpendicular directions with no shear stresses.

Note that the normal stresses acting on the 45° element (Fig. 9-27b) correspond to an element subjected to shear stresses τ acting in the directions shown in Fig. 9-27a. If the shear stresses acting on the element of Fig. 9-27a are reversed in direction, the normal stresses acting on the 45° planes also will change directions.

FIGURE 9-26

Graph of normal stresses σ_θ and shear stresses τ_θ versus angle θ of the inclined plane

FIGURE 9-27

Stress elements oriented at $\theta = 0$ and $\theta = 45°$ for pure shear

(a)   (b)

FIGURE 9-28

Torsion failure of a brittle material by tension cracking along a 45° helical surface

If a stress element is oriented at an angle other than 45°, both normal and shear stresses will act on the inclined faces [see Eqs. (9-29a and b) and Fig. 9-26]. Stress elements subjected to these more general conditions are discussed in detail in Chapter 11.

The equations derived in this section are valid for a stress element in pure shear regardless of whether the element is cut from a bar in torsion or from some other structural element. Also, since Eqs. (9-29) were derived from equilibrium only, they are valid for any material, whether or not it behaves in a linearly elastic manner.

The existence of maximum tensile stresses on planes at 45° to the x axis (Fig. 9-27b) explains why bars in torsion that are made of materials that are brittle and weak in tension fail by cracking along a 45° helical surface (Fig. 9-28). As mentioned in Section 9.3, this type of failure is readily demonstrated by twisting a piece of classroom chalk.

Strains in Pure Shear

Now consider the strains that exist in an element in pure shear. For instance, consider the element in pure shear shown in Fig. 9-27a. The corresponding shear strains are shown in Fig. 9-29a, where the deformations are highly exaggerated. The shear strain γ is the change in angle between two lines that were originally perpendicular to each other, as discussed previously in Section 7.6. Thus, the decrease in the angle at the lower left-hand corner of the element is the shear strain γ (measured in radians). This same change in angle occurs at the upper right-hand corner (where the angle decreases) and at the other two corners (where the angles increase). However, the lengths of the sides of the element, including the thickness perpendicular to the plane of the paper, do not change when these shear deformations occur. Therefore, the element changes its shape from a rectangular parallelepiped (Fig. 9-27a) to an oblique parallelepiped (Fig. 9-29a). This change in shape is called a **shear distortion**.

If the material is linearly elastic, the shear strain for the element oriented at $\theta = 0$ (Fig. 9-29a) is related to the shear stress by Hooke's law in shear:

$$\gamma = \frac{\tau}{G} \tag{9-30}$$

where, as usual, the symbol G represents the shear modulus of elasticity.

Next, consider the strains that occur in an element oriented at $\theta = 45°$ (Fig. 9-29b). The tensile stresses acting at 45° tend to elongate the element in that direction. Because of the Poisson effect, they also tend to shorten it in the

FIGURE 9-29

Strains in pure shear: (a) shear distortion of an element oriented at $\theta = 0$ and (b) distortion of an element oriented at $\theta = 45°$

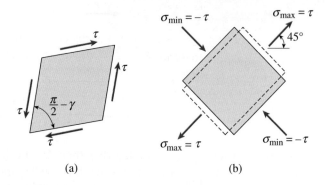

(a)   (b)

perpendicular direction (the direction where $\theta = 135°$ or $-45°$). Similarly, the compressive stresses acting at $135°$ tend to shorten the element in that direction and elongate it in the $45°$ direction. These dimensional changes are shown in Fig. 9-29b, where the dashed lines show the deformed element. Since there are no shear distortions, the element remains a rectangular parallelepiped even though its dimensions have changed.

If the material is linearly elastic and follows Hooke's law, you can obtain an equation relating strain to stress for the element at $\theta = 45°$ (Fig. 9-29b). The tensile stress σ_{max} acting at $\theta = 45°$ produces a positive normal strain in that direction equal to σ_{max}/E. Since $\sigma_{max} = \tau$, this strain also can be expressed as τ/E. The stress σ_{max} also produces a negative strain in the perpendicular direction equal to $-\nu\tau/E$, where ν is Poisson's ratio. Similarly, the stress $\sigma_{min} = -\tau$ (at $\theta = 135°$) produces a negative strain equal to $-\tau/E$ in that direction and a positive strain in the perpendicular direction (the $45°$ direction) equal to $\nu\tau/E$. Therefore, the normal strain in the $45°$ direction is

$$\varepsilon_{max} = \frac{\tau}{E} + \frac{\nu\tau}{E} = \frac{\tau}{E}(1 + \nu) \tag{9-31}$$

which is positive, representing elongation. The strain in the perpendicular direction is a negative strain of the same amount. In other words, pure shear produces elongation in the $45°$ direction and shortening in the $135°$ direction. These strains are consistent with the shape of the deformed element of Fig. 9-29a, because the $45°$ diagonal has lengthened and the $135°$ diagonal has shortened.

The next section uses the geometry of the deformed element to relate the shear strain γ (Fig. 9-29a) to the normal strain ε_{max} in the $45°$ direction (Fig. 9-29b). In so doing, the following relationship is derived:

$$\varepsilon_{max} = \frac{\gamma}{2} \tag{9-32}$$

This equation, in conjunction with Eq. (9-30), can be used to calculate the maximum shear strains and maximum normal strains in pure torsion when the shear stress τ is known.

Example 9-6

A circular tube with an outside diameter of 80 mm and an inside diameter of 60 mm is subjected to a torque $T = 4.0$ kN·m (Fig. 9-30). The tube is made of aluminum alloy 7075-T6.

(a) Determine the maximum shear, tensile, and compressive stresses in the tube and show these stresses on sketches of properly oriented stress elements.

(b) Determine the corresponding maximum strains in the tube and show these strains on sketches of the deformed elements.

(c) What is the maximum permissible torque T_{max} if the allowable normal strain is $\varepsilon_a = 0.9 \times 10^{-3}$?

(d) If $T = 4.0$ kN·m and $\varepsilon_a = 0.9 \times 10^{-3}$, what new outer diameter is required so that the tube can carry the required torque T (assume that the inner diameter of the tube remains at 60 mm)?

FIGURE 9-30

Example 9-6: Circular tube in torsion

$T = 4.0$ kN·m

|—60—|
mm
|—80—|
mm

Solution:

The four-step problem-solving approach is used in the solution; only essential steps are shown in each part.

Part (a): Maximum stresses.

1, 2. *Conceptualize, Categorize*: The maximum values of all three stresses (shear, tensile, and compressive) are equal numerically, although they act on different planes. Their magnitudes are found from the torsion formula:

$$\tau_{max} = \frac{Tr}{I_p} = \frac{(4000 \text{ N} \cdot \text{m})(0.040 \text{ m})}{\dfrac{\pi}{32}\left[(0.080 \text{ m})^4 - (0.060 \text{ m})^4\right]} = 58.2 \text{ MPa} \qquad \leftarrow$$

3. *Analyze*: The maximum shear stresses act on cross-sectional and longitudinal planes, as shown by the stress element in Fig. 9-31a, where the x axis is parallel to the longitudinal axis of the tube.

The maximum tensile and compressive stresses are

$$\sigma_t = 58.2 \text{ MPa} \qquad \sigma_c = -58.2 \text{ MPa} \qquad \leftarrow$$

FIGURE 9-31

Stress and strain elements for the tube of Example 9-6: (a) maximum shear stresses, (b) maximum tensile and compressive stresses, (c) maximum shear strains, and (d) maximum tensile and compressive strains

(a)

(b)

(c)

(d)

4. *Finalize*: These stresses act on planes at 45° to the axis (Fig. 9-31b).

Part (b): Maximum strains.

1, 2. *Conceptualize, Categorize*: The maximum shear strain in the tube is obtained from Eq. (9-30). The shear modulus of elasticity is obtained from Table F-2, Appendix F, as $G = 27$ GPa.

3. *Analyze*: Therefore, the maximum shear strain is

$$\gamma_{max} = \frac{\tau_{max}}{G} = \frac{58.2 \text{ MPa}}{27 \text{ GPa}} = 0.0022 \text{ rad}$$

The deformed element is shown by the dashed lines in Fig. 9-31c.

The magnitude of the maximum normal strains [from Eq. (9-32)] is

$$\varepsilon_{max} = \frac{\gamma_{max}}{2} = 0.0011$$

Thus, the maximum tensile and compressive strains are

$$\varepsilon_t = 0.0011 \qquad \varepsilon_c = -0.0011$$

4. *Finalize*: The deformed element is shown by the dashed lines in Fig. 9-31d for an element with sides of unit length.

Part (c): Maximum permissible torque.

1, 2. *Conceptualize, Categorize*: The tube is in *pure shear*, so the allowable shear strain is twice the allowable normal strain [see Eq. (9-32)]:

$$\gamma_a = 2\varepsilon_a = 2(0.9 \times 10^{-3}) = 1.8 \times 10^{-3}$$

3. *Analyze*: From the shear formula [Eq. (9-13)],

$$\tau_{max} = \frac{T\left(\dfrac{d_2}{2}\right)}{I_p} \quad \text{so} \quad T_{max} = \frac{\tau_a I_p}{\left(\dfrac{d_2}{2}\right)} = \frac{2(G\gamma_a)I_p}{d_2}$$

where d_2 is the outer diameter. Substituting numerical values gives

$$T_{max} = \frac{2(27 \text{ GPa})(1.8 \times 10^{-3})\left[\dfrac{\pi}{32}[(0.08 \text{ m})^4 - (0.06 \text{ m})^4]\right]}{0.08 \text{ m}}$$

$$= 3.34 \text{ kN} \cdot \text{m}$$

4. *Finalize*: The allowable shear strain is less than γ_{max} in part (b), so T_{max} is less than 4.0 kN·m, as expected.

Part (d): New outer diameter of tube.

1, 2. *Conceptualize, Categorize*: Use the previous equation but with $T = 4.0$ kN · m to find the required outer diameter d_2:

$$\frac{I_p}{d_2} = \frac{T}{2G\gamma_a} \quad \text{or} \quad \frac{d_2^4 - (0.06 \text{ m})^4}{d_2} = \frac{\left(\dfrac{32}{\pi}\right)4 \text{ kN} \cdot \text{m}}{2(27 \text{ GPa})(1.8 \times 10^{-3})} = 4.192 \times 10^{-4} \text{ m}^3$$

3. *Analyze*: Solving for the required outer diameter d_2 numerically gives

$$d_2 = 83.2 \text{ mm}$$

4. *Finalize*: Outer diameter d_2 is larger than that in part (c), as expected, since the applied torque here is larger than T_{max} in part (c).

9.6 Relationship Between Moduli of Elasticity *E* and *G*

An important relationship between the moduli of elasticity *E* and *G* can be obtained from the equations derived in the preceding section. For this purpose, consider the stress element *abcd* shown in Fig. 9-32a. The front face of the element is assumed to be square with the length of each side denoted as *h*. When this element is subjected to pure shear by stresses τ, the front face distorts into a rhombus (Fig. 9-32b) with sides of length *h* and with shear strain $\gamma = \tau/G$. Because of the distortion, diagonal *bd* is lengthened, and diagonal *ac* is shortened. The length of diagonal *bd* is equal to its initial length $\sqrt{2}h$ times the factor $1 + \varepsilon_{max}$, where ε_{max} is the normal strain in the 45° direction; thus,

$$L_{bd} = \sqrt{2}h(1 + \varepsilon_{max}) \tag{9-33}$$

This length can be related to the shear strain γ by considering the geometry of the deformed element.

FIGURE 9-32

Geometry of deformed element
in pure shear

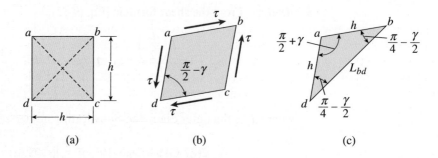

(a)   (b)   (c)

To obtain the required geometric relationships, consider triangle *abd* (Fig. 9-32c), which represents one-half of the rhombus pictured in Fig. 9-32b. Side *bd* of this triangle has a length L_{bd} [Eq. (9-33)], and the other sides each have a length *h*. Angle *adb* of the triangle is equal to one-half of angle *adc* of the rhombus, or $\pi/4 - \gamma/2$. The angle *abd* in the triangle is the same. Therefore, angle *dab* of the triangle equals $\pi/2 + \gamma$. Now use the law of cosines (see Appendix A) for triangle *abd* to get

$$L_{bd}^2 = h^2 + h^2 - 2h^2 \cos\left(\frac{\pi}{2} + \gamma\right)$$

Substitute for L_{bd} from Eq. (9-33) and simplify to get

$$(1 + \varepsilon_{max})^2 = 1 - \cos\left(\frac{\pi}{2} + \gamma\right)$$

Expand the term on the left-hand side, and also observe that $\cos(\pi/2 + \gamma) = -\sin\gamma$ to obtain

$$1 + 2\varepsilon_{max} + \varepsilon_{max}^2 = 1 + \sin\gamma$$

Because ε_{max} and γ are very small strains, disregard ε_{max}^2 in comparison with $2\varepsilon_{max}$ and replace $\sin\gamma$ by γ. The resulting expression is

$$\varepsilon_{max} = \frac{\gamma}{2} \tag{9-34}$$

which establishes the relationship already presented in Section 9.5 as Eq. (9-32).

The shear strain γ appearing in Eq. (9-34) is equal to τ/G by Hooke's law [Eq. (9-30)], and the normal strain ε_{max} is equal to $\tau(1 + v)/E$ by Eq. (9-31). Making both of these substitutions in Eq. (9-34) yields

$$G = \frac{E}{2(1 + v)} \tag{9-35}$$

So *E*, *G*, and *v* are not independent properties of a linearly elastic material. Instead, if any two of them are known, the third can be calculated from Eq. (9-35).

Typical values of *E*, *G*, and *v* are listed in Table F-2, Appendix F.

9.7 Transmission of Power by Circular Shafts

The most important use of circular shafts is to transmit mechanical power from one device or machine to another, as in the drive shaft of an automobile, the propeller shaft of a ship, or the axle of a bicycle. The power is transmitted through the rotary motion of the shaft, and the amount of power transmitted depends upon the magnitude of the torque and the speed of rotation. A common design problem is to determine the required size of a shaft so that it will transmit a specified amount of power at a specified rotational speed without exceeding the allowable stresses for the material.

FIGURE 9-33

Shaft transmitting a constant torque T at an angular speed ω

Suppose that a motor-driven shaft (Fig. 9-33) is rotating at an angular speed ω, measured in radians per second (rad/s). The shaft transmits a torque T to a device (not shown in the figure) that is performing useful work. The torque applied by the shaft to the external device has the same sense as the angular speed ω, that is, its vector points to the left. However, the torque shown in the figure is the torque exerted *on the shaft* by the device, and so its vector points in the opposite direction.

In general, the work W done by a torque of constant magnitude is equal to the product of the torque and the angle through which it rotates; that is,

$$W = T\psi \tag{9-36}$$

where ψ is the angle of rotation in radians.

Power is the *rate* at which work is done, or

$$P = \frac{dW}{dt} = T\frac{d\psi}{dt} \tag{9-37}$$

in which P is the symbol for power and t represents time. The rate of change $d\psi/dt$ of the angular displacement ψ is the angular speed ω, and therefore, the preceding equation becomes

$$P = T\omega \quad (\omega = \text{rad/s}) \tag{9-38}$$

This formula, which is familiar from elementary physics, gives the power transmitted by a rotating shaft transmitting a constant torque T.

The **units** to be used in Eq. (9-38) are as follows. If the torque T is expressed in newton meters, then the power is expressed in watts (W). One watt is equal to one newton meter per second (or one joule per second). If T is expressed in pound-feet, then the power is expressed in foot-pounds per second.

Angular speed is often expressed as the frequency f of rotation, which is the number of revolutions per unit of time. The unit of frequency is the hertz (Hz), which is equal to one revolution per second (s^{-1}). One revolution equals 2π radians, so

$$\omega = 2\pi f \quad (\omega = \text{rad/s}, f = \text{Hz} = s^{-1}) \tag{9-39}$$

The expression for power [Eq. (9-3)] then becomes

$$P = 2\pi fT \quad (f = \text{Hz} = s^{-1}) \tag{9-40}$$

Another commonly used unit is the number of revolutions per minute (rpm), denoted by the letter n. It follows that

$$n = 60 f \tag{9-41}$$

and

$$P = \frac{2\pi nT}{60} \quad (n = \text{rpm})$$ (9-42)

In Eqs. (9-40) and (9-42), the quantities P and T have the same units as in Eq. (9-38); that is, P has units of watts if T has units of newton meters, and P has units of foot-pounds per second if T has units of pound-feet.

In U.S. engineering practice, power is sometimes expressed in horsepower (hp), which is a unit equal to 550 ft-lb/s. Therefore, the horsepower H being transmitted by a rotating shaft is

$$H = \frac{2\pi nT}{60(550)} = \frac{2\pi nT}{33,000} \quad (n = \text{rpm}, \; T = \text{lb-ft}, \; H = \text{hp})$$ (9-43)

One horsepower is approximately 746 watts.

The preceding equations relate the torque acting in a shaft to the power transmitted by the shaft. Once the torque is known, you can determine the shear stresses, shear strains, angles of twist, and other desired quantities by the methods described in Sections 9.2 through 9.5.

The following examples illustrate some of the procedures for analyzing rotating shafts.

Example 9-7

A motor driving a solid circular steel shaft transmits 40 hp to a gear at B (Fig. 9-34). The allowable shear stress in the steel is 6000 psi.

(a) What is the required diameter d of the shaft if it is operated at 500 rpm?

(b) What is the required diameter d if it is operated at 3000 rpm?

FIGURE 9-34

Example 9-7: Steel shaft in torsion

Solution:

Use the four-step problem-solving approach. Combine steps as needed for an efficient solution.

Part (a): Motor operating at 500 rpm.

1, 2. *Conceptualize, Categorize*: Knowing the horsepower and the speed of rotation, the torque T acting on the shaft is found from Eq. (9-43). Solve that equation for T to get

$$T = \frac{33,000H}{2\pi n} = \frac{33,000(40 \text{ hp})}{2\pi(500 \text{ rpm})} = 420.2 \text{ lb-ft} = 5042 \text{ lb-in.}$$

This torque is transmitted by the shaft from the motor to the gear.

The maximum shear stress in the shaft can be obtained from the modified torsion formula [Eq. (9-14)]:

$$\tau_{max} = \frac{16T}{\pi d^3}$$

3. *Analyze*: Solve that equation for the diameter d, and also substitute τ_{allow} for τ_{max} to get

$$d^3 = \frac{16T}{\pi \tau_{allow}} = \frac{16(5042 \text{ lb-in.})}{\pi(6000 \text{ psi})} = 4.280 \text{ in}^3$$

from which

$$d = 1.62 \text{ in.} \quad \leftarrow$$

4. *Finalize*: The diameter of the shaft must be at least this large if the allowable shear stress is not to be exceeded.

Part (b): Motor operating at 3000 rpm.

1, 2. *Conceptualize, Categorize*: Follow the same procedure as in part (a).

3. *Analyze*: Torque T and diameter d are now

$$T = \frac{33,000H}{2\pi n} = \frac{33,000(40 \text{ hp})}{2\pi(3000 \text{ rpm})} = 70.03 \text{ lb-ft} = 840.3 \text{ lb-in.}$$

$$d^3 = \frac{16T}{\pi \tau_{allow}} = \frac{16(840.3 \text{ lb-in.})}{\pi(6000 \text{ psi})} = 0.7133 \text{ in}^3$$

$$d = 0.89 \text{ in.} \quad \leftarrow$$

which is less than the diameter found in part (a).

4. *Finalize*: This example illustrates that the higher the speed of rotation, the smaller the required size of the shaft (for the same power and the same allowable stress).

Example 9-8

A solid steel shaft ABC with a 50 mm diameter (Fig. 9-35a) is driven at A by a motor that transmits 50 kW to the shaft at 10 Hz. The gears at B and C drive machinery requiring power equal to 35 kW and 15 kW, respectively.

Compute the maximum shear stress τ_{max} in the shaft and the angle of twist ϕ_{AC} between the motor at A and the gear at C. (Use $G = 80 \text{ GPa}$.)

Solution:

Use the four-step problem-solving approach.

1. *Conceptualize*:

   Torques acting on the shaft: Begin the analysis by determining the torques applied to the shaft by the motor and the two gears. Since the motor supplies

FIGURE 9-35

Example 9-8: Steel shaft in torsion

(c) TMD

(d) TDD

50 kW at 10 Hz, it creates a torque T_A at end A of the shaft (Fig. 9-35b) calculated from Eq. (9-40):

$$T_A = \frac{P}{2\pi f} = \frac{50 \text{ kW}}{2\pi(10 \text{ Hz})} = 796 \text{ N} \cdot \text{m}$$

In a similar manner, calculate the torques T_B and T_C applied by the gears to the shaft:

$$T_B = \frac{P}{2\pi f} = \frac{35 \text{ kW}}{2\pi(10 \text{ Hz})} = 557 \text{ N} \cdot \text{m}$$

$$T_C = \frac{P}{2\pi f} = \frac{15 \text{ kW}}{2\pi(10 \text{ Hz})} = 239 \text{ N} \cdot \text{m}$$

These torques are shown in the free-body diagram of the shaft (Fig. 9-35b). Note that the torques applied by the gears are opposite in direction to the torque applied by the motor. (Think of T_A as the "load" applied to the shaft by the motor, then the torques T_B and T_C are the "reactions" of the gears.)

The internal torques in the two segments of the shaft are now found (by inspection) from the free-body diagram of Fig. 9-35b:

$$T_{AB} = 796 \text{ N} \cdot \text{m} \qquad T_{BC} = 239 \text{ N} \cdot \text{m}$$

The TMD is shown in Fig. 9-35c.

2. *Categorize*: Both internal torques act in the same direction; therefore, the angles of twist in segments AB and BC are additive when finding the total angle of twist. (To be specific, both torques are positive according to the sign convention adopted in Section 9.4.)

3. *Analyze*:

Shear stresses and angles of twist: The shear stress and angle of twist in segment AB of the shaft are found in the usual manner from Eqs. (9-14) and (9-17):

$$\tau_{AB} = \frac{16T_{AB}}{\pi d^3} = \frac{16(796 \text{ N} \cdot \text{m})}{\pi(50 \text{ mm})^3} = 32.4 \text{ MPa}$$

$$\phi_{AB} = \frac{T_{AB}L_{AB}}{GI_p} = \frac{(796 \text{ N} \cdot \text{m})(1.0 \text{ m})}{(80 \text{ GPa})\left(\dfrac{\pi}{32}\right)(50 \text{ mm})^4} = 0.0162 \text{ rad}$$

The corresponding quantities for segment BC are

$$\tau_{BC} = \frac{16T_{BC}}{\pi d^3} = \frac{16(239 \text{ N} \cdot \text{m})}{\pi(50 \text{ mm})^3} = 9.7 \text{ MPa}$$

$$\phi_{BC} = \frac{T_{BC}L_{BC}}{GI_p} = \frac{(239 \text{ N} \cdot \text{m})(1.2 \text{ m})}{(80 \text{ GPa})\left(\dfrac{\pi}{32}\right)(50 \text{ mm})^4} = 0.0058 \text{ rad}$$

4. *Finalize*: Thus, the maximum shear stress in the shaft occurs in segment AB and is

$$\tau_{\max} = 32.4 \text{ MPa}$$

Also, the total angle of twist between the motor at A and the gear at C is

$$\phi_{AC} = \phi_{AB} + \phi_{BC} = 0.0162 \text{ rad} + 0.0058 \text{ rad} = 0.0220 \text{ rad} = 1.26°$$

The TDD is shown in Fig. 9-35d. As explained previously, both parts of the shaft twist in the same direction; therefore, the angles of twist are added.

9.8   Statically Indeterminate Torsional Members

The bars and shafts described in the preceding sections of this chapter are *statically determinate* because all internal torques and all reactions can be obtained from free-body diagrams and equations of equilibrium. However, if additional restraints, such as fixed supports, are added to the bars, the equations of equilibrium will no longer be adequate for determining the torques. The bars are then classified as **statically indeterminate**. Torsional members of this kind can be analyzed by supplementing the equilibrium equations with compatibility equations pertaining to the rotational displacements. Thus, the general method for analyzing statically indeterminate torsional members is the same as described in Section 8.4 for statically indeterminate bars with axial loads.

The first step in the analysis is to write **equations of equilibrium**, obtained from free-body diagrams of the given physical situation. The unknown quantities in the equilibrium equations are torques, either internal torques or reaction torques.

FIGURE 9-36

Statically indeterminate bar in torsion

(a)

(b)

(c)

(d)

(e)

The second step in the analysis is to formulate **equations of compatibility** based upon physical conditions pertaining to the angles of twist. As a consequence, the compatibility equations contain angles of twist as unknowns.

The third step is to relate the angles of twist to the torques by **torque–displacement relations**, such as $\phi = TL/GI_p$. After introducing these relations into the compatibility equations, they too become equations containing torques as unknowns. Therefore, the last step is to obtain the unknown torques by solving simultaneously the equations of equilibrium and compatibility.

To illustrate the method of solution, analyze the composite bar AB shown in Fig. 9-36a. The bar is attached to a fixed support at end A and loaded by a torque T at end B. Furthermore, the bar consists of two parts: a solid bar and a tube (Figs. 9-36b and c) with both the solid bar and the tube joined to a rigid end plate at B.

For convenience, identify the solid bar and tube (and their properties) by the numerals 1 and 2, respectively. For instance, the diameter of the solid bar is denoted d_1 and the outer diameter of the tube is denoted d_2. A small gap exists between the bar and the tube; therefore, the inner diameter of the tube is slightly larger than the diameter d_1 of the bar.

When the torque T is applied to the composite bar, the end plate rotates through a small angle ϕ (Fig. 9-36c), and torques T_1 and T_2 are developed in the solid bar and the tube, respectively (Figs. 9-36d and e). From equilibrium, the sum of these torques equals the applied load, so the *equation of equilibrium* is

$$T_1 + T_2 = T \tag{9-44}$$

Because this equation contains two unknowns (T_1 and T_2), the composite bar is statically indeterminate.

To obtain a second equation, consider the rotational displacements of both the solid bar and the tube. Note that the angle of twist of the solid bar (Fig. 9-36d) is ϕ_1, and the angle of twist of the tube is ϕ_2 (Fig. 9-36e). These angles of twist must be equal because the bar and tube are securely joined to the end plate and rotate with it; consequently, the *equation of compatibility* is

$$\phi_1 = \phi_2 \tag{9-45}$$

The angles ϕ_1 and ϕ_2 are related to the torques T_1 and T_2 by the *torque–displacement relations*, which in the case of linearly elastic materials are obtained from the equation $\phi = TL/GI_p$. Thus,

$$\phi_1 = \frac{T_1L}{G_1I_{p1}} \qquad \phi_2 = \frac{T_2L}{G_2I_{p2}} \tag{9-46a,b}$$

in which G_1 and G_2 are the shear moduli of elasticity of the materials and I_{p1} and I_{p2} are the polar moments of inertia of the cross sections.

When the preceding expressions for ϕ_1 and ϕ_2 are substituted into Eq. (9-45), the equation of compatibility becomes

$$\frac{T_1L}{G_1I_{p1}} = \frac{T_2L}{G_2I_{p2}} \tag{9-47}$$

The two equations [Eqs. (9-44) and (9-47)] have two unknowns, so solving for the torques T_1 and T_2 gives

$$T_1 = T\left(\frac{G_1 I_{p1}}{G_1 I_{p1} + G_2 I_{p2}}\right) \qquad T_2 = T\left(\frac{G_2 I_{p2}}{G_1 I_{p1} + G_2 I_{p2}}\right) \quad \textbf{(9-48a,b)}$$

With these torques known, the essential part of the statically indeterminate analysis is completed. All other quantities, such as stresses and angles of twist, now can be found from the torques.

The preceding discussion illustrates the general methodology for analyzing a statically indeterminate system in torsion. In the following example, this same approach is used to analyze a bar that is fixed against rotation at both ends. In the example and in the problems, assume that the bars are made of linearly elastic materials. However, the general methodology is also applicable to bars of nonlinear materials—the only change is in the torque–displacement relations.

Example 9-9

The bar ACB shown in Figs. 9-37a and b is fixed at both ends and loaded by a torque T_0 at point C. Segments AC and CB of the bar have diameters d_A and d_B, lengths L_A and L_B, and polar moments of inertia I_{PA} and I_{pB}, respectively. The material of the bar is the same throughout both segments.

Obtain formulas for (a) the reactive torques T_A and T_B at the ends, (b) the maximum shear stresses τ_{AC} and τ_{CB} in each segment of the bar, and (c) the angle of rotation ϕ_C at the cross section where the load T_0 is applied.

Solution:

Use the four-step problem-solving approach. Combine steps as needed for an efficient solution.

1, 2. *Conceptualize, Categorize*:

Equation of equilibrium: The load T_0 produces reactions T_A and T_B at the fixed ends of the bar, as shown in Figs. 9-37a and b. Equilibrium of the bar gives

$$T_A + T_B = T_0 \tag{a}$$

Because there are two unknowns in this equation (and no other useful equations of equilibrium), the bar is statically indeterminate.

Equation of compatibility: Now separate the bar from its support at end B and obtain a bar that is fixed at end A and free at end B (Figs. 9-37c and d). When the load T_0 acts alone (Fig. 9-37c), it produces an angle of twist at end B denoted as ϕ_1. Similarly, when the reactive torque T_B acts alone, it produces an angle ϕ_2 (Fig. 9-37d). The angle of twist at end B in the original bar, which is equal to the sum of ϕ_1 and ϕ_2, is zero. Therefore, the equation of compatibility is

$$\phi_1 + \phi_2 = 0 \tag{b}$$

Note that ϕ_1 and ϕ_2 are assumed to be positive in the direction shown in the figure.

FIGURE 9-37

Example 9-9: Statically indeterminate bar in torsion

(a)

(b)

(c)

(d)

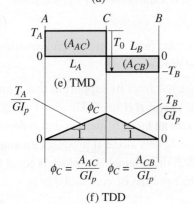

(e) TMD

(f) TDD

Torque–displacement equations: The angles of twist ϕ_1 and ϕ_2 can be expressed in terms of the torques T_0 and T_B by referring to Figs. 9-37c and d and using the equation $\phi = TL/GI_p$. The equations are

$$\phi_1 = \frac{T_0 L_A}{GI_{pA}} \qquad \phi_2 = -\frac{T_B L_A}{GI_{pA}} - \frac{T_B L_B}{GI_{pB}} \qquad \textbf{(c,d)}$$

The minus signs appear in Eq. (d) because T_I produces a rotation that is opposite in direction to the positive direction of ϕ_2 (Fig. 9-37d).

Now substitute the angles of twist [Eqs. (c) and (d)] into the compatibility equation [Eq. (b)] and obtain

$$\frac{T_0 L_A}{GI_{pA}} - \frac{T_B L_A}{GI_{pA}} - \frac{T_B L_B}{GI_{pB}} = 0$$

or

$$\frac{T_B L_A}{I_{pA}} + \frac{T_B L_B}{I_{pB}} = \frac{T_0 L_A}{I_{pA}} \qquad \textbf{(e)}$$

3. *Analyze*:

Solution of equations: The preceding equation can be solved for the torque T_B, which then can be substituted into the equation of equilibrium [Eq. (a)] to obtain the torque T_A. The results are

$$T_A = T_0 \left(\frac{L_B I_{pA}}{L_B I_{pA} + L_A I_{pB}} \right)$$

$$\qquad\qquad\qquad\qquad\qquad\qquad\qquad \longleftarrow \textbf{(9-49a,b)}$$

$$T_B = T_0 \left(\frac{L_A I_{pB}}{L_B I_{pA} + L_A I_{pB}} \right)$$

Thus, the reactive torques at the ends of the bar have been found, and the statically indeterminate part of the analysis is completed.

As a special case, note that if the bar is prismatic ($I_{pA} = I_{pB} = I_p$), the preceding results simplify to

$$T_A = \frac{T_0 L_B}{L} \qquad T_B = \frac{T_0 L_A}{L} \qquad \textbf{(9-50a,b)}$$

where L is the total length of the bar. These equations are analogous to those for the reactions of an axially loaded bar with fixed ends [see Eqs. (8-12a and b)]. The TMD for this special case is shown in Fig. 9-37e.

Maximum shear stresses: The maximum shear stresses in each part of the bar are obtained directly from the torsion formula:

$$\tau_{AC} = \frac{T_A d_A}{2I_{pA}} \qquad \tau_{CB} = \frac{T_B d_B}{2I_{pB}}$$

Substituting from Eqs. (9-49a and b) gives

$$\tau_{AC} = \frac{T_0 L_B d_A}{2(L_B I_{pA} + L_A I_{pB})} \qquad \tau_{CB} = \frac{T_0 L_A d_B}{2(L_B I_{pA} + L_A I_{pB})} \quad \text{(9-51a,b)}$$

By comparing the product $L_B d_A$ with the product $L_A d_B$, you can immediately determine which segment of the bar has the larger stress.

Angle of rotation: The angle of rotation ϕ_C at section C is equal to the angle of twist of either segment of the bar, since both segments rotate through the same angle at section C. Therefore, the angle of rotation is

$$\phi_C = \frac{T_A L_A}{G I_{pA}} = \frac{T_B L_B}{G I_{pB}} = \frac{T_0 L_A L_B}{G(L_B I_{pA} + L_A I_{pB})} \quad \text{(9-52)}$$

In the special case of a prismatic bar ($I_{pA} = I_{pB} = I_p$), the angle of rotation at the section where the load is applied is

$$\phi_C = \frac{T_0 L_A L_B}{G L I_p} \qquad \text{(9-53)}$$

The TDD for the case of a prismatic bar is shown in Fig. 9-37f.

4. *Finalize*: This example illustrates not only the analysis of a statically indeterminate bar but also the techniques for finding stresses and angles of rotation. In addition, note that the results obtained in this example are valid for a bar consisting of either solid or tubular segments.

9.9 Torsion of Noncircular Prismatic Shafts

Sections 9.1 through 9.8 of this chapter covered the twisting of circular shafts. Shafts with circular cross sections (either solid or hollow) do not *warp* when torsional moments are applied. Plane cross sections remain plane (as shown in Fig. 9-38), and shearing stresses and strains vary linearly with a distance ρ from the longitudinal axis of the shaft to the outer surface ($\rho = r$). Now consider prismatic shafts of length L acted upon by torsional moments T at either end but having *noncircular* cross sections. These cross sections could be solid (such as the elliptical, triangular, and rectangular shapes shown in Fig. 9-38), or they could be thin-walled *open* cross sections such as the I-beam, channel, and Z-shaped cross sections depicted in Fig. 9-39.

These noncircular cross sections *warp* under the action of torsional moments, and this warping alters the shear stress and strain distributions in the cross section. The simple *torsion formula* of Eq. 9-13 can no longer be used to compute shear stresses, and the *torque–displacement relation* of Eq. 9-17 cannot be used to find the angle of twist of the shaft. For example, warping

(a)   (b)

(c)

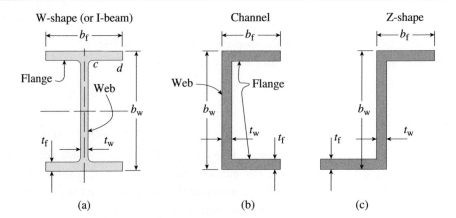

(a)   (b)   (c)

distortions of a rectangular bar of length L acted upon by torques T at either end are shown in Fig. 9-40a; the cross sections remain rectangular, but a grid on the surface of the bar is distorted as shown and $+/-x$ displacements represent the out-of-plane warping of the cross sections. The torsional shear stress distribution in the rectangular cross section is shown in Fig. 9-40b. The shear stresses at the corners are zero, and the maximum shear stress occurs at the midpoint of the longer side (point A in Figs. 9-38 and 9-40b). A more advanced theory, developed by Saint-Venant, is required to formulate expressions for torsion of shafts of noncircular cross section. Simple formulas for maximum shear stress and angle of twist for the cross-sectional shapes shown in Figs. 9-38 and 9-39 will be presented next and then used in calculations in Examples 9-10 and 9-11. However, derivation of these formulas is beyond the scope of this text; future coursework on the *theory of elasticity* and, perhaps, the *finite element method* will provide more detailed analysis.

Shear Stress Distribution and Angle of Twist

The following discussion presents only the basic relations between applied torsional moment T and three key items of interest for a variety of noncircular cross sections.

FIGURE 9-40

(a) Torsion of a bar of rectangular cross section and (b) shear stress distribution for a bar of rectangular cross section acted upon by torsional moment T

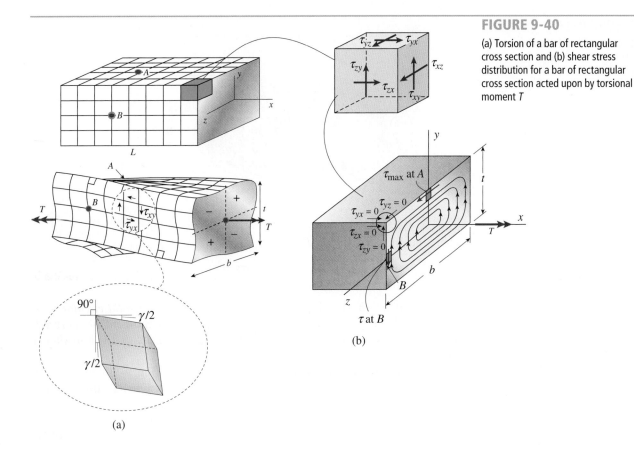

The three key items of interest are:

1. The location and value of the maximum shear stress τ_{max} in the cross section
2. The torsional rigidity GJ
3. The angle of twist ϕ of a prismatic bar of length L

Constant G is the shearing modulus of elasticity of the material, and variable J is the *torsion constant* for the cross section. Note that only for a circular cross section does torsion constant J become the polar moment of inertia I_p.

Elliptical, Triangular, and Rectangular Cross Sections

The shear stress distribution for a bar with an *elliptical cross section* ($2a$ along major axis, $2b$ along minor axis, area $A = \pi ab$) is shown in Fig. 9-41. The maximum shear stress is at the ends of the *minor axis* and may be computed using

$$\tau_{max} = \frac{2T}{\pi ab^2} \qquad \textbf{(9-54)}$$

where a is greater than or equal to b. The angle of twist ϕ of a prismatic shaft of length L with an elliptical cross section is expressed as

$$\phi = \frac{TL}{GJ_e}$$

FIGURE 9-41

Shear stress distribution in an
elliptical cross section

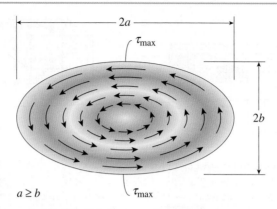

where the torsion constant J_e is

$$J_e = \frac{\pi a^3 b^3}{a^2 + b^2}$$ (9-55a,b)

Note that if $a = b$, the elliptical cross section becomes a *solid circular cross section*, and the expression for J_e becomes the polar moment of inertia I_p [Eq. (9-12)] and Eqs. (9-54) and (9-55a) reduce to Eqs. (9-13) and (9-17), respectively.

   Next consider an *equilateral triangular cross section* (Fig. 9-38b) for the shaft of length L acted upon by torques T at each end. Each side has dimension b_t, and the triangle height is h_t. The torsion constant J_t is

$$J_t = \frac{h_t^4}{15\sqrt{3}}$$ (9-56)

The maximum shear stress occurs on the surface at the *midpoint of each side* (points A in Fig. 9-38b). The maximum shear stress and the angle of twist ϕ of a prismatic shaft of length L with an *equilateral triangular cross section* are expressed as

$$\tau_{\text{max}} = \frac{T\left(\dfrac{h_t}{2}\right)}{J_t} = \frac{15\sqrt{3}T}{2h_t^3}$$ (9-57)

where

$$J_t = \frac{h_t^4}{15\sqrt{3}}$$

and

$$\phi = \frac{TL}{GJ_t} = \frac{15\sqrt{3}TL}{Gh_t^4}$$ (9-58)

Finally, consider a *rectangular cross section* ($b \times t$, $b/t \geq 1$) (see Figs. 9-38c and 9-40). Theory of elasticity solutions provide expressions for *maximum shear stress at point A* in the cross section and the angle of twist for a variety of aspect ratios b/t as

$$\tau_{\text{max}} = \frac{T}{k_1 b t^2}$$ (9-59)

$$\phi = \frac{TL}{(k_2 b t^3)G} = \frac{TL}{GJ_r}$$ (9-60)

b/t	1.00	1.50	1.75	2.00	2.50	3.00	4	6	8	10	∞	**Table 9-1**
k_1	0.208	0.231	0.239	0.246	0.258	0.267	0.282	0.298	0.307	0.312	0.333	Dimensionless coefficients for
k_2	0.141	0.196	0.214	0.229	0.249	0.263	0.281	0.298	0.307	0.312	0.333	rectangular bars

where

$$J_r = k_2 b t^3$$

and dimensionless coefficients k_1 and k_2 are listed in Table 9-1.

It is especially important to note that, for the elliptical, triangular, and rectangular sections considered here, *maximum shear stress does not occur at the largest distance from the axis of the shaft like it does for circular sections.* Instead, maximum shear strain and stress occur at the midpoints of the sides for each section. In fact, the shear stresses are zero in the corners of the triangular and rectangular sections (as indicated by the appearance of zero shear strain at the corners of the rectangular section in Fig. 9-40a, for example).

Thin-Walled Open Cross Sections: I-beam, Angle, Channel, and Z-shape

Metal structural shapes of open cross section (see Fig. 9-39) can be represented as assemblages of rectangles for purposes of computing their torsional properties and response to applied torsional moments. Torsion constants for typical structural steel shapes are tabulated in the American Institute of Steel Construction (AISC) manual and may be up to 10% higher than properties based on use of rectangles to represent flanges and web. Hence, maximum shear stress values and twist angles computed using the formulas presented here may be somewhat conservative.

The total torque is assumed to be equal to the sum of the torques carried by the flanges and web. First compute the flange b_f/t_f ratio (see Fig. 9-39 for cross-sectional dimensions). Then find constant k_2 from Table 9-1 (interpolation between values may be necessary). For the web, use the ratio $(b_w - 2t_f)/t_w$ in Table 9-1 to find a new constant k_2 for the web. The separate torsion constants for both flanges and the web are expressed as

$$J_f = k_2 b_f t_f^3 \quad J_w = k_2 (b_w - 2t_f)\left(t_w^3\right) \tag{9-61a,b}$$

The total torsion constant for the thin, open cross section is obtained (assuming two flanges) as

$$J = J_w + 2J_f \tag{9-62}$$

The maximum shear stress and angle of twist then can be computed as

$$\tau_{max} = \frac{2T\left(\dfrac{t}{2}\right)}{J} \quad \text{and} \quad \phi = \frac{TL}{GJ} \tag{9-63a,b}$$

where the *larger* of t_f and t_w is used in the formula for τ_{max}.

Examples 9-10 and 9-11 illustrate the application of these formulas to obtain maximum shear stress and angle of twist values for prismatic bars with noncircular cross sections, such as those presented in Figs. 9-38 and 9-39.

Example 9-10

A shaft with a length $L = 6$ ft is subjected to torques $T = 45$ kip-in. at either end (Fig. 9-42). Segment AB ($L_1 = 3$ ft) is made of brass ($G_b = 6000$ ksi) and has a square cross section ($a = 3$ in.). Segment BC ($L_2 = 3$ ft) is made of steel ($G_s = 10{,}800$ ksi) and has a circular cross section ($d = a = 3$ in.). Ignore stress concentrations near B.

(a) Find the maximum shear stress and angle of twist for each segment of the shaft.

(b) Find a new value for the dimension a of bar AB if the maximum shear stress in AB and BC are to be equal.

(c) Repeat part (b) if the angles of twist of segments AB and BC are to be equal.

(d) If dimension a is reset to $a = 3$ in. and bar BC is now a hollow pipe with an outer diameter $d_2 = a$, find the inner diameter d_1 so that the angles of twist of segments AB and BC are equal.

FIGURE 9-42

Example 9-10: Torsion of shaft with noncircular cross section

Solution:

Use the four-step problem-solving approach. Combine steps as needed for an efficient solution.

Part (a): Maximum shear stress and angles of twist for each segment.

1, 2. *Conceptualize, Categorize*: Both segments of the shaft have internal torque equal to the applied torque T. For square segment AB, obtain torsion coefficients k_1 and k_2 from Table 9-1.

3. *Analyze*: Use Eqs. (9-59) and (9-60) to compute the maximum shear stress and angle of twist as

$$\tau_{max1} = \frac{T}{k_1 bt^2} = \frac{T}{k_1 a^3} = \frac{(45 \text{ kip-in.})}{0.208(3 \text{ in.})^3} = 8.01 \text{ ksi} \quad \Leftarrow \text{(a)}$$

$$\phi_1 = \frac{TL_1}{(k_2 bt^3)G_b} = \frac{TL_1}{k_2 a^4 G_b} = \frac{(45 \text{ kip-in.})(36 \text{ in.})}{0.141(3 \text{ in.})^4 (6000 \text{ ksi})}$$
$$= 2.36 \times 10^{-2} \text{ radians} \quad \Leftarrow \text{(b)}$$

The maximum shear stress in AB occurs at the midpoint of each side of the square cross section.

Segment BC is a solid, circular cross section, so use Eqs. (9-14) and (9-17) to compute the maximum shear stress and angle of twist for segment BC:

$$\tau_{max2} = \frac{16T}{\pi d^3} = \frac{16(45 \text{ kip-in.})}{\pi(3 \text{ in.})^3} = 8.49 \text{ ksi} \quad \Leftarrow \text{(c)}$$

$$\phi_2 = \frac{TL_2}{G_s I_p} = \frac{(45 \text{ kip-in.})(36 \text{ in.})}{10{,}800 \text{ ksi}\left[\dfrac{\pi}{32}(3 \text{ in.})^4\right]} = 1.886 \times 10^{-2} \quad \Leftarrow \text{(d)}$$

4. *Finalize*: Compare the shear stress and angle of twist values for square segment *AB* and circular segment *BC*. Steel pipe *BC* has 6% greater maximum shear stress but 20% less twist rotation than the brass bar *AB*.

Part (b): New value for dimension *a* of bar *AB* so that maximum shear stresses in *AB* and *BC* are equal.

1, 2. *Conceptualize, Categorize*: Equate expressions for τ_{max1} and τ_{max2} in Eqs. (a) and (c) and solve for the required new value of dimension *a* of bar *AB*.

3. *Analyze*:

$$\tau_{max1} = \tau_{max2} \quad \text{so} \quad \frac{16}{\pi d^3} = \frac{1}{k_1 a_{new}^3} \quad \text{or} \quad a_{new} = \left(\frac{\pi d^3}{16 k_1}\right)^{\frac{1}{3}} = 2.94 \text{ in.} \quad \blacktriangleleft \text{(e)}$$

4. *Finalize*: The diameter of bar *BC* remains unchanged at $d = 3$ in., so a slight reduction in dimension *a* for bar *AB* leads to the same maximum shear stress of 8.49 ksi [Eq. (c)] in the two bar segments.

Part (c): New value for dimension *a* of bar *AB* so that twist rotations in *AB* and *BC* are equal.

1, 2. *Conceptualize, Categorize*: Now, equate expressions for ϕ_1 and ϕ_2 in Eqs. (b) and (d) and solve for the required new value of dimension *a* of bar *AB*.

3. *Analyze*:

$$\phi_1 = \phi_2$$

so

$$\frac{L_1}{k_2 a_{new}^4 G_b} = \frac{L_2}{G_s I_p} \tag{f}$$

or

$$a_{new} = \left[\frac{L_1}{L_2}\left(\frac{G_s I_p}{k_2 G_b}\right)\right]^{\frac{1}{4}} = 3.17 \text{ in.} \quad \blacktriangleleft$$

4. *Finalize*: The diameter of bar *BC* remains unchanged at $d = 3$ in., so a slight increase in dimension *a* for brass bar *AB* leads to the same twist rotation of 0.01886 radians, as in Eq. (d) in each of the two bar segments.

Part (d): Change segment *BC* to hollow pipe; find inner diameter d_1 so that twist rotations in *AB* and *BC* are equal.

1, 2. *Conceptualize, Categorize*: Side dimension *a* of square segment *AB* is equal to 3 in., and outer diameter $d_2 = 3$ in. (Fig. 9-43). Using Eq. (9-19) for the polar moment of inertia of segment *BC*, twist angle ϕ_2 is

$$\phi_2 = \frac{TL_2}{G_s\left[\frac{\pi}{32}\left(d_2^4 - d_1^4\right)\right]} \tag{g}$$

FIGURE 9-43

Hollow pipe cross section for segment *BC*

3. *Analyze*: Once again, equate expressions for ϕ_1 and ϕ_2 but now use Eqs. (b) and (g). Solve for d_1 to get

$$d_1 = \left[d_2^4 - 32 \left(\frac{L_2}{L_1} \right) \left(\frac{G_b}{G_s} \right) \left(\frac{a^4 k_2}{\pi} \right) \right]^{\frac{1}{4}}$$

$$= \left[(3 \text{ in.})^4 - 32 \left(\frac{3 \text{ ft}}{3 \text{ ft}} \right) \left(\frac{6000 \text{ ksi}}{10,800 \text{ ksi}} \right) \left[\frac{(3 \text{ in.})^4 (0.141)}{\pi} \right] \right]^{\frac{1}{4}}$$ **(h)**

$$= 2.01 \text{ in.}$$

4. *Finalize*: So the square, solid brass pipe AB ($a \times a$, $a = 3$ in.) and hollow steel pipe BC ($d_2 = 3$ in., $d_1 = 2.01$ in.) are each 3 ft in length and have the same twist rotation (0.0236 radians) due to applied torque T. However, additional calculations will show that the maximum shear stress in segment BC is now increased from 8.49 ksi [Eq. (c)] to 10.64 ksi by using a hollow rather than solid bar for BC.

Note that by deriving the formula for inner diameter d_1 in Eq. (h) (rather than finding a numerical solution alone), you can also investigate other solutions of possible interest using different values of the key variables. For example, if bar AB is increased in length to $L_1 = 3.5$ ft, inner diameter d_1 for BC can be increased to 2.43 in., and the angles of twist for AB and BC will be the same.

Example 9-11

FIGURE 9-44

Example 9-11: W-shape and angle steel sections in torsion

A steel angle, L178 × 102 × 19, and a steel wide-flange beam, W 360 × 39, each of length $L = 3.5$ m, are subjected to torque T (see Fig. 9-44). The allowable shear stress is 45 MPa, and the maximum permissible twist rotation is 5°. Find the value of the maximum torque T than can be applied to each section. Assume that $G = 80$ GPa and ignore stress concentration effects. [See Tables C-1(b) and C-5(b) for cross-sectional properties and dimensions.]

Solution:

Use the four-step problem-solving approach. Combine steps as needed for an efficient solution.

The angle and wide-flange steel shapes have the same cross-sectional area [$A = 4960$ mm²; see Tables C-1(b) and C-5(b)] but the thicknesses of flange and web components of each section are quite different. First, consider the angle section.

Part (a): Steel angle section.

1,2. *Conceptualize, Categorize*: Approximate the unequal leg angle as one long rectangle with length $b_L = 280$ mm and constant thickness $t_L = 19$ mm, so $b_L / t_L = 14.7$. From Table 9-1, estimate coefficients $k_1 = k_2$ to be approximately 0.319.

3. *Analyze*: The maximum allowable torques can be obtained from Eqs. (9-59) and (9-60) based on the given allowable shear stress and allowable twist rotation, respectively, as

$$T_{\max 1} = \tau_a k_1 b_L t_L^2 = 45 \text{ MPa } (0.319)(280 \text{ mm})[(19 \text{ mm})^2] = 1451 \text{ N} \cdot \text{m} \qquad \textbf{(a)}$$

$$T_{\max 2} = \phi_a(k_2 b_L t_L^3)\frac{G}{L} = \left(\frac{5\pi}{180} \text{ rad}\right)(0.319)(280 \text{ mm})[(19 \text{ mm})^3]\frac{80 \text{ GPa}}{3500 \text{ mm}}$$

$$= 1222 \text{ N} \cdot \text{m} \qquad\qquad\qquad \longleftarrow \textbf{(b)}$$

Alternatively, compute the torsion constant for the angle J_L as

$$J_L = k_2 b_L t_L^3 = 6.128 \times 10^5 \text{ mm}^4 \qquad \textbf{(c)}$$

then use Eqs. (9-61) and (9-63) to find the maximum allowable torque values. From Eq. (9-63a), find $T_{\max 1}$, and from Eq. (9-63b), obtain $T_{\max 2}$:

$$T_{\max 1} = \frac{\tau_a J_L}{t_L} = 1451 \text{ N} \cdot \text{m} \quad \text{and} \quad T_{\max 2} = \frac{G J_L}{L}\phi_a = 1222 \text{ N} \cdot \text{m}$$

4. *Finalize*: For the angle, the lesser value controls, so $T_{\max} = 1222$ N·m.

Part (b): Steel W shape.

1,2. *Conceptualize, Categorize*: The two flanges and the web are separate rectangles that together resist the applied torsional moment. However, the dimensions (b, t) of each of these rectangles are different: for a W360 × 39, each flange has a width of $b_f = 128$ mm and a thickness of $t_f = 10.7$ mm [see Table C-1(b)]. The web has thickness $t_w = 6.48$ mm [Table C-1(b)] and, conservatively, $b_w = (d_w - 2t_f) = (353 \text{ mm} - 2(10.7 \text{ mm})) = 331.6$ mm. Based on the b/t ratios, find separate coefficients k_2 for the flanges and web from Table 9-1, then compute the torsion constants J for each component using Eqs. (9-61) as

For the flanges:

$$\frac{b_f}{t_f} = 11.963$$

so an estimated value for $k_{2f} = 0.316$. Thus,

$$J_f = k_{2f} b_f t_f^3 = 0.316(128 \text{ mm})[(10.7 \text{ mm})^3] = 4.955 \times 10^4 \text{ mm}^4 \qquad \textbf{(d)}$$

For the web:

$$\frac{d_w - 2t_f}{t_w} = 51.173$$

and k_{2w} is estimated as $k_{2w} = 0.329$, so

$$J_w = k_{2w}(d_w - 2t_f)(t_w^3) = 0.329[353\ \text{mm} - 2(10.7\ \text{mm})][(6.48\ \text{mm})^3]$$
$$= 2.968 \times 10^4\ \text{mm}^4 \tag{e}$$

The torsion constant for the entire W360 × 39 section is obtained by adding web and flange contributions [Eqs. (d) and (e)]:

$$J_W = 2J_f + J_w = [2(4.955) + 2.968](10^4)\ \text{mm}^4 = 1.288 \times 10^5\ \text{mm}^4 \tag{f}$$

3. *Analyze*: Now, use Eq. (9-63a) and the allowable shear stress τ_a to compute the maximum allowable torque based on both flange and web maximum shear stresses:

$$T_{\max f} = \tau_a \frac{J_W}{t_f} = 45\ \text{MPa}\left(\frac{1.288 \times 10^5\ \text{mm}^4}{10.7\ \text{mm}}\right) = 542\ \text{N} \cdot \text{m} \tag{g}$$

$$T_{\max w} = \tau_a \frac{J_W}{t_w} = 45\ \text{MPa}\left(\frac{1.288 \times 10^5\ \text{mm}^4}{6.48\ \text{mm}}\right) = 894\ \text{N} \cdot \text{m} \tag{h}$$

Note that since the flanges have greater thickness than the web, the maximum shear stress will be in the flanges. So a calculation of $T_{\max}$ based on the maximum web shear stress using Eq. (h) is not necessary.

Finally, use Eq. (9-63b) to compute $T_{\max}$ based on the allowable angle of twist:

$$T_{\max \phi} = \frac{GJ_W}{L}\phi_a = \frac{80\ \text{GPa}(1.288 \times 10^5\ \text{mm}^4)}{3500\ \text{mm}}\left(\frac{5\pi}{180}\ \text{rad}\right)$$
$$= 257\ \text{N} \cdot \text{m} \qquad \Longleftarrow \text{(i)}$$

4. *Finalize*: For the W shape, the most restrictive requirement is the allowable twist rotation, so $T_{\max} = 257\ \text{N} \cdot \text{m}$ governs [Eq. (i)].

It is interesting to note that, even though both angle and W shapes have the same cross-sectional area, the W shape is considerably weaker in torsion, because its component rectangles are much thinner ($t_w = 6.48\ \text{mm}$, $t_f = 10.7\ \text{mm}$) than the angle section ($t_L = 19\ \text{mm}$). However, Chapter 10 shows that, although weak in torsion, the W shape has a considerable advantage in resisting bending and *transverse* shear stresses.

Chapter 9 discussed the behavior of bars and hollow tubes acted on by concentrated torques or distributed torsional moments as well as prestrain effects. Torque-displacement relations were derived for use in computing angles of twist of bars under both uniform and nonuniform conditions. Then, equilibrium and compatibility equations were developed for statically indeterminate structures in a superposition procedure leading to solution for all unknown torques, rotational displacements and, stresses. Equations were developed for normal and shear stresses on inclined sections. Here are the major concepts presented in this chapter.

1. For circular bars and tubes, the **shearing stress** (τ) and **strain** (γ) vary linearly with radial distance from the center of the cross-section.

$$\tau = (\rho/r)\tau_{\max} \quad \gamma = (\rho/r)\gamma_{\max}$$

2. The **torsion formula** defines the relation between shear stress and torsional moment. Maximum shear stress $\tau_{\max}$ occurs on the outer surface of the bar or tube and depends on torsional moment T, radial distance r, and polar moment of inertia I_p for circular cross sections. Thin-walled tubes are more efficient in torsion because the available material is more uniformly stressed than solid circular bars.

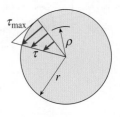

$$\tau_{\max} = \frac{Tr}{I_p}$$

3. The angle of twist ϕ of prismatic circular bars subjected to torsional moment(s) is proportional to both the torque T and the length of the bar L, and it is inversely proportional to the torsional rigidity (GI_p) of the bar; this relationship is called the **torque–displacement relation**.

$$\phi = \frac{TL}{GI_p}$$

4. The angle of twist per unit length of a bar is referred to as its **torsional flexibility** (f_T), and the inverse relationship is the **torsional stiffness** ($k_T = 1/f_T$) of the bar or shaft.

$$k_T = \frac{GI_p}{L} \qquad f_T = \frac{L}{GI_p}$$

5. The summation of the twisting deformations of the individual segments of a nonprismatic shaft equals the twist of the entire bar (ϕ). Free-body diagrams are used to find the torsional moments (T_i) in each segment i.

$$\phi = \sum_{i=1}^{n} \phi_i = \sum_{i=1}^{n} \frac{T_i L_i}{G_i (I_p)_i}$$

If torsional moments and/or cross sectional properties (I_p) vary continuously, an integral expression is required.

$$\phi = \int_0^L d\phi = \int_0^L \frac{T(x)dx}{GI_p(x)}$$

A graphical display of the internal torsional moments T_i is called a **torsional-moment diagram** (TMD). The TMD can be used along with the *torque–displacement relation* to create the displaced shape of the shaft referred to as the **torsional-displacement diagram** (TDD). These diagrams are useful visual aids that help to identify locations of maximum moment and shear stress, twist displacement, and shear strain.

6. If the bar structure is **statically indeterminate**, additional equations are required to solve for unknown moments. **Compatibility equations** are used to relate bar rotations to support conditions and thereby generate additional relationships among the unknowns. It is convenient to use a **superposition** of "released" (or statically determinate) structures to represent the actual statically indeterminate bar structure.

7. **Misfits** and **prestrains** induce torsional moments only in statically indeterminate bars or shafts.

8. A circular shaft is subjected to **pure shear** due to torsional moments. **Maximum normal** and **shear stresses** are obtained by considering an inclined stress element. The maximum shear stress occurs on an element aligned with the axis of the bar, but the maximum normal stress occurs at an inclination of 45° to the axis of the bar, and the maximum normal stress is equal to the maximum shear stress

$$\sigma_{max} = \tau$$

The relationship between the maximum shear and normal strains for the case of pure shear is

$$\varepsilon_{max} = \gamma_{max}/2$$

9. Circular shafts are commonly used to transmit mechanical power from one device or machine to another. If the torque T is expressed in newton meters and n is the shaft rpm, the power P is expressed in watts as

$$P = \frac{2\pi n T}{60}$$

In U.S. customary units, torque T is given in ft-lb and power is given in horsepower (hp), H, as

$$H = \frac{2\pi n T}{33,000}$$

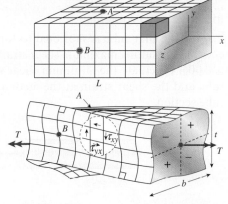

10. The torsion formula and torque–displacement relation do not apply to prismatic **noncircular shafts.** Shafts with elliptical, triangular, and rectangular cross sections *warp* under the action of torsional moments, and the *maximum shear stress does not occur at the largest distance from the axis of the shaft like it does for circular sections.*

PROBLEMS

9.2 Torsional Deformations of a Circular Bar

Introductory Problems

9.2-1 A circular tube is subjected to torque T at its ends. The resulting maximum shear strain in the tube is 0.005. Calculate the minimum shear strain in the tube and the shear strain at the median line of the tube section.

Median line

2.5 in.

3 in.

PROBLEM 9.2-1

9.2-2 A plastic bar of diameter $d = 56$ mm is to be twisted by torques T (see figure) until the angle of rotation between the ends of the bar is 4.0°.

(a) If the allowable shear strain in the plastic is 0.012 rad, what is the minimum permissible length of the bar?

(b) If the length of the bar is 200 mm, what is the maximum permissible diameter of the bar?

9.2-3 A copper rod of length $L = 18.0$ in. is to be twisted by torques T (see figure) until the angle of rotation between the ends of the rod is 3.0°.

(a) If the allowable shear strain in the copper is 0.0006 rad, what is the maximum permissible diameter of the rod?

(b) If the rod diameter is 0.5 in., what is the minimum permissible length of the rod?

d

L

PROBLEMS 9.2-2 and 9.2-3

Representative Problems

9.2-4 A circular steel tube of length $L = 1.0$ m is loaded in torsion by torques T (see figure).

(a) If the inner radius of the tube is $r_1 = 45$ mm and the measured angle of twist between the ends is 0.5°, what is the shear strain γ_1 (in radians) at the inner surface?

(b) If the maximum allowable shear strain is 0.0004 rad and the angle of twist is to be kept at 0.45° by adjusting the torque T, what is the maximum permissible outer radius $(r_2)_{max}$?

9.2-5 Solve the preceding problem if the length $L = 56$ in., the inner radius $r_1 = 1.25$ in., the angle of twist is 0.5°, and the allowable shear strain is 0.0004 rad.

9.2-6 A circular aluminum tube subjected to pure torsion by torques T (see figure) has an outer radius r_2 equal to 1.5 times the inner radius r_1.

(a) If the maximum shear strain in the tube is measured as 400×10^{-6} rad, what is the shear strain γ_1 at the inner surface?

(b) If the maximum allowable rate of twist is 0.125 °/m and the maximum shear strain is to be kept at 400×10^{-6} rad by adjusting the torque T, what is the minimum required outer radius $(r_2)_{min}$?

L

r_2

r_1

PROBLEMS 9.2-4, 9.2-5, and 9.2-6

9.3 Circular Bars of Linearly Elastic Materials

Introductory Problems

9.3-1 A solid steel bar of circular cross section has diameter $d = 2.5$ in., $L = 60$ in., and shear modulus of elasticity $G = 11.5 \times 10^6$ psi. The bar is subjected to torques $T = 300$ lb-ft at the ends. Calculate the angle of twist between the ends. What is the maximum shear stress and the shear stress at a distance $r_A = 1.0$ in. measured from the center of the bar?

$d = 2.5$ in.

T　　　T

$L = 60$ in.

$r_A = 1$ in.

1.25 in.

PROBLEM 9.3-1

9.3-2 A solid copper bar of circular cross section has length $L = 1.25$ m and shear modulus of elasticity $G = 45$ GPa. The bar is designed to carry a 250 N · m torque acting at the ends. If the allowable shear stress is 30 MPa and the allowable angle of twist between the ends is 2.5°, what is the minimum required diameter?

T　　　T

$L = 1.25$ m

PROBLEM 9.3-2

9.3-3 Repeat Problem 9.3-1, but now use a circular tube with outer diameter $d_o = 2.5$ in. and inner diameter $d_i = 1.5$ in.

T　　　T

1.5 in.

2.5 in.

PROBLEM 9.3-3

9.3-4 A copper tube with circular cross section has length $L = 1.25$ m, thickness $t = 2$ mm, and shear modulus of elasticity $G = 45$ GPa. The bar is designed to carry a 300 N · m torque acting at the ends. If the allowable shear stress is 25 MPa and the allowable angle of twist between the ends is 2.5°, what is the minimum required outer diameter d?

T　　　T

t

d

PROBLEM 9.3-4

Representative Problems

9.3-5 A prospector uses a hand-powered winch (see figure) to raise a bucket of ore in his mine shaft. The axle of the winch is a steel rod of diameter $d = 0.625$ in. Also, the distance from the center of the axle to the center of the lifting rope is $b = 4.0$ in.

(a) If the weight of the loaded bucket is $W = 100$ lb, what is the maximum shear stress in the axle due to torsion?

(b) If the maximum bucket load is 125 lb and the allowable shear stress in the axle is 9250 psi, what is the minimum permissible axle diameter?

P

W

d

b

W

PROBLEM 9.3-5

9.3-6 When drilling a hole in a table leg, a furniture maker uses a hand-operated drill (see figure) with a bit of diameter $d = 4.0$ mm.

(a) If the resisting torque supplied by the table leg is equal to 0.3 N · m, what is the maximum shear stress in the drill bit?

(b) If the allowable shear stress in the drill bit is 32 MPa, what is the maximum resisting torque before the drill binds up?

(c) If the shear modulus of elasticity of the steel is $G = 75$ GPa, what is the rate of twist of the drill bit (degrees per meter)?

PROBLEM 9.3-6

9.3-7 While removing a wheel to change a tire, a driver applies forces $P = 25$ lb at the ends of two of the arms of a lug wrench (see figure). The wrench is made of steel with shear modulus of elasticity $G = 11.4 \times 10^6$ psi. Each arm of the wrench is 9.0 in. long and has a solid circular cross section of diameter $d = 0.5$ in.

PROBLEM 9.3-7

698

(a) Determine the maximum shear stress in the arm that is turning the lug nut (arm A).

(b) Determine the angle of twist (in degrees) of this same arm.

9.3-8 An aluminum bar of solid circular cross section is twisted by torques T acting at the ends (see figure). The dimensions and shear modulus of elasticity are $L = 1.4$ m, $d = 32$ mm, and $G = 28$ GPa.

(a) Determine the torsional stiffness of the bar.

(b) If the angle of twist of the bar is 5°, what is the maximum shear stress? What is the maximum shear strain (in radians)?

(c) If a hole of diameter $d/2$ is drilled longitudinally through the bar, what is the ratio of the torsional stiffnesses of the hollow and solid bars? What is the ratio of their maximum shear stresses if both are acted on by the same torque?

(d) If the hole diameter remains at $d/2$, what new outside diameter d_2 will result in equal stiffnesses of the hollow and solid bars?

PROBLEM 9.3-8

9.3-9 A high-strength steel drill rod used for boring a hole in the earth has a diameter of 0.5 in. (see figure). The allowable shear stress in the steel is 40 ksi and the shear modulus of elasticity is 11,600 ksi.

(a) What is the minimum required length of the rod so that one end of the rod can be twisted 30° with respect to the other end without exceeding the allowable stress?

(b) If the shear strain in part (a) is limited to 3.2×10^{-3}, what is the minimum required length of the drill rod?

PROBLEM 9.3-9

9.3-10 The steel shaft of a socket wrench has a diameter of 8.0 mm and a length of 200 mm (see figure).

If the allowable stress in shear is 60 MPa, what is the maximum permissible torque T_{max} that may be exerted with the wrench?

Through what angle ϕ (in degrees) will the shaft twist under the action of the maximum torque? (Assume $G = 78$ GPa and disregard any bending of the shaft.)

PROBLEM 9.3-10

9.3-11 A circular tube of aluminum is subjected to torsion by torques T applied at the ends (see figure). The bar is 24 in. long, and the inside and outside diameters are 1.25 in. and 1.75 in., respectively. It is determined by measurement that the angle of twist is 4° when the torque is 6200 lb-in.

(a) Calculate the maximum shear stress τ_{max} in the tube, the shear modulus of elasticity G, and the maximum shear strain γ_{max} (in radians).

(b) If the maximum shear strain in the tube is limited to 2.5×10^{-3} and the inside diameter is increased to 1.375 in., what is the maximum permissible torque?

PROBLEM 9.3-11

9.3-12 A propeller shaft for a small yacht is made of a solid steel bar 104 mm in diameter. The allowable stress in shear is 48 MPa, and the allowable rate of twist is 2.0° in 3.5 meters.

(a) Assuming that the shear modulus of elasticity is $G = 80$ GPa, determine the maximum torque T_{max} that can be applied to the shaft.

(b) Repeat part (a) if the shaft is now hollow with an inner diameter of $5d/8$. Compare T_{max} values to corresponding values from part (a).

PROBLEM 9.3-12

9.3-13 Three identical circular disks A, B, and C are welded to the ends of three identical solid circular bars (see figure). The bars lie in a common plane and the disks lie in planes perpendicular to the axes of the bars. The bars are welded at their intersection D to form a rigid connection. Each bar has diameter $d_1 = 0.5$ in. and each disk has diameter $d_2 = 3.0$ in.

Forces P_1, P_2, and P_3 act on disks A, B, and C, respectively, thus subjecting the bars to torsion. If $P_1 = 28$ lb, what is the maximum shear stress τ_{max} in any of the three bars?

PROBLEM 9.3-13

9.3-14 The steel axle of a large winch on an ocean liner is subjected to a torque of 1.65 kN · m (see figure).

(a) What is the minimum required diameter d_{min} if the allowable shear stress is 48 MPa and the allowable rate of twist is 0.75°/m? (Assume that the shear modulus of elasticity is 80 GPa.)

(b) Repeat part (a) if the shaft is now hollow with an inner diameter of $5d/8$. Compare d_{min} values to corresponding values from part (a).

PROBLEM 9.3-14

699

9.3-15 A hollow steel shaft used in a construction auger has an outer diameter $d_2 = 6.0$ in. and inner diameter $d_1 = 4.5$ in. (see figure). The steel has a shear modulus of elasticity $G = 11.0 \times 10^6$ psi.

For an applied torque of 150 kip-in., determine the following quantities:

(a) Shear stress τ_2 at the outer surface of the shaft

(b) Shear stress τ_1 at the inner surface

(c) Rate of twist θ (degrees per unit of length)

Also, draw a diagram showing how the shear stresses vary in magnitude along a radial line in the cross section.

PROBLEMS 9.3-15 and 9.3-16

9.3-16 Solve the preceding problem if the shaft has an outer diameter $d_2 = 150$ mm and inner diameter $d_1 = 100$ mm. Also, the steel has a shear modulus of elasticity $G = 75$ GPa, and the applied torque is 16 kN · m.

9.3-17 A vertical pole of solid, circular cross section is twisted by horizontal forces $P = 1100$ lb acting at the ends of a rigid horizontal arm AB (see figure part a). The distance from the outside of the pole to the line of action of each force is $c = 5.0$ in. (see figure part b) and the pole height is $L = 14$ in.

(a) If the allowable shear stress in the pole is 4500 psi, what is the minimum required diameter d_{min} of the pole?

(b) Find the torsional stiffness of the pole (kip-in./rad). Assume that $G = 10{,}800$ ksi.

(c) If two translational springs, each with stiffness $k = 33$ kips/in., are added at $2c/5$ from A and B (see figure part c), repeat part (a) to find d_{min}. *Hint:* Consider the pole and pair of springs as "springs in parallel."

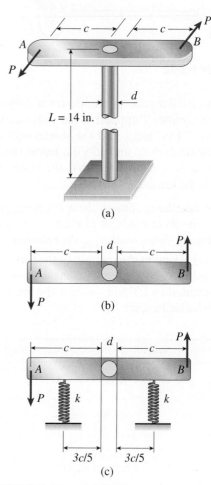

(a)

(b)

(c)

PROBLEMS 9.3-17 and 9.3-18

9.3-18 A vertical pole of solid, circular cross section is twisted by horizontal forces $P = 5$ kN acting at the ends of a rigid horizontal arm AB (see figure part a). The distance from the outside of the pole to the line of action of each force is $c = 125$ mm (see figure part b) and the pole height $L = 350$ mm.

(a) If the allowable shear stress in the pole is 30 MPa, what is the minimum required diameter d_{min} of the pole?

(b) What is the torsional stiffness of the pole (kN · m/rad)? Assume that $G = 28$ GPa.

(c) If two translational springs, each with stiffness $k = 2550$ kN/m, are added at $2c/5$ from A and B (see figure part c), repeat part (a) to find d_{min}. *Hint*: Consider the pole and pair of springs as "springs in parallel."

9.3-19 A solid brass bar of diameter $d = 1.25$ in. is subjected to torques T_1, as shown in part a of the figure. The allowable shear stress in the brass is 12 ksi.

(a) What is the maximum permissible value of the torques T_1?

(b) If a hole of diameter 0.625 in. is drilled longitudinally through the bar, as shown in part b of the figure, what is the maximum permissible value of the torques T_2?

(c) What is the percent decrease in torque and the percent decrease in weight due to the hole?

(a)

(b)

PROBLEM 9.3-19

9.3-20 A hollow aluminum tube used in a roof structure has an outside diameter $d_2 = 104$ mm and an inside $d_1 = 82$ mm (see figure). The tube is 2.75 m long, and the aluminum has shear modulus $G = 28$ GPa.

(a) If the tube is twisted in pure torsion by torques acting at the ends, what is the angle of twist (in degrees) when the maximum shear stress is 48 MPa?

(b) What diameter d is required for a solid shaft (see figure) to resist the same torque with the same maximum stress?

(c) What is the ratio of the weight of the hollow tube to the weight of the solid shaft?

PROBLEM 9.3-20

9.3-21 A circular tube of inner radius r_1 and outer radius r_2 is subjected to a torque produced by forces $P = 900$ lb (see figure part a). The forces have their lines of action at a distance $b = 5.5$ in. from the outside of the tube.

(a) If the allowable shear stress in the tube is 6300 psi and the inner radius $r_1 = 1.2$ in., what is the minimum permissible outer radius r_2?

(b) If a torsional spring of stiffness $k_R = 450$ kip-in./rad is added at the end of the tube (see figure part b), what is the maximum value of forces P if the allowable shear stress is not to be exceeded? Assume that the tube has a length of $L = 18$ in., outer radius of $r_2 = 1.45$ in., and shear modulus $G = 10,800$ ksi. *Hint*: Consider the tube and torsional spring as "springs in parallel."

(a)

(b)

(c)

PROBLEM 9.3-21

701

9.4 Nonuniform Torsion

Introductory Problems

9.4-1 A stepped shaft ABC consisting of two solid circular segments is subjected to torques T_1 and T_2 acting in opposite directions, as shown in the figure. The larger segment of the shaft has a diameter of $d_1 = 2.25$ in. and length $L_1 = 30$ in.; the smaller segment has a diameter of $d_2 = 1.75$ in. and length of $L_2 = 20$ in. The material is steel with shear modulus $G = 11 \times 10^6$ psi, and the torques are $T_1 = 20,000$ lb-in. and $T_2 = 8000$ lb-in.

(a) Calculate the maximum shear stress τ_{max} in the shaft and the angle of twist ϕ_c (in degrees) at end C.

(b) If the maximum shear stress in BC must be the same as that in AB, what is the required diameter of segment BC? What is the resulting twist at end C?

PROBLEM 9.4-1

9.4-2 A circular tube of outer diameter $d_3 = 70$ mm and inner diameter $d_2 = 60$ mm is welded at the right-hand end to a fixed plate and at the left-hand end to a rigid end plate (see figure). A solid, circular bar with a diameter of $d_1 = 40$ mm is inside of, and concentric

PROBLEM 9.4-2

with, the tube. The bar passes through a hole in the fixed plate and is welded to the rigid end plate.

The bar is 1.0 m long and the tube is half as long as the bar. A torque $T = 1000$ N·m acts at end A of the bar. Also, both the bar and tube are made of an aluminum alloy with a shear modulus of elasticity $G = 27$ GPa.

(a) Determine the maximum shear stresses in both the bar and tube.

(b) Determine the angle of twist (in degrees) at end A of the bar.

9.4-3 A stepped shaft $ABCD$ consisting of solid circular segments is subjected to three torques, as shown in the figure. The torques have magnitudes of 12.5 kip-in., 9.8 kip-in., and 9.2 kip-in. The length of each segment is 25 in. and the diameters of the segments are 3.5 in., 2.75 in., and 2.5 in. The material is steel with shear modulus of elasticity $G = 11.6 \times 10^3$ ksi.

(a) Calculate the maximum shear stress τ_{max} in the shaft and the angle of twist ϕ_D (in degrees) at end D.

(b) If each segment must have the same shear stress, find the required diameter of each segment in part (a) so that all three segments have shear stress τ_{max} from part (a). What is the resulting angle of twist at D?

PROBLEM 9.4-3

9.4-4 A solid, circular bar ABC consists of two segments, as shown in the figure. One segment has a diameter of $d_1 = 56$ mm and length of $L_1 = 1.45$ m; the other segment has a diameter of $d_2 = 48$ mm and length of $L_2 = 1.2$ m.

What is the allowable torque T_{allow} if the shear stress is not to exceed 30 MPa and the angle of twist between the ends of the bar is not to exceed 1.25°? (Assume $G = 80$ GPa.)

PROBLEM 9.4-4

9.4-5 A hollow tube *ABCDE* constructed of monel metal is subjected to five torques acting in the directions shown in the figure. The magnitudes of the torques are $T_1 = 1000$ lb-in., $T_2 = T_4 = 500$ lb-in., and $T_3 = T_5 = 800$ lb-in. The tube has an outside diameter of $d_2 = 1.0$ in. The allowable shear stress is 12,000 psi and the allowable rate of twist is 2.0°/ft.

Determine the maximum permissible inside diameter d_1 of the tube.

$T_1 =$ $T_2 =$ $T_3 =$ $T_4 =$ $T_5 =$
1000 lb-in. 500 lb-in. 800 lb-in. 500 lb-in. 800 lb-in.

A B C D E

$d_2 = 1.0$ in.

PROBLEM 9.4-5

9.4-6 A shaft with a solid, circular cross section consisting of two segments is shown in part a of the figure. The left-hand segment has a diameter of 80 mm and length of 1.2 m; the right-hand segment has a diameter of 60 mm and length of 0.9 m.

Shown in part b of the figure is a hollow shaft made of the same material and having the same length. The thickness t of the hollow shaft is $d/10$, where d is the outer diameter. Both shafts are subjected to the same torque.

(a) If the hollow shaft is to have the same torsional stiffness as the solid shaft, what should be its outer diameter d?

(b) If torque T is applied at either end of both shafts and the hollow shaft is to have the same maximum shear stress as the solid shaft, what should be its outer diamteter d?

80 mm   60 mm

1.2 m   0.9 m

(a)

d $t = \dfrac{d}{10}$

2.1 m

(b)

PROBLEM 9.4-6

Representative Problems

9.4-7 A solid steel shaft *ABCDE* turns freely in bearings at points *A* and *E*. The shaft is driven by the gear at *C*, which applies a torque $T_2 = 325$ lb-ft. Gears at *B* and *D* are driven by the shaft and have resisting torques $T_1 = 200$ lb-ft and $T_3 = 125$ lb-ft, respectively. Segments *BC* and *CD* have lengths $L_{BC} = 20$ in. and $L_{CD} = 15$ in. and the shear modulus $G = 11,600$ ksi. Determine the minimum required diameter (*d*) of the shaft if the allowable shear stress $\tau_a = 6$ ksi. Also calculate the angle of twist between gears *B* and *D*.

T_1 T_2 T_3
        d

A
        B C D E

L_{BC} L_{CD}

PROBLEM 9.4-7

9.4-8 Two sections of steel drill pipe, joined by bolted flange plates at *B*, are being tested to assess the adequacy of both the pipes. In the test, the pipe structure is fixed at *A*, a concentrated torque of 500 kN · m is applied at $x = 0.5$ m, and uniformly distributed torque intensity $t_0 = 250$ kN · m/m is applied on pipe *BC*. Both pipes have the same inner diameter $d = 200$ mm. Pipe *AB* has thickness $t_{AB} = 15$ mm, while pipe *BC* has thickness $t_{BC} = 12$ mm. Find the maximum shear stress and maximum twist of the pipe and their locations along the pipe. Assume $G = 75$ GPa.

A 500 kN · m   250 kN · m/m

x
        2 m B C
0.5 m   1.5 m

PROBLEM 9.4-8

9.4-9 Four gears are attached to a circular shaft and transmit the torques shown in the figure. The allowable shear stress in the shaft is 10,000 psi.

(a) What is the required diameter d of the shaft if it has a solid cross section?

(b) What is the required outside diameter d if the shaft is hollow with an inside diameter of 1.0 in.?

8000 lb-in.

19,000 lb-in.

4000 lb-in.

7000 lb-in.

A

B

C

D

PROBLEM 9.4-9

9.4-10 A tapered bar AB with a solid circular cross section is twisted by torques T (see figure). The diameter of the bar varies linearly from d_A at the left-hand end to d_B at the right-hand end.

(a) Confirm that the angle of twist of the tapered bar is

$$\phi = \frac{32TL}{3\pi G(d_B - d_A)}\left(\frac{1}{d_A^3} - \frac{1}{d_B^3}\right)$$

(b) For what ratio d_B/d_A will the angle of twist of the tapered bar be one-half the angle of twist of a prismatic bar of diameter d_A? (The prismatic bar is made of the same material, has the same length, and is subjected to the same torque as the tapered bar.)

T A B T

L

d_A d_B

PROBLEMS 9.4-10, 9.4-11, and 9.4-12

9.4-11 A tapered bar AB with a solid circular cross section is twisted by torques $T = 36,000$ lb-in. (see figure). The diameter of the bar varies linearly from d_A at the left-hand end to d_B at the right-hand end. The bar has length $L = 4.0$ ft and is made of an aluminum alloy having shear modulus of elasticity $G = 3.9 \times 10^6$ psi. The allowable shear stress in the bar is 15,000 psi and the allowable angle of twist is 3.0°.

704

If the diameter at end B is 1.5 times the diameter at end A, what is the minimum required diameter d_A at end A?

9.4-12 The bar shown in the figure is tapered linearly from end A to end B and has a solid circular cross section. The diameter at the smaller end of the bar is $d_A = 25$ mm and the length is $L = 300$ mm. The bar is made of steel with shear modulus of elasticity $G = 82$ GPa.

If the torque $T = 180$ N $\cdot$ m and the allowable angle of twist is 0.3°, what is the minimum allowable diameter d_B at the larger end of the bar?

9.4-13 The nonprismatic, cantilever circular bar shown has an internal cylindrical hole from 0 to x, so the net polar moment of inertia of the cross section for segment 1 is $(7/8)I_p$. Torque T is applied at x and torque $T/2$ is applied at $x = L$. Assume that G is constant.

(a) Find the reaction moment R_1.

(b) Find internal torsional moments T_i in segments 1 and 2.

(c) Find x required to obtain twist at joint 3 of $\phi_3 = TL/GI_p$.

(d) What is the rotation at joint 2, ϕ_2?

(e) Draw the torsional moment (TMD: $T(x)$, $0 \leq x \leq L$) and displacement (TDD: $\phi(x), 0 \leq x \leq L$) diagrams.

Segment 1 — Segment 2

x

R_1

$\frac{7}{8}I_p$

I_p

$\frac{T}{2}$

T

1

2

3

x

$L - x$

PROBLEM 9.4-13

9.4-14 A uniformly tapered tube AB with a hollow circular cross section is shown in the figure. The tube has constant wall thickness t and length L. The average diameters at the ends are d_A and $d_B = 2d_A$. The polar moment of inertia may be represented by the approximate formula $I_p \approx \pi d^3 t/4$ [see Eq. (9-21)].

A B

T T

L

t

d_A

t

$d_B = 2d_A$

PROBLEM 9.4-14

Derive a formula for the angle of twist ϕ of the tube when it is subjected to torques T acting at the ends.

9.4-15 A uniformly tapered aluminum-alloy tube AB with a circular cross section and length L is shown in the figure. The outside diameters at the ends are d_A and $d_B = 2d_A$. A hollow section of length $L/2$ and constant thickness $t = d_A/10$ is cast into the tube and extends from B halfway toward A.

(a) Find the angle of twist ϕ of the tube when it is subjected to torques T acting at the ends. Use numerical values: $d_A = 2.5$ in., $L = 48$ in., $G = 3.9 \times 10^6$, and $T = 40,000$ in.-lb.

(b) Repeat part (a) if the hollow section has constant diameter d_A (see figure part b).

(a)

(b)

PROBLEM 9.4-15

9.4-16 For the *thin* nonprismatic steel pipe of constant thickness t and variable diameter d shown with applied torques at joints 2 and 3, determine the following.

(a) Find the reaction moment R_1.

(b) Find an expression for twist rotation ϕ_3 at joint 3. Assume that G is constant.

(c) Draw the torsional-moment diagram (TMD: $T(x), 0 \le x \le L$).

PROBLEM 9.4-16

9.4-17 A mountain-bike rider going uphill applies torque $T = Fd$ ($F = 15$ lb, $d = 4$ in.) to the end of the handlebars $ABCD$ (Fig. a) by pulling on the handlebar extenders DE (Fig. b). Consider the right half of the handlebar assembly only (assume the bars are fixed at the fork at A). Segments AB and CD are prismatic with lengths $L_1 = 2$ in. and $L_3 = 8.5$ in. and with outer diameters and thicknesses $d_{01} = 1.25$ in., $t_{01} = 0.125$ in. and $d_{03} = 0.87$ in., $t_{03} = 0.115$ in., respectively, as shown. Segment BC of length $L_2 = 1.2$ in. however, is tapered, and outer diameter and thickness vary linearly between dimensions at B and C.

Consider torsion effects only. Assume $G = 4000$ ksi is constant.

Derive an integral expression for the angle of twist ϕ_D of half of the handlebar tube when it is subjected to torque $T = Fd$ acting at the end. Evaluate ϕ_D for the given numerical values.

(a)

(b)

PROBLEM 9.4-17

9.4-18 A prismatic bar AB of length L and solid circular cross section (diameter d) is loaded by a distributed torque of constant intensity t per unit distance (see figure).

(a) Determine the maximum shear stress τ_{max} in the bar.

(b) Determine the angle of twist ϕ between the ends of the bar.

PROBLEM 9.4-18

9.4-19 A prismatic bar AB with a solid circular cross section (diameter d) is loaded by a distributed torque (see figure). The intensity of the torque, that is, the torque per unit distance, is denoted $t(x)$ and varies linearly from a maximum value t_A at end A to zero at end B. Also, the length of the bar is L and the shear modulus of elasticity of the material is G.

(a) Determine the maximum shear stress τ_{max} in the bar.

(b) Determine the angle of twist ϕ between the ends of the bar.

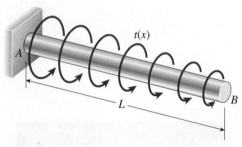

PROBLEM 9.4-19

9.4-20 A magnesium-alloy wire of diameter $d = 4\,mm$ and length L rotates inside a flexible tube in order to open or close a switch from a remote location (see figure). A torque T is applied manually (either clockwise or counterclockwise) at end B, thus twisting the wire inside the tube. At the other end A, the rotation of the wire operates a handle that opens or closes the switch.

A torque $T_0 = 0.2\,N \cdot m$ is required to operate the switch. The torsional stiffness of the tube, combined with friction between the tube and the wire, induces a distributed torque of constant intensity $t = 0.04\,N \cdot m/m$ (torque per unit distance) acting along the entire length of the wire.

(a) If the allowable shear stress in the wire is $\tau_{allow} = 30\,MPa$, what is the longest permissible length L_{max} of the wire?

(b) If the wire has length $L = 4.0\,m$ and the shear modulus of elasticity for the wire is $G = 15\,GPa$, what is the angle of twist ϕ (in degrees) between the ends of the wire?

PROBLEM 9.4-20

9.4-21 A nonprismatic bar ABC with a solid circular cross section is loaded by distributed torques (see figure). The intensity of the torques, that is, the torque per unit distance, is denoted $t(x)$ and varies linearly from zero at A to a maximum value T_0/L at B. Segment BC has linearly distributed torque of intensity $t(x) = T_0/3L$ of opposite sign to that applied along AB. Also, the polar moment of inertia of AB is twice that of BC, and the shear modulus of elasticity of the material is G.

(a) Find the reaction torque R_A.

(b) Find internal torsional moments $T(x)$ in segments AB and BC.

(c) Find the rotation ϕ_C.

(d) Find the maximum shear stress τ_{max} and its location along the bar.

(e) Draw the torsional-moment diagram (TMD $: T(x), 0 \le x \le L$).

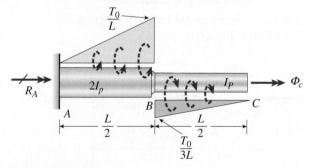

PROBLEM 9.4-21

9.4-22 Two tubes (AB, BC) of the same material are connected by three pins (pin diameter $= d_p$) just left of B as shown in the figure. Properties and dimensions for each tube are given in the figure. Torque $2T$ is applied at $x = 2L/5$ and uniformly distributed torque intensity $t_0 = 3T/L$ is applied on tube BC.

(a) Find the maximum value of load variable $T(\text{N·m})$ based on allowable shear (τ_a) and bearing (σ_{ba}) stresses in the three pins which connect the two tubes at B. Use the following numerical properties: $L = 1.5\,\text{m}$, $E = 74\,\text{GPa}$, $v = 0.33$, $d_p = 18\,\text{mm}$, $\tau_a = 45\,\text{MPa}$, $\sigma_{ba} = 90\,\text{MPa}$, $d_1 = 85\,\text{mm}$, $d_2 = 73\,\text{mm}$, and $d_3 = 60\,\text{mm}$.

(b) What is the maximum shear stress in the tubes for the applied torque in part (a)?

PROBLEM 9.4-22

9.5 Stresses and Strains in Pure Shear

Introductory Problems

9.5-1 A circular copper bar with diameter $d = 3$ in. is subjected to torques $T = 30$ kip-in. at its ends. Find the maximum shear, tensile, and compressive stresses in the tube and their corresponding strains. Assume that $G = 6000$ ksi.

PROBLEM 9.5-1

9.5-2 A circular steel tube with an outer diameter of 75 mm and inner diameter of 65 mm is subjected to torques T at its ends. Calculate the maximum permissible torque T_{max} if the allowable normal strain is $\varepsilon_a = 5 \times 10^{-4}$. Assume that $G = 75$ GPa.

PROBLEM 9.5-2

9.5-3 A hollow aluminum shaft (see figure) has an outside diameter $d_2 = 4.0$ in. and inside diameter $d_1 = 2.0$ in. When twisted by torques T, the shaft has an angle of twist per unit distance equal to $0.54°/\text{ft}$. The shear modulus of elasticity of the aluminum is $G = 4.0 \times 10^6$ psi.

(a) Determine the maximum tensile stress σ_{max} in the shaft.

(b) Determine the magnitude of the applied torques T.

PROBLEMS 9.5-3, 9.5-4, and 9.5-5

9.5-4 A hollow steel bar ($G = 80$ GPa) is twisted by torques T (see figure). The twisting of the bar produces a maximum shear strain $\gamma_{max} = 640 \times 10^{-6}$ rad. The bar has outside and inside diameters of 150 mm and 120 mm, respectively.

(a) Determine the maximum tensile strain in the bar.

(b) Determine the maximum tensile stress in the bar.

(c) What is the magnitude of the applied torques T?

9.5-5 A tubular bar with outside diameter $d_2 = 4.0$ in. is twisted by torques $T = 70.0$ kip-in. (see figure). Under the action of these torques, the maximum tensile stress in the bar is found to be 6400 psi.

(a) Determine the inside diameter d_1 of the bar.

(b) If the bar has length $L = 48.0$ in. and is made of aluminum with shear modulus $G = 4.0 \times 10^6$ psi, what is the angle of twist ϕ (in degrees) between the ends of the bar?

(c) Determine the maximum shear strain γ_{max} (in radians)?

Representative Problems

9.5-6 A solid circular bar of diameter $d = 50$ mm (see figure) is twisted in a testing machine until the applied torque reaches the value $T = 500$ N · m. At this value of torque, a strain gage oriented at 45° to the axis of the bar gives a reading $\varepsilon = 339 \times 10^{-6}$.

What is the shear modulus G of the material?

PROBLEM 9.5-6

9.5-7 A steel tube ($G = 11.5 \times 10^6$ psi) has an outer diameter $d_2 = 2.0$ in. and an inner diameter $d_1 = 1.5$ in. When twisted by a torque T, the tube develops a maximum normal strain of 170×10^{-6}.

What is the magnitude of the applied torque T?

9.5-8 A solid circular bar of steel ($G = 78$ GPa) transmits a torque $T = 360$ N · m. The allowable stresses in tension, compression, and shear are 90 MPa, 70 MPa, and 40 MPa, respectively. Also, the allowable tensile strain is 220×10^{-6}.

(a) Determine the minimum required diameter d of the bar.

(b) If the bar diameter $d = 40$ mm, what is T_{max}?

9.5-9 The normal strain in the 45° direction on the surface of a circular tube (see figure) is 880×10^{-6} when the torque $T = 750$ lb-in. The tube is made of copper alloy with $G = 6.2 \times 10^6$ psi and $\nu = 0.35$.

(a) If the outside diameter d_2 of the tube is 0.8 in., what is the inside diameter d_1?

(b) If the allowable normal stress in the tube is 14 ksi, what is the maximum permissible inside diameter d_1?

PROBLEM 9.5-9

9.5-10 An aluminum tube has inside diameter $d_1 = 50$ mm, shear modulus of elasticity $G = 27$ GPa, $\nu = 0.33$, and torque $T = 4.0$ kN · m. The allowable shear stress in the aluminum is 50 MPa, and the allowable normal strain is 900×10^{-6}.

(a) Determine the required outside diameter d_2.

(b) Re-compute the required outside diameter d_2 if allowable normal stress is 62 MPa and allowable shear strain is 1.7×10^{-3}.

9.5-11 A solid steel bar ($G = 11.8 \times 10^6$ psi) of diameter $d = 2.0$ in. is subjected to torques $T = 8.0$ kip-in. acting in the directions shown in the figure.

(a) Determine the maximum shear, tensile, and compressive stresses in the bar and show these stresses on sketches of properly oriented stress elements.

(b) Determine the corresponding maximum strains (shear, tensile, and compressive) in the bar and show these strains on sketches of the deformed elements.

PROBLEM 9.5-11

9.5-12 A solid aluminum bar ($G = 27$ GPa) of diameter $d = 40$ mm is subjected to torques $T = 300$ N · m acting in the directions shown in the figure.

(a) Determine the maximum shear, tensile, and compressive stresses in the bar and show these stresses on sketches of properly oriented stress elements.

(b) Determine the corresponding maximum strains (shear, tensile, and compressive) in the bar and show these strains on sketches of the deformed elements.

PROBLEM 9.5-12

9.5-13 Two circular aluminum pipes of equal length $L = 24$ in. are loaded by torsional moments T (see figure). Pipe 1 has outside and inside diameters $d_2 = 3$ in. and $d_1 = 2.5$ in., respectively. Pipe 2 has a constant outer diameter of d_2 along its entire length L and an inner diameter of d_1 but has an increased inner diameter of $d_3 = 2.65$ in. over the middle third.

Assume that $E = 10,400$ ksi, $v = 0.33$, and allowable shear stress $\tau_a = 6500$ psi.

(a) Find the maximum acceptable torques that can be applied to Pipe 1; repeat for Pipe 2.

(b) If the maximum twist ϕ of Pipe 2 cannot exceed 5/4 of that of Pipe 1, what is the maximum acceptable length of the middle segment? Assume both pipes have total length L and the same applied torque T.

(c) Find the new value of inner diameter d_3 of Pipe 2 if the maximum torque carried by Pipe 2 is to be 7/8 of that for Pipe 1.

(d) If the maximum normal strain in each pipe is known to be $\varepsilon_{max} = 811 \times 10^{-6}$, what is the applied torque on each pipe? Also, what is the maximum twist of each pipe? Use the original properties and dimensions.

(a)

(b)

PROBLEM 9.5-13

9.7 Transmission of Power by Circular Shafts

Introductory Problems

9.7-1 A generator shaft in a small hydroelectric plant turns at 120 rpm and delivers 50 hp (see figure).

(a) If the diameter of the shaft is $d = 3.0$ in., what is the maximum shear τ_{max} in the shaft?

(b) If the shear stress is limited to 4000 psi, what is the minimum permissible diameter d_{min} of the shaft?

PROBLEM 9.7-1

9.7-2 A motor drives a shaft at 12 Hz and delivers 20 kW of power (see figure).

(a) If the shaft has a diameter of 30 mm, what is the maximum shear stress τ_{max} in the shaft?

(b) If the maximum allowable shear stress is 40 MPa, what is the maximum permissible diameter d_{min} of the shaft?

PROBLEM 9.7-2

9.7-3 A motor driving a solid circular steel shaft with diameter $d = 1.5$ in. transmits 50 hp to a gear at B. The allowable shear stress in the steel is 6000 psi. Calculate the required speed of rotation (number of revolutions per minute) so that the shear stress in the shaft does not exceed the allowable limit.

PROBLEM 9.7-3

9.7-4 A solid steel shaft ABC with diameter $d = 40$ mm is driven at A by a motor that transmits 75 kW to the shaft at 15 Hz. The gears at B and C drive machinery requiring power equal to 50 kW and 25 kW, respectively. Compute the maximum shear stress and angle of twist in the shaft between the motor at A and the gear at C. Assume that $G = 75$ GPa.

PROBLEM 9.7-4

Representative Problems

9.7-5 The propeller shaft of a large ship has an outside diameter 18 in. and inside diameter 12 in., as shown in the figure. The shaft is rated for a maximum shear stress of 4500 psi.

(a) If the shaft is turning at 100 rpm, what is the maximum horsepower that can be transmitted without exceeding the allowable stress?

(b) If the rotational speed of the shaft is doubled but the power requirements remain unchanged, what happens to the shear stress in the shaft?

PROBLEM 9.7-5

9.7-6 The drive shaft for a truck (outer diameter 60 mm and inner diameter 40 mm) is running at 2500 rpm (see figure).

(a) If the shaft transmits 150 kW, what is the maximum shear stress in the shaft?

(b) If the allowable shear stress is 30 MPa, what is the maximum power that can be transmitted?

PROBLEM 9.7-6

9.7-7 A hollow circular shaft for use in a pumping station is being designed with an inside diameter equal to 0.75 times the outside diameter. The shaft must transmit 400 hp at 400 rpm without exceeding the allowable shear stress of 6000 psi.

Determine the minimum required outside diameter d.

9.7-8 A tubular shaft being designed for use on a construction site must transmit 120 kW at 1.75 Hz. The inside diameter of the shaft is to be one-half of the outside diameter.

If the allowable shear stress in the shaft is 45 MPa, what is the minimum required outside diameter d?

9.7-9 A propeller shaft of solid circular cross section and diameter d is spliced by a collar of the same material (see figure). The collar is securely bonded to both parts of the shaft.

What should be the minimum outer diameter d_1 of the collar in order that the splice can transmit the same power as the solid shaft?

PROBLEM 9.7-9

9.7-10 What is the maximum power that can be delivered by a hollow propeller shaft (outside diameter 50 mm, inside diameter 40 mm, and shear modulus of elasticity 80 GPa) turning at 600 rpm if the allowable shear stress is 100 MPa and the allowable rate of twist is 3.0°/m?

9.7-11 A motor delivers 275 hp at 1000 rpm to the end of a shaft (see figure). The gears at B and C take out 125 and 150 hp, respectively.

Determine the required diameter d of the shaft if the allowable shear stress is 7500 psi and the angle of twist between the motor and gear C is limited to 1.5°. (Assume $G = 11.5 \times 10^6$ psi, $L_1 = 6$ ft, and $L_2 = 4$ ft.)

PROBLEMS 9.7-11 and 9.7-12

9.7-12 The shaft ABC shown in the figure is driven by a motor that delivers 300 kW at a rotational speed of 32 Hz. The gears at B and C take out 120 kW and 180 kW, respectively. The lengths of the two parts of the shaft are $L_1 = 1.5$ m and $L_2 = 0.9$ m.

Determine the required diameter d of the shaft if the allowable shear stress is 50 MPa, the allowable angle of twist between points A and C is 4.0°, and $G = 75$ GPa.

9.8 Statically Indeterminate Torsional Members

Introductory Problems

9.8-1 A solid circular bar $ABCD$ with fixed supports is acted upon by torques T_0 and $2T_0$ at the locations shown in the figure.

(a) Obtain a formula for the maximum angle of twist ϕ_{max} of the bar.

(b) What is ϕ_{max} if the applied torque T_0 at B is reversed in direction?

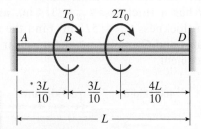

PROBLEM 9.8-1

9.8-2 A solid circular bar $ABCD$ with fixed supports at ends A and D is acted upon by two equal and oppositely directed torques T_0, as shown in the figure. The torques are applied at points B and C, each of which is located at distance x from one end of the bar. (The distance x may vary from zero to $L/2$.)

(a) For what distance x will the angle of twist at points B and C be a maximum?

(b) What is the corresponding angle of twist ϕ_{max}?

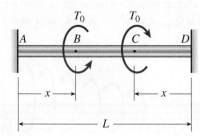

PROBLEM 9.8-2

9.8-3 A solid circular shaft AB of diameter d is fixed against rotation at both ends (see figure). A circular disk is attached to the shaft at the location shown.

What is the largest permissible angle of rotation ϕ_{max} of the disk if the allowable shear stress in the shaft is τ_{allow}? [Assume that $a > b$. Also, use Eqs. (9-50a and b) of Example 9-9 to obtain the reactive torques.]

PROBLEM 9.8-3

9.8-4 A hollow steel shaft ACB of outside diameter 50 mm and inside diameter 40 mm is held against rotation at ends A and B (see figure). Horizontal forces P are applied at the ends of a vertical arm that is welded to the shaft at point C.

Determine the allowable value of the forces P if the maximum permissible shear stress in the shaft is 45 MPa.

PROBLEM 9.8-4

9.8-5 A stepped shaft ACB having solid circular cross sections with two different diameters is held against rotation at the ends (see figure).

(a) If the allowable shear stress in the shaft is 6000 psi, what is the maximum torque $(T_0)_{max}$ that may be applied at section C?

(b) Find $(T_0)_{max}$ if the maximum angle of twist is limited to $0.55°$. Let $G = 10,600$ ksi.

PROBLEM 9.8-5

9.8-6 A stepped shaft ACB having solid circular cross sections with two different diameters is held against rotation at the ends (see figure).

(a) If the allowable shear stress in the shaft is 43 MPa, what is the maximum torque $(T_0)_{max}$ that may be applied at section C?

(b) Find $(T_0)_{max}$ if the maximum angle of twist is limited to 1.85°. Let $G = 28$ GPa.

PROBLEM 9.8-6

9.8-7 A stepped shaft ACB is held against rotation at ends A and B and subjected to a torque T_0 acting at section C (see figure). The two segments of the shaft (AC and CB) have diameters d_A and d_B, respectively, and polar moments of inertia I_{pA} and I_{pB}, respectively. The shaft has length L and segment AC has length a.

(a) For what ratio a/L will the maximum shear stresses be the same in both segments of the shaft?

(b) For what ratio a/L will the internal torques be the same in both segments of the shaft?

PROBLEM 9.8-7

Representative Problems

9.8-8 A solid circular aluminum bar AB is fixed at both ends and loaded by a uniformly distributed torque 150 N $\cdot$ m/m. The bar has diameter $d = 30$ mm. Calculate the reactive torques at the supports and the angle of twist at midspan. Assume that $G = 28$ GPa.

712

PROBLEM 9.8-8

9.8-9 Two sections of steel drill pipe, joined by bolted flange plates at B, are subjected to a concentrated torque 4000 kip-in. at $x = 3$ ft, and a uniformly distributed torque $t_0 = 50$ kip-ft/ft is applied on pipe BC. Let $G = 11,800$ ksi and assume that pipes AB and BC have the same inner diameter, $d = 12$ in. Pipe AB has a thickness $t_{AB} = 3/4$ in., and pipe BC has a thickness $t_{BC} = 5/8$ in. Find the reactive torques at A and C and the maximum shear stresses in each segment.

PROBLEM 9.8-9

9.8-10 A circular bar AB of length L is fixed against rotation at the ends and loaded by a distributed torque $t(x)$ that varies linearly in intensity from zero at end A to t_0 at end B (see figure).

(a) Obtain formulas for the fixed-end torques T_A and T_B.

(b) Find an expression for the angle of twist $\phi(x)$. What is ϕ_{max}, and where does it occur along the bar?

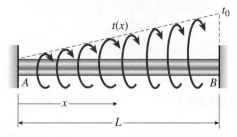

PROBLEM 9.8-10

9.8-11 A circular bar AB with ends fixed against rotation has a hole extending for half of its length (see figure). The outer diameter of the bar is $d_2 = 3.0$ in., and the diameter of the hole is $d_1 = 2.4$ in. The total length of the bar is $L = 50$ in.

(a) At what distance x from the left-hand end of the bar should a torque T_0 be applied so that the reactive torques at the supports will be equal?

(b) Based on the solution for x in part (a), what is ϕ_{max}, and where does it occur? Assume that $T_0 = 87.4$ kip-in. and $G = 10,600$ ksi.

PROBLEM 9.8-11

9.8-12 A solid steel bar of diameter $d_1 = 25.0$ mm is enclosed by a steel tube of outer diameter $d_3 = 37.5$ mm and inner diameter $d_2 = 30.0$ mm (see figure). Both bar and tube are held rigidly by a support at end A and joined securely to a rigid plate at end B. The composite bar, which has a length $L = 550$ mm, is twisted by a torque $T = 400$ N · m acting on the end plate.

(a) Determine the maximum shear stresses τ_1 and τ_2 in the bar and tube, respectively.

(b) Determine the angle of rotation ϕ (in degrees) of the end plate, assuming that the shear modulus of the steel is $G = 80$ GPa.

(c) Determine the torsional stiffness k_T of the composite bar.

PROBLEMS 9.8-12 and 9.8-13

9.8-13 A solid steel bar of diameter $d_1 = 1.50$ in. is enclosed by a steel tube of outer diameter $d_3 = 2.25$ in. and inner diameter $d_2 = 1.75$ in. (see figure). Both bar and tube are held rigidly by a support at end A and joined securely to a rigid plate at end B. The composite bar, which has length $L = 30.0$ in., is twisted by a torque $T = 5000$ lb-in. acting on the end plate.

(a) Determine the maximum shear stresses τ_1 and τ_2 in the bar and tube, respectively.

(b) Determine the angle of rotation ϕ (in degrees) of the end plate, assuming that the shear modulus of the steel is $G = 11.6 \times 10^6$ psi.

(c) Determine the torsional stiffness k_T of the composite bar.

9.8-14 The composite shaft shown in the figure is manufactured by shrink-fitting a steel sleeve over a brass core so that the two parts act as a single solid bar in torsion. The outer diameters of the two parts are $d_1 = 40$ mm for the brass core and $d_2 = 50$ mm for the steel sleeve. The shear moduli of elasticity are $G_b = 36$ GPa for the brass and $G_s = 80$ GPa for the steel.

(a) Assuming that the allowable shear stresses in the brass and steel are $\tau_b = 48$ MPa and $\tau_s = 80$ MPa, respectively, determine the maximum permissible torque T_{max} that may be applied to the shaft.

(b) If the applied torque $T = 2500$ kN · m, find the required diameter d_2 so that allowable shear stress τ_s is reached in the steel.

PROBLEMS 9.8-14 and 9.8-15

9.8-15 The composite shaft shown in the figure is manufactured by shrink-fitting a steel sleeve over a brass core so that the two parts act as a single solid bar in torsion. The outer diameters of the two parts are $d_1 = 1.6$ in. for the brass core and $d_2 = 2.0$ in. for the steel sleeve. The shear moduli of elasticity are $G_b = 5400$ ksi for the brass and $G_s = 12,000$ ksi for the steel.

(a) Assuming that the allowable shear stresses in the brass and steel are $\tau_b = 4500$ psi and $\tau_s = 7500$ psi, respectively, determine the maximum permissible torque T_{max} that may be applied to the shaft.

(b) If the applied torque $T = 15$ kip-in., find the required diameter d_2 so that allowable shear stress τ_s is reached in the steel.

9.8-16 A steel shaft ($G_s = 80$ GPa) of total length $L = 3.0$ m is encased for one-third of its length by a brass sleeve ($G_b = 40$ GPa) that is securely bonded to the steel (see figure). The outer diameters of the shaft and sleeve are $d_1 = 70$ mm and $d_2 = 90$ mm, respectively.

PROBLEM 9.8-16

(a) Determine the allowable torque T_1 that may be applied to the ends of the shaft if the angle of twist between the ends is limited to 8.0°.

(b) Determine the allowable torque T_2 if the shear stress in the brass is limited to $\tau_b = 70$ MPa.

(c) Determine the allowable torque T_3 if the shear stress in the steel is limited to $\tau_s = 110$ MPa.

(d) What is the maximum allowable torque T_{max} if all three of the preceding conditions must be satisfied?

9.8-17 A uniformly tapered aluminum-alloy tube AB of circular cross section and length L is fixed against rotation at A and B, as shown in the figure. The outside diameters at the ends are d_A and $d_B = 2d_A$. A hollow section of length $L/2$ and constant thickness $t = d_A/10$ is cast into the tube and extends from B halfway toward A. Torque T_0 is applied at $L/2$.

(a) Find the reactive torques at the supports T_A and T_B. Use numerical values as follows: $d_A = 2.5$ in., $L = 48$ in., $G = 3.9 \times 10^6$ psi, and $T_0 = 40,000$ in.-lb.

(b) Repeat part (a) if the hollow section has constant diameter d_A.

(a)

(b)

PROBLEM 9.8-17

9.8-18 Two pipes ($L_1 = 2.5$ m and $L_2 = 1.5$ m) are joined at B by flange plates (thickness $t_f = 14$ mm) with five bolts ($d_{bf} = 13$ mm) arranged in a circular pattern (see figure). Also, each pipe segment is attached to a wall (at A and C, see figure) using a base plate ($t_b = 15$ mm) and four bolts ($d_{bb} = 16$ mm). All bolts are tightened until just snug. Assume $E_1 = 110$ GPa, $E_2 = 73$ GPa, $v_1 = 0.33$, and $v_2 = 0.25$. Neglect the self-weight of the pipes, and assume the pipes are in a stress-free state initially. The cross-sectional areas of the pipes are $A_1 = 1500$ mm^2 and $A_2 = (3/5)A_1$. The outer diameter of Pipe 1 is 60 mm. The outer diameter of Pipe 2 is equal to the inner diameter of Pipe 1. The bolt radius $r = 64$ mm for both base and flange plates.

(a) If torque T is applied at $x = L_1$, find an *expression* for reactive torques R_1 and R_2 in terms of T.

(b) Find the maximum load variable T (i.e., T_{max}) if allowable torsional stress in the two pipes is $\tau_{allow} = 65$ MPa.

(c) Draw torsional-moment (TMD) and torsional-displacement (TDD) diagrams. Label all key ordinates. What is ϕ_{max}?

(d) Find T_{max} if allowable shear and bearing stresses in the base plate and flange bolts cannot be exceeded. Assume allowable stresses in shear and bearing for all bolts are $\tau_{allow} = 45$ MPa and $\sigma_{allow} = 90$ MPa.

(e) Remove torque T at $x = L_1$. Now assume the flange-plate bolt holes are misaligned by some angle β (see figure). Find the expressions for reactive torques R_1 and R_2 if the pipes are twisted to align the flange-plate bolt holes, bolts are then inserted, and the pipes released.

(f) What is the maximum permissible misalignment angle β_{max} if allowable stresses in shear and bearing for all bolts [from part (d)] are not to be exceeded?

A Base plate
(4 bolts)   Flange plate 2

x T

E_1 B E_2
                              C
d_{bb} d_{bf}

L_1 L_2

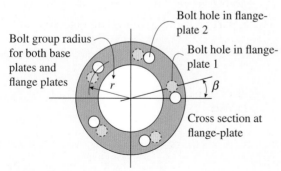

Bolt group radius
for both base
plates and
flange plates

Bolt hole in flange-plate 2

Bolt hole in flange-plate 1

r β

Cross section at flange-plate

PROBLEM 9.8-18

Stresses in Beams (Basic Topics)

Beams are essential load carrying components in a wide variety of modern structures.

CHAPTER OBJECTIVES

- Develop a relationship between moment and curvature for a beam loaded by transverse applied loads and bending moments.

- Define the *flexure formula*, which shows that *normal stresses* vary linearly over the depth of a beam and are proportional to the bending moment and inversely proportional to the moment of inertia of the cross section.

- Define the *section modulus* of a beam and use it to design beams made of steel, wood, or other materials based upon an allowable stress for the material.

- Investigate *shear stresses* in beams of different shapes and study the variation of shear stress over the depth of a beam using a shear formula.

- Develop a modified flexure formula for computation of normal stresses in beams made up of two different materials, referred to as *composite beams*.

- Develop a general procedure for analyzing an *unsymmetric beam* subjected to any bending moment *M* resolved into components along the principal centroidal axes of the cross section.

Chapter Outline

10.1 Introduction

In the preceding chapter, you saw how the loads acting on a beam create internal actions (or *stress resultants*) in the form of shear forces and bending moments. This chapter goes one step further and investigates the *stresses* and *strains* associated with those shear forces and bending moments. Knowing the stresses and strains, you will analyze and design beams subjected to a variety of loading conditions.

The loads acting on a beam cause the beam to bend (or *flex*), thereby deforming its axis into a curve. As an example, consider a cantilever beam AB subjected to a load P at the free end (Fig. 10-1a). The initially straight axis is bent into a curve (Fig. 10-1b), called the **deflection curve** of the beam.

For reference purposes, a system of **coordinate axes** (Fig. 10-1b) is constructed with the origin located at a suitable point on the longitudinal axis of the beam. In this illustration, the origin is placed at the fixed support. The positive x axis is directed to the right, and the positive y axis is directed upward. The z axis, not shown in the figure, is directed outward (that is, toward the viewer), so that the three axes form a right-handed coordinate system.

The beams considered in this chapter are assumed to be symmetric about the x-y plane, which means that the y axis is an axis of symmetry of the cross section. In addition, all loads must act in the x-y plane. As a consequence, the bending deflections occur in this same plane, known as the **plane of bending**. Thus, the deflection curve shown in Fig. 10-1b is a plane curve lying in the plane of bending.

The **deflection** of the beam at any point along its axis is the *displacement* of that point from its original position measured in the y direction. Denote the deflection by the letter v to distinguish it from the coordinate y itself (see Fig. 10-1b).[1]

FIGURE 10-1

Bending of a cantilever beam: (a) beam with load and (b) deflection curve

(a)

(b)

10.2 Pure Bending and Nonuniform Bending

When analyzing beams, it is often necessary to distinguish between pure bending and nonuniform bending. **Pure bending** refers to the flexure of a beam under a constant bending moment. Therefore, pure bending occurs only in regions of a beam where the shear force is zero (because $V = dM/dx$). In contrast, **nonuniform bending** refers to flexure in the presence of shear forces, which means that the bending moment changes as you move along the axis of the beam.

As an example of pure bending, consider a simple beam AB loaded by two couples M_1 having the same magnitude but acting in opposite directions (Fig. 10-2a). These loads produce a constant bending moment $M = M_1$ throughout the length of the beam, as shown by the bending moment diagram in (Fig. 10-2b). Note that the shear force V is zero at all cross sections of the beam.

Figure 10-3a shows pure bending, where the cantilever beam AB is subjected to a clockwise couple M_2 at the free end. There are no shear forces in this beam, and the bending moment M is constant throughout its length. The bending moment is negative $(M = -M_2)$, as shown by the bending moment diagram in Fig. 10-3b.

The symmetrically loaded simple beam of Fig. 10-4a is an example of a beam that is partly in pure bending and partly in nonuniform bending, as seen from the shear-force and bending-moment diagrams (Figs. 10-4b and c).

FIGURE 10-2

Simple beam in pure bending $(M = M_1)$

(a)

(b)

---

[1] In applied mechanics, the traditional symbols for displacements in the x, y, and z directions are u, v, and w, respectively.

The central region of the beam is in pure bending because the shear force is zero and the bending moment is constant. The parts of the beam near the ends are in nonuniform bending because shear forces are present and the bending moments vary.

In the following two sections, the strains and stresses in beams subjected only to pure bending are investigated. Fortunately, the results obtained for pure bending can be used even when shear forces are present, as explained in Section 10.7.

10.3 Curvature of a Beam

When loads are applied to a beam, its longitudinal axis is deformed into a curve, as illustrated previously in Fig. 10-1. The resulting strains and stresses in the beam are directly related to the **curvature** of the deflection curve.

To illustrate the concept of curvature, consider again a cantilever beam subjected to a load P acting at the free end (see Fig. 10-5a on the next page). The deflection curve of this beam is shown in Fig. 10-5b. For purposes of analysis, identify two points m_1 and m_2 on the deflection curve. Point m_1 is selected at an arbitrary distance x from the y axis, and point m_2 is located a small distance ds further along the curve. At each of these points, draw a line normal to the *tangent* to the deflection curve, that is, normal to the curve itself. These normals intersect at point O', which is the **center of curvature** of the deflection curve. Because most beams have very small deflections and nearly flat deflection curves, point O' is usually located much farther from the beam than is indicated in the figure.

The distance m_1O' from the curve to the center of curvature is called the **radius of curvature** ρ (rho), and the **curvature** κ (kappa) is defined as the reciprocal of the radius of curvature. Thus,

$$\kappa = \frac{1}{\rho} \tag{10-1}$$

Curvature is a measure of how sharply a beam is bent. If the load on a beam is small, the beam will be nearly straight, the radius of curvature will be very large, and the curvature will be very small. If the load is increased, the amount of bending will increase—the radius of curvature will become smaller, and the curvature will become larger.

The geometry of triangle $O'm_1m_2$ (Fig. 10-5b) leads to

$$\rho d\theta = ds \tag{10-2}$$

FIGURE 10-3

Cantilever beam in pure bending $(M = -M_2)$

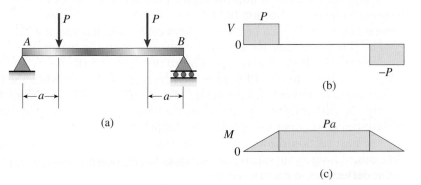

(a)

(b)

(c)

FIGURE 10-4

Simple beam with central region in pure bending and end regions in nonuniform bending

FIGURE 10-5

Curvature of a bent beam:
(a) beam with load and
(b) deflection curve

(a)   (b)

in which $d\theta$ (measured in radians) is the infinitesimal angle between the normals and ds is the infinitesimal distance along the curve between points m_1 and m_2. Combine Eq. (10-2) with Eq. (10-1) to get

$$\kappa = \frac{1}{\rho} = \frac{d\theta}{ds} \tag{10-3}$$

This equation for **curvature** is derived in textbooks on calculus and holds for any curve, regardless of the amount of curvature. If the curvature is *constant* throughout the length of a curve, the radius of curvature also will be constant, and the curve will be an arc of a circle.

The deflections of a beam are usually very small compared to its length (consider, for instance, the deflections of the structural frame of an automobile or a beam in a building). Small deflections mean that the deflection curve is nearly flat. Consequently, the distance ds along the curve may be set equal to its horizontal projection dx (see Fig. 10-5b). Under these special conditions of **small deflections**, the equation for the curvature becomes

$$\kappa = \frac{1}{\rho} = \frac{d\theta}{dx} \tag{10-4}$$

Both the curvature and the radius of curvature are functions of the distance x measured along the x axis. It follows that the position O' of the center of curvature also depends upon the distance x.

The curvature at a particular point on the axis of a beam depends upon the bending moment at that point and upon the properties of the beam itself (shape of cross section and type of material). Therefore, if the beam is prismatic and the material is homogeneous, the curvature varies only with the bending moment (see Section 10.5). Consequently, a beam in *pure bending* has constant curvature, and a beam in *nonuniform bending* has varying curvature.

The **sign convention for curvature** depends upon the orientation of the coordinate axes. If the x axis is positive to the right and the y axis is positive upward, as shown in Fig. 10-6, then the curvature is positive when the beam is bent concave upward and the center of curvature is above the beam. Conversely, the curvature is negative when the beam is bent concave downward, and the center of curvature is below the beam.

The next section shows how the longitudinal strains in a bent beam are determined from its curvature, and Chapter 13 covers how curvature is related to the deflections of beams.

FIGURE 10-6

Sign convention for curvature

(a)

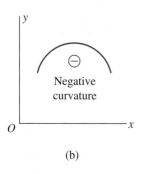

(b)

10.4 Longitudinal Strains in Beams

The longitudinal strains in a beam can be found by analyzing the curvature of the beam and the associated deformations. For this purpose, consider a portion *AB* of a beam in pure bending subjected to positive bending moments *M* (Fig. 10-7a). Assume that the beam initially has a straight longitudinal axis (the *x* axis in the figure) and that its cross section is symmetric about the *y* axis, as shown in Fig. 10-7b.

Under the action of the bending moments, the beam deflects in the *x-y* plane (the plane of bending) and its longitudinal axis is bent into a circular curve (curve *s–s* in Fig. 10-7c). The beam is bent concave upward, which is positive curvature (Fig. 10-6a).

Cross sections of the beam, such as sections *mn* and *pq* in Fig. 10-7a, remain plane and normal to the longitudinal axis (Fig. 10-7c). The fact that cross sections of a beam in pure bending remain plane is so fundamental to beam theory that it is often called an assumption. However, it also could be called a theorem because it can be proved rigorously using only rational arguments based upon symmetry. The basic point is that the symmetry of the beam and its loading (Figs. 10-7a and b) means that all elements of the beam (such as element *mpqn*) must deform in an identical manner, which is possible only if cross sections remain plane during bending (Fig. 10-7c). This conclusion is valid for beams of any material, whether the material is elastic or inelastic, linear or nonlinear. Of course, the material properties, like the dimensions, must be symmetric about the plane of bending. *Note:* Even though a plane cross section in pure bending remains plane, there still may be deformations in the plane itself. Such deformations are due to the effects of Poisson's ratio, as explained at the end of this discussion.

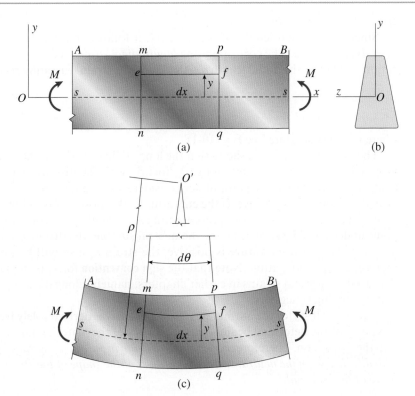

FIGURE 10-7

Deformations of a beam in pure bending: (a) side view of beam, (b) cross section of beam, and (c) deformed beam

Because of the bending deformations shown in Fig. 10-7c, cross sections *mn* and *pq* rotate with respect to each other about axes perpendicular to the *x*-*y* plane. Longitudinal lines on the lower part of the beam are elongated, whereas those on the upper part are shortened. Thus, the lower part of the beam is in tension and the upper part is in compression. Somewhere between the top and bottom of the beam is a surface in which longitudinal lines do not change in length. This surface, indicated by the dashed line *s–s* in Figs. 10-7a and c, is called the **neutral surface** of the beam. Its intersection with any cross-sectional plane is called the **neutral axis** of the cross section; for instance, the *z* axis is the neutral axis for the cross section of Fig. 10-7b.

The planes containing cross sections *mn* and *pq* in the deformed beam (Fig. 10-7c) intersect in a line through the center of curvature *O'*. The angle between these planes is denoted $d\theta$, and the distance from *O'* to the neutral surface *s–s* is the radius of curvature ρ. The initial distance dx between the two planes (Fig. 10-7a) is unchanged at the neutral surface (Fig. 10-7c), hence $\rho d\theta = dx$. However, all other longitudinal lines between the two planes either lengthen or shorten, thereby creating **normal strains** ε_x.

To evaluate these normal strains, consider a typical longitudinal line *ef* located within the beam between planes *mn* and *pq* (Fig. 10-7a). Identify line *ef* by its distance *y* from the neutral surface in the initially straight beam. Now assume that the *x* axis lies along the neutral surface of the *undeformed* beam. Of course, when the beam deflects, the neutral surface moves with the beam, but the *x* axis remains fixed in position. Nevertheless, the longitudinal line *ef* in the deflected beam (Fig. 10-7c) is still located at the same distance *y* from the neutral surface. Thus, the length L_1 of line *ef* after bending takes place is

$$L_1 = (\rho - y)d\theta = dx - \frac{y}{\rho}dx$$

after substitution of $d\theta = dx/\rho$.

Since the original length of line *ef* is dx, it follows that its elongation is $L_1 - dx$, or $-ydx/\rho$. The corresponding *longitudinal strain* is equal to the elongation divided by the initial length dx; therefore, the **strain–curvature relation** is

$$\varepsilon_x = -\frac{y}{\rho} = -\kappa y \tag{10-5}$$

where κ is the curvature [see Eq. (10-1)].

The preceding equation shows that the longitudinal strains in the beam are proportional to the curvature and vary linearly with the distance *y* from the neutral surface. When the point under consideration is above the neutral surface, the distance *y* is positive. If the curvature is also positive (as in Fig. 10-7c), then ε_x will be a negative strain, representing a shortening. By contrast, if the point under consideration is below the neutral surface, the distance *y* will be negative and, if the curvature is positive, the strain ε_x also will be positive, representing an elongation. Note that the **sign convention** for ε_x is the same as that used for normal strains in earlier chapters, namely, elongation is positive and shortening is negative.

Equation (10-5) for the normal strains in a beam was derived solely from the geometry of the deformed beam—the properties of the material did not enter into the discussion. Therefore, *the strains in a beam in pure bending vary linearly with distance from the neutral surface regardless of the shape of the stress-strain curve of the material.*

The next step in the analysis, namely, finding the stresses from the strains, requires the use of the *stress-strain curve*. This step is described in the next section for linearly elastic materials.

The longitudinal strains in a beam are accompanied by *transverse strains* (that is, normal strains in the y and z directions) because of the effects of Poisson's ratio. However, there are no accompanying transverse stresses because beams are free to deform laterally. This stress condition is analogous to that of a prismatic bar in tension or compression, and therefore, *longitudinal elements in a beam in pure bending are in a state of uniaxial stress.*

Example 10-1

FIGURE 10-8

Example 10-1: Beam in pure bending: (a) beam with loads and (b) deflection curve

(a)

(b)

A simply supported steel beam AB (Fig. 10-8a) of a length $L = 8.0$ ft and height $h = 6.0$ in. is bent by couples M_0 into a circular arc with a downward deflection δ at the midpoint (Fig. 10-8b). The longitudinal normal strain (elongation) on the bottom surface of the beam is 0.00125, and the distance from the neutral surface to the bottom surface of the beam is 3.0 in.

Determine the radius of curvature ρ, the curvature κ, and the deflection δ of the beam.

Note: This beam has a relatively large deflection because its length is large compared to its height ($L/h = 16$), and the strain of 0.00125 is also large. (This is about the same as the yield strain for ordinary structural steel.)

Solution:

Use the four-step problem-solving approach. Combine steps as needed for an efficient solution.

Part (a): Curvature.

1, 2. *Conceptualize* [*hypothesize, sketch*], *Categorize* [*simplify, classify*]: Since the longitudinal strain at the bottom surface of the beam ($\varepsilon_x = 0.00125$) and the distance from the neutral surface to the bottom surface ($y = -3.0$ in.) are known, use Eq. (10-5) to calculate both the radius of curvature and the curvature.

3. *Analyze* [*evaluate; select relevant equations, carry out mathematical solution*]: Rearrange Eq. (10-5) and substitute numerical values to get

$$\rho = -\frac{y}{\varepsilon_x} = -\frac{-3.0 \text{ in.}}{0.00125} = 2400 \text{ in.} = 200 \text{ ft} \qquad \kappa = \frac{1}{\rho} = 0.0050 \text{ ft}^{-1}$$

4. *Finalize* [*conclude; examine answer—Does it make sense? Are units correct? How does it compare to similar problem solutions?*]: These results show that the radius of curvature is extremely large compared to the length of the beam even when the strain in the material is large. If, as usual, the strain is less, the radius of curvature is even larger.

Part (b): Deflection.

1, 2. *Conceptualize, Categorize:* As pointed out in Section 10.3, a constant bending moment (pure bending) produces constant curvature throughout the length of a beam. Therefore, the deflection curve is a circular arc. From Fig. 10-8b, the distance from the center of curvature O' to the midpoint C' of the deflected beam is the radius of curvature r, and the distance from O' to point C on the x axis is $r \cos u$, where u is angle $BO'C$. This leads to the expression for the deflection at the midpoint of the beam:

$$\delta = \rho(1 - \cos\theta) \tag{10-6}$$

For a nearly flat curve, assume that the distance between supports is the same as the length of the beam itself. Therefore, from triangle $BO'C$,

$$\sin\theta = \frac{L/2}{\rho} \tag{10-7}$$

3. *Analyze:* Substitute numerical values to obtain

$$\sin\theta = \frac{(8.0 \text{ ft})(12 \text{ in/ft})}{2(2400 \text{ in.})} = 0.0200$$

and

$$\theta = 0.0200 \text{ rad} = 1.146°$$

For practical purposes, consider $\sin\theta$ and θ (radians) to be equal numerically because θ is a very small angle.

Now substitute into Eq. (10-6) for the deflection and obtain

$$\delta = \rho(1 - \cos\theta) = (2400 \text{ in.})(1 - 0.999800) = 0.480 \text{ in.} \quad \longleftarrow$$

4. *Finalize:* This deflection is very small compared to the length of the beam, as shown by the ratio of the span length to the deflection:

$$\frac{L}{\delta} = \frac{(8.0 \text{ ft})(12 \text{ in/ft})}{0.480 \text{ in.}} = 200$$

This confirms that the deflection curve is nearly flat in spite of the large strains. Of course, in Fig. 10-8b, the deflection of the beam is highly exaggerated for clarity.

Note: The purpose of this example is to show the relative magnitudes of the radius of curvature, length of the beam, and deflection of the beam. However, the method used for finding the deflection has little practical value because it is limited to pure bending, which produces a circular deflected shape. More useful methods for finding beam deflections are presented in Chapter 13.

10.5 Normal Stresses in Beams (Linearly Elastic Materials)

Longitudinal strains ε_x in a beam in pure bending were investigated in the preceding section [see Eq. (10-5) and Fig. 10-7]. Since longitudinal elements of a beam are subjected only to tension or compression, now use the **stress-strain curve** for the material to determine the stresses from the strains. The stresses act over the entire cross section of the beam and vary in intensity, depending upon the shape of the stress-strain diagram and the dimensions of the cross section. Since the x direction is longitudinal (Fig. 10-7a), use the symbol σ_x to denote these stresses.

The most common stress-strain relationship encountered in engineering is the equation for a **linearly elastic material**. For such materials, substitute Hooke's law for uniaxial stress ($\sigma = E\varepsilon$) into Eq. (10-5) and obtain

$$\sigma_x = E\varepsilon_x = -\frac{Ey}{\rho} = -E\kappa y \qquad \textbf{(10-8)}$$

This equation shows that the normal stresses acting on the cross section vary linearly with the distance y from the neutral surface. This stress distribution is pictured in Fig. 10-9a for the case in which the bending moment M is positive and the beam bends with positive curvature.

When the curvature is positive, the stresses σ_x are negative (compression) above the neutral surface and positive (tension) below it. In the figure, compressive stresses are indicated by arrows pointing *toward* the cross section and tensile stresses are indicated by arrows pointing *away* from the cross section.

In order for Eq. (10-8) to be of practical value, locate the origin of the coordinates so that you can determine the distance y. In other words, locate the neutral axis of the cross section. You also need to obtain a relationship between the curvature and the bending moment—so that you can substitute into Eq. (10-8) and obtain an equation relating the stresses to the bending moment. These two objectives can be accomplished by determining the resultant of the stresses σ_x acting on the cross section.

In general, the **resultant of the normal stresses** consists of two stress resultants: (1) a force acting in the x direction and (2) a bending couple acting about the z axis. However, the axial force is zero when a beam is in pure bending. Therefore, write the following equations of statics: (1) The resultant force in the x direction is equal to zero, and (2) the resultant moment is equal to the bending moment M. The first equation gives the location of the neutral axis, and the second gives the moment–curvature relationship.

FIGURE 10-9

Normal stresses in a beam of linearly elastic material:
(a) side view of beam showing distribution of normal stresses and
(b) cross section of beam showing the z axis as the neutral axis of the cross section

(a)

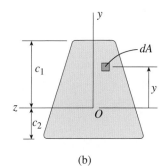

(b)

Location of Neutral Axis

To obtain the first equation of statics, consider an element of area dA in the cross section (Fig. 10-9b). The element is located at a distance y from the neutral axis; therefore, the stress σ_x acting on the element is given by Eq. (10-8). The *force* acting on the element is equal to $\sigma_x dA$ and is compressive when y is positive. Because there is no resultant force acting on the cross section, the integral of $\sigma_x dA$ over the area A of the entire cross section must vanish; thus, the *first equation of statics* is

$$\int_A \sigma_x \, dA = -\int_A E\kappa y \, dA = 0 \tag{10-9a}$$

Because the curvature κ and modulus of elasticity E are nonzero constants at any given cross section of a bent beam, they are not involved in the integration over the cross-sectional area. Therefore, drop them from the equation and obtain

$$\int_A y \, dA = 0 \tag{10-9b}$$

This equation states that the first moment of the area of the cross section, when evaluated with respect to the z axis, is zero. In other words, the z axis must pass through the centroid of the cross section.[2]

The z axis is also the neutral axis, so

The neutral axis passes through the centroid of the cross-sectional area when the material follows Hooke's law and there is no axial force acting on the cross section.

This observation makes it relatively simple to determine the position of the neutral axis.

As explained in Section 10.1, this discussion is limited to beams for which the y axis is an axis of symmetry. Consequently, the y axis also passes through the centroid. Therefore,

The origin O of coordinates (Fig. 10-9b) is located at the centroid of the cross-sectional area.

Because the y axis is an axis of symmetry of the cross section, the y axis is a *principal axis* (see Section 5.8 for a discussion of principal axes). Since the z axis is perpendicular to the y axis, it too is a principal axis. Thus, when a beam of linearly elastic material is subjected to pure bending, *the y and z axes are principal centroidal axes.*

Moment–Curvature Relationship

The *second equation of statics* expresses the fact that the moment resultant of the normal stresses σ_x acting over the cross section is equal to the bending moment M (Fig. 10-9a). The element of force $\sigma_x dA$ acting on the element of area dA (Fig. 10-9b) is in the positive direction of the x axis when σ_x is positive and in the negative direction when σ_x is negative. Since the element dA is located above the neutral axis, a positive stress σ_x acting on that element produces an element of moment equal to $\sigma_x y \, dA$. This element of moment acts opposite in direction to the positive bending moment M shown in Fig. 10-9a. Therefore, the elemental moment is

$$dM = -\sigma_x y \, dA$$

---

[2]Centroids and first moments of areas are discussed in Sections 5.2 and 5.3.

The integral of all such elemental moments over the entire cross-sectional area A must equal the bending moment:

$$M = -\int_A \sigma_x y\, dA \qquad (10\text{-}10a)$$

or, upon substituting for σ_x from Eq. (10-9),

$$M = \int_A \kappa E y^2\, dA = \kappa E \int_A y^2\, dA \qquad (10\text{-}10b)$$

This equation relates the curvature of the beam to the bending moment M.

Since the integral in the preceding equation is a property of the cross-sectional area, it is convenient to rewrite the equation as

$$M = \kappa E I \qquad (10\text{-}11)$$

in which

$$I = \int_A y^2\, dA \qquad (10\text{-}12)$$

This integral is the **moment of inertia** of the cross-sectional area with respect to the z axis (that is, with respect to the neutral axis). Moments of inertia are always positive and have dimensions of length to the fourth power; for instance, typical USCS units are in^4 and typical SI units are mm^4 when performing beam calculations.[3]

Equation (10-11) now can be rearranged to express the *curvature* in terms of the bending moment in the beam:

$$\kappa = \frac{1}{\rho} = \frac{M}{EI} \qquad (10\text{-}13)$$

Known as the **moment–curvature equation**, Eq. (10-13) shows that the curvature is directly proportional to the bending moment M and inversely proportional to the quantity EI, which is called the **flexural rigidity** of the beam. Flexural rigidity is a measure of the resistance of a beam to bending, that is, the larger the flexural rigidity, the smaller the curvature for a given bending moment.

Comparing the **sign convention** for bending moments with that for curvature (Fig. 10-6), note that *a positive bending moment produces positive curvature and a negative bending moment produces negative curvature* (see Fig. 10-10).

Flexure Formula

Now that the neutral axis has been located and the moment–curvature relationship has been derived, determine the stresses in terms of the bending moment. Substitute the expression for curvature [Eq. (10-13)] into the expression for the stress σ_x [Eq. (10-8)] to get

$$\sigma_x = -\frac{My}{I} \qquad (10\text{-}14)$$

This equation, called the **flexure formula**, shows that the stresses are directly proportional to the bending moment M and inversely proportional to the moment of inertia I of the cross section. Also, the stresses vary linearly with the distance y from the neutral axis, as previously observed. Stresses calculated from the flexure formula are called **bending stresses** or **flexural stresses**.

FIGURE 10-10

Relationships between signs of bending moments and signs of curvatures

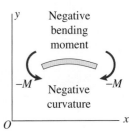

---

[3]Moments of inertia of areas are discussed in Section 5.6.

FIGURE 10-11

Relationships between signs of bending moments and directions of normal stresses: (a) positive bending moment and (b) negative bending moment

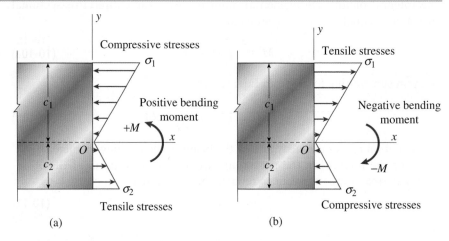

If the bending moment in the beam is positive, the bending stresses will be positive (tension) over the part of the cross section where y is negative, that is, over the lower part of the beam. The stresses in the upper part of the beam will be negative (compression). If the bending moment is negative, the stresses will be reversed. These relationships are shown in Fig. 10-11.

Maximum Stresses at a Cross Section

The maximum tensile and compressive bending stresses acting at any given cross section occur at points located farthest from the neutral axis. Denote by c_1 and c_2 the distances from the neutral axis to the extreme elements in the positive and negative y directions, respectively (see Fig. 10-9b and Fig. 10-11). Then the corresponding **maximum normal stresses** σ_1 and σ_2 (from the flexure formula) are

$$\sigma_1 = -\frac{Mc_1}{I} = -\frac{M}{S_1} \qquad \sigma_2 = \frac{Mc_2}{I} = \frac{M}{S_2} \qquad \textbf{(10-15a,b)}$$

in which

$$S_1 = \frac{I}{c_1} \qquad S_2 = \frac{I}{c_2} \qquad \textbf{(10-16a,b)}$$

The quantities S_1 and S_2 are known as the **section moduli** of the cross-sectional area. From [Eqs. (10-16a and b)], note that each section modulus has dimensions of a length to the third power (for example, in³ or mm³). Also note that the distances c_1 and c_2 to the top and bottom of the beam are always taken as positive quantities.

The advantage of expressing the maximum stresses in terms of section moduli arises from the fact that each section modulus combines the beam's relevant cross-sectional properties into a single quantity. Then this quantity can be listed in tables and handbooks as a property of the beam, which is a convenience to designers. (Design of beams using section moduli is explained in the next section.)

Doubly Symmetric Shapes

If the cross section of a beam is symmetric with respect to the z axis as well as the y axis (*doubly symmetric cross section*), then $c_1 = c_2 = c$, and the maximum tensile and compressive stresses are equal numerically:

$$\sigma_1 = -\sigma_2 = -\frac{Mc}{I} = -\frac{M}{S} \qquad \text{or} \qquad \sigma_{max} = \frac{M}{S} \qquad \textbf{(10-17a,b)}$$

in which

$$S = \frac{I}{c} \qquad (10\text{-}18)$$

is the only section modulus for the cross section.

For a beam of **rectangular cross section** with width b and height h (Fig. 10-12a), the moment of inertia and section modulus are

$$I = \frac{bh^3}{12} \qquad S = \frac{bh^2}{6} \qquad (10\text{-}19\text{a,b})$$

For a **circular cross section** of diameter d (Fig. 10-12b), these properties are

$$I = \frac{\pi d^4}{64} \qquad S = \frac{\pi d^3}{32} \qquad (10\text{-}20\text{a,b})$$

Properties of other doubly symmetric shapes, such as hollow tubes (either rectangular or circular) and wide-flange shapes, can be readily obtained from the preceding formulas.

Properties of Beam Cross Sections

Moments of inertia of many plane figures are listed in Appendix B for convenient reference. Also, the dimensions and properties of standard sizes of steel and wood beams are listed in Appendixes C and D and in many engineering handbooks, as explained in more detail in the next section.

For other cross-sectional shapes, determine the location of the neutral axis, the moment of inertia, and the section moduli by direct calculation, using the techniques described in Chapter 5. This procedure is illustrated later in Example 10-4.

Limitations

The analysis presented in this section is for the pure bending of prismatic beams composed of homogeneous, linearly elastic materials. If a beam is subjected to nonuniform bending, the shear forces will produce *warping* (or out-of-plane distortion) of the cross sections. Thus, a cross section that was plane before bending is no longer plane after bending. Warping due to shear deformations greatly complicates the behavior of the beam. However, detailed investigations show that the normal stresses calculated from the flexure formula are not significantly altered by the presence of shear stresses and the associated warping. Thus, you may justifiably use the theory of pure bending for calculating normal stresses in beams subjected to nonuniform bending.[4]

The flexure formula gives results that are accurate only in regions of the beam where the stress distribution is not disrupted by changes in the shape of the beam or by discontinuities in loading. For instance, the flexure formula is not applicable near the supports of a beam or close to a concentrated load. Such irregularities produce localized stresses, or *stress concentrations*, that are much greater than the stresses obtained from the flexure formula.

---

[4]Beam theory began with Galileo Galilei (1564–1642), who investigated the behavior of various types of beams. His work in mechanics of materials is described in his famous book *Two New Sciences*, first published in 1638. Although Galileo made many important discoveries regarding beams, he did not obtain the stress distribution used today. Further progress in beam theory was made by Mariotte, Jacob Bernoulli, Euler, Parent, Saint-Venant, and others.

FIGURE 10-12

Doubly symmetric cross-sectional shapes

(a)

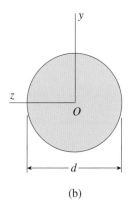

(b)

Example 10-2

A high-strength steel wire with a diameter d is bent around a cylindrical drum of radius R_0 (Fig. 10-13).

Determine the bending moment M and maximum bending stress σ_{max} in the wire, assuming $d = 4$ mm and $R_0 = 0.5$ m. (The steel wire has a modulus of elasticity $E = 200$ GPa and a proportional limit $\sigma_{pl} = 1200$ MPa.)

FIGURE 10-13

Example 10-2: Wire bent around a drum

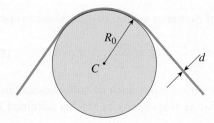

Solution:

Use the four-step problem-solving approach.

1. *Conceptualize*: The first step in this example is to determine the radius of curvature ρ of the bent wire. Knowing ρ, then find the bending moment and maximum stresses.

2. *Categorize*:

   Radius of curvature: The radius of curvature of the bent wire is the distance from the center of the drum to the neutral axis of the cross section of the wire:

   $$\rho = R_0 + \frac{d}{2} \tag{10-21}$$

   Bending moment: The bending moment in the wire may be found from the moment–curvature relationship (Eq. 10-13):

   $$M = \frac{EI}{\rho} = \frac{2EI}{2R_0 + d} \tag{10-22}$$

   in which I is the moment of inertia of the cross-sectional area of the wire. Substitute for I in terms of the diameter d of the wire [Eq. (10-20a)] to get

   $$M = \frac{\pi E d^4}{32(2R_0 + d)} \tag{10-23}$$

   This result was obtained without regard to the *sign* of the bending moment, since the direction of bending is obvious from the figure.

   Maximum bending stresses: The maximum tensile and compressive stresses, which are equal numerically, are obtained from the flexure formula as given by Eq. (10-17b):

   $$\sigma_{max} = \frac{M}{S}$$

in which S is the section modulus for a circular cross section. Substitute for M from Eq. (10-23) and for S from Eq. (10-20b) to get

$$\sigma_{max} = \frac{Ed}{2R_0 + d} \qquad \textbf{(10-24)}$$

This same result can be obtained directly from Eq. (10-8) by replacing y with $d/2$ and substituting for ρ from Eq. (10-21).

Inspection of Fig. 10-13 reveals that the stress is compressive on the lower (or inner) part of the wire and tensile on the upper (or outer) part.

3. *Analyze*:

Numerical results: Now substitute the given numerical data into Eqs. (10-23) and (10-24) and obtain

$$M = \frac{\pi Ed^4}{32(2R_0 + d)} = \frac{\pi(200 \text{ GPa})(4 \text{ mm})^4}{32[2(0.5 \text{ m}) + 4 \text{ mm}]} = 5.01 \text{ N} \cdot \text{m} \qquad \leftarrow$$

$$\sigma_{max} = \frac{Ed}{2R_0 + d} = \frac{(200 \text{ GPa})(4 \text{ mm})}{2(0.5 \text{ m}) + 4 \text{ mm}} = 797 \text{ MPa} \qquad \leftarrow$$

4. *Finalize*: Maximum stress σ_{max} is less than the proportional limit of the steel wire; therefore, the calculations are valid.

Note: Because the radius of the drum is large compared to the diameter of the wire, d in comparison with $2R_0$ in the denominators of the expressions for M and σ_{max} can be safely disregarded. Then Eqs. (10-23) and (10-24) give

$$M = 5.03 \text{ N} \cdot \text{m} \qquad \sigma_{max} = 800 \text{ MPa}$$

These results are on the conservative side and differ by less than 1% from the more precise values.

Example 10-3

A simple beam with an overhang (from Example 6-5) is shown in Fig. 10-14. A uniform load with an intensity $q = 400$ lb/ft acts throughout the length of the beam, and a concentrated load $P = 2400$ lb acts at a point 9 ft from the left-hand support. Uniform load q includes the weight of the beam. The beam is constructed of structural glued and laminated timber, has a cross section width of $b = 5$ in., and has a height of $h = 22$ in. (Fig. 10-15).

(a) Determine the maximum tensile and compressive stresses in the beam due to bending.

(b) If load q is unchanged, find the maximum permissible value of load P if the allowable normal stress in tension and compression is $\sigma_a = 1875$ psi.

FIGURE 10-14

Beam with an overhang and uniform and concentrated loads

FIGURE 10-15

Beam cross section

Solution:

Use the four-step problem-solving approach. Combine steps as needed for an efficient solution.

Part (a): Maximum normal stresses.

1, 2. *Conceptualize, Categorize:* Begin the analysis by drawing the shear-force and bending-moment diagrams (Fig. 10-16); then determine the maximum bending moment, which occurs under the concentrated load. This is detailed in Example 6-5, and the resulting diagrams are shown in Fig. 10-16. The moment diagram shows that $M_{max} = 37,800$ lb-ft at 9 ft to the right of support A. The maximum bending stresses in the beam occur at the cross section of the maximum moment.

Section modulus: The section modulus for the rectangular cross-sectional area in Fig. 10-15 is from Eq. (10-19b):

$$S = \frac{bh^2}{6} = \frac{1}{6}(5 \text{ in.})(22 \text{ in.})^2 = 403.3 \text{ in}^3 \qquad \textbf{(a)}$$

3. *Analyze:*

Maximum stresses: The maximum tensile and compressive stresses are obtained from Eq. (10-17):

$$\sigma_t = \frac{M_{max}}{S} = \frac{(37,800 \text{ lb-ft})(12 \text{ in/ft})}{403.3 \text{ in}^3} = 1125 \text{ psi}$$

$$\sigma_c = -\frac{M_{max}}{S} = -1125 \text{ psi}$$

⟵ **(b)**

FIGURE 10-16

(a, b) Shear and moment diagrams (from Example 6-5)

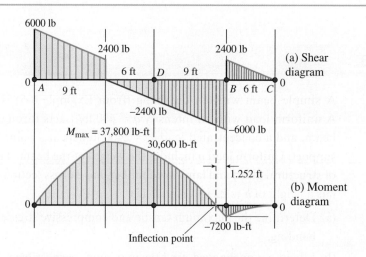

(a) Shear diagram

(b) Moment diagram

4. *Finalize*: The moment diagram is plotted on the compression side of the beam, so most of span AB has compressive stress on the top and tension stress on the bottom of the beam. The reverse is true for the portion of the beam to the right of the inflection point, which includes overhang segment BC.

Part (b): Maximum permissible load P.

1, 2. *Conceptualize, Categorize*: The normal stresses in Eq. (b) at the location of the maximum moment are well below the allowable value of 1875 psi, so the beam can carry a much larger value of load P than that applied in part (a). Let the distance from support A to load P be $a = 9$ ft, span AB length $L = 24$ ft, and the uniform load be unchanged at $q = 400$ lb/ft.

3. *Analyze*: Apply concentrated load P and uniform load q and solve for the reaction at A:

$$R_A = P\left(\frac{L-a}{L}\right) + \frac{15}{32}qL \tag{c}$$

The maximum moment is at distance a from support A and is written as

$$M_{max} = R_A a - \frac{qa^2}{2} \tag{d}$$

Equate M_{max} to $(\sigma_a)(S) = 63{,}016$ lb-ft, insert numerical values in Eqs. (c) and (d), and solve for $P_{max} = 6883$ lb.

Alternate solution: Apply additional load ΔP to increase the maximum moment from 37,800 lb-ft to 63,016 lb-ft, that is, $\Delta M = 25{,}216$ lb-ft. Now compute the required additional load ΔP, which gives the moment at the location of a concentrated load:

$$\Delta P = \frac{L}{a(L-a)}\Delta M = \frac{24 \text{ ft}}{9 \text{ ft}(24 \text{ ft} - 9 \text{ ft})}(25{,}216 \text{ lb-ft}) = 4483 \text{ lb} \tag{e}$$

Add ΔP to the load $P = 2400$ lb from part (a) to get

$$P_{max} = P + \Delta P = 2400 \text{ lb} + 4483 \text{ lb} = 6883 \text{ lb} \tag{f}$$

4. *Finalize*: Check that the maximum permissible value of P produces normal stresses at the allowable level at the point of maximum moment. Substitute P_{max} into Eqs. (c) and (d) to find that $R_A = 8802$ lb and $M_{max} = 63{,}016$ lb-ft. Using these values, the stresses at the point of load P_{max} application are

$$\sigma_t = -\sigma_c = \frac{M_{max}}{S} = \frac{(63{,}016 \text{ lb-ft})(12 \text{ in/ft})}{403.3 \text{ in}^3} = 1875 \text{ psi} \tag{g}$$

Example 10-4

The beam ABC shown in Fig. 10-17a has simple supports at A and B and an overhang from B to C. The length of the span is $L = 3.0$ m, and the length of the overhang is $L/2 = 1.5$ m. A uniform load of intensity $q = 3.2$ kN/m acts throughout the entire length of the beam (4.5 m).

The beam has a cross section of channel shape with a width of $b = 300$ mm and height of $h = 80$ mm (Fig. 10-18). The web thickness is $t = 12$ mm, and the average thickness of the sloping flanges is the same. For the purpose of calculating the properties of the cross section, assume that the cross section consists of three rectangles, as shown in Fig. 10-18b.

(a) Determine the maximum tensile and compressive stresses in the beam due to the uniform load.

(b) Find the maximum permissible value of uniform load q (in kN/m) if allowable stresses in tension and compression are $\sigma_{aT} = 110$ MPa and $\sigma_{aC} = 92$ MPa, respectively.

FIGURE 10-17

Example 10-4: Stresses in a beam with an overhang

(a) Beam with an overhang

(b) Shear diagram

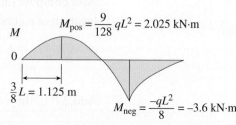

(c) Moment diagram

Solution:

Use the four-step problem-solving approach. Combine steps as needed for an efficient solution.

Part (a): Maximum tensile and compressive stresses.

1, 2. *Conceptualize, Categorize*: Reactions, shear forces, and bending moments are computed in the analysis of this beam. First, find the reactions at supports A and B using statics, as described in Chapter 6. The results are

$$R_A = \frac{3}{8}qL = 3.6 \text{ kN} \qquad R_B = \frac{9}{8}qL = 10.8 \text{ kN}$$

From these values, construct the shear-force diagram (Fig. 10-17b).

Note that the shear force changes sign and is equal to zero at two locations: (i) at a distance of 1.125 m from the left-hand support and (ii) at the right-hand reaction.

Next, draw the bending-moment diagram shown in Fig. 10-17c. Both the maximum positive and maximum negative bending moments occur at the cross sections where the shear force changes sign. These maximum moments are

$$M_{\text{pos}} = \frac{9}{128}qL^2 = 2.025 \text{ kN} \cdot \text{m} \qquad M_{\text{neg}} = \frac{-qL^2}{8} = -3.6 \text{ kN} \cdot \text{m}$$

respectively.

Neutral axis of the cross section (Fig. 10-18b): The origin O of the y-z coordinates is placed at the centroid of the cross-sectional area; therefore, the z axis becomes the neutral axis of the cross section. The centroid is located by using the techniques described in Section 5.3, as follows.

First, divide the area into three rectangles (A_1, A_2, and A_3). Second, establish a reference axis Z–Z across the upper edge of the cross section, and let y_1 and y_2 be the distances from the Z–Z axis to the centroids of areas A_1 and A_2,

FIGURE 10-18

Cross section of beam discussed in Example 10-4: (a) actual shape and (b) idealized shape for use in analysis (the thickness of the beam is exaggerated for clarity)

(a)   (b)

respectively. Then the calculations for locating the centroid of the entire channel section (distances c_1 and c_2) are

Area 1: $y_1 = t/2 = 6$ mm

$$A_1 = (b - 2t)(t) = (276 \text{ mm})(12 \text{ mm}) = 3312 \text{ mm}^2$$

Area 2: $y_2 = h/2 = 40$ mm

$$A_2 = ht = (80 \text{ mm})(12 \text{ mm}) = 960 \text{ mm}^2$$

Area 3: $y_3 = y_2 \quad A_3 = A_2$

$$c_1 = \frac{\Sigma y_i A_i}{\Sigma A_i} = \frac{y_1 A_1 + 2y_2 A_2}{A_1 + 2A_2}$$

$$= \frac{(6 \text{ mm})(3312 \text{ mm}^2) + 2(40 \text{ mm})(960 \text{ mm}^2)}{3312 \text{ mm}^2 + 2(960 \text{ mm}^2)} = 18.48 \text{ mm}$$

$$c_2 = h - c_1 = 80 \text{ mm} - 18.48 \text{ mm} = 61.52 \text{ mm}$$

Thus, the position of the neutral axis (the z axis) is determined.

Moment of inertia: In order to calculate the stresses from the flexure formula, determine the moment of inertia of the cross-sectional area with respect to the neutral axis. These calculations require the use of the parallel axis theorem (see Section 5.6).

Beginning with area A_1, obtain its moment of inertia $(I_z)_1$ about the z axis from the equation

$$(I_z)_1 = (I_c)_1 + A_1 d_1^2 \tag{a}$$

In this equation, $(I_c)_1$ is the moment of inertia of area A_1 about its own centroidal axis:

$$(I_c)_1 = \frac{1}{12}(b - 2t)(t)^3 = \frac{1}{12}(276 \text{ mm})(12 \text{ mm})^3 = 39{,}744 \text{ mm}^4$$

and d_1 is the distance from the centroidal axis of area A_1 to the z axis:

$$d_1 = c_1 - t/2 = 18.48 \text{ mm} - 6 \text{ mm} = 12.48 \text{ mm}$$

Therefore, the moment of inertia of area A_1 about the z axis [from Eq. (a)] is

$$(I_z)_1 = 39{,}744 \text{ mm}^4 + (3312 \text{ mm}^2)(12.48 \text{ mm})^2 = 555{,}600 \text{ mm}^4$$

Proceed in the same manner for areas A_2 and A_3 to get

$$(I_z)_2 = (I_z)_3 = 956{,}600 \text{ mm}^4$$

Thus, the centroidal moment of inertia I_z of the entire cross-sectional area is

$$I_z = (I_z)_1 + (I_z)_2 + (I_z)_3 = 2.469 \times 10^6 \text{ mm}^4$$

Section moduli: The section moduli for the top and bottom of the beam, respectively, are

$$S_1 = \frac{I_z}{c_1} = 133,600 \text{ mm}^3 \qquad S_2 = \frac{I_z}{c_2} = 40,100 \text{ mm}^3$$

[see Eqs. (10-16a and b)]. With the cross-sectional properties determined, now calculate the maximum stresses from Eqs. (10-15a and b).

3. *Analyze:*

Maximum stresses: At the cross section of maximum positive bending moment, the largest tensile stress occurs at the bottom of the beam (σ_2) and the largest compressive stress occurs at the top (σ_1). Thus, from Eqs. (10-15b) and (10-15a), respectively, you get

$$\sigma_t = \sigma_2 = \frac{M_{\text{pos}}}{S_2} = \frac{2.025 \text{ kN} \cdot \text{m}}{40,100 \text{ mm}^3} = 50.5 \text{ MPa}$$

$$\sigma_c = \sigma_1 = -\frac{M_{\text{pos}}}{S_1} = -\frac{2.025 \text{ kN} \cdot \text{m}}{133,600 \text{ mm}^3} = -15.2 \text{ MPa}$$

Similarly, the largest stresses at the section of maximum negative moment are

$$\sigma_t = \sigma_1 = -\frac{M_{\text{neg}}}{S_1} = -\frac{-3.6 \text{ kN} \cdot \text{m}}{133,600 \text{ mm}^3} = 26.9 \text{ MPa}$$

$$\sigma_c = \sigma_2 = \frac{M_{\text{neg}}}{S_2} = \frac{-3.6 \text{ kN} \cdot \text{m}}{40,100 \text{ mm}^3} = -89.8 \text{ MPa}$$

4. *Finalize:* A comparison of these four stresses shows that the largest tensile stress in the beam is 50.5 MPa and occurs at the bottom of the beam at the cross section of maximum positive bending moment; thus,

$$(\sigma_t)_{\text{max}} = 50.5 \text{ MPa}$$

The largest compressive stress is −89.8 MPa and occurs at the bottom of the beam at the section of maximum negative moment:

$$(\sigma_c)_{\text{max}} = -89.8 \text{ MPa}$$

Recall that these are the maximum bending stresses due to the uniform load acting on the beam.

Part (b): Maximum permissible value of uniform load q.

1, 2. *Conceptualize, Categorize:* Next, find q_{max} based on the given allowable normal stresses, which are different for tension and compression. The allowable compression stress is σ_{aC} lower than that for tension, σ_{aT}, to account for the possibility of local buckling of the flanges of the C shape (if they are in compression).

Use the flexure formula to compute potential values of q_{max} at four locations: at the top and bottom of the beam at the location of the maximum positive moment (M_{pos}) and at the top and bottom of the beam at the location of the maximum negative moment (M_{neg}). In each case, be sure to use the proper value of allowable stress. Assume that the C shape is used in the orientation shown in Fig. 10-18 (flanges downward), so at the location of M_{pos}, the top of the beam is in compression and the bottom is in tension, while the opposite is true at point B. Using the expressions for M_{pos} and M_{neg} and equating each to the appropriate product of allowable stress and section modulus, solve for possible values of q_{max} as given here.

3. *Analyze:* In beam segment AB at the *top* of beam,

$$M_{pos} = \frac{9}{128}q_1L^2 = \sigma_{aC}S_1 \quad \text{so} \quad q_1 = \frac{128}{9L^2}(\sigma_{aC}S_1) = 19.42 \text{ kN/m}$$

In beam segment AB at the *bottom* of beam,

$$M_{pos} = \frac{9}{128}q_2L^2 = \sigma_{aT}S_2 \quad \text{so} \quad q_2 = \frac{128}{9L^2}(\sigma_{aT}S_2) = 6.97 \text{ kN/m}$$

At joint B at the top of beam,

$$M_{pos} = \frac{1}{8}q_3L^2 = \sigma_{aT}S_1 \quad \text{so} \quad q_3 = \frac{8}{L^2}(\sigma_{aT}S_1) = 13.06 \text{ kN/m}$$

At joint B at *bottom* of the beam,

$$M_{pos} = \frac{1}{8}q_4L^2 = \sigma_{aC}S_2 \quad \text{so} \quad q_4 = \frac{8}{L^2}(\sigma_{aC}S_2) = 3.28 \text{ kN/m}$$

4. *Finalize:* From these calculations, the bottom of the beam near joint B (where the flange tips are in compression) does indeed control the maximum permissible value of uniform load q. Hence,

$$q_{max} = 3.28 \text{ kN/m} \quad \longleftarrow$$

10.6 Design of Beams for Bending Stresses

The process of designing a beam requires that many factors be considered, including the type of structure (airplane, automobile, bridge, building, or whatever), the materials to be used, the loads to be supported, the environmental conditions to be encountered, and the costs to be paid. However, from the standpoint of strength, the task eventually reduces to selecting a shape and size of beam such that the actual stresses in the beam do not exceed the allowable stresses for the material. This section considers only the bending stresses [that is, the stresses obtained from the flexure formula, Eq. (10-14)].

When designing a beam to resist bending stresses, begin by calculating the **required section modulus**. For instance, if the beam has a doubly symmetric cross section and the allowable stresses are the same for both tension and compression, calculate the required modulus by dividing the maximum bending moment by the allowable bending stress for the material [see Eq. (10-17)]:

$$S = \frac{M_{max}}{\sigma_{allow}} \tag{10-25}$$

The allowable stress is based upon the properties of the material and the desired factor of safety. To ensure that this stress is not exceeded, choose a beam that provides a section modulus at least as large as that obtained from Eq. (10-25).

If the cross section is not doubly symmetric, or if the allowable stresses are different for tension and compression, it may be necessary to determine two required section moduli—one based upon tension and the other based upon compression. Then provide a beam that satisfies both criteria.

To minimize weight and save material, select a beam that has the least cross-sectional area while still providing the required section moduli (and also meeting any other design requirements that may be imposed).

Beams are constructed in a great variety of shapes and sizes to suit a myriad of purposes. For instance, very large steel beams are fabricated by welding (Fig. 10-19), aluminum beams are extruded as round or rectangular tubes, wood beams are cut and glued to fit special requirements, and reinforced concrete beams are cast in any desired shape by proper construction of the forms.

In addition, beams of steel, aluminum, plastic, and wood can be ordered in **standard shapes and sizes** from catalogs supplied by dealers and manufacturers. Readily available shapes include wide-flange beams, I-beams, angles, channels, rectangular beams, and tubes.

FIGURE 10-19

Welder fabricating a large wide flange steel beam

B. Franklin/Shutterstock.com

Beams of Standardized Shapes and Sizes

The dimensions and properties of many kinds of beams are listed in engineering handbooks. For instance, in the United States, the shapes and sizes of structural-steel beams are standardized by the American Institute of Steel Construction (AISC), which publishes manuals giving their properties in both USCS and SI units. The tables in these manuals list cross-sectional dimensions and properties such as weight, cross-sectional area, moment of inertia, and section modulus.

Properties of aluminum and wood beams are tabulated in a similar manner and are available in publications of the Aluminum Association and the American Forest and Paper Association.

Abridged tables of steel beams and wood beams are given later in this book for use in solving problems using both USCS and SI units (see Appendixes C and D).

Structural-steel sections are given a designation such as W 30×211 in USCS units, which means that the section is of W shape (also called a wide-flange shape) with a nominal depth of 30 in. and a weight of 211 lb per ft of length (see Table C-1(a), Appendix C). The corresponding properties for each W shape are also given in SI units in Table C-1(b). For example, in SI units, the W 30×211 is listed as W 760×314 with a nominal depth of 760 millimeters and mass of 314 kilograms per meter of length.

Similar designations are used for S shapes (also called I-beams) and C shapes (also called channels), as shown in Tables C-2(a) and C-3(a) in USCS units and in Tables C-2(b) and C-3(b) in SI units. Angle sections, or L shapes, are designated by the lengths of the two legs and the thickness (see Tables C-4 and C-5). For example, L $8 \times 6 \times 1$ [see Table C-5(a)] denotes an angle with unequal legs, one of length 8 in. and the other of length 6 in., with a thickness of 1 in. The corresponding label in SI units for this unequal leg angle is L 203 × 152 × 25.4 [see Table C-5(b)].

The standardized steel sections described here are manufactured by *rolling*, a process in which a billet of hot steel is passed back and forth between rolls until it is formed into the desired shape.

Aluminum structural sections are usually made by the process of *extrusion*, in which a hot billet is pushed, or extruded, through a shaped die. Since dies are relatively easy to make and the material is workable; aluminum beams can be extruded in almost any desired shape. Standard shapes of wide-flange beams, I-beams, channels, angles, tubes, and other sections are listed in the *Aluminum Design Manual*. In addition, custom-made shapes can be ordered.

Most **wood beams** have rectangular cross sections and are designated by nominal dimensions, such as 4 × 8 inches. These dimensions represent the rough-cut size of the lumber. The net dimensions (or actual dimensions) of a wood beam are smaller than the nominal dimensions if the sides of the rough lumber have been planed, or *surfaced*, to make them smooth. Thus, a 4 × 8 wood beam has actual dimensions of 3.5 × 7.25 in. after it has been surfaced. Of course, the net dimensions of surfaced lumber should be used in all engineering computations. Therefore, net dimensions and the corresponding properties (in USCS units) are given in Appendix D. Similar tables are available in SI units.

Relative Efficiency of Various Beam Shapes

One of the objectives in designing a beam is to use the material as efficiently as possible within the constraints imposed by function, appearance, manufacturing costs, and the like. From the standpoint of strength alone, efficiency in bending depends primarily upon the shape of the cross section. In particular, the most efficient beam is one in which the material is located as far as practical from the neutral axis. The farther a given amount of material is from the neutral axis, the larger the section modulus becomes—and the larger the section modulus, the larger the bending moment that can be resisted (for a given allowable stress).

As an illustration, consider a cross section in the form of a **rectangle** of width b and height h (Fig. 10-20a). The section modulus [from Eq. (10-19b)] is

$$S = \frac{bh^2}{6} = \frac{Ah}{6} = 0.167 Ah \tag{10-26}$$

where A denotes the cross-sectional area. This equation shows that a rectangular cross section of given area becomes more efficient as the height h is increased (and the width b is decreased to keep the area constant). Of course, there is a practical limit to the increase in height, because the beam becomes laterally

---

FIGURE 10-20

Cross-sectional shapes of beams

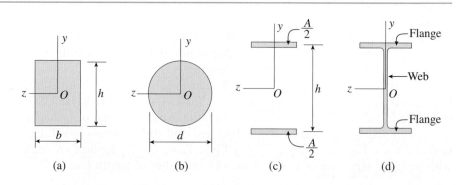

(a)   (b)   (c)   (d)

unstable when the ratio of height to width becomes too large. Thus, a beam of very narrow rectangular section will fail due to lateral (sideways) buckling rather than to insufficient strength of the material.

Next, compare a **solid circular cross section** of diameter d (Fig. 10-20b) with a square cross section of the same area. The side h of a square having the same area as the circle is $h = (d/2)\sqrt{\pi}$. The corresponding section moduli [from Eqs. (10-19b) and (10-20b)] are

$$S_{\text{square}} = \frac{h^3}{6} = \frac{\pi\sqrt{\pi}d^3}{48} = 0.1160d^3 \qquad \textbf{(10-27a)}$$

$$S_{\text{circle}} = \frac{\pi d^3}{32} = 0.0982d^3 \qquad \textbf{(10-27b)}$$

which gives

$$\frac{S_{\text{square}}}{S_{\text{circle}}} = 1.18 \qquad \textbf{(10-28)}$$

This result shows that a beam of square cross section is more efficient in resisting bending than is a circular beam of the same area. The reason, of course, is that a circle has a relatively larger amount of material located near the neutral axis. This material is less highly stressed; therefore, it does not contribute as much to the strength of the beam.

The **ideal cross-sectional shape** for a beam of given cross-sectional area A and height h would be obtained by placing one-half of the area at a distance $h/2$ above the neutral axis and the other half at distance $h/2$ below the neutral axis, as shown in Fig. 10-20c. For this ideal shape, obtain

$$I = 2\left(\frac{A}{2}\right)\left(\frac{h}{2}\right)^2 = \frac{Ah^2}{4} \qquad S = \frac{I}{h/2} = 0.5Ah \qquad \textbf{(10-29a,b)}$$

These theoretical limits are approached in practice by wide-flange sections and I-sections, which have most of their material in the flanges (Fig. 10-20d). For standard wide-flange beams, the section modulus is approximately

$$S \approx 0.35Ah \qquad \textbf{(10-30)}$$

which is less than the ideal but much larger than the section modulus for a rectangular cross section of the same area and height [see Eq. (10-26)].

Another desirable feature of a wide-flange beam is its greater width; hence, its greater stability with respect to sideways buckling when compared to a rectangular beam of the same height and section modulus. On the other hand, there are practical limits to how thin the web can be for a wide-flange beam. A web that is too thin is susceptible to localized buckling or it may be overstressed in shear (see Section 10.9).

The following four examples illustrate the process of selecting a beam on the basis of the allowable stresses. Only the effects of bending stresses (obtained from the flexure formula) are considered.

Note: When solving examples and problems that require the selection of a steel or wood beam from the tables in the appendixes, use the following rule: *If several choices are available in a table, select the lightest beam that provides the required section modulus.*

Example 10-5

A simply supported wood beam with a span length $L = 12$ ft carries a uniform load $q = 420$ lb/ft (Fig. 10-21). The allowable bending stress is 1800 psi, the wood weighs 35 lb/ft^3, and the beam is supported laterally against sideways buckling and tipping.

Select a suitable size for the beam from the table in Appendix D.

FIGURE 10-21

Example 10-5: Design of a simply supported wood beam

$q = 420$ lb/ft

$L = 12$ ft

Solution:

Use the four-step problem-solving approach. Combine steps as needed for an efficient solution.

1, 2. *Conceptualize, Categorize*: Since the beam weight is not known in advance, proceed by trial-and-error:

  i. Calculate the required section modulus based upon the given uniform load.

  ii. Select a trial size for the beam.

  iii. Add the weight of the beam to the uniform load and calculate a new required section modulus.

  Check to see that the selected beam is still satisfactory. If it is not, select a larger beam and repeat the process.

3. *Analyze*:

  i. The maximum bending moment in the beam occurs at the midpoint:

$$M_{max} = \frac{qL^2}{8} = \frac{(420 \text{ lb/ft})(12 \text{ ft})^2(12 \text{ in./ft})}{8} = 90{,}720 \text{ lb-in.}$$

  The required section modulus [Eq. (10-25)] is

$$S = \frac{M_{max}}{\sigma_{allow}} = \frac{90{,}720 \text{ lb-in.}}{1800 \text{ psi}} = 50.40 \text{ in}^3$$

  ii. From the table in Appendix D, the lightest beam that supplies a section modulus of at least 50.40 in^3 about axis 1–1 is a 3 × 12 in. beam (nominal dimensions). This beam has a section modulus equal to 52.73 in^3 and weighs 6.8 lb/ft. (Note that Appendix D gives weights of beams based upon a density of 35 lb/ft^3.)

iii. The uniform load on the beam now becomes 426.8 lb/ft, and the corresponding required section modulus is

$$S = (50.40 \text{ in}^3)\left(\frac{426.8 \text{ lb/ft}}{420 \text{ lb/ft}}\right) = 51.22 \text{ in}^3$$

4. *Finalize:* The previously selected beam has a section modulus of 52.73 in^3, which is larger than the required modulus of 51.22 in^3.

Therefore, a 3 × 12 in. beam is satisfactory.

Note: If the weight density of the wood is other than 35 lb/ft^3, compute the weight of the beam per linear foot by multiplying the value in the last column in Appendix D by the ratio of the actual weight density to 35 lb/ft^3.

Example 10-6

A vertical post 2.5-meters high must support a lateral load $P = 12$ kN at its upper end (Fig. 10-22). Two plans are proposed—a solid wood post and a hollow aluminum tube.

(a) What is the minimum required diameter d_1 of the wood post if the allowable bending stress in the wood is 15 MPa?

(b) What is the minimum required outer diameter d_2 of the aluminum tube if its wall thickness is to be one-eighth of the outer diameter and the allowable bending stress in the aluminum is 50 MPa?

FIGURE 10-22

Example 10-6: (a) Solid wood post and (b) aluminum tube

(a)   (b)

Solution:

Use the four-step problem-solving approach. Combine steps as needed for an efficient solution.

1. *Conceptualize*:

   Maximum bending moment: The maximum moment occurs at the base of the post and is equal to the load P times the height h; thus,

   $$M_{max} = Ph = (12 \text{ kN})(2.5 \text{ m}) = 30 \text{ kN} \cdot \text{m}$$

Part (a): Wood post.

2, 3. *Categorize, Analyze*: The required section modulus S_1 for the wood post [see Eqs. (10-20b and 10-25)] is

$$S_1 = \frac{\pi d_1^3}{32} = \frac{M_{max}}{\sigma_{allow}} = \frac{30 \text{ kN} \cdot \text{m}}{15 \text{ MPa}} = 0.0020 \text{ m}^3 = 2 \times 10^6 \text{ mm}^3$$

Solving for the diameter gives

$$d_1 = 273 \text{ mm} \qquad \leftarrow$$

4. *Finalize*: The diameter selected for the wood post must be equal to or larger than 273 mm if the allowable stress is not to be exceeded.

Part (b): Aluminum tube.

2, 3. *Categorize, Analyze*: To determine the section modulus S_2 for the tube, first find the moment of inertia I_2 of the cross section. The wall thickness of the tube is $d_2/8$; therefore, the inner diameter is $d_2 - d_2/4$, or $0.75d_2$. Thus, the moment of inertia [see Eq. (10-20a)] is

$$I_2 = \frac{\pi}{64}[d_2^4 - (0.75d_2)^4] = 0.03356d_2^4$$

The section modulus of the tube is now obtained from Eq. (10-18) as

$$S_2 = \frac{I_2}{c} = \frac{0.03356d_2^4}{d_2/2} = 0.06712d_2^3$$

The required section modulus is obtained from Eq. (10-25):

$$S_2 = \frac{M_{max}}{\sigma_{allow}} = \frac{30 \text{ kN} \cdot \text{m}}{50 \text{ MPa}} = 0.0006 \text{ m}^3 = 600 \times 10^3 \text{ mm}^3$$

Equate the two preceding expressions for the section modulus, then solve for the required outer diameter:

$$d_2 = \left(\frac{600 \times 10^3 \text{ mm}^3}{0.06712}\right)^{1/3} = 208 \text{ mm} \qquad \leftarrow$$

4. *Finalize*: The corresponding inner diameter is 0.75(208 mm), or 156 mm.

Example 10-7

A simple beam AB of span length 21 ft must support a uniform load $q = 2000$ lb/ft distributed along the beam in the manner shown in Fig. 10-23a.

Considering both the uniform load and the weight of the beam, and also using an allowable bending stress of 18,000 psi, select a structural steel beam of wide-flange shape to support the loads.

FIGURE 10-23

Example 10-7: Design of a simple beam with partial uniform loads

(a)   (b)

Solution:

Use the four-step problem-solving approach. Combine steps as needed for an efficient solution.

1, 2. *Conceptualize, Categorize:* In this example, proceed as follows:

   i. Find the maximum bending moment in the beam due to the uniform load.

   ii. Knowing the maximum moment, find the required section modulus.

   iii. Select a trial wide-flange beam from Table C-1(a) in Appendix C and obtain the weight of the beam.

   iv. With the weight known, calculate a new value of the bending moment and a new value of the section modulus.

Determine whether the selected beam is still satisfactory. If it is not, select a new beam size and repeat the process until a satisfactory size of beam has been found.

Maximum bending moment: To assist in locating the cross section of maximum bending moment, construct the shear-force diagram (Fig. 10-23b) using the methods described in Chapter 6. As part of that process, determine the reactions at the supports:

$$R_A = 18,860 \text{ lb} \qquad R_B = 17,140 \text{ lb}$$

The distance x_1 from the left-hand support to the cross section of zero shear force is obtained from

$$V = R_A - qx_1 = 0$$

which is valid in the range $0 \leq x \leq 12$ ft. Solve for x_1 to get

$$x_1 = \frac{R_A}{q} = \frac{18,860 \text{ lb}}{2000 \text{ lb/ft}} = 9.430 \text{ ft}$$

which is less than 12 ft; therefore, the calculation is valid.

The maximum bending moment occurs at the cross section where the shear force is zero; therefore,

$$M_{max} = R_A x_1 - \frac{q x_1^2}{2} = 88,920 \text{ lb-ft}$$

3. *Analyze*:

Required section modulus: The required section modulus (based only upon the load q) is obtained from Eq. (10-25):

$$S = \frac{M_{max}}{\sigma_{allow}} = \frac{(88,920 \text{ lb-ft})(12 \text{ in/ft})}{18,000 \text{ psi}} = 59.3 \text{ in}^3$$

Trial beam: Now turn to Table C-1(a) and select the lightest wide-flange beam having a section modulus greater than 59.3 in^3. The lightest beam that provides this section modulus is W 12×50 with $S = 64.7$ in^3. This beam weighs 50 lb/ft. (Recall that the tables in Appendix C are abridged, so a lighter beam may actually be available.)

Now recalculate the reactions, maximum bending moment, and required section modulus with the beam loaded by both the uniform load q and its own weight. Under these combined loads, the reactions are

$$R_A = 19,380 \text{ lb} \qquad R_B = 17,670 \text{ lb}$$

and the distance to the cross section of zero shear becomes

$$x_1 = \frac{19,380 \text{ lb}}{2050 \text{ lb/ft}} = 9.454 \text{ ft}$$

The maximum bending moment increases to 91,610 lb-ft, and the new required section modulus is

$$S = \frac{M_{max}}{\sigma_{allow}} = \frac{(91,610 \text{ lb-ft})(12 \text{ in/ft})}{18,000 \text{ psi}} = 61.1 \text{ in}^3$$

4. *Finalize*: Thus, the W 12×50 beam with section modulus $S = 64.7$ in^3 is still satisfactory.

Note: If the new required section modulus exceeded that of the W 12×50 beam, a new beam with a larger section modulus would be selected and the process repeated.

Example 10-8

FIGURE 10-24

Example 10-8: Wood dam with horizontal planks A supported by vertical posts B

(a) Top view   (b) Side view

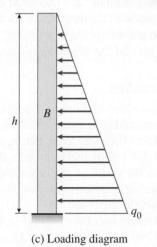

(c) Loading diagram

A temporary wood dam is constructed of horizontal planks A supported by vertical wood posts B that are sunk into the ground so that they act as cantilever beams (Fig. 10-24). The posts are of square cross section (dimensions $b \times b$) and spaced at distance $s = 0.8$ m, center to center. Assume that the water level behind the dam is at its full height $h = 2.0$ m.

Determine the minimum required dimension b of the posts if the allowable bending stress in the wood is $\sigma_{\text{allow}} = 8.0$ MPa.

Solution:

Use the four-step problem-solving approach.

1. *Conceptualize*:

Loading diagram: Each post is subjected to a triangularly distributed load produced by the water pressure acting against the planks. Consequently, the loading diagram for each post is triangular (Fig. 10-24c). The maximum intensity q_0 of the load on the posts is equal to the water pressure at depth h times the spacing s of the posts:

$$q_0 = \gamma h s \tag{a}$$

where γ is the specific weight of water. Note that q_0 has units of force per unit distance, γ has units of force per unit volume, and both h and s have units of length.

2. *Categorize*:

Section modulus: Since each post is a cantilever beam, the maximum bending moment occurs at the base and is given by

$$M_{\text{max}} = \frac{q_0 h}{2}\left(\frac{h}{3}\right) = \frac{\gamma h^3 s}{6} \tag{b}$$

Therefore, the required section modulus [Eq. (10-25)] is

$$S = \frac{M_{\text{max}}}{\sigma_{\text{allow}}} = \frac{\gamma h^3 s}{6\sigma_{\text{allow}}} \tag{c}$$

3. *Analyze*: For a beam of square cross section, the section modulus is $S = b^3/6$ [see Eq. (10-19b)]. Substitute this expression for S into Eq. (c) to get a formula for the cube of the minimum dimension b of the posts:

$$b^3 = \frac{\gamma h^3 s}{\sigma_{\text{allow}}} \tag{d}$$

Numerical values: Now substitute numerical values into Eq. (d) and obtain

$$b^3 = \frac{(9.81 \text{ kN/m}^3)(2.0 \text{ m})^3(0.8 \text{ m})}{8.0 \text{ MPa}} = 0.007848 \text{ m}^3 = 7.848 \times 10^6 \text{ mm}^3$$

from which

$$b = 199 \text{ mm}$$

4. *Finalize*: Thus, the minimum required dimension b of the posts is 199 mm. Any larger dimension, such as 200 mm, ensures that the actual bending stress is less than the allowable stress.

10.7 Shear Stresses in Beams of Rectangular Cross Section

When a beam is in *pure bending*, the only stress resultants are the bending moments and the only stresses are the normal stresses acting on the cross sections. However, most beams are subjected to loads that produce both bending moments and shear forces (*nonuniform bending*). In these cases, both normal and shear stresses are developed in the beam. The normal stresses are calculated from the flexure formula (see Section 10.5), provided the beam is constructed of a linearly elastic material. The shear stresses are discussed in this and the following two sections.

FIGURE 10-25

Shear stresses in a beam of rectangular cross section

(a)

(b)   (c)

Vertical and Horizontal Shear Stresses

Consider a beam of rectangular cross section (width b and height h) subjected to a positive shear force V (Fig. 10-25a). It is reasonable to assume that the shear stresses τ acting on the cross section are parallel to the shear force, that is, parallel to the vertical sides of the cross section. It is also reasonable to assume that the shear stresses are uniformly distributed across the width of the beam, although they may vary over the height. Using these two assumptions, you can determine the intensity of the shear stress at any point on the cross section.

For purposes of analysis, isolate a small element mn of the beam (Fig. 10-25a) by cutting between two adjacent cross sections and between two horizontal planes. Assume the shear stresses τ acting on the front face of this element are vertical and uniformly distributed from one side of the beam to the other. Also, from the discussion of shear stresses in Section 7.6, shear stresses acting on one side of an element are accompanied by shear stresses of equal magnitude acting on perpendicular faces of the element (Figs. 10-25b and c). Thus, there are horizontal shear stresses acting between horizontal layers of the beam as well as vertical shear stresses acting on the cross sections. At any point in the beam, these complementary shear stresses are equal in magnitude.

The equality of the horizontal and vertical shear stresses acting on an element leads to an important conclusion regarding the shear stresses at the top and bottom of the beam. If you imagine that the element mn (Fig. 10-25a) is located at either the top or the bottom, it follows that the horizontal shear stresses must vanish, because there are no stresses on the outer surfaces of the beam. It follows that the vertical shear stresses must also vanish at those locations; in other words, $\tau = 0$ where $y = \pm h/2$.

The existence of horizontal shear stresses in a beam can be demonstrated by a simple experiment. Place two identical rectangular beams on simple supports and load them by a force P, as shown in Fig. 10-26a. If friction between the beams is small, the beams bend independently (Fig. 10-26b). Each beam is in compression above its own neutral axis and in tension below its neutral axis; therefore, the bottom surface of the upper beam slides with respect to the top surface of the lower beam.

Now suppose that the two beams are glued along the contact surface, so they become a single solid beam. When this beam is loaded, horizontal shear stresses must develop along the glued surface in order to prevent the sliding shown in Fig. 10-26b. Because of the presence of these shear stresses, the single solid beam is much stiffer and stronger than the two separate beams.

Derivation of Shear Formula

Now derive a formula for the shear stresses τ in a rectangular beam. However, instead of evaluating the vertical shear stresses acting on a cross section, it is easier to evaluate the horizontal shear stresses acting between layers of the beam. Of course, the vertical shear stresses have the same magnitudes as the horizontal shear stresses.

Now consider a beam in nonuniform bending (Fig. 10-27a). Take two adjacent cross sections mn and m_1n_1 at a distance dx apart, and consider the **element** mm_1n_1n. The bending moment and shear force acting on the left-hand face of this element are denoted M and V, respectively. Since both the bending moment and shear force may change when moving along the axis of the beam, the corresponding quantities on the right-hand face (Fig. 10-27a) are denoted $M + dM$ and $V + dV$.

Because of the presence of the bending moments and shear forces, the element shown in Fig. 10-27a is subjected to normal and shear stresses on both cross-sectional faces. However, only the normal stresses are needed in the following derivation, so only the normal stresses are shown in Fig. 10-27b. On cross sections mn and m_1n_1, the normal stresses are, respectively,

$$\sigma_1 = -\frac{My}{I} \quad \text{and} \quad \sigma_2 = -\frac{(M + dM)y}{I} \qquad \textbf{(10-31a,b)}$$

as given by the flexure formula [Eq. (10-14)]. In these expressions, y is the distance from the neutral axis and I is the moment of inertia of the cross-sectional area about the neutral axis.

Next, isolate a **subelement** mm_1p_1p by passing a horizontal plane pp_1 through element mm_1n_1n (Fig. 10-27b). The plane pp_1 is at distance y_1 from the neutral surface of the beam. The subelement is shown separately in Fig. 10-27c. Note that its top face is part of the upper surface of the beam and thus is free from stress. Its bottom face (which is parallel to the neutral surface and distance y_1 from it) is acted upon by the horizontal shear stresses τ existing at this level in the beam. Its cross-sectional faces mp and m_1p_1 are acted upon by the bending stresses σ_1 and σ_2, respectively, which are produced by the bending moments. Vertical shear stresses also act on the cross-sectional faces; however, these stresses do not affect the equilibrium of the subelement in the horizontal direction (the x direction), so they are not shown in Fig. 10-27c.

If the bending moments at cross sections mn and m_1n_1 (Fig. 10-27b) are equal (that is, if the beam is in pure bending), the normal stresses σ_1 and σ_2

FIGURE 10-26

Bending of two separate beams

(a)

(b)

FIGURE 10-27

Shear stresses in a beam
of rectangular cross section

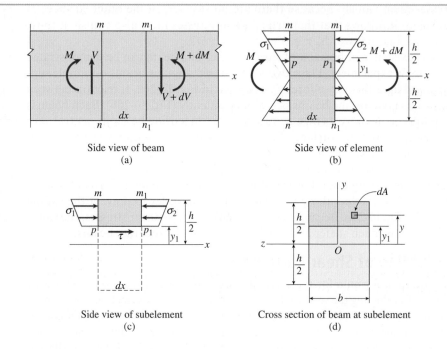

Side view of beam
(a)

Side view of element
(b)

Side view of subelement
(c)

Cross section of beam at subelement
(d)

acting over the sides mp and m_1p_1 of the subelement (Fig. 10-27c) also are equal. Under these conditions, the subelement is in equilibrium under the action of the normal stresses alone; therefore, the shear stresses τ acting on the bottom face pp_1 vanish. This conclusion is obvious inasmuch as a beam in pure bending has no shear force and hence no shear stresses.

If the bending moments vary along the x axis (nonuniform bending), the shear stress τ acting on the bottom face of the subelement (Fig. 10-27c) can be determined by considering the equilibrium of the subelement in the x direction.

Begin by identifying an element of area dA in the *cross section* at a distance y from the neutral axis (Fig. 10-27d). The force acting on this element is σdA, where σ is the normal stress obtained from the flexure formula. If the element of area is located on the left-hand face mp of the subelement (where the bending moment is M), the normal stress is given by Eq. (10-31a); therefore, the element of force is

$$\sigma_1 dA = \frac{My}{I} dA$$

Note that only absolute values are used in this equation because the directions of the stresses are obvious from Fig. 10-27. Summing these elements of force over the area of face mp of the subelement (Fig. 10-27c) gives the total horizontal force F_1 acting on that face:

$$F_1 = \int \sigma_1 dA = \int \frac{My}{I} dA \qquad \text{(10-32a)}$$

Note that this integration is performed over the area of the shaded part of the cross section shown in Fig. 10-27d, that is, over the area of the cross section from $y = y_1$ to $y = h/2$.

The force F_1 is shown in Fig. 10-28 on a partial free-body diagram of the subelement. (Vertical forces have been omitted.)

In a similar manner, the total force F_2 acting on the right-hand face m_1p_1 of the subelement (Fig. 10-28 and Fig. 10-27c) is

$$F_2 = \int \sigma_2 dA = \int \frac{(M + dM)y}{I} dA \qquad \text{(10-32b)}$$

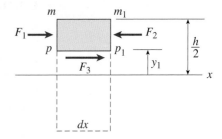

Knowing the forces F_1 and F_2, now determine the horizontal force F_3 acting on the bottom face of the subelement.

Since the subelement is in equilibrium, sum forces in the x direction and obtain

$$F_3 = F_2 - F_1 \qquad \text{(10-32c)}$$

or

$$F_3 = \int \frac{(M + dM)y}{I} dA - \int \frac{My}{I} dA = \int \frac{(dM)y}{I} dA$$

The quantities dM and I in the last term can be moved outside the integral sign because they are constants at any given cross section and are not involved in the integration. Thus, the expression for the force F_3 becomes

$$F_3 = \frac{dM}{I} \int y dA \qquad \text{(10-33)}$$

If the shear stresses τ are uniformly distributed across the width b of the beam, the force F_3 is also equal to

$$F_3 = \tau b dx \qquad \text{(10-34)}$$

in which $b dx$ is the area of the bottom face of the subelement.

Combine Eqs. (10-33) and (10-34) and solve for the shear stress τ to get

$$\tau = \frac{dM}{dx} \left(\frac{1}{Ib} \right) \int y dA \qquad \text{(10-35)}$$

The quantity dM/dx is equal to the shear force V, so the preceding expression becomes

$$\tau = \frac{V}{Ib} \int y dA \qquad \text{(10-36)}$$

The integral in this equation is evaluated over the shaded part of the cross section (Fig. 10-27d), as already explained. Thus, the integral is the first moment of the shaded area with respect to the neutral axis (the z axis). In other words, *the integral is the first moment of the cross-sectional area above the level at which the shear stress τ is being evaluated.* This first moment is usually denoted by the symbol Q:

$$Q = \int y dA \qquad \text{(10-37)}$$

With this notation, the equation for the shear stress becomes

$$\tau = \frac{VQ}{Ib} \qquad \text{(10-38)}$$

This equation, known as the **shear formula**, can be used to determine the shear stress τ at any point in the cross section of a rectangular beam. Note that for a specific cross section, the shear force V, moment of inertia I, and width b are constants. However, the first moment Q (and hence the shear stress τ) varies with the distance y_1 from the neutral axis.

Calculation of the First Moment Q

If the level where the shear stress is to be determined is above the neutral axis, as shown in Fig. 10-27d, it is natural to obtain Q by calculating the first moment of the cross-sectional area *above* that level (the shaded area in the figure). However, as an alternative, you could calculate the first moment of the remaining cross-sectional area, that is, the area *below* the shaded area. Its first moment is equal to the negative of Q.

The explanation lies in the fact that the first moment of the entire cross-sectional area with respect to the neutral axis is equal to zero (because the neutral axis passes through the centroid). Therefore, the value of Q for the area below the level y_1 is the negative of Q for the area above that level. Use the area above the level y_1 when the point where the shear stress is computed is in the upper part of the beam, and use the area below the level y_1 when the point is in the lower part of the beam.

Furthermore, don't bother with sign conventions for V and Q. Instead, treat all terms in the shear formula as positive quantities and determine the direction of the shear stresses by inspection, since the stresses act in the same direction as the shear force V itself. This procedure for determining shear stresses is illustrated in Example 10-9.

FIGURE 10-29

Distribution of shear stresses in a beam of rectangular cross section: (a) cross section of beam and (b) diagram showing the parabolic distribution of shear stresses over the height of the beam

Distribution of Shear Stresses in a Rectangular Beam

Now find the distribution of the shear stresses in a beam of rectangular cross section (Fig. 10-29). Obtain the first moment Q of the shaded part of the cross-sectional area by multiplying the area by the distance from its own centroid to the neutral axis:

$$Q = b\left(\frac{h}{2} - y_1\right)\left(y_1 + \frac{h/2 - y_1}{2}\right) = \frac{b}{2}\left(\frac{h^2}{4} - y_1^2\right) \quad \text{(10-39a)}$$

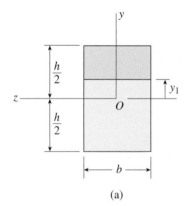

(a)

This same result can be obtained by integration using Eq. (10-37):

$$Q = \int y\,dA = \int_0^{h/2} ybdy = \frac{b}{2}\left(\frac{h^2}{4} - y_1^2\right) \quad \text{(10-39b)}$$

Substitute the expression for Q into the shear formula [Eq. (10-38)] to get

$$\tau = \frac{V}{2I}\left(\frac{h^2}{4} - y_1^2\right) \quad \text{(10-40)}$$

This equation shows that the shear stresses in a rectangular beam vary quadratically with the distance y_1 from the neutral axis. Thus, when plotted along the height of the beam, τ varies as shown in Fig. 10-29b. Note that the shear stress is zero when $y_1 = \pm h/2$.

The maximum value of the shear stress occurs at the neutral axis ($y_1 = 0$) where the first moment Q has its maximum value. Substitute $y_1 = 0$ into Eq. (10-40) to get

(b)

$$\tau_{\max} = \frac{Vh^2}{8I} = \frac{3V}{2A} \quad \text{(10-41)}$$

where $A = bh$ is the cross-sectional area. Thus, the maximum shear stress in a beam of rectangular cross section is 50% larger than the average shear stress V/A.

Note again that the preceding equations for the shear stresses can be used to calculate either the vertical shear stresses acting on the cross sections or the horizontal shear stresses acting between horizontal layers of the beam.[5]

Limitations

The formulas for shear stresses in this section are subject to the same restrictions as the flexure formula from which they are derived. Thus, they are valid only for beams of linearly elastic materials with small deflections.

In the case of rectangular beams, the accuracy of the shear formula depends upon the height-to-width ratio of the cross section. The formula may be considered as exact for very narrow beams (height h much larger than the width b). However, it becomes less accurate as b increases relative to h. For instance, when the beam is square ($b = h$), the true maximum shear stress is about 13% larger than the value given by Eq. (10-41).

A common error is to apply the shear formula [(Eq. (10-38)] to cross-sectional shapes for which it is not applicable. For instance, it is not applicable to sections of triangular or semicircular shapes. To avoid misusing the formula, keep in mind the following assumptions that underlie the derivation: (1) The edges of the cross section must be parallel to the y axis (so that the shear stresses act parallel to the y axis), and (2) the shear stresses must be uniform across the width of the cross section. These assumptions are fulfilled only in certain cases, such as those discussed in this and the next two sections.

Finally, the shear formula applies only to prismatic beams. If a beam is nonprismatic (for instance, if the beam is tapered), the shear stresses are quite different from those predicted by the formulas given here.

Effects of Shear Strains

Because the shear stress τ varies parabolically over the height of a rectangular beam, it follows that the shear strain $\gamma = \tau/G$ also varies parabolically. As a result of these shear strains, cross sections of the beam that were originally plane surfaces become warped. This warping is shown in Fig. 10-30, where cross sections mn and pq, originally plane, have become curved surfaces m_1n_1 and p_1q_1, with the maximum shear strain occurring at the neutral surface. At points m_1, p_1, n_1, and q_1, the shear strain is zero, and therefore the curves m_1n_1 and p_1q_1 are perpendicular to the upper and lower surfaces of the beam.

If the shear force V is constant along the axis of the beam, warping is the same at every cross section. Therefore, stretching and shortening of longitudinal elements due to the bending moments is unaffected by the shear strains, and the distribution of the normal stresses is the same as in pure bending. Moreover, detailed investigations using advanced methods of analysis show that the warping of cross sections due to shear strains does not substantially affect the longitudinal strains even when the shear force varies continuously along the length. Thus, under most conditions, it is justifiable to use the flexure formula [Eq. (10-14)] for nonuniform bending, even though the formula was derived for pure bending.

FIGURE 10-30

Warping of the cross sections of a beam due to shear strains

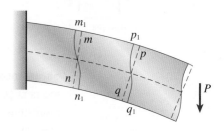

---

[5]The shear-stress analysis presented in this section was developed by the Russian engineer D. J. Jourawski.

Example 10-9

FIGURE 10-31

Example 10-9: (a) Simple beam with uniform load, (b) cross section of beam, and (c) stress element showing the normal and shear stresses at point C

(a)

(b)

(c)

A metal beam with a span $L = 3$ ft is simply supported at points A and B (Fig. 10-31a). The uniform load on the beam (including its own weight) is $q = 160$ lb/in. The cross section of the beam is rectangular (Fig. 10-31b) with width $b = 1$ in. and height $h = 4$ in. The beam is adequately supported against sideways buckling.

Determine the normal stress σ_C and shear stress τ_C at point C, which is located 1 in. below the top of the beam and 8 in. from the right-hand support. Show these stresses on a sketch of a stress element at point C.

Solution:

Use the four-step problem-solving approach. Combine steps as needed for an efficient solution.

1, 2. *Conceptualize, Categorize:*

Shear force and bending moment: The shear force V_C and bending moment M_C at the cross section through point C are found as described in Section 6.4. The results are

$$M_C = 17{,}920 \text{ lb-in.} \qquad V_C = -1600 \text{ lb}$$

The signs of these quantities are based upon the standard sign conventions for bending moments and shear forces (see Section 6.4).

Moment of inertia: The moment of inertia of the cross-sectional area about the neutral axis (the z axis in Fig. 10-31b) is

$$I = \frac{bh^3}{12} = \frac{1}{12}(1.0 \text{ in.})(4.0 \text{ in.})^3 = 5.333 \text{ in}^4$$

3. *Analyze:*

Normal stress at point C: The normal stress at point C is found from the flexure formula [Eq. (10-14)] with the distance y from the neutral axis equal to 1.0 in.; thus,

$$\sigma_C = -\frac{My}{I} = -\frac{(17{,}920 \text{ lb-in.})(1.0 \text{ in.})}{5.333 \text{ in}^4} = -3360 \text{ psi} \quad \longleftarrow$$

The minus sign indicates that the stress is compressive, as expected.

Shear stress at point C: To obtain the shear stress at point C, evaluate the first moment Q_C of the cross-sectional area above point C (Fig. 10-31b). This first moment is equal to the product of the area and its centroidal distance (denoted y_C) from the z axis; thus,

$$A_C = (1.0 \text{ in.})(1.0 \text{ in.}) = 1.0 \text{ in}^2 \qquad y_C = 1.5 \text{ in.} \qquad Q_C = A_C y_C = 1.5 \text{ in}^3$$

Now substitute numerical values into the shear formula [Eq. (10-41)] and obtain the magnitude of the shear stress:

$$\tau_C = \frac{V_C Q_C}{Ib} = \frac{(1600 \text{ lb})(1.5 \text{ in}^3)}{(5.333 \text{ in}^4)(1.0 \text{ in.})} = 450 \text{ psi}$$

4. *Finalize*: The direction of this stress can be established by inspection because it acts in the same direction as the shear force. In this example, the shear force acts upward on the part of the beam to the left of point C and downward on the part of the beam to the right of point C. The best way to show the directions of both the normal and shear stresses is to draw a stress element.

Stress element at point C: The stress element, shown in Fig. 10-31c, is cut from the side of the beam at point C (Fig. 10-31a). Compressive stresses $\sigma_C = 3360$ psi act on the cross-sectional faces of the element and shear stresses $\tau_C = 450$ psi act on the top and bottom faces as well as the cross-sectional faces.

Example 10-10

FIGURE 10-32

Example 10-10: Wood beam with concentrated loads

(a)

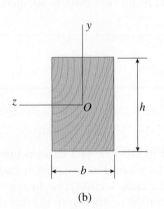

(b)

A wood beam AB supporting two concentrated loads P (Fig. 10-32a) has a rectangular cross section of width $b = 100$ mm and height $h = 150$ mm (Fig. 10-32b). The distance from each end of the beam to the nearest load is $a = 0.5$ m.

Determine the maximum permissible value P_{max} of the loads if the allowable stress in bending is $\sigma_{allow} = 11$ MPa (for both tension and compression) and the allowable stress in horizontal shear is $\tau_{allow} = 1.2$ MPa. (Disregard the weight of the beam itself.)

Note: Wood beams are much weaker in *horizontal shear* (shear parallel to the longitudinal fibers in the wood) than in *cross-grain shear* (shear on the cross sections). Consequently, the allowable stress in horizontal shear is usually considered in design.

Solution:

Use the four-step problem-solving approach.

1. *Conceptualize*: The maximum shear force occurs at the supports, and the maximum bending moment occurs throughout the region between the loads. Their values are

$$V_{max} = P \qquad M_{max} = Pa$$

Also, the section modulus S and cross-sectional area A are

$$S = \frac{bh^2}{6} \qquad A = bh$$

2. *Categorize*: The maximum normal and shear stresses in the beam are obtained from the flexure and shear formulas [Eqs. (10-17) and (10-41)]:

$$\sigma_{max} = \frac{M_{max}}{S} = \frac{6Pa}{bh^2} \qquad \tau_{max} = \frac{3V_{max}}{2A} = \frac{3P}{2bh}$$

Therefore, the maximum permissible values of the load P in bending and shear, respectively, are

$$P_{bending} = \frac{\sigma_{allow}bh^2}{6a} \qquad P_{shear} = \frac{2\tau_{allow}bh}{3}$$

3. *Analyze*: Substitute numerical values into these formulas to get

$$P_{bending} = \frac{(11\ \text{MPa})(100\ \text{mm})(150\ \text{mm})^2}{6(0.5\ \text{m})} = 8.25\ \text{kN}$$

$$P_{shear} = \frac{2(1.2\ \text{MPa})(100\ \text{mm})(150\ \text{mm})}{3} = 12.0\ \text{kN}$$

Thus, the bending stress governs the design, and the maximum permissible load is

$$P_{max} = 8.25\ \text{kN}$$

4. *Finalize*: A more complete analysis of this beam would require that the weight of the beam be taken into account, thus reducing the permissible load.

Notes:

   i. In this example, the maximum normal stresses and maximum shear stresses do not occur at the same locations in the beam—the normal stress is maximum in the middle region of the beam at the top and bottom of the cross section, and the shear stress is maximum near the supports at the neutral axis of the cross section.

   ii. For most beams, the bending stresses (not the shear stresses) control the allowable load, as in this example.

   iii. Although wood is not a homogeneous material and often departs from linearly elastic behavior, approximate results still can be obtained from the flexure and shear formulas. These approximate results are usually adequate for designing wood beams.

10.8 Shear Stresses in Beams of Circular Cross Section

When a beam has a **circular cross section** (Fig. 10-33), you can no longer assume that the shear stresses act parallel to the y axis. For instance, it is easy to prove that at point m (on the boundary of the cross section) the shear stress τ must act *tangent* to the boundary. This observation follows from the fact that the outer surface of the beam is free of stress, and the shear stress acting on the cross section can have no component in the radial direction.

Although there is no simple way to find the shear stresses acting throughout the entire cross section, the shear stresses at the neutral axis (where the stresses are the largest) are found by making some reasonable assumptions about the stress distribution. Assume that the stresses act parallel to the y axis and have a constant intensity across the width of the beam (from point p to point q in Fig. 10-33). Since these assumptions are the same as those used in deriving the shear formula $\tau = VQ/Ib$ [Eq. (10-38)], use the shear formula to calculate the stresses at the neutral axis.

For use in the shear formula, the following properties pertaining to a circular cross section having radius r are needed:

$$I = \frac{\pi r^4}{4} \quad Q = A\bar{y} = \left(\frac{\pi r^2}{2}\right)\left(\frac{4r}{3\pi}\right) = \frac{2r^3}{3} \quad b = 2r \quad \textbf{(10-42a,b)}$$

The expression for the moment of inertia I is taken from Case 9 of Appendix B, and the expression for the first moment Q is based upon the formulas for a semicircle (Case 10, Appendix B). Substitute these expressions into the shear formula to obtain

$$\tau_{max} = \frac{VQ}{Ib} = \frac{V(2r^3/3)}{(\pi r^4/4)(2r)} = \frac{4V}{3\pi r^2} = \frac{4V}{3A} \quad \textbf{(10-43)}$$

in which $A = \pi r^2$ is the area of the cross section. This equation shows that the maximum shear stress in a circular beam is equal to 4/3 times the average vertical shear stress V/A.

For a beam with a **hollow circular cross section** (Fig. 10-34), again assume with reasonable accuracy that the shear stresses at the neutral axis are parallel to the y axis and uniformly distributed across the section. Consequently, the shear formula is used to find the maximum stresses. The required properties for a hollow circular section are

$$I = \frac{\pi}{4}(r_2^4 - r_1^4) \quad Q = \frac{2}{3}(r_2^3 - r_1^3) \quad b = 2(r_2 - r_1) \, \textbf{(10-44a,b,c)}$$

where r_1 and r_2 are the inner and outer radii of the cross section, respectively. Therefore, the maximum stress is

$$\tau_{max} = \frac{VQ}{Ib} = \frac{4V}{3A}\left(\frac{r_2^2 + r_2 r_1 + r_1^2}{r_2^2 + r_1^2}\right) \quad \textbf{(10-45)}$$

where

$$A = \pi(r_2^2 - r_1^2)$$

is the area of the cross section. Note that if $r_1 = 0$, Eq. (10-45) reduces to Eq. (10-43) for a solid circular beam.

Although the preceding theory for shear stresses in beams of circular cross section is approximate, it gives results differing by only a few percent from those obtained using the exact theory of elasticity. Consequently, Eqs. (10-43) and (10-45) can be used to determine the maximum shear stresses in circular beams under ordinary circumstances.

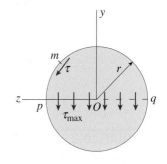

FIGURE 10-33

Shear stresses acting on the cross section of a circular beam

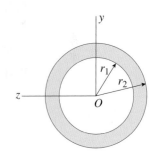

FIGURE 10-34

Hollow circular cross section

Example 10-11

A vertical pole consisting of a circular tube of outer diameter $d_2 = 4.0$ in. and inner diameter $d_1 = 3.2$ in. is loaded by a horizontal force $P = 1500$ lb (Fig. 10-35a).

(a) Determine the maximum shear stress in the pole.

(b) For the same load P and the same maximum shear stress, what is the diameter d_0 of a solid circular pole (Fig. 10-35b)?

FIGURE 10-35

Example 10-11: Shear stresses in beams of circular cross section

(a)   (b)

Solution:

Use the four-step problem-solving approach. Combine steps as needed for an efficient solution.

Part (a): Maximun shear stress.

1, 2. *Conceptualize, Categorize*: For the pole having a hollow circular cross section (Fig. 10-35a), use Eq. (10-45) with the shear force V replaced by the load P and the cross-sectional area A replaced by the expression $\pi(r_2^2 - r_1^2)$; thus,

$$\tau_{max} = \frac{4P}{3\pi}\left(\frac{r_2^2 + r_2 r_1 + r_1^2}{r_2^4 - r_1^4}\right) \tag{a}$$

3, 4. *Analyze, Finalize*: Next, substitute numerical values, namely,

$$P = 1500 \text{ lb} \qquad r_2 = d_2/2 = 2.0 \text{ in.} \qquad r_1 = d_1/2 = 1.6 \text{ in.}$$

to obtain

$$\tau_{max} = 658 \text{ psi}$$

which is the maximum shear stress in the pole.

Part (b): Diameter of solid circular pole.

1, 2. *Conceptualize, Categorize*: For the pole having a solid circular cross section (Fig. 10-33b), use Eq. (10-43) with V replaced by P and r replaced by $d_0/2$:

$$\tau_{max} = \frac{4P}{3\pi(d_0/2)^2} \qquad \text{(b)}$$

3. *Analyze*: Solve for d_0 to obtain

$$d_0^2 = \frac{16P}{3\pi\tau_{max}} = \frac{16(1500\ lb)}{3\pi(658\ psi)} = 3.87\ in.^2$$

that produces

$$d_0 = 1.97\ in.$$

4. *Finalize*: In this particular example, the solid circular pole has a diameter approximately one-half that of the tubular pole.

   Note: Shear stresses rarely govern the design of either circular or rectangular beams made of metals such as steel and aluminum. In these kinds of materials, the allowable shear stress is usually in the range 25 to 50% of the allowable tensile stress. In the case of the tubular pole in this example, the maximum shear stress is only 658 psi. In contrast, the maximum bending stress obtained from the flexure formula is 9700 psi for a relatively short pole of length 24 in. Thus, as the load increases, the allowable tensile stress will be reached long before the allowable shear stress is reached.

   The situation is quite different for materials that are weak in shear, such as wood. For a typical wood beam, the allowable stress in horizontal shear is in the range of 4 to 10% of the allowable bending stress. Consequently, even though the maximum shear stress is relatively low in value, it sometimes governs the design.

10.9 Shear Stresses in the Webs of Beams with Flanges

When a beam of wide-flange shape (Fig. 10-36a) is subjected to shear forces as well as bending moments (nonuniform bending), both normal and shear stresses are developed on the cross sections. The distribution of the shear stresses in a wide-flange beam is more complicated than in a rectangular beam. For instance, the shear stresses in the flanges of the beam act in both vertical and horizontal directions (the y and z directions), as shown by the small arrows in Fig. 10-36b. The horizontal shear stresses are much larger than the vertical shear stresses in the flanges.

FIGURE 10-36

(a) Beam of wide-flange shape
and (b) directions of the shear
stresses acting on a cross section

(a)

(b)

The shear stresses in the web of a wide-flange beam act only in the vertical direction and are larger than the stresses in the flanges. These stresses can be found by the same techniques used for finding shear stresses in rectangular beams.

Shear Stresses in the Web

Begin the analysis by determining the shear stresses at line ef in the web of a wide-flange beam (Fig. 10-37a). Make the same assumptions as those made for a rectangular beam; that is, assume that the shear stresses act parallel to the y axis and are uniformly distributed across the thickness of the web. Then the shear formula $\tau = VQ/Ib$ will still apply. However, the width b is now the thickness t of the web, and the area used in calculating the first moment Q is the area between line ef and the top edge of the cross section (indicated by the shaded area of Fig. 10-37a).

When finding the first moment Q of the shaded area, disregard the effects of the small fillets at the juncture of the web and flange (points b and c in Fig. 10-37a). The error in ignoring the areas of these fillets is very small. Then divide the shaded area into two rectangles. The first rectangle is the upper flange itself, which has the area

$$A_1 = b\left(\frac{h}{2} - \frac{h_1}{2}\right) \tag{10-46a}$$

in which b is the width of the flange, h is the overall height of the beam, and h_1 is the distance between the insides of the flanges. The second rectangle is the part of the web between ef and the flange, that is, rectangle $efcb$, which has the area

$$A_2 = t\left(\frac{h_1}{2} - y_1\right) \tag{10-46b}$$

in which t is the thickness of the web and y_1 is the distance from the neutral axis to line ef.

The first moments of areas A_1 and A_2, evaluated about the neutral axis, are obtained by multiplying these areas by the distances from their respective

FIGURE 10-37

Shear stresses in the web of
a wide-flange beam:
(a) cross section of beam
and (b) distribution of vertical
shear stresses in the web

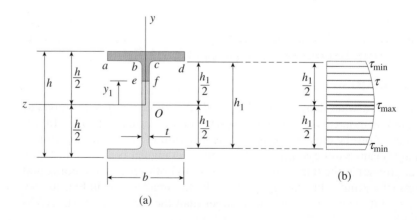

(a)

(b)

centroids to the z axis. Adding these first moments gives the first moment Q of the combined area:

$$Q = A_1\left(\frac{h_1}{2} + \frac{h/2 - h_1/2}{2}\right) + A_2\left(y_1 + \frac{h_1/2 - y_1}{2}\right)$$

Substituting for A_1 and A_2 from Eqs. (10-46a and b) and then simplifying gives

$$Q = \frac{b}{8}(h^2 - h_1^2) + \frac{t}{8}(h_1^2 - 4y_1^2) \qquad \textbf{(10-47)}$$

Therefore, the shear stress τ in the web of the beam at distance y_1 from the neutral axis is

$$\tau = \frac{VQ}{It} = \frac{V}{8It}\left[b(h^2 - h_1^2) + t(h_1^2 - 4y_1^2)\right] \qquad \textbf{(10-48)}$$

where the moment of inertia of the cross section is

$$I = \frac{bh^3}{12} - \frac{(b - t)h_1^3}{12} = \frac{1}{12}(bh^3 - bh_1^3 + th_1^3) \qquad \textbf{(10-49)}$$

Since all quantities in Eq. (10-48) are constants except y_1, note that τ varies quadratically throughout the height of the web, as shown by the graph in Fig. 10-37b. The graph is drawn only for the web and does not include the flanges. The reason is simple enough—Eq. (10-48) cannot be used to determine the vertical shear stresses in the flanges of the beam (see the discussion titled "Limitations" later in this section).

Maximum and Minimum Shear Stresses

The maximum shear stress in the web of a wide-flange beam occurs at the neutral axis where $y_1 = 0$. The minimum shear stress occurs where the web meets the flanges ($y_1 = \pm h_1/2$). These stresses, found from Eq. (10-48), are

$$\tau_{\max} = \frac{V}{8It}(bh^2 - bh_1^2 + th_1^2) \qquad \tau_{\min} = \frac{Vb}{8It}(h^2 - h_1^2) \quad \textbf{(10-50a,b)}$$

Both $\tau_{\max}$ and $\tau_{\min}$ are labeled on the graph of Fig. 10-37b. For typical wide-flange beams, the maximum stress in the web is from 10 to 60% greater than the minimum stress.

Although it may not be apparent from the preceding discussion, the stress $\tau_{\max}$ given by Eq. (10-50a) not only is the largest shear stress in the web but also is the largest shear stress anywhere in the cross section.

Shear Force in the Web

The vertical shear force carried by the web alone may be determined by multiplying the area of the shear-stress diagram (Fig. 10-37b) by the thickness t of the web. The shear-stress diagram consists of two parts: a rectangle of area $h_1\tau_{\min}$ and a parabolic segment of area

$$\frac{2}{3}(h_1)(\tau_{\max} - \tau_{\min})$$

FIGURE 10-37 (Repeated)

Shear stresses in the web of a wide-flange beam: (a) cross section of beam and (b) distribution of vertical shear stresses in the web

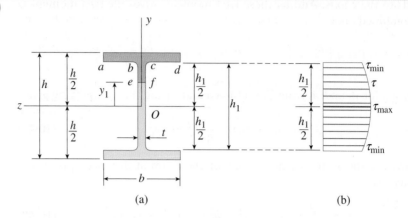

(a)   (b)

Adding these two areas, multiplying by the thickness t of the web, and then combining terms gives the total shear force in the web:

$$V_{web} = \frac{th_1}{3}(2\tau_{max} + \tau_{min}) \tag{10-51}$$

For beams of typical proportions, the shear force in the web is 90 to 98% of the total shear force V acting on the cross section; the remainder is carried by shear in the flanges.

Since the web resists most of the shear force, designers often calculate an approximate value of the maximum shear stress by dividing the total shear force by the area of the web. The result is the average shear stress in the web, assuming that the web carries *all* of the shear force:

$$\tau_{aver} = \frac{V}{th_1} \tag{10-52}$$

For typical wide-flange beams, the average stress calculated in this manner is within 10% (plus or minus) of the maximum shear stress calculated from Eq. (10-50a). Thus, Eq. (10-52) provides a simple way to estimate the maximum shear stress.

Limitations

The elementary shear theory presented in this section is suitable for determining the vertical shear stresses in the web of a wide-flange beam. However, when investigating vertical shear stresses in the flanges, you can no longer assume that the shear stresses are constant across the width of the section, that is, across the width b of the flanges (Fig. 10-37a). Hence, you cannot use the shear formula to determine these stresses.

To emphasize this point, consider the junction of the web and upper flange ($y_1 = h_1/2$), where the width of the section changes abruptly from t to b. The shear stresses on the free surfaces ab and cd (Fig. 10-37a) must be zero, whereas the shear stress across the web at line bc is τ_{min}. These observations indicate that the distribution of shear stresses at the junction of the web and the flange is quite complex and cannot be investigated by elementary methods. The stress analysis is further complicated by the use of fillets at the re-entrant corners (corners b and c). The fillets are necessary to prevent the stresses from becoming dangerously large, but they also alter the stress distribution across the web.

Thus, the shear formula cannot be used to determine the vertical shear stresses in the flanges. However, the shear formula does give good results for the shear stresses acting *horizontally* in the flanges (Fig. 10-36b).

This method for determining shear stresses in the webs of wide-flange beams also can be used for other sections having thin webs. For instance, Example 10-13 illustrates the procedure for a T-beam.

Example 10-12

A beam of wide-flange shape (Fig. 10-38a) is subjected to a vertical shear force $V = 45$ kN. The cross-sectional dimensions of the beam are $b = 165$ mm, $t = 7.5$ mm, $h = 320$ mm, and $h_1 = 290$ mm.

Determine the maximum shear stress, minimum shear stress, and total shear force in the web. (Disregard the areas of the fillets when making calculations.)

FIGURE 10-38

Example 10-12: Shear stresses in the web of a wide-flange beam

(a)

(b)

Solution:

Use the four-step problem-solving approach. Combine steps as needed for an efficient solution.

1, 2. *Conceptualize, Categorize*:

Maximum and minimum shear stresses: The maximum and minimum shear stresses in the web of the beam are given by Eqs. (10-50a and b). Before substituting into those equations, calculate the moment of inertia of the cross-sectional area from Eq. (10-49):

$$I = \frac{1}{12}(bh^3 - bh_1^3 + th_1^3) = 130.45 \times 10^6 \text{ mm}^4$$

3. *Analyze*: Now substitute this value for I, as well as the numerical values for the shear force V and the cross-sectional dimensions, into Eqs. (10-50a and b):

$$\tau_{max} = \frac{V}{8It}(bh^2 - bh_1^2 + th_1^2) = 21.0 \text{ MPa}$$

$$\tau_{min} = \frac{Vb}{8It}(h^2 - h_1^2) = 17.4 \text{ MPa}$$

In this case, the ratio of τ_{max} to τ_{min} is 1.21, that is, the maximum stress in the web is 21% larger than the minimum stress. The variation of the shear stresses over the height h_1 of the web is shown in Fig. 10-38b.

Total shear force: The shear force in the web is calculated from Eq. (10-51) as

$$V_{web} = \frac{th_1}{3}(2\tau_{max} + \tau_{min}) = 43.0 \text{ kN}$$

4. *Finalize*: From this result, note that the web of this particular beam resists 96% of the total shear force.

     Note: The average shear stress in the web of the beam [from Eq. (10-52)] is

$$\tau_{aver} = \frac{V}{th_1} = 20.7 \text{ MPa}$$

which is only 1% less than the maximum stress.

Example 10-13

A beam having a T-shaped cross section (Fig. 10-39a) is subjected to a vertical shear force $V = 10,000$ lb. The cross-sectional dimensions are $b = 4$ in., $t = 1.0$ in., $h = 8.0$ in., and $h_1 = 7.0$ in.

     Determine the shear stress τ_1 at the top of the web (level *n–n*) and the maximum shear stress τ_{max}. (Disregard the areas of the fillets.)

Solution:

Use the four-step problem-solving approach. Combine steps as needed for an efficient solution.

1, 2. *Conceptualize, Categorize*:

     Location of neutral axis: The neutral axis of the T-beam is located by calculating the distances c_1 and c_2 from the top and bottom of the beam to the

FIGURE 10-39

Example 10-13: Shear stresses in web of T-shaped beam

(a)   (b)

centroid of the cross section (Fig. 10-39a). First, divide the cross section into two rectangles: the flange and the web (see the dashed line in Fig. 10-39a). Then calculate the first moment Q_{aa} of these two rectangles with respect to line aa at the bottom of the beam. The distance c_2 is equal to Q_{aa} divided by the area A of the entire cross section (see Section 5.3 for methods for locating centroids of composite areas). The calculations are

$$A = \Sigma A_i = b(h - h_1) + th_1 = 11.0 \text{ in}^2$$

$$Q_{aa} = \Sigma y_i A_i = \left(\frac{h + h_1}{2}\right)(b)(h - h_1) + \frac{h_1}{2}(th_1) = 54.5 \text{ in}^3$$

$$c_2 = \frac{Q_{aa}}{A} = \frac{54.5 \text{ in}^3}{11.0 \text{ in}^2} = 4.955 \text{ in.} \quad c_1 = h - c_2 = 3.045 \text{ in.}$$

Moment of inertia: Find the moment of inertia I of the entire cross-sectional area (with respect to the neutral axis) by determining the moment of inertia I_{aa} about line $a–a$ at the bottom of the beam and then use the parallel-axis theorem (see Section 5.6):

$$I = I_{aa} - Ac_2^2$$

The calculations are

$$I_{aa} = \frac{bh^3}{3} - \frac{(b - t)h_1^3}{3} = 339.67 \text{ in}^4 \qquad Ac_2^2 = 270.02 \text{ in}^4 \qquad I = 69.65 \text{ in}^4$$

3. *Analyze*:

Shear stress at top of web: To find the shear stress τ_1 at the top of the web (along line n–n) calculate the first moment Q_1 of the area above level n–n. This first moment is equal to the area of the flange times the distance from the neutral axis to the centroid of the flange:

$$Q_1 = b(h - h_1)\left(c_1 - \frac{h - h_1}{2}\right)$$

$$= (4 \text{ in.})(1 \text{ in.})(3.045 \text{ in.} - 0.5 \text{ in.}) = 10.18 \text{ in}^3$$

You get the same result if you calculate the first moment of the area below level n–n:

$$Q_1 = th_1\left(c_2 - \frac{h_1}{2}\right) = (1 \text{ in.})(7 \text{ in.})(4.955 \text{ in.} - 3.5 \text{ in.}) = 10.18 \text{ in}^3$$

Substitute into the shear formula to find

$$\tau_1 = \frac{VQ_1}{It} = \frac{(10{,}000 \text{ lb})(10.18 \text{ in}^3)}{(69.65 \text{ in}^4)(1 \text{ in.})} = 1460 \text{ psi}$$

This stress exists both as a vertical shear stress acting on the cross section and as a horizontal shear stress acting on the horizontal plane between the flange and the web.

Maximum shear stress: The maximum shear stress occurs in the web at the neutral axis. Therefore, calculate the first moment Q_{max} of the cross-sectional area below the neutral axis:

$$Q_{max} = tc_2\left(\frac{c_2}{2}\right) = (1 \text{ in.})(4.955 \text{ in.})\left(\frac{4.955 \text{ in.}}{2}\right) = 12.28 \text{ in}^3$$

The same result is obtained if the first moment of the area above the neutral axis is computed, but those calculations would be slighter longer.

Substitute into the shear formula to obtain

$$\tau_{max} = \frac{VQ_{max}}{It} = \frac{(10{,}000 \text{ lb})(12.28 \text{ in}^3)}{(69.65 \text{ in}^4)(1 \text{ in.})} = 1760 \text{ psi}$$

which is the maximum shear stress in the beam.

4. *Finalize*: The parabolic distribution of shear stresses in the web is shown in Fig. 10-39b.

10.10 Composite Beams

The next topic in the study of bending of beams is composite beams; beams made of more than one material. It will be necessary to modify the flexure formula to account for the material properties of the different segments bonded together to form the beam. First, a general theory of flexure for composite beams is presented. This is followed by an alternate approach in which the composite beam is transformed into an equivalent beam of one material to simplify the calculations.

Later, Chapter 13 presents two additional subjects of fundamental importance in beam design—deflections of beams and statically indeterminate beams.

General Theory for Composite Beams

Beams that are fabricated of more than one material are called **composite beams**. Examples are bimetallic beams (such as those used in thermostats), plastic-coated pipes, and wood beams with steel reinforcing plates (see Fig. 10-40).

Many other types of composite beams have been developed in recent years, primarily to save material and reduce weight. For instance, **sandwich beams** are widely used in the aviation and aerospace industries, where light weight plus high strength and rigidity are required. Such familiar objects as skis, doors, wall panels, book shelves, and cardboard boxes are also manufactured in sandwich style.

A typical sandwich beam (Fig. 10-41) consists of two thin *faces* of relatively high-strength material (such as aluminum) separated by a thick *core* of lightweight, low-strength material. Since the faces are at the greatest distance from the neutral axis (where the bending stresses are highest), they function somewhat like the flanges of an I-beam. The core serves as a filler and provides support for the faces, stabilizing them against wrinkling or buckling. Lightweight plastics and foams, as well as honeycombs and corrugations, are often used for cores.

Strains and Stresses

The strains in composite beams are determined from the same basic axiom used for finding the strains in beams of one material, namely, cross sections remain plane during bending. This axiom is valid for pure bending regardless of the nature of the material (see Section 10.4). Therefore, the longitudinal strains ε_x in a composite beam vary linearly from top to bottom of the beam, as expressed by Eq. (10-5), which is repeated here:

$$\varepsilon_x = -\frac{y}{\rho} = -\kappa y \qquad \text{(10-53)}$$

In this equation, y is the distance from the neutral axis, ρ is the radius of curvature, and κ is the curvature.

Begin with the linear strain distribution represented by Eq. (10-53) to determine the strains and stresses in any composite beam. To show how this is accomplished, consider the composite beam shown in Fig. 10-42. This beam consists of two materials, labeled 1 and 2 in the figure, which are securely bonded so that they act as a single solid beam.

FIGURE 10-40

Examples of composite beams: (a) bimetallic beam, (b) plastic-coated steel pipe, and (c) wood beam reinforced with a steel plate

(a)

(b)

(c)

Roof structure: composite timber-steel I-beam and rafters

© Can Stock Photo Inc./tin

FIGURE 10-41

Sandwich beams: (a) plastic core, (b) honeycomb core, and (c) corrugated core (© Barry Goodno)

(a)

(b)

(c)

As in previous discussions of beams, assume that the x-y plane is a plane of symmetry and that the x-z plane is the neutral plane of the beam. However, the neutral axis (the z axis in Fig. 10-42) does *not* pass through the centroid of the cross-sectional area when the beam is made of two different materials.

If the beam is bent with positive curvature, the strains ϵ_x will vary as shown in Fig. 10-42, where ε_A is the compressive strain at the top of the beam, ε_B is the tensile strain at the bottom, and ε_C is the strain at the contact surface of the two materials. Of course, the strain is zero at the neutral axis (the z axis).

The normal stresses acting on the cross section can be obtained from the strains by using the stress–strain relationships for the two materials. Assume that both materials behave in a linearly elastic manner so that Hooke's law for uniaxial stress is valid. Then the stresses in the materials are obtained by multiplying the strains by the appropriate modulus of elasticity.

Denoting the moduli of elasticity for materials 1 and 2 as E_1 and E_2, respectively, and also assuming that $E_2 > E_1$, obtain the stress diagram shown in Fig. 10-42. The compressive stress at the top of the beam is $\sigma_A = E_1\varepsilon_A$, and the tensile stress at the bottom is $\sigma_B = E_2\varepsilon_B$.

At the contact surface (C) the stresses in the two materials are different because their moduli are different. In material 1, the stress is $\sigma_{1C} = E_1\varepsilon_C$, and in material 2, it is $\sigma_{2C} = E_2\varepsilon_C$.

Using Hooke's law and Eq. (10-53), the normal stresses at distance y from the neutral axis in terms of the curvature are

$$\sigma_{x1} = -E_1\kappa y \qquad \sigma_{x2} = -E_2\kappa y \qquad \textbf{(10-54a,b)}$$

where σ_{x1} is the stress in material 1 and σ_{x2} is the stress in material 2. With the aid of these equations, the next step is to locate the neutral axis and obtain the moment–curvature relationship.

FIGURE 10-42

(a) Composite beam of two materials, (b) cross section of beam, (c) distribution of strains ε_x throughout the height of the beam, and (d) distribution of stresses σ_x in the beam for the case where $E_2 > E_1$

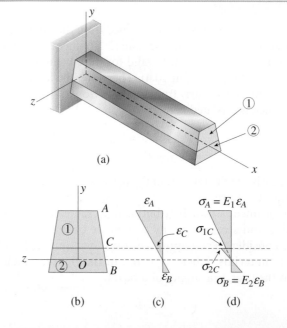

Neutral Axis

The position of the neutral axis (the z axis) is found from the condition that the resultant axial force acting on the cross section is zero (see Section 10.5); therefore,

$$\int_1 \sigma_{x1}\,dA + \int_2 \sigma_{x2}\,dA = 0 \qquad \text{(10-55)}$$

where it is understood that the first integral is evaluated over the cross-sectional area of material 1 and the second integral is evaluated over the cross-sectional area of material 2. Replace σ_{x1} and σ_{x2} in the preceding equation by their expressions from Eqs. (10-54a and b) to get

$$-\int_1 E_1\kappa y\,dA - \int_2 E_2\kappa y\,dA = 0$$

Since the curvature is a constant at any given cross section, it is not involved in the integrations and can be cancelled from the equation; thus, the equation for locating the **neutral axis** becomes

$$E_1\int_1 y\,dA + E_2\int_2 y\,dA = 0 \qquad \text{(10-56)}$$

FIGURE 10-43

Doubly symmetric cross section

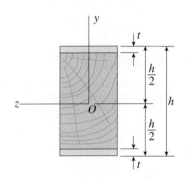

The integrals in this equation represent the first moments of the two parts of the cross-sectional area with respect to the neutral axis. (If there are more than two materials—a rare condition—additional terms are required in the equation.)

Equation (10-56) is a generalized form of the analogous equation for a beam of one material (Eq. 10-9). The details of the procedure for locating the neutral axis with the aid of Eq. (10-56) are illustrated later in Example 10-15.

If the cross section of a beam is **doubly symmetric**, as in the case of a wood beam with steel cover plates on the top and bottom (Fig. 10-43), the neutral axis is located at the mid-height of the cross section and Eq. (10-56) is not needed.

Moment–Curvature Relationship

The moment–curvature relationship for a composite beam of two materials (Fig. 10-42) may be determined from the condition that the moment resultant of the bending stresses is equal to the bending moment M acting at the cross section. Follow the same steps as for a beam of one material [see Eqs. (10-10) through (10-13)], and also use Eqs. (10-54a and b) to obtain

$$M = -\int_A \sigma_x y\,dA = -\int_1 \sigma_{x1} y\,dA - \int_2 \sigma_{x2} y\,dA$$

$$= \kappa E_1 \int_1 y^2\,dA + \kappa E_2 \int_2 y^2\,dA \qquad \text{(10-57a)}$$

This equation can be written in the simpler form

$$M = \kappa(E_1 I_1 + E_2 I_2) \qquad \text{(10-57b)}$$

where I_1 and I_2 are the moments of inertia about the neutral axis (the z axis) of the cross-sectional areas of materials 1 and 2, respectively. Note that $I = I_1 + I_2$, where I is the moment of inertia of the *entire* cross-sectional area about the neutral axis.

Equation (10-57b) now can be solved for the curvature in terms of the bending moment:

$$\kappa = \frac{1}{\rho} = \frac{M}{E_1 I_1 + E_2 I_2} \qquad \text{(10-58)}$$

This equation is the **moment–curvature relationship** for a beam of two materials [compare with Eq. (10-13) for a beam of one material]. The denominator on the right-hand side is the **flexural rigidity** of the composite beam.

Normal Stresses (Flexure Formulas)

The normal stresses (or bending stresses) in the beam are obtained by substituting the expression for curvature [Eq. (10-58)] into the expressions for σ_{x1} and σ_{x2} [Eqs. (10-54a and b)]; thus,

$$\sigma_{x1} = -\frac{MyE_1}{E_1I_1 + E_2I_2} \qquad \sigma_{x2} = -\frac{MyE_2}{E_1I_1 + E_2I_2} \qquad \textbf{(10-59a,b)}$$

These expressions, known as the **flexure formulas for a composite beam**, give the normal stresses in materials 1 and 2, respectively. If the two materials have the same modulus of elasticity ($E_1 = E_2 = E$), both equations reduce to the flexure formula for a beam of one material [Eq. (10-14)].

The analysis of composite beams, using Eqs. (10-56) through (10-59), is illustrated in Examples 10-14 and 10-15 at the end of this section.

Approximate Theory for Bending of Sandwich Beams

Sandwich beams having doubly symmetric cross sections and composed of two linearly elastic materials (Fig. 10-44) can be analyzed for bending using Eqs. (10-58) and (10-59), as described previously. However, an approximate theory for bending of sandwich beams can be developed by introducing some simplifying assumptions.

If the material of the faces (material 1) has a much larger modulus of elasticity than does the material of the core (material 2), it is reasonable to disregard the normal stresses in the core and assume that the faces resist all of the longitudinal bending stresses. This assumption is equivalent to saying that the modulus of elasticity E_2 of the core is zero. Under these conditions, the flexure formula for material 2 [Eq. (10-59b)] gives $\sigma_{x2} = 0$ (as expected), and the flexure formula for material 1 [Eq. (10-59a)] gives

$$\sigma_{x1} = -\frac{My}{I_1} \qquad \textbf{(10-60)}$$

which is similar to the ordinary flexure formula [Eq. (10-14)]. The quantity I_1 is the moment of inertia of the two faces evaluated with respect to the neutral axis; thus,

$$I_1 = \frac{b}{12}\left(h^3 - h_c^3\right) \qquad \textbf{(10-61)}$$

in which b is the width of the beam, h is the overall height of the beam, and h_c is the height of the core. Note that $h_c = h - 2t$ where t is the thickness of the faces.

The maximum normal stresses in the sandwich beam occur at the top and bottom of the cross section where $y = h/2$ and $-h/2$, respectively. Thus, from Eq. (10-60),

$$\sigma_{\text{top}} = -\frac{Mh}{2I_1} \qquad \sigma_{\text{bottom}} = \frac{Mh}{2I_1} \qquad \textbf{(10-62a,b)}$$

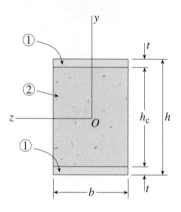

FIGURE 10-44

Cross section of a sandwich beam having two axes of symmetry (doubly symmetric cross section)

If the bending moment M is positive, the upper face is in compression and the lower face is in tension. [These equations are conservative because they give stresses in the faces that are higher than those obtained from Eqs. (10-59a and b).]

If the faces are thin compared to the thickness of the core (that is, if t is small compared to h_c), disregard the shear stresses in the faces and assume that the core carries all of the shear stresses. Under these conditions the average shear stress and average shear strain in the core are, respectively,

$$\tau_{aver} = \frac{V}{bh_c} \qquad \gamma_{aver} = \frac{V}{bh_cG_c} \qquad \textbf{(10-63a,b)}$$

in which V is the shear force acting on the cross section and G_c is the shear modulus of elasticity for the core material. (Although the maximum shear stress and maximum shear strain are larger than the average values, the average values are often used for design purposes.)

Limitations

The preceding discussion of composite beams assumed that both materials followed Hooke's law and that the two parts of the beam were adequately bonded so that they acted as a single unit. Thus, the analysis is highly idealized and represents only a first step in understanding the behavior of composite beams and composite materials. Methods for dealing with nonhomogeneous and nonlinear materials, bond stresses between the parts, shear stresses on the cross sections, buckling of the faces, and other such matters are treated in reference books dealing specifically with composite construction.

Reinforced concrete beams are one of the most complex types of composite construction (Fig. 10-45), and their behavior differs significantly from that of the composite beams discussed in this section. Concrete is strong in compression but extremely weak in tension. Consequently, its tensile strength is usually disregarded entirely. Under those conditions, *the formulas given in this section do not apply.* Working stress design in which the portion of the beam in tension is removed from the composite beam cross section is used in evaluating deflections of reinforced concrete beams, so an allowable stress approach is presented in Example 10-16 to show the general analysis procedure. Example 10-16 uses a "cracked section analysis" to demonstrate this reinforced-concrete analysis procedure.

Note that most reinforced concrete beams are not designed on the basis of linearly elastic behavior—instead, more realistic design methods based upon load-carrying capacity instead of allowable stresses are used. The design of reinforced concrete members is highly specialized and is presented in courses and textbooks devoted solely to that subject.

FIGURE 10-45

Reinforced concrete beam with longitudinal reinforcing bars and vertical stirrups

Example 10-14

A sandwich beam having aluminum-alloy faces enclosing a plastic core (Fig. 10-46) is subjected to a bending moment $M = 3.0$ kN $\cdot$ m. The thickness of the faces is $t = 5$ mm, and their modulus of elasticity is $E_1 = 72$ GPa. The height of the plastic core is $h_c = 150$ mm, and its modulus of elasticity is $E_2 = 800$ MPa. The overall dimensions of the beam are $h = 160$ mm and $b = 200$ mm.

Determine the maximum tensile and compressive stresses in the faces and the core using (a) the general theory for composite beams and (b) the approximate theory for sandwich beams.

FIGURE 10-46

Example 10-14: Cross section of sandwich beam having aluminum-alloy faces and a plastic core

Solution:

Use the four-step problem-solving approach. Combine steps as needed for an efficient solution.

1. *Conceptualize*: Use the general theory of flexure for composite beams.

2. *Categorize*:

   Neutral axis: Because the cross section is doubly symmetric, the neutral axis (the z axis in Fig. 10-46) is located at mid-height.

   Moments of inertia: The moment of inertia I_1 of the cross-sectional areas of the faces (about the z axis) is

   $$I_1 = \frac{b}{12}(h^3 - h_c^3) = \frac{200 \text{ mm}}{12}\left[(160 \text{ mm})^3 - (150 \text{ mm})^3\right]$$
   $$= 12.017 \times 10^6 \text{ mm}^4$$

   and the moment of inertia I_2 of the plastic core is

   $$I_2 = \frac{b}{12}(h_c^3) = \frac{200 \text{ mm}}{12}(150 \text{ mm})^3 = 56.250 \times 10^6 \text{ mm}^4$$

   As a check on these results, note that the moment of inertia of the entire cross-sectional area about the z axis ($I = bh^3/12$) is equal to the sum of I_1 and I_2.

3, 4. *Analyze, Finalize*:

   Part (a): Normal stresses calculated from the general theory for composite beams.
   To calculate these stresses, use Eqs. (10-59a and b).

   As a preliminary matter, evaluate the term in the denominator of those equations (that is, the flexural rigidity of the composite beam):

   $$E_1 I_1 + E_2 I_2 = (72 \text{ GPa})(12.017 \times 10^6 \text{ mm}^4) + (800 \text{ MPa})(56.250 \times 10^6 \text{ mm}^4)$$
   $$= 910,200 \text{ N} \cdot \text{m}^2$$

   The maximum tensile and compressive stresses in the aluminum faces are found from Eq. (10-59a):

   $$(\sigma_1)_{\max} = \pm\frac{M(h/2)(E_1)}{E_1 I_1 + E_2 I_2}$$
   $$= \pm\frac{(3.0 \text{ kN} \cdot \text{m})(80 \text{ mm})(72 \text{ GPa})}{910,200 \text{ N} \cdot \text{m}^2} = \pm 19.0 \text{ MPa}$$

The corresponding quantities for the plastic core [from Eq. (10-59b)] are

$$(\sigma_2)_{max} = \pm \frac{M(h_c/2)(E_2)}{E_1 I_1 + E_2 I_2}$$

$$= \pm \frac{(3.0 \text{ kN} \cdot \text{m})(75 \text{ mm})(800 \text{ MPa})}{910,200 \text{ N} \cdot \text{m}^2} = \pm 0.198 \text{ MPa} \quad \blacktriangleleft$$

The maximum stresses in the faces are 96 times greater than the maximum stresses in the core, primarily because the modulus of elasticity of the aluminum is 90 times greater than that of the plastic.

Part (b): Normal stresses calculated from the approximate theory for sandwich beams.

The approximate theory disregards the normal stresses in the core and assumes that the faces transmit the entire bending moment. Then the maximum tensile and compressive stresses in the faces is found from Eqs. (10-62a and b), as

$$(\sigma_1)_{max} = \pm \frac{Mh}{2I_1} = \pm \frac{(3.0 \text{ kN} \cdot \text{m})(80 \text{ mm})}{12.017 \times 10^6 \text{ mm}^4} = \pm 20.0 \text{ MPa} \quad \blacktriangleleft$$

As expected, the approximate theory gives slightly higher stresses in the faces than does the general theory for composite beams.

Example 10-15

FIGURE 10-47

Example 10-15: Cross section of a composite beam of wood and steel

A composite beam (Fig. 10-47) is constructed from a wood beam (4.0 in. × 6.0 in. actual dimensions) and a steel reinforcing plate (4.0 in. wide and 0.5 in. thick). The wood and steel are securely fastened to act as a single beam. The beam is subjected to a positive bending moment $M = 60$ kip-in.

Calculate the largest tensile and compressive stresses in the wood (material 1) and the maximum and minimum tensile stresses in the steel (material 2) if $E_1 = 1500$ ksi and $E_2 = 30,000$ ksi.

Solution:

Use the four-step problem-solving approach.

1. *Conceptualize*: Use the general theory of flexure for composite beams.

2. *Categorize*:

   Neutral axis: The first step in the analysis is to locate the neutral axis of the cross section. For that purpose, denote the distances from the neutral axis to the top and bottom of the beam as h_1 and

h_2, respectively. To obtain these distances, use Eq. (10-56). Evaluate the integrals in that equation by taking the first moments of areas 1 and 2 about the z axis, as

$$\int_1 y \, dA = \bar{y}_1 A_1 = (h_1 - 3 \text{ in.})(4 \text{ in.} \times 6 \text{ in.}) = (h_1 - 3 \text{ in.})(24 \text{ in}^2)$$

$$\int_2 y \, dA = \bar{y}_2 A_2 = -(6.25 \text{ in.} - h_1)(4 \text{ in.} \times 0.5 \text{ in.}) = (h_1 - 6.25 \text{ in.})(2 \text{ in}^2)$$

where A_1 and A_2 are the areas of parts 1 and 2 of the cross section, $\bar{y}_1$ and $\bar{y}_2$ are the y coordinates of the centroids of the respective areas, and h_1 has units of inches.

Substitute the preceding expressions into Eq. (10-56) to get the equation for locating the neutral axis as

$$E_1 \int_1 y \, dA + E_2 \int_2 y \, dA = 0$$

or

$$(1500 \text{ ksi})(h_1 - 3 \text{ in.})(24 \text{ in}^2) + (30{,}000 \text{ ksi})(h_1 - 6.25 \text{ in.})(2 \text{ in}^2) = 0$$

Solve this equation to obtain the distance h_1 from the neutral axis to the top of the beam:

$$h_1 = 5.031 \text{ in.}$$

Also, the distance h_2 from the neutral axis to the bottom of the beam is

$$h_2 = 6.5 \text{ in.} - h_1 = 1.469 \text{ in.}$$

Thus, the position of the neutral axis is established.

Moments of inertia: The moments of inertia I_1 and I_2 of areas A_1 and A_2 with respect to the neutral axis are found by using the parallel-axis theorem (see Section 5.6). Beginning with area 1 (Fig. 10-46),

$$I_1 = \frac{1}{12}(4 \text{ in.})(6 \text{ in.})^3 + (4 \text{ in.})(6 \text{ in.})(h_1 - 3 \text{ in.})^2 = 171.0 \text{ in}^4$$

Similarly, for area 2,

$$I_2 = \frac{1}{12}(4 \text{ in.})(0.5 \text{ in.})^3 + (4 \text{ in.})(0.5 \text{ in.})(h_2 - 0.25 \text{ in.})^2 = 3.01 \text{ in}^4$$

To check these calculations, compute the moment of inertia I of the entire cross-sectional area about the z axis as

$$I = \frac{1}{3}(4 \text{ in.})h_1^3 + \frac{1}{3}(4 \text{ in.})h_2^3 = 169.8 \text{ in}^4 + 4.2 \text{ in}^4 = 174.0 \text{ in}^4$$

which agrees with the sum of I_1 and I_2.

3. *Analyze*:

Normal stresses: The stresses in materials 1 and 2 are calculated from the flexure formulas for composite beams [Eqs. (10-59a and b)]. The largest compressive stress in material 1 occurs at the top of the beam (A) where $y = h_1 = 5.031$ in. Denoting this stress by σ_{1A} and using Eq. (10-59a) gives

$$\sigma_{1A} = -\frac{Mh_1E_1}{E_1I_1 + E_2I_2}$$

$$= -\frac{(60 \text{ kip-in.})(5.031 \text{ in.})(1500 \text{ ksi})}{(1500 \text{ ksi})(171.0 \text{ in}^4) + (30,000 \text{ ksi})(3.01 \text{ in}^4)} = -1310 \text{ psi}$$

The largest tensile stress in material 1 occurs at the contact plane between the two materials (C) where $y = -(h_2 - 0.5 \text{ in.}) = -0.969$ in. Proceed as in the previous calculation to obtain

$$\sigma_{1C} = -\frac{(60 \text{ kip-in.})(-0.969 \text{ in.})(1500 \text{ ksi})}{(1500 \text{ ksi})(171.0 \text{ in}^4) + (30,000 \text{ ksi})(3.01 \text{ in}^4)} = 251 \text{ psi}$$

These are the largest compressive and tensile stresses in the wood.

The steel plate (material 2) is located below the neutral axis; therefore, it is entirely in tension. The maximum tensile stress occurs at the bottom of the beam (B) where $y = -h_2 = -1.469$ in. Hence, from Eq. (10-59b),

$$\sigma_{2B} = -\frac{M(-h_2)E_2}{E_1I_1 + E_2I_2}$$

$$= -\frac{(60 \text{ kip-in.})(-1.469 \text{ in.})(30,000 \text{ ksi})}{(1500 \text{ ksi})(171.0 \text{ in}^4) + (30,000 \text{ ksi})(3.01 \text{ in}^4)} = 7620 \text{ psi}$$

The minimum tensile stress in material 2 occurs at the contact plane (C) where $y = -0.969$ in. Thus,

$$\sigma_{2C} = -\frac{(60 \text{ kip-in.})(-0.969 \text{ in.})(30,000 \text{ ksi})}{(1500 \text{ ksi})(171.0 \text{ in}^4) + (30,000 \text{ ksi})(3.01 \text{ in}^4)} = 5030 \text{ psi}$$

These stresses are the maximum and minimum tensile stresses in the steel.

4. *Finalize*: The stress distribution over the cross section of the composite wood-steel beam is shown in Fig. 10-48.

Note: At the contact plane, the ratio of the stress in the steel to the stress in the wood is

$$\sigma_{2C}/\sigma_{1C} = 5030 \text{ psi}/251 \text{ psi} = 20$$

which is equal to the ratio E_2/E_1 of the moduli of elasticity (as expected). Although the strains in the steel and wood are equal at the contact plane, the stresses are different because of the different moduli.

FIGURE 10-48

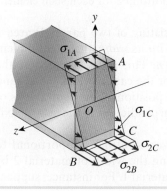

Transformed-Section Method

The transformed-section method is an alternative procedure for analyzing the bending stresses in a composite beam. The method is based upon the theories and equations developed in the preceding section; therefore, it is subject to the same limitations (for instance, it is valid only for linearly elastic materials) and gives the same results. Although the transformed-section method does not reduce the calculating effort, many designers find that it provides a convenient way to visualize and organize the calculations.

The method consists of transforming the cross section of a composite beam into an equivalent cross section of an imaginary beam that is composed of only one material. This new cross section is called the **transformed section**. Then the imaginary beam with the transformed section is analyzed in the customary manner for a beam of one material. As a final step, the stresses in the transformed beam are converted to those in the original beam.

Neutral Axis and Transformed Section

FIGURE 10-49

Composite beam of two materials:
(a) actual cross section and
(b) transformed section consisting
only of material 1

If the transformed beam is to be equivalent to the original beam, *its neutral axis must be located in the same place and its moment-resisting capacity must be the same*. To show how these two requirements are met, consider again a composite beam of two materials (Fig. 10-49a). The **neutral axis** of the cross section is obtained from Eq. (10-56), which is repeated here:

$$E_1 \int_1 y \, dA + E_2 \int_2 y \, dA = 0 \qquad \text{(10-64)}$$

In this equation, the integrals represent the first moments of the two parts of the cross section with respect to the neutral axis.

Now introduce the notation

$$n = \frac{E_2}{E_1} \qquad \text{(10-65)}$$

where n is the **modular ratio**. With this notation, rewrite Eq. (10-64) in the form

$$\int_1 y \, dA + \int_2 y n \, dA = 0 \qquad \text{(10-66)}$$

(a)

(b)

Since Eqs. (10-64) and (10-66) are equivalent, the preceding equation shows that the neutral axis is unchanged if each element of area dA in material 2 is multiplied by the factor n, provided that the y coordinate for each such element of area is not changed.

Therefore, create a new cross section consisting of two parts: (1) area 1 with its dimensions unchanged and (2) area 2 with its *width* (that is, its dimension parallel to the neutral axis) multiplied by n. This new cross section (the transformed section) is shown in Fig. 10-49b for the case where $E_2 > E_1$ (and therefore $n > 1$). Its neutral axis is in the same position as the neutral axis of the original beam. (Note that all dimensions perpendicular to the neutral axis remain the same.)

Since the stress in the material (for a given strain) is proportional to the modulus of elasticity ($\sigma = E\varepsilon$), multiplying the width of material 2 by $n = E_2/E_1$ is equivalent to transforming it to material 1. For instance, suppose that $n = 10$. Then the area of part 2 of the cross section is now 10 times wider than before. If you imagine that this part of the beam is now material 1, it will carry the same force as before because its modulus is *reduced* by a factor of 10 (from E_2 to E_1) at the same time that its area is *increased* by a factor of 10. Thus, the new section (the transformed section) consists only of material 1.

Moment–Curvature Relationship

The *moment–curvature relationship* for the transformed beam must be the same as for the original beam. To show that this is indeed the case, note that the stresses in the transformed beam (since it consists only of material 1) are given by Eq. (10-8) of Section 10.5:

$$\sigma_x = -E_1 \kappa y$$

Using this equation, and also following the same procedure as for a beam of one material (see Section 10.5), the moment–curvature relation for the transformed beam is

$$M = -\int_A \sigma_x y \, dA = -\int_1 \sigma_x y \, dA - \int_2 \sigma_x y \, dA$$

$$= E_1 \kappa \int_1 y^2 \, dA + E_1 \kappa \int_2 y^2 \, dA = \kappa(E_1 I_1 + E_1 n I_2)$$

or

$$M = \kappa(E_1 I_1 + E_2 I_2) \tag{10-67}$$

This equation is the same as Eq. (10-57b), thereby demonstrating that the moment–curvature relationship for the transformed beam is the same as for the original beam.

Normal Stresses

Since the transformed beam consists of only one material, the *normal stresses* (or *bending stresses*) can be found from the standard flexure formula [Eq. (10-14)]. Thus, the normal stresses in the beam transformed to material 1 (Fig. 10-49b) are

$$\sigma_{x1} = -\frac{My}{I_T} \tag{10-68}$$

where I_T is the moment of inertia of the transformed section with respect to the neutral axis. By substituting into this equation, you can calculate the stresses at any point in the *transformed* beam. (As explained later, the stresses in the transformed beam match those in the part of the original beam consisting of material 1; however, in the part of the original beam consisting of material 2, the stresses are different from those in the transformed beam.)

Equation (10-68) can be verified by noting that the moment of inertia of the transformed section (Fig. 10-49b) is related to the moment of inertia of the original section (Fig. 10-49a) by the relation:

$$I_T = I_1 + n I_2 = I_1 + \frac{E_2}{E_1} I_2 \tag{10-69}$$

Substituting this expression for I_T into Eq. (10-68) gives

$$\sigma_{x1} = -\frac{My E_1}{E_1 I_1 + E_2 I_2} \tag{10-70a}$$

which is the same as Eq. (10-59a), demonstrating that the stresses in material 1 in the original beam are the same as the stresses in the corresponding part of the transformed beam.

As mentioned previously, the stresses in material 2 in the original beam are *not* the same as the stresses in the corresponding part of the transformed beam. Instead, the stresses in the transformed beam [Eq. (10-68)] must be multiplied by the modular ratio n to obtain the stresses in material 2 of the original beam:

$$\sigma_{x2} = -\frac{My}{I_T}n \qquad (10\text{-}70b)$$

Verify this formula by substituting Eq. (10-69) for I_T into Eq. (10-70b) to get

$$\sigma_{x2} = -\frac{MynE_1}{E_1I_1 + E_2I_2} = -\frac{MyE_2}{E_1I_1 + E_2I_2} \qquad (10\text{-}70c)$$

which is the same as Eq. (10-59b).

General Comments

In this discussion of the transformed-section method, the original beam was transformed to a beam consisting entirely of material 1. It is also possible to transform the beam to material 2. In that case, the stresses in the original beam in material 2 will be the same as the stresses in the corresponding part of the transformed beam. However, the stresses in material 1 in the original beam must be obtained by multiplying the stresses in the corresponding part of the transformed beam by the modular ratio n, which in this case is defined as $n = E_1/E_2$.

It is also possible to transform the original beam into a material having any arbitrary modulus of elasticity E, in which case all parts of the beam must be transformed to the fictitious material. Of course, the calculations are simpler if you transform to one of the original materials. Finally, with a little ingenuity, it is possible to extend the transformed-section method to composite beams of more than two materials.

Example 10-16

FIGURE 10-50

Example 10-16: Cross section of a singly reinforced concrete inverted T-beam

b_w

Neutral axis

y

d

t_f

b

An inverted precast concrete T-beam is used to support precast double-tee floor beams in a parking deck (see Fig. 10-50 and photo). Beam dimensions are $b = 500$ mm, $b_w = 300$ mm, $d = 600$ mm, and $t_f = 100$ mm. Steel reinforcement consists of four bars each with a 25-mm diameter. The modulus of elasticity for the concrete is $E_c = 25$ GPa, while that of the steel is $E_s = 200$ GPa. Allowable stresses for concrete and steel are $\sigma_{ac} = 9.3$ MPa and $\sigma_{as} = 137$ MPa, respectively.

(a) Use the transformed section in Fig. 10-51 (in which the concrete in tension is neglected and the steel reinforcing bars are converted to the equivalent concrete) to find the maximum permissible moment that can be applied to this beam.

(b) Repeat part (a) if the beam is rotated 180°, as shown in Fig. 10-52, and if the steel reinforcement remains in the bottom tension zone.

FIGURE 10-51

Transformed section for singly reinforced concrete inverted T-beam

FIGURE 10-52

Transformed section for singly reinforced concrete T-beam

Solution:

Use the four-step problem-solving approach. Combine steps as needed for an efficient solution.

Part (a): Inverted T-beam.

1, 2. *Conceptualize, Categorize*: Start by finding the *neutral axis* (at some distance y down from the top of the beam) for the transformed section shown in Fig. 10-51. Equating the first moments of areas of concrete in compression ($b_w \times y$) and the transformed area of steel in tension ($n \times A_s$) leads to a quadratic equation. The solution for y gives the position of the neutral axis as

$$b_w y \frac{y}{2} - nA_s(d - y) = 0 \quad \text{where} \quad n = \frac{E_s}{E_c} = 8 \tag{a}$$

$$y = \sqrt{\left(\frac{nA_s}{b_w}\right)^2 + 2d\left(\frac{nA_s}{b_w}\right)} - \left(\frac{nA_s}{b_w}\right) = 0.204 \text{ m} \tag{b}$$

Now use Eq. (10-69) to compute the moment of inertia of the transformed section:

$$I_T = \frac{b_w y^3}{3} + nA_s[(d - y)^2] = 3.312 \times 10^{-3} \text{ m}^4 \tag{c}$$

3, 4. *Analyze, Finalize*: Finally, the moment capacity of the beam is found by solving Eqs. (10-68) (allowable stress in concrete controls) and Eq. (10-70b) (allowable stress in steel controls) for M, where the lower value based on allowable stress in the steel governs:

$$M_c = \frac{\sigma_{ac}}{y} I_T = \left(\frac{9.3 \text{ MPa}}{0.204 \text{ m}}\right)(3.312 \times 10^{-3} \text{ m}^4) = 151 \text{ kN} \cdot \text{m} \tag{d}$$

$$M_s = \frac{\sigma_{as}}{n(d - y)} I_T = \frac{137 \text{ MPa}}{8(0.6 \text{ m} - 0.204 \text{ m})}(3.312 \times 10^{-3} \text{ m}^4)$$
$$= 143.2 \text{ kN} \cdot \text{m} \tag{e}$$

Part (b): T-beam.

1, 2. *Conceptualize, Categorize*: Now the flange of the T-beam with a thickness of t_f is on top, so start by assuming that neutral axis location distance y is greater than t_f. Divide the compression area of the concrete for the transformed section (Fig. 10-52) into three rectangles, then equate the first moments of areas of the concrete in compression and transformed area of the steel ($n \times A_s$) to get a quadratic equation for distance y. The solution for y gives the position of the neutral axis as

$$(b - b_w)t_f\left(y - \frac{t_f}{2}\right) + b_w y \frac{y}{2} - nA_s(d - y) = 0 \tag{f}$$

Solve Eq. (f) for y to get

$$y = 0.1702 \text{ m}$$

which exceeds t_f as assumed. The moment of inertia of the transformed section is now

$$I_T = \frac{b_w y^3}{3} + \frac{(b - b_w)t_f^3}{12} + (b - b_w)t_f\left(y - \frac{t_f}{2}\right)^2 + nA_s(d - y)^2$$

$$= 3.7 \times 10^{-3} \text{ m}^4$$

3. *Analyze*: Finally, repeat the solutions for maximum permissible moment M in Eqs. (d) and (e) as

$$M_c = \frac{\sigma_{ac}}{y}I_T = \frac{(9.3 \text{ MPa})}{0.1702 \text{ m}}(3.7 \times 10^{-3} \text{ m}^4) = 202 \text{ kN} \cdot \text{m} \tag{g}$$

$$M_s = \frac{\sigma_{as}}{n(d - y)}I_T = \frac{137 \text{ MPa}}{8(0.6 \text{ m} - 0.1702 \text{ m})}(3.7 \times 10^{-3} \text{ m}^4) \tag{h}$$

$$= 147.4 \text{ kN} \cdot \text{m}$$

4. *Finalize*: Once again, the lower value of moment M based on allowable stress in the steel governs. Since the allowable stress in the reinforcing steel bars controls both beams, their moment capacities [Eqs. (e) and (h)] are approximately the same.

Example 10-17

The composite beam shown in Fig. 10-50a is formed of a wood beam (4.0 in. × 6.0 in. actual dimensions) and a steel reinforcing plate (4.0 in. wide and 0.5 in. thick). The beam is subjected to a positive bending moment $M = 60$ kip-in.

Using the transformed-section method, calculate the largest tensile and compressive stresses in the wood (material 1) and the maximum and minimum tensile stresses in the steel (material 2) if $E_1 = 1500$ ksi and $E_2 = 30,000$ ksi.

Note: This same beam was analyzed previously in Example 10-15.

FIGURE 10-53

Example 10-17: Composite beam of Example 10-15 analyzed by the transformed-section method: (a) cross section of original beam and (b) transformed section (material 1)

(a)   (b)

Solution:

Use the four-step problem-solving approach. Combine steps as needed for an efficient solution.

1, 2. *Conceptualize, Categorize*: Use the transformed-section approach and compare results to those found in Example 10-15.

3. *Analyze*:

Transformed section: Transform the original beam into a beam of material 1, which means that the modular ratio is defined as

$$n = \frac{E_2}{E_1} = \frac{30{,}000 \text{ ksi}}{1500 \text{ ksi}} = 20$$

The part of the beam made of wood (material 1) is not altered, but the part made of steel (material 2) has its width multiplied by the modular ratio. Thus, the width of this part of the beam becomes

$$n(4 \text{ in.}) = 20(4 \text{ in.}) = 80 \text{ in.}$$

in the transformed section (Fig. 10-53b).

Neutral axis: Because the transformed beam consists of only one material, the neutral axis passes through the centroid of the cross-sectional area. Therefore, with the top edge of the cross section serving as a reference line and the distance y_i measured positive downward, calculate the distance h_1 to the centroid as

$$h_1 = \frac{\Sigma y_i A_i}{\Sigma A_i} = \frac{(3 \text{ in.})(4 \text{ in.})(6 \text{ in.}) + (6.25 \text{ in.})(80 \text{ in.})(0.5 \text{ in.})}{(4 \text{ in.})(6 \text{ in.}) + (80 \text{ in.})(0.5 \text{ in.})}$$

$$= \frac{322.0 \text{ in}^3}{64.0 \text{ in}^2} = 5.031 \text{ in.}$$

Also, the distance h_2 from the lower edge of the section to the centroid is

$$h_2 = 6.5 \text{ in.} - h_1 = 1.469 \text{ in.}$$

Thus, the location of the neutral axis is determined.

Moment of inertia of the transformed section: Using the parallel-axis theorem (see Section 5.6), calculate the moment of inertia I_T of the entire cross-sectional area with respect to the neutral axis as

$$
I_T = \frac{1}{12}(4 \text{ in.})(6 \text{ in.})^3 + (4 \text{ in.})(6 \text{ in.})(h_1 - 3 \text{ in.})^2
$$

$$
+ \frac{1}{12}(80 \text{ in.})(0.5 \text{ in.})^3 + (80 \text{ in.})(0.5 \text{ in.})(h_2 - 0.25 \text{ in.})^2
$$

$$
= 171.0 \text{ in}^4 + 60.3 \text{ in}^4 = 231.3 \text{ in}^4
$$

Normal stresses in the wood (material 1): The stresses in the transformed beam (Fig. 10-53b) at the top of the cross section (A) and at the contact plane between the two parts (C) are the same as in the original beam (Fig. 10-53a). These stresses can be found from the flexure formula in Eq. (10-68) as

$$
\sigma_{1A} = -\frac{My}{I_T} = -\frac{(60 \text{ kip-in.})(5.031 \text{ in.})}{231.3 \text{ in}^4} = -1310 \text{ psi} \qquad \Leftarrow
$$

$$
\sigma_{1C} = -\frac{My}{I_T} = -\frac{(60 \text{ kip-in.})(-0.969 \text{ in.})}{231.3 \text{ in}^4} = 251 \text{ psi} \qquad \Leftarrow
$$

These are the largest tensile and compressive stresses in the wood (material 1) in the original beam. The stress σ_{1A} is compressive and the stress σ_{1C} is tensile.

Normal stresses in the steel (material 2): The maximum and minimum stresses in the steel plate are found by multiplying the corresponding stresses in the transformed beam by the modular ratio n in Eq. (10-70b). The maximum stress occurs at the lower edge of the cross section (B) and the minimum stress occurs at the contact plane (C):

$$
\sigma_{2B} = -\frac{My}{I_T}n = -\frac{(60 \text{ kip-in.})(-1.469 \text{ in.})}{231.3 \text{ in}^4}(20) = 7620 \text{ psi} \qquad \Leftarrow
$$

$$
\sigma_{2C} = -\frac{My}{I_T}n = -\frac{(60 \text{ kip-in.})(-0.969 \text{ in.})}{231.3 \text{ in}^4}(20) = 5030 \text{ psi} \qquad \Leftarrow
$$

Both of these stresses are tensile.

4. *Finalize*: Note that the stresses calculated by the transformed-section method agree with those found in Example 10-15 by direct application of the formulas for a composite beam.

Balanced design: As a final evaluation of the wood-steel composite beam considered here and in Example 10-15, note that neither wood nor steel has reached typical allowable stress levels. Perhaps some redesigning of this beam would be of interest; consider only the steel plate here (you could also re-size the wood beam).

A *balanced design* is one in which wood and steel reach their allowable stress values at the same time under the design moment; this could be regarded as a more efficient design of this beam. *First*, holding the steel plate thickness at $t_s = 0.5$ in., find the required width b_s of the steel plate, so the wood and steel reach allowable stress values simultaneously under design moment M_D. Then setting $b_s = 4$ in., repeat the previous process but also find required plate thickness t_s to achieve the same objective. Assume that the allowable stress values for wood and steel are $\sigma_{aw} = 1850$ psi and $\sigma_{as} = 14$ ksi, respectively. Also assume that the wood beam dimensions are unchanged.

Using the transformed-section approach, write the expressions for the stresses at the top of the wood and bottom of the steel. Equate each to its allowable value as

$$\sigma_{aw} = \frac{-M_D h_1}{I_T} \quad \text{and} \quad \sigma_{as} = \frac{-M_D h_2 n}{I_T} \tag{a,b}$$

Next, solve each of Eqs. (a) and (b) for ratio M_D/I_T; then equate the two expressions to find the h_1/h_2 ratio for which allowable stress levels are reached in both materials:

$$\frac{h_1}{h_2} = n\frac{\sigma_{aw}}{\sigma_{as}} \tag{c}$$

Expressions for h_1 and h_2 can be obtained in terms of the transformed-section dimensions $b, h, b_s,$ and t_s (Figs. 10-54a and b) by taking first moments about the z axis to get

$$h_1 = \frac{h}{2} + \frac{(b_s n t_s^2) + (b_s h n t_s)}{(2bh) + (2b_s n t_s)} \quad \text{and} \quad h_2 = \frac{n b_s t_s \left(\frac{t_s}{2}\right) + bh\left(t_s + \frac{h}{2}\right)}{(n b_s t_s) + (bh)} \tag{d}$$

FIGURE 10-54

Example 10-17: Balanced design of composite beam: (a) original beam and (b) transformed beam

(a)

(b)

With some effort (and perhaps with computer assistance), rewrite Eq. (c) as

$$\frac{bh^2 + (2b_s nht_s) + (b_s nt_s^2)}{bh^2 + (2bht_s) + (b_s nt_s^2)} = n\frac{\sigma_{aw}}{\sigma_{as}} \tag{e}$$

Collecting terms and solving for the required width of b_s for the steel plate (with thickness t_s unchanged) then substituting numerical values, width b_s (instead of the original width of 4 in.) is

$$b = 4 \text{ in.} \quad h = 6 \text{ in.} \quad t_s = 0.5 \text{ in.} \quad n = 20 \quad \sigma_{aw} = 1850 \text{ psi} \quad \sigma_{as} = 14 \text{ ksi}$$

$$b_s = \frac{\left(n\dfrac{\sigma_{aw}}{\sigma_{as}}\right)(bh^2 + 2bht_s) - bh^2}{(2nht_s) + nt_s^2\left(1 - n\dfrac{\sigma_{aw}}{\sigma_{as}}\right)} = 2.68 \text{ in.} \tag{f}$$

So for a 4 in. $\times$ 6 in. wood beam reinforced by a 2.68 in. $\times$ 0.5 in. steel plate (Fig. 10-51a) under any applied moment M that is less than or equal to M_D, the stress ratio σ_{1A}/σ_{2B} will be equal to σ_{aw}/σ_{as}. If $M = M_D$, then $\sigma_{1A} = \sigma_{aw}$ and $\sigma_{2B} = \sigma_{as}$.

Alternatively, reformulate Eq. (e) to get a quadratic equation for the steel plate thickness t_s (with the original width $b_s = 4$ in.) to obtain

$$t_s^2\left[nb_s\left(1 - n\frac{\sigma_{aw}}{\sigma_{as}}\right)\right] + t_s\left\{2h\left[nb_s - b\left(n\frac{\sigma_{aw}}{\sigma_{as}}\right)\right]\right\} + bh^2\left(1 - n\frac{\sigma_{aw}}{\sigma_{as}}\right) = 0 \tag{g}$$

The solution of Eq. (g) results in a reduced steel plate thickness leading to a *balanced design* of the wood-steel composite beam as $t_s = 0.298$ in. Once again, for $b_s = 4$ in. and $t_s = 0.298$ in., the stress ratio $\sigma_{1A}/\sigma_{2B} = \sigma_{aw}/\sigma_{as}$ for applied moments M which are less than or equal to M_D.

10.11 Bending of Unsymmetric Beams

In previous discussions of bending, the beams were assumed to have cross sections with at least one axis of symmetry. Now that restriction is removed, and beams may have unsymmetric cross sections. As in earlier discussions, it is assumed that the beams are made of linearly elastic materials.

Suppose that a beam having an unsymmetric cross section is subjected to a bending moment M acting at the end cross section (Fig. 10-55a). Of interest here are the stresses in the beam and the position of the neutral axis. Unfortunately, at this stage of the analysis, there is no direct way of determining these quantities. Therefore, an indirect approach must be used—instead of starting with a bending moment and trying to find the neutral axis, start with an assumed neutral axis and find the associated bending moment.

Neutral Axis

Begin by constructing two perpendicular axes (the y and z axes) at an arbitrarily selected point in the plane of the cross section (Fig. 10-55b). The axes may have any orientation, but for convenience, orient them horizontally and vertically. Next, *assume* that the beam is bent in such a manner that the z axis is the neutral axis of the cross section. Consequently, the beam deflects in the x-y plane, which becomes the plane of bending. Under these conditions, the normal stress acting on an element of area dA located at distance y from the neutral axis [see Fig. 10-55b] is

$$\sigma_x = -E\kappa_y y \qquad \text{(10-71)}$$

The minus sign is needed because the part of the beam above the z axis (the neutral axis) is in compression when the curvature is positive. (The sign convention for curvature when the beam is bent in the x-y plane is shown in Fig. 10-56a.)

The force acting on the element of area dA is $\sigma_x dA$, and the resultant force acting on the entire cross section is the integral of this elemental force over the cross-sectional area A. Since the beam is in pure bending, the resultant force must be zero; hence,

$$\int_A \sigma_x dA = -\int_A E\kappa_y y\, dA = 0$$

The modulus of elasticity and the curvature are constants at any given cross section, and therefore

$$\int_A y\, dA = 0 \qquad \text{(10-72)}$$

This equation shows that the z axis (the neutral axis) passes through the centroid C of the cross section.

Now assume that the beam is bent in such a manner that the y axis is the neutral axis and the x-z plane is the plane of bending. Then the normal stress acting on the element of area dA (Fig. 10-55b) is

$$\sigma_x = -E\kappa_z z \qquad \text{(10-73)}$$

The sign convention for the curvature κ_z in the x-z plane is shown in Fig. 10-56b. The minus sign is needed in Eq. (10-73) because positive curvature in the x-z plane produces compression on the element dA. The resultant force for this case is

$$\int_A \sigma_x dA = -\int_A E\kappa_z z\, dA = 0$$

which gives

$$\int_A z\, dA = 0 \qquad \text{(10-74)}$$

and again note that the neutral axis must pass through the centroid. Thus, it follows that *the origin of the y and z axes for an unsymmetric beam must be placed at the centroid C.*

Now consider the moment resultant of the stresses σ_x. Once again assume that bending takes place with the z axis as the neutral axis, in which case the

FIGURE 10-55
Unsymmetric beam subjected to a bending moment M

(a)

(b)

Unsymmetric composite beam made up from channel section and old wood beam

Franz Pfluegl/Shutterstock.com

Positive
curvature κ_y

(a)

Positive
curvature κ_z

(b)

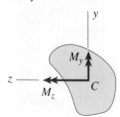

stresses σ_x are given by Eq. (10-71). The corresponding bending moments M_z and M_y about the z and y axes, respectively (Fig. 10-57), are

$$M_z = -\int_A \sigma_x y \, dA = \kappa_y E \int_A y^2 \, dA = \kappa_y E I_z \qquad \text{(10-75a)}$$

$$M_y = \int_A \sigma_x z \, dA = -\kappa_y E \int_A yz \, dA = -\kappa_y E I_{yz} \qquad \text{(10-75b)}$$

In these equations, I_z is the moment of inertia of the cross-sectional area with respect to the z axis and I_{yz} is the *product of inertia* with respect to the y and z axes.[6]

Equations (10-75a and b) lead to the following conclusions: (1) If the z axis is selected in an arbitrary direction through the centroid, it will be the neutral axis *only* if moments M_y and M_z act about the y and z axes and *only* if these moments are in the ratio established by Eqs. (10-75a and b). (2) If the z axis is selected as a *principal axis*, then the product of inertia I_{yz} equals zero and the only bending moment is M_z. In that case, the z axis is the neutral axis, bending takes place in the x-y plane, and the moment M_z acts in that same plane. Thus, bending occurs in a manner analogous to that of a symmetric beam.

In summary, an unsymmetric beam bends in the same general manner as a symmetric beam provided the z axis is a *principal centroidal axis* and the only bending moment is the moment M_z acting about that same axis.

Now assume that the y axis is the neutral axis to arrive at similar conclusions. The stresses σ_x are given by Eq. (10-73) and the bending moments are

$$M_y = \int_A \sigma_x z \, dA = -\kappa_z E \int_A z^2 \, dA = -\kappa_z E I_y \qquad \text{(10-76a)}$$

$$M_z = -\int_A \sigma_x y \, dA = \kappa_z E \int_A yz \, dA = \kappa_z E I_{yz} \qquad \text{(10-76b)}$$

in which I_y is the moment of inertia with respect to the y axis. Again observe that if the neutral axis (the y axis in this case) is oriented arbitrarily, moments M_y and M_z must exist. However, if the y axis is a principal axis, the only moment is M_y and there is ordinary bending in the x-z plane. Therefore, an unsymmetric beam bends in the same general manner as a symmetric beam when the y axis is a *principal centroidal axis* and the only bending moment is the moment M_y acting about that same axis.

One further observation—since the y and z axes are orthogonal, if *either* axis is a principal axis, then the other axis is automatically a principal axis.

This discussion leads to the following important conclusion: *When an unsymmetric beam is in pure bending, the plane in which the bending moment acts is perpendicular to the neutral surface only if the y and z axes are principal centroidal axes of the cross section and the bending moment acts in one of the two principal planes (the x-y plane or the x-z plane).* In such a case, the principal plane in which the bending moment acts becomes the plane of bending, and the usual bending theory (including the flexure formula) is valid.

This conclusion leads to a direct method for finding the stresses in an unsymmetric beam subjected to a bending moment acting in an arbitrary direction.

---

[6]Products of inertia are discussed in Section 5.6.

Procedure for Analyzing an Unsymmetric Beam

A general procedure is now presented for analyzing an unsymmetric beam subjected to any bending moment M (Fig. 10-58). Begin by locating the centroid C of the cross section and constructing a set of principal axes at that point (the y and z axes in the figure).[7] Next, the bending moment M is resolved into components M_y and M_z, which are positive in the directions shown in the figure. These components are

$$M_y = M \sin\theta \qquad M_z = M \cos\theta \qquad \text{(10-77a,b)}$$

where θ is the angle between the moment vector M and the z axis (Fig. 10-58). Since each component acts in a principal plane, it produces pure bending in that same plane. Thus, the usual formulas for pure bending apply, and the stresses are computed using the moments M_y and M_z acting separately. The bending stresses obtained from the moments acting separately are then superposed to obtain the stresses produced by the original bending moment M. (Note that this general procedure is similar to that described in the preceding section for analyzing doubly symmetric beams with inclined loads.)

The superposition of the bending stresses in order to obtain the resultant stress at any point in the cross section is given by Eq. (10-19):

$$\sigma_x = \frac{M_y z}{I_y} - \frac{M_z y}{I_z} = \frac{(M \sin\theta)z}{I_y} - \frac{(M \cos\theta)y}{I_z} \qquad \text{(10-78)}$$

where y and z are the coordinates of the point under consideration.

Also, the equation of the neutral axis $n–n$ (Fig. 10-58) is obtained by setting σ_x equal to zero and simplifying:

$$\frac{\sin\theta}{I_y} z - \frac{\cos\theta}{I_z} y = 0 \qquad \text{(10-79)}$$

The angle β between the neutral axis and the z axis can be obtained from the preceding equation, as

$$\tan\beta = \frac{y}{z} = \frac{I_z}{I_y} \tan\theta \qquad \text{(10-80)}$$

This equation shows that in general the angles β and θ are not equal; hence the neutral axis is generally not perpendicular to the plane in which the applied couple M acts.

This section has focused attention on unsymmetric beams. Of course, symmetric beams are special cases of unsymmetric beams; therefore, the discussions of this section also apply to symmetric beams. If a beam is singly symmetric, the axis of symmetry is one of the centroidal principal axes of the cross section; the other principal axis is perpendicular to the axis of symmetry at the centroid. If a beam is doubly symmetric, the two axes of symmetry are centroidal principal axes.

In a strict sense, the discussions of this section apply only to pure bending, which means that no shear forces act on the cross sections. *When shear forces*

---

[7]Principal axes are discussed in Sections 5.8.

FIGURE 10-58

Unsymmetric cross section with the bending moment M resolved into components M_y and M_z acting about the principal centroidal axes *do exist, the possibility arises that the beam will twist about the longitudinal axis.* However, twisting is avoided when the shear forces act through the *shear center.*

Example 10-18 illustrates the analysis of a beam having one axis of symmetry. The calculations for an unsymmetric beam having no axes of symmetry proceed in the same general manner, except that the determination of the various cross-sectional properties is much more complex.

Alternate Procedure for Analyzing an Unsymmetric Beam

FIGURE 10-59

Unsymmetric Z section with nonprincipal centroidal axes

In the preceding discussion, an unsymmetric beam was analyzed by locating the principal centroidal axes of the cross section and then resolving the bending moment into components in those directions. The advantage of this method is that all standard formulas for stresses and deflections apply because bending takes place in a principal plane. However, the method is inconvenient when the orientation of the principal axes cannot be obtained by inspection (as when an axis of symmetry exists) or from tables (as in the case of standard angle sections). If the orientation of the principal axes and the magnitudes of the principal moments of inertia must be calculated, such as for the Z section in Fig. 10-59, it may be easier to work with the **nonprincipal centroidal axes** that are aligned with the sides of the cross section.

To derive the equations of a generalized bending theory for nonprincipal axes, consider the unsymmetric cross section of Fig. 10-60. The y and z axes have their origin at the centroid, but they are not principal axes. Bending moments M_y and M_z act on the cross section, and bending of the beam occurs in both the x-y and x-z planes, neither of which is a principal plane. If the curvatures in the x-y and x-z planes are κ_y and κ_z, respectively (see Fig. 10-56), the normal stress at point A is

$$\sigma_x = -\kappa_y E y - \kappa_z E z \tag{10-81}$$

FIGURE 10-60

Unsymmetric cross section with the bending moment M resolved into components M_y and M_z that act about the nonpricipal centroidal axes

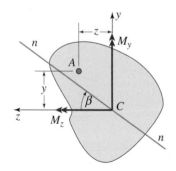

The resultant (axial) force acting on the cross section in the x direction is equated to zero to obtain

$$\int \sigma_x dA = 0 \quad \text{or} \quad \kappa_y E \int y dA + \kappa_z E \int z dA = 0 \tag{10-82}$$

This equation is satisfied automatically because the origin of the axes is at the centroid of the cross section.

The moment M_y is the moment stress resultant about the y axis:

$$M_y = \int \sigma_x z dA = -\kappa_y E \int yz dA - \kappa_z E \int z^2 dA \tag{10-83a}$$

or

$$M_y = -\kappa_y E I_{yz} - \kappa_z E I_y \tag{10-83b}$$

in which I_{yz} is the product of inertia of the cross-sectional area with respect to the y and z axes. In a similar way, the expression for moment about the z axis is

$$M_z = -\int \sigma_x y dA = \kappa_y E \int y^2 dA + \kappa_z E \int yz dA \tag{10-84a}$$

or

$$M_z = \kappa_y E I_z + \kappa_z E I_{yz} \tag{10-84b}$$

Solve Eqs. (10-83) and (10-84) simultaneously to obtain the expressions for curvatures in terms of the bending moments:

$$\kappa_y = \frac{M_z I_y + M_y I_{yz}}{E(I_y I_z - I_{yz}^2)} \quad \text{and} \quad \kappa_z = -\frac{M_y I_z + M_z I_{yz}}{E(I_y I_z - I_{yz}^2)} \qquad \textbf{(10-85)}$$

Now substitute these expressions for curvatures into Eq. (10-81) to obtain the *normal stress* σ_x at any point in an unsymmetric beam when moments M_y and M_z are known:

$$\sigma_x = \frac{(M_y I_z + M_z I_{yz})z - (M_z I_y + M_y I_{yz})y}{I_y I_z - I_{yz}^2} \qquad \textbf{(10-86)}$$

Equation (10-86) is the *generalized flexure formula* for an unsymmetric beam acted on by moments about perpendicular centroidal axes that are not necessarily principal axes. Note that if the y and z axes are principal centroidal axes, Eq. (10-86) reduces to Eq. (10-78), since $I_{yz} = 0$ for principal axes.

The orientation of the *neutral axis nn* (Fig. 10-57) is obtained by equating σ_x to zero to find an expression for angle β between the z axis and the neutral axis:

$$\tan \beta = \frac{y}{z} = \frac{M_y I_z + M_z I_{yz}}{M_z I_y + M_y I_{yz}} \qquad \textbf{(10-87)}$$

The plane of bending is perpendicular to the neutral axis.

Example 10-18

FIGURE 10-61

Example 10-18: Channel section subjected to a bending moment M acting at an angle θ to the z axis

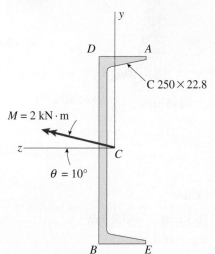

A channel section C 250 × 22.8 is subjected to a bending moment $M = 2$ kN · m oriented at an angle $\theta = 10°$ to the z axis (Fig. 10-61).

Calculate the bending stresses σ_A and σ_B at points A and B, respectively, and determine the position of the neutral axis.

Solution:

Use the four-step problem-solving approach. Combine steps as needed for an efficient solution.

1, 2. *Conceptualize, Categorize*:

> **Properties of the cross section:** The centroid C is located on the axis of symmetry (the z axis) at a distance
>
> $$c = 16.1 \text{ mm}$$
>
> from the back of the channel (Fig. 10-62).[8] The y and z axes are principal centroidal axes with moments of inertia:
>
> $$I_y = 94.5 \text{ cm}^4 \quad I_z = 2800 \text{ cm}^4$$

---

[8]See Table C-3(b), Appendix C, for dimensions and properties of channel sections.

FIGURE 10-62

Solution to Example 10-18:
(a) Cross section properties
and (b) normal stress
distribution in channel section

(a)

(b)

Also, the coordinates of points A, B, D, and E are

$$y_A = 127 \text{ mm} \quad z_A = -66 \text{ mm} + 16.1 \text{ mm} = -49.9 \text{ mm}$$

$$y_B = -127 \text{ mm} \quad z_B = 16.1 \text{ mm}$$

$$y_D = y_A, \quad\quad\quad z_D = z_B$$

$$y_E = y_B, \quad\quad\quad z_E = z_A$$

Bending moments: The bending moments about the y and z axes (Fig. 10-62a) are

$$M_y = M \sin \theta = (2 \text{ kN} \cdot \text{m})(\sin 10°) = 0.347 \text{ kN} \cdot \text{m}$$

$$M_z = M \cos \theta = (2 \text{ kN} \cdot \text{m})(\cos 10°) = 1.970 \text{ kN} \cdot \text{m}$$

3. *Analyze:*

Bending stresses: Now calculate the stress at point A from Eq. (10-78):

$$\sigma_A = \frac{M_y z_A}{I_y} - \frac{M_z y_A}{I_z}$$

$$= \frac{(0.347 \text{ kN} \cdot \text{m})(-0.0499 \text{ m})}{0.945 \times 10^{-6} \text{ m}^4} - \frac{(1.970 \text{ kN} \cdot \text{m})(0.127 \text{ m})}{28.0 \times 10^{-6} \text{ m}^4}$$

$$= -18.34 \text{ MPa} - 8.93 \text{ MPa} = -27.27 \text{ MPa}$$

By a similar calculation, the stress at point B is

$$\sigma_B = \frac{M_y z_B}{I_y} - \frac{M_z y_B}{I_z}$$

$$= \frac{(0.347 \text{ kN} \cdot \text{m})(0.0161 \text{ m})}{0.945 \times 10^{-6} \text{ m}^4} - \frac{(1.970 \text{ kN} \cdot \text{m})(-0.127 \text{ m})}{28.0 \times 10^{-6} \text{ m}^4}$$

$$= 5.92 \text{ MPa} + 8.93 \text{ MPa} = 14.85 \text{ MPa}$$

These stresses are the maximum compressive and tensile stresses in the beam.

The normal stresses at points D and E also can be computed using the procedure shown. Thus,

$$\sigma_D = -3.02 \text{ MPa} \quad \sigma_E = -9.41 \text{ MPa}$$

The normal stresses acting on the cross section are shown in Fig. 10-62b.

Neutral axis: The angle β that locates the neutral axis [Eq. (10-80)] is found as

$$\tan \beta = \frac{I_z}{I_y} \tan \theta = \frac{2800 \text{ cm}^4}{94.5 \text{ cm}^4} \tan 10° = 5.225 \quad \beta = 79.16°$$

The neutral axis $n–n$ is shown in Fig. 10-62, and note that points A and B are located at the farthest distances from the neutral axis, thus confirming that σ_A and σ_B are the largest stresses in the beam.

4. *Finalize:* In this example, the angle β between the z axis and the neutral axis is much larger than the angle θ (Fig. 10-62) because the ratio I_z/I_y is large. The angle β varies from 0 to 79.16° as the angle θ varies from 0 to 10°. Beams with large I_z/I_y ratios are very sensitive to the direction of loading. Thus, beams of this kind should be provided with lateral support to prevent excessive lateral deflections.

Chapter 10 covered the behavior of beams with loads applied and bending occurring in the *x-y* plane: a plane of symmetry in the beam cross section. Both pure bending and nonuniform bending were considered. The normal stresses (σ) were seen to vary linearly from the neutral surface in accordance with the **flexure formula**. Horizontal and vertical shear stresses (τ) were computed using the **shear formula** for the case of nonuniform bending of beams with either rectangular or circular cross sections. The special case of shear in beams with flanges also was considered. Finally, the bending of beams made up of more than one material, referred to as **composite beams**, and the bending of unsymmetric beams were discussed.

Here are some of the major concepts and findings presented in this chapter.

1. If the *x-y* plane is a plane of symmetry of a beam cross section and applied loads act in the *x-y* plane, the bending deflections occur in this same plane, known as the **plane of bending**.

2. A beam in pure bending has constant curvature κ, and a beam in nonuniform bending has varying curvature. Longitudinal strains (ε_x) in a bent beam are proportional to its curvature, and the strains in a beam in pure bending vary linearly with distance from the neutral surface, regardless of the shape of the stress-strain curve of the material, as

$$\varepsilon_x = -\kappa y$$

3. The neutral axis passes through the centroid of the cross-sectional area when the material follows Hooke's law and there is no axial force acting on the cross section. When a beam of linearly elastic material is subjected to pure bending, the *y* and *z* axes are **principal centroidal** axes.

4. If the material of a beam is linearly elastic and follows Hooke's law, the **moment–curvature equation** shows that the curvature is directly proportional to the bending moment *M* and inversely proportional to the quantity *EI*, the **flexural rigidity** of the beam. The moment–curvature relation is

$$\kappa = \frac{M}{EI}$$

5. The **flexure formula** shows that the normal stresses σ_x are directly proportional to the bending moment *M* and inversely proportional to the moment of inertia *I* of the cross section:

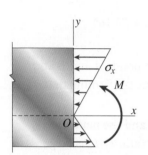

$$\sigma_x = -\frac{My}{I}$$

The maximum tensile and compressive bending stresses acting at any given cross section occur at points located farthest from the neutral axis. Thus,

$$(y = c_1, \; y = -c_2)$$

6. The normal stresses calculated from the flexure formula are not significantly altered by the presence of shear stresses and the associated warping of the cross section for the case of nonuniform bending. However, the flexure formula is not applicable near the supports of a beam or close to a concentrated load, because such irregularities produce **stress concentrations** that are much greater than the stresses obtained from the flexure formula.

7. To **design** a beam to resist bending stresses, calculate the required **section modulus** S from the maximum moment and allowable normal stress as

$$S = \frac{M_{max}}{\sigma_{allow}}$$

To minimize weight and save material, select a beam from a material design manual (see sample tables in Appendixes C and D for steel and wood) that has the least cross-sectional area while still providing the required section modulus; wide-flange sections and I-sections have most of their material in the flanges, and the width of their flanges helps to reduce the likelihood of sideways buckling.

8. Beams subjected to loads that produce both bending moments (M) and shear forces (V) (**nonuniform bending**) develop both normal and shear stresses in the beam. Normal stresses are calculated from the **flexure formula** (provided the beam is constructed of a linearly elastic material), and shear stresses are computed using the **shear formula**

$$\tau = \frac{VQ}{Ib}$$

Shear stress varies parabolically over the height of a rectangular beam, and shear strain also varies parabolically; these shear strains cause cross sections of the beam that were originally plane surfaces to become warped. The maximum values of the shear stress and strain (τ_{max}, γ_{max}) occur at the neutral axis, and the shear stress and strain are zero on the top and bottom surfaces of the beam.

9. The shear formula applies only to prismatic beams and is valid only for beams of linearly elastic materials with small deflections; also, the edges of the cross section must be **parallel** to the y axis. For **rectangular** beams, the accuracy of the shear formula depends upon the height-to-width ratio of the cross section: The formula may be considered exact for very narrow beams but becomes less accurate as width b increases relative to height h. Use the shear formula to calculate the shear stresses only at the neutral axis of a beam of **circular** cross section.

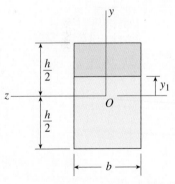

For rectangular cross sections,

$$\tau_{max} = \frac{3}{2}\frac{V}{A}$$

and for solid circular cross sections,

$$\tau_{max} = \frac{4}{3}\frac{V}{A}$$

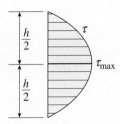

10. Shear stresses rarely govern the design of either circular or rectangular beams made of metals such as steel and aluminum for which the allowable shear stress is usually in the range 25 to 50% of the allowable tensile stress. However, for **materials that are weak in shear,** such as wood, the allowable stress in horizontal shear is in the range of 4 to 10% of the allowable bending stress and so may govern the design.

11. Shear stresses in the flanges of **wide-flange beams** act in both vertical and horizontal directions. The horizontal shear stresses are much larger than the vertical shear stresses in the flanges. The shear stresses in the **web of a wide-flange beam** act only in the vertical direction, are larger than the stresses in the flanges, and may be computed using the shear formula. The maximum shear stress in the web of a wide-flange beam occurs at the neutral axis, and the minimum shear stress occurs where the web meets the flanges. For beams of typical proportions, the shear force in the web is 90 to 98% of the total shear force V acting on the cross section; the remainder is carried by shear in the flanges.

12. In the introduction to **composite beams**, specialized moment–curvature relationship and flexure formulas for composite beams of two materials were developed:

$$\kappa = \frac{1}{\rho} = \frac{M}{E_1 I_1 + E_2 I_2}$$

$$\sigma_{x1} = -\frac{M_Y E_1}{E_1 I_1 + E_2 I_2} \qquad \sigma_{x2} = -\frac{M_Y E_2}{E_1 I_1 + E_2 I_2}$$

Both materials must follow Hooke's law, and the two parts of the beam must be adequately bonded so that they act as a single unit. The formulas here do not apply to reinforced concrete beams, which are not designed on the basis of linearly elastic behavior. However, a transformed-section approach (see Example 10-16) can be used as part of a cracked section analysis of reinforced concrete beams.

13. The **transformed-section method** offers a convenient way of transforming the cross section of a composite beam into an equivalent cross section of an imaginary beam that is composed of only one material. The ratio of the modulus of elasticity of material 2 to that of material 1 is known as the **modular ratio**, $n = E_2/E_1$. The moment of inertia of the transformed section is defined as

$$I_T = I_1 + nI_2 = I_1 + \frac{E_2}{E_1}I_2$$

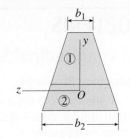

Normal stresses in the beam transformed to material 1 are computed using the simplified flexure formula:

$$\sigma_{x1} = -\frac{My}{I_T}$$

while those in material 2 are computed as

$$\sigma_{x2} = -\frac{My}{I_T}n$$

14. When the restriction of symmetry about at least one axis of the cross section is removed for pure bending, the plane in which the bending moment acts is perpendicular to the neutral surface only if the y and z axes are principal centroidal axes of the cross section, and the bending moment acts in one of the two principal planes (the x-y plane or the x-z plane). A **general procedure for computing normal stresses in** *unsymmetric beams* acted on by any moment M is (a) find the centroid and (b) superpose results of the flexure formula about the two principal centroidal axes to find normal stresses. An alternate approach is to use the **generalized flexure formula** with the original centroidal axes aligned with the beam cross section.

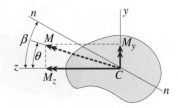

PROBLEMS

10.4 Longitudinal Strains in Beams

Introductory Problems

10.4-1 A steel wire with a diameter of $d = 1/16$ in. is bent around a cylindrical drum with a radius of $R = 36$ in. (see figure).

(a) Determine the maximum normal strain ε_{max}.
(b) What is the minimum acceptable radius of the drum if the maximum normal strain must remain below yield? Assume $E = 30,000$ ksi and $\sigma_Y = 100$ ksi.
(c) If $R = 36$ in., what is the maximum acceptable diameter of the wire if the maximum normal strain must remain below yield?

PROBLEM 10.4-1

10.4-2 A copper wire having a diameter of $d = 4$ mm is bent into a circle and held with the ends just touching (see figure).

(a) If the maximum permissible strain in the copper is $\varepsilon_{max} = 0.0024$, what is the shortest length L of wire that can be used?
(b) If $L = 5.5$ m, what is the maximum acceptable diameter of the wire if the maximum normal strain must remain below yield? Assume $E = 120$ GPa and $\sigma_Y = 300$ MPa.

d = diameter

L = length

PROBLEM 10.4-2

10.4-3 A 4.75-in. outside diameter polyethylene pipe designed to carry chemical waste is placed in a trench and bent around a quarter-circular 90° bend (see figure). The bent section of the pipe is 52 ft long.

(a) Determine the maximum compressive strain ε_{max} in the pipe.
(b) If the normal strain cannot exceed 6.1×10^{-3}, what is the maximum diameter of the pipe?
(c) If $d = 4.75$ in., what is the minimum acceptable length of the bent section of the pipe?

90°

PROBLEM 10.4-3

Representative Problems

10.4-4 A cantilever beam AB is loaded by a couple M_0 at its free end (see figure). The length of the beam is $L = 2.0$ m, and the longitudinal normal strain at the top surface is $\varepsilon = 0.0010$. The distance from the top surface of the beam to the neutral surface is $c = 85$ mm.

(a) Calculate the radius of curvature ρ, the curvature κ, and the vertical deflection δ at the end of the beam.
(b) If allowable strain $\varepsilon_a = 0.0008$, what is the maximum acceptable depth of the beam? [Assume that the curvature is unchanged from part(a)].
(c) If allowable strain $\varepsilon_a = 0.0008$, $c = 85$ mm, and $L = 4$ m, what is deflection δ?

A

B

L

M_0

PROBLEM 10.4-4

10.4-5 A thin strip of steel with a length of $L = 19$ in. and thickness of $t = 0.275$ in. is bent by couples M_0 (see figure). The deflection at the midpoint of the strip (measured from a line joining its end points) is found to be 0.30 in.

(a) Determine the longitudinal normal strain ε at the top surface of the strip.
(b) If allowable strain $\varepsilon_a = 0.0008$, what is the maximum acceptable thickness of the strip?
(c) If allowable strain $\varepsilon_a = 0.0008$, $t = 0.275$ in., and $L = 32$ in., what is deflection δ?
(d) If allowable strain $\varepsilon_a = 0.0008$, $t = 0.275$ in., and the deflection cannot exceed 1.0 in., what is the maximum permissible length of the strip?

PROBLEM 10.4-5

10.4-6 A bar of rectangular cross section is loaded and supported as shown in the figure. The distance between supports is $L = 1.75$ m, and the height of the bar is $h = 140$ mm. The deflection at the midpoint is measured as 2.5 mm.

(a) What is the maximum normal strain ε at the top and bottom of the bar?
(b) If allowable strain $\varepsilon_a = 0.0006$ and the deflection cannot exceed 4.3 mm, what is the maximum permissible length of the bar?

PROBLEM 10.4-6

10.4-7 A simply supported beam with a length $L = 10$ ft and height 7 in. is bent by couples M_0 into a circular arc with downward deflection δ at the midpoint. If the curvature of the beam is 0.003 ft^{-1}, calculate the deflection, δ, at the mid-span of the beam and the longitudinal strain at the bottom fiber given that the distance between the neutral surface and the bottom surface is 3.5 in.

PROBLEM 10.4-7

10.4-8 A cantilever beam is subjected to a concentrated moment at B. The length of the beam $L = 3$ m and the height $h = 600$ mm. The longitudinal strain at the top of the beam is 0.0005 and the distance from the neutral surface to the bottom surface of the beam is 300 mm. Find the radius of curvature, the curvature, and the deflection of the beam at B.

PROBLEM 10.4-8

10.5 Normal Stresses in Beams (Linearly Elastic Materials)

Introductory Problems

10.5-1 A thin strip of hard copper ($E = 16,000$ ksi) having length $L = 90$ in. and thickness $t = 3/32$ in. is bent into a circle and held with the ends just touching (see figure).

(a) Calculate the maximum bending stress σ_{max} in the strip.
(b) By what percent does the stress increase or decrease if the thickness of the strip is increased by 1/32 in.?
(c) Find the new length of the strip so that the stress in part (b) ($t = 1/8$ in. and $L = 90$ in.) is equal to that in part (a) ($t = 3/32$ in. and $L = 90$ in.).

PROBLEM 10.5-1

10.5-2 A steel wire ($E = 200$ GPa) of a diameter $d = 1.25$ mm is bent around a pulley of a radius $R_0 = 500$ mm (see figure).

(a) What is the maximum stress σ_{max} in the wire?

(b) By what percent does the stress increase or decrease if the radius of the pulley is increased by 25%?

(c) By what percent does the stress increase or decrease if the diameter of the wire is increased by 25% while the pulley radius remains at $R_0 = 500$ mm?

PROBLEM 10.5-2

10.5-3 A thin, high-strength steel rule ($E = 30 \times 10^6$ psi) having a thickness $t = 0.175$ in. and length $L = 48$ in. is bent by couples M_0 into a circular arc subtending a central angle $\alpha = 40°$ (see figure).

(a) What is the maximum bending stress σ_{max} in the rule?

(b) By what percent does the stress increase or decrease if the central angle is increased by 10%?

(c) What percent increase or decrease in rule thickness will result in the maximum stress reaching the allowable value of 42 ksi?

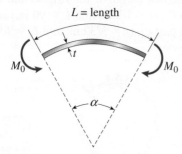

PROBLEM 10.5-3

Representative Problems

10.5-4 A simply supported wood beam AB with a span length $L = 4$ m carries a uniform load of intensity $q = 5.8$ kN/m (see figure).

(a) Calculate the maximum bending stress σ_{max} due to the load q if the beam has a rectangular cross section with width $b = 140$ mm and height $h = 240$ mm.

(b) Repeat part (a) but use the trapezoidal distributed load shown in the figure part b.

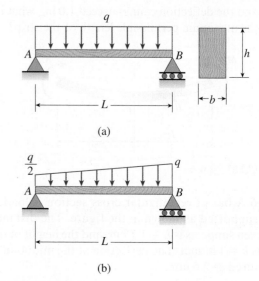

(a)

(b)

PROBLEM 10.5-4

10.5-5 Beam ABC has simple supports at A and B and an overhang from B to C. The beam is constructed from a steel W 16 × 31. The beam must carry its own weight in addition to uniform load $q = 150$ lb/ft. Determine the maximum tensile and compressive stresses in the beam.

PROBLEM 10.5-5

10.5-6 A simply supported beam is subjected to a linearly varying distributed load $q(x) = \dfrac{x}{L}q_0$ with maximum intensity q_0 at B. The beam has a length $L = 4$ m and rectangular cross section with a width of 200 mm and height of 300 mm. Determine the maximum permissible value for the maximum intensity, q_0, if the allowable normal stresses in tension and compression are 120 MPa.

PROBLEM 10.5-6

10.5-7 Each girder of the lift bridge (see figure) is 180 ft long and simply supported at the ends. The design load for each girder is a uniform load of intensity 1.6 kips/ft. The girders are fabricated by welding three steel plates to form an I-shaped cross section (see figure) having section modulus $S = 3600$ in^3.

What is the maximum bending stress σ_{max} in a girder due to the uniform load?

PROBLEM 10.5-7

10.5-8 A freight-car axle AB is loaded approximately as shown in the figure, with the forces P representing the car loads (transmitted to the axle through the axle boxes) and the forces R representing the rail loads (transmitted to the axle through the wheels). The diameter of the axle is $d = 82$ mm, the distance between centers of the rails is L, and the distance between the forces P and R is $b = 220$ mm.

Calculate the maximum bending stress σ_{max} in the axle if $P = 50$ kN.

PROBLEM 10.5-8

10.5-9 A seesaw weighing 3 lb/ft of length is occupied by two children, each weighing 90 lb (see figure). The center of gravity of each child is 8 ft from the fulcrum. The board is 19 ft long, 8 in. wide, and 1.5 in. thick.

What is the maximum bending stress in the board?

PROBLEM 10.5-9

10.5-10 During construction of a highway bridge, the main girders are cantilevered outward from one pier toward the next (see figure). Each girder has a cantilever length of 48 m and an I-shaped cross section with dimensions shown in the figure. The load on each girder (during construction) is assumed to be 9.5 kN/m, which includes the weight of the girder.

Determine the maximum bending stress in a girder due to this load.

PROBLEM 10.5-10

799

10.5-11 The horizontal beam *ABC* of an oil-well pump has the cross section shown in the figure. If the vertical pumping force acting at end *C* is 9 kips and if the distance from the line of action of that force to point *B* is 16 ft, what is the maximum bending stress in the beam due to the pumping force?

Horizontal beam transfers loads as part of oil well pump

PROBLEM 10.5-11

10.5-12 A railroad tie (or *sleeper*) is subjected to two rail loads, each of magnitude $P = 175$ kN, acting as shown in the figure. The reaction q of the ballast is assumed to be uniformly distributed over the length of the tie, which has cross-sectional dimensions $b = 300$ mm and $h = 250$ mm.

Calculate the maximum bending stress σ_{max} in the tie due to the loads *P*, assuming the distance $L = 1500$ mm and the overhang length $a = 500$ mm.

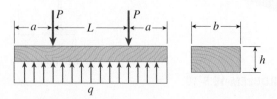

PROBLEM 10.5-12

800

10.5-13 A fiberglass pipe is lifted by a sling, as shown in the figure. The outer diameter of the pipe is 6.0 in., its thickness is 0.25 in., and its weight density is 0.053 lb/in^3. The length of the pipe is $L = 36$ ft and the distance between lifting points is $s = 11$ ft.

(a) Determine the maximum bending stress in the pipe due to its own weight.

(b) Find the spacing *s* between lift points which minimizes the bending stress. What is the minimum bending stress?

(c) What spacing *s* leads to maximum bending stress? What is that stress?

PROBLEM 10.5-13

10.5-14 A small dam of height $h = 2.0$ m is constructed of vertical wood beams *AB* of thickness $t = 120$ mm, as shown in the figure. Consider the beams to be simply supported at the top and bottom.

Determine the maximum bending stress σ_{max} in the beams, assuming that the weight density of water is $\gamma = 9.81$ kN/m^3.

PROBLEM 10.5-14

10.5-15 Determine the maximum tensile stress σ_t (due to pure bending about a horizontal axis through *C* by positive bending moments *M*) for beams having cross sections as follows (see figure).

(a) A semicircle of diameter *d*.

(b) An isosceles trapezoid with bases $b_1 = b$ and $b_2 = 4b/3$ and altitude *h*.

(c) A circular sector with $\alpha = \pi/3$ and $r = d/2$

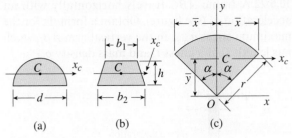

(a)   (b)   (c)

PROBLEM 10.5-15

10.5-16 Determine the maximum bending stress σ_{max} (due to pure bending by a moment M) for a beam having a cross section in the form of a circular core (see figure). The circle has diameter d and the angle $\beta = 60°$. *Hint:* Use the formulas given in Appendix B, Cases 9 and 15.

PROBLEM 10.5-16

10.5-17 A simple beam AB of a span length $L = 24$ ft is subjected to two wheel loads acting at a distance $d = 5$ ft apart (see figure). Each wheel transmits a load $P = 3.0$ kips, and the carriage may occupy any position on the beam.

(a) Determine the maximum bending stress σ_{max} due to the wheel loads if the beam is an I-beam having section modulus $S = 16.2$ in.³.

(b) If $d = 5$ ft, find the required span length L to reduce the maximum stress in part (a) to 18 ksi.

(c) If $L = 24$ ft, find the required wheel spacing s to reduce the maximum stress in part (a) to 18 ksi.

PROBLEM 10.5-17

10.5-18 Determine the maximum tensile stress σ_t and maximum compressive stress σ_c due to the load P acting on the simple beam AB (see figure).

(a) Data are $P = 6.2$ kN, $L = 3.2$ m, $d = 1.25$ m, $b = 80$ mm, $t = 25$ mm, $h = 120$ mm, and $h_1 = 90$ mm.

(b) Find the value of d for which tensile and compressive stresses are the largest. What are these stresses?

PROBLEM 10.5-18

10.5-19 A cantilever beam AB, loaded by a uniform load and a concentrated load (see figure), is constructed of a channel section.

(a) Find the maximum tensile stress σ_t and maximum compressive stress σ_c if the cross section has the dimensions indicated and the moment of inertia about the z axis (the neutral axis) is $I = 3.36$ in.⁴. *Note:* The uniform load represents the weight of the beam.

(b) Find the maximum value of the concentrated load if the maximum tensile stress cannot exceed 4 ksi and the maximum compressive stress is limited to 14.5 ksi.

(c) How far from A can load $P = 250$ lb be positioned if the maximum tensile stress cannot exceed 4 ksi and the maximum compressive stress is limited to 14.5 ksi?

PROBLEM 10.5-19

10.5-20 A cantilever beam AB of an isosceles trapezoidal cross section has a length $L = 0.8$ m, dimensions $b_1 = 80$ mm and $b_2 = 90$ mm, and height $h = 110$ mm (see figure). The beam is made of brass weighing 85 kN/m³.

(a) Determine the maximum tensile stress σ_t and maximum compressive stress σ_c due to the beam's own weight.
(b) If the width b_1 is doubled, what happens to the stresses?
(c) If the height h is doubled, what happens to the stresses?

PROBLEM 10.5-20

10.5-21 A cantilever beam, a $C\,12 \times 30$ section, is subjected to its own weight and a point load at B. Find the maximum permissible value of load P at B (kips) if the allowable stress in tension and compression is $\sigma_a = 18$ ksi.

PROBLEM 10.5-21

10.5-22 A frame ABC travels horizontally with an acceleration a_0 (see figure). Obtain a formula for the maximum stress σ_{max} in the vertical arm AB, which has length L, thickness t, and mass density ρ.

PROBLEM 10.5-22

10.5-23 A beam ABC with an overhang from B to C supports a uniform load of 200 lb/ft throughout its length (see figure). The beam is a channel section with dimensions as shown in the figure. The moment of inertia about the z axis (the neutral axis) equals 8.13 in⁴.

(a) Calculate the maximum tensile stress σ_t and maximum compressive stress σ_c due to the uniform load.
(b) Find the required span length a that results in the ratio of larger to smaller compressive stress being equal to the ratio of larger to smaller tensile stress for the beam. Assume that the total length $L = a + b = 18$ ft remains unchanged.

PROBLEM 10.5-23

10.5-24 A cantilever beam AB with a rectangular cross section has a longitudinal hole drilled throughout its length (see figure). The beam supports a load $P = 600$ N. The cross section is 25 mm wide and 50 mm high, and the hole has a diameter of 10 mm.

Find the bending stresses at the top of the beam, at the top of the hole, and at the bottom of the beam.

PROBLEM 10.5-24

10.5-25 A beam with a T-section is supported and loaded as shown in the figure. The cross section has width $b = 2\ 1/2$ in., height $h = 3$ in., and thickness $t = 3/8$ in.

(a) Determine the maximum tensile and compressive stresses in the beam.

(b) If the allowable stresses in tension and compression are 18 ksi and 12 ksi, respectively, what is the required depth h of the beam? Assume that thickness t remains at 3/8 in. and that flange width $b = 2.5$ in.

(c) Find the new values of loads P and q so that the allowable tension (18 ksi) and compression

(12 ksi) stresses are reached simultaneously for the beam. Use the beam cross section in part (a) (see figure) and assume that L_1, L_2, and L_3 are unchanged.

PROBLEM 10.5-25

10.5-26 Consider the compound beam with segments AB and BCD joined by a pin connection (moment release) just right of B (see figure part a). The beam cross section is a double-T made up from three 50 mm × 150 mm wood members (actual dimensions, see figure part b).

(a) Find the centroid C of the double-T cross section (c_1, c_2); then compute the moment of inertia, $[I_z(\text{mm}^4)]$.

(b) Find the maximum *tensile* normal stress σ_t and maximum *compressive* normal stress σ_c (kPa) for the loading shown. (Ignore the weight of the beam.)

(a)

(b)

PROBLEM 10.5-26

10.5-27 A small dam of a height $h = 6$ ft is constructed of vertical wood beams AB, as shown in the figure. The wood beams, which have a thickness $t = 2.5$ in., are simply supported by horizontal steel beams at A and B.

Construct a graph showing the maximum bending stress σ_{max} in the wood beams versus the depth d of the water above the lower support at B. Plot the stress σ_{max} (psi) as the ordinate and the depth d (ft) as the abscissa. *Note:* The weight density γ of water equals 62.4 lb/ft^3.

Steel beam
Wood beam
Steel beam   Wood beam

Side view   Top view

PROBLEM 10.5-27

10.5-28 A foot bridge on a hiking trail is constructed using two timber logs each having a diameter $d = 0.5$ m (see figure a). The bridge is simply supported and has

(a)

(b)

Section *x–x*
(c)

PROBLEM 10.5-28

a length $L = 4$ m. The top of each log is trimmed to form the walking surface (see Fig. b). A simplified model of the bridge is shown in Fig. c. *Each log* must carry its own weight $w = 1.2$ kN/m and the weight ($P = 850$ N) of a person at mid-span (see Fig. b).

(a) Determine the maximum tensile and compressive stresses in the beam (Fig. b) due to bending.

(b) If load w is unchanged, find the maximum permissible value of load P_{max} if the allowable normal stress in tension and compression is 2.5 MPa.

10.5-29 A steel post ($E = 30 \times 10^6$ psi) having thickness $t = 1/8$ in. and height $L = 72$ in. supports a stop sign (see figure), where $s = 12.5$ in. The height of the

L

Section A–A

5/8 in.   Circular cut-out, $d = 0.375$ in.
Post, $t = 0.125$ in.

z ------ 1.5 in.

C

c_1
c_2

Stop sign

0.5 in. 1.0 in. 1.0 in. 0.5 in.

Wind load

Numerical properties of post
$A = 0.578$ in.^{2}, $c_1 = 0.769$ in., $c_2 = 0.731$ in., $I_y = 0.44867$ in.^{4}, $I_z = 0.16101$ in.^{4}

A A

Elevation
view of post

PROBLEM 10.5-29

post L is measured from the base to the centroid of the sign. The stop sign is subjected to wind pressure $p = 20$ lb/ft^2 normal to its surface. Assume that the post is fixed at its base.

(a) What is the resultant load on the sign? (See Appendix B, Case 25, for properties of an octagon, $n = 8$.)

(b) What is the maximum bending stress σ_{max} in the post?

(c) Repeat part (b) if the circular cut-outs are eliminated over the height of the post.

10.5-30 Beam $ABCDE$ has a moment release just right of joint B and has concentrated moment loads at D and E. In addition, a cable with tension P is attached at F and runs over a pulley at C (Fig. a). The beam is constructed using two steel plates, which are welded to form a T cross section (see Fig. b). *Consider flexural stresses only.* Find the maximum permissible value of load *variable P* if the allowable bending stress is 130 MPa. Ignore the self-weight of the frame members and let length *variable L = 0.75 m*.

(a)

PROBLEM 10.5-30   (b)

10.6 Design of Beams for Bending Stresses

Introductory Problems

10.6-1 A simply supported wood beam having a span length $L = 12$ ft is subjected to unsymmetrical point loads, as shown in the figure. Select a suitable size for the beam from the table in Appendix D. The allowable bending stress is 1800 psi and the wood weighs 35 lb/ft^3.

PROBLEM 10.6-1

10.6-2 A simply supported beam ($L = 4.5$ m) must support mechanical equipment represented as a distributed load with intensity $q = 30$ kN/m acting over the middle segment of the beam (see figure). Select the most economical W-shape steel beam from Table C-1(b) to support the loads. Consider both the distributed force q and the weight of the beam. Use an allowable bending stress of 140 MPa.

PROBLEM 10.6-2

10.6-3 The cross section of a narrow-gage railway bridge is shown in part a of the figure. The bridge is constructed with longitudinal steel girders that support the wood cross ties. The girders are restrained against lateral buckling by diagonal bracing, as indicated by the dashed lines.

The spacing of the girders is $s_1 = 50$ in. and the spacing of the rails is $s_2 = 30$ in. The load transmitted by each rail to a single tie is $P = 1500$ lb. The cross section of a tie, shown in part b of the figure, has a width $b = 5.0$ in. and depth d.

Determine the minimum value of d based upon an allowable bending stress of 1125 psi in the wood tie. (Disregard the weight of the tie itself.)

PROBLEM 10.6-3

10.6-4 A fiberglass bracket $ABCD$ with a solid circular cross section has the shape and dimensions shown in the figure. A vertical load $P = 40$ N acts at the free end D.

(a) Determine the minimum permissible diameter d_{min} of the bracket if the allowable bending stress in the material is 30 MPa and $b = 37$ mm. *Note:* Disregard the weight of the bracket itself.

(b) If $d = 10$ mm, $b = 37$ mm, and $\sigma_{allow} = 30$ MPa, what is the maximum value of load P if vertical load P at D is replaced with horizontal loads P at B and D (see figure part b)?

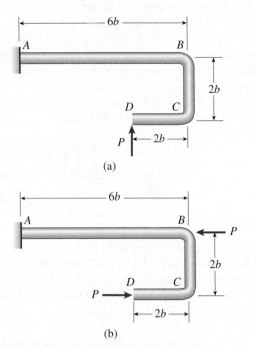

(a)

(b)

PROBLEM 10.6-4

Representative Problems

10.6-5 A cantilever beam AB is loaded by a uniform load q and a concentrated load P, as shown in the figure.

(a) Select the most economical steel C shape from Table C-3(a) in Appendix C; use $q = 20$ lb/ft and $P = 300$ lb (assume allowable normal stress is $\sigma_a = 18$ ksi).

(b) Select the most economical steel S shape from Table C-2(a) in Appendix C; use $q = 45$ lb/ft and $P = 2000$ lb (assume allowable normal stress is $\sigma_a = 20$ ksi).

(c) Select the most economical steel W shape from Table C-1(a) in Appendix C; use $q = 45$ lb/ft and $P = 2000$ lb (assume allowable normal stress is $\sigma_a = 20$ ksi). *However*, assume that the design requires that the W shape must be used in weak axis bending, i.e., it must bend about the 2–2 (or y) axis of the cross section.

Note: For parts (a), (b), and (c), revise your initial beam selection as needed to include the distributed weight of the beam in addition to uniform load q.

PROBLEM 10.6-5

10.6-6 A simple beam of length $L = 5$ m carries a uniform load of intensity $q = 5.8$ kN/m and a concentrated load 22.5 kN (see figure).

(a) Assuming $\sigma_{allow} = 110$ MPa, calculate the required section modulus S. Then select the most economical wide-flange beam (W shape) from Table C-1(b) in Appendix C, and recalculate S, taking into account the weight of the beam. Select a new beam if necessary.

(b) Repeat part (a), but now assume that the design requires that the W shape must be used in weak axis bending (i.e., it must bend about the 2–2 (or y) axis of the cross section).

PROBLEM 10.6-6

10.6-7 A simple beam AB is loaded as shown in the figure.

(a) Calculate the required section modulus S if $\sigma_{allow} = 18,000$ psi, $L = 32$ ft, $P = 2900$ lb, and $q = 450$ lb/ft. Then select a suitable I-beam (S shape) from Table C-2(a), Appendix C, and recalculate S taking into account the weight of the beam. Select a new beam size if necessary.

(b) What is the maximum load P that can be applied to your final beam selection in part (a)?

PROBLEM 10.6-7

10.6-8 A pontoon bridge (see figure) is constructed of two longitudinal wood beams, known as *balks*, that span between adjacent pontoons and support the transverse floor beams, which are called *chesses*. For purposes of design, assume that a uniform floor load of 7.5 kPa acts over the chesses. (This load includes an allowance for the weights of the chesses and balks.) Also, assume that the chesses are 2.5 m long and that the balks are simply supported with a span of 3.0 m. The allowable bending stress in the wood is 15 MPa.

(a) If the balks have a square cross section, what is their minimum required width b_{min}?

(b) Repeat part (a) if the balk width is $1.5 b$ and the balk depth is b; compare the cross-sectional areas of the two designs.

PROBLEM 10.6-8

10.6-9 A floor system in a small building consists of wood planks supported by 2-in. (nominal width) joists spaced at distance s and measured from center to center (see figure). The span length L of each joist is 12 ft, the spacing s of the joists is 16 in., and the allowable bending stress in the wood is 1250 psi. The uniform floor load is 120 lb/ft², which includes an allowance for the weight of the floor system itself.

(a) Calculate the required section modulus S for the joists, and then select a suitable joist size (surfaced lumber) from Appendix D, assuming that each joist may be represented as a simple beam carrying a uniform load.

(b) What is the maximum floor load that can be applied to your final beam selection in part (a)?

PROBLEMS 10.6-9 and 10.6-10

10.6-10 The wood joists supporting a plank floor (see figure) are 38 mm × 220 mm in cross section (actual dimensions) and have a span length of $L = 4.0$ m. The floor load is 5.0 kPa, which includes the weight of the joists and the floor.

(a) Calculate the maximum permissible spacing s of the joists if the allowable bending stress is 14 MPa. (Assume that each joist may be represented as a simple beam carrying a uniform load.)

(b) If spacing $s = 406$ mm, what is the required depth h of the joist? Assume all other variables remain unchanged.

10.6-11 A beam ABC with an overhang from B to C is constructed of a C 10 × 30 channel section with flanges facing upward (see figure). The beam supports its own weight (30 lb/ft) plus a *triangular* load of maximum intensity q_0 acting on the overhang. The allowable stresses in tension and compression are 18 ksi and 12 ksi, respectively.

(a) Determine the allowable *triangular* load intensity $q_{0,allow}$ if the distance L equals 4 ft.

(b) What is the allowable triangular load intensity $q_{0,allow}$ if the beam is rotated 180° about its longitudinal centroidal axis so that the flanges are downward?

PROBLEM 10.6-11

10.6-12 A "trapeze bar" in a hospital room provides a means for patients to exercise while in bed (see figure). The bar is 2.1 m long and has a cross section in the shape of a regular octagon. The design load is 1.2 kN applied at the midpoint of the bar, and the allowable bending stress is 200 MPa.

Determine the minimum height h of the bar. (Assume that the ends of the bar are simply supported and that the weight of the bar is negligible.)

PROBLEM 10.6-12

10.6-13 A two-axle carriage that is part of an overhead traveling crane in a testing laboratory moves slowly across a simple beam AB (see figure). The load transmitted to the beam from the front axle is 2200 lb and from the rear axle is 3800 lb. The weight of the beam itself may be disregarded.

(a) Determine the minimum required section modulus S for the beam if the allowable bending stress is 17.0 ksi, the length of the beam is 18 ft, and the wheelbase of the carriage is 5 ft.

(b) Select the most economical I-beam (S shape) from Table C-2(a), Appendix C.

PROBLEM 10.6-13

10.6-14 A cantilever beam AB with a circular cross section and length $L = 750$ mm supports a load $P = 800$ N acting at the free end (see figure). The beam is made of steel with an allowable bending stress of 120 MPa.

(a) Determine the required diameter d_{min} (figure part a) of the beam, considering the effect of the beam's own weight.

(b) Repeat part (a) if the beam is hollow with wall thickness $t = d/8$ (figure part b); compare the cross-sectional areas of the two designs.

PROBLEM 10.6-14

10.6-15 A propped cantilever beam ABC (see figure) has a shear release just right of the mid-span.

(a) Select the most economical wood beam from the table in Appendix D; assume $q = 55$ lb/ft, $L = 16$ ft, $\sigma_{aw} = 1750$ psi, and $\tau_{aw} = 375$ psi. Include the self-weight of the beam in your design.

(b) If a C 10 × 25 steel beam is now used for beam ABC, what is the maximum permissible value of load variable q? Assume $\sigma_{as} = 16$ ksi and $L = 10$ ft. Include the self-weight of the beam in your analysis.

PROBLEM 10.6-15

10.6-16 A small balcony constructed of wood is supported by three identical cantilever beams (see figure). Each beam has length $L_1 = 2.1$ m, width b, and height $h = 4b/3$. The dimensions of the balcony floor are $L_1 \times L_2$, where $L_2 = 2.5$ m. The design load is 5.5 kPa acting over the entire floor area. (This load accounts for all loads except the weights of the cantilever beams, which have a weight density $\gamma = 5.5$ kN/m^3.) The allowable bending stress in the cantilevers is 15 MPa.

Assuming that the middle cantilever supports 50% of the load and each outer cantilever supports 25% of the load, determine the required dimensions b and h.

PROBLEM 10.6-16

10.6-17 A beam having a cross section in the form of an unsymmetric wide-flange shape (see figure) is subjected to a negative bending moment acting about the z axis.

Determine the width b of the top flange in order that the stresses at the top and bottom of the beam will be in the ratio 4:3, respectively.

PROBLEM 10.6-17

10.6-18 A beam having a cross section in the form of a channel (see figure) is subjected to a bending moment acting about the z axis.

Calculate the thickness t of the channel in order that the bending stresses at the top and bottom of the beam will be in the ratio 7:3, respectively.

PROBLEM 10.6-18

10.6-19 Determine the ratios of the weights of four beams that have the same length, are made of the same material, are subjected to the same maximum bending moment, and have the same maximum bending stress if their cross sections are (1) a rectangle with height equal to twice the width, (2) a square, (3) a circle, and (4) a pipe with outer diameter d and wall thickness $t = d/8$ (see figures).

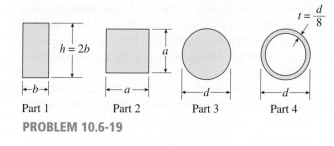

PROBLEM 10.6-19

10.6-20 A horizontal shelf AD of length $L = 1215$ mm, width $b = 305$ mm, and thickness $t = 22$ mm is supported by brackets at B and C (see part a of the figure). The brackets are adjustable and may be placed in any desired positions between the ends of the shelf. A uniform load of intensity q, which includes the weight of the shelf itself, acts on the shelf (see part b of the figure).

(a) Determine the maximum permissible value of the load q if the allowable bending stress in the shelf is $\sigma_{allow} = 8.5$ MPa and the position of the supports is adjusted for maximum load carrying capacity.

(b) The bookshelf owner decides to reinforce the shelf with a bottom wood plate $b/2 \times t/2$ along its entire length (see figure part c). Find the new maximum permissible value of the load q if the allowable bending stress in the shelf remains at $\sigma_{allow} = 8.5$ MPa.

PROBLEM 10.6-20

10.6-21 A steel plate (called a *cover plate*) having cross-sectional dimensions 6.0 in. × 0.5 in. is welded along the full length of the bottom flange of a W 12 × 50 wide-flange beam (see figure, which shows the beam cross section).

What is the percent increase in the smaller section modulus (as compared to the wide-flange beam alone)?

W 12 × 50

6.0 × 0.5 in. cover plate

PROBLEM 10.6-21

10.6-22 A steel beam *ABC* is simply supported at *A* and *B* and has an overhang *BC* of length $L = 150$ mm (see figure). The beam supports a uniform load of intensity $q = 4.0$ kN/m over its entire span *AB* and 1.5q over *BC*. The cross section of the beam is rectangular with width b and height $2b$. The allowable bending stress in the steel is $\sigma_{allow} = 60$ MPa, and its weight density is $\gamma = 77.0$ kN/m^3.

(a) Disregarding the weight of the beam, calculate the required width b of the rectangular cross section.

(b) Taking into account the weight of the beam, calculate the required width b.

1.5 q

q

A

B

C

2b

2L

L

b

PROBLEM 10.6-22

10.6-23 A retaining wall 6 ft high is constructed of horizontal wood planks 2.5 in. thick (actual dimension) that are supported by vertical wood piles of a 12 in. diameter (actual dimension), as shown in the figure. The lateral earth pressure is $p_1 = 125$ lb/ft^2 at the top of the wall and $p_2 = 425$ lb/ft^2 at the bottom.

(a) Assuming that the allowable stress in the wood is 1175 psi, calculate the maximum permissible spacing s of the piles.

(b) Find the required diameter of the wood piles so that piles and planks ($t = 2.5$ in.) reach the allowable stress at the same time.

Hint: Observe that the spacing of the piles may be governed by the load-carrying capacity of either the planks or the piles. Consider the piles to act as cantilever beams subjected to a trapezoidal distribution of load, and consider the planks to act as simple beams between the piles. To be on the safe side, assume that the pressure on the bottom plank is uniform and equal to the maximum pressure.

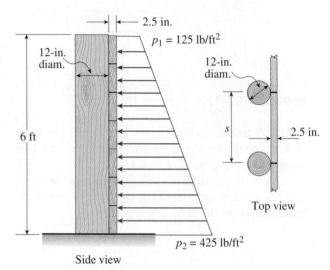

2.5 in.

$p_1 = 125$ lb/ft^2

12-in. diam.

12-in. diam.

6 ft

s

2.5 in.

Top view

$p_2 = 425$ lb/ft^2

Side view

PROBLEM 10.6-23

10.6-24 A retaining wall (Fig. a) is constructed using steel W-shape columns and concrete panel infill (Fig. b). Each column is subjected to lateral soil pressure with peak intensity q_0 (Figs. b and c). The tensile and compressive strength of the beam is 600 MPa. Select the most economical W 360 section from Table C-1(b) based on safety factor of 3.0.

10.6-25 A beam of square cross section (a = length of each side) is bent in the plane of a diagonal (see figure). By removing a small amount of material at the top and bottom corners, as shown by the shaded triangles in the figure, you can increase the section modulus and obtain a stronger beam, even though the area of the cross section is reduced.

(a) Determine the ratio β defining the areas that should be removed in order to obtain the strongest cross section in bending.

(b) By what percent is the section modulus increased when the areas are removed?

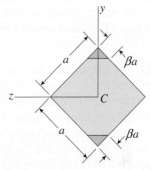

PROBLEM 10.6-25

10.6-26 The cross section of a rectangular beam having a width b and height h is shown in part a of the figure. For reasons unknown to the beam designer, it is planned to add structural projections of width $b/9$ and height d to the top and bottom of the beam (see part b of the figure).

For what values of d is the bending-moment capacity of the beam increased? For what values is it decreased?

(a)

Concrete

Steel pile

(b)

4 m

$q_0 = 80$ kN/m

(c)

PROBLEM 10.6-24

(a)   (b)

PROBLEM 10.6-26

10.7 Shear Stresses in Beams of Rectangular Cross Section

Introductory Problems

10.7-1 The shear stresses τ in a rectangular beam are given by Eq. (10-40):

$$\tau = \frac{V}{2I}\left(\frac{h^2}{4} - y_1^2\right)$$

in which V is the shear force, I is the moment of inertia of the cross-sectional area, h is the height of the beam, and y_1 is the distance from the neutral axis to the point where the shear stress is being determined (Fig. 10-29).

By integrating over the cross-sectional area, show that the resultant of the shear stresses is equal to the shear force V.

10.7-2 Calculate the maximum shear stress τ_{max} and the maximum bending stress σ_{max} in a wood beam (see figure) carrying a uniform load of 22.5 kN/m (which includes the weight of the beam) if the length is 1.95 m and the cross section is rectangular with width 150 mm and height 300 mm, and the beam is either (a) simply supported as in the figure part a or (b) has a sliding support at right as in the figure part b.

(a)

(b)

PROBLEM 10.7-2

10.7-3 A simply supported wood beam is subjected to uniformly distributed load q. The width of the beam is 6 in. and the height is 8 in. Determine the normal stress and the shear stress at point C. Show these stresses on a sketch of a stress element at point C.

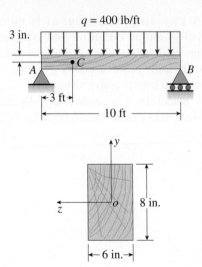

PROBLEM 10.7-3

10.7-4 A simply supported wood beam with over-hang is subjected to uniformly distributed load q. The beam has a rectangular cross section with width $b = 200$ mm and height $h = 250$ mm. Determine the maximum permissible value q if the allowable bending stress is $\sigma_{all} = 11$ MPa, and the allowable shear stress is $\tau_{all} = 1.2$ MPa.

PROBLEM 10.7-4

10.7-5 Two wood beams, each of rectangular cross section (3.0 in. × 4.0 in., actual dimensions), are glued together to form a solid beam with dimensions

812

6.0 in. × 4.0 in. (see figure). The beam is simply supported with a span of 8 ft.

(a) What is the maximum moment M_{max} that may be applied at the left support if the allowable shear stress in the glued joint is 200 psi? (Include the effects of the beam's own weight, assuming that the wood weighs 35 lb/ft^3.)

(b) Repeat part (a) if M_{max} is based on allowable bending stress of 2500 psi.

PROBLEM 10.7-5

10.7-6 A cantilever beam of length $L = 2$ m supports a load $P = 8.0$ kN (see figure). The beam is made of wood with cross-sectional dimensions 120 mm × 200 mm.

Calculate the shear stresses due to the load P at points located 25 mm, 50 mm, 75 mm, and 100 mm from the top surface of the beam. From these results, plot a graph showing the distribution of shear stresses from top to bottom of the beam.

PROBLEM 10.7-6

10.7-7 A steel beam of length $L = 16$ in. and cross-sectional dimensions $b = 0.6$ in. and $h = 2$ in. (see figure) supports a uniform load of intensity $q = 240$ lb/in., which includes the weight of the beam.

Calculate the shear stresses in the beam (at the cross section of maximum shear force) at points located 1/4 in., 1/2 in., 3/4 in., and 1 in. from the top surface of the beam. From these calculations, plot a graph showing the distribution of shear stresses from top to bottom of the beam.

PROBLEM 10.7-7

Representative Problems

10.7-8 A beam of rectangular cross section (width b and height h) supports a uniformly distributed load along its entire length L. The allowable stresses in bending and shear are σ_{allow} and τ_{allow}, respectively.

(a) If the beam is simply supported, what is the span length L_0 below which the shear stress governs the allowable load and above which the bending stress governs?

(b) If the beam is supported as a cantilever, what is the length L_0 below which the shear stress governs the allowable load and above which the bending stress governs?

10.7-9 A laminated wood beam on simple supports (figure part a) is built up by gluing together four 2 in. × 4 in. boards (actual dimensions) to form a solid beam 4 in. × 8 in. in cross section, as shown in the figure part b. The allowable shear stress in the glued joints is 62 psi, the allowable shear stress in the wood is 175 psi, and the allowable bending stress in the wood is 1650 psi.

(a) If the beam is 12 ft long, what is the allowable load P acting at the one-third point along the beam, as shown? (Include the effects of the beam's own weight, assuming that the wood weighs 35 lb/ft^3.)

(b) Repeat part (a) if the beam is assembled by gluing together two 3 in. × 4 in. boards and a 2 in. × 4 in. board (see figure part c).

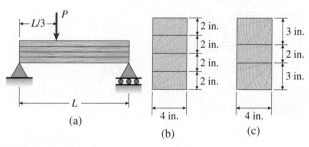

PROBLEM 10.7-9

10.7-10 A laminated plastic beam of square cross section is built up by gluing together three strips, each 10 mm × 30 mm in cross section (see figure). The beam has a total weight of 3.6 N and is simply supported with span length $L = 360$ mm.

Considering the weight of the beam (q), calculate the maximum permissible CCW moment M that may be placed at the right support.

(a) The allowable shear stress in the glued joints is 0.3 MPa.

(b) The allowable bending stress in the plastic is 8 MPa.

PROBLEM 10.7-10

10.7-11 A wood beam AB on simple supports with span length equal to 10 ft is subjected to a uniform load of intensity 125 lb/ft acting along the entire length of the beam, a concentrated load of magnitude 7500 lb acting at a point 3 ft from the right-hand support, and a moment at A of 18,500 ft-lb (see figure). The allowable stresses in bending and shear, respectively, are 2250 psi and 160 psi.

(a) From the table in Appendix D, select the lightest beam that will support the loads (disregard the weight of the beam).

(b) Taking into account the weight of the beam (weight density = 35 lb/ft^3), verify that the selected beam is satisfactory, or if it is not, select a new beam.

PROBLEM 10.7-11

10.7-12 A simply supported wood beam of rectangular cross section and span length 1.2 m carries a concentrated load P at midspan in addition to its own weight (see figure). The cross section has width 140 mm and height 240 mm. The weight density of the wood is 5.4 kN/m^3.

Calculate the maximum permissible value of the load P if (a) the allowable bending stress is 8.5 MPa and (b) the allowable shear stress is 0.8 MPa.

PROBLEM 10.7-12

10.7-13 A square wood platform is 8 ft × 8 ft in area and rests on masonry walls (see figure). The deck of the platform is constructed of 2-in. nominal thickness tongue-and-groove planks (actual thickness 1.5 in.; see Appendix D) supported on two 8-ft long beams. The beams have 4 in. × 6 in. nominal dimensions (actual dimensions 3.5 in. × 5.5 in.).

The planks are designed to support a uniformly distributed load w (lb/ft^2) acting over the entire top surface of the platform. The allowable bending stress for the planks is 2400 psi and the allowable shear stress is 100 psi. When analyzing the planks, disregard their weights and assume that their reactions are uniformly distributed over the top surfaces of the supporting beams.

(a) Determine the allowable platform load w_1 (lb/ft^2) based upon the bending stress in the planks.

(b) Determine the allowable platform load w_2 (lb/ft^2) based upon the shear stress in the planks.

(c) Which of the preceding values becomes the allowable load w_{allow} on the platform?

Hints: Use care in constructing the loading diagram for the planks, noting especially that the reactions are distributed loads instead of concentrated loads. Also, note that the maximum shear forces occur at the inside faces of the supporting beams.

PROBLEM 10.8-13

10.7-14 A wood beam ABC with simple supports at A and B and an overhang BC has height $h = 300$ mm (see figure). The length of the main span of the beam is $L = 3.6$ m and the length of the overhang is $L/3 = 1.2$ m. The beam supports a concentrated load $3P = 18$ kN at the midpoint of the main span and a moment $PL/2 = 10.8$ kN · m at the free end of the overhang. The wood has a weight density $\gamma = 5.5$ kN/m^3.

(a) Determine the required width b of the beam based upon an allowable bending stress of 8.2 MPa.

(b) Determine the required width based upon an allowable shear stress of 0.7 MPa.

PROBLEM 10.7-14

10.8 Shear Stresses in Beams of Circular Cross Section

Introductory Problems

10.8-1 A wood pole with a solid circular cross section (d = diameter) is subjected to a triangular distributed horizontal force of peak intensity $q_0 = 20$ lb/in. (see figure). The length of the pole is $L = 6$ ft, and the

allowable stresses in the wood are 1900 psi in bending and 120 psi in shear.

Determine the minimum required diameter of the pole based upon (a) the allowable bending stress and (b) the allowable shear stress.

PROBLEM 10.8-1

10.8-2 A simple log bridge in a remote area consists of two parallel logs with planks across them (see figure). The logs are Douglas fir with an average diameter 300 mm. A truck moves slowly across the bridge, which spans 2.5 m. Assume that the weight of the truck is equally distributed between the two logs.

Because the wheelbase of the truck is greater than 2.5 m, only one set of wheels is on the bridge at a time. Thus, the wheel load on one log is equivalent to a concentrated load W acting at any position along the span. In addition, the weight of one log and the planks it supports is equivalent to a uniform load of 850 N/m acting on the log.

Determine the maximum permissible wheel load W based upon (a) an allowable bending stress of 7.0 MPa and (b) an allowable shear stress of 0.75 MPa.

PROBLEM 10.8-2

815

Representative Problems

10.8-3 A vertical pole consisting of a circular tube of outer diameter 5 in. and inner diameter 4.5 in. is loaded by a linearly varying distributed force with maximum intensity of q_0. Find the maximum shear stress in the pole.

$q_0 = 400$ lb/ft

PROBLEM 10.8-3

10.8-4 A circular pole is subjected to linearly varying distributed force with maximum intensity q_0. Calculate the diameter d_0 of the pole if the maximum allowable shear stress for the pole is 75 MPa.

$q_0 = 100$ kN/m

PROBLEM 10.8-4

10.8-5 A sign for an automobile service station is supported by two aluminum poles of hollow circular cross section, as shown in the figure. The poles are being designed to resist a wind pressure of 75 lb/ft^2 against the full area of the sign. The dimensions of the poles and sign are $h_1 = 20$ ft, $h_2 = 5$ ft, and $b = 10$ ft. To prevent buckling of the walls of the poles, the thickness t is specified as one-tenth the outside diameter d.

(a) Determine the minimum required diameter of the poles based upon an allowable bending stress of 7500 psi in the aluminum.

(b) Determine the minimum required diameter based upon an allowable shear stress of 2000 psi.

$t = \dfrac{d}{10}$

PROBLEM 10.8-5

10.8-6 A steel pipe is subjected to a quadratic distributed load over its height with the peak intensity q_0 at the base (see figure). Assume the following pipe properties and dimensions: height L, outside diameter $d = 200$ mm, and wall thickness $t = 10$ mm. Allowable stresses for flexure and shear are $\sigma_a = 125$ MPa and $\tau_a = 30$ MPa.

(a) If $L = 2.6$ m, find $q_{0,\text{max}}$ (kN/m), assuming that allowable flexure and shear stresses in the pipe are not to be exceeded.

(b) If $q_0 = 60$ kN/m, find the maximum height L_{max} (m) of the pipe if the allowable flexure and shear stresses in the pipe are not to be exceeded.

$q(x) = q_0[1-(x/L)^2]$

PROBLEM 10.8-6

816

10.9 Shear Stresses in the Webs of Beams with Flanges

Introductory Problems

10.9-1 through 10.9-6 A wide-flange beam (see figure) is subjected to a shear force V. Using the dimensions of the cross section, calculate the moment of inertia and then determine the following quantities:

(a) The maximum shear stress τ_{max} in the web.

(b) The minimum shear stress τ_{min} in the web.

(c) The average shear stress τ_{aver} (obtained by dividing the shear force by the area of the web) and the ratio τ_{max}/τ_{aver}.

(d) The shear force V_{web} carried in the web and the ratio V_{web}/V.

Note: Disregard the fillets at the junctions of the web and flanges and determine all quantities, including the moment of inertia, by considering the cross section to consist of three rectangles.

10.9-1 Dimensions of cross section: $b = 6$ in., $t = 0.5$ in., $h = 12$ in., $h_1 = 10.5$ in., and $V = 30$ k.

10.9-2 Dimensions of cross section: $b = 180$ mm, $t = 12$ mm, $h = 420$ mm, $h_1 = 380$ mm, and $V = 125$ kN.

10.9-3 Wide-flange shape, W 8×28 (see Table C-1(a), Appendix C); $V = 10$ k.

10.9-4 Dimensions of cross section: $b = 220$ mm, $t = 12$ mm, $h = 600$ mm, $h_1 = 570$ mm, and $V = 200$ kN.

10.9-5 Wide-flange shape, W 18×71 (see Table C-1(a), Appendix C); $V = 21$ k.

10.9-6 Dimensions of cross section: $b = 120$ mm, $t = 7$ mm, $h = 350$ mm, $h_1 = 330$ mm, and $V = 60$ kN.

PROBLEMS 10.9-1 through 10.9-6

Representative Problems

10.9-7 A cantilever beam AB of length $L = 6.5$ ft supports a trapezoidal distributed load of peak intensity q, and minimum intensity $q/2$, that includes the weight of the beam (see figure). The beam is a steel W 12×14 wide-flange shape (see Table C-1(a), Appendix C).

Calculate the maximum permissible load q based upon (a) an allowable bending stress $\sigma_{allow} = 18$ ksi and (b) an allowable shear stress $\tau_{allow} = 7.5$ ksi. *Note:* Obtain the moment of inertia and section modulus of the beam from Table C-1(a).

PROBLEM 10.9-7

10.9-8 A bridge girder AB on a simple span of length $L = 14$ m supports a distributed load of maximum intensity q at mid-span and minimum intensity $q/2$ at supports A and B that includes the weight of the girder (see figure). The girder is constructed of three plates welded to form the cross section shown.

Determine the maximum permissible load q based upon (a) an allowable bending stress $\sigma_{allow} = 110$ MPa and (b) an allowable shear stress $\tau_{allow} = 50$ MPa.

PROBLEM 10.9-8

817

10.9-9 A simple beam with an overhang supports a uniform load of intensity $q = 1200$ lb/ft and a concentrated $P = 3000$ lb load at 8 ft to the right of A and also at C (see figure). The uniform load includes an allowance for the weight of the beam. The allowable stresses in bending and shear are 18 ksi and 11 ksi, respectively.

Select from Table C-2(a), Appendix C, the lightest I-beam (S shape) that will support the given loads.

Hint: Select a beam based upon the bending stress and then calculate the maximum shear stress. If the beam is overstressed in shear, select a heavier beam and repeat.

PROBLEM 10.9-9

10.9-10 A hollow steel box beam has the rectangular cross section shown in the figure. Determine the maximum allowable shear force V that may act on the beam if the allowable shear stress is 36 MPa.

PROBLEM 10.9-10

10.9-11 A hollow aluminum box beam has the square cross section shown in the figure. Calculate the maximum and minimum shear stresses τ_{max} and τ_{min} in the webs of the beam due to a shear force $V = 28$ k.

PROBLEM 10.9-11

818

10.9-12 The T-beam shown in the figure has cross-sectional dimensions: $b = 210$ mm, $t = 16$ mm, $h = 300$ mm, and $h_1 = 280$ mm. The beam is subjected to a shear force $V = 68$ kN.

Determine the maximum shear stress τ_{max} in the web of the beam.

10.9-13 Calculate the maximum shear stress τ_{max} in the web of the T-beam shown in the figure if $b = 10$ in., $t = 0.5$ in., $h = 7$ in., $h_1 = 6.2$ in., and the shear force $V = 5300$ lb.

PROBLEMS 10.9-12 and 10.9-13

10.10 Composite Beams

To solve the problems for Section 10.10, assume that the component parts of the beams are securely bonded by adhesives or connected by fasteners. Also, use the general theory for composite beams.

Introductory Problems

10.10-1 A composite beam is constructed using a steel plate (0.5 in. × 6 in.) with two wood beams (3 in. × 6 in.) on either side. The wood and steel are securely fastened to act as a single beam. The beam is subjected to a positive bending moment $M_z = 60$ kip-in. Calculate the maximum tensile and compressive stresses in the wood and steel if $E_w = 1500$ ksi and $E_s = 29,500$ ksi.

PROBLEM 10.10-1

10.10-2 A wood beam is strengthened using two steel plates as shown in Fig. a. The beam has simple supports and an overhang and is subjected to a point load and a uniform load as shown in Fig. b. Calculate the maximum tensile and compressive stresses of the beam. Assume that $E_w = 11$ GPa and $E_s = 200$ GPa.

(a)

(b)

PROBLEM 10.10-2

10.10-3 A composite beam consisting of fiberglass faces and a core of particle board has the cross section shown in the figure. The width of the beam is 2.0 in., the thickness of the faces is 0.10 in., and the thickness of the core is 0.50 in. The beam is subjected to a bending moment of 250 lb-in. acting about the z axis.

Find the maximum bending stresses σ_f and σ_c in the faces and the core, respectively, if their respective moduli of elasticity are 4×10^6 psi and 1.5×10^6 psi.

PROBLEM 10.10-3

10.10-4 A wood beam with cross-sectional dimensions 200 mm × 300 mm is reinforced on its sides by steel plates 12 mm thick (see figure). The moduli of elasticity for the steel and wood are $E_s = 190$ GPa and $E_w = 11$ GPa, respectively. Also, the corresponding allowable stresses are $\sigma_s = 110$ MPa and $\sigma_w = 7.5$ MPa.

(a) Calculate the maximum permissible bending moment M_{max} when the beam is bent about the z axis.

(b) Repeat part (a) if the beam is now bent about its y axis.

(c) Find the required thickness of the steel plates on the beam bent about the y axis so that M_{max} is the same for both beam orientations.

(a)   (b)

PROBLEM 10.10-4

10.10-5 A hollow box beam is constructed with webs of Douglas-fir plywood and flanges of pine, as shown in the figure in a cross-sectional view. The plywood is 1 in. thick and 12 in. wide; the flanges are 2 in. × 4 in. (nominal size). The modulus of elasticity for the plywood is 1,800,000 psi and for the pine is 1,400,000 psi.

(a) If the allowable stresses are 2000 psi for the plywood and 1750 psi for the pine, find the allowable bending moment M_{max} when the beam is bent about the z axis.

(b) Repeat part (a) if the beam is now bent about its y axis.

(a)   (b)

PROBLEM 10.10-5

Representative Problems

10.10-6 A round *titanium* tube of outside diameter d_2 and a *copper* core of diameter d_1 are bonded to form a composite beam, as shown in the figure.

(a) Derive formulas for the allowable bending moment M that can be carried by the beam based upon an allowable stress σ_{Ti} in the titanium and an allowable stress σ_{Cu} in the copper (Assume that the moduli of elasticity for the titanium and copper are E_{Ti} and E_{Cu}, respectively.)

(b) If $d_2 = 40$ mm, $d_1 = 36$ mm, $E_{Ti} = 120$ GPa, $E_{Cu} = 110$ GPa, $\sigma_{Ti} = 840$ MPa, and $\sigma_{Cu} = 700$ MPa, what is the maximum bending moment M?

(c) What new value of copper diameter d_1 will result in a balanced design? (i.e., a balanced design is that in which titanium and copper reach allowable stress values at the same time).

PROBLEM 10.10-6

10.10-7 A beam with a guided support and 10-ft span supports a distributed load of intensity $q = 660$ lb/ft over its first half (see figure part a) and a moment $M_0 = 300$ ft-lb at joint B. The beam consists of a wood member (nominal dimensions 6 in. × 12 in. and actual dimensions 5.5 in. × 11.5 in. in cross section, as shown in the figure part b) that is reinforced by 0.25-in.-thick steel plates on top and bottom. The moduli of elasticity for the steel and wood are $E_s = 30 \times 10^6$ psi and $E_w = 1.5 \times 10^6$ psi, respectively.

(a) Calculate the maximum bending stresses σ_s in the steel plates and σ_w in the wood member due to the applied loads.

(b) If the allowable bending stress in the steel plates is $\sigma_{as} = 14,000$ psi and that in the wood is $\sigma_{aw} = 900$ psi, find q_{max}. (Assume that the moment at B, M_0, remains at 300 ft-lb.)

(c) If $q = 660$ lb/ft and allowable stress values in part (b) apply, what is $M_{0,max}$ at B?

PROBLEM 10.10-7

10.10-8 A plastic-lined steel pipe has the cross-sectional shape shown in the figure. The steel pipe has an outer diameter $d_3 = 100$ mm and an inner diameter $d_2 = 94$ mm. The plastic liner has an inner diameter $d_1 = 82$ mm. The modulus of elasticity of the steel is 75 times the modulus of the plastic.

(a) Determine the allowable bending moment M_{allow} if the allowable stress in the steel is 35 MPa and in the plastic is 600 kPa.

(b) If pipe and liner diameters remain unchanged, what new value of allowable stress for the steel pipe will result in the steel pipe and plastic liner reaching their allowable stress values under the same maximum moment (i.e., a balanced design)? What is the new maximum moment?

PROBLEM 10.10-8

10.10-9 The cross section of a sandwich beam consisting of aluminum alloy faces and a foam core is shown in the figure. The width b of the beam is 8.0 in., the thickness t of the faces is 0.25 in., and the height h_c of the core is 5.5 in. (total height $h = 6.0$ in.). The moduli of elasticity are 10.5×10^6 psi for the aluminum faces and 12,000 psi for the foam core. A bending moment $M = 40$ kip-in. acts about the z axis.

Determine the maximum stresses in the faces and the core using (a) the general theory for composite beams and (b) the approximate theory for sandwich beams.

10.10-10 The cross section of a sandwich beam consisting of fiberglass faces and a lightweight plastic core is shown in the figure. The width b of the beam is 50 mm, the thickness t of the faces is 4 mm, and the height h_c of the core is 92 mm (total height $h = 100$ mm). The moduli of elasticity are 75 GPa for the fiberglass and 1.2 GPa for the plastic. A bending moment $M = 275$ N $\cdot$ m acts about the z axis.

Determine the maximum stresses in the faces and the core using (a) the general theory for composite beams and (b) the approximate theory for sandwich beams.

PROBLEMS 10.10-9 and 10.10-10

10.10-11 A bimetallic beam used in a temperature-control switch consists of strips of aluminum and copper bonded together as shown in the figure, which is a cross-sectional view. The width of the beam is 1.0 in., and each strip has a thickness of 1/16 in.

Under the action of a bending moment $M = 12$ lb-in. acting about the z axis, what are the maximum stresses σ_a and σ_c in the aluminum and copper, respectively? (Assume $E_{Al} = 10.5 \times 10^6$ psi and $E_{Cu} = 16.8 \times 10^6$ psi.)

PROBLEM 10.10-11

10.10-12 A simply supported composite beam 3 m long carries a uniformly distributed load of intensity $q = 30$ kN/m (see figure). The beam is constructed of a wood member, 100 mm wide by 150 mm deep, and is reinforced on its lower side by a steel plate 8 mm thick and 100 mm wide.

(a) Find the maximum bending stresses σ_w and σ_s in the wood and steel, respectively, due to the uniform load if the moduli of elasticity are $E_w = 10$ GPa for the wood and $E_s = 210$ GPa for the steel.

(b) Find the required thickness of the steel plate so that the steel plate and wood reach their allowable stress values, $\sigma_{as} = 100$ MPa and $\sigma_{aw} = 8.5$ MPa, simultaneously under the maximum moment.

PROBLEM 10.10-12

10.10-13 A simply supported wooden I-beam with a 12-ft span supports a distributed load of intensity $q = 90$ lb/ft over its length (see figure part a). The beam is constructed with a web of Douglas-fir plywood and flanges of pine glued to the web, as shown in the figure part b. The plywood is 3/8 in. thick; the flanges are 2 in. × 2 in. (actual size). The modulus of elasticity for the plywood is 1,600,000 psi and for the pine is 1,200,000 psi.

(a) Calculate the maximum bending stresses in the pine flanges and in the plywood web.

(b) What is q_{max} if allowable stresses are 1600 psi in the flanges and 1200 psi in the web?

PROBLEM 10.10-13

10.10-14 A simply supported composite beam with a 3.6 m span supports a triangularly distributed load of peak intensity q_0 at mid-span (see figure part a). The beam is constructed of two wood joists, each 50 mm × 280 mm, fastened to two steel plates, one of dimensions 6 mm × 80 mm and the lower plate of dimensions 6 mm × 120 mm (see figure part b). The modulus of elasticity for the wood is 11 GPa and for the steel is 210 GPa.

If the allowable stresses are 7 MPa for the wood and 120 MPa for the steel, find the allowable peak load intensity $q_{0,\max}$ when the beam is bent about the z axis. Neglect the weight of the beam.

(a) (b)

PROBLEM 10.10-14

10.10-15 A composite beam is constructed from a wood beam (3 in. × 6 in.) and a steel plate (3 in. wide). The wood and the steel are securely fastened to act as a single beam. The beam is subjected to a positive bending moment $M_z = 75$ kip-in. Calculate the required thickness of the steel plate based on the following limit states:

(a) Allowable compressive stress in the wood = 2 ksi
(b) Allowable tensile stress in the wood = 2 ksi
(c) Allowable tensile stress in the steel plate = 16 ksi
Assume that $E_w = 1,500$ ksi and $E_s = 30,000$ ksi.

PROBLEM 10.10-15

10.10-16 A wood beam in a historic theater is reinforced with two angle sections at the outside lower corners (see figure). If the allowable stress in the wood is 12 MPa and that in the steel is 140 MPa, what is ratio of the maximum permissible moments for the beam before and after reinforcement with the angle sections? See Table C-5(b) of Appendix C for angle section properties. Assume that $E_w = 12$ GPa and $E_s = 210$ GPa.

PROBLEM 10.10-16

10.10-17 Repeat Problem 10.10-1 but now assume that the steel plate is smaller (0.5 in. × 5 in.) and is aligned with the top of the beam as shown in the figure.

PROBLEM 10.10-17

Transformed-Section Method

To solve the problems for this section, assume that the component parts of the beams are securely bonded by adhesives or connected by fasteners. Also, use the transformed-section method in the solutions.

Introductory Problems

10.10-18 A sandwich beam having steel faces enclosing a plastic core is subjected to a bending moment $M = 5$ kN · m. The thickness of each steel face is $t = 3$ mm with modulus of elasticity $E_s = 200$ GPa. The height of the plastic core is $h_p = 140$ mm, and its modulus of elasticity is $E_p = 800$ MPa. The overall dimensions of the beam are $h = 146$ mm and $b = 175$ mm.

Using the transformed-section method, determine the maximum tensile and compressive stresses in the faces and the core.

PROBLEM 10.10-18

10.10-19 A wood beam 8 in. wide and 12 in. deep (nominal dimensions) is reinforced on top and bottom by 0.25-in.-thick steel plates (see figure part a).

(a) Find the allowable bending moment M_{max} about the z axis if the allowable stress in the wood is 1100 psi and in the steel is 15,000 psi. (Assume that the ratio of the moduli of elasticity of steel and wood is 20.)

(b) Compare the moment capacity of the beam in part a with that shown in the figure part b, which has two 4 in. × 12 in. joists (nominal dimensions) attached to a 1/4 in. × 11.0 in. steel plate.

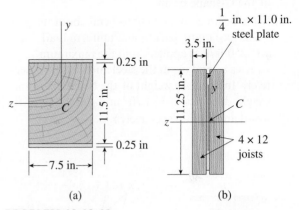

PROBLEM 10.10-19

10.10-20 A simple beam of span length 3.2 m carries a uniform load of intensity 48 kN/m. The cross section of the beam is a hollow box with wood flanges and steel side plates, as shown in the figure. The wood flanges are 75 mm × 100 mm in cross section, and the steel plates are 300 mm deep.

What is the required thickness t of the steel plates if the allowable stresses are 120 MPa for the steel and 6.5 MPa for the wood? (Assume that the moduli of elasticity for the steel and wood are 210 GPa and 10 GPa, respectively, and disregard the weight of the beam.)

PROBLEM 10.10-20

10.10-21 A simple beam that is 18 ft long supports a uniform load of intensity q. The beam is constructed of two C 8×11.5 sections (channel sections or C shapes) on either side of a 4×8 (actual dimensions) wood beam (see the cross section shown in the figure part a). The modulus of elasticity of the steel ($E_s = 30,000$ ksi) is 20 times that of the wood (E_w).

(a) If the allowable stresses in the steel and wood are 12,000 psi and 900 psi, respectively, what is the allowable load q_{allow}? *Note:* Disregard the weight of the beam, and see Table C-3(a) of Appendix C for the dimensions and properties of the C-shape beam.

(b) If the beam is rotated 90° to bend about its y axis (see figure part b) and uniform load $q = 250$ lb/ft is applied, find the maximum stresses σ_s and σ_w in the steel and wood, respectively. Include the weight of the beam. (Assume weight densities of 35 lb/ft^3 and 490 lb/ft^3 for the wood and steel, respectively.)

C 8×11.5 Wood beam   Wood beam

(a)   (b)

PROBLEM 10.10-21

10.10-22 The composite beam shown in the figure is simply supported and carries a total uniform load of 40 kN/m on a span length of 4.0 m. The beam is built of a southern pine wood member having cross-sectional dimensions of 150 mm × 250 mm and two brass plates of cross-sectional dimensions 30 mm × 150 mm.

(a) Determine the maximum stresses σ_B and σ_w in the brass and wood, respectively, if the moduli of elasticity are $E_B = 96$ GPa and $E_w = 14$ GPa. (Disregard the weight of the beam.)

(b) Find the required thickness of the brass plates so that the plate and wood reach their allowable stress values of $\sigma_{aB} = 70$ MPa and $\sigma_{aw} = 8.5$ MPa simultaneously under the maximum moment. What is the maximum moment?

PROBLEM 10.10-22

Representative Problems

10.10-23 The cross section of a beam made of thin strips of aluminum separated by a lightweight plastic is shown in the figure. The beam has width $b = 3.0$ in., the aluminum strips have thickness $t = 0.1$ in., and the plastic segments have heights $d = 1.2$ in. and $3d = 3.6$ in. The total height of the beam is $h = 6.4$ in.

The moduli of elasticity for the aluminum and plastic are $E_{Al} = 11 \times 10^6$ psi and $E_p = 440 \times 10^3$ psi, respectively.

Determine the maximum stresses σ_{Al} and σ_p in the aluminum and plastic, respectively, due to a bending moment of 6.0 kip-in.

10.10-24 Consider the preceding problem if the beam has width $b = 75$ mm, the aluminum strips have thickness $t = 3$ mm, the plastic segments have heights $d = 40$ mm and $3d = 120$ mm, and the total height of the beam is $h = 212$ mm. Also, the moduli of elasticity are $E_{AI} = 75$ GPa and $E_p = 3$ GPa, respectively.

Determine the maximum stresses σ_{AI} and σ_p in the aluminum and plastic, respectively, due to a bending moment of 1.0 kN · m.

PROBLEMS 10.10-23 and 10.10-24

10.10-25 A simple beam that is 18 ft long supports a uniform load of intensity q. The beam is constructed of two angle sections, each L $6 \times 4 \times 1/2$, on either side of a 2 in. × 8 in. (actual dimensions) wood beam (see the cross section shown in the figure part a). The modulus of elasticity of the steel is 20 times that of the wood.

(a) If the allowable stresses in the steel and wood are 12,000 psi and 900 psi, respectively, what is the allowable load q_{allow}? *Note:* Disregard the weight of the beam, and see Table C-5(a) of Appendix C for the dimensions and properties of the angles.

(b) Repeat part (a) if a 1 in. × 10 in. wood flange (actual dimensions) is added (see figure part b).

(a)

PROBLEM 10.10-25   (b)

10.10-26 The cross section of a composite beam made of aluminum and steel is shown in the figure. The moduli of elasticity are $E_{Al} = 75$ GPa and $E_s = 200$ GPa.

(a) Under the action of a bending moment that produces a maximum stress of 50 MPa in the aluminum, what is the maximum stress σ_s in the steel?

(b) If the height of the beam remains at 120 mm and allowable stresses in steel and aluminum are defined as 94 MPa and 40 MPa, respectively,

PROBLEM 10.10-26

what heights h_{Al} and h_s are required for aluminum and steel, respectively, so that both steel and aluminum reach their allowable stress values under the maximum moment?

10.10-27 A beam is constructed of two angle sections, each L 5 × 3 × 1/2, that reinforce a 2 × 8 (actual dimensions) wood plank (see the cross section shown in the figure). The modulus of elasticity for the wood is $E_w = 1.2 \times 10^6$ psi and for the steel is $E_s = 30 \times 10^6$ psi.

Find the allowable bending moment M_{allow} for the beam if the allowable stress in the wood is $\sigma_w = 1100$ psi and in the steel is $\sigma_s = 12,000$ psi. *Note:* Disregard the weight of the beam, and see Table C-5(a) of Appendix C for the dimensions and properties of the angles.

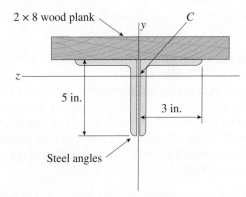

PROBLEM 10.10-27

10.10-28 The cross section of a bimetallic strip is shown in the figure. Assuming that the moduli of elasticity for metals A and B are $E_A = 168$ GPa and $E_B = 90$ GPa, respectively, determine the smaller of the two section moduli for the beam. (Recall that section modulus is equal to bending moment divided by maximum bending stress.) In which material does the maximum stress occur?

PROBLEM 10.10-28

825

10.10-29 A W 12 × 50 steel wide-flange beam and a segment of a 4-inch thick concrete slab (see figure) jointly resist a positive bending moment of 95 kip-ft. The beam and slab are joined by shear connectors that are welded to the steel beam. (These connectors resist the horizontal shear at the contact surface.) The moduli of elasticity of the steel and the concrete are in the ratio 12 to 1.

Determine the maximum stresses σ_s and σ_c in the steel and concrete, respectively. *Note:* See Table C-1(a) of Appendix C for the dimensions and properties of the steel beam.

PROBLEM 10.10-29

10.10-30 A reinforced concrete beam (see figure) is acted on by a positive bending moment of $M = 160$ kN · m. Steel reinforcement consists of 4 bars of 28 mm diameter. The modulus of elasticity for the concrete is $E_c = 25$ GPa while that of the steel is $E_s = 200$ GPa.

(a) Find the maximum stresses in steel and concrete.

(b) If *allowable* stresses for concrete and steel are $\sigma_{ac} = 9.2$ MPa and $\sigma_{as} = 135$ MPa, respectively, what is the maximum permissible positive bending moment?

(c) What is the required area of steel reinforcement, A_s, if a balanced condition must be achieved? What is the allowable positive bending moment? (Recall that in a balanced design, both steel and concrete reach allowable stress values simultaneously under the design moment.)

PROBLEM 10.10-30

10.10-31 A reinforced concrete T-beam (see figure) is acted on by a positive bending moment of $M = 175$ kip-ft. Steel reinforcement consists of four bars of 1.41-inch diameter. The modulus of elasticity for the concrete is $E_c = 3000$ ksi while that of the steel is $E_s = 29,000$ ksi. Let $b = 48$ in., $t_f = 4$ in., $b_w = 15$ in., and $d = 24$ in.

(a) Find the maximum stresses in steel and concrete.

(b) If *allowable* stresses for concrete and steel are $\sigma_{ac} = 1400$ psi and $\sigma_{as} = 18$ ksi, respectively, what is the maximum permissible positive bending moment?

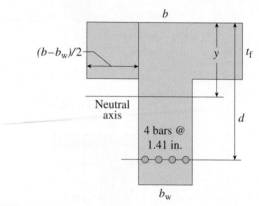

PROBLEM 10.10-31

10.10-32 A reinforced concrete slab (see figure) is reinforced with 13-mm bars spaced 160 mm apart at $d = 105$ mm from the top of the slab. The modulus of elasticity for the concrete is $E_c = 25$ GPa, while that of the steel is $E_s = 200$ GPa. Assume that *allowable* stresses for concrete and steel are $\sigma_{ac} = 9.2$ MPa and $\sigma_{as} = 135$ MPa.

(a) Find the maximum permissible positive bending moment for a 1-m wide strip of the slab.

(b) What is the required area of steel reinforcement, A_s, if a *balanced condition* must be achieved? What is the allowable positive bending moment? (Recall that in a balanced design, both steel and concrete reach allowable stress values simultaneously under the design moment.)

PROBLEM 10.10-32

10.10-33 A wood beam reinforced using two channels is subjected to a positive bending moment $M_z = 25$ kip-ft. Calculate the largest tensile and compressive stresses in the wood and steel if $E_w = 1500$ ksi and $E_s = 30,000$ ksi.

PROBLEM 10.10-33

10.10-34 A wood beam reinforced by an aluminum channel section is shown in the figure. The beam has a cross section of dimensions 150 mm × 250 mm, and the channel has a uniform thickness of 6.5 mm. If the allowable stresses in the wood and aluminum are 8 MPa and 38 MPa, respectively, and if their moduli of elasticity are in the ratio 1 to 6, what is the maximum allowable bending moment for the beam?

PROBLEM 10.10-34

10.11 Bending of Unsymmetric Beams

To solve the problems for Section 10.11, draw a sketch of the cross section showing the orientation of the neutral axis and the locations of the points where the stresses are being found.

10.11-1 A beam with a channel section is subjected to a bending moment M having its vector at an angle θ to the z axis (see figure).

Determine the orientation of the neutral axis and calculate the maximum tensile stress σ_t and maximum compressive stress σ_c in the beam.

Use the following data: C 8 × 11.5 section, $M = 20$ kip-in., $\tan \theta = 1/3$. See Table C-3(a) of Appendix C for the dimensions and properties of the channel section.

10.11-2 A beam with a channel section is subjected to a bending moment M having its vector at an angle θ to the z axis (see figure).

Determine the orientation of the neutral axis and calculate the maximum tensile stress σ_t and maximum compressive stress σ_c in the beam. Use a C 200 × 20.5 channel section with $M = 0.75$ kN · m and $\theta = 20°$.

PROBLEMS 10.11-1 and 10.11-2

10.11-3 An angle section with equal legs is subjected to a bending moment M having its vector directed along the 1–1 axis, as shown in the figure.

Determine the orientation of the neutral axis and calculate the maximum tensile stress σ_t and maximum compressive stress σ_c if the angle is an L 6 × 6 × 3/4 section and $M = 20$ kip-in. See Table C-4(a) of Appendix C for the dimensions and properties of the angle section.

10.11-4 An angle section with equal legs is subjected to a bending moment M having its vector directed along the 1–1 axis, as shown in the figure.

Determine the orientation of the neutral axis and calculate the maximum tensile stress σ_t and maximum compressive stress σ_c if the section is an L 152 × 152 × 12.7 section and $M = 2.5$ kN · m. See Table C-4(b) of Appendix C for the dimensions and properties of the angle section.

PROBLEMS 10.11-3 and 10.11-4

10.11-5 A beam made up of two unequal leg angles is subjected to a bending moment M having its vector at an angle θ to the z axis (see figure part a).

(a)

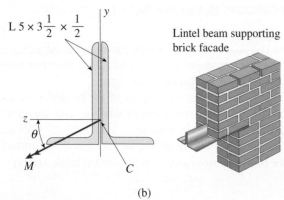

(b)

PROBLEM 10.11-5

828

(a) For the position shown in the figure, determine the orientation of the neutral axis and calculate the maximum tensile stress σ_t and maximum compressive stress σ_c in the beam. Assume that $\theta = 30°$ and $M = 30$ kip-in.

(b) The two angles are now inverted and attached back-to-back to form a lintel beam that supports two courses of brick façade (see figure part b). Find the new orientation of the neutral axis and calculate the maximum tensile stress σ_t and maximum compressive stress σ_c in the beam using $\theta = 30°$ and $M = 30$ kip-in.

10.11-6 The Z-section in the figure is subjected to $M = 5$ kN · m, as shown.

Determine the orientation of the neutral axis and calculate the maximum tensile stress σ_t and maximum compressive stress σ_c in the beam. Use the following numerical data: height $h = 200$ mm, width $b = 90$ mm, constant thickness $t = 15$ mm, and $\theta_p = 19.2°$. Use $I_1 = 32.6 \times 10^6$ mm^4 and $I_2 = 2.4 \times 10^6$ mm^4.

PROBLEM 10.11-6

10.11-7 The cross section of a steel beam is constructed of a W 18 × 71 wide-flange section with a 6 in. × 1/2 in. cover plate welded to the top flange and a C 10 × 30 channel section welded to the bottom flange. This beam is subjected to a bending moment M having its vector at an angle θ to the z axis (see figure).

PROBLEM 10.11-7

Determine the orientation of the neutral axis and calculate the maximum tensile stress σ_t and maximum compressive stress σ_c in the beam. Assume that $\theta = 30°$ and $M = 75$ kip-in.

10.11-8 The cross section of a steel beam is shown in the figure. This beam is subjected to a bending moment M having its vector at an angle θ to the z axis.

Determine the orientation of the neutral axis and calculate the maximum tensile stress σ_t and maximum compressive stress σ_c in the beam. Assume that $\theta = 22.5°$ and $M = 4.5$ kN · m. Use cross-sectional properties $I_{x_1} = 93.14 \times 10^6$ mm^4, $I_{y_1} = 152.7 \times 10^6$ mm^4, and $\theta_p = 27.3°$.

PROBLEM 10.11-8

10.11-9 A beam with a semicircular cross section of radius r is subjected to a bending moment M having its vector at an angle θ to the z axis (see figure).

Derive formulas for the maximum tensile stress σ_t and the maximum compressive stress σ_c in the beam for $\theta = 0, 45°$, and $90°$. Express the results in the form $\alpha\, M/r^3$, where α is a numerical value.

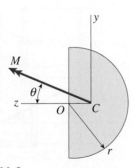

PROBLEM 10.11-9

10.11-10 A built-up beam supporting a condominium balcony is made up of a structural T (one half of a W 200 × 31.3) for the top flange and web and two angles (2 L 102 × 76 × 6.4, long legs back-to-back) for the bottom flange and web, as shown. The beam is subjected to a bending moment M having its vector at an angle θ to the z axis (see figure).

Determine the orientation of the neutral axis and calculate the maximum tensile stress σ_t and maximum compressive stress σ_c in the beam. Assume that $\theta = 30°$ and $M = 15$ kN · m.

Use the numerical properties: $c_1 = 4.111$ mm, $c_2 = 4.169$ mm, $b_f = 134$ mm, $L_s = 76$ mm, $A = 4144$ mm^2, $I_y = 3.88 \times 10^6$ mm^4, and $I_z = 34.18 \times 10^6$ mm^4.

Built-up beam

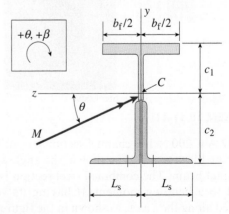

PROBLEM 10.11-10

829

10.11-11 A steel post ($E = 30 \times 10^6$ psi) having thickness $t = 1/8$ in. and height $L = 72$ in. supports a stop sign (see figure). The stop sign post is subjected to a bending moment M having its vector at an angle θ to the z axis.

Determine the orientation of the neutral axis and calculate the maximum tensile stress σ_t and maximum compressive stress σ_c in the beam. Assume that $\theta = 30°$ and $M = 5.0$ kip-in.

Use the following numerical properties for the post: $A = 0.578$ in.^{2}, $c_1 = 0.769$ in., $c_2 = 0.731$ in., $I_y = 0.44867$ in.^{4}, and $I_z = 0.16101$ in.^{4}.

Section A–A

5/8 in.

Circular cut-out, $d = 0.375$ in.

Post, $t = 0.125$ in.

$z \leftarrow 1.5$ in.

c_1

c_2

C

θ

Stop sign M

0.5 in. 1.0 in. 1.0 in. 0.5 in.

STOP

L

A A

Elevation view of post

© Barry Goodno

Steel post

PROBLEM 10.11-11

10.11-12 A C 200 × 17.1 channel section has an angle with equal legs attached as shown; the angle serves as a lintel beam. The combined steel section is subjected to a bending moment M having its vector directed along the z axis, as shown in the figure. The centroid C of the combined section is located at distances x_c and y_c from the centroid (C_1) of the channel alone. Principal axes x_1 and y_1 are also shown in the figure and properties I_{x1}, I_{y1}, and θ_p are given.

Find the orientation of the neutral axis and calculate the maximum tensile stress σ_t and maximum compressive stress σ_c if the angle is an L 76 × 76 × 6.4 section and $M = 3.5$ kN · m. Use the following properties for principal axes for the combined section: $I_{x1} = 18.49 \times 10^6$ mm^4, $I_{y1} = 1.602 \times 10^6$ mm^4, $\theta_p = 7.448°$ (CW), $x_c = 10.70$ mm, and $y_c = 24.07$ mm.

y_1

y

C

C_1 y_c

z

M

θ_p

x_c

x_1

L 76 × 76 × 6.4 lintel

C 200 × 17.1

PROBLEM 10.11-12

10.11-13 A cold-formed steel section is made by folding a steel plate to form a structural section such as that shown in the figure. This beam is subjected to bending moment $M = 2$ kip-in. at angle $\theta = 10°$ to the z axis. Find the centroid and the orientation of the neutral axis. Find flexural normal stresses at points A and B (see figure). Use the flexure formula based on principal axes and then confirm your solution using the general flexure formula.

Hint: The plate is thin, so use centerline dimensions.

y

A

$t = \dfrac{3}{32}$ in.

$\dfrac{5}{8}$ in.

$\dfrac{1}{2}$ in.

M

1.75 in.

$\dfrac{1}{2}$ in.

z

θ

B

$\dfrac{7}{8}$ in. $\dfrac{7}{8}$ in.

PROBLEM 10.11-13

Analysis of Stress and Strain

Two polarizing filters are used at different angles to show stress patterns in a clear plastic machine part under load.

CHAPTER OBJECTIVES

- Define the state of stress at a point on a body using a plane stress element.

- Develop *plane stress transformation equations* that give equivalent descriptions of the state of stress at a point for various rotated positions of the stress element.

- Use the transformation equations to find the maximum and minimum normal stresses, known as *principal normal stresses*, and also the *maximum shear stress* at any point of interest.

- Display the transformation equations in graphical form on a plot known as *Mohr's circle*, which shows the relationship between

normal and shear stresses acting on inclined planes at a point in a stressed body.

- Define *Hooke's law for plane stress*, which provides the relationship between normal strains and stresses for homogeneous and isotropic materials.

- Study states of stress in the three dimensions and define triaxial states of stress known as *spherical stress* and *hydrostatic stress*.

- Develop transformation equations for *plane strain* for use in evaluation of experimental measurements made with strain gages; use Mohr's circle for plane strain transformations.

Chapter Outline

11.1 Introduction

Normal and shear stresses in beams, shafts, and bars can be calculated from the basic formulas discussed in the preceding chapters. For instance, the stresses in a beam are given by the flexure and shear formulas ($\sigma = My/I$ and $\tau = VQ/Ib$), and the stresses in a shaft are given by the torsion formula $\tau = T\rho/I_p$. The stresses calculated from these formulas act on cross sections of the members, but larger stresses may occur on **inclined sections**. Therefore, the analysis of stresses and strains begins with a discussion of methods for finding the normal and shear stresses acting on inclined sections cut through a member.

Expressions for the normal and shear stresses acting on inclined sections in both *uniaxial stress* and *pure shear* were presented in Sections 8.6 and 9.5, respectively. In the case of uniaxial stress, the maximum shear stresses occur on planes inclined at 45° to the axis, whereas the maximum normal stresses occur on the cross sections. In the case of pure shear, the maximum tensile and compressive stresses occur on 45° planes. In an analogous manner, the stresses on inclined sections cut through a beam may be larger than the stresses acting on a cross section. To calculate such stresses, determine the stresses acting on inclined planes under a more general stress state known as **plane stress** (Section 11.2).

In this discussion of plane stress, **stress elements** represent the state of stress at a point in a body. Stress elements were discussed previously in a specialized context (see Sections 8.6 and 9.5), but now they are used in a more formalized manner. Begin the analysis by considering an element on which the stresses are known and then derive the **transformation equations** that give the stresses acting on the sides of an element oriented in a different direction.

When working with stress elements, keep in mind that only one intrinsic **state of stress** exists at a point in a stressed body, regardless of the orientation of the element being used to portray that state of stress. When there are two elements with different orientations at the same point in a body, the stresses acting on the faces of the two elements are different, but they still represent the same state of stress, namely, the stress at the point under consideration. This situation is analogous to the representation of a force vector by its components—although the components are different when the coordinate axes are rotated to a new position, the force itself is the same.

Furthermore, keep in mind that stresses are *not* vectors. This fact can sometimes be confusing, because engineers customarily represent stresses by arrows just as they represent force vectors by arrows. *Although the arrows used to represent stresses have magnitude and direction, they are not vectors because they do not combine according to the parallelogram law of addition.* Instead, stresses are much more complex quantities than are vectors, and in mathematics, they are called **tensors**. Other tensor quantities in mechanics are strains and moments of inertia.

11.2 Plane Stress

The stress conditions encountered in earlier chapters when analyzing bars in tension and compression, shafts in torsion, and beams in bending are examples of a state of stress called **plane stress**. To explain plane stress, consider the stress element shown in Fig. 11-1a. This element is infinitesimal in size and can be sketched either as a cube or as a rectangular parallelepiped. The *xyz* axes are parallel to the edges of the element, and the faces of the element are designated by the directions of their outward normals, as explained previously in Section 7.6. For instance, the right-hand face of the element is referred to as the positive *x* face, and the left-hand face (hidden from the viewer) is referred to as the negative *x* face. Similarly, the top face is the positive *y* face, and the front face is the positive *z* face.

When the material is in plane stress in the *x-y* plane, only the *x* and *y* faces of the element are subjected to stresses, and all stresses act parallel to the *x* and *y* axes, as shown in Fig. 11-1a. This stress condition is very common because it exists at the surface of any stressed body, except at points where external loads act on the surface. When the element shown in Fig. 11-1a is located at the free surface of a body, the *z* axis is normal to the surface and the *z* face is in the plane of the surface.

The symbols for the stresses shown in Fig. 11-1a have the following meanings. A **normal stress** σ has a subscript that identifies the face on which the stress acts; for instance, the stress σ_x acts on the *x* face of the element and the stress σ_y acts on the *y* face of the element. Since the element is infinitesimal in size, equal normal stresses act on the opposite faces. The **sign convention for normal stresses** is the familiar one, namely, tension is positive and compression is negative.

A **shear stress** τ has two subscripts—the first subscript denotes the face on which the stress acts, and the second gives the direction on that face. Thus, the stress τ_{xy} acts on the *x* face in the direction of the *y* axis (Fig. 11-1a), and the stress τ_{yx} acts on the *y* face in the direction of the *x* axis.

The **sign convention for shear stresses** is as follows. A shear stress is positive when it acts on a positive face of an element in the positive direction of an axis, and it is negative when it acts on a positive face of an element in the negative direction of an axis. Therefore, the stresses τ_{xy} and τ_{yx} shown on the positive *x* and *y* faces in Fig. 11-1a are positive shear stresses. Similarly, on a negative face of the element, a shear stress is positive when it acts in the negative direction of an axis. Hence, the stresses τ_{xy} and τ_{yx} shown on the negative *x* and *y* faces of the element are also positive.

This sign convention for shear stresses is easy to remember if stated as follows:

A shear stress is positive when the directions associated with its subscripts are plus-plus or minus-minus; the stress is negative when the directions are plus-minus or minus-plus.

The preceding sign convention for shear stresses is consistent with the equilibrium of the element, because shear stresses on opposite faces of an infinitesimal element must be equal in magnitude and opposite in direction. Hence, according to this sign convention, a positive stress τ_{xy} acts upward on the positive face (Fig. 11-1a) and downward on the negative face. In a similar manner, the stresses τ_{yx} acting on the top and bottom faces of the element are positive—although they have opposite directions.

Shear stresses on perpendicular planes are equal in magnitude and have directions such that both stresses point toward, or both point away from, the line of intersection of the faces. Inasmuch as τ_{xy} and τ_{yx} are positive in the directions shown in the figure, they are consistent with this observation. Therefore, note that

$$\tau_{xy} = \tau_{yx} \qquad (11\text{-}1)$$

FIGURE 11-1

Elements in plane stress: (a) three-dimensional view of an element oriented to the *xyz* axes, (b) two-dimensional view of the same element, and (c) two-dimensional view of an element oriented to the $x_1y_1z_1$ axes

(a)

(b)

(c)

FIGURE 11-1 (Repeated)

(a)

(b)

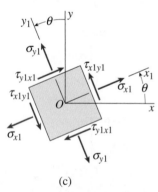

(c)

This relationship was derived previously from equilibrium of the element (see Section 7.6).

For convenience in sketching plane-stress elements, draw only a two-dimensional view of the element, as shown in Fig. 11-1b. Although a figure of this kind is adequate for showing all stresses acting on the element, keep in mind that the element is a solid body with a thickness perpendicular to the plane of the figure.

Stresses on Inclined Sections

Now consider the stresses acting on inclined sections, assuming that the stresses σ_x, σ_y, and τ_{xy} (Figs. 11-1a and b) are known. To portray the stresses acting on an inclined section, consider a new stress element (Fig. 11-1c) that is located at the same point in the material as the original element (Fig. 11-1b). However, the new element has faces that are parallel and perpendicular to the inclined direction. Associated with this new element are axes x_1, y_1, and z_1, such that the z_1 axis coincides with the z axis and the x_1y_1 axes are rotated counterclockwise through an angle θ with respect to the x-y axes.

The normal and shear stresses acting on this new element are denoted σ_{x1}, σ_{y1}, τ_{x1y1}, and τ_{y1x1}, using the same subscript designations and sign conventions described previously for the stresses acting on the x-y element. The previous conclusions regarding the shear stresses still apply, so that

$$\tau_{x1y1} = \tau_{y1x1} \tag{11-2}$$

From this equation and the equilibrium of the element, observe that *the shear stresses acting on all four side faces of an element in plane stress are known if you determine the shear stress acting on any one of those faces.*

The stresses acting on the inclined x_1y_1 element (Fig. 11-1c) can be expressed in terms of the stresses on the x-y element (Fig. 11-1b) by using equations of equilibrium. For this purpose, choose a **wedge-shaped stress element** (Fig. 11-2a) having an inclined face that is the same as the x_1 face of the inclined element shown in Fig. 11-1c. The other two side faces of the wedge are parallel to the x and y axes.

In order to write equations of equilibrium for the wedge, construct a free-body diagram showing the forces acting on the faces. Denote the area of the left-hand side face (that is, the negative x face) as A_0. Then the normal and shear forces acting on that face are $\sigma_x A_0$ and $\tau_{xy} A_0$, as shown in the free-body diagram of Fig. 11-2b. The area of the bottom face (or negative y face) is $A_0 \tan \theta$, and the area of the inclined face (or positive x_1 face) is $A_0 \sec \theta$. Thus, the normal and shear forces acting on these faces have the magnitudes and directions shown in Fig. 11-2b.

The forces acting on the left-hand and bottom faces can be resolved into orthogonal components acting in the x_1 and y_1 directions. Then obtain two equations of equilibrium by summing forces in those directions. The first equation, obtained by summing forces in the x_1 direction, is

$$\sigma_{x1} A_0 \sec \theta - \sigma_x A_0 \cos \theta - \tau_{xy} A_0 \sin \theta$$
$$- \sigma_y A_0 \tan \theta \sin \theta - \tau_{yx} A_0 \tan \theta \cos \theta = 0$$

In the same manner, summation of forces in the y_1 direction gives

$$\tau_{x1y1} A_0 \sec \theta + \sigma_x A_0 \sin \theta - \tau_{xy} A_0 \cos \theta$$
$$- \sigma_y A_0 \tan \theta \cos \theta + \tau_{yx} A_0 \tan \theta \sin \theta = 0$$

Use the relationship $\tau_{xy} = \tau_{yx}$ and also simplify and rearrange to obtain the following two equations:

$$\sigma_{x1} = \sigma_x \cos^2\theta + \sigma_y \sin^2\theta + 2\tau_{xy}\sin\theta\cos\theta \qquad \text{(11-3a)}$$

$$\tau_{x1y1} = -(\sigma_x - \sigma_y)\sin\theta\cos\theta + \tau_{xy}(\cos^2\theta - \sin^2\theta) \qquad \text{(11-3b)}$$

FIGURE 11-2

Wedge-shaped stress element in plane stress: (a) stresses acting on the element and (b) forces acting on the element (free-body diagram)

Equations (11-3a and b) give the normal and shear stresses acting on the x_1 plane in terms of the angle θ and the stresses σ_x, σ_y, and τ_{xy} acting on the x and y planes.

For the special case when $\theta = 0$, note that Eqs. (11-3a and b) give $\sigma_{x1} = \sigma_x$ and $\tau_{x1y1} = \tau_{xy}$, as expected. Also, when $\theta = 90°$, the equations give $\sigma_{x1} = \sigma_y$ and $\tau_{x1y1} = -\tau_{xy} = -\tau_{yx}$. In the latter case, since the x_1 axis is vertical when $\theta = 90°$, the stress τ_{x1y1} will be positive when it acts to the left. However, the stress τ_{yx} acts to the right, so $\tau_{x1y1} = -\tau_{yx}$.

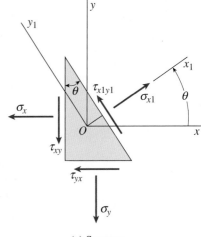

(a) Stresses

Transformation Equations for Plane Stress

Equations (11-3a and b) for the stresses on an inclined section can be expressed in a more convenient form by introducing the following trigonometric identities (see Appendix A):

$$\cos^2\theta = \frac{1}{2}(1 + \cos 2\theta) \qquad \sin^2\theta = \frac{1}{2}(1 - \cos$$

$$\sin\theta\cos\theta = \frac{1}{2}\sin 2\theta$$

When these substitutions are made, the equations become

$$\sigma_{x1} = \frac{\sigma_x + \sigma_y}{2} + \frac{\sigma_x - \sigma_y}{2}\cos 2\theta + \tau_{xy}\sin 2\theta \qquad \text{(11-4a)}$$

$$\tau_{x1y1} = -\frac{\sigma_x - \sigma_y}{2}\sin 2\theta + \tau_{xy}\cos 2\theta \qquad \text{(11-4b)}$$

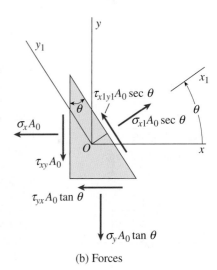

(b) Forces

These equations are usually called the **transformation equations for plane stress** because they transform the stress components from one set of axes to another. However, the intrinsic state of stress at the point under consideration is the same whether represented by stresses acting on the xy element (Fig. 11-1b) or by stresses acting on the inclined x_1y_1 element (Fig. 11-1c).

Since the transformation equations were derived solely from equilibrium of an element, they are applicable to stresses in any kind of material, whether linear or nonlinear, elastic or inelastic.

An important observation concerning the normal stresses can be obtained from the transformation equations. As a preliminary matter, note that the normal stress σ_{y1} acting on the y_1 face of the inclined element (Fig. 11-1c) can be obtained from Eq. (11-4a) by substituting $\theta + 90°$ for θ. The result is the following equation for σ_{y1}:

$$\sigma_{y1} = \frac{\sigma_x + \sigma_y}{2} - \frac{\sigma_x - \sigma_y}{2}\cos 2\theta - \tau_{xy}\sin 2\theta \qquad \text{(11-5)}$$

FIGURE 11-3

Graph of normal stress σ_{x1} and shear stress τ_{x1y1} versus the angle θ (for $\sigma_y = 0.2\sigma_x$ and $\tau_{xy} = 0.8\sigma_x$)

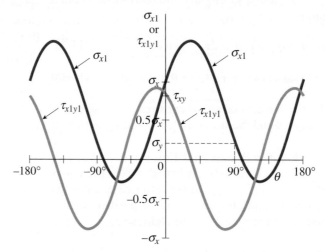

Summing the expressions for σ_{x1} and σ_{y1} [Eqs. (11-4a) and (11-5)] produces the following equation for plane stress:

$$\sigma_{x1} + \sigma_{y1} = \sigma_x + \sigma_y \tag{11-6}$$

This equation shows that the sum of the normal stresses acting on perpendicular faces of plane-stress elements (at a given point in a stressed body) is constant and independent of the angle θ.

The manner in which the normal and shear stresses vary is shown in Fig. 11-3, which is a graph of σ_{x1} and τ_{x1y1} versus the angle θ [from Eqs. (11-4a and b)]. The graph is plotted for the particular case of $\sigma_y = 0.2\sigma_x$ and $\tau_{xy} = 0.8\sigma_x$. The plot shows that the stresses vary continuously as the orientation of the element is changed. At certain angles, the normal stress reaches a maximum or minimum value; at other angles, it becomes zero. Similarly, the shear stress has maximum, minimum, and zero values at certain angles. A detailed investigation of these maximum and minimum values is made in Section 11.3.

FIGURE 11-4

Element in uniaxial stress

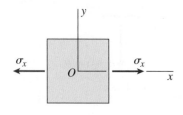

Special Cases of Plane Stress

The general case of plane stress reduces to simpler states of stress under special conditions. For instance, if all stresses acting on the xy element (Fig. 11-1b) are zero except for the normal stress σ_x, then the element is in **uniaxial stress** (Fig. 11-4). The corresponding transformation equations, obtained by setting σ_y and τ_{xy} equal to zero in Eqs. (11-4a and b), are

$$\sigma_{x1} = \frac{\sigma_x}{2}(1 + \cos 2\theta) \qquad \tau_{x1y1} = -\frac{\sigma_x}{2}(\sin 2\theta) \tag{11-7a,b}$$

FIGURE 11-5

Element in pure shear

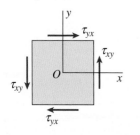

These equations agree with the equations derived previously in Section 8.6 [see Eqs. (8-29a and b)], except that now a more generalized notation for the stresses acting on an inclined plane is used.

Another special case is **pure shear** (Fig. 11-5), for which the transformation equations are obtained by substituting $\sigma_x = 0$ and $\sigma_y = 0$ into Eqs. (11-4a and b):

$$\sigma_{x1} = \tau_{xy} \sin 2\theta \qquad \tau_{x1y1} = \tau_{xy} \cos 2\theta \tag{11-8a,b}$$

Again, these equations correspond to those derived earlier [see Eqs. (9-29a and b) in Section 9.5].

Finally, note the special case of **biaxial stress**, in which the xy element is subjected to normal stresses in both the x and y directions but without any shear stresses (Fig. 11-6). The equations for biaxial stress are obtained from Eqs. (11-4a and b) simply by dropping the terms containing τ_{xy}, as

FIGURE 11-6

Element in biaxial stress

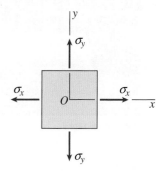

$$\sigma_{x1} = \frac{\sigma_x + \sigma_y}{2} + \frac{\sigma_x - \sigma_y}{2} \cos 2\theta \qquad \textbf{(11-9a)}$$

$$\tau_{x1y1} = -\frac{\sigma_x - \sigma_y}{2} \sin 2\theta \qquad \textbf{(11-9b)}$$

Biaxial stress occurs in many kinds of structures, including thin-walled pressure vessels (see Sections 12.2 and 12.3).

Example 11-1

FIGURE 11-7

Example 11-1: (a) Cylindrical pressure vessel with stress element at C, (b) element C in plane stress, and (c) element C inclined at an angle $\theta = 45°$

(a)

(b)

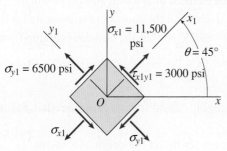

(c)

A cylindrical pressure vessel rests on simple supports at A and B (see Fig. 11-7). The vessel is under internal pressure resulting in longitudinal stress $\sigma_x = 6,000$ psi and circumferential stress $\sigma_y = 12,000$ psi on a stress element at point C on the wall of the vessel. In addition, differential settlement after an earthquake has caused the support at B to rotate, which applies a torsional moment to the vessel leading to shear stress $\tau_{xy} = 2500$ psi. Find the stresses acting on the element at C when rotated through angle $\theta = 45°$.

Solution:

Use the four-step problem-solving approach. Combine steps as needed for an efficient solution.

1, 2. *Conceptualize [hypothesize, sketch]*, *Categorize*, *[simplify, classify]*:

Transformation equations: To determine the stresses acting on an inclined element, use the transformation equations given in Eqs. (11-4a and b). From the given numerical data, obtain the following values for substitution into those equations:

$$\frac{\sigma_x + \sigma_y}{2} = 9000 \text{ psi} \qquad \frac{\sigma_x - \sigma_y}{2} = -3000 \text{ psi} \qquad \tau_{xy} = 2500 \text{ psi}$$

$$\sin 2\theta = \sin 90° = 1 \qquad \cos 2\theta = \cos 90° = 0$$

Fuel storage tanks

© Barry Goodno

3. **Analyze** [*evaluate; select relevant equations, carry out mathematical solution*]: Substitute these values into Eqs. (11-4a and b) to get

$$\sigma_{x1} = \frac{\sigma_x + \sigma_y}{2} + \frac{\sigma_x - \sigma_y}{2} \cos 2\theta + \tau_{xy} \sin 2\theta$$

$$= 9000 \text{ psi} + (-3000 \text{ psi})(0) + (2500 \text{ psi})(1) = 11{,}500 \text{ psi}$$

$$\tau_{x1y1} = -\frac{\sigma_x - \sigma_y}{2} \sin 2\theta + \tau_{xy} \cos 2\theta$$

$$= -(-3000 \text{ psi})(1) + (2500 \text{ psi})(0) = 3000 \text{ psi}$$

In addition, the stress σ_{y1} may be obtained from Eq. (11-5):

$$\sigma_{y1} = \frac{\sigma_x + \sigma_y}{2} - \frac{\sigma_x - \sigma_y}{2} \cos 2\theta - \tau_{xy} \sin 2\theta$$

$$= 9000 \text{ psi} - (-3000 \text{ psi})(0) - (2500 \text{ psi})(1) = 6500 \text{ psi}$$

4. **Finalize** [*conclude; examine answer—Does it make sense? Are units correct? How does it compare to similar problem solutions?*]:

Stress elements: From these results, you can readily obtain the stresses acting on all sides of an element oriented at $\theta = 45°$, as shown in Fig. 11-7c. The arrows show the true directions in which the stresses act. Note especially the directions of the shear stresses, all of which have the same magnitude. Also, observe that the sum of the normal stresses remains constant and equal to 18,000 psi [see Eq. (11-6)].

Note: The stresses shown in Fig. 11-7c represent the same intrinsic state of stress as do the stresses shown in Fig. 11-7b. However, the stresses have different values, because the elements on which they act have different orientations.

Example 11-2

A cylindrical pressure vessel rests on simple supports at A and B (see Fig. 11-8). The vessel has a helical weld joint oriented at $\theta = 35°$ to the longitudinal axis. The vessel is under internal pressure and also has some torsional shear stress due to differential settlement of the support at B. The state of stress on the element at D along and perpendicular to the weld seam is known and is given in Fig. 11-8b. Find the equivalent stress state for the element at D when rotated through angle $\theta = -35°$ so that the element is aligned with the longitudinal axis of the vessel.

Solution:

Use the four-step problem-solving approach. Combine steps as needed for an efficient solution.

1, 2. *Conceptualize, Categorize*: The stresses acting on the original element (Fig. 11-8b) have the values:

$$\sigma_x = 40 \text{ MPa} \qquad \sigma_y = 80 \text{ MPa} \qquad \tau_{xy} = -22 \text{ MPa}$$

FIGURE 11-8

Example 11-2: (a) Cylindrical pressure vessel with stress element at *D*, (b) element *D* in plane stress, and (c) element *D* inclined at an angle $\theta = -35°$

(a)

(b)

(c)

Fuel storage tank supported on pedestals

© Barry Goodno

An element oriented at a clockwise angle of $-35°$ is shown in Fig. 11-8c, where the x_1 axis is at an angle $\theta = -35°$ with respect to the x axis.

Stress transformation equations: You can readily calculate the stresses on the x_1 face of the element oriented at $\theta = -35°$ by using the transformation equations given in Eqs. (11-4a and b). The calculation results are

$$\frac{\sigma_x + \sigma_y}{2} = 60 \text{ MPa} \qquad \frac{\sigma_x - \sigma_y}{2} = -20 \text{ MPa}$$

$$\sin 2\theta = \sin(-70°) = -0.94 \quad \cos 2\theta = \cos(-70°) = 0.342$$

3. *Analyze*: Substitute into the transformation equations to get

$$\sigma_{x1} = \frac{\sigma_x + \sigma_y}{2} + \frac{\sigma_x - \sigma_y}{2}\cos(2\theta) + \tau_{xy}\sin(2\theta)$$
$$= 60 \text{ MPa} + (-20 \text{ MPa})(0.342) + (-22 \text{ MPa})(-0.94)$$
$$= 73.8 \text{ MPa}$$

$$\tau_{x1y1} = -\left(\frac{\sigma_x - \sigma_y}{2}\right)\sin(2\theta) + \tau_{xy}\cos(2\theta)$$
$$= -(-20 \text{ MPa})(-0.94) + (-22 \text{ MPa})(0.342) = -26.3 \text{ MPa}$$

The normal stress acting on the y_1 face [see Eq. (11-5)] is

$$\sigma_{y1} = \frac{\sigma_x + \sigma_y}{2} - \frac{\sigma_x - \sigma_y}{2}\cos(2\theta) - \tau_{xy}\sin(2\theta)$$
$$= 60 \text{ MPa} - (-20 \text{ MPa})(0.342) - (-22 \text{ MPa})(-0.94)$$
$$= 46.2 \text{ MPa}$$

4. *Finalize*: As a check on the results, note that $\sigma_{x1} + \sigma_{y1} = \sigma_x + \sigma_y$.

The stresses acting on the rotated element are shown in Fig. 11-8c, where the arrows indicate the true directions of the stresses. Note that both stress elements shown in Fig. 11-8 represent the same state of stress.

Frans Lemmens/The Image Bank/ Getty Images

Courtesy Eann Patterson

FIGURE 11-9

Photoelastic fringe pattern displays principal stresses in a model of a crane-hook: (a) photo of a crane-hook, (b) photoelastic fringe pattern

(a)

(b)

11.3 Principal Stresses and Maximum Shear Stresses

The transformation equations for plane stress show that the normal stresses σ_{x1} and the shear stresses τ_{x1y1} vary continuously as the axes are rotated through the angle θ. This variation is pictured in Fig. 11-3 for a particular combination of stresses. From the figure, both the normal and shear stresses reach maximum and minimum values at 90° intervals. Not surprisingly, these maximum and minimum values are usually needed for design purposes. For instance, fatigue failures of structures such as machines and aircraft are often associated with the maximum stresses, and hence their magnitudes and orientations should be determined as part of the design process (see Fig. 11-9).

Principal Stresses

The maximum and minimum normal stresses, called the **principal stresses**, can be found from the transformation equation for the normal stress σ_{x1} [(Eq. 11-4a)]. Take the derivative of σ_{x1} with respect to θ and set it equal to zero to obtain an equation that gives the values of θ at which σ_{x1} is a maximum or a minimum. The equation for the derivative is

$$\frac{d\sigma_{x1}}{d\theta} = -(\sigma_x - \sigma_y)\sin 2\theta + 2\tau_{xy}\cos 2\theta = 0 \tag{11-10}$$

which now gives

$$\tan 2\theta_p = \frac{2\tau_{xy}}{\sigma_x - \sigma_y} \tag{11-11}$$

The subscript p indicates that the angle θ_p defines the orientation of the **principal planes**, that is, the planes on which the principal stresses act.

Two values of the angle $2\theta_p$ in the range from 0 to 360° can be obtained from Eq. (11-11). These values differ by 180°, with one value between 0 and 180° and the other between 180° and 360°. Therefore, the angle θ_p has two values that differ by 90°, one value between 0 and 90° and the other between 90° and 180°. The two values of θ_p are known as the **principal angles**. For one of these angles, the normal stress σ_{x1} is a *maximum* principal stress; for the other, it is a *minimum* principal stress. Because the principal angles differ by 90°, *the principal stresses occur on mutually perpendicular planes*.

The principal stresses can be calculated by substituting each of the two values of θ_p into the first stress-transformation equation [Eq. (11-4a)] and solving for σ_{x1}. By determining the principal stresses in this manner, you not only obtain the values of the principal stresses but you also learn which principal stress is associated with which principal angle.

You can also obtain general formulas for the principal stresses. To do so, refer to the right triangle in Fig. 11-10, which is constructed from Eq. (11-11). Note that the hypotenuse of the triangle, obtained from the Pythagorean theorem, is

FIGURE 11-10

Geometric representation of Eq. (11-11)

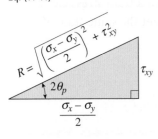

$$R = \sqrt{\left(\frac{\sigma_x - \sigma_y}{2}\right)^2 + \tau_{xy}^2} \tag{11-12}$$

The quantity R is always a positive number and, like the other two sides of the triangle, has units of stress. Two additional relations from the triangle are

$$\cos 2\theta_p = \frac{\sigma_x - \sigma_y}{2R} \qquad \sin 2\theta_p = \frac{\tau_{xy}}{R} \qquad \text{(11-13a,b)}$$

Now substitute these expressions for $\cos 2\theta_p$ and $\sin 2\theta_p$ into Eq. (11-4a) and obtain the algebraically larger of the two principal stresses, denoted by σ_1:

$$\sigma_1 = \sigma_{x1} = \frac{\sigma_x + \sigma_y}{2} + \frac{\sigma_x - \sigma_y}{2} \cos 2\theta_p + \tau_{xy} \sin 2\theta_p$$

$$= \frac{\sigma_x + \sigma_y}{2} + \frac{\sigma_x - \sigma_y}{2}\left(\frac{\sigma_x - \sigma_y}{2R}\right) + \tau_{xy}\left(\frac{\tau_{xy}}{R}\right)$$

Substitute for R from Eq. (11-12) and perform some algebraic manipulations to obtain

$$\sigma_1 = \frac{\sigma_x + \sigma_y}{2} + \sqrt{\left(\frac{\sigma_x - \sigma_y}{2}\right)^2 + \tau_{xy}^2} \qquad \text{(11-14)}$$

The smaller of the principal stresses, denoted by σ_2, may be found from the condition that the sum of the normal stresses on perpendicular planes is constant [see Eq. (11-6)]:

$$\sigma_1 + \sigma_2 = \sigma_x + \sigma_y \qquad \text{(11-15)}$$

Substitute the expression for σ_1 into Eq. (11-15) and solve for σ_2 to get

$$\sigma_2 = \sigma_x + \sigma_y - \sigma_1$$

$$= \frac{\sigma_x + \sigma_y}{2} - \sqrt{\left(\frac{\sigma_x - \sigma_y}{2}\right)^2 + \tau_{xy}^2} \qquad \text{(11-16)}$$

This equation has the same form as the equation for σ_1 but differs by the presence of the minus sign before the square root.

The preceding formulas for σ_1 and σ_2 can be combined into a single formula for the **principal stresses**:

$$\sigma_{1,2} = \frac{\sigma_x + \sigma_y}{2} \pm \sqrt{\left(\frac{\sigma_x - \sigma_y}{2}\right)^2 + \tau_{xy}^2} \qquad \text{(11-17)}$$

The plus sign gives the algebraically larger principal stress, and the minus sign gives the algebraically smaller principal stress.

Principal Angles

Now denote the two angles defining the principal planes as θ_{p1} and θ_{p2}, corresponding to the principal stresses σ_1 and σ_2, respectively. Both angles can be determined from the equation for $\tan 2\theta_p$ [Eq. (11-11)]. However, you cannot tell from that equation which angle is θ_{p1} and which is θ_{p2}. A simple procedure for making this determination is to take one of the values and substitute it into the equation for σ_{x1} [Eq. (11-4a)]. The resulting value of σ_{x1} will be

FIGURE 11-11

Elements in uniaxial and biaxial stress

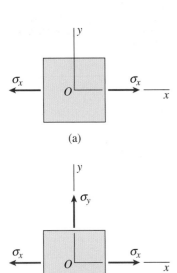

(a)

(b)

recognized as either σ_1 or σ_2 [assuming you have already found σ_1 and σ_2 from Eq. (11-17)], thus correlating the two principal angles with the two principal stresses.

Another method for correlating the principal angles and principal stresses is to use Eqs. (11-13a and b) to find θ_p, since the only angle that satisfies *both* of those equations is θ_{p1}. Thus, rewrite those equations as

$$\cos 2\theta_{p1} = \frac{\sigma_x - \sigma_y}{2R} \qquad \sin 2\theta_{p1} = \frac{\tau_{xy}}{R} \qquad \textbf{(11-18a,b)}$$

Only one angle exists between 0 and 360° that satisfies both of these equations. Thus, the value of θ_{p1} can be determined uniquely from Eqs. (11-18a and b). The angle θ_{p2}, corresponding to σ_2, defines a plane that is perpendicular to the plane defined by θ_{p1}. Therefore, θ_{p2} can be taken as 90° larger or 90° smaller than θ_{p1}.

Shear Stresses on the Principal Planes

An important characteristic of the principal planes can be obtained from the transformation equation for the shear stresses [Eq. (11-4b)]. If the shear stress τ_{x1y1} is set equal to zero, the resulting equation is the same as Eq. (11-10). Therefore, if you solve that equation for the angle 2θ, you get the same expression for $\tan 2\theta$ as before [Eq. (11-11)]. In other words, the angles to the planes of zero shear stress are the same as the angles to the principal planes.

This leads to the following important observation: *The shear stresses are zero on the principal planes.*

FIGURE 11-12

(a) Element in pure shear and (b) principal stresses

(a)

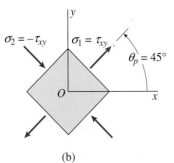

(b)

Special Cases

The principal planes for elements in **uniaxial stress** and **biaxial stress** are the x and y planes themselves (Fig. 11-11) because $\tan 2\theta_p = 0$ [see Eq. (11-11)] and the two values of θ_p are 0 and 90°. The x and y planes are the principal planes based on the fact that the shear stresses are zero on those planes.

For an element in **pure shear** (Fig. 11-12a), the principal planes are oriented at 45° to the x axis (Fig. 11-12b), because $\tan 2\theta_p$ is infinite, and the two values of θ_p are 45° and 135°. If τ_{xy} is positive, the principal stresses are $\sigma_1 = \tau_{xy}$ and $\sigma_2 = -\tau_{xy}$ (see Section 9.5 for a discussion of pure shear).

The Third Principal Stress

The preceding discussion of principal stresses refers only to rotation of axes in the x-y plane, that is, rotation about the z axis (Fig. 11-13a). Therefore, the two principal stresses determined from Eq. (11-17) are called the **in-plane principal stresses**. However, the stress element is actually three-dimensional and has three (not two) principal stresses acting on three mutually perpendicular planes.

A more complete three-dimensional analysis shows that the three principal planes for a plane-stress element are the two principal planes already described plus the z face of the element. These principal planes are shown in Fig. 11-13b, where a stress element has been oriented at the principal angle θ_{p1}, which corresponds to the principal stress σ_1. The principal stresses σ_1 and σ_2 are given by Eq. (11-17), and the third principal stress (σ_3) equals zero.

FIGURE 11-13

Elements in plane stress:
(a) original element and
(b) element oriented to the three
principal planes and three principal
stresses

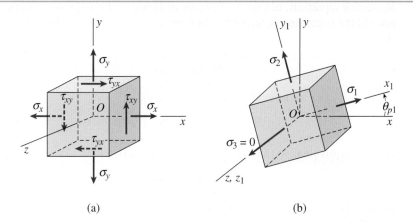

(a)   (b)

By definition, σ_1 is algebraically larger than σ_2, but σ_3 may be algebraically larger than, between, or smaller than σ_1 and σ_2. Of course, it is also possible for some or all of the principal stresses to be equal. Note again that there are no shear stresses on any of the principal planes.[1]

Maximum Shear Stresses

Having found the principal stresses and their directions for an element in plane stress, now consider the determination of the maximum shear stresses and the planes on which they act. The shear stresses τ_{x1y1} acting on inclined planes are given by the second transformation equation [Eq. (11-4b)]. Take the derivative of τ_{x1y1} with respect to θ and set it equal to zero to obtain

$$\frac{d\tau_{x1y1}}{d\theta} = -(\sigma_x - \sigma_y)\cos 2\theta - 2\tau_{xy}\sin 2\theta = 0 \qquad \textbf{(11-19)}$$

from which

$$\tan 2\theta_s = -\frac{\sigma_x - \sigma_y}{2\tau_{xy}} \qquad \textbf{(11-20)}$$

The subscript s indicates that the angle θ_s defines the orientation of the planes of maximum positive and negative shear stresses.

Equation (11-20) yields one value of θ_s between 0 and 90° and another between 90° and 180°. Furthermore, these two values differ by 90°; therefore, the maximum shear stresses occur on perpendicular planes. Because shear stresses on perpendicular planes are equal in absolute value, the maximum positive and negative shear stresses differ only in sign.

Comparing Eq. (11-20) for θ_s with Eq. (11-11) for θ_p shows that

$$\tan 2\theta_s = -\frac{1}{\tan 2\theta_p} = -\cot 2\theta_p \qquad \textbf{(11-21)}$$

---

[1]The determination of principal stresses is an example of a type of mathematical analysis known as *eigenvalue analysis*, which is described in books on matrix algebra. The stress-transformation equations and the concept of principal stresses are due to the French mathematicians A. L. Cauchy (1789–1857) and Barré de Saint-Venant (1797–1886) and to the Scottish scientist and engineer W. J. M. Rankine (1820–1872).

From this equation, obtain a relationship between the angles θ_s and θ_p. First, rewrite the preceding equation in the form

$$\frac{\sin 2\theta_s}{\cos 2\theta_s} + \frac{\cos 2\theta_p}{\sin 2\theta_p} = 0$$

Multiply by the terms in the denominator to get

$$\sin 2\theta_s \sin 2\theta_p + \cos 2\theta_s \cos 2\theta_p = 0$$

which is equivalent to the expression (see Appendix A):

$$\cos(2\theta_s - 2\theta_p) = 0$$

Therefore,

$$2\theta_s - 2\theta_p = \pm 90°$$

and

$$\theta_s = \theta_p \pm 45° \tag{11-22}$$

This equation shows that *the planes of maximum shear stress occur at 45° to the principal planes.*

The plane of the maximum positive shear stress τ_{max} is defined by the angle θ_{s1}, for which the following equations apply:

$$\cos 2\theta_{s1} = \frac{\tau_{xy}}{R} \qquad \sin 2\theta_{s1} = -\frac{\sigma_x - \sigma_y}{2R} \tag{11-23a,b}$$

where R is given by Eq. (11-12). Also, the angle θ_{s1} is related to the angle θ_{p1} [see Eqs. (11-18a and b)] as

$$\theta_{s1} = \theta_{p1} - 45° \tag{11-24}$$

The corresponding maximum shear stress is obtained by substituting the expressions for $\cos 2\theta_{s1}$ and $\sin 2\theta_{s1}$ into the second transformation equation [Eq. (11-4b)], yielding

$$\tau_{max} = \sqrt{\left(\frac{\sigma_x - \sigma_y}{2}\right)^2 + \tau_{xy}^2} \tag{11-25}$$

The maximum negative shear stress τ_{min} has the same magnitude but opposite sign.

Another expression for the maximum shear stress can be obtained from the principal stresses σ_1 and σ_2, both of which are given by Eq. (11-17). Subtract the expression for σ_2 from that for σ_1, and then compare with Eq. (11-25) to get

$$\tau_{max} = \frac{\sigma_1 - \sigma_2}{2} \tag{11-26}$$

Thus, *the maximum shear stress is equal to one-half the difference of the principal stresses.*

The planes of maximum shear stress also contain normal stresses. The **normal stress** acting on the planes of maximum positive shear stress can be determined by substituting the expressions for the angle θ_{s1} [Eqs. (11-23a and b)] into the equation for σ_{x1} [Eq. (11-4a)]. The resulting stress is equal to the average of the normal stresses on the x and y planes:

$$\sigma_{\text{aver}} = \frac{\sigma_x + \sigma_y}{2} \qquad \text{(11-27)}$$

This same normal stress acts on the planes of maximum negative shear stress.

In the particular cases of **uniaxial stress** and **biaxial stress** (Fig. 11-11), the planes of maximum shear stress occur at 45° to the x and y axes. In the case of **pure shear** (Fig. 11-12), the maximum shear stresses occur on the x and y planes.

In-Plane and Out-of-Plane Shear Stresses

The preceding analysis of shear stresses dealt only with **in-plane shear stresses**, that is, stresses acting in the x-y plane. To obtain the maximum in-plane shear stresses [Eqs. (11-25) and (11-26)], elements were considered that were obtained by rotating the xyz axes about the z axis, which is a principal axis (Fig. 11-13a). The result was that maximum shear stresses occur on planes at 45° to the principal planes. The principal planes for the element of Fig. 11-13a are shown in Fig. 11-13b, where σ_1 and σ_2 are the principal stresses. Therefore, the maximum in-plane shear stresses are found on an element obtained by rotating the $x_1 y_1 z_1$ axes (Fig. 11-13b) about the z_1 axis through an angle of 45°. These stresses are given by Eq. (11-25) or Eq. (11-26).

Maximum shear stresses also can be obtained by 45° rotations about the other two principal axes (the x_1 and y_1 axes in Fig. 11-13b). As a result, three sets of **maximum positive** and **maximum negative shear stresses** are found [compare with Eq. (11-26)]:

$$(\tau_{\text{max}})_{x1} = \pm\frac{\sigma_2}{2} \qquad (\tau_{\text{max}})_{y1} = \pm\frac{\sigma_1}{2}$$

$$(\tau_{\text{max}})_{z1} = \pm\frac{\sigma_1 - \sigma_2}{2} \qquad \text{(11-28a,b,c)}$$

where the subscripts indicate the principal axes about which the 45° rotations take place. The stresses obtained by rotations about the x_1 and y_1 axes are called **out-of-plane shear stresses**.

The algebraic values of σ_1 and σ_2 determine which of the preceding expressions gives the numerically largest shear stress. If σ_1 and σ_2 have the same sign, then one of the first two expressions is numerically largest; if they have opposite signs, the last expression is largest.

FIGURE 11-13 (Repeated)

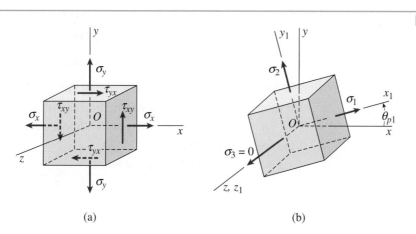

(a)   (b)

Example 11-3

FIGURE 11-14

Example 11-3: (a) Beam structure, (b) element at C in plane stress, (c) principal stresses, and (d) maximum shear stresses

(a)

(b)

(c)

(d)

A simply supported, wide-flange beam has a concentrated load P applied at mid-span (Fig. 11-14a). The state of stress in the beam web at element C is known (Fig. 11-14b) to be $\sigma_x = 12{,}300$ psi, $\sigma_y = -4200$ psi, and $\tau_{xy} = -4700$ psi.

(a) Determine the principal stresses and show them on a sketch of a properly oriented element.

(b) Determine the maximum shear stresses and show them on a sketch of a properly oriented element. (Consider only the in-plane stresses.)

Solution:

Use the four-step problem-solving approach. Combine steps as needed for an efficient solution.

Part (a): Principal stresses.

1, 2. *Conceptualize, Categorize*: The normal and shear stresses for the stress element at C (Fig. 11-14b) are aligned with the axis of the beam. These stresses were computed using the flexure and shear formulas presented in Sections 10.5 and 10.9, respectively. First, find the orientation of principal axes; then use stress-transformation formulas to find the principal normal stresses. Next, find the orientation angle for the plane on which maximum shear stress acts and the value of that shear stress using Eq. (11-25).

3. *Analyze*: The principal angles θ_p that locate the principal planes can be obtained from Eq. (11-11):

$$\tan 2\theta_p = \frac{2\tau_{xy}}{\sigma_x - \sigma_y}$$

$$= \frac{2(-4700 \text{ psi})}{12{,}300 \text{ psi} - (-4200 \text{ psi})} = -0.5697$$

Solve for the angles to get the two sets of values

$$2\theta_p = 150.3° \text{ and } \theta_p = 75.2°$$

$$2\theta_p = 330.3° \text{ and } \theta_p = 165.2°$$

The principal stresses are obtained by substituting the two values of $2\theta_p$ into the transformation equation for σ_{x1} from Eq. (11-4a). As a preliminary calculation, determine the following quantities:

$$\frac{\sigma_x + \sigma_y}{2} = \frac{12{,}300 \text{ psi} - 4200 \text{ psi}}{2} = 4050 \text{ psi}$$

$$\frac{\sigma_x - \sigma_y}{2} = \frac{12{,}300 \text{ psi} + 4200 \text{ psi}}{2} = 8250 \text{ psi}$$

Now substitute the first value of $2\theta_p$ into Eq. (11-4a) and obtain

$$\sigma_{x1} = \frac{\sigma_x + \sigma_y}{2} + \frac{\sigma_x - \sigma_y}{2} \cos 2\theta + \tau_{xy} \sin 2\theta$$
$$= 4050 \text{ psi} + (8250 \text{ psi})(\cos 150.3°) - (4700 \text{ psi})(\sin 150.3°)$$
$$= -5440 \text{ psi}$$

In a similar manner, substitute the second value of $2\theta_p$ and obtain $\sigma_{x1} = 13,540$ psi. Thus, the principal stresses and their corresponding principal angles are

$$\sigma_1 = 13,540 \text{ psi} \quad \text{and} \quad \theta_{p1} = 165.2°$$
$$\sigma_2 = -5440 \text{ psi} \quad \text{and} \quad \theta_{p2} = 75.2°$$

4. *Finalize*: Note that θ_{p1} and θ_{p2} differ by 90° and that $\sigma_1 + \sigma_2 = \sigma_x + \sigma_y$.

The principal stresses are shown on a properly oriented element in Fig. 11-14c. Of course, no shear stresses act on the principal planes.

Alternative solution for the principal stresses: The principal stresses also may be calculated directly from Eq. (11-17):

$$\sigma_{1,2} = \frac{\sigma_x + \sigma_y}{2} \pm \sqrt{\left(\frac{\sigma_x - \sigma_y}{2}\right)^2 + \tau_{xy}^2}$$
$$= 4050 \text{ psi} \pm \sqrt{(8250 \text{ psi})^2 + (-4700 \text{ psi})^2}$$
$$\sigma_{1,2} = 4050 \text{ psi} \pm 9490 \text{ psi}$$

Therefore,

$$\sigma_1 = 13,540 \text{ psi} \qquad \sigma_2 = -5440 \text{ psi}$$

The angle θ_{p1} to the plane on which σ_1 acts is obtained from Eqs. (11-18a and b):

$$\cos 2\theta_{p1} = \frac{\sigma_x - \sigma_y}{2R} = \frac{8250 \text{ psi}}{9490 \text{ psi}} = 0.869$$

$$\sin 2\theta_{p1} = \frac{\tau_{xy}}{R} = \frac{-4700 \text{ psi}}{9490 \text{ psi}} = -0.495$$

where R is given by Eq. (11-12) and is equal to the square-root term in the preceding calculation for the principal stresses σ_1 and σ_2.

The only angle between 0 and 360° having the specified sine and cosine is $2\theta_{p1} = 330.3°$, hence, $\theta_{p1} = 165.2°$. This angle is associated with the algebraically larger principal stress $\sigma_1 = 13,540$ psi. The other angle is 90° larger or smaller than θ_{p1}; hence, $\theta_{p2} = 75.2°$. This angle corresponds to the smaller principal stress $\sigma_2 = -5440$ psi. Note that these results for the principal stresses and principal angles agree with those found previously.

FIGURE 11-14 (Repeated)

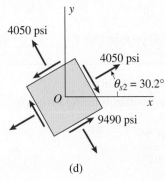

(d)

Part (b): Maximum shear stresses.

2, 3. *Categorize, Analyze*: The maximum in-plane shear stresses are given by Eq. (11-25):

$$\tau_{max} = \sqrt{\left(\frac{\sigma_x - \sigma_y}{2}\right)^2 + \tau_{xy}^2}$$

$$= \sqrt{(8250 \text{ psi})^2 + (-4700 \text{ psi})^2} = 9490 \text{ psi} \quad \leftarrow$$

The angle θ_{s1} to the plane having the maximum positive shear stress is calculated from Eq. (11-24):

$$\theta_{s1} = \theta_{p1} - 45° = 165.2° - 45° = 120.2° \quad \leftarrow$$

It follows that the maximum negative shear stress acts on the plane for which $\theta_{s2} = 120.2° - 90° = 30.2°$.

The normal stresses acting on the planes of maximum shear stresses are calculated from Eq. (11-27):

$$\sigma_{aver} = \frac{\sigma_x + \sigma_y}{2} = 4050 \text{ psi} \quad \leftarrow$$

4. *Finalize*: Finally, the maximum shear stresses and associated normal stresses are shown on the stress element of Fig. 11-14d.

As an alternative approach to finding the maximum shear stresses, use Eq. (11-20) to determine the two values of the angles θ_s, and then use the second transformation equation (Eq. 11-4b) to obtain the corresponding shear stresses.

11.4 Mohr's Circle for Plane Stress

The transformation equations for plane stress can be represented in graphical form by a plot known as **Mohr's circle**. This graphical representation is extremely useful because it displays the relationships between the normal and shear stresses acting on various inclined planes at a point in a stressed body. It also provides a means for calculating principal stresses, maximum shear stresses, and stresses on inclined planes. Furthermore, Mohr's circle is valid not only for stresses but also for other quantities of a similar mathematical nature, including strains and moments of inertia.[2]

Equations of Mohr's Circle

The equations of Mohr's circle can be derived from the transformation equations for plane stress in Eqs. (11-4a and b). The two equations are repeated here, but with a slight rearrangement of the first equation:

$$\sigma_{x1} - \frac{\sigma_x + \sigma_y}{2} = \frac{\sigma_x - \sigma_y}{2} \cos 2\theta + \tau_{xy} \sin 2\theta \qquad \textbf{(11-29a)}$$

---

[2]Mohr's circle is named after the famous German civil engineer Otto Christian Mohr (1835–1918), who developed the circle in 1882.

$$\tau_{x1y1} = -\frac{\sigma_x - \sigma_y}{2} \sin 2\theta + \tau_{xy} \cos 2\theta \qquad \textbf{(11-29b)}$$

From analytic geometry, you might recognize that these two equations are the equations of a circle in parametric form. The angle 2θ is the parameter and the stresses σ_{x1} and τ_{x1y1} are the coordinates. However, it is not necessary to recognize the nature of the equations at this stage—if you eliminate the parameter, the significance of the equations will become apparent.

To eliminate the parameter 2θ, square both sides of each equation and then add the two equations. The equation that results is

$$\left(\sigma_{x1} - \frac{\sigma_x + \sigma_y}{2}\right)^2 + \tau_{x1y1}^2 = \left(\frac{\sigma_x - \sigma_y}{2}\right)^2 + \tau_{xy}^2 \qquad \textbf{(11-30)}$$

This equation can be written in simpler form by using the following notation from Section 11.3 [see Eqs. (11-27) and (11-12), respectively]:

$$\sigma_{aver} = \frac{\sigma_x + \sigma_y}{2} \qquad R = \sqrt{\left(\frac{\sigma_x - \sigma_y}{2}\right)^2 + \tau_{xy}^2} \qquad \textbf{(11-31a,b)}$$

Equation (11-30) now becomes

$$(\sigma_{x1} - \sigma_{aver})^2 + \tau_{x1y1}^2 = R^2 \qquad \textbf{(11-32)}$$

which is the equation of a circle in standard algebraic form. The coordinates are σ_{x1} and τ_{x1y1}, the radius is R, and the center of the circle has coordinates $\sigma_{x1} = \sigma_{aver}$ and $\tau_{x1y1} = 0$.

Two Forms of Mohr's Circle

Mohr's circle can be plotted from Eqs. (11-29) and (11-32) in either of two forms. In the first form of Mohr's circle, plot the normal stress σ_{x1} positive to the right and the shear stress τ_{x1y1} positive downward, as shown in Fig. 11-15a. The advantage of plotting shear stresses positive downward is that the angle 2θ on Mohr's circle will be positive when counterclockwise, which agrees with the positive direction of 2θ in the derivation of the transformation equations (see Figs. 11-1 and 11-2).

In the second form of Mohr's circle, τ_{x1y1} is plotted positive upward but the angle 2θ is now positive clockwise (Fig. 11-15b), which is opposite to its usual positive direction.

Both forms of Mohr's circle are mathematically correct, and either one can be used. However, it is easier to visualize the orientation of the stress element if the positive direction of the angle 2θ is the same in Mohr's circle as it is for the element itself. Furthermore, a counterclockwise rotation agrees with the customary right-hand rule for rotation.

Therefore, the first form of Mohr's circle (Fig. 11-15a) is used here, where *a positive shear stress is plotted downward and a positive angle 2θ is plotted counterclockwise.*

Construction of Mohr's Circle

Mohr's circle can be constructed in a variety of ways, depending upon which stresses are known and which are to be found. To show the basic properties of the circle, now assume that you know the stresses σ_x, σ_y, and τ_{xy} acting on the x and y planes of an element in plane stress (Fig. 11-16a). This information

FIGURE 11-15

Two forms of Mohr's circle: (a) τ_{x1y1} is positive downward and the angle 2θ is positive counterclockwise and (b) τ_{x1y1} is positive upward and the angle 2θ is positive clockwise (*Note:* The first form is used in this book)

(a)

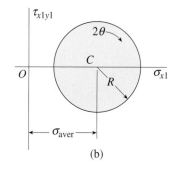

(b)

FIGURE 11-16

Construction of Mohr's circle
for plane stress

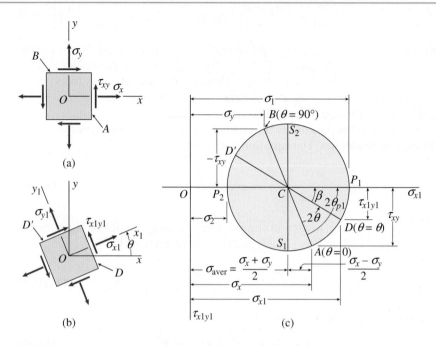

(a)

(b)

(c)

is sufficient to construct the circle. Then, with the circle drawn, determine the stresses σ_{x1}, σ_{y1}, and τ_{x1y1} acting on an inclined element (Fig. 11-16b). You can also obtain the principal stresses and maximum shear stresses from the circle.

With σ_x, σ_y, and τ_{xy} known, the **procedure for constructing Mohr's circle** is as follows (see Fig. 11-16c).

1. Draw a set of coordinate axes with σ_{x1} as abscissa (positive to the right) and τ_{x1y1} as ordinate (positive downward).

2. Locate the center C of the circle at the point having coordinates and $\sigma_{x1} = \sigma_{aver}$ and $\tau_{x1y1} = 0$ [see Eqs. (11-31a and b)].

3. Locate point A, representing the stress conditions on the x face of the element shown in Fig. 11-16a, by plotting its coordinates $\sigma_{x1} = \sigma_x$ and $\tau_{x1y1} = \tau_{xy}$. Note that point A on the circle corresponds to $\theta = 0$. Also, note that the x face of the element (Fig. 11-16a) is labeled "A" to show its correspondence with point A on the circle.

4. Locate point B, representing the stress conditions on the y face of the element shown in Fig. 11-16a, by plotting its coordinates $\sigma_{x1} = \sigma_y$ and $\tau_{x1y1} = -\tau_{xy}$. Note that point B on the circle corresponds to $\theta = 90°$. In addition, the y face of the element (Fig. 11-16a) is labeled "B" to show its correspondence with point B on the circle.

5. Draw a line from point A to point B. This line is a diameter of the circle and passes through the center C. Points A and B, representing the stresses on planes at $90°$ to each other (Fig. 11-16a), are at opposite ends of the diameter (and therefore are $180°$ apart on the circle).

6. Using point C as the center, draw Mohr's circle through points A and B. The circle drawn in this manner has radius R (Eq. 11-31b), as shown in the next paragraph.

With the circle drawn, verify by geometry that lines CA and CB are radii and have lengths equal to R. Note that the abscissas of points C and

A are $(\sigma_x + \sigma_y)/2$ and σ_x, respectively. The difference in these abscissas is $(\sigma_x - \sigma_y)/2$, as dimensioned in the figure. Also, the ordinate to point A is τ_{xy}. Therefore, line CA is the hypotenuse of a right triangle having one side of length $(\sigma_x - \sigma_y)/2$ and the other side of length τ_{xy}. Taking the square root of the sum of the squares of these two sides gives the radius R:

$$R = \sqrt{\left(\frac{\sigma_x - \sigma_y}{2}\right)^2 + \tau_{xy}^2}$$

which is the same as Eq. (11-31b). By a similar procedure, you can show that the length of line CB is also equal to the radius R of the circle.

Stresses on an Inclined Element

Now consider the stresses σ_{x1}, σ_{y1}, and τ_{x1y1} acting on the faces of a plane-stress element oriented at an angle θ from the x axis (Fig. 11-16b). If the angle θ is known, these stresses can be determined from Mohr's circle. The procedure is as follows.

On the circle (Fig. 11-16c), measure an angle 2θ counterclockwise from radius CA, because point A corresponds to $\theta = 0$ and is the reference point from which angles are measured. The angle 2θ locates point D on the circle, which (as shown in the next paragraph) has coordinates σ_{x1} and τ_{x1y1}. Therefore, point D represents the stresses on the x_1 face of the element of Fig. 11-16b. Consequently, this face of the element is labeled "D" in Fig. 11-16b.

Note that an angle 2θ on Mohr's circle corresponds to an angle θ on a stress element. For instance, point D on the circle is at an angle 2θ from point A, but the x_1 face of the element shown in Fig. 11-16b (the face labeled "D") is at an angle θ from the x face of the element shown in Fig. 11-16a (the face labeled "A"). Similarly, points A and B are 180° apart on the circle, but the corresponding faces of the element (Fig. 11-16a) are 90° apart.

To show that the coordinates σ_{x1} and τ_{x1y1} of point D on the circle are indeed given by the stress-transformation equations given in Eqs. (11-4a and b), again use the geometry of the circle. Let β be the angle between the radial line CD and the σ_{x1} axis. Then, from the geometry of the figure, the following expressions are obtained for the coordinates of point D:

$$\sigma_{x1} = \frac{\sigma_x + \sigma_y}{2} + R\cos\beta \qquad \tau_{x1y1} = R\sin\beta \qquad \textbf{(11-33a,b)}$$

Note that the angle between the radius CA and the horizontal axis is $2\theta + \beta$, so

$$\cos(2\theta + \beta) = \frac{\sigma_x - \sigma_y}{2R} \qquad \sin(2\theta + \beta) = \frac{\tau_{xy}}{R}$$

Expanding the cosine and sine expressions (see Appendix A) gives

$$\cos 2\theta \cos\beta - \sin 2\theta \sin\beta = \frac{\sigma_x - \sigma_y}{2R} \qquad \textbf{(11-34a)}$$

$$\sin 2\theta \cos\beta + \cos 2\theta \sin\beta = \frac{\tau_{xy}}{R} \qquad \textbf{(11-34b)}$$

Multiply the first of these equations by $\cos 2\theta$ and the second by $\sin 2\theta$ and then add to obtain

$$\cos\beta = \frac{1}{R}\left(\frac{\sigma_x - \sigma_y}{2}\cos 2\theta + \tau_{xy}\sin 2\theta\right) \qquad \textbf{(11-34c)}$$

FIGURE 11-16 (Repeated)

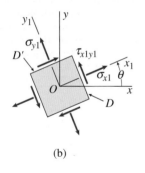

(b)

Also, multiply Eq. (11-34a) by sin 2θ and Eq. (11-34b) by cos 2θ and then subtract to get

$$\sin \beta = \frac{1}{R}\left(\frac{\sigma_x - \sigma_y}{2}\sin 2\theta + \tau_{xy}\cos 2\theta\right) \tag{11-34d}$$

Substitute these expressions for cos β and sin β into Eqs. (11-33a and b) to obtain the stress-transformation equations for σ_{x1} and τ_{x1y1} [Eqs. (11-4a and b)]. Thus, point D on Mohr's circle, defined by the angle 2θ, represents the stress conditions on the x_1 face of the stress element defined by the angle θ (Fig. 11-16b).

Point D', which is diametrically opposite point D on the circle, is located by an angle 2θ (measured from line CA) that is 180° greater than the angle 2θ to point D. Therefore, point D' on the circle represents the stresses on a face of the stress element (Fig. 11-16b) at 90° from the face represented by point D. Thus, point D' on the circle gives the stresses σ_{y1} and $-\tau_{x1y1}$ on the y_1 face of the stress element (the face labeled "D'" in Fig. 11-16b).

The stresses represented by points on Mohr's circle are related to the stresses acting on an element. The stresses on an inclined plane defined by the angle θ (Fig. 11-16b) are found on the circle at the point where the angle from the reference point (point A) is 2θ. Thus, as the $x_1 y_1$ axes rotate counterclockwise through an angle θ (Fig. 11-16b), the point on Mohr's circle corresponding to the x_1 face moves counterclockwise through an angle 2θ. Similarly, if the axes rotate clockwise through an angle, the point on the circle moves clockwise through an angle twice as large.

Principal Stresses

The determination of principal stresses is probably the most important application of Mohr's circle. Note that as you move around Mohr's circle (Fig. 11-16c), you encounter point P_1 where the normal stress reaches its algebraically largest value and the shear stress is zero. Hence, point P_1 represents a **principal stress** and a **principal plane**. The abscissa σ_1 of point P_1 gives the algebraically larger principal stress, and its angle $2\theta_{p1}$ from the reference point A (where $\theta = 0$) gives the orientation of the principal plane. The other principal plane, associated with the algebraically smallest normal stress, is represented by point P_2, which is diametrically opposite point P_1.

From the geometry of the circle, the algebraically larger principal stress is

$$\sigma_1 = OC + \overline{CP_1} = \frac{\sigma_x + \sigma_y}{2} + R$$

which, upon substitution of the expression for R [Eq. (11-31b)], agrees with the earlier equation for this stress [Eq. (11-14)]. In a similar manner, you can verify the expression for the algebraically smaller principal stress σ_2.

The principal angle θ_{p1} between the x axis (Fig. 11-16a) and the plane of the algebraically larger principal stress is one-half the angle $2\theta_{p1}$, which is the angle on Mohr's circle between radii CA and CP_1. The cosine and sine of the angle $2\theta_{p1}$ can be obtained by inspection from the circle:

$$\cos 2\theta_{p1} = \frac{\sigma_x - \sigma_y}{2R} \qquad \sin 2\theta_{p1} = \frac{\tau_{xy}}{R}$$

These equations agree with Eqs. (11-18a and b), and once again the geometry of the circle matches the equations derived earlier. On the circle, the angle $2\theta_{p2}$ to the other principal point (point P_2) is 180° larger than $2\theta_{p1}$; hence, $\theta_{p2} = \theta_{p1} + 90°$, as expected.

Maximum Shear Stresses

Points S_1 and S_2, representing the planes of maximum positive and maximum negative shear stresses, respectively, are located at the bottom and top of Mohr's circle (Fig. 11-16c). These points are at angles $2\theta = 90°$ from points P_1 and P_2, which agrees with the fact that the planes of maximum shear stress are oriented at 45° to the principal planes.

The maximum shear stresses are numerically equal to the radius R of the circle [compare Eq. (11-31b) for R with Eq. (11-25) for τ_{max}]. Also, the normal stresses on the planes of maximum shear stress are equal to the abscissa of point C, which is the average normal stress σ_{aver} [see Eq. (11-31a)].

Alternative Sign Convention for Shear Stresses

An alternative sign convention for shear stresses is sometimes used when constructing Mohr's circle. In this convention, the direction of a shear stress acting on an element of the material is indicated by the sense of the rotation that it tends to produce (Figs. 11-17a and b). If the shear stress τ tends to rotate the stress element clockwise, it is called a *clockwise shear stress*, and if it tends to rotate it counterclockwise, it is called a *counterclockwise stress*. Then, when constructing Mohr's circle, clockwise shear stresses are plotted upward and counterclockwise shear stresses are plotted downward (Fig. 11-17c).

It is important to realize that *the alternative sign convention produces a circle that is identical to the circle already described* (Fig. 11-16c). The reason is that a positive shear stress τ_{x1y1} is also a counterclockwise shear stress, and both are plotted downward. Also, a negative shear stress τ_{x1y1} is a clockwise shear stress, and both are plotted upward.

Thus, the alternative sign convention merely provides a different point of view. Instead of thinking of the vertical axis as having negative shear stresses plotted upward and positive shear stresses plotted downward (which is a bit awkward), you can think of the vertical axis as having clockwise shear stresses plotted upward and counterclockwise shear stresses plotted downward (Fig. 11-17c).

General Comments about the Circle

From the preceding discussions in this section, it is apparent that the stresses acting on any inclined plane, as well as the principal stresses and maximum shear stresses, can be found from Mohr's circle. However, only rotations of axes in the x-y plane (that is, rotations about the z axis) are considered; therefore, *all stresses on Mohr's circle are in-plane stresses.*

For convenience, the circle of Fig. 11-16 was drawn with σ_x, σ_y, and τ_{xy} as positive stresses, but the same procedures may be followed if one or more of the stresses is negative. If one of the normal stresses is negative, part or all of the circle will be located to the left of the origin, as illustrated in Example 11-6 that follows.

Point A in Fig. 11-16c, representing the stresses on the plane $\theta = 0$, may be situated anywhere around the circle. However, the angle 2θ is always measured counterclockwise from the radius CA, regardless of where point A is located.

In the special cases of *uniaxial stress*, *biaxial stress*, and *pure shear*, the construction of Mohr's circle is simpler than in the general case of plane stress. These special cases are illustrated in Example 11-4 and in Problems 11.4-1 through 11.4-9.

Besides using Mohr's circle to obtain the stresses on inclined planes when the stresses on the x and y planes are known, you can also use the circle in the opposite manner. If you know the stresses σ_{x1}, σ_{y1}, and τ_{x1y1} acting on an inclined element oriented at a known angle θ, you can easily construct the circle and

FIGURE 11-17

Alternative sign convention for shear stresses: (a) clockwise shear stress, (b) counterclockwise shear stress, and (c) axes for Mohr's circle (Note that clockwise shear stresses are plotted upward and counterclockwise shear stresses are plotted downward)

(a)

(b)

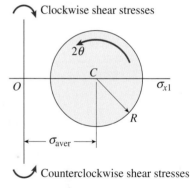

(c)

determine the stresses σ_x, σ_y, and τ_{xy} for the angle $\theta = 0$. The procedure is to locate points D and D' from the known stresses and then draw the circle using line DD' as a diameter. Measure the angle 2θ in a negative sense from radius CD to locate point A, corresponding to the x face of the element. Then locate point B by constructing a diameter from A. Finally, determine the coordinates of points A and B, and thereby obtain the stresses acting on the element for which $\theta = 0$.

If desired, construct Mohr's circle to scale and measure values of stress from the drawing. However, it is usually preferable to obtain the stresses by numerical calculations—either directly from the various equations or by using trigonometry and the geometry of the circle.

Mohr's circle makes it possible to visualize the relationships between stresses acting on planes at various angles, and it also serves as a simple memory device for calculating stresses. Although many graphical techniques are no longer used in engineering work, Mohr's circle remains valuable because it provides a simple and clear picture of an otherwise complicated analysis.

Mohr's circle is also applicable to the transformations for plane strain and moments of inertia of plane areas because these quantities follow the same transformation laws as do stresses (see Section 11.7, and Sections 5.7 and 5.8).

Example 11-4

At a point on the surface of a hydraulic ram on a piece of construction equipment (Fig. 11-18a), the material is subjected to biaxial stresses $\sigma_x = 90$ MPa and $\sigma_y = 20$ MPa, as shown on the stress element of Fig. 11-18b. Using Mohr's circle, determine the stresses acting on an element inclined at an angle $\theta = 30°$. (Consider only the in-plane stresses, and show the results on a sketch of a properly oriented element.)

Solution:

Use the four-step problem-solving approach. Combine steps as needed for an efficient solution.

1, 2. *Conceptualize, Categorize*:

Construction of Mohr's circle: Begin by setting up the axes for the normal and shear stresses, with σ_{x1} positive to the right and τ_{x1y1} positive downward, as shown in Fig. 11-18c. Then place the center C of the circle on the σ_{x1} axis at the point where the stress equals the average normal stress [Eq. (11-31a)]:

$$\sigma_{\text{aver}} = \frac{\sigma_x + \sigma_y}{2} = \frac{90 \text{ MPa} + 20 \text{ MPa}}{2} = 55 \text{ MPa}$$

Point A, representing the stresses on the x face of the element ($\theta = 0$), has coordinates

$$\sigma_{x1} = 90 \text{ MPa} \qquad \tau_{x1y1} = 0$$

Similarly, the coordinates of point B, representing the stresses on the y face ($\theta = 90°$), are

$$\sigma_{x1} = 20 \text{ MPa} \qquad \tau_{x1y1} = 0$$

FIGURE 11-18

Example 11-4: (a) Hydraulic cylinder on construction equipment, (b) element on hydraulic ram in plane stress, and (c) the corresponding Mohr's circle (*Note:* All stresses on the circle have units of MPa)

(a)

(b)

(c)

Now draw the circle through points A and B with the center at C and radius R [see Eq. (11-31b)] equal to

$$R = \sqrt{\left(\frac{\sigma_x - \sigma_y}{2}\right)^2 + \tau_{xy}^2} = \sqrt{\left(\frac{90 \text{ MPa} - 20 \text{ MPa}}{2}\right)^2 + 0}$$
$$= 35 \text{ MPa}$$

3. *Analyze:*

   Stresses on an element inclined at $\theta = 30°$: The stresses acting on a plane oriented at an angle $\theta = 30°$ are given by the coordinates of point D, which is at an angle $2\theta = 60°$ from point A (Fig. 11-18c). By inspection of the circle, note that the coordinates of point D are

   (Point D) $\sigma_{x1} = \sigma_{aver} + R \cos 60°$

   $\qquad\qquad = 55 \text{ MPa} + (35 \text{ MPa})(\cos 60°) = 72.5 \text{ MPa}$ ◄

   $\qquad \tau_{x1y1} = -R \sin 60° = -(35 \text{ MPa})(\sin 60°)$

   $\qquad\qquad = -30.3 \text{ MPa}$ ◄

   In a similar manner, find the stresses represented by point D', which corresponds to an angle $\theta = 120°$ (or $2\theta = 240°$):

   (Point D') $\sigma_{x1} = \sigma_{aver} - R \cos 60°$

   $\qquad\qquad = 55 \text{ MPa} - (35 \text{ MPa})(\cos 60°) = 37.5 \text{ MPa}$ ◄

   $\qquad \tau_{x1y1} = R \sin 60° = (35 \text{ MPa})(\sin 60°) = 30.3 \text{ MPa}$

4. *Finalize:* These results are shown in Fig. 11-19 on a sketch of an element oriented at an angle $\theta = 30°$ with all stresses shown in their true directions. Note that the sum of the normal stresses on the inclined element is equal to $\sigma_x + \sigma_y$ or 110 MPa.

FIGURE 11-19

Example 11-4: Stresses acting on an element oriented at an angle $\theta = 30°$

Example 11-5

FIGURE 11-20

Example 11-5: (a) Oil drilling pumps, (b) element in plane stress, and (c) the corresponding Mohr's circle (*Note:* All stresses on the circle have units of psi)

(a)

(b)

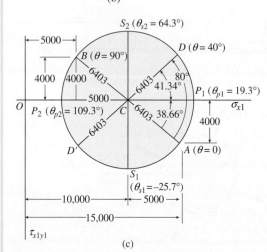

(c)

An element in plane stress on the surface of an oil-drilling pump arm (Fig. 11-20a) is subjected to stresses $\sigma_x = 15,000$ psi, $\sigma_y = 5000$ psi, and $\tau_{xy} = 4000$ psi, as shown in Fig. 11-20b.

Using Mohr's circle, determine (a) the stresses acting on an element inclined at an angle $\theta = 40°$, (b) the principal stresses, and (c) the maximum shear stresses. (Consider only the in-plane stresses, and show all results on sketches of properly oriented elements.)

Solution:

Use the four-step problem-solving approach. Combine steps as needed for an efficient solution.

1, 2. *Conceptualize, Categorize:*

Construction of Mohr's circle: The first step in the solution is to set up the axes for Mohr's circle, with σ_{x1} positive to the right and τ_{x1y1} positive downward (Fig. 11-20c). The center C of the circle is located on the σ_{x1} axis at the point where σ_{x1} equals the average normal stress [Eq. (11-31a)]:

$$\sigma_{aver} = \frac{\sigma_x + \sigma_y}{2} = \frac{15,000 \text{ psi} + 5000 \text{ psi}}{2} = 10,000 \text{ psi}$$

Point A, representing the stresses on the x face of the element ($\theta = 0$), has coordinates

$$\sigma_{x1} = 15,000 \text{ psi} \qquad \tau_{x1y1} = 4000 \text{ psi}$$

Similarly, the coordinates of point B, representing the stresses on the y face ($\theta = 90°$), are

$$\sigma_{x1} = 5000 \text{ psi} \qquad \tau_{x1y1} = -4000 \text{ psi}$$

The circle is now drawn through points A and B with center at C. The radius of the circle, from Eq. (11-31b), is

$$R = \sqrt{\left(\frac{\sigma_x - \sigma_y}{2}\right)^2 + \tau_{xy}^2}$$

$$= \sqrt{\left(\frac{15,000 \text{ psi} - 5000 \text{ psi}}{2}\right)^2 + (4000 \text{ psi})^2} = 6403 \text{ psi}$$

Part (a): Stresses on an element inclined at $\theta = 40°$.

3. *Analyze:*

The stresses acting on a plane oriented at an angle $\theta = 40°$ are given by the coordinates of point D, which is at an angle $2\theta = 80°$ from point A (Fig. 11-20c).

FIGURE 11-21

Example 11-5: (a) Stresses acting on an element oriented at $\theta = 40°$, (b) principal stresses, and (c) maximum shear stresses

(a)

(b)

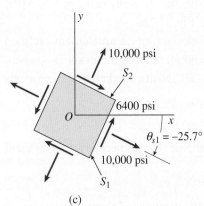

(c)

To evaluate these coordinates, you need to know the angle between line CD and the σ_{x1} axis (that is, angle DCP_1), which in turn requires that you know the angle between line CA and the σ_{x1} axis (angle ACP_1). These angles are found from the geometry of the circle, as

$$\tan \overline{ACP_1} = \frac{4000 \text{ psi}}{5000 \text{ psi}} = 0.8 \quad \overline{ACP_1} = 38.66°$$

$$\overline{DCP_1} = 80° - \overline{ACP_1} = 80° - 38.66° = 41.34°$$

Knowing these angles, determine the coordinates of point D directly from Fig. 11-21c:

(Point D) $\quad \sigma_{x1} = 10,000 \text{ psi} + (6403 \text{ psi})(\cos 41.34°)$

$$= 14,810 \text{ psi}$$

$$\tau_{x1y1} = -(6403 \text{ psi})(\sin 41.34°) = -4230 \text{ psi}$$

In an analogous manner, find the stresses represented by point D', which corresponds to a plane inclined at an angle $\theta = 130°$ (or $2\theta = 260°$):

(Point D') $\quad \sigma_{x1} = 10,000 \text{ psi} - (6403 \text{ psi})(\cos 41.34°)$

$$= 5190 \text{ psi}$$

$$\tau_{x1y1} = (6403 \text{ psi})(\sin 41.34°) = 4230 \text{ psi}$$

4. *Finalize*: These stresses are shown in Fig. 11-21a on a sketch of an element oriented at an angle $\theta = 40°$ (all stresses are shown in their true directions). Also, note that the sum of the normal stresses is equal to $\sigma_x + \sigma_y$ or 20,000 psi.

Part (b): Principal stresses.

3. *Analyze*: The principal stresses are represented by points P_1 and P_2 on Mohr's circle (Fig. 11-20c). The algebraically larger principal stress (point P_1) is

$$\sigma_1 = 10,000 \text{ psi} + 6400 \text{ psi} = 16,400 \text{ psi}$$

as seen by inspection of the circle. The angle $2\theta_{p1}$ to point P_1 from point A is the angle ACP_1 on the circle, that is,

$$\overline{ACP_1} = 2\theta_{p1} = 38.66° \qquad \theta_{p1} = 19.3°$$

Thus, the plane of the algebraically larger principal stress is oriented at an angle $\theta_{p1} = 19.3°$, as shown in Fig. 11-21b.

The algebraically smaller principal stress (represented by point P_2) is obtained from the circle in a similar manner:

$$\sigma_2 = 10,000 \text{ psi} - 6400 \text{ psi} = 3600 \text{ psi}$$

4. *Finalize*: The angle $2\theta_{p2}$ to point P_2 on the circle is $38.66° + 180° = 218.66°$; thus, the second principal plane is defined by the angle $\theta_{p2} = 109.3°$.

The principal stresses and principal planes are shown in Fig. 11-21b. Note that the sum of the normal stresses is equal to 20,000 psi.

Part (c): Maximum shear stresses.

3. *Analyze*: The maximum shear stresses are represented by points S_1 and S_2 on Mohr's circle; therefore, the maximum in-plane shear stress (equal to the radius of the circle) is

$$\tau_{max} = 6400 \text{ psi}$$

The angle ACS_1 from point A to point S_1 is $90° - 38.66° = 51.34°$; therefore, the angle $2\theta_{s1}$ for point S_1 is

$$2\theta_{s1} = -51.34°$$

4. *Finalize*: This angle is negative because it is measured clockwise on the circle. The corresponding angle θ_{s1} to the plane of the maximum positive shear stress is one-half that value, or $\theta_{s1} = -25.7°$, as shown in Figs. 11-20c and 11-21c. The maximum negative shear stress (point S_2 on the circle) has the same numerical value as the maximum positive stress (6400 psi).

The normal stresses acting on the planes of maximum shear stress are equal to σ_{aver}, which is the abscissa of the center C of the circle (10,000 psi). These stresses are also shown in Fig. 11-21c. Note that the planes of maximum shear stress are oriented at 45° to the principal planes.

Example 11-6

At a point on the surface of a metal-working lathe, the stresses are $\sigma_x = -50$ MPa, $\sigma_y = 10$ MPa, and $\tau_{xy} = -40$ MPa, as shown in Fig. 11-22a.

Using Mohr's circle, determine (a) the stresses acting on an element inclined at an angle $\theta = 45°$, (b) the principal stresses, and (c) the maximum shear stresses. (Consider only the in-plane stresses, and show all results on sketches of properly oriented elements.)

Solution:

Use the four-step problem-solving approach. Combine steps as needed for an efficient solution.

1, 2. *Conceptualize, Categorize*:

Construction of Mohr's circle: The axes for the normal and shear stresses are shown in Fig. 11-22b, with σ_{x1} positive to the right and τ_{x1y1} positive downward. The center C of the circle is located on the σ_{x1} axis at the point where the stress equals the average normal stress [Eq. (11-31a)]:

$$\sigma_{aver} = \frac{\sigma_x + \sigma_y}{2} = \frac{-50 \text{ MPa} + 10 \text{ MPa}}{2} = -20 \text{ MPa}$$

FIGURE 11-22

Example 11-6: (a) Element in plane stress and (b) the corresponding Mohr's circle (*Note*: All stresses on the circle have units of MPa)

(a)

(b)

Point A, representing the stresses on the x face of the element ($\theta = 0$), has coordinates

$$\sigma_{x1} = -50 \text{ MPa} \qquad \tau_{x1y1} = -40 \text{ MPa}$$

Similarly, the coordinates of point B, representing the stresses on the y face ($\theta = 90°$), are

$$\sigma_{x1} = 10 \text{ MPa} \qquad \tau_{x1y1} = 40 \text{ MPa}$$

The circle is now drawn through points A and B with center at C and radius R equal to [from Eq. (11-31b)]

$$R = \sqrt{\left(\frac{\sigma_x - \sigma_y}{2}\right)^2 + \tau_{xy}^2}$$

$$= \sqrt{\left(\frac{-50 \text{ MPa} - 10 \text{ MPa}}{2}\right)^2 + (-40 \text{ MPa})^2} = 50 \text{ MPa}$$

Part (a): Stresses on an element inclined at $\theta = 45°$.

3. *Analyze*: The stresses acting on a plane oriented at an angle $\theta = 45°$ are given by the coordinates of point D, which is at an angle $2\theta = 90°$ from point A (Fig. 11-22b). To evaluate these coordinates, you need to know the angle between line CD and the negative σ_{x1} axis (that is, angle DCP_2), which in turn requires that you know the angle between line CA and the negative σ_{x1} axis (angle ACP_2). These angles are found from the geometry of the circle

$$\tan \overline{ACP}_2 = \frac{40 \text{ MPa}}{30 \text{ MPa}} = \frac{4}{3} \qquad \overline{ACP}_2 = 53.13°$$

$$\overline{DCP}_2 = 90° - \overline{ACP}_2 = 90° - 53.13° = 36.87°$$

Knowing these angles, obtain the coordinates of point D directly from Fig. 11-22b:

$$\text{(Point } D) \quad \sigma_{x1} = -20 \text{ MPa} - (50 \text{ MPa})(\cos 36.87°) = -60 \text{ MPa}$$

$$\tau_{x1y1} = (50 \text{ MPa})(\sin 36.87°) = 30 \text{ MPa}$$

In an analogous manner, find the stresses represented by point D', which corresponds to a plane inclined at an angle $\theta = 135°$ (or $2\theta = 270°$):

$$\text{(Point } D') \quad \sigma_{x1} = -20 \text{ MPa} + (50 \text{ MPa})(\cos 36.87°) = 20 \text{ MPa}$$

$$\tau_{x1y1} = (-50 \text{ MPa})(\sin 36.87°) = -30 \text{ MPa}$$

4. *Finalize*: These stresses are shown in Fig. 11-23a on a sketch of an element oriented at an angle $\theta = 45°$ (all stresses are shown in their true directions). Also, note that the sum of the normal stresses is equal to $\sigma_x + \sigma_y$ or -40 MPa.

FIGURE 11-23

Example 11-6: (a) Stresses acting on an element oriented at $\theta = 45°$, (b) principal stresses, and (c) maximum shear stresses

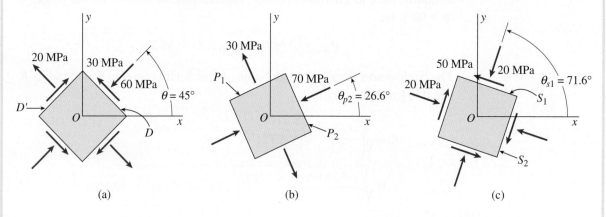

(a)   (b)   (c)

Part (b): Principal stresses.

3. *Analyze*: The principal stresses are represented by points P_1 and P_2 on Mohr's circle (Fig. 11-22b). The algebraically larger principal stress (represented by point P_1) is

$$\sigma_1 = -20 \text{ MPa} + 50 \text{ MPa} = 30 \text{ MPa}$$

as seen by inspection of the circle. The angle $2\theta_{p1}$ to point P_1 from point A is the angle ACP_1 measured counterclockwise on the circle, that is,

$$\overline{ACP_1} = 2\theta_{p1} = 53.13° + 180° = 233.13° \quad \theta_{p1} = 116.6°$$

Thus, the plane of the algebraically larger principal stress is oriented at an angle $\theta_{p1} = 116.16°$.

The algebraically smaller principal stress (point P_2) is obtained from the circle in a similar manner:

$$\sigma_2 = -20 \text{ MPa} - 50 \text{ MPa} = -70 \text{ MPa} \qquad \leftarrow$$

The angle $2\theta_{p2}$ to point P_2 on the circle is 53.13°; thus, the second principal plane is defined by the angle $\theta_{p2} = 26.6°$.

4. *Finalize*: The principal stresses and principal planes are shown in Fig. 11-23b. Note that the sum of the normal stresses is equal to $\sigma_x + \sigma_y$ or −40 MPa.

Part (c): Maximum shear stresses.

3. *Analyze*: The maximum positive and negative shear stresses are represented by points S_1 and S_2 on Mohr's circle (Fig. 11-22b). Their magnitudes, equal to the radius of the circle, are

$$\tau_{\max} = 50 \text{ MPa} \qquad \leftarrow$$

The angle ACS_1 from point A to point S_1 is $90° + 53.13° = 143.13°$; therefore, the angle $2\theta_{s1}$ for point S_1 is

$$2\theta_{s1} = 143.13°$$

4. *Finalize*: The corresponding angle θ_{s1} to the plane of the maximum positive shear stress is one-half that value, or $\theta_{s1} = 71.6°$, as shown in Fig. 11-23c. The maximum negative shear stress (point S_2 on the circle) has the same numerical value as the positive stress (50 MPa).

   The normal stresses acting on the planes of maximum shear stress are equal to σ_{aver}, which is the coordinate of the center C of the circle (−20 MPa). These stresses are also shown in Fig. 11-23c. Note that the planes of maximum shear stress are oriented at 45° to the principal planes.

11.5 Hooke's Law for Plane Stress

FIGURE 11-24

Element of material in plane stress ($\sigma_z = 0$)

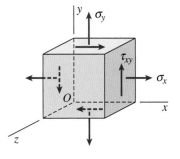

The stresses acting on inclined planes when the material is subjected to plane stress (Fig. 11-24) were discussed in Sections 11.2, 11.3, and 11.4. The stress-transformation equations derived in those discussions were obtained solely from equilibrium, so the properties of the materials were not needed. Now, in this section, the *strains* in the material are investigated, which means that the material properties must be considered. However, the discussion is limited to materials that meet two important conditions. First, *the material is uniform throughout the body and has the same properties in all directions* (homogeneous and isotropic material). Second, *the material follows Hooke's law* (linearly elastic material). Under these conditions, the relationships between the stresses and strains in the body are readily obtained.

FIGURE 11-25

Element of material subjected to normal strains ε_x, ε_y, and ε_z

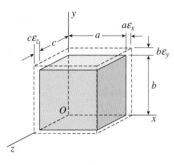

Begin by considering the **normal strains** ε_x, ε_y, and ε_z in plane stress. The effects of these strains are pictured in Fig. 11-25, which shows the changes in dimensions of a small element having edges of lengths a, b, and c. All three strains are shown positive (elongation) in the figure. The strains can be expressed in terms of the stresses (Fig. 11-24) by superimposing the effects of the individual stresses.

For instance, the strain ε_x in the x direction due to the stress σ_x is equal to σ_x/E, where E is the modulus of elasticity. Also, the strain ε_x due to the stress σ_y is equal to $-\nu\sigma_y/E$, where ν is Poisson's ratio (see Section 7.5). Of course, the shear stress τ_{xy} produces no normal strains in the x, y, or z directions. Thus, the resultant strain in the x direction is

$$\varepsilon_x = \frac{1}{E}(\sigma_x - \nu\sigma_y) \tag{11-35a}$$

In a similar manner, the strains in the y and z directions are obtained:

$$\varepsilon_y = \frac{1}{E}(\sigma_y - \nu\sigma_x) \qquad \varepsilon_z = -\frac{\nu}{E}(\sigma_x + \sigma_y) \tag{11-35b,c}$$

These equations may be used to find the normal strains (in plane stress) when the stresses are known.

The shear stress τ_{xy} (Fig. 11-24) causes a distortion of the element such that each z face becomes a rhombus (Fig. 11-26). The **shear strain** γ_{xy} is the *decrease* in angle between the x and y faces of the element and is related to the shear stress by Hooke's law in shear, as

FIGURE 11-26

Shear strain γ_{xy}

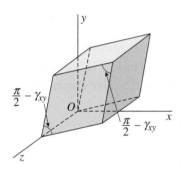

$$\gamma_{xy} = \frac{\tau_{xy}}{G} \tag{11-36}$$

where G is the shear modulus of elasticity. Note that the normal stresses σ_x and σ_y have no effect on the shear strain γ_{xy}. Consequently, Eqs. (11-35) and (11-36) give the strains (in plane stress) when all stresses (σ_x, σ_y, and τ_{xy}) act simultaneously.

The first two equations (Eqs. 11-35a and b) give the strains ε_x and ε_y in terms of the stresses. These equations can be solved simultaneously for the stresses in terms of the strains:

$$\sigma_x = \frac{E}{1 - \nu^2}(\varepsilon_x + \nu\varepsilon_y) \qquad \sigma_y = \frac{E}{1 - \nu^2}(\varepsilon_y + \nu\varepsilon_x) \tag{11-37a,b}$$

In addition, the following equation expresses the shear stress in terms of the shear strain:

$$\tau_{xy} = G\gamma_{xy} \tag{11-38}$$

Equations (11-37) and (11-38) may be used to find the stresses (in plane stress) when the strains are known. Of course, the normal stress σ_z in the z direction is equal to zero.

Equations (11-35) through (11-38) are known collectively as **Hooke's law for plane stress**. They contain three material constants (E, G, and ν), but only two are independent because of the relationship

$$G = \frac{E}{2(1 + \nu)} \tag{11-39}$$

that was derived previously in Section 9.6.

Special Cases of Hooke's Law

FIGURE 11-11 (Repeated)

In the special case of **biaxial stress** (Fig. 11-11b), the shear stress $\tau_{xy} = 0$; therefore, Hooke's law for plane stress simplifies to

$$\varepsilon_x = \frac{1}{E}(\sigma_x - \nu\sigma_y) \qquad \varepsilon_y = \frac{1}{E}(\sigma_y - \nu\sigma_x)$$

$$\varepsilon_z = -\frac{\nu}{E}(\sigma_x + \sigma_y) \qquad \textbf{(11-40a,b,c)}$$

$$\sigma_x = \frac{E}{1-\nu^2}(\varepsilon_x + \nu\varepsilon_y) \qquad \sigma_y = \frac{E}{1-\nu^2}(\varepsilon_y + \nu\varepsilon_x) \quad \textbf{(11-41a,b)}$$

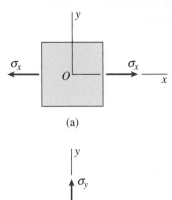

(a)

These equations are the same as Eqs. (11-35) and (11-37) because the effects of normal and shear stresses are independent of each other.

For **uniaxial stress**, with $\sigma_y = 0$ (Fig. 11-11a), the equations of Hooke's law simplify even further:

$$\varepsilon_x = \frac{\sigma_x}{E} \qquad \varepsilon_y = \varepsilon_z = -\frac{\nu\sigma_x}{E} \qquad \sigma_x = E\varepsilon_x \qquad \textbf{(11-42a,b,c)}$$

Finally, consider **pure shear** (Fig. 11-12a), which means that $\sigma_x = \sigma_y = 0$, so

$$\varepsilon_x = \varepsilon_y = \varepsilon_z = 0 \qquad \gamma_{xy} = \frac{\tau_{xy}}{G} \qquad \textbf{(11-43a,b)}$$

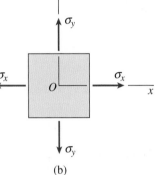

(b)

In all three of these special cases, the normal stress σ_z is equal to zero.

Volume Change

When a solid object undergoes strains, both its dimensions and its volume will change. The change in volume can be determined if the normal strains in three perpendicular directions are known. To show how this is accomplished, consider the small element of material shown in Fig. 11-25. The original element is a rectangular parallelepiped having sides of lengths a, b, and c in the x, y, and z directions, respectively. The strains ε_x, ε_y, and ε_z produce the changes in dimensions shown by the dashed lines. Thus, the increases in the lengths of the sides are $a\varepsilon_x$, $b\varepsilon_y$, and $c\varepsilon_z$.

The original volume of the element is

$$V_0 = abc \qquad \textbf{(11-44a)}$$

and its final volume is

$$V_1 = (a + a\varepsilon_x)(b + b\varepsilon_y)(c + c\varepsilon_z)$$
$$= abc(1 + \varepsilon_x)(1 + \varepsilon_y)(1 + \varepsilon_z) \qquad \textbf{(11-44b)}$$

By referring to Eq. (11-44a), express the final volume of the element [Eq. (11-44b)] in the form

$$V_1 = V_0(1 + \varepsilon_x)(1 + \varepsilon_y)(1 + \varepsilon_z) \qquad \textbf{(11-45a)}$$

Expand the terms on the right-hand side to obtain the following equivalent expression:

$$V_1 = V_0(1 + \varepsilon_x + \varepsilon_y + \varepsilon_z + \varepsilon_x\varepsilon_y + \varepsilon_x\varepsilon_z + \varepsilon_y\varepsilon_z + \varepsilon_x\varepsilon_y\varepsilon_z) \quad \textbf{(11-45b)}$$

The preceding equations for V_1 are valid for both large and small strains.

Limiting the discussion to structures having only very *small strains* (as is usually the case), you can disregard the terms in Eq. (11-45b) that consist of products of small strains. Such products are themselves small in comparison to the individual strains ε_x, ε_y, and ε_z. Then the expression for the final volume simplifies to

$$V_1 = V_0(1 + \varepsilon_x + \varepsilon_y + \varepsilon_z) \tag{11-46}$$

and the **volume change** is

$$\Delta V = V_1 - V_0 = V_0(\varepsilon_x + \varepsilon_y + \varepsilon_z) \tag{11-47}$$

This expression can be used for any volume of material *provided the strains are small and remain constant throughout the volume*. Note also that the material does not have to follow Hooke's law. Furthermore, the expression is not limited to plane stress, but is valid for any stress conditions. (As a final note, shear strains produce no change in volume.)

The **unit volume change** e, also known as the **dilatation**, is defined as the change in volume divided by the original volume; thus,

$$e = \frac{\Delta V}{V_0} = \varepsilon_x + \varepsilon_y + \varepsilon_z \tag{11-48}$$

By applying this equation to a differential element of volume and then integrating, you can obtain the change in volume of a body even when the normal strains vary throughout the body.

The preceding equations for volume changes apply to both tensile and compressive strains, inasmuch as the strains ε_x, ε_y, and ε_z are algebraic quantities (positive for elongation and negative for shortening). With this sign convention, positive values for ΔV and e represent increases in volume, and negative values represent decreases.

Now return to materials that follow **Hooke's law** and are subjected only to **plane stress** (Fig. 11-24). In this case, the strains ε_x, ε_y, and ε_z are given by Eqs. (11-35a, b, and c). Substitute those relationships into Eq. (11-48) to obtain the unit volume change in terms of stresses:

$$e = \frac{\Delta V}{V_0} = \frac{1 - 2\nu}{E}(\sigma_x + \sigma_y) \tag{11-49}$$

Note that this equation also applies to **biaxial stress**.

In the case of a prismatic bar in tension, that is, **uniaxial stress**, Eq. (11-49) simplifies to

$$e = \frac{\Delta V}{V_0} = \frac{\sigma_x}{E}(1 - 2\nu) \tag{11-50}$$

From this equation, note that the maximum possible value of Poisson's ratio for common materials is 0.5, because a larger value means that the volume decreases when the material is in tension, which is contrary to ordinary physical behavior.[3]

FIGURE 11-24 (Repeated)

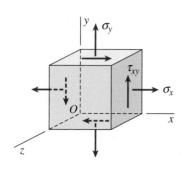

---

[3]See discussion in Section 7.5 about materials that have a negative Poisson's ratio.

Example 11-7

FIGURE 11-27

Example 11-7: Rectangular aluminum plate with strain gages *A* and *B*

Strain gages *A* and *B* (oriented in the *x* and *y* directions, respectively) are attached to a rectangular aluminum plate with a thickness of $t = 0.275$ in. The plate is subjected to uniform normal stresses σ_x and σ_y, as shown in Fig. 11-27, and the gage readings for normal strains are $\varepsilon_x = -0.00075$ (shortening, gage *A*) and $\varepsilon_y = 0.00125$ (elongation, gage *B*). The modulus of elasticity is $E = 10,600$ ksi, and Poisson's ratio is $\nu = 0.33$. Find the stresses σ_x and σ_y and the change Δt in the thickness of the plate. Also, find the unit volume change (or dilatation) *e* for the plate.

Solution:

Use the four-step problem-solving approach. Combine steps as needed for an efficient solution.

1, 2, 3. *Conceptualize, Categorize, Analyze*: For a plate in *biaxial stress*, use Eqs. (11-41a and b) to find the normal stresses σ_x and σ_y in the *x* and *y* directions, respectively, based upon the measured normal strains ε_x and ε_y:

$$\sigma_x = \frac{E}{1 - \nu^2}(\varepsilon_x + \nu\varepsilon_y) = \frac{10,600 \text{ ksi}}{1 - 0.33^2}[-0.00075 + (0.33)(0.00125)]$$
$$= -4.01 \text{ ksi}$$

$$\sigma_y = \frac{E}{1 - \nu^2}(\varepsilon_y + \nu\varepsilon_x) = \frac{10,600 \text{ ksi}}{1 - 0.33^2}[0.00125 + (0.33)(-0.00075)]$$
$$= 11.93 \text{ ksi}$$

The normal strain in the *z* direction is then computed from Eq. (11-40c) as

$$\varepsilon_z = \frac{-\nu}{E}(\sigma_x + \sigma_y) = \frac{-(0.33)}{10,600 \text{ ksi}}(-4.01 \text{ ksi} + 11.93 \text{ ksi})$$
$$= -2.466 \times 10^{-4}$$

The change (here a decrease) in the thickness of the plate is then

$$\Delta t = \varepsilon_z t = [-2.466(10^{-4})](0.275 \text{ in.}) = -6.78 \times 10^{-5} \text{ in.}$$

Use Eq. (11-49) to find the dilatation or unit volume change *e* of the plate as

$$e = \frac{1 - 2\nu}{E}(\sigma_x + \sigma_y) = 2.537 \times 10^{-4}$$

4. *Finalize*: The positive sign for *e* means that the plate under biaxial stress increases in volume (although the increase is very small) while decreasing in thickness.

FIGURE 11-28

Element in triaxial stress

(a)

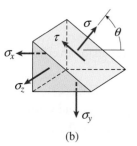

(b)

11.6 Triaxial Stress

An element of material subjected to normal stresses σ_x, σ_y, and σ_z acting in three mutually perpendicular directions is said to be in a state of **triaxial stress** (Fig. 11-28a). Since there are no shear stresses on the x, y, and z faces, the stresses σ_x, σ_y, and σ_z are the *principal stresses* in the material.

If an inclined plane parallel to the z axis is cut through the element (Fig. 11-28b), the only stresses on the inclined face are the normal stress σ and shear stress τ, both of which act parallel to the x-y plane. These stresses are analogous to the stresses σ_{x1} and τ_{x1y1} encountered in earlier discussions of plane stress (see, for instance, Fig. 11-2a). Because the stresses σ and τ (Fig. 11-28b) are found from equations of force equilibrium in the x-y plane, they are independent of the normal stress σ_z. Therefore, use the transformation equations of plane stress, as well as Mohr's circle for plane stress, when determining the stresses σ and τ in triaxial stress. The same general conclusion holds for the normal and shear stresses acting on inclined planes cut through the element parallel to the x and y axes.

Maximum Shear Stresses

From the previous discussions of plane stress, the maximum shear stresses occur on planes oriented at 45° to the principal planes. Therefore, for a material in triaxial stress (Fig. 11-28a), the maximum shear stresses occur on elements oriented at angles of 45° to the x, y, and z axes. For example, consider an element obtained by a 45° rotation about the z axis. The maximum positive and negative shear stresses acting on this element are

$$(\tau_{\max})_z = \pm\frac{\sigma_x - \sigma_y}{2} \qquad \textbf{(11-51a)}$$

Similarly, by rotating about the x and y axes through angles of 45°, the following maximum shear stresses are obtained:

$$(\tau_{\max})_x = \pm\frac{\sigma_y - \sigma_z}{2} \qquad (\tau_{\max})_y = \pm\frac{\sigma_x - \sigma_z}{2} \qquad \textbf{(11-51b,c)}$$

The absolute maximum shear stress is the numerically largest of the stresses determined from Eqs. (11-51a, b, and c). It is equal to one-half the difference between the algebraically largest and algebraically smallest of the three principal stresses.

The stresses acting on elements oriented at various angles to the x, y, and z axes can be visualized with the aid of **Mohr's circles**. For elements oriented by rotations about the z axis, the corresponding circle is labeled A in Fig. 11-29. Note that this circle is drawn for the case in which $\sigma_x > \sigma_y$ and both σ_x and σ_y are tensile stresses.

In a similar manner, construct circles B and C for elements oriented by rotations about the x and y axes, respectively. The radii of the circles represent the maximum shear stresses given by Eqs. (11-51a, b, and c), and the absolute maximum shear stress is equal to the radius of the largest circle. The normal stresses acting on the planes of maximum shear stresses have magnitudes given by the abscissas of the centers of the respective circles.

The preceding discussion of triaxial stress only considered stresses acting on planes obtained by rotating about the x, y, and z axes. Thus, every plane considered is parallel to one of the axes. For instance, the inclined plane of Fig. 11-28b is parallel to the z axis, and its normal is parallel to the x-y plane. Of course, you can also cut through the element in **skew directions**, so that the resulting inclined

FIGURE 11-29

Mohr's circles for an element in triaxial stress

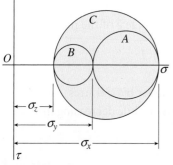

planes are skew to all three coordinate axes. The normal and shear stresses acting on such planes can be obtained by a more complicated three-dimensional analysis. However, the normal stresses acting on skew planes are intermediate in value between the algebraically maximum and minimum principal stresses, and the shear stresses on those planes are smaller (in absolute value) than the absolute maximum shear stress obtained from Eqs. (11-51a, b, and c).

Hooke's Law for Triaxial Stress

If the material follows Hooke's law, you can obtain the relationships between the normal stresses and normal strains by using the same procedure as for plane stress (see Section 11.5). The strains produced by the stresses σ_x, σ_y, and σ_z acting independently are superimposed to obtain the resultant strains. Thus, the following equations pertain to the **strains in triaxial stress**:

$$\varepsilon_x = \frac{\sigma_x}{E} - \frac{\nu}{E}(\sigma_y + \sigma_z) \tag{11-52a}$$

$$\varepsilon_y = \frac{\sigma_y}{E} - \frac{\nu}{E}(\sigma_z + \sigma_x) \tag{11-52b}$$

$$\varepsilon_z = \frac{\sigma_z}{E} - \frac{\nu}{E}(\sigma_x + \sigma_y) \tag{11-52c}$$

In these equations, the standard sign conventions are used; that is, tensile stress σ and extensional strain ε are positive.

The preceding equations can be solved simultaneously for the **stresses in terms of the strains**:

$$\sigma_x = \frac{E}{(1 + \nu)(1 - 2\nu)}[(1 - \nu)\varepsilon_x + \nu(\varepsilon_y + \varepsilon_z)] \tag{11-53a}$$

$$\sigma_y = \frac{E}{(1 + \nu)(1 - 2\nu)}[(1 - \nu)\varepsilon_y + \nu(\varepsilon_z + \varepsilon_x)] \tag{11-53b}$$

$$\sigma_z = \frac{E}{(1 + \nu)(1 - 2\nu)}[(1 - \nu)\varepsilon_z + \nu(\varepsilon_x + \varepsilon_y)] \tag{11-53c}$$

Equations (11-52) and (11-53) represent **Hooke's law for triaxial stress**.

In the special case of **biaxial stress** (Fig. 11-11b), the equations of Hooke's law are obtained by substituting $\sigma_z = 0$ into the preceding equations. The resulting equations reduce to Eqs. (11-40) and (11-41) of Section 11.5.

Unit Volume Change

The unit volume change (or *dilatation*) for an element in triaxial stress is obtained in the same manner as for plane stress (see Section 11.5). If the element is subjected to strains ε_x, ε_y, and ε_z, use Eq. (11-48) for the unit volume change:

$$e = \varepsilon_x + \varepsilon_y + \varepsilon_z \tag{11-54}$$

This equation is valid for any material provided the strains are small.

If Hooke's law holds for the material, substitute for the strains ε_x, ε_y, and ε_z from Eqs. (11-52a, b, and c) and obtain

$$e = \frac{1 - 2\nu}{E}(\sigma_x + \sigma_y + \sigma_z) \tag{11-55}$$

Equations (11-54) and (11-55) give the unit volume change in triaxial stress in terms of the strains and stresses, respectively.

FIGURE 11-30

Element in spherical stress

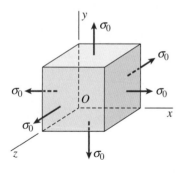

Spherical Stress

A special type of triaxial stress, called **spherical stress**, occurs whenever all three normal stresses are equal (Fig. 11-30):

$$\sigma_x = \sigma_y = \sigma_z = \sigma_0 \tag{11-56}$$

Under these stress conditions, *any* plane cut through the element will be subjected to the same normal stress σ_0 and will be free of shear stress. Thus, there are equal normal stresses in every direction and no shear stresses anywhere in the material. Every plane is a principal plane, and the three Mohr's circles shown in Fig. 11-29 reduce to a single point.

The normal strains in spherical stress are also the same in all directions, provided the material is homogeneous and isotropic. If Hooke's law applies, the normal strains are

$$\varepsilon_0 = \frac{\sigma_0}{E}(1 - 2\nu) \tag{11-57}$$

as obtained from Eqs. (11-52a, b, and c).

Since there are no shear strains, an element in the shape of a cube changes in size but remains a cube. In general, any body subjected to spherical stress maintains its relative proportions but expands or contracts in volume depending upon whether σ_0 is tensile or compressive.

The expression for the unit volume change can be obtained from Eq. (11-54) by substituting for the strains from Eq. (11-57). The result is

$$e = 3\varepsilon_0 = \frac{3\sigma_0(1 - 2\nu)}{E} \tag{11-58}$$

Equation (11-58) is usually expressed in more compact form by introducing a new quantity K called the **volume modulus of elasticity**, or **bulk modulus of elasticity**, which is defined as

$$K = \frac{E}{3(1 - 2\nu)} \tag{11-59}$$

With this notation, the expression for the unit volume change becomes

$$e = \frac{\sigma_0}{K} \tag{11-60}$$

and the volume modulus is

$$K = \frac{\sigma_0}{e} \tag{11-61}$$

Thus, the volume modulus can be defined as the ratio of the spherical stress to the volumetric strain, which is analogous to the definition of the modulus E in uniaxial stress. Note that the preceding formulas for e and K are based upon the assumptions that *the strains are small and Hooke's law holds for the material.*

Equation (11-61) for K shows that if Poisson's ratio ν equals 1/3, the moduli K and E are numerically equal. If $\nu = 0$, then K has the value $E/3$, and if $\nu = 0.5$, K becomes infinite, which corresponds to a rigid material having no change in volume (that is, the material is incompressible).

The preceding formulas for spherical stress were derived for an element subjected to uniform tension in all directions, but of course, the formulas also apply to an element in uniform compression. In the case of uniform compression, the stresses and strains have negative signs. Uniform compression occurs when the material is subjected to uniform pressure in all directions; for example, an object submerged in water or rock deep within the earth. This state of stress is often called **hydrostatic stress**.

Although uniform compression is relatively common, a state of uniform tension is difficult to achieve. It can be realized by suddenly and uniformly heating the outer surface of a solid metal sphere, so that the outer layers are at a higher temperature than the interior. The tendency of the outer layers to expand produces uniform tension in all directions at the center of the sphere.

11.7 Plane Strain

The strains at a point in a loaded structure vary according to the orientation of the axes in a manner similar to that for stresses. This section presents the derivation of the transformation equations that relate the strains in inclined directions to the strains in the reference directions. These transformation equations are widely used in laboratory and field investigations involving measurements of strains.

Strains are customarily measured by *strain gages*; for example, gages are placed in aircraft to measure structural behavior during flight, and gages are placed in buildings to measure the effects of earthquakes. Since each gage measures the strain in one particular direction, it is usually necessary to calculate the strains in other directions by means of the transformation equations.

Plane Strain versus Plane Stress

Consider what is meant by plane strain and how it relates to plane stress. A small element of material has sides of lengths a, b, and c in the x, y, and z directions, respectively (Fig. 11-31a). If the only deformations are those in the x-y plane, then three strain components may exist—the normal strain ε_x in the x direction (Fig. 11-31b), the normal strain ε_y in the y direction (Fig. 11-31c), and the shear strain γ_{xy} (Fig. 11-31d). An element of material subjected to these strains (and *only* these strains) is said to be in a state of **plane strain**.

It follows that an element in plane strain has no normal strain ε_z in the z direction and no shear strains γ_{xz} and γ_{yz} in the x-z and y-z planes, respectively. Thus, plane strain is defined by the following conditions:

$$\varepsilon_z = 0 \qquad \gamma_{xz} = 0 \qquad \gamma_{yz} = 0 \qquad \textbf{(11-62a,b,c)}$$

The remaining strains (ε_x, ε_y, and γ_{xy}) may have nonzero values.

The preceding definition shows that plane strain occurs when the front and rear faces of an element of material (Fig. 11-31a) are fully restrained against displacement in the z direction—an idealized condition that is seldom reached in actual structures. However, this does not mean that the transformation equations of plane strain are not useful. It turns out that they are extremely useful because they also apply to the strains in plane stress, as explained in the following paragraphs.

FIGURE 11-31

Strain components ε_x, ε_y, and γ_{xy} in the x-y plane (plane strain)

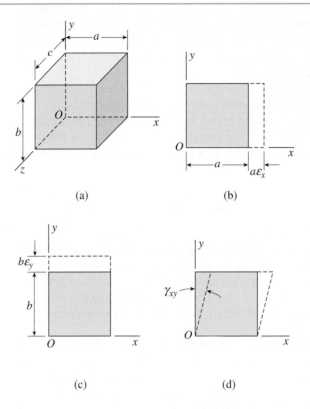

(a)

(b)

(c)

(d)

vThe definition of plane strain [Eqs. (11-62a, b, and c)] is analogous to that for plane stress. In plane stress, the following stresses must be zero:

$$\sigma_z = 0 \qquad \tau_{xz} = 0 \qquad \tau_{yz} = 0 \qquad \text{(11-63a,b,c)}$$

whereas the remaining stresses (σ_x, σ_y, and τ_{xy}) may have nonzero values. A comparison of the stresses and strains in plane stress and plane strain is given in Fig. 11-32.

It should not be inferred from the similarities in the definitions of plane stress and plane strain that both occur simultaneously. In general, an element in plane stress undergoes a strain in the z direction (Fig. 11-32); hence, it is *not* in plane strain. Also, an element in plane strain usually has stresses σ_z acting on it because of the requirement that $\varepsilon_z = 0$; therefore, it is *not* in plane stress. Thus, under ordinary conditions plane stress and plane strain do not occur simultaneously.

An exception occurs when an element in plane stress is subjected to equal and opposite normal stresses (that is, when $\sigma_x = -\sigma_y$) and Hooke's law holds for the material. In this special case, there is no normal strain in the z direction, as shown by Eq. (11-35c); therefore, the element is in a state of plane strain as well as plane stress. Another special case, albeit a hypothetical one, is when a material has Poisson's ratio equal to zero ($\nu = 0$); then every plane stress element is also in plane strain because $\varepsilon_z = 0$ [Eq. (11-35c)].[4]

---

[4]The discussions of this chapter omit the effects of temperature changes and prestrains, both of which produce additional deformations that may alter some of these conclusions.

FIGURE 11-32

Comparison of plane stress and plane strain

Plane stress	Plane strain
(diagram of cube with σ_y, τ_{xy}, σ_x, axes x, y, z, origin O)	*(diagram of cube with ε_y, γ_{xy}, ε_x, axes x, y, z, origin O)*

	Plane stress	Plane strain
Stresses	$\sigma_z = 0 \qquad \tau_{xz} = 0 \qquad \tau_{yz} = 0$ σ_x, σ_y, and τ_{xy} may have nonzero values	$\tau_{xz} = 0 \qquad \tau_{yz} = 0$ σ_x, σ_y, σ_z, and τ_{xy} may have nonzero values
Strains	$\gamma_{xz} = 0 \qquad \gamma_{yz} = 0$ ε_x, ε_y, ε_z, and γ_{xy} may have nonzero values	$\varepsilon_z = 0 \qquad \gamma_{xz} = 0 \qquad \gamma_{yz} = 0$ ε_x, ε_y, and γ_{xy} may have nonzero values

Application of the Transformation Equations

The stress-transformation equations derived for plane stress in the x-y plane [Eqs. (11-4a and b)] are valid even when a normal stress σ_z is present. The explanation lies in the fact that the stress σ_z does not enter the equations of equilibrium used in deriving Eqs. (11-4a and b). Therefore, *the transformation equations for plane stress also can be used for the stresses in plane strain.*

An analogous situation exists for plane strain. Although the strain-transformation equations are derived for the case of plane strain in the x-y plane, the equations are valid even when a strain ε_z exists. The reason is simple enough—the strain ε_z does not affect the geometric relationships used in the derivations. Therefore, *the transformation equations for plane strain also can be used for the strains in plane stress.*

Finally, recall that the transformation equations for plane stress are derived solely from equilibrium and therefore are valid for any material, whether linearly elastic or not. The same conclusion applies to the transformation equations for plane strain—since they are derived solely from geometry, *they are independent of the material properties.*

Transformation Equations for Plane Strain

In the derivation of the transformation equations for plane strain, use the coordinate axes shown in Fig. 11-33. Assume that the normal strains ε_x and ε_y and the shear strain γ_{xy} associated with the xy axes are known (Fig. 11-31). The objectives of this analysis are to determine the normal strain ε_{x1} and the shear strain γ_{x1y1} associated with the x_1y_1 axes, which are rotated counterclockwise through an angle θ from the xy axes. (It is not necessary to derive a separate equation for the normal strain ε_{y1} because it can be obtained from the equation for ε_{x1} by substituting $\theta + 90°$ for θ.)

FIGURE 11-33

Axes x_1 and y_1 rotated through an angle θ from the x-y axes

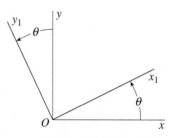

Normal Strain ε_{x1}. To determine the normal strain ε_{x1} in the x_1 direction, consider a small element of material selected so that the x_1 axis is along a diagonal of the z face of the element and the x and y axes are along the sides of the element (Fig. 11-34a). The figure shows a two-dimensional view of the element with the z axis toward the viewer. Of course, the element is actually three-dimensional, as in Fig. 11-31a, with a dimension in the z direction.

Consider first the strain ε_x in the x direction (Fig. 11-34a). This strain produces an elongation in the x direction equal to $\varepsilon_x\,dx$, where dx is the length of the corresponding side of the element. As a result of this elongation, the diagonal of the element increases in length by an amount

$$\varepsilon_x\,dx\cos\theta \tag{11-64a}$$

as shown in Fig. 11-34a.

Next, consider the strain ε_y in the y direction (Fig. 11-34b). This strain produces an elongation in the y direction equal to $\varepsilon_y\,dy$, where dy is the length of the side of the element parallel to the y axis. As a result of this elongation, the diagonal of the element increases in length by an amount

$$\varepsilon_y\,dy\sin\theta \tag{11-64b}$$

which is shown in Fig. 11-34b.

Finally, consider the shear strain γ_{xy} in the x-y plane (Fig. 11-34c). This strain produces a distortion of the element such that the angle at the lower-left corner of the element decreases by an amount equal to the shear strain. Consequently, the upper face of the element moves to the right (with respect to the lower face) by an amount $\gamma_{xy}\,dy$. This deformation results in an increase in the length of the diagonal equal to

$$\gamma_{xy}\,dy\cos\theta \tag{11-64c}$$

as shown in Fig. 11-34c.

FIGURE 11-34

Deformations of an element in plane strain due to (a) normal strain ε_x, (b) normal strain ε_y, and (c) shear strain γ_{xy}

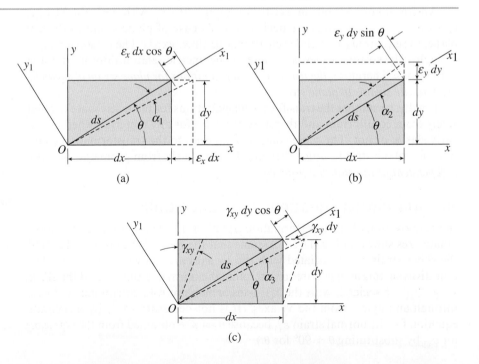

The total increase Δd in the length of the diagonal is the sum of the preceding three expressions; thus,

$$\Delta d = \varepsilon_x\, dx \cos \theta + \varepsilon_y\, dy \sin \theta + \gamma_{xy}\, dy \cos \theta \qquad \textbf{(11-65)}$$

The normal strain ε_{x1} in the x_1 direction is equal to this increase in length divided by the initial length ds of the diagonal:

$$\varepsilon_{x1} = \frac{\Delta d}{ds} = \varepsilon_x \frac{dx}{ds} \cos \theta + \varepsilon_y \frac{dy}{ds} \sin \theta + \gamma_{xy} \frac{dy}{ds} \cos \theta \qquad \textbf{(11-66)}$$

The two relations $dx/ds = \cos \theta$ and $dy/ds = \sin \theta$ lead to the following equation for the **normal strain**:

$$\varepsilon_{x1} = \varepsilon_x \cos^2 \theta + \varepsilon_y \sin^2 \theta + \gamma_{xy} \sin \theta \cos \theta \qquad \textbf{(11-67)}$$

Thus, Eq. (11-67) represents an expression for the normal strain in the x_1 direction in terms of the strains ε_x, ε_y, and γ_{xy} associated with the x-y axes.

As mentioned previously, the normal strain ε_{y1} in the y_1 direction is obtained from the preceding equation by substituting $\theta + 90°$ for θ.

Shear Strain γ_{x1y1}. Now turn to the shear strain γ_{x1y1} associated with the $x_1 y_1$ axes. This strain is equal to the decrease in angle between lines in the material that were initially along the x_1 and x_1 axes. To clarify this idea, consider Fig. 11-35, which shows both the xy and $x_1 y_1$ axes, with the angle θ between them. Let line Oa represent a line in the material that *initially* was along the x_1 axis (that is, along the diagonal of the element in Fig. 11-34). The deformations caused by the strains ε_x, ε_y, and γ_{xy} (Fig. 11-34) cause line Oa to rotate through a counterclockwise angle α from the x_1 axis to the position shown in Fig. 11-35. Similarly, line Ob was originally along the y_1 axis, but because of the deformations, it rotates through a clockwise angle β. The shear strain γ_{x1y1} is the decrease in angle between the two lines that originally were at right angles; therefore,

$$\gamma_{x1y1} = \alpha + \beta \qquad \textbf{(11-68)}$$

Thus, to find the shear strain γ_{x1y1}, find the angles α and β.

The angle α can be found from the deformations pictured in Fig. 11-34 as follows. The strain ε_x (Fig. 11-34a) produces a clockwise rotation of the diagonal of the element. Denote this angle of rotation as α_1. The angle α_1 is equal to the distance $\varepsilon_x\, dx \sin \theta$ divided by the length ds of the diagonal:

$$\alpha_1 = \varepsilon_x \frac{dx}{ds} \sin \theta \qquad \textbf{(11-69a)}$$

Similarly, the strain ε_y produces a counterclockwise rotation of the diagonal through an angle α_2 (Fig. 11-34b). This angle is equal to the distance $\varepsilon_y\, dy \cos \theta$ divided by ds:

$$\alpha_2 = \varepsilon_y \frac{dy}{ds} \cos \theta \qquad \textbf{(11-69b)}$$

Finally, the strain γ_{xy} produces a clockwise rotation through an angle α_3 (Fig. 11-34c), which is equal to the distance $\gamma_{xy}\, dy \sin \theta$ divided by ds:

$$\alpha_3 = \gamma_{xy} \frac{dy}{ds} \sin \theta \qquad \textbf{(11-69c)}$$

FIGURE 11-35

Shear strain γ_{x1y1} associated with the $x_1 y_1$ axes

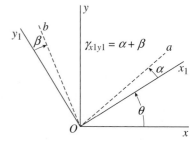

Therefore, the resultant counterclockwise rotation of the diagonal (Fig. 11-34), which is equal to the angle α shown in Fig. 11-35, is

$$\alpha = -\alpha_1 + \alpha_2 - \alpha_3$$

$$= -\varepsilon_x \frac{dx}{ds} \sin\theta + \varepsilon_y \frac{dy}{ds} \cos\theta - \gamma_{xy} \frac{dy}{ds} \sin\theta \tag{11-70}$$

Again observe that $dx/ds = \cos\theta$ and $dy/ds = \sin\theta$ so

$$\alpha = -(\varepsilon_x - \varepsilon_y)\sin\theta\cos\theta - \gamma_{xy}\sin^2\theta \tag{11-71}$$

The rotation of line Ob (Fig. 11-35), which initially was at 90° to line Oa, can be found by substituting $\theta + 90°$ for θ in the expression for α. The resulting expression is counterclockwise when positive (because α is counterclockwise when positive); hence, it is equal to the negative of the angle β (because β is positive when clockwise). Thus,

$$\beta = (\varepsilon_x - \varepsilon_y)\sin(\theta + 90°)\cos(\theta + 90°) + \gamma_{xy}\sin^2(\theta + 90°)$$

$$= -(\varepsilon_x - \varepsilon_y)\sin\theta\cos\theta + \gamma_{xy}\cos^2\theta \tag{11-72}$$

Adding α and β gives the shear strain γ_{x1y1} [see Eq. (11-68)]:

$$\gamma_{x1y1} = -2(\varepsilon_x - \varepsilon_y)\sin\theta\cos\theta + \gamma_{xy}(\cos^2\theta - \sin^2\theta) \tag{11-73}$$

To put the equation in a more useful form, divide each term by 2:

$$\frac{\gamma_{x1y1}}{2} = -(\varepsilon_x - \varepsilon_y)\sin\theta\cos\theta + \frac{\gamma_{xy}}{2}(\cos^2\theta - \sin^2\theta) \tag{11-74}$$

The result is an expression for the **shear strain** γ_{x1y1} associated with the x_1–y_1 axes in terms of the strains ε_x, ε_y, and γ_{xy} associated with the x-y axes.

Transformation Equations for Plane Strain.

The equations for plane strain [Eqs. (11-67) and (11-74)] can be expressed in terms of the angle 2θ by using the following trigonometric identities:

$$\cos^2\theta = \frac{1}{2}(1 + \cos 2\theta) \quad \sin^2\theta = \frac{1}{2}(1 - \cos 2\theta)$$

$$\sin\theta\cos\theta = \frac{1}{2}\sin 2\theta$$

Thus, the transformation equations for plane strain become

$$\varepsilon_{x1} = \frac{\varepsilon_x + \varepsilon_y}{2} + \frac{\varepsilon_x - \varepsilon_y}{2}\cos 2\theta + \frac{\gamma_{xy}}{2}\sin 2\theta \tag{11-75a}$$

and

$$\frac{\gamma_{x1y1}}{2} = -\frac{\varepsilon_x - \varepsilon_y}{2}\sin 2\theta + \frac{\gamma_{xy}}{2}\cos 2\theta \tag{11-75b}$$

These equations are the counterparts of Eqs. (11-4a and b) for plane stress.

When comparing the two sets of equations, note that ε_{x1} corresponds to σ_{x1}, $\gamma_{x1y1}/2$ corresponds to τ_{x1y1}, ε_x corresponds to σ_x, ε_y corresponds to σ_y, and

$\gamma_{xy}/2$ corresponds to τ_{xy}. The corresponding variables in the two sets of transformation equations are listed in Table 11-1.

The analogy between the transformation equations for plane stress and those for plane strain shows that all of the observations made in Sections 11.2, 11.3, and 11.4 concerning plane stress, principal stresses, maximum shear stresses, and Mohr's circle have their counterparts in plane strain. For instance, the sum of the normal strains in perpendicular directions is a constant [compare with Eq. (11-6)]:

$$\varepsilon_{x1} + \varepsilon_{y1} = \varepsilon_x + \varepsilon_y \tag{11-76}$$

This equality can be verified easily by substituting the expressions for ε_{x1} [from Eq. (11-75a)] and ε_{y1} [from Eq. (11-75a) with θ replaced by $\theta + 90°$].

Principal Strains

Principal strains exist on perpendicular planes with the principal angles θ_p calculated from the following equation [compare with Eq. (11-11)]:

$$\tan 2\theta_p = \frac{\gamma_{xy}}{\varepsilon_x - \varepsilon_y} \tag{11-77}$$

The principal strains can be calculated from the equation

$$\varepsilon_{1,2} = \frac{\varepsilon_x + \varepsilon_y}{2} \pm \sqrt{\left(\frac{\varepsilon_x - \varepsilon_y}{2}\right)^2 + \left(\frac{\gamma_{xy}}{2}\right)^2} \tag{11-78}$$

which corresponds to Eq. (11-17) for the principal stresses. The two principal strains (in the x-y plane) can be correlated with the two principal directions using the technique described in Section 11.3 for the principal stresses. (This technique is illustrated later in Example 11-8.) Finally, note that in plane strain the third principal strain is $\varepsilon_z = 0$. Also, the shear strains are zero on the principal planes.

Maximum Shear Strains

The maximum shear strains in the x-y plane are associated with axes at 45° to the directions of the principal strains. The algebraically maximum shear strain (in the x-y plane) is given by the following equation [compare with Eq. (11-25)]:

$$\frac{\gamma_{max}}{2} = \sqrt{\left(\frac{\varepsilon_x - \varepsilon_y}{2}\right)^2 + \left(\frac{\gamma_{xy}}{2}\right)^2} \tag{11-79}$$

The minimum shear strain has the same magnitude but is negative. In the directions of maximum shear strain, the normal strains are

$$\varepsilon_{aver} = \frac{\varepsilon_x + \varepsilon_y}{2} \tag{11-80}$$

which is analogous to Eq. (11-27) for stresses. The maximum out-of-plane shear strains, that is, the shear strains in the x-z and y-z planes, can be obtained from equations analogous to Eq. (11-79).

Table 11-1

Corresponding variables in the transformation equations for plane stress [Eqs. (11-4a and b)] and plane strain [Eqs. (11-75a and b)]

Stresses	Strains
σ_x	ε_x
σ_y	ε_y
τ_{xy}	$\gamma_{xy}/2$
σ_{x1}	ε_{x1}
τ_{x1y1}	$\gamma_{x1y1}/2$

An element in plane stress that is oriented to the principal directions of stress (see Fig. 11-13b) has no shear stresses acting on its faces. Therefore, the shear strain γ_{x1y1} for this element is zero. It follows that the normal strains in this element are the principal strains. Thus, at a given point in a stressed body, *the principal strains and principal stresses occur in the same directions.*

Mohr's Circle for Plane Strain

Mohr's circle for plane strain is constructed in the same manner as the circle for plane stress, as illustrated in Fig. 11-36. Normal strain ε_{x1} is plotted as the abscissa (positive to the right), and one-half the shear strain ($\gamma_{x1y1}/2$) is plotted as the ordinate (positive downward). The center C of the circle has an abscissa equal to ε_{aver} [Eq. (11-80)].

Point A, representing the strains associated with the x direction ($\theta = 0$), has coordinates ε_x and $\gamma_{xy}/2$. Point B, at the opposite end of a diameter from A, has coordinates ε_y and $-\gamma_{xy}/2$, representing the strains associated with a pair of axes rotated through an angle $\theta = 90°$.

The strains associated with axes rotated through an angle θ are given by point D, which is located on the circle by measuring an angle 2θ counterclockwise from radius CA. The principal strains are represented by points P_1 and P_2, and the maximum shear strains by points S_1 and S_2. All of these strains can be determined from the geometry of the circle or from the transformation equations.

Strain Measurements

An electrical-resistance **strain gage** is a device for measuring normal strains on the surface of a stressed object. These gages are quite small, with lengths typically in the range from one-eighth to one-half of an inch. The gages are bonded securely to the surface of the object so that they change in length in proportion to the strains in the object itself.

FIGURE 11-36

Mohr's circle for plane strain

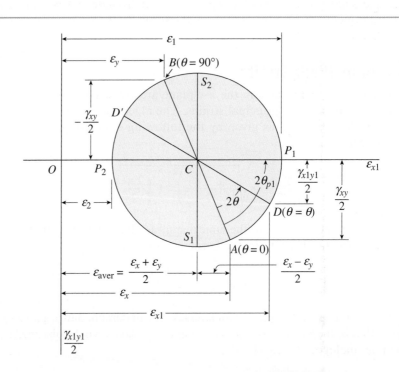

FIGURE 11-37

Three electrical-resistance strain gages arranged as a 45° strain rosette (magnified view) (Courtesy of Micro-Measurements Division of Vishay Precision Group, Raleigh, NC, USA)

(a) 45° strain-gage three-element rosette

(b) Three-element strain-gage rosettes prewired

Each gage consists of a fine metal grid that is stretched or shortened when the object is strained at the point where the gage is attached. The grid is equivalent to a continuous wire that goes back and forth from one end of the grid to the other, thereby effectively increasing its length (Fig. 11-37). The electrical resistance of the wire is altered when it stretches or shortens—then this change in resistance is converted into a measurement of strain. The gages are extremely sensitive and can measure strains as small as 1×10^{-6}.

Since each gage measures the normal strain in only one direction, and since the directions of the principal stresses are usually unknown, it is necessary to use three gages in combination, with each gage measuring the strain in a different direction. From three such measurements, it is possible to calculate the strains in any direction, as illustrated in Example 11-9.

A group of three gages arranged in a particular pattern is called a **strain rosette**. Because the rosette is mounted on the surface of the body, where the material is in plane stress, use the transformation equations for plane strain to calculate the strains in various directions. (As explained earlier in this section, the transformation equations for plane strain can also be used for the strains in plane stress.)

Calculation of Stresses from the Strains

The strain equations presented in this section are derived solely from geometry, as already pointed out. Therefore, the equations apply to any material, whether linear or nonlinear, elastic or inelastic. However, if it is desired to determine the stresses from the strains, the material properties must be taken into account.

If the material follows Hooke's law, find the stresses using the appropriate stress-strain equations from either Section 11.5 (for plane stress) or Section 11.6 (for triaxial stress).

As a first example, suppose that the material is in plane stress and that are known the strains ε_x, ε_y, and γ_{xy}, perhaps from strain-gage measurements. Then use the stress-strain equations for plane stress [Eqs. (11-37) and (11-38)] to obtain the stresses in the material.

Now consider a second example. Suppose you have determined the three principal strains ε_1, ε_2, and ε_3 for an element of material (if the element is in plane strain, then $\varepsilon_3 = 0$). Knowing these strains, you can find the principal stresses using Hooke's law for triaxial stress [see Eqs. (11-53a, b, and c)]. Once the principal stresses are known, the stresses on inclined planes are computed using the transformation equations for plane stress (see the discussion at the beginning of Section 11.6).

Example 11-8

An element of material in plane strain undergoes the following strains:

$$\varepsilon_x = 340 \times 10^{-6} \qquad \varepsilon_y = 110 \times 10^{-6} \qquad \gamma_{xy} = 180 \times 10^{-6}$$

These strains are shown highly exaggerated in Fig. 11-38a, which shows the deformations of an element of unit dimensions. Since the edges of the element have unit lengths, the changes in linear dimensions have the same magnitudes as the normal strains ε_x and ε_y. The shear strain γ_{xy} is the decrease in angle at the lower-left corner of the element.

Determine the following quantities: (a) the strains for an element oriented at an angle $\theta = 30°$, (b) the principal strains, and (c) the maximum shear strains. (Consider only the in-plane strains, and show all results on sketches of properly oriented elements.)

Solution:

Use the four-step problem-solving approach. Combine steps as needed for an efficient solution.

FIGURE 11-38

Example 11-8: Element of material in plane strain: (a) element oriented to the x and y axes, (b) element oriented at an angle $\theta = 30°$, (c) principal strains, and (d) maximum shear strains (*Note:* The edges of the elements have unit lengths)

(a)

(b)

(c)

(d)

Part (a): Element oriented at an angle $\theta = 30°$.

1, 2, 3. *Conceptualize, Categorize, Analyze*: The strains for an element oriented at an angle θ to the x axis can be found from the transformation equations of Eqs. (11-75a and b). As a preliminary matter, make the following calculations:

$$\frac{\varepsilon_x + \varepsilon_y}{2} = \frac{(340 + 110)10^{-6}}{2} = 225 \times 10^{-6}$$

$$\frac{\varepsilon_x - \varepsilon_y}{2} = \frac{(340 - 110)10^{-6}}{2} = 115 \times 10^{-6}$$

$$\frac{\gamma_{xy}}{2} = 90 \times 10^{-6}$$

Now substitute into Eqs. (11-75a and b) to get

$$\varepsilon_{x1} = \frac{\varepsilon_x + \varepsilon_y}{2} + \frac{\varepsilon_x - \varepsilon_y}{2} \cos 2\theta + \frac{\gamma_{xy}}{2} \sin 2\theta$$
$$= (225 \times 10^{-6}) + (115 \times 10^{-6})(\cos 60°) + (90 \times 10^{-6})(\sin 60°)$$
$$= 360 \times 10^{-6}$$

$$\frac{\gamma_{x1y1}}{2} = -\frac{\varepsilon_x - \varepsilon_y}{2} \sin 2\theta + \frac{\gamma_{xy}}{2} \cos 2\theta$$
$$= -(115 \times 10^{-6})(\sin 60°) + (90 \times 10^{-6})(\cos 60°)$$
$$= -55 \times 10^{-6}$$

Therefore, the shear strain is

$$\gamma_{x1y1} = -110 \times 10^{-6}$$

The strain ε_{y1} can be obtained from Eq. (11-76), as

$$\varepsilon_{y1} = \varepsilon_x + \varepsilon_y - \varepsilon_{x1} = (340 + 110 - 360)10^{-6} = 90 \times 10^{-6}$$

4. *Finalize*: The strains ε_{x1}, ε_{y1}, and γ_{x1y1} are shown in Fig. 11-38b for an element oriented at $\theta = 30°$. Note that the angle at the lower-left corner of the element increases because γ_{x1y1} is negative.

Part (b): Principal strains.

1, 2, 3. *Conceptualize, Categorize, Analyze*: The principal strains are readily determined from Eq. (11-78), as

$$\varepsilon_{1,2} = \frac{\varepsilon_x + \varepsilon_y}{2} \pm \sqrt{\left(\frac{\varepsilon_x - \varepsilon_y}{2}\right)^2 + \left(\frac{\gamma_{xy}}{2}\right)^2}$$
$$= 225 \times 10^{-6} \pm \sqrt{(115 \times 10^{-6})^2 + (90 \times 10^{-6})^2}$$
$$= 225 \times 10^{-6} \pm 146 \times 10^{-6}$$

Thus, the principal strains are

$$\varepsilon_1 = 370 \times 10^{-6} \qquad \varepsilon_2 = 80 \times 10^{-6}$$

where ε_1 denotes the algebraically larger principal strain and ε_2 denotes the algebraically smaller principal strain. (Recall that only in-plane strains are considered in this example.)

The angles to the principal directions can be obtained from Eq. (11-77):

$$\tan 2\theta_p = \frac{\gamma_{xy}}{\varepsilon_x - \varepsilon_y} = \frac{180}{340 - 110} = 0.7826$$

The values of $2\theta_p$ between 0 and 360° are 38.0° and 218.0°; therefore, the angles to the principal directions are

$$\theta_p = 19.0° \text{ and } 109.0°$$

To determine the value of θ_p associated with each principal strain, substitute $\theta_p = 19.0°$ into the first transformation equation [Eq. (11-75a)] and solve for the strain:

$$\varepsilon_{x1} = \frac{\varepsilon_x + \varepsilon_y}{2} + \frac{\varepsilon_x + \varepsilon_y}{2} \cos 2\theta + \frac{\gamma_{xy}}{2} \sin 2\theta$$
$$= (225 \times 10^{-6}) + (115 \times 10^{-6})(\cos 38.0°) + (90 \times 10^{-6})(\sin 38.0°)$$
$$= 370 \times 10^{-6}$$

4. *Finalize*: This result shows that the larger principal strain ε_1 is at the angle $\theta_{p1} = 19.0°$. The smaller strain ε_2 acts at 90° from that direction ($\theta_{p2} = 109.0°$). Thus,

$$\varepsilon_1 = 370 \times 10^{-6} \quad \text{and} \quad \theta_{p1} = 19.0°$$
$$\varepsilon_2 = 80 \times 10^{-6} \quad \text{and} \quad \theta_{p2} = 109.0°$$

Note that $\varepsilon_1 + \varepsilon_2 = \varepsilon_x + \varepsilon_y$.

The principal strains are portrayed in Fig. 11-38c. There are, of course, no shear strains on the principal planes.

FIGURE 11-38c,d (Repeated)

(c)   (d)

Part (c): Maximum shear strain.

1, 2, 3. *Conceptualize, Categorize, Analyze*: The maximum shear strain is calculated from Eq. (11-79):

$$\frac{\gamma_{max}}{2} = \sqrt{\left(\frac{\varepsilon_x - \varepsilon_y}{2}\right)^2 + \left(\frac{\gamma_{xy}}{2}\right)^2} = 146 \times 10^{-6} \qquad \gamma_{max} = 290 \times 10^{-6}$$

The element having the maximum shear strains is oriented at 45° to the principal directions; therefore, $\theta_s = 19.0° + 45° = 64.0°$ and $2\theta_s = 128.0°$. Substitute this value of $2\theta_s$ into the second transformation equation [Eq. (11-75b)] to determine the sign of the shear strain associated with this direction. The calculations are

$$\frac{\gamma_{x1y1}}{2} = \frac{\varepsilon_x - \varepsilon_y}{2} \sin 2\theta + \frac{\gamma_{xy}}{2} \cos 2\theta$$
$$= -(115 \times 10^{-6})(\sin 128.0°) + (90 \times 10^{-6})(\cos 128.0°)$$
$$= -146 \times 10^{-6}$$

4. *Finalize*: This result shows that an element oriented at an angle $\theta_{s2} = 64.0°$ has the maximum negative shear strain.

You can arrive at the same result by observing that the angle θ_{s1} to the direction of maximum positive shear strain is always 45° less than θ_{p1}. Hence,

$$\theta_{s1} = \theta_{p1} - 45° = 19.0° - 45° = -26.0°$$
$$\theta_{s2} = \theta_{s1} + 90° = 64.0°$$

The shear strains corresponding to θ_{s1} and θ_{s2} are $\gamma_{max} = 290 \times 10^{-6}$ and $\gamma_{min} = -290 \times 10^{-6}$, respectively.

The normal strains on the element having the maximum and minimum shear strains are

$$\varepsilon_{aver} = \frac{\varepsilon_x + \varepsilon_y}{2} = 225 \times 10^{-6}$$

A sketch of the element having the maximum in-plane shear strains is shown in Fig. 11-38d.

In this example, the strains were computed by using the transformation equations. However, all of the results can be obtained just as easily from Mohr's circle, as illustrated in Fig. 11-39.

FIGURE 11-39

Example 11-8: Mohr's circle for element of material in plane strain

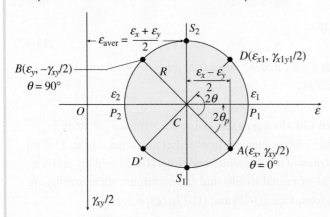

Example 11-9

FIGURE 11-40

Example 11-9: (a) 45° strain rosette and (b) element oriented at an angle θ to the x-y axes

(a)

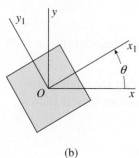

(b)

A 45° strain rosette (also called a *rectangular rosette*) consists of three electrical-resistance strain gages arranged to measure strains in two perpendicular directions and also at a 45° angle between them, as shown in Fig. 11-40a. The rosette is bonded to the surface of the structure before it is loaded. Gages A, B, and C measure the normal strains ε_a, ε_b, and ε_c in the directions of lines Oa, Ob, and Oc, respectively.

Explain how to obtain the strains ε_{x1}, ε_{y1}, and γ_{x1y1} associated with an element oriented at an angle θ to the xy axes (Fig. 11-40b).

Solution:

Use the four-step problem-solving approach. Combine steps as needed for an efficient solution.

1, 2. *Conceptualize, Categorize*: At the surface of the stressed object, the material is in plane stress. Since the strain-transformation equations [Eqs. (11-75a and b)] apply to plane stress as well as to plane strain, use those equations to determine the strains in any desired direction.

3. *Analyze*:

Strains associated with the x-y axes: Begin by determining the strains associated with the x-y axes. Because gages A and C are aligned with the x and y axes, respectively, they give the strains ε_x and ε_y directly:

$$\varepsilon_x = \varepsilon_a \qquad \varepsilon_y = \varepsilon_c \qquad \text{(11-81a,b)}$$

To obtain the shear strain γ_{xy}, use the transformation equation for normal strains [Eq. (11-75a)]:

$$\varepsilon_{x1} = \frac{\varepsilon_x + \varepsilon_y}{2} + \frac{\varepsilon_x - \varepsilon_y}{2}\cos 2\theta + \frac{\gamma_{xy}}{2}\sin 2\theta$$

For an angle $\theta = 45°$, $\varepsilon_{x1} = \varepsilon_b$ (Fig. 11-40a); therefore, the preceding equation gives

$$\varepsilon_b = \frac{\varepsilon_a + \varepsilon_c}{2} + \frac{\varepsilon_a - \varepsilon_c}{2}(\cos 90°) + \frac{\gamma_{xy}}{2}(\sin 90°)$$

Solve for γ_{xy} to get

$$\gamma_{xy} = 2\varepsilon_b - \varepsilon_a - \varepsilon_c \qquad \text{(11-82)}$$

Thus, the strains ε_x, ε_y, and γ_{xy} are easily determined from the given strain-gage readings.

4. *Finalize*:

Strains associated with the x_1y_1 axes: Using the strains ε_x, ε_y, and γ_{xy}, calculate the strains for an element oriented at any angle θ (Fig. 11-40b) from the strain-transformation equations [Eqs. (11-75a and b)] or from Mohr's circle. The principal strains and the maximum shear strains are computed using from Eqs. (11-78) and (11-79), respectively.

CHAPTER SUMMARY AND REVIEW

Chapter 11 discussed the **state of stress** at a point on a stressed body and then displayed it on a stress element. In two dimensions, **plane stress** was defined and expressed in terms of transformation equations that gave different, but equivalent, expressions of the state of normal and shear stresses at that point. **Principal normal stresses** and **maximum shear stress**, and their orientations, were seen to be the most important information for design. A graphical representation of the transformation equations, **Mohr's circle**, was shown to be a convenient way of exploring various representations of the state of stress at a point, including those orientations of the stress element where principal stresses and maximum shear stress occur. Later, strains were introduced and **Hooke's law for plane stress** was derived for homogeneous and isotropic materials and then specialized to obtain stress-strain relationships for **biaxial stress**, uniaxial stress, and pure shear. The stress state in three dimensions, referred to as **triaxial stress**, was introduced along with Hooke's law for triaxial stress. **Spherical stress** and **hydrostatic stress** were defined as special cases of triaxial stress. Finally, **plane strain** was defined for use in experimental stress analysis and compared to **plane stress**. Here are the major concepts presented in this chapter.

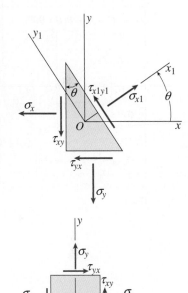

1. The **stresses on inclined sections** cut through a body, such as a beam, may be larger than the stresses acting on a stress element aligned with the cross section.

2. Stresses are tensors, not vectors, so equilibrium of a wedge element was used to transform the stress components from one set of axes to another. The transformation equations are applicable to stresses in any kind of material, whether linear, nonlinear, elastic, or inelastic. The **transformation equations for plane stress** are

$$\sigma_{x1} = \frac{\sigma_x + \sigma_y}{2} + \frac{\sigma_x - \sigma_y}{2} \cos 2\theta + \tau_{xy} \sin 2\theta$$

$$\tau_{x1y1} = -\frac{\sigma_x - \sigma_y}{2} \sin 2\theta + \tau_{xy} \cos 2\theta$$

$$\sigma_{y1} = \frac{\sigma_x + \sigma_y}{2} - \frac{\sigma_x - \sigma_y}{2} \cos 2\theta - \tau_{xy} \sin 2\theta$$

3. If two elements with different orientations are used to display the **state of plane stress** at the same point in a body, the stresses acting on the faces of the two elements are different, but they still represent the same intrinsic state of stress at that point.

4. The shear stresses acting on all four side faces of a stress element in plane stress are known if the shear stress acting on any one of those faces is known.

5. The sum of the normal stresses acting on perpendicular faces of plane-stress elements at a given point in a stressed body is constant and independent of the angle θ:

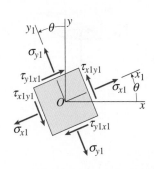

$$\sigma_{x1} + \sigma_{y1} = \sigma_x + \sigma_y$$

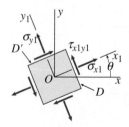

6. The maximum and minimum normal stresses, called the **principal stresses** σ_1 and σ_2, can be found from the transformation equation for normal stress as

$$\sigma_{1,2} = \frac{\sigma_x + \sigma_y}{2} \pm \sqrt{\left(\frac{\sigma_x - \sigma_y}{2}\right)^2 + \tau_{xy}^2}$$

The principal planes on which they act are at orientation angle θ_p. The shear stresses are zero on the principal planes, the planes of maximum shear stress occur at 45° to the principal planes, and the maximum shear stress is equal to one-half the difference of the principal stresses. Maximum shear stress can be computed from the normal and shear stresses on the original element or from the principal stresses as

$$\tau_{max} = \sqrt{\left(\frac{\sigma_x - \sigma_y}{2}\right)^2 + \tau_{xy}^2}$$

$$\tau_{max} = \frac{\sigma_1 - \sigma_2}{2}$$

7. The transformation equations for plane stress are represented in graphical form as **Mohr's circle**, which displays the relationship between normal and shear stresses acting on various inclined planes at a point in a stressed body. It also is used for calculating principal stresses, maximum shear stresses, and the orientations of the elements on which they act.

8. **Hooke's law for plane stress** provides the relationships between normal strains and stresses for homogeneous and isotropic materials which follow Hooke's law. These relationships contain three material constants (E, G, and v). When the normal stresses in plane stress are known, the normal strains in the x, y, and z directions are

$$\varepsilon_x = \frac{1}{E}(\sigma_x - v\sigma_y)$$

$$\varepsilon_y = \frac{1}{E}(\sigma_y - v\sigma_x)$$

$$\varepsilon_z = -\frac{v}{E}(\sigma_x + \sigma_y)$$

Solution of these equations gives the x and y normal stresses in terms of the strains:

$$\sigma_x = \frac{E}{1 - v^2}(\varepsilon_x + v\varepsilon_y)$$

$$\sigma_y = \frac{E}{1 - v^2}(\varepsilon_y + v\varepsilon_x)$$

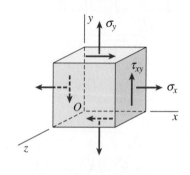

9. The **unit volume change** e, or the **dilatation** of a solid body, is defined as the change in volume divided by the original volume and is equal to the sum of the normal strains in three perpendicular directions:

$$e = \frac{\Delta V}{V_0} = \varepsilon_x + \varepsilon_y + \varepsilon_z$$

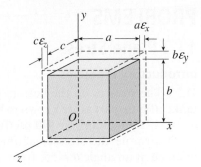

10. A state of **triaxial stress** exists in an element if it is subjected to normal stresses in three mutually perpendicular directions and there are no shear stresses on the faces of the element; the stresses are the principal stresses in the material. A special type of triaxial stress called **spherical stress** occurs when all three normal stresses are equal and tensile. If all three stresses are equal and compressive, the triaxial stress state is referred to as **hydrostatic stress**.

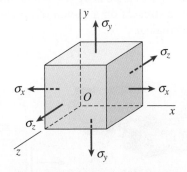

11. **Transformation equations for plane strain** are derived for use in the interpretation of experimental measurements made with strain gages. Plane strains at any orientation are represented in graphical form using **Mohr's circle for plane strain**. Plane stress and plane strain are compared in Fig. 11-32, and under ordinary conditions they do not occur simultaneously. The transformation equations for plane strain are independent of the material properties. At a given point in a stressed body, the **principal strains and principal stresses** occur in the same directions. Last, the transformation equations for plane stress also can be used for the stresses in plane strain, and the transformation equations for plane strain can be used for the strains in plane stress. The transformation equations for plane strain are

$$\varepsilon_{x1} = \frac{\varepsilon_x + \varepsilon_y}{2} + \frac{\varepsilon_x - \varepsilon_y}{2} \cos 2\theta + \frac{\gamma_{xy}}{2} \sin 2\theta$$

$$\frac{\gamma_{x1y1}}{2} = -\frac{\varepsilon_x - \varepsilon_y}{2} \sin 2\theta + \frac{\gamma_{xy}}{2} \cos 2\theta$$

PROBLEMS

11.2 Plane Stress

Introductory Problems

11.2-1 The stresses on the bottom surface of a fuel tanker (figure part a) are known to be $\sigma_x = 7750$ psi, $\sigma_y = 1175$ psi, and $\tau_{xy} = 940$ psi (figure part b).

Determine the stresses acting on an element oriented at an angle $\theta = 55°$ from the x axis, where the angle θ is positive when counterclockwise. Show these stresses on a sketch of an element oriented at the angle θ.

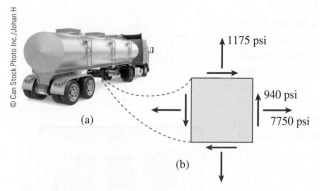

(a)

(b)

PROBLEM 11.2-1

11.2-2 Solve the preceding problem for an element in *plane stress* on the bottom surface of a fuel tanker (figure part a). Stresses are $\sigma_x = 105$ MPa, $\sigma_y = 75$ MPa, and $\tau_{xy} = 25$ MPa.

Determine the stresses acting on an element oriented at an angle $\theta = 40°$ from the x axis, where the angle θ is positive when counterclockwise. Show these stresses on a sketch of an element oriented at the angle θ.

PROBLEM 11.2-2

11.2-3 The stresses on an element are $\sigma_x = 1000$ psi, $\sigma_y = 500$ psi, and $\tau_{xy} = 350$ psi. Find the stresses acting on an element oriented at an angle $\theta = 25°$. Show these stresses on the rotated element.

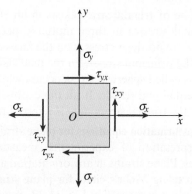

PROBLEM 11.2-3

11.2-4 The stresses on an element are known to be $\sigma_x = 120$ MPa, $\sigma_y = 100$ MPa, and $\tau_{xy} = 75$ MPa. Find the stresses on an inclined section through the element at an angle $\theta = 45°$.

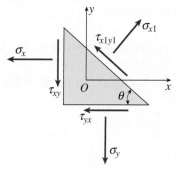

PROBLEM 11.2-4

11.2-5 The stresses acting on element A on the web of a train rail (see figure part a) are found to be 6500 psi tension in the horizontal direction and 18,500 psi compression in the vertical direction (see figure part b). Also, shear stresses with a magnitude of 3800 psi act in the directions shown.

Determine the stresses acting on an element oriented at a counterclockwise angle of 30° from the horizontal. Show these stresses on a sketch of an element oriented at this angle.

(a)

PROBLEMS 11.2-5, 11.3-6 and 11.3-7

(b)

11.2-6 Solve the preceding problem if the stresses acting on element A on the web of a train rail (see figure part a of Problem 11.2-5) are found to be 40 MPa in tension in the horizontal direction and 160 MPa in compression in the vertical direction. Also, shear stresses of magnitude 54 MPa act in the directions shown in the figure.

Determine the stresses acting on an element oriented at a counterclockwise angle of 52° from the horizontal. Show these stresses on a sketch of an element oriented at this angle.

160 MPa

40 MPa

54 MPa

A

PROBLEM 11.2-6

11.2-7 The stresses acting on element B on the web of a train rail (see figure part a of Problem 11.2-5) are found to be 5700 psi in compression in the horizontal direction and 2300 psi in compression in the vertical direction (see figure). Also, shear stresses of magnitude 2500 psi act in the directions shown.

Determine the stresses acting on an element oriented at a counterclockwise angle of 50° from the horizontal. Show these stresses on a sketch of an element oriented at this angle.

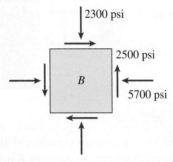

2300 psi

2500 psi

B

5700 psi

PROBLEM 11.2-7

11.2-8 An element in *plane stress* on the fuselage of an airplane (figure part a) is subjected to compressive stresses with a magnitude of 42 MPa in the horizontal direction and tensile stresses with a magnitude of 9.5 MPa in the vertical direction (see figure part b). Also, shear stresses with a magnitude of 15.5 MPa act in the directions shown.

Determine the stresses acting on an element oriented at a clockwise angle of 40° from the horizontal. Show these stresses on a sketch of an element oriented at this angle.

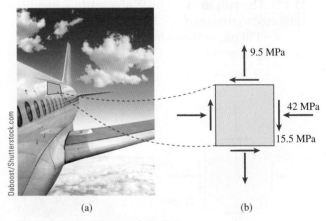

9.5 MPa

42 MPa

15.5 MPa

(a)

(b)

PROBLEMS 11.2-8 and 11-3-8

11.2-9 The stresses acting on element B (see figure part a) on the web of a wide-flange beam are found to be 14,500 psi in compression in the horizontal direction and 2530 psi in compression in the vertical direction (see figure part b). Also, shear stresses with a magnitude of 3500 psi act in the directions shown.

Determine the stresses acting on an element oriented at a counterclockwise angle of 38° from the horizontal. Show these stresses on a sketch of an element oriented at this angle.

(b)

(a)

PROBLEM 11.2-9

11.2-10 Solve the preceding problem if the normal and shear stresses acting on element B are 56 MPa, 17 MPa, and 27 MPa (in the directions shown in the figure) and the angle is 40° (clockwise).

PROBLEM 11.2-10

Representative Problems

11.2-11 The polyethylene liner of a settling pond is subjected to stresses $\sigma_x = 350$ psi, $\sigma_y = 112$ psi, and $\tau_{xy} = -120$ psi, as shown by the plane-stress element in the figure part a.

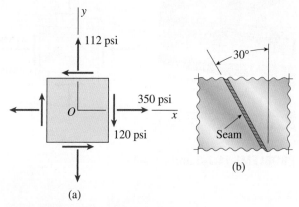

(b)

(a)

PROBLEM 11.2-11

Determine the normal and shear stresses acting on a seam oriented at an angle of 30° to the element, as shown in the figure part b. Show these stresses on a sketch of an element having its sides parallel and perpendicular to the seam.

11.2-12 Solve the preceding problem if the normal and shear stresses acting on the element are $\sigma_x = 2100$ kPa, $\sigma_y = 300$ kPa, and $\tau_{xy} = -560$ kPa, and the seam is oriented at an angle of 22.5° to the element.

PROBLEM 11.2-12

11.2-13 Two steel rods are welded together (see figure); the seam is oriented at angle $\theta = 50°$. The stresses on the rotated element are $\sigma_{x1} = 10$ ksi, $\sigma_{y1} = -12$ ksi, and $\tau_{x1y1} = -5$ ksi. Find the state of plane stress on the element if it is rotated clockwise to align the x_1 axis with the longitudinal axis of the rods.

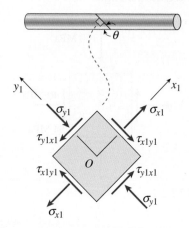

PROBLEMS 11.2-13 and 11.2-14

11.2-14 Repeat the previous problem using $\theta = 50°$ and stresses on the rotated element: $\sigma_{x1} = 70$ MPa, $\sigma_{y1} = -82$ MPa, and $\tau_{x1y1} = -35$ MPa.

11.2-15 A rectangular plate of dimensions 3.0 in. × 5.0 in. is formed by welding two triangular plates (see figure). The plate is subjected to a tensile stress of 500 psi in the long direction and a compressive stress of 350 psi in the short direction.

Determine the normal stress σ_w acting perpendicular to the line of the weld and the shear stress τ_w acting parallel to the weld. (Assume that the normal stress σ_w is positive when it acts in tension against the weld and the shear stress τ_w is positive when it acts counterclockwise against the weld.)

PROBLEM 11.2-15

11.2-16 Solve the preceding problem for a plate of dimensions 100 mm × 250 mm subjected to a compressive stress of 2.5 MPa in the long direction and a tensile stress of 12.0 MPa in the short direction (see figure).

PROBLEM 11.2-16

11.2-17 A simply supported beam is subjected to point load P at mid-span. The normal stress on an element at mid-span is known to be $\sigma_x = 1.5$ ksi.

PROBLEMS 11.2-17 and 11.2-18

Determine the element stresses if it is rotated through angle $\theta = -45°$. Show these stresses on a sketch of an element oriented at that angle.

11.2-18 Repeat the previous problem using $\sigma_x = 12$ MPa.

11.2-19 At a point on the surface of an elliptical exercise machine, the material is in *biaxial stress* with $\sigma_x = 1400$ psi and $\sigma_y = -900$ psi, as shown in the figure part a. The figure part b shows an inclined plane *aa* cut through the same point in the material but oriented at an angle θ.

Determine the value of the angle θ between zero and 90° such that no normal stress acts on plane *aa*. Sketch a stress element having plane *aa* as one of its sides and show all stresses acting on the element.

PROBLEM 11.2-19

11.2-20 Solve the preceding problem for $\sigma_x = 11$ MPa and $\sigma_y = -20$ MPa (see figure).

PROBLEM 11.2-20

11.2-21 An element in *plane stress* from the frame of a racing car is oriented at a known angle θ (see figure). On this inclined element, the normal and shear stresses have the magnitudes and directions shown in the figure.

Determine the normal and shear stresses acting on an element whose sides are parallel to the xy axes, that is, determine σ_x, σ_y, and τ_{xy}. Show the results on a sketch of an element oriented at $\theta = 0°$.

PROBLEM 11.2-21

11.2-22 Solve the preceding problem for the element shown in the figure.

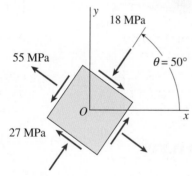

PROBLEM 11.2-22

11.2-23 A gusset plate on a truss bridge is in *plane stress* with normal stresses σ_x and σ_y and *shear stress* τ_{xy}, as shown in the figure. At counterclockwise angles $\theta = 32°$ and $\theta = 78°$ from the x axis, the normal stress is 4200 psi in tension.

If the stress σ_x equals 2650 psi in tension, what are the stresses σ_y and τ_{xy}?

11.2-24 The surface of an airplane wing is subjected to plane stress with normal stresses σ_x and σ_y and shear stress τ_{xy}, as shown in the figure. At a counterclockwise angle $\theta = 32°$ from the x axis, the normal stress is 29 MPa in tension, and at an angle $\theta = 46°$, it is 17 MPa in compression.

If the stress σ_x equals 105 MPa in tension, what are the stresses σ_y and τ_{xy}?

PROBLEM 11.2-24

11.2-25 At a point on the web of a girder on an overhead bridge crane in a manufacturing facility, the stresses are known to be $\sigma_x = -4300$ psi, $\sigma_y = 1700$ psi, and $\tau_{xy} = 3100$ psi (the sign convention for these stresses is shown in Fig. 11-1). A stress element located at the same point in the structure (but oriented at a counterclockwise angle θ_1 with respect to the x axis) is subjected to the stresses shown in the figure (σ_b, τ_b, and 2100 psi).

Assuming that the angle θ_1 is between zero and 90°, calculate the normal stress σ_b, the shear stress τ_b, and the angle θ_1.

PROBLEM 11.2-23

PROBLEM 11.2-25

11.2-26 A rectangular plate of dimensions 125 mm × 75 mm is subjected to tensile stress $\sigma_x = 67$ kPa and compressive stress σ_y. If it is known that the normal stress along the diagonal t–t is $\sigma_t = -6.57$ kPa, find stress σ_y on element A.

PROBLEM 11.2-26

11.2-27 A square plate with side dimension of 2 in. is subjected to compressive stress σ_x and tensile stress σ_y. The stresses on element A oriented at angle $\theta = 45°$ are $\sigma_{x1} = 75$ psi, $\sigma_{y1} = 75$ psi, and $\tau_{x1y1} = 275$ psi. Find the state of stress on the element if it is rotated clockwise to align the x_1 axis with the horizontal x axis.

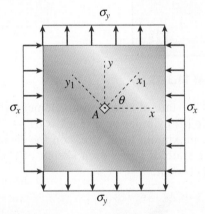

PROBLEM 11.2-27

11.3 Principal Stresses and Maximum Shear Stresses

To solve the problems for Section 11.3, consider only the in-plane stresses (the stresses in the x-y plane).

Introductory Problems

11.3-1 The stresses acting on an element are $\sigma_x = 750$ psi, $\sigma_y = 600$ psi, and $\tau_{xy} = 400$ psi. Determine the principal stresses and show them on a sketch of a properly oriented element.

11.3-2 Repeat the preceding problem using $\sigma_x = 5.5$ MPa, $\sigma_y = 4$ MPa, and $\tau_{xy} = 3.2$ MPa.

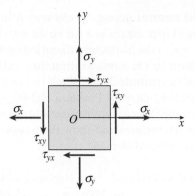

PROBLEMS 11.3-1 through 11.3-5

11.3-3 An element in *plane stress* is subjected to stresses $\sigma_x = 5750$ psi, $\sigma_y = 1100$ psi, and $\tau_{xy} = 750$ psi.
Determine the principal stresses and show them on a sketch of a properly oriented element.

11.3-4 An element in *plane stress* is subjected to stresses $\sigma_x = 105$ MPa, $\sigma_y = 75$ MPa, and $\tau_{xy} = 25$ MPa.
Determine the principal stresses and show them on a sketch of a properly oriented element.

11.3-5 An element in *plane stress* is subjected to stresses $\sigma_x = -5500$ psi, $\sigma_y = -2000$ psi, and $\tau_{xy} = 1900$ psi.
Determine the principal stresses and show them on a sketch of a properly oriented element.

11.3-6 The stresses acting on element A in the web of a train rail are found to be 40 MPa tension in the horizontal direction and 160 MPa compression in the vertical direction. Also, shear stresses of magnitude 54 MPa act in the directions shown (see the figure b for Problem 11.2-5).
Determine the principal stresses and show them on a sketch of a properly oriented element.

11.3-7 The normal and shear stresses acting on element A are 6500 psi, 17,300 psi, and 2900 psi (see the figure b for Problem 11.2-5).
Determine the maximum shear stresses and associated normal stresses and show them on a sketch of a properly oriented element.

11.3-8 An element in *plane stress* from the fuselage of an airplane is subjected to compressive stresses of magnitude 35 MPa in the horizontal direction and tensile stresses of magnitude 6.5 MPa in the vertical direction. Also, shear stresses of magnitude 12.5 MPa act in the directions shown (see the figure for Problem 11.2-8).
Determine the maximum shear stresses and associated normal stresses and show them on a sketch of a properly oriented element.

11.3-9 The stresses acting on element B in the web of a wide-flange beam are found to be 14,000 psi compression in the horizontal direction and 2600 psi compression in the vertical direction. Also, shear stresses of magnitude 3800 psi act in the directions shown (see the figure for Problem 11.2-9).

Determine the maximum shear stresses and associated normal stresses and show them on a sketch of a properly oriented element.

11.3-10 The normal and shear stresses acting on element B are $\sigma_x = -46$ MPa, $\sigma_y = -13$ MPa, and $\tau_{xy} = 21$ MPa (see figure for Problem 11.2-10).

Determine the maximum shear stresses and associated normal stresses and show them on a sketch of a properly oriented element.

Representative Problems

11.3-11 The stresses on an element are $\sigma_x = -300$ psi and $\sigma_y = -600$ psi. Find the maximum shear stresses on the element and show them on a sketch of a properly oriented element.

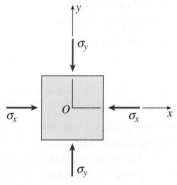

PROBLEM 11.3-11

11.3-12 A simply supported beam is subjected to two point loads as shown in the figure. The stresses on

PROBLEM 11.3-12

element A are $\tau_{xy} = -20$ kPa. Find the principal stresses on element A and show them on a sketch of a properly oriented element.

11.3-13 A shear wall in a reinforced concrete building is subjected to a vertical uniform load of intensity q and a horizontal force H, as shown in the first part of the figure. (The force H represents the effects of wind and earthquake loads.) As a consequence of these loads, the stresses at point A on the surface of the wall have the values shown in the second part of the figure (compressive stress equal to 1100 psi and shear stress equal to 480 psi).

(a) Determine the principal stresses and show them on a sketch of a properly oriented element.

(b) Determine the maximum shear stresses and associated normal stresses and show them on a sketch of a properly oriented element.

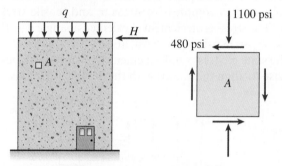

PROBLEM 11.3-13

11.3-14 The state of stress on an element along the hydraulic lift cylinder on a truck is $\sigma_y = -5$ MPa. Find the maximum shear stress on the element and show the state of stress on a sketch of a properly oriented element.

PROBLEMS 11.3-14 and 11.3-15

11.3-15 Repeat the preceding problem using $\sigma_y = -750$ psi.

11.3-16 A propeller shaft subjected to combined torsion and axial thrust is designed to resist a shear stress of 57 MPa and a compressive stress of 105 MPa (see figure).

(a) Determine the principal stresses and show them on a sketch of a properly oriented element.

(b) Determine the maximum shear stresses and associated normal stresses and show them on a sketch of a properly oriented element.

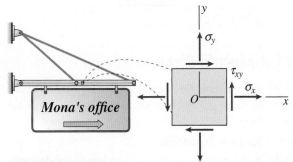

PROBLEM 11.3-16

11.3-17 The stresses at a point along a beam supporting a sign (see figure) are $\sigma_x = 2250$ psi, $\sigma_y = 1175$ psi, and $\tau_{xy} = -820$ psi.

(a) Find the principal stresses. Show them on a sketch of a properly oriented element.

(b) Find the maximum shear stresses and associated normal stresses. Show them on a sketch of a properly oriented element.

PROBLEM 11.3-17

11.3-18 through 11.3-22 An element in *plane stress* (see figure) is subjected to stresses σ_x, σ_y, and τ_{xy}.

(a) Determine the principal stresses and show them on a sketch of a properly oriented element.

(b) Determine the maximum shear stresses and associated normal stresses and show them on a sketch of a properly oriented element.

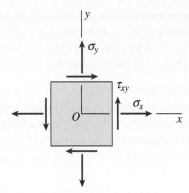

PROBLEMS 11.3-18 through 11.3-22

11.3-18 $\sigma_x = 2150$ kPa, $\sigma_y = 375$ kPa, $\tau_{xy} = -460$ kPa

11.3-19 $\sigma_x = 14{,}500$ psi, $\sigma_y = 1070$ psi, $\tau_{xy} = 1900$ psi

11.3-20 $\sigma_x = 16.5$ MPa, $\sigma_y = -91$ MPa, $\tau_{xy} = -39$ MPa

11.3-21 $\sigma_x = -3300$ psi, $\sigma_y = -11{,}000$ psi, $\tau_{xy} = 4500$ psi

11.3-22 $\sigma_x = -108$ MPa, $\sigma_y = 58$ MPa, $\tau_{xy} = -58$ MPa

11.3-23 At a point on the web of a girder on a gantry crane, the stresses acting on the x face of a stress element are $\sigma_x = 6250$ psi and $\tau_{xy} = 1425$ psi (see figure).

What is the allowable range of values for the stress σ_y if the maximum shear stress is limited to $\tau_0 = 2150$ psi?

PROBLEM 11.3-23

11.3-24 The stresses acting on a stress element on the arm of a power excavator (see figure) are $\sigma_x = 52$ MPa and $\tau_{xy} = 33$ MPa (see figure).

What is the allowable range of values for the stress σ_y if the maximum shear stress is limited to $\tau_0 = 37$ MPa?

PROBLEM 11.3-24

893

11.3-25 The stresses at a point on the down tube of a bicycle frame are $\sigma_x = 4800$ psi and $\tau_{xy} = -1950$ psi (see figure). It is known that one of the principal stresses equals 6375 psi in tension.

(a) Determine the stress σ_y.
(b) Determine the other principal stress and the orientation of the principal planes, then show the principal stresses on a sketch of a properly oriented element.

PROBLEM 11.3-25

11.3-26 An element in *plane stress* on the surface of an automobile drive shaft (see figure) is subjected to stresses of $\sigma_x = -45$ MPa and $\tau_{xy} = 39$ MPa (see figure). It is known that one of the principal stresses equals 41 MPa in tension.

(a) Determine the stress σ_y.
(b) Determine the other principal stress and the orientation of the principal planes, then show the principal stresses on a sketch of a properly oriented element.

PROBLEM 11.3-26

11.3-27 A simply supported wood beam is subjected to point load P at mid-span. The stresses on element C are known to be $\sigma_x = -92$ psi and $\tau_{xy} = -7$ psi. Find the principal stresses on the element and show them on a sketch of a properly oriented element.

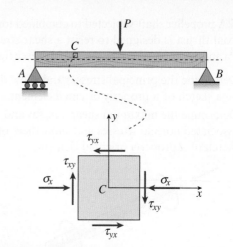

PROBLEM 11.3-27

11.3-28 A simply supported wood beam is subjected to point load P at mid-span. The normal stress on element C is known to be $\sigma_x = 12$ MPa. Find the maximum shear stress on the element and show the state of stress on a sketch of a properly oriented element.

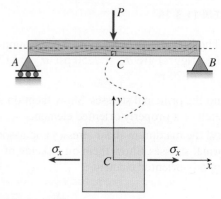

PROBLEM 11.3-28

11.4 Mohr's Circle

Solve the problems for Section 11.4 using Mohr's circle. Consider only the in-plane stresses (the stresses in the x-y plane).

Introductory Problems

11.4-1 An element in *uniaxial stress* is subjected to tensile stresses $\sigma_x = 14,250$ psi, as shown in the figure. Using Mohr's circle, determine the following.

(a) The stresses acting on an element oriented at a counterclockwise angle $\theta = 29°$ from the x axis.
(b) The maximum shear stresses and associated normal stresses.

Show all results on sketches of properly oriented elements.

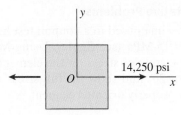

PROBLEM 11.4-1

11.4-2 An element in *uniaxial stress* is subjected to tensile stresses $\sigma_x = 57$ MPa, as shown in the figure. Using Mohr's circle, determine the following.

(a) The stresses acting on an element oriented at an angle $\theta = -33°$ from the x axis (minus means clockwise).

(b) The maximum shear stresses and associated normal stresses.

Show all results on sketches of properly oriented elements.

PROBLEM 11.4-2

11.4-3 An element on the gusset plate in Problem 11.2-23 in *uniaxial stress* is subjected to compressive stresses of magnitude 6750 psi, as shown in the figure. Using Mohr's circle, determine the following.

(a) The stresses acting on an element oriented at a slope of 1 on 2 (see figure).

(b) The maximum shear stresses and associated normal stresses.

Show all results on sketches of properly oriented elements.

PROBLEM 11.4-3

11.4-4 An element on the top surface of the fuel tanker in Problem 11.2-1 is in *biaxial stress* and is subjected to stresses $\sigma_x = -48$ MPa and $\sigma_y = 19$ MPa, as shown in the figure. Using Mohr's circle, determine the following.

(a) The stresses acting on an element oriented at a counterclockwise angle $\theta = 25°$ from the x axis.

(b) The maximum shear stresses and associated normal stresses.

Show all results on sketches of properly oriented elements.

PROBLEM 11.4-4

11.4-5 An element on the top surface of the fuel tanker in Problem 11.2-1 is in *biaxial stress* and is subjected to stresses $\sigma_x = 6250$ psi and $\sigma_y = -1750$ psi, as shown in the figure. Using Mohr's circle, determine the following.

(a) The stresses acting on an element oriented at a counterclockwise angle $\theta = 55°$ from the x axis.

(b) The maximum shear stresses and associated normal stresses.

Show all results on sketches of properly oriented elements.

PROBLEM 11.4-5

11.4-6 An element in *biaxial stress* is subjected to stresses $\sigma_x = -29$ MPa and $\sigma_y = 57$ MPa, as shown in the figure. Using Mohr's circle, determine the following.

(a) The stresses acting on an element oriented at a slope of 1 on 2.5 (see figure).

(b) The maximum shear stresses and associated normal stresses.

Show all results on sketches of properly oriented elements.

PROBLEM 11.4-6

11.4-7 An element on the surface of a drive shaft is in *pure shear* and is subjected to stresses $\tau_{xy} = 2700$ psi, as shown in the figure. Using Mohr's circle, determine the following.

(a) The stresses acting on an element oriented at a counterclockwise angle $\theta = 52°$ from the x axis.

(b) The principal stresses.

Show all results on sketches of properly oriented elements.

PROBLEM 11.4-7

Representative Problems

11.4-8 A specimen used in a coupon test has normal stress $\sigma_y = 15$ MPa (see figure). Using Mohr's circle, find the state of stress on the element oriented at angle $\theta = 20°$ and show the full stress state on a sketch of a properly oriented element.

PROBLEM 11.4-8

11.4-9 A specimen used in a coupon test is shown in the figure. The stresses on element A are known to be $\sigma_y = -1500$ psi. Use Mohr's circle to:

(a) Find the stresses acting on the element oriented at an angle $\theta = -35°$.

(b) Find maximum normal and shear stresses and show them on sketches of properly oriented elements.

PROBLEM 11.4-9

11.4-10 The rotor shaft of a helicopter (see figure part a) drives the rotor blades that provide the lifting force and is subjected to a combination of torsion and axial loading (see figure part b).

It is known that normal stress $\sigma_y = 68$ MPa and shear stress $\tau_{xy} = -100$ MPa. Using Mohr's circle, determine the following.

(a) The stresses acting on an element oriented at a counterclockwise angle $\theta = 22.5°$ from the x axis.

(b) Find the maximum tensile stress, maximum compressive stress, and maximum shear stress in the shaft.

Show all results on sketches of properly oriented elements.

(a)

(b)

PROBLEM 11.4-10

11.4-11 An element in *pure shear* is subjected to stresses $\tau_{xy} = 3750$ psi, as shown in the figure. Using Mohr's circle, determine the following.

(a) The stresses acting on an element oriented at a slope of 3 on 4 (see figure).

(b) The principal stresses.

Show all results on sketches of properly oriented elements.

PROBLEM 11.4-11

11.4-12 through 11.4-17 An element in *plane stress* is subjected to stresses σ_x, σ_y, and τ_{xy} (see figure).

Using Mohr's circle, determine the stresses acting on an element oriented at an angle θ from the x axis.

Show these stresses on a sketch of an element oriented at the angle θ. *Note:* The angle θ is positive when counterclockwise and negative when clockwise.

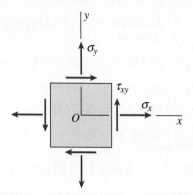

PROBLEMS 11.4-12 through 11.4-17

11.4-12 $\sigma_x = 27$ MPa, $\sigma_y = 14$ MPa, $\tau_{xy} = 6$ MPa, $\theta = 40°$

11.4-13 $\sigma_x = 3500$ psi, $\sigma_y = 12,200$ psi, $\tau_{xy} = -3300$ psi, $\theta = -51°$

11.4-14 $\sigma_x = -47$ MPa, $\sigma_y = -186$ MPa, $\tau_{xy} = -29$ MPa, $\theta = -33°$

11.4-15 $\sigma_x = -1720$ psi, $\sigma_y = -680$ psi, $\tau_{xy} = 320$ psi, $\theta = 14°$

11.4-16 $\sigma_x = 33$ MPa, $\sigma_y = -9$ MPa, $\tau_{xy} = 29$ MPa, $\theta = 35°$

11.4-17 $\sigma_x = -5700$ psi, $\sigma_y = 950$ psi, $\tau_{xy} = -2100$ psi, $\theta = 65°$

11.4-18 through 11.4-25 An element in *plane stress* is subjected to stresses σ_x, σ_y, and τ_{xy} (see figure).

Using Mohr's circle, determine (a) the principal stresses and (b) the maximum shear stresses and associated normal stresses. Show all results on sketches of properly oriented elements.

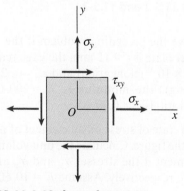

PROBLEMS 11.4-18 through 11.4-25

897

11.4-18 $\sigma_x = 2900\,\text{kPa}, \sigma_y = 9100\,\text{kPa}, \tau_{xy} = -3750\,\text{kPa}$

11.4-19 $\sigma_x = 800\,\text{psi}, \sigma_y = -2200\,\text{psi}, \tau_{xy} = 2900\,\text{psi}$

11.4-20 $\sigma_x = -3.3\,\text{MPa}, \sigma_y = 8.9\,\text{MPa}, \tau_{xy} = -14.1\,\text{MPa}$

11.4-21 $\sigma_x = -11,500\,\text{psi}, \sigma_y = -18,250\,\text{psi},$
$\tau_{xy} = -7200\,\text{psi}$

11.4-22 $\sigma_x = -29.5\,\text{MPa}, \sigma_y = 29.5\,\text{MPa}, \tau_{xy} = 27\,\text{MPa}$

11.4-23 $\sigma_x = 2050\,\text{psi}, \sigma_y = 6100\,\text{psi}, \tau_{xy} = 2750\,\text{psi}$

11.4-24 $\sigma_x = 0\,\text{MPa}, \sigma_y = -23.4\,\text{MPa}, \tau_{xy} = -9.6\,\text{MPa}$

11.4-25 $\sigma_x = 7300\,\text{psi}, \sigma_y = 0\,\text{psi}, \tau_{xy} = 1300\,\text{psi}$

11.5 Hooke's Law for Plane Stress

To solve the problems for Section 11.5, assume that the material is linearly elastic with modulus of elasticity E and Poisson's ratio ν.

Introductory Problems

11.5-1 A rectangular steel plate with thickness $t = 5/8$ in. is subjected to uniform normal stresses σ_x and σ_y, as shown in the figure. Strain gages A and B, oriented in the x and y directions, respectively, are attached to the plate. The gage readings give normal strains $\varepsilon_x = 0.00065$ (elongation) and $\varepsilon_y = 0.00040$ (elongation).

Knowing that $E = 30 \times 10^6\,\text{psi}$ and $\nu = 0.3$, determine the stresses σ_x and σ_y and the change Δt in the thickness of the plate.

PROBLEM 11.5-1 and 11.5-2

11.5-2 Solve the preceding problem if the thickness of the steel plate is $t = 12$ mm, the gage readings are $\varepsilon_x = 530 \times 10^{-6}$ (elongation) and $\varepsilon_y = -210 \times 10^{-6}$ (shortening), the modulus is $E = 200$ GPa, and Poisson's ratio is $\nu = 0.30$.

11.5-3 The state of stress on an element of material is shown in the figure. Calculate the unit volume change of the element if the stresses σ_x and σ_y are –20 ksi and 10 ksi, respectively. Assume $E = 10,600$ ksi and $\nu = 0.33$.

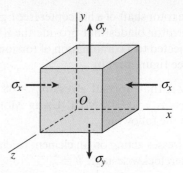

PROBLEM 11.5-3

11.5-4 An element of a material is subjected to plane stresses as shown in the figure. The stresses σ_x, σ_y, and τ_{xy} are 10 MPa, –15 MPa, and 5 MPa, respectively. Assume $E = 200$ GPa and $\nu = 0.3$. Calculate the normal strain in the x, y, and z directions and the shear strain.

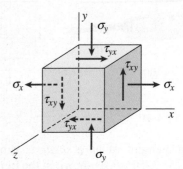

PROBLEM 11.5-4

11.5-5 Assume that the normal strains ε_x and ε_y for an element in plane stress (see figure) are measured with strain gages.

(a) Obtain a formula for the normal strain ε_z in the z direction in terms of ε_x, ε_y, and Poisson's ratio ν.
(b) Obtain a formula for the dilatation e in terms of ε_x, ε_y, and Poisson's ratio ν.

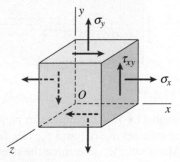

PROBLEM 11.5-5

11.5-6 A cast-iron plate in *biaxial stress* is subjected to tensile stresses $\sigma_x = 31$ MPa and $\sigma_y = 17$ MPa (see figure). The corresponding strains in the plate are $\varepsilon_x = 240 \times 10^{-6}$ and $\varepsilon_y = 85 \times 10^{-6}$.

Determine Poisson's ratio ν and the modulus of elasticity E for the material.

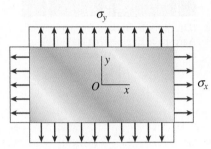

PROBLEMS 11.5-6 through 11.5-9

11.5-7 Solve the preceding problem for a steel plate with $\sigma_x = 11,600$ psi (tension), $\sigma_y = -5700$ psi (compression), $\varepsilon_x = 450 \times 10^{-6}$ (elongation), and $\varepsilon_y = -310 \times 10^{-6}$ (shortening).

11.5-8 A rectangular plate in *biaxial stress* (see figure) is subjected to normal stresses $\sigma_x = 67$ MPa (tension) and $\sigma_x = -23$ MPa (compression). The plate has dimensions $400 \times 550 \times 20$ mm and is made of steel with $E = 200$ GPa and $\nu = 0.30$.

(a) Determine the maximum in-plane shear strain γ_{max} in the plate.
(b) Determine the change Δt in the thickness of the plate.
(c) Determine the change ΔV in the volume of the plate.

11.5-9 Solve the preceding problem for an aluminum plate with $\sigma_x = 12,000$ psi (tension), $\sigma_y = -3000$ psi (compression), dimensions $20 \times 30 \times 0.5$ in., $E = 10.5 \times 10^6$ psi, and $\nu = 0.33$.

11.5-10 A brass cube of 48 mm on each edge is compressed in two perpendicular directions by forces $P = 160$ kN (see figure).

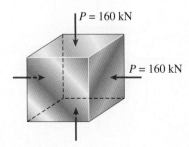

PROBLEM 11.5-10

(a) Calculate the change ΔV in the volume of the cube, assuming $E = 100$ GPa and $\nu = 0.34$.
(b) Repeat part (a) if the cube is made of an aluminum alloy with $E = 73$ GPa and $\nu = 0.33$.

Representative Problems

11.5-11 A 4.0 in. cube of concrete ($E = 4.5 \times 10^6$ psi, $\nu = 0.2$) is compressed in *biaxial stress* by means of a framework that is loaded as shown in the figure.

Assuming that each load F equals 25 kips, determine the change ΔV in the volume of the cube.

PROBLEM 11.5-11

11.5-12 A square plate of a width b and thickness t is loaded by normal forces P_x and P_y and by shear forces V, as shown in the figure. These forces produce uniformly distributed stresses acting on the side faces of the plate.

(a) Calculate the change ΔV in the volume of the plate if the dimensions are $b = 600$ mm and $t = 40$ mm; the plate is made of magnesium with $E = 41$ GPa and $\nu = 0.35$; and the forces are $P_x = 420$ kN, $P_y = 210$ kN, and $V = 96$ kN.

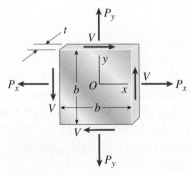

PROBLEMS 11.5-12 and 11.5-13

(b) Find the minimum width b of the square plate of thickness $t = 40$ mm when the change in volume of the plate cannot exceed 0.018% of the original volume.

11.5-13 Solve the preceding problem for an aluminum plate with $b = 10$ in., $t = 0.75$ in., $E = 10,600$ ksi, $\nu = 0.33$, $P_x = 96$ kips, $P_y = 24$ kips, and $V = 18$ kips.

For part (b) of Problem 11.5-12, assume that the change in volume cannot exceed 0.05%.

11.5-14 A circle of a diameter $d = 200$ mm is etched on a brass plate (see figure). The plate has dimensions of $400 \times 400 \times 20$ mm. Forces are applied to the plate, producing uniformly distributed normal stresses $\sigma_x = 59$ MPa and $\sigma_y = -17$ MPa. Calculate the following quantities: (a) the change in length Δac of diameter ac; (b) the change in length Δbd of diameter bd; (c) the change Δt in the thickness of the plate; (d) the change ΔV in the volume of the plate; and (e) the maximum permissible value of normal stress σ_x when the change in volume of the plate cannot exceed 0.015% of the original volume. (Assume $E = 100$ GPa and $\nu = 0.34$.)

(Dr. Mark E. Barkey, Department of Aerospace Engineering and Mechanics, University of Alabama)

PROBLEM 11.5-15

11.5-16 A rubber sheet in biaxial stress is subjected to tensile stresses $\sigma_x = 270$ Pa and $\sigma_y = 144$ Pa. The corresponding strains in the sheet are $\varepsilon_x = 0.0002$ and $\varepsilon_y = 0.000015$. Determine Poisson's ratio and the modulus elasticity of the material.

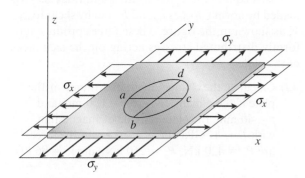

PROBLEM 11.5-14

11.5-15 The normal stress on an elastomeric rubber pad in a test machine is $\sigma_y = -100$ psi (see figure). Assume $E = 312$ psi and shear modulus $G = 105$ psi.

(a) Calculate the strains in the pad in the x, y, and z directions.

(b) Calculate the unit volume change of the rubber.

(Courtesy of TestResources, http://www.testresources.net)

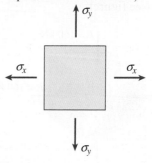

PROBLEM 11.5-16

11.6 Triaxial Stress

To solve the problems for Section 11.6, assume that the material is linearly elastic with modulus of elasticity E and Poisson's ratio ν.

Introductory Problems

11.6-1 An element of aluminum is subjected to triaxial stresses. Calculate the strains in the element in x, y, and z directions if the stresses σ_x, σ_y, and σ_z are –2000 psi, 3500 psi, and 2700 psi, respectively. Assume $E = 10,000$ ksi and $\nu = 0.33$. Also, find the unit volume change of the element.

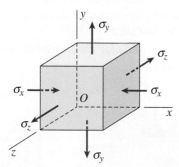

PROBLEM 11.6-1

11.6-2 An element of aluminum is subjected to triaxial stresses. Calculate the strains in the element in x, y, and z directions if the stresses σ_x, σ_y, and σ_z are –20 MPa, 28 MPa, and –18 MPa, respectively. Assume $E = 70$ GPa and $\nu = 0.33$.

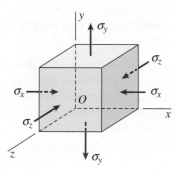

PROBLEM 11.6-2

11.6-3 An element of aluminum in the form of a rectangular parallelepiped (see figure) of dimensions $a = 5.5$ in., $b = 4.5$ in., and $c = 3.5$ in. is subjected to *triaxial stresses* $\sigma_x = 12,500$ psi, $\sigma_y = -5000$ psi, and $\sigma_z = -1400$ psi acting on the x, y, and z faces, respectively.

Determine the following quantities: (a) the maximum shear stress τ_{max} in the material; (b) the changes Δa, Δb, and Δc in the dimensions of the element;

(c) the change ΔV in the volume; and (d) the maximum value of σ_x when the change in volume must be limited to 0.021%. (Assume $E = 10,400$ ksi and $\nu = 0.33$.)

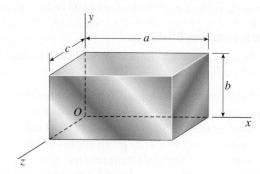

PROBLEMS 11.6-3 and 11.6-4

11.6-4 Solve the preceding problem if the element is steel ($E = 200$ GPa, $\nu = 0.30$) with dimensions $a = 300$ mm, $b = 150$ mm, and $c = 150$ mm and with the stresses $\sigma_x = -62$ MPa, $\sigma_y = -45$ MPa, and $\sigma_z = -45$ MPa.

For part (d) of Problem 11.6-3, find the maximum value of σ_x if the change in volume must be limited to -0.028%.

11.6-5 A cube of cast iron with sides of length $a = 4.0$ in. (see figure) is tested in a laboratory under *triaxial stress*. Gages mounted on the testing machine show that the compressive strains in the material are $\varepsilon_x = -225 \times 10^{-6}$ and $\varepsilon_y = \varepsilon_z = -37.5 \times 10^{-6}$.

Determine the following quantities: (a) the normal stresses σ_x, σ_y, and σ_z acting on the x, y, and z faces of the cube; (b) the maximum shear stress τ_{max} in the material; (c) the change ΔV in the volume of the cube; and (d) the maximum value of σ_x when the change in volume must be limited to 0.028%. (Assume $E = 14,000$ ksi and $\nu = 0.25$.)

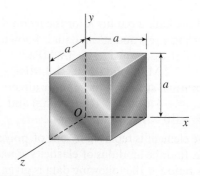

PROBLEMS 11.6-5 and 11.6-6

11.6-6 Solve the preceding problem if the cube is granite ($E = 80$ GPa, $\nu = 0.25$) with dimensions $a = 89$ mm and compressive strains $\varepsilon_x = 690 \times 10^{-6}$ and $\varepsilon_y = \varepsilon_z = 255 \times 10^{-6}$. For part (d) of Problem 11.6-5, find the maximum value of σ_x when the change in volume must be limited to 0.11%.

Representative Problems

11.6-7 An element of aluminum is subjected to *triaxial* stress (see figure).

(a) Find the bulk modulus K for the aluminum if the following stress and strain data are known: normal stresses are $\sigma_x = 5200$ psi (tension), $\sigma_y = -4750$ psi (compression); and $\sigma_z = -3090$ psi (compression) and normal strains in the x and y directions are $\varepsilon_x = 713.8 \times 10^{-6}$ (elongation) and $\varepsilon_y = -502.3 \times 10^{-6}$ (shortening).

(b) If the element is replaced by one of magnesium, find the modulus of elasticity E and Poisson's ratio ν if the following data is given: bulk modulus $K = 6.8 \times 10^{6}$ psi; normal stresses are $\sigma_x = 4550$ psi (tension), $\sigma_y = -1700$ psi (compression), and $\sigma_z = -1090$ psi (compression); and normal strain in the x direction is $\varepsilon_x = 900 \times 10^{-6}$ (elongation).

PROBLEMS 11.6-7 and 11.6-8

11.6-8 Solve the preceding problem if the material is nylon.

(a) Find the bulk modulus K for the nylon if the following stress and strain data is known: normal stresses are $\sigma_x = -3.9$ MPa, $\sigma_y = -3.2$ MPa, and $\sigma_z = -1.8$ MPa; and normal strains in the x and y directions are $\varepsilon_x = -640 \times 10^{-6}$ (shortening) and $\varepsilon_y = -310 \times 10^{-6}$ (shortening).

(b) If the element is replaced by one of polyethylene, find the modulus of elasticity E and Poisson's ratio ν if the following data is given: bulk

modulus $K = 2162$ MPa; normal stresses are $\sigma_x = -3.6$ MPa (compression), $\sigma_y = -2.1$ MPa (compression), and $\sigma_z = -2.1$ MPa (compression); and normal strain in the x direction is $\varepsilon_x = -1480 \times 10^{-6}$ (shortening).

11.6-9 A rubber cylinder R of length L and cross-sectional area A is compressed inside a steel cylinder S by a force F that applies a uniformly distributed pressure to the rubber (see figure).

(a) Derive a formula for the lateral pressure p between the rubber and the steel. (Disregard friction between the rubber and the steel, and assume that the steel cylinder is rigid when compared to the rubber.)

(b) Derive a formula for the shortening δ of the rubber cylinder.

PROBLEM 11.6-9

11.6-10 A block R of rubber is confined between plane parallel walls of a steel block S (see figure). A uniformly distributed pressure p_0 is applied to the top of the rubber block by a force F.

(a) Derive a formula for the lateral pressure p between the rubber and the steel. (Disregard friction between the rubber and the steel, and assume that the steel block is rigid when compared to the rubber.)

(b) Derive a formula for the dilatation e of the rubber.

PROBLEM 11.6-10

11.6-11 A rubber cube R of a side $L = 3$ in. and cross-sectional area $A = 9$ in^2 is compressed inside a steel cube S by a force $F = 5$ lb that applies uniformly distributed pressure to the rubber. Assume $E = 0.3$ ksi and $v = 0.45$.

(a) Calculate the lateral pressure between the rubber and steel (disregard friction between the rubber and the steel, and assume that the steel block is rigid when compared to the rubber).

(b) Calculate the change in volume of the rubber.

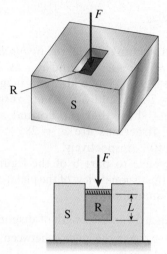

PROBLEM 11.6-11

11.6-12 A copper bar with a square cross section is inserted into a square rigid tube as shown in the figure. The length of the copper bar is 1.2 m and the area of the cross section is 300 mm². The bar is subjected to a force P that applies a uniformly distributed pressure to the copper. Calculate the force P if the longitudinal displacement of the bar is 2 mm. Assume that the modulus of elasticity of the bar is $E = 110$ GPa and Poisson's ratio is $v = 0.33$. Disregard friction between the copper and the rigid tube.

PROBLEM 11.6-12

11.6-13 A solid spherical ball of magnesium alloy $(E = 6.5 \times 10^{-6}\,\text{psi}, v = 0.35)$ is lowered into the ocean to a depth of 8000 ft. The diameter of the ball is 9.0 in.

(a) Determine the decrease Δd in diameter and the decrease ΔV in volume of the ball.

(b) At what depth will the volume change be equal to 0.0324% of the original volume?

11.6-14 A solid steel sphere $(E = 210$ GPa, $v = 0.3)$ is subjected to hydrostatic pressure p such that its volume is reduced by 0.4%.

(a) Calculate the pressure p.

(b) Calculate the volume modulus of elasticity K for the steel.

11.6-15 A solid bronze sphere (volume modulus of elasticity $K = 14.5 \times 10^6$ psi) is suddenly heated around its outer surface. The tendency of the heated part of the sphere to expand produces uniform tension in all directions at the center of the sphere.

If the stress at the center is 12,000 psi, what is the strain? Also, calculate the unit volume change e.

11.7 Plane Strain

To solve the problems for Section 11.7, consider only the in-plane strains (the strains in the x-y plane) unless stated otherwise. Use the transformation equations of plane strain except when Mohr's circle is specified.

Introductory Problems

11.7-1 An element of material in plain strain has the following strains: $\varepsilon_x = -0.001$ and $\varepsilon_y = 0.0015$.

(a) Determine the strains for an element oriented at an angle $\theta = 25°$.

(b) Find the principal strains of the element.

Confirm the solution using Mohr's circle for plane strain.

PROBLEM 11.7-1

903

11.7-2 An element of material in plain strain has the following strains: $\varepsilon_x = 0.002$ and $\varepsilon_y = 0.0015$.

(a) Determine the principal strains of the element.

(b) Determine the maximum shear strain of the element.

Confirm the solution using Mohr's circle for plane strain.

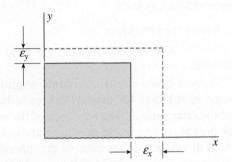

PROBLEM 11.7-2

11.7-3 An element of material in plain strain is subjected to shear strain $\gamma_{xy} = 0.0003$.

(a) Determine the strains for an element oriented at an angle $\theta = 30°$.

(b) Determine the principal strains of the element.

Confirm the solution using Mohr's circle for plane strain.

PROBLEM 11.7-3

11.7-4 An element of material in plain strain is subjected to strains $\varepsilon_x = 0.0015$, $\varepsilon_y = -0.0002$, and $\gamma_{xy} = 0.0003$.

(a) Determine the strains for an element oriented at an angle $\theta = 20°$.

(b) Determine the principal strains of the element.

Confirm the solution using Mohr's circle for plane strain.

PROBLEM 11.7-4

11.7-5 A thin rectangular plate in *biaxial stress* is subjected to stresses σ_x and σ_y, as shown in part a of the figure. The width and height of the plate are $b = 7.5$ in. and $h = 2.5$ in., respectively. Measurements show that the normal strains in the x and y directions are $\varepsilon_x = 285 \times 10^{-6}$ and $\varepsilon_y = -190 \times 10^{-6}$, respectively.

With reference to part b of the figure, which shows a two-dimensional view of the plate, determine the following quantities.

(a) The increase Δd in the length of diagonal Od.

(b) The change $\Delta\phi$ in the angle ϕ between diagonal Od and the x axis.

(c) The change $\Delta\psi$ in the angle ψ between diagonal Od and the y axis.

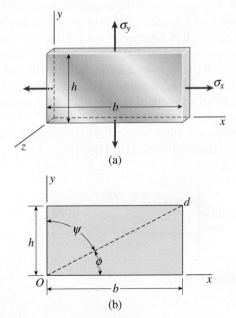

PROBLEMS 11.7-5 and 11.7-6

11.7-6 Solve the preceding problem if $b = 180$ mm and $h = 70$ mm, respectively. Measurements show that the normal strains in the x and y directions are $\varepsilon_x = 390 \times 10^{-6}$ and $\varepsilon_y = -240 \times 10^{-6}$, respectively.

11.7-7 A thin square plate in *biaxial stress* is subjected to stresses σ_x and σ_y, as shown in part a of the figure. The width of the plate is $b = 12.0$ in. Measurements show that the normal strains in the x and y directions are $\varepsilon_x = 427 \times 10^{-6}$ and $\varepsilon_y = 113 \times 10^{-6}$, respectively.

With reference to part b of the figure, which shows a two-dimensional view of the plate, determine the following quantities.

(a) The increase Δd in the length of diagonal Od.
(b) The change $\Delta\phi$ in the angle ϕ between diagonal Od and the x axis.
(c) The shear strain γ associated with diagonals Od and cf (that is, find the decrease in angle ced).

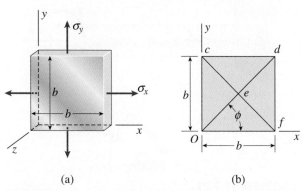

(a)   (b)

PROBLEMS 11.7-7 and 11.7-8

11.7-8 Solve the preceding problem if $b = 225$ mm, $\varepsilon_x = 845 \times 10^{-6}$, and $\varepsilon_y = 211 \times 10^{-6}$.

11.7-9 An element of material subjected to *plane strain* (see figure) has strains of $\varepsilon_x = 280 \times 10^{-6}$, $\varepsilon_y = 420 \times 10^{-6}$, and $\gamma_{xy} = 150 \times 10^{-6}$.

Calculate the strains for an element oriented at an angle $\theta = 35°$. Show these strains on a sketch of a properly oriented element.

PROBLEMS 11.7-9 through 11.7-14

11.7-10 Solve the preceding problem for the following data: $\varepsilon_x = 190 \times 10^{-6}$, $\varepsilon_y = -230 \times 10^{-6}$, $\gamma_{xy} = 160 \times 10^{-6}$, and $\theta = 40°$.

11.7-11 The strains for an element of material in *plane strain* (see figure) are as follows: $\varepsilon_x = 480 \times 10^{-6}$, $\varepsilon_y = 140 \times 10^{-6}$, and $\gamma_{xy} = -350 \times 10^{-6}$.

Determine the principal strains and maximum shear strains, and show these strains on sketches of properly oriented elements.

11.7-12 Solve the preceding problem for the following strains: $\varepsilon_x = 120 \times 10^{-6}$, $\varepsilon_y = -450 \times 10^{-6}$, and $\gamma_{xy} = -360 \times 10^{-6}$.

Representative Problems

11.7-13 An element of material in *plane strain* (see figure) is subjected to strains $\varepsilon_x = 480 \times 10^{-6}$, $\varepsilon_y = 70 \times 10^{-6}$, and $\gamma_{xy} = 420 \times 10^{-6}$.

Determine the following quantities: (a) the strains for an element oriented at an angle $\theta = 75°$, (b) the principal strains, and (c) the maximum shear strains. Show the results on sketches of properly oriented elements.

11.7-14 Solve the preceding problem for the following data: $\varepsilon_x = -1120 \times 10^{-6}$, $\varepsilon_y = -430 \times 10^{-6}$, $\gamma_{xy} = 780 \times 10^{-6}$, and $\theta = 45°$.

11.7-15 A brass plate with a modulus of elasticity $E = 16 \times 10^6$ psi and Poisson's ratio $\nu = 0.34$ is loaded in *biaxial stress* by normal stresses σ_x and σ_y (see figure). A strain gage is bonded to the plate at an angle $\phi = 35°$.

If the stress σ_x is 10,700 psi and the strain measured by the gage is $\varepsilon = 390 \times 10^{-6}$, what is the maximum in-plane shear stress $(\tau_{max})_{xy}$ and shear strain $(\gamma_{max})_{xy}$? What is the maximum shear strain $(\gamma_{max})_{xz}$ in the x-z plane? What is the maximum shear strain $(\gamma_{max})_{yz}$ in the y-z plane?

PROBLEMS 11.7-15 and 11.7-16

11.7-16 Solve the preceding problem if the plate is made of aluminum with $E = 72$ GPa and Poisson's ratio $\nu = 0.33$. The plate is loaded in *biaxial stress* with normal stress $\sigma_x = 79$ MPa, angle $\phi = 18°$, and the strain measured by the gage is $\varepsilon = 925 \times 10^{-6}$.

11.7-17 An element in *plane stress* is subjected to stresses $\sigma_x = -8400$ psi, $\sigma_y = 1100$ psi, and $\tau_{xy} = -1700$ psi (see figure). The material is aluminum with modulus of elasticity $E = 10,000$ ksi and Poisson's ratio $\nu = 0.33$.

Determine the following quantities: (a) the strains for an element oriented at an angle $\theta = 30°$, (b) the principal strains, and (c) the maximum shear strains. Show the results on sketches of properly oriented elements.

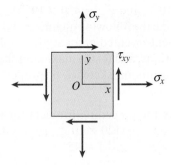

PROBLEMS 11.7-17 and 11.7-18

11.7-18 Solve the preceding problem for the following data: $\sigma_x = -150$ MPa, $\sigma_y = -210$ MPa, $\tau_{xy} = -16$ MPa, and $\theta = 50°$. The material is brass with $E = 100$ GPa and $\nu = 0.34$.

11.7-19 During a test of an airplane wing, the strain gage readings from a 45° rosette (see figure) are as follows: gage A, 520×10^{-6}; gage B, 360×10^{-6}; and gage C, -80×10^{-6}.

Determine the principal strains and maximum shear strains, and show them on sketches of properly oriented elements.

PROBLEMS 11.7-19 and 11.7-20

11.7-20 A 45° strain rosette (see figure) mounted on the surface of an automobile frame gives the following readings: gage A, 310×10^{-6}; gage B, 180×10^{-6}; and gage C, -160×10^{-6}.

Determine the principal strains and maximum shear strains, and show them on sketches of properly oriented elements.

11.7-21 A solid circular bar with a diameter of $d = 1.25$ in. is subjected to an axial force P and a torque T (see figure). Strain gages A and B mounted on the surface of the bar give readings $\varepsilon_A = 140 \times 10^{-6}$ and $\varepsilon_B = -60 \times 10^{-6}$. The bar is made of steel having $E = 30 \times 10^6$ psi and $\nu = 0.29$.

(a) Determine the axial force P and the torque T.

(b) Determine the maximum shear strain γ_{max} and the maximum shear stress τ_{max} in the bar.

PROBLEM 11.7-21

11.7-22 A cantilever beam with a rectangular cross section (width $b = 20$ mm, height $h = 175$ mm) is loaded by a force P that acts at the mid-height of the beam and is inclined at an angle α to the vertical (see figure). Two strain gages are placed at point C, which also is at the mid-height of the beam. Gage A

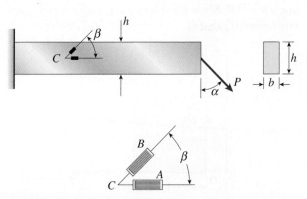

PROBLEMS 11.7-22 and 11.7-23

measures the strain in the horizontal direction, and gage B measures the strain at an angle $\beta = 60°$ to the horizontal. The measured strains are $\varepsilon_A = 145 \times 10^{-6}$ and $\varepsilon_B = -165 \times 10^{-6}$.

Determine the force P and the angle α, assuming the material is steel with $E = 200$ GPa and $\nu = 1/3$.

11.7-23 Solve the preceding problem if the cross-sectional dimensions are $b = 1.5$ in. and $h = 5.0$ in., the gage angle is $\beta = 75°$, the measured strains are $\varepsilon_A = 209 \times 10^{-6}$ and $\varepsilon_B = -110 \times 10^{-6}$, and the material is a magnesium alloy with modulus $E = 6.0 \times 10^6$ psi and Poisson's ratio $\nu = 0.35$.

11.7-24 A 60° strain rosette, or *delta rosette*, consists of three electrical-resistance strain gages arranged as shown in the figure. Gage A measures the normal strain ε_a in the direction of the x axis. Gages B and C measure the strains ε_b and ε_c in the inclined directions shown.

Obtain the equations for the strains ε_x, ε_y, and γ_{xy} associated with the x-y axes.

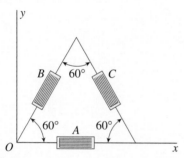

PROBLEM 11.7-24

11.7-25 On the surface of a structural component in a space vehicle, the strains are monitored by means of three strain gages arranged as shown in the figure. During a certain maneuver, the following strains were recorded: $\varepsilon_a = 1100 \times 10^{-6}$, $\varepsilon_b = 200 \times 10^{-6}$, and $\varepsilon_c = 200 \times 10^{-6}$.

Determine the principal strains and principal stresses in the material, which is a magnesium alloy

PROBLEM 11.7-25

for which $E = 6000$ ksi and $\nu = 0.35$. Show the principal strains and principal stresses on sketches of properly oriented elements.

11.7-26 The strains on the surface of an experimental device made of pure aluminum ($E = 70$ GPa, $\nu = 0.33$) and tested in a space shuttle were measured by means of strain gages. The gages were oriented as shown in the figure, and the measured strains were $\varepsilon_a = 1100 \times 10^{-6}$, $\varepsilon_b = 1496 \times 10^{-6}$, and $\varepsilon_c = -39.44 \times 10^{-6}$.

What is the stress σ_x in the x direction?

PROBLEM 11.11-26

11.7-27 Solve Problem 11.7-9 by using Mohr's circle for plane strain.

11.7-28 Solve Problem 11.7-10 by using Mohr's circle for plane strain.

11.7-29 Solve Problem 11.7-11 by using Mohr's circle for plane strain.

11.7-30 Solve Problem 11.7-12 by using Mohr's circle for plane strain.

11.7-31 Solve Problem 11.7-13 by using Mohr's circle for plane strain.

11.7-32 Solve Problem 11.7-14 by using Mohr's circle for plane strain.

PROBLEMS 11.7-27 through 11.7-32

907

Applications of Plane Stress
(Pressure Vessels and Combined Loadings)

Thin-walled spherical pressure vessels are used for storage of natural gas in this refinery.

muratart/Shutterstock.com

CHAPTER OBJECTIVES

- Study the condition of stress known as *plane stress* and its application in such structures as *spherical and cylindrical pressure vessels.*

- Evaluate *combined loadings* such as axial, torsional, shear, bending, and possibly internal pressure at key locations on

  structures to define the state of plane stress at different points of interest.

- Present a detailed approach for analysis of critical points in a structure acted on by combined loadings.

Chapter Outline

12.1 Introduction

This chapter presents the analysis of some practical examples of structures and components in states of plane stress or strain, building upon the concepts presented in Chapter 11. First, stresses and strains in the walls of thin pressure vessels are examined. Then structures acted upon by combined loadings are evaluated to find the maximum normal and shear stresses that govern their design.

12.2 Spherical Pressure Vessels

Pressure vessels are closed structures containing liquids or gases under pressure. Familiar examples include tanks, pipes, and pressurized cabins in aircraft and space vehicles. When pressure vessels have walls that are thin in comparison to their overall dimensions, they are included within a more general category known as **shell structures**. Examples of shell structures are roof domes, airplane wings, and submarine hulls.

This section considers thin-walled pressure vessels of spherical shape, like the compressed-air tank shown in Fig. 12-1. The term **thin-walled** is not precise, but as a general rule, pressure vessels are considered to be thin-walled when the ratio of radius r to wall thickness t (Fig. 12-2) is greater than 10. When this condition is met, the stresses in the walls can be determined with reasonable accuracy using statics alone.

Assume in the following discussions that the internal pressure p (Fig. 12-2) exceeds the pressure acting on the outside of the shell. Otherwise, the vessel may collapse inward due to buckling.

A sphere is the theoretically ideal shape for a vessel that resists internal pressure. You only need to contemplate the familiar soap bubble to recognize that a sphere is the "natural" shape for this purpose. To determine the stresses in a spherical vessel, cut through the sphere on a vertical diametral plane (Fig. 12-3a) and isolate half of the shell *and its fluid contents* as a single free body (Fig. 12-3b). Acting on this free body are the tensile stresses σ in the wall of the vessel and the fluid pressure p. This pressure acts horizontally against the plane circular area of fluid remaining inside the hemisphere. Since the pressure is uniform, the resultant pressure force P (Fig. 12-3b) is

$$P = p(\pi r^2) \tag{12-1}$$

where r is the inner radius of the sphere.

Note that the pressure p is not the absolute pressure inside the vessel but is the net internal pressure, or the **gage pressure**. Gage pressure is the internal pressure *above* the pressure acting on the outside of the vessel. If the internal and external pressures are the same, no stresses are developed in the wall of the vessel—only the excess of internal pressure over external pressure has any effect on these stresses.

Because of the symmetry of the vessel and its loading (Fig. 12-3b), the tensile stress σ is uniform around the circumference. Furthermore, since the wall is thin, you can assume with good accuracy that the stress is uniformly distributed across the thickness t. The accuracy of this approximation increases as the shell becomes thinner and decreases as it becomes thicker.

Thin-walled spherical pressure vessel used for storage of propane in this oil refinery

FIGURE 12-1

Spherical pressure vessel

Welded seam

FIGURE 12-2

Cross section of spherical pressure vessel showing inner radius r, wall thickness t, and internal pressure p

The resultant of the tensile stresses σ in the wall is a horizontal force equal to the stress σ times the area over which it acts, or

$$\sigma(2\pi r_m t)$$

where t is the thickness of the wall and r_m is its mean radius:

$$r_m = r + \frac{t}{2} \tag{12-2}$$

Thus, equilibrium of forces in the horizontal direction (Fig. 12-3b) gives

$$\Sigma F_{\text{horiz}} = 0 \qquad \sigma(2\pi r_m t) - p(\pi r^2) = 0 \tag{12-3}$$

from which the *tensile stresses* in the wall of the vessel are derived:

$$\sigma = \frac{pr^2}{2r_m t} \tag{12-4}$$

Since this analysis is valid only for thin shells, disregard the small difference between the two radii appearing in Eq. (12-4) and replace r by r_m or replace r_m by r. While either choice is satisfactory for this approximate analysis, the stresses are closer to the theoretically exact stresses if the inner radius r is used instead of the mean radius r_m. Therefore, the following formula is used for calculating the **tensile stresses in the wall of a spherical shell**:

$$\sigma = \frac{pr}{2t} \tag{12-5}$$

As is evident from the symmetry of a spherical shell, you obtain the same equation for the tensile stresses when you cut a plane through the center of the sphere in any direction whatsoever. Thus, *the wall of a pressurized spherical vessel is subjected to uniform tensile stresses σ in all directions*. This stress condition is represented in Fig. 12-3c by the small stress element with stresses σ acting in mutually perpendicular directions.

Stresses that act tangentially to the curved surface of a shell, such as the stresses σ shown in Fig. 12-3c, are known as **membrane stresses**. The name arises from the fact that these are the only stresses that exist in true membranes, such as soap films.

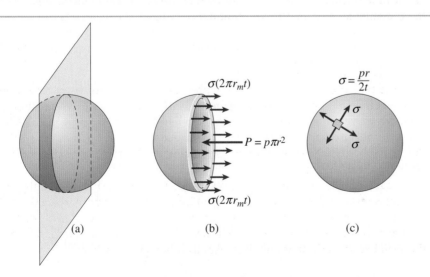

FIGURE 12-3

Tensile stresses σ in the wall of a spherical pressure vessel

$\sigma(2\pi r_m t)$

$\sigma = \dfrac{pr}{2t}$

$P = p\pi r^2$

σ

σ

$\sigma(2\pi r_m t)$

(a)   (b)   (c)

FIGURE 12-4

Stresses in a spherical pressure
vessel at (a) the outer surface and
(b) the inner surface

(a)

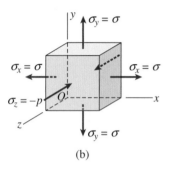

(b)

Stresses at the Outer Surface

The outer surface of a spherical pressure vessel is usually free of any loads. Therefore, the element shown in Fig. 12-3c is in *biaxial stress*. To aid in analyzing the stresses acting on this element, it appears again in Fig. 12-4a, where a set of coordinate axes is oriented parallel to the sides of the element. The x and y axes are tangential to the surface of the sphere, and the z axis is perpendicular to the surface. Thus, the normal stresses σ_x and σ_y are the same as the membrane stresses σ, and the normal stress σ_z is zero. No shear stresses act on the sides of this element.

Analyzing the element of Fig. 12-4a by using the transformation equations for plane stress [see Fig. 11-1 and Eqs. (11-4a and b) of Section 11.2] gives

$$\sigma_{x1} = \sigma \quad \text{and} \quad \tau_{x1y1} = 0$$

as expected. In other words, when you consider elements obtained by rotating the axes about the z axis, the normal stresses remain constant and there are no shear stresses. *Every plane is a principal plane and every direction is a principal direction.* Thus, the **principal stresses** for the element are

$$\sigma_1 = \sigma_2 = \frac{pr}{2t} \qquad \sigma_3 = 0 \tag{12-6a,b}$$

The stresses σ_1 and σ_2 lie in the x-y plane, and the stress σ_3 acts in the z direction.

To obtain the **maximum shear stresses**, consider out-of-plane rotations, that is, rotations about the x and y axes (because all in-plane shear stresses are zero). Elements oriented by making 45° rotations about the x and y axes have maximum shear stresses equal to $\sigma/2$ and normal stresses equal to $\sigma/2$. Therefore,

$$\tau_{max} = \frac{\sigma}{2} = \frac{pr}{4t} \tag{12-7}$$

These stresses are the largest shear stresses in the element.

Stresses at the Inner Surface

At the inner surface of the wall of a spherical vessel, a stress element (Fig. 12-4b) has the same membrane stresses σ_x and σ_y as does an element at the outer surface (Fig. 12-4a). In addition, a compressive stress σ_z equal to the pressure p acts in the z direction (Fig. 12-4b). This compressive stress decreases from p at the inner surface of the sphere to zero at the outer surface.

The element shown in Fig. 12-4b is in triaxial stress with principal stresses

$$\sigma_1 = \sigma_2 = \frac{pr}{2t} \qquad \sigma_3 = -p \tag{12-8a,b}$$

The in-plane shear stresses are zero, but the maximum out-of-plane shear stress (obtained by a 45° rotation about either the x or y axis) is

$$\tau_{max} = \frac{\sigma + p}{2} = \frac{pr}{4t} + \frac{p}{2} = \frac{p}{2}\left(\frac{r}{2t} + 1\right) \tag{12-9}$$

When the vessel is thin-walled and the ratio r/t is large, disregard the number 1 in comparison with the term $r/2t$. In other words, the principal stress σ_3 in the z direction is small when compared with the principal stresses σ_1 and σ_2. Consequently, consider the stress state at the inner surface to be the same as at the outer surface (biaxial stress). This approximation is consistent with the approximate nature of thin-shell theory, so use Eqs. (12-5), (12-6a,b), and (12-7) to obtain the stresses in the wall of a spherical pressure vessel.

General Comments

Pressure vessels usually have openings in their walls (to serve as inlets and outlets for the fluid contents) as well as fittings and supports that exert forces on the shell (Fig. 12-1). These features result in nonuniformities in the stress distribution, or *stress concentrations*, that cannot be analyzed by the elementary formulas given here. Instead, more advanced methods of analysis are needed. Other factors that affect the design of pressure vessels include corrosion, accidental impacts, and temperature changes.

Some of the limitations of thin-shell theory as applied to pressure vessels are listed here.

1. The wall thickness must be small in comparison to the other dimensions (the ratio r/t should be 10 or more).
2. The internal pressure must exceed the external pressure (to avoid inward buckling).
3. The analysis presented in this section is based only on the effects of internal pressure (the effects of external loads, reactions, the weight of the contents, and the weight of the structure are not considered).
4. The formulas derived in this section are valid throughout the wall of the vessel *except* near points of stress concentrations.

The following example illustrates how the principal stresses and maximum shear stresses are used in the analysis of a spherical shell.

Example 12-1

A compressed-air tank having an inner diameter of 18 feet and a wall thickness of 1.75 inches is formed by welding two steel hemispheres (Fig. 12-5).

(a) If the allowable tensile stress in the steel is 13.5 ksi, what is the maximum permissible air pressure p_a in the tank?

(b) If the allowable shear stress in the steel is 6200 psi, what is the maximum permissible pressure p_b?

(c) If the normal strain at the outer surface of the tank is not to exceed 0.0003, what is the maximum permissible pressure p_c? (Assume that Hooke's law is valid and that the modulus of elasticity for the steel is 29×10^6 psi and Poisson's ratio is 0.28.)

(d) Tests on the welded seam show that failure occurs when the tensile load on the welds exceeds 42 kip/in. of weld. If the required factor of safety against failure of the weld is 2.5, what is the maximum permissible pressure p_d?

(e) Considering the four preceding factors, what is the allowable pressure p_{allow} in the tank?

FIGURE 12-5

Example 12-1: Spherical pressure vessel
(Attachments and supports are shown in photo.)

Spherical tanks at oil refinery

Solution:

Use the four-step problem-solving approach. Combine steps as needed for an efficient solution.

1. *Conceptualize [hypothesize, sketch]*: The compressed-air tank is a thin-walled pressure vessel. Evaluate membrane stresses at the outer surface of the tank. Tensile stresses in the wall are evaluated using Eq. (12-5). Computed stresses will be closer to theoretically exact values if the inner radius of the shell rather than the mean radius is used.

2,3. *Categorize [simplify, classify], Analyze [evaluate; select relevant equations, carry out mathematical solution]*:

Part (a): Allowable pressure based upon the tensile stress in the steel.

The maximum tensile stress in the wall of the tank is given by the formula $\sigma = pr/2t$ [see Eq. (12-5)]. Solve this equation for the pressure in terms of the allowable stress to get

$$p_a = \frac{2t\sigma_{\text{allow}}}{r} = \frac{2(1.75 \text{ in.})(13{,}500 \text{ psi})}{108 \text{ in.}} = 437.5 \text{ psi} \quad \Leftarrow$$

Thus, the maximum allowable pressure based upon tension in the wall of the tank is $p_a = 437$ psi. (Note that in a calculation of this kind, you round downward, not upward.)

Part (b): Allowable pressure based upon the shear stress in the steel.

The maximum shear stress in the wall of the tank is given by Eq. (12-7), from which the pressure is obtained:

$$p_b = \frac{4t\tau_{\text{allow}}}{r} = \frac{4(1.75 \text{ in.})(6200 \text{ psi})}{108 \text{ in.}} = 401.9 \text{ psi} \quad \Leftarrow$$

Therefore, the allowable pressure based upon shear is $p_b = 401$ psi.

Part (c): Allowable pressure based upon the normal strain in the steel.

The normal strain is obtained from Hooke's law for biaxial stress [Eq. (11-40a)]:

$$\varepsilon_x = \frac{1}{E}(\sigma_x - \nu\sigma_y) \tag{a}$$

Substitute $\sigma_x = \sigma_y = \sigma = pr/2t$ (see Fig. 12-4a) to obtain

$$\varepsilon_x = \frac{\sigma}{E}(1 - \nu) = \frac{pr}{2tE}(1 - \nu) \tag{12-10}$$

This equation can be solved for the pressure p_c:

$$p_c = \frac{2tE\varepsilon_{allow}}{r(1-\nu)} = \frac{2(1.75 \text{ in.})(29,000 \text{ ksi})(0.0003)}{108 \text{ in.}(1-0.28)} = 391.6 \text{ psi}$$

Thus, the allowable pressure based upon the normal strain in the wall is $p_c = 391$ psi.

Part (d): Allowable pressure based upon the tension in the welded seam.

The allowable tensile load on the welded seam is equal to the failure load divided by the factor of safety:

$$T_{allow} = \frac{T_{failure}}{n} = \frac{42 \text{ kip/in.}}{2.5} = 16,800 \text{ lb/in.}$$

The corresponding allowable tensile stress is equal to the allowable load on a one-inch length of weld divided by the cross-sectional area of a one-inch length of weld:

$$\sigma_{allow} = \frac{T_{allow}(1.0 \text{ in.})}{(1.0 \text{ in.})(t)} = \frac{16,800 \text{ lb/in.}(1.0 \text{ in.})}{(1.0 \text{ in.})(1.75 \text{ in.})} = 9600 \text{ psi}$$

Finally, solve for the internal pressure by using Eq. (12-5):

$$p_d = \frac{2t\sigma_{allow}}{r} = \frac{2(1.75 \text{ in.})(9600 \text{ psi})}{108 \text{ in.}} = 311.1 \text{ psi}$$

This result gives the allowable pressure based upon tension in the welded seam.

Part (e): Allowable pressure.

Comparing the preceding results for p_a, p_b, p_c, and p_d, note that tension in the welded seam governs and the allowable pressure in the tank is

$$p_{allow} = 311 \text{ psi}$$

4. *Finalize* [conclude; examine answer—Does it make sense? Are units correct? How does it compare to similar problem solutions?]: This example illustrates how various stresses and strains enter into the design of a spherical pressure vessel.

   Note: When the internal pressure is at its maximum allowable value (311 psi), the tensile stresses in the shell are

$$\sigma = \frac{pr}{2t} = \frac{311 \text{ psi } (108 \text{ in.})}{2(1.75 \text{ in.})} = 9597 \text{ psi}$$

Thus, at the inner surface of the shell (Fig. 12-4b), the ratio of the principal stress in the z direction (311 psi) to the in-plane principal stresses (9597 psi) is only 0.032. Therefore, the earlier assumption that the principal stress σ_3 in the z direction can be disregarded and the entire shell to be in biaxial stress is justified.

FIGURE 12-6

Cylindrical pressure vessels with circular cross sections

(a)

(b)

Cylindrical storage tanks in a petrochemical plant

12.3 Cylindrical Pressure Vessels

Cylindrical pressure vessels with a circular cross section (Fig. 12-6) are found in industrial settings (compressed air tanks and rocket motors), in homes (fire extinguishers and spray cans), and in the countryside (propane tanks and grain silos). Pressurized pipes, such as water-supply pipes and penstocks, are also classified as cylindrical pressure vessels.

The analysis of cylindrical vessels begins by determining the normal stresses in a *thin-walled circular tank AB* subjected to internal pressure (Fig. 12-7a). A *stress element* with its faces parallel and perpendicular to the axis of the tank is shown on the wall of the tank. The normal stresses σ_1 and σ_2 acting on the side faces of this element are the membrane stresses in the wall. No shear stresses act on these faces because of the symmetry of the vessel and its loading. Therefore, the stresses σ_1 and σ_2 are principal stresses.

Because of their directions, the stress σ_1 is called the **circumferential stress** or the **hoop stress**, and the stress σ_2 is called the **longitudinal stress** or the **axial stress**. Each of these stresses can be calculated from equilibrium by using appropriate free-body diagrams.

Circumferential Stress

To determine the circumferential stress σ_1, make two cuts (*mn* and *pq*) perpendicular to the longitudinal axis and a distance *b* apart (Fig. 12-7a). Then make a third cut in a vertical plane through the longitudinal axis of the tank, resulting in the free body shown in Fig. 12-7b. This free body consists not only of the half-circular piece of the tank but also of the fluid contained within the cuts. Acting on the longitudinal cut (plane *mpqn*) are the circumferential stresses σ_1 and the internal pressure *p*.

Stresses and pressures also act on the left-hand and right-hand faces of the free body. However, these stresses and pressures are not shown in the figure because they do not enter the equation of equilibrium used here. As in the analysis of a spherical vessel, disregard the weight of the tank and its contents.

FIGURE 12-7

Stresses in a circular cylindrical pressure vessel

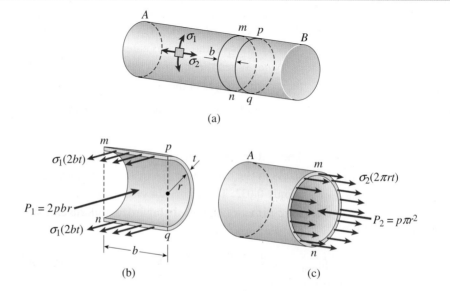

(a)

(b)   (c)

The circumferential stresses σ_1 acting in the wall of the vessel have a resultant equal to $\sigma_1(2bt)$, where t is the thickness of the wall. Also, the resultant force P_1 of the internal pressure is equal to $2pbr$, where r is the inner radius of the cylinder. The resulting equation of equilibrium is

$$\sigma_1(2bt) - 2pbr = 0$$

This equation leads to the formula for the *circumferential stress in a pressurized cylinder*:

$$\sigma_1 = \frac{pr}{t} \tag{12-11}$$

This stress is uniformly distributed over the thickness of the wall, provided the thickness is small compared to the radius.

Longitudinal Stress

The longitudinal stress σ_2 is obtained from the equilibrium of a free body of the part of the vessel to the left of cross section *mn* (Fig. 12-7c). Again, the free body includes not only part of the tank but also its contents. The stresses σ_2 act longitudinally and have a resultant force equal to $\sigma_2(2\pi rt)$. Note that the inner radius of the shell is used in place of the mean radius, as explained in Section 12.2.

The resultant force P_2 of the internal pressure is a force equal to $p\pi r^2$. Thus, the equation of equilibrium for the free body is

$$\sigma_2(2\pi rt) - p\pi r^2 = 0$$

Solve this equation for σ_2 to obtain the formula for the *longitudinal stress* in a cylindrical pressure vessel:

$$\sigma_2 = \frac{pr}{2t} \tag{12-12}$$

This stress is equal to the membrane stress in a spherical vessel [Eq. (12-5)].

Compare Eqs. (12-11) and (12-12) to see that the circumferential stress in a cylindrical vessel is equal to twice the longitudinal stress:

$$\sigma_1 = 2\sigma_2 \tag{12-13}$$

From this result, note that a longitudinal welded seam in a pressurized tank must be twice as strong as a circumferential seam.

Stresses at the Outer Surface

The principal stresses σ_1 and σ_2 at the outer surface of a cylindrical vessel are shown on the stress element of Fig. 12-8a. Since the third principal stress (acting in the z direction) is zero, the element is in *biaxial stress.*

The maximum *in-plane shear stresses* occur on planes that are rotated 45° about the z axis; these stresses are

$$(\tau_{max})_z = \frac{\sigma_1 - \sigma_2}{2} = \frac{\sigma_1}{4} = \frac{pr}{4t} \tag{12-14}$$

(a)

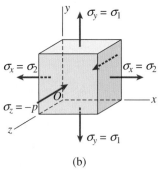

(b)

The maximum *out-of-plane shear stresses* are obtained by 45° rotations about the x and y axes, respectively; thus,

$$(\tau_{max})_x = \frac{\sigma_1}{2} = \frac{pr}{2t} \qquad (\tau_{max})_y = \frac{\sigma_2}{2} = \frac{pr}{4t} \qquad \textbf{(12-15a,b)}$$

The preceding results show that the *absolute maximum shear stress* is

$$\tau_{max} = \frac{\sigma_1}{2} = \frac{pr}{2t} \qquad \textbf{(12-16)}$$

This stress occurs on a plane that has been rotated 45° about the x axis.

Stresses at the Inner Surface

The stress conditions at the inner surface of the wall of the vessel are shown in Fig. 12-8b. The principal stresses are

$$\sigma_1 = \frac{pr}{t} \qquad \sigma_2 = \frac{pr}{2t} \qquad \sigma_3 = -p \qquad \textbf{(12-17a,b,c)}$$

The three maximum shear stresses, obtained by 45° rotations about the x, y, and z axes, are

$$(\tau_{max})_x = \frac{\sigma_1 - \sigma_3}{2} = \frac{pr}{2t} + \frac{p}{2} \qquad \textbf{(12-18a)}$$

$$(\tau_{max})_y = \frac{\sigma_2 - \sigma_3}{2} = \frac{pr}{4t} + \frac{p}{2} \qquad \textbf{(12-18b)}$$

$$(\tau_{max})_z = \frac{\sigma_1 - \sigma_2}{2} = \frac{pr}{4t} \qquad \textbf{(12-18c)}$$

The first of these three stresses is the largest. However, as explained in the discussion of shear stresses in a spherical shell, you may disregard the additional term $p/2$ in Eqs. (12-18a and b) when the shell is thin-walled. Equations (12-18a, b, and c) then become the same as Eqs. (12-15) and (12-14), respectively.

Therefore, in all examples and problems pertaining to cylindrical pressure vessels, *disregard the presence of the compressive stress in the z direction.* (This compressive stress varies from p at the inner surface to zero at the outer surface.) With this approximation, the stresses at the inner surface become the same as the stresses at the outer surface (biaxial stress). As explained in the discussion of spherical pressure vessels, this procedure is satisfactory when you consider the numerous other approximations in this theory.

General Comments

The preceding formulas for stresses in a circular cylinder are valid in parts of the cylinder away from any discontinuities that cause stress concentrations, as discussed previously for spherical shells. An obvious discontinuity exists at the ends of the cylinder where the heads are attached, because the geometry of the structure changes abruptly. Other stress concentrations occur at openings, at points of support, and wherever objects or fittings are attached to the cylinder. The stresses at such points cannot be determined solely from equilibrium equations; instead, more advanced methods of analysis (such as shell theory and finite-element analysis) must be used.

Some of the limitations of the elementary theory for thin-walled shells are listed in Section 12.2.

Example 12-2

FIGURE 12-9

Example 12-2: Cylindrical pressure vessel with a helical weld

Helical weld

Cylindrical pressure vessel on simple supports

A cylindrical pressure vessel is constructed from a long, narrow steel plate by wrapping the plate around a mandrel and then welding along the edges of the plate to make a helical joint (Fig. 12-9). The helical weld makes an angle $\alpha = 55°$ with the longitudinal axis. The vessel has inner radius $r = 1.8$ m and wall thickness $t = 20$ mm. The material is steel with modulus $E = 200$ GPa and Poisson's ratio $\nu = 0.30$. The internal pressure p is 800 kPa.

Calculate the following quantities for the cylindrical part of the vessel: (a) the circumferential and longitudinal stresses σ_1 and σ_2, respectively; (b) the maximum in-plane and out-of-plane shear stresses; (c) the circumferential and longitudinal strains ε_1 and ε_2, respectively; and (d) the normal stress σ_w and shear stress τ_w acting perpendicular and parallel, respectively, to the welded seam.

Solution:

Use the four-step problem-solving approach. Combine steps as needed for an efficient solution.

1. *Conceptualize*: The circular cylindrical tank is a thin-walled pressure vessel. Evaluate membrane stresses at the outer surface of the tank. Evaluate circumferential stresses in the wall using Eq. (12-11) and longitudinal stresses using Eq. (12-12). Computed stresses will be closer to theoretically exact values if you use the inner radius of the shell rather than the mean radius.

2, 3. *Categorize, Analyze*:

Part (a): Circumferential and longitudinal stresses.

The circumferential and longitudinal stresses σ_1 and σ_2, respectively, are pictured in Fig. 12-10a, where they are shown acting on a stress element at point A on the wall of the vessel. Calculate the magnitudes of the stresses from Eqs. (12-11) and (12-12):

$$\sigma_1 = \frac{pr}{t} = \frac{(800 \text{ kPa})(1.8 \text{ m})}{20 \text{ mm}} = 72 \text{ MPa} \qquad \sigma_2 = \frac{pr}{2t} = \frac{\sigma_1}{2} = 36 \text{ MPa}$$

The stress element at point A is shown again in Fig. 12-10b, where the x axis is in the longitudinal direction of the cylinder and the y axis is in the circumferential direction. Since there is no stress in the z direction ($\sigma_3 = 0$), the element is in biaxial stress.

Note that the ratio of the internal pressure (800 kPa) to the smaller in-plane principal stress (36 MPa) is 0.022. Therefore, the assumption that any stresses in the z direction can be disregarded and all elements in the cylindrical shell, even those at the inner surface, are in biaxial stress is justified.

FIGURE 12-10

Solution to Example 12-2

(a)

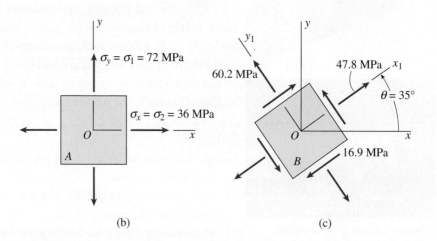

(b)   (c)

Part (b): Maximum shear stresses.

The largest in-plane shear stress is obtained from Eq. (12-14):

$$(\tau_{max})_z = \frac{\sigma_1 - \sigma_2}{2} = \frac{\sigma_1}{4} = \frac{pr}{4t} = 18 \text{ MPa} \quad \Longleftarrow$$

Because the normal stress in the z direction is disregarded, the largest out-of-plane shear stress is obtained from Eq. (12-15a):

$$\tau_{max} = \frac{\sigma_1}{2} = \frac{pr}{2t} = 36 \text{ MPa} \quad \Longleftarrow$$

This last stress is the absolute maximum shear stress in the wall of the vessel.

Part (c): Circumferential and longitudinal strains.

Since the largest stresses are well below the yield stress of steel (see Table F-3, Appendix F), assume that Hooke's law applies to the wall of the vessel. Then obtain the strains in the x and y directions (Fig. 12-10b) from Eqs. (11-40a and b) for biaxial stress:

$$\varepsilon_x = \frac{1}{E}(\sigma_x - \nu\sigma_y) \qquad \varepsilon_y = \frac{1}{E}(\sigma_y - \nu\sigma_x) \qquad \textbf{(a,b)}$$

Note that the strain ε_x is the same as the principal strain ε_2 in the longitudinal direction and that the strain ε_y is the same as the principal strain ε_1 in the circumferential direction. Also, the stress σ_x is the same as the stress σ_2, and the stress σ_y is the same as the stress σ_1. Therefore, the preceding two equations can be written in the forms:

$$\varepsilon_2 = \frac{\sigma_2}{E}(1 - 2\nu) = \frac{pr}{2tE}(1 - 2\nu) \qquad \text{(12-19a)}$$

$$\varepsilon_1 = \frac{\sigma_1}{2E}(2 - \nu) = \frac{pr}{2tE}(2 - \nu) \qquad \text{(12-19b)}$$

Substituting numerical values:

$$\varepsilon_2 = \frac{\sigma_2}{E}(1 - 2\nu) = \frac{(36 \text{ MPa})[1 - 2(0.30)]}{200 \text{ GPa}} = 72 \times 10^{-6}$$

$$\varepsilon_1 = \frac{\sigma_1}{2E}(2 - \nu) = \frac{(72 \text{ MPa})(2 - 0.30)}{2(200 \text{ GPa})} = 306 \times 10^{-6}$$

These are the longitudinal and circumferential strains in the cylinder.

Part (d): Normal and shear stresses acting on the welded seam.

The stress element at point B in the wall of the cylinder (Fig. 12-10a) is oriented so that its sides are parallel and perpendicular to the weld. The angle θ for the element is

$$\theta = 90° - \alpha = 35°$$

as shown in Fig. 12-10c. Either the stress-transformation equations or Mohr's circle may be used to obtain the normal and shear stresses acting on the side faces of this element.

Stress-transformation equations: The normal stress σ_{x1} and the shear stress τ_{x1y1} acting on the x_1 face of the element (Fig. 12-10c) are obtained from Eqs. (11-4a and b), which are repeated here:

$$\sigma_{x1} = \frac{\sigma_x + \sigma_y}{2} + \frac{\sigma_x - \sigma_y}{2}\cos 2\theta + \tau_{xy}\sin 2\theta \qquad \text{(12-20a)}$$

$$\tau_{x1y1} = -\frac{\sigma_x - \sigma_y}{2}\sin 2\theta + \tau_{xy}\cos 2\theta \qquad \text{(12-20b)}$$

Substitute $\sigma_x = \sigma_2 = pr/2t$, $\sigma_y = \sigma_1 = pr/t$, and $\tau_{xy} = 0$, to obtain

$$\sigma_{x1} = \frac{pr}{4t}(3 - \cos 2\theta) \qquad \tau_{x1y1} = \frac{pr}{4t}\sin 2\theta \qquad \text{(12-21a,b)}$$

These equations give the normal and shear stresses acting on an inclined plane oriented at an angle θ with the longitudinal axis of the cylinder.

Substitute $pr/4t = 18$ MPa and $\theta = 35°$ into Eqs. (12-21a and b) to obtain

$$\sigma_{x1} = 47.8 \text{ MPa} \qquad \tau_{x1y1} = 16.9 \text{ MPa}$$

These stresses are shown on the stress element of Fig. 12-10c.

FIGURE 12-11

Mohr's circle for the biaxial stress element of Figure 12-10b (*Note:* All stresses on the circle have units of MPa)

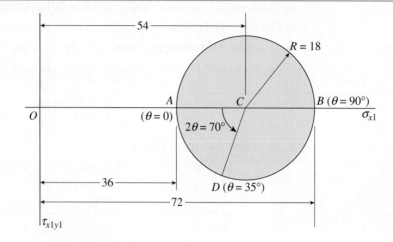

To complete the stress element, calculate the normal stress σ_{y1} acting on the y_1 face of the element from the sum of the normal stresses on perpendicular faces [Eq. (11-6)]:

$$\sigma_1 + \sigma_2 = \sigma_{x1} + \sigma_{y1} \tag{12-22}$$

Substitute numerical values to get

$$\sigma_{y1} = \sigma_1 + \sigma_2 - \sigma_{x1} = 72 \text{ MPa} + 36 \text{ MPa} - 47.8 \text{ MPa} = 60.2 \text{ MPa}$$

as shown in Fig. 12-10c.

From the figure, the normal and shear stresses acting perpendicular and parallel, respectively, to the welded seam are

$$\sigma_w = 47.8 \text{ MPa} \qquad \tau_w = 16.9 \text{ MPa}$$

4. *Finalize*:

Mohr's circle: The Mohr's circle construction for the biaxial stress element of Fig. 12-10b is shown in Fig. 12-11. Point A represents the stress $\sigma_2 = 36$ MPa on the x face ($\theta = 0$) of the element, and point B represents the stress $\sigma_1 = 72$ MPa on the y face ($\theta = 90°$). The center C of the circle is at a stress of 54 MPa, and the radius of the circle is

$$R = \frac{72 \text{ MPa} - 36 \text{ MPa}}{2} = 18 \text{ MPa}$$

A counterclockwise angle $2\theta = 70°$ (measured on the circle from point A) locates point D, which corresponds to the stresses on the x_1 face ($\theta = 35°$)

of the element. The coordinates of point D (from the geometry of the circle) are

$$\sigma_{x1} = 54 \text{ MPa} - R \cos 70° = 54 \text{ MPa} - (18 \text{ MPa})(\cos 70°) = 47.8 \text{ MPa}$$
$$\tau_{x1y1} = R \sin 70° = (18 \text{ MPa})(\sin 70°) = 16.9 \text{ MPa}$$

These results are the same as those found earlier from the stress-transformation equations.

Note: When seen in a side view, a **helix** follows the shape of a sine curve (Fig. 12-12). The pitch of the helix is

$$p = \pi d \tan \theta \tag{12-23}$$

FIGURE 12-12

Side view of a helix

where d is the diameter of the circular cylinder and θ is the angle between a normal to the helix and a longitudinal line. The width of the flat plate that wraps into the cylindrical shape is

$$w = \pi d \sin \theta \tag{12-24}$$

Thus, if the diameter of the cylinder and the angle θ are given, both the pitch and the plate width are established. For practical reasons, the angle θ is usually in the range from 20° to 35°.

12.4 Combined Loadings

Previous chapters presented the analysis of structural members subjected to a single type of loading. For instance, axially loaded bars were analyzed in Chapters 7 and 8, shafts in torsion in Chapter 9, and beams in bending in Chapter 10. Analysis of pressure vessels was presented earlier in this chapter. For each type of loading, methods were developed for finding stresses, strains, and deformations.

However, in many structures, the members are required to resist more than one kind of loading. For example, a beam may be subjected to the simultaneous action of bending moments and axial forces (Fig. 12-13a), a pressure vessel may be supported so that it also functions as a beam (Fig. 12-13b), or a shaft in torsion may carry a bending load (Fig. 12-13c). Known as **combined loadings**, situations similar to those shown in Fig. 12-13 occur in a great variety of machines, buildings, vehicles, tools, equipment, and many other kinds of structures.

A structural member subjected to combined loadings often can be analyzed by superimposing the stresses and strains caused by each load acting separately. However, superposition of both stresses and strains is permissible only under certain conditions, as explained in earlier chapters. One requirement is that the stresses and strains must be linear functions of the applied loads, which in turn requires that the material follow Hooke's law and the displacements remain small.

A second requirement is that there must be no interaction between the various loads, that is, the stresses and strains due to one load must not be affected by the presence of the other loads. Most ordinary structures satisfy these two conditions; therefore, the use of superposition is very common in engineering work.

FIGURE 12-13

Example of structures subjected to combined loading: (a) wide-flange beam supported by a cable (combined bending and axial load), (b) cylindrical pressure vessel supported as a beam, and (c) shaft in combined torsion and bending

Cable

Beam

(a)

Pressure vessel

(b)

Gear

Shaft

Cam

(c)

Method of Analysis

While there are many ways to analyze a structure subjected to more than one type of load, the procedure usually includes the following steps:

1. Select a point in the structure where the stresses and strains are to be determined. The point is usually selected at a cross section where the stresses are large, such as at a cross section where the bending moment has its maximum value.

2. For each load on the structure, determine the stress resultants at the cross section containing the selected point. The possible stress resultants are an axial force, a twisting moment, a bending moment, and a shear force.

3. Calculate the normal and shear stresses at the selected point due to each of the stress resultants. If the structure is a pressure vessel, determine the stresses due to the internal pressure. The stresses are found from the stress formulas derived previously; for instance, $\sigma = P/A$, $\tau = T\rho/I_p$, $\sigma = My/I$, $\tau = VQ/Ib$, and $\sigma = pr/t$.

4. Combine the individual stresses to obtain the resultant stresses at the selected point. In other words, obtain the stresses σ_x, σ_y, and τ_{xy} acting on a stress element at the point. (Note that this chapter deals only with elements in plane stress.)

5. Determine the principal stresses and maximum shear stresses at the selected point using either the stress-transformation equations or Mohr's circle. If required, determine the stresses acting on other inclined planes.

6. Determine the strains at the point with the aid of Hooke's law for plane stress.

7. Select additional points and repeat the process. Continue until enough stress and strain information is available to satisfy the purposes of the analysis.

Illustration of the Method

To illustrate the procedure for analyzing a member subjected to combined loadings, consider the stresses in the cantilever bar of circular cross section shown in Fig. 12-14a. This bar is subjected to two types of load—a torque T and a vertical load P, both acting at the free end of the bar.

Begin by arbitrarily selecting two points A and B for investigation (Fig. 12-14a). Point A is located at the top of the bar and point B is located on the side. Both points are located at the same cross section.

The stress resultants acting at the cross section (Fig. 12-14b) are a twisting moment equal to the torque T, a bending moment M equal to the load P times the distance b from the free end of the bar to the cross section, and a shear force V equal to the load P.

The stresses acting at points A and B are shown in Fig. 12-14c. The twisting moment T produces torsional shear stresses

$$\tau_1 = \frac{Tr}{I_p} = \frac{2T}{\pi r^3} \tag{12-25}$$

where r is the radius of the bar and $I_p = \pi r^4/2$ is the polar moment of inertia of the cross-sectional area. The stress τ_1 acts horizontally to the left at point A and vertically downward at point B, as shown in the figure.

The bending moment M produces a tensile stress at point A:

$$\sigma_A = \frac{Mr}{I} = \frac{4M}{\pi r^3} \qquad \textbf{(12-26)}$$

in which $I = \pi r^4/4$ is the moment of inertia about the neutral axis. However, the bending moment produces no stress at point B, because B is located on the neutral axis.

The shear force V produces no shear stress at the top of the bar (point A), but at point B, the shear stress is [see Eq. (10-43) in Chapter 10]:

$$\tau_2 = \frac{4V}{3A} = \frac{4V}{3\pi r^2} \qquad \textbf{(12-27)}$$

where $A = \pi r^2$ is the cross-sectional area.

The stresses σ_A and τ_1 acting at point A (Fig. 12-14c) are shown acting on a stress element in Fig. 12-15a. This element is cut from the top of the bar at point A. A two-dimensional view of the element, obtained by looking vertically downward on the element, is shown in Fig. 12-15b. For the purpose of determining the principal stresses and maximum shear stresses, construct x and y axes through the element. The x axis is parallel to the longitudinal axis of the circular bar (Fig. 12-14a) and the y axis is horizontal. Note that the element is in plane stress with $\sigma_x = \sigma_A$, $\sigma_y = 0$, and $\tau_{xy} = -\tau_1$.

A stress element at point B (also in plane stress) is shown in Fig. 12-16a. The only stresses acting on this element are the shear stresses, equal to $\tau_1 + \tau_2$ (see Fig. 12-14c). A two-dimensional view of the stress element is shown in Fig. 12-16b, with the x axis parallel to the longitudinal axis of the bar and the y axis in the vertical direction. The stresses acting on the element are $\sigma_x = \sigma_y = 0$ and $\tau_{xy} = -(\tau_1 + \tau_2)$.

Now that the stresses acting at points A and B have been determined and shown on the corresponding stress elements, you can use the transformation equations of plane stress (Sections 11.2 and 11.3) or Mohr's circle (Section 11.4) to determine principal stresses, maximum shear stresses, and stresses acting in inclined directions. You can also use Hooke's law (Section 11.5) to determine the strains at points A and B.

The procedure described previously for analyzing the stresses at points A and B (Fig. 12-14a) can be used at other points in the bar. Of particular interest are the points where the stresses calculated from the flexure and shear formulas have maximum or minimum values, called **critical points**. For instance, the normal stresses due to bending are largest at the cross section of maximum bending moment, which is at the support. Therefore, points C and D at the top and bottom of the beam at the fixed end (Fig. 12-14a) are critical points where the stresses should be calculated. Another critical point is point B itself because the shear stresses are a maximum at this point. (Note that, in this example, the shear stresses do not change if point B is moved along the bar in the longitudinal direction.)

As a final step, the principal stresses and maximum shear stresses at the critical points can be compared with one another in order to determine the absolute maximum normal and shear stresses in the bar.

This example illustrates the general procedure for determining the stresses produced by combined loadings. Note that no new theories are involved—only applications of previously derived formulas and concepts. Since the variety of

FIGURE 12-14

Cantilever bar subjected to combined torsion and bending: (a) loads acting on the bar, (b) stress resultants at a cross section, and (c) stresses at point A and B

(a)

(b)

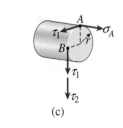

(c)

FIGURE 12-15

Stress element at point A

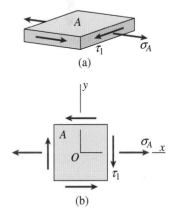

(a)

(b)

FIGURE 12-16

Stress element at point B

(a)

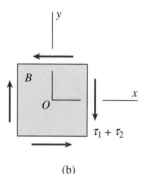

(b)

practical situations seems to be endless, it is not practical to derive general formulas for calculating the maximum stresses. Instead, treat each structure as a special case.

Selection of Critical Points

If the objective of the analysis is to determine the largest stresses *anywhere* in the structure, then the critical points should be selected at cross sections where the stress resultants have their largest values. Furthermore, within those cross sections, the points should be selected where either the normal stresses or the shear stresses have their largest values. By using good judgment in the selection of the points, you often can be reasonably certain of obtaining the absolute maximum stresses in the structure.

However, it is sometimes difficult to recognize in advance where the maximum stresses in the member are to be found. Then it may be necessary to investigate the stresses at a large number of points, perhaps even using trial-and-error in the selection of points. Other strategies may also prove fruitful—such as deriving equations specific to the problem at hand or making simplifying assumptions to facilitate an otherwise difficult analysis.

The following examples illustrate the methods used to calculate stresses in structures subjected to combined loadings.

Example 12-3

The hollow pipe casing for a production oil well (see Fig. 12-17) is 8 in. in outer diameter and 0.75 in. in thickness. The internal pressure due to gas and oil is 2175 psi. At some point above the blowout preventer, the compressive force in the pipe (due to the weight of the pipe) is 40 kips, and the torque is 10 kip-ft. Determine the maximum tensile, compressive, and shear stresses in the pipe casing.

Solution:

Use the four-step problem-solving approach. Combine steps as needed for an efficient solution.

1. *Conceptualize*: The stresses in the well casing are produced by the combined action of the axial force P, the torque T, and *internal pressure p* (Fig. 12-17b). Therefore, the stresses at any point on the surface of the shaft at some depth consist of circumferential stress σ_x, longitudinal stress σ_y, and shear stresses τ_{xy}, as shown on the stress element on the surface of the casing in Fig. 12-17b. Note that the y axis is parallel to the longitudinal axis of the casing.

2. *Categorize*: The circumferential stress σ_x is due to the internal pressure of oil and gas and is computed using Eq. (12-11) as

$$\sigma_x = \frac{pr}{t} = \frac{[2175 \text{ psi} \times (4 \text{ in.})]}{0.75 \text{ in.}} = 11{,}600 \text{ psi}$$

The longitudinal stress σ_y is caused by the axial compressive force P (due to self-weight) and is divided by casing cross-sectional area A. The longitudinal tensile

FIGURE 12-17

Example 12-3: Production oil well casing (combined torsion and axial force and internal pressure) (Courtesy of EMNRD)

Production oil well

(a)   (b)

stress σ_L is due to internal pressure [see Eq. (12-12) to find σ_L, which is nonzero when the well is capped and not operational]. You can assume that oil and gas are flowing, so σ_L is zero and σ_y is computed as

$$\sigma_y = \frac{-P}{A} = \frac{-40 \text{ kips}}{\pi\left[r^2 - (r-t)^2\right]} = -2342 \text{ psi}$$

Find the shear stress τ_{xy} from the torsion formula [see Eq. (9-13) of Section 9.3]:

$$\tau_{xy} = \frac{Tr}{I_p} = \frac{(10 \text{ kip-ft}) \times (4 \text{ in.})}{226.88 \text{ in}^4} = 2116 \text{ psi}$$

The shear stress is positive in accordance with the sign convention established in Section 7.6.

3. *Analyze:* Since you know the stresses σ_x, σ_y, and τ_{xy}, compute the principal stresses and maximum shear stresses by the methods described in Section 11.3. Obtain the principal stresses from Eq. (11-17):

$$\sigma_{1,2} = \frac{\sigma_x + \sigma_y}{2} \pm \sqrt{\left(\frac{\sigma_x - \sigma_y}{2}\right)^2 + \tau_{xy}^2}$$

Substitute $\sigma_x = 11{,}600$ psi, $\sigma_y = -2342$ psi, and $\tau_{xy} = 2116$ psi to get

$$\sigma_{1,2} = 4629 \text{ psi} \pm 7285 \text{ psi or } \sigma_1 = 11{,}914 \text{ psi} \quad \sigma_2 = -2656 \text{ psi}$$ ←

These are the maximum tensile and compressive stresses in the drill casing. The maximum in-plane shear stresses from Eq. (11-25) are

$$\tau_{max} = \sqrt{\left(\frac{\sigma_x - \sigma_y}{2}\right)^2 + \tau_{xy}^2} = 7285 \text{ psi}$$ ←

4. *Finalize*: Because the principal stresses σ_1 and σ_2 have opposite signs, the maximum in-plane shear stresses are larger than the maximum out-of-plane shear stresses [see Eqs. (11-28a, b, and c) and the accompanying discussion]. Therefore, the maximum shear stress in the drill casing is 7285 psi.

Example 12-4

FIGURE 12-18

Example 12-4: Wind pressure against a sign (combined bending, torsion, and shear of the pole)

A sign with dimensions of 2.0 m × 1.2 m is supported by a hollow circular pole having an outer diameter of 220 mm and an inner diameter of 180 mm (Fig. 12-18). The sign is offset 0.5 m from the centerline of the pole, and its lower edge is 6.0 m above the ground.

Determine the principal stresses and maximum shear stresses at points A and B at the base of the pole due to a wind pressure of 2.0 kPa against the sign.

Solution:

Use the four-step problem-solving approach. Combine steps as needed for an efficient solution to determine the circular pole stress resultants.

1. *Conceptualize*: The wind pressure against the sign produces a resultant force W that acts at the midpoint of the sign (Fig. 12-19a) and is equal to the pressure p times the area A over which it acts:

$$W = pA = (2.0 \text{ kPa})(2.0 \text{ m} \times 1.2 \text{ m}) = 4.8 \text{ kN}$$

The line of action of this force is at height $h = 6.6$ m above the ground and at distance $b = 1.5$ m from the centerline of the pole.

The wind force acting on the sign is statically equivalent to a lateral force W and a torque T acting on the pole (Fig. 12-19b). The torque is equal to the force W times the distance b:

$$T = Wb = (4.8 \text{ kN})(1.5 \text{ m}) = 7.2 \text{ kN·m}$$

The stress resultants at the base of the pole (Fig. 12-19c) consist of a bending moment M, a torque T, and a shear force V. Their magnitudes are

$$M = Wh = (4.8 \text{ kN})(6.6 \text{ m}) = 31.68 \text{ kN·m}$$
$$T = 7.2 \text{ kN·m} \qquad V = W = 4.8 \text{ kN}$$

Examination of these stress resultants shows that maximum bending stresses occur at point A and maximum shear stresses at point B. Therefore, A and B

FIGURE 12-19

Solution to Example 12-4

(a)   (b)   (c)

(d)   (e)   (f)

are critical points where the stresses should be determined. (Another critical point is diametrically opposite point A, as explained at the end of this example.)

2,3. Categorize, Analyze:

Stresses at points A and B: The bending moment M produces a tensile stress σ_A at point A (Fig. 12-19d) but no stress at point B (which is located on the neutral axis). The stress σ_A is obtained from the flexure formula:

$$\sigma_A = \frac{M(d_2/2)}{I}$$

in which d_2 is the outer diameter (220 mm) and I is the moment of inertia of the cross section. The moment of inertia is

$$I = \frac{\pi}{64}(d_2^4 - d_1^4) = \frac{\pi}{64}\left[(220 \text{ mm})^4 - (180 \text{ mm})^4\right] = 63.46 \times 10^{-6} \text{ m}^4$$

in which d_1 is the inner diameter. Therefore, the stress σ_A is

$$\sigma_A = \frac{Md_2}{2I} = \frac{(31.68 \text{ kN·m})(220 \text{ mm})}{2(63.46 \times 10^{-6} \text{ m}^4)} = 54.91 \text{ MPa}$$

The torque T produces shear stresses τ_1 at points A and B (Fig. 12-19d). Calculate these stresses from the torsion formula:

$$\tau_1 = \frac{T(d_2/2)}{I_p}$$

in which I_p is the polar moment of inertia:

$$I_p = \frac{\pi}{32}(d_2^4 - d_1^4) = 2I = 126.92 \times 10^{-6} \text{ m}^4$$

Thus,

$$\tau_1 = \frac{Td_2}{2I_p} = \frac{(7.2 \text{ kN · m})(220 \text{ mm})}{2(126.92 \times 10^{-6} \text{ m}^4)} = 6.24 \text{ MPa}$$

Finally, calculate the shear stresses at points A and B due to the shear force V. The shear stress at point A is zero, and the shear stress at point B (denoted τ_2 in Fig. 12-19d) is obtained from the shear formula for a circular tube [Eq. (10-45) of Section 10.8]:

$$\tau_2 = \frac{4V}{3A}\left(\frac{r_2^2 + r_2 r_1 + r_1^2}{r_2^2 + r_1^2}\right) \tag{a}$$

in which r_2 and r_1 are the outer and inner radii, respectively, and A is the cross-sectional area:

$$r_2 = \frac{d_2}{2} = 110 \text{ mm} \qquad r_1 = \frac{d_1}{2} = 90 \text{ mm}$$

$$A = \pi(r_2^2 - r_1^2) = 12{,}570 \text{ mm}^2$$

Substitute numerical values into Eq. (a) to obtain

$$\tau_2 = 0.76 \text{ MPa}$$

The stresses acting on the cross section at points A and B now have been calculated.

Stress elements: The next step is to show these stresses on stress elements (Figs. 12-19e and f). For both elements, the y axis is parallel to the longitudinal axis of the pole, and the x axis is horizontal. At point A, the stresses acting on the element are

$$\sigma_x = 0 \qquad \sigma_y = \sigma_A = 54.91 \text{ MPa} \qquad \tau_{xy} = \tau_1 = 6.24 \text{ MPa}$$

At point B, the stresses are

$$\sigma_x = \sigma_y = 0 \qquad \tau_{xy} = \tau_1 + \tau_2 = 6.24 \text{ MPa} + 0.76 \text{ MPa} = 7.00 \text{ MPa}$$

Since there are no normal stresses acting on the element, point B is in pure shear.

Now that all stresses acting on the stress elements (Figs. 12-19e and f) are known, use the equations given in Section 11.3 to determine the principal stresses and maximum shear stresses.

Principal stresses and maximum shear stresses at point A: The principal stresses are obtained from Eq. (11-17), which is repeated here:

$$\sigma_{1,2} = \frac{\sigma_x + \sigma_y}{2} \pm \sqrt{\left(\frac{\sigma_x - \sigma_y}{2}\right)^2 + \tau_{xy}^2} \qquad \text{(b)}$$

Substitute $\sigma_x = 0$, $\sigma_y = 54.91 \text{ MPa}$, and $\tau_{xy} = 6.24 \text{ MPa}$ to get

$$\sigma_{1,2} = 27.5 \text{ MPa} \pm 28.2 \text{ MPa}$$

or

$$\sigma_1 = 55.7 \text{ MPa} \qquad \sigma_2 = -0.7 \text{ MPa} \qquad \longleftarrow$$

The maximum in-plane shear stresses may be obtained from Eq. (11-25):

$$\tau_{\max} = \sqrt{\left(\frac{\sigma_x - \sigma_y}{2}\right)^2 + \tau_{xy}^2} \qquad \text{(c)}$$

This term was evaluated previously, so it follows that

$$\tau_{\max} = 28.2 \text{ MPa} \qquad \longleftarrow$$

Because the principal stresses σ_1 and σ_2 have opposite signs, the maximum in-plane shear stresses are larger than the maximum out-of-plane shear stresses [see Eqs. (11-28a, b, and c) and the accompanying discussion]. Therefore, the maximum shear stress at point A is 28.2 MPa.

Principal stresses and maximum shear stresses at point B: The stresses at this point are $\sigma_x = 0$, $\sigma_y = 0$, and $\tau_{xy} = 7.0$ MPa. Since the element is in pure shear, the principal stresses are

$$\sigma_1 = 7.0 \text{ MPa} \qquad \sigma_2 = -7.0 \text{ MPa}$$

and the maximum in-plane shear stress is

$$\tau_{max} = 7.0 \text{ MPa}$$

The maximum out-of-plane shear stresses are half this value.

4. *Finalize*: If the largest stresses anywhere in the pole are needed, then also determine the stresses at the critical point diametrically opposite point *A*, because at that point the compressive stress due to bending has its largest value. The principal stresses at that point are

$$\sigma_1 = 0.7 \text{ MPa} \qquad \sigma_2 = -55.7 \text{ MPa}$$

and the maximum shear stress is 28.2 MPa. Therefore, the largest tensile stress in the pole is 55.7 MPa, the largest compressive stress is −55.7 MPa, and the largest shear stress is 28.2 MPa. (Keep in mind that only the effects of the wind pressure are considered in this analysis. Other loads, such as the weight of the structure, also produce stresses at the base of the pole.)

Example 12-5

The cylindrical pressure vessel from Example 12-2 (see photo) is now placed on simple supports and is acted on by a uniformly distributed load $q = 10,500$ lb/ft, which includes the weight of the tank and its contents. The 20-ft-long tank has an inner radius of $r = 4$ ft and a wall thickness of $t = 0.75$ in. The material is steel with a modulus of $E = 29,000$ ksi and the internal pressure $p = 105$ psi.

Example 12-2 investigated the longitudinal and circumferential stresses and strains, as well as the maximum in-plane and out-of-plane shear stresses. Now, the effect of distributed load q is considered to *find states of stress at element locations A and B* (see Fig. 12-20) due to the combined effects of internal pressure and transverse shear and bending moment (shear-force and bending-moment diagrams are given in Figs. 12-20c and d). (a) Element *A* is on the outer surface of the vessel, just to the right of the left-hand support. (b) Element *B* is located on the bottom surface of the tank at the mid-span.

Solution:

Use the four-step problem-solving approach. Combine steps as needed for an efficient solution.

1. *Conceptualize*: The stresses in the wall of the pressure vessel are caused by the *combined action* of internal pressure and transverse shear and bending.

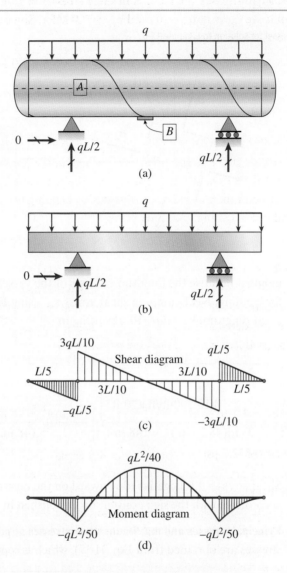

FIGURE 12-20

Example 12-5: Cylindrical pressure vessel subjected to combined internal pressure p and transverse load q

Part (a): State of stress at point A.

2,3. *Categorize, Analyze*: Isolate a stress element similar to that shown in Fig. 12-21a. The x axis is parallel to the longitudinal axis of the pressure vessel, and the y axis is circumferential. There are shear stresses acting on element A due to load q (assume that element A is a sufficient distance from the support so that any stress concentration effects are negligible). The stresses are computed as

$$\sigma_x = \sigma_L = \frac{pr}{2t} = \frac{105 \text{ psi} \times (48 \text{ in.})}{2(0.75 \text{ in.})} = 3360 \text{ psi}$$

$$\sigma_y = \sigma_r = \frac{pr}{t} = \frac{105 \text{ psi} \times (48 \text{ in.})}{0.75 \text{ in.}} = 6720 \text{ psi}$$

where σ_L is the longitudinal stress and σ_r is the circumferential (or radial) stress due to internal pressure p. There are no normal stresses due to bending

FIGURE 12-21

Stresses in a cylindrical pressure vessel for solution to Example 12-5

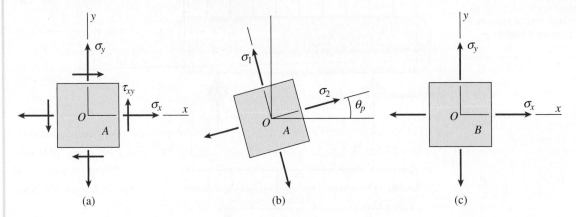

(a)   (b)   (c)

moment because the longitudinal axis of the vessel lies in the neutral plane for bending. Next compute shear stress τ_{xy} using Eq. (10-45), where, from the shear diagram, $V = 3qL/10$, resulting in

$$\tau_{xy} = \frac{-4}{3}\frac{V}{A}\left(\frac{r_1^2 + r_1 r_2 + r_2^2}{r_1^2 + r_2^2}\right)$$

$$\tau_{xy} = \frac{-4}{3}\frac{\left[\dfrac{3}{10}(10{,}500\ \text{lb/ft})(20\ \text{ft})\right]}{\pi[(48.75\ \text{in.})^2 - (48\ \text{in.})^2]}\left[\frac{(48\ \text{in.})^2 + 48\ \text{in.}(48.75\ \text{in.}) + (48.75\ \text{in.})^2}{(48\ \text{in.})^2 + (48.75\ \text{in.})^2}\right]$$

$$= -552.7\ \text{psi}$$

Shear stress τ_{xy} is negative (downward on the positive x face of the element) in accordance with the sign convention established in Section 7.6.

Principal stresses and maximum shear stresses at point A: The principal stresses are obtained from Eq. (11-17), which is repeated here:

$$\sigma_{1,2} = \frac{\sigma_x + \sigma_y}{2} \pm \sqrt{\left(\frac{\sigma_x - \sigma_y}{2}\right)^2 + \tau_{xy}^2}$$

so

$$\sigma_1 = 5040\ \text{psi} + 1769\ \text{psi} = 6809\ \text{psi}$$

$$\sigma_2 = 5040\ \text{psi} - 1769\ \text{psi} = 3271\ \text{psi}$$

4. *Finalize*: The principal stresses at point A are shown on an element rotated through $\theta_p = 9.11°$ in Fig. 12-21b.

The *maximum in-plane shear stress* is computed using Eq. (11-28c):

$$\tau_{\max} = \frac{\sigma_1 - \sigma_2}{2} = 1769\ \text{psi}$$

but the *maximum out-of-plane shear stress* controls using Eqs. (11-28b):

$$\tau_{max} = \frac{\sigma_1}{2} = 3405 \text{ psi}$$

Because the principal stresses have the same signs, one of the out-of-plane shear stresses is the largest shear stress [see the discussion following Eqs. (11-28a, b, and c)].

Part (b): State of stress at point *B*.

2,3. Categorize, Analyze: The stress element is located on the bottom surface of the vessel and (as you look up at it from the bottom of the tank) is oriented as shown in Fig. 12-21c. The *x* axis is parallel to the longitudinal axis of the pressure vessel, and the *y* axis is circumferential. There are no shear stresses acting on element *B* due to load *q*, because element *B* is on the bottom free surface, but normal tensile stress is maximum due to bending. The stresses are computed as

$$\sigma_x = \sigma_L + \frac{Mr}{I_z}$$

where I_z for the vessel is

$$I_z = \frac{\pi}{4}\left[(r+t)^4 - r^4\right] = 12.864 \text{ ft}^4$$

so

$$\sigma_x = \frac{pr}{2t} + \frac{\left(\dfrac{qL^2}{40}\right)(r+t)}{I_z}$$

$$= \frac{105 \text{ psi}(48 \text{ in.})}{2(0.75 \text{ in.})} + \frac{\left[10{,}500 \text{ lb/ft} \dfrac{(20 \text{ ft})^2}{40}\right](48.75 \text{ in.})}{12.864 \text{ ft}^4}$$

$$\sigma_x = 3360 \text{ psi} + 230 \text{ psi} = 3590 \text{ psi}$$

$$\sigma_y = \sigma_r = \frac{pr}{t} = \frac{105 \text{ psi}(48 \text{ in.})}{0.75 \text{ in.}} = 6720 \text{ psi}$$

4. Finalize: Because there are no shear stresses acting at *B*, normal stresses σ_x and σ_y are the principal normal stresses, (i.e., $\sigma_x = \sigma_2$ and $\sigma_y = \sigma_1$). The maximum in-plane and out-of-plane shear stresses can be found from Eqs. (11-28a, b, and c).

The *maximum in-plane shear stress* is computed using Eq. (11-28c) as

$$\tau_{max} = \frac{\sigma_1 - \sigma_2}{2} = 1565 \text{ psi}$$

but the *maximum out-of-plane shear stress* controls using Eqs. (11-28b) for

$$\tau_{max} = \frac{\sigma_1}{2} = 3360 \text{ psi}$$

Example 12-6

A jib crane at a marina carries a load at the end of the jib beam with components 12 kN in the x direction, 22 kN in the $-y$ direction, and 18 kN in the z direction (Fig. 12-22a). The crane is fixed at the base at O. Find the state of plane stress on the surface of the vertical post at points such as A and B just above the base (see Fig. 12-22b). The post has an outer radius $r_2 = 165$ mm and wall thickness $t = 20$ mm. Neglect the weight of the crane.

Solution:

Use the four-step problem-solving approach. Combine steps as needed for an efficient solution.

1. *Conceptualize*: Use statics to find reaction forces and moments at support point O or to find the load resultants just above the base of the post (Fig. 12-22b) (the resultants are the negatives of the reactions). Use the load resultants to find normal and shear stresses on each plane stress element. View each element by walking around the base of the post; in each case, the local x axis for the element (at A, B, or elsewhere) is to your right, and the y axis is upward (parallel to the y axis for the entire crane structure).

2. *Categorize*: Start with axial force resultant F_y at the base (Fig. 12-22b) and find the axial compressive normal stress σ_y acting on each stress element. Next, use force resultants F_x and F_z in the *shear formula* for a hollow circular cross section [Eq. (10-45)] to find the transverse shear stress τ_V on the $+y$ face of each stress element. Use torsional moment M_y in the *torsion formula* [Eq. (9-13)] to compute the shear stress τ_T on the $+y$ face due to torsion. Finally, use moments M_x and

FIGURE 12-22

(a) Load components at end of beam on jib crane; (b) load resultants and plane stress elements A and B on surface of post at base

baitong333/Shutterstock.com

(a)

(b)

M_z in the *flexure formula* [Eq. (10-14)] to find additional normal stresses σ_y on the $+y$ face of each element. Note that elements not aligned with the x and z axes are subjected to *biaxial bending* and that every axis through the centroid of the cross section of the post is a *principal axis*. Also observe that the normal stress σ_x in the x direction on each element is zero. Assume that all stress elements are a sufficient distance from support O so that stress-concentration effects are negligible.

Cross-sectional properties of post: Compute the cross-sectional area A, polar moment of inertia I_p, and moment of inertia for flexure I for use in stress calculations. Use the inner and outer radii and thickness of the post, $r_2 = 165$ mm, $t = 20$ mm, and $r_1 = r_2 - t = 145$ mm, as

$$A = \pi(r_2^2 - r_1^2) = 1.948 \times 10^4 \text{ mm}^2$$

$$I_p = \frac{\pi}{2}(r_2^4 - r_1^4) = 4.699 \times 10^8 \text{ mm}^4 \quad \text{and} \quad I = I_p/2 = 2.350 \times 10^8 \text{ mm}^4 \quad \textbf{(a)}$$

In addition, the following term is needed in the calculation of maximum shear stress at the neutral axis for a hollow circular cross section [see Eq. (10-45)]:

$$\alpha = \frac{r_2^2 + r_2 r_1 + r_1^2}{r_2^2 + r_1^2} = 1.496 \qquad \textbf{(b)}$$

Stress resultants (forces, moments) at base of post: Summing forces in x, y, and z directions and also moments about the x, y, and z axes leads to the values for the force and moment resultants shown in Fig. 12-22b:

$$F_x = 12 \text{ kN} \quad F_y = -22 \text{ kN} \quad F_z = 18 \text{ kN}$$

$$M_x = (22 \text{ kN})(3.5 \text{ m}) + (18 \text{ kN})(4 \text{ m}) = 149 \text{ kN} \cdot \text{m}$$

$$M_y = (12 \text{ kN})(3.5 \text{ m}) = 42 \text{ kN} \cdot \text{m} \qquad \textbf{(c)}$$

$$M_z = -(12 \text{ kN})(4 \text{ m}) = -48 \text{ kN} \cdot \text{m}$$

The negatives of these values are the support reactions at base point O.

3. *Analyze:* Compute the states of plane stress for elements at points A and B and at other locations of interest, as follows.

Element A: Element A is located at coordinates $(r_2, 0, 0)$. The normal stress σ_x is zero in the element local x direction, as noted previously. Normal stress σ_y is the sum of axial compression due to force F_y and flexural normal stress due to M_z:

$$\sigma_y = \frac{F_y}{A} + \frac{M_z r_2}{I} = -1.129 \text{ MPa} - 33.71 \text{ MPa} = -34.8 \text{ MPa} \quad \leftarrow$$

The shear stress on the $+y$ face of the element is the sum of transverse shear stress τ_V due to force F_z and torsional shear stress τ_T due to torsional moment M_y:

$$\tau_V = -\frac{4}{3}\left[\frac{F_z}{A}\right]\alpha = -1.843 \text{ MPa} \qquad \tau_T = \frac{M_y r_2}{I_p} = 14.75 \text{ MPa}$$

so $\qquad \tau_A = \tau_V + \tau_T = 12.91 \text{ MPa} \quad \leftarrow$

The state of plane stress at element A is shown in Fig. 12-23a. Use the plane stress-transformation equations [Eqs. (11-11), (11-17), and (11-25)] or Mohr's circle (see Fig. 12-23d) to find the principal normal and maximum shear stress states shown in Figs. 12-23b and c.

Element B: Element B is located at coordinates $(0, 0, r_2)$. Normal stress σ_y is the sum of axial compression due to force F_y and flexural normal stress due to M_x:

$$\sigma_y = \frac{F_y}{A} - \frac{M_x r_2}{I} = -1.129 \text{ MPa} - 104.6 \text{ MPa} = -105.7 \text{ MPa}$$

Similar to element A, the shear stress on the $+y$ face of the element is the sum of transverse shear stress τ_V due to force F_x and torsional shear stress τ_T due to torsional moment M_y:

$$\tau_B = \tau_V + \tau_T = \frac{4}{3}\left[\frac{F_x}{A}\right]\alpha + \frac{M_y r_2}{I_p} = 1.229 \text{ MPa} + 14.75 \text{ MPa} = 15.98 \text{ MPa}$$

FIGURE 12-23

(a) Plane stress state at element A; (b) principal normal stresses; (c) maximum shear stress; and (d) Mohr's circle for element A

FIGURE 12-24

(a) Plane stress state at element *B*; (b) principal normal stresses; and (c) maximum shear stress

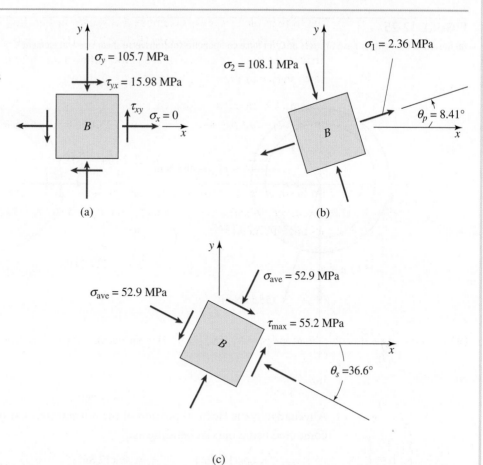

(a)

(b)

(c)

The state of plane stress on element *B* is shown in Fig. 12-24a. The plane stress-transformation equations (or Mohr's circle) provide the principal normal and maximum shear stress states and rotated element orientations shown in Figs. 12-24b and c, respectively.

Element *C*: Since $I_x = I_z$, the neutral axis *nn* for flexure only is defined by the angle between moment components M_x and M_z, that is,

$$\beta = \tan^{-1}\left[\frac{M_z}{M_x}\right] = 17.86°, \text{ as shown in Fig. 12-25. Now the neutral axis no}$$

longer passes through the centroid of the cross section when both axial and flexural effects are considered. As a result, the maximum compressive normal stress is at point *C* and is the sum of uniform axial compressive stress due to force F_y and flexural normal stress due to the resultant of moment components M_x and M_z. Hence $\sigma_x = 0$, and the maximum normal stress σ_y is

$$\sigma_y = \frac{F_y}{A} - \frac{\left(\sqrt{M_x^2 + M_z^2}\right)r_2}{I} = \frac{-22 \text{ kN}}{A} - \frac{(156.5 \text{ kN} \cdot \text{m})r_2}{I}$$
$$= -1.129 \text{ MPa} - 109.9 \text{ MPa} = -111.1 \text{ MPa}$$

FIGURE 12-25

(a) Resultant moment for flexural stress at C; (b) force components for transverse shear stress at element C

(a)   (b)

Alternatively, the flexural portion of the normal stress at point C can be computed using biaxial bending as

$$(-M_x)\frac{r_2 \cos(17.86°)}{I} + (M_z)\frac{r_2 \sin(17.86°)}{I} = -99.6 \text{ MPa} - 10.3 \text{ MPa}$$

$$= -109.9 \text{ MPa}$$

Additional analysis at point C leads to shear stress τ_V due to transverse shear force. Use components of forces F_x and F_z (Fig. 12-25b) to find transverse shear stress τ_V and then add uniform torsional shear stress as

$$\tau_C = \tau_V + \tau_T = \frac{4}{3}\left[\frac{F_x \cos(17.86°) - F_z \sin(17.86°)}{A}\right]\alpha + \frac{M_y r_2}{I_p}$$

$$= 0.60 \text{ MPa} + 14.75 \text{ MPa} = 15.35 \text{ MPa}$$

Finally, use plane stress transformations to find principal normal stresses and the maximum shear stress at point C for points A and B. The principal normal stresses and maximum shear stress at point C are $\sigma_1 = 2.08$ MPa, $\sigma_2 = -113.2$ MPa, and $\tau_{max} = 57.6$ MPa.

4. *Finalize*: As expected, the normal stress σ_y at point C is the largest axial compressive stress at the base of the post for this loading. If the weight of the crane itself is considered in this analysis, load resultants F_y and M_x will increase, resulting in larger normal compressive stress values at the base.

Example 12-7

FIGURE 12-26

Example 12-7: Loads on a post (combined axial load, bending, and shear)

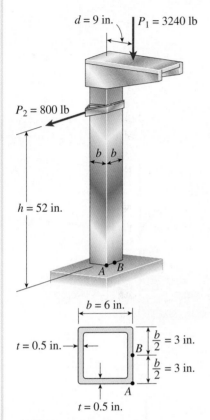

A tubular post of square cross section supports a horizontal platform (Fig. 12-26). The tube has an outer dimension $b = 6$ in. and wall thickness $t = 0.5$ in. The platform has dimensions 6.75 in. × 24.0 in. and supports a uniformly distributed load of 20 psi acting over its upper surface. The resultant of this distributed load is a vertical force P_1:

$$P_1 = (20 \text{ psi})(6.75 \text{ in.} \times 24.0 \text{ in.}) = 3240 \text{ lb}$$

This force acts at the midpoint of the platform, which is at distance $d = 9$ in. from the longitudinal axis of the post. A second load $P_2 = 800$ lb acts horizontally on the post at a height $h = 52$ in. above the base.

Determine the principal stresses and maximum shear stresses at points A and B at the base of the post due to the loads P_1 and P_2.

Solution:

Use the four-step problem-solving approach. Combine steps as needed for an efficient solution.

1. *Conceptualize*:

    Stress resultants: The force P_1 acting on the platform (Fig. 12-26) is statically equivalent to a force P_1 and a moment $M_1 = P_1 d$ acting at the centroid of the cross section of the post (Fig. 12-27a). The load P_2 is also shown in this figure.

    The stress resultants at the base of the post due to the loads P_1 and P_2 and the moment M_1 are shown in Fig. 12-27b. These stress resultants are

    i. An axial compressive force $P_1 = 3240$ lb.
    ii. A bending moment M_1 produced by the force P_1:

    $$M_1 = P_1 d = (3240 \text{ lb})(9 \text{ in.}) = 29{,}160 \text{ lb-in.}$$

    iii. A shear force $P_2 = 800$ lb.
    iv. A bending moment M_2 produced by the force P_2:

    $$M_2 = P_2 h = (800 \text{ lb})(52 \text{ in.}) = 41{,}600 \text{ lb-in.}$$

    Examination of these stress resultants (Fig. 12-27b) shows that both M_1 and M_2 produce maximum compressive stresses at point A and the shear force produces maximum shear stresses at point B. Therefore, A and B are critical points where the stresses should be determined. (Another critical point is diagonally opposite point A, as explained in the note at the end of this example.)

FIGURE 12-27

Solution to Example 12-7

(a)

(b)   (c)

$\sigma_A = \sigma_{P1} + \sigma_{M1} + \sigma_{M2} = 4090$ psi

$\sigma_B = \sigma_{P1} + \sigma_{M1} = 1860$ psi

$\tau_{P2} = 160$ psi

(d)   (e)

2,3. *Categorize, Analyze*:

Stresses at points *A* and *B*:

    i. The axial force P_1 (Fig. 12-27b) produces uniform compressive stresses through-out the post. These stresses are

$$\sigma_{P1} = \frac{P_1}{A}$$

in which A is the cross-sectional area of the post:

$$A = b^2 - (b - 2t)^2 = 4t(b - t)$$
$$= 4(0.5 \text{ in.})(6 \text{ in.} - 0.5 \text{ in.}) = 11.00 \text{ in}^2$$

Therefore, the axial compressive stress is

$$\sigma_{P1} = \frac{P_1}{A} = \frac{3240 \text{ lb}}{11.00 \text{ in}^2} = 295 \text{ psi}$$

The stress σ_{P1} is shown acting at points A and B in Fig. 12-27c.

ii. The bending moment M_1 (Fig. 12-27b) produces compressive stresses σ_{M1} at points A and B (Fig. 12-27c). These stresses are obtained from the flexure formula:

$$\sigma_{M1} = \frac{M_1(b/2)}{I} = \frac{M_1 b}{2I}$$

where I is the moment of inertia of the cross-sectional area:

$$I = \frac{b^4}{12} - \frac{(b - 2t)^4}{12} = \frac{1}{12}[(6 \text{ in.})^4 - (5 \text{ in.})^4] = 55.92 \text{ in}^4$$

Thus, the stress σ_{M1} is

$$\sigma_{M1} = \frac{M_1 b}{2I} = \frac{(29{,}160 \text{ lb-in.})(6 \text{ in.})}{2(55.92 \text{ in}^4)} = 1564 \text{ psi}$$

iii. The shear force P_2 (Fig. 12-27b) produces a shear stress at point B but not at point A. From the discussion of shear stresses in the webs of beams with flanges (Section 10.9), an approximate value of the shear stress can be obtained by dividing the shear force by the web area (see Eq. 10-52 in Section 10.9). Thus, the shear stress produced at point B by the force P_2 is

$$\tau_{P2} = \frac{P_2}{A_{\text{web}}} = \frac{P_2}{2t(b - 2t)} = \frac{800 \text{ lb}}{2(0.5 \text{ in.})(6 \text{ in.} - 1 \text{ in.})} = 160 \text{ psi}$$

The stress τ_{P2} acts at point B in the direction shown in Fig. 12-27c.

Calculate the shear stress τ_{P2} from the more accurate formula of Eq. (10-50a) in Section 10.9. The result of that calculation is $\tau_{P2} = 163$ psi, which shows that the shear stress obtained from the approximate formula is satisfactory.

iv. The bending moment M_2 (Fig. 12-27b) produces a compressive stress at point A but no stress at point B. The stress at A is

$$\sigma_{M2} = \frac{M_2(b/2)}{I} = \frac{M_2 b}{2I} = \frac{(41{,}600 \text{ lb-in.})(6 \text{ in.})}{2(55.92 \text{ in}^4)} = 2232 \text{ psi}$$

This stress is also shown in Fig. 12-27c.

Stress elements: The next step is to show the stresses acting on stress elements at points A and B (Figs. 12-27d and e). Each element is oriented so that the y axis is vertical (that is, parallel to the longitudinal axis of the

FIGURE 12-27 (Repeated)

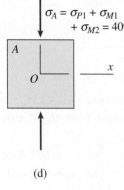

(d)

post) and the x axis is horizontal. At point A, the only stress is a compressive stress σ_A in the y direction (Fig. 12-27d):

$$\sigma_A = \sigma_{P1} + \sigma_{M1} + \sigma_{M2}$$
$$= 295 \text{ psi} + 1564 \text{ psi} + 2232 \text{ psi} = 4090 \text{ psi (compression)}$$

Thus, this element is in uniaxial stress.

At point B, the compressive stress in the y direction (Fig. 12-27e) is

$$\sigma_B = \sigma_{P1} + \sigma_{M1} = 295 \text{ psi} + 1564 \text{ psi} = 1860 \text{ psi (compression)}$$

and the shear stress is

$$\tau_{P2} = 160 \text{ psi}$$

The shear stress acts leftward on the top face of the element and downward on the x face of the element.

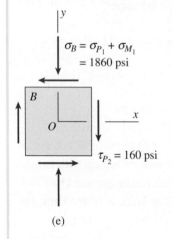

(e)

Principal stresses and maximum shear stresses at point A: Using the standard notation for an element in plane stress (Fig. 12-28), write the stresses for element A (Fig. 12-27d) as

$$\sigma_x = 0 \qquad \sigma_y = -\sigma_A = -4090 \text{ psi} \qquad \tau_{xy} = 0$$

Since the element is in uniaxial stress, the principal stresses are

$$\sigma_1 = 0 \qquad \sigma_2 = -4090 \text{ psi}$$

and the maximum in-plane shear stress [Eq. (11-26)] is

$$\tau_{max} = \frac{\sigma_1 - \sigma_2}{2} = \frac{4090 \text{ psi}}{2} = 2050 \text{ psi}$$

The maximum out-of-plane shear stress [Eq. (11-28a)] has the same magnitude.

FIGURE 12-28

Notation for an element in plane stress

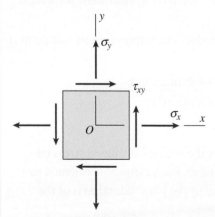

Principal stresses and maximum shear stresses at point B: Again using the standard notation for plane stress (Fig. 12-28), the stresses at point B (Fig. 12-27e) are

$$\sigma_x = 0 \qquad \sigma_y = -\sigma_B = -1860 \text{ psi} \qquad \tau_{xy} = -\tau_{P2} = -160 \text{ psi}$$

To obtain the principal stresses, use Eq. (11-17), which is repeated here:

$$\sigma_{1,2} = \frac{\sigma_x + \sigma_y}{2} \pm \sqrt{\left(\frac{\sigma_x - \sigma_y}{2}\right)^2 + \tau_{xy}^2} \qquad \text{(a)}$$

Substitute for σ_x, σ_y, and τ_{xy} to get

$$\sigma_{1,2} = -930 \text{ psi} \pm 944 \text{ psi}$$

or

$$\sigma_1 = 14 \text{ psi} \qquad \sigma_2 = -1870 \text{ psi}$$

The maximum in-plane shear stresses may be obtained from Eq. (11-25):

$$\tau_{max} = \sqrt{\left(\frac{\sigma_x - \sigma_y}{2}\right)^2 + \tau_{xy}^2} \qquad \text{(b)}$$

This term was evaluated previously, so

$$\tau_{max} = 944 \text{ psi} \qquad \blacktriangleleft$$

Because the principal stresses σ_1 and σ_2 have opposite signs, the maximum in-plane shear stresses are larger than the maximum out-of-plane shear stresses [see Eqs. (11-28a, b, and c) and the accompanying discussion]. Therefore, the maximum shear stress at point B is 944 psi.

4. *Finalize:* If the largest stresses anywhere at the base of the post are needed, then you must also determine the stresses at the critical point diagonally opposite point A (Fig. 12-27c) because at that point each bending moment produces the maximum tensile stress. Thus, the tensile stress acting at that point is

$$\sigma_y = -\sigma_{P1} + \sigma_{M1} + \sigma_{M2} = -295 \text{ psi} + 1564 \text{ psi} + 2232 \text{ psi} = 3500 \text{ psi}$$

The stresses acting on a stress element at that point (see Fig. 12-28) are

$$\sigma_x = 0 \qquad \sigma_y = 3500 \text{ psi} \qquad \tau_{xy} = 0$$

and therefore the principal stresses and maximum shear stress are

$$\sigma_1 = 3500 \text{ psi} \qquad \sigma_2 = 0 \qquad \tau_{max} = 1750 \text{ psi}$$

Thus, the largest tensile stress anywhere at the base of the post is 3500 psi, the largest compressive stress is 4090 psi, and the largest shear stress is 2050 psi. (Keep in mind that only the effects of the loads P_1 and P_2 are considered in this analysis. Other loads, such as the weight of the structure, also produce stresses at the base of the post.)

Chapter 12 presented some practical examples of structures in states of plane stress, building upon the material presented in Sections 11.2 through 11.5 in the previous chapter. The **stresses in thin-walled spherical and cylindrical vessels**, such as storage tanks containing compressed gases or liquids, were discussed. The maximum normal and shear stresses were found at various points in structures or components acted upon by **combined loadings**. Here are the major concepts and findings presented in this chapter.

1. **Plane stress** is a common stress condition that exists in all ordinary structures, such as in the walls of pressure vessels, in the webs and/or flanges of beams of various shapes, and in a wide variety of structures subject to the combined effects of axial, shear, and bending loads, as well as internal pressure.

2. The wall of a pressurized **thin-walled spherical vessel** is in a state of plane stress—specifically, biaxial stress—with uniform tensile stresses known as membrane stresses σ acting in all directions. The tensile stresses σ in the wall of a spherical shell are calculated as

$$\sigma = \frac{pr}{2t}$$

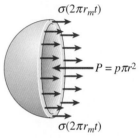

Only the excess of internal pressure over external pressure, or gage pressure, has any effect on these stresses. Additional important considerations for more detailed analysis or design of spherical vessels include: stress concentrations around openings, effects of external loads and self-weight, including contents and influence of corrosion, impacts, and temperature changes.

3. The walls of **thin-walled cylindrical pressure vessels** with circular cross sections are also in a state of biaxial stress. The circumferential stress σ_1 is referred to as the hoop stress, and the stress parallel to the axis of the tank is called the longitudinal stress or the axial stress σ_2. The circumferential stress is equal to twice the longitudinal stress. Both are principal stresses. The formulas for σ_1 and σ_2 are

$$\sigma_1 = \frac{pr}{t} \qquad \sigma_2 = \frac{pr}{2t}$$

These formulas were derived using elementary theory for thin-walled shells and are only valid in parts of the cylinder away from any discontinuities that cause stress concentrations.

4. A structural member subjected to **combined loadings** often is analyzed by superimposing the stresses and strains caused by each load acting separately. However, the stresses and strains must be linear functions of the applied loads, which in turn require that the material follow Hooke's law and the displacements remain small. There must be no interaction between the various loads, that is, the stresses and strains due to one load must not be affected by the presence of the other loads.

5. A **detailed approach for analysis** of critical points in a structure or component subjected to more than one type of load was presented in Section 12.4.

PROBLEMS

12.2 Spherical Pressure Vessels

To solve the problems for Section 12.2, assume that the given radius or diameter is an inside dimension and that all internal pressures are gage pressures.

Introductory Problems

12.2-1 A spherical balloon is filled with a gas. The outer diameter of the balloon is 20 in. and the thickness is 0.012 in. Calculate the maximum permissible pressure in the balloon if the allowable tensile stress and the allowable shear stress in the balloon are 1 ksi and 0.3 ksi, respectively.

12.2-2 A spherical balloon with an outer diameter of 500 mm and thickness 0.3 mm is filled with a gas. Calculate maximum permissible pressure in the balloon if the allowable normal strain at the outer surface of the balloon is 0.1. Assume $E = 4$ MPa and $\nu = 0.45$.

12.2-3 A large spherical tank (see figure) contains gas at a pressure of 420 psi. The tank is 45 ft in diameter and is constructed of high-strength steel having a yield stress in tension of 80 ksi.

(a) Determine the required thickness (to the nearest 1/4 inch) of the wall of the tank if a factor of safety of 3.5 with respect to yielding is required.

(b) If the tank wall thickness is 2.25 in., what is the maximum permissible internal pressure?

PROBLEMS 12.2-3 and 12.2-4

12.2-4 Solve the preceding problem if the internal pressure is 3.85 MPa, the diameter is 20 m, the yield stress is 590 MPa, and the factor of safety is 3.0.

(a) Determine the required thickness to the nearest millimeter.

(b) If the tank wall thickness is 85 mm, what is the maximum permissible internal pressure?

Representative Problems

12.2-5 A hemispherical window (or *viewport*) in a decompression chamber (see figure) is subjected to an internal air pressure of 85 psi. The window is attached to the wall of the chamber by 14 bolts.

(a) Find the tensile force F in each bolt and the tensile stress σ in the viewport if the radius of the hemisphere is 14 in. and its thickness is 1.25 in.

(b) If the yield stress for each of the 14 bolts is 50 ksi and the factor of safety is 3.0, find the required bolt diameter.

(c) If the stress in the viewport is limited to 500 psi, find the required radius of the hemisphere.

PROBLEM 12.2-5

12.2-6 A rubber ball (see figure) is inflated to a pressure of 65 kPa. At that pressure, the diameter of the ball is 240 mm and the wall thickness is 1.25 mm. The rubber has a modulus of elasticity $E = 3.7$ MPa and Poisson's ratio $\nu = 0.48$.

(a) Determine the maximum stress and strain in the ball.

(b) If the strain must be limited to 0.425, find the minimum required wall thickness of the ball.

PROBLEMS 12.2-6 and 12.2-7

12.2-7 (a) Solve part (a) of the preceding problem if the pressure is 8.5 psi, the diameter is 10 in., the wall thickness is 0.05 in., the modulus of elasticity is 200 psi, and Poisson's ratio is 0.48.

(b) If the strain must be limited to 1.01, find the maximum acceptable inflation pressure.

12.2-8 A spherical steel pressure vessel (diameter 500 mm, thickness 10 mm) is coated with brittle lacquer that cracks when the strain reaches 150×10^{-6} (see figure).

(a) What internal pressure p will cause the lacquer to develop cracks? (Assume $E = 205$ GPa and $\nu = 0.30$.)

(b) If the strain is measured at 125×10^{-6}, what is the internal pressure at that point?

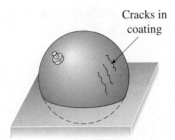

Cracks in coating

PROBLEM 12.2-8

12.2-9 A spherical tank of diameter 48 in. and wall thickness 1.75 in. contains compressed air at a pressure of 2200 psi. The tank is constructed of two hemispheres joined by a welded seam (see figure).

(a) What is the tensile load f (lb per in. of length of weld) carried by the weld?

(b) What is the maximum shear stress τ_{max} in the wall of the tank?

(c) What is the maximum normal strain ε in the wall? (For steel, assume $E = 30 \times 10^6$ psi and $\nu = 0.29$.)

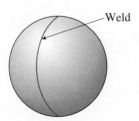

Weld

PROBLEMS 12.2-9 and 12.2-10

12.2-10 Solve the preceding problem for the following data: diameter 1.0 m, thickness 48 mm, pressure 22 MPa, modulus 210 GPa, and Poisson's ratio 0.29.

12.2-11 A spherical stainless-steel tank having a diameter of 26 in. is used to store propane gas at a pressure of 2075 psi. The properties of the steel are as follows: yield stress in tension, 140,000 psi; yield stress in shear, 65,000 psi; modulus of elasticity, 30×10^6 psi; and Poisson's ratio, 0.28. The desired factor of safety with respect to yielding is 2.8. Also, the normal strain must not exceed 1250×10^{-6}.

(a) Determine the minimum permissible thickness t_{min} of the tank.

(b) If the tank thickness is 0.30 in. and normal strain is measured at 990×10^{-6}, what is the internal pressure in the tank at that point?

12.2-12 Solve the preceding problem if the diameter is 480 mm, the pressure is 20 MPa, the yield stress in tension is 975 MPa, the yield stress in shear is 460 MPa, the factor of safety is 2.75, the modulus of elasticity is 210 GPa, Poisson's ratio is 0.28, and the normal strain must not exceed 1190×10^{-6}. For part (b), assume that the tank thickness is 8 mm and the measured normal strain is 990×10^{-6}.

12.2-13 A hollow, pressurized sphere having a radius $r = 4.8$ in. and wall thickness $t = 0.4$ in. is lowered into a lake (see figure). The compressed air in the tank is at a pressure of 24 psi (gage pressure when the tank is out of the water).

At what depth D_0 will the wall of the tank be subjected to a compressive stress of 90 psi?

D_0

PROBLEM 12.2-13

12.3 Cylindrical Pressure Vessels

To solve the problems for Section 12.3, assume that the given radius or diameter is an inside dimension and that all internal pressures are gage pressures.

Introductory Problems

12.3-1 A fire extinguisher tank is designed for an internal pressure of 825 psi. The tank has an outer diameter of 4.5 in. and thickness of 0.08 in. Calculate the longitudinal stress, the circumferential stress, and the maximum shear stresses (out-of-plane and in-plane) at the outer surface of the tank.

PROBLEMS 12.3-1 and 12.3-2

12.3-2 Repeat Problem 12.3-1 for a fire extinguisher tank with an internal pressure of 1.8 MPa, diameter of 130 mm, and thickness 1.5 mm.

12.3-3 A scuba tank (see figure) is being designed for an internal pressure of 2640 psi with a factor of safety of 2.0 with respect to yielding. The yield stress of the steel is 65,000 psi in tension and 32,000 psi in shear.

(a) If the diameter of the tank is 7.0 in., what is the minimum required wall thickness?
(b) If the wall thickness is 0.25 in., what is the maximum acceptable internal pressure?

PROBLEM 12.3-3

12.3-4 A tall standpipe with an open top (see figure) has diameter $d = 2.2$ m and wall thickness $t = 20$ mm.

(a) What height h of water will produce a circumferential stress of 12 MPa in the wall of the standpipe?
(b) What is the axial stress in the wall of the tank due to the water pressure?

PROBLEM 12.3-4

Representative Problems

12.3-5 An inflatable structure used by a traveling circus has the shape of a half-circular cylinder with closed ends (see figure). The fabric and plastic structure is inflated by a small blower and has a radius of 40 ft when fully inflated. A longitudinal seam runs the entire length of the "ridge" of the structure.

If the longitudinal seam along the ridge tears open when it is subjected to a tensile load of 540 pounds per inch of seam, what is the factor of safety n against tearing when the internal pressure is 0.5 psi and the structure is fully inflated?

Longitudinal seam

PROBLEM 12.3-5

12.3-6 A thin-walled cylindrical pressure vessel of a radius r is subjected simultaneously to internal gas pressure p and a compressive force F acting at the ends (see figure).

(a) What should be the magnitude of the force F in order to produce pure shear in the wall of the cylinder?

(b) If force $F = 190$ kN, internal pressure $p = 12$ MPa, inner diameter $= 200$ mm, and allowable normal and shear stresses are 110 MPa and 60 MPa, respectively, what is the required thickness of the vessel?

PROBLEM 12.3-6

12.3-7 A strain gage is installed in the longitudinal direction on the surface of an aluminum beverage can (see figure). The radius-to-thickness ratio of the can is 200. When the lid of the can is popped open, the strain changes by $\varepsilon_0 = 187 \times 10^{-6}$.

(a) What was the internal pressure p in the can? (Assume $E = 10 \times 10^6$ psi and $\nu = 0.33$.)

(b) What is the change in strain in the radial direction when the lid is opened?

PROBLEM 12.3-7

12.3-8 A circular cylindrical steel tank (see figure) contains a volatile fuel under pressure. A strain gage at point A records the longitudinal strain in the tank and transmits this information to a control room. The ultimate shear stress in the wall of the tank is 98 MPa, and a factor of safety of 2.8 is required.

(a) At what value of the strain should the operators take action to reduce the pressure in the tank? (Data for the steel are modulus of elasticity $E = 210$ GPa and Poisson's ratio $\nu = 0.30$.)

(b) What is the associated strain in the radial direction?

PROBLEM 12.3-8

12.3-9 A cylinder filled with oil is under pressure from a piston, as shown in the figure. The diameter d of the piston is 1.80 in. and the compressive force F is 3500 lb. The maximum allowable shear stress τ_{allow} in the wall of the cylinder is 5500 psi.

What is the minimum permissible thickness t_{min} of the cylinder wall? (See figure.)

PROBLEMS 12.3-9 and 12.3-10

12.3-10 Solve the preceding problem if $d = 90$ mm, $F = 42$ kN, and $\tau_{allow} = 40$ MPa.

12.3-11 A standpipe in a water-supply system (see figure) is 12 ft in diameter and 6 in. thick. Two horizontal pipes carry water out of the standpipe; each is 2 ft in diameter and 1 in. thick. When the system is shut down and water fills the pipes but is not moving, the hoop stress at the bottom of the standpipe is 130 psi.

(a) What is the height h of the water in the standpipe?

(b) If the bottoms of the pipes are at the same elevation as the bottom of the standpipe, what is the hoop stress in the pipes?

PROBLEM 12.3-11

12.3-12 A cylindrical tank with hemispherical heads is constructed of steel sections that are welded circumferentially (see figure). The tank diameter is 1.25 m, the wall thickness is 22 mm, and the internal pressure is 1750 kPa.

(a) Determine the maximum tensile stress σ_h in the heads of the tank.

(b) Determine the maximum tensile stress σ_c in the cylindrical part of the tank.

(c) Determine the tensile stress σ_w acting perpendicular to the welded joints.

(d) Determine the maximum shear stress τ_h in the heads of the tank.

(e) Determine the maximum shear stress τ_c in the cylindrical part of the tank.

Welded seams

PROBLEMS 12.3-12 and 12.3-13

12.3-13 A cylindrical tank with diameter $d = 18$ in. is subjected to internal gas pressure $p = 450$ psi. The tank is constructed of steel sections that are welded circumferentially (see figure). The heads of the tank are hemispherical. The allowable tensile and shear stresses are 8200 psi and 3000 psi, respectively. Also, the allowable tensile stress perpendicular to a weld is 6250 psi.

Determine the minimum required thickness t_{min} of (a) the cylindrical part of the tank and (b) the hemispherical heads.

12.3-14 A pressurized steel tank is constructed with a helical weld that makes an angle $\alpha = 55°$ with the longitudinal axis (see figure). The tank has radius $r = 0.6$ m, wall thickness $t = 18$ mm, and internal pressure $p = 2.8$ MPa. Also, the steel has modulus of elasticity $E = 200$ GPa and Poisson's ratio $\nu = 0.30$.

Determine the following quantities for the cylindrical part of the tank.

(a) The circumferential and longitudinal stresses.

(b) The maximum in-plane and out-of-plane shear stresses.

(c) The circumferential and longitudinal strains.

(d) The normal and shear stresses acting on planes parallel and perpendicular to the weld (show these stresses on a properly oriented stress element).

Helical weld

PROBLEMS 12.3-14 and 12.3-15

12.3-15 Solve the preceding problem for a welded tank with $\alpha = 62°$, $r = 19$ in., $t = 0.65$ in., $p = 240$ psi, $E = 30 \times 10^6$ psi, and $\nu = 0.30$.

12.4 Combined Loadings

Solve the problems for Section 12.4 assuming that the structures behave linearly elastically and that the stresses caused by two or more loads may be superimposed to obtain the resultant stresses acting at a point. Consider both in-plane and out-of-plane shear stresses unless otherwise specified.

Introductory Problems

12.4-1 A W12×35 steel cantilever beam is subjected to an axial load $P = 10$ kips and a transverse load $V = 15$ kips. The beam has length $L = 6$ ft. (a) Calculate the principal normal stresses and the maximum shear stress for an element located at C near the fixed support. Neglect the weight of the beam. (b) Repeat part (a) for point D, which is 4 in. above point C (see figure). See Table C-1(a), Appendix C, for beam properties.

PROBLEM 12.4-1

12.4-2 A W 310×52 steel beam is subjected to a point load $P = 45$ kN and a transverse load $V = 20$ kN at B. The beam has length $L = 2$ m. (a) Calculate the principal normal stresses and the maximum shear stress on element D located on the web right below the top flange and near the fixed support. Neglect the weight of the beam. (b) Repeat part (a) at centroid C (see figure). See Table C-1(b), Appendix C, for beam properties.

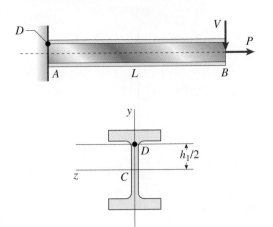

PROBLEM 12.4-2

12.4-3 A solid circular bar is fixed at point A. The bar is subjected to transverse load $V = 70$ lb and torque $T = 300$ lb-in. at point B. The bar has a length $L = 60$ in. and diameter $d = 3$ in. Calculate the principal normal stresses and the maximum shear stress at element 1 located on the bottom surface of the bar at fixed end A (see figure).

Assume that element 1 is a sufficient distance from support A so that stress concentration effects are negligible.

PROBLEM 12.4-3

12.4-4 Solve the preceding problem using transverse load $V = 300$ N and torque $T = 3.5$ N·m applied at point B. The bar has length $L = 1.5$ m and diameter $d = 8$ mm. Calculate the principal stresses and the maximum shear stress for element 2 located at the side of the bar at fixed end A (see figure).

Assume that element 2 is a sufficient distance from support A so that stress concentration effects are negligible.

PROBLEM 12.4-4

12.4-5 A cylindrical tank having a diameter $d = 2.5$ in. is subjected to internal gas pressure $p = 600$ psi and an external tensile load $T = 1000$ lb (see figure).

Determine the minimum thickness t of the wall of the tank based upon an allowable shear stress of 3000 psi.

PROBLEM 12.4-5

12.4-6 A cylindrical tank subjected to internal pressure p is simultaneously compressed by an axial force $F = 72$ kN (see figure). The cylinder has diameter $d = 100$ mm and wall thickness $t = 4$ mm.

Calculate the maximum allowable internal pressure p_{max} based upon an allowable shear stress in the wall of the tank of 60 MPa.

PROBLEM 12.4-6

12.4-7 A cylindrical pressure vessel having a radius $r = 14$ in. and wall thickness $t = 0.5$ in. is subjected to internal pressure $p = 375$ psi. In addition, a torque $T = 90$ kip-ft acts at each end of the cylinder (see figure).

(a) Determine the maximum tensile stress σ_{max} and the maximum in-plane shear stress τ_{max} in the wall of the cylinder.

(b) If the allowable in-plane shear stress is 4.5 ksi, what is the maximum allowable torque T?

(c) If $T = 150$ kip-ft and allowable in-plane shear and allowable normal stresses are 4.5 ksi and 11.5 ksi, respectively, what is the minimum required wall thickness?

PROBLEM 12.4-7

12.4-8 A pressurized cylindrical tank with flat ends is loaded by torques T and tensile forces P (see figure). The tank has a radius of $r = 125$ mm and wall thickness $t = 6.5$ mm. The internal pressure $p = 7.25$ MPa and the torque $T = 850$ N·m.

(a) What is the maximum permissible value of the forces P if the allowable tensile stress in the wall of the cylinder is 160 MPa?

(b) If forces $P = 400$ kN, what is the maximum acceptable internal pressure in the tank?

PROBLEM 12.4-8

12.4-9 A cylindrical pressure vessel with flat ends is subjected to a torque T and a bending moment M (see figure). The outer radius is 12.0 in. and the wall thickness is 1.0 in. The loads are $T = 800$ kip-in., $M = 1000$ kip-in., and the internal pressure $p = 900$ psi.

Determine the maximum tensile stress σ_t, maximum compressive stress σ_c, and maximum shear stress τ_{max} in the wall of the cylinder.

PROBLEM 12.4-9

12.4-10 The torsional pendulum shown in the figure consists of a horizontal circular disk of a mass $M = 60$ kg suspended by a vertical steel wire ($G = 80$ GPa) of a length $L = 2$ m and diameter $d = 4$ mm.

Calculate the maximum permissible angle of rotation ϕ_{max} of the disk (that is, the maximum amplitude of torsional vibrations) so that the stresses in the wire do not exceed 100 MPa in tension or 50 MPa in shear.

PROBLEM 12.4-10

Representative Problems

12.4-11 The hollow drill pipe for an oil well (see figure) is 6.2 in. in outer diameter and 0.75 in. in thickness. Just above the bit, the compressive force in the pipe (due to the weight of the pipe) is 62 kips and the torque (due to drilling) is 185 kip-in.

Determine the maximum tensile, compressive, and shear stresses in the drill pipe.

PROBLEM 12.4-11

12.4-12 A segment of a generator shaft is subjected to a torque T and an axial force P, as shown in the figure. The shaft is hollow (outer diameter $d_2 = 300$ mm and inner diameter $d_1 = 250$ mm) and delivers 1800 kW at 4.0 Hz.

If the compressive force $P = 540$ kN, what are the maximum tensile, compressive, and shear stresses in the shaft?

PROBLEMS 12.4-12 AND 12.4-13

12.4-13 A segment of a generator shaft with a hollow circular cross section is subjected to a torque $T = 240$ kip-in. (see figure). The outer and inner diameters of the shaft are 12.0 in. and 6.25 in., respectively.

What is the maximum permissible compressive load P that can be applied to the shaft if the allowable in-plane shear stress is $\tau_{allow} = 6250$ psi?

12.4-14 A post having a hollow, circular cross section supports a $P = 3.2$ kN load acting at the end of an arm that is $b = 1.5$ m long (see figure). The height of the post is $L = 9$ m, and its section modulus is $S = 2.65 \times 10^5$ mm³. Assume that the outer radius of the post is $r_2 = 123$ mm, and the inner radius is $r_1 = 117$ mm.

(a) Calculate the maximum tensile stress σ_{max} and maximum in-plane shear stress τ_{max} at point A on the outer surface of the post along the x axis due to the load P. Load P acts at B along line BC.

(b) If the maximum tensile stress and maximum in-plane shear stress at point A are limited to 90 MPa and 38 MPa, respectively, what is the largest permissible value of the load P?

PROBLEM 12.4-14

12.4-15 A sign is supported by a pole of hollow circular cross section, as shown in the figure. The outer and inner diameters of the pole are 10.5 in. and 8.5 in., respectively. The pole is 42 ft high and weighs 4 kips. The sign has dimensions 8 ft × 3 ft and weighs 500 lb. Note that its center of gravity is 53.25 in. from the axis of the pole. The wind pressure against the sign is 35 lb/ft².

(a) Determine the stresses acting on a stress element at point A, which is on the outer surface of the pole at the "front" of the pole, that is, the part of the pole nearest to the viewer.

(b) Determine the maximum tensile, compressive, and shear stresses at point A.

PROBLEM 12.4-15

12.4-16 A sign is supported by a pipe (see figure) having an outer diameter 110 mm and inner diameter 90 mm. The dimensions of the sign are 2.0 m × 1.0 m, and its lower edge is 3.0 m above the base. Note that the center of gravity of the sign is 1.05 m from the axis of the pipe. The wind pressure against the sign is 1.5 kPa.

Determine the maximum in-plane shear stresses due to the wind pressure on the sign at points A, B, and C, located on the outer surface at the base of the pipe.

PROBLEM 12.4-16

12.4-17 A traffic light and signal pole is subjected to the weight of each traffic signal $W_S = 45$ lb and the weight of the road lamp $W_L = 55$ lb. The pole is fixed at the base. Find the principal normal stresses and the maximum shear stress on element B located 19 ft above the base (see figure). Assume that the weight of the pole and lateral arms is included in the signal and lamp weights.

PROBLEM 12.4-17

12.4-18 Repeat the preceding problem but now find the stress state on Element A at the base. Let $W_S = 240$ N, $W_L = 250$ N, $t = 5$ mm, $d = 360$ mm. See the figure for the locations of element A and all loads.

PROBLEM 12.4-18

12.4-19 A bracket $ABCD$ having a hollow circular cross section consists of a vertical arm $AB (L = 6$ ft), a horizontal arm BC parallel to the x_0 axis, and a horizontal arm CD parallel to the z_0 axis (see figure). The arms BC and CD have lengths $b_1 = 3.6$ ft and $b_2 = 2.2$ ft, respectively. The outer and inner diameters of the bracket are $d_2 = 7.5$ in. and $d_1 = 6.8$ in. An inclined load $P = 2200$ lb acts at point D along line DH. Determine the maximum tensile, compressive, and shear stresses in the vertical arm.

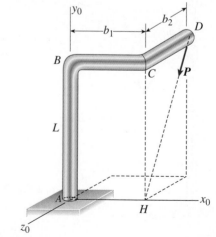

PROBLEM 12.4-19

956

12.4-20 A gondola on a ski lift is supported by two bent arms, as shown in the figure. Each arm is offset by the distance $b = 180$ mm from the line of action of the weight force W. The allowable stresses in the arms are 100 MPa in tension and 50 MPa in shear.

If the loaded gondola weighs 12 kN, what is the mininum diameter d of the arms?

PROBLEM 12.4-20

12.4-21 Determine the maximum tensile, compressive, and shear stresses at points A and B on the bicycle pedal crank shown in the figure.

The pedal and crank are in a horizontal plane and points A and B are located on the top of the crank. The

Top view

PROBLEM 12.4-21

load $P = 160$ lb acts in the vertical direction and the distances (in the horizontal plane) between the line of action of the load and points A and B are $b_1 = 5.0$ in., $b_2 = 2.5$ in., and $b_3 = 1.0$ in. Assume that the crank has a solid circular cross section with diameter $d = 0.6$ in.

12.4-22 A double-decker bicycle rack made up of *square* steel tubing is fixed at A (figure a). The weight of a bicycle is represented as a point load applied at B on a plane frame model of the rack (figure b).

(a) Find the state of plane stress on an element C located on the surface at the left side of the vertical tube at the base A. *Include* the weight of the framing system. (Assume weight density $\gamma = 77$ kN/m^3.)

(b) Find the maximum *shear* stresses on an element at C and show them on a sketch of a properly oriented element.

Assume that element C is a sufficient distance from support A so that stress concentration effects are negligible.

(a)

(b)

PROBLEM 12.4-22

12.5-23 A semicircular bar AB lying in a horizontal plane is supported at B (see figure part a). The bar has a centerline radius R and weight q per unit of length (total weight of the bar equals $\pi q R$). The cross section of the bar is circular with diameter d.

(a) Obtain formulas for the maximum tensile stress σ_t, maximum compressive stress σ_c, and maximum in-plane shear stress τ_{max} at the top of the bar at the support due to the weight of the bar.

(b) Repeat part (a) if the bar is a quarter-circular segment (see figure part b) but has the same total weight as the semicircular bar.

(a)   (b)

PROBLEM 12.4-23

12.4-24 Repeat Problem 12.4-22 but replace the square tube column with a *circular* tube having a wall thickness $t = 5$ mm and the same cross-sectional area (3900 mm^2) as that of the square tube in figure b in Problem 12.4-22. Also, add force $P_z = 120$ N at B.

(a) Find the state of plane stress at C.

(b) Find maximum *normal* stresses and show them on a sketch of a properly oriented element.

(c) Find maximum *shear* stresses and show them on a sketch of a properly oriented element.

PROBLEM 12.4-24

12.4-25 An L-shaped bracket lying in a horizontal plane supports a load $P = 150$ lb (see figure). The bracket has a hollow rectangular cross section with thickness $t = 0.125$ in. and outer dimensions $b = 2.0$ in. and $h = 3.5$ in. The centerline lengths of the arms are $b_1 = 20$ in. and $b_2 = 30$ in.

Considering only the load P, calculate the maximum tensile stress σ_t, maximum compressive stress σ_c, and maximum shear stress τ_{max} at point A, which is located on the top of the bracket at the support.

PROBLEM 12.4-25

12.4-26 A horizontal bracket ABC consists of two perpendicular arms AB of a length 0.75 m and BC of a length 0.5 m. The bracket has a solid, circular cross section with a diameter equal to 65 mm. The bracket is inserted in a frictionless sleeve at A (which is slightly larger in diameter), so it is free to rotate about the z_0 axis at A and is supported by a pin at C. Moments are applied at point C, where $M_1 = 1.5$ kN·m in the x direction and $M_2 = 1.0$ kN·m acts in the $-z$ direction.

Cross section at A

PROBLEM 12.4-26

Considering only the moments M_1 and M_2, calculate the maximum tensile stress σ_t, the maximum compressive stress σ_c, and the maximum in-plane shear stress τ_{max} at point p, which is located at support A on the side of the bracket at mid-height.

12.4-27 An arm ABC lying in a horizontal plane and supported at A (see figure) is made of two identical solid steel bars AB and BC welded together at a right angle. Each bar is 22 in. long.

(a) Knowing that the maximum tensile stress (principal stress) at the top of the bar at support A due solely to the weights of the bars is 1025 psi, determine the diameter d of the bars.

(b) If the allowable tensile stress is 1475 psi and each bar has a diameter $d = 2.0$ in., what is the maximum downward load P that can be applied at C (in addition to self-weight)?

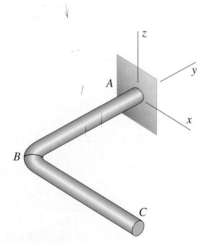

PROBLEM 12.4-27

12.4-28 A crank arm consists of a solid segment of length b_1 and diameter d, a segment of length b_2, and a segment of length b_3, as shown in the figure. Two loads P act as shown: one parallel to $-x$ and another parallel to $-y$. Each load P equals 1.2 kN. The crankshaft dimensions are $b_1 = 75$ mm, $b_2 = 125$ mm, and $b_3 = 35$ mm. The diameter of the upper shaft is $d = 22$ mm.

(a) Determine the maximum tensile, compressive, and shear stresses at point A, which is located on the surface of the shaft at the z axis.

(b) Determine the maximum tensile, compressive, and shear stresses at point B, which is located on the surface of the shaft at the y axis.

PROBLEM 12.4-28

12.4-29 A moveable steel stand supports an automobile engine weighing $W = 750$ lb, as shown in the figure part a. The stand is constructed of 2.5 in. × 2.5 in. × 1/8 in.-thick steel tubing. Once in position, the stand is restrained by pin supports at B and C. Of interest are stresses at point A at the base of the vertical post; point A has coordinates ($x = 1.25$, $y = 0$, $z = 1.25$) in inches. Neglect the weight of the stand.

(a) Initially, the engine weight acts in the $-z$ direction through point Q, which has coordinates (24, 0, 1.25) inches. Find the maximum tensile, compressive, and shear stresses at point A.

(b) Repeat part (a) assuming now that, during repair, the engine is rotated about its own longitudinal axis (which is parallel to the x axis) so that W acts through Q' [with coordinates (24, 6, 1.25) in inches] and force $F_y = 200$ lb is applied parallel to the y axis at distance $d = 30$ in.

(b) Top view

PROBLEM 12.4-29

12.4-30 A mountain bike rider going uphill applies a force $P = 65$ N to each end of the handlebars $ABCD$, made of aluminum alloy 7075-T6, by pulling on the handlebar extenders (DF on right handlebar segment). Consider the right half of the handlebar assembly only (assume the bars are fixed at the fork at A). Segments AB and CD are prismatic with lengths L_1 and L_3 and with outer diameters and thicknesses d_{01}, t_{01} and d_{03}, t_{03}, respectively, as shown. Segment BC of length L_2, however, is tapered, and outer diameter and thickness vary linearly between dimensions at B and C. Consider shear, torsion, and bending effects only for segment AD; assume DF is rigid.

Find the maximum tensile, compressive, and shear stresses adjacent to support A. Show where each maximum stress value occurs.

(a)

(b) Section D–F

PROBLEM 12.4-30

12.4-31 Determine the maximum tensile, compressive, and shear stresses acting on the cross section of the tube at point A of the hitch bicycle rack shown in the figure.

The rack is made up of 2 in. × 2 in. steel tubing which is 1/8 in. thick. Assume that the weight of each of four bicycles is distributed evenly between the two support arms so that the rack can be represented as a cantilever beam ($ABCDEF$) in the xy plane. The overall weight of the rack alone is $W = 60$ lb directed through C, and the weight of each bicycle is $P = 30$ lb.

PROBLEM 12.4-31

12.4-32 Consider the mountain bike shown in the figure. To account for impact, crashes, and other loading uncertainties, a *design load P* = 5000 N is used to design the seat post. The length of the seat post is $L = 254$ mm.

(a) Find the required diameter of the seat post if it is to be constructed using an *aluminum alloy* with the ultimate stress $\sigma_U = 550$ MPa and a factor of safety of 2.8. Consider only axial and flexural normal stresses in your design.

(b) Repeat part (a) if a *titanium alloy* is used instead. Assume the ultimate stress $\sigma_U = 900$ MPa and a factor of safety of 2.5.

PROBLEM 12.4-32

12.4-33 A plumber's valve wrench is used to replace valves in plumbing fixtures. A simplified model of the wrench (see figure part a) consists of pipe AB (length L, outer diameter d_2, inner diameter d_1), which is fixed at A and has holes of a diameter d_b on either side of the pipe at B. A solid, cylindrical bar CBD (length a, diameter d_b) is inserted into the holes at B and *only one* force $F = 55$ lb is applied in the $-Z$ direction at C to loosen the fixture valve at A (see figure part c). Let $G = 11{,}800$ ksi, $\nu = 0.30$, $L = 4$ in., $a = 4.5$ in., $d_2 = 1.25$ in., $d_1 = 1$ in., and $d_b = 0.25$ in.

Find the state of plane stress on the top of the pipe near A (at coordinates $X = 0, Y = 0, Z = d_2/2$), and show all stresses on a plane stress element (see figure part b). Compute the principal stresses and maximum shear stress, and show them on properly rotated stress elements.

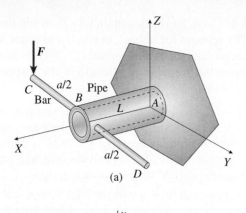

(a)

(b) Repeat part (a) for a plane stress element located at mid-height of the beam at A.

(c) If the maximum tensile stress and maximum in-plane shear stress at point A are limited to 90 MPa and 42 MPa, respectively, what is the largest permissible value of the cable force P?

View looking down on stress element on top of pipe at A

(b)

View of cross section

(c)

PROBLEM 12.4-33

12.4-34 A compound beam $ABCD$ has a cable with force P anchored at C. The cable passes over a pulley at D, and force P acts in the $-x$ direction. There is a moment release just left of B. Neglect the self-weight of the beam and cable. Cable force $P = 450$ N and dimension variable $L = 0.25$ m. The beam has a rectangular cross section ($b = 20$ mm, $h = 50$ mm).

(a) Calculate the maximum normal stresses and maximum in-plane shear stress on the bottom surface of the beam at support A.

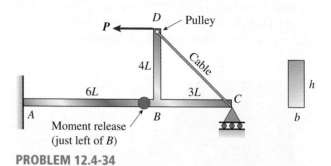

PROBLEM 12.4-34

12.4-35 A steel hanger bracket $ABCD$ has a solid, circular cross section with a diameter of $d = 2$ in. The dimension variable is $b = 6$ in. (see figure). Load $P = 1200$ lb is applied at D along a line DH; the coordinates of point H are $(8b, -5b, 3b)$. Find normal and shear stresses on a plane stress element on the surface of the bracket at A. Then find the principal stresses and maximum shear stress. Show each stress state on properly rotated elements.

PROBLEM 12.4-35

Deflections of Beams

©Kristiina Paul

The roof structure for the atrium of this shopping mall is a complex assemblage of beam elements. Calculation of the forces, moments and deflections of these beams is an important step in the overall roof design.

CHAPTER OBJECTIVES

- Study several different methods for computing deflections of beams, including *integration* of the differential equation of the deflection curve and the *method of superposition.*
- Investigate the behavior of different types of *statically indeterminate beams.*

- Determine the *degree of indeterminacy* of statically indeterminate beams.
- Develop a superposition approach (known as the *force* or *flexibility method*) for solving for redundant forces and moments in statically indeterminate beams.

Chapter Outline

13.1 Introduction

When a beam with a straight longitudinal axis is loaded by lateral forces, the axis is deformed into a curve, called the **deflection curve** of the beam. In Chapter 10, the curvature of the bent beam was used to find the normal strains and stresses in the beam. However, a method for finding the deflection curve itself was not presented. In this chapter, the equation of the deflection curve and also deflections at specific points along the axis of the beam will be determined.

The calculation of deflections is an important part of structural analysis and design. For example, finding deflections is an essential ingredient in the analysis of statically indeterminate structures (Section 13.6). Deflections are also important in dynamic analyses, as when investigating the vibrations of aircraft or the response of buildings to earthquakes.

Deflections are sometimes calculated in order to verify that they are within tolerable limits. For instance, specifications for the design of buildings usually place upper limits on the deflections. Large deflections in buildings are unsightly (and even unnerving) and can cause cracks in ceilings and walls. In the design of machines and aircraft, specifications may limit deflections in order to prevent undesirable vibrations.

13.2 Differential Equations of the Deflection Curve

Most procedures for finding beam deflections are based on the differential equations of the deflection curve and their associated relationships. Consequently, the discussion here begins by deriving the basic equation for the deflection curve of a beam.

Consider a cantilever beam with a concentrated load acting upward at the free end (Fig. 13-1a). Under the action of this load, the axis of the beam deforms into a curve, as shown in Fig. 13-1b. The reference axes have their origin at the fixed end of the beam, with the x axis directed to the right and the y axis directed upward. The z axis is directed outward from the figure (toward the viewer).

As in previous discussions of beam bending in Chapter 10, assume that the x-y plane is a plane of symmetry of the beam, and assume that all loads act in this plane (the *plane of bending*).

The **deflection** v is the displacement in the y direction of any point on the axis of the beam (Fig. 13-1b). Because the y axis is positive upward, the deflections are also positive when upward.[1]

To obtain the equation of the deflection curve, express the deflection v as a function of the coordinate x. The deflection v at any point m_1 on the deflection curve is shown in Fig. 13-2a. Point m_1 is located at distance x from the origin (measured along the x axis). A second point m_2, located at distance $x + dx$ from the origin, is also shown. The deflection at this second point is $v + dv$, where dv is the increment in deflection as you move along the curve from m_1 to m_2.

FIGURE 13-1

Deflection curve of a cantilever beam

(a)

(b)

---

[1]As mentioned in Section 10.1, the traditional symbols for displacements in the x, y, and z directions are u, v, and w, respectively. The advantage of this notation is that it emphasizes the distinction between a *coordinate* and a *displacement*.

When the beam is bent, there is not only a deflection at each point along the axis but also a rotation. The **angle of rotation** θ of the axis of the beam is the angle between the x axis and the tangent to the deflection curve, as shown for point m_1 in the enlarged view of Fig. 13-2b. For the choice of axes (x positive to the right and y positive upward), the angle of rotation is positive when counterclockwise. (Other names for the angle of rotation are *angle of inclination* and *angle of slope*.)

The angle of rotation at point m_2 is $\theta + d\theta$, where $d\theta$ is the increase in angle as you move from point m_1 to point m_2. It follows that if you construct lines normal to the tangents (Figs. 13-2a and b), the angle between these normals is $d\theta$. Also, as discussed earlier in Section 10.3, the point of intersection of these normals is the **center of curvature** O' (Fig. 13-2a) and the distance from O' to the curve is the **radius of curvature** ρ. From Fig. 13-2a, it follows that

$$\rho \, d\theta = ds \qquad\qquad \textbf{(13-1)}$$

where $d\theta$ is in radians and ds is the distance along the deflection curve between points m_1 and m_2. Therefore, the **curvature** κ (equal to the reciprocal of the radius of curvature) is given by the equation

$$\kappa = \frac{1}{\rho} = \frac{d\theta}{ds} \qquad\qquad \textbf{(13-2)}$$

The **sign convention** for curvature is pictured in Fig. 13-3, which is repeated from Fig. 10-6 of Section 10.3. Note that curvature is positive when the angle of rotation increases as you move along the beam in the positive x direction.

The **slope of the deflection curve** is the first derivative dv/dx of the expression for the deflection v. In geometric terms, the slope is the increment dv in the deflection (as you go from point m_1 to point m_2 in Fig. 13-2) divided by the increment dx in the distance along the x axis. Since dv and dx are infinitesimally small, the slope dv/dx is equal to the tangent of the angle of rotation θ (Fig. 13-2b). Thus,

$$\frac{dv}{dx} = \tan\theta \qquad\qquad \theta = \arctan\frac{dv}{dx} \qquad\qquad \textbf{(13-3a,b)}$$

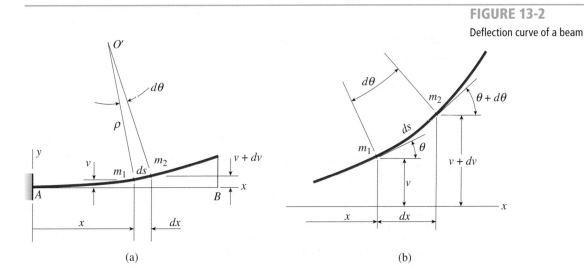

FIGURE 13-2

Deflection curve of a beam

(a)   (b)

FIGURE 13-3

Sign convention for curvature

Positive
curvature

(a)

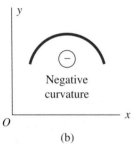

Negative
curvature

(b)

In a similar manner, the following relationships are obtained:

$$\cos\theta = \frac{dx}{ds} \qquad \sin\theta = \frac{dv}{ds} \qquad \text{(13-4a,b)}$$

Note that when the x and y axes have the directions shown in Fig. 13-2a, the slope dv/dx is positive when the tangent to the curve slopes upward to the right.

Equations (13-2) through (13-4) are based only upon geometric considerations; therefore, they are valid for beams of any material. Furthermore, there are no restrictions on the magnitudes of the slopes and deflections.

Beams with Small Angles of Rotation

The structures encountered in everyday life, such as buildings, automobiles, aircraft, and ships, undergo relatively small changes in shape while in service. The changes are so small as to be unnoticed by a casual observer. Consequently, the deflection curves of most beams and columns have very small angles of rotation, very small deflections, and very small curvatures. Under these conditions, some mathematical approximations that greatly simplify beam analysis can be made.

Consider, for instance, the deflection curve shown in Fig. 13-2. If the angle of rotation θ is a very small quantity (and hence the deflection curve is nearly horizontal), the distance ds along the deflection curve is practically the same as the increment dx along the x axis. This same conclusion can be obtained directly from Eq. (13-4a). Since $\cos \approx 1$ when the angle θ is small, Eq. (13-4a) gives

$$ds \approx dx \qquad \text{(13-5)}$$

With this approximation, the curvature becomes [see Eq. (13-2)]

$$\kappa = \frac{1}{\rho} = \frac{d\theta}{dx} \qquad \text{(13-6)}$$

Also, since $\tan\theta \approx \theta$ when θ is small, the following approximation to Eq. (13-3a) can be made:

$$\theta \approx \tan\theta = \frac{dv}{dx} \qquad \text{(13-7)}$$

Thus, if the rotations of a beam are small, assume that the angle of rotation θ and the slope dv/dx are equal. (Note that the angle of rotation must be measured in radians.)

Take the derivative of θ with respect to x in Eq. (13-7) to get

$$\frac{d\theta}{dx} = \frac{d^2v}{dx^2} \qquad \text{(13-8)}$$

Combine this equation with Eq. (13-6) to obtain a relation between the **curvature** of a beam and its deflection:

$$\kappa = \frac{1}{\rho} = \frac{d^2v}{dx^2} \qquad \text{(13-9)}$$

This equation is valid for a beam of any material, provided the rotations are small quantities.

If the material of a beam is **linearly elastic** and follows Hooke's law, the curvature [from Eq. 10-13, Chapter 10] is

$$\kappa = \frac{1}{\rho} = \frac{M}{EI} \tag{13-10}$$

where M is the bending moment and EI is the flexural rigidity of the beam. Equation (13-10) shows that a positive bending moment produces positive curvature and a negative bending moment produces negative curvature, as shown earlier in Fig. 10-10.

Combining Eq. (13-9) with Eq. (13-10) yields the basic **differential equation of the deflection curve** of a beam:

$$\frac{d^2v}{dx^2} = \frac{M}{EI} \tag{13-11}$$

This equation can be integrated in each particular case to find the deflection v, provided the bending moment M and flexural rigidity EI are known as functions of x.

As a reminder, the **sign conventions** to be used with the preceding equations are repeated here: (1) The x and y axes are positive to the right and upward, respectively; (2) the deflection v is positive upward; (3) the slope dv/dx and angle of rotation θ are positive when counterclockwise with respect to the positive x axis; (4) the curvature κ is positive when the beam is bent concave upward; and (5) the bending moment M is positive when it produces compression in the upper part of the beam.

Additional equations can be obtained from the relations between bending moment M, shear force V, and intensity q of distributed load. In Section 6.4 of Chapter 6, the following equations between M, V, and q [see Eqs. (6-1) and (6-4)] were derived:

$$\frac{dV}{dx} = -q \qquad \frac{dM}{dx} = V \tag{13-12a,b}$$

The sign conventions for these quantities are shown in Fig. 13-4. Differentiate Eq. (13-11) with respect to x; then substitute the preceding equations for shear force and load to obtain additional equations. In so doing, consider two cases: nonprismatic beams and prismatic beams.

Nonprismatic Beams

In the case of a nonprismatic beam, the flexural rigidity EI is variable; therefore, Eq. (13-11) is now written in the form

$$EI_x \frac{d^2v}{dx^2} = M \tag{13-13a}$$

where the subscript x is inserted as a reminder that the flexural rigidity may vary with x. Differentiate both sides of this equation and use Eqs. (13-12a and b) to obtain

$$\frac{d}{dx}\left(EI_x \frac{d^2v}{dx^2}\right) = \frac{dM}{dx} = V \tag{13-13b}$$

$$\frac{d^2}{dx^2}\left(EI_x \frac{d^2v}{dx^2}\right) = \frac{dV}{dx} = -q \tag{13-13c}$$

FIGURE 13-4

Sign conventions for bending moment M, shear force V, and intensity q of distributed load

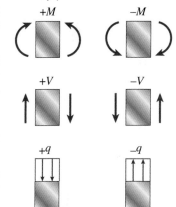

The deflection of a nonprismatic beam can be found by solving (either analytically or numerically) any one of the three preceding differential equations. The choice usually depends upon which equation provides the most efficient solution.

Prismatic Beams

In the case of a prismatic beam (constant EI), the differential equations become

$$EI\frac{d^2v}{dx^2} = M \qquad EI\frac{d^3v}{dx^3} = V \qquad EI\frac{d^4v}{dx^4} = -q \qquad \text{(13-14a,b,c)}$$

To simplify the writing of these and other equations, **primes** are often used to denote differentiation:

$$v' \equiv \frac{dv}{dx} \qquad v'' \equiv \frac{d^2v}{dx^2} \qquad v''' \equiv \frac{d^3v}{dx^3} \qquad v'''' \equiv \frac{d^4v}{dx^4} \qquad \text{(13-15)}$$

Using this notation produces the differential equations for a prismatic beam in the forms:

$$EIv'' = M \qquad EIv''' = V \qquad EIv'''' = -q \qquad \text{(13-16a,b,c)}$$

Refer to these equations as the **bending-moment equation**, the **shear-force equation**, and the **load equation**, respectively.

In the next two sections, the preceding equations are used to find deflections of beams. The general procedure consists of integrating the equations and then evaluating the constants of integration from boundary and other conditions pertaining to the beam.

When deriving the differential equations [Eqs. (13-13), (13-14), and (13-16)], it is assumed that the material followed Hooke's law and that the slopes of the deflection curve are very small. Also any shear deformations are assumed to be negligible; consequently, only the deformations due to pure bending are considered. All of these assumptions are satisfied by most beams in common use.

Exact Expression for Curvature

If the deflection curve of a beam has large slopes, you cannot use the approximations given by Eqs. (13-5) and (13-7). Instead, the exact expressions for curvature and angle of rotation must be used [see Eqs. (13-2) and (13-3b)]. Combine those expressions to get

$$\kappa = \frac{1}{\rho} = \frac{d\theta}{ds} = \frac{d(\arctan v')}{dx}\frac{dx}{ds} \qquad \text{(13-17)}$$

From Fig. 13-2, note that

$$ds^2 = dx^2 + dv^2 \qquad \text{or} \qquad ds = [dx^2 + dv^2]^{1/2} \qquad \text{(13-18a,b)}$$

Dividing both sides of Eq. (13-18b) by dx gives

$$\frac{ds}{dx} = \left[1 + \left(\frac{dv}{dx}\right)^2\right]^{1/2} = [1 + (v')^2]^{1/2} \qquad \text{or} \qquad \frac{dx}{ds} = \frac{1}{[1 + (v')^2]^{1/2}} \qquad \text{(13-18c,d)}$$

Also, differentiation of the arctangent function (see Appendix A) gives

$$\frac{d}{dx}(\arctan v') = \frac{v''}{1 + (v')^2} \qquad \text{(13-18e)}$$

Substitution of expressions Eqs. (13-18d and e) into the equation for curvature from Eq. (13-17) yields

$$\kappa = \frac{1}{\rho} = \frac{v''}{[1 + (v')^2]^{3/2}} \qquad \textbf{(13-19)}$$

Comparing this equation with Eq. (13-9), note that the assumption of small rotations is equivalent to disregarding $(v')^2$ in comparison to one. Equation (13-19) should be used for the curvature whenever the slopes are large.[2]

13.3 Deflections by Integration of the Bending-Moment Equation

To solve the differential equations of the deflection curve and obtain deflections of beams, the first equation you will use is the bending-moment equation (Eq. 13-16a). Since this equation is of second order, two integrations are required. The first integration produces the slope $v' = dv / dx$, and the second produces the deflection v.

Begin the analysis by writing the equation (or equations) for the bending moments in the beam. Since statically determinate beams are considered first in this chapter, obtain the bending moments from free-body diagrams and equations of equilibrium, using the procedures described in Chapter 6. In some cases, a single bending-moment expression holds for the entire length of the beam, as illustrated in Examples 13-1 and 13-2. In other cases, the bending moment changes abruptly at one or more points along the axis of the beam. Then write separate bending-moment expressions for each region of the beam between points where changes occur, as illustrated in Example 13-3.

Regardless of the number of bending-moment expressions, the general procedure for solving the differential equations is as follows. For each region of the beam, substitute the expression for M into the differential equation and integrate to obtain the slope v'. Each such integration produces one constant of integration. Next, integrate each slope equation to obtain the corresponding deflection v. Again, each integration produces a new constant. Thus, there are two constants of integration for each region of the beam. These constants are evaluated from known conditions pertaining to the slopes and deflections. The conditions fall into three categories: (1) boundary conditions, (2) continuity conditions, and (3) symmetry conditions.

Boundary conditions pertain to the deflections and slopes at the supports of a beam. For example, at a simple support (either a pin or a roller), the deflection is zero (Fig. 13-5), and at a fixed support, both the deflection and the slope are zero (Fig. 13-6). Each such boundary condition supplies one equation that can be used to evaluate the constants of integration.

FIGURE 13-5

Boundary conditions at simple supports

$v_A = 0$ $\qquad\qquad v_B = 0$

FIGURE 13-6

Boundary conditions at a fixed support

$v_A = 0$
$v'_A = 0$

---

[2]The basic relationship stating that the curvature of a beam is proportional to the bending moment (Eq. 13-10) was first obtained by Jacob Bernoulli, although he obtained an incorrect value for the constant of proportionality. The relationship was used later by Euler, who solved the differential equation of the deflection curve for both large deflections (using Eq. 13-19) and small deflections (using Eq. 13-11).

FIGURE 13-7

Continuity conditions at point C

At point C: $(v)_{AC} = (v)_{CB}$
$(v')_{AC} = (v')_{CB}$

Continuity conditions occur at points where the regions of integration meet, such as at point C in the beam of Fig. 13-7. The deflection curve of this beam is physically continuous at point C; therefore, the deflection at point C as determined for the left-hand part of the beam must be equal to the deflection at point C as determined for the right-hand part. Similarly, the slopes found for each part of the beam must be equal at point C. Each of these continuity conditions supplies an equation for evaluating the constants of integration.

Symmetry conditions also may be available. For instance, if a simple beam supports a uniform load throughout its length, the slope of the deflection curve at the midpoint must be zero. This condition supplies an additional equation, as illustrated in Example 13-1.

Each boundary, continuity, and symmetry condition leads to an equation containing one or more of the constants of integration. Since the number of *independent* conditions always matches the number of constants of integration, you can always solve these equations for the constants. (The boundary and continuity conditions alone are always sufficient to determine the constants. Any symmetry conditions provide additional equations, but they are not independent of the other equations. The choice of which conditions to use is a matter of convenience.)

Once the constants are evaluated, they can be substituted back into the expressions for slopes and deflections, thus yielding the final equations of the deflection curve. These equations then can be used to obtain the deflections and angles of rotation at particular points along the axis of the beam.

The preceding method for finding deflections is sometimes called the **method of successive integrations**. The following examples illustrate this method in detail.

Note: Sketches of deflection curves, such as those shown in the following examples and in Figs. 13-5, 13-6, and 13-7, are greatly exaggerated for clarity. However, it always should be kept in mind that the actual deflections are very small quantities.

Example 13-1

Determine the equation of the deflection curve for a simple beam AB supporting a uniform load of intensity q acting throughout the span of the beam (Fig. 13-8a).

Also, determine the maximum deflection δ_{max} at the midpoint of the beam and the angles of rotation θ_A and θ_B at the supports (Fig. 13-8b). *Note:* The beam has length L and constant flexural rigidity EI.

Solution:

Use the four-step problem-solving approach.

1. *Conceptualize* [*hypothesize, sketch*]: The beam is statically determinate. Begin by finding reaction R_A. The left-hand free-body diagram in Fig. 13-9 is then used to obtain an expression for internal moment $M(x)$.

   Bending moment in the beam: The bending moment at a cross-section a distance x from the left-hand support is obtained from the free-body diagram of

FIGURE 13-8

Example 13-1: Deflections of a simple beam
with a uniform load

(a)

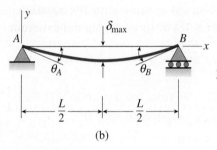

(b)

FIGURE 13-9

Example 13-1: Free-body diagram used in
determining the bending moment M

Fig. 13-9. Since the reaction at the support is $qL/2$, the equation for the bending moment is

$$M = \frac{qL}{2}(x) - qx\left(\frac{x}{2}\right) = \frac{qLx}{2} - \frac{qx^2}{2} \qquad \textbf{(13-20)}$$

2. *Categorize* [*simplify, classify*]: Use the expression for internal moment $M(x)$ in the bending-moment equation [Eq. (13-16a)] to find expressions for slopes and deflections for this beam.

Differential equation of the deflection curve: Substitute the expression for the bending moment (Eq. 13-20) into the differential equation (Eq. 13-16a) to obtain

$$EIv'' = \frac{qLx}{2} - \frac{qx^2}{2} \qquad \textbf{(13-21)}$$

3. *Analyze* [*evaluate; select relevant equations; carry out mathematical solution*]: This equation can now be integrated to obtain the slope and deflection of the beam.

Slope of the beam: Multiply both sides of the differential equation by dx to get

$$EIv'' \, dx = \frac{qLx}{2} dx - \frac{qx^2}{2} dx$$

Integrate each term to obtain

$$EI \int v'' dx = \int \frac{qLx}{2} dx - \int \frac{qx^2}{2} dx$$

or

$$EIv' = \frac{qLx^2}{4} - \frac{qx^3}{6} + C_1 \qquad \textbf{(a)}$$

in which C_1 is a constant of integration.

To evaluate the constant C_1, observe from the symmetry of the beam and its load that the slope of the deflection curve at mid-span is equal to zero. Thus, the following symmetry condition applies here:

$$v' = 0 \quad \text{when} \quad x = \frac{L}{2}$$

This condition may be expressed more succinctly as

$$v'\left(\frac{L}{2}\right) = 0$$

Applying this condition to Eq. (a) gives

$$0 = \frac{qL}{4}\left(\frac{L}{2}\right)^2 - \frac{q}{6}\left(\frac{L}{2}\right)^3 + C_1 \quad \text{or} \quad C_1 = -\frac{qL^3}{24}$$

The equation for the slope of the beam [Eq. (a)] then becomes

$$EIv' = \frac{qLx^2}{4} - \frac{qx^3}{6} - \frac{qL^3}{24} \tag{b}$$

or

$$v' = -\frac{q}{24EI}(L^3 - 6Lx^2 + 4x^3) \tag{13-22}$$

As expected, the slope is negative (that is, clockwise) at the left-hand end of the beam ($x = 0$), positive at the right-hand end ($x = L$), and equal to zero at the midpoint ($x = L/2$).

Deflection of the beam: The deflection is obtained by integrating the equation for the slope. Thus, multiply both sides of Eq. (b) by dx and integrate to obtain

$$EIv = \frac{qLx^3}{12} - \frac{qx^3}{24} - \frac{qL^3x}{24} + C_2 \tag{c}$$

The constant of integration C_2 may be evaluated from the condition that the deflection of the beam at the left-hand support is equal to zero; that is, $v = 0$ when $x = 0$, or

$$v(0) = 0$$

Applying this condition to Eq. (c) yields $C_2 = 0$; hence, the equation for the deflection curve is

$$EIv = \frac{qLx^3}{12} - \frac{qx^4}{24} - \frac{qL^3x}{24} \tag{d}$$

or

$$v = -\frac{qx}{24EI}(L^3 - 2Lx^2 + x^3) \tag{13-23}$$

This equation gives the deflection at any point along the axis of the beam. Note that the deflection is zero at both ends of the beam ($x = 0$ and $x = L$) and negative elsewhere (recall that downward deflections are negative).

Maximum deflection: From symmetry, the maximum deflection occurs at the midpoint of the span (Fig. 13-8b). Thus, set x equal to $L/2$ in Eq. (13-23) to obtain

$$v\left(\frac{L}{2}\right) = -\frac{5qL^4}{384EI}$$

in which the negative sign means that the deflection is downward (as expected). Variable δ_{max} represents the magnitude of this deflection and is expressed as

$$\delta_{max} = \left|v\left(\frac{L}{2}\right)\right| = \frac{5qL^4}{384EI} \tag{13-24}$$

Angles of rotation: The maximum angles of rotation occur at the supports of the beam. At the left-hand end of the beam, the angle θ_A, which is a clockwise angle (Fig. 13-8b), is equal to the negative of the slope v'. Thus, substitute $x = 0$ into Eq. (13-22) to find

$$\theta_A = -v'(0) = \frac{qL^3}{24EI} \qquad \text{◀ (13-25)}$$

In a similar manner, obtain the angle of rotation θ_B at the right-hand end of the beam. Since θ_B is a counterclockwise angle, it is equal to the slope at the end:

$$\theta_B = v'(L) = \frac{qL^3}{24EI} \qquad \text{◀ (13-26)}$$

4. *Finalize [conclude; examine answer—Does it make sense? Are units correct? How does it compare to similar problem solutions?]:* Because the beam and loading are symmetric about the midpoint, the angles of rotation at the ends are equal.

   This example illustrates the process of setting up and solving the differential equation of the deflection curve. It also illustrates the process of finding slopes and deflections at selected points along the axis of a beam.

   Note: Now that formulas for the maximum deflection and maximum angles of rotation have been derived [see Eqs. (13-24), (13-25), and (13-26)], you can evaluate those quantities numerically and observe that the deflections and angles are indeed small, as the theory requires.

   Consider a steel beam on simple supports with a span length $L = 6\ \text{ft}$. The cross section is rectangular with width $b = 3\ \text{in.}$ and height $h = 6\ \text{in.}$ The intensity of uniform load is $q = 8000\ \text{lb/ft}$, which is relatively large because it produces a stress in the beam of 24,000 psi. (Thus, the deflections and slopes are larger than would normally be expected.)

   Substitute into Eq. (13-24), and use $E = 30 \times 10^6$ psi to find that the maximum deflection is $\delta_{max} = 0.144\ \text{in.}$, which is only 1/500 of the span length. Also, from Eq. (13-25), the maximum angle of rotation is computed as $\theta_A = 0.0064$ radians, or $0.37°$, which is a very small angle.

   Thus, the assumption that the slopes and deflections are small is validated.

Example 13-2

Determine the equation of the deflection curve for a cantilever beam AB subjected to a uniform load of intensity q (Fig. 13-10a).

Also, determine the angle of rotation θ_B and the deflection δ_B at the free end (Fig. 13-10b). *Note:* The beam has length L and constant flexural rigidity EI.

FIGURE 13-10

Example 13-2: Deflections of a cantilever beam with a uniform load

(a)

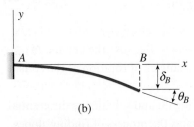

(b)

FIGURE 13-11

Example 13-2: Free-body diagram used in determining the bending moment M

Solution:

Use the four-step problem-solving approach.

1. *Conceptualize*: The beam is statically determinate. Begin by finding reaction force R_A and reaction moment M_A. The left-hand free-body diagram in Fig. 13-11 is then used to obtain an expression for internal moment $M(x)$.

   Bending moment in the beam: The bending moment at distance x from the fixed support is obtained from the free-body diagram of Fig. 13-11. Note that the vertical reaction at the support is equal to qL and the moment reaction is equal to $qL^2/2$. Consequently, the expression for the bending moment M is

   $$M = -\frac{qL^2}{2} + qLx - \frac{qx^2}{2} \tag{13-27}$$

2. *Categorize*: Use the expression for internal moment $M(x)$ in the bending moment equation (Eq. 13-14a) to find expressions for slopes and deflections for this beam.

   Differential equation of the deflection curve: When the preceding expression for the bending moment is substituted into the differential equation (Eq. 13-16a), the following equation is obtained:

   $$EIv'' = -\frac{qL^2}{2} + qLx - \frac{qx^2}{2} \tag{13-28}$$

3. *Analyze*: Now integrate both sides of this equation to obtain the slopes and deflections.

Slope of the beam: The first integration of Eq. (13-28) gives the following equation for the slope:

$$EIv' = -\frac{qL^2 x}{2} + \frac{qLx^2}{2} - \frac{qx^3}{6} + C_1 \tag{a}$$

The constant of integration C_1 can be found from the boundary condition that the slope of the beam is zero at the support; which is expressed as

$$v'(0) = 0$$

When this condition is applied to Eq. (a), the result is $C_1 = 0$. Therefore, Eq. (a) becomes

$$EIv' = -\frac{qL^2 x}{2} + \frac{qLx^2}{2} - \frac{qx^3}{6} \tag{b}$$

and the slope is

$$v' = -\frac{qx}{6EI}(3L^2 - 3Lx + x^2) \tag{13-29}$$

As expected, the slope obtained from this equation is zero at the support ($x = 0$) and negative (i.e., clockwise) throughout the length of the beam.

Deflection of the beam: Integration of the slope equation [Eq. (b)] yields

$$EIv = -\frac{qL^2x^2}{4} + \frac{qLx^3}{6} - \frac{qx^4}{24} + C_2 \qquad \text{(c)}$$

The constant C_2 is found from the boundary condition that the deflection of the beam is zero at the support:

$$v(0) = 0$$

When this condition is applied to Eq. (c), the result is $C_2 = 0$. Therefore, the equation for the deflection v is

$$v = -\frac{qx^2}{24EI}(6L^2 - 4Lx + x^2) \qquad \text{◀ (13-30)}$$

As expected, the deflection obtained from this equation is zero at the support ($x = 0$) and negative (that is, downward) elsewhere.

Angle of rotation at the free end of the beam: The clockwise angle of rotation θ_B at end B of the beam (Fig. 13-10b) is equal to the negative of the slope at that point. Thus, use Eq. (13-29) to get

$$\theta_B = -v'(L) = \frac{qL^3}{6EI} \qquad \text{◀ (13-31)}$$

This angle is the maximum angle of rotation for the beam.

Deflection at the free end of the beam: Since the deflection δ_B is downward (Fig. 13-10b), it is equal to the negative of the deflection obtained from Eq. (13-30):

$$\delta_B = -v(L) = \frac{qL^4}{8EI} \qquad \text{◀ (13-32)}$$

This deflection is the maximum deflection of the beam.

4. *Finalize*: Equations (13-29) to (13-32) are listed as Case 1 in Table E-1, Appendix E.

Example 13-3

A simple beam AB supports a concentrated load P acting at distances a and b from the left-hand and right-hand supports, respectively (Fig. 13-12a).

Determine the equations of the deflection curve, the angles of rotation θ_A and θ_B at the supports, the maximum deflection $\delta_{\max}$, and the deflection δ_C at the midpoint C of the beam (Fig. 13-12b). *Note:* The beam has length L and constant flexural rigidity EI.

FIGURE 13-12

Example 13-3: Deflections of a simple beam with a concentrated load

(a)

(b)

FIGURE 13-13

Example 13-3: Free-body diagrams used in determining the bending moments

(a)

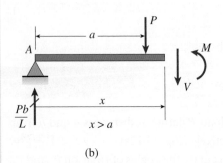

(b)

Solution:

Use the four-step problem-solving approach.

1. *Conceptualize*: Internal shear $V(x)$ and moment $M(x)$ are discontinuous at load point P. Two free-body diagrams are required to obtain moment expressions over the entire length of the beam (see Figs. 13-13a and b).

Bending moments in the beam: In this example, the bending moments are expressed by two equations—one for each part of the beam. Use the free-body diagrams of Fig. 13-13 to arrive at

$$M = \frac{Pbx}{L} \qquad (0 \le x \le a) \qquad \textbf{(13-33a)}$$

$$M = \frac{Pbx}{L} - P(x - a) \qquad (a \le x \le L) \qquad \textbf{(13-33b)}$$

2. *Categorize*:

Differential equations of the deflection curve: The differential equations for the two parts of the beam are obtained by substituting the bending-moment expressions [Eqs. (13-33a and b)] into Eq. (13-16a). The results are

$$EIv'' = \frac{Pbx}{L} \qquad (0 \le x \le a) \qquad \textbf{(13-34a)}$$

$$EIv'' = \frac{Pbx}{L} - P(x - a) \qquad (a \le x \le L) \qquad \textbf{(13-34b)}$$

3. *Analyze*:

Slopes and deflections of the beam: The first integrations of the two differential equations yield the following expressions for the slopes:

$$EIv' = \frac{Pbx^2}{2L} + C_1 \qquad (0 \le x \le a) \qquad \textbf{(a)}$$

$$EIv' = \frac{Pbx^2}{2L} - \frac{P(x - a)^2}{2} + C_2 \qquad (a \le x \le L) \qquad \textbf{(b)}$$

in which C_1 and C_2 are constants of integration. A second pair of integrations gives the deflections:

$$EIv = \frac{Pbx^3}{6L} + C_1x + C_3 \qquad (0 \le x \le a) \qquad \textbf{(c)}$$

$$EIv = \frac{Pbx^3}{6L} - \frac{P(x - a)^3}{6} + C_2x + C_4 \qquad (a \le x \le L) \qquad \textbf{(d)}$$

These equations contain two additional constants of integration, making a total of four constants to be evaluated.

Constants of integration: The four constants of integration can be found from the following four conditions:

  i. At $x = a$, the slopes v' for the two parts of the beam are the same.
 ii. At $x = a$, the deflections v for the two parts of the beam are the same.
iii. At $x = 0$, the deflection v is zero.
 iv. At $x = L$, the deflection v is zero.

The first two conditions are continuity conditions based upon the fact that the axis of the beam is a continuous curve. Conditions (iii) and (iv) are boundary conditions that must be satisfied at the supports.

Condition (i) means that the slopes determined from Eqs. (a) and (b) must be equal when $x = a$; therefore,

$$\frac{Pba^2}{2L} + C_1 = \frac{Pba^2}{2L} + C_2 \quad \text{or} \quad C_1 = C_2$$

Condition (ii) means that the deflections found from Eqs. (c) and (d) must be equal when $x = a$; therefore,

$$\frac{Pba^3}{6L} + C_1 a + C_3 = \frac{Pba^3}{6L} + C_2 a + C_4$$

Inasmuch as $C_1 = C_2$, this equation gives $C_3 = C_4$.

Next, apply condition (iii) to Eq. (c) and obtain $C_3 = 0$; therefore,

$$C_3 = C_4 = 0 \tag{e}$$

Finally, apply condition (iv) to Eq. (d) and obtain

$$\frac{PbL^2}{6} - \frac{Pb^3}{6} + C_2 L = 0$$

Therefore,

$$C_1 = C_2 = -\frac{Pb(L^2 - b^2)}{6L} \tag{f}$$

Equations of the deflection curve: Now substitute the constants of integration [Eqs. (e) and (f)] into the equations for the deflections [Eqs. (c) and (d)] and obtain the deflection equations for the two parts of the beam. The resulting equations, after a slight rearrangement, are

$$v = -\frac{Pbx}{6LEI}(L^2 - b^2 - x^2) \qquad (0 \le x \le a) \quad \Longleftarrow \textbf{(13-35a)}$$

$$v = -\frac{Pbx}{6LEI}(L^2 - b^2 - x^2) - \frac{P(x - a)^3}{6EI} \qquad (a \le x \le L) \quad \Longleftarrow \textbf{(13-35b)}$$

Equation (13-35a) gives the deflection curve for the part of the beam to the left of the load P, and Eq. (13-35b) gives the deflection curve for the part of the beam to the right of the load.

The slopes for the two parts of the beam can be found either by substituting the values of C_1 and C_2 into Eqs. (a) and (b) or by taking the first derivatives of the deflection equations [Eqs. (13-35a and b)]. The resulting equations are

$$v' = -\frac{Pb}{6LEI}(L^2 - b^2 - 3x^2) \qquad (0 \le x \le a) \qquad \textbf{(13-36a)}$$

$$v' = -\frac{Pb}{6LEI}(L^2 - b^2 - 3x^2) - \frac{P(x-a)^2}{2EI} \qquad (a \le x \le L) \qquad \textbf{(13-36b)}$$

The deflection and slope at any point along the axis of the beam can be calculated from Eqs. (13-35) and (13-36).

Angles of rotation at the supports: To obtain the angles of rotation θ_A and θ_B at the ends of the beam (Fig. 13-12b), substitute $x = 0$ into Eq. (13-36a) and $x = L$ into Eq. (13-36b):

$$\theta_A = -v'(0) = \frac{Pb(L^2 - b^2)}{6LEI} = \frac{Pab(L + b)}{6LEI} \qquad \Longleftarrow \textbf{(13-37a)}$$

$$\theta_B = v'(L) = \frac{Pb(2L^2 - 3bL + b^2)}{6LEI} = \frac{Pab(L + a)}{6LEI} \qquad \Longleftarrow \textbf{(13-37b)}$$

FIGURE (13-12b Repeated)

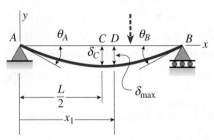

Note that the angle θ_A is clockwise and the angle θ_B is counterclockwise, as shown in Fig. 13-12b.

The angles of rotation are functions of the position of the load and reach their largest values when the load is located near the midpoint of the beam. In the case of the angle of rotation θ_A, the maximum value of the angle is

$$(\theta_A)_{max} = \frac{PL^2\sqrt{3}}{27EI} \qquad \textbf{(13-38)}$$

and occurs when $b = L/\sqrt{3} = 0.577L$ (or $a = 0.423L$). This value of b is obtained by taking the derivative of θ_A with respect to b [using the first of the two expressions for θ_A in Eq. (13-37a)] and then setting it equal to zero.

Maximum deflection of the beam: The maximum deflection δ_{max} occurs at point D (Fig. 13-12b) where the deflection curve has a horizontal tangent. If the load is to the right of the midpoint, that is, if $a > b$, point D is in the part of the beam to the left of the load. Locate this point by equating the slope v' from Eq. (13-36a) to zero and solving for the distance x, which is now denoted as x_1. In this manner, obtain the following formula for x_1:

$$x_1 = \sqrt{\frac{L^2 - b^2}{3}} \qquad (a \ge b) \qquad \textbf{(13-39)}$$

From this equation, note that, as the load P moves from the middle of the beam ($b = L/2$) to the right-hand end ($b = 0$), the distance x_1 varies from $L/2$ to $L/\sqrt{3} = 0.577L$. Thus, the maximum deflection occurs at a point very close to the midpoint of the beam, and this point is always between the midpoint of the beam and the load.

The maximum deflection δ_{max} is found by substituting x_1 [from Eq. (13-39)] into the deflection equation [Eq. (13-35a)] and then inserting a minus sign:

$$\delta_{max} = -(v)_{x=x1} = \frac{Pb(L^2 - b^2)^{3/2}}{9\sqrt{3}\, LEI} \quad (a \geq b) \quad \longleftarrow \textbf{(13-40)}$$

The minus sign is needed because the maximum deflection is downward (Fig. 13-12b), whereas the deflection v is positive upward.

The maximum deflection of the beam depends on the position of the load P, that is, on the distance b. The maximum value of the maximum deflection (the "max-max" deflection) occurs when $b = L/2$ and the load is at the midpoint of the beam. This maximum deflection is equal to $PL^3/48EI$.

Deflection at the midpoint of the beam: The deflection δ_C at the midpoint C when the load is acting to the right of the midpoint (Fig. 13-12b) is obtained by substituting $x = L/2$ into Eq. (13-35a), as

$$\delta_C = -v\left(\frac{L}{2}\right) = \frac{Pb(3L^2 - 4b^2)}{48EI} \quad (a \geq b) \quad \textbf{(13-41)}$$

4. *Finalize*: Because the maximum deflection always occurs near the midpoint of the beam, Eq. (13-41) yields a close approximation to the maximum deflection. In the most unfavorable case (when b approaches zero), the difference between the maximum deflection and the deflection at the midpoint is less than 3% of the maximum deflection, as demonstrated in Problem 13.3-9.

Special case (load at the midpoint of the beam): An important special case occurs when the load P acts at the midpoint of the beam ($a = b = L/2$). Then the following results are obtained from Eqs. (13-36a), (13-35a), (13-37), and (13-40), respectively:

$$v' = -\frac{P}{16EI}(L^2 - 4x^2) \quad \left(0 \leq x \leq \frac{L}{2}\right) \quad \textbf{(13-42)}$$

$$v = -\frac{Px}{48EI}(3L^2 - 4x^2) \quad \left(0 \leq x \leq \frac{L}{2}\right) \quad \textbf{(13-43)}$$

$$\theta_A = \theta_B = \frac{PL^2}{16EI} \quad \textbf{(13-44)}$$

$$\delta_{max} = \delta_C = \frac{PL^3}{48EI} \quad \textbf{(13-45)}$$

Since the deflection curve is symmetric about the midpoint of the beam, the equations for v' and v are given only for the left-hand half of the beam in Eqs. (13-42) and (13-43). If needed, the equations for the right-hand half can be obtained from Eqs. (13-36b) and (13-35b) by substituting $a = b = L/2$.

13.4 Deflections by Integration of the Shear-Force and Load Equations

The equations of the deflection curve in terms of the shear force V and the load q [Eqs. (13-16b and c), respectively] also may be integrated to obtain slopes and deflections. Since the loads are usually known quantities, whereas the bending moments must be determined from free-body diagrams and equations of equilibrium, many analysts prefer to start with the load equation. For this same reason, most computer programs for finding deflections begin with the load equation and then perform numerical integrations to obtain the shear forces, bending moments, slopes, and deflections.

The procedure for solving either the load equation or the shear-force equation is similar to that for solving the bending-moment equation, except that more integrations are required. For instance, with the load equation, four integrations are needed in order to arrive at the deflections. Thus, four constants of integration are introduced for each load equation that is integrated. As before, these constants are found from boundary, continuity, and symmetry conditions. However, these conditions now include conditions on the shear forces and bending moments as well as conditions on the slopes and deflections.

Conditions on the shear forces are equivalent to conditions on the third derivative (because $EIv''' = V$). In a similar manner, conditions on the bending moments are equivalent to conditions on the second derivative (because $EIv'' = M$). When the shear-force and bending-moment conditions are added to those for the slopes and deflections, there are always enough independent conditions to solve for the constants of integration.

The following examples illustrate the techniques of analysis in detail. The first example begins with the load equation and the second begins with the shear-force equation.

Example 13-4

Determine the equation of the deflection curve for a cantilever beam AB supporting a triangularly distributed load of maximum intensity q_0 (Fig. 13-14a).

Also, determine the deflection δ_B and angle of rotation θ_B at the free end (Fig. 13-14b). Use the fourth-order differential equation of the deflection curve (the load equation). *Note:* The beam has length L and constant flexural rigidity EI.

Solution:

Use the four-step problem-solving approach.

1. *Conceptualize:* The cantilever beam is statically determinate. Use the load equation to find the deflection curve of the beam. An expression for load intensity $q(x)$ can be obtained as shown in Fig. 13-14a.

   Differential equation of the deflection curve: The intensity of the distributed load is given by (see Fig. 13-14a):

   $$q(x) = \frac{q_0(L - x)}{L} \tag{13-46}$$

Cantilever portion of roof structure (Courtesy of the National Information Service for Earthquake Engineering EERC, University of California, Berkeley)

FIGURE 13-14

Example 13-4: Deflections of a cantilever beam with a triangular load

(a)

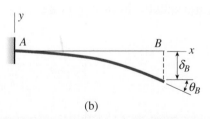

(b)

2. *Categorize*: Consequently, the fourth-order differential equation [Eq. (13-16c)] becomes

$$EIv'''' = -q(x) = -\frac{q_0(L-x)}{L} \tag{a}$$

3. *Analyze*:

Shear force in the beam: The first integration of Eq. (a) gives

$$EIv''' = \frac{q_0}{2L}(L - x)^2 + C_1 \tag{b}$$

The right-hand side of this equation represents the shear force V [see Eq. (13-16b)]. Because the shear force is zero at $x = L$, the boundary condition is

$$v'''(L) = 0$$

Use this condition with Eq. (b), to get $C_1 = 0$. Therefore, Eq. (b) simplifies to

$$EIv''' = \frac{q_0}{2L}(L - x)^2 \tag{c}$$

and the shear force in the beam is

$$V = EIv''' = \frac{q_0}{2L}(L - x)^2 \tag{13-47}$$

Bending moment in the beam: Integrate a second time to obtain the following equation from Eq. (c):

$$EIv'' = -\frac{q_0}{6L}(L - x)^3 + C_2 \tag{d}$$

This equation is equal to the bending moment M [see Eq. (13-16a)]. The bending moment is zero at the free end of the beam, so the boundary condition is

$$v''(L) = 0$$

Apply this condition to Eq. (d) to obtain $C_2 = 0$, so the bending moment is

$$M = EIv'' = -\frac{q_0}{6L}(L - x)^3 \tag{13-48}$$

Slope and deflection of the beam: The third and fourth integrations yield

$$EIv' = \frac{q_0}{24L}(L - x)^4 + C_3 \tag{e}$$

$$EIv = -\frac{q_0}{120L}(L - x)^5 + C_3 x + C_4 \tag{f}$$

The boundary conditions at the fixed support, where both the slope and deflection are equal to zero, are

$$v'(0) = 0 \qquad v(0) = 0$$

Apply these conditions to Eqs. (e) and (f), respectively, to find

$$C_3 = -\frac{q_0 L^3}{24} \qquad C_4 = \frac{q_0 L^4}{120}$$

Substitute these expressions for the constants into Eqs. (e) and (f) to obtain the equations for the slope and deflection of the beam:

$$v' = -\frac{q_0 x}{24 LEI}(4L^3 - 6L^2 x + 4Lx^2 - x^3) \tag{13-49}$$

$$v = -\frac{q_0 x^2}{120 LEI}(10L^3 - 10L^2 x + 5Lx^2 - x^3) \tag{13-50}$$

Angle of rotation and deflection at the free end of the beam: The angle of rotation θ_B and deflection δ_B at the free end of the beam (Fig. 13-14b) are obtained from Eqs. (13-49) and (13-50), respectively, by substituting $x = L$. The results are

$$\theta_B = -v'(L) = \frac{q_0 L^3}{24 EI} \qquad \delta_B = -v(L) = \frac{q_0 L^4}{30 EI} \tag{13-51a,b}$$

4. *Finalize:* The required slopes and deflections of the beam were found by solving the fourth-order differential equation of the deflection curve. This was shown to be an efficient approach for use with the linearly varying distributed load.

Example 13-5

A simple beam AB with an overhang BC supports a concentrated load P at the end of the overhang (Fig. 13-15a). The main span of the beam has length L and the overhang has length $L/2$.

Determine the equations of the deflection curve and the deflection δ_C at the end of the overhang (Fig. 13-15b). Use the third-order differential equation of the

FIGURE 13-15

Example 13-5: Deflections of a beam with an overhang

(a)   (b)

Bridge girder with overhang during transport to the construction site

deflection curve (the shear-force equation). *Note:* The beam has constant flexural rigidity *EI*.

Solution:

Use the four-step problem-solving approach.

1. *Conceptualize*:

    Differential equations of the deflection curve: Because reactive forces act at supports *A* and *B*, write separate differential equations for parts *AB* and *BC* of the beam. Begin by finding the shear forces in each part of the beam.

    The downward reaction at support *A* is equal to *P/2*, and the upward reaction at support *B* is equal to 3*P/2* (see Fig. 13-15a). It follows that the shear forces in parts *AB* and *BC* are

$$V = -\frac{P}{2} \quad (0 < x < L) \tag{13-52a}$$

$$V = P \quad \left(L < x < \frac{3L}{2}\right) \tag{13-52b}$$

where *x* is measured from end *A* of the beam (Fig. 13-15b).

2. *Categorize*: The third-order differential equations for the beam now become [see Eq. (13-16b)]:

$$EIv''' = -\frac{P}{2} \quad (0 < x < L) \tag{a}$$

$$EIv''' = P \quad \left(L < x < \frac{3L}{2}\right) \tag{b}$$

3. *Analyze*:

    Bending moments in the beam: Integration of the preceding two equations yields the bending-moment equations:

$$M = EIv'' = -\frac{Px}{2} + C_1 \quad (0 \leq x \leq L) \tag{c}$$

$$M = EIv'' = Px + C_2 \quad \left(L \leq x \leq \frac{3L}{2}\right) \tag{d}$$

The bending moments at points *A* and *C* are zero; hence, the following boundary conditions apply to this beam:

$$v''(0) = 0 \qquad v''\left(\frac{3L}{2}\right) = 0$$

Use these conditions with Eqs. (c) and (d) to get

$$C_1 = 0 \qquad C_2 = -\frac{3PL}{2}$$

Therefore, the bending moments are

$$M = EIv'' = -\frac{Px}{2} \qquad (0 \leq x \leq L) \tag{13-53a}$$

$$M = EIv'' = -\frac{P(3L - 2x)}{2} \quad \left(L \leq x \leq \frac{3L}{2}\right) \tag{13-53b}$$

These equations can be verified by determining the bending moments from free-body diagrams and equations of equilibrium.

Slopes and deflections of the beam: The next integrations yield the slopes:

$$EIv' = -\frac{Px^2}{4} + C_3 \qquad (0 \leq x \leq L)$$

$$EIv' = -\frac{Px(3L - x)}{2} + C_4 \qquad \left(L \leq x \leq \frac{3L}{2}\right)$$

The only condition on the slopes is the continuity condition at support B. According to this condition, the slope at point B as found for part AB of the beam is equal to the slope at the same point as found for part BC of the beam. Therefore, substitute $x = L$ into each of the two preceding equations for the slopes and obtain

$$-\frac{PL^2}{4} + C_3 = -PL^2 + C_4$$

This equation eliminates one constant of integration because C_4 can be expressed in terms of C_3:

$$C_4 = C_3 + \frac{3PL^2}{4} \qquad \text{(e)}$$

The third and last integrations give

$$EIv = -\frac{Px^3}{12} + C_3 x + C_5 \qquad (0 \leq x \leq L) \qquad \text{(f)}$$

$$EIv = -\frac{Px^2(9L - 2x)}{12} + C_4 x + C_6 \qquad \left(L \leq x \leq \frac{3L}{2}\right) \qquad \text{(g)}$$

For part AB of the beam (Fig. 13-15a), there are two boundary conditions on the deflections, namely, the deflection is zero at points A and B:

$$v(0) = 0 \quad \text{and} \quad v(L) = 0$$

Apply these conditions to Eq. (f) to obtain

$$C_5 = 0 \qquad C_3 = \frac{PL^2}{12} \qquad \text{(h,i)}$$

Substitute the preceding expression for C_3 in Eq. (e) to get

$$C_4 = \frac{5PL^2}{6} \qquad \text{(j)}$$

For part BC of the beam, the deflection is zero at point B. Therefore, the boundary condition is

$$v(L) = 0$$

Apply this condition to Eq. (g), and also substitute Eq. (j) for C_4 to get

$$C_6 = -\frac{PL^3}{4} \tag{k}$$

All constants of integration have now been evaluated.

The deflection equations are obtained by substituting the constants of integration (Eqs. h, i, j, and k) into Eqs. (f) and (g). The results are

$$v = \frac{Px}{12EI}(L^2 - x^2) \qquad\qquad (0 \le x \le L) \qquad \Longleftarrow \textbf{(13-54a)}$$

$$v = -\frac{P}{12EI}(3L^3 - 10L^2x + 9Lx^2 - 2x^3) \quad \left(L \le x \le \frac{3L}{2}\right) \quad \Longleftarrow \textbf{(13-54b)}$$

Note that the deflection is always positive (upward) in part AB of the beam [Eq. (13-54a)] and always negative (downward) in the overhang BC [Eq. (13-54b)].

Deflection at the end of the overhang: Find the deflection δ_C at the end of the overhang (Fig. 13-15b) by substituting $x = 3L/2$ in Eq. (13-54b):

$$\delta_C = -v\left(\frac{3L}{2}\right) = \frac{PL^3}{8EI} \qquad \Longleftarrow \textbf{(13-55)}$$

4. *Finalize:* The required deflections of the overhanging beam [Eqs. (13-54) and (13-55)] were obtained by solving the third-order differential equation of the deflection curve. The deflection curve is shown in Fig. 13-15b.

FIGURE 13-15 (Repeated)

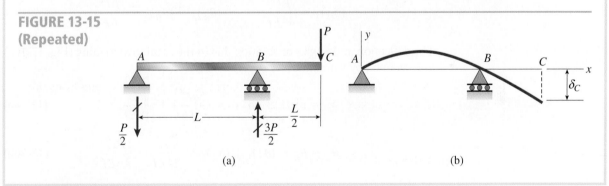

(a)   (b)

13.5 Deflections by Method of Superposition

The **method of superposition** is a practical and commonly used technique for obtaining deflections and angles of rotation of beams. The underlying concept is quite simple and may be stated as follows:

Under suitable conditions, the deflection of a beam produced by several different loads acting simultaneously can be found by superposing the deflections produced by the same loads acting separately.

For instance, if v_1 represents the deflection at a particular point on the axis of a beam due to a load q_1, and if v_2' represents the deflection at that same point due to a different load q_2, then the deflection at that point due to loads q_1 and q_2

FIGURE 13-16

Simple beam with two loads

(a)

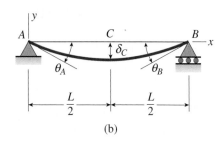

(b)

acting simultaneously is $v_1 + v_2$. (The loads q_1 and q_2 are independent loads and each may act anywhere along the axis of the beam.)

The justification for superposing deflections lies in the nature of the differential equations of the deflection curve [Eqs. (13-16a, b, and c)]. These equations are *linear* differential equations, because all terms containing the deflection v and its derivatives are raised to the first power. Therefore, the solutions of these equations for several loading conditions may be added algebraically, or *superposed*. (The conditions for superposition to be valid are described later in the subsection "Principle of Superposition.")

As an **illustration** of the superposition method, consider the simple beam *ACB* shown in Fig. 13-16a. This beam supports two loads: (1) a uniform load of intensity q acting throughout the span, and (2) a concentrated load P acting at the midpoint. Suppose you wish to find the deflection δ_C at the midpoint and the angles of rotation θ_A and θ_B at the ends (Fig. 13-16b). Using the method of superposition, first obtain the effects of each load acting separately and then combine the results.

For the uniform load acting alone, the deflection at the midpoint and the angles of rotation are obtained from the formulas of Example 13-1 [see Eqs. (13-24), (13-25), and (13-26)]:

$$(\delta_C)_1 = \frac{5qL^4}{384EI} \qquad (\theta_A)_1 = (\theta_B)_1 = \frac{qL^3}{24EI}$$

where EI is the flexural rigidity of the beam and L is its length.

For the load P acting alone, the corresponding quantities are obtained from the formulas of Example 13-3 [see Eqs. (13-44) and (13-45)]:

$$(\delta_C)_2 = \frac{PL^3}{48EI} \qquad (\theta_A)_2 = (\theta_B)_2 = \frac{PL^2}{16EI}$$

The deflection and angles of rotation due to the combined loading (Fig. 13-16a) are obtained by summation:

$$\delta_C = (\delta_C)_1 + (\delta_C)_2 = \frac{5qL^4}{384EI} + \frac{PL^3}{48EI} \qquad \textbf{(13-56a)}$$

$$\theta_A = \theta_B = (\theta_A)_1 + (\theta_A)_2 = \frac{qL^3}{24EI} + \frac{PL^2}{16EI} \qquad \textbf{(13-56b)}$$

The deflections and angles of rotation at other points on the beam axis can be found by this same procedure. However, the method of superposition is not limited to finding deflections and angles of rotation at single points. The method also may be used to obtain general equations for the slopes and deflections of beams subjected to more than one load.

Tables of Beam Deflections

The method of superposition is useful only when formulas for deflections and slopes are readily available. To provide convenient access to such formulas, tables for both cantilever and simple beams are given in Appendix E. Similar tables can be found in engineering handbooks. Use these tables and the method of superposition to find deflections and angles of rotation for many different loading conditions, as illustrated in the examples at the end of this section.

Distributed Loads

Sometimes a distributed load of interest is not included in a table of beam deflections. In such cases, superposition still may be useful. Consider an element of the distributed load as though it were a concentrated load, and then find the required deflection by integrating throughout the region of the beam where the load is applied.

FIGURE 13-17

Simple beam with a triangular load

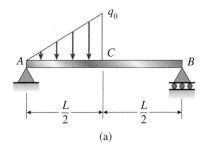

(a)

To illustrate this process of integration, consider a simple beam ACB with a triangular load acting on the left-hand half (Fig. 13-17a). The goal is to obtain the deflection δ_C at the midpoint C and the angle of rotation θ_A at the left-hand support (Fig. 13-17c).

Begin by visualizing an element qdx of the distributed load as a concentrated load (Fig. 13-17b). Note that the load acts to the left of the midpoint of the beam. The deflection at the midpoint due to this concentrated load is obtained from Case 5 of Table E-2, Appendix E. The formula given there for the midpoint deflection (for the case in which $a \leq b$) is

$$\frac{Pa}{48EI}(3L^2 - 4a^2)$$

In this example (Fig. 13-17b), substitute qdx for P and x for a:

$$\frac{(qdx)(x)}{48EI}(3L^2 - 4x^2) \qquad \textbf{(13-57)}$$

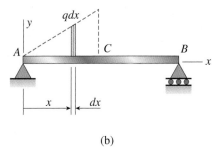

(b)

This expression gives the deflection at point C due to the element qdx of the load.

Next, note that the intensity of the uniform load (Figs. 13-17a and b) is

$$q = \frac{2q_0 x}{L} \qquad \textbf{(13-58)}$$

where q_0 is the maximum intensity of the load. With this substitution for q, the formula for the deflection [Eq. (13-57)] becomes

$$\frac{q_0 x^2}{24LEI}(3L^2 - 4x^2)dx$$

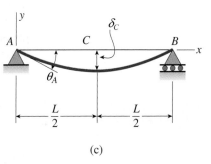

(c)

Now, integrate throughout the region of the load to obtain the deflection δ_C at the midpoint of the beam due to the entire triangular load:

$$\delta_C = \int_0^{L/2} \frac{q_0 x^2}{24LEI}(3L^2 - 4x^2)dx \qquad \textbf{(13-59)}$$

$$= \frac{q_0}{24LEI}\int_0^{L/2}(3L^2 - 4x^2)x^2 dx = \frac{q_0 L^4}{240EI}$$

By a similar procedure, you can calculate the angle of rotation θ_A at the left-hand end of the beam (Fig. 13-17c). The expression for this angle due to a concentrated load P (see Case 5 of Table E-2) is

$$\frac{Pab(L + b)}{6LEI}$$

Replace P with $2q_0 xdx/L$, a with x, and b with $L - x$ to obtain

$$\frac{2q_0 x^2(L - x)(L + L - x)}{6L^2 EI}dx \quad \text{or} \quad \frac{q_0}{3L^2 EI}(L - x)(2L - x)x^2 dx$$

Finally, integrate throughout the region of the load:

$$\theta_A = \int_0^{L/2} \frac{q_0}{3L^2 EI}(L-x)(2L-x)x^2 dx = \frac{41q_0 L^3}{2880 EI} \quad \textbf{(13-60)}$$

This is the angle of rotation produced by the triangular load.

This illustrates how to use superposition and integration to find deflections and angles of rotation produced by distributed loads of almost any kind. If the integration cannot be performed easily by analytical means, numerical methods can be used.

Principle of Superposition

The method of superposition for finding beam deflections is an example of a more general concept known in mechanics as the **principle of superposition**. This principle is valid whenever the quantity to be determined is a linear function of the applied loads. When that is the case, the desired quantity may be found due to each load acting separately, and these results may be superposed to obtain the desired quantity due to all loads acting simultaneously. In ordinary structures, the principle is usually valid for stresses, strains, bending moments, and many other quantities besides deflections.

In the particular case of **beam deflections**, the principle of superposition is valid under the following conditions: (1) Hooke's law holds for the material, (2) the deflections and rotations are small, and (3) the presence of the deflections does not alter the actions of the applied loads. These requirements ensure that the differential equations of the deflection curve are linear.

The following examples provide additional illustrations where the principle of superposition is used to calculate deflections and angles of rotation of beams.

Example 13-6

A cantilever beam AB supports a uniform load of intensity q acting over part of the span and a concentrated load P acting at the free end (Fig. 13-18a).

Determine the deflection δ_B and angle of rotation θ_B at end B of the beam (Fig. 13-18b). *Note:* The beam has length L and constant flexural rigidity EI.

Solution:

Use the four-step problem-solving approach.

1. *Conceptualize*: Obtain the deflection and angle of rotation at end B of the beam by combining the effects of the loads acting separately.

2. *Categorize*: If the uniform load acts alone, the deflection and angle of rotation (obtained from Case 2 of Table E-1, Appendix E) are

$$(\delta_B)_1 = \frac{qa^3}{24EI}(4L-a) \qquad (\theta_B)_1 = \frac{qa^3}{6EI}$$

FIGURE 13-18

Example 13-6: Cantilever beam with a uniform load and a concentrated load

(a)

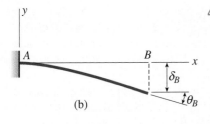

(b)

If the load P acts alone, the corresponding quantities (from Case 4, Table E-1) are

$$(\delta_B)_2 = \frac{PL^3}{3EI} \qquad (\theta_B)_2 = \frac{PL^2}{2EI}$$

3. *Analyze:* Therefore, the deflection and angle of rotation due to the combined loading (Fig. 13-18a) are

$$\delta_B = (\delta_B)_1 + (\delta_B)_2 = \frac{qa^3}{24EI}(4L - a) + \frac{PL^3}{3EI} \qquad \text{(13-61)}$$

$$\theta_B = (\theta_B)_1 + (\theta_B)_2 = \frac{qa^3}{6EI} + \frac{PL^2}{2EI} \qquad \text{(13-62)}$$

4. *Finalize:* The required quantities were found by using tabulated formulas and the method of superposition. Note that you would have to integrate over two separate regions $(0 \leq x \leq a, a \leq x \leq L)$ to find the deflection curve shown in Fig. 13-18b.

Example 13-7

A cantilever beam AB with a uniform load of intensity q acting on the right-hand half of the beam is shown in Fig. 13-19a.

Obtain formulas for the deflection δ_B and angle of rotation θ_B at the free end (Fig. 13-19c). *Note:* The beam has length L and constant flexural rigidity EI.

Solution:

Use the four-step problem-solving approach.

1. *Conceptualize:* In this example, determine the deflection and angle of rotation by treating an element of the uniform load as a concentrated load and then integrating (see Fig. 13-19b).

2. *Categorize:* The element of the load has a magnitude $q\,dx$ and is located at distance x from the support. The resulting differential deflection $d\delta_B$ and differential angle of rotation $d\theta_B$ at the free end are found from the corresponding formulas in Case 5 of Table E-1, Appendix E, by replacing P with $q\,dx$ and a with x; thus,

$$d\delta_B = \frac{(q\,dx)(x^2)(3L - x)}{6EI} \qquad d\theta_B = \frac{(q\,dx)(x^2)}{2EI}$$

Example 13-7: Cantilever
beam with a uniform load
acting on the right-hand half
of the beam

(a)

(b)

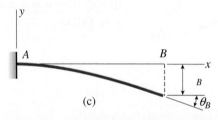

(c)

3. *Analyze*: Integrate over the loaded region to get

$$\delta_B = \int d\delta_B = \frac{q}{6EI} \int_{L/2}^{L} x^2(3L - x)dx = \frac{41qL^4}{384EI} \qquad\text{(13-63)}$$

$$\theta_B = \int d\theta_B = \frac{q}{2EI} \int_{L/2}^{L} x^2 dx = \frac{7qL^3}{48EI} \qquad\text{(13-64)}$$

4. *Finalize*: These same results can be obtained by using the formulas in Case 3 of
   Table E-1 and substituting $a = b = L/2$. Also, note that subtraction of Case 2 from
   Case 1 in Table E-1 (with $a = b = L/2$) leads to the same results.

Example 13-8

A compound beam ABC has a roller support at A, an internal hinge (that is,
moment release) at B, and a fixed support at C (Fig. 13-20a). Segment AB has a
length of a, and segment BC has a length of b. A concentrated load P acts at a distance $2a/3$ from support A, and a uniform load of intensity q acts between points
B and C.

Determine the deflection δ_B at the hinge and the angle of rotation θ_A at support
A (Fig. 13-20d). *Note:* The beam has constant flexural rigidity EI.

FIGURE 13-20

Example 13-8: Compound beam with a hinge

(a)

(b)

(c)

(d)

Solution:

Use the four-step problem-solving approach.

1. *Conceptualize*: For purposes of analysis, consider the compound beam to consist of two individual beams: a simple beam AB of length a and a cantilever beam BC of length b. The two beams are linked together by a pin connection at B.

2. *Categorize*: If you separate beam AB from the rest of the structure (Fig. 13-20b), there is a vertical force F at end B equal to $2P/3$. This same force acts downward at end B of the cantilever (Fig. 13-20c). Consequently, the cantilever beam BC is subjected to two loads: a uniform load and a concentrated load.

3. *Analyze*: The deflection at the end of this cantilever (which is the same as the deflection δ_B of the hinge) is readily found from Cases 1 and 4 of Table E-1, Appendix E:

$$\delta_B = \frac{qb^4}{8EI} + \frac{Fb^3}{3EI}$$

or, since $F = 2P/3$,

$$\delta_B = \frac{qb^4}{8EI} + \frac{2Pb^3}{9EI} \qquad \blacktriangleleft (13\text{-}65)$$

The angle of rotation θ_A at support A (Fig. 13-20d) consists of two parts: *an angle BAB′* produced by the downward displacement of the hinge and *an additional angle of rotation* produced by the bending of beam AB (or beam $AB′$) as a simple beam. *The angle BAB′* is

$$(\theta_A)_1 = \frac{\delta_B}{a} = \frac{qb^4}{8aEI} + \frac{2Pb^3}{9aEI}$$

The angle of rotation at the end of a simple beam with a concentrated load is obtained from Case 5 of Table E-2. The formula given there is

$$\frac{Pab(L + b)}{6LEI}$$

where L is the length of the simple beam, a is the distance from the left-hand support to the load, and b is the distance from the right-hand support to the load. Thus, in the notation of this example (Fig. 13-20a), the angle of rotation is

$$(\theta_A)_2 = \frac{P\left(\dfrac{2a}{3}\right)\left(\dfrac{a}{3}\right)\left(a + \dfrac{a}{3}\right)}{6aEI} = \frac{4Pa^2}{81EI}$$

Combine the two angles to obtain the total angle of rotation at support A:

$$\theta_A = (\theta_A)_1 + (\theta_A)_2 = \frac{qb^4}{8aEI} + \frac{2Pb^3}{9aEI} + \frac{4Pa^2}{81EI} \qquad \text{(13-66)}$$

4. *Finalize*: This example illustrates how the method of superposition can be adapted to handle a seemingly complex situation in a relatively simple manner.

Example 13-9

A simple beam AB of a span length L has an overhang BC of length a (Fig. 13-21a). The beam supports a uniform load of intensity q throughout its length.

Obtain a formula for the deflection δ_C at the end of the overhang (Fig. 13-21c).

Note: The beam has constant flexural rigidity EI.

FIGURE 13-21

Example 13-9: Simple beam with an overhang

Beam with overhang loaded by gravity uniform load (Courtesy of the National Information Service for Earthquake Engineering EERC, University of California, Berkeley)

Solution:

Use the four-step problem-solving approach.

1. *Conceptualize*: Find the deflection of point C by imagining the overhang BC (Fig. 13-21a) to be a cantilever beam subjected to two actions. The first action is the rotation of the support of the cantilever through an angle θ_B, which is the angle of rotation of beam ABC at support B (Fig. 13-21c). (Assume that a clockwise angle θ_B is positive.) This angle of rotation causes a rigid-body rotation of the overhang BC, resulting in a downward displacement δ_1 of point C.

The second action is the bending of BC as a cantilever beam supporting a uniform load. This bending produces an additional downward displacement δ_2 (Fig. 13-21c). The superposition of these two displacements gives the total displacement δ_C at point C.

2. *Categorize*:

Deflection δ_1: Begin by finding the deflection δ_1 caused by the angle of rotation θ_B at point B. To find this angle, observe that part AB of the beam is in the same condition as a simple beam (Fig. 13-21b) subjected to the following loads: a uniform load of intensity q, a couple M_B (equal to $qa^2/2$), and a vertical load P (equal to qa). Only the loads q and M_B produce angles of rotation at end B of this simple beam. These angles are found from Cases 1 and 7 of Table E-2, Appendix E. Thus, the angle θ_B is

$$\theta_B = -\frac{qL^3}{24EI} + \frac{M_B L}{3EI} = -\frac{qL^3}{24EI} + \frac{qa^2 L}{6EI} = \frac{qL(4a^2 - L^2)}{24EI} \quad \textbf{(13-67)}$$

in which a clockwise angle is positive, as shown in Fig. 13-21c.

The downward deflection δ_1 of point C, due solely to the angle of rotation θ_B, is equal to the length of the overhang times the angle (Fig. 13-21c):

$$\delta_1 = a\theta_B = \frac{qaL(4a^2 - L^2)}{24EI} \quad \textbf{(a)}$$

Deflection δ_2: Bending of the overhang BC produces an additional downward deflection δ_2 at point C. This deflection is equal to the deflection of a cantilever beam of length a subjected to a uniform load of intensity q (see Case 1 of Table E-1):

$$\delta_2 = \frac{qa^4}{8EI} \quad \textbf{(b)}$$

3. *Analyze*:

Deflection δ_C: The total downward deflection of point C is the algebraic sum of δ_1 and δ_2:

$$\delta_C = \delta_1 + \delta_2 = \frac{qaL(4a^2 - L^2)}{24EI} + \frac{qa^4}{8EI} = \frac{qa}{24EI}[L(4a^2 - L^2) + 3a^3]$$

or

$$\delta_C = \frac{qa}{24EI}(a + L)(3a^2 + aL - L^2) \qquad \text{(13-68)}$$

4. *Finalize*: In the preceding equation, the deflection δ_C may be upward or downward, depending upon the relative magnitudes of the lengths L and a. If a is relatively large, the last term in the equation (the three-term expression in parentheses) is positive, and the deflection δ_C is downward. If a is relatively small, the last term is negative, and the deflection is upward. The deflection is zero when the last term is equal to zero:

$$3a^2 + aL - L^2 = 0$$

or

$$a = \frac{L(\sqrt{13} - 1)}{6} = 0.4343L \qquad \text{(c)}$$

From this result, note that if a is greater than $0.4343L$, the deflection of point C is downward; if a is less than $0.4343L$, the deflection is upward.

Deflection curve: The shape of the deflection curve for the beam in this example is shown in Fig. 13-21c for the case where a is large enough ($a > 0.4343L$) to produce a downward deflection at C and small enough ($a < L$) to ensure that the reaction at A is upward. Under these conditions, the beam has a positive bending moment between support A and a point such as D. The deflection curve in region AD is concave upward (positive curvature). From D to C, the bending moment is negative; therefore, the deflection curve is concave downward (negative curvature).

Point of inflection: At point D, the curvature of the deflection curve is zero because the bending moment is zero. A point such as D, where the curvature and bending moment *change signs*, is called a **point of inflection** (or *point of contraflexure*). The bending moment M and the second derivative d^2v/dx^2 always vanish at an inflection point.

However, a point where M and d^2v/dx^2 equal zero is not necessarily an inflection point because it is possible for those quantities to be zero without changing signs at that point; for example, they could have maximum or minimum values.

13.6 Statically Indeterminate Beams

Introduction

This chapter discusses the analysis of beams in which the number of reactions exceeds the number of independent equations of equilibrium. Since the reactions of such beams cannot be determined by statics alone, the beams are said to be **statically indeterminate**.

The analysis of statically indeterminate beams is quite different from that of statically determinate beams. When a beam is statically determinate, all reactions, shear forces, and bending moments can be obtained from free-body diagrams and equations of equilibrium. Then, knowing the shear forces and bending moments, the stresses and deflections are computed.

However, when a beam is statically indeterminate, the equilibrium equations are not sufficient, and additional equations are needed. The most fundamental method for analyzing a statically indeterminate beam is to solve the differential equations of the deflection curve, as described later in this section. Although this method serves as a good starting point in the analysis, it is practical for only the simplest types of statically indeterminate beams.

Therefore, the focus here is on the method of superposition, which is a method that is applicable to a wide variety of structures. In the method of superposition, the equilibrium equations are supplemented with compatibility equations and force–displacement equations. (This same method was described earlier in Section 8.4, where statically indeterminate bars subjected to tension and compression were analyzed.) Throughout this chapter, the beams are assumed to be made of **linearly elastic materials**.

Although only statically indeterminate beams are discussed in this section, the fundamental ideas have much wider application. Most structures, including automobile frames, buildings, and aircraft, are statically indeterminate. However, they are much more complex than beams and must be designed using very sophisticated analytical techniques. Many of these techniques rely on the concepts described here; therefore, this section may be viewed as an introduction to the analysis of statically indeterminate structures of all kinds.

Types of Statically Indeterminate Beams

Statically indeterminate beams are usually identified by the arrangement of their supports. For instance, a beam that is fixed at one end and simply supported at the other (Fig. 13-22a) is called a **propped cantilever beam**. The reactions of the beam shown in the figure consist of horizontal and vertical forces at support A,

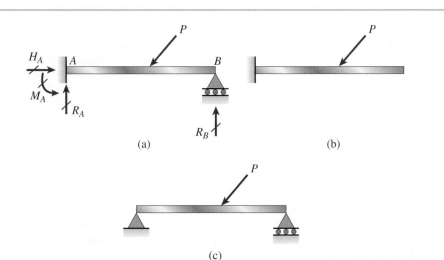

FIGURE 13-22

Propped cantilever beam: (a) beam with load and reactions; (b) released structure when the reaction at end B is selected as the redundant; and (c) released structure when the moment reaction at end A is selected as the redundant a moment at support A, and a vertical force at support B. Because there are only three independent equations of equilibrium for this beam, it is not possible to calculate all four of the reactions from equilibrium alone. The number of reactions in *excess* of the number of equilibrium equations is called the **degree of static indeterminacy**. Thus, a propped cantilever beam is statically indeterminate to the first degree.

The excess reactions are called **static redundants** and must be selected in each particular case. For example, the reaction R_B of the propped cantilever beam shown in Fig. 13-22a may be selected as the redundant reaction. Since this reaction is in excess of those needed to maintain equilibrium, it can be released from the structure by removing the support at B. When support B is removed, the resulting structure is a cantilever beam (Fig. 13-22b). The structure that remains when the redundants are released is called the **released structure** or the **primary structure**. The released structure must be stable (so that it is capable of carrying loads), and it must be statically determinate (so that all force quantities can be determined by equilibrium alone).

Another possibility for the analysis of the propped cantilever beam of Fig. 13-22a is to select the reactive moment M_A as the redundant. Then, when the moment restraint at support A is removed, the released structure is a simple beam with a pin support at one end and a roller support at the other (Fig. 13-22c).

A special case arises if all loads acting on the beam are vertical (Fig. 13-23). Then the horizontal reaction at support A vanishes, and three reactions remain. However, only two independent equations of equilibrium are now available; therefore, the beam is still statically indeterminate to the first degree. If the reaction R_B is chosen as the redundant, the released structure is a cantilever beam; if the moment M_A is chosen, the released structure is a simple beam.

Another type of statically indeterminate beam, known as a **fixed-end beam**, is shown in Fig. 13-24a. This beam has fixed supports at both ends, resulting in a total of six unknown reactions (two forces and a moment at each support). Because there are only three equations of equilibrium, the beam is statically indeterminate to the third degree. (Other names for this type of beam are *clamped beam* and *built-in beam*.)

FIGURE 13-23

Propped cantilever beam with vertical loads only

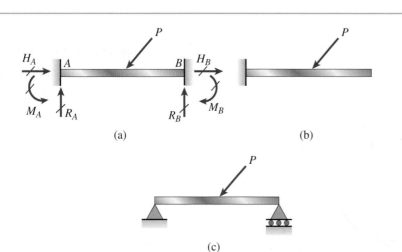

(a)   (b)

(c)

FIGURE 13-24

Fixed-end beam: (a) beam with load and reactions; (b) released structure when the three reactions at end B are selected as the redundants; and (c) released structure when the two moment reactions and the horizontal reaction at end B are selected as the redundants

If the three reactions at end B of the beam are selected as the redundants and if the corresponding restraints are removed, a cantilever beam is left as the released structure (Fig. 13-24b). If the two fixed-end moments and one horizontal reaction are removed instead, the released structure is a simple beam (Fig. 13-24c).

Again considering the special case of only vertical loads (Fig. 13-25), the fixed-end beam now has only four nonzero reactions (one force and one moment at each support). The number of available equilibrium equations is two, so the beam is statically indeterminate to the second degree. If the two reactions at end B are selected as the redundants, the released structure is a cantilever beam; if the two moment reactions are selected, the released structure is a simple beam.

The beam shown in Fig. 13-26a is an example of a **continuous beam**, so called because it has more than one span and is continuous over an interior support. This particular beam is statically indeterminate to the first degree because there are four reactive forces and only three equations of equilibrium.

If the reaction R_B at the interior support is selected as the redundant, and if the corresponding support is removed from the beam, there remains a released structure in the form of a statically determinate simple beam (Fig. 13-26b). If the reaction R_C is selected as the redundant, the released structure is a simple beam with an overhang (Fig. 13-26c).

The following sections discuss two methods for analyzing statically indeterminate beams. The objective in each case is to determine the redundant reactions. Once they are known, all remaining reactions (plus the shear forces and bending moments) can be found from equations of equilibrium. In effect, the structure has become statically determinate. Therefore, as the final step in the analysis, the stresses and deflections can be found by the methods described in preceding chapters.

Long-span bridges are often constructed using continuous beams

FIGURE 13-25

Fixed-end beam with vertical loads only

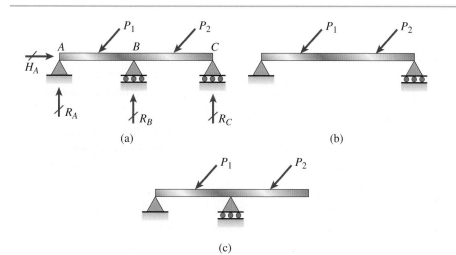

(a)   (b)

(c)

FIGURE 13-26

Example of a continuous beam: (a) beam with loads and reactions; (b) released structure when the reaction at support B is selected as the redundant; and (c) released structure when the reaction at end C is selected as the redundant

Analysis by the Differential Equations of the Deflection Curve

Statically indeterminate beams may be analyzed by solving any one of the three differential equations of the deflection curve: (1) the second-order equation in terms of the bending moment [Eq. (13-13a) or Eq. (13-16a) if prismatic], (2) the third-order equation in terms of the shear force [Eq. (13-13b) or Eq. (13-16b)], or (3) the fourth-order equation in terms of the intensity of distributed load [Eq. (13-13c) or Eq. (13-16c)].

The procedure is essentially the same as that for a statically determinate beam (see Sections 13.2, 13.3, and 13.4) and consists of writing the differential equation, integrating to obtain its general solution, and then applying boundary and other conditions to evaluate the unknown quantities. The unknowns consist of the redundant reactions as well as the constants of integration.

The differential equation for a beam may be solved in symbolic terms only when the beam and its loading are relatively simple and uncomplicated. The resulting solutions are in the form of general-purpose formulas. However, in more complex situations, the differential equations must be solved numerically, using computer programs intended for that purpose. In such cases, the results apply only to specific numerical problems.

The following examples illustrate the analysis of statically indeterminate beams by solving the differential equations in symbolic terms.

Example 13-10

FIGURE 13-27

Example 13-10: Propped cantilever beam with a uniform load

A propped cantilever beam AB of a length L supports a uniform load of intensity q (Fig. 13-27). Analyze this beam by solving the second-order differential equation of the deflection curve (the bending-moment equation). Determine the reactions, shear forces, bending moments, slopes, and deflections of the beam.

Solution:

Use the four-step problem-solving approach.

1. *Conceptualize*: Because the load on this beam acts in the vertical direction (Fig. 13-27), there is no horizontal reaction at the fixed support. Therefore, the beam has three unknown reactions (M_A, R_A, and R_B). Only two equations of equilibrium are available for determining these reactions; therefore, the beam is statically indeterminate to the first degree.

2. *Categorize*: To analyze this beam by solving the bending-moment equation, begin with a general expression for the moment. This expression is in terms of both the load and the selected redundant.

Redundant reaction: Choose the reaction R_B at the simple support as the redundant. Then, by considering the equilibrium of the entire beam, express the other two reactions in terms of R_B:

$$R_A = qL - R_B \qquad M_A = \frac{qL^2}{2} - R_B L \tag{a,b}$$

Bending moment: The bending moment M at distance x from the fixed support is expressed in terms of the reactions as

$$M = R_A x - M_A - \frac{qx^2}{2} \tag{c}$$

This equation is obtained by constructing a free-body diagram of part of the beam and solving an equation of equilibrium.

Substitute into Eq. (c) from Eqs. (a) and (b) to obtain the bending moment in terms of the load and the redundant reaction:

$$M = qLx - R_B x - \frac{qL^2}{2} + R_B L - \frac{qx^2}{2} \tag{d}$$

3. *Analyze*:

Differential equation: The second-order differential equation of the deflection curve [Eq. (13-16a)] now becomes

$$EIv'' = M = qLx - R_B x - \frac{qL^2}{2} + R_B L - \frac{qx^2}{2} \tag{e}$$

After two successive integrations, the following equations are obtained for the slopes and deflections of the beam:

$$EIv' = \frac{qLx^2}{2} - \frac{R_B x^2}{2} - \frac{qL^2 x}{2} + R_B Lx - \frac{qx^3}{6} + C_1 \tag{f}$$

$$EIv = \frac{qLx^3}{6} - \frac{R_B x^3}{6} - \frac{qL^2 x^2}{4} + \frac{R_B Lx^2}{2} - \frac{qx^4}{24} + C_1 x + C_2 \tag{g}$$

These equations contain three unknown quantities (C_1, C_2, and R_B).

Boundary conditions: Three boundary conditions pertaining to the deflections and slopes of the beam are apparent from an inspection of Fig. 13-27. These conditions are (1) the deflection at the fixed support is zero, (2) the slope at the fixed support is zero, and (3) the deflection at the simple support is zero. Thus,

$$v(0) = 0 \qquad v'(0) = 0 \qquad v(L) = 0$$

Apply these conditions to the equations for slopes and deflections given in Eqs. (f) and (g) to find that $C_1 = 0, C_2 = 0$, and

$$R_B = \frac{3qL}{8} \qquad \qquad \blacktriangleleft \text{(13-69)}$$

Thus, the redundant reaction R_B is now known.

Reactions: With the value of the redundant established, the remaining reactions from Eqs. (a) and (b) are

$$R_A = \frac{5qL}{8} \qquad M_A = \frac{qL^2}{8} \qquad \text{(13-70a,b)}$$

Knowing these reactions, the shear forces and bending moments in the beam now can be found.

Shear forces and bending moments: These quantities are obtained by the usual techniques involving free-body diagrams and equations of equilibrium. The results are

$$V = R_A - qx = \frac{5qL}{8} - qx \qquad \text{(13-71)}$$

$$M = R_A x - M_A - \frac{qx^2}{2} = \frac{5qLx}{8} - \frac{qL^2}{8} - \frac{qx^2}{2} \qquad \text{(13-72)}$$

Shear-force and bending-moment diagrams for the beam are drawn with the aid of these equations (see Fig. 13-28).

From the diagrams, note that the maximum shear force occurs at the fixed support and is equal to

$$V_{\text{max}} = \frac{5qL}{8} \qquad \text{(13-73)}$$

Also, the maximum positive and negative bending moments are

$$M_{\text{pos}} = \frac{9qL^2}{128} \qquad M_{\text{neg}} = -\frac{qL^2}{8} \qquad \text{(13-74a,b)}$$

Finally, note that the bending moment is equal to zero at distance $x = L/4$ from the fixed support.

Slopes and deflections of the beam: Return to Eqs. (f) and (g) for the slopes and deflections; now substitute the values of the constants of integration ($C_1 = 0$ and $C_2 = 0$) as well as the expression for the redundant R_B [Eq. (13-69)] and obtain

$$v' = \frac{qx}{48EI}(-6L^2 + 15Lx - 8x^2) \qquad \text{(13-75)}$$

$$v = -\frac{qx^2}{48EI}(3L^2 - 5Lx + 2x^2) \qquad \text{(13-76)}$$

FIGURE 13-28

Shear-force and bending-moment diagrams for the propped cantilever beam of Fig. 13-27

FIGURE 13-29

Deflection curve for the propped cantilever beam of Fig. 13-27

The deflected shape of the beam as obtained from Eq. (13-76) is shown in Fig. 13-29.

To determine the maximum deflection of the beam, set the slope [Eq. (13-75)] equal to zero and solve for the distance x_1 to the point where this deflection occurs:

$$v' = 0 \quad \text{or} \quad -6L^2 + 15Lx - 8x^2 = 0$$

from which

$$x_1 = \frac{15 - \sqrt{33}}{16} L = 0.5785L \tag{13-77}$$

Substitute this value of x into the equation for the deflection [Eq. (13-76)] and also change the sign to get the maximum deflection:

$$\delta_{max} = -v(x_1) = \frac{qL^4}{65,536\,EI}(39 + 55\sqrt{33})$$

$$= \frac{qL^4}{184.6\,EI} = 0.005416\frac{qL^4}{EI} \tag{13-78}$$

The point of inflection is located where the bending moment is equal to zero, that is, where $x = L/4$. The corresponding deflection δ_0 of the beam [from Eq. (13-76)] is

$$\delta_0 = -v(L/4) = \frac{5qL^4}{2048\,EI} = 0.002441\frac{qL^4}{EI} \tag{13-79}$$

Note that when $x < L/4$, both the curvature and the bending moment are negative, and when $x > L/4$, the curvature and bending moment are positive.

To determine the angle of rotation θ_B at the simply supported end of the beam, use Eq. (13-75), as

$$\theta_B = v'(L) = \frac{qL^3}{48\,EI} \tag{13-80}$$

Slopes and deflections at other points along the axis of the beam can be obtained by similar procedures.

4. *Finalize*: This example presents the analysis of the beam by taking the reaction R_B (Fig. 13-27) as the redundant reaction. An alternative approach is to take the reactive moment M_A as the redundant. Then express the bending moment M in terms of M_A, substitute the resulting expression into the second-order differential equation, and solve as before. Still another approach is to begin with the fourth-order differential equation, as illustrated in the next example.

Example 13-11

FIGURE 13-30

Example 13-11: Fixed-end beam with a concentrated load at the midpoint

The fixed-end beam ACB shown in Fig. 13-30 supports a concentrated load P at the midpoint. Analyze this beam by solving the fourth-order differential equation of the deflection curve (the load equation). Determine the reactions, shear forces, bending moments, slopes, and deflections of the beam.

Solution:

Use the four-step problem-solving approach.

1. *Conceptualize*: Because the load on this beam acts only in the vertical direction, there are no horizontal reactions at the supports. Therefore, the beam has four unknown reactions—two at each support. Since only two equations of equilibrium are available, the beam is statically indeterminate to the second degree.

2. *Categorize*: The analysis can be simplified by observing from the symmetry of the beam and its loading that the forces and moments at supports A and B are equal, that is,

$$R_A = R_B \quad \text{and} \quad M_A = M_B$$

The vertical reactions at the supports are equal, so from equilibrium of forces in the vertical direction, each force is equal to $P/2$:

$$R_A = R_B = \frac{P}{2} \qquad \text{(13-81)}$$

Thus, the only unknown quantities that remain are the moment reactions M_A and M_B. For convenience, select the moment M_A as the redundant quantity.

3. *Analyze*:

Differential equation: Because there is no load acting on the beam between points A and C, the fourth-order differential equation [Eq. (13-19c)] for the left-hand half of the beam is

$$EIv'''' = -q = 0 \qquad (0 < x < L/2) \qquad \text{(a)}$$

Successive integrations of this equation yield the following equations, which are valid for the left-hand half of the beam:

$$EIv''' = C_1 \qquad \text{(b)}$$

$$EIv'' = C_1 x + C_2 \qquad \text{(c)}$$

$$EIv' = \frac{C_1 x^2}{2} + C_2 x + C_3 \qquad \text{(d)}$$

$$EIv = \frac{C_1 x^3}{6} + \frac{C_2 x^2}{2} + C_3 x + C_4 \qquad \text{(e)}$$

These equations contain four unknown constants of integration. There are now five unknowns ($C_1, C_2, C_3, C_4,$ and M_A), so five boundary conditions are needed.

Boundary conditions: The boundary conditions applicable to the left-hand half of the beam are

i. The shear force in the left-hand segment of the beam is equal to R_A, or $P/2$. Therefore, from Eq. (13-16b)

$$EIv''' = V = \frac{P}{2}$$

Combine this equation with Eq. (b) to obtain $C_1 = P/2$.

ii. The bending moment at the left-hand support is equal to $-M_A$. Therefore, from Eq. (13-16a),

$$EIv'' = M = -M_A \quad \text{at} \quad x = 0$$

Combining this equation with Eq. (c) leads to $C_2 = -M_A$.

iii. The slope of the beam at the left-hand support ($x = 0$) is equal to zero. Therefore, Eq. (d) yields $C_3 = 0$.

iv. The slope of the beam at the midpoint ($x = L/2$) is also equal to zero (from symmetry). Therefore, from Eq. (d),

$$M_A = M_B = \frac{PL}{8} \qquad \text{(13-82)}$$

Thus, the reactive moments at the ends of the beam have been determined.

v. The deflection of the beam at the left-hand support ($x = 0$) is equal to zero. Therefore, from Eq. (e), $C_4 = 0$.

In summary, the four constants of integration are

$$C_1 = \frac{P}{2} \qquad C_2 = -M_A = -\frac{PL}{8} \qquad C_3 = 0 \qquad C_4 = 0 \quad \text{(f,g,h,i)}$$

Shear forces and bending moments: The shear forces and bending moments are found by substituting the appropriate constants of integration into Eqs. (b) and (c). The results are

$$EIv''' = V = \frac{P}{2} \qquad (0 < x < L/2) \qquad \text{(13-83)}$$

$$EIv'' = M = \frac{Px}{2} - \frac{PL}{8} \quad (0 \le x \le L/2) \qquad \text{(13-84)}$$

Since the reactions of the beam are known, these expressions also can be obtained directly from free-body diagrams and equations of equilibrium.

The shear-force and bending moment diagrams are shown in Fig. 13-31.

FIGURE 13-31

Shear-force and bending-moment diagrams for the fixed-end beam of Fig. 13-30

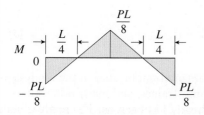

FIGURE 13-32

Deflection curve for the fixed-end beam of Fig. 13-30

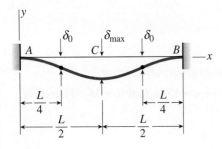

FIGURE 13-32

Deflection curve for the fixed-end beam of Fig. 13-30

Slopes and deflections: The slopes and deflections in the left-hand half of the beam can be found from Eqs. (d) and (e) by substituting the expressions for the constants of integration. The resulting expressions are

$$v' = -\frac{Px}{8EI}(L - 2x) \qquad (0 \le x \le L/2) \qquad \text{(13-85)}$$

$$v = -\frac{Px^2}{48EI}(3L - 4x) \quad (0 \le x \le L/2) \qquad \text{(13-86)}$$

The deflection curve of the beam is shown in Fig. 13-32.

To find the maximum deflection δ_{max}, set x equal to $L/2$ in Eq. (13-86) and change the sign; thus,

$$\delta_{max} = -v(L/2) = \frac{PL^3}{192EI} \qquad \text{(13-87)}$$

The point of inflection in the left-hand half of the beam occurs where the bending moment M is equal to zero, that is, where $x = L/4$ [see Eq. (13-84)]. The corresponding deflection δ_0 [from Eq. (13-86)] is

$$\delta_0 = -v(L/4) = \frac{PL^3}{384EI} \qquad \text{(13-88)}$$

which is equal numerically to one-half of the maximum deflection. A second point of inflection occurs in the right-hand half of the beam at distance $L/4$ from end B.

4. *Finalize:* This example demonstrates that the number of boundary and other conditions is always sufficient to evaluate not only the constants of integration but also the redundant reactions.

Note: Sometimes it is necessary to set up differential equations for more than one region of the beam and use conditions of continuity between regions, as illustrated in Example 13-3 of Section 13.3 and Example 13-5 of Section 13.4 for statically determinate beams. Such analyses are likely to be long and tedious because of the large number of conditions that must be satisfied. However, if deflections and angles of rotation are needed at only one or two specific points, the method of superposition may be useful, as shown in next section.

Method of Superposition

The method of superposition is of fundamental importance in the analysis of statically indeterminate bars, trusses, beams, frames, and many other kinds of structures. The superposition method already has been used to analyze statically indeterminate structures composed of bars in tension and compression

(Section 8.4) and shafts in torsion (Section 9.8). In this section, the method is applied to beams.

Begin the analysis by noting the degree of static indeterminacy and selecting the redundant reactions. Then, having identified the redundants, write **equations of equilibrium** that relate the other unknown reactions to the redundants and the loads.

Next, assume that both the original loads and the redundants act upon the released structure. Then find the deflections in the released structure by superposing the separate deflections due to the loads and the redundants. The sum of these deflections must match the deflections in the original beam. However, the deflections in the original beam (at the points where restraints were removed) are either zero or have known values. Therefore, **equations of compatibility** (or *equations of superposition*) now can be written expressing the fact that the deflections of the released structure (at the points where restraints were removed) are the same as the deflections in the original beam (at those same points).

The released structure is statically determinate, so determine its deflections by using the techniques described in Sections 13.3 through 13.5. The relationships between the loads and the deflections of the released structure are called **force–displacement relations**. When these relations are substituted into the equations of compatibility, you obtain equations in which the redundants are the unknown quantities. Therefore, solving those equations gives the redundant reactions. Then, with the redundants known, all other reactions are found using the equations of equilibrium. Furthermore, the shear forces and bending moments are also obtained from equilibrium.

The preceding steps can be made clearer by considering a particular case, namely, a propped cantilever beam supporting a uniform load (Fig. 13-33a). Two analyses are presented here: the first with the force reaction R_B selected as the redundant and the second with the moment reaction M_A as the redundant. (This same beam was analyzed in Example 13-10 by solving the differential equation of the deflection curve.)

Analysis with R_B as Redundant

For the first case, select the reaction R_B at the simple support (Fig. 13-33a) as the redundant. Then the *equations of equilibrium* that express the other unknown reactions in terms of the redundant are

$$R_A = qL - R_B \qquad M_A = \frac{qL^2}{2} - R_B L \qquad \textbf{(13-89a,b)}$$

These equations come from equations of equilibrium that apply to the entire beam taken as a free body (Fig. 13-33a).

The next step is to remove the restraint corresponding to the redundant (in this case, remove the support at end B). The *released structure* that remains is a cantilever beam (Fig. 13-33b). The uniform load q and the redundant force R_B are now applied as loads on the released structure (Figs. 13-33c and d).

The deflection at end B of the released structure due solely to the uniform load is denoted $(\delta_B)_1$, and the deflection at the same point due solely to the redundant is denoted $(\delta_B)_2$. The deflection δ_B at point B in the original

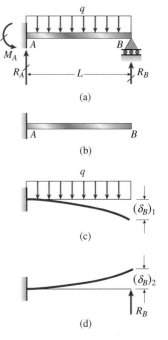

FIGURE 13-33

Analysis of a propped cantilever beam by the method of superposition with the reaction R_B selected as the redundant structure is obtained by superposing these two deflections. The deflection in the original beam is equal to zero, so the *equation of compatibility* is

$$\delta_B = (\delta_B)_1 - (\delta_B)_2 = 0 \tag{13-90}$$

The minus sign appears in this equation because $(\delta_B)_1$ is positive downward whereas $(\delta_B)_2$ is positive upward.

The *force–displacement relations* that give the deflections $(\delta_B)_1$ and $(\delta_B)_2$ in terms of the uniform load q and the redundant R_B, respectively, are found with the aid of Table E-1 in Appendix E (see Cases 1 and 4). Use the formulas given there to get

$$(\delta_B)_1 = \frac{qL^4}{8EI} \qquad (\delta_B)_2 = \frac{R_B L^3}{3EI} \tag{13-91a,b}$$

Substitute these force–displacement relations into the equation of compatibility to find

$$\delta_B = \frac{qL^4}{8EI} - \frac{R_B L^3}{3EI} = 0 \tag{13-91c}$$

which can be solved for the *redundant reaction*:

$$R_B = \frac{3qL}{8} \tag{13-92}$$

Note that this equation gives the redundant in terms of the loads acting on the original beam.

The remaining reactions (R_A and M_A) can be found from the equilibrium equations [Eqs. (13-89a and b)]; the results are

$$R_A = \frac{5qL}{8} \qquad M_A = \frac{qL^2}{8} \tag{13-93a,b}$$

Knowing all reactions, you can now obtain the shear forces and bending moments throughout the beam and plot the corresponding diagrams (see Fig. 13-28 for these diagrams).

The *deflections and slopes* of the original beam also can be detemined using the principle of superposition. The procedure consists of superposing the deflections of the released structure when acted upon by the loads shown in Figs. 13-33c and d. For instance, the equations of the deflection curves for those two loading systems are obtained from Cases 1 and 4, respectively, of Table E-1, Appendix E:

$$v_1 = -\frac{qx^2}{24EI}(6L^2 - 4Lx + x^2)$$

$$v_2 = \frac{R_B x^2}{6EI}(3L - x)$$

Substituting for R_B from Eq. (13-92) and then adding the deflections v_1 and v_2 leads to the following equation for the deflection curve of the original statically indeterminate beam (Fig. 13-33a):

$$v = v_1 + v_2 = -\frac{qx^2}{48EI}(3L^2 - 5Lx + 2x^2)$$

This equation agrees with Eq. (13-76) of Example 13-10. Other deflection quantitites can be found in an analogous manner.

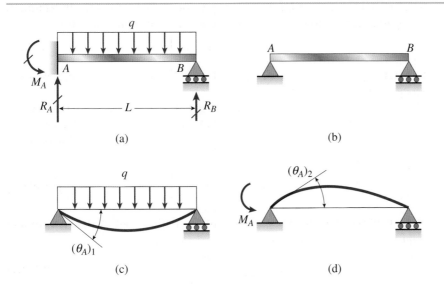

FIGURE 13-34

Analysis of a propped cantilever beam by the method of superposition with the moment reaction M_A selected as the redundant

(a)   (b)

(c)   (d)

Analysis with M_A as Redundant

Now analyze the same propped cantilever beam by selecting the moment reaction M_A as the redundant (Fig. 13-34). In this case, the released structure is a simple beam (Fig. 13-34b). The equations of equilibrium for the reactions R_A and R_B in the original beam are

$$R_A = \frac{qL}{2} + \frac{M_A}{L} \qquad R_B = \frac{qL}{2} - \frac{M_A}{L} \qquad \textbf{(13-94a,b)}$$

The equation of compatibility expresses the fact that the angle of rotation θ_A at the fixed end of the original beam is equal to zero. Since this angle is obtained by superposing the angles of rotation $(\theta_A)_1$ and $(\theta_A)_2$ in the released structure (Figs. 13-34c and d), the *compatibility equation* becomes

$$\theta_A = (\theta_A)_1 - (\theta_A)_2 = 0 \qquad \textbf{(13-95a)}$$

In this equation, the angle $(\theta_A)_1$ is assumed to be positive when clockwise, and the angle $(\theta_A)_2$ is assumed to be positive when counterclockwise.

The angles of rotation in the released structure are obtained from the formulas given in Table E-2 of Appendix E (see Cases 1 and 7). Thus, the *force–displacement relations* are

$$(\theta_A)_1 = \frac{qL^3}{24EI} \qquad (\theta_A)_2 = \frac{M_A L}{3EI}$$

Substituting into the compatibility equation (Eq. 13-95a) gives

$$\theta_A = \frac{qL^3}{24EI} - \frac{M_A L}{3EI} = 0 \qquad \textbf{(13-95b)}$$

Solve this equation for the redundant to get $M_A = qL^2/8$, which agrees with the previous result [Eq. (13-93b)]. Also, the equations of equilibrium [Eqs. (13-94a and b)] yield the same results as before for the reactions R_A and R_B [see Eqs. (13-93a) and (13-92), respectively].

Now that all reactions have been found, you can determine the shear forces, bending moments, slopes, and deflections by the techniques already described.

General Comments

The method of superposition described in this section is also called the *flexibility method* or the *force method*. The latter name arises from the use of force quantities (forces and moments) as the redundants; the former name is used because the coefficients of the unknown quantities in the compatibility equation [terms such as $L^3/3EI$ and $L/3EI$ in Eq. (13-95b)] are *flexibilities* (that is, deflections or angles produced by a unit load).

Since the method of superposition involves the superposition of deflections, it is applicable only to linearly elastic structures. (Recall that this same limitation applies to all topics discussed in this chapter.)

The following examples, and the problems at the end of the chapter, are concerned primarily with finding the reactions, since this is the key step in the solutions.

Example 13-12

A two-span continuous beam ABC supports a uniform load of intensity q, as shown in Fig. 13-35a. Each span of the beam has a length L. Using the method of superposition, determine all reactions for this beam.

FIGURE 13-35

Example 13-12: Two-span continuous beam with a uniform load

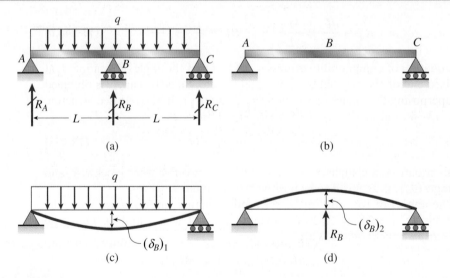

Solution:

Use the four-step problem-solving approach.

1. *Conceptualize*: This beam has three unknown reactions (R_A, R_B, and R_C). Since there are two equations of equilibrium for the beam as a whole, it is statically indeterminate to the first degree. For convenience, select the reaction R_B at the middle support as the redundant.

2. *Categorize*:

Equations of equilibrium: Express the reactions R_A and R_C in terms of the redundant R_B by means of two equations of equilibrium. The first equation, which is for equilibrium of moments about point B, shows that R_A and R_C are equal. The second equation, which is for equilibrium in the vertical direction, yields

$$R_A = R_C = qL - \frac{R_B}{2} \tag{a}$$

Equation of compatibility: Because the reaction R_B is selected as the redundant, the released structure is a simple beam with supports at A and C (Fig. 13-35b). The deflections at point B in the released structure due to the uniform load q and the redundant R_B are shown in Figs. 13-35c and d, respectively. Note that the deflections are denoted $(\delta_B)_1$ and $(\delta_B)_2$. The superposition of these deflections must produce the deflection δ_B in the original beam at point B. Since the latter deflection is equal to zero, the equation of compatibility is

$$\delta_B = (\delta_B)_1 - (\delta_B)_2 = 0 \tag{b}$$

where the deflection $(\delta_B)_1$ is positive downward and the deflection $(\delta_B)_2$ is positive upward.

Force–displacement relations: The deflection $(\delta_B)_1$ caused by the uniform load acting on the released structure (Fig. 13-35c) is obtained from Table E-2, Case 1, as

$$(\delta_B)_1 = \frac{5q(2L)^4}{384EI} = \frac{5qL^4}{24EI}$$

where $2L$ is the length of the released structure. The deflection $(\delta_B)_2$ produced by the redundant (Fig. 13-35d) is

$$(\delta_B)_2 = \frac{R_B(2L)^3}{48EI} = \frac{R_B L^3}{6EI}$$

as obtained from Table E-2, Case 4.

3. *Analyze*:

Reactions: The equation of compatibility pertaining to the vertical deflection at point B [Eq. (b)] now becomes

$$\delta_B = \frac{5qL^4}{24EI} - \frac{R_B L^3}{6EI} = 0 \tag{c}$$

from which the reaction at the middle support is

$$R_B = \frac{5qL}{4} \tag{13-96}$$

The other reactions are obtained from Eq. (a):

$$R_A = R_C = \frac{3qL}{8} \qquad \text{(13-97)}$$

With the reactions known, the shear forces, bending moments, stresses, and deflections can be found without difficulty.

4. *Finalize*: The purpose of this example is to demonstrate the method of superposition, so all steps were described in the analysis. However, this particular beam (Fig. 13-35a) can be analyzed by inspection because of the symmetry of the beam and its loading.

From symmetry, the slope of the beam at the middle support must be zero; therefore, each half of the beam is in the same condition as a propped cantilever beam with a uniform load (see, for instance, Fig. 13-27). Consequently, all of the previous results for a propped cantilever beam with a uniform load [Eqs. (13-69) to (13-80)] can be adapted immediately to the continuous beam of Fig. 13-35a.

Example 13-13

A fixed-end beam AB (Fig. 13-36a) is loaded by a force P acting at an intermediate point D. Find the reactive forces and moments at the ends of the beam using the method of superposition. Also, determine the deflection at point D where the load is applied.

Solution:

Use the four-step problem-solving approach.

1. *Conceptualize*: This beam has four unknown reactions (a force and a moment at each support) but only two independent equations of equilibrium are available. Therefore, the beam is statically indeterminate to the second degree. In this example, select the reactive moments M_A and M_B as the redundants.

2. *Categorize*:

Equations of equilibrium: The two unknown force reactions (R_A and R_B) are expressed in terms of the redundants (M_A and M_B) with the aid of two equations of equilibrium. The first equation is for moments about point B, and the second is for moments about point A. The resulting expressions are

$$R_A = \frac{Pb}{L} + \frac{M_A}{L} - \frac{M_B}{L} \qquad R_B = \frac{Pa}{L} - \frac{M_A}{L} + \frac{M_B}{L} \qquad \text{(a,b)}$$

Equations of compatibility: When both redundants are released by removing the rotational restraints at the ends of the beam, a simple beam remains as

FIGURE 13-36

Example 13-13: Fixed-end beam with a concentrated load

(a)

(b)

(c)

(d)

the released structure (Figs. 13-36b, c, and d). The angles of rotation at the ends of the released structure due to the concentrated load P are denoted $(\theta_A)_1$ and $(\theta_B)_1$, as shown in Fig. 13-36b. In a similar manner, the angles at the ends due to the redundant M_A are denoted $(\theta_A)_2$ and $(\theta_B)_2$, and the angles due to the redundant M_B are denoted $(\theta_A)_3$ and $(\theta_B)_3$.

Since the angles of rotation at the supports of the original beam are equal to zero, the two equations of compatibility are

$$\theta_A = (\theta_A)_1 - (\theta_A)_2 - (\theta_A)_3 = 0 \qquad \textbf{(c)}$$

$$\theta_B = (\theta_B)_1 - (\theta_B)_2 - (\theta_B)_3 = 0 \qquad \textbf{(d)}$$

in which the signs of the various terms are determined by inspection from the figures.

Force–displacement relations: The angles at the ends of the beam due to the load P (Fig. 13-36b) are obtained from Case 5 of Table E-2:

$$(\theta_A)_1 = \frac{Pab(L + b)}{6LEI} \qquad (\theta_B)_1 = \frac{Pab(L + a)}{6LEI}$$

where a and b are the distances from the supports to point D where the load is applied.

Also, the angles at the ends due to the redundant moment M_A are (see Case 7 of Table E-2):

$$(\theta_A)_2 = \frac{M_A L}{3EI} \qquad (\theta_B)_2 = \frac{M_A L}{6EI}$$

Similarly, the angles due to the moment M_B are

$$(\theta_A)_3 = \frac{M_B L}{6EI} \qquad (\theta_B)_3 = \frac{M_B L}{3EI}$$

3. *Analyze:*

Reactions: When the preceding expressions for the angles are substituted into the equations of compatibility [Eqs. (c) and (d)], two simultaneous equations containing M_A and M_B as unknowns are obtained:

$$\frac{M_A L}{3EI} + \frac{M_B L}{6EI} = \frac{Pab(L + b)}{6LEI} \qquad \textbf{(e)}$$

$$\frac{M_A L}{6EI} + \frac{M_B L}{3EI} = \frac{Pab(L + a)}{6LEI} \qquad \textbf{(f)}$$

Solving these equations for the redundants gives

$$M_A = \frac{Pab^2}{L^2} \qquad M_B = \frac{Pa^2b}{L^2}$$ ◀ **(13-98a,b)**

Substitute these expressions for M_A and M_B into the equations of equilibrium [Eqs. (a) and (b)] to obtain the vertical reactions:

$$R_A = \frac{Pb^2}{L^3}(L + 2a) \qquad R_B = \frac{Pa^2}{L^3}(L + 2b)$$ ◀ **(13-99a,b)**

Thus, all reactions for the fixed-end beam have been determined.

The reactions at the supports of a beam with fixed ends are commonly referred to as **fixed-end moments** and **fixed-end forces**. They are widely used in structural analysis, and formulas for these quantities are listed in engineering handbooks.

Deflection at point D: To obtain the deflection at point D in the original fixed-end beam (Fig. 13-36a), again use the principle of superposition. The deflection at point D is equal to the sum of three deflections: (1) the downward deflection $(\delta_D)_1$ at point D in the released structure due to the load P (Fig. 13-36b); (2) the upward deflection $(\delta_D)_2$ at the same point in the released structure due to the redundant M_A (Fig. 13-36c); and (3) the upward deflection $(\delta_D)_3$ at the same point in the released structure due to the redundant M_B (Fig. 13-36d). This superposition of deflections is expressed by

$$\delta_D = (\delta_D)_1 - (\delta_D)_2 - (\delta_D)_3$$ **(g)**

where δ_D is the downward deflection in the original beam.

The deflections appearing in Eq. (g) are obtained from the formulas given in Table E-2 of Appendix E (see Cases 5 and 7) by making the appropriate substitutions and algebraic simplifications. The results of these manipulations are

$$(\delta_D)_1 = \frac{Pa^2b^2}{3LEI} \qquad (\delta_D)_2 = \frac{M_A ab}{6LEI}(L + b) \qquad (\delta_D)_3 = \frac{M_B ab}{6LEI}(L + a)$$

Substitute the expressions for M_A and M_B from Eqs. (13-98a and b) into the last two expressions to get

$$(\delta_D)_2 = \frac{Pa^2b^3}{6L^3EI}(L + b) \qquad (\delta_D)_3 = \frac{Pa^3b^2}{6L^3EI}(L + a)$$

Therefore, the deflection at point D in the original beam, obtained by substituting $(\delta_D)_1$, $(\delta_D)_2$, and $(\delta_D)_3$ into Eq. (g) and simplifying, is

$$\delta_D = \frac{Pa^3b^3}{3L^3EI}$$ ◀ **(13-100)**

4. *Finalize:* The method described in this example for finding the deflection δ_D can be used not only to find deflections at individual points, but also to find the equations of the deflection curve.

Concentrated load acting at the midpoint of the beam:
When the load P acts at the midpoint C (Fig. 13-37), the reactions of the beam [from Eqs. (13-98) and (13-99) with $a = b = L/2$] are

$$M_A = M_B = \frac{PL}{8} \qquad R_A = R_B = \frac{P}{2} \qquad \textbf{(13-101a,b)}$$

Also, the deflection at the midpoint [from Eq. (13-100)] is

$$\delta_C = \frac{PL^3}{192EI} \qquad \textbf{(13-102)}$$

This deflection is only one-fourth of the deflection at the midpoint of a simple beam with the same load, which shows the stiffening effect of clamping the ends of the beam.

The preceding results for the reactions at the ends and the deflection at the middle [Eqs. (13-101) and (13-102)] agree with those found in Example 13-11 by solving the differential equation of the deflection curve [see Eqs. (13-81), (13-82), and (13-87)].

FIGURE 13-37

Fixed-end beam with a concentrated load acting at the midpoint

Example 13-14

A fixed-end beam AB supports a uniform load of intensity q acting over part of the span (Fig. 13-38a).

Determine the reactions of this beam (that is, find the fixed-end moments and fixed-end forces).

FIGURE 13-38

Example 13-14:
(a) Fixed-end beam with a uniform load over part of the span and (b) reactions produced by an element qdx of the uniform load

(a) (b)

Solution:

Use the four-step problem-solving approach.

1. *Conceptualize:* Find the reactions of this beam by using the principle of superposition together with the results obtained in Example 13-13. In that example, reactions

were found for a fixed-end beam subjected to a concentrated load P acting at distance a from the left-hand end [see Fig. 13-36a and Eqs. (13-98) and (13-99)].

In order to apply those results to the uniform load of Fig. 13-38a, treat an element of the uniform load as a concentrated load of magnitude qdx acting at distance x from the left-hand end (Fig. 13-38b). Then, using the formulas derived in Example 13-13, find the reactions caused by this element of load. Finally, by integrating over the length a of the uniform load, obtain the reactions due to the entire uniform load.

2. *Categorize*:

Fixed-end moments: Begin with the moment reactions using Eqs. (13-98a and b) of Example 13-13. To obtain the moments caused by the element qdx of the uniform load (compare Fig. 13-38b with Fig. 13-36a), replace P with qdx, a with x, and b with $L - x$. Thus, the fixed-end moments due to the element of load (Fig. 13-38b) are

$$dM_A = \frac{qx(L-x)^2\,dx}{L^2} \qquad dM_B = \frac{qx^2(L-x)\,dx}{L^2}$$

3. *Analyze*: Integrate over the loaded part of the beam to get the fixed-end moments due to the entire uniform load:

$$M_A = \int dM_A = \frac{q}{L^2}\int_0^a x(L-x)^2\,dx = \frac{qa^2}{12L^2}(6L - 8aL + 3a^2) \quad \Longleftarrow \textbf{(13-103a)}$$

$$M_B = \int dM_B = \frac{q}{L^2}\int_0^a x^2(L-x)\,dx = \frac{qa^3}{12L^2}(4L - 3a) \quad \Longleftarrow \textbf{(13-103b)}$$

Fixed-end forces: Proceed in a similar manner as for the fixed-end moments, but use Eqs. (13-99a and b) to obtain the following expressions for the fixed-end forces due to the element qdx of load:

$$dR_A = \frac{q(L-x)^2(L+2x)\,dx}{L^3} \qquad dR_B = \frac{qx^2(3L-2x)\,dx}{L^3}$$

Integration gives

$$R_A = \int dR_A = \frac{q}{L^3}\int_0^a (L-x)^2(L+2x)\,dx = \frac{qa}{2L^3}(2L^3 - 2La^2 + a^3) \quad \Longleftarrow \textbf{(13-104a)}$$

$$R_B = \int dR_B = \frac{q}{L^3}\int_0^a x^2(3L-2x)\,dx = \frac{qa^3}{2L^3}(2L - a) \quad \Longleftarrow \textbf{(13-104b)}$$

Thus, all reactions (fixed-end moments and fixed-end forces) have been found.

4. *Finalize*:

Uniform load acting over the entire length of the beam: When the load acts over the entire span (Fig. 13-39), obtain the reactions by substituting $a = L$ into the preceding equations, yielding

$$M_A = M_B = \frac{qL^2}{12} \qquad R_A = R_B = \frac{qL}{2} \qquad \textbf{(13-105a,b)}$$

FIGURE 13-39

Fixed-end beam with a uniform load

The deflection at the midpoint of a uniformly loaded beam is also of interest. The simplest way to obtain this deflection is to use the method of superposition. The first step is to remove the moment restraints at the supports and obtain a released structure in the form of a simple beam. The downward deflection at the midpoint of a simple beam due to a uniform load (from Case 1, Table E-2) is

$$(\delta_C)_1 = \frac{5qL^4}{384EI} \tag{a}$$

and the upward deflection at the midpoint due to the end moments (from Case 10, Table E-2) is

$$(\delta_C)_2 = \frac{M_A L^2}{8EI} = \frac{(qL^2/12)L^2}{8EI} = \frac{qL^4}{96EI} \tag{b}$$

Thus, the final downward deflection of the original fixed-end beam (Fig. 13-39) is

$$\delta_C = (\delta_C)_1 - (\delta_C)_2$$

Substitute for the deflections from Eqs. (a) and (b) to get

$$\delta_C = \frac{qL^4}{384EI} \tag{13-106}$$

This deflection is one-fifth of the deflection at the midpoint of a simple beam with a uniform load [Eq. (a)], again illustrating the stiffening effect of fixity at the ends of the beam.

Example 13-15

A beam ABC (Fig. 13-40a) rests on simple supports at points A and B and is supported by a cable at point C. The beam has total length $2L$ and supports a uniform load of intensity q. Prior to the application of the uniform load, there is no force in the cable nor is there any slack in the cable.

When the uniform load is applied, the beam deflects downward at point C and a tensile force T develops in the cable.

Find the magnitude of this force.

Solution:

Use the four-step problem-solving approach.

1. *Conceptualize*:

   Redundant force: The structure $ABCD$, consisting of the beam and cable, has three vertical reactions (at points A, B, and D). However, only two equations

FIGURE 13-40

Example 13-15:
Beam *ABC* with
one end supported
by a cable

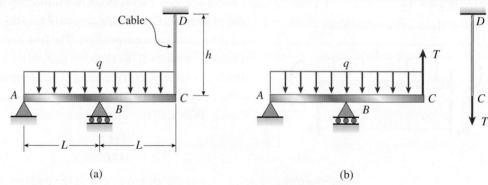

(a)   (b)

of equilibrium are available from a free-body diagram of the entire structure. Therefore, the structure is statically indeterminate to the first degree, and one redundant quantity must be selected for purposes of analysis.

The tensile force T in the cable is a suitable choice for the redundant. Release this force by removing the connection at point C, thereby cutting the structure into two parts (Fig. 13-40b). The released structure consists of the beam ABC and the cable CD as separate elements, with the redundant force T acting upward on the beam and downward on the cable.

2. *Categorize*:

Equation of compatibility: The deflection at point C of beam ABC (Fig. 13-40b) consists of two parts: a downward deflection $(\delta_C)_1$ due to the uniform load and an upward deflection $(\delta_C)_2$ due to the force T. At the same time, the lower end C of cable CD displaces downward by an amount $(\delta_C)_3$, which is equal to the elongation of the cable due to the force T. Therefore, the *equation of compatibility*, which expresses the fact that the downward deflection of end C of the beam is equal to the elongation of the cable, is

$$(\delta_C)_1 - (\delta_C)_2 = (\delta_C)_3 \tag{a}$$

Having formulated this equation, now turn to the task of evaluating all three displacements.

3. *Analyze*:

Force–displacement relations: The deflection $(\delta_C)_1$ at the end of the overhang (point C in beam ABC) due to the uniform load can be found from the results given in Example 13-9 of Section 13.5 (see Fig. 13-21). Use Eq. (13-68) of that example and substitute $a = L$ to get

$$(\delta_C)_1 = \frac{qL^4}{4E_bI_b} \tag{b}$$

where E_bI_b is the flexural rigidity of the beam.

The deflection of the beam at point C due to the force T is the deflection $(\delta_C)_2$ at the end of the overhang when the length of the overhang is a:

$$(\delta_C)_2 = \frac{Ta^2(L+a)}{3E_bI_b}$$

Now substitute $a = L$ to obtain the desired deflection:

$$(\delta_C)_2 = \frac{2TL^3}{3E_bI_b} \tag{c}$$

Finally, use the force–displacement relation (Eq. 8-3) to find the elongation of the cable

$$(\delta_C)_3 = \frac{Th}{E_cA_c} \tag{d}$$

where h is the length of the cable and E_cA_c is its axial rigidity.

Force in the cable: Substitute the three displacements [Eqs. (b), (c), and (d)] into the equation of compatibility [Eq. (a)] to get

$$\frac{qL^4}{4E_bI_b} - \frac{2TL^3}{3E_bI_b} = \frac{Th}{E_cA_c}$$

Solve for the force T to find

$$T = \frac{3qL^4E_cA_c}{8L^3E_cA_c + 12hE_bI_b} \qquad \Longleftarrow \textbf{(13-107)}$$

4. *Finalize*: With the force T known, all reactions, shear forces, and bending moments can be found by means of free-body diagrams and equations of equilibrium.

This example illustrates how an internal force quantity (instead of an external reaction) can be used as the redundant.

Chapter 13 discussed the linear elastic, small displacement behavior of beams of different types, with different support conditions, and acted upon by a wide variety of loadings. Methods based on integration of the second-, third-, or fourth-order differential equations of the deflection curve were described. Equations were derived for the deflected shape of the entire beam, and translational and rotational displacements were computed at specific points along a beam. The principle of superposition was used to solve more complicated beams and loadings by combining the simpler standard solutions.

This chapter also discussed the behavior of statically indeterminate beams acted on by concentrated and distributed loads. Two analysis approaches were presented: (1) **integration** of the equation of the elastic curve and (2) the more general approach based on **superposition**. In the superposition procedure, the **equilibrium** equations from statics were used along with **compatibility** equations and the **force–displacement relations** to generate the additional equations needed to solve the problem. The number of additional equations required is dependent on the **degree of static indeterminacy** of the structure. The superposition approach is limited to structures made of linearly elastic materials. Here are the major concepts presented in this chapter.

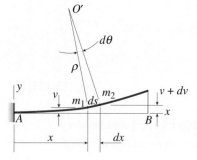

1. Combining expressions for linear curvature ($\kappa = d^2v / dx^2$) and the moment-curvature relation ($\kappa = M / EI$) resulted in the **ordinary differential equation of the deflection curve** for a beam, which is valid only for linear elastic behavior.

$$EI \frac{d^2v}{dx^2} = M$$

2. The differential equation of the deflection curve can be differentiated once to obtain a third-order equation relating shear force V and first derivative of moment dM/dx or twice to obtain a fourth-order equation relating intensity of distributed load q and first derivative of shear dV/dx.

$$EI \frac{d^3v}{dx^3} = V$$

$$EI \frac{d^4v}{dx^4} = -q$$

The choice of second-, third-, or fourth-order differential equations depends on which is most efficient for a particular beam support case and applied loading.

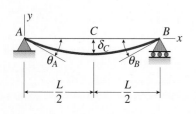

3. Write expressions for either moment (M), shear (V), or load intensity (q) for each separate region of the beam whenever q, V, M, or EI vary. Then apply **boundary**, **continuity**, or **symmetry conditions** (as appropriate) to solve for unknown constants of integration. The beam deflection equation, $v(x)$, can be evaluated at a particular value of x to find the translational displacement at that point; evaluation of dv/dx at that same point provides the slope of the deflection equation.

4. The **method of superposition** can be used to solve for displacements and rotations for more complicated beams and loadings. Superposition is only applicable to beams undergoing small displacements and behaving in a linear elastic manner.

5. Several types of statically indeterminate beam structures, such as propped cantilever, fixed-end, and continuous beams were discussed. The **degree of static indeterminacy** was determined for each beam type, and a **released** structure was defined by removing a sufficient number of **redundant** reaction forces.

6. The released structure must be statically determinate and **stable** under the action of the applied loadings. It is also possible to insert **internal releases** on axial force, shear, and moment to create a released structure.

7. For simple, statically indeterminate beam structures, the **differential equation of the elastic curve** can be written as a second-, third-, or fourth-order equation in terms of moment, shear force, and distributed load, respectively. Applying boundary and other conditions leads to a solution for the constants of integration and the redundant reactions.

8. A more general solution approach for more complex beam and other structures is the **method of superposition** (also known as the **force** or **flexibility** method). Here, additional equations that describe the **compatibility** of displacements and incorporate the appropriate **force–displacement relations** for beams are used to supplement the **equilibrium** equations.

PROBLEMS

13.2 Differential Equations of the Deflection Curve

The beams described in the problems for Section 13.2 have constant flexural rigidity EI.

Introductory Problems

13.2-1 The equation of the deflection curve for a cantilever beam is

$$v(x) = \frac{M_0 x^2}{2EI}$$

(a) Describe the loading acting on the beam.

(b) Draw the moment diagram for the beam.

13.2-2 The equation of the deflection curve for a simply supported beam is

$$v(x) = \frac{q_0}{24EI}(2Lx^3 - x^4 - L^3 x)$$

(a) Derive the slope equation of the beam.

(b) Derive the bending-moment equation of the beam.

(c) Derive the shear-force equation of the beam.

(d) Describe the loading acting on the beam.

13.2-3 The deflection curve for a simple beam AB (see figure) is given by

$$v = -\frac{q_0 x}{360LEI}(7L^4 - 10L^2 x^2 + 3x^4)$$

Describe the load acting on the beam.

PROBLEMS 13.2-3 and 13.2-4

13.2-4 The deflection curve for a simple beam AB (see figure) is given by

$$v = -\frac{q_0 L^4}{\pi^4 EI}\sin\frac{\pi x}{L}$$

(a) Describe the load acting on the beam.

(b) Determine the reactions R_A and R_B at the supports.

(c) Determine the maximum bending moment M_{max}.

Representative Problems

13.2-5 The deflection curve for a cantilever beam AB (see figure) is given by

$$v = -\frac{q_0 x^2}{120LEI}(10L^3 - 10L^2 x + 5Lx^2 - x^3)$$

Describe the load acting on the beam.

PROBLEMS 13.2-5 and 13.2-6

13.2-6 The deflection curve for a cantilever beam AB (see figure) is given by

$$v = -\frac{q_0 x^2}{360L^2 EI}(45L^4 - 40L^3 x + 15L^2 x^2 - x^4)$$

(a) Describe the load acting on the beam.

(b) Determine the reactions R_A and M_A at the support.

13.3 Deflections by Integration of the Bending-Moment Equation

Solve problems in this section using available deflection formulas. *All beams have constant flexural rigidity EI.*

Introductory Problems

13.3-1 A simply supported beam is loaded with a point load, as shown in the figure. The beam is a steel wide flange shape (W 12 × 35) in strong axis bending.

PROBLEM 13.3-1

Calculate the maximum deflection of the beam and the rotation at joint A if $L = 10$ ft, $a = 7$ ft, $b = 3$ ft, and $P = 10$ kips. Neglect the weight of the beam.

13.3-2 A 1-meter-long, simply supported copper beam ($E = 117$ GPa) carries uniformly distributed load q. The maximum deflection is measured as 1.5 mm.

(a) Calculate the magnitude of the distributed load q if the beam has a rectangular cross section (width $b = 20$ mm, height $h = 40$ mm).

(b) If instead the beam has circular cross section and $q = 500$ N/m, calculate the radius r of the cross section.

Neglect the weight of the beam.

PROBLEM 13.3-2

13.3-3 A wide-flange beam (W 12 × 35) supports a uniform load on a simple span of length $L = 14$ ft (see figure).

Calculate the maximum deflection δ_{max} at the midpoint and the angles of rotation θ at the supports if $q = 1.8$ kip/ft and $E = 30 \times 10^6$ psi. (Use the formulas of Example 13-1.)

PROBLEMS 13.3-3, 13.3-4, and 13.3-5

13.3-4 A uniformly loaded, steel wide-flange beam with simple supports (see figure) has a downward

deflection of 10 mm at the midpoint and angles of rotation equal to 0.01 radians at the ends.

Calculate the height h of the beam if the maximum bending stress is 90 MPa and the modulus of elasticity is 200 GPa. (Use the formulas of Example 13-1.)

13.3-5 What is the span length L of a uniformly loaded, simple beam of wide-flange cross section (see figure) if the maximum bending stress is 12,000 psi, the maximum deflection is 0.1 in., the height of the beam is 12 in., and the modulus of elasticity is 30×10^6 psi? (Use the formulas of Example 13-1.)

13.3-6 Calculate the maximum deflection δ_{max} of a uniformly loaded simple beam (see figure) if the span length $L = 2.0$ m, the intensity of the uniform load $q = 2.0$ kN/m, and the maximum bending stress $\sigma = 60$ MPa.

The cross section of the beam is square, and the material is aluminum having modulus of elasticity $E = 70$ GPa. (Use the formulas of Example 13-1.)

PROBLEM 13.3-6

Representative Problems

13.3-7 A cantilever beam with a uniform load (see figure) has a height h equal to 1/8 of the length L. The beam is a steel wide-flange section with $E = 28 \times 10^6$ psi and an allowable bending stress of 17,500 psi in both tension and compression. Calculate the ratio δ/L of the deflection at the free end to the length, assuming that the beam carries the maximum allowable load. (Use the formulas of Example 13-2.)

PROBLEM 13.3-7

13.3-8 A gold-alloy microbeam attached to a silicon wafer behaves like a cantilever beam subjected to a uniform load (see figure). The beam has a length $L = 27.5\,\mu m$ and rectangular cross section of a width $b = 4.0\,\mu m$ and thickness $t = 0.88\,\mu m$. The total load on the beam is 17.2 μN. If the deflection at the end of the beam is 2.46 μm, what is the modulus of elasticity E_g of the gold alloy? (Use the formulas of Example 13-2.)

PROBLEM 13.3-8

13.3-9 Obtain a formula for the ratio δ_C / δ_{max} of the deflection at the midpoint to the maximum deflection for a simple beam supporting a concentrated load P (see figure).

From the formula, plot a graph of δ_C / δ_{max} versus the ratio a/L that defines the position of the load ($0.5 < a/L < 1$). What conclusion do you draw from the graph? (Use the formulas of Example 13-3.)

PROBLEM 13.3-9

13.3-10 A cantilever beam model is often used to represent micro-electrical-mechanical systems (MEMS) (see figure). The cantilever beam is made of polysilicon ($E = 150$ GPa) and is subjected to an electrostatic moment M applied at the end of the cantilever beam. If dimensions are $b = 2\,\mu m$, $h = 4\,\mu m$, and $L = 320\,\mu m$, find expressions for the tip deflection and rotation of the cantilever beam in terms of moment M.

Based on Gary O'Brien, David J. Monk and Liwei Lin, "MEMS cantilever beam electrostatic pull-in model," Proc. SPIE 4593, Design, Characterization, and Packaging for MEMS and Microelectronics II, 31 (November 19, 2001).

PROBLEM 13.3-10

13.3-11 Beams AB and CDE are connected using rigid link DB with hinges (or moment releases) at ends D and B (see figure a). Beam AB is fixed at joint A and beam CDE is pin-supported at joint E. Load $P = 150$ lb is applied at C.

(a) Calculate the deflections of joints B and joint C. Assume $L = 9$ ft and $EI = 127,000$ kip-in^2.

(b) Repeat part (a) if rigid link DB is replaced by a linear spring with $k = 20$ kip/in. (see figure b).

(a)

(b)

PROBLEM 13.3-11

Solve Problems 13.3-12 through 13.3-23 by integrating the second-order differential equation of the deflection curve (the bending-moment equation). The origin of coordinates is at the left-hand end of each beam, and all beams have constant flexural rigidity EI.

Introductory Problems

13.3-12 Derive the equation of the deflection curve for a cantilever beam AB supporting a load P at the free end (see figure). Also, determine the deflection δ_B and angle of rotation θ_B at the free end. Use the second-order differential equation of the deflection curve.

PROBLEM 13.3-12

13.3-13 Derive the equation of the deflection curve for a simple beam AB loaded by a couple M_0 at the left-hand support (see figure). Also, determine the maximum deflection δ_{max}. Use the second-order differential equation of the deflection curve.

PROBLEM 13.3-13

13.3-14 A cantilever beam AB supporting a triangularly distributed load of maximum intensity q_0 is shown in the figure.

Derive the equation of the deflection curve and then obtain formulas for the deflection δ_B and angle of rotation θ_B at the free end. Use the second-order differential equation of the deflection curve.

PROBLEM 13.3-14

13.3-15 A cantilever beam has a length $L = 12$ ft and a rectangular cross section ($b = 16$ in., $h = 24$ in.). A linearly varying distributed load with peak intensity q_0 acts on the beam.

(a) Find peak intensity q_0 if the deflection at joint B is known to be 0.18 in. Assume that modulus $E = 30,000$ ksi.

(b) Find the location and magnitude of the maximum rotation of the beam.

PROBLEM 13.3-15

13.3-16 A simple beam with an overhang is subjected to a point load $P = 6$ kN. If the maximum allowable deflection at point C is 0.5 mm, select the lightest W360 section from Table C-1(b) that can be used for the beam. Assume that $L = 3$ m and ignore the distributed weight of the beam.

PROBLEM 13.3-16

Representative Problems

13.3-17 A cantilever beam AB is acted upon by a uniformly distributed moment (bending moment, not torque) of intensity m per unit distance along the axis of the beam (see figure).

Derive the equation of the deflection curve and then obtain formulas for the deflection δ_B and angle of rotation θ_B at the free end. Use the second-order differential equation of the deflection curve.

PROBLEM 13.3-17

13.3-18 The beam shown in the figure has a sliding support at A and a spring support at B. The sliding support permits vertical movement but no rotation. Derive the equation of the deflection curve and determine the deflection δ_B at end B due to the uniform load of intensity q. Use the second-order differential equation of the deflection curve.

PROBLEM 13.3-18

13.3-19 Derive the equations of the deflection curve for a simple beam AB loaded by a couple M_0 acting at distance a from the left-hand support (see figure). Also, determine the deflection δ_0 at the point where the load is applied. Use the second-order differential equation of the deflection curve.

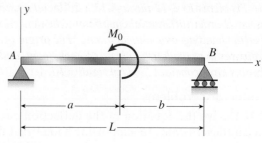

PROBLEM 13.3-19

13.3-20 Derive the equations of the deflection curve for a cantilever beam AB carrying a uniform load of intensity q over part of the span (see figure). Also, determine the deflection δ_B at the end of the beam. Use the second-order differential equation of the deflection curve.

PROBLEM 13.3-20

13.3-21 Derive the equations of the deflection curve for a cantilever beam AB supporting a distributed load of peak intensity q_0 acting over one-half of the length (see figure). Also, obtain formulas for the deflections δ_B and δ_C at points B and C, respectively. Use the second-order differential equation of the deflection curve.

PROBLEM 13.3-21

13.3-22 Derive the equations of the deflection curve for a simple beam AB with a distributed load of peak intensity q_0 acting over the left-hand half of the span (see figure). Also, determine the deflection δ_C at the midpoint of the beam. Use the second-order differential equation of the deflection curve.

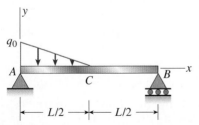

PROBLEM 13.3-22

13.3-23 The beam shown in the figure has a sliding support at A and a roller support at B. The sliding support permits vertical movement but no rotation. Derive the equation of the deflection curve and determine the deflection δ_A at end A and also δ_C at point C due to the uniform load of intensity $q = P/L$ applied over segment CB and load P at $x = L/3$. Use the second-order differential equation of the deflection curve.

PROBLEM 13.3-23

13.4 Deflections by Integration of the Shear-Force and Load Equations

The beams described in the problems for Section 13.4 have constant flexural rigidity EI. Also, the origin of coordinates is at the left-hand end of each beam.

Introductory Problems

13.4-1 Derive the equation of the deflection curve for a cantilever beam AB when a couple M_0 acts counterclockwise at the free end (see figure). Also, determine the deflection δ_B and slope θ_B at the free end. Use the third-order differential equation of the deflection curve (the shear-force equation).

PROBLEM 13.4-1

13.4-2 A simple beam AB is subjected to a distributed load of intensity $q(x) = q_0 \sin \pi x / L$, where q_0 is the maximum intensity of the load (see figure).

Derive the equation of the deflection curve, and then determine the deflection δ_{max} at the midpoint of the beam. Use the fourth-order differential equation of the deflection curve (the load equation).

PROBLEM 13.4-2

13.4-3 The simple beam AB shown in the figure has moments $2M_0$ and M_0 acting at the ends.

Derive the equation of the deflection curve, and then determine the maximum deflection δ_{max}. Use the third-order differential equation of the deflection curve (the shear-force equation).

PROBLEM 13.4-3

Representative Problems

13.4-4 A beam with a uniform load has a sliding support at one end and spring support at the other. The spring has a stiffness $k = 48EI / L^3$. Derive the equation of the deflection curve by starting with the third-order differential equation (the shear-force equation). Also, determine the angle of rotation θ_B at support B.

PROBLEM 13.4-4

13.4-5 The distributed load acting on a cantilever beam AB has an intensity $q(x)$ given by the expression $q_0 \cos \pi x / 2L$, where q_0 is the maximum intensity of the load (see figure).

Derive the equation of the deflection curve, and then determine the deflection δ_B at the free end. Use the fourth-order differential equation of the deflection curve (the load equation).

PROBLEM 13.4-5

13.4-6 A cantilever beam AB is subjected to a parabolically varying load of intensity $q(x) = q_0(L^2 - x^2)/L^2$, where q_0 is the maximum intensity of the load (see figure).

Derive the equation of the deflection curve, and then determine the deflection δ_B and angle of rotation θ_B at the free end. Use the fourth-order differential equation of the deflection curve (the load equation).

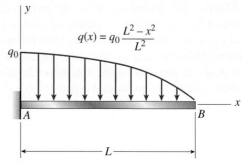

PROBLEM 13.4-6

13.4-7 A beam on simple supports is subjected to a parabolically distributed load of intensity $q(x) = 4q_0 x(L - x)/L^2$, where q_0 is the maximum intensity of the load (see figure).

Derive the equation of the deflection curve, and then determine the maximum deflection δ_{max}. Use the fourth-order differential equation of the deflection curve (the load equation).

PROBLEM 13.4-7

13.4-8 Derive the equation of the deflection curve for beam AB with sliding support at A and roller at B, carrying a triangularly distributed load of maximum intensity q_0 (see figure). Also, determine the maximum deflection δ_{max} of the beam. Use the fourth-order differential equation of the deflection curve (the load equation).

PROBLEM 13.4-8

13.4-9 Derive the equations of the deflection curve for beam ABC with sliding support at A and roller support at B, supporting a uniform load of intensity q acting on the overhang portion of the beam (see figure). Also, determine deflection δ_C and angle of rotation θ_C. Use the fourth-order differential equation of the deflection curve (the load equation).

PROBLEM 13.4-9

13.4-10 Derive the equations of the deflection curve for beam AB with sliding support at A and roller support at B, supporting a distributed load of maximum intensity q_0 acting on the right-hand half of the beam (see figure). Also, determine deflection δ_A, angle of rotation θ_B, and deflection δ_C at the midpoint. Use the fourth-order differential equation of the deflection curve (the load equation).

PROBLEM 13.4-10

13.5 Method of Superposition

Solve the problems for Section 13.5 by the method of superposition. All beams have constant flexural rigidity EI.

Introductory Problems

13.5-1 A simply supported beam ($E = 1600$ ksi) is loaded by a triangular distributed load from A to C (see figure). The load has a peak intensity $q_0 = 10$ lb / ft, and the deflection is known to be 0.01 in. at point C. The length of the beam is 12 ft, and the ratio of the height to the width of the cross section is ($h{:}b$) 2:1. Find the height h and width b of the cross section of the beam.

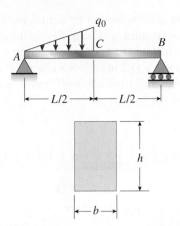

PROBLEM 13.5-1

13.5-2 A simply supported beam ($E = 12$ GPa) carries a uniformly distributed load $q = 125$ N/m, and a point load $P = 200$ N at mid-span. The beam has a rectangular cross section ($b = 75$ mm, $h = 200$ mm) and a length of 3.6 m. Calculate the maximum deflection of the beam.

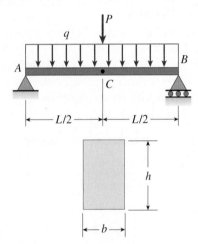

PROBLEM 13.5-2

13.5-3 Copper beam AB has circular cross section with a radius of 0.25 in. and length $L = 3$ ft. The beam is subjected to a uniformly distributed load $w = 3.5$ lb/ft. Calculate the required load P at joint B so that the total deflection at joint B is zero. Assume that $E = 16,000$ ksi.

PROBLEM 13.5-3

13.5-4 Beam ABC is loaded by a uniform load q and point load P at joint C. Using the method of superposition, calculate the deflection at joint C. Assume that $L = 4$ m, $a = 2$ m, $q = 15$ kN/m, $P = 7.5$ kN, $E = 200$ GPa, and $I = 70.8 \times 10^6$ mm^4.

PROBLEM 13.5-4

13.5-5 A cantilever beam of a length $L = 2.5$ ft has a rectangular cross section ($b = 4$ in., $h = 8$ in.) and modulus $E = 10,000$ ksi. The beam is subjected to a linearly varying distributed load with a peak intensity $q_0 = 900$ lb/ft. Use the method of superposition and Cases 1 and 9 in Table E-1 to calculate the deflection and rotation at B.

PROBLEM 13.5-5

13.5-6 A cantilever beam carries a trapezoidal distributed load (see figure). Let $w_B = 2.5$ kN/m, $w_A = 5.0$ kN/m, and $L = 2.5$ m. The beam has a modulus $E = 45$ GPa and a rectangular cross section with width $b = 200$ mm and depth $h = 300$ mm. Use the method of superposition and Cases 1 and 8 in Table E-1 to calculate the deflection and rotation at B.

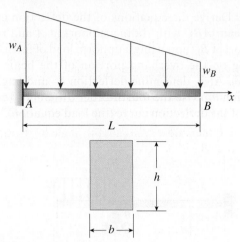

PROBLEM 13.5-6

13.5-7 A cantilever beam AB carries three equally spaced concentrated loads, as shown in the figure. Obtain formulas for the angle of rotation θ_B and deflection δ_B at the free end of the beam.

PROBLEM 13.5-7

13.5-8 A simple beam AB supports five equally spaced loads P (see figure).

(a) Determine the deflection δ_1 at the midpoint of the beam.

(b) If the same total load ($5P$) is distributed as a uniform load on the beam, what is the deflection δ_2 at the midpoint?

(c) Calculate the ratio of δ_1 to δ_2.

PROBLEM 13.5-8

Representative Problems

13.5-9 The cantilever beam AB shown in the figure has an extension BCD attached to its free end. A force P acts at the end of the extension.

(a) Find the ratio a/L so that the vertical deflection of point B will be zero.

(b) Find the ratio a/L so that the angle of rotation at point B will be zero.

PROBLEM 13.5-9

13.5-10 Beam ACB hangs from two springs, as shown in the figure. The springs have stiffnesses k_1 and k_2 and the beam has flexural rigidity EI.

(a) What is the downward displacement of point C, which is at the midpoint of the beam, when the moment M_0 is applied? Data for the structure are as follows: $M_0 = 10.0$ kN·m, $L = 1.8$ m, $EI = 216$ kN·m^2, $k_1 = 250$ kN/m, and $k_2 = 160$ kN/m.

(b) Repeat part (a), but remove M_0 and apply a uniform load $q = 3.5$ kN/m to the entire beam.

$q = 3.5$ kN/m (for part (b) only)

PROBLEM 13.5-10

13.5-11 What must be the equation $y = f(x)$ of the axis of the slightly curved beam AB (see figure) *before* the load is applied in order that the load P, moving along the bar, always stays at the same level?

PROBLEM 13.5-11

13.5-12 Determine the angle of rotation θ_B and deflection δ_B at the free end of a cantilever beam AB having a uniform load of intensity q acting over the middle third of its length (see figure).

PROBLEM 13.5-12

13.5-13 The cantilever beam ACB shown in the figure has flexural rigidity $EI = 2.1 \times 10^6$ kip-in^2. Calculate the downward deflections δ_C and δ_B at points C and B, respectively, due to the simultaneous action of the moment of 35 kip-in. applied at point C and the concentrated load of 2.5 kips applied at the free end B.

PROBLEM 13.5-13

13.5-14 A cantilever beam is subjected to load P at mid-span and counterclockwise moment M at B (see figure).

(a) Find an *expression* for moment M in terms of the load P so that the reaction moment M_A at A is equal to zero.

(b) Find an *expression* for moment M in terms of the load P so that the deflection is $\delta_B = 0$; *also*, what is rotation θ_B?

(c) Find an *expression* for moment M in terms of the load P so that the rotation $\theta_B = 0$; *also*, what is deflection δ_B?

PROBLEM 13.5-14

13.5-15 Use the method of superposition to find the angles of rotation θ_A and θ_B at the supports, and the maximum deflection δ_{max} for a simply supported beam subjected to *symmetric* loads P at distance a from each support. Assume that EI is constant, total beam length is L and $a = L/3$. *Hint:* Use the formulas of Example 13-3.

PROBLEM 13.5-15

13.5-16 Repeat Problem 13.5-15 for the *anti-symmetric* loading shown in the figure.

PROBLEM 13.5-16

13.5-17 A cantilever beam is subjected to a quadratic distributed load $q(x)$ over the length of the beam (see figure). Find an *expression* for moment M in terms of the peak distributed load intensity q_0 so that the deflection is $\delta_B = 0$.

PROBLEM 13.5-17

13.5-18 A beam $ABCD$ consisting of a simple span BD and an overhang AB is loaded by a force P acting at the end of the bracket CEF (see figure).

(a) Determine the deflection δ_A at the end of the overhang.

(b) Under what conditions is this deflection upward? Under what conditions is it downward?

PROBLEM 13.5-18

13.5-19 A horizontal load P acts at end C of the bracket ABC shown in the figure.

(a) Determine the deflection δ_C of point C.

(b) Determine the maximum upward deflection δ_{max} of member AB.

Note: Assume that the flexural rigidity EI is constant throughout the frame. Also, disregard the effects of axial deformations and consider only the effects of bending due to the load P.

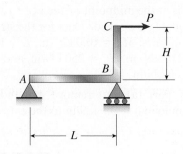

PROBLEM 13.5-19

13.5-20 A beam ABC having flexural rigidity $EI = 75 \text{ kN} \cdot \text{m}^2$ is loaded by a force $P = 800 \text{ N}$ at end C and tied down at end A by a wire having axial rigidity $EA = 900 \text{ kN}$ (see figure).

What is the deflection at point C when the load P is applied?

PROBLEM 13.5-20

13.5-21 Determine the angle of rotation θ_B and deflection δ_B at the free end of a cantilever beam AB supporting a parabolic load defined by the equation $q(x) = q_0 x^2 / L^2$ (see figure).

PROBLEM 13.5-21

13.5-22 A simple beam AB supports a uniform load of intensity q acting over the middle region of the span (see figure).

Determine the angle of rotation θ_A at the left-hand support and the deflection δ_{max} at the midpoint.

PROBLEM 13.5-22

13.5-23 The overhanging beam $ABCD$ supports two concentrated loads P and Q (see figure).

(a) For what ratio P/Q will the deflection at point B be zero?

(b) For what ratio will the deflection at point D be zero?

(c) If Q is replaced by a uniform load with intensity q (on the overhang), repeat parts (a) and (b), but find ratio $P/(qa)$.

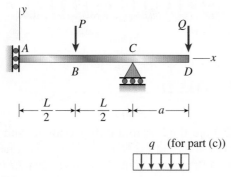

PROBLEM 13.5-23

13.5-24 A thin metal strip of total weight W and length L is placed across the top of a flat table of width $L/3$ as shown in the figure.

What is the clearance δ between the strip and the middle of the table? (The strip of metal has flexural rigidity EI.)

PROBLEM 13.5-24

13.5-25 An overhanging beam ABC with flexural rigidity $EI = 15$ kip-in^2 is supported by a sliding support at A and by a spring of stiffness k at point B (see figure). Span AB has a length $L = 30$ in. and carries a uniform load. The overhang BC has a length $b = 15$ in. For what stiffness k of the spring will the uniform load produce no deflection at the free end C?

PROBLEM 13.5-25

13.5-26 A beam $ABCD$ rests on simple supports at B and C (see figure). The beam has a slight initial curvature so that end A is 18 mm above the elevation of the supports and end D is 12 mm above. What moments M_1 and M_2, acting at points A and D, respectively, will move points A and D downward to the level of the supports? (The flexural rigidity EI of the beam is 2.5×10^6 N·m^2 and $L = 2.5$m.)

PROBLEM 13.5-26

13.5-27 The compound beam ABC shown in the figure has a sliding support at A and a fixed support at C. The beam consists of two members joined by a pin connection (i.e., moment release) at B. Find the deflection δ under the load P.

PROBLEM 13.5-27

13.5-28 A compound beam $ABCDE$ (see figure) consists of two parts (ABC and CDE) connected by a hinge (i.e., moment release) at C. The elastic support at B has stiffness $k = EI/b^3$. Determine the deflection δ_E at the free end E due to the load P acting at that point.

PROBLEM 13.5-28

13.5-29 A steel beam ABC is simply supported at A and held by a high-strength steel wire at B (see figure). A load $P = 240$ lb acts at the free end C. The wire has axial rigidity $EA = 1500 \times 10^3$ lb, and the beam has flexural rigidity $EI = 36 \times 10^6$ lb-in^2.

What is the deflection δ_C of point C due to the load P?

PROBLEM 13.5-29

13.5-30 Calculate the deflection at point C of a beam subjected to uniformly distributed load $w = 275$ N/m on span AB and point load $P = 10$ kN at C. Assume that $L = 5$ m and $EI = 1.50 \times 10^7$ N·m^2.

PROBLEM 13.5-30

13.5-31 Compound beam ABC is loaded by point load $P = 1.5$ kips at distance $2a/3$ from point A and a triangularly distributed load on segment BC with peak intensity $q_0 = 0.5$ kip/ft. If length $a = 5$ ft and length $b = 10$ ft, find the deflection at B and rotation at A. Assume that $E = 29,000$ ksi and $I = 53.8$ in^4.

PROBLEM 13.5-31

13.5-32 The compound beam shown in the figure consists of a cantilever beam AB (length L) that is pin-connected to a simple beam BD (length $2L$). After the beam is constructed, a clearance c exists between the beam and a support at C, midway between points B and D. Subsequently, a uniform load is placed along the entire length of the beam.

What intensity q of the load is needed to close the gap at C and bring the beam into contact with the support?

PROBLEM 13.5-32

13.5-33 Find the horizontal deflection δ_h and vertical deflection δ_v at the free end C of the frame ABC shown in the figure. (The flexural rigidity EI is constant throughout the frame.)

PROBLEM 13.5-33

Note: Disregard the effects of axial deformations and consider only the effects of bending due to the load *P*.

13.5-34 The frame *ABCD* shown in the figure is squeezed by two collinear forces *P* acting at points *A* and *D*. What is the decrease δ in the distance between points *A* and *D* when the loads *P* are applied? (The flexural rigidity *EI* is constant throughout the frame.)
Note: Disregard the effects of axial deformations and consider only the effects of bending due to the loads *P*.

PROBLEM 13.5-34

13.5-35 A framework *ABCD* is acted on by counter-clockwise moment *M* at *A* (see figure). Assume that *EI* is constant.

(a) Find expressions for reactions at supports *B* and *C*.

(b) Find expressions for angles of rotation at *A*, *B*, *C*, and *D*.

(c) Find expressions for horizontal deflections δ_A and δ_D.

(d) If length $L_{AB} = L/2$, find length L_{CD} in terms of *L* for the absolute value of the ratio $\left| \delta_A / \delta_D \right| = 1$.

PROBLEM 13.5-35

13.5-36 A framework *ABCD* is acted on by force *P* at $2L/3$ from *B* (see figure). Assume that *EI* is constant.

(a) Find expressions for reactions at supports *B* and *C*.

(b) Find expressions for angles of rotation at *A*, *B*, *C*, and *D*.

(c) Find expressions for horizontal deflections δ_A and δ_D.

(d) If length $L_{AB} = L/2$, find length L_{CD} in terms of *L* for the absolute value of the ratio $\left| \delta_A / \delta_D \right| = 1$.

PROBLEM 13.5-36

13.5-37 A beam *ABCDE* has simple supports at *B* and *D* and symmetrical overhangs at each end (see figure). The center span has length *L* and each overhang has length *b*. A uniform load of intensity *q* acts on the beam.

(a) Determine the ratio *b/L* so that the deflection δ_C at the midpoint of the beam is equal to the deflections δ_A and δ_E at the ends.

(b) For this value of *b/L*, what is the deflection δ_C at the midpoint?

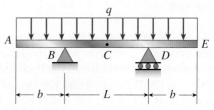

PROBLEM 13.5-37

13.5-38 A frame *ABC* is loaded at point *C* by a force *P* acting at an angle α to the horizontal (see figure). Both members of the frame have the same length and the same flexural rigidity.

Determine the angle α so that the deflection of point *C* is in the same direction as the load. (Disregard the effects of axial deformations and consider only the effects of bending due to the load *P*.)

Note: A direction of loading such that the resulting deflection is in the same direction as the load is called a *principal direction*. For a given load on a planar structure, there are two principal directions that are perpendicular to each other.

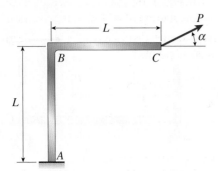

PROBLEM 13.5-38

13.5-39 The wing of a large commercial jet is represented by a simplified prismatic cantilever beam model with uniform load *w* and concentrated loads *P* at the two engine locations (see figure). Find expressions for the tip deflection and rotation at *D* in terms of *w*, *P*, *L*, and *EI*.

PROBLEM 13.5-39

13.5-40 The wing of a small plane is represented by a simplified prismatic cantilever beam model acted on by the distributed loads shown in the figure. Assume constant $EI = 1200 \text{ kN} \cdot \text{m}^2$. Find the tip deflection and rotation at *B*.

PROBLEM 13.5-40

13.5-41 Find an expression for required moment M_A (in terms of *q* and *L*) that will result in rotation $\theta_B = 0$ due to M_A and *q* loadings applied at the same time. Also, what is the resulting net rotation at support *A*?

PROBLEM 13.5-41

13.5-42 Find an expression for required moment M_A (in terms of *q* and *L*) that will result in rotation $\theta_B = 0$ due to M_A and *q* loadings applied at the same time. Also, what is the resulting net rotation at support *A*?

PROBLEM 13.5-42

13.5-43 Find required distance d (in terms of L) so that rotation $\theta_B = 0$ is due to M and q loadings applied at the same time. Also, what is the resulting net rotation θ_A at support A? Moment M is applied at distance d from joint B.

PROBLEM 13.5-43

13.5-44 A cantilever beam has two triangular loads as shown in the figure.

(a) Find an expression for beam deflection δ_C using superposition.

(b) Find the required magnitude of load intensity q_2 in terms of q_1 so that the deflection at C is zero.

(c) Find an expression for the deflection at C if both load intensities, q_1 and q_2, are equal to q_0.

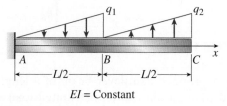

PROBLEM 13.5-44

13.6 Statically Indeterminate Beams

Differential Equations of the Deflection Curve

Solve Problems 13.6-1 though 13.6-16 by integrating the differential equations of the deflection curve. All beams have constant flexural rigidity EI. When drawing shear-force and bending-moment diagrams, label all critical ordinates, including maximum and minimum values.

Introductory Problems

13.6-1 A propped cantilever steel beam is constructed from a W12 × 35 section. The beam is loaded by its self-weight with intensity q. The length of the beam is 11.5 ft. Let $E = 30,000$ ksi.

(a) Calculate the reactions at joints A and B.

(b) Find the location of zero moment within span AB.

(c) Calculate the maximum deflection of the beam and the rotation at joint B.

PROBLEM 13.6-1

13.6-2 A fixed-end beam is subjected to a point load at mid-span. The beam has a rectangular cross section (assume that the h/b ratio is 2) and is made of wood ($E = 11$ GPa).

(a) Find height h of the cross section if the maximum displacement of the beam is 2 mm.

(b) Calculate the displacement of the beam at the inflection points.

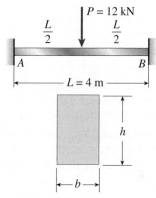

PROBLEM 13.6-2

13.6-3 A propped cantilever beam AB of a length L is loaded by a counterclockwise moment M_0 acting at support B (see figure).

Beginning with the second-order differential equation of the deflection curve (the bending-moment equation), obtain the reactions, shear forces, bending moments, slopes, and deflections of the beam. Construct the shear-force and bending-moment diagrams, labeling all critical ordinates.

PROBLEM 13.6-3

13.6-4 A fixed-end beam AB of a length L supports a uniform load of intensity q (see figure).

Beginning with the second-order differential equation of the deflection curve (the bending-moment equation), obtain the reactions, shear forces, bending moments, slopes, and deflections of the beam. Construct the shear-force and bending-moment diagrams, labeling all critical ordinates.

PROBLEM 13.6-4

Representative Problems

13.6-5 A cantilever beam AB of a length L has a fixed support at A and a roller support at B (see figure). The support at B is moved downward through a distance δ_B.

Using the fourth-order differential equation of the deflection curve (the load equation), determine the reactions of the beam and the equation of the deflection curve. *Note:* Express all results in terms of the imposed displacement δ_B.

PROBLEM 13.6-5

13.6-6 A cantilever beam of a length L and loaded by a uniform load of intensity q has a fixed support at A and spring support at B with rotational stiffness k_R. A rotation θ_B at B results in a reaction moment $M_B = k_R \times \theta_B$.

Find rotation θ_B and displacement δ_B at end B. Use the second-order differential equation of the deflection curve to solve for displacements at end B.

PROBLEM 13.6-6

13.6-7 A cantilever beam has a length L and is loaded by a triangularly distributed load of maximum intensity q_0 at B.

Use the fourth-order differential equation of the deflection curve to solve for reactions at A and B and also the equation of the deflection curve.

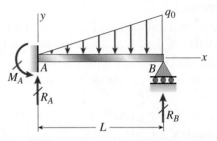

PROBLEM 13.6-7

13.6-8 A propped cantilever beam of a length L is loaded by a parabolically distributed load with a maximum intensity q_0 at B.

(a) Use the fourth-order differential equation of the deflection curve to solve for reactions at A and B and also the equation of the deflection curve.

(b) Repeat part (a) if the parabolic load is replaced by $q_0 \sin(\pi x / 2L)$.

PROBLEM 13.6-8

13.6-9 A propped cantilever beam of a length L is loaded by a parabolically distributed load with a maximum intensity q_0 at A.

(a) Use the fourth-order differential equation of the deflection curve to solve for reactions at A and B and also the equation of the deflection curve.

(b) Repeat part (a) if the parabolic load is replaced by $q_0 \cos(\pi x/2L)$.

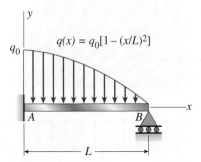

$q(x) = q_0[1 - (x/L)^2]$

PROBLEM 13.6-9

13.6-10 A fixed-end beam of a length L is loaded by a distributed load in the form of a cosine curve with a maximum intensity q_0 at A.

(a) Use the fourth-order differential equation of the deflection curve to solve for reactions at A and B and also the equation of the deflection curve.

(b) Repeat part (a) using the distributed load $q_0 \sin(\pi x/L)$.

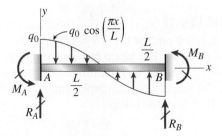

PROBLEM 13.6-10

13.6-11 A fixed-end beam of a length L is loaded by a distributed load in the form of a cosine curve with a maximum intensity q_0 at A.

(a) Use the fourth-order differential equation of the deflection curve to solve for reactions at A and B and also the equation of the deflection curve.

(b) Repeat part (a) if the distributed load is now $q_0(1 - x^2/L^2)$.

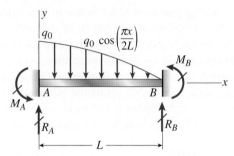

PROBLEM 13.6-11

13.6-12 A fixed-end beam of a length L is loaded by triangularly distributed load of a maximum intensity q_0 at B.

Use the fourth-order differential equation of the deflection curve to solve for reactions at A and B and also the equation of the deflection curve.

PROBLEM 13.6-12

13.6-13 A counterclockwise moment M_0 acts at the midpoint of a fixed-end beam ACB of length L (see figure).

Beginning with the second-order differential equation of the deflection curve (the bending-moment equation), determine all reactions of the beam and obtain the equation of the deflection curve for the left-hand half of the beam.

Then construct the shear-force and bending-moment diagrams for the entire beam, labeling all critical ordinates. Also, draw the deflection curve for the entire beam.

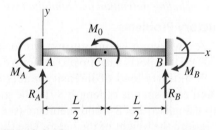

PROBLEM 13.6-13

1037

13.6-14 A propped cantilever beam of a length L is loaded by a concentrated moment M_0 at midpoint C. Use the second-order differential equation of the deflection curve to solve for reactions at A and B. Draw shear-force and bending-moment diagrams for the entire beam. Also find the equations of the deflection curves for both halves of the beam, and draw the deflection curve for the entire beam.

PROBLEM 13.6-14

13.6-15 A propped cantilever beam is subjected to uniform load q. The beam has flexural rigidity $EI = 2000 \text{ kip-ft}^2$ and the length of the beam is 10 ft. Find the intensity q of the distributed load if the maximum displacement of the beam is $\delta_{max} = 0.125$ in.

PROBLEMS 13.6-15 and 13.6-16

13.6-16 Repeat Problem 13.6-15 using $L = 3.5$ m, $\delta_{max} = 3$ mm, and $EI = 800 \text{ kN·m}^2$.

Method of Superposition

Solve Problems 13.6-17 through 13.6-59 by the method of superposition. All beams have constant flexural rigidity EI unless otherwise stated. When drawing shear-force and bending-moment diagrams, label all critical ordinates, including maximum and minimum values.

Introductory Problems

13.6-17 A two-span, continuous wood girder ($E = 1700$ ksi) supports a roof patio structure (figure part a). A uniform load of intensity q acts on the girder, and each span is of length 8 ft. The girder is made up using two 2×8 wood members (see figure part b). Ignore the weight of the beam. Use the nominal dimensions of the beam in your calculations.

1038

(a) Find the reactions at A, B, and C.

(b) Use the method of superposition to calculate the displacement of the beam at the mid-span of segment AB. *Hint*: See Figs. 13-35c and 13-35d in Example 13-12.

(a)

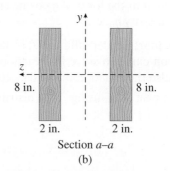

Section *a–a*
(b)

PROBLEM 13.6-17

13.6-18 A fixed-end beam AB carries point load P acting at point C. The beam has a rectangular cross section ($b = 75$ mm, $h = 150$ mm). Calculate the reactions of the beam and the displacement at point C. Assume that $E = 190$ GPa.

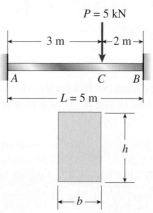

PROBLEM 13.6-18

13.6-19 A fixed-end beam AB supports a uniform load of intensity $q = 75$ lb/ft acting over part of the span. Assume that $EI = 300$ kip-ft^2.

(a) Calculate the reactions at A and B.

(b) Find the maximum displacement and its location.

(c) Repeat part (a) if the distributed load is applied from A to B.

PROBLEM 13.6-19

13.6-20 A cantilever beam is supported at B by cable BC. The beam carries a uniform load $q = 200$ N/m. If the length of the beam is $L = 3$ m, find the force in the cable and the reactions at A. Ignore the axial flexibility of the cable.

PROBLEM 13.6-20

13.6-21 A propped cantilever beam AB of a length L carries a concentrated load P acting at the position shown in the figure.

Determine the reactions R_A, R_B, and M_A for this beam. Also, draw the shear-force and bending-moment diagrams, labeling all critical ordinates.

PROBLEM 13.6-21

Representative Problems

13.6-22 A beam with a sliding support at B is loaded by a uniformly distributed load with intensity q. Use the method of superposition to solve for all reactions. Also draw shear-force and bending-moment diagrams, labeling all critical ordinates.

PROBLEM 13.6-22

13.6-23 A propped cantilever beam of a length $2L$ with a support at B is loaded by a uniformly distributed load with intensity q. Use the method of superposition to solve for all reactions. Also draw shear-force and bending-moment diagrams, labeling all critical ordinates.

PROBLEM 13.6-23

13.6-24 The continuous frame ABC has a pin support at A, roller supports at B and C, and a rigid corner connection at B (see figure). Members AB and BC each have flexural rigidity EI. A moment M_0 acts counterclockwise at B. *Note:* Disregard axial deformations in member AB and consider only the effects of bending.

(a) Find all reactions of the frame.

(b) Find joint rotations θ at A, B, and C.

(c) Find the required new length of member BC in terms of L, so that θ_B in part (b) is doubled in size.

PROBLEM 13.6-24

13.6-25 The continuous frame *ABC* has a pin support at *A*, roller supports at *B* and *C*, and a rigid corner connection at *B* (see figure). Members *AB* and *BC* each have flexural rigidity *EI*. A moment M_0 acts counterclockwise at *A*. *Note:* Disregard axial deformations in member *AB* and consider only the effects of bending.

(a) Find all reactions of the frame.

(b) Find joint rotations θ at *A*, *B*, and *C*.

(c) Find the required new length of member *AB* in terms of *L*, so that θ_A in part (b) is doubled in size.

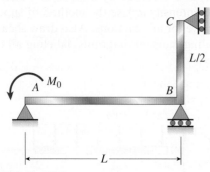

PROBLEM 13.6-25

13.6-26 Beam *AB* has a pin support at *A* and a roller support at *B*. Joint *B* is also restrained by a linearly elastic rotational spring with stiffness k_R, which provides a resisting moment M_B due to rotation at *B*. Member *AB* has flexural rigidity *EI*. A moment M_0 acts counterclockwise at *B*.

(a) Use the method of superposition to solve for all reactions.

(b) Find an expression for joint rotation θ_A in terms of spring stiffness k_R. What is θ_A when $k_R \to 0$? What is θ_A when $k_R \to \infty$? What is θ_A when $k_R = 6EI/L$?

PROBLEM 13.6-26

13.6-27 The continuous frame *ABCD* has a pin support at *B*; roller supports at *A*, *C*, and *D*; and rigid corner connections at *B* and *C* (see figure). Members *AB*, *BC*, and *CD* each have flexural rigidity *EI*. Moment M_0 acts counterclockwise at *B* and clockwise at *C*. *Note:* Disregard axial deformations in member *AB* and consider only the effects of bending.

(a) Find all reactions of the frame.

(b) Find joint rotations θ at *A*, *B*, *C*, and *D*.

(c) Repeat parts (a) and (b) if both moments M_0 are counterclockwise.

PROBLEM 13.6-27

13.6-28 Two flat beams *AB* and *CD*, lying in horizontal planes, cross at right angles and jointly support a vertical load *P* at their midpoints (see figure). Before the load *P* is applied, the beams just touch each other. Both beams are made of the same material and have the same widths. Also, the ends of both beams are simply supported. The lengths of beams *AB* and *CD* are L_{AB} and L_{CD}, respectively.

What should be the ratio t_{AB}/t_{CD} of the thicknesses of the beams if all four reactions are to be the same?

PROBLEM 13.6-28

13.6-29 A propped cantilever beam of a length $2L$ is loaded by a uniformly distributed load with intensity q. The beam is supported at B by a linearly elastic spring with stiffness k. Use the method of superposition to solve for all reactions. Also draw shear-force and bending-moment diagrams, labeling all critical ordinates. Let $k = 6EI/L^3$.

PROBLEM 13.6-29

13.6-30 A propped cantilever beam of a length $2L$ is loaded by a uniformly distributed load with intensity q. The beam is supported at B by a linearly elastic rotational spring with stiffness k_R, which provides a resisting moment M_B due to rotation θ_B. Use the method of superposition to solve for all reactions. Also draw shear-force and bending-moment diagrams, labeling all critical ordinates. Let $k_R = EI/L$.

PROBLEM 13.6-30

13.6-31 Determine the fixed-end moments (M_A and M_B) and fixed-end forces (R_A and R_B) for a beam of length L supporting a triangular load of maximum intensity q_0 (see figure). Then draw the shear-force and bending-moment diagrams, labeling all critical ordinates.

PROBLEM 13.6-31

13.6-32 A continuous beam ABC with two unequal spans, one of length L and one of length $2L$, supports a uniform load of intensity q (see figure).

Determine the reactions R_A, R_B, and R_C for this beam. Also, draw the shear-force and bending-moment diagrams, labeling all critical ordinates.

PROBLEM 13.6-32

13.6-33 Beam ABC is fixed at support A and rests (at point B) upon the midpoint of beam DE (see part a of the figure). Thus, beam ABC may be represented as a propped cantilever beam with an overhang BC and a linearly elastic support of stiffness k at point B (see part b of the figure).

The distance from A to B is $L = 10$ ft, the distance from B to C is $L/2 = 5$ ft, and the length of beam DE is $L = 10$ ft. Both beams have the same flexural rigidity EI. A concentrated load $P = 1700$ lb acts at the free end of beam ABC.

Determine the reactions R_A, R_B, and M_A for beam ABC. Also, draw the shear-force and bending-moment diagrams for beam ABC, labeling all critical ordinates.

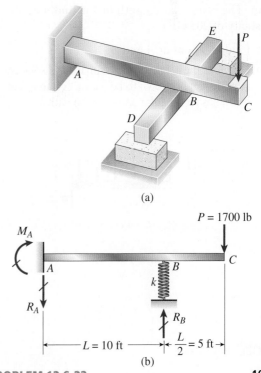

PROBLEM 13.6-33

13.6-34 A propped cantilever beam has flexural rigidity $EI = 4.5\,MN \cdot m^2$. When the loads shown are applied to the beam, it settles at joint B by 5 mm. Find the reaction at joint B.

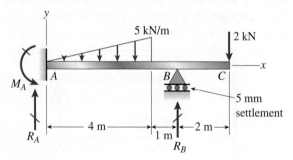

PROBLEM 13.6-34

13.6-35 A triangularly distributed load with a maximum intensity of $q_0 = 10$ lb/ft acts on propped cantilever beam AB. If the length L of the beam is 10 ft, find the reactions at A and B.

PROBLEM 13.6-35

13.6-36 A fixed-end beam is loaded by a uniform load $q = 15$ kN/m and a point load $P = 30$ kN at midspan. The beam has a length of 4 m and modulus of elasticity of 205 GPa.

(a) Find reactions at A and B.

(b) Calculate the height of the beam if the displacement at mid-span is known to be 3 mm. Assume that the beam has rectangular cross section with $h/b = 2$.

PROBLEM 13.6-36

13.6-37 Uniform load $q = 10$ lb/ft acts over part of the span of fixed-end beam AB (see figure). Upward load $P = 250$ lb is applied 9 ft to the right of joint A. Find the reactions at A and B.

PROBLEM 13.6-37

13.6-38 A propped cantilever beam with a length $L = 4$ m is subjected to a trapezoidal load with intensities $q_0 = 10$ kN/m and $q_1 = 15$ kN/m. Find the reactions at A and B. *Hint:* The loading is the sum of uniform and triangular loads.

PROBLEM 13.6-38

13.6-39 A cantilever beam is supported by a tie rod at B as shown. Both the tie rod and the beam are steel with $E = 30 \times 10^6$ psi. The tie rod is just taut before the distributed load $q = 200$ lb/ft is applied.

(a) Find the tension force in the tie rod.

(b) Draw shear-force and bending-moment diagrams for the beam, labeling all critical ordinates.

PROBLEM 13.6-39

13.6-40 The figure shows a nonprismatic, propped cantilever beam AB with flexural rigidity $2EI$ from A to C and EI from C to B.

Determine all reactions of the beam due to the uniform load of intensity q.

PROBLEM 13.6-40

13.6-41 A beam ABC is fixed at end A and supported by beam DE at point B (see figure). Both beams have the same cross section and are made of the same material.

(a) Determine all reactions due to the load P.

(b) What is the numerically largest bending moment in either beam?

PROBLEM 13.6-41

13.6-42 A three-span continuous beam $ABCD$ with three equal spans supports a uniform load of intensity q (see figure).

Determine all reactions of this beam and draw the shear-force and bending-moment diagrams, labeling all critical ordinates.

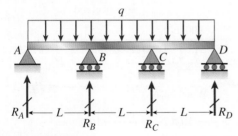

PROBLEM 13.6-42

13.6-43 A beam rests on supports at A and B and is loaded by a distributed load with intensity q as shown. A small gap Δ exists between the unloaded beam and the support at C. Assume that span length $L = 40$ in. and flexural rigidity of the beam $EI = 0.4 \times 10^9$ lb-in^2. Plot a graph of the bending moment at B as a function of load intensity q.

Hint: See Example 13-9 for guidance on computing the deflection at C.

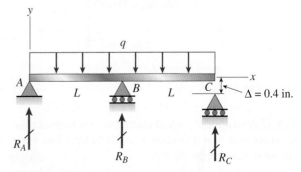

PROBLEM 13.6-43

13.6-44 A propped cantilever beam is subjected to two triangularly distributed loads, each with a peak load intensity equal to q_0 (see figure). Find the expressions for reactions at A and C using superposition. Plot shear and moment diagrams.

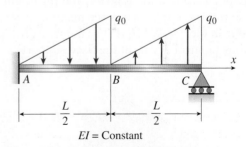

$EI = $ Constant

PROBLEM 13.6-44

13.6-45 A propped cantilever beam is loaded by a triangular distributed load from A to C (see figure). The load has a peak intensity $q_0 = 10$ lb/ft. The length of the beam is 12 ft. Find support reactions at A and B.

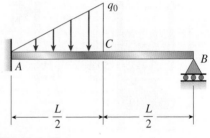

PROBLEM 13.6-45

13.6-46 A fixed-end beam AB of a length L is subjected to a moment M_0 acting at the position shown in the figure.

(a) Determine all reactions for this beam.

(b) Draw shear-force and bending-moment diagrams for the special case in which $a = b = L/2$.

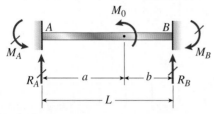

PROBLEM 13.6-46

13.6-47 A temporary wood flume serving as a channel for irrigation water is shown in the figure. The vertical boards forming the sides of the flume are sunk in the ground, which provides a fixed support. The top of the flume is held by tie rods that are tightened so that there is no deflection of the boards at that point. Thus, the vertical boards may be modeled as a beam AB, supported and loaded as shown in the last part of the figure.

Assuming that the thickness t of the boards is 1.5 in., the depth d of the water is 40 in., and the height h to the tie rods is 50 in., what is the maximum bending stress σ in the boards? *Hint:* The numerically largest bending moment occurs at the fixed support.

PROBLEM 13.6-47

13.6-48 Two identical, simply supported beams AB and CD are placed so that they cross each other at their midpoints (see figure). Before the uniform load is applied, the beams just touch each other at the crossing point.

Determine the maximum bending moments $(M_{AB})_{max}$ and $(M_{CD})_{max}$ in beams AB and CD, respectively, due to the uniform load if the intensity of the load is $q = 6.4$ kN/m and the length of each beam is $L = 4$ m.

PROBLEM 13.6-48

13.6-49 The cantilever beam AB shown in the figure is an S6×12.5 steel I-beam with $E = 30 \times 10^6$ psi. The simple beam DE is a wood beam 4 in. $\times$ 12 in. (nominal dimensions) in cross section with $E = 1.5 \times 10^6$ psi. A steel rod AC of diameter 0.25 in., length 10 ft, and $E = 30 \times 10^6$ psi serves as a hanger joining the two beams. The hanger fits snugly between the beams before the uniform load is applied to beam DE.

Determine the tensile force F in the hanger and the maximum bending moments M_{AB} and M_{DE} in the two beams due to the uniform load, which has an intensity $q = 400$ lb/ft. *Hint:* To aid in obtaining the maximum bending moment in beam DE, draw the shear-force and bending-moment diagrams.

PROBLEM 13.6-49

13.6-50 The beam AB shown in the figure is simply supported at A and B and supported on a spring of stiffness k at its midpoint C. The beam has flexural rigidity EI and length $2L$.

What should be the stiffness k of the spring in order that the maximum bending moment in the beam (due to the uniform load) will have the smallest possible value?

PROBLEM 13.6-51

PROBLEM 13.6-50

13.6-51 The continuous frame ABC has a fixed support at A, a roller support at C, and a rigid corner connection at B (see figure). Members AB and BC each have length L and flexural rigidity EI. A horizontal force P acts at mid-height of member AB.

(a) Find all reactions of the frame.

(b) What is the largest bending moment $M_{\max}$ in the frame? *Note:* Disregard axial deformations in member AB and consider only the effects of bending.

13.6-52 The continuous frame ABC has a pinned support at A, a sliding support at C, and a rigid corner connection at B (see figure). Members AB and BC each have length L and flexural rigidity EI. A horizontal force P acts at mid-height of member AB.

(a) Find all reactions of the frame.

(b) What is the largest bending moment $M_{\max}$ in the frame? *Note:* Disregard axial deformations in members AB and BC and consider only the effects of bending.

Ship container handling cranes made up of two plane frames (Courtesy of the National Information Service for Earthquake Engineering EERC, University of California Berkeley)

PROBLEM 13.6-52

13.6-53 A wide-flange beam *ABC* rests on three identical spring supports at points *A*, *B*, and *C* (see figure). The flexural rigidity of the beam is $EI = 6912 \times 10^6$ lb-in^2, and each spring has stiffness $k = 62{,}500$ lb/in. The length of the beam is $L = 16$ ft.

If the load *P* is 6000 lb, what are the reactions R_A, R_B, and R_C? Also, draw the shear-force and bending-moment diagrams for the beam, labeling all critical ordinates.

PROBLEM 13.6-53

13.6-54 A fixed-end beam *AB* of a length *L* is subjected to a uniform load of intensity *q* acting over the middle region of the beam (see figure).

(a) Obtain a formula for the fixed-end moments M_A and M_B in terms of the load *q*, the length *L*, and the length *b* of the loaded part of the beam.

(b) Plot a graph of the fixed-end moment M_A versus the length *b* of the loaded part of the beam. For convenience, plot the graph in the following non-dimensional form:

$$\frac{M_A}{qL^2/12} \quad \text{versus} \quad \frac{b}{L}$$

with the ratio *b/L* varying between its extreme values of 0 and 1.

(c) For the special case in which $a = b = L/3$, draw the shear-force and bending-moment diagrams for the beam, labeling all critical ordinates.

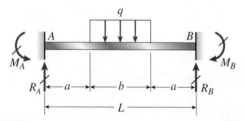

PROBLEM 13.6-54

13.6-55 A beam supporting a uniform load of intensity *q* throughout its length rests on pistons at points *A*, *C*, and *B* (see figure). The cylinders are filled with oil and are connected by a tube so that the oil pressure on each piston is the same. The pistons at *A* and *B* have diameter d_1, and the piston at *C* has diameter d_2.

(a) Determine the ratio of d_2 to d_1 so that the largest bending moment in the beam is as small as possible.

(b) Under these optimum conditions, what is the largest bending moment M_{max} in the beam?

(c) What is the difference in elevation between point *C* and the end supports?

PROBLEM 13.6-55

13.6-56 A thin steel beam *AB* used in conjunction with an electromagnet in a high-energy physics experiment is securely bolted to rigid supports (see figure). A magnetic field produced by coils *C* results in a force acting on the beam. The force is trapezoidally distributed with maximum intensity $q_0 = 18$ kN/m. The length of the beam between supports is $L = 200$ mm, and the dimension *c* of the trapezoidal load is 50 mm. The beam has a rectangular cross section with width $b = 60$ and height $h = 20$ mm.

Determine the maximum bending stress σ_{max} and the maximum deflection δ_{max} for the beam. (Disregard any effects of axial deformations and consider only the effects of bending. Use $E = 200$ GPa.)

PROBLEM 13.6-56

13.6-57 Find an expression for required moment M_A (in terms of q and L) that will result in rotation $\theta_A = 0$ due to M_A and q loadings applied at the same time.

PROBLEM 13.6-57

13.6-58 Repeat Problem 13.6-57 for the loading shown in the figure.

PROBLEM 13.6-58

13.6-59 A propped cantilever beam is loaded by two different load patterns (see figures a and b). Assume that EI is constant and the total beam length is L. Find expressions for reactions at A and B for each beam. Plot shear and moment diagrams. Assume that $a = L/3$.

PROBLEM 13.6-59

Critical load-carrying elements in structures such as columns and other slender compression members are susceptible to buckling failure.

CHAPTER OBJECTIVES

- Study basic concepts of elastic buckling of slender columns.

- Derive formulas for the *Euler buckling load* of slender columns having a variety of different support conditions.

- Study the buckled *mode shapes* and compute the *effective lengths* of columns with different support conditions.

- Study the *secant formula*, which gives maximum compressive stresses in slender columns acted on by eccentric axial compressive loads.

Chapter Outline

(a)   (b)

14.1 Introduction

Load-carrying structures may fail in a variety of ways, depending upon the type of structure, the conditions of support, the kinds of loads, and the materials used. For instance, an axle in a vehicle may fracture suddenly from repeated cycles of loading, or a beam may deflect excessively, so that the structure is unable to perform its intended functions. These kinds of failures are prevented by designing structures so that the maximum stresses and maximum displacements remain within tolerable limits. Thus, **strength** and **stiffness** are important factors in design, as discussed throughout the preceding chapters.

Another type of failure is **buckling**, which is the subject matter of this chapter. The focus here is the buckling of **columns**, which are long, slender structural members loaded axially in compression (Fig. 14-1a). If a compression member is relatively slender, it may deflect laterally and fail by bending (Fig. 14-1b) rather than failing by direct compression of the material. You can demonstrate this behavior by compressing a plastic ruler or other slender object. When lateral bending occurs, the column has *buckled*. Under an increasing axial load, the lateral deflections will increase too, and eventually the column will collapse completely.

The phenomenon of buckling is not limited to columns. Buckling can occur in many kinds of structures and can take many forms. When you step on the top of an empty aluminum can, the thin cylindrical walls buckle under your weight and the can collapses. When a large bridge collapsed a few years ago, investigators found that failure was caused by the buckling of a thin steel plate that wrinkled under compressive stresses. Buckling is one of the major causes of failures in structures; therefore, the possibility of buckling should always be considered in design.

14.2 Buckling and Stability

To illustrate the fundamental concepts of buckling and stability, consider the **idealized structure**, or **buckling model**, shown in Fig. 14-2a. This hypothetical structure consists of two rigid bars AB and BC, each of a length $L/2$. They are joined at B by a pin connection and held in a vertical position by a rotational spring having a stiffness β_R.[1]

This idealized structure is analogous to the column of Fig. 14-1a, because both structures have simple supports at the ends and are compressed by an axial load P. However, the elasticity of the idealized structure is "concentrated" in the rotational spring, whereas a real column can bend throughout its length (Fig. 14-1b).

In the idealized structure, the two bars are perfectly aligned and the axial load P has its line of action along the longitudinal axis (Fig. 14-2a). Consequently, the spring is initially unstressed, and the bars are in direct compression.

Now suppose that the structure is disturbed by some external force that causes point B to move a small distance laterally (Fig. 14-2b). The rigid bars

---

[1]The general relationship for a rotational spring is $M = \beta_R\theta$, where M is the moment acting on the spring, β_R is the rotational stiffness of the spring, and θ is the angle through which the spring rotates. Thus, rotational stiffness has units of moment divided by angle, such as lb-in/rad or N · m/rad. The analogous relationship for a translational spring is $F = \beta\delta$, where F is the force acting on the spring, β is the translational stiffness of the spring (or spring constant), and δ is the change in length of the spring. Thus, translational stiffness has units of force divided by length, such as lb/in. or N/m.

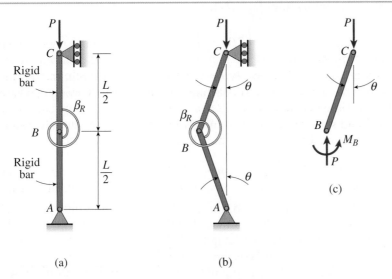

(a) (b) (c)

rotate through small angles θ, and a moment develops in the spring. The direction of this moment is such that it tends to return the structure to its original straight position; therefore, it is called a **restoring moment**. At the same time, however, the tendency of the axial compressive force is to increase the lateral displacement. Thus, these two actions have opposite effects—the restoring moment tends to *decrease* the displacement and the axial force tends to *increase* it.

Now consider what happens when the disturbing force is removed. If the axial force P is relatively small, the action of the restoring moment will predominate over the action of the axial force, and the structure will return to its initial straight position. Under these conditions, the structure is said to be **stable**. However, if the axial force P is large, the lateral displacement of point B will increase, and the bars will rotate through larger and larger angles until the structure collapses. Under these conditions, the structure is **unstable** and fails by lateral buckling.

Critical Load

The transition between the stable and unstable conditions occurs at a special value of the axial force known as the **critical load** (denoted by the symbol P_{cr}). The critical load of this buckling model can be determined by considering the structure in the disturbed position (Fig. 14-2b) and investigating its equilibrium.

First, consider the entire structure as a free body and sum moments about support A. This step leads to the conclusion that there is no horizontal reaction at support C. Second, consider bar BC as a free body (Fig. 14-2c) and note that it is subjected to the action of the axial forces P and the moment M_B in the spring. The moment M_B is equal to the rotational stiffness β_R times the angle of rotation 2θ of the spring; thus,

$$M_B = 2\beta_R\theta \qquad \textbf{(14-1a)}$$

Since the angle θ is a small quantity, the lateral displacement of point B is $\theta L/2$. Therefore, the equation of equilibrium is obtained by summing moments about point B for bar BC (Fig. 14-2c):

$$M_B - P\left(\frac{\theta L}{2}\right) = 0 \qquad \textbf{(14-1b)}$$

or, upon substituting from Eq. (14-1a),

$$\left(2\beta_R - \frac{PL}{2}\right)\theta = 0 \qquad (14\text{-}2)$$

One solution of this equation is $\theta = 0$, which is a trivial solution and merely means that the structure is in equilibrium when it is perfectly straight, regardless of the magnitude of the force P.

A second solution is obtained by setting the term in parentheses equal to zero and solving for the load P, which is the *critical load*:

$$P_{cr} = \frac{4\beta_R}{L} \qquad (14\text{-}3)$$

At the critical value of the load, the structure is in equilibrium regardless of the magnitude of the angle θ [provided the angle remains small, because that assumption was made when deriving Eq. (14-1b)].

From the preceding analysis, note that the critical load is the *only* load for which the structure will be in equilibrium in the disturbed position. At this value of the load, the restoring effect of the moment in the spring just matches the buckling effect of the axial load. Therefore, the critical load represents the boundary between stable and unstable conditions.

If the axial load is less than P_{cr}, the effect of the moment in the spring predominates, and the structure returns to the vertical position after a slight disturbance; if the axial load is larger than P_{cr}, the effect of the axial force predominates and the structure buckles:

If $P < P_{cr}$, the structure is *stable*.

If $P > P_{cr}$, the structure is *unstable*.

Equation (14-3) shows that the stability of the structure is increased either by *increasing its stiffness* or by *decreasing its length*. Later in this chapter, these same observations apply when critical loads are determined for various types of columns.

Summary

In summary, the idealized structure (Fig. 14-2a) behaves as follows as the axial load P increases from zero to a large value.

When the axial load is less than the critical load ($0 < P < P_{cr}$), the structure is in equilibrium when it is perfectly straight. Because the equilibrium is **stable**, the structure returns to its initial position after being disturbed. Thus, the structure is in equilibrium *only* when it is perfectly straight ($\theta = 0$).

When the axial load is greater than the critical load ($P > P_{cr}$), the structure is still in equilibrium when $\theta = 0$ (because it is in direct compression and there is no moment in the spring), but the equilibrium is **unstable** and cannot be maintained. The slightest disturbance will cause the structure to buckle.

At the critical load ($P = P_{cr}$), the structure is in equilibrium even when point B is displaced laterally by a small amount. In other words, the structure is in equilibrium for *any* small angle θ, including $\theta = 0$. However, the structure is neither stable nor unstable—it is at the boundary between stability and instability. This condition is referred to as **neutral equilibrium**.

The three equilibrium conditions for the idealized structure are shown in the graph of axial load P versus angle of rotation θ (Fig. 14-3). The two heavy lines—one vertical and one horizontal—represent the equilibrium conditions. Point B, where the equilibrium diagram branches, is called a *bifurcation point*.

The horizontal line for neutral equilibrium extends to the left and right of the vertical axis because the angle θ may be clockwise or counterclockwise. The line extends only a short distance, however, because this analysis is based upon the assumption that θ is a small angle. (This assumption is quite valid, because θ is indeed small when the structure first departs from its vertical position. If buckling continues and θ becomes large, the line labeled "Neutral equilibrium" curves upward, as shown later in Fig. 14-12.)

The three equilibrium conditions represented by the diagram of Fig. 14-3 are analogous to those of a ball placed upon a smooth surface (Fig. 14-4). If the surface is concave upward, like the inside of a dish, the equilibrium is stable, and the ball always returns to the low point when disturbed. If the surface is convex upward, like a dome, the ball theoretically can be in equilibrium on top of the surface, but the equilibrium is unstable and in reality the ball rolls away. If the surface is perfectly flat, the ball is in neutral equilibrium and remains wherever it is placed.

In the next section, the behavior of an ideal elastic column is shown to be analogous to that of the buckling model shown in Fig. 14-2. Furthermore, many other kinds of structural and mechanical systems fit this model.

FIGURE 14-3

Equilibrium diagram for buckling of an idealized structure

FIGURE 14-4

Ball in stable, unstable, and neutral equilibrium

Example 14-1

Two idealized columns are shown in Fig. 14-5. Both columns are initially straight and vertical with load P applied at joint A.

(a) The first column (Column 1, Fig. 14-5a) consists of a rigid bar ABC with a roller support at B and is connected to bar CD by an elastic connection with rotational stiffness $\beta_R = \beta L^2$.

(b) The second column (Column 2, Fig. 14-5b) is a single rigid bar $ABCD$ that has a roller support at B and a spring support at D.

Column 1 has a sliding support at D that cannot rotate, while Column 2 has a roller support at D with a translational spring with stiffness β. Find an expression for the critical load P_{cr} for each column.

Solution:

Use the four-step problem-solving approach. Combine steps as needed for an efficient solution.

Part (a): Column 1.

1. *Conceptualize* [*hypothesize, sketch*]: Begin by considering the equilibrium of Column 1 in a displaced position caused by some external load P applied at A causing *small* lateral displacement Δ_A (Fig. 14-5a). Sum lateral forces for

FIGURE 14-5

Example 14-1: Two idealized columns made up of rigid bars: (a) Column 1 with rotational elastic connection at *C* and, (b) Column 2 supported by a translational elastic support at *D*, (c) free-body diagram of bar *ABC* in Column 1, (d) free-body diagram of Column 2

the entire column to find that reaction $H_B = 0$. Next, draw the free-body diagram of bar *ABC* (Fig. 14-5c). Note that $\Delta_C = \Delta_A$ and that bar *ABC* rotation is $\Delta_A/(L/2) = 2\Delta_A/L$.

2. *Categorize* [*simplify, classify*]: The rotation of bar *ABC* can be expressed as $2\Delta_A/L$ because the rotation is small, so the elastic connection resisting moment at *C* is equal to rotational stiffness β_R times the total relative rotation of the spring or $M_C = (\beta_R)(2\Delta_A/L)$.

3. *Analyze* [*evaluate; select relevant equations; carry out mathematical solution*]: Sum moments about *C* in Fig. 14-5c and solve for *P* to obtain the critical load P_{cr1} for Column 1:

$$P_{cr1} = \frac{M_C}{2\Delta_A} = \frac{\beta_R\left(\dfrac{2\Delta_A}{L}\right)}{2\Delta_A} = \frac{\beta_R}{L} = \beta L \qquad \text{(a)}$$

4. *Finalize* [*conclude; examine answer—Does it make sense? Are units correct? How does it compare to similar problem solutions?*]: The buckled *mode shape* for Column 1 is the displaced position shown in Fig. 14-5a.

Part (b): Column 2.

1, 2. *Conceptualize, Categorize*: Investigate the equilibrium of Column 2 in a *displaced* position, once again defined by lateral displacement Δ_A, as shown in Fig. 14-5b. Use a free-body diagram of the entire column (Fig. 14-5d) and sum lateral forces to find that

$$H_B = -(\beta)(3\Delta_A) \qquad \text{(b)}$$

3. *Analyze*: Sum moments about D in Fig. 14-5d to find the critical load:

$$P_{cr2} = \frac{-H_B\left(\dfrac{3L}{2}\right)}{4\Delta_A} = \frac{9}{8}\beta L \qquad \qquad \longleftarrow \text{(c)}$$

4. *Finalize*: The buckled *mode shape* for Column 2 is the displaced position shown in Fig. 14-5b.

Combined model and analysis.

1. *Conceptualize*: Now create a more advanced structure model by combining the features of Column 1 and Column 2 into a single column as shown in Fig. 14-5e. This idealized structure has elastic rotational springs at both C and D with rotational stiffnesses β_{R1} and β_{R2}, respectively. The roller support remains at B, and the sliding support at D is now restrained by an elastic translational spring with stiffness β. Small lateral displacements Δ_C and Δ_D are selected as the degrees of freedom that define the possible displaced positions of the column. (Alternatively, rotation angles such as θ_B and θ_D could be used as degrees of freedom to uniquely describe any arbitrary position of the displaced structure.) Hence, the *combined* structure has *two degrees of freedom*; therefore, it has two possible buckled mode shapes and two different critical loads—each of which causes the associated buckling mode. In contrast, Columns 1 and 2 are *single degree-of-freedom* structures because only Δ_A is needed to define the buckled shape of each structure depicted in Figs. 14-5a and b.

2. *Categorize*: Observe that if rotational spring β_{R2} becomes infinitely stiff and translational spring stiffness $\beta = 0$ in the combined structure (Fig. 14-5e) (while β_{R1} remains finite), the two degree of freedom (2DOF) combined model reduces to the single degree-of-freedom (SDOF) model of Fig. 14-5a. Similarly, if rotational spring stiffness β_{R1} goes to infinity and spring $\beta_{R2} = 0$ in Fig. 14-5e (while translational spring stiffness β remains finite), the model becomes that shown in Fig. 14-5b. It follows *that the solutions for P_{cr} for Columns 1 and 2 [Eqs. (a) and (c)] are simply two special case solutions of the general combined model in* Fig. 14-5e.

3. *Analyze*: The goal now is to find a general solution for the 2DOF model in Fig. 14-5e and then to show that solutions for P_{cr} for Columns 1 and 2 can be obtained from this general solution.

   First, consider the equilibrium of the entire 2DOF model in the displaced position shown in Fig. 14-5f. Sum horizontal forces to find that $H_B = \beta\Delta_D$. Rotation angles θ_B and θ_D in Fig. 14-5f can be expressed in terms of translations Δ_C and Δ_D as

$$\theta_D = \frac{\Delta_D - \Delta_C}{L} \qquad \theta_B = \frac{2}{L}\Delta_C \qquad \qquad \text{(d)}$$

FIGURE 14-5 (*continued*)

(e) combined model of Columns 1 and 2, (f) free-body diagram of combined model, and (g) free-body diagram of bar *ABC* in combined model

Two degree-of-freedom model

Elastic springs at *D*

(e)   (f)   (g)

Sum moments about *B* in free-body diagram *ABC* (Fig. 14-5g), noting that the moment at *C* is equal to rotational spring stiffness β_{R1} times the *relative* rotation $(\theta_D - \theta_B)$ at *C*, to get

$$2P\Delta_C - \beta\Delta_D \frac{L}{2} + \beta_{R1}(\theta_D - \theta_B) = 0$$

Substituting expressions for θ_D and θ_B from Eqs. (d) produces the following equation in terms of unknown displacements Δ_C and Δ_D:

$$\left(2P - \frac{3\beta_{R1}}{L}\right)\Delta_C + \left(\frac{\beta_{R1}}{L} - \frac{L\beta}{2}\right)\Delta_D = 0 \qquad \textbf{(e)}$$

Obtain a second equation that describes the equilibrium of the displaced structure from the free-body diagram of bar *CD* alone. Summing moments about *C* for bar *CD* alone gives

$$P(\Delta_D - \Delta_C) - \beta_{R1}(\theta_D - \theta_B) - \beta_{R2}\theta_D - \beta\Delta_D L = 0$$

which can be rewritten as

$$\left(\frac{3\beta_{R1}}{L} - P + \frac{\beta_{R2}}{L}\right)\Delta_C + \left(P - \frac{\beta_{R1}}{L} - \frac{\beta_{R2}}{L} - L\beta\right)\Delta_D = 0 \qquad \textbf{(f)}$$

There are now two algebraic equations [Eqs. (e) and (f)] and two unknowns (Δ_C and Δ_D). Expressing Eqs. (e) and (f) in matrix form gives

FIGURE 14-5
(continued)

(h) buckled mode
shapes for combined
model

Mode 1

Mode 2

(h)

$$
\begin{pmatrix}
2P - \dfrac{3\beta_{R1}}{L} & \dfrac{\beta_{R1}}{L} - \dfrac{L\beta}{2} \\
\dfrac{3\beta_{R1}}{L} - P + \dfrac{\beta_{R2}}{L} & P - \dfrac{\beta_{R1}}{L} - \dfrac{\beta_{R2}}{L} - L\beta
\end{pmatrix}
\begin{pmatrix}
\Delta_C \\
\Delta_D
\end{pmatrix}
=
\begin{pmatrix}
0 \\
0
\end{pmatrix}
\tag{g}
$$

These homogeneous equations have a nonzero (nontrivial) solution only if the determinant of the coefficient matrix in Eq. (g) is equal to zero. If both elastic connections have the same rotational stiffness, $\beta_{R1} = \beta_{R2} = \beta L^2$, and Eq. (g) becomes

$$
\begin{pmatrix}
2P - 3L\beta & \dfrac{L\beta}{2} \\
4L\beta - P & P - 3L\beta
\end{pmatrix}
\begin{pmatrix}
\Delta_C \\
\Delta_D
\end{pmatrix}
=
\begin{pmatrix}
0 \\
0
\end{pmatrix}
\tag{h}
$$

So the determinant of the coefficient matrix in Eq. (h) (known as the characteristic equation) is

$$
P^2 - \left(\frac{17}{4}\beta L\right)P + \frac{7}{2}(\beta L)^2 = 0
\tag{i}
$$

Solving Eq. (i) using the quadratic formula results in two values of the critical load:

$$
P_{cr1} = \beta L\left(\frac{17 - \sqrt{65}}{8}\right) = 1.117\beta L
$$

$$
P_{cr2} = \beta L\left(\frac{17 + \sqrt{65}}{8}\right) = 3.13\beta L
$$

4. *Finalize*: These are the *eigenvalues* of the combined 2DOF system when elastic connection stiffnesses are defined as $\beta_{R1} = \beta_{R2} = \beta L^2$. Usually the lower value of the critical load is of more interest because the structure will buckle first at this lower load value. Substitute P_{cr1} and P_{cr2} back into Eq. (e) or (f) to find the buckled mode shape (*eigenvector*) associated with each critical load. The resulting eigenvectors are given here and are shown in Fig. 14-5h:

$$
\Phi_1 = \begin{pmatrix} 0.653 \\ 1 \end{pmatrix} \quad \Phi_2 = \begin{pmatrix} -0.153 \\ 1 \end{pmatrix}
$$

Application of combined model to Columns 1 and 2.

If the rotational spring stiffness β_{R2} goes to infinity and translational spring stiffness β goes to zero while rotational stiffness β_{R1} remains finite, the combined model (Fig. 14-5e) reduces to Column 1, and the critical load obtained from the solution of Eq. (g) is that given in Eq. (a). [Alternatively, equating Δ_C and Δ_D and setting $\beta = 0$ in Eq. (e) confirms P_{cr1} in Eq. (a)]. If, instead, rotational spring stiffness β_{R1} goes to infinity and rotational stiffness β_{R2} goes to zero while translational stiffness β remains finite, the solution of Eq. (g) gives the critical load for Column 2 in Eq. (c).

14.3 Columns with Pinned Ends

The investigation of the stability behavior of columns begins by analyzing a slender column with pinned ends (Fig. 14-6a). The column is loaded by a vertical force P that is applied through the centroid of the end cross section. The column itself is perfectly straight and is made of a linearly elastic material that follows Hooke's law. Since the column is assumed to have no imperfections, it is referred to as an **ideal column**.

For purposes of analysis, construct a coordinate system with its origin at support A and with the x axis along the longitudinal axis of the column. The y axis is directed to the left in the figure, and the z axis (not shown) comes out of the plane of the figure toward the viewer. Assume that the x-y plane is a plane of symmetry of the column and that any bending takes place in that plane (Fig. 14-6b). The coordinate system is identical to the one used in our previous discussions of beams, as can be seen by rotating the column clockwise through an angle of 90°.

When the axial load P has a small value, the column remains perfectly straight and undergoes direct axial compression. The only stresses are the uniform compressive stresses obtained from the equation $\sigma = P/A$. The column is in **stable equilibrium**, which means that it returns to the straight position after a disturbance. For instance, if a small lateral load is applied and the column bends, the deflection will disappear and the column will return to its original position when the lateral load is removed.

As the axial load P is gradually increased, a condition of **neutral equilibrium** is reached in which the column may have a bent shape. The corresponding value of the load is the **critical load** P_{cr}. At this load, the column may undergo small lateral deflections with no change in the axial force. For instance, a small lateral load will produce a bent shape that does not disappear when the lateral load is removed. Thus, the critical load can maintain the column in equilibrium *either* in the straight position or in a slightly bent position.

At higher values of the load, the column is **unstable** and may collapse by buckling, that is, by excessive bending. For the ideal case, the column will be

FIGURE 14-6

Column with pinned ends: (a) ideal column, (b) buckled shape, and (c) axial force P and bending moment M acting at a cross section

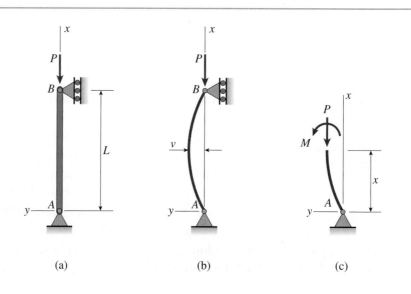

(a)   (b)   (c)

in equilibrium in the straight position even when the axial force P is greater than the critical load. However, since the equilibrium is unstable, the smallest imaginable disturbance will cause the column to deflect sideways. Once that happens, the deflections will immediately increase, and the column will fail by buckling. The behavior is similar to that described in the preceding section for the idealized buckling model (Fig. 14-2).

The behavior of an ideal column compressed by an axial load P (Figs. 14-6a and b) may be summarized as follows.

If $P < P_{cr}$, the column is in stable equilibrium in the straight position.

If $P = P_{cr}$, the column is in neutral equilibrium in either the straight or a slightly bent position.

If $P > P_{cr}$, the column is in unstable equilibrium in the straight position and will buckle under the slightest disturbance.

Of course, a real column does not behave in this idealized manner because imperfections are always present. For instance, the column is not *perfectly* straight, and the load is not *exactly* at the centroid. Nevertheless, this discussion begins by studying ideal columns because they provide insight into the behavior of real columns.

Differential Equation for Column Buckling

To determine the critical loads and corresponding deflected shapes for an ideal pin-ended column (Fig. 14-6a), use one of the differential equations of the deflection curve of a beam [see Eqs. (13-16a, b, and c) in Section 13.2]. These equations are applicable to a buckled column because the column bends as though it were a beam (Fig. 14-6b).

Although both the fourth-order differential equation (the load equation) and the third-order differential equation (the shear-force equation) are suitable for analyzing columns, use the second-order equation (the bending-moment equation) because its general solution is usually the simplest. The **bending-moment equation** (Eq. 13-16a) is

$$EIv'' = M \qquad \textbf{(14-4)}$$

where M is the bending moment at any cross section, v is the lateral deflection in the y direction, and EI is the flexural rigidity for bending in the x-y plane.

The bending moment M at distance x from end A of the buckled column is shown acting in its positive direction in Fig. 14-6c. Note that the bending-moment sign convention is the same as that used in earlier chapters, namely, positive bending moment produces positive curvature (see Figs. 13-3 and 13-4).

The axial force P acting at the cross section is also shown in Fig. 14-6c. Since there are no horizontal forces acting at the supports, there are no shear forces in the column. Therefore, equilibrium of moments about point A gives

$$M + Pv = 0 \quad \text{or} \quad M = -Pv \qquad \textbf{(14-5)}$$

where v is the deflection at the cross section.

This same expression for the bending moment is obtained if it is assumed that the column buckles to the right instead of to the left (Fig. 14-7a). When

FIGURE 14-7

Column with pinned ends
(alternative direction of buckling)

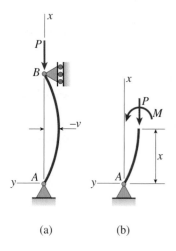

(a)   (b)

the column deflects to the right, the deflection itself is $-v$, but the moment of the axial force about point A also changes sign. Thus, the equilibrium equation for moments about point A (see Fig. 14-7b) is

$$M - P(-v) = 0$$

which gives the same expression for the bending moment M as before.

The **differential equation of the deflection curve** (Eq. 14-4) now becomes

$$EIv'' + Pv = 0 \qquad \text{(14-6)}$$

By solving this equation, which is a *homogeneous, linear, differential equation of second order with constant coefficients*, you can determine the magnitude of the critical load and the deflected shape of the buckled column.

Note that the buckling of columns involves solving the same basic differential equation as the one solved in Chapter 13 when finding beam deflections. However, there is a fundamental difference in the two types of analysis. In the case of beam deflections, the bending moment M appearing in Eq. (14-4) is a function of the loads only—it does not depend upon the deflections of the beam. In the case of buckling, the bending moment is a function of the deflections themselves (Eq. 14-5).

Thus, a new aspect of bending analysis is revealed. In the previous work, the deflected shape of the structure was not considered, and the equations of equilibrium were based upon the geometry of the *undeformed* structure. Now, however, the geometry of the *deformed* structure is taken into account when writing equations of equilibrium.

Solution of the Differential Equation

For convenience in writing the solution of the differential equation of Eq. (14-6), use the notation

$$k^2 = \frac{P}{EI} \qquad \text{or} \qquad k = \sqrt{\frac{P}{EI}} \qquad \text{(14-7a,b)}$$

where k is always taken as a positive quantity. Note that k has units of the reciprocal of length; therefore, quantities such as kx and kL are nondimensional.

Using this notation, rewrite Eq. (14-6) in the form

$$v'' + k^2v = 0 \qquad \text{(14-8)}$$

From mathematics, you know that the **general solution** of this equation is

$$v = C_1 \sin kx + C_2 \cos kx \qquad \text{(14-9)}$$

in which C_1 and C_2 are constants of integration (to be evaluated from the boundary conditions, or end conditions, of the column). Note that the number of arbitrary constants in the solution (two in this case) agrees with the order of the differential equation. Also, note that the solution can be verified by substituting the expression for v [Eq. (14-9)] into the differential equation [Eq. (14-8)] and reducing it to an identity.

To evaluate the **constants of integration** appearing in the solution [Eq. (14-9)], use the boundary conditions at the ends of the column; namely, the deflection is zero when $x = 0$ and $x = L$ (see Fig. 14-6b):

$$v(0) = 0 \quad \text{and} \quad v(L) = 0 \qquad \text{(14-10a,b)}$$

The first condition gives $C_2 = 0$, so

$$v = C_1 \sin kx \qquad \text{(14-10c)}$$

The second condition gives

$$C_1 \sin kL = 0 \qquad \text{(14-10d)}$$

From this equation, either $C_1 = 0$ or $\sin kL = 0$. Both of these possibilities are now considered.

Case 1. If the constant C_1 equals zero, the deflection v is also zero [see Eq. (14-10c)]; therefore, the column remains straight. In addition, note that when C_1 equals zero, Eq. (14-10d) is satisfied for *any* value of the quantity kL. Consequently, the axial load P also may have any value [see Eq. (14-7b)]. This solution of the differential equation (known in mathematics as the *trivial solution*) is represented by the vertical axis of the load–deflection diagram (Fig. 14-8). It gives the behavior of an ideal column that is in equilibrium (either stable or unstable) in the straight position (no deflection) under the action of the compressive load P.

Case 2. The second possibility for satisfying Eq. (14-10d) is given by the equation known as the **buckling equation**:

$$\sin kL = 0 \qquad \text{(14-11)}$$

This equation is satisfied when $kL = 0, \pi, 2\pi, \ldots$. However, since $kL = 0$ means that $P = 0$, this solution is not of interest. Therefore, the solutions considered here are

$$kL = n\pi \qquad n = 1, 2, 3, \ldots \qquad \text{(14-12)}$$

or [see Eq. (14-7a)]:

$$P = \frac{n^2 \pi^2 EI}{L^2} \qquad n = 1, 2, 3, \ldots \qquad \text{(14-13)}$$

This formula gives the values of P that satisfy the buckling equation and provide solutions (other than the trivial solution) to the differential equation.

The equation of the **deflection curve** from Eqs. (14-10c) and (14-12) is

$$v = C_1 \sin kx = C_1 \sin \frac{n\pi x}{L} \qquad n = 1, 2, 3, \ldots \qquad \text{(14-14)}$$

Only when P has one of the values given by Eq. (14-13) is it theoretically possible for the column to have a bent shape [given by Eq. (14-14)]. For all other values of P, the column is in equilibrium only if it remains straight. Therefore, the values of P given by Eq. (14-13) are the **critical loads** for this column.

Critical Loads

The lowest critical load for a column with pinned ends (Fig. 14-9a) is obtained when $n = 1$:

$$P_{cr} = \frac{\pi^2 EI}{L^2} \qquad \text{(14-15)}$$

The corresponding buckled shape (sometimes called a *mode shape*—see Example 14-1) is

$$v = C_1 \sin \frac{\pi x}{L} \qquad \text{(14-16)}$$

FIGURE 14-8

Load–deflection diagram for an ideal, linearly elastic column

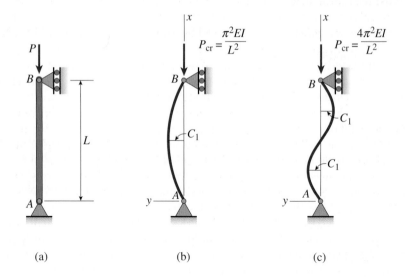

(a)　　　　　　　　　　(b)　　　　　　　　　　(c)

as shown in Fig. 14-9b. The constant C_1 represents the deflection at the midpoint of the column and may have any small value, either positive or negative. Therefore, the part of the load–deflection diagram corresponding to P_{cr} is a horizontal straight line (Fig. 14-8). Thus, the deflection at the critical load is *undefined*, although it must remain small for our equations to be valid. Above the bifurcation point B, the equilibrium is unstable, and below point B it is stable.

Buckling of a pinned-end column in the first mode is called the **fundamental case** of column buckling.

The type of buckling described in this section is called **Euler buckling**, and the critical load for an ideal elastic column is often called the **Euler load**. The famous mathematician Leonhard Euler (1707–1783), generally recognized as the greatest mathematician of all time, was the first person to investigate the buckling of a slender column and determine its critical load (Euler published his results in 1744).

Taking higher values of the index n in Eqs. (14-13) and (14-14) leads to an infinite number of critical loads and corresponding mode shapes. The mode shape for $n = 2$ has two half-waves, as pictured in Fig. 14-9c. The corresponding critical load is four times larger than the critical load for the fundamental case. The magnitudes of the critical loads are proportional to the square of n, and the number of half-waves in the buckled shape is equal to n.

Buckled shapes for the **higher modes** are often of no practical interest because the column buckles when the axial load P reaches its lowest critical value. The only way to obtain modes of buckling higher than the first is to provide lateral support of the column at intermediate points, such as at the midpoint of the column shown in Fig. 14-9 (see Example 14-2 at the end of this section).

General Comments

From Eq. (14-15), note that the critical load of a column is proportional to the flexural rigidity EI and inversely proportional to the square of the length. Of particular interest is the fact that the *strength* of the material itself, as represented by a quantity such as the proportional limit or the yield stress, does not appear in the equation for the critical load. Therefore, increasing a strength property does not raise the critical load of a slender column. It only can be raised by increasing the flexural rigidity, reducing the length, or providing additional lateral support.

The *flexural rigidity* can be increased by using a "stiffer" material (that is, a material with larger modulus of elasticity E) or by distributing the material in such a way as to increase the moment of inertia I of the cross section—just as a beam can be made stiffer by increasing the moment of inertia. The moment of inertia is increased by distributing the material farther from the centroid of the cross section. Hence, a hollow tubular member is generally more economical for use as a column than a solid member having the same cross-sectional area.

Reducing the *wall thickness* of a tubular member and increasing its lateral dimensions (while keeping the cross-sectional area constant) also increases the critical load because the moment of inertia is increased. This process has a practical limit, however, because eventually the wall itself will become unstable. When that happens, localized buckling occurs in the form of small corrugations or wrinkles in the walls of the column. Thus, a distinction must be made between *overall buckling* of a column, which is discussed in this chapter, and *local buckling* of its parts. The latter requires more detailed investigations and is beyond the scope of this book.

The preceding analysis (see Fig. 14-9) assumed that the x-y plane was a plane of symmetry of the column and that buckling took place in that plane. The latter assumption will be met if the column has lateral supports perpendicular to the plane of the figure, so that the column is constrained to buckle in the x-y plane. If the column is supported only at its ends and is free to buckle in *any* direction, the bending will occur about the principal centroidal axis having the smaller moment of inertia.

For instance, consider the rectangular and wide-flange cross sections shown in Fig. 14-10. In each case, the moment of inertia I_1 is greater than the moment of inertia I_2; hence, the column will buckle in the 1–1 plane, and the smaller moment of inertia I_2 should be used in the formula for the critical load. If the cross section is square or circular, all centroidal axes have the same moment of inertia, and buckling may occur in any longitudinal plane.

Critical Stress

After finding the critical load for a column, calculate the corresponding **critical stress** by dividing the load by the cross-sectional area. For the fundamental case of buckling (Fig. 14-9b), the critical stress is

$$\sigma_{cr} = \frac{P_{cr}}{A} = \frac{\pi^2 EI}{AL^2} \qquad (14\text{-}17)$$

in which I is the moment of inertia for the principal axis about which buckling occurs. This equation can be written in a more useful form by introducing the notation

$$r = \sqrt{\frac{I}{A}} \qquad (14\text{-}18)$$

where r is the **radius of gyration** of the cross section in the plane of bending.[2] Then the equation for the critical stress becomes

$$\sigma_{cr} = \frac{\pi^2 E}{(L/r)^2} \qquad (14\text{-}19)$$

---

[2]Radius of gyration is described in Section 5.6.

FIGURE 14-10

Cross sections of columns showing principal centroidal axes with $I_1 > I_2$

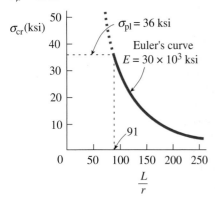

in which L/r is a nondimensional ratio called the **slenderness ratio**:

$$\text{Slenderness ratio} = \frac{L}{r} \qquad (14\text{-}20)$$

Note that the slenderness ratio depends only on the dimensions of the column. A column that is long and slender will have a high slenderness ratio and therefore a low critical stress. A column that is short and stubby will have a low slenderness ratio and will buckle at a high stress. Typical values of the slenderness ratio for actual columns are between 30 and 150.

The critical stress is the average compressive stress on the cross section at the instant the load reaches its critical value. You can plot a graph of this stress as a function of the slenderness ratio and obtain a curve known as **Euler's curve** (Fig. 14-11). The curve shown in the figure is plotted for a structural steel with $E = 30 \times 10^3$ ksi. The curve is valid only when the critical stress is less than the proportional limit of the steel, because the equations were derived using Hooke's law. Therefore, draw a horizontal line on the graph at the proportional limit of the steel (assumed to be 36 ksi) and terminate Euler's curve at that level of stress.[3]

Effects of Large Deflections, Imperfections, and Inelastic Behavior

The equations for critical loads were derived for ideal columns, that is, columns for which the loads are precisely applied, the construction is perfect, and the material follows Hooke's law. As a consequence, the magnitudes of the small deflections at buckling were undefined.[4] Thus, when $P = P_{cr}$, the column may have any small deflection, which is a condition represented by the horizontal line labeled A in the load–deflection diagram of Fig. 14-12. (This figure shows only the right-hand half of the diagram, but the two halves are symmetric about the vertical axis.)

The theory for ideal columns is limited to small deflections because the second derivative v'' for the curvature was used. A more exact analysis, based upon the exact expression for curvature [Eq. (13-19) in Section 13.2], shows that there is no indefiniteness in the magnitudes of the deflections at buckling. Instead, for an ideal, linearly elastic column, the load–deflection diagram goes upward in accord with curve B of Fig. 14-12. Thus, after a linearly elastic column begins to buckle, an increasing load is required to cause an increase in the deflections.

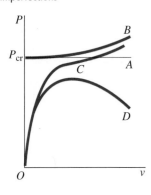

Now suppose that the column is not constructed perfectly; for instance, the column might have an imperfection in the form of a small initial curvature, so that the unloaded column is not perfectly straight. Such imperfections produce deflections from the onset of loading, as shown by curve C in Fig. 14-12. For small deflections, curve C approaches line A as an asymptote. However, as the deflections become large, it approaches curve B. The larger the imperfections,

---

[3]Euler's curve is not a common geometric shape. It is sometimes mistakenly called a hyperbola, but hyperbolas are plots of polynomial equations of the second degree in two variables, whereas Euler's curve is a plot of an equation of the third degree in two variables.

[4]In mathematical terminology, you solved a *linear eigenvalue problem* (see Example 14-1). The critical load is an *eigenvalue*, and the corresponding buckled mode shape is an *eigenfunction*.

the further curve *C* moves to the right, away from the vertical line. Conversely, if the column is constructed with considerable accuracy, curve *C* approaches the vertical axis and the horizontal line labeled *A*. By comparing lines *A, B,* and *C*, note that for practical purposes the critical load represents the maximum load-carrying capacity of an elastic column, because large deflections are not acceptable in most applications.

Finally, consider what happens when the stresses exceed the proportional limit and the material no longer follows Hooke's law. Of course, the load–deflection diagram is unchanged up to the level of load when the proportional limit is reached. Then the curve for inelastic behavior (curve *D*) departs from the elastic curve, continues upward, reaches a maximum, and turns downward.

The precise shapes of the curves in Fig. 14-12 depend upon the material properties and column dimensions, but the general nature of the behavior is typified by the curves shown.

Only extremely slender columns remain elastic up to the critical load. Stockier columns behave inelastically and follow a curve such as *D*. Thus, the maximum load that can be supported by an inelastic column may be considerably less than the Euler load for that same column. Furthermore, the descending part of curve *D* represents sudden and catastrophic collapse because it takes smaller and smaller loads to maintain larger and larger deflections. By contrast, the curves for elastic columns are quite stable because they continue upward as the deflections increase; therefore, it takes larger and larger loads to cause an increase in deflection.

Optimum Shapes of Columns

Compression members usually have the same cross sections throughout their lengths, so only prismatic columns are analyzed in this chapter. However, prismatic columns are not the optimum shape if minimum weight is desired. The critical load of a column consisting of a given amount of material may be increased by varying the shape so that the column has larger cross sections in those regions where the bending moments are larger. Consider, for instance, a column of solid circular cross section with pinned ends. A column shaped as shown in Fig. 14-13a will have a larger critical load than a prismatic column made from the same volume of material. As a means of approximating this optimum shape, prismatic columns are sometimes reinforced over part of their lengths (Fig. 14-13b).

Now consider a prismatic column with pinned ends that is free to buckle in *any* lateral direction (Fig. 14-14a). Also, assume that the column has a solid cross section, such as a circle, square, triangle, rectangle, or hexagon (Fig. 14-14b). An interesting question arises: For a given cross-sectional area, which of these shapes makes the most efficient column? Or, in more precise terms, which cross section gives the largest critical load? Of course, the assumption is that the critical load is calculated from the Euler formula $P_{cr} = \pi^2 EI/L^2$ using the smallest moment of inertia for the cross section.

While a common answer to this question is "the circular shape," you can readily demonstrate that a cross section in the shape of an equilateral triangle gives a 21% higher critical load than does a circular cross section of the same area (see Prob. 14.3-17). The critical load for an equilateral triangle is also higher than the loads obtained for the other shapes; hence, an equilateral triangle is the optimum cross section (based only upon theoretical considerations).

FIGURE 14-13

Nonprismatic columns

(a)   (b)

FIGURE 14-14

Which cross-sectional shape is the optimum shape for a prismatic column?

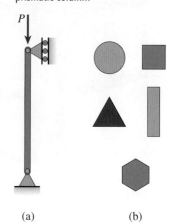

(a)   (b)

Example 14-2

FIGURE 14-15

Example 14-2: Euler buckling of a slender column

(a)   (b)

Section X–X

Slender steel column with lateral support near mid-height

A long, slender column *ABC* is pin-supported at the ends and compressed by an axial load *P* (Fig. 14-15). Lateral support is provided at the midpoint *B* in the plane of the figure. However, lateral support perpendicular to the plane of the figure is provided only at the ends.

The column is constructed of a steel I-beam section S 200 × 34 having modulus of elasticity $E = 200$ GPa and proportional limit $\sigma_{pl} = 300$ MPa. The total length of the column is $L = 8$ m.

Determine the allowable load P_{allow} using a factor of safety $n = 2.5$ with respect to Euler buckling of the column.

Solution:

Use the four-step problem-solving approach.

1. *Conceptualize*: Because of the manner in which it is supported, this column may buckle in either of the two principal planes of bending. As one possibility, it may buckle in the plane of the figure, in which case the distance between lateral supports is $L/2 = 4$ m and bending occurs about axis 2–2 (see Fig. 14-9c for the mode shape of buckling).

As a second possibility, the column may buckle perpendicular to the plane of the figure with bending about axis 1–1. Because the only lateral support in this direction is at the ends, the distance between lateral supports is $L = 8$ m (see Fig. 14-9b for the mode shape of buckling).

Column properties: From Table C-2(b), obtain the following moments of inertia and cross-sectional area for a S 200 × 34 column:

$$I_1 = 26.9 \times 10^6 \text{ mm}^4 \qquad I_2 = 1.78 \times 10^6 \text{ mm}^4 \qquad A = 4360 \text{ mm}^2$$

2. *Categorize*:

Critical loads: If the column buckles in the plane of the figure, the critical load is

$$P_{cr} = \frac{\pi^2 EI_2}{(L/2)^2} = \frac{4\pi^2 EI_2}{L^2}$$

3. *Analyze*: Substitute numerical values to obtain

$$P_{cr} = \frac{4\pi^2 EI_2}{L^2} = \frac{4\pi^2 (200 \text{ GPa})(1.78 \times 10^6 \text{ mm}^4)}{(8 \text{ m})^2} = 220 \text{ kN}$$

If the column buckles perpendicular to the plane of the figure, the critical load is

$$P_{cr} = \frac{\pi^2 E I_1}{L^2} = \frac{\pi^2 (200 \text{ GPa})(26.9 \times 10^6 \text{ mm}^4)}{(8 \text{ m})^2} = 830 \text{ kN}$$

Therefore, the critical load for the column (the smaller of the two preceding values) is

$$P_{cr} = 220 \text{ kN}$$

and buckling occurs in the plane of the figure.

Critical stresses: Since the calculations for the critical loads are valid only if the material follows Hooke's law, verify that the critical stresses do not exceed the proportional limit of the material. For the larger critical load, the critical stress is

$$\sigma_{cr} = \frac{P_{cr}}{A} = \frac{830 \text{ kN}}{4360 \text{ mm}^2} = 190.4 \text{ MPa}$$

Since this stress is less than the proportional limit ($\sigma_{pl} = 300$ MPa), both critical-load calculations are satisfactory.

4. *Finalize*:

Allowable load: The allowable axial load for the column, based on Euler buckling, is

$$P_{allow} = \frac{P_{cr}}{n} = \frac{220 \text{ kN}}{2.5} = 88 \text{ kN}$$

in which $n = 2.5$ is the desired factor of safety.

14.4 Columns with Other Support Conditions

Buckling of a column with pinned ends (described in the preceding section) is usually considered as the most basic case of buckling. However, in practice, engineers encounter many other end conditions, such as fixed ends, free ends, and elastic supports. The critical loads for columns with various kinds of support conditions can be determined from the differential equation of the deflection curve by following the same procedure that was used when analyzing a pinned-end column.

The procedure is as follows. First, with the column assumed to be in the buckled state, obtain an expression for the bending moment in the column. Second, set up the differential equation of the deflection curve using the bending-moment equation ($EIv'' = M$). Third, solve the equation and obtain its general solution, which contains two constants of integration plus any other unknown quantities. Fourth, apply boundary conditions pertaining to the deflection v and the slope v' and obtain a set of simultaneous equations. Finally, solve those equations to obtain the critical load and the deflected shape of the buckled column.

This straightforward mathematical procedure is illustrated in the following discussion of three types of columns.

Slender concrete columns fixed at the base and free at the top during construction

Column Fixed at the Base and Free at the Top

The first case to consider is an ideal column that is fixed at the base, free at the top, and subjected to an axial load P (Fig. 14-16a).[5] The deflected shape of the buckled column is shown in Fig. 14-16b. From this figure, note that the bending moment at distance x from the base is

$$M = P(\delta - v) \tag{14-21}$$

where δ is the deflection at the free end of the column. The **differential equation** of the deflection curve then becomes

$$EIv'' = M = P(\delta - v) \tag{14-22}$$

in which I is the moment of inertia for buckling in the x-y plane.

Using the notation $k^2 = P/EI$ from Eq. (14-7a), rearrange Eq. (14-22) into the form

$$v'' + k^2 v = k^2 \delta \tag{14-23}$$

which is a linear differential equation of second order with constant coefficients. However, it is a more complicated equation than the equation for a column with pinned ends [see Eq. (14-8)] because it has a nonzero term on the right-hand side.

The **general solution** of Eq. (14-23) consists of two parts: (1) the *homogeneous solution*, which is the solution of the homogeneous equation obtained by replacing the right-hand side with zero, and (2) the *particular solution*, which is the solution of Eq. (14-23) that produces the term on the right-hand side.

The homogeneous solution (also called the *complementary solution*) is the same as the solution of Eq. (14-8); hence,

$$v_H = C_1 \sin kx + C_2 \cos kx \tag{14-24a}$$

where C_1 and C_2 are constants of integration. Note that when v_H is substituted into the left-hand side of the differential equation of Eq. (14-23), it produces zero.

FIGURE 14-16

Ideal column fixed at the base and free at the top: (a) initially straight column, (b) buckled shape for $n = 1$, (c) buckled shape for $n = 3$, and (d) buckled shape for $n = 5$

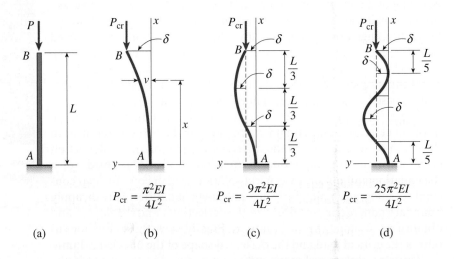

$$P_{cr} = \frac{\pi^2 EI}{4L^2}$$

$$P_{cr} = \frac{9\pi^2 EI}{4L^2}$$

$$P_{cr} = \frac{25\pi^2 EI}{4L^2}$$

(a)   (b)   (c)   (d)

[5]This column is of special interest because it is the one first analyzed by Euler in 1744.

The particular solution of the differential equation is

$$v_P = \delta \qquad \textbf{(14-24b)}$$

When v_P is substituted into the left-hand side of the differential equation, it produces the right-hand side, that is, it produces the term $k^2\delta$. Consequently, the *general solution* of the equation, equal to the sum of v_H and v_P, is

$$v = C_1 \sin kx + C_2 \cos kx + \delta \qquad \textbf{(14-25)}$$

This equation contains three unknown quantities (C_1, C_2, and δ); therefore, three **boundary conditions** are needed to complete the solution.

At the base of the column, the deflection and slope are each equal to zero. Therefore, the boundary conditions for this case are

$$v(0) = 0 \qquad v'(0) = 0$$

Apply the first condition to Eq. (14-25) to find

$$C_2 = -\delta \qquad \textbf{(14-26)}$$

To apply the second condition, first differentiate Eq. (14-25) to obtain the slope:

$$v' = C_1 k \cos kx - C_2 k \sin kx \qquad \textbf{(14-27)}$$

Apply the second condition to this equation to find $C_1 = 0$.

Now substitute the expressions for C_1 and C_2 into the general solution of Eq. (14-25) and obtain the **equation of the deflection curve** for the buckled column:

$$v = \delta(1 - \cos kx) \qquad \textbf{(14-28)}$$

Note that this equation gives only the *shape* of the deflection curve—the amplitude δ remains undefined. Thus, when the column buckles, the deflection given by Eq. (14-28) may have any arbitrary magnitude, except that it must remain small (because the differential equation is based upon small deflections).

The third boundary condition applies to the upper end of the column, where the deflection v is equal to δ:

$$v(L) = \delta$$

Use this condition with Eq. (14-28) to get

$$\delta \cos kL = 0 \qquad \textbf{(14-29)}$$

It follows from this equation that either $\delta = 0$ or $\cos kL = 0$. If $\delta = 0$, there is no deflection of the bar [see Eq. (14-28)] and you have the *trivial solution*—the column remains straight and buckling does not occur. In that case, Eq. (14-29) will be satisfied for any value of the quantity kL, that is, for any value of the load P. This conclusion is represented by the vertical line in the load–deflection diagram of Fig. 14-8.

The other possibility for solving Eq. (14-29) is

$$\cos kL = 0 \qquad \textbf{(14-30)}$$

which is the **buckling equation**. In this case, Eq. (14-29) is satisfied regardless of the value of the deflection δ. Thus, as already observed, δ is undefined and may have any small value.

The equation $\cos kL = 0$ is satisfied when

$$kL = \frac{n\pi}{2} \qquad n = 1, 3, 5, \ldots \qquad \textbf{(14-31)}$$

Use the expression $k^2 = P/EI$ to obtain the following formula for the **critical loads**:

$$P_{cr} = \frac{n^2\pi^2 EI}{4L^2} \qquad n = 1, 3, 5, \ldots \qquad \textbf{(14-32)}$$

Also, the **buckled mode shapes** are obtained from Eq. (14-28):

$$v = \delta\left(1 - \cos\frac{n\pi x}{2L}\right) \qquad n = 1, 3, 5, \ldots \qquad \textbf{(14-33)}$$

The lowest critical load is obtained by substituting $n = 1$ in Eq. (14-32):

$$P_{cr} = \frac{\pi^2 EI}{4L^2} \qquad \textbf{(14-34)}$$

The corresponding buckled shape (from Eq. 14-33) is

$$v = \delta\left(1 - \cos\frac{\pi x}{2L}\right) \qquad \textbf{(14-35)}$$

and is shown in Fig. 14-16b.

Taking higher values of the index n leads to an infinite number of critical loads from Eq. (14-32). The corresponding buckled mode shapes have additional waves in them. For instance, when $n = 3$, the buckled column has the shape shown in Fig. 14-16c, and P_{cr} is nine times larger than for $n = 1$. Similarly, the buckled shape for $n = 5$ has even more waves (Fig. 14-16d), and the critical load is 25 times larger.

Effective Lengths of Columns

The critical loads for columns with various support conditions can be related to the critical load of a pinned-end column through the concept of an **effective length**. To demonstrate this idea, consider the deflected shape of a column fixed at the base and free at the top (Fig. 14-17a). This column buckles in a curve that is one-quarter of a complete sine wave. If the deflection curve is extended (Fig. 14-17b), it becomes one-half of a complete sine wave, which is the deflection curve for a pinned-end column.

The effective length L_e for any column is the length of the equivalent pinned-end column, that is, it is the length of a pinned-end column having a deflection curve that exactly matches all or part of the deflection curve of the original column.

Another way of expressing this idea is to say that the effective length of a column is the distance between points of inflection (that is, points of zero moment) in its deflection curve, assuming that the curve is extended (if necessary) until points of inflection are reached. Thus, for a fixed-free column (Fig. 14-17), the effective length is

$$L_e = 2L \qquad \textbf{(14-36)}$$

Because the effective length is the length of an equivalent pinned-end column, a general formula for critical loads can be written as

$$P_{cr} = \frac{\pi^2 EI}{L_e^2} \qquad \textbf{(14-37)}$$

FIGURE 14-17

Deflection curves showing the effective length L_e for a column fixed at the base and free at the top

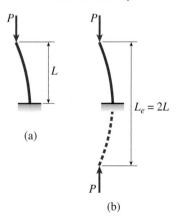

If the effective length of a column is known (no matter how complex the end conditions may be), you can substitute into the preceding equation and determine the critical load. For instance, in the case of a fixed-free column, substitute $L_e = 2L$ and obtain Eq. (14-34).

The effective length is often expressed in terms of an **effective-length factor** K:

$$L_e = KL \qquad \textbf{(14-38)}$$

where L is the actual length of the column. Thus, the critical load is

$$P_{cr} = \frac{\pi^2 EI}{(KL)^2} \qquad \textbf{(14-39)}$$

The factor K equals 2 for a column fixed at the base and free at the top and equals 1 for a pinned-end column. The effective-length factor is often included in design formulas for columns.

Column with Both Ends Fixed Against Rotation

Next, consider a column with both ends fixed against rotation (Fig. 14-18a). Note that in this figure the standard symbol for the fixed support is used at the base of the column. However, since the column is free to shorten under an axial load, a new symbol is introduced at the top of the column. This new symbol shows a rigid block that is constrained in such a manner that rotation and horizontal displacement are prevented but vertical movement can occur. (As a convenience when drawing sketches, this more accurate symbol is replaced with the standard symbol for a fixed support—see Fig. 14-18b—with the understanding that the column is free to shorten.)

The buckled shape of the column in the first mode is shown in Fig. 14-18c. Note that the deflection curve is symmetrical (with zero slope at the midpoint) and has zero slope at the ends. Because rotation at the ends is prevented, reactive moments M_0 develop at the supports. These moments, as well as the reactive force at the base, are shown in the figure.

From the previous solutions of the differential equation, the equation of the deflection curve involves sine and cosine functions. Also, the curve is symmetric about the midpoint. Therefore, the curve must have inflection points at distances $L/4$ from the ends. It follows that the middle portion of the deflection curve has the same shape as the deflection curve for a pinned-end column. Thus, the effective length of a column with fixed ends, equal to the distance between inflection points, is

$$L_e = \frac{L}{2} \qquad \textbf{(14-40)}$$

Substituting into Eq. (14-37) gives the critical load:

$$P_{cr} = \frac{4\pi^2 EI}{L^2} \qquad \textbf{(14-41)}$$

This formula shows that the critical load for a column with fixed ends is four times that for a column with pinned ends. As a check, this result may be verified by solving the differential equation of the deflection curve (see Prob. 14.4-11).

FIGURE 14-18

Buckling of a column with both ends fixed against rotation

(a)   (b)

(c)

Column Fixed at the Base and Pinned at the Top

The critical load and buckled mode shape for a column that is fixed at the base and pinned at the top (Fig. 14-19a) can be determined by solving the differential equation of the deflection curve. When the column buckles (Fig. 14-19b), a reactive moment M_0 develops at the base because there can be no rotation at that point. Then from the equilibrium of the entire column, there must be horizontal reactions R at each end such that

$$M_0 = RL \tag{14-42}$$

The bending moment in the buckled column, at distance x from the base, is

$$M = M_0 - Pv - Rx = -Pv + R(L - x) \tag{14-43}$$

therefore, the **differential equation** is

$$EIv'' = M = -Pv + R(L - x) \tag{14-44}$$

Again substitute $k^2 = P/EI$ and rearrange to get

$$v'' + k^2 v = \frac{R}{EI}(L - x) \tag{14-45}$$

The **general solution** of this equation is

$$v = C_1 \sin kx + C_2 \cos kx + \frac{R}{P}(L - x) \tag{14-46}$$

in which the first two terms on the right-hand side constitute the homogeneous solution and the last term is the particular solution. This solution can be verified by substitution into the differential equation of Eq. (14-44).

Since the solution contains three unknown quantities (C_1, C_2, and R), the following three **boundary conditions** are required:

$$v(0) = 0 \qquad v'(0) = 0 \qquad v(L) = 0$$

FIGURE 14-19

Column fixed at the base and pinned at the top

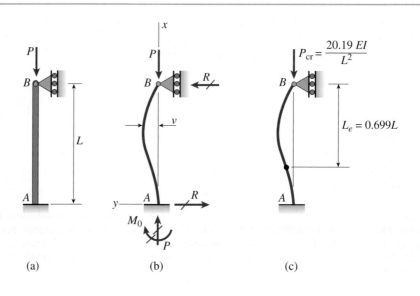

(a)   (b)   (c)

Applying these conditions to Eq. (14-46) yields

$$C_2 + \frac{RL}{P} = 0 \qquad C_1 k - \frac{R}{P} = 0$$

$$C_1 \tan kL + C_2 = 0$$

(14-47a,b,c)

All three equations are satisfied if $C_1 = C_2 = R = 0$, in which case the trivial solution results and the deflection is zero.

To obtain the solution for buckling, solve Eqs. (14-47a, b, and c) in a more general manner. One method of solution is to eliminate R from the first two equations, which yields

$$C_1 kL + C_2 = 0 \quad \text{or} \quad C_2 = -C_1 kL \qquad \textbf{(14-47d)}$$

Next, substitute this expression for C_2 into Eq. (14-47c) and obtain the **buckling equation**:

$$kL = \tan kL \qquad \textbf{(14-48)}$$

The solution of this equation gives the critical load.

Since the buckling equation is a transcendental equation, it cannot be solved explicitly.[6] Nevertheless, the values of kL that satisfy the equation can be determined numerically by using a computer program for finding roots of equations. The smallest nonzero value of kL that satisfies Eq. (14-48) is

$$kL = 4.4934 \qquad \textbf{(14-49)}$$

The corresponding **critical load** is

$$P_{cr} = \frac{20.19\,EI}{L^2} = \frac{2.046\pi^2 EI}{L^2} \qquad \textbf{(14-50)}$$

which (as expected) is higher than the critical load for a column with pinned ends and lower than the critical load for a column with fixed ends [see Eqs. (14-15) and (14-41)].

The **effective length** of the column may be obtained by comparing Eqs. (14-50) and (14-37); thus,

$$L_e = 0.699\,L \approx 0.7\,L \qquad \textbf{(14-51)}$$

This length is the distance from the pinned end of the column to the point of inflection in the buckled shape (Fig. 14-19c).

The equation of the **buckled mode shape** is obtained by substituting $C_2 = -C_1 kL$ [Eq. (14-47d)] and $R/P = C_1 k$ [Eq. (14-47b)] into the general solution [Eq. (14-46)]:

$$v = C_1[\sin kx - kL \cos kx + k(L - x)] \qquad \textbf{(14-52)}$$

in which $k = 4.4934/L$. The term in brackets gives the mode shape for the deflection of the buckled column. However, the amplitude of the deflection curve is undefined because C_1 may have any value (within the usual limitation that the deflections must remain small).

---

[6]In a transcendental equation, the variables are contained within transcendental functions. A transcendental function cannot be expressed by a finite number of algebraic operations; hence, trigonometric, logarithmic, exponential, and other such functions are transcendental.

FIGURE 14-20

Critical loads, effective lengths, and effective-length factors for ideal columns

(a) Pinned-pinned column	(b) Fixed-free column	(c) Fixed-fixed column	(d) Fixed-pinned column
$P_{cr} = \dfrac{\pi^2 EI}{L^2}$	$P_{cr} = \dfrac{\pi^2 EI}{4L^2}$	$P_{cr} = \dfrac{4\pi^2 EI}{L^2}$	$P_{cr} = \dfrac{2.046\,\pi^2 EI}{L^2}$
$L_e = L$	$L_e = 2L$	$L_e = 0.5L$	$L_e = 0.699L$
$K = 1$	$K = 2$	$K = 0.5$	$K = 0.699$

Limitations

In addition to the requirement of small deflections, the Euler buckling theory used in this section is valid only if the column is perfectly straight before the load is applied, the column and its supports have no imperfections, and the column is made of a linearly elastic material that follows Hooke's law. These limitations were explained previously in Section 14.3.

Summary of Results

The lowest critical loads and corresponding effective lengths for the four columns analyzed here are summarized in Fig. 14-20.

Example 14-3

A viewing platform in a wild-animal park (Fig. 14-21a) is supported by a row of aluminum pipe columns having a length $L = 10$ ft 8 in. and an outer diameter $d = 4$ in. The bases of the columns are set in concrete footings, and the tops of the columns are supported laterally by the platform. The columns are being designed to support compressive loads where $P = 22.5$ kips.

Determine the minimum required thickness t of the columns (Fig. 14-21b) if a factor of safety $n = 3$ is required with respect to Euler buckling. (For the aluminum, use 10,400 ksi for the modulus of elasticity and use 70 ksi for the proportional limit.)

FIGURE 14-21

Example 14-3: Aluminum
pipe column

(a)

(b)

Solution:

Use the four-step problem-solving approach. Combine steps as needed for an efficient solution.

1, 2. *Conceptualize, Categorize*:

Critical load: Because of the manner in which the columns are constructed, each column is modeled as a fixed-pinned column (see Fig. 14-20d). Therefore, the critical load is

$$P_{cr} = \frac{2.046\pi^2 EI}{L^2} \tag{a}$$

where I is the moment of inertia of the tubular cross section:

$$I = \frac{\pi}{64}[d^4 - (d - 2t)^4] \tag{b}$$

Substitute $d = 4$ in. to get

$$I = \frac{\pi}{64}[(4 \text{ in.})^4 - (4 \text{ in.} - 2t)^4] \tag{c}$$

in which t is expressed in inches.

3. *Analyze*:

Required thickness of the columns: Since the load per column is 22.5 kips and the factor of safety is 3, each column must be designed for the critical load:

$$P_{cr} = nP = 3(22.5 \text{ kips}) = 67.5 \text{ kips}$$

Substitute this value for P_{cr} in Eq. (a), and also replace I with its expression from Eq. (c) to obtain

$$67.5 \text{ kips} = \frac{2.046\,\pi^2(10,400 \text{ ksi})}{(128 \text{ in.})^2}\left[\frac{\pi}{64}[(4 \text{ in.})^4 - (4 \text{ in.} - 2t)^4]\right]$$

Note that all terms in this equation are expressed in units of kips and inches.

Solve the previous equation to find that the minimum required thickness of the column to meet the specified conditions is

$$t_{min} = 0.254 \text{ in.}$$

4. *Finalize*:

Supplementary calculations: Knowing the diameter and thickness of the column, now calculate its moment of inertia, cross-sectional area, and radius of gyration. Using the minimum thickness of 0.254 in. gives

$$I = \frac{\pi}{64}[d^4 - (d - 2t)^4] = 5.267 \text{ in}^4$$

$$A = \frac{\pi}{4}[d^2 - (d - 2t)^2] = 2.989 \text{ in}^2 \qquad r = \sqrt{\frac{I}{A}} = 1.327 \text{ in.}$$

The slenderness ratio L/r of the column is approximately 96.5, which is in the customary range for slender columns. The diameter-to-thickness ratio d/t is approximately 16, which should be adequate to prevent local buckling of the walls of the column.

The critical stress in the column must be less than the proportional limit of the aluminum if the formula for the critical load of Eq. (a) is to be valid. The critical stress is

$$\sigma_{cr} = \frac{P_{cr}}{A} = \frac{67.5 \text{ kips}}{2.989 \text{ in}^2} = 22.6 \text{ ksi}$$

which is less than the proportional limit (70 ksi). Therefore, the calculation for the critical load using the Euler buckling theory is satisfactory.

14.5 Columns with Eccentric Axial Loads

Sections 14.3 and 14.4 covered the analysis of ideal columns in which the axial loads acted through the centroids of the cross sections. Under these conditions, the columns remain straight until the critical loads are reached, after which bending may occur.

Now assume that a column is compressed by loads P that are applied with a small eccentricity e measured from the axis of the column (Fig. 14-22a). Each eccentric axial load is equivalent to a centric load P and a couple of moment $M_0 = Pe$ (Fig. 14-22b). This moment exists from the instant the load is first applied; therefore, the column begins to deflect at the onset of loading. The deflection then becomes steadily larger as the load increases.

To analyze the pin-ended column shown in Fig. 14-22, make the same assumptions as in previous sections; namely, the column is initially perfectly straight, the material is linearly elastic, and the x-y plane is a plane of symmetry. The bending moment in the column at distance x from the lower end (Fig. 14-22b) is

$$M = M_0 + P(-v) = Pe - Pv \qquad \text{(14-53)}$$

where v is the deflection of the column (positive when in the positive direction of the y axis). Note that the deflections of the column are negative when the eccentricity of the load is positive.

The **differential equation** of the deflection curve is

$$EIv'' = M = Pe - Pv \qquad (14\text{-}54)$$

or

$$v'' + k^2v = k^2e \qquad (14\text{-}55)$$

in which $k^2 = P/EI$, as before. The general solution of this equation is

$$v = C_1 \sin kx + C_2 \cos kx + e \qquad (14\text{-}56)$$

where C_1 and C_2 are constants of integration in the homogeneous solution and e is the particular solution. As usual, verify the solution by substituting it into the differential equation.

The **boundary conditions** for determining the constants C_1 and C_2 are obtained from the deflections at the ends of the column (Fig. 14-22b):

$$v(0) = 0 \qquad v(L) = 0$$

These conditions yield

$$C_2 = -e \qquad C_1 = -\frac{e(1 - \cos kL)}{\sin kL} = -e \tan \frac{kL}{2}$$

Therefore, the **equation of the deflection curve** is

$$v = -e \left(\tan \frac{kL}{2} \sin kx + \cos kx - 1 \right) \qquad (14\text{-}57)$$

For a column with known loads P and known eccentricity e, use this equation to calculate the deflection at any point along the x axis.

The behavior of a column with an eccentric load is quite different from that of a centrally loaded column, as can be seen by comparing Eq. (14-57) with Eqs. (14-16), (14-33), and (14-52). Equation (14-57) shows that each value of the eccentric load P produces a definite value of the deflection, just as each value of the load on a beam produces a definite deflection. In contrast, the deflection equations for centrally loaded columns give the buckled mode shape (when $P = P_{cr}$) but with the amplitude undefined.

Because the column shown in Fig. 14-22 has pinned ends, its critical load (when centrally loaded) is

$$P_{cr} = \frac{\pi^2 EI}{L^2} \qquad (14\text{-}58)$$

Use this formula as a reference quantity in some of the equations that follow.

Maximum Deflection

The maximum deflection δ produced by the eccentric loads occurs at the midpoint of the column (Fig. 14-23) and is obtained by setting x equal to $L/2$ in Eq. (14-57):

$$\delta = -v\left(\frac{L}{2}\right) = e\left(\tan \frac{kL}{2} \sin \frac{kL}{2} + \cos \frac{kL}{2} - 1 \right)$$

FIGURE 14-22

Column with eccentric axial loads

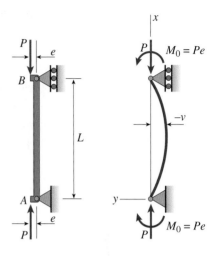

(a)　　　　(b)

FIGURE 14-23

Maximum deflection δ of a column with eccentric axial loads or, after simplifying,

$$\delta = e\left(\sec\frac{kL}{2} - 1\right) \tag{14-59}$$

This equation can be written in a slightly different form by replacing the quantity k with its equivalent in terms of the critical load [see Eq. (14-58)]:

$$k = \sqrt{\frac{P}{EI}} = \sqrt{\frac{P\pi^2}{P_{cr}L^2}} = \frac{\pi}{L}\sqrt{\frac{P}{P_{cr}}} \tag{14-60}$$

Thus, the nondimensional term kL becomes

$$kL = \pi\sqrt{\frac{P}{P_{cr}}} \tag{14-61}$$

and Eq. (14-59) for the **maximum deflection** becomes

$$\delta = e\left[\sec\left(\frac{\pi}{2}\sqrt{\frac{P}{P_{cr}}}\right) - 1\right] \tag{14-62}$$

FIGURE 14-24

Load–deflection diagram for a column with eccentric axial loads [see Fig. 14-23 and Eq. (14-62)]

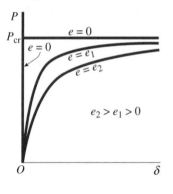

As special cases, note: (1) The deflection δ is zero when the eccentricity e is zero and P is not equal to P_{cr}, (2) the deflection is zero when the axial load P is zero, and (3) the deflection becomes infinitely large as P approaches P_{cr}. These characteristics are shown in the **load–deflection diagram** of Fig. 14-24.

To plot the load–deflection diagram, select a particular value e_1 of the eccentricity and then calculate δ for various values of the load P. The resulting curve is labeled $e = e_1$ in Fig. 14-24. Note that the deflection δ increases as P increases, but the relationship is nonlinear. Therefore, the principle of superposition cannot be used for calculating deflections due to more than one load, even though the material of the column is linearly elastic. As an example, the deflection due to an axial load $2P$ is *not* equal to twice the deflection caused by an axial load P.

Additional curves, such as the curve labeled $e = e_2$, are plotted in a similar manner. Since the deflection δ is linear with e in Eq. (14-62), the curve for $e = e_2$ has the same *shape* as the curve for $e = e_1$, but the abscissas are larger by the ratio e_2/e_1.

As the load P approaches the critical load, the deflection δ increases without limit, and the horizontal line corresponding to $P = P_{cr}$ becomes an asymptote for the curves. In the limit, as e approaches zero, the curves on the diagram approach two straight lines: one vertical and one horizontal (compare with Fig. 14-8). Thus, as expected, an ideal column with a centrally applied load ($e = 0$) is the limiting case of a column with an eccentric load ($e > 0$).

Although the curves plotted in Fig. 14-24 are mathematically correct, keep in mind that the differential equation is valid only for small deflections. Therefore, when the deflections become large, the curves are no longer physically valid and must be modified to take into account the presence of large deflections and (if the proportional limit of the material is exceeded) inelastic bending effects (see Fig. 14-12).

The reason for the nonlinear relationship between loads and deflections, even when the deflections are small and Hooke's law holds, can be understood if you observe once again that the axial loads P are equivalent to centrally applied loads P plus couples Pe acting at the ends of the column (Fig. 14-22b). The couples Pe, if acting alone, would produce bending deflections of the column in the same manner as for a beam. In a beam, the presence of the deflections does not change the action of the loads, and the bending moments are the same whether the deflections exist or not. However, when an axial load is applied to the member, the existence of deflections increases the bending moments (the increases are equal to the product of the axial load and the deflections). When the bending moments increase, the deflections are further increased—hence, the moments increase even more, and so on. Thus, the bending moments in a column depend upon the deflections, which in turn depend upon the bending moments. This type of behavior results in a nonlinear relationship between the axial loads and the deflections.

In general, a straight structural member subjected to both bending loads and axial compressive loads is called a **beam-column**. In the case of a column with eccentric loads (Fig. 14-22), the bending loads are the moments $M_0 = Pe$ and the axial loads are the forces P.

Maximum Bending Moment

The maximum bending moment in an eccentrically loaded column occurs at the midpoint where the deflection is a maximum (Fig. 14-23):

$$M_{max} = P(e + \delta) \tag{14-63}$$

Substitute for δ from Eqs. (14-59) and (14-62) to obtain

$$M_{max} = Pe \sec\frac{kL}{2} = Pe \sec\left(\frac{\pi}{2}\sqrt{\frac{P}{P_{cr}}}\right) \tag{14-64}$$

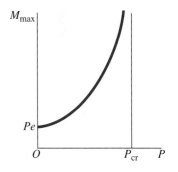

The manner in which M_{max} varies as a function of the axial load P is shown in Fig. 14-25.

When P is small, the maximum moment is equal to Pe, which means that the effect of the deflections is negligible. As P increases, the bending moment grows nonlinearly and theoretically becomes infinitely large as P approaches the critical load. However, as explained before, these equations are valid only when the deflections are small, and they cannot be used when the axial load approaches the critical load. Nevertheless, the preceding equations and accompanying graphs indicate the general behavior of beam-columns.

Other End Conditions

The equations given in this section were derived for a pinned-end column, as shown in Figs. 14-22 and 14-23. If a column is fixed at the base and free at the top (Fig. 14-20b), use Eqs. (14-59) and (14-64) by replacing the actual length L with the equivalent length $2L$ (see Prob. 14.5-11). However, the equations do not apply to a column that is fixed at the base and pinned at the top (Fig. 14-20d). The use of an equivalent length equal to $0.699L$ gives erroneous results; instead, you must return to the differential equation and derive a new set of equations.

In the case of a column with both ends fixed against rotation (Fig. 14-20c), the concept of an eccentric axial load acting at the end of the column has no meaning. Any moment applied at the end of the column is resisted directly by the supports and produces no bending of the column itself.

Example 14-4

FIGURE 14-26

Example 14-4: Brass bar with an eccentric axial load

Brass bar

A brass bar AB projecting from the side of a large machine is loaded at end B by a force $P = 7$ kN acting with an eccentricity $e = 11$ mm (Fig. 14-26). The bar has a rectangular cross section with height $h = 30$ mm and width $b = 15$ mm.

What is the longest permissible length L_{max} of the bar if the deflection at the end is limited to 3 mm? (For the brass, use $E = 110$ GPa.)

Solution:

Use the four-step problem-solving approach.

1. *Conceptualize*:

   Critical load: This bar is a slender column that is fixed at end A and free at end B. Therefore, the critical load (see Fig. 14-20b) is

   $$P_{cr} = \frac{\pi^2 EI}{4L^2} \tag{a}$$

   The moment of inertia for the axis about which bending occurs is

   $$I = \frac{hb^3}{12} = \frac{(30 \text{ mm})(15 \text{ mm})^3}{12} = 8.44 \times 10^3 \text{ mm}^4$$

   Therefore, the expression for the critical load becomes

   $$P_{cr} = \frac{\pi^2 (110 \text{ GPa})(8.44 \times 10^3 \text{ mm}^4)}{4L^2} = \frac{2.29 \text{ kN} \cdot \text{m}^2}{L^2} \tag{b}$$

   in which P_{cr} has units of kN and L has units of meters.

2. *Categorize*:

   Deflection: The deflection at the end of the bar is given by Eq. (14-62), which applies to a fixed-free column as well as a pinned-end column:

   $$\delta = e \left[\sec \left(\frac{\pi}{2} \sqrt{\frac{P}{P_{cr}}} \right) - 1 \right] \tag{c}$$

   In this equation, P_{cr} is given by Eq. (a).

3. *Analyze*:

   Length: To find the maximum permissible length of the bar, substitute for δ its limiting value of 3 mm. Also, substitute $e = 11$ mm and $P = 7$ kN, and substitute for P_{cr} from Eq. (b). Thus,

   $$3 \text{ mm} = (11 \text{ mm}) \left[\sec \left(\frac{\pi}{2} \sqrt{\frac{7 \text{ kN}}{2.29/L^2}} \right) - 1 \right]$$

The only unknown in this equation is the length L (meters). To solve for L, perform the various arithmetic operations in the equation and then rearrange the terms. The result is

$$0.2727 = \sec{(2.746L)} - 1$$

Use radians and solve this equation to get $L = 0.243$ m. Thus, the maximum permissible length of the bar is

$$L_{max} = 0.243 \text{ m}$$

4. *Finalize*: If a longer bar is used, the deflection will exceed the allowable value of 3 mm.

14.6 The Secant Formula for Columns

In the preceding section, the maximum deflection and maximum bending moment were found for a pin-ended column subjected to eccentric axial loads. This section discusses the maximum stresses in the column and a special formula for calculating them.

The maximum stresses in a column with eccentric axial loads occur at the cross section where the deflection and bending moment have their largest values, that is, at the midpoint (Fig. 14-27a). Acting at this cross section are the compressive force P and the bending moment M_{max} (Fig. 14-27b). The stresses due to the force P are equal to P/A, where A is the cross-sectional area of the column, and the stresses due to the bending moment M_{max} are obtained from the flexure formula.

FIGURE 14-27

Column with eccentric axial loads

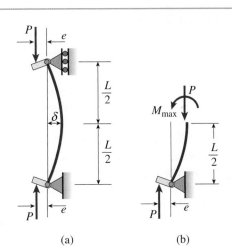

(a)   (b)

Thus, the maximum compressive stress, which occurs on the concave side of the column, is

$$\sigma_{max} = \frac{P}{A} + \frac{M_{max}c}{I} \tag{14-65}$$

where I is the moment of inertia in the plane of bending and c is the distance from the centroidal axis to the extreme point on the concave side of the column. Note that in this equation compressive stresses are considered to be positive, since these are the important stresses in a column.

The bending moment M_{max} is obtained from Eq. (14-64), which is repeated here:

$$M_{max} = Pe \sec\left(\frac{\pi}{2}\sqrt{\frac{P}{P_{cr}}}\right)$$

Since $P_{cr} = \pi^2 EI/L^2$ for a pinned-end column and $I = Ar^2$, where r is the radius of gyration in the plane of bending, the preceding equation becomes

$$M_{max} = Pe \sec\left(\frac{L}{2r}\sqrt{\frac{P}{EA}}\right) \tag{14-66}$$

Substitute into Eq. (14-65) to obtain the following formula for the **maximum compressive stress**:

$$\sigma_{max} = \frac{P}{A} + \frac{Pec}{I}\sec\left(\frac{L}{2r}\sqrt{\frac{P}{EA}}\right)$$

or

$$\sigma_{max} = \frac{P}{A}\left[1 + \frac{ec}{r^2}\sec\left(\frac{L}{2r}\sqrt{\frac{P}{EA}}\right)\right] \tag{14-67}$$

Equation (14-67) is commonly known as the **secant formula** for an eccentrically loaded column with pinned ends.

The secant formula gives the maximum compressive stress in the column as a function of the average compressive stress P/A, the modulus of elasticity E, and two nondimensional ratios—the slenderness ratio L/r from Eq. (14-20) and the **eccentricity ratio**:

$$\text{Eccentricity ratio} = \frac{ec}{r^2} \tag{14-68}$$

As the name implies, the eccentricity ratio is a measure of the eccentricity of the load as compared to the dimensions of the cross section. Its numerical value depends upon the position of the load, but typical values are in the range from 0 to 3 with the most common values being less than 1.

When analyzing a column, use the secant formula to calculate the maximum compressive stress whenever the axial load P and its eccentricity e are known. Then the maximum stress can be compared with the allowable stress to determine if the column is adequate to support the load.

The secant formula also can be used in the reverse manner, that is, if the allowable stress is known calculate the corresponding value of the load P. However, because the secant formula is transcendental, it is not practical to derive a formula for the load P. Instead, solve Eq. (14-67) numerically in each individual case.

A **graph of the secant formula** is shown in Fig. 14-28. The abscissa is the slenderness ratio L/r, and the ordinate is the average compressive stress P/A. The graph is plotted for a steel column with modulus of elasticity $E = 30 \times 10^3$ ksi and maximum stress $\sigma_{max} = 36$ ksi. Curves are plotted for several values of the eccentricity ratio ec/r^2. These curves are valid only when the maximum stress is less than the proportional limit of the material, because the secant formula was derived using Hooke's law.

A special case arises when the eccentricity of the load disappears ($e = 0$), because then you have an ideal column with a centrally applied load. Under these conditions, the maximum load is the critical load ($P_{cr} = \pi^2 EI/L^2$), and the corresponding maximum stress is the critical stress [see Eqs. (14-17) and (14-19)]:

$$\sigma_{cr} = \frac{P_{cr}}{A} = \frac{\pi^2 EI}{AL^2} = \frac{\pi^2 E}{(L/r)^2} \qquad \textbf{(14-69)}$$

Since this equation gives the stress P/A in terms of the slenderness ratio L/r, you can plot it on the graph of the secant formula (Fig. 14-28) as **Euler's curve**.

Now assume that the proportional limit of the material is the same as the selected maximum stress, that is, 36 ksi. Construct a horizontal line on the graph at a value of 36 ksi, and terminate Euler's curve at that stress. The horizontal line and Euler's curve represent the limits of the secant-formula curves as the eccentricity e approaches zero.

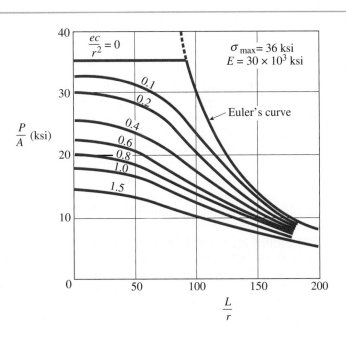

FIGURE 14-28

Graph of the secant formula (Eq. 14-67) for $\sigma_{max} = 36$ ksi and $E = 30 \times 10^3$ ksi

Discussion of the Secant Formula

The graph of the secant formula shows that the load-carrying capacity of a column decreases significantly as the slenderness ratio L/r increases, especially in the intermediate region of L/r values. Thus, long slender columns are much less stable than short, stocky columns. The graph also shows that the load-carrying capacity decreases with increasing eccentricity e; furthermore, this effect is relatively greater for short columns than for long ones.

The secant formula was derived for a column with pinned ends, but it also can be used for a column that is fixed at the base and free at the top. All that is required is to replace the length L in the secant formula with the equivalent length $2L$. However, because it is based upon Eq. (14-64), the secant formula is not valid for the other end conditions discussed.

Now consider an actual column that inevitably differs from an ideal column because of imperfections, such as initial curvature of the longitudinal axis, imperfect support conditions, and nonhomogeneity of the material. Furthermore, even when the load is supposed to be centrally applied, there will be unavoidable eccentricities in its direction and point of application. The extent of these imperfections varies from one column to another, so there is considerable scatter in the results of laboratory tests performed with actual columns.

All imperfections have the effect of producing bending in addition to direct compression. Therefore, it is reasonable to assume that the behavior of an imperfect, centrally loaded column is similar to that of an ideal, eccentrically loaded column. In such cases, the secant formula can be used by choosing an approximate value of the eccentricity ratio ec/r^2 to account for the combined effects of the various imperfections. For instance, a commonly used value of the eccentricity ratio for pinned-end columns in structural-steel design is $ec/r^2 = 0.25$. The use of the secant formula in this manner for columns with centrally applied loads provides a rational means of accounting for the effects of imperfections, rather than accounting for them simply by increasing the factor of safety. (For further discussions of the secant formula and the effects of imperfections, see books on buckling and stability.)

The procedure for analyzing a centrally loaded column by means of the secant formula depends upon the particular conditions. For instance, if the objective is to determine the allowable load, the procedure is as follows. Assume a value of the eccentricity ratio ec/r^2 based upon test results, code values, or practical experience. Substitute this value into the secant formula, along with the values of L/r, A, and E for the actual column. Assign a value to σ_{max}, such as the yield stress σ_Y or the proportional limit σ_{pl} of the material. Then solve the secant formula for the load P_{max} that produces the maximum stress. (This load always will be less than the critical load P_{cr} for the column.) The allowable load on the column equals the load P_{max} divided by the factor of safety n.

The following example illustrates how the secant formula may be used to determine the maximum stress in a column when the load is known, and also how to determine the load when the maximum stress is given.

Example 14-5

FIGURE 14-29

Example 14-5: Column with an
eccentrically applied axial load

(a)

(b)

(c)

A steel wide-flange column of a W 14 × 82 shape (Fig. 14-29a) is
pin-supported at the ends and has a length of 25 ft. The column
supports a centrally applied load $P_1 = 320$ kips and an eccentrically
applied load $P_2 = 40$ kips (Fig. 14-29b). Bending takes place about
axis 1–1 of the cross section, and the eccentric load acts on axis 2–2
at a distance of 13.5 in. from the centroid C.

(a) Using the secant formula and assuming $E = 30,000$ ksi, calculate the maximum compressive stress in the column.

(b) If the yield stress for the steel is $\sigma_Y = 42$ ksi, what is the factor of safety with respect to yielding?

Solution:

Use the four-step problem-solving approach.

Part (a): Maximum compressive stress.

1. *Conceptualize*: The two loads P_1 and P_2 acting as shown
in Fig. 14-29b are statically equivalent to a single load
$P = 360$ kips acting with an eccentricity $e = 1.5$ in.
(Fig. 14-29c). Since the column is now loaded by a single force
P having an eccentricity e, use the secant formula to find the
maximum stress.

The required properties of the W 14 × 82 wide-flange
shape are obtained from Table C-1(a) in Appendix C:

$$A = 24.0 \text{ in}^2 \qquad r = 6.05 \text{ in.} \qquad c = \frac{14.3 \text{ in.}}{2} = 7.15 \text{ in.}$$

2. *Categorize*: The required terms in the secant formula of
Eq. (14-67) are calculated as

$$\frac{P}{A} = \frac{360 \text{ kips}}{24 \text{ in}^2} = 15 \text{ ksi}$$

$$\frac{ec}{r^2} = \frac{(1.5 \text{ in.})(7.15 \text{ in.})}{(6.05 \text{ in.})^2} = 0.293$$

$$\frac{L}{r} = \frac{(25 \text{ ft})(12 \text{ in/ft})}{6.05 \text{ in.}} = 49.59$$

$$\frac{P}{EA} = \frac{360 \text{ kips}}{(30,000 \text{ ksi})(24 \text{ in}^2)} = 500 \times 10^{-6}$$

3. *Analyze*: Substitute these values into the secant formula to get

$$\sigma_{max} = \frac{P}{A}\left[1 + \frac{ec}{r^2}\sec\left(\frac{L}{2r}\sqrt{\frac{P}{EA}}\right)\right]$$

$$= (15\text{ ksi})(1 + 0.345) = 20.1\text{ ksi}$$

4. *Finalize*: This compressive stress occurs at mid-height of the column on the concave side (the right-hand side in Fig. 14-29b).

Part (b): Factor of safety with respect to yielding.

1. *Conceptualize*: To find the factor of safety, determine the value of the load P, acting at the eccentricity e, that will produce a maximum stress equal to the yield stress $\sigma_Y = 42$ ksi. Since this value of the load is just sufficient to produce initial yielding of the material, denote it as P_Y.

2. *Categorize*: Note that force P_Y cannot be determined by multiplying the load P (equal to 360 kips) by the ratio σ_Y/σ_{max}. The reason is that there is a nonlinear relationship between load and stress. Instead, substitute $\sigma_{max} = \sigma_Y = 42$ ksi in the secant formula and then solve for the corresponding load P, which becomes P_Y. In other words, find the value of P_Y that satisfies

$$\sigma_Y = \frac{P_Y}{A}\left[1 + \frac{ec}{r^2}\sec\left(\frac{L}{2r}\sqrt{\frac{P_Y}{EA}}\right)\right] \qquad \text{(14-70)}$$

3. *Analyze*: Substitute numerical values to obtain

$$42\text{ ksi} = \frac{P_Y}{24.0\text{ in}^2}\left[1 + 0.293\sec\left(\frac{49.59}{2}\sqrt{\frac{P_Y}{(30,000\text{ ksi})(24.0\text{ in}^2)}}\right)\right]$$

or

$$1008\text{ kips} = P_Y[1 + 0.293\sec(0.02916\sqrt{P_Y})]$$

in which P_Y has units of kips. Solving this equation numerically gives

$$P_Y = 714\text{ kips}$$

This load will produce yielding of the material (in compression) at the cross section of maximum bending moment.

Since the actual load is $P = 360$ kips, the factor of safety against yielding is

$$n = \frac{P_Y}{P} = \frac{714\text{ kips}}{360\text{ kips}} = 1.98$$

4. *Finalize*: This example illustrates two of the many ways the secant formula may be used. Other types of analysis are illustrated in the problems at the end of the chapter.

Chapter 14 discussed the elastic behavior of axially loaded members known as columns. First, the concepts of **buckling and stability** of these slender compression elements were discussed using the equilibrium of simple column models made up of rigid bars and elastic springs. Then elastic columns with pinned ends acted on by centroidal compressive loads were considered, and the differential equation of the deflection curve was solved to obtain the **buckling load** (P_{cr}) and **buckled mode shape**; linear elastic behavior was assumed. Three additional support cases were investigated, and the buckling load for each case was expressed in terms of the **column's effective length**. Behavior of pinned-end columns with **eccentric axial loads** was discussed, and the **secant formula** that defines the maximum stress in these columns was derived.

Here are the major concepts presented in this chapter.

1. Buckling instability of slender columns is an important mode of failure that must be considered in their design (in addition to strength and stiffness).

2. A slender column with pinned ends and length L acted on by a compressive load at the centroid of the cross section and restricted to linear elastic behavior buckles at the **Euler buckling load**

$$P_{cr} = \pi^2 EI / L^2$$

in the fundamental mode; hence, the buckling load depends on the flexural rigidity (EI) and length (L) but not the strength of the material.

3. Changing the support conditions, or providing additional lateral supports, changes the critical buckling load. However, P_{cr} for these **other support cases** may be obtained by replacing the actual column length (L) by the **effective length** (L_e) in the formula for P_{cr} above. Three additional support cases are shown in the table below. The effective length L_e can be expressed in terms of an effective-length factor K as

$$L_e = KL$$

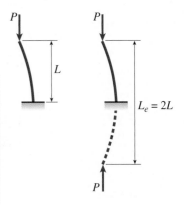

(a) Pinned-pinned column	(b) Fixed-free column	(c) Fixed-fixed column	(d) Fixed-pinned column
$P_{cr} = \dfrac{\pi^2 EI}{L^2}$	$P_{cr} = \dfrac{\pi^2 EI}{4L^2}$	$P_{cr} = \dfrac{4\pi^2 EI}{L^2}$	$P_{cr} = \dfrac{2.046\,\pi^2 EI}{L^2}$
$L_e = L$	$L_e = 2L$	$L_e = 0.5L$	$L_e = 0.699L$
$K = 1$	$K = 2$	$K = 0.5$	$K = 0.699$

where $K = 1$ for a pinned-end column and $K = 2$ for a column fixed at its base. The critical load P_{cr} then is expressed as

$$P_{cr} = \frac{\pi^2 EI}{(KL)^2}$$

Effective-length factor K is often used in column design formulas.

4. Columns with **eccentric axial loads** behave quite differently from those with centroidal loads. The maximum compressive stress in pinned-end columns with load P applied at eccentricity e is defined by the **secant formula**; a graph of this formula (see graph below) shows that column load-carrying capacity decreases with increasing eccentricity. The secant formula gives the maximum compressive stress σ_{max} in an eccentrically loaded, pinned-end column in terms of average compressive stress P/A, modulus of elasticity E, slenderness ratio L/r, and eccentricity ratio ec/r^2 as

$$\sigma_{max} = \frac{P}{A}\left[1 + \frac{ec}{r^2}\sec\left(\frac{L}{2r}\sqrt{\frac{P}{EA}}\right)\right]$$

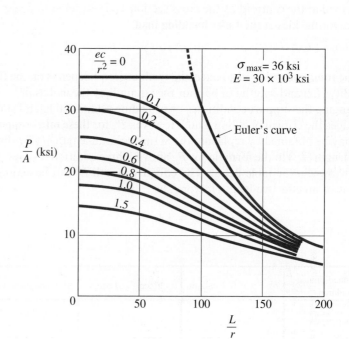

PROBLEMS

14.2 Buckling and Stability

Introductory Problems

14.2-1 A rigid bar of length L is supported by a linear elastic rotational spring with rotational stiffness β_R at A.

Determine the critical load P_{cr} for the structure.

PROBLEM 14.2-1

14.2-2 The figure shows an idealized structure consisting of a rigid bar with pinned connections and linearly elastic springs. Rotational stiffness is denoted β_R, and translational stiffness is denoted β.

(a) Determine the critical load P_{cr} for the structure from the figure part a.

(b) Find P_{cr} if another rotational spring is added at B from the figure part b.

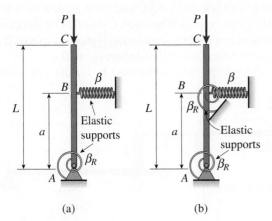

(a)   (b)

PROBLEM 14.2-2

14.2-3 Two rigid bars are connected with a rotational spring, as shown in the figure. Assume that the elastic rotational spring constant is $\beta_R = 75$ kip-in./rad. Calculate the critical load P_{cr} of the system. Assume that $L = 6$ ft.

PROBLEMS 14.2-3 and 14.2-4

14.2-4 Repeat Problem 14.2-3 assuming that $\beta_R = 10$ kN·m/rad and $L = 2$ m.

14.2-5 The figure shows an idealized structure consisting of two **rigid bars** with pinned connections and linearly elastic rotational springs. Rotational stiffness is denoted β_R.

Determine the critical load P_{cr} for the structure.

PROBLEM 14.2-5

Representative Problems

14.2-6 An idealized column consists of rigid bar $ABCD$ with a roller support at B and a roller and spring support at D. The spring constant at D is $\beta = 750$ N/m. Find the critical load P_{cr} of the column.

PROBLEM 14.2-6

14.2-7 An idealized column is made up of rigid segments ABC and CD that are joined by an elastic connection at C with rotational stiffness $\beta_R = 100$ kip-in./rad. The column has a roller support at B and a sliding support at D. Calculate the critical load P_{cr} of the column.

PROBLEM 14.2-7

14.2-8 The figure shows an idealized structure consisting of bars AB and BC that are connected using a hinge at B and linearly elastic springs at A and B. Rotational stiffness is denoted β_R and translational stiffness is denoted β.

(a) Determine the critical load P_{cr} for the structure from the figure part a.

(b) Find P_{cr} if an elastic connection is now used to connect bar segments AB and BC from the figure part b.

(a)   (b)

PROBLEM 14.2-8

14.2-9 The figure shows an idealized structure consisting of two rigid bars joined by an elastic connection with rotational stiffness β_R. Determine the critical load P_{cr} for the structure.

PROBLEM 14.2-9

14.2-10 The figure shows an idealized structure consisting of rigid bars ABC and DEF joined by a linearly elastic spring β between C and D. The structure is also supported by translational elastic support β at B and rotational elastic support β_R at E.

Determine the critical load P_{cr} for the structure.

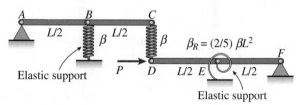

PROBLEM 14.2-10

14.2-11 The figure shows an idealized structure consisting of an L-shaped rigid bar structure supported by linearly elastic springs at A and C. Rotational

1090

stiffness is denoted β_R, and translational stiffness is denoted β.

Determine the critical load P_{cr} for the structure.

PROBLEM 14.2-11

14.2-12 Rigid column $ABCD$ has an elastic support at B with translational stiffness β. Find an expression for the critical load P_{cr} of the column.

PROBLEM 14.2-12

14.2-13 An idealized column is made up of rigid bars ABC and CD that are joined by a rotational elastic connection at C with stiffness β_R. The column has a roller support at B and a pin support at D. Find an expression for the critical load P_{cr} of the column.

PROBLEM 14.2-13

14.2-14 An idealized column is composed of rigid bars ABC and CD joined by an elastic connection with rotational stiffness β_{R1} at C. There is a roller support at B and an elastic support at D with translational spring stiffness β and rotational stiffness β_{R2}. Find the critical buckling loads for *each of the two buckling modes* of the column. Assume that $L = 3$ m, $\beta = 9$ kN/m, and $\beta_{R1} = \beta_{R2} = \beta L^2$. Sketch the buckled mode shapes.

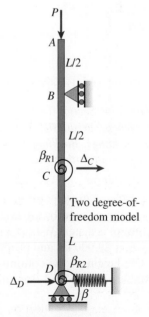

Elastic springs at D

PROBLEM 14.2-14 and 14.2-15

14.2-15 Repeat Problem 14.2-14 using $L = 12$ ft, $\beta = 0.25$ kips/in., $\beta_{R1} = 1.5\beta L^2$, and $\beta_{R2} = 2\beta_{R1}$.

14.2-16 An idealized column is composed of rigid bars ABC and CD joined by an elastic connection with rotational stiffness β_R at C. There is an elastic support at B with translational spring stiffness β and a pin support at D. Find the critical buckling loads for *each of the two buckling modes* of the column in terms of βL. Assume that $\beta_R = \beta L^2$. Sketch the buckled mode shapes.

PROBLEM 14.2-16

14.3 Columns with Pinned Ends

Solve the problems for Section 14.3 using the assumptions of ideal, slender, prismatic, and linearly elastic columns (Euler buckling). Buckling occurs in the plane of the figure unless stated otherwise.

Introductory Problems

14.3-1 Column AB has a pin support at A, a roller support at B, and is compressed by an axial load P (see figure). The column is a steel W 12×35 with modulus of elasticity $E = 29,000$ ksi and proportional limit $\sigma_{pl} = 50$ ksi. The height of the column is $L = 12$ ft. Find the allowable value of load P assuming a factor of safety $n = 2.5$.

PROBLEM 14.3-1

14.3-2 Slender column ABC is supported at A and C and is subjected to axial load P. Lateral support is provided at mid-height B but only in the plane of the figure; lateral support perpendicular to the plane of the figure is provided only at ends A and C. The column is a steel W shape with modulus of elasticity $E = 200$ GPa and proportional limit $\sigma_{pl} = 400$ MPa. The total length of the column $L = 9$ m. If the allowable load is 150 kN and the factor of safety is 2.5, determine the lightest W 200 section that can be used for the column. (See Table C-1(b), Appendix C).

PROBLEM 14.3-2

14.3-3 Calculate the critical load P_{cr} for a W 8×35 steel column (see figure) having a length $L = 24$ ft and $E = 30 \times 10^6$ psi under the following conditions:

(a) The column buckles by bending about its strong axis (axis 1–1).

(b) The column buckles by bending about its weak axis (axis 2–2).

In both cases, assume that the column has pinned ends.

PROBLEMS 14.3-3 through 14.3-5

14.3-4 Solve the preceding problem for a W 250 × 89 steel column having a length $L = 10$ m. Let $E = 200$ GPa.

14.3-5 Solve Problem 14.3-3 for a W 10 × 45 steel column having a length $L = 28$ ft.

Representative Problems

14.3-6 A horizontal beam AB is pin-supported at end A and carries a clockwise moment M at joint B, as shown in the figure. The beam is also supported at C by a pinned-end column of length L; the column is restrained laterally at $0.6L$ from the base at D. Assume the column can only buckle in the plane of the frame. The column is a solid steel bar ($E = 200$ GPa) of square cross section having length $L = 2.4$ m and side dimensions $b = 70$ mm. Let dimensions $d = L/2$. Based upon the critical load of the column, determine the allowable moment M if the factor of safety with respect to buckling is $n = 2.0$.

PROBLEM 14.3-6

14.3-7 A column ABC is supported at ends A and C and compressed by an axial load P (figure a). Lateral support is provided at point B but only in the plane of the figure; lateral support perpendicular to the plane of the figure is provided only at A and C. The column is constructed of two channel sections

(C 6 × 8.2) back to back (see figure b). The modulus of elasticity of the column is $E = 29,500$ ksi and the proportional limit is 50 ksi. The height of the column is $L = 15$ ft. Find the allowable value of load P using a factor of safety of 2.5.

(a)   (b)

PROBLEM 14.3-7

14.3-8 Find the controlling buckling load (kN) for the steel column shown in the figure. The column is pinned at top and bottom and is made up of two C 150 × 12.2 shapes that act together. Assume that $E = 205$ GPa and $L = 6$ m.

PROBLEM 14.3-8

14.3-9 A column, pinned at top and bottom, is made up of two C 6 × 13 steel shapes (see figure) that act together.

(a) Find the buckling load (kips) if the gap is zero.

(b) Find required separation distance d (inches) so that the buckling load is the same in y and z directions.

Assume that $E = 30,000$ ksi and $L = 18$ ft. Note that distance d is measured between the centroids of the two C shapes.

PROBLEM 14.3-9 and 14.3-10

14.3-10 Repeat Problem 14.3-9. Use two C 150 × 12.2 steel shapes and assume that $E = 205$ GPa and $L = 6$ m.

14.3-11 A horizontal beam AB is pin-supported at end A and carries a load Q at joint B, as shown in the figure. The beam is also supported at C by a pinned-end column of length L; the column is restrained laterally at $0.6L$ from the base at D. Assume the column can only buckle in the plane of the frame. The column is a solid aluminum bar ($E = 10 \times 10^6$ psi) of square cross section having a length $L = 30$ in. and side dimensions $b = 1.5$ in. Let dimension $d = L/2$. Based upon the critical load of the column, determine the allowable force Q if the factor of safety with respect to buckling is $n = 1.8$.

(a)

(b)

PROBLEM 14.3-12

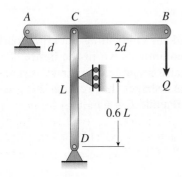

PROBLEM 14.3-11

14.3-12 A horizontal beam AB is supported at end A and carries a load Q at joint B, as shown in the figure part a. The beam is also supported at C by a pinned-end column of length L. The column has flexural rigidity EI.

(a) For the case of a sliding support at A (figure part a), what is the critical load Q_{cr}? (In other words, at what load Q_{cr} does the system collapse because of Euler buckling of the column DC?)

(b) Repeat part (a) if the sliding support at A is replaced by column AF with a length $3L/2$ and flexural rigidity EI (see figure part b).

14.3-13 A horizontal beam AB has a sliding support at end A and carries a load Q at end B, as shown in the figure part a. The beam is supported at C and D by two identical pinned-end columns of length L. Each column has flexural rigidity EI.

(a) Find an expression for the critical load Q_{cr}. (In other words, at what load Q_{cr} does the system collapse because of Euler buckling of the columns?)

(b) Repeat part (a), but assume a pin support at A. Find an expression for the critical moment M_{cr} (i.e., find the moment M at B at which the system collapses because of Euler buckling of the columns).

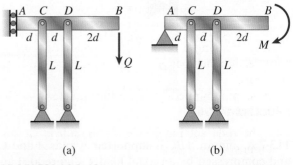

(a)   (b)

PROBLEM 14.3-13

14.3-14 A slender bar AB with pinned ends and length L is held between immovable supports (see figure).

What increase ΔT in the temperature of the bar will produce buckling at the Euler load?

PROBLEM 14.3-14

14.3-15 A rectangular column with cross-sectional dimensions b and h is pin-supported at ends A and C (see figure). At mid-height, the column is restrained in the plane of the figure but is free to deflect perpendicularly to the plane of the figure.

Determine the ratio h/b such that the critical load is the same for buckling in the two principal planes of the column.

PROBLEM 14.3-15

14.3.16 Three identical, solid circular rods, each of radius r and length L, are placed together to form a compression member (see the cross section shown in the figure).

Assuming pinned-end conditions, determine the critical load P_{cr}.

(a) The rods act independently as individual columns.

(b) The rods are bonded by epoxy throughout their lengths so that they function as a single member.

What is the effect on the critical load when the rods act as a single member?

PROBLEM 14.3-16

14.3-17 Three pinned-end columns of the same material have the same length and the same cross-sectional area (see figure). The columns are free to buckle in any direction. The columns have cross sections as: (a) a circle, (b) a square, and (c) an equilateral triangle.

Determine the ratios $P_a : P_b : P_c$ of the critical loads for these columns.

(a)   (b)   (c)

PROBLEM 14.3-17

14.3-18 A long slender column ABC is pinned at ends A and C and compressed by an axial force P (see figure). At the midpoint B, lateral support is provided to prevent deflection in the plane of the figure. The column is a steel wide-flange section (W 250 × 67) with $E = 200$ GPa. The distance between lateral supports is $L = 5.5$ m.

Calculate the allowable load P using a factor of safety $n = 2.4$, taking into account the possibility of Euler buckling about either principal centroidal axis (i.e., axis 1–1 or axis 2–2).

PROBLEM 14.3-18

14.3-19 The roof over a concourse at an airport is supported by the use of pretensioned cables. At a typical joint in the roof structure, a strut AB is compressed by the action of tensile forces F in a cable that makes an angle $\alpha = 75°$ with the strut (see figure and photo). The strut is a circular tube of steel ($E = 30,000$ ksi) with outer diameter $d_2 = 2.5$ in. and inner diameter $d_1 = 2.0$ in. The strut is 5.75 ft long and is assumed to be pin-connected at both ends.

Using a factor of safety $n = 2.5$ with respect to the critical load, determine the allowable force F in the cable.

PROBLEM 14.3-19

Cable and strut at typical joint of airport concourse roof

14.3-20 The hoisting arrangement for lifting a large pipe is shown in the figure. The spreader is a steel tubular section with outer diameter 70 mm and inner diameter 57 mm. Its length is 2.6 m, and its modulus of elasticity is 200 GPa.

Based upon a factor of safety of 2.25 with respect to Euler buckling of the spreader, what is the maximum weight of pipe that can be lifted? (Assume pinned conditions at the ends of the spreader.)

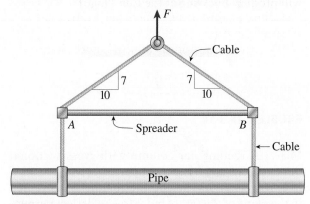

PROBLEM 14.3-20

14.3-21 A pinned-end strut of aluminum ($E = 10,400$ ksi) with a length $L = 6$ ft is constructed of circular tubing with an outside diameter $d = 2$ in. (see figure). The strut must resist an axial load $P = 4$ kips with a factor of safety $n = 2.0$ with respect to the critical load.

Determine the required thickness t of the tube.

$d = 2$ in.

PROBLEM 14.3-21

14.3-22 The cross section of a column built up of two steel I-beams (S 150 × 25.7 sections) is shown in the figure. The beams are connected by spacer bars, or *lacing*, to ensure that they act together as a single column. (The lacing is represented by dashed lines in the figure.)

The column is assumed to have pinned ends and may buckle in any direction. Assuming $E = 200$ GPa and $L = 8.5$ m, calculate the critical load P_{cr} for the column.

S 150 × 25.7

100 mm

PROBLEM 14.3-22

14.3-23 The truss ABC shown in the figure supports a vertical load W at joint B. Each member is a slender circular steel pipe ($E = 30,000$ ksi) with an outside diameter of 4 in. and wall thickness 0.25 in. The distance between supports is 23 ft. Joint B is restrained against displacement perpendicular to the plane of the truss.

Determine the critical value W_{cr} of the load.

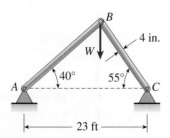

PROBLEM 14.3-23

14.3-24 A truss ABC supports a load W at joint B, as shown in the figure. The length L_1 of member AB is fixed, but the length of strut BC varies as the angle θ is changed. Strut BC has a solid circular cross section. Joint B is restrained against displacement perpendicular to the plane of the truss.

Assuming that collapse occurs by Euler buckling of the strut, determine the angle θ for minimum weight of the strut.

PROBLEM 14.3-24

14.3-25 An $S6 \times 12.5$ steel cantilever beam AB is supported by a steel tie rod at B as shown. The tie rod is just taut when a roller support is added at C at a distance s to the left of B, then the distributed load q is applied to beam segment AC. Assume $E = 30 \times 10^6$ psi and neglect the self-weight of the beam and tie rod. See Table C-2(a) in Appendix C for the properties of the S-shape beam.

(a) What value of uniform load q will, if exceeded, result in buckling of the tie rod if $L_1 = 6$ ft, $s = 2$ ft, $H = 3$ ft, and $d = 0.25$ in.?

(b) What minimum beam moment of inertia I_b is required to prevent buckling of the tie rod if $q = 200$ lb/ft, $L_1 = 6$ ft, $H = 3$ ft, $d = 0.25$ in., and $s = 2$ ft?

(c) For what distance s will the tie rod be just on the verge of buckling if $q = 200$ lb/ft, $L_1 = 6$ ft, $H = 3$ ft, and $d = 0.25$ in.?

PROBLEM 14.3-25

14.3-26 The plane truss shown in the figure supports vertical loads F at joint D, $2F$ at joint C, and $3F$ at joint B. Each member is a slender circular pipe ($E = 70$ GPa) with an outside diameter of 60 mm and wall thickness of 5 mm. Joint B is restrained against displacement perpendicular to the plane of the truss. Determine the critical value of load variable F (kN) at which member BF fails by Euler buckling.

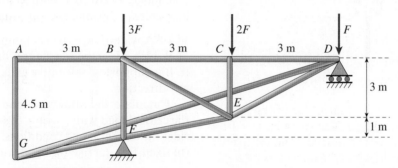

PROBLEM 14.3-26

14.3-27 A space truss is restrained at joints O, A, B, and C, as shown in the figure. Load P is applied at joint A and load $2P$ acts downward at joint C. Each member is a slender, circular pipe ($E = 10,600$ ksi) with an outside diameter of 3.5 in. and wall thickness of 0.25 in. Length variable $L = 11$ ft. Determine the critical value of load variable P (kips) at which member OB fails by Euler buckling.

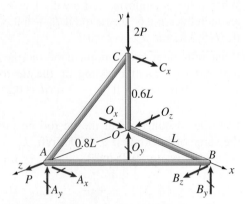

PROBLEM 14.3-27

14.4 Columns with Other Support Conditions

Solve the problems for Section 14.4 using the assumptions of ideal, slender, prismatic, and linearly elastic columns (Euler buckling). Buckling occurs in the plane of the figure unless stated otherwise.

Introductory Problems

14.4-1 A fixed-end column with circular cross section is acted on by compressive axial load P. The 18-ft-long-column has an outer diameter of 5 in., a thickness of 0.5 in., and is made of aluminum with a modulus of elasticity of 10,000 ksi. Find the buckling load of the column.

PROBLEM 14.4-1

14.4-2 A cantilever aluminum column has a square tube cross section with an outer dimension of 150 mm. The column has a length $L = 4$ m and is designed to support an axial load of 45 kN. Find the minimum required thickness of the section if the factor of safety $n = 2.5$ with respect to buckling. Assume that the modulus of elasticity is 72 GPa and the proportional limit is 480 MPa.

PROBLEM 14.4-2

14.4-3 An aluminum pipe column ($E = 10,400$ ksi) with a length $L = 10.0$ ft has inside and outside diameters $d_1 = 5.0$ in. and $d_2 = 6.0$ in., respectively (see figure). The column is supported only at the ends and may buckle in any direction.

Calculate the critical load P_{cr} for the following end conditions: (a) pinned-pinned, (b) fixed-free, (c) fixed-pinned, and (d) fixed-fixed.

PROBLEMS 14.4-3 and 14.4-4

14.4-4 Solve the preceding problem for a steel pipe column ($E = 210$ GPa) with length $L = 1.2$ m, inner diameter $d_1 = 36$ mm, and outer diameter $d_2 = 40$ mm.

14.4-5 A wide-flange steel column ($E = 30 \times 10^6$ psi) of W 12×87 shape (see figure) has a length $L = 28$ ft. It is supported only at the ends and may buckle in any direction.

Calculate the allowable load P_{allow} based upon the critical load with a factor of safety $n = 2.5$. Consider the following end conditions: (a) pinned-pinned, (b) fixed-free, (c) fixed-pinned, and (d) fixed-fixed.

PROBLEMS 14.4-5 and 14.4-6

14.4-6 Solve the preceding problem for a W 250 × 89 shape with length $L = 7.5$ m and $E = 200$ GPa.

Representative Problems

14.4-7 The upper end of a W 8 × 21 wide-flange steel column ($E = 30 \times 10^3$ ksi) is supported laterally between two pipes (see figure). The pipes are not attached to the column, and friction between the pipes and the column is unreliable. The base of the column provides a fixed support, and the column is 13 ft long.

Determine the critical load for the column, considering Euler buckling in the plane of the web and also perpendicular to the plane of the web.

W 8 × 21

PROBLEM 14.4-7

14.4-8 A vertical post AB is embedded in a concrete foundation and held at the top by two cables (see figure). The post is a hollow steel tube with modulus of elasticity 200 GPa, outer diameter of 40 mm, and thickness of 5 mm. The cables are tightened equally by turnbuckles.

If a factor of safety of 3.0 against Euler buckling in the plane of the figure is desired, what is the maximum allowable tensile force T_{allow} in the cables?

PROBLEM 14.4-8

14.4-9 The horizontal beam ABC shown in the figure is supported by columns BD and CE. The beam is prevented from moving horizontally by the pin support at end A. Each column is pinned at its upper end to the beam, but at the lower ends, support D is a sliding support and support E is pinned. Both columns are solid steel bars ($E = 30 \times 10^6$ psi) of square cross section with width equal to 0.625 in. A load Q acts at distance a from column BD.

(a) If the distance $a = 12$ in., what is the critical value Q_{cr} of the load?

(b) If the distance a can be varied between 0 and 40 in., what is the maximum possible value of Q_{cr}? What is the corresponding value of the distance a?

PROBLEM 14.4-9

14.4-10 The roof beams of a warehouse are supported by pipe columns (see figure) having an outer diameter $d_2 = 100$ mm and inner diameter $d_1 = 90$ mm. The columns have a length $L = 4.0$ m, modulus $E = 210$ GPa, and fixed supports at the base.

Calculate the critical load P_{cr} of one of the columns using the following assumptions: (a) the upper end is pinned and the beam prevents horizontal displacement; (b) the upper end is fixed against rotation and the beam prevents horizontal displacement; (c) the upper end is pinned, but the beam is free to move horizontally; and (d) the upper end is fixed against rotation, but the beam is free to move horizontally.

Roof beam

Pipe column

d_2

L

PROBLEM 14.4-10

14.4-11 Determine the critical load P_{cr} and the equation of the buckled shape for an ideal column with ends fixed against rotation (see figure) by solving the differential equation of the deflection curve. (See also Fig. 14-18.)

P

B

L

A

PROBLEM 14.4-11

14.4-12 A fixed-pinned column is a $W\,310 \times 21$ steel shape and is designed to carry an axial load of 125 kN. Determine the maximum permissible height L of the column if a factor of safety $n = 2.5$ is required with respect to the buckling of the column. Use $E = 205$ GPa and assume that the proportional limit is 340 MPa. The column may buckle about either axis of the cross section.

P

L

PROBLEM 14.4-12

14.4-13 Find the controlling buckling load (kips) for the steel column shown in the figure. The column is fixed at the base and free at the top and is made up of two $C\,6 \times 10.5$ shapes that act together. Assume that $E = 30,000$ ksi and $L = 14$ ft.

y

z

PROBLEM 14.4-13

14.4-14 A column, fixed at the base and free at the top, is made up of two $C\,100 \times 10.8$ steel shapes (see figure) that act together.

(a) Find the buckling load (kN) if the gap is zero.

(b) Find the required separation distance d (mm) so that the buckling load is the same in y and z directions.

Assume that $E = 205$ GPa and $L = 6$ m. Note that distance d is measured between the centroids of the two C shapes.

Gap y

z

d

PROBLEM 14.4-14

14.4-15 A rigid L-shaped frame is supported by a steel pipe column AB (see figure) and is subjected to a horizontal load $F = 140$ kips. If the pipe has an outside diameter $d = 4$ in. and a factor of safety of 2.5 is required with respect to Euler buckling, what is the minimum acceptable thickness t of the pipe? Assume that $E = 30,000$ ksi for AB.

PROBLEM 14.4-15

14.4-16 An aluminum tube AB with a circular cross section has a sliding support at the base and is pinned at the top to a horizontal beam supporting a load $Q = 200$ kN (see figure).

Determine the required thickness t of the tube if its outside diameter d is 200 mm and the desired factor of safety with respect to Euler buckling is $n = 3.0$. (Assume $E = 72$ GPa.)

PROBLEM 14.4-16

14.4-17 The frame ABC consists of two members AB and BC that are rigidly connected at joint B, as shown in part a of the figure. The frame has pin supports at A and C. A concentrated load P acts at joint B, thereby placing member AB in direct compression.

To assist in determining the buckling load for member AB, represent it as a pinned-end column, as shown in part b of the figure. At the top of the column, a rotational spring of stiffness β_R represents the restraining action of the horizontal beam BC on the column (note that the horizontal beam provides resistance to rotation of joint B when the column buckles). Also, consider only bending effects in the analysis (i.e., disregard the effects of axial deformations).

(a) By solving the differential equation of the deflection curve, derive the buckling equation for this column:

$$\frac{\beta_R L}{EI}(kL \cot kL - 1) - k^2 L^2 = 0$$

where L is the length of the column and EI is its flexural rigidity.

(b) For the particular case when member BC is identical to member AB, the rotational stiffness β_R equals $3EI/L$ (see Case 7, Table E-2, Appendix E). For this special case, determine the critical load P_{cr}.

(a)   (b)

PROBLEM 14.4-17

14.5 Columns with Eccentric Axial Loads

To solve the problems for Section 14.5, assume that bending occurs in the principal plane containing the eccentric axial load.

Introductory Problems

14.5-1 An aluminum bar having a rectangular cross section (2.0 in. $\times 1.0$ in.) and length $L = 30$ in. is compressed by axial loads that have a resultant $P = 2800$ lb acting at the midpoint of the long side of the cross section (see figure).

PROBLEM 14.5-1

Assuming that the modulus of elasticity E is equal to 10×10^6 psi and that the ends of the bar are pinned, calculate the maximum deflection δ and the maximum bending moment M_{max}.

14.5-2 A steel bar having a square cross section (50 mm × 50 mm) and length $L = 2.0$ m is compressed by axial loads that have a resultant $P = 60$ kN acting at the midpoint of one side of the cross section (see figure).

Assuming that the modulus of elasticity E is equal to 210 GPa and that the ends of the bar are pinned, calculate the maximum deflection δ and the maximum bending moment M_{max}.

PROBLEM 14.5-2

14.5-3 A simply supported slender column is subjected to axial load $P = 175$ kips applied at distance $e = 0.5$ in. from joints A and B (see figure). The column has a circular cross section with an outer diameter of 10 in. and wall thickness of 0.5 in. Calculate the height of the column if the maximum deflection caused by the axial loads is 0.1 in. Assume that $E = 29,000$ ksi.

PROBLEM 14.5-3

1102

14.5-4 A brass bar of a length $L = 0.4$ m is loaded at end B by force $P = 10$ kN with an eccentricity $e = 6$ mm. The bar has a rectangular cross section with an h/b ratio of 1.5. Find the dimensions of the bar if the deflection at the end is limited to 4 mm. Assume that $E = 110$ GPa.

PROBLEM 14.5-4

Representative Problems

14.5-5 Determine the bending moment M in the pinned-end column with eccentric axial loads shown in the figure. Then plot the bending-moment diagram for an axial load $P = 0.3P_{cr}$.

Note: Express the moment as a function of the distance x from the end of the column, and plot the diagram in nondimensional form with M/Pe as ordinate and x/L as abscissa.

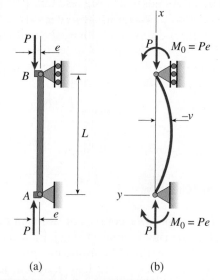

(a)   (b)

PROBLEMS 14.5-5 through 14.5-7

14.5-6 Plot the load–deflection diagram for a pinned-end column with eccentric axial loads (see figure) if the eccentricity e of the load is 5 mm and the column has a length $L = 3.6$ m, moment of inertia $I = 9.0 \times 10^6$ mm^4, and modulus of elasticity $E = 210$ GPa.

Note: Plot the axial load as ordinate and the deflection at the midpoint as abscissa.

14.5-7 Solve the preceding problem for a column with $e = 0.20$ in., $L = 12$ ft, $I = 21.7$ in^4, and $E = 30 \times 10^6$ psi.

14.5-8 A wide-flange member (W 200 × 22.5) is compressed by axial loads that have a resultant P acting at the point shown in the figure. The member has modulus of elasticity $E = 200$ GPa and pinned conditions at the ends. Lateral supports prevent any bending about the weak axis of the cross section.

If the length of the member is 6.2 m and the deflection is limited to 6.5 mm, what is the maximum allowable load P_{allow}?

PROBLEM 14.5-8

14.5-9 A wide-flange member (W 10 × 30) is compressed by axial loads that have a resultant $P = 20$ kips acting at the point shown in the figure. The material is steel with modulus of elasticity $E = 29,000$ ksi. Assuming pinned-end conditions, determine the maximum permissible length L_{max} if the deflection is not to exceed 1/400th of the length.

PROBLEMS 14.5-9 and 14.5-10

14.5-10 Solve the preceding problem (W 250 × 44.8) if the resultant force P equals 110 kN and $E = 200$ GPa.

14.5-11 The column shown in the figure is fixed at the base and free at the upper end. A compressive load P acts at the top of the column with an eccentricity e from the axis of the column.

Beginning with the differential equation of the deflection curve, derive formulas for the maximum deflection δ of the column and the maximum bending moment M_{max} in the column.

(a)   (b)

PROBLEM 14.5-11

14.5-12 An aluminum box column with a square cross section is fixed at the base and free at the top (see figure). The outside dimension b of each side is 100 mm and the thickness t of the wall is 8 mm. The resultant of the compressive loads acting on the top of the column is a force $P = 50$ kN acting at the outer edge of the column at the midpoint of one side.

What is the longest permissible length L_{max} of the column if the deflection at the top is not to exceed 30 mm? (Assume $E = 73$ GPa.)

Section A–A

PROBLEMS 14.5-12 and 14.5-13

14.5-13 Solve the preceding problem for an aluminum column with $b = 6.0$ in., $t = 0.5$ in., $P = 30$ kips, and $E = 10.6 \times 10^3$ ksi. The deflection at the top is limited to 2.0 in.

14.5-14 A steel post *AB* with a hollow circular cross section is fixed at the base and free at the top (see figure). The inner and outer diameters are $d_1 = 96$ mm and $d_2 = 110$ mm, respectively, and the length is $L = 4.0$ m.

A cable *CBD* passes through a fitting that is welded to the side of the post. The distance between the plane of the cable (plane *CBD*) and the axis of the post is $e = 100$ mm, and the angles between the cable and the ground are $\alpha = 53.13°$. The cable is pretensioned by tightening the turnbuckles.

If the deflection at the top of the post is limited to $\delta = 20$ mm, what is the maximum allowable tensile force *T* in the cable? (Assume $E = 205$ GPa.)

PROBLEM 14.5-14

14.5-15 A frame *ABCD* is constructed of steel wide-flange members (W 8 × 21; $E = 30 \times 10^6$ psi) and subjected to triangularly distributed loads of maximum intensity q_0 acting along the vertical members (see figure). The distance between supports is $L = 20$ ft and the height of the frame is $h = 4$ ft. The members are rigidly connected at *B* and *C*.

(a) Calculate the intensity of load q_0 required to produce a maximum bending moment of 80 kip-in. in the horizontal member *BC*.

PROBLEM 14.5-15   Section *E–E*

1104

(b) If the load q_0 is reduced to one-half of the value calculated in part (a), what is the maximum bending moment in member *BC*? What is the ratio of this moment to the moment of 80 kip-in. in part (a)?

14.6 The Secant Formula for Columns

To solve the problems for Section 14.6, assume that bending occurs in the principal plane containing the eccentric axial load.

Introductory Problems

14.6-1 A steel bar has a square cross section of width $b = 2.0$ in. (see figure). The bar has pinned supports at the ends and is 3.0 ft long. The axial forces acting at the end of the bar have a resultant $P = 20$ kips located at distance $e = 0.75$ in. from the center of the cross section. Also, the modulus of elasticity of the steel is 29,000 ksi.

(a) Determine the maximum compressive stress σ_{max} in the bar.
(b) If the allowable stress in the steel is 18,000 psi, what is the maximum permissible length L_{max} of the bar?

PROBLEMS 14.6-1 through 14.6-3

14.6-2 A brass bar ($E = 100$ GPa) with a square cross section is subjected to axial forces having a resultant *P* acting at distance *e* from the center (see figure). The bar is pin supported at the ends and is 0.6 m in length. The side dimension *b* of the bar is 30 mm and the eccentricity *e* of the load is 10 mm.

If the allowable stress in the brass is 150 MPa, what is the allowable axial force P_{allow}?

14.6-3 A square aluminum bar with pinned ends carries a load $P = 25$ kips acting at distance $e = 2.0$ in. from the center (see figure). The bar has a length $L = 54$ in. and modulus of elasticity $E = 10,600$ ksi.

If the stress in the bar is not to exceed 6 ksi, what is the minimum permissible width b_{min} of the bar?

14.6-4 A pinned-end column of a length $L = 2.1$ m is constructed of steel pipe ($E = 210$ GPa) having an inside diameter $d_1 = 60$ mm and outside diameter $d_2 = 68$ mm (see figure). A compressive load $P = 10$ kN acts with eccentricity $e = 30$ mm.

(a) What is the maximum compressive stress σ_{max} in the column?

(b) If the allowable stress in the steel is 50 MPa, what is the maximum permissible length L_{max} of the column?

PROBLEMS 14.6-4 through 14.6-6

14.6-5 A pinned-end strut of a length $L = 5.2$ ft is constructed of steel pipe ($E = 30 \times 10^3$ ksi) having an inside diameter $d_1 = 2.0$ in. and outside diameter $d_2 = 2.2$ in. (see figure). A compressive load $P = 2.0$ kips is applied with eccentricity $e = 1.0$ in.

(a) What is the maximum compressive stress σ_{max} in the strut?

(b) What is the allowable load P_{allow} if a factor of safety $n = 2$ with respect to yielding is required? (Assume that the yield stress σ_Y of the steel is 42 ksi.)

14.6-6 A circular aluminum tube with pinned ends supports a load $P = 18$ kN acting at a distance $e = 50$ mm from the center (see figure). The length of the tube is 3.5 m, and its modulus of elasticity is 73 GPa.

If the maximum permissible stress in the tube is 20 MPa, what is the required outer diameter d_2 if the ratio of diameters is to be $d_1 / d_2 = 0.9$?

Representative Problems

14.6-7 A steel W 12 × 35 column is pin-supported at the ends. The column carries an axial load $P = 150$ kips with eccentricity $e = 3$ in. (see figure). Find the length of the column if the maximum stress is restricted to the proportional limit $\sigma_{pl} = 50$ ksi and a factor of safety of 2.0 is assumed. Use modulus of elasticity $E = 29,000$ ksi.

PROBLEM 14.6-7

14.6-8 A steel W 310 × 52 column is pin-supported at the ends and has a length $L = 4$ m. The column supports two eccentrically applied loads $P_1 = 750$ kN and $P_2 = 500$ kN (see figure). Bending occurs about axis 1–1 of the cross section, and the eccentric loads are applied on axis 2–2 at distances of 200 mm from centroid C. Calculate the maximum compressive stress in the column. Assume that modulus elasticity $E = 200$ GPa.

PROBLEM 14.6-8

14.6-9 A steel column ($E = 30 \times 10^3$ ksi) with pinned ends is constructed of a W 10×60 wide-flange shape (see figure). The column is 24 ft long. The resultant of the axial loads acting on the column is a force P acting with an eccentricity $e = 2.0$ in.

(a) If $P = 120$ kips, determine the maximum compressive stress σ_{max} in the column.

(b) Determine the allowable load P_{allow} if the yield stress is $\sigma_Y = 42$ ksi and the factor of safety with respect to yielding of the material is $n = 2.5$.

$e = 2.0$ in.

W 10×60

PROBLEM 14.6-9

14.6-10 A W 410×85 steel column is compressed by a force $P = 340$ kN acting with an eccentricity $e = 38$ mm, as shown in the figure. The column has pinned ends and a length L. Also, the steel has a modulus of elasticity $E = 200$ GPa and yield stress $\sigma_Y = 250$ MPa.

(a) If the length $L = 3$ m, what is the maximum compressive stress σ_{max} in the column?

(b) If a factor of safety $n = 2.0$ is required with respect to yielding, what is the longest permissible length L_{max} of the column?

$P = 340$ kN $e = 38$ mm

W 410×85

PROBLEM 14.6-10

14.6-11 A steel column ($E = 30 \times 10^3$ ksi) that is fixed at the base and free at the top is constructed of a W 8×35 wide-flange member (see figure). The column is 9.0 ft long. The force P acting at the top of the column has an eccentricity $e = 1.25$ in.

(a) If $P = 40$ kips, what is the maximum compressive stress in the column?

(b) If the yield stress is 36 ksi and the required factor of safety with respect to yielding is 2.1, what is the allowable load P_{allow}?

P e

A A P e

L

Section A–A

PROBLEMS 14.6-11 and 14.6-12

14.6-12 A W 310×74 wide-flange steel column with length $L = 3.8$ m is fixed at the base and free at the top (see figure). The load P acting on the column is intended to be centrally applied, but because of unavoidable discrepancies in construction, an eccentricity ratio of 0.25 is specified. Also, the following data are supplied: $E = 200$ GPa, $\sigma_Y = 290$ MPa, and $P = 310$ kN.

(a) What is the maximum compressive stress σ_{max} in the column?

(b) What is the factor of safety n with respect to yielding of the steel?

14.6-13 A pinned-end column with a length $L = 18$ ft is constructed from a W 12×87 wide-flange shape (see figure). The column is subjected to a centrally applied load $P_1 = 180$ kips and an eccentrically applied load $P_2 = 75$ kips. The load P_2 acts at a distance $s = 5.0$ in. from the centroid of the cross section. The properties of the steel are $E = 29,000$ ksi and $\sigma_Y = 36$ ksi.

(a) Calculate the maximum compressive stress in the column.

(b) Determine the factor of safety with respect to yielding.

Wide-flange column

PROBLEMS 14.6-13 and 14.6-14

L

A A

Section A–A

PROBLEMS 14.6-15 and 14.6-16

14.6-14 The wide-flange, pinned-end column shown in the figure carries two loads: a force $P_1 = 450$ kN acting at the centroid and a force $P_2 = 270$ kN acting at a distance $s = 100$ mm from the centroid. The column is a W 250 × 67 shape with $L = 4.2$ m, $E = 200$ GPa, and $\sigma_Y = 290$ MPa.

(a) What is the maximum compressive stress in the column?

(b) If the load P_1 remains at 450 kN, what is the largest permissible value of the load P_2 in order to maintain a factor of safety of 2.0 with respect to yielding?

14.6-15 A W 14 × 53 wide-flange column of a length $L = 15$ ft is fixed at the base and free at the top (see figure). The column supports a centrally applied load $P_1 = 120$ kips and a load $P_2 = 40$ kips supported on a bracket. The distance from the centroid of the column to the load P_2 is $s = 12$ in. Also, the modulus of elasticity is $E = 29,000$ ksi, and the yield stress is $\sigma_Y = 36$ ksi.

(a) Calculate the maximum compressive stress in the column.

(b) Determine the factor of safety with respect to yielding.

14.6-16 A wide-flange column with a bracket is fixed at the base and free at the top (see figure). The column supports a load $P_1 = 340$ kN acting at the centroid and a load $P_2 = 110$ kN acting on the bracket at a distance $s = 250$ mm from the load P_1. The column is a W 310 × 52 shape with $L = 5$ m, $E = 200$ GPa, and $\sigma_Y = 290$ MPa.

(a) What is the maximum compressive stress in the column?

(b) If the load P_1 remains at 340 kN, what is the largest permissible value of the load P_2 in order to maintain a factor of safety of 1.8 with respect to yielding?

Mathematical Formulas

Vector Operations

Properties

A *vector* is a quantity that has both magnitude and direction. Examples of vector quantities are force, displacement, velocity, and acceleration. A *scalar,* on the other hand, has only magnitude. Scalars include such items as mass, volume, density, time, and speed. In this text, a vector is shown in a bold italic font (V) while the magnitude of the vector is shown as V using a regular (non-bold italic) font. The vector magnitude is a scalar quantity and can also be shown as $|V|$.

Addition, Subtraction

Parallelogram Law: The *sum* of two vectors can be represented by the vector equation

$$V_s = V_1 + V_2 = V_2 + V_1$$

The order of the vector addition does not affect the final result. Resultant vector V (see figure below) is the diagonal of the *parallelogram* formed from vectors V_1 and V_2. The *difference* between two vectors is obtained using vector subtraction as follows:

$$V_d = V_1 - V_2 = V_1 + (-V_2)$$

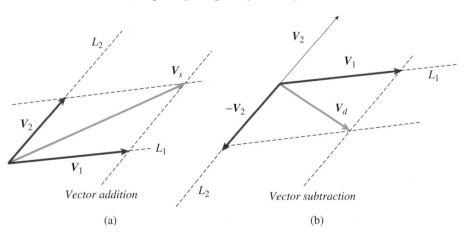

Vector addition   (a)   *Vector subtraction*   (b)

Triangle Law: An alternative way to represent vector addition and subtraction is with the triangle law. Vector V_2 is moved to the head of vector V_1 (since both are free vectors) to find the resulting equivalent vector (V_s or V_d for addition and subtraction, respectively) as shown in the figure below.

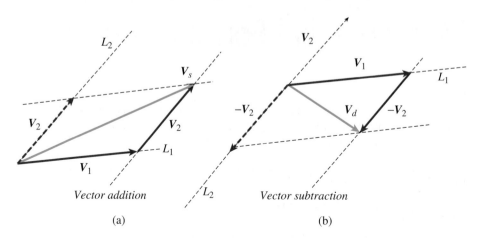

Vector addition

(a)

Vector subtraction

(b)

Commutative and Associative Laws:

$$\text{Commutative: } V_1 + V_2 = V_2 + V_1 \qquad \text{and}$$
$$\text{Associative: } (V_1 + V_2) + V_3 = V_1 + (V_2 + V_3)$$

Vector Components: In three dimensions, vector V has three components (see figure below), one along each coordinate axis (x, y, and z), and vector addition is carried out as follows:

$$V = V_x + V_y + V_z$$

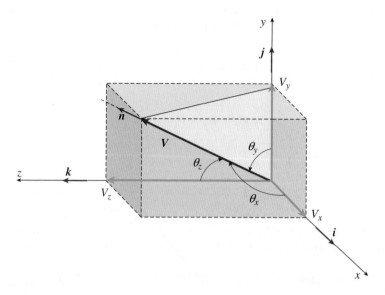

Unit Vectors

Vector V can also be expressed mathematically as the product of vector magnitude $V = |V|$ and unit vector n_V as follows: $V = |V|n_V = Vn_V$, where the magnitude of V is $|V| = V$ and $V = \sqrt{V_x^2 + V_y^2 + V_z^2}$. Hence, $n_V = \dfrac{V}{|V|} = \dfrac{V}{V}$ and unit vector $\boldsymbol{n}_V$ has unit magnitude [$|\boldsymbol{n}_V| = 1$], is dimensionless, and has the same direction as $\boldsymbol{V}$. Unit vectors along x, y, and z axis directions are defined as $\boldsymbol{i}, \boldsymbol{j},$ and $\boldsymbol{k}$ so vector $\boldsymbol{V}$ can be expressed in terms of its rectangular components, in three dimensions, as follows:

$$\boldsymbol{V} = V_x\boldsymbol{i} + V_y\boldsymbol{j} + V_z\boldsymbol{k} = \begin{pmatrix} V_x \\ V_y \\ V_z \end{pmatrix} \text{ where } \boldsymbol{V}_x = V_x\boldsymbol{i}, \boldsymbol{V}_y = V_y\boldsymbol{j}, \text{ and } \boldsymbol{V}_z = V_z\boldsymbol{k}.$$

Direction Cosines

The *direction cosines* of a vector $\boldsymbol{V}$ are the cosines of the angles between the vector and the three coordinate axes (see figure above). The direction cosines of $\boldsymbol{V}$ can be defined as follows:

$$l = \cos(\theta_x) = \frac{V_x}{V}, \; m = \cos(\theta_y) = \frac{V_y}{V} \text{ and } n = \cos(\theta_z) = \frac{V_z}{V}$$

Using the direction cosines gives

$$\boldsymbol{V} = V(l\boldsymbol{i} + m\boldsymbol{j} + n\boldsymbol{k}) = V\begin{pmatrix} l \\ m \\ n \end{pmatrix}$$

It follows that

$$l^2 + m^2 + n^2 = \frac{V_x^2 + V_y^2 + V_z^2}{V^2} = 1$$

Dot or Scalar Product

The dot or scalar product of two vectors, $\boldsymbol{F}$ and $\boldsymbol{P}$, is defined as the product of their magnitudes (F times P) times the cosine of the angle (θ) between them: $\boldsymbol{F} \cdot \boldsymbol{P} = FP\cos(\theta)$. Two vectors are perpendicular if $\boldsymbol{F} \cdot \boldsymbol{P} = \boldsymbol{P} \cdot \boldsymbol{F} = 0$.

The dot product can be used to find the angle θ between two vectors $\boldsymbol{F}$ and $\boldsymbol{P}$ as $\theta = \cos^{-1}\left[\dfrac{\boldsymbol{F} \cdot \boldsymbol{P}}{F\,P}\right]$. The dot product is also used to find components of a force along ($\boldsymbol{F}_{//}$) and perpendicular to ($\boldsymbol{F}_\perp$) a line of action of interest (defined by unit vector $\boldsymbol{n}_P$ in the figure above). The component of $\boldsymbol{F}$ along (or parallel to) the direction defined by $\boldsymbol{n}_P$ is $\boldsymbol{F}_{//} = (\boldsymbol{F} \cdot \boldsymbol{n}_P)\boldsymbol{n}_P$ and the perpendicular component is $\boldsymbol{F}_\perp = \boldsymbol{F} - \boldsymbol{F}_{//}$.

Finally, the *distributive law* gives: $\boldsymbol{F} \cdot (\boldsymbol{P} + \boldsymbol{Q}) = \boldsymbol{F} \cdot \boldsymbol{P} + \boldsymbol{F} \cdot \boldsymbol{Q}$

Cross or Vector Product

The cross product of two vectors $\boldsymbol{F}$ and $\boldsymbol{P}$ is a vector which is perpendicular to the plane formed by $\boldsymbol{F}$ and $\boldsymbol{P}$ (see figure) with a direction defined by the right hand rule. As a result, $\boldsymbol{F} \times \boldsymbol{P} = -\boldsymbol{P} \times \boldsymbol{F}$. The magnitude of the cross product is defined as $|\boldsymbol{F} \times \boldsymbol{P}| = FP\sin\theta$. The *distributive law* gives $\boldsymbol{F} \times (\boldsymbol{P} + \boldsymbol{Q}) = \boldsymbol{F} \times \boldsymbol{P} + \boldsymbol{F} \times \boldsymbol{Q}$.

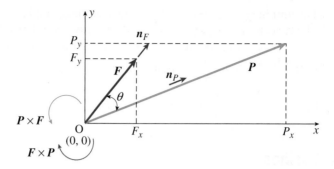

The cross product of vectors F and P can be found in three dimensions using a determinant as follows.

$$F \times P = \begin{vmatrix} i & j & k \\ F_x & F_y & F_z \\ P_x & P_y & P_z \end{vmatrix}$$

Determinant expansion by minors using the cofactors of the first row gives

$$F \times P = (F_y P_z - F_z P_y)i + (F_z P_x - F_x P_z)j + (F_x P_y - F_y P_x)k$$

Cross products of unit vectors in a right-handed Cartesian coordinate system are summarized as follows.

$$i \times j = k \quad j \times k = i \quad k \times i = j$$
$$j \times i = -k \quad k \times j = -i \quad i \times k = -j$$
$$i \times i = j \times j = k \times k = 0$$

Triple Scalar Product; Triple Vector Product

The *triple scalar product* is defined as $F \times P \cdot Q = (F \times P) \cdot Q = Q \cdot (F \times P)$. The result is a scalar quantity; parentheses are not needed because vector F cannot be crossed with scalar $P \cdot Q$. The triple scalar product can be found using any of the following determinants:

$$F \times P \cdot Q = \begin{vmatrix} F_x & F_y & F_z \\ P_x & P_y & P_z \\ Q_x & Q_y & Q_z \end{vmatrix} = \begin{vmatrix} Q_x & Q_y & Q_z \\ F_x & F_y & F_z \\ P_x & P_y & P_z \end{vmatrix} = \begin{vmatrix} P_x & P_y & P_z \\ Q_x & Q_y & Q_z \\ F_x & F_y & F_z \end{vmatrix}$$

The *triple vector product* is defined as $(F \times P) \times Q = -Q \times (F \times P) = Q \times (P \times F)$, which is equivalent to $(F \times P) \times Q = Q \cdot FP - Q \cdot PF$. Note that $Q \cdot FP$ means that the scalar product of Q and F is then multiplied by vector P. Also note that $(F \times P) \times Q \neq F \times (P \times Q)$.

Mathematical Constants

$\pi = 3.14159\ldots \quad e = 2.71828\ldots \quad 2\pi \text{ radians} = 360 \text{ degrees}$

$1 \text{ radian} = \dfrac{180}{\pi} \text{ degrees} = 57.2958°$

$1 \text{ degree} = \dfrac{\pi}{180} \text{ radians} = 0.0174533 \text{ rad}$

Conversions: Multiply degrees by $\dfrac{\pi}{180}$ to obtain radians

Multiply radians by $\dfrac{180}{\pi}$ to obtain degrees

Exponents

$$A^n A^m = A^{n+m} \qquad \frac{A^m}{A^n} = A^{m-n} \qquad (A^m)^n = A^{mn} \qquad A^{-m} = \frac{1}{A^m}$$

$$(AB)^n = A^n B^n \qquad \left(\frac{A}{B}\right)^n = \frac{A^n}{B^n} \qquad A^{m/n} = \sqrt[n]{A^m} \qquad A^0 = 1 (A \neq 0)$$

Logarithms

$\log \equiv$ common logarithm (logarithm to the base 10)

$10^x = y \qquad \log y = x$

$\ln \equiv$ natural logarithm (logarithm to the base e) $\qquad e^x = y \qquad \ln y = x$

$e^{\ln A} = A \qquad 10^{\log A} = A \qquad \ln e^A = A \qquad \log 10^A = A$

$$\log AB = \log A + \log B \qquad \log \frac{A}{B} = \log A - \log B \qquad \log \frac{1}{A} = -\log A$$

$\log A^n = n \log A \qquad \log 1 = \ln 1 = 0 \qquad \log 10 = 1 \qquad \ln e = 1$

$\ln A = (\ln 10)(\log A) = 2.30259 \log A$

$\log A = (\log e)(\ln A) = 0.434294 \ln A$

Trigonometric Functions

$$\tan x = \frac{\sin x}{\cos x} \qquad \cot x = \frac{\cos x}{\sin x} \qquad \sec x = \frac{1}{\cos x} \qquad \csc x = \frac{1}{\sin x}$$

$\sin^2 x + \cos^2 x = 1 \qquad \tan^2 x + 1 = \sec^2 x \qquad \cot^2 x + 1 = \csc^2 x$

$\sin(-x) = -\sin x \qquad \cos(-x) = \cos x \qquad \tan(-x) = -\tan x$

$\sin(x \pm y) = \sin x \cos y \pm \cos x \sin y$

$\cos(x \pm y) = \cos x \cos y \mp \sin x \sin y$

$\sin 2x = 2 \sin x \cos x \qquad \cos 2x = \cos^2 x - \sin^2 x$

$$\tan 2x = \frac{2 \tan x}{1 - \tan^2 x}$$

$$\tan x = \frac{1 - \cos 2x}{\sin 2x} = \frac{\sin 2x}{1 + \cos 2x}$$

$$\sin^2 x = \frac{1}{2}(1 - \cos 2x) \qquad \cos^2 x = \frac{1}{2}(1 + \cos 2x)$$

For any triangle with sides a, b, c and opposite angles A, B, C:

Law of sines $\quad \dfrac{a}{\sin A} = \dfrac{b}{\sin B} = \dfrac{c}{\sin C}$

Law of cosines $\quad c^2 = a^2 + b^2 - 2ab \cos C$

Quadratic Equation and Quadratic Formula

$$ax^2 + bx + c = 0 \qquad x = \frac{-b \pm \sqrt{b^2 - 4ac}}{2a}$$

Infinite Series

$$\frac{1}{1 + x} = 1 - x + x^2 - x^3 + \cdots \qquad (-1 < x < 1)$$

$$\sqrt{1 + x} = 1 + \frac{x}{2} - \frac{x^2}{8} + \frac{x^3}{16} - \cdots \qquad (-1 < x < 1)$$

$$\frac{1}{\sqrt{1 + x}} = 1 - \frac{x}{2} + \frac{3x^2}{8} - \frac{5x^3}{16} + \cdots \qquad (-1 < x < 1)$$

$$e^x = 1 + x + \frac{x^2}{2!} + \frac{x^3}{3!} + \cdots \qquad (-\infty < x < \infty)$$

$$\sin x = x - \frac{x^3}{3!} + \frac{x^5}{5!} - \frac{x^7}{7!} + \cdots \qquad (-\infty < x < \infty)$$

$$\cos x = 1 - \frac{x^2}{2!} + \frac{x^4}{4!} - \frac{x^6}{6!} + \cdots \qquad (-\infty < x < \infty)$$

Note: If x is very small compared to 1, only the first few terms in the series are needed.

Derivatives

$$\frac{d}{dx}(ax) = a \qquad\qquad \frac{d}{dx}(x^n) = nx^{n-1} \qquad\qquad \frac{d}{dx}(au) = a\frac{du}{dx}$$

$$\frac{d}{dx}(uv) = u\frac{dv}{dx} + v\frac{du}{dx} \qquad\qquad \frac{d}{dx}\left(\frac{u}{v}\right) = \frac{v(du/dx) - u(dv/dx)}{v^2}$$

$$\frac{d}{dx}(u^n) = nu^{n-1}\frac{du}{dx} \qquad\qquad \frac{dy}{dx} = \frac{dy}{du}\frac{du}{dx} \qquad\qquad \frac{du}{dx} = \frac{1}{dx / du}$$

$$\frac{d}{dx}(\sin u) = \cos u\frac{du}{dx} \qquad\qquad \frac{d}{dx}(\cos u) = -\sin u\frac{du}{dx}$$

$$\frac{d}{dx}(\tan u) = \sec^2 u\frac{du}{dx} \qquad\qquad \frac{d}{dx}(\cot u) = -\csc^2 u\frac{du}{dx}$$

$$\frac{d}{dx}(\sec u) = \sec u \tan u\frac{du}{dx} \qquad\qquad \frac{d}{dx}(\csc u) = -\csc u \cot u\frac{du}{dx}$$

$$\frac{d}{dx}(\arctan u) = \frac{1}{1 + u^2}\frac{du}{dx} \qquad \frac{d}{dx}(\log u) = \frac{\log e}{u}\frac{du}{dx} \qquad \frac{d}{dx}(\ln u) = \frac{1}{u}\frac{du}{dx}$$

$$\frac{d}{dx}(a^u) = a^u \ln a \frac{du}{dx} \qquad\qquad \frac{d}{dx}(e^u) = e^u \frac{du}{dx}$$

Indefinite Integrals

Note: A constant must be added to the result of every integration

$$\int a\,dx = ax \qquad\qquad\qquad \int u\,dv = uv - \int v\,du \quad \text{(integration by parts)}$$

$$\int x^n dx = \frac{x^{n+1}}{n+1} \quad (n \neq -1) \qquad \int \frac{dx}{x} = \ln|x| \quad (x \neq 0)$$

$$\int \frac{dx}{x^n} = \frac{x^{1-n}}{1-n} \quad (n \neq 1) \qquad \int (a+bx)^n dx = \frac{(a+bx)^{n+1}}{b(n+1)} \quad (n \neq -1)$$

$$\int \frac{dx}{a+bx} = \frac{1}{b}\ln(a+bx) \qquad \int \frac{dx}{(a+bx)^2} = \frac{1}{b(a+bx)}$$

$$\int \frac{dx}{(a+bx)^n} = -\frac{1}{(n-1)(b)(a+bx)^{n-1}} \quad (n \neq 1)$$

$$\int \frac{dx}{a^2 + b^2 x^2} = \frac{1}{ab}\tan^{-1}\frac{bx}{a} \qquad (x \text{ in radians}) \quad (a > 0, b > 0)$$

$$\int \frac{dx}{a^2 - b^2 x^2} = \frac{1}{2ab}\ln\left(\frac{a+bx}{a-bx}\right) \quad (x \text{ in radians}) \quad (a > 0, b > 0)$$

$$\int \frac{x\,dx}{a+bx} = \frac{1}{b^2}[bx - a\ln(a+bx)]$$

$$\int \frac{x\,dx}{(a+bx)^2} = \frac{1}{b^2}\left[\frac{a}{a+bx} + \ln(a+bx)\right]$$

$$\int \frac{x\,dx}{(a+bx)^3} = -\frac{a+2bx}{2b^2(a+bx)^2} \qquad \int \frac{x\,dx}{(a+bx)^4} = -\frac{a+3bx}{6b^2(a+bx)^3}$$

$$\int \frac{x^2 dx}{a+bx} = \frac{1}{2b^3}[(a+bx)(-3a+bx) + 2a^2\ln(a+bx)]$$

$$\int \frac{x^2 dx}{(a+bx)^2} = \frac{1}{b^3}\left[\frac{bx(2a+bx)}{a+bx} - 2a\ln(a+bx)\right]$$

$$\int \frac{x^2 dx}{(a+bx)^3} = \frac{1}{b^3}\left[\frac{a(3a+4bx)}{2(a+bx)^2} + \ln(a+bx)\right]$$

$$\int \frac{x^2 dx}{(a+bx)^4} = -\frac{a^2 + 3abx + 3b^2 x^2}{3b^3(a+bx)^3}$$

$$\int \sin ax\,dx = -\frac{\cos ax}{a} \qquad\qquad \int \cos ax\,dx = \frac{\sin ax}{a}$$

$$\int \tan ax \, dx = \frac{1}{a} \ln(\sec ax) \qquad \int \cot ax \, dx = \frac{1}{a} \ln(\sin ax)$$

$$\int \sec ax \, dx = \frac{1}{a} \ln(\sec ax + \tan ax)$$

$$\int \csc ax \, dx = \frac{1}{a} \ln(\csc ax - \cot ax)$$

$$\int \sin^2 ax \, dx = \frac{x}{2} - \frac{\sin 2ax}{4a}$$

$$\int \cos^2 ax \, dx = \frac{x}{2} + \frac{\sin 2ax}{4a} \qquad (x \text{ in radians})$$

$$\int x \sin ax \, dx = \frac{\sin ax}{a^2} - \frac{x \cos ax}{a} \qquad (x \text{ in radians})$$

$$\int x \cos ax \, dx = \frac{\cos ax}{a^2} + \frac{x \sin ax}{a} \qquad (x \text{ in radians})$$

$$\int e^{ax} dx = \frac{e^{ax}}{a} \qquad \int x e^{ax} dx = \frac{e^{ax}}{a^2}(ax - 1)$$

$$\int \ln ax \, dx = x(\ln ax - 1)$$

$$\int \frac{dx}{1 + \sin ax} = -\frac{1}{a} \tan\left(\frac{\pi}{4} - \frac{ax}{2}\right)$$

$$\int \sqrt{a + bx} \, dx = \frac{2}{3b}(a + bx)^{3/2}$$

$$\int \sqrt{a^2 + b^2 x^2} \, dx = \frac{x}{2}\sqrt{a^2 + b^2 x^2} + \frac{a^2}{2b} \ln\left(\frac{bx}{a} + \sqrt{1 + \frac{b^2 x^2}{a^2}}\right)$$

$$\int \frac{dx}{\sqrt{a^2 + b^2 x^2}} = \frac{1}{b} \ln\left(\frac{bx}{a} + \sqrt{1 + \frac{b^2 x^2}{a^2}}\right)$$

$$\int \sqrt{a^2 - b^2 x^2} \, dx = \frac{x}{2}\sqrt{a^2 - b^2 x^2} + \frac{a^2}{2b} \sin^{-1}\frac{bx}{a}$$

Definite Integrals

$$\int_a^b f(x)dx = -\int_b^a f(x)dx \qquad \int_a^b f(x)dx = \int_a^c f(x)dx + \int_c^b f(x)dx$$

Properties of Plane Areas

Notation:

A = area

$\bar{x}, \bar{y}$ = distances to centroid C

I_x, I_y = moments of inertia with respect to the x and y axes, respectively

I_{xy} = product of inertia with respect to the x and y axes

$I_P = I_x + I_y$ = polar moment of inertia with respect to the origin of the x and y axes

I_{BB} = moment of inertia with respect to axis B–B

1

Rectangle (Origin of axes at centroid)

$$A = bh \qquad \bar{x} = \frac{b}{2} \qquad \bar{y} = \frac{h}{2}$$

$$I_x = \frac{bh^3}{12} \qquad I_y = \frac{hb^3}{12} \qquad I_{xy} = 0 \qquad I_P = \frac{bh}{12}(h^2 + b^2)$$

2

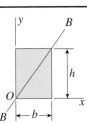

Rectangle (Origin of axes at corner)

$$I_x = \frac{bh^3}{3} \qquad I_y = \frac{hb^3}{3} \qquad I_{xy} = \frac{b^2h^2}{4} \qquad I_P = \frac{bh}{3}(h^2 + b^2)$$

$$I_{BB} = \frac{b^3h^3}{6(b^2 + h^2)}$$

(see Problem 5.6-5)

3

Triangle (Origin of axes at centroid)

$$A = \frac{bh}{2} \qquad \bar{x} = \frac{b + c}{3} \qquad \bar{y} = \frac{h}{3}$$

$$I_x = \frac{bh^3}{36} \qquad I_y = \frac{bh}{36}(b^2 - bc + c^2)$$

$$I_{xy} = \frac{bh^2}{72}(b - 2c) \qquad I_P = \frac{bh}{36}(h^2 + b^2 - bc + c^2)$$

4

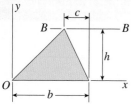

Triangle (Origin of axes at vertex)

$$I_x = \frac{bh^3}{12} \qquad I_y = \frac{bh}{12}(3b^2 - 3bc + c^2)$$

$$I_{xy} = \frac{bh^2}{24}(3b - 2c) \qquad I_{BB} = \frac{bh^3}{4}$$

(see Problem 5.6-1)

5

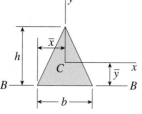

Isosceles triangle (Origin of axes at centroid)

$$A = \frac{bh}{2} \qquad \bar{x} = \frac{b}{2} \qquad \bar{y} = \frac{h}{3}$$

$$I_x = \frac{bh^3}{36} \qquad I_y = \frac{hb^3}{48} \qquad I_{xy} = 0$$

$$I_P = \frac{bh}{144}(4h^2 + 3b^2) \qquad I_{BB} = \frac{bh^3}{12}$$

(*Note:* For an equilateral triangle, $h = \sqrt{3}\,b/2$.)
(see Problem 5.6-22)

6

Right triangle (Origin of axes at centroid)

$$A = \frac{bh}{2} \qquad \bar{x} = \frac{b}{3} \qquad \bar{y} = \frac{h}{3}$$

$$I_x = \frac{bh^3}{36} \qquad I_y = \frac{hb^3}{36} \qquad I_{xy} = -\frac{b^2h^2}{72}$$

$$I_P = \frac{bh}{36}(h^2 + b^2) \qquad I_{BB} = \frac{bh^3}{12}$$

(see Problems 5.2-1, 5.6-25)

7

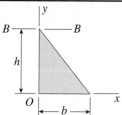

Right triangle (Origin of axes at vertex)

$$I_x = \frac{bh^3}{12} \qquad I_y = \frac{hb^3}{12} \qquad I_{xy} = \frac{b^2h^2}{24}$$

$$I_P = \frac{bh}{12}(h^2 + b^2) \qquad I_{BB} = \frac{bh^3}{4}$$

8

Trapezoid (Origin of axes at centroid)

$$A = \frac{h(a + b)}{2} \qquad \bar{y} = \frac{h(2a + b)}{3(a + b)}$$

$$I_x = \frac{h^3(a^2 + 4ab + b^2)}{36(a + b)} \qquad I_{BB} = \frac{h^3(3a + b)}{12}$$

(see Problems 5.2-2, 5.6-2)

9

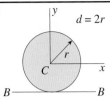

Circle (Origin of axes at center)

$$A = \pi r^2 = \frac{\pi d^2}{4} \qquad I_x = I_y = \frac{\pi r^4}{4} = \frac{\pi d^4}{64}$$

$$I_{xy} = 0 \qquad I_P = \frac{\pi r^4}{2} = \frac{\pi d^4}{32} \qquad I_{BB} = \frac{5\pi r^4}{4} = \frac{5\pi d^4}{64}$$

(see Problem 5.6-4)

10

Semicircle (Origin of axes at centroid)

$$A = \frac{\pi r^2}{2} \qquad \bar{y} = \frac{4r}{3\pi}$$

$$I_x = \frac{(9\pi^2 - 64)r^4}{72\pi} \approx 0.1098 r^4 \qquad I_y = \frac{\pi r^4}{8} \qquad I_{xy} = 0 \qquad I_{BB} = \frac{\pi r^4}{8}$$

(see Problem 5.2-3)

11

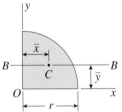

Quarter circle (Origin of axes at center of circle)

$$A = \frac{\pi r^2}{4} \qquad \bar{x} = \bar{y} = \frac{4r}{3\pi}$$

$$I_x = I_y = \frac{\pi r^4}{16} \qquad I_{xy} = \frac{r^4}{8} \qquad I_{BB} = \frac{(9\pi^2 - 64)r^4}{144\pi} \approx 0.05488 r^4$$

12

Quarter-circular spandrel (Origin of axes at point of tangency)

$$A = \left(1 - \frac{\pi}{4}\right)r^2 \qquad \bar{x} = \frac{2r}{3(4 - \pi)} \approx 0.7766r \qquad \bar{y} = \frac{(10 - 3\pi)r}{3(4 - \pi)} \approx 0.2234r$$

$$I_x = \left(1 - \frac{5\pi}{16}\right)r^4 \approx 0.01825 r^4 \qquad I_y = I_{BB} = \left(\frac{1}{3} - \frac{\pi}{16}\right)r^4 \approx 0.1370 r^4$$

(see Problems 5.6-26, 5.6-30)

13

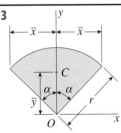

Circular sector (Origin of axes at center of circle)

$$\alpha = \text{angle in radians} \qquad (\alpha \le \pi/2)$$

$$A = \alpha r^2 \qquad \bar{x} = r \sin\alpha \qquad \bar{y} = \frac{2r\sin\alpha}{3\alpha}$$

$$I_x = \frac{r^4}{4}(\alpha + \sin\alpha\cos\alpha) \qquad I_y = \frac{r^4}{4}(\alpha - \sin\alpha\cos\alpha)$$

$$I_{xy} = 0 \qquad I_P = \frac{\alpha r^4}{2}$$

(see Problem 5.6-24)

14

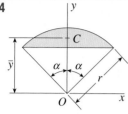

Circular segment (Origin of axes at center of circle)

α = angle in radians $(\alpha \le \pi/2)$

$$A = r^2(\alpha - \sin\alpha\cos\alpha) \qquad \bar{y} = \frac{2r}{3}\left(\frac{\sin^3\alpha}{\alpha - \sin\alpha\cos\alpha}\right)$$

$$I_x = \frac{r^4}{4}(\alpha - \sin\alpha\cos\alpha + 2\sin^3\alpha\cos\alpha) \qquad I_{xy} = 0$$

$$I_y = \frac{r^4}{12}(3\alpha - 3\sin\alpha\cos\alpha - 2\sin^3\alpha\cos\alpha)$$

15

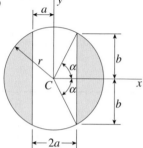

Circle with core removed (Origin of axes at center of circle)

α = angle in radians $(\alpha \le \pi/2)$

$$\alpha = \arccos\frac{a}{r} \qquad b = \sqrt{r^2 - a^2} \qquad A = 2r^2\left(\alpha - \frac{ab}{r^2}\right)$$

$$I_x = \frac{r^4}{6}\left(3\alpha - \frac{3ab}{r^2} - \frac{2ab^3}{r^4}\right) \qquad I_y = \frac{r^4}{2}\left(\alpha - \frac{ab}{r^2} + \frac{2ab^3}{r^4}\right) \qquad I_{xy} = 0$$

16

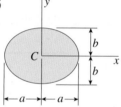

Ellipse (Origin of axes at centroid)

$$A = \pi ab \qquad I_x = \frac{\pi ab^3}{4} \qquad I_y = \frac{\pi ba^3}{4}$$

$$I_{xy} = 0 \qquad I_P = \frac{\pi ab}{4}(b^2 + a^2)$$

$$\text{Circumference} \approx \pi[1.5(a + b) - \sqrt{ab}] \quad (a/3 \le b \le a)$$

$$\approx 4.17b^2/a + 4a \qquad (0 \le b \le a/3)$$

17

Parabolic semisegment (Origin of axes at corner)

$$y = f(x) = h\left(1 - \frac{x^2}{b^2}\right)$$

$$A = \frac{2bh}{3} \qquad \bar{x} = \frac{3b}{8} \qquad \bar{y} = \frac{2h}{5}$$

$$I_x = \frac{16bh^3}{105} \qquad I_y = \frac{2hb^3}{15} \qquad I_{xy} = \frac{b^2h^2}{12}$$

(see Problem 5.6-29)

18

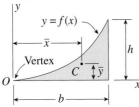

Parabolic spandrel (Origin of axes at vertex)

$$y = f(x) = \frac{hx^2}{b^2}$$

$$A = \frac{bh}{3} \qquad \bar{x} = \frac{3b}{4} \qquad \bar{y} = \frac{3h}{10}$$

$$I_x = \frac{bh^3}{21} \qquad I_y = \frac{hb^3}{5} \qquad I_{xy} = \frac{b^2h^2}{12}$$

(see Problems 5.2-4, 5.6-3)

19

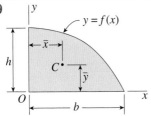

Semisegment of nth degree (Origin of axes at corner)

$$y = f(x) = h\left(1 - \frac{x^n}{b^n}\right) \quad (n > 0)$$

$$A = bh\left(\frac{n}{n+1}\right) \qquad \bar{x} = \frac{b(n+1)}{2(n+2)} \qquad \bar{y} = \frac{hn}{2n+1}$$

$$I_x = \frac{2bh^3n^3}{(n+1)(2n+1)(3n+1)} \qquad I_y = \frac{hb^3n}{3(n+3)} \qquad I_{xy} = \frac{b^2h^2n^2}{4(n+1)(n+2)}$$

(see Problem 5.2-5)

20

Spandrel of nth degree (Origin of axes at point of tangency)

$$y = f(x) = \frac{hx^n}{b^n} \quad (n > 0)$$

$$A = \frac{bh}{n+1} \qquad \bar{x} = \frac{b(n+1)}{n+2} \qquad \bar{y} = \frac{h(n+1)}{2(2n+1)}$$

$$I_x = \frac{bh^3}{3(3n+1)} \qquad I_y = \frac{hb^3}{n+3} \qquad I_{xy} = \frac{b^2h^2}{4(n+1)}$$

21

Sine wave (Origin of axes at centroid)

$$A = \frac{4bh}{\pi} \qquad \bar{y} = \frac{\pi h}{8}$$

$$I_x = \left(\frac{8}{9\pi} - \frac{\pi}{16}\right)bh^3 \approx 0.08659bh^3 \qquad I_y = \left(\frac{4}{\pi} - \frac{32}{\pi^3}\right)hb^3 \approx 0.2412hb^3$$

$$I_{xy} = 0 \qquad I_{BB} = \frac{8bh^3}{9\pi}$$

22

Thin circular ring (Origin of axes at center) Approximate formulas for case when t is small

$$A = 2\pi rt = \pi dt \qquad I_x = I_y = \pi r^3 t = \frac{\pi d^3 t}{8}$$

$$I_{xy} = 0 \qquad I_P = 2\pi r^3 t = \frac{\pi d^3 t}{4}$$

23

Thin circular arc (Origin of axes at center of circle) Approximate formulas for case when t is small

β = angle in radians (*Note:* For a semicircular arc, $\beta = \pi/2$.)

$$A = 2\beta rt \qquad \bar{y} = \frac{r \sin \beta}{\beta}$$

$$I_x = r^3 t(\beta + \sin \beta \cos \beta) \qquad I_y = r^3 t(\beta - \sin \beta \cos \beta)$$

$$I_{xy} = 0 \qquad I_{BB} = r^3 t \left(\frac{2\beta + \sin 2\beta}{2} - \frac{1 - \cos 2\beta}{\beta} \right)$$

24

Thin rectangle (Origin of axes at centroid) Approximate formulas for case when t is small

$$A = bt$$

$$I_x = \frac{tb^3}{12} \sin^2 \beta \qquad I_y = \frac{tb^3}{12} \cos^2 \beta \qquad I_{BB} = \frac{tb^3}{3} \sin^2 \beta$$

25

Regular polygon with n sides (Origin of axes at centroid)

C = centroid (at center of polygon)

n = number of sides ($n \geq 3$) b = length of a side

β = central angle for a side α = interior angle (or vertex angle)

$$\beta = \frac{360°}{n} \qquad \alpha = \left(\frac{n-2}{n} \right) 180° \qquad \alpha + \beta = 180°$$

R_1 = radius of circumscribed circle (line CA)

R_2 = radius of inscribed circle (line CB)

$$R_1 = \frac{b}{2} \csc \frac{\beta}{2} \qquad R_2 = \frac{b}{2} \cot \frac{\beta}{2} \qquad A = \frac{nb^2}{4} \cot \frac{\beta}{2}$$

I_c = moment of inertia about any axis through C (the centroid C is a principal point and every axis through C is a principal axis)

$$I_c = \frac{nb^4}{192} \left(\cot \frac{\beta}{2} \right) \left(3 \cot^2 \frac{\beta}{2} + 1 \right) \qquad I_P = 2I_c$$

Properties of Structural-Steel Shapes

In the following tables, the properties of a few structural-steel shapes are presented as an aid in solving problems in the text. These tables were compiled from the extensive tables in the *Manual of Steel Construction*, published by the American Institute of Steel Construction, Inc.

Notation:

I = moment of inertia

S = section modulus

$r = \sqrt{I/A}$ = radius of gyration

Table C-1(a)

Properties of Wide-Flange Sections (W Shapes)—USCS Units (Abridged List)

Designation	Weight per Foot	Area	Depth	Web Thickness	Flange		Axis 1–1			Axis 2–2		
					Width	Thickness	I	S	r	I	S	r
	lb	in²	in.	in.	in.	in.	in⁴	in³	in.	in⁴	in³	in.
W 30 × 211	211	62.2	30.9	0.775	15.1	1.32	10300	665	12.9	757	100	3.49
W 30 × 132	132	38.9	30.3	0.615	10.5	1.00	5770	380	12.2	196	37.2	2.25
W 24 × 162	162	47.7	25.0	0.705	13.0	1.22	5170	414	10.4	443	68.4	3.05
W 24 × 94	94.0	27.7	24.3	0.515	9.07	0.875	2700	222	9.87	109	24.0	1.98
W 18 × 119	119	35.1	19.0	0.655	11.3	1.06	2190	231	7.90	253	44.9	2.69
W 18 × 71	71.0	20.8	18.5	0.495	7.64	0.810	1170	127	7.50	60.3	15.8	1.70
W 16 × 100	100	29.5	17.0	0.585	10.4	0.985	1490	175	7.10	186	35.7	2.51
W 16 × 77	77.0	22.6	16.5	0.455	10.3	0.760	1110	134	7.00	138	26.9	2.47
W 16 × 57	57.0	16.8	16.4	0.430	7.12	0.715	758	92.2	6.72	43.1	12.1	1.60
W 16 × 31	31.0	9.13	15.9	0.275	5.53	0.440	375	47.2	6.41	12.4	4.49	1.17
W 14 × 120	120	35.3	14.5	0.590	14.7	0.940	1380	190	6.24	495	67.5	3.74
W 14 × 82	82.0	24.0	14.3	0.510	10.1	0.855	881	123	6.05	148	29.3	2.48
W 14 × 53	53.0	15.6	13.9	0.370	8.06	0.660	541	77.8	5.89	57.7	14.3	1.92
W 14 × 26	26.0	7.69	13.9	0.255	5.03	0.420	245	35.3	5.65	8.91	3.55	1.08
W 12 × 87	87.0	25.6	12.5	0.515	12.1	0.810	740	118	5.38	241	39.7	3.07
W 12 × 50	50.0	14.6	12.2	0.370	8.08	0.640	391	64.2	5.18	56.3	13.9	1.96
W 12 × 35	35.0	10.3	12.5	0.300	6.56	0.520	285	45.6	5.25	24.5	7.47	1.54
W 12 × 14	14.0	4.16	11.9	0.200	3.97	0.225	88.6	14.9	4.62	2.36	1.19	0.753
W 10 × 60	60.0	17.6	10.2	0.420	10.1	0.680	341	66.7	4.39	116	23.0	2.57
W 10 × 45	45.0	13.3	10.1	0.350	8.02	0.620	248	49.1	4.32	53.4	13.3	2.01
W 10 × 30	30.0	8.84	10.5	0.300	5.81	0.510	170	32.4	4.38	16.7	5.75	1.37
W 10 × 12	12.0	3.54	9.87	0.190	3.96	0.210	53.8	10.9	3.90	2.18	1.10	0.785
W 8 × 35	35.0	10.3	8.12	0.310	8.02	0.495	127	31.2	3.51	42.6	10.6	2.03
W 8 × 28	28.0	8.24	8.06	0.285	6.54	0.465	98.0	24.3	3.45	21.7	6.63	1.62
W 8 × 21	21.0	6.16	8.28	0.250	5.27	0.400	75.3	18.2	3.49	9.77	3.71	1.26
W 8 × 15	15.0	4.44	8.11	0.245	4.01	0.315	48.0	11.8	3.29	3.41	1.70	0.876

Note: Axes 1–1 and 2–2 are principal centroidal axes.

Table C-1(b)

Properties of Wide-Flange Sections (W Shapes)—SI Units (Abridged List)

Designation	Mass per Meter	Area	Depth	Web Thick-ness	Flange		Axis 1–1			Axis 2–2		
					Width	Thick-ness	I	S	r	I	S	r
	kg	mm²	mm	mm	mm	mm	$\times 10^6$ mm⁴	$\times 10^3$ mm³	mm	$\times 10^6$ mm⁴	$\times 10^3$ mm³	mm
W 760 × 314	314	40100	785	19.7	384	33.5	4290	10900	328	315	1640	88.6
W 760 × 196	196	25100	770	15.6	267	25.4	2400	6230	310	81.6	610	57.2
W 610 × 241	241	30800	635	17.9	330	31.0	2150	6780	264	184	1120	77.5
W 610 × 140	140	17900	617	13.1	230	22.2	1120	3640	251	45.4	393	50.3
W 460 × 177	177	22600	483	16.6	287	26.9	912	3790	201	105	736	68.3
W 460 × 106	106	13400	470	12.6	194	20.6	487	2080	191	25.1	259	43.2
W 410 × 149	149	19000	432	14.9	264	25.0	620	2870	180	77.4	585	63.8
W 410 × 114	114	14600	419	11.6	262	19.3	462	2200	178	57.4	441	62.7
W 410 × 85	85.0	10800	417	10.9	181	18.2	316	1510	171	17.9	198	40.6
W 410 × 46.1	46.1	5890	404	6.99	140	11.2	156	773	163	5.16	73.6	29.7
W 360 × 179	179	22800	368	15.0	373	23.9	574	3110	158	206	1110	95.0
W 360 × 122	122	15500	363	13.0	257	21.7	367	2020	154	61.6	480	63.0
W 360 × 79	79.0	10100	353	9.40	205	16.8	225	1270	150	24.0	234	48.8
W 360 × 39	39.0	4960	353	6.48	128	10.7	102	578	144	3.71	58.2	27.4
W 310 × 129	129	16500	318	13.1	307	20.6	308	1930	137	100	651	78.0
W 310 × 74	74.0	9420	310	9.40	205	16.3	163	1050	132	23.4	228	49.8
W 310 × 52	52.0	6650	318	7.62	167	13.2	119	747	133	10.2	122	39.1
W 310 × 21	21.0	2680	302	5.08	101	5.72	36.9	244	117	0.982	19.5	19.1
W 250 × 89	89.0	11400	259	10.7	257	17.3	142	1090	112	48.3	377	65.3
W 250 × 67	67.0	8580	257	8.89	204	15.7	103	805	110	22.2	218	51.1
W 250 × 44.8	44.8	5700	267	7.62	148	13.0	70.8	531	111	6.95	94.2	34.8
W 250 × 17.9	17.9	2280	251	4.83	101	5.33	22.4	179	99.1	0.907	18.0	19.9
W 200 × 52	52.0	6650	206	7.87	204	12.6	52.9	511	89.2	17.7	174	51.6
W 200 × 41.7	41.7	5320	205	7.24	166	11.8	40.8	398	87.6	9.03	109	41.1
W 200 × 31.3	31.3	3970	210	6.35	134	10.2	31.3	298	88.6	4.07	60.8	32.0
W 200 × 22.5	22.5	2860	206	6.22	102	8.00	20.0	193	83.6	1.42	27.9	22.3

Note: Axes 1–1 and 2–2 are principal centroidal axes.

Table C-2(a)

Properties of I-Beam Sections (S Shapes)—USCS Units (Abridged List)

Designation	Weight per Foot	Area	Depth	Web Thickness	Flange		Axis 1–1			Axis 2–2		
					Width	Average Thickness	I	S	r	I	S	r
	lb	in²	in.	in.	in.	in.	in⁴	in³	in.	in⁴	in³	in.
S 24 × 100	100	29.3	24.0	0.745	7.25	0.870	2380	199	9.01	47.4	13.1	1.27
S 24 × 80	80.0	23.5	24.0	0.500	7.00	0.870	2100	175	9.47	42.0	12.0	1.34
S 20 × 96	96.0	28.2	20.3	0.800	7.20	0.920	1670	165	7.71	49.9	13.9	1.33
S 20 × 75	75.0	22.0	20.0	0.635	6.39	0.795	1280	128	7.62	29.5	9.25	1.16
S 18 × 70	70.0	20.5	18.0	0.711	6.25	0.691	923	103	6.70	24.0	7.69	1.08
S 18 × 54.7	54.7	16.0	18.0	0.461	6.00	0.691	801	89.0	7.07	20.7	6.91	1.14
S 15 × 50	50.0	14.7	15.0	0.550	5.64	0.622	485	64.7	5.75	15.6	5.53	1.03
S 15 × 42.9	42.9	12.6	15.0	0.411	5.50	0.622	446	59.4	5.95	14.3	5.19	1.06
S 12 × 50	50.0	14.6	12.0	0.687	5.48	0.659	303	50.6	4.55	15.6	5.69	1.03
S 12 × 35	35.0	10.2	12.0	0.428	5.08	0.544	228	38.1	4.72	9.84	3.88	0.980
S 10 × 35	35.0	10.3	10.0	0.594	4.94	0.491	147	29.4	3.78	8.30	3.36	0.899
S 10 × 25.4	25.4	7.45	10.0	0.311	4.66	0.491	123	24.6	4.07	6.73	2.89	0.950
S 8 × 23	23.0	6.76	8.00	0.441	4.17	0.425	64.7	16.2	3.09	4.27	2.05	0.795
S 8 × 18.4	18.4	5.40	8.00	0.271	4.00	0.425	57.5	14.4	3.26	3.69	1.84	0.827
S 6 × 17.2	17.3	5.06	6.00	0.465	3.57	0.359	26.2	8.74	2.28	2.29	1.28	0.673
S 6 × 12.5	12.5	3.66	6.00	0.232	3.33	0.359	22.0	7.34	2.45	1.80	1.08	0.702
S 4 × 9.5	9.50	2.79	4.00	0.326	2.80	0.293	6.76	3.38	1.56	0.887	0.635	0.564
S 4 × 7.7	7.70	2.26	4.00	0.193	2.66	0.293	6.05	3.03	1.64	0.748	0.562	0.576

Note: Axes 1–1 and 2–2 are principal centroidal axes.

Table C-2(b)

Properties of I-Beam Sections (S Shapes)—SI Units (Abridged List)

Designation	Mass per Meter	Area	Depth	Web Thick-ness	Flange		Axis 1–1			Axis 2–2		
					Width	Thickness	I	S	r	I	S	r
	kg	mm²	mm	mm	mm	mm	$\times 10^6$ mm⁴	$\times 10^3$ mm³	mm	$\times 10^6$ mm⁴	$\times 10^3$ mm³	mm
S 610 × 149	149	18900	610	18.9	184	22.1	991	3260	229	19.7	215	32.3
S 610 × 119	119	15200	610	12.7	178	22.1	874	2870	241	17.5	197	34.0
S 510 × 143	143	18200	516	20.3	183	23.4	695	2700	196	20.8	228	33.8
S 510 × 112	112	14200	508	16.1	162	20.2	533	2100	194	12.3	152	29.5
S 460 × 104	104	13200	457	18.1	159	17.6	384	1690	170	10.0	126	27.4
S 460 × 81.4	81.4	10300	457	11.7	152	17.6	333	1460	180	8.62	113	29.0
S 380 × 74	74.0	9480	381	14.0	143	15.8	202	1060	146	6.49	90.6	26.2
S 380 × 64	64.0	8130	381	10.4	140	15.8	186	973	151	5.95	85.0	26.9
S 310 × 74	74.0	9420	305	17.4	139	16.7	126	829	116	6.49	93.2	26.2
S 310 × 52	52.0	6580	305	10.9	129	13.8	94.9	624	120	4.10	63.6	24.9
S 250 × 52	52.0	6650	254	15.1	125	12.5	61.2	482	96.0	3.45	55.1	22.8
S 250 × 37.8	37.8	4810	254	7.90	118	12.5	51.2	403	103	2.80	47.4	24.1
S 200 × 34	34.0	4360	203	11.2	106	10.8	26.9	265	78.5	1.78	33.6	20.2
S 200 × 27.4	27.4	3480	203	6.88	102	10.8	23.9	236	82.8	1.54	30.2	21.0
S 150 × 25.7	25.7	3260	152	11.8	90.7	9.12	10.9	143	57.9	0.953	21.0	17.1
S 150 × 18.6	18.6	2360	152	5.89	84.6	9.12	9.16	120	62.2	0.749	17.7	17.8
S 100 × 14.1	14.1	1800	102	8.28	71.1	7.44	2.81	55.4	39.6	0.369	10.4	14.3
S 100 × 11.5	11.5	1460	102	4.90	67.6	7.44	2.52	49.7	41.7	0.311	9.21	14.6

Note: Axes 1–1 and 2–2 are principal centroidal axes.

Table C-3(a)

Properties of Channel Sections (C Shapes)—USCS Units (Abridged List)

Designation	Weight per Foot	Area	Depth	Web Thickness	Flange		Axis 1–1			Axis 2–2			
					Width	Average Thickness	I	S	r	I	S	r	c
	lb	in²	in.	in.	in.	in.	in⁴	in³	in.	in⁴	in³	in.	in.
C 15 × 50	50.0	14.7	15.0	0.716	3.72	0.650	404	53.8	5.24	11.0	3.77	0.865	0.799
C 15 × 40	40.0	11.8	15.0	0.520	3.52	0.650	348	46.5	5.45	9.17	3.34	0.883	0.778
C 15 × 33.9	33.9	10.0	15.0	0.400	3.40	0.650	315	42.0	5.62	8.07	3.09	0.901	0.788
C 12 × 30	30.0	8.81	12.0	0.510	3.17	0.501	162	27.0	4.29	5.12	2.05	0.762	0.674
C 12 × 25	25.0	7.34	12.0	0.387	3.05	0.501	144	24.0	4.43	4.45	1.87	0.779	0.674
C 12 × 20.7	20.7	6.08	12.0	0.282	2.94	0.501	129	21.5	4.61	3.86	1.72	0.797	0.698
C 10 × 30	30.0	8.81	10.0	0.673	3.03	0.436	103	20.7	3.42	3.93	1.65	0.668	0.649
C 10 × 25	25.0	7.34	10.0	0.526	2.89	0.436	91.1	18.2	3.52	3.34	1.47	0.675	0.617
C 10 × 20	20.0	5.87	10.0	0.379	2.74	0.436	78.9	15.8	3.66	2.80	1.31	0.690	0.606
C 10 × 15.3	15.3	4.48	10.0	0.240	2.60	0.436	67.3	13.5	3.87	2.27	1.15	0.711	0.634
C 8 × 18.7	18.7	5.51	8.00	0.487	2.53	0.390	43.9	11.0	2.82	1.97	1.01	0.598	0.565
C 8 × 13.7	13.7	4.04	8.00	0.303	2.34	0.390	36.1	9.02	2.99	1.52	0.848	0.613	0.554
C 8 × 11.5	11.5	3.37	8.00	0.220	2.26	0.390	32.5	8.14	3.11	1.31	0.775	0.623	0.572
C 6 × 13	13.0	3.81	6.00	0.437	2.16	0.343	17.3	5.78	2.13	1.05	0.638	0.524	0.514
C 6 × 10.5	10.5	3.08	6.00	0.314	2.03	0.343	15.1	5.04	2.22	0.860	0.561	0.529	0.500
C 6 × 8.2	8.20	2.39	6.00	0.200	1.92	0.343	13.1	4.35	2.34	0.687	0.488	0.536	0.512
C 4 × 7.2	7.20	2.13	4.00	0.321	1.72	0.296	4.58	2.29	1.47	0.425	0.337	0.447	0.459
C 4 × 5.4	5.40	1.58	4.00	0.184	1.58	0.296	3.85	1.92	1.56	0.312	0.277	0.444	0.457

Notes: 1. Axes 1–1 and 2–2 are principal centroidal axes.

2. The distance c is measured from the centroid to the back of the web.

3. For axis 2–2, the tabulated value of S is the smaller of the two section moduli for this axis.

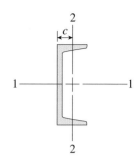

Table C-3(b)

Properties of Channel Sections (C Shapes)—SI Units (Abridged List)

Designation	Mass per Meter	Area	Depth	Web Thick-ness	Flange Width	Flange Thick-ness	Axis 1–1 I	Axis 1–1 S	Axis 1–1 r	Axis 2–2 I	Axis 2–2 S	Axis 2–2 r	c
	kg	mm²	mm	mm	mm	mm	$\times 10^6\,\text{mm}^4$	$\times 10^3\,\text{mm}^3$	mm	$\times 10^6\,\text{mm}^4$	$\times 10^3\,\text{mm}^3$	mm	mm
C 380 × 74	74.0	9480	381	18.2	94.5	16.5	168	882	133	4.58	61.8	22.0	20.3
C 380 × 60	60.0	7610	381	13.2	89.4	16.5	145	762	138	3.82	54.7	22.4	19.8
C 380 × 50.4	50.4	6450	381	10.2	86.4	16.5	131	688	143	3.36	50.6	22.9	20.0
C 310 × 45	45.0	5680	305	13.0	80.5	12.7	67.4	442	109	2.13	33.6	19.4	17.1
C 310 × 37	37.0	4740	305	9.83	77.5	12.7	59.9	393	113	1.85	30.6	19.8	17.1
C 310 × 30.8	30.8	3920	305	7.16	74.7	12.7	53.7	352	117	1.61	28.2	20.2	17.7
C 250 × 45	45.0	5680	254	17.1	77.0	11.1	42.9	339	86.9	1.64	27.0	17.0	16.5
C 250 × 37	37.0	4740	254	13.4	73.4	11.1	37.9	298	89.4	1.39	24.1	17.1	15.7
C 250 × 30	30.0	3790	254	9.63	69.6	11.1	32.8	259	93.0	1.17	21.5	17.5	15.4
C 250 × 22.8	22.8	2890	254	6.10	66.0	11.1	28.0	221	98.3	0.945	18.8	18.1	16.1
C 200 × 27.9	27.9	3550	203	12.4	64.3	9.91	18.3	180	71.6	0.820	16.6	15.2	14.4
C 200 × 20.5	20.5	2610	203	7.70	59.4	9.91	15.0	148	75.9	0.633	13.9	15.6	14.1
C 200 × 17.1	17.1	2170	203	5.59	57.4	9.91	13.5	133	79.0	0.545	12.7	15.8	14.5
C 150 × 19.3	19.3	2460	152	11.1	54.9	8.71	7.20	94.7	54.1	0.437	10.5	13.3	13.1
C 150 × 15.6	15.6	1990	152	7.98	51.6	8.71	6.29	82.6	56.4	0.358	9.19	13.4	12.7
C 150 × 12.2	12.2	1540	152	5.08	48.8	8.71	5.45	71.3	59.4	0.286	8.00	13.6	13.0
C 100 × 10.8	10.8	1370	102	8.15	43.7	7.52	1.91	37.5	37.3	0.177	5.52	11.4	11.7
C 100 × 8	8.00	1020	102	4.67	40.1	7.52	1.60	31.5	39.6	0.130	4.54	11.3	11.6

Notes: 1. Axes 1–1 and 2–2 are principal centroidal axes.

2. The distance c is measured from the centroid to the back of the web.

3. For axis 2–2, the tabulated value of S is the smaller of the two section moduli for this axis.

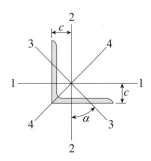

Table C-4(a)

Properties of Angle Sections with Equal Legs (L Shapes)—USCS Units (Abridged List)

Designation	Weight per Foot	Area	Axis 1–1 and Axis 2–2				Axis 3–3
			I	S	r	c	r_{min}
in.	lb	in²	in⁴	in³	in.	in.	in.
L 8 × 8 × 1	51.0	15.0	89.1	15.8	2.43	2.36	1.56
L 8 × 8 × 3/4	38.9	11.4	69.9	12.2	2.46	2.26	1.57
L 8 × 8 × 1/2	26.4	7.75	48.8	8.36	2.49	2.17	1.59
L 6 × 6 × 1	37.4	11.0	35.4	8.55	1.79	1.86	1.17
L 6 × 6 × 3/4	28.7	8.46	28.1	6.64	1.82	1.77	1.17
L 6 × 6 × 1/2	19.6	5.77	19.9	4.59	1.86	1.67	1.18
L 5 × 5 × 7/8	27.2	7.98	17.8	5.16	1.49	1.56	0.971
L 5 × 5 × 1/2	16.2	4.75	11.3	3.15	1.53	1.42	0.980
L 5 × 5 × 3/8	12.3	3.61	8.76	2.41	1.55	1.37	0.986
L 4 × 4 × 3/4	18.5	5.44	7.62	2.79	1.18	1.27	0.774
L 4 × 4 × 1/2	12.8	3.75	5.52	1.96	1.21	1.18	0.776
L 4 × 4 × 3/8	9.80	2.86	4.32	1.50	1.23	1.13	0.779
L 3 − 1/2 × 3 − 1/2 × 3/8	8.50	2.48	2.86	1.15	1.07	1.00	0.683
L 3 − 1/2 × 3 − 1/2 × 1/4	5.80	1.69	2.00	0.787	1.09	0.954	0.688
L 3 × 3 × 1/2	9.40	2.75	2.20	1.06	0.895	0.929	0.580
L 3 × 3 × 1/4	4.90	1.44	1.23	0.569	0.926	0.836	0.585

Notes:
1. Axes 1–1 and 2–2 are centroidal axes parallel to the legs.
2. The distance c is measured from the centroid to the back of the legs.
3. For axes 1–1 and 2–2, the tabulated value of S is the smaller of the two section moduli for those axes.
4. Axes 3–3 and 4–4 are principal centroidal axes.
5. The moment of inertia for axis 3–3, which is the smaller of the two principal moments of inertia, can be found from the equation $I_{33} = Ar_{min}^2$.
6. The moment of inertia for axis 4–4, which is the larger of the two principal moments of inertia, can be found from the equation $I_{44} + I_{33} = I_{11} + I_{22}$.

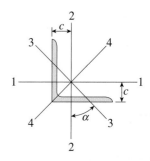

Table C-4(b)

Properties of Angle Sections with Equal Legs (L Shapes)—SI Units (Abridged List)

Designation	Mass per Meter	Area	Axis 1–1 and Axis 2–2				Axis 3–3
			I	S	r	c	r_{min}
mm	kg	mm²	×10⁶ mm⁴	×10³ mm³	mm	mm	mm
L 203 × 203 × 25.4	75.9	9680	37.1	259	61.7	59.9	39.6
L 203 × 203 × 19	57.9	7350	29.1	200	62.5	57.4	39.9
L 203 × 203 × 12.7	39.3	5000	20.3	137	63.2	55.1	40.4
L 152 × 152 × 25.4	55.7	7100	14.7	140	45.5	47.2	29.7
L 152 × 152 × 19	42.7	5460	11.7	109	46.2	45.0	29.7
L 152 × 152 × 12.7	29.2	3720	8.28	75.2	47.2	42.4	30.0
L 127 × 127 × 22.2	40.5	5150	7.41	84.6	37.8	39.6	24.7
L 127 × 127 × 12.7	24.1	3060	4.70	51.6	38.9	36.1	24.9
L 127 × 127 × 9.5	18.3	2330	3.65	39.5	39.4	34.8	25.0
L 102 × 102 × 19	27.5	3510	3.17	45.7	30.0	32.3	19.7
L 102 × 102 × 12.7	19.0	2420	2.30	32.1	30.7	30.0	19.7
L 102 × 102 × 9.5	14.6	1850	1.80	24.6	31.2	28.7	19.8
L 89 × 89 × 9.5	12.6	1600	1.19	18.8	27.2	25.4	17.3
L 89 × 89 × 6.4	8.60	1090	0.832	12.9	27.7	24.2	17.5
L 76 × 76 × 12.7	14.0	1770	0.916	17.4	22.7	23.6	14.7
L 76 × 76 × 6.4	7.30	929	0.512	9.32	23.5	21.2	14.9

Notes:
1. Axes 1–1 and 2–2 are centroidal axes parallel to the legs.
2. The distance c is measured from the centroid to the back of the legs.
3. For axes 1–1 and 2–2, the tabulated value of S is the smaller of the two section moduli for those axes.
4. Axes 3–3 and 4–4 are principal centroidal axes.
5. The moment of inertia for axis 3–3, which is the smaller of the two principal moments of inertia, can be found from the equation $I_{33} = Ar_{min}^2$.
6. The moment of inertia for axis 4–4, which is the larger of the two principal moments of inertia, can be found from the equation $I_{44} + I_{33} = I_{11} + I_{22}$.

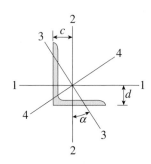

Table C-5(a)

Properties of Angle Sections with Unequal Legs (L Shapes)—USCS Units (Abridged List)

Designation	Weight per Foot	Area	Axis 1–1				Axis 2–2				Axis 3–3	
			I	S	r	d	I	S	r	c	r_{min}	$\tan \alpha$
in.	lb	in²	in⁴	in³	in.	in.	in⁴	in³	in.	in.	in.	
L 8 × 6 × 1	44.2	13.0	80.9	15.1	2.49	2.65	38.8	8.92	1.72	1.65	1.28	0.542
L 8 × 6 × 1/2	23.0	6.75	44.4	8.01	2.55	2.46	21.7	4.79	1.79	1.46	1.30	0.557
L 7 × 4 × 3/4	26.2	7.69	37.8	8.39	2.21	2.50	9.00	3.01	1.08	1.00	0.855	0.324
L 7 × 4 × 1/2	17.9	5.25	26.6	5.79	2.25	2.40	6.48	2.10	1.11	0.910	0.866	0.334
L 6 × 4 × 3/4	23.6	6.94	24.5	6.23	1.88	2.07	8.63	2.95	1.12	1.07	0.856	0.428
L 6 × 4 × 1/2	16.2	4.75	17.3	4.31	1.91	1.98	6.22	2.06	1.14	0.981	0.864	0.440
L 5 × 3 − 1/2 × 3/4	19.8	5.81	13.9	4.26	1.55	1.74	5.52	2.20	0.974	0.993	0.744	0.464
L 5 × 3 − 1/2 × 1/2	13.6	4.00	10.0	2.97	1.58	1.65	4.02	1.55	1.00	0.901	0.750	0.479
L 5 × 3 × 1/2	12.8	3.75	9.43	2.89	1.58	1.74	2.55	1.13	0.824	0.746	0.642	0.357
L 5 × 3 × 1/4	6.60	1.94	5.09	1.51	1.62	1.64	1.41	0.600	0.853	0.648	0.652	0.371
L 4 × 3 − 1/2 × 1/2	11.9	3.50	5.30	1.92	1.23	1.24	3.76	1.50	1.04	0.994	0.716	0.750
L 4 × 3 − 1/2 × 1/4	6.20	1.81	2.89	1.01	1.26	1.14	2.07	0.794	1.07	0.897	0.723	0.759
L 4 × 3 × 1/2	11.1	3.25	5.02	1.87	1.24	1.32	2.40	1.10	0.858	0.822	0.633	0.542
L 4 × 3 × 3/8	8.50	2.48	3.94	1.44	1.26	1.27	1.89	0.851	0.873	0.775	0.636	0.551
L 4 × 3 × 1/4	5.80	1.69	2.75	0.988	1.27	1.22	1.33	0.585	0.887	0.725	0.639	0.558

Notes: 1. Axes 1–1 and 2–2 are centroidal axes parallel to the legs.

2. The distances c and d are measured from the centroid to the backs of the legs.

3. For axes 1–1 and 2–2, the tabulated value of S is the smaller of the two section moduli for those axes.

4. Axes 3–3 and 4–4 are principal centroidal axes.

5. The moment of inertia for axis 3–3, which is the smaller of the two principal moments of inertia, can be found from the equation $I_{33} = Ar_{min}^2$.

6. The moment of inertia for axis 4–4, which is the larger of the two principal moments of inertia, can be found from the equation $I_{44} + I_{33} = I_{11} + I_{22}$.

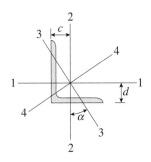

Table C-5(b)

Properties of Angle Sections with Unequal Legs (L Shapes)—SI Units (Abridged List)

Designation	Mass per Meter	Area	Axis 1–1				Axis 2–2				Axis 3–3	
			I	S	r	d	I	S	r	c	r_{min}	$\tan \alpha$
mm	kg	mm²	$\times 10^6$ mm⁴	$\times 10^3$ mm³	mm	mm	$\times 10^6$ mm⁴	$\times 10^3$ mm³	mm	mm	mm	
L 203 × 152 × 25.4	65.5	8390	33.7	247	63.2	67.3	16.1	146	43.7	41.9	32.5	0.542
L 203 × 152 × 12.7	34.1	4350	18.5	131	64.8	62.5	9.03	78.5	45.5	37.1	33.0	0.557
L 178 × 102 × 19	38.8	4960	15.7	137	56.1	63.5	3.75	49.3	27.4	25.4	21.7	0.324
L 178 × 102 × 12.7	26.5	3390	11.1	94.9	57.2	61.0	2.70	34.4	28.2	23.1	22.0	0.334
L 152 × 102 × 19	35.0	4480	10.2	102	47.8	52.6	3.59	48.3	28.4	27.2	21.7	0.428
L 152 × 102 × 12.7	24.0	3060	7.20	70.6	48.5	50.3	2.59	33.8	29.0	24.9	21.9	0.440
L 127 × 89 × 19	29.3	3750	5.79	69.8	39.4	44.2	2.30	36.1	24.7	25.2	18.9	0.464
L 127 × 89 × 12.7	20.2	2580	4.15	48.7	40.1	41.9	1.67	25.4	25.4	22.9	19.1	0.479
L 127 × 76 × 12.7	19.0	2420	3.93	47.4	40.1	44.2	1.06	18.5	20.9	18.9	16.3	0.357
L 127 × 76 × 6.4	9.80	1250	2.12	24.7	41.1	41.7	0.587	9.83	21.7	16.5	16.6	0.371
L 102 × 89 × 12.7	17.6	2260	2.21	31.5	31.2	31.5	1.57	24.6	26.4	25.2	18.2	0.750
L 102 × 89 × 6.4	9.20	1170	1.20	16.6	32.0	29.0	0.862	13.0	27.2	22.8	18.4	0.759
L 102 × 76 × 12.7	16.4	2100	2.09	30.6	31.5	33.5	0.999	18.0	21.8	20.9	16.1	0.542
L 102 × 76 × 9.5	12.6	1600	1.64	23.6	32.0	32.3	0.787	13.9	22.2	19.7	16.2	0.551
L 102 × 76 × 6.4	8.60	1090	1.14	16.2	32.3	31.0	0.554	9.59	22.5	18.4	16.2	0.558

Notes: **1.** Axes 1–1 and 2–2 are centroidal axes parallel to the legs.

   2. The distances c and d are measured from the centroid to the backs of the legs.

   3. For axes 1–1 and 2–2, the tabulated value of S is the smaller of the two section moduli for those axes.

   4. Axes 3–3 and 4–4 are principal centroidal axes.

   5. The moment of inertia for axis 3–3, which is the smaller of the two principal moments of inertia, can be found from the equation
   $I_{33} = Ar_{min}^2$.

   6. The moment of inertia for axis 4–4, which is the larger of the two principal moments of inertia, can be found from the equation
   $I_{44} + I_{33} = I_{11} + I_{22}$.

Properties of Structural Lumber

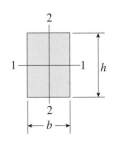

Properties of Surfaced Lumber (Abridged List)

Nominal Dimensions $b \times h$	Net Dimensions $b \times h$	Area $A = bh$	Axis 1–1		Axis 2–2		Weight per Linear Foot (weight density $= 35\,\text{lb}/\text{ft}^3$)
			Moment of Inertia $I_1 = \dfrac{bh^3}{12}$	Section Modulus $S_1 = \dfrac{bh^2}{6}$	Moment of Inertia $I_2 = \dfrac{hb^3}{12}$	Section Modulus $S_2 = \dfrac{hb^2}{6}$	
in.	in.	in²	in⁴	in³	in⁴	in³	lb
2 × 4	1.5 × 3.5	5.25	5.36	3.06	0.98	1.31	1.3
2 × 6	1.5 × 5.5	8.25	20.80	7.56	1.55	2.06	2.0
2 × 8	1.5 × 7.25	10.88	47.63	13.14	2.04	2.72	2.6
2 × 10	1.5 × 9.25	13.88	98.93	21.39	2.60	3.47	3.4
2 × 12	1.5 × 11.25	16.88	177.98	31.64	3.16	4.22	4.1
3 × 4	2.5 × 3.5	8.75	8.93	5.10	4.56	3.65	2.1
3 × 6	2.5 × 5.5	13.75	34.66	12.60	7.16	5.73	3.3
3 × 8	2.5 × 7.25	18.13	79.39	21.90	9.44	7.55	4.4
3 × 10	2.5 × 9.25	23.13	164.89	35.65	12.04	9.64	5.6
3 × 12	2.5 × 11.25	28.13	296.63	52.73	14.65	11.72	6.8
4 × 4	3.5 × 3.5	12.25	12.51	7.15	12.51	7.15	3.0
4 × 6	3.5 × 5.5	19.25	48.53	17.65	19.65	11.23	4.7
4 × 8	3.5 × 7.25	25.38	111.15	30.66	25.90	14.80	6.2
4 × 10	3.5 × 9.25	32.38	230.84	49.91	33.05	18.89	7.9
4 × 12	3.5 × 11.25	39.38	415.28	73.83	40.20	22.9	9.6
6 × 6	5.5 × 5.5	30.25	76.3	27.7	76.3	27.7	7.4
6 × 8	5.5 × 7.5	41.25	193.4	51.6	104.0	37.8	10.0
6 × 10	5.5 × 9.5	52.25	393.0	82.7	131.7	47.9	12.7
6 × 12	5.5 × 11.5	63.25	697.1	121.2	159.4	58.0	15.4
8 × 8	7.5 × 7.5	56.25	263.7	70.3	263.7	70.3	13.7
8 × 10	7.5 × 9.5	71.25	535.9	112.8	334.0	89.1	17.3
8 × 12	7.5 × 11.5	86.25	950.5	165.3	404.3	107.8	21.0

Note: Axes 1–1 and 2–2 are principal centroidal axes.

Deflections and Slopes of Beams

Table E-1

Deflections and Slopes of Cantilever Beams

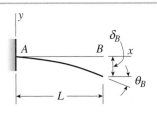

Notation:

v = deflection in the y direction (positive upward)

$v' = dv/dx$ = slope of the deflection curve

$\delta_B = -v(L)$ = deflection at end B of the beam (positive downward)

$\theta_B = -v'(L)$ = angle of rotation at end B of the beam (positive clockwise)

EI = constant

1

$$v = -\frac{qx^2}{24EI}(6L^2 - 4Lx + x^2) \qquad v' = \frac{qx}{6EI}(3L^2 - 3Lx + x^2)$$

$$\delta_B = \frac{qL^4}{8EI} \qquad \theta_B = \frac{qL^3}{6EI}$$

2

$$v = -\frac{qx^2}{24EI}(6a^2 - 4ax + x^2) \qquad (0 \le x \le a)$$

$$v' = -\frac{qx}{6EI}(3a^2 - 3ax + x^2) \qquad (0 \le x \le a)$$

$$v = -\frac{qa^3}{24EI}(4x - a) \qquad v' = -\frac{qa^3}{6EI} \qquad (a \le x \le L)$$

At $x = a$:
$$v = -\frac{qa^4}{8EI} \qquad\qquad v' = -\frac{qa^3}{6EI}$$

$$\delta_B = \frac{qa^3}{24EI}(4L - a) \qquad \theta_B = \frac{qa^3}{6EI}$$

(Continues)

Table E-1 (*Continued*)

3

$$v = -\frac{qbx^2}{12EI}(3L + 3a - 2x) \qquad (0 \le x \le a)$$

$$v' = -\frac{qbx}{2EI}(L + a - x) \qquad (0 \le x \le a)$$

$$v = -\frac{q}{24EI}(x^4 - 4Lx^3 + 6L^2x^2 - 4a^3x + a^4) \qquad (a \le x \le L)$$

$$v' = -\frac{q}{6EI}(x^3 - 3Lx^2 + 3L^2x - a^3) \qquad (a \le x \le L)$$

At $x = a$: $\quad v = -\dfrac{qa^2b}{12EI}(3L + a) \qquad\qquad v' = -\dfrac{qabL}{2EI}$

$$\delta_B = \frac{q}{24EI}(3L^4 - 4a^3L + a^4) \qquad \theta_B = \frac{q}{6EI}(L^3 - a^3)$$

4

$$v = -\frac{Px^2}{6EI}(3L - x) \qquad v' = -\frac{Px}{2EI}(2L - x)$$

$$\delta_B = \frac{PL^3}{3EI} \qquad\qquad \theta_B = \frac{PL^2}{2EI}$$

5

$$v = -\frac{Px^2}{6EI}(3a - x) \qquad v' = -\frac{Px}{2EI}(2a - x) \qquad (0 \le x \le a)$$

$$v = -\frac{Pa^2}{6EI}(3x - a) \qquad v' = -\frac{Pa^2}{2EI} \qquad\qquad (a \le x \le L)$$

At $x = a$: $\quad v = -\dfrac{Pa^3}{3EI} \qquad\qquad v' = -\dfrac{Pa^2}{2EI}$

$$\delta_B = \frac{Pa^2}{6EI}(3L - a) \qquad \theta_B = \frac{Pa^2}{2EI}$$

6

$$v = -\frac{M_0x^2}{2EI} \qquad v' = -\frac{M_0x}{EI}$$

$$\delta_B = \frac{M_0L^2}{2EI} \qquad \theta_B = \frac{M_0L}{EI}$$

Table E-1 (*Continued*)

7

$$v = -\frac{M_0 x^2}{2EI} \qquad v' = -\frac{M_0 x}{EI} \qquad (0 \le x \le a)$$

$$v = -\frac{M_0 a}{2EI}(2x - a) \qquad v' = -\frac{M_0 a}{EI} \qquad (a \le x \le L)$$

At $x = a$: $v = -\dfrac{M_0 a^2}{2EI} \qquad v' = -\dfrac{M_0 a}{EI}$

$$\delta_B = \frac{M_0 a}{2EI}(2L - a) \qquad \theta_B = \frac{M_0 a}{EI}$$

8

$$v = -\frac{q_0 x^2}{120 LEI}(10L^3 - 10L^2 x + 5Lx^2 - x^3)$$

$$v' = -\frac{q_0 x}{24 LEI}(4L^3 - 6L^2 x + 4Lx^2 - x^3)$$

$$\delta_B = \frac{q_0 L^4}{30EI} \qquad \theta_B = \frac{q_0 L^3}{24EI}$$

9

$$v = -\frac{q_0 x^2}{120 LEI}(20L^3 - 10L^2 x + x^3)$$

$$v' = -\frac{q_0 x}{24 LEI}(8L^3 - 6L^2 x + x^3)$$

$$\delta_B = \frac{11 q_0 L^4}{120EI} \qquad \theta_B = \frac{q_0 L^3}{8EI}$$

10 $q = q_0 \cos\dfrac{\pi x}{2L}$

$$v = -\frac{q_0 L}{3\pi^4 EI}\left(48L^3 \cos\frac{\pi x}{2L} - 48L^3 + 3\pi^3 Lx^2 - \pi^3 x^3\right)$$

$$v' = -\frac{q_0 L}{\pi^3 EI}\left(2\pi^2 Lx - \pi^2 x^2 - 8L^2 \sin\frac{\pi x}{2L}\right)$$

$$\delta_B = \frac{2q_0 L^4}{3\pi^4 EI}(\pi^3 - 24) \qquad \theta_B = \frac{q_0 L^3}{\pi^3 EI}(\pi^2 - 8)$$

Table E-2

Deflections and Slopes of Simple Beams

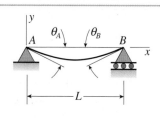

Notation:

v = deflection in the y direction (positive upward)

$v' = dv/dx$ = slope of the deflection curve

$\delta_C = -v(L/2)$ = deflection at midpoint C of the beam (positive downward)

x_1 = distance from support A to point of maximum deflection

$\delta_{max} = -v_{max}$ = maximum deflection (positive downward)

$\theta_A = -v'(0)$ = angle of rotation at left-hand end of the beam (positive clockwise)

$\theta_B = v'(L)$ = angle of rotation at right-hand end of the beam (positive counterclockwise)

EI = constant

1

$$v = -\frac{qx}{24EI}(L^3 - 2Lx^2 + x^3)$$

$$v' = -\frac{q}{24EI}(L^3 - 6Lx^2 + 4x^3)$$

$$\delta_C = \delta_{max} = \frac{5qL^4}{384EI} \qquad \theta_A = \theta_B = \frac{qL^3}{24EI}$$

2

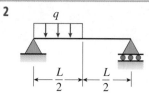

$$v = -\frac{qx}{384EI}(9L^3 - 24Lx^2 + 16x^3) \qquad \left(0 \le x \le \frac{L}{2}\right)$$

$$v' = -\frac{q}{384EI}(9L^3 - 72Lx^2 + 64x^3) \qquad \left(0 \le x \le \frac{L}{2}\right)$$

$$v = -\frac{qL}{384EI}(8x^3 - 24Lx^2 + 17L^2x - L^3) \qquad \left(\frac{L}{2} \le x \le L\right)$$

$$v' = -\frac{qL}{384EI}(24x^2 - 48Lx + 17L^2) \qquad \left(\frac{L}{2} \le x \le L\right)$$

$$\delta_C = \frac{5qL^4}{768EI} \qquad \theta_A = \frac{3qL^3}{128EI} \qquad \theta_B = \frac{7qL^3}{384EI}$$

3

$$v = -\frac{qx}{24LEI}(a^4 - 4a^3L + 4a^2L^2 + 2a^2x^2 - 4aLx^2 + Lx^3) \qquad (0 \le x \le a)$$

$$v' = -\frac{q}{24LEI}(a^4 - 4a^3L + 4a^2L^2 + 6a^2x^2 - 12aLx^2 - 4Lx^3) \quad (0 \le x \le a)$$

$$v = -\frac{qa^2}{24LEI}(-a^2L + 4L^2x + a^2x - 6Lx^2 + 2x^3) \qquad (a \le x \le L)$$

$$v' = -\frac{qa^2}{24LEI}(4L^2 + a^2 - 12Lx + 6x^2) \qquad (a \le x \le L)$$

$$\theta_A = \frac{qa^2}{24LEI}(2L - a)^2 \qquad \theta_B = \frac{qa^2}{24LEI}(2L^2 - a^2)$$

Table E-2 (*Continued*)

4

$$v = -\frac{Px}{48EI}(3L^2 - 4x^2) \quad v' = -\frac{P}{16EI}(L^2 - 4x^2) \quad \left(0 \le x \le \frac{L}{2}\right)$$

$$\delta_C = \delta_{\max} = \frac{PL^3}{48EI} \qquad \theta_A = \theta_B = \frac{PL^2}{16EI}$$

5

$$v = -\frac{Pbx}{6LEI}(L^2 - b^2 - x^2) \quad v' = -\frac{Pb}{6LEI}(L^2 - b^2 - 3x^2) \quad (0 \le x \le a)$$

$$\theta_A = \frac{Pab(L+b)}{6LEI} \qquad \theta_B = \frac{Pab(L+a)}{6LEI}$$

$$\text{If } a \ge b, \ \delta_C = \frac{Pb(3L^2 - 4b^2)}{48EI} \qquad \text{If } a \le b, \ \delta_C = \frac{Pa(3L^2 - 4a^2)}{48EI}$$

$$\text{If } a \ge b, \ x_1 = \sqrt{\frac{L^2 - b^2}{3}} \qquad \text{and} \qquad \delta_{\max} = \frac{Pb(L^2 - b^2)^{3/2}}{9\sqrt{3}LEI}$$

6

$$v = -\frac{Px}{6EI}(3aL - 3a^2 - x^2) \quad v' = -\frac{P}{2EI}(aL - a^2 - x^2) \quad (0 \le x \le a)$$

$$v = -\frac{Pa}{6EI}(3Lx - 3x^2 - a^2) \quad v' = -\frac{Pa}{2EI}(L - 2x) \qquad (a \le x \le L - a)$$

$$\delta_C = \delta_{\max} = \frac{Pa}{24EI}(3L^2 - 4a^2) \quad \theta_A = \theta_B = \frac{Pa(L-a)}{2EI}$$

7

$$v = -\frac{M_0 x}{6LEI}(2L^2 - 3Lx + x^2) \quad v' = -\frac{M_0}{6LEI}(2L^2 - 6Lx + 3x^2)$$

$$\delta_C = \frac{M_0 L^2}{16EI} \qquad \theta_A = \frac{M_0 L}{3EI} \qquad \theta_B = \frac{M_0 L}{6EI}$$

$$x_1 = L\left(1 - \frac{\sqrt{3}}{3}\right) \quad \text{and} \quad \delta_{\max} = \frac{M_0 L^2}{9\sqrt{3}EI}$$

8

$$v = -\frac{M_0 x}{24LEI}(L^2 - 4x^2) \quad v' = -\frac{M_0}{24LEI}(L^2 - 12x^2) \quad \left(0 \le x \le \frac{L}{2}\right)$$

$$\delta_C = 0 \qquad \theta_A = \frac{M_0 L}{24EI} \qquad \theta_B = -\frac{M_0 L}{24EI}$$

9

$$v = -\frac{M_0 x}{6LEI}(6aL - 3a^2 - 2L^2 - x^2) \quad (0 \le x \le a)$$

$$v' = -\frac{M_0}{6LEI}(6aL - 3a^2 - 2L^2 - 3x^2) \quad (0 \le x \le a)$$

$$\text{At } x = a: \quad v = -\frac{M_0 ab}{3LEI}(2a - L) \qquad v' = -\frac{M_0}{3LEI}(3aL - 3a^2 - L^2)$$

$$\theta_A = \frac{M_0}{6LEI}(6aL - 3a^2 - 2L^2) \quad \theta_B = \frac{M_0}{6LEI}(3a^2 - L^2)$$

(*Continues*)

Table E-2 (Continued)

10

$$v = -\frac{M_0 x}{2EI}(L - x) \quad v' = -\frac{M_0}{2EI}(L - 2x)$$

$$\delta_C = \delta_{max} = \frac{M_0 L^2}{8EI} \quad \theta_A = \theta_B = \frac{M_0 L}{2EI}$$

11

$$v = -\frac{q_0 x}{360 LEI}(7L^4 - 10L^2 x^2 + 3x^4)$$

$$v' = -\frac{q_0}{360 LEI}(7L^4 - 30L^2 x^2 + 15x^4)$$

$$\delta_C = \frac{5q_0 L^4}{768 EI} \quad \theta_A = \frac{7q_0 L^3}{360 EI} \quad \theta_B = \frac{q_0 L^3}{45 EI}$$

$$x_1 = 0.5193L \quad \delta_{max} = 0.00652 \frac{q_0 L^4}{EI}$$

12

$$v = -\frac{q_0 x}{960 LEI}(5L^2 - 4x^2)^2 \qquad \left(0 \le x \le \frac{L}{2}\right)$$

$$v' = -\frac{q_0}{192 LEI}(5L^2 - 4x^2)(L^2 - 4x^2) \qquad \left(0 \le x \le \frac{L}{2}\right)$$

$$\delta_C = \delta_{max} = \frac{q_0 L^4}{120 EI} \quad \theta_A = \theta_B = \frac{5q_0 L^3}{192 EI}$$

13

$$q = q_0 \sin\frac{\pi x}{L}$$

$$v = -\frac{q_0 L^4}{\pi^4 EI}\sin\frac{\pi x}{L} \quad v' = -\frac{q_0 L^3}{\pi^3 EI}\cos\frac{\pi x}{L}$$

$$\delta_C = \delta_{max} = \frac{q_0 L^4}{\pi^4 EI} \quad \theta_A = \theta_B = \frac{q_0 L^3}{\pi^3 EI}$$

Properties of Materials

Notes:

1. Properties of materials vary greatly depending upon manufacturing processes, chemical composition, internal defects, temperature, previous loading history, age, dimensions of test specimens, and other factors. The tabulated values are typical but should never be used for specific engineering or design purposes. Manufacturers and materials suppliers should be consulted for information about a particular product.

2. Except when compression or bending is indicated, the modulus of elasticity E, yield stress σ_Y, and ultimate stress σ_U are for materials in tension.

Table F-1

Weights and Mass Densities

Material	Weight Density γ		Mass Density ρ	
	lb/ft³	kN/m³	slugs/ft³	kg/m³
Aluminum alloys	160–180	26–28	5.2–5.4	2600–2800
2014-T6, 7075-T6	175	28	5.4	2800
6061-T6	170	26	5.2	2700
Brass	520–540	82–85	16–17	8400–8600
Bronze	510–550	80–86	16–17	8200–8800
Cast iron	435–460	68–72	13–14	7000–7400
Concrete				
Plain	145	23	4.5	2300
Reinforced	150	24	4.7	2400
Lightweight	70–115	11–18	2.2–3.6	1100–1800
Copper	556	87	17	8900
Glass	150–180	24–28	4.7–5.4	2400–2800
Magnesium alloys	110–114	17–18	3.4–3.5	1760–1830
Monel (67% Ni, 30% Cu)	550	87	17	8800
Nickel	550	87	17	8800
Plastics				
Nylon	55–70	8.6–11	1.7–2.2	880–1100
Polyethylene	60–90	9.4–14	1.9–2.8	960–1400
Rock				
Granite, marble, quartz	165–180	26–28	5.1–5.6	2600–2900
Limestone, sandstone	125–180	20–28	3.9–5.6	2000–2900
Rubber	60–80	9–13	1.9–2.5	960–1300
Sand, soil, gravel	75–135	12–21	2.3–4.2	1200–2200
Steel	490	77.0	15.2	7850
Titanium	280	44	8.7	4500
Tungsten	1200	190	37	1900
Water, fresh	62.4	9.81	1.94	1000
sea	63.8	10.0	1.98	1020
Wood (air dry)				
Douglas fir	30–35	4.7–5.5	0.9–1.1	480–560
Oak	40–45	6.3–7.1	1.2–1.4	640–720
Southern pine	35–40	5.5–6.3	1.1–1.2	560–640

Table F-2

Moduli of Elasticity and Poisson's Ratios

Material	Modulus of Elasticity E		Shear Modulus of Elasticity G		Poisson's Ratio ν
	ksi	GPa	ksi	GPa	
Aluminum alloys	10,000–11,400	70–79	3800–4300	26–30	0.33
2014-T6	10,600	73	4000	28	0.33
6061-T6	10,000	70	3800	26	0.33
7075-T6	10,400	72	3900	27	0.33
Brass	14,000–16,000	96–110	5200–6000	36–41	0.34
Bronze	14,000–17,000	96–120	5200–6300	36–44	0.34
Cast iron	12,000–25,000	83–170	4600–10,000	32–69	0.2–0.3
Concrete (compression)	2500–4500	17–31			0.1–0.2
Copper and copper alloys	16,000–18,000	110–120	5800–6800	40–47	0.33–0.36
Glass	7000–12,000	48–83	2700–5100	19–35	0.17–0.27
Magnesium alloys	6000–6500	41–45	2200–2400	15–17	0.35
Monel (67% Ni, 30% Cu)	25,000	170	9500	66	0.32
Nickel	30,000	210	11,400	80	0.31
Plastics					
Nylon	300–500	2.1–3.4			0.4
Polyethylene	100–200	0.7–1.4			0.4
Rock (compression)					
Granite, marble,	6000–14,000	40–100			0.2–0.3
quartz	3000–10,000	20–70			0.2–0.3
Limestone, sandstone					
Rubber	0.1–0.6	0.0007–0.004	0.03–0.2	0.0002–0.001	0.45–0.50
Steel	28,000–30,000	190–210	10,800–11,800	75–80	0.27–0.30
Titanium alloys	15,000–17,000	100–120	5600–6400	39–44	0.33
Tungsten	50,000–55,000	340–380	21,000–23,000	140–160	0.2
Wood (bending)					
Douglas fir	1600–1900	11–13			
Oak	1600–1800	11–12			
Southern pine	1600–2000	11–14			

Table F-3

Mechanical Properties

Material	Yield Stress σ_Y		Ultimate Stress σ_U		Percent Elongation (2-in. gage length)
	ksi	MPa	ksi	MPa	
Aluminum alloys	5–70	35–500	15–80	100–550	1–45
2014-T6	60	410	70	480	13
6061-T6	40	270	45	310	17
7075-T6	70	480	80	550	11
Brass	10–80	70–550	30–90	200–620	4–60
Bronze	12–100	82–690	30–120	200–830	5–60
Cast iron (tension)	17–42	120–290	10–70	69–480	0–1
Cast iron (compression)			50–200	340–1400	
Concrete (compression)			1.5–10	10–70	
Copper and copper alloys	8–110	55–760	33–120	230–830	4–50
Glass			5–150	30–1000	0
Plate glass			10	70	
Glass fibers			1000–3000	7000–20,000	
Magnesium alloys	12–40	80–280	20–50	140–340	2–20
Monel (67% Ni, 30% Cu)	25–160	170–1100	65–170	450–1200	2–50
Nickel	15–90	100–620	45–110	310–760	2–50
Plastics					
Nylon			6–12	40–80	20–100
Polyethylene			1–4	7–28	15–300
Rock (compression)					
Granite, marble, quartz			8–40	50–280	
Limestone, sandstone			3–30	20–200	
Rubber	0.2–1.0	1–7	1–3	7–20	100–800
Steel					
High-strength	50–150	340–1000	80–180	550–1200	5–25
Machine	50–100	340–700	80–125	550–860	5–25
Spring	60–240	400–1600	100–270	700–1900	3–15
Stainless	40–100	280–700	60–150	400–1000	5–40
Tool	75	520	130	900	8
Steel, structural	30–100	200–700	50–120	340–830	10–40
ASTM-A36	36	250	60	400	30
ASTM-A572	50	340	70	500	20
ASTM-A514	100	700	120	830	15

Table F-3 (Continued)

Material	Yield Stress σ_Y		Ultimate Stress σ_U		Percent Elongation (2-in. gage length)
	ksi	MPa	ksi	MPa	
Steel wire	40–150	280–1000	80–200	550–1400	5–40
Titanium alloys	110–150	760–1000	130–170	900–1200	10
Tungsten			200–600	1400–4000	0–4

Wood (bending)
    Douglas fir
    Oak
    Southern pine
	5–8
	6–9
	6–9
		30–50
		40–60
		40–60
			8–12
			8–14
			8–14
				50–80
				50–100
				50–100	

Wood (compression parallel to grain)
    Douglas fir
    Oak
    Southern pine
	4–8
	4–6
	4–8
		30–50
		30–40
		30–50
			6–10
			5–8
			6–10
				40–70
				30–50
				40–70	

Table F-4

Coefficients of Thermal Expansion

Material	Coefficient of Thermal Expansion α		Material	Coefficient of Thermal Expansion α	
	$10^{-6}/°F$	$10^{-6}/°C$		$10^{-6}/°F$	$10^{-6}/°C$

Aluminum alloys	13	23	Plastics
			    Nylon
			    Polyethylene
				
				40–80
				80–160
					
					70–140
					140–290

Brass	10.6–11.8	19.1–21.2			
Bronze	9.9–11.6	18–21	Rock	3–5	5–9
Cast iron	5.5–6.6	9.9–12	Rubber	70–110	130–200

Concrete	4–8	7–14	Steel
			    High-strength
			    Stainless
			    Structural
				5.5–9.9
				8.0
				9.6
				6.5
					10–18
					14
					17
					12

Copper and copper alloys	9.2–9.8	16.6–17.6			
Glass	3–6	5–11			
Magnesium alloys	14.5–16.0	26.1–28.8	Titanium alloys	4.5–6.0	8.1–11
Monel (67% Ni, 30% Cu)	7.7	14	Tungsten	2.4	4.3
Nickel	7.2	13			

CHAPTER 1

1.2-1 (a) $F_{plane} = 155.1$ kips
(b) $W_{plane} = 155.5$ kips

1.2-2 (a) $F_{moonE} = 2.05 \times 10^{17}$ kN
(b) $F_{JupiterE} = 2.185 \times 10^{15}$ kN
(c) $F_{MarsE} = 8.244 \times 10^{13}$ kN
(d) $F_{SaturnE} = 1.586 \times 10^{14}$ kN
The moon has the largest gravitational interaction with earth.

1.2-3 29.8 lb

1.2-4 $F_2 = 1.461$ kN $> F_1 = 0.483$ kN

1.3-1 $l = 0.555, \quad m = 0.832, \quad n = 0$
$n_V = (0.555)i + (0.832)j$

1.3-2 $V_s = V_1 + V_2 = \begin{pmatrix} 18 \\ 3 \end{pmatrix}$

$V_d = V_1 - V_2 = \begin{pmatrix} 2 \\ 27 \end{pmatrix}$

1.3-3 $V_s = V_1 + V_2 = \begin{pmatrix} -2 \\ 3 \end{pmatrix}$

$V_d = V_1 - V_2 = \begin{pmatrix} -18 \\ 27 \end{pmatrix}$

1.3-4 $V_1 = \begin{pmatrix} -2.31 \\ -2.82 \\ 0 \end{pmatrix}, \quad V_2 = \begin{pmatrix} 6.31 \\ -5.18 \\ 7 \end{pmatrix}$

1.3-5 $V_y = 1$

1.3-6 $d_{CA} = \begin{pmatrix} -5 \\ 9 \\ 4 \end{pmatrix}$ m, $n_{CA} = \begin{pmatrix} -0.453 \\ 0.815 \\ 0.362 \end{pmatrix}$

$F = \begin{pmatrix} -8.15 \\ 14.67 \\ 6.52 \end{pmatrix}$ N

$d_{BC} = \begin{pmatrix} 5 \\ 0 \\ -4 \end{pmatrix}$ m, $d_{BD} = \begin{pmatrix} 5 \\ 9 \\ -4 \end{pmatrix}$ m

1.3-7 $l = -0.453, \quad m = 0.815, \quad n = 0.362$

1.3-8 $F = \dfrac{|F|}{\sqrt{3}}(i + j + k)$

1.3-9 $F_{1x} = 11.26$ lb, $F_{1y} = -6.5$ lb
$F_{2x} = 10.61$ lb, $F_{2y} = 10.61$ lb

1.3-10 $\theta_x = 64.1°, \quad \theta_y = 150.8°,$
$\theta_z = 77.4°$

1.3-11 (a) $n_s = \begin{pmatrix} 0.983 \\ 0.185 \end{pmatrix}, \quad n_d = \begin{pmatrix} 0.038 \\ -0.999 \end{pmatrix}$

(b) $r_{OP} = \begin{pmatrix} -10 \text{ in.} \\ 12 \text{ in.} \end{pmatrix}, \quad |r_{OP}| = 15.62$ in.

$n_{OP} = \dfrac{r_{OP}}{|r_{OP}|} = \begin{pmatrix} -0.64 \\ 0.768 \end{pmatrix}$

1.3-12 $F_s = \begin{pmatrix} 0 \\ 143 \end{pmatrix}$ N $\quad F_d = \begin{pmatrix} 230 \\ -50 \end{pmatrix}$ N

1.4-1 $m_{USCS} = 3500$ lb/g $= 108.8$ slugs
$m_{SI} = 14.5939(108.8) = 1588$ kg

1.4-2 $m_c = 4.93$ slugs, $W_c = m_c\, g = 158.7$ lb

1.4-3 $V = \dfrac{1}{3}\pi r^2 h = 4.16$ in^3, $\quad L^2 L \rightarrow L^3$

1.4-4 $E_{AL} = 7.308 \times 10^{10}$ Pa
$E_{AL} = 7.308 \times 10^7$ kPa
$E_{AL} = 7.308 \times 10^4$ MPa
$E_{AL} = 73.08$ GPa
$\gamma_{AL} = 2.75 \times 10^4$ N/m^3, $\gamma_{AL} = 27.5$ kN/m^3
$V = \dfrac{4}{3}\pi r^3 = 113.1$ in^3

$m_{USCS} = \dfrac{\gamma_{AL}}{g}V = 0.356$ slugs

$m_{SI} = \dfrac{\gamma_{AL}}{g}V = 5.2$ kg

1.4-5 $m_{USCS} = \dfrac{W}{g} = 0.466$ slugs

$m_{SI} = \dfrac{W}{g} = 6.8$ kg

1.5-1 For $\alpha_a = 5°$, sine, cosine, and tangent errors are 0.127%, 91.2%, and 0.254%, respectively
For $\alpha_d = 20°$, sine, cosine, and tangent errors are 2.06%, 62.8%, and 4.1%, respectively

1.5-2 Error$(\alpha_a) = -1.28\%$
Error$(\alpha_b) = -5.33\%$
Error$(\alpha_c) = -12.77\%$
Error$(\alpha_d) = -25.02\%$

1.5-3 $P + Q = 10.66$, $P - Q = -5.20$, $\dfrac{P}{Q} = 0.344$

$(P + Q)^2 = 113.6$, $\sqrt{-P + Q} = 2.28$

CHAPTER 2

2.1-1 (a) $\alpha = -8$, (b) $\alpha = 1.934$, (c) $\varphi = 0$

2.1-2 $\theta = 40.5°$, $\beta = 27.7°$

2.1-3 (a) $F_x = 250$ lb, $F_y = 433$ lb
(b) $F_\alpha = 577$ lb, $F_\beta = 289$ lb
(c) $F_{\alpha proj} = 433$ lb, $F_{\beta proj} = 0$

2.1-4 (a) $F_x = -1.768$ kN, $F_y = 1.768$ kN
(b) $F_n = -2.415$ kN, $F_t = -0.647$ kN

2.1-5 $F_2 = 21.2$ lb, $F_4 = -49.5$ lb

2.1-6 (a) $R_{12} = \begin{pmatrix} 130 \\ 52 \end{pmatrix}$ kN, $|R_{12}| = 140$ kN

$\beta_{12} = 21.79°$

(b) $R_{34} = \begin{pmatrix} -70 \\ -52 \end{pmatrix}$ kN, $|R_{34}| = 87.2$ kN

$\beta_{34} = 217°$

(c) $R = R_{12} + R_{34} = \begin{pmatrix} 60 \\ 0 \end{pmatrix}$ kN, $|R| = 60$ kN

2.1-7 $R = 154.5$ lb, $\alpha = 15°$

2.1-8 (a) $R = \begin{pmatrix} 0 \\ 401 \\ 0 \end{pmatrix}$ N   at $90°$ to x axis

(b) $R_n = 347$ N $= N_2$, $R_t = 200$ N $= N_1$

2.1-9 $R = \begin{pmatrix} 1120 \\ 61 \\ 0 \end{pmatrix}$ lb, $\theta_x = 3.11°$

2.1-10 $R = \begin{pmatrix} -785 \\ 554 \end{pmatrix}$ N, $\theta_x = 144.8°$

2.1-11 $R = 8572(-j)$ lb   or $R = \begin{pmatrix} 0 \\ -8572 \end{pmatrix}$ lb

$\beta_1 = 31°, \beta_2 = 31°$

2.1-12 $R = \begin{pmatrix} 0 \\ -180 \end{pmatrix}$ N

2.1-13 (a) $F_{EB} = \begin{pmatrix} -781 \\ 4432 \\ 0 \end{pmatrix}$ lb

(b) $F_{CA} = 4229$ lb

2.1-14 $R_a = \begin{pmatrix} -3 \\ 6 \\ 2 \end{pmatrix}$ N, $R_b = \begin{pmatrix} -5 \\ -6 \\ 6 \end{pmatrix}$ N,

$R_c = \begin{pmatrix} 1.45 \\ 4.37 \\ -4.37 \end{pmatrix}$ N

2.1-15 $T = \begin{pmatrix} 385 \\ 1538 \\ 2308 \end{pmatrix}$ lb

2.1-16 $T = 6.54$ kN

2.1-17 $T_{CE} = -265\,i + 159\,j - 212\,k$ lb
T_{CE} projection $= 212$ lb

2.1-18 $R = \begin{pmatrix} 0 \\ 1442 \\ -1923 \end{pmatrix}$ N

2.1-19 (a) $R = \begin{pmatrix} 265 \\ -339 \\ 452 \end{pmatrix}$ lb, $|R| = 624$ lb

(b) $R_{EB} = 565$ lb, $R_{EC} = 587$ lb

2.1-20 (a) $F_s = \begin{pmatrix} 0 \\ 145.7 \\ 49.8 \end{pmatrix}$ N

(b) $F_{AD} = 61.9$ N, $\theta = 66.3°$

2.1-21 (a) $F = \begin{pmatrix} -95.9 \\ 55.3 \\ 332 \end{pmatrix}$ lb

(b) $F_{BC} = 55.3$ lb, $\theta = 80.9°$

2.1-22 $n_{BC} = \begin{pmatrix} 0.183 \\ 0.365 \\ 0.913 \end{pmatrix}$, $T_B = \begin{pmatrix} 676 \\ 1351 \\ 3378 \end{pmatrix}$ N

2.1-23 (a) $T_{AB} = \begin{pmatrix} -1.493 \\ -7.463 \\ 12.926 \end{pmatrix}$ lb,

(b) $T_{CD} = -1.493$ lb, (c) $T_{CB} = 6.72$ lb

2.1-24 (a) $P = \begin{pmatrix} 1.54 \\ 0 \\ 2.57 \end{pmatrix}$ kN, $Q = \begin{pmatrix} 1.59 \\ 2.54 \\ 2.65 \end{pmatrix}$ kN

(b) $R = \begin{pmatrix} 3.13 \\ 2.54 \\ 5.22 \end{pmatrix}$ kN, $\theta_y = 67.3°$

(c) $P_{CApar} = 1.091$ kN, $P_{CAper} \begin{pmatrix} 0.77 \\ 0.77 \\ 2.57 \end{pmatrix}$ kN

(d) $Q_{BApar} = -0.749$ kN

$Q_{BAper} = \begin{pmatrix} 2.12 \\ 2.54 \\ 2.12 \end{pmatrix}$ kN

(e) $\theta = 39.5°$

2.1-25 $n_{BQ} = \begin{pmatrix} -0.631 \\ 0.451 \\ 0.631 \end{pmatrix}$, $T_2 = \begin{pmatrix} -3282 \\ 2344 \\ 3282 \end{pmatrix}$ lb

$T_{BD} = 3679$ lb

2.1-26 (a) $F = \begin{pmatrix} 12.86 \\ 12.86 \\ 17.15 \end{pmatrix}$ kN, (b) $n_{OA} = \begin{pmatrix} 0.51 \\ 0.51 \\ 0.69 \end{pmatrix}$

(c) $F_{x'} = 4.71$ kN, $F_{y'} = 17.57$ kN,
$F_{z'} = 17.15$ kN

2.1-27 (a) $\boldsymbol{P} = \begin{pmatrix} 5.66 \\ 7.07 \\ -4.24 \end{pmatrix}$ kips

(b) $P_{HC\mathrm{par}} = 7.07$ kips, $\boldsymbol{P}_{HC\mathrm{per}} = \begin{pmatrix} 0 \\ 7.07 \\ 0 \end{pmatrix}$ kips

(c) $P_{DG} = 1.98$ kips, $\theta = 78.6°$

2.1-28 (a) $\boldsymbol{P} = \begin{pmatrix} 30 \\ 25 \\ -75 \end{pmatrix}$ kN, (b) 83.3 kN, (c) $\theta = 9.88°$

2.1-29 (a) $\boldsymbol{T}_{BD} = \begin{pmatrix} 7.28 \\ -4.85 \\ 4.85 \end{pmatrix}$ kips

$\boldsymbol{T}_{CD} = \begin{pmatrix} 7.28 \\ -4.85 \\ -4.85 \end{pmatrix}$ kips

$\boldsymbol{T}_{DE} = \begin{pmatrix} 0 \\ -10.61 \\ -10.61 \end{pmatrix}$ kips

$\boldsymbol{T}_{DF} = \begin{pmatrix} 0 \\ -10.61 \\ 10.61 \end{pmatrix}$ kips

(b) $\boldsymbol{R} = \begin{pmatrix} 14.55 \\ -30.91 \\ 0 \end{pmatrix}$ kips

2.1-30 (a) $\boldsymbol{W} = \begin{pmatrix} 0 \\ 0 \\ -4413 \end{pmatrix}$ N, $\boldsymbol{P} = \begin{pmatrix} 4350 \\ 5801 \\ -18,639 \end{pmatrix}$ N

(b) $\boldsymbol{R} = \begin{pmatrix} 4350 \\ 5801 \\ -23,052 \end{pmatrix}$ N

2.2-1 $\boldsymbol{R}_A = \begin{pmatrix} 7.07 \\ -17.07 \\ 0 \end{pmatrix}$ kips

$\boldsymbol{M}_A = \begin{pmatrix} 0 \\ 0 \\ -1329 \end{pmatrix}$ kip-in.

$x = 77.8$ in.

2.2-2 $\boldsymbol{R}_A = \begin{pmatrix} 70.7 \\ -108.9 \\ 0 \end{pmatrix}$ kN

$\boldsymbol{M}_A = \begin{pmatrix} 0 \\ 0 \\ -545 \end{pmatrix}$ kN $\cdot$ m

$y = 7.71$ m

2.2-3 $\boldsymbol{R}_A = \begin{pmatrix} -17.5 \\ -110 \\ 0 \end{pmatrix}$ kips, $\boldsymbol{M}_A = \begin{pmatrix} 0 \\ 0 \\ -955 \end{pmatrix}$ kip-ft

2.2-4 $\boldsymbol{R}_O = \begin{pmatrix} 1 \\ 1 \\ 0 \end{pmatrix} F$, $M_{Oz} = -\left(2 + \sqrt{3}\right) Fr$

2.2-5 $\boldsymbol{M}_C = \begin{pmatrix} 0 \\ 0 \\ -25,811 \end{pmatrix}$ lb-ft, $M_{Cz} = -25,811$ lb-ft

2.2-6 $\boldsymbol{M}_O = \begin{pmatrix} 0 \\ 0 \\ 139.5 \end{pmatrix}$ kN $\cdot$ m

2.2-7 (a) $\boldsymbol{M}_D = \begin{pmatrix} -830 \\ 0 \\ -240 \end{pmatrix}$ lb-ft

(b) $M_{Dx} = -830$ lb-ft, $M_{Dy} = 0$, $M_{Dz} = -240$ lb-ft

(c) $\boldsymbol{M}_{BD\mathrm{par}} = \begin{pmatrix} 0 \\ 110.6 \\ -165.9 \end{pmatrix}$ lb-ft

$\boldsymbol{M}_{BD\mathrm{per}} = \begin{pmatrix} -830.1 \\ -110.6 \\ -73.7 \end{pmatrix}$ lb-ft

2.2-8 (a) $\boldsymbol{M}_B = \begin{pmatrix} 225 \\ 0 \\ 90 \end{pmatrix}$ kN $\cdot$ m

(b) $M_{Bx} = 225$ kN $\cdot$ m, $M_{By} = 0$, $M_{Bz} = 90$ kN $\cdot$ m

(c) $\boldsymbol{M}_{BC\mathrm{par}} = \begin{pmatrix} 8.8 \\ 2.93 \\ -17.61 \end{pmatrix}$ kN $\cdot$ m

$\boldsymbol{M}_{BC\mathrm{per}} = \begin{pmatrix} 216.2 \\ -2.9 \\ 107.6 \end{pmatrix}$ kN $\cdot$ m

(d) $\boldsymbol{M}_O = \begin{pmatrix} -150 \\ 180 \\ 0 \end{pmatrix}$ kN $\cdot$ m

$\boldsymbol{M}_{OB\mathrm{par}} = \begin{pmatrix} 0 \\ 36 \\ 72 \end{pmatrix}$ kN $\cdot$ m

$\boldsymbol{M}_{OB\mathrm{per}} = \begin{pmatrix} -150 \\ 144 \\ -72 \end{pmatrix}$ kN $\cdot$ m

2.2-9 (a) $x_D = 6.67$ ft, $y_D = 13.33$ ft

(b) $\left| \boldsymbol{M}_O \right| = 268$ kip-ft, $\left| \boldsymbol{M}_C \right| = 268$ kip-ft

(c) $d = 11.95$ ft

(d) $M_{Ox} = -240$ kip-ft, $M_{Oy} = 120$ kip-ft

(e) $z_Q = 25$ ft

2.2-10 $\boldsymbol{R}_O = \begin{pmatrix} -84.6 \\ 0 \\ 0 \end{pmatrix}$ kN, $\boldsymbol{M}_{R\mathrm{par}} = \begin{pmatrix} -150 \\ 0 \\ 0 \end{pmatrix}$ kN $\cdot$ m,

$y = 4$ m, $z = -2.13$ m

2.2-11   (a) $M_B = \begin{pmatrix} 1245 \\ 719 \\ 240 \end{pmatrix}$ lb-ft

(b) $M_{Bx} = 1245$ lb-ft, $M_{By} = 719$ lb-ft,
$M_{Bz} = 240$ lb-ft

(c) $M_{BD\,par} = \begin{pmatrix} 0 \\ 110.6 \\ -165.9 \end{pmatrix}$ lb-ft

$M_{BD\,per} = \begin{pmatrix} -830.1 \\ -110.6 \\ -73.7 \end{pmatrix}$ lb-ft

2.2-12   (a), (b) $M_B = \begin{pmatrix} 1695 \\ 0 \\ 2119 \end{pmatrix}$ N·m

(c) $M_{BD\,par} = \begin{pmatrix} 0 \\ 0 \\ 0 \end{pmatrix}$ N·m

$M_{BD\,per} = \begin{pmatrix} 1695 \\ 0 \\ 2119 \end{pmatrix}$ N·m

(d) $M_O = \begin{pmatrix} 0 \\ -942 \\ 2119 \end{pmatrix}$ N·m

$M_{OD\,par} = \begin{pmatrix} -400 \\ -720 \\ 0 \end{pmatrix}$ N·m

$M_{OD\,per} = \begin{pmatrix} 400 \\ -222 \\ 2119 \end{pmatrix}$ N·m

2.2-13   (a), (b) $M_O = \begin{pmatrix} 0 \\ -77.6 \\ -44.8 \end{pmatrix}$ in-lb

(c) $M_{CD\,par} = \begin{pmatrix} -259 \\ 0 \\ 0 \end{pmatrix}$ in-lb

$M_{CD\,per} = \begin{pmatrix} 0 \\ -77.6 \\ -74.6 \end{pmatrix}$ in-lb

2.2-14   (a), (b) $M_O = \begin{pmatrix} 0 \\ 14.72 \\ 0 \end{pmatrix}$ N·m

2.2-15   (a) $R_O = \begin{pmatrix} 0 \\ 0 \\ -40 \end{pmatrix}$ lb

$M_O = \begin{pmatrix} -130 \\ 153.2 \\ 0 \end{pmatrix}$ in-lb

(b) $M_{AB} = \begin{pmatrix} -46.9 \\ 0 \\ -39.4 \end{pmatrix}$ in-lb

2.2-16   (a), (b) $M_O = \begin{pmatrix} -4.44 \\ 13.25 \\ 3.81 \end{pmatrix}$ N·m

(c) $M_{OC\,par} = \begin{pmatrix} 3.65 \\ 2.84 \\ 0 \end{pmatrix}$ N·m

$M_{OC\,per} = \begin{pmatrix} -8.09 \\ 10.41 \\ 3.81 \end{pmatrix}$ N·m

(d) $d = 3.62$ m

2.2-17 $M_A = \begin{pmatrix} 0 \\ -27 \\ 0 \end{pmatrix}$ lb-ft

$M_D = M_A$, $M_O = M_A$

2.2-18   (a) $M_C = \begin{pmatrix} -3600 \\ 900 \\ -1200 \end{pmatrix}$ N·m

(b) $M_{CD} = \begin{pmatrix} -2541 \\ 1694 \\ -1906 \end{pmatrix}$ N·m

(c) $d = 7.21$ m

2.2-19   (a) $M_A = \begin{pmatrix} -37.1 \\ -60 \\ -22.3 \end{pmatrix}$ in-lb

(b) $M_{AC} = \begin{pmatrix} -14.4 \\ -19.2 \\ 24 \end{pmatrix}$ in-lb

(c) $d = 1.697$ in.

2.2-20 $M_{SE} = \begin{pmatrix} 0.832 \\ 0.894 \\ 0.002 \end{pmatrix} M_0$

2.2-21   (a) $M_{EF1} = \begin{pmatrix} 0 \\ -28.8 \\ -38.4 \end{pmatrix}$ ft-lb

(b) $M_{EF2} = \begin{pmatrix} 43.2 \\ 0 \\ 57.6 \end{pmatrix}$ ft-lb

(c) $M_{EC} = \begin{pmatrix} 7.2 \\ -7.2 \\ 9.6 \end{pmatrix}$ ft-lb

(d) $M_{EG} = \begin{pmatrix} -2.81 \\ -2.81 \\ 2.11 \end{pmatrix}$ ft-lb

(e) $M_{EB} = \begin{pmatrix} 0 \\ -36 \\ 0 \end{pmatrix}$ ft-lb

2.2-22 (a) $R = \begin{pmatrix} 0 \\ 355 \\ 0 \end{pmatrix}$ N

(b) $M_B = \begin{pmatrix} -485 \\ 0 \\ 913 \end{pmatrix}$ N·m

(c) Apply R in y-dir at $x = 2.57$ m, $z = 1.366$ m

2.2-23 (a), (b) $R_A = \begin{pmatrix} -5 \\ -1 \\ 0 \end{pmatrix}$ lb

$M_A = \begin{pmatrix} 12 \\ -20 \\ -1 \end{pmatrix}$ in-lb

(c) $y = -0.2$ in., $z = 4.31$ in.

2.2-24 $R = \begin{pmatrix} 29.5 \\ 36.4 \\ -81.4 \end{pmatrix}$ N, $M_{R\,\mathrm{par}} = \begin{pmatrix} 42.8 \\ 52.7 \\ -117.8 \end{pmatrix}$ N·m

$x = 3.1$ m, $y = 7.26$ m

2.2-25 $R_O = \begin{pmatrix} 0 \\ 0 \\ 0 \end{pmatrix}$, $M_O = \begin{pmatrix} -440 \\ 120 \\ 0 \end{pmatrix}$ kip-ft

2.2-26 $R_A = \begin{pmatrix} -140 \\ 80 \\ -120 \end{pmatrix}$ N, $M_A = \begin{pmatrix} -160 \\ 325 \\ 1380 \end{pmatrix}$ N·m

$M_{R\,\mathrm{par}} = \begin{pmatrix} 406 \\ -232 \\ 348 \end{pmatrix}$ N·m

$y_P = 7.37$ m, $z_P = -3.98$ m

2.2-27 $R_O = \begin{pmatrix} 0 \\ -875 \\ 0 \end{pmatrix}$ lb, $M_O = \begin{pmatrix} 3500 \\ 0 \\ -4850 \end{pmatrix}$ lb-ft

2.2-28 $R_D = \begin{pmatrix} 34.4 \\ 41.7 \\ -10 \end{pmatrix}$ kN, $M_D = \begin{pmatrix} -140 \\ 75 \\ 0 \end{pmatrix}$ kN·m

$M_{DF\,\mathrm{par}} = \begin{pmatrix} -14.8 \\ -18.5 \\ -11.1 \end{pmatrix}$ kN·m

$M_{DF\,\mathrm{per}} = \begin{pmatrix} -125.2 \\ 93.5 \\ 11.1 \end{pmatrix}$ kN·m

$x = 9.82$ m, $y = 12.08$ m

2.2-29 (a) $R_O = \begin{pmatrix} -15 \\ 0 \\ -25 \end{pmatrix}$ lb, $M_O = \begin{pmatrix} -187.5 \\ 31.25 \\ 0 \end{pmatrix}$ in-lb

(b) $x = 1.25$ in., $y = 5.51$ in.

2.2-30 $x = \dfrac{L}{2}$, $y = \dfrac{\sqrt{3}}{10}L$

2.2-31 $R_O = \begin{pmatrix} -10 \\ 2 \\ 0 \end{pmatrix}$ lb, $M_O = \begin{pmatrix} 18 \\ -120 \\ -3.5 \end{pmatrix}$ in-lb

2.2-32 $R_O = \begin{pmatrix} 0 \\ -600 \\ 0 \end{pmatrix}$ N, $M_O = \begin{pmatrix} -1300 \\ 0 \\ 0 \end{pmatrix}$ N·m

2.2-33 (a) $R_O = \begin{pmatrix} -23.64 \\ -1.97 \\ -7.88 \end{pmatrix}$ kips

$M_O = \begin{pmatrix} -183.7 \\ 0 \\ 851.2 \end{pmatrix}$ kip-in.

(b) $M_{OA\,\mathrm{par}} = \begin{pmatrix} -183.7 \\ 0 \\ 0 \end{pmatrix}$ kip-in.

$M_{OA\,\mathrm{per}} = \begin{pmatrix} 0 \\ 0 \\ 851 \end{pmatrix}$ kip-in.

2.2-34 (a) $R_O = \begin{pmatrix} 3.13 \\ 2.54 \\ 5.22 \end{pmatrix}$ kN

$M_O = \begin{pmatrix} 2.95 \\ 0 \\ -1.77 \end{pmatrix}$ kN·m

(b) $M_{AD\,\mathrm{par}} = \begin{pmatrix} 0 \\ 0 \\ -9.4 \end{pmatrix}$ kN·m

$M_{AD\,\mathrm{per}} = \begin{pmatrix} 2.95 \\ 15.66 \\ 0 \end{pmatrix}$ kN·m

2.2-35 (a) $R_D = \begin{pmatrix} 5.66 \\ 7.07 \\ -4.24 \end{pmatrix}$ kips

$M_D = \begin{pmatrix} -21.21 \\ 16.97 \\ 0 \end{pmatrix}$ kip-in.

(b) $M_{DG\,\mathrm{par}} = \begin{pmatrix} -13.58 \\ 0 \\ -10.18 \end{pmatrix}$ kip-in.

$M_{DG\,\mathrm{per}} = \begin{pmatrix} -7.64 \\ 16.97 \\ 10.18 \end{pmatrix}$ kip-in.

(c) $d = 3.3$ in.

2.2-36   (a) $R_A = \begin{pmatrix} -35.4 \\ 21.2 \\ -28.3 \end{pmatrix}$ kN

$M_A = \begin{pmatrix} -84.9 \\ -141.4 \\ 0 \end{pmatrix}$ kN·m

(b) $M_{AD\text{par}} = \begin{pmatrix} -51.7 \\ 0 \\ -41.4 \end{pmatrix}$ kN·m

$M_{AD\text{per}} = \begin{pmatrix} -33.1 \\ -141.4 \\ 41.4 \end{pmatrix}$ kN·m

2.2-37   (a) $R_A = \begin{pmatrix} 14.55 \\ -30.91 \\ 0 \end{pmatrix}$ kips

$M_A = \begin{pmatrix} 0 \\ 0 \\ -121.8 \end{pmatrix}$ kip-ft

(b) Zero

2.2-38   (a) $R_A = \begin{pmatrix} 0 \\ -50.4 \\ 49.8 \end{pmatrix}$ N

$M_A = \begin{pmatrix} 119.6 \\ 0 \\ -33.3 \end{pmatrix}$ N·m

(b) $M_{CD} = -0.067$ N·m

2.2-39 $R_O = \begin{pmatrix} 0.98 \\ 1.31 \\ -5.17 \end{pmatrix}$ kips

$M_O = \begin{pmatrix} -178.2 \\ 37.7 \\ -26.4 \end{pmatrix}$ kip-ft

CHAPTER 3

3.3-1 $R_x = -1120$ lb, $R_y = -60.8$ lb

3.3-2 $R_x = -672$ N, $R_y = -180$ N

3.3-3 $N_1 = 45$ lb, $N_2 = 77.9$ lb

3.3-4 $N_1 = 2.01\ W$, $F = 1.521\ W$

3.3-5 $N_1 = 1.48\ W$, $F = 0.448\ W$

3.3-6 $N_1 = 1.777\ W$, $F = 1.326\ W$

3.3-7 $N_1 = 1.777\ W$, $F = 0.53\ W$

3.3-8 $T = 34.3$ N, $L_1 = 2.33$ m, $L_2 = 1.167$ m

3.3-9 $T_{AB} = 0.763\ W$, $T_{BC} = 0.214\ W$

3.3-10 $T_{AB} = 0.863\ W$, $T_{BC} = 0.643\ W$

3.3-11 $T_{AB} = 0.621\ W$, $T_{BC} = 0.532\ W$

3.3-12 $T = 1.526$ kN, $A_x = -1.27$ kN,
$A_y = 3.1$ kN, $D_x = 0$, $D_y = -1.692$ kN

3.3-13 $L = 6.5$ ft, $F = 139$ lb

3.3-14 $T = 1019$ N, $R_{Ox} = 1559$ N,
$R_{Oy} = 1815$ N
$A_x = -1559$ N, $A_y = 1185$ N

3.3-15 $A_x = -4.12$ lb, $A_y = 12.64$ lb,
$R_C = -21.2$ lb

3.3-16 $A_x = -76.6$ N, $A_y = 33$ N, $R_F = 89$ N

3.3-17 $R_{Ax} = P$, $R_{Ay} = P$, $M_{Az} = 6Pb$

3.3-18 $T = 6.54$ kN
$R_{Ox} = 0$, $R_{Oy} = 33.87$ kN
$R_{Ax} = -7.12$ kN, $R_{Ay} = -6.83$ kN
$R_{Bx} = 7.12$ kN, $R_{By} = -7.04$ kN

3.3-19 $A_x = 0$, $A_y = 23.3$ kips, $B_y = 16.22$ kips

3.3-20 $T = 25$ N

3.3-21 $F = 200$ lb

3.3-22 $T_1 = 13.68$ kN, $T_2 = 15.41$ kN

3.3-23 $A_y = 5$ lb, $B_y = -5$ lb
$C_x = 50$ lb, $C_y = 0$

3.3-24 $A_x = 0$, $M_A = 0$, $C_y = 253.3$ N
$D_y = -93.3$ N

3.3-25 $C_x = -86.6$ lb, $C_y = 51.7$ lb
$F_y = 62.2$ lb

3.3-26 $A_x = 0$, $A_y = 37.5$ kN, $E_y = 37.5$ kN

3.3-27 $A_x = -15.2$ kips, $A_y = 8.36$ kips
$E_x = 25.1$ kips, $E_y = 0$

3.3-28 $A_x = -P$, $A_y = \dfrac{5P}{8}$, $C_y = \dfrac{3P}{8}$

3.3-29 $A_x = -0.894P$, $A_y = -0.25P$,
$C_y = 0.697P$

3.3-30 $H_x = 0$, $H_y = 1060$ N,
$M_H = -4120$ N·m

3.3-31 $A_x = -1050$ lb, $A_y = -437$ lb
$D_x = 1050$ lb, $D_y = 788$ lb
$C_x = 1050$ lb, $C_y = 788$ lb

3.3-32 $A_x = -65$ kN, $A_y = -35.8$ kN
$M_A = 152.7$ kN·m, $C_y = 35.8$ kN

3.3-33 $A_x = -540$ lb, $A_y = -55.6$ lb

   $M_A = 4320$ lb-ft, $C_y = 55.6$ lb

3.3-34 $A_x = 280$ N, $A_y = 8.91$ N

   $M_A = -667$ N·m, $D_y = 151.1$ N

3.3-35 $B_x = -10$ kips, $B_y = -60$ kips,

   $C_y = 60$ kips

3.3-36 $A_x = 320$ N, $A_y = -240$ N

   $C_y = 192$ N, $E_y = -192$ N

3.3-37 $A_x = 30$ lb, $A_y = 140$ lb

   $C_x = -30$ lb, $C_y = 60$ lb

3.3-38 $A_x = -93.9$ kN, $A_y = -187.8$ kN

   $B_x = 93.9$ kN, $B_y = 176.7$ kN

   $C_y = 221$ kN

3.3-39 $A_y = -1250$ lb, $E_x = 0$, $E_y = 1750$ lb

3.3-40 $B_x = -T$, $B_y = -T$, $G_y = -\dfrac{7\,T}{4}$

3.3-41 $B_x = -T$, $B_y = -T$, $G_y = -\dfrac{5\,T}{4}$

3.3-42 $B_x = -T$, $B_y = -T$, $G_y = -\dfrac{T}{2}$

3.3-43 $R_{Ox} = -25$ lb, $R_{Oy} = 0$, $R_{Oz} = 25$ lb

   $M_{Ox} = 0$, $M_{Oy} = -150$ in-lb,

   $M_{Oz} = -121.9$ in-lb

3.3-44 $A_x = -45$ N, $A_y = 390$ N, $F = 312$ N

3.3-45 $A_x = -50$ lb, $A_y = 0$, $A_z = -75$ lb

   $M_{Ax} = 750$ in-lb, $M_{Ay} = 1500$ in-lb,

   $M_{Az} = -500$ in-lb

3.3-46 $T_A = 4577$ N, $T_B = 2328$ N,

   $T_D = 2328$ N

3.3-47 $T_1 = 5877$ lb, $T_2 = 4679$ lb, $T_3 = 7159$ lb

3.3-48   Fig. (a) $R_O = \begin{pmatrix} 3 \\ -6 \\ -2 \end{pmatrix}$ N, $M_O = \begin{pmatrix} 12 \\ -12 \\ -18 \end{pmatrix}$ N·m

   Fig. (b) $R_O = \begin{pmatrix} 5 \\ 6 \\ -6 \end{pmatrix}$ N

   $M_O = \begin{pmatrix} -12 \\ 28 \\ 18 \end{pmatrix}$ N·m

   Fig. (c) $R_O = \begin{pmatrix} -1.452 \\ -4.366 \\ 4.366 \end{pmatrix}$ N

   $M_O = \begin{pmatrix} 8.73 \\ -16 \\ -13.1 \end{pmatrix}$ N·m

3.3-49 $A_x = 0$, $A_y = 0$, $A_z = 0$

   $M_{Ax} = -120$ in-lb, $M_{Ay} = 0$,

   $M_{Az} = -95$ in-lb

3.3-50 $A_x = 0$, $A_y = -120$ N, $A_z = -150$ N

   $M_{Ax} = -263$ N·m, $M_{Ay} = -7.5$ N·m,

   $M_{Az} = -180$ N·m

   $D_x = -60$ N, $D_y = 120$ N, $D_z = 30$ N

3.3-51   (a) $A_x = -5.77$ lb, $A_y = -47.3$ lb,

   $A_z = 2.31$ lb, $M_{Az} = -200$ in-lb

   (b) $T_{CD} = -3.81$ lb, $T_{CE} = -6.79$ lb

3.3-52 $C_x = 120$ N, $C_y = -160$ N, $C_z = 506$ N

   $D_z = 466$ N, $H_y = 320$ N, $H_z = 499$ N

3.3-53 $A_x = -250$ lb, $A_y = 0$

   $B_y = 0$, $B_z = -200$ lb

   $C_x = 0$, $O_x = 250$ lb, $O_y = 400$ lb,

   $O_z = 0$

3.3-54 $F_s = 154$ N, $A_x = 0$,

   $A_y = -145.7$ N, $A_z = -49.8$ N

3.3-55 $R_A = \begin{pmatrix} -102.9 \\ 77.2 \\ -77.2 \end{pmatrix}$ lb, $M_A = \begin{pmatrix} 771.7 \\ 1543.5 \\ 514.5 \end{pmatrix}$ in-lb

3.3-56 $F = 1215$ N

3.3-57   (a) $T = 366$ lb

   (b) $A_x = 0$, $A_y = 375$ lb, $A_z = 500$ lb

   $G_y = 375$ lb, $G_z = 500$ lb

3.3-58   (a) $T = 1.598$ kN

   (b) $A_x = 0$, $A_y = 1.5$ kN,

   $A_z = 2.5$ kN, $G_y = 1.5$ kN

3.3-59 $T = 396$ lb, $A_x = 0$, $A_y = 425$ lb,

   $A_z = 567$ lb, $G_y = 425$ lb, $G_z = 567$ lb

3.3-60   (a) $P = 1500$ N   (b) $P = 3005$ N

3.3-61   (a) $\delta_C = -1.877$ in., $A_x = 0$

   $A_y = 5.63$ lb, $R_C = 5.63$ lb

   (b) Compression $M = 0.149$ slugs

   Tension $M = 0.314$ slugs

3.3-62   (a) $A_x = 0$, $A_y = 49$ N

   $M_C = 63.7$ N·m

   (b) M_C is CCW, $M = 0.711$ kg

   M_C is CW, $M = 5.9$ kg

3.4-1 $\mu_{smin} = 0.42$

3.4-2 $\beta = 56.9°$

3.4-3 $L/s = 10.94$

3.4-4 $\mu_s = 0.333$

3.4-5   (a) $x_{max} = 1.5$ in.

   (b) $\mu_{min} = 0.624$

   (c) $W_{min} = 14.77$ lb

3.4-6   (a) $\theta_{max} = 27.2°$,   (b) $\theta_{max} = 31°$

3.4-7   (a) $P = 11.34$ kips

   (b) 3.41 kips $\leq P \leq 19.27$ kips

3.4-8   (a) $W_{1,min} = 29.7$ N

   (b) $W_{1,max} = 173.7$ N

3.4-9 $\mu_s = 0.315$

3.4-10 (a) $\theta = 56.6°$ (b) $\theta = 61.9°$

3.4-11 $\mu_{min} = 0.746$

3.4-12 (a) $\mu_1 = 0.322$ (b) $\mu_2 = 0.638$

3.4-13 (a) $\mu_{min} = 0.229$ (b) $\mu_{min} = 0.417$

3.4-14 $x_{max} = 152$ mm

3.4-15 $x_{max} = 1.98$ in.

3.4-16 $\mu_{min} = 0.512$

3.4-17 $\mu_{min} = 0.451$

3.4-18 2.81 N $\leq P \leq 11.85$ N

3.4-19 (a) $\mu = 0.563$ (b) $\mu = 0594$

3.4-20 (a) $x_{max} = 0.301$ m
(b) $\mu_{min} = 0.429$
(c) $W_{min} = 34.5$ N

3.4-21 (a) $\mu_{min} = 0.452$ (b) $\mu_{min} = 0.216$

3.4-22 (a) At B when $P = 1.767$ W
(b) At B when $P = 4.71$ W

3.4-23 $P = 206$ lb

3.4-24 (a) $H = 1778$ N (b) $H_{min} = 112$ N

3.4-25 $M = 39.3$ lb-ft

3.4-26 $M = 12.62$ kg

3.4-27 (a) $P_{min} = 86.5$ lb (b) $P_{min} = 94.3$ lb

3.4-28 (a) $W_{1,min} = 20.6$ N (b) $W_{1,max} = 250$ N

3.4-29 $P_{min} = 2.35$ kips

3.4-30 $P = 210$ N

3.4-31 (a) $P = 82$ lb (b) $P = 34.5$ lb

3.4-32 (a) $P = 343$ N (b) $P = 77$ N

3.4-33 (a) $F_T = 4.64$ lb (b) $F_L = 0.495$ lb

3.4-34 $F_r = 19.14$ N
$F_L = 5.3$ N

3.4-35 $d_m = 0.997$ in.

3.4-36 $P = 251$ N, $M_L = 230$ N $\cdot$ mm

CHAPTER 4

4.2-1 $A_x = 105$ kips, $A_y = 13.5$ kips,
$B_y = -48.5$ kips
$AC = -15.6$ kips, $AB = -97.2$ kips,
$BC = 78.9$ kips

4.2-2 $E_x = -80$ kN, $E_y = 30$ kN,
$F_x = 80$ kN
$BC = 60$ kN, $BH = -63.2$ kN,
$CH = 10$ kN, $DG = 0$,
$EG = 6$ kN, $CG = -18$ kN,
$DE = 75$ kN, $DC = 75$ kN,
$GH = -63.2$ kN, $EF = 26.7$ kN,
$FG = -84.3$ kN

4.2-3 Reaction forces: $F_x = -15$ kips,
$F_y = -18.75$ kips, $H_y = 18.75$ kips
Bar forces (kips):
$AB = -5$, $CD = 2.50$, $DE = -7.50$,
$FG = 15$, $GH = 0$,
$AC = 7.50$, $BE = -7.50$,
$CF = 18.75$, $EH = -18.75$,
$AD = -9.01$, $BD = 9.01$,
$CG = -13.52$, $EG = 13.52$

4.2-4 $A_x = -1000$ N, $D_x = 1000$ N,
$D_y = 1000$ N,
$AB = 1000$ N, $BC = 1414$ N,
$BD = -1000$ N, $CD = -1000$ N

4.2-5 $AB = BC = 450$ lb, $AD = EC = -636$ lb,
$DB = BE = -212$ lb,
$FB = 300$ lb, $DF = FE = -424$ lb

4.2-6 $AB = 0$, $AE = 0$, $BE = 0$,
$ED = 0$,
$BD = 0$, $BC = 0$, $CD = 50$ kN

4.2-7 $A_x = -450$ lb, $A_y = 0$, $J_x = -450$ lb
$BC = -225$ lb, $FE = 0$,
$FG = 0$, $IJ = 318$ lb

4.2-8 $A_x = 0$, $A_y = 60$ kN, $E_y = 60$ kN
$AB = DE = -90$ kN, $BC = CD = -90$ kN,
$AF = HE = 108.2$ kN,
$BF = DH = 0$, $CG = 0$,
$FC = CH = 0$, $FG = GH = 94.9$ kN

4.2-9 $A_x = 0$, $A_y = 520$ lb, $E_y = 680$ lb
$GH = 2746$ lb, $CD = -1556$ lb

4.2-10 $R_{3x} = 9$ kN, $R_{3y} = 2$ kN, $R_{4x} = -9$ kN
$M_6 = 2.25$ kN, $M_{10} = -9.02$ kN,
$M_{11} = -1.25$ kN

4.2-11 $A_x = 0$, $A_y = 1$ kip, $E_y = 5$ kips
$FE = 1.898$ kips

4.2-12 $D_y = 6$ kN, $F_x = 0$, $F_y = 12$ kN
$FE = 0$

4.2-13 $A_x = -20$ kips, $A_y = 18$ kips,
$G_x = 20$ kips, $AE = -22.4$ kips

4.2-14 $A_x = -20$ kN, $A_y = 18$ kN,
$G_x = 20$ kN, $AE = -63.7$ kN

4.2-15 $AG = 6.17P$, $FD = 4.51P$

4.2-16 $AE = -4.03P$, $FC = -3.33P$

4.2-17 $FC = 1.667P$, $FD = -4.51P$

4.2-18 (a) $GB = 1.146P$, (b) $GB = 0.208P$

4.2-19   (a) $FD = 2.85P$,   (b) $GA = 1.424P$

4.2-20 $BG = 3.33P$, $FG = -7P$

4.2-21 $A_x = 0$, $A_y = \dfrac{P}{2}$, $G_y = -2.5P$

$AG = -3.08P$, $FD = 1.187P$

4.2-22 $R_{3x} = 40$ N, $R_{3y} = -25$ N,

$R_{5x} = 20$ N

$F_{11} = 0$, $F_{13} = 28.3$ N

4.2-23 $R_{5x} = -15$ lb, $R_{5y} = 4.41$ lb,

$R_{8y} = 10.59$ lb

$F_{11} = 7.35$ lb, $F_{13} = 19.09$ lb

4.2-24 $R_{1x} = 22.5$ N, $R_{1y} = 15$ N,

$R_{5x} = -37.5$ N

$F_{11} = 21.2$ N, $F_{13} = 0$

4.2-25 $A_x = -732$, $A_y = 404$

$T = 384$ (all units lb)

$AB = -74.9$, $BC = -475$,

$AG = 902$, $BG = -300$,

$GF = 192.1$, $CG = 960$, $EG = 0$,

$CE = -430$, $EF = -430$

4.2-26   (1) $DB = 0$, $BC = -P$, $AB = \dfrac{-2}{3}P$,

$AC = \dfrac{5}{4}P$, $DC = \dfrac{-3}{4}P$, $AD = 0$

(2) $AC = 0$, $DC = P$, $BC = \dfrac{-2}{3}P$

$DB = \dfrac{5}{6}P$, $AB = \dfrac{-P}{2}$, $AD = \dfrac{-2}{3}P$

4.2-27 $A_x = -15.2$ kips, $A_y = 8.36$ kips,

$E_x = 25.1$ kips

$AB = 15.2$ kips, $AE = 8.36$ kips,

$BC = 15.2$ kips, $BG = 0$,

$BF = 0$, $BE = 0$

$CD = 15.2$ kips, $CG = 0$, $EF = -26.4$ kips

$FG = -26.4$ kips, $GD = -26.4$ kips

4.2-28 $GC = 1.299$ kN, $CF = -2.5$ kN,

4.2-29 $CD = 9.35$ lb, $DH = -13.09$ lb,

$HI = -1000$ lb

4.2-30 $S_x = 0$, $S_y = -200$ kN,

$T_y = 250$ kN

$FG = 65.1$ kN, $NO = -151.5$ kN,

$LM = -44.7$ kN

4.2-31   (a) $FD = -17.09$ kips, $GB = 6.67$ kips
(b) $FD = -21.4$ kips, $GB = -0.833$ kips
(c) $FD = 14.2$ kips, $GB = 6.67$ kips

4.2-32   (a) $FD = -64.1$ kN, $GB = -31.3$ kN
(b) $FD = -42.7$ kN, $GB = -46.9$ kN
(c) $FD = -10.68$ kN, $GB = -37.5$ kN

4.2-33 $AB = 0$, $CD = 0$, $EF = -6.3$ kips

4.2-34 $A_y = 200$ N, $C_x = -260$ N,

$C_y = 200$ N, $BF = 28.9$ N

4.2-35 $A_x = 33.2$ kips, $A_y = 6.36$ kips,

$B_x = -28.7$ kips

$IJ = 11.18$ kips, $EF = 9.74$ kips

4.2-36 $FG = -7$ kN, $AB = 4.5$ kN,

$AG = 3.54$ kN

4.2-37 $BC = -6.25$ kips, $GC = -0.354$ kips,

$GH = 12.5$ kips

4.2-38 $BC = -35$ kN, $HI = 30$ kN,

$IJ = 40$ kN, $JK = 40$ kN

4.2-39 $EI = -0.156$ kips, $BF = 0.786$ kips,

$AF = 0.221$ kips, $CF = 0.423$ kips

4.2-40 $BC = -14.77$ kN, $CH = -6.96$ kN,

$HI = 18.83$ kN

4.3-1 $A_x = 0$, $A_y = -11.91$ kips,

$A_z = 24.2$ kips,

$B_y = -23.1$ kips, $B_z = 0$,

$C_z = -24.2$ kips

$AB = 0$, $AC = 27$ kips,

$BC = 73.7$ kips

4.3-2 $A_x = 4.01$ kN, $A_y = -0.34$ kN,

$C_z = -0.69$ kN

$B_x = -2.01$ kN, $B_y = -0.66$ kN,

$B_z = 2.69$ kN

$AB = -4.83$ kN, $AC = 0.77$ kN,

$BC = 2.11$ kN

4.3-3   (a) $B_x = -0.8P$, $B_z = 2.0P$,
$O_z = -1.25P$;   (b) $F_{AC} = 0.960P$

4.3-4 $A_y = -10$ kN, $A_z = -10$ kN

$B_x = 1.25$ kN, $B_y = 2.5$ kN

$C_x = 3.75$ kN, $C_y = 7.5$ kN

$AB = 11.18$ kN, $AC = 10$ kN,

$BC = -11.25$ kN

4.3-5 $A_y = 13.33$ kips, $A_z = -10$ kips

$B_x = 10$ kips, $B_y = -14.17$ kips

$C_x = -5$ kips, $C_y = 0.833$ kips

$AB = -17.95$ kips, $AC = 6.01$ kips,

$BC = -1.179$ kips

4.3-6 $A_y = 48$ kN, $A_z = -36$ kN

$B_x = 63$ kN, $B_y = -102$ kN

$C_x = -45$ kN, $C_y = 54$ kN

$AB = -60$ kN, $AC = 40.2$ kN,

$BC = -66.1$ kN

4.3-7 $A_y = 64$ kips, $A_z = -24$ kips

$B_x = 48$ kips, $B_y = -100$ kips

$C_x = -36$ kips, $C_y = 36$ kips

$AB = -80$ kips, $AC = 43.3$ kips,

$BC = -40.2$ kips

4.3-8   (a) $A_x = -1.25P$, $B_y = 0$, $B_z = -P$;
(b) $F_{AB} = 1.601P$

4.3-9 (a) $A_y = 4.67\ P$, $A_z = -4.0\ P$;
(b) $F_{AB} = -8.33P$

4.3-10 (a) $A_z = 0$, $B_x = -3.75$ kN;
(b) $F_{AB} = 6.73$ kN

4.3-11 $BC = -33.9$ kips, $BD = 29.4$ kips,
$A_x = 12$ kips, $F_y = -11.09$ kips,

4.3-12 $A_x = -3.13$ kN, $A_y = -5$ kN,
$A_z = 0$
$B_x = -3.75$ kN, $C_x = 1.875$ kN,
$C_y = 5$ kN
$AB = 6.73$ kN, $CO = 0$

4.3-13 $A_x = -106.2$ lb, $A_y = -850$ lb,
$A_z = 0$
$B_x = -1488$ lb, $C_x = 744$ lb,
$C_y = 850$ lb
$AB = 2670$ lb, $CO = 1603$ lb

4.3-14 $A_x = -0.875$ kN, $A_y = -7$ kN,
$A_z = 0$
$B_x = -12.25$ kN, $C_x = 6.13$ kN,
$C_y = 7$ kN
$AB = -3.14$ kN, $CO = -14.94$ kN

4.3-15 $A_x = 438$ lb, $A_y = -500$ lb,
$A_z = 0$
$B_x = -875$ lb, $C_x = -62.5$ lb,
$C_y = 500$ lb,
$AB = 1571$ lb, $CO = 667$ lb

4.3-16 $F_{AB} = 50$ N, $F_{AC} = -5.59$ N,
$F_{AE} = -5.59$ N
$F_{BC} = -114.6$ N, $F_{BD} = 50.3$ N,
$F_{BF} = -5.59$ N

4.3-17 $A_y = -9.33$ kips, $B_x = 5.25$ kips,
$C_z = -7$ kips
$O_x = -12.25$ kips, $O_y = 2.33$ kips,
$O_z = 0$
$OA = 12.25$ kips, $OB = 0$,
$OC = 2.33$ kips
$AB = -8.75$ kips, $AC = -11.67$ kips,
$BC = 9.9$ kips

4.3-18 $A_x = 0$, $A_y = -5.63$ kN,
$A_z = 0$
$B_x = -15$ kN, $B_y = 11.25$ kN,
$B_z = 0$
$C_x = 0$, $C_y = -5.63$ kN, $D_z = 0$
$AO = 7.5$ kN, $BO = 0$,
$CO = 7.5$ kN, $DO = 0$, $OF = 0$
$AF = -9.38$ kN, $BF = 18.75$ kN,
$CF = -9.38$ kN, $DF = 0$

4.3-19 $A_x = 0$, $B_z = 0$, $C_x = 0$,
$C_y = 0$,
$D_y = 21$ kips, $D_z = 0$
$F_x = -12$ kips, $F_y = -33$ kips,
$F_z = 0$
$AO = 0$, $BO = -16$ kips,
$CO = 0$, $DO = -16$ kips,
$OF = 0$, $AF = 0$, $BF = 20$ kips,
$CF = 0$ $DF = 35$ kips

4.3-20 $A_y = -3.33$ kN, $C_x = 0$, $D_z = 0$
$O_x = -10$ kN, $O_y = 3.33$ kN, $O_z = 0$
$OA = 2.5$ kN, $OB = 10.61$ kN,
$OC = -7.5$ kN, $AB = 0$,
$BC = 0$, $OD = 3.33$ kN,
$CD = 12.5$ kN, $BD = -14.58$ kN,
$AD = -4.17$ kN

4.3-21 $A_x = 300$ lb, $A_z = -25$ lb,
$B_z = -475$ lb, $D_x = 0$, $E_y = 0$,
$F_y = 600$ lb, $F_{CB} = 593.7$ lb,
$F_{DA} = 0$, $F_{DF} = 0$

4.3-22 $A_x = -50$ N, $A_z = 125$ N, $B_x = 0$,
$C_x = 50$ N, $C_y = 50$ N, $D_z = -25$ N,
$F_{AE} = 75$ N, $F_{BE} = 0$

4.3-23 $DA = 500$ lb, $DB = -976$ lb,
$DC = -976$ lb

4.4-1 $A_x = 0$, $A_y = 622$ lb, $C_y = 775$ lb

4.4-2 $A_x = -9.26$ kN, $A_y = 0$
$B_x = 18.98$ kN, $B_y = 12.31$ kN
$C_x = 9.26$ kN, $C_y = 0$

4.4-3 (1) $A_x = -P$, $A_y = \dfrac{-4}{9}P$, $D_y = \dfrac{4}{9}P$
(2) $A_x = \dfrac{-3}{5}P$, $A_y = 0$, $D_y = \dfrac{4}{5}P$

4.4-4 $A_x = 450$ N, $A_y = 600$ N,
$M_A = 900$ N·m, $C_y = -600$ N

4.4-5 $A_x = -28.9$ lb, $A_y = 50$ lb,
$B_x = -65$ lb, $T = 71.6$ lb

4.4-6 $B_y = -172.5$ N, $B_z = -80.8$ N,
$C_z = 80.8$ N,
$D_x = 0$, $D_y = -182.5$ N, $D_z = 0$

4.4-7 $A_x = -115.6$ lb, $A_y = 0$,
$B_x = 393$ lb, $B_y = 75$ lb
$D_x = -884$ lb, $D_y = -168.8$ lb

4.4-8 $BD = 11.4$ kN, $C_x = 3.49$ kN,
$C_y = -9.38$ kN, $C_{res} = 10.01$ kN

4.4-9 $A_x = 0$, $A_y = 5.46$ kips,
$A_z = -0.72$ kips, $F_k = 0.9$ kips,
$M_{Ax} = 0$, $M_{Ay} = 8.64$ kip-ft,
$M_{Az} = 47.5$ kip-ft

4.4-10 $F_{BE} = -47.1$ kN

4.4-11 $F_{EB} = 5604$ lb, $C_x = 973$ lb,

           $C_y = 9619$ lb, $p_{EB} = 446$ psi

4.4-12   (a) $F_{BC} = 93.6$ kN, $F_{DE} = -93.4$ kN,

           $F_{GH} = 16.74$ kN

           (b) $p_{GH} = 2.36$ MPa

4.4-13   (a) $F_{30} = 10.78$ kips   (b) $p = 549$ psi

CHAPTER 5

5.2-6 $x_L = 1.903$, $y_L = 0.381$

5.2-7 $x_L = 1.273$, $y_L = 1.096$

5.2-8 $x_L = 2.486$, $y_L = 0.313$

5.2-9 $x_L = 0.38$, $y_L = -0.235$

5.2-10 $x_L = 0$, $y_L = 1.823$ m

5.2-11 $\bar{x} = \bar{z} = 0$ due to symmetry.

         $\bar{y} = \dfrac{h}{4} = 2$ in.

5.2-12 $\bar{x} = 0$, $\bar{y} = \dfrac{h}{2} = 50$ mm

         $\bar{z} = \dfrac{-4r}{3\pi} = -12.73$ mm

5.2-13 $\bar{x} = 3.53$ in., $\bar{y} = 0$, $\bar{z} = 2.36$ in.

5.2-14 $\bar{z} = \bar{x} = 0$, $\bar{y} = 1.6$ m

5.2-15   (a) $\bar{x} = 0.75$, $\bar{y} = 0.75$, $\bar{z} = 0.75$

         (b) $\bar{x} = 1$, $\bar{y} = 1.25$, $\bar{z} = 1.5$

5.2-16 $\bar{x} = 1.248$, $\bar{y} = 1$, $\bar{z} = 1.248$

5.2-17   (a) $\bar{x} = 0$ $\bar{y} = 1$ in., $\bar{z} = 0$

         (b) $\bar{x} = 0$, $\bar{y} = 0.964$ in., $\bar{z} = 0$

5.2-18 $\bar{x} = \dfrac{67}{148}h$, $\bar{y} = \bar{z} = 0$

5.3-2 $\bar{x} = \bar{y} = 5a/12$

5.3-3 $\bar{y} = 1.10$ in.

5.3-4 $2c^2 = ab$

5.3-5 $\bar{y} = 13.94$ in.

5.3-6 $\bar{y} = 52.5$ mm

5.3-7 $\bar{x} = 0.99$ in., $\bar{y} = 1.99$ in.

5.3-8 $\bar{x} = 137$ mm, $\bar{y} = 132$ mm

5.3-9 $\bar{x} = 0.714$ in., $\bar{y} = 3.43$ in.

5.3-10 $\bar{x} = 0$, $\bar{y} = 36.1$ mm

5.3-11 $h = 5.29$ in.

5.3-12 $\bar{x} = \dfrac{11L}{10}$, $\bar{y} = \dfrac{L}{10}$, $\bar{z} = \dfrac{L}{2}$

5.3-13 $\bar{x} = 3.08$ in., $\bar{y} = 0.5$ in.,

         $\bar{z} = 0.808$ in.

5.3-14   Area: $\bar{x} = 0$, $\bar{y} = 70$ mm

         Lines: $x_L = 0$, $y_L = 60.3$ mm

5.3-15   Area: $\bar{x} = 0.187r$, $\bar{y} = 0.813r$

         Lines: $\bar{x} = 0.355r$, $\bar{y} = 0.749r$

5.3-16 $\bar{x} = 3.66$ m, $\bar{y} = 1.127$ m,

         $\bar{z} = -1.092$ m

5.3-17 $\bar{x} = 3.87$ in., $\bar{y} = 2$ in., $\bar{z} = 0.919$ in.

5.3-18 $\bar{x} = \dfrac{55}{92}h$, $\bar{y} = 0$, $\bar{z} = 0$

5.3-19 $\bar{x} = \dfrac{43}{100}h$, $\bar{y} = 0$, $\bar{z} = 0$

5.3-20   For both: $\bar{x} = 0$, $\bar{z} = 0$,

         $y_a = 8.91$ mm, $y_b = 9.22$ mm

5.3-21 $\bar{x} = 0$, $\bar{y} = 29.4$ mm, $\bar{z} = 0$

5.4-1 $x_{cm} = 1$ in., $y_{cm} = 3.2$ in.

5.4-2 $x_{cm} = 2L/3 = 300$ mm

5.4-3 $x_{cm} = 11L/12$, $y_{cm} = L/6$, $z_{cm} = 5L/12$

5.4-4 $x_{cm} = 0$, $y_{cm} = 40.4$ mm

5.4-5 $x_{cm} = 55/18$ in., $y_{cm} = 19/18$ in.,

        $z_{cm} = 7/12$ in.

5.4-6 $x_{cm} = -14.01$ mm, $y_{cm} = 70.0$ mm

5.4-7 $x_{cm} = 0.292r$, $y_{cm} = 0.583r$

5.4-8 $x_{cg} = 46.7$ mm, $y_{cg} = 32.3$ mm

5.4-9 $x_{cg} = 5.85$ in., $y_{cg} = 5.0$ in.

5.4-10 $x_{cg} = 3.1$ m, $y_{cg} = 0.567$ m,

        $z_{cg} = 1.05$ m

5.4-11 $x_{cm} = 73h/164$, $y_{cm} = 0$, $z_{cm} = 0$

5.4-12 $x_{cm} = 0$, $y_{cm} = 27.8$ mm, $z_{cm} = 0$

5.5-1 $A_s = 2\pi r^2 = 157.1$ ft^2

5.5-2 $A_s = 2\pi y_L L = 25.1$

5.5-3 $A_S = \left(\dfrac{4}{9}\pi\right)x_L L = 17.87$

5.5-4 $A_S = 2\pi(x_L)L = 34.7$

5.5-5 $A_S = 2\pi(y_L)L = 111.6$

5.5-6 $A_S = 2\pi(x_L)L = 64.8$

5.5-7 $V = 12.31$ ft^3

5.5-8 $A_S = 6.032 \times 10^4$ mm^2

        $V = 4.524 \times 10^5$ mm^3

5.5-9 $V = 1.161 \times 10^5$ ft^3

5.5-10 $V = 2.367 \times 10^4$ mm^3

5.5-11 $V = 1414$ ft^3

5.5-12   (a) $V_y = 3.815 \times 10^4$ mm^3

        (b) $V_x = 7.948 \times 10^3$ mm^3

5.5-13 $V = \dfrac{4\pi r^3}{3}$

5.6-6 $I_x = 518 \times 10^3$ mm^4

5.6-7 $I_x = 36.1$ in^4, $I_y = 10.9$ in^4

5.6-8 $I_x = I_y = 194.6 \times 10^6$ mm^4,

$r_x = r_y = 80.1$ mm

5.5-9 $I_1 = 1480$ in^4, $I_2 = 186$ in^4, $r_1 = 7.10$ in.,

$r_2 = 2.52$ in.

5.6-10 $I_x = \dfrac{65\pi b^4}{648} = 4.084 \times 10^6$ mm^4

$I_y = \dfrac{245\pi b^4}{648} = 1.539 \times 10^7$ mm^4

$r_x = \sqrt{\dfrac{I_x}{A}} = 36.1$ mm, $r_y = \sqrt{\dfrac{I_y}{A}} = 70$ mm

5.6-11 $I_x = \dfrac{bh^3}{3} = 1458$ in^4

$I_y = \dfrac{22b^3 h}{27} = 1584$ in^4

$r_x = \sqrt{\dfrac{I_x}{A}} = 5.2$ in., $r_y = \sqrt{\dfrac{I_y}{A}} = 5.42$ in.

5.6-12 $I_c = 11a^4/192$

5.6-13 $I_b = 940$ in^4

5.6-14 $I_2 = 405 \times 10^3$ mm^4

5.6-15 $I_{xc} = 7.23$ in^4

5.6-16 $I_{xc} = 106 \times 10^6$ mm^4

5.6-17 $I_{xc} = 6050$ in^4, $I_{yc} = 475$ in^4

5.6-18 $b = 250$ mm

5.6-19 $I_{xc} = 17.40$ in^4, $I_{yc} = 6.27$ in^4

5.6-20 $I_{xc} = 8.155 \times 10^5$ mm^4

5.6-21 $I_{xc} = 365$ in^4

5.6-22 $I_P = bh(b^2 + 12h^2)/48$

5.6-23 $I_P = 233$ in^4

5.6-24 $(I_P)_C = r^4(9\alpha^2 - 8\sin^2\alpha)/18\alpha$

5.6-25 $I_P = bh(b^2 + h^2)/24$

5.6-26 $(I_P)_C = r^4(176 - 84\pi + 9\pi^2)/[72(4 - \pi)]$

5.6-27 $I_{PC} = 855$ in^4

5.6-28 $I_{PC} = 1.294 \times 10^{-5}$ m^4

5.6-30 $I_{xy} = r^4/24$

5.6-31 $b = 2r$

5.6-32 $I_{xy} = t^2(2b^2 - t^2)/4$

5.6-33 $I_{12} = -20.5$ in^4

5.6-34 $I_{xy} = 24.3 \times 10^6$ mm^4

5.6-35 $I_{xcyc} = -6.079$ in^4

5.7-1 $I_{x1} = I_{y1} = b^4/12$, $I_{x1y1} = 0$

5.7-2 $I_{x1} = \dfrac{b^3 h^3}{6(b^2 + h^2)}$, $I_{y1} = \dfrac{bh(b^4 + h^4)}{12(b^2 + h^2)}$,

$I_{x1y1} = \dfrac{b^2 h^2(h^2 - b^2)}{12(b^2 + h^2)}$

5.7-3 $I_d = 159$ in^4

5.7-4 $I_{x1} = 12.44 \times 10^6$ mm^4,

$I_{y1} = 9.68 \times 10^6$ mm^4,

$I_{x1y1} = 6.03 \times 10^6$ mm^4

5.7-5 $I_{x1} = 13.50$ in^4, $I_{y1} = 3.84$ in^4,

$I_{x1y1} = 4.76$ in^4

5.7-6 $I_{x1} = 8.75 \times 10^6$ mm^4,

$I_{y1} = 1.02 \times 10^6$ mm^4,

$I_{x1y1} = -0.356 \times 10^6$ mm^4

5.8-1 (a) $c = \sqrt{a^2 - b^2}/2$; (b) $a/b = \sqrt{5}$;

(c) $1 \le a/b < \sqrt{5}$

5.8-2 Shows that two different sets of principal axes exist at each point.

5.8-3 $\theta_{p1} = -29.87°$, $\theta_{p2} = 60.13°$,

$I_1 = 311.1$ in^4, $I_2 = 88.9$ in^4

5.8-4 $\theta_{p1} = -8.54°$, $\theta_{p2} = 81.46°$,

$I_1 = 17.24 \times 10^6$ mm^4,

$I_2 = 4.88 \times 10^6$ mm^4

5.8-5 $\theta_{p1} = 37.73°$, $\theta_{p2} = 127.73°$,

$I_1 = 15.45$ in^4, $I_2 = 1.89$ in^4

5.8-6 $\theta_{p1} = 32.63°$, $\theta_{p2} = 122.63°$,

$I_1 = 8.76 \times 10^6$ mm^4,

$I_2 = 1.00 \times 10^6$ mm^4

5.8-7 $\theta_{p1} = 16.85°$, $\theta_{p2} = 106.85°$,

$I_1 = 0.2390b^4$, $I_2 = 0.0387b^4$

5.8-8 $\theta_{p1} = 74.08°$, $\theta_{p2} = -15.92°$,

$I_1 = 8.29 \times 10^6$ mm^4,

$I_2 = 1.00 \times 10^6$ mm^4

5.8-9 $\theta_{p1} = 75.73°$, $\theta_{p2} = -14.27°$,

$I_1 = 20.07$ in^4, $I_2 = 2.12$ in^4

CHAPTER 6

6.2-1 $N_{AB} = 1600$ lb, $N_{BC} = -100$ lb,

$N_{CD} = -1300$ lb

6.2-2 $N_{AB} = -(P_1 + P_2) = -1120$ kN

$N_{BC} = -P_1 = -400$ kN

6.2-3 $N_{AB}(x) = qx$ $N_{BC} = P$

6.2-4 (a) $N_a(y) = -P\dfrac{y}{L}$;

(b) $N_b(y) = \dfrac{P}{L^2}y^2 + \left(-\dfrac{2P}{L}\right)y$

6.2-5 $N(x) = \dfrac{(L - x)^3}{L^3}W$

6.2-6 $N_{AB} = P - qL_2 = 14$ kN

$N_{BC}(x) = -q(L_1 + L_2 - x)$

6.2-7 (a) $N_{AB}(x) = \dfrac{q_0(3L^2 - 4x^2)}{4L}$

$N_{BC}(x) = q_0(L - x)$

(b) $N_{AB}(x) = -\dfrac{Lq_0}{4} - \dfrac{q_0 x^2}{L}$

$N_{BC}(x) = -q_0(L - x)$

6.2-8 $N_{AB}(x) = 2.25$ kN

$N_{BC}(x) = 3.0x - 0.25x^2 - 6.75$ kN

6.2-9 $N_{AB}(x) = -\dfrac{125x^2}{6}$ lb

$N_{BC}(x) = -(x - 12)\left(\dfrac{125x}{6} - 250\right)$ lb

6.2-10 $N(x) = -\dfrac{2\sqrt{2}(8-x)^{\frac{3}{2}}}{3}$ kN

6.3-1 $T_{AB} = T_1 + T_2 - T_3$, $T_{BC} = T_2 - T_3$,

$T_{CD} = -T_3$

6.3-2 (a) $T_{AC} = T_b + T_a$, $T_{CB} = T_a$

(b) $T_{AC} = -T_b + T_a$, $T_{CB} = T_a$

6.3-3 $T_{AB} = 31.5$ kip-in., $T_{BC} = 19$ kip-in.,

$T_{CD} = 9.2$ kip-in.

6.3-4 $R_A = -(tL)$, $T(x) = tL\left(1 - \dfrac{x}{L}\right)$

6.3-5 $T_{AB} = -1000$, $T_{BC} = -500$,

$T_{CD} = -1300$, $T_{DE} = -800$ (all in lb-in units)

6.3-6 $R_A = \dfrac{Lt_0}{6}$, $T(x) = \dfrac{-1}{6}t_0 L + \dfrac{t_0 x^2}{2L}$

6.3-7 $R_1 = -T_0$, $T_{12} = T_0$, $T_{23} = \dfrac{3}{2}T_0$

6.3-8 $R_A = -480$ N·m, $T_{AC} = 480$ N·m,

$T_{CB} = 720$ N·m

6.3-9 $T_{AB}(x) = 300 - \dfrac{75x^2}{8}$ lb-ft

$T_{BC}(x) = -(x - 12)\left(\dfrac{75x}{4} - 225\right)$ lb-ft

6.3-10 $R_A = -\dfrac{2Lt_0}{3}$,

$T(x) = \dfrac{2Lt_0}{3} - t_0 x + \dfrac{t_0 x^3}{3L^2}$

6.4-1 $V = 333$ lb, $M = 34{,}000$ in-lb

6.4-2 $V = -0.938$ kN, $M = 5.06$ kN·m

6.4-3 (a) $V_{\text{mid}} = 0$, $M_{\text{mid}} = 0$;

(b) $V_{\text{mid}} = 0$, $M_{\text{mid}} = 0$

6.4-4 $V = 3$ kN, $M = -6$ kN·m

6.4-5 (a) $V = -190$ lb, $M = 16{,}580$ ft-lb;

(b) $q = 370.4$ lb/ft (upward)

6.4-6 (a) $V = -1.0$ kN, $M = -7$ kN·m;

(b) $P_2 = 4$ kN; (c) $P_1 = -8$ kN (acts to right)

6.4-7 (a) $b/L = 1/2$; (b) $b/L = 1$

6.4-8 $M = 108$ N·m

6.4-9 $N = P\sin\theta$, $V = P\cos\theta$, $M = Pr\sin\theta$

6.4-10 $V = -3172$ N, $M = 5553$ N·m

6.4-11 (a) $P = 1200$ lb; (b) $P = 133.3$ lb

6.4-12 $V = -4.17$ kN, $M = 75$ kN·m

6.4-13 (a) $V_B = 6000$ lb, $M_B = 9000$ lb-ft;

(b) $V_m = 0$, $M_m = 21{,}000$ lb-ft

6.4-14 (a) $V_C = -15.5$ kN, $M_C = -44.8$ kN·m

(b) $M_A = -81$ kN·m, $V_C = -25$ kN,

$M_C = -30.5$ kN·m

6.4-15 $V_{\text{mid}} = \dfrac{2}{45}q_0 L$, $M_{\text{mid}} = \dfrac{13}{180}q_0 L^2$

6.4-16 (a) $V_{23} = \dfrac{-5}{36}q_0 L$, $M_{23} = \dfrac{23}{324}q_0 L^2$

(b) $V_{\text{mid}} = \dfrac{5}{144}q_0 L$, $M_{\text{mid}} = \dfrac{23}{288}q_0 L^2$

6.4-17 (a) $V_{23} = \dfrac{-1}{36}q_0 L$, $M_{23} = \dfrac{-5}{324}q_0 L^2$

(b) $V_{23} = \dfrac{-5}{72}q_0 L$, $M_{23} = \dfrac{35}{648}q_0 L^2$

6.4-18 $V\left(\dfrac{L}{2}\right) = \dfrac{\sqrt{2}-4}{6}q_0 L$, $M\left(\dfrac{L}{2}\right) = \dfrac{\sqrt{2}+2}{30}q_0 L^2$

6.4-19 $V\left(\dfrac{2L}{3}\right) = \dfrac{-19}{81}q_0 L$, $M\left(\dfrac{2L}{3}\right) = \dfrac{43}{972}q_0 L^2$

6.4-20 $V_{\text{mid}} = \dfrac{-q_0 L}{6}$, $M_{\text{mid}} = 0$

6.4-21 $V_{\text{mid}} = \dfrac{P}{2}(\sqrt{2} - 1)$, $M_{\text{mid}} = \dfrac{-PL}{4}$

6.4-22 (a) $V_{\text{mid}} = \dfrac{41}{216}q_0 L$, $M_{\text{mid}} = \dfrac{59}{432}q_0 L^2$

(b) $V_{\text{mid}} = \dfrac{-4}{27}q_0 L$, $M_{\text{mid}} = \dfrac{2}{27}q_0 L^2$

6.4-23 $V\left(\dfrac{L}{3}\right) = \dfrac{-P}{\sqrt{5}}$, $M\left(\dfrac{L}{3}\right) = \dfrac{2\sqrt{5}}{15}PL$

6.4-24 $V\left(\dfrac{L}{3}\right) = \dfrac{11P}{36}$, $M\left(\dfrac{L}{3}\right) = \dfrac{-43}{81}PL$

6.4-25 $N = \dfrac{-P}{2}$, $V = 0$, $M = 0$

6.4-26 (a) $N = 21.6$ kN (compression), $V = 7.2$ kN,

$M = 50.4$ kN·m;

(b) $N = 21.6$ kN (compression), $V = -5.4$ kN, $M = 0$

(at moment release)

6.4-27 (a) $V_{\text{max}} = P$, $M_{\text{max}} = Pa$;

(b) $A_y = -B_y = P\left(2\dfrac{a}{L} - 1\right)$

6.4-28 (a) $V_{\text{max}} = \dfrac{M_0}{L}$, $M_{\text{max}} = M_0\dfrac{a}{L}$;

(b) $V_{\text{max}} = 2\dfrac{M_0}{L}$, $M_{\text{max}} = 2M_0\dfrac{a}{L}$

6.4-29 $V_{\text{max}} = qL/2$, $M_{\text{max}} = -3qL^2/8$

6.4-30 $V_{\text{max}} = P$, $M_{\text{max}} = PL/4$

6.4-31 (a) $q_1 = 3q$; (b) $q_0 = 4.5q$

6.4-32 $V_{\text{max}} = -2P/3$, $M_{\text{max}} = PL/9$

6.4-33 $V_{\text{max}} = 2M_1/L$, $M_{\text{max}} = 7M_1/3$

6.4-34   (a) $V_{max} = \dfrac{P}{2}$ (on AB),

$$M_{max} = R_C\left(\dfrac{3L}{4}\right) = \dfrac{3LP}{8} \text{ (just right of } B);$$

(b) $N_{max} = P$ (tension on AB), $V_{max} = \dfrac{P}{5}$,

$$M_{max} = \dfrac{-P}{5}\left(\dfrac{3L}{4}\right) = -\dfrac{3LP}{20} \text{ (just right of B)}$$

6.4-35 $N_{max} = P, V_{max} = P, M_{max} = -\dfrac{2}{3}PL$

6.4-36   (a) $V_{max} = P, M_{max} = -Pa$;

(b) $M = 3Pa$ (CCW); $V_{max} = 2P, M_{max} = 2Pa$

6.4-37   (a) $V_{max} = \dfrac{1}{2}qL, M_{max} = \dfrac{5}{72}qL^2$;

(b) $V_{max} = \dfrac{7}{20}qL, M_{max} = \dfrac{5}{54}qL^2$

6.4-38   (a) $V_{max} = -q_0L/2, M_{max} = -q_0L^2/6$;

(b) $V_{max} = -\dfrac{2Lq_0}{3}, M_{max} = -\dfrac{4L^2q_0}{15}$ (at B)

6.4-39 $V_{max} = -173.3$ lb, $M_{max} = 4303$ lb-in.

6.4-40   (a) $V_{max} = 1200$ N, $M_{max} = 960$ N·m;

(b) $V_{max} = 675$ N, $M_{max} = 640$ N·m

6.4-41 $V_{max} = 200$ lb, $M_{max} = -1600$ lb-ft

6.4-42 $V_{max} = 4.5$ kN, $M_{max} = -12.67$ kN·m

6.4-43 $V_{max} = -1900$ lb, $M_{max} = -28{,}800$ lb-in.

6.4-44 $V_{max} = 13.15$ kN, $M_{max} = 11.21$ kN·m

6.4-45   (a) $V_{max} = \dfrac{4}{5}P, M_{max} = -\dfrac{17}{20}PL$

(b) $V_{max} = -\dfrac{7}{5}P, M_{max} = \dfrac{1}{2}PL$

6.4-46   (a) $N_{max} = \dfrac{-3}{5}P, V_{max} = \dfrac{-12}{5}P, M_{max} = PL$;

(b) $N_{max} = \dfrac{-3}{5}P, V_{max} = \dfrac{-4}{5}P, M_{max} = \dfrac{-4}{5}PL$;

(c) $N_{max} = \dfrac{-3}{5}P, V_{max} = \dfrac{4}{5}P, M_{max} = \dfrac{-6}{5}PL$

6.4-47 $V_{max} = 900$ lb, $M_{max} = -900$ lb-ft

6.4-48 $V_{max} = -7$ kN, $M_{max} = -12$ kN·m

6.4-49   Two cases have the same maximum moment: (PL).

6.4-50 $V_{max} = 31.5$ kN, $M_{max} = -73.5$ kN·m, $M_{mid} = -72.6$ kN·m

6.4-51   (a) $V_{max} = 1440$ lb, $M_{max} = 8640$ lb-ft;

(b) $V_{max} = 1440$ lb, $M_{max} = -8640$ lb-ft

6.4-52 $M_{Az} = -PL$ (CW), $A_x = 0, A_y = 0,$

$$C_y = \dfrac{1}{12}P \text{ (upward)}, D_y = \dfrac{1}{6}P \text{ (upward)}, V_{max} = P/12,$$

$$M_{max} = PL$$

6.4-53   (a) $V_{max} = -640$ lb, $M_{max} = 1688$ lb-ft;

(b) $a = 5.74$ ft, $V_{max} = -565$ lb, $M_{max} = 1838$ lb-ft;

(c) $a = 4.7$ ft, $M_{max} = 1873$ lb-ft

6.4-54   (a) $V_{max} = -2.4$ kN, $M_{max} = 1.266$ kN·m

(b) $M_0 = 0.523$ kN·m (CW)

6.4-55   (a) $a = 0.586L, V_{max} = 0.293qL, M_{max} = 0.0214qL^2$;

(b) $a = 0.404L, V_{max} = 0.161qL, M_{max} = 0.00882qL^2$

6.4-56 $V_{max} = -2.25$ kN, $M_{max} = 3.5$ kN·m

6.4-57   (a) $V_{max} = -\dfrac{1}{2}q_0L, M_{max} = \dfrac{1}{3}q_0L^2$

(b) $V_{max} = -\dfrac{2}{3}q_0L, M_{max} = \dfrac{2}{5}q_0L^2$

6.4-58 $A_y = 6.25$ kN, $B_y = 8.75$ kN, $M_{max} = 8.75$ kN·m   at $x = 3$ m

6.4-59 $A_y = 600$ lb at $x = 4$ ft, $B_y = 400$ lb at $x = 20$ ft, $M_{max} = 1425$ lb-ft at $x = 13$ ft

6.4-60 $V_{max} = -w_0L/3, M_{max} = -w_0L^2/12$

6.4-61 $M_A = -\dfrac{w_0}{30}L^2$ (CW), $A_x = -3w_0L/10$ (leftward),

$A_y = -3w_0L/20$ (downward), $C_y = w_0L/12$ (upward),

$D_y = w_0L/6$ (upward), $V_{max} = w_0L/4,$

$M_{max} = -w_0L^2/24$ at B

6.4-62   (a) $x = 0$ and $V_{max} = 56.8$ kN;

(b) $x_m = 7.1$ m and $M_{max} = 202$ kN·m

6.4-63   (a) $A_x = 50.38$ lb (right), $A_y = 210$ lb (upward),

$B_x = -50.38$ lb (left), $N_{max} = -214.8$ lb,

$V_{max} = -47.5$ lb, $M_{max} = 270$ lb-ft;

(b) $A_x = 0, A_y = 67.5$ lb, $B_x = 0,$

$B_y = 142.5$ lb, $N_{max} = 134.4$ lb, $V_{max} = -47.5$ lb,

$M_{max} = 270$ lb-ft

6.4-64   (a) $N_{max} = \dfrac{1}{2}q_0L, V_{max} = \dfrac{19}{18}q_0L,$

$$M_{max} = q_0L^2; \text{ (b) } N_{max} = \dfrac{-1}{2}q_0L,$$

$$V_{max} = \dfrac{1}{2}q_0L, M_{max} = q_0L^2$$

6.4-65   (a) $N_{AB} = 0, V_{AB} = 0, M_{AB\,max} = \dfrac{1}{6}q_0L^2$

$$N_{BC} = 0, V_{BC\,max} = \dfrac{-1}{2}q_0L, M_{BC\,max} = \dfrac{1}{6}q_0L^2;$$

(b) $N_{AB} = 0, V_{AB\,max} = \dfrac{4}{3}q_0L, M_{AB\,max} = \dfrac{-9}{10}q_0L^2,$

$$N_{BC} = 0, V_{BC\,max} = \dfrac{-1}{2}q_0L, M_{BC\,max} = \dfrac{1}{6}q_0L^2$$

6.4-66 $M_A = 0, A_x = 0,$

$A_y = -18.41$ kN (downward), $M_D = 0,$

$D_x = -63.0$ kN (leftward), $D_y = 62.1$ kN

(upward), $N_{max} = -62.1$ kN, $V_{max} = 63.0$ kN,

$M_{max} = 756$ kN·m

CHAPTER 7

7.2-1 (a) $\sigma_{AB} = 1443$ psi; (b) $P_2 = 1487.5$ lb; (c) $t_{BC} = 0.5$ in.

7.2-2 (a) $\sigma_1 = 23.2$ MPa, $\sigma_2 = 10.61$ MPa; (b) $\varepsilon_1 = 6.8 \times 10^{-3}$, $\varepsilon_2 = 3.13 \times 10^{-3}$

7.2-3 $\sigma_{max} = 0.566$ ksi

7.2-4 (a) $\sigma = 130.2$ MPa; (b) $\varepsilon = 4.652 \times 10^{-4}$

7.2-5 (a) $R_B = 90$ lb (cantilever), 191.3 lb (V-brakes); $\sigma_{pad} = 144$ psi (cantilever), 306 psi (V-brakes); (b) $\sigma_{cable} = 26,946$ psi (both)

7.2-6 (a) $\varepsilon_s = 3.101 \times 10^{-4}$; (b) $\delta = 0.1526$ mm; (c) $P_{max} = 89.5$ kN

7.2-7 (a) $\sigma_C = 2.46$ ksi; (b) $x_C = 19.56$ in., $y_C = 19.56$ in.

7.2-8 (a) $\sigma_t = 132.7$ MPa; (b) $\alpha_{max} = 34.4°$

7.2-9 (a) $\sigma_1 = 34.4$ ksi, $\sigma_2 = 30.6$ ksi; (b) $d_{1\,new} = 3.18 \times 10^{-2}$ in.; (c) $\sigma_1 = 19.6$ ksi, $\sigma_2 = 18.78$ ksi, $\sigma_3 = 22.7$ ksi

7.2-10 $\sigma_C = 5.21$ MPa

7.2-11 (a) $T = 184$ lb, $\sigma = 10.8$ ksi; (b) $\varepsilon_{cable} = 5 \times 10^{-4}$

7.2-12 (a) $T = 819$ N, $\sigma = 74.5$ MPa; (b) $\varepsilon_{cable} = 4.923 \times 10^{-4}$

7.2-13 (a) $T = \begin{Bmatrix} 5877 \\ 4679 \\ 7159 \end{Bmatrix}$ lb;

(b) $\sigma = \begin{Bmatrix} 48,975 \\ 38,992 \\ 59,658 \end{Bmatrix}$ psi;

(c) $T = \begin{Bmatrix} 4278 \\ 6461 \\ 3341 \\ 4278 \end{Bmatrix}$ lb, $\sigma = \begin{Bmatrix} 35,650 \\ 53,842 \\ 27,842 \\ 35,650 \end{Bmatrix}$ psi

7.2-14 (a) $\sigma_x = \gamma \omega^2 (L^2 - x^2)/2g$; (b) $\sigma_{max} = \gamma \omega^2 L^2 / 2g$

7.2-15 (a) $T_{AB} = 1620$ lb, $T_{BC} = 1536$ lb, $T_{CD} = 1640$ lb; (b) $\sigma_{AB} = 13,501$ psi, $\sigma_{BC} = 12,799$ psi, $\sigma_{CD} = 13,667$ psi

7.2-16 (a) $T_{AQ} = T_{BQ} = 50.5$ kN; (b) $\sigma = 166$ MPa

7.2-17 (a) $T_A = 1278$ lb, $T_B = 1046$ lb; (b) $\sigma_A = 2.71$ ksi, $\sigma_B = 2.22$ ksi

7.3-1 (a) $L_{max} = 11,800$ ft; (b) $L_{max} = 13,500$ ft

7.3-2 (a) $L_{max} = 7143$ m; (b) $L_{max} = 8209$ m

7.3-3 % elongation = 6.5, 24.0, 39.0; % reduction = 8.1, 37.9, 74.9; Brittle, ductile, ductile

7.3-4 11.9×10^3 m; 12.7×10^3 m; 6.1×10^3 m; 6.5×10^3 m; 23.9×10^3 m

7.3-5 $\sigma = 52.3$ ksi

7.3-6 $\sigma_{pl} \approx 47$ MPa, Slope ≈ 2.4 GPa, $\sigma_\gamma \approx 53$ MPa; Brittle

7.3-7 $\sigma_{pl} \approx 65,000$ psi, Slope $\approx 30 \times 10^6$ psi, $\sigma_Y \approx 69,000$ psi, $\sigma_U \approx 113,000$ psi; Elongation = 6%, Reduction = 31%

7.4-1 (a) 0.097 in. longer (b) $\sigma = 45.3$ ksi, No permanent set

7.4-2 4.0 mm longer

7.4-3 0.13 in. longer

7.4-4 (a) $\delta_{pset} = 4.28$ mm; (b) $\sigma_B = 65.6$ MPa

7.4-5 (a) $\delta_{pset} = 1.596$ in.; (b) $\sigma_B = 30$ ksi

7.4-6 (a) $\delta = 11.23$ mm; (b) No permanent set

7.4-7 (b) 0.71 in.; (c) 0.58 in.; (d) 49 ksi

7.5-1 $P_{max} = 157$ kips

7.5-2 $P = 27.4$ kN (tension)

7.5-3 $P = -15.708$ kips

7.5-4 (a) $L_p = 405.25$ mm; (b) $\dfrac{V_i}{V_f} = 1.003$

7.5-5 $P = 175$ kips, Gap = 0.1762 in.

7.5-6 (a) $P = 74.1$ kN; (b) Shortening: $\delta = \varepsilon L = 0.469$ mm, $\dfrac{A_f - A}{A} = +0.052\%$, $\Delta V_1 = V_{1f} - Vol_1 = -207$ mm^3; (c) $d_3 = 65.4$ mm

7.5-7 $\Delta d = -1.56 \times 10^{-4}$ in., $P = 2.154$ kips

7.5-8 (a) $E = 104$ GPa; (b) $\nu = 0.34$

7.5-9 (a) $\Delta d_{BC\,inner} = 8 \times 10^{-4}$ in.; (b) $\nu_{brass} = 0.34$; (c) $\Delta t_{AB} = 2.732 \times 10^{-4}$ in., $\Delta d_{AB\,inner} = 1.366 \times 10^{-4}$ in.

7.5-10 (a) $\Delta L_1 = 12.66$ mm; $\Delta L_2 = 5.06$ mm; $\Delta L_3 = 3.8$ mm; (b) $\Delta Vol_1 = 21,548$ mm^3; $\Delta Vol_2 = 21,601$ mm^3; $\Delta Vol_3 = 21,610$ mm^3

7.6-1 $\sigma_b = 7.04$ ksi, $\tau_{ave} = 10.756$ ksi

7.6-2 $\sigma_b = 139.86$ MPa; $P_{ult} = 144.45$ kN

7.6-3 (a) $\tau = 12.732$ ksi; (b) $\sigma_{bf} = 20$ ksi, $\sigma_{bg} = 26.667$ ksi

7.6-4 (a) $B_x = -252.8$ N, $A_x = -B_x$, $A_y = 1150.1$ N; (b) $A_{resultant} = 1178$ N; (c) $\tau = 5.86$ MPa, $\sigma_{b,rail} = 18.4$ MPa

7.6-5 (a) $\tau_{max} = 2979$ psi; (b) $\sigma_{b\,max} = 936$ psi

7.6-6 $T_1 = 13.176$ kN, $T_2 = 10.772$ kN, $\tau_{1ave} = 25.888$ MPa, $\tau_{2ave} = 21.166$ MPa, $\sigma_{b1} = 9.15$ MPa, $\sigma_{b2} = 7.48$ MPa

7.6-7 (a) Resultant = 1097 lb; (b) $\sigma_b = 4999$ psi; (c) $\tau_{nut} = 2793$ psi, $\tau_{pl} = 609$ psi

7.6-8 $G = 2.5$ MPa

7.6-9 (a) $\gamma_{aver} = 0.004$; (b) $V = 89.6$ kips

7.6-10 $\tau_{pl} = 348$ MPa, $\sigma_U = 154.7$ MPa, $\sigma_L = 348$ MPa

7.6-11 $d = 8.33 \times 10^{-3}$ in., $\gamma = 0.0167$

7.6-12 $\tau_{ave} = 0.299$ MPa

7.6-13 (a) $\sigma_b = 69.5$ ksi, $\sigma_{brg} = 39.1$ ksi, $\tau_f = 6.7$ ksi; (b) $\sigma_b = 60.4$ ksi, $\sigma_{brg} = 34$ ksi, $\tau_f = 5.83$ ksi

7.6-14 (a) $\gamma_{aver} = 0.50$; (b) $\delta = 4.92$ mm

7.6-15   (a) $A_x = 0$, $A_y = 170$ lb, $M_A = 4585$ in.-lb;
(b) $B_x = 253.6$ lb, $B_y = 160$ lb, $B_{res} = 299.8$ lb,
$C_x = -B_x$; (c) $\tau_B = 3054$ psi, $\tau_C = 1653$ psi;
(d) $\sigma_{bB} = 4797$ psi, $\sigma_{bC} = 3246$ psi

7.6-16 $\tau_{aver} = 42.9$ MPa

7.6-17   (a) $\tau = \dfrac{P}{2\pi rh}$; (b) $\delta = \dfrac{P}{2\pi hG} \ln \dfrac{b}{d}$

7.6-18   (a) $\tau_1 = 2.95$ MPa, $\tau_4 = 0$;
(b) $\sigma_{b1} = 1.985$ MPa, $\sigma_{b4} = 0$;
(c) $\sigma_{b4} = 41$ MPa; (d) $\tau = 10.62$ MPa;
(e) $\sigma_3 = 75.1$ MPa

7.6-19   (a) $O_x = 12.68$ lb, $O_y = 1.294$ lb, $O_{res} = 12.74$ lb;
(b) $\tau_O = 519$ psi, $\sigma_{bO} = 816$ psi;
(c) $\tau = 362$ psi

7.6-20   (a) $F_x = 153.9$ N, $\sigma = 3.06$ MPa;
(b) $\tau_{ave} = 1.96$ MPa; (c) $\sigma_b = 1.924$ MPa

7.6-21   (a) $P = 395$ lb; (b) $C_x = 374$ lb, $C_y = -237$ lb,
$C_{res} = 443$ lb;
(c) $\tau = 18.04$ ksi, $\sigma_{bC} = 4.72$ ksi

7.6-22   (a) $R_x = -145.4$ kN, $R_y = -229$ kN,
$R_z = -51.5$ kN, $M_x = -23.2$ kN $\cdot$ m, $M_y = 0$,
$M_z = 65.4$ kN $\cdot$ m; (b) $\tau_{ave} = 42.6$ MPa

7.6-23   (a) $R_O = \left\{ \begin{array}{c} 0 \\ 206 \\ -70.8 \end{array} \right\}$ lb, $M_O = \left\{ \begin{array}{c} 1027 \\ -1185 \\ -3858 \end{array} \right\}$ lb-in.;
(b) $\tau = 1594$ psi

7.6-24   For a bicycle with $L/R = 1.8$: (a) $T = 1440$ N;
(b) $\tau_{aver} = 147$ MPa

7.7-1 $P_{allow} = 3140$ lb

7.7-2 $T_{max} = 33.4$ kN $\cdot$ m

7.7-3 $P_{allow} = 607$ lb

7.7-4   (a) $P_{allow} = 8.74$ kN; (b) $P_{allow} = 8.69$ kN;
(c) $P_{allow} = 21.2$ kN, $P_{allow} = 8.69$ kN
(shear controls)

7.7-5 $P = 294$ kips

7.7-6 $d = 56.8$ mm

7.7-7   Air: $n = 198$, Water: $n = 228$

7.7-8   (a) $F = 1.171$ kN;
(b) Shear: $F_{allow} = 2.86$ kN

7.7-9 $W_{max} = 5110$ lb

7.7-10   (a) $F_A = \sqrt{2}T$, $F_B = 2T$, $F_C = T$;
(b) Shear at A: $W_{max} = 66.5$ kN

7.7-11 $P_{allow} = 10.21$ kips

7.7-12 $C_{ult} = 5739$ N: $P_{max} = 445$ N

7.7-13 $W_{max} = 0.305$ kips

7.7-14   Shear in rivets in CG & CD controls: $P_{allow} = 45.8$ kN

7.7-15   (a) $P_{allow} = \sigma_{allow}(0.587d^2)$; (b) $P_{allow} = 21.6$ kips

7.7-16 $P_{allow} = 96.5$ kN

7.7-17 $p_{max} = 11.98$ psf

7.7-18   (a) $P_{allow} = \sigma_c(\pi d^2 / 4)\sqrt{1 - (R/L)^2}$;
(b) $P_{allow} = 9.77$ kN

7.8-1   (a) $d_{min} = 3.75$ in.; (b) $d_{min} = 4.01$ in.

7.8-2   (a) $d_{min} = 164.6$ mm; (b) $d_{min} = 170.9$ mm

7.8-3   (a) $d_{min} = 0.704$ in.; (b) $d_{min} = 0.711$ in.

7.8-4 $d_{min} = 63.3$ mm

7.8-5 $d_{pin} = 1.029$ in.

7.8-6 $A_{cable} = 75$ mm^2, $d_O = 7.71$ mm, $d_B = d_D = 8.64$ mm

7.8-7   (a) $A_{cable} = 0.0675$ in^2; (b) $d_O = 0.412$ in.,
$d_A = d_B = 0.359$ in., $d_D = 0.504$ in.

7.8-8   (b) $A_{min} = 435$ mm^2

7.8-9 $d_{min} = 0.372$ in.

7.8-10 $d_{min} = 5.96$ mm

7.8-11 $n = 11.6$, use 12 bolts

7.8-12 $(d_2)_{min} = 131$ mm

7.8-13 $A_c = 1.498$ in^2

7.8-14   (a) $t_{min} = 18.8$ mm, use $t = 20$ mm;
(b) $D_{min} = 297$ mm

7.8-15   (a) $\sigma_{DF} = 10.38$ ksi $< \sigma_{allow}$,
$\sigma_{bF} = 378$ psi $< \sigma_{ba}$; (b) new $\sigma_{BC} = 25$ ksi,
so increase rod BC to $\dfrac{1}{4}$-in. diameter; required
diameter of washer $= 1 + \dfrac{5}{16}$ in. $= 1.312$ in.

7.8-16   (a) $d_m = 24.7$ mm; (b) $P_{max} = 49.4$ kN

7.8-17   (a) $P_{max} = 2719$ lb;
(b) $d_A = d_B = 0.486$ in., $d_D = 0.677$ in.

7.8-18 $d_A = d_B = 8.32$ mm, $d_D = 11.60$ mm

7.8-19 $\theta = \arccos 1/\sqrt{3} = 54.7°$

CHAPTER 8

8.2-1 $\delta_A = 0.833$ in.

8.2-2 $\delta_{rod} = 0.1471$ mm
$\delta_B = 0.1912$ mm

8.2-3   (a) $\delta = \dfrac{6W}{5k}$; (b) $\delta = \dfrac{4W}{5k}$

8.2-4   (a) $\delta = 12.5$ mm; (b) $n = 5.8$

8.2-5   (a) $\dfrac{\delta_a}{\delta_s} = \dfrac{E_s}{E_a} = \dfrac{30}{11}$; (b) $\dfrac{d_a}{d_s} = \sqrt{\dfrac{E_s}{E_a}} = 1.651$;
(c) $\dfrac{L_a}{L_s} = 1.5\dfrac{E_a}{E_s} = 0.55$;
(d) $E_1 = \dfrac{E_s}{1.7} = 17,647$ ksi (cast iron or copper alloy)
(see App. F)

8.2-6 $h = 13.4$ mm

8.2-7 $\Delta_{DC} = 2.8 \times 10^{-3}$ in., $\delta_B = 10.5 \times 10^{-3}$ in.

8.2-8 $\Delta_A = 0.542$ mm

8.2-9 $h = L - \pi\rho_{max}d^2/4k$

8.2-10   (a) $x = 102.6$ mm; (b) $x = 205$ mm;
         (c) $P_{max} = 12.51$ N; (d) $\theta_{init} = 1.325°$;
         (e) $P = 20.4$ N

8.2-11 $W = 2.63$ lb

8.2-12 $n = 8.9$

8.2-13   (a) $\delta_4 = \dfrac{26P}{3k}$; (b) $\delta_4 = \dfrac{104P}{45k}$, Ratio $= \dfrac{15}{4} = 3.75$

8.2-14   (a) $\delta_B = 1.827$ mm; (b) $P_{max} = 390$ kN;
         (c) $\delta_{Bx} = 6.71$ mm, $P_{max} = 106.1$ kN

8.2-15 $P_{max} = 72.3$ lb

8.2-16   (a) $x = 134.7$ mm; (b) $k_1 = 0.204$ N/mm;
         (c) $b = 74.1$ mm; (d) $k_3 = 0.638$ N/mm

8.2-17   (a) $t_{c,\ min} = 0.021$ in.; (b) $\delta_r = 0.031$ in.;
         (c) $h_{min} = 0.051$ in.

8.2-18 $\delta_A = 0.200$ mm, $\delta_D = 0.880$ mm

8.2-19   (a) $\delta_D = \dfrac{P}{16}(28f_2 - 9f_1)$; (b) $\dfrac{L_1}{L_2} = \dfrac{27}{16}$;
         (c) $\dfrac{d_1}{d_2} = 1.225$; (d) $x = \dfrac{365L}{236}$

8.2-20   (a) $\theta = 35.1°$, $\delta = 44.6$ mm, $R_A = 25$ N,
         $R_C = 25$ N; (b) $\theta = 43.3°$, $\delta = 8.19$ mm,
         $R_A = 31.5$ N, $R_C = 18.5$ N, $M_A = 1.882$ N · m

8.2-21   (a) $\theta = 35.1°$, $\delta = 1.782$ in., $R_A = 5$ lb, $R_C = 5$ lb;
         (b) $\theta = 43.3°$, $\delta = 0.327$ in., $R_A = 6.3$ lb, $R_C = 3.71$ lb,
         $M_A = 1.252$ lb-ft

8.3-1   (a) $\delta = 0.0276$ in.; (b) $d_B = 1.074$ in.

8.3-2   (a) $\delta = 0.675$ mm; (b) $P_{max} = 267$ kN

8.3-3   (a) $\delta = 0.01125$ in. (elongation);
         (b) So new value of P_3 is 1690 lb, an increase of 390 lb;
         (c) $A_{AB} = 0.78$ in^2

8.3-4 $\delta_B = 2.29 \times 10^{-4}$ mm, $\delta_D = 4.67 \times 10^{-4}$ mm,
         $\delta_E = 6.85 \times 10^{-4}$ mm

8.3-5 $\delta_B = -1.255 \times 10^{-5}$ in., $\delta_C = -2.09 \times 10^{-5}$ in.,
         $\delta_D = -5.44 \times 10^{-5}$ in.

8.3-6 $\delta_B = 4.22 \times 10^{-4}$ mm, $\delta_D = 7.84 \times 10^{-4}$ mm,
         $\delta_E = 10.51 \times 10^{-4}$ mm

8.3-7 $\delta_B = 1.135 \times 10^{-6}$ in., $\delta_C = 2.54 \times 10^{-6}$ in.,
         $\delta_D = -23.1 \times 10^{-6}$ in.

8.3-8   (a) $\delta = \dfrac{7PL}{6Ebt}$; (b) $\delta = 0.5$ mm; (c) $L_{slot} = 244$ mm

8.3-9   (a) $\delta = \dfrac{7PL}{6Ebt}$; (b) $\delta = 0.021$ in.; (c) $L_{slot} = 10$ in.

8.3-10   (a) $\delta_{AC} = 3.72$ mm; (b) $P_0 = 44.2$ kN

8.3-11   (a) $\delta_a = 0.0589$ in.; (b) $\delta_b = 0.0501$ in.;
         (c) $\dfrac{\delta_c}{\delta_a} = 0.846$, $\dfrac{\delta_c}{\delta_b} = 0.993$

8.3-12   (a) $d_{max} = 23.9$ mm; (b) $b = 4.16$ mm;
         (c) $x = 183.3$ mm

8.3-13   (a) $\delta = \dfrac{PL}{2EA}$; (b) $\sigma(y) = \dfrac{P}{A}\left(\dfrac{y}{L}\right)$;
         (c) $\delta = \dfrac{PL}{EA}\left(\dfrac{2}{3}\right)$, $\sigma(y) = \dfrac{P}{A}\left[\dfrac{y}{L}\left(2 - \dfrac{y}{L}\right)\right]$

8.3-14   (a) $\delta_{2-4} = 0.024$ mm; (b) $P_{max} = 8.15$ kN;
         (c) $L_2 = 9.16$ mm

8.3-15   (a) $R_1 = -3P/2$; (b) $N_1 = 3P/2$ (tension), $N_2 = P/2$
         (tension); (c) $x = L/3$;
         (d) $\delta_2 = 2PL/3EA$; (e) $\beta = 1/11$

8.3-16   (a) $\delta_C = W(L^2 - h^2)/2EAL$;
         (b) $\delta_B = WL/2EA$; (c) $\beta = 3$;
         (d) $\delta = \dfrac{WL}{2EA} = 359$ mm (in sea water);
         $\delta = \dfrac{WL}{2EA} = 412$ mm (in air)

8.3-17   (b) $\delta = 0.010$ in.

8.3-18 $\delta_B = 0.937$ mm, $\delta_C = 2.95$ mm

8.3-19 $\delta_B = 0.087$ in., $\delta_C = 0.261$ in.

8.3-20 $P_1 = 415$ kN to left, $\delta_B = -1.389$ mm

8.3-21 $L = 9.68$ ft

8.3-22 $\delta = 4.14$ mm

8.3-23 $\delta = 2WL/\pi d^2 E$

8.3-24 $\delta = 2PH/3Eb^2$

8.3-25   (b) $\delta = 11.14$ ft

8.3-26   (a) $\delta = 2.18$ mm; (b) $\delta = 6.74$ mm

8.3-27 $\delta_C = 0.027$ in. (downward)

8.3-28 $\delta_B = -4\dfrac{PL}{EA}$, $\delta_F = -\dfrac{11}{3}\dfrac{PL}{EA}$

8.3-29 $\delta_B = -\dfrac{3}{4}\dfrac{PL}{EA}$, $\delta_D = -\dfrac{5}{4}\dfrac{PL}{EA}$

8.3-30 $\delta_B = \delta_D = -\dfrac{1}{2}\dfrac{PL}{EA}$

8.4-1   (a) $P = 1330$ lb; (b) $P_{allow} = 1300$ lb

8.4-2   (a) $P = 104$ kN; (b) $P_{max} = 116$ kN

8.4-3 $A_y = -1.667$ kips,
         $B_y = -0.833$ kips
         (both downward),
         $\delta_C = 2.59 \times 10^{-4}$ in. (upward)

8.4-4 $A = 70.3$ mm^2

8.4-5 $P_{allow} = 709$ kips

8.4-6   (a) $P_B/P = 3/11$; (b) $\sigma_B/\sigma_A = 1/2$;
         (c) Ratio $= 1$

8.4-7   (a) If $x \leq L/2$, $R_A = (-3PL)/(2(x + 3L))$,
         $R_B = -P(2x + 3L)/(2(x + 3L))$.
         If $x \geq L/2$, $R_A = (-P(x + L))/(x + 3L)$,
         $R_B = (-2PL)/(x + 3L)$; (b) If $x \leq L/2$,
         $\delta = PL(2x + 3L)/[(x + 3L)E\pi d^2]$.
         If $x \geq L/2$, $\delta = 8PL(x + L)/[3(x + 3L)E\pi d^2]$;
         (c) $x = 3L/10$ or $x = 2L/3$;
         (d) $R_B = \rho g\pi d^2 L/8$, $R_A = 3\rho g\pi d^2 L/32$

8.4-8 $R_A = -80$ kN, $R_C = -120$ kN, $\delta_B = 0.466$ mm

8.4-9 $R_A = -15.55$ kips, $R_C = -4.45$ kips,
         $\delta_B = 1.542 \times 10^{-2}$ in.

8.4-10 (a) $\delta = 1.91$ mm; (b) $\delta = 1.36$ mm;
(c) $\delta = 2.74$ mm

8.4-11 (a) 41.7%; (b) $\sigma_M = 32.7$ ksi, $\sigma_O = 51.4$ ksi

8.4-12 (a) $R_A = 10.5$ kN (to the left),
$R_D = 2.0$ kN (to the right);
(b) $F_{BC} = 15.0$ kN (compression)

8.4-13 (a) $R_A = 2P/3$, $R_E = -5P/3$;
(b) $\delta_B = -\dfrac{LP}{6EA}$, $\delta_C = \dfrac{LP}{6EA}$, $\delta_D = \dfrac{5LP}{6EA}$;
(c) $\delta_{max} = \dfrac{5LP}{6EA}$ (to the right), $\delta_A = \delta_E = 0$;
(d) $P_{max} = 12.37$ kips

8.4-14 (a) $P = 13.73$ kN, $R_1 = 9.07$ kN, $R_2 = 4.66$ kN,
$\sigma_2 = 7$ MPa;
(b) $\delta_{cap} = 190.9$ mm, Axial Force Diagram:
$N(x) = -R_2$ if $x \le L_2$, $N(x) = R_1$ if $x > L_2$; Axial Displacement Diagram: $\delta(x) = \left[\dfrac{-R_2}{EA_2}(x)\right]$ if $x \le L_2$,
$\delta(x) = \left[\dfrac{-R_2 L_2}{EA_2} + \dfrac{R_1}{EA_1}(x - L_2)\right]$ if $x > L_2$;
(c) $q = 1.552$ kN/m

8.4-15 (b) $\sigma_a = 1610$ psi (compression), $\sigma_s = 9350$ psi
(tension)

8.4-16 (a) $P_{allow} = 1504$ N; (b) $P_{allow} = 820$ N;
(c) $P_{allow} = 703$ N

8.4-17 (a) $P_1 = PE_1/(E_1 + E_2)$;
(b) $e = b(E_2 - E_1)/[2(E_2 + E_1)]$;
(c) $\sigma_1/\sigma_2 = E_1/E_2$

8.4-18 (a) $A_x = -41.2$ kN, $A_y = -71.4$ kN, $B_x = -329$ kN,
$B_y = 256$ kN;
(b) $P_{max} = 233$ kN

8.4-19 $d_2 = 0.338$ in., $L_2 = 48.0$ in.

8.4-20 $P_{max} = 1800$ N

8.4-21 (a) $\sigma_C = 10{,}000$ psi, $\sigma_D = 12{,}500$ psi;
(b) $\delta_B = 0.0198$ in.

8.4-22 $M_A = \dfrac{2}{3}PL$, $A_x = 0$, $D_x = 0$, $D_y = \dfrac{2}{3}P$, $R_F = \dfrac{4}{3}P$

8.4-23 $\sigma_s = 3.22$ ksi, $\sigma_b = 1.716$ ksi, $\sigma_c = 1.93$ ksi

8.4-24 $A_x = 0$, $A_y = -6P$, $D_x = 0$, $D_y = \dfrac{18}{5}P$, $R_F = \dfrac{27}{5}P$

8.5-1 $\sigma = 11{,}700$ psi

8.5-2 $T = 40.3°C$

8.5-3 $\Delta T = 185°F$

8.5-4 (a) $\Delta T = 24°$ C, $\sigma_{rod} = 57.6$ MPa;
(b) Clevis: $\sigma_{bc} = 42.4$ MPa,
Washer: $\sigma_{bw} = 74.1$ MPa; (c) $d_b = 10.68$ mm

8.5-5 (a) $\sigma_c = E\alpha(\Delta T_B)/4$;
(b) $\sigma_c = E\alpha(\Delta T_B)/[4(EA/kL + 1)]$

8.5-6 $\sigma_T = -1.285$ MPa, $\delta_B = 1.601$ mm

8.5-7 $\sigma_T = -3.77$ksi

8.5-8 (a) $N = 51.8$ kN, max. $\sigma_c = 26.4$ MPa,
$\delta_C = -0.314$ mm;
(b) $N = 31.2$ kN, max. $\sigma_c = 15.91$ MPa,
$\delta_C = -0.546$ mm

8.5-9 $R_A = -R_C = 1.429$ kips, $\delta_B = 2.66 \times 10^{-4}$ in. to right

8.5-10 $R_A = -R_C = 117.4$ kN, $\delta_B = 0$

8.5-11 $\delta = 0.123$ in.

8.5-12 Ends $\sigma_T = -22.7$ MPa,
Middle $\sigma_T = -27.0$ MPa

8.5-13 $\tau = 15.0$ ksi

8.5-14 $\Delta T = 34°$ C

8.5-15 (a) $T_A = 400$ lb, $T_B = 200$ lb;
(b) $T_A = 454$ lb, $T_B = 92$ lb;
(c) $\Delta T = 153°$ F

8.5-16 $P_{allow} = 39.5$ kN

8.5-17 (a) $\sigma = -957$ psi; (b) $F_k = 3006$ lb (C);
(c) $\sigma = -2560$ psi

8.5-18 (a) $\sigma = 98$ MPa; (b) $T = 35°$ C

8.5-19 $F_{tube} = -227$ kips, $F_{each-cable} = 113.6$ kips,
$\delta_{tube} = -0.0953$ in.

8.5-20 $s = PL/6EA$

8.5-21 (a) $P_1 = 231$ kips, $R_A = -55.2$ kips,
$R_B = 55.2$ kips; (b) $P_2 = 145.1$ kips,
$R_A = -55.2$ kips, $R_B = 55.2$ kips; (c) For P_1,
$\tau_{max} = 13.39$ ksi, for P_2, $\tau_{max} = 19.44$ ksi;
(d) $\Delta T = 65.8°$ F, $R_A = 0$, $R_B = 0$;
(e) $R_A = -55.2$ kips, $R_B = 55.2$ kips

8.5-22 (a) $R_A = [-s + \alpha\Delta T(L_1 + L_2)]/[(L_1/EA_1)$
$+ (L_2/EA_2) + (L/k_3)]$, $R_D = -R_A$;
(b) $\delta_B = \alpha\Delta T(L_1) - R_A(L_1/EA_1)$,
$\delta_C = \alpha\Delta T(L_1 + L_2) - R_A[(L_1/EA_1) + (L_2/EA_2)]$

8.5-23 $T_B = 660$ lb, $T_C = 780$ lb

8.5-24 $P_{allow} = 1.8$ MN

8.5-25 (a) $\sigma_p = -0.196$ ksi, $\sigma_r = 3.42$ ksi;
(b) $\sigma_b = 2.74$ ksi, $\tau_c = 0.285$ ksi

8.5-26 $\sigma_p = 25.0$ MPa

8.5-27 $\sigma_p = 2400$ psi

8.5-28 (a) $P_B = 25.4$ kN, $P_s = -P_B$;
(b) $s_{reqd} = 25.7$ mm; (c) $\delta_{final} = 0.35$ mm

8.5-29 (a) $F_k = -0.174$ kips; (b) $F_t = 0.174$ kips;
(c) $L_f = 12.01$ in.; (d) $\Delta T = 141.9°$ F

8.5-30 $\sigma_s = 500$ MPa (tension),
$\sigma_c = 10$ MPa (compression)

8.5-31 (a) $F_k = 0.174$ kips; (b) $F_t = -0.174$ kips;
(c) $L_f = 11.99$ in.; (d) $\Delta T = -141.6°$ F

8.6-1 $P_{max} = 42{,}600$ lb

8.6-2 $d_{min} = 6.81$ mm

8.6-3 $P_{max} = 24{,}000$ lb

8.6-4 (a) $\Delta T_{max} = -46°$ C; (b) $\Delta T = +9.93°$ C

8.6-5 (a) $\tau_{max} = 10{,}800$ psi; (b) $\Delta T_{max} = -49.9°$ F;
(c) $\Delta T = +75.9°$ F

8.6-6 (a) $\sigma_x = 84$ MPa; (b) $\tau_{max} = 42$ MPa;
(c) On rotated x face: $\sigma_{x1} = 42$ MPa,
$\tau_{x1y1} = 42$ MPa; On rotated y face: $\sigma_{y1} = 42$ MPa;
(d) On rotated x face: $\sigma_{x1} = 71.7$ MPa,
$\tau_{x1y1} = -29.7$ MPa; On rotated y face: $\sigma_{y1} = 12.3$ MPa

8.6-7   (a) $\sigma_{max} = 18{,}000$ psi; (b) $\tau_{max} = 9000$ psi

8.6-8   (a) Element A: $\sigma_x = 105$ MPa (compression), Element B: $\tau_{max} = 52.5$ MPa; (b) $\theta = 33.1°$

8.6-9 $\sigma_\theta = -0.839$ ksi, $\sigma_{\theta + \pi/2} = -0.411$ ksi, $\tau_\theta = 0.587$ ksi

8.6-10 $\sigma_\theta = -30$ MPa, $\sigma_{\theta + \pi/2} = -30$ MPa, $\tau_\theta = 30$ MPa

8.6-11   (a) $\tau_{max\,AC} = \dfrac{\sigma_{AC}}{2} = 1.859$ ksi,

$\tau_{max\,AB} = \dfrac{\sigma_{AB}}{2} = 7.42$ ksi, $\tau_{max\,BC} = \dfrac{\sigma_{BC}}{2} = -9.41$ ksi;

(b) $P_{max} = 36.5$ kips

8.6-12   (a) (1) $\sigma_x = -945$ kPa; (2) $\sigma_\theta = -807$ kPa, $\tau_\theta = 334$ kPa; (3) $\sigma_\theta = -472$ kPa, $\tau_\theta = 472$ kPa, $\sigma_{max} = -945$ kPa, $\tau_{max} = -472$ kPa; (b) $\sigma_{max} = -378$ kPa, $\tau_{max} = -189$ kPa

8.6-13   (a) $\tau_{pq} = 1154$ psi; (b) $\sigma_{pq} = -1700$ psi, $\sigma(pq + \pi/2) = -784$ psi; (c) $P_{max} = 14{,}688$ lb

8.6-14   (a) $\Delta T_{max} = 31.3°$ C;

(b) $\sigma_{pq} = -21.0$ MPa (compression), $\tau_{pq} = 30$ MPa (CCW);

(c) $\beta = 0.62$

8.6-15 $N_{AC} = 5.77$ kips; $d_{min} = 1.08$ in.

8.6-16   (a) $\sigma_\theta = 0.57$ MPa, $\tau_\theta = -1.58$ MPa; (b) $\alpha = 33.3°$; (c) $\alpha = 26.6°$

8.6-17   (a) $\theta = 35.26°$, $\tau_0 = -7070$ psi; (b) $\sigma_{max} = 15{,}000$ psi, $\tau_{max} = 7500$ psi

8.6-18 $\sigma_{\theta 1} = 54.9$ MPa, $\sigma_{\theta 2} = 18.3$ MPa, $\tau_\theta = -31.7$ MPa

8.6-19 $\sigma_{max} = 10{,}000$ psi, $\tau_{max} = 5000$ psi

8.6-20   (a) $\theta = 30.96°$; (b) $P_{max} = 1.53$ kN

8.6-21   (a) $\tau_\theta = 348$ psi, $\theta = 20.1°$;

(b) $\sigma_{x1} = -950$ psi, $\sigma_{y1} = -127.6$ psi;

(c) $k_{max} = 15{,}625$ lb/in.; (d) $L_{max} = 1.736$ ft;

(e) $\Delta T_{max} = 92.8°$ F

8.6-22 $P_{max} = 5.63$ kN

CHAPTER 9

9.2-1 $\gamma_{min} = 4.17 \times 10^{-3}$ rad., $\gamma_{med} = 4.58 \times 10^{-3}$ rad.

9.2-2   (a) $L_{min} = 162.9$ mm; (b) $d_{max} = 68.8$ mm

9.2-3   (a) $d_{max} = 0.413$ in.; (b) $L_{min} = 21.8$ in.

9.2-4   (a) $\gamma_1 = 393 \times 10^{-6}$ rad; (b) $r_{2,max} = 50.9$ mm

9.2-5   (a) $\gamma_1 = 195 \times 10^{-6}$ rad; (b) $r_{2,max} = 2.57$ in.

9.2-6   (a) $\gamma_1 = 267 \times 10^{-6}$ rad; (b) $r_{2,min} = 183.3$ mm

9.3-1 $\phi = 4.9 \times 10^{-3}$ rad, $\tau_{max} = 1.173$ ksi, $\tau_A = 0.939$ ksi

9.3-2 $d_{min} = 35.7$ mm

9.3-3 $\phi = 5.63 \times 10^{-3}$ rad, $\tau_{max} = 1.348$ ksi, $\tau_A = 1.079$ ksi

9.3-4 $d_{min} = 64.7$ mm

9.3-5   (a) $\tau_{max} = 8344$ psi; (b) $d_{min} = 0.651$ in.

9.3-6   (a) $\tau_{max} = 23.8$ MPa; (b) $T_{max} = 0.402$ N·m; (c) $\theta = 9.12°/m$

9.3-7   (a) $\tau_{max} = 18{,}300$ psi; (b) $\phi = 3.32°$

9.3-8   (a) $k_T = 2059$ N·m/rad;

(b) $\tau_{max} = 27.9$ MPa, $\gamma_{max} = 997 \times 10^{-6}$ radians;

(c) $\dfrac{k_{Thollow}}{k_{Tsolid}} = 0.938$, $\dfrac{\tau_{max\,H}}{\tau_{max\,S}} = 1.067$;

(d) $d_2 = 32.5$ mm

9.3-9   (a) $L_{min} = 38.0$ in.; (b) $L_{min} = 40.9$ in.

9.3-10 $T_{max} = 6.03$ N·m, $\phi = 2.20°$

9.3-11   (a) $\tau_{max} = 7965$ psi; $\gamma_{max} = 0.00255$ radians, $G = 3.13 \times 10^6$ psi; (b) $T_{max} = 5096$ lb-in.

9.3-12   (a) $T_{max} = 9164$ N·m; (b) $T_{max} = 7765$ N·m;

9.3-13 $\tau_{max} = 4840$ psi

9.3-14   (a) $d_{min} = 63.3$ mm; (b) $d_{min} = 66$ mm (4.2% increase in diameter)

9.3-15   (a) $\tau_2 = 5170$ psi; (b) $\tau_1 = 3880$ psi; (c) $\theta = 0.00898°/in.$

9.3-16   (a) $\tau_2 = 30.1$ MPa; (b) $\tau_1 = 20.1$ MPa; (c) $\theta = 0.306°/m$

9.3-17   (a) $d_{min} = 2.50$ in.; (b) $k_T = 2941 \dfrac{\text{in-kip}}{\text{rad}}$; (c) $d_{min} = 1.996$ in.

9.3-18   (a) $d_{min} = 64.4$ mm;

(b) $k_T = 134.9$ kN·m/rad; (c) $d_{min} = 50$ mm

9.3-19   (a) $T_{1,max} = 4.60$ in-kip; (b) $T_{1,max} = 4.31$ in-kip; (c) Torque: 6.25%, Weight: 25%

9.3-20   (a) $\phi = 5.19°$; (b) $d = 88.4$ mm; (c) Ratio = 0.524

9.3-21   (a) $r_2 = 1.399$ in.; (b) $P_{max} = 1387$ lb

9.4-1   (a) $\tau_{max} = \tau_{BC} = 7602$ psi, $\phi_C = 0.16°$; (b) $d_{BC} = 1.966$ in., $\phi_C = -0.177°$

9.4-2   (a) $\tau_{bar} = 79.6$ MPa, $\tau_{tube} = 32.3$ MPa; (b) $\phi_A = 9.43°$

9.4-3   (a) $\tau_{max} = \tau_{BC} = 4653$ psi, $\phi_D = 0.978°$; (b) $d_{AB} = 3.25$ in., $d_{BC} = 2.75$ in., $d_{CD} = 2.16$ in., $\phi_D = 1.303°$

9.4-4 $T_{allow} = 459$ N·m

9.4-5 $d_1 = 0.818$ in.

9.4-6   (a) $d = 77.5$ mm; (b) $d = 71.5$ mm

9.4-7 $d_{min} = 1.268$ in., $\phi_{BD} = -0.497°$

9.4-8 $\tau_{max\,AB} = -367$ MPa, $\tau_{max\,BC} = -466$ MPa, $\phi_{max} = -5.63°$

9.4-9   (a) $d = 1.78$ in.; (b) $d = 1.83$ in.

9.4-10 $d_B/d_A = 1.45$

9.4-11   Minimum $d_A = 2.52$ in.

9.4-12   Minimum $d_B = 48.6$ mm

9.4-13   (a) $R_1 = -3T/2$; (b) $T_1 = 1.5T$, $T_2 = 0.5T$; (c) $x = 7L/17$; (d) $\phi_2 = (12/17)(TL/GI_p)$

9.4-14 $\phi = 3TL/2\pi Gtd_A^3$

9.4-15 (a) $\phi = 2.79°$; (b) $\phi = 2.21°$

9.4-16 (a) $R_1 = \dfrac{-T}{2}$; (b) $\phi_3 = \dfrac{19}{8}\dfrac{TL}{\pi Gtd^3}$

9.4-17 $\phi_D = \dfrac{4Fd}{\pi G}\left\{\dfrac{L_1}{t_{01}d_{01}^3}\right.$

$+\displaystyle\int_0^{L_2}\dfrac{L_2^4}{(d_{01}L_2 - d_{01}x + d_{03}x)^3(t_{01}L_2 - t_{01}x + t_{03}x)}$

$\left. dx + \dfrac{L_3}{t_{03}d_{03}^3}\right\}$,

$\phi_D = 0.142°$

9.4-18 (a) $\tau_{max} = 16tL/\pi d^3$; (b) $\phi = 16tL^2/\pi Gd^4$

9.4-19 (a) $\tau_{max} = 8t_A L/\pi d^3$; (b) $\phi = 16t_A L^2/3\pi Gd^4$

9.4-20 (a) $L_{max} = 4.42$ m; (b) $\phi = 170°$

9.4-21 (a) $R_A = \dfrac{T_0}{6}$;

(b) $T_{AB}(x) = \left(\dfrac{T_0}{6} - \dfrac{x^2}{L^2}T_0\right),\ 0 \le x \le \dfrac{L}{2}$,

$T_{BC}(x) = -\left[\left(\dfrac{x-L}{L}\right)^2\dfrac{T_0}{3}\right],\ \dfrac{L}{2} \le x \le L$;

(c) $\phi_c = \dfrac{T_0 L}{144GI_p}$; (d) $\tau_{max} = \dfrac{8}{3\pi}\dfrac{T_0}{d_{AB}^3}$

9.4-22 (a) $T_{max} = 875$ N·m; (b) $\tau_{max} = 25.3$ MPa

9.5-1 $\tau_{max} = 5.66$ ksi, $\gamma_{max} = 9.43 \times 10^{-4}$ rad, $\sigma_{max} = 5.66$ ksi, $\sigma_{min} = -5.66$ ksi, $\varepsilon_{max} = 4.72 \times 10^{-4}$

9.5-2 $T_{max} = 2.71$ kN·m

9.5-3 (a) $\sigma_{max} = 6280$ psi; (b) $T = 74{,}000$ lb-in.

9.5-4 (a) $\varepsilon_{max} = 320 \times 10^{-6}$; (b) $\sigma_{max} = 51.2$ MPa; (c) $T = 20.0$ kN·m

9.5-5 (a) $d_1 = 2.40$ in.; (b) $\phi = 2.20°$; (c) $\gamma_{max} = 1600 \times 10^{-6}$ rad

9.5-6 $G = 30.0$ GPa

9.5-7 $T = 4200$ lb-in.

9.5-8 (a) $d_{min} = 37.7$ mm; (b) $T_{max} = 431$ N·m

9.5-9 (a) $d_1 = 0.6$ in.; (b) $d_{1max} = 0.661$ in.

9.5-10 (a) $d_2 = 79.3$ mm; (b) $d_2 = 80.5$ mm

9.5-11 (a) $\tau_{max} = 5090$ psi; (b) $\gamma_{max} = 432 \times 10^{-6}$ rad

9.5-12 (a) $\tau_{max} = 23.9$ MPa; (b) $\gamma_{max} = 884 \times 10^{-6}$ rad

9.5-13 (a) $T_{1allow} = 17.84$ kip-in., $T_{2allow} = 13.48$ kip-in., (b) $L_{mid} = 18.54$ in.; (c) $d_{3new} = 2.58$ in.; (d) $T_{max1} = 17.41$ kip-in., $T_{max2} = 13.15$ kip-in., $\phi_{max1} = 1.487°$, $\phi_{max2} = 1.245°$

9.7-1 (a) $\tau_{max} = 4950$ psi; (b) $d_{min} = 3.22$ in.

9.7-2 (a) $\tau_{max} = 50.0$ MPa; (b) $d_{min} = 32.3$ mm

9.7-3 $n = 793$ rpm

9.7-4 $\tau_{max} = 63.3$ MPa, $\phi_{AC} = 0.0535$ rad

9.7-5 (a) $H = 6560$ hp; (b) Shear stress is halved.

9.7-6 (a) $\tau_{max} = 16.8$ MPa; (b) $P_{max} = 267$ kW

9.7-7 $d_{min} = 4.28$ in.

9.7-8 $d_{min} = 110$ mm

9.7-9 Minimum $d_1 = 1.221d$

9.7-10 $P_{max} = 91.0$ kW

9.7-11 $d = 2.75$ in.

9.7-12 $d = 53.4$ mm

9.8-1 (a) $\phi_{max} = 3T_0 L/5GI_P$; (b) $\phi_{max} = \dfrac{9LT_0}{25GI_p}$

9.8-2 (a) $x = L/4$; (b) $\phi_{max} = T_0 L/8GI_P$

9.8-3 $\phi_{max} = 2b\tau_{allow}/Gd$

9.8-4 $P_{allow} = 2717$ N

9.8-5 (a) $T_{0,max} = 3678$ lb-in.; (b) $T_{0,max} = 3898$ lb-in.

9.8-6 (a) $T_{0,max} = 150$ N·m; (b) $T_{0,max} = 140$ N·m

9.8-7 (a) $a/L = d_A/(d_A + d_B)$; (b) $a/L = d_A^4/(d_A^4 + d_B^4)$

9.8-8 $T_A = T_B = 225$ N·m, $\phi_{L/2} = -4.34°$

9.8-9 $T_A = -196.8$ kip-ft, $T_C = 63.5$ kip-ft, $\tau_{AB\,max} = 13.01$ ksi, $\tau_{BC\,max} = -10.96$ ksi

9.8-10 (a) $T_A = \dfrac{Lt_0}{6}, T_B = \dfrac{Lt_0}{3}$; (b) $\phi_{max} = \phi\left(\dfrac{L}{\sqrt{3}}\right) = -\dfrac{\sqrt{3}L^2 t_0}{27GI_P}$

9.8-11 (a) $x = 30.12$ in.; (b) $\phi_{max} = -1°$ (at $x = 30.12$ in.)

9.8-12 (a) $\tau_1 = 32.7$ MPa, $\tau_2 = 49.0$ MPa; (b) $\phi = 1.030°$; (c) $k_T = 22.3$ kN·m/rad

9.8-13 (a) $\tau_1 = 1790$ psi, $\tau_2 = 2690$ psi; (b) $\phi = 0.354°$; (c) $k_T = 809$ kip-in./rad

9.8-14 (a) $T_{max} = 1.521$ kN·m; (b) $d_2 = 56.9$ mm

9.8-15 (a) $T_{max} = 9.13$ kip-in.; (b) $d_2 = 2.27$ in.

9.8-16 (a) $T_{1,allow} = 7.14$ kN·m; (b) $T_{2,allow} = 6.35$ kN·m; (c) $T_{3,allow} = 7.41$ kN·m; (d) $T_{max} = 6.35$ kN·m

9.8-17 (a) $T_A = 15{,}292$ in-lb, $T_B = 24{,}708$ in-lb; (b) $T_A = 8734$ in-lb, $T_B = 31{,}266$ in-lb

9.8-18 (a) $R_1 = -0.77T$, $R_2 = -0.23T$; (b) $T_{max} = 2.79$ kN·m; (c) $\phi_{max} = 7.51°$; (d) $T_{max} = 2.48$ kN·m (shear in flange plate bolts controls); (e) $R_2 = \dfrac{\beta}{f_{T1} + f_{T2}}$, $R_1 = -R_2$, with $f_{T1} = \dfrac{L_1}{G_1 I_{p1}}$, $f_{T2} = \dfrac{L_2}{G_2 I_{p2}}$; (f) $\beta_{max} = 29.1°$

CHAPTER 10

10.4-1 (a) $\varepsilon_{max} = 8.67 \times 10^{-4}$; (b) $R_{min} = 9.35$ in.;
(c) $d_{max} = 0.24$ in.

10.4-2 (a) $L_{min} = 5.24$ m; (b) $d_{max} = 4.38$ mm

10.4-3 (a) $\varepsilon_{max} = 5.98 \times 10^{-3}$; (b) $d_{max} = 4.85$ in.;
(c) $L_{min} = 51$ ft

10.4-4 (a) $\rho = 85$ m, $\kappa = 0.0118\dfrac{1}{m}$, $\delta = 23.5$ mm;
(b) $h_{max} = 136$ mm; (c) $\delta = 75.3$ mm

10.4-5 (a) $\varepsilon = 9.14 \times 10^{-4}$; (b) $t_{max} = 0.241$ in.;
(c) $\delta = 0.744$ in.; (d) $L_{max} = 37.1$ in.

10.4-6 (a) $\varepsilon = 4.57 \times 10^{-4}$; (b) $L_{max} = 2$ m

10.4-7 $\delta = 0.45$ in., $\varepsilon_x = 8.75 \times 10^{-4}$

10.4-8 $\rho = 600$ m, $\kappa = 1.667 \times 10^{-3}\dfrac{1}{m}$, $\delta_B = 7.5$ mm

10.5-1 (a) $\sigma_{max} = 52.4$ ksi; (b) 33.3%;
(c) $L_{new} = 120$ in.

10.5-2 (a) $\sigma_{max} = 250$ MPa; (b) -19.98%; (c) $+25\%$

10.5-3 (a) $\sigma_{max} = 38.2$ ksi; (b) $+10\%$; (c) $+10\%$

10.5-4 (a) $\sigma_{max} = 8.63$ MPa; (b) $\sigma_{max} = 6.49$ MPa

10.5-5 $\sigma_{max} = 575$ psi

10.5-6 $q_{0,max} = 351\dfrac{kN}{m}$

10.5-7 $\sigma_{max} = 21.6$ ksi

10.5-8 $\sigma_{max} = 203$ MPa

10.5-9 $\sigma_{max} = 3420$ psi

10.5-10 $\sigma_{max} = 101$ MPa

10.5-11 $\sigma_{max} = 9.53$ ksi

10.5-12 $\sigma_{max} = 7.0$ MPa

10.5-13 (a) $\sigma_{max} = 432$ psi;
(b) $s = 0.58579\,L$, $\sigma_{min} = 153.7$ psi;
(c) $s = 0$ or L, $\sigma_{max} = 896$ psi

10.5-14 $\sigma_{max} = 2.10$ MPa

10.5-15 (a) $\sigma_t = 30.93\ M/d^3$; (b) $\sigma_t = 360\ M/(73bh^2)$;
(c) $\sigma_t = 85.24\ M/d^3$

10.5-16 $\sigma_{max} = 10.965\ M/d^3$

10.5-17 (a) $\sigma_{max} = 21.4$ ksi; (b) $L = 20.9$ ft;
(c) $d = 8.56$ ft

10.5-18 (a) $\sigma_t = 35.4$ MPa, $\sigma_c = 61$ MPa;
(b) $d_{max} = \dfrac{L}{2}$, $\sigma_t = 37.1$ MPa, $\sigma_c = 64.1$ MPa

10.5-19 (a) $\sigma_t = 4.34$ ksi, $\sigma_c = 15.96$ ksi;
(b) $P_{max} = 214$ lb; (c) 4.28 ft

10.5-20 (a) $\sigma_c = 1.456$ MPa, $\sigma_t = 1.514$ MPa;
(b) $\sigma_c = 1.666$ MPa $(+14\%)$, $\sigma_t = 1.381$ MPa (-9%);
(c) $\sigma_c = 0.728$ MPa (-50%), $\sigma_t = 0.757$ MPa (-50%)

10.5-21 $P_{max} = 0.207$ kips

10.5-22 $\sigma_{max} = 3\rho L^2 a_0/t$

10.5-23 (a) $\sigma_t = 7810$ psi, $\sigma_c = 13,885$ psi;
(b) $a = 12.73$ ft

10.5-24 (a) $\sigma = 25.1$ MPa, 17.8 MPa, -23.5 MPa

10.5-25 (a) $\sigma_t = 20,360$ psi, $\sigma_c = 13,188$ psi;
(b) $h = 3.20$ in.;
(c) $q = 97.2$ lb/ft, $P = 675$ lb

10.5-26 (a) $c_1 = 91.7$ mm, $c_2 = 108.3$ mm,
$I_z = 7.969 \times 10^7$ mm^4;
(b) $\sigma_t = 4659$ kPa (top of beam at C), $\sigma_c = 5506$ kPa
(bottom of beam at C)

10.5-27 $d = 3$ ft, $\sigma_{max} = 171$ psi, $d = 6$ ft, $\sigma_{max} = 830$ psi

10.5-28 (a) $\sigma_t = 0.335$ MPa, $\sigma_c = 0.288$ MPa
(b) $P_{max} = 21.9$ kN

10.5-29 (a) $F_{res} = 104.8$ lb;
(b) $\sigma_{max} = 36.0$ ksi (compression at base);
(c) $\sigma_{max} = 32.4$ ksi (tension at base)

10.5-30 $P_{max} = 675$ N

10.6-1 6×10

10.6-2 W 360×39

10.6-3 $d_{min} = 4.00$ in.

10.6-4 (a) $d_{min} = 12.62$ mm; (b) $P_{max} = 39.8$ N

10.6-5 (a) C 15×33.9; (b) S 8×18.4; (c) W 8×35

10.6-6 (a) W 360×39; (b) W 250×89

10.6-7 (a) S 10×35; (b) $P_{max} = 3152$ lb

10.6-8 (a) $b_{min} = 161.6$ mm; (b) $b_{min} = 141.2$ mm,
area$_{(b)}$/area$_{(a)} = 1.145$

10.6-9 (a) 2×12; (b) $w_{max} = 137.3$ lb/ft^2

10.6-10 (a) $s_{max} = 429$ mm; (b) $h_{min} = 214$ mm

10.6-11 (a) $q_{0,allow} = 419$ lb/ft; (b) $q_{0,allow} = 264$ lb/ft

10.6-12 $h_{min} = 30.6$ mm

10.6-13 (a) $S_{reqd} = 15.37$ in^3; (b) S 8×23

10.6-14 (a) $d_{min} = 37.6$ mm; (b) $d_{min} = 42.4$ mm,
area$_{(b)}$/area$_{(a)} = 0.557$

10.6-15 (a) 4×12; (b) $q_{max} = 14.2$ lb/ft

10.6-16 $b = 152$ mm, $h = 202$ mm

10.6-17 $b = 10.25$ in.

10.6-18 $t = 13.61$ mm

10.6-19 $W_1 : W_2 : W_3 : W_4 = 1 : 1.260 : 1.408 : 0.794$

10.6-20 (a) $q_{max} = 6.61$ kN/m; (b) $q_{max} = 9.37$ kN/m

10.6-21 6.57%

10.6-22 (a) $b_{min} = 11.91$ mm; (b) $b_{min} = 11.92$ mm

10.6-23 (a) $s_{max} = 49.2$ in.; (b) $d = 12.65$ in.

10.6-24 W 360×79

10.6-25 (a) $\beta = 1/9$; (b) 5.35%

10.6-26 Increase when $d/h > 0.6861$; decrease when $d/h < 0.6861$

10.7-2 (a) $\tau_{max} = 731$ kPa, $\sigma_{max} = 4.75$ MPa;
(b) $\tau_{max} = 1462$ kPa, $\sigma_{max} = 19.01$ MPa

10.7-3 $\sigma_x = -196.9$ psi, $\tau_{xy} = -23.4$ psi

10.7-4 $q_{max} = 24$ kN/m

10.7-5 (a) $M_{max} = 25.4$ kip-ft; (b) $M_{max} = 1.746$ kip-ft

10.7-6 $\tau_{max} = 500$ kPa

10.7-7 $\tau_{max} = 2400$ psi

10.7-8 (a) $L_0 = h(\sigma_{allow}/\tau_{allow})$;
(b) $L_0 = (h/2)(\sigma_{allow}/\tau_{allow})$

10.7-9 (a) $P_{max} = 1.914$ kips; (b) $P_{max} = 2.05$ kips

10.7-10 (a) $M_{max} = 72.2$ N·m; (b) $M_{max} = 36.0$ N·m

10.7-11 (a) 8×12-in. beam; (b) 8×12-in. beam

10.7-12 (a) $P = 38.0$ kN; (b) $P = 35.6$ kN

10.7-13 (a) $w_1 = 121$ lb/ft^2; (b) $w_2 = 324$ lb/ft^2;
(c) $w_{allow} = 121$ lb/ft^2

10.7-14 (a) $b = 89.3$ mm; (b) $b = 87.8$ mm

10.8-1 $d_{min} = 5.70$ in.

10.8-2 (a) $W = 28.6$ kN; (b) $W = 38.7$ kN

10.8-3 $\tau_{max} = 1.07$ ksi

10.8-4 $d_0 = 47.6$ mm

10.8-5 (a) $d = 10.52$ in.; (b) $d = 2.56$ in.

10.8-6 (a) $q_{0,max} = 55.7$ kN/m; (b) $L_{max} = 2.51$ m

10.9-1 (a) $\tau_{max} = 5795$ psi; (b) $\tau_{min} = 4555$ psi;
(c) $\tau_{aver} = 5714$ psi; (d) $V_{web} = 28.25$ kips

10.9-2 (a) $\tau_{max} = 28.43$ MPa; (b) $\tau_{min} = 21.86$ MPa;
(c) $\tau_{aver} = 27.41$ MPa; (d) $V_{web} = 119.7$ kN

10.9-3 (a) $\tau_{max} = 4861$ psi; (b) $\tau_{min} = 4202$ psi;
(c) $\tau_{aver} = 4921$ psi; (d) $V_{web} = 9.432$ kips

10.9-4 (a) $\tau_{max} = 32.28$ MPa; (b) $\tau_{min} = 21.45$ MPa;
(c) $\tau_{aver} = 29.24$ MPa; (d) $V_{web} = 196.1$ kN

10.9-5 (a) $\tau_{max} = 2634$ psi; (b) $\tau_{min} = 1993$ psi;
(c) $\tau_{aver} = 2518$ psi; (d) $V_{web} = 20.19$ kips

10.9-6 (a) $\tau_{max} = 28.40$ MPa; (b) $\tau_{min} = 19.35$ MPa;
(c) $\tau_{aver} = 25.97$ MPa; (d) $V_{web} = 58.63$ kN

10.9-7 $q_{max} = 1270$ lb/ft

10.9-8 (a) $q_{max} = 184.7$ kN/m; (b) $q_{max} = 247$ kN/m

10.9-9 S 8×23

10.9-10 $V = 273$ kN

10.9-11 $\tau_{max} = 1.42$ ksi, $\tau_{min} = 1.03$ ksi

10.9-12 $\tau_{max} = 19.7$ MPa

10.9-13 $\tau_{max} = 2221$ psi

10.10-1 $\sigma_w = 632$ psi, $\sigma_s = 12{,}421$ psi

10.10-2 $\sigma_w = 4.66$ MPa, $\sigma_s = 91.5$ MPa

10.10-3 $\sigma_{face} = \pm 1980$ psi, $\sigma_{core} = \pm 531$ psi

10.10-4 (a) $M_{max} = 58.7$ kN·m;
(b) $M_{max} = 90.9$ kN·m; (c) $t = 7.08$ mm

10.10-5 (a) $M_{max} = 172$ kip-in.; (b) $M_{max} = 96$ kip-in.

10.10-6 (a) $M_{allowTi} = \dfrac{\sigma_{Ti}(E_{Ti}I_{Ti} + E_{Cu}I_{Cu})}{E_{Ti}\left(\dfrac{d_2}{2}\right)}$,

$M_{allowCu} = \sigma_{Cu}\dfrac{(E_{Ti}I_{Ti} + E_{Cu}I_{Cu})}{E_{Cu}\left(\dfrac{d_1}{2}\right)}$;

(b) $M_{maxTi} = 4989$ N·m; (c) $d_1 = 36.4$ mm

10.10-7 (a) $\sigma_w = 666$ psi, $\sigma_s = 13{,}897$ psi;
(b) $q_{max} = 665$ lb/ft; (c) $M_{0,max} = 486$ lb-ft

10.10-8 (a) $M_{allow} = 768$ N·m; (b) $\sigma_{sa} = 47.9$ MPa, $M_{max} = 1051$ N·m

10.10-9 (a) $\sigma_{face} = 3610$ psi, $\sigma_{core} = 4$ psi;
(b) $\sigma_{face} = 3630$ psi, $\sigma_{core} = 0$

10.10-10 (a) $\sigma_{face} = 14.1$ MPa, $\sigma_{core} = 0.214$ MPa;
(b) $\sigma_{face} = 14.9$ MPa, $\sigma_{core} = 0$

10.10-11 $\sigma_a = 4120$ psi, $\sigma_c = 5230$ psi

10.10-12 (a) $\sigma_w = 5.1$ MPa (compression), $\sigma_s = 37.6$ MPa (tension); (b) $t_s = 3.09$ mm

10.10-13 (a) $\sigma_{plywood} = 1131$ psi, $\sigma_{pine} = 969$ psi
(b) $q_{max} = 95.5$ lb/ft

10.10-14 $q_{0,max} = 15.53$ kN/m

10.10-15 (a) $t_s = 0.755$ in.; (b) $t_s = 0.079$ in.;
(c) $t_s = 0.356$ in.

10.10-16 Ratio $= 0.755$

10.10-17 Top of beam: $\sigma_w = -754$ psi, $\sigma_s = -14{,}838$ psi
Bottom of beam: $\sigma_w = 915$ psi
Bottom of steel: $\sigma_s = 12{,}524$ psi

10.10-18 Top of core: $\sigma_p = -0.253$ MPa
Top of beam: $\sigma_s = -66$ MPa

10.10-19 (a) $M_{max} = 442$ kip-in.; (b) $M_{max} = 189$ kip-in.

10.10-20 $t_{min} = 15.0$ mm

10.10-21 (a) $q_{allow} = 454$ lb/ft;
(b) $\sigma_{wood} = 277$ psi, $\sigma_{steel} = 11{,}782$ psi

10.10-22 (a) $\sigma_B = 60.3$ MPa, $\sigma_w = 7.09$ MPa;
(b) $t_B = 25.1$ mm, $M_{max} = 80$ kN·m

10.10-23 $\sigma_{AL} = 1860$ psi, $\sigma_p = 72$ psi

10.10-24 $\sigma_{AL} = 12.14$ MPa, $\sigma_p = 0.47$ MPa

10.10-25 (a) $q_{allow} = 264$ lb/ft; (b) $q_{allow} = 280$ lb/ft

10.10-26 (a) $\sigma_s = 93.5$ MPa;
(b) $h_s = 5.08$ mm, $h_{AL} = 114.92$ mm

10.10-27 $M_{max} = 81.1$ kip-in.

10.10-28 $S_A = 50.6$ mm^3, Metal A

10.10-29 $\sigma_s = 13{,}400$ psi (tension), $\sigma_c = 812$ psi (compression).

10.10-30 (a) $\sigma_c = 8.51$ MPa, $\sigma_s = 118.3$ MPa;
(b) $M_{max} = M_c = 172.9$ kN·m;
(c) $A_s = 2254$ mm^2, $M_{allow} = 167.8$ kN·m

10.10-31 (a) $\sigma_c = 649$ psi, $\sigma_s = 15{,}246$ psi;
(b) $M_{allow} = M_s = 207$ kip-ft

10.10-32 (a) $M_{max} = M_s = 10.59$ kN·m;
(b) $A_s = 1262$ mm^2, $M_{allow} = 15.79$ kN·m

10.10-33 Top of beam: $\sigma_w = -0.829$ ksi,
$\sigma_s = -16.58$ ksi, Bottom of wood: $\sigma_w = 0.463$ ksi,
Bottom of steel: $\sigma_s = 17.89$ ksi

10.10-34 $M_{allow} = M_w = 16.39$ kN $\cdot$ m

10.11-1 $\beta = 83.1°, \sigma_t = 5060$ psi, $\sigma_c = -10,420$ psi

10.11-2 $\beta = 83.4°, \sigma_t = 10.5$ MPa, $\sigma_c = -23.1$ MPa

10.11-3 $\beta = 75.6°, \sigma_t = 3080$ psi, $\sigma_c = -3450$ psi

10.11-4 $\beta = 75.8°, \sigma_t = 31.7$ MPa, $\sigma_c = -39.5$ MPa

10.11-5 (a) $\beta = -28.7°, \sigma_t = 4975$ psi, $\sigma_c = -4903$ psi;
(b) $\beta = -38.5°, \sigma_t = 5756$ psi, $\sigma_c = -4868$ psi

10.11-6 $\beta = 78.1°, \sigma_t = 40.7$ MPa, $\sigma_c = -40.7$ MPa

10.11-7 $\beta = 82.3°, \sigma_t = 1397$ psi, $\sigma_c = -1157$ psi

10.11-8 $\beta = 2.93°, \sigma_t = 6.56$ MPa, $\sigma_c = -6.54$ MPa

10.11-9 For $\theta = 0$: $\sigma_t = -\sigma_c = 2.546 M/r^3$; For $\theta = 45°$:
$\sigma_t = 4.535 M/r^3, \sigma_c = -3.955 M/r^3$;
For $\theta = 90°$: $\sigma_t = 3.867 M/r^3, \sigma_c = -5.244 M/r^3$

10.11-10 $\beta = -78.9°, \sigma_t = 131.1$ MPa, $\sigma_t = -148.5$ MPa

10.11-11 $\beta = -11.7°, \sigma_t = 28.0$ ksi, $\sigma_c = -24.2$ ksi

10.11-12 $\beta = -56.5°, \sigma_t = 31.0$ MPa, $\sigma_c = -29.0$ MPa

10.11-13 $\sigma_A = -13.70$ ksi, $\sigma_B = 10.17$ ksi

CHAPTER 11

11.2-1 For $\theta = 55°$: $\sigma_{x1} = 4221$ psi, $\sigma_{y1} = 4704$ psi,
$\tau_{x1y1} = -3411$ psi

11.2-2 For $\theta = 40°$: $\sigma_{x1} = 117.2$ MPa, $\sigma_{y1} = 62.8$ MPa,
$\tau_{x1y1} = -10.43$ MPa

11.2-3 $\sigma_{x1} = 1179$ psi, $\sigma_{y1} = 321$ psi, $\tau_{x1y1} = 33$ psi

11.2-4 $\sigma_{x1} = 185$ MPa, $\sigma_{y1} = 35$ MPa, $\tau_{x1y1} = -10$ MPa

11.2-5 For $\theta = 30°$: $\sigma_{x1} = -3041$ psi, $\sigma_{y1} = -8959$ psi,
$\tau_{x1y1} = -12,725$ psi

11.2-6 For $\theta = 52°$: $\sigma_{x1} = -136.6$ MPa, $\sigma_{y1} = 16.6$ MPa,
$\tau_{x1y1} = -84$ MPa

11.2-7 For $\theta = 50°$: $\sigma_{x1} = -1243$ psi, $\sigma_{y1} = -6757$ psi,
$\tau_{x1y1} = 1240$ psi

11.2-8 For $\theta = -40°$: $\sigma_{x1} = -5.5$ MPa, $\sigma_{y1} = -27$ MPa,
$\tau_{x1y1} = -28.1$ MPa

11.2-9 For $\theta = 38°$: $\sigma_{x1} = -13,359$ psi, $\sigma_{y1} = -3671$ psi,
$\tau_{x1y1} = 4960$ psi

11.2-10 For $\theta = -40°$: $\sigma_{x1} = -66.5$ MPa, $\sigma_{y1} = -6.52$ MPa,
$\tau_{x1y1} = -14.52$ MPa

11.2-11 Normal stress on seam, 187 psi tension. Shear stress,
163 psi (clockwise).

11.2-12 Normal stress on seam, 1440 kPa tension. Shear stress,
1030 kPa (clockwise).

11.2-13 $\sigma_{x1} = 5.83$ ksi, $\sigma_{y1} = -7.83$ ksi, $\tau_{x1y1} = 9.96$ ksi

11.2-14 $\sigma_{x1} = 41.7$ MPa, $\sigma_{y1} = -53.7$ MPa, $\tau_{x1y1} = 68.8$ MPa

11.2-15 $\sigma_w = -125$ psi, $\tau_w = 375$ psi

11.2-16 $\sigma_w = 10.0$ MPa, $\tau_w = -5.0$ MPa

11.2-17 $\sigma_{x1} = 0.75$ ksi, $\sigma_{y1} = 0.75$ ksi, $\tau_{x1y1} = 0.75$ ksi

11.2-18 $\sigma_{x1} = 6$ MPa, $\sigma_{y1} = 6$ MPa, $\tau_{x1y1} = 6$ MPa

11.2-19 $\theta = 51.3°, \sigma_{y1} = 500$ psi, $\tau_{x1y1} = -1122$ psi

11.2-20 $\theta = 36.6°, \sigma_{y1} = -9$ MPa, $\tau_{x1y1} = -14.83$ MPa

11.2-21 For $\theta = -36°$: $\sigma_{x1} = -12,068$ psi, $\sigma_{y1} = -4732$ psi,
$\tau_{x1y1} = -4171$ psi

11.2-22 For $\theta = -50°$: $\sigma_{x1} = +51.4$ MPa, $\sigma_{y1} = -14.4$ MPa,
$\tau_{x1y1} = -31.3$ MPa

11.2-23 $\sigma_y = 3673$ psi, $\tau_{xy} = 1405$ psi

11.2-24 $\sigma_y = -77.7$ MPa, $\tau_{xy} = -27.5$ MPa

11.2-25 $\sigma_b = -4700$ psi, $\tau_b = 2655$ psi, $\theta_1 = 48.04°$

11.2-26 $\sigma_y = -211$ kPa

11.2-27 $\sigma_{x1} = -200$ psi, $\sigma_{y1} = 350$ psi, $\tau_{x1y1} = 0$

11.3-1 $\sigma_1 = 1082$ psi, $\sigma_2 = 268$ psi, $\tau_{max} = 407$ psi

11.3-2 $\sigma_1 = 8.04$ MPa, $\sigma_2 = 1.463$ MPa, $\tau_{max} = 3.29$ MPa

11.3-3 $\sigma_1 = 5868$ psi, $\sigma_2 = 982$ psi, $\theta_{p1} = 8.94°$

11.3-4 $\sigma_1 = 119.2$ MPa, $\sigma_2 = 60.8$ MPa, $\theta_{p1} = 29.52°$

11.3-5 $\sigma_1 = -6333$ psi, $\sigma_2 = -1167$ psi, $\theta_{p1} = -23.68°$

11.3-6 $\sigma_1 = 53.6$ MPa, $\theta_{p1} = -14.2°$

11.3-7 $\sigma_1 = 5771$ psi, $\sigma_2 = 18,029$ psi,
$\tau_{max} = \dfrac{\sigma_1 - \sigma_2}{2} = -6129$ psi, $\theta_{p1} = -14.12°$

11.3-8 $\tau_{max} = 24.2$ MPa, $\sigma_{x1} = -14.25$ MPa,
$\sigma_{y1} = -14.25$ MPa, $\theta_{s1} = 60.53°$

11.3-9 $\tau_{max} = 6851$ psi, $\theta_{s1} = 61.8°$

11.3-10 $\tau_{max} = 26.7$ MPa, $\theta_{s1} = 19.08°$

11.3-11 $(\tau_{max})_{z1} = 150$ psi, $(\tau_{max})_{x1} = -300$ psi,
$(\tau_{max})_{y1} = -150$ psi

11.3-12 $\sigma_1 = 20$ kPa, $\sigma_2 = -20$ kPa, $\tau_{max} = 20$ kPa

11.3-13 (a) $\sigma_1 = 180$ psi, $\theta_{p1} = -20.56°$;
(b) $\tau_{max} = 730$ psi, $\theta_{s1} = -65.56°$

11.3-14 $\sigma_1 = 0, \sigma_2 = -5$ MPa, $\tau_{max} = 2.5$ MPa

11.3-15 $\sigma_1 = 0, \sigma_2 = -750$ psi, $\tau_{max} = 375$ psi

11.3-16 (a) $\sigma_1 = 25$ MPa, $\sigma_2 = -130$ MPa;
(b) $\tau_{max} = 77.5$ MPa, $\sigma_{ave} = -52.5$ MPa

11.3-17 (a) $\sigma_1 = 2693$ psi, $\sigma_2 = 732$ psi;
(b) $\tau_{max} = 980$ psi, $\sigma_{ave} = 1713$ psi

11.3-18 (a) $\sigma_1 = 2262$ kPa, $\theta_{p1} = -13.70°$
(b) $\tau_{max} = 1000$ kPa, $\theta_{s1} = -58.7°$

11.3-19 (a) $\sigma_1 = 14,764$ psi, $\theta_{p1} = 7.90°$
(b) $\tau_{max} = 6979$ psi, $\theta_{s1} = -37.1°$

11.3-20 (a) $\sigma_1 = 29.2$ MPa, $\theta_{p1} = -17.98°$
(b) $\tau_{max} = 66.4$ MPa, $\theta_{s1} = -63.0°$

11.3-21 (a) $\sigma_1 = -1228$ psi, $\theta_{p1} = 24.7°$
(b) $\tau_{max} = 5922$ psi, $\theta_{s1} = -20.3°$

11.3-22 (a) $\sigma_1 = 76.3$ MPa, $\theta_{p1} = 107.5°$
(b) $\tau_{max} = 101.3$ MPa, $\theta_{s1} = -62.5°$

11.3-23 $3030 \text{ psi} \le \sigma_y \le 9470 \text{ psi}$

11.3-24 $18.5 \text{ MPa} \le \sigma_y \le 85.5 \text{ MPa}$

11.3-25 (a) $\sigma_y = 3961 \text{ psi}$; (b) $\theta_{p1} = -38.93°$, $\sigma_1 = 6375 \text{ psi}$, $\theta_{p2} = 51.07°$, $\sigma_2 = 2386 \text{ psi}$

11.3-26 (a) $\sigma_y = 23.3 \text{ MPa}$;
(b) $\theta_{p1} = 65.6°$, $\sigma_1 = 41 \text{ MPa}$, $\theta_{p2} = -24.4°$,
$\sigma_2 = -62.7 \text{ MPa}$

11.3-27 $\sigma_1 = 0.53 \text{ psi}$, $\sigma_2 = -92.5 \text{ psi}$, $\tau_{max} = 46.5 \text{ psi}$

11.3-28 $\sigma_1 = 12 \text{ MPa}$, $\sigma_2 = 0$, $\tau_{max} = 6 \text{ MPa}$

11.4-1 (a) $\sigma_{x1} = 10,901 \text{ psi}$, $\sigma_{y1} = 3349 \text{ psi}$, $\tau_{x1y1} = -6042 \text{ psi}$;
(b) $\tau_{max} = 7125 \text{ psi}$, $\sigma_{ave} = 7125 \text{ psi}$

11.4-2 (a) $\sigma_{x1} = 40.1 \text{ MPa}$, $\sigma_{y1} = 16.91 \text{ MPa}$, $\tau_{x1y1} = 26 \text{ MPa}$;
(b) $\tau_{max} = 28.5 \text{ MPa}$, $\sigma_{ave} = 28.5 \text{ MPa}$

11.4-3 (a) $\sigma_{x1} = -5400 \text{ psi}$, $\sigma_{y1} = -1350 \text{ psi}$, $\tau_{x1y1} = 2700 \text{ psi}$;
(b) $\tau_{max} = -3375 \text{ psi}$, $\sigma_{aver} = -3375 \text{ psi}$

11.4-4 For $\theta = 25°$: (a) $\sigma_{x1} = -36.0 \text{ MPa}$, $\tau_{x1y1} = 25.7 \text{ MPa}$;
(b) $\tau_{max} = 33.5 \text{ MPa}$, $\theta_{s1} = 45.0°$

11.4-5 For $\theta = 55°$: (a) $\sigma_{x1} = 882 \text{ psi}$, $\tau_{x1y1} = -3759 \text{ psi}$;
(b) $\tau_{max} = 4000 \text{ psi}$, $\theta_{s1} = -45.0°$, $\sigma_{aver} = 2250 \text{ psi}$

11.4-6 For $\theta = 21.80°$: (a) $\sigma_{x1} = -17.1 \text{ MPa}$,
$\tau_{x1y1} = 29.7 \text{ MPa}$; (b) $\tau_{max} = 43.0 \text{ MPa}$, $\theta_{s1} = 45.0°$

11.4-7 For $\theta = 52°$: (a) $\sigma_{x1} = 2620 \text{ psi}$, $\tau_{x1y1} = -653 \text{ psi}$;
(b) $\sigma_1 = 2700 \text{ psi}$, $\theta_{p1} = 45.0°$

11.4-8 $\sigma_{x1} = 1.75 \text{ MPa}$, $\sigma_{y1} = 13.25 \text{ MPa}$, $\tau_{x1y1} = 4.82 \text{ MPa}$

11.4-9 (a) $\sigma_{x1} = -493 \text{ psi}$, $\sigma_{y1} = -1007 \text{ psi}$, $\tau_{x1y1} = 705 \text{ psi}$;
(b) $\sigma_1 = 0$, $\sigma_2 = -1500 \text{ psi}$, $\tau_{max} = 750 \text{ psi}$

11.4-10 (a) $\sigma_{x1} = -60.8 \text{ MPa}$, $\sigma_{y1} = 128.8 \text{ MPa}$,
$\tau_{x1y1} = -46.7 \text{ MPa}$; (b) $\sigma_1 = 139.6 \text{ MPa}$,
$\sigma_2 = -71.6 \text{ MPa}$, $\tau_{max} = 105.6 \text{ MPa}$

11.4-11 For $\theta = 36.87°$: (a) $\sigma_{x1} = 3600 \text{ psi}$, $\tau_{x1y1} = 1050 \text{ psi}$;
(b) $\sigma_1 = 3750 \text{ psi}$, $\theta_{p1} = 45.0°$

11.4-12 For $\theta = 40°$: $\sigma_{x1} = 27.5 \text{ MPa}$, $\tau_{x1y1} = -5.36 \text{ MPa}$

11.4-13 For $\theta = -51°$: $\sigma_{x1} = 11,982 \text{ psi}$, $\tau_{x1y1} = -3569 \text{ psi}$

11.4-14 For $\theta = -33°$: $\sigma_{x1} = -61.7 \text{ MPa}$, $\tau_{x1y1} = 51.7 \text{ MPa}$,
$\sigma_{y1} = -171.3 \text{ MPa}$

11.4-15 For $\theta = 14°$: $\sigma_{x1} = -1509 \text{ psi}$, $\tau_{x1y1} = 527 \text{ psi}$,
$\sigma_{y1} = -891 \text{ psi}$

11.4-16 For $\theta = 35°$: $\sigma_{x1} = 46.4 \text{ MPa}$, $\tau_{x1y1} = -9.81 \text{ MPa}$

11.4-17 For $\theta = 65°$: $\sigma_{x1} = -1846 \text{ psi}$, $\tau_{x1y1} = 3897 \text{ psi}$

11.4-18 (a) $\sigma_1 = 10,865 \text{ kPa}$, $\theta_{p1} = 115.2°$;
(b) $\tau_{max} = 4865 \text{ kPa}$, $\theta_{s1} = 70.2°$

11.4-19 (a) $\sigma_1 = 2565 \text{ psi}$, $\theta_{p1} = 31.3°$;
(b) $\tau_{max} = 3265 \text{ psi}$, $\theta_{s1} = -13.70°$

11.4-20 (a) $\sigma_1 = 18.2 \text{ MPa}$, $\theta_{p1} = 123.3°$;
(b) $\tau_{max} = 15.4 \text{ MPa}$, $\theta_{s1} = 78.3°$

11.4-21 (a) $\sigma_1 = -6923 \text{ psi}$, $\theta_{p1} = -32.4°$;
(b) $\tau_{max} = 7952 \text{ psi}$, $\theta_{s1} = 102.6°$

11.4-22 (a) $\sigma_1 = 40.0 \text{ MPa}$, $\theta_{p1} = 68.8°$;
(b) $\tau_{max} = 40.0 \text{ MPa}$, $\theta_{s1} = 23.8°$

11.4-23 (a) $\sigma_1 = 7490 \text{ psi}$, $\theta_{p1} = 63.2°$;
(b) $\tau_{max} = 3415 \text{ psi}$, $\theta_{s1} = -18.20°$

11.4-24 (a) $\sigma_1 = 3.43 \text{ MPa}$, $\theta_{p1} = -19.68°$;
(b) $\tau_{max} = 15.13 \text{ MPa}$, $\theta_{s1} = -64.7°$

11.4-25 (a) $\sigma_1 = 7525 \text{ psi}$, $\theta_{p1} = 9.80°$;
(b) $\tau_{max} = 3875 \text{ psi}$, $\theta_{s1} = -35.2°$

11.5-1 $\sigma_x = 25,385 \text{ psi}$, $\sigma_y = 19,615 \text{ psi}$,
$\Delta t = -2.81 \times 10^{-4} \text{ in.}$

11.5-2 $\sigma_x = 102.6 \text{ MPa}$, $\sigma_y = -11.21 \text{ MPa}$,
$\Delta t = -1.646 \times 10^{-3} \text{ mm}$

11.5-3 $e = -3.21 \times 10^{-4}$

11.5-4 $\epsilon_x = 7.25 \times 10^{-5}$, $\epsilon_y = -9 \times 10^{-5}$, $\epsilon_z = 7.5 \times 10^{-6}$,
$\gamma_{xy} = 6.5 \times 10^{-5}$;

11.5-5 (a) $\epsilon_z = -v(\epsilon_x + \epsilon_y)/(1 - v)$;
(b) $e = (1 - 2v)(\epsilon_x + \epsilon_y)/(1 - v)$

11.5-6 $v = 0.24$, $E = 112.1 \text{ GPa}$

11.5-7 $v = 0.3$, $E = 29,560 \text{ ksi}$

11.5-8 (a) $\gamma_{max} = 5.85 \times 10^{-4}$;
(b) $\Delta t = -1.32 \times 10^{-3} \text{ mm}$;
(c) $\Delta V = 387 \text{ mm}^3$

11.5-9 (a) $\gamma_{max} = 1900 \times 10^{-6}$;
(b) $\Delta t = -141 \times 10^{-6} \text{ in.}$ (decrease);
(c) $\Delta V = 0.0874 \text{ in}^3$ (increase)

11.5-10 (a) $\Delta V_b = -49.2 \text{ mm}^3$, (b) $\Delta V_a = -71.5 \text{ mm}^3$

11.5-11 $\Delta V = -0.0377 \text{ in}^3$

11.5-12 (a) $\Delta V = 2766 \text{ mm}^3$, (b) $b_{min} = 640 \text{ mm}$

11.5-13 (a) $\Delta V = 0.0385 \text{ in}^3$, (b) $b_{min} = 10.26 \text{ in.}$

11.5-14 (a) $\Delta ac = \epsilon_x d = 0.1296 \text{ mm}$ (increase);
(b) $\Delta bc = \epsilon_y d = -0.074 \text{ mm}$ (decrease);
(c) $\Delta t = \epsilon_z t = -2.86 \times 10^{-3} \text{ mm}$ (decrease);
(d) $\Delta V = eV_0 = 430 \text{ mm}^3$;
(e) $\sigma_{x max} = 63.9 \text{ MPa}$

11.5-15 (a) $\epsilon_x = 0.1557$, $\epsilon_y = -0.321$, $\epsilon_z = 0.1557$;
(b) $e = -9.16 \times 10^{-3}$

11.5-16 $\nu = 0.477$, $E = 1.006 \text{ MPa}$

11.6-1 $\epsilon_x = -4.05 \times 10^{-4}$, $\epsilon_y = 3.27 \times 10^{-4}$, $\epsilon_z = 2.2 \times 10^{-4}$,
$e = 1.428 \times 10^{-4}$

11.6-2 $\epsilon_x = -3.33 \times 10^{-4}$, $\epsilon_y = 5.79 \times 10^{-4}$,
$\epsilon_z = -2.95 \times 10^{-4}$

11.6-3 (a) $\tau_{max} = \dfrac{\sigma_1 - \sigma_3}{2} = 8750 \text{ psi}$;

(b) $\Delta a = a\epsilon_x = 7.73 \times 10^{-3} \text{ in.}$,
$\Delta b = \epsilon_y b = -3.75 \times 10^{-3} \text{ in.}$,
$\Delta c = \epsilon_z c = -1.3 \times 10^{-3} \text{ in.}$;
(c) $\Delta V = eV_0 = 0.0173 \text{ in}^3$;
(d) $\sigma_{x max} = 12,824 \text{ psi}$

11.6-4 (a) $\tau_{max} = \dfrac{\sigma_1 - \sigma_3}{2} = 8.5 \text{ MPa}$;

(b) $\Delta a = a\epsilon_x = -0.0525 \text{ mm}$,
$\Delta b = \epsilon_y b = -9.67 \times 10^{-3} \text{ mm}$,
$\Delta c = \epsilon_z c = -9.67 \times 10^{-3} \text{ mm}$;
(c) $\Delta V = eV_0 = -2.052 \times 10^3 \text{ mm}^3$;
(d) $\sigma_{x max} = -50 \text{ MPa}$

11.6-5 (a) $\sigma_x = -4200$ psi, $\sigma_y = -2100$ psi, $\sigma_z = -2100$ psi;

(b) $\tau_{max} = \dfrac{\sigma_1 - \sigma_3}{2} = 1050$ psi;

(c) $\Delta V = eV_0 = -0.0192$ in^3;

(d) $\sigma_{x\,max} = -3864$ psi

11.6-6 (a) $\sigma_x = -82.6$ MPa, $\sigma_y = -54.7$ MPa, $\sigma_z = -54.7$ MPa;

(b) $\tau_{max} = \dfrac{\sigma_1 - \sigma_3}{2} = 13.92$ MPa;

(c) $\Delta V = eV_0 = -846$ mm^3;

(d) $\sigma_{x\,max} = -73$ MPa

11.6-7 (a) $K_{AL} = 1 \times 10^7$ psi; (b) $E = 6139$ ksi, $v = 0.35$

11.6-8 (a) $K = 4.95$ GPa; (b) $E = 1.297$ GPa, $v = 0.40$

11.6-9 (a) $p = vF/[A(1 - v)]$;
(b) $\delta = FL(1 + v)(1 - 2v)/EA(1 - v)$

11.6-10 (a) $p = vp_0$; (b) $e = -p_0(1 + v)(1 - 2v)/E$

11.6-11 (a) $p = -0.455$ psi; (b) $\Delta V = -0.01318$ in^3

11.6-12 $P = -81.5$ kN

11.6-13 (a) $\Delta d = 1.472 \times 10^{-3}$ in., $\Delta V = 0.187$ in^3,
(b) $h = 5282$ ft

11.6-14 (a) $p = 700$ MPa; (b) $K = 175$ GPa

11.6-15 $\epsilon_0 = 276 \times 10^{-6}$, $e = 828 \times 10^{-6}$

11.7-1 (a) $\epsilon_{x1} = -5.53 \times 10^{-4}$, $\epsilon_{y1} = 1.053 \times 10^{-3}$,
$\gamma_{x1y1} = 1.915 \times 10^{-3}$; (b) $\epsilon_1 = 1.5 \times 10^{-3}$,
$\epsilon_2 = -1.0 \times 10^{-3}$, $\gamma_{max} = 2.5 \times 10^{-3}$

11.7-2 (a) $\epsilon_1 = 2.0 \times 10^{-3}$, $\epsilon_2 = 1.5 \times 10^{-3}$;
(b) $\gamma_{max} = 5.0 \times 10^{-4}$

11.7-3 (a) $\epsilon_{x1} = 1.299 \times 10^{-4}$, $\epsilon_{y1} = -1.299 \times 10^{-4}$,
$\gamma_{x1y1} = 1.5 \times 10^{-4}$; (b) $\epsilon_1 = 1.5 \times 10^{-4}$,
$\epsilon_2 = -1.5 \times 10^{-4}$, $\gamma_{max} = 3 \times 10^{-4}$

11.7-4 (a) $\epsilon_{x1} = 1.398 \times 10^{-3}$, $\epsilon_{y1} = -9.76 \times 10^{-5}$,
$\gamma_{x1y1} = -8.63 \times 10^{-4}$; (b) $\epsilon_1 = 1.513 \times 10^{-3}$,
$\epsilon_2 = -2.13 \times 10^{-4}$, $\gamma_{max} = 1.726 \times 10^{-3}$

11.7-5 (a) $\Delta d = 1.878 \times 10^{-3}$ in.;
(b) $\Delta\phi = -\alpha = 1.425 \times 10^{-4}$ (decrease, radians);
(c) $\Delta\psi = -\alpha = 1.425 \times 10^{-4}$ (increase, radians)

11.7-6 (a) $\Delta d = \epsilon_{x1}L_d = 0.062$ mm;
(b) $\Delta\phi = -\alpha = 1.89 \times 10^{-4}$ (decrease, radians);
(c) $\Delta\psi = -\alpha = 1.89 \times 10^{-4}$ (increase, radians)

11.7-7 (a) $\Delta d = 0.00458$ in. (increase);
(b) $\Delta\phi = 157 \times 10^{-6}$ rad (decrease);
(c) $\gamma = -314 \times 10^{-6}$ rad (angle *ced* increases)

11.7-8 (a) $\Delta d = 0.168$ mm (increase);
(b) $\Delta\phi = 317 \times 10^{-6}$ rad (decrease);
(c) $\gamma = -634 \times 10^{-6}$ rad (angle *ced* increases)

11.7-9 $\epsilon_{x1} = 3.97 \times 10^{-4}$, $\epsilon_{y1} = 3.03 \times 10^{-4}$,
$\gamma_{x1y1} = 1.829 \times 10^{-4}$

11.7-10 $\epsilon_{x1} = 9.53 \times 10^{-5}$, $\epsilon_{y1} = -1.353 \times 10^{-4}$,
$\gamma_{x1y1} = -3.86 \times 10^{-4}$

11.7-11 $\epsilon_1 = 554 \times 10^{-6}$, $\theta_{p1} = -22.9°$, $\gamma_{max} = 488 \times 10^{-6}$

11.7-12 $\epsilon_1 = 172 \times 10^{-6}$, $\theta_{p1} = 163.9°$, $\gamma_{max} = 674 \times 10^{-6}$

11.7-13 For $\theta = 75°$: (a) $\epsilon_{x1} = 202 \times 10^{-6}$,
$\gamma_{x1y1} = -569 \times 10^{-6}$;
(b) $\epsilon_1 = 568 \times 10^{-6}$, $\theta_{p1} = 22.8°$;
(c) $\gamma_{max} = 587 \times 10^{-6}$

11.7-14 For $\theta = 45°$: (a) $\epsilon_{x1} = -385 \times 10^{-6}$,
$\gamma_{x1y1} = 690 \times 10^{-6}$;
(b) $\varepsilon_1 = -254 \times 10^{-6}$, $\theta_{p1} = 65.7°$;
(c) $\gamma_{max} = 1041 \times 10^{-6}$

11.7-15 $\tau_{max\,xy} = \dfrac{\sigma_x - \sigma_y}{2} = 4076$ psi,

$\gamma_{xy\,max} = 2\sqrt{\left(\dfrac{\varepsilon_x - \varepsilon_y}{2}\right)^2 + \left(\dfrac{\gamma_{xy}}{2}\right)^2}$
$= 6.83 \times 10^{-4}$,

$\gamma_{xz\,max} = 2\sqrt{\left(\dfrac{\varepsilon_x - \varepsilon_z}{2}\right)^2 + \gamma_{xz}^2} = 8.96 \times 10^{-4}$,

$\gamma_{yz\,max} = 2\sqrt{\left(\dfrac{\varepsilon_y - \varepsilon_z}{2}\right)^2 + \gamma_{yz}^2} = 2.13 \times 10^{-4}$

11.7-16 $\tau_{max\,xy} = \dfrac{\sigma_x - \sigma_y}{2} = 33.7$ MPa,

$\gamma_{xy\,max} = 2\sqrt{\left(\dfrac{\varepsilon_x - \varepsilon_y}{2}\right)^2 + \left(\dfrac{\gamma_{xy}}{2}\right)^2}$,
$= 1.244 \times 10^{-3}$,

$\gamma_{xz\,max} = 2\sqrt{\left(\dfrac{\varepsilon_x - \varepsilon_z}{2}\right)^2 + \gamma_{xz}^2} = 1.459 \times 10^{-3}$,

$\gamma_{yz\,max} = 2\sqrt{\left(\dfrac{\varepsilon_y - \varepsilon_z}{2}\right)^2 + \gamma_{yz}^2} = 2.15 \times 10^{-4}$

11.7-17 For $\theta = 30°$: (a) $\varepsilon_{x1} = -756 \times 10^{-6}$,
$\gamma_{x1y1} = 868 \times 10^{-6}$; (b) $\varepsilon_1 = 426 \times 10^{-6}$, $\theta_{p1} = 99.8°$;
(c) $\gamma_{max} = 1342 \times 10^{-6}$

11.7-18 For $\theta = 50°$: (a) $\varepsilon_{x1} = -1469 \times 10^{-6}$,
$\gamma_{x1y1} = -717 \times 10^{-6}$;
(b) $\varepsilon_1 = -732 \times 10^{-6}$, $\theta_{p1} = 166.0°$;
(c) $\gamma_{max} = 911 \times 10^{-6}$

11.7-19 $\epsilon_1 = 551 \times 10^{-6}$, $\theta_{p1} = 12.5°$, $\gamma_{max} = 662 \times 10^{-6}$

11.7-20 $\epsilon_1 = 332 \times 10^{-6}$, $\theta_{p1} = 12.0°$, $\gamma_{max} = 515 \times 10^{-6}$

11.7-21 (a) $P = 5154$ lb, $T = -978$ in-lb;
(b) $\gamma_{max} = 2.84 \times 10^{-4}$, $\tau_{max} = 3304$ psi

11.7-22 $P = 121.4$ kN, $\alpha = 56.7°$

11.7-23 $P = 9726$ lb, $\alpha = 75.2°$

11.7-24 $\epsilon_x = \epsilon_a$, $\epsilon_y = (2\epsilon_b + 2\epsilon_c - \epsilon_a)/3$, $\gamma_{xy} = 2(\epsilon_b - \epsilon_c)/\sqrt{3}$

11.7-25 For $\theta_{p1} = 30°$: $\epsilon_1 = 1550 \times 10^{-6}$, $\epsilon_2 = -250 \times 10^{-6}$,
$\sigma_1 = 10{,}000$ psi, $\sigma_2 = 2000$ psi

11.7-26 $\sigma_x = 91.6$ MPa

11.7-27 $\epsilon_{x1} = 3.97 \times 10^{-4}$, $\epsilon_{y1} = 3.03 \times 10^{-4}$,
$\gamma_{x1y1} = 1.829 \times 10^{-4}$

11.7-28 $\epsilon_{x1} = 9.53 \times 10^{-5}$, $\epsilon_{y1} = -1.353 \times 10^{-4}$,
$\gamma_{x1y1} = -3.86 \times 10^{-4}$

11.7-29 $\epsilon_1 = 554 \times 10^{-6}$, $\theta_{p1} = 157.1°$, $\gamma_{max} = 488 \times 10^{-6}$

11.7-30 $\epsilon_1 = 172 \times 10^{-6}$, $\theta_{p1} = 163.9°$, $\gamma_{max} = 674 \times 10^{-6}$

11.7-31 For $\theta = 75°$: (a) $\epsilon_{x1} = 202 \times 10^{-6}$, $\gamma_{x1y1} = -569 \times 10^{-6}$; (b) $\epsilon_1 = 568 \times 10^{-6}$, $\theta_{p1} = 22.8°$; (c) $\gamma_{max} = 587 \times 10^{-6}$

11.7-32 For $\theta = 45°$: (a) $\epsilon_{x1} = -385 \times 10^{-6}$, $\gamma_{x1y1} = 690 \times 10^{-6}$; (b) $\epsilon_1 = -254 \times 10^{-6}$, $\theta_{p1} = 65.7°$; (c) $\gamma_{max} = 1041 \times 10^{-6}$

CHAPTER 12

12.2-1 $p_\sigma = 2.4$ psi, $p_\tau = 1.44$ psi

12.2-2 $p_a = 1.748$ kPa

12.2-3 (a) Use $t = 2.5$ in.; (b) $p_{max} = 381$ psi

12.2-4 (a) Use $t = 98$ mm; (b) $p_{max} = 3.34$ MPa

12.2-5 (a) $F = 3738$ lb, $\sigma = 476$ psi; (b) $d_b = 0.534$ in.; (c) $r = 14.71$ in.

12.2-6 (a) $\sigma_{max} = 3.12$ MPa, $\epsilon_{max} = 0.438$; (b) $t_{reqd} = 1.29$ mm

12.2-7 (a) $\sigma_{max} = 425$ psi, $\epsilon_{max} = 1.105$; (b) $p_{max} = 7.77$ psi

12.2-8 (a) $p_{max} = 3.51$ MPa; (b) $p_{max} = 2.93$ MPa

12.2-9 (a) $f = 26.4$ kip/in.; (b) $\tau_{max} = 7543$ psi; (c) $\epsilon_{max} = 3.57 \times 10^{-4}$

12.2-10 (a) $f = 5.5$ MN/m; (b) $\tau_{max} = 57.3$ MPa; (c) $\epsilon_{max} = 3.87 \times 10^{-4}$

12.2-11 (a) $t_{min} = 0.291$ in.; (b) $p = 1904$ psi

12.2-12 (a) $t_{min} = 7.17$ mm; (b) $p = 19.25$ MPa

12.2-13 $D_0 = 90$ ft

12.3-1 $\sigma_1 = 22.4$ ksi, $\sigma_2 = 11.19$ ksi, $\tau_{maxIn} = 5.6$ ksi, $\tau_{maxOut} = 11.19$ ksi

12.3-2 $\sigma_1 = 76.2$ MPa, $\sigma_2 = 38.1$ MPa, $\tau_{maxIn} = 19.05$ MPa, $\tau_{maxOut} = 38.1$ MPa

12.3-3 (a) $t_{min} = 0.289$ in.; (b) $p_{max} = 2286$ psi

12.3-4 (a) $h = 22.2$ m; (b) zero

12.3-5 $n = 2.25$

12.3-6 (a) $F = 3\pi pr^2$; (b) $t_{reqd} = 10.91$ mm

12.3-7 (a) $p = 55$ psi; (b) $\epsilon_r = 9.18 \times 10^{-4}$

12.3-8 (a) $\epsilon_{max} = 6.67 \times 10^{-5}$; (b) $\epsilon_r = 2.83 \times 10^{-4}$

12.3-9 $t_{min} = 0.113$ in.

12.3-10 $t_{min} = 3.71$ mm

12.3-11 (a) $h = 25$ ft; (b) $\sigma_1 \approx 125$ psi

12.3-12 (a) $\sigma_h = 24.9$ MPa; (b) $\sigma_c = 49.7$ MPa; (c) $\sigma_w = 24.9$ MPa; (d) $\tau_h = 12.43$ MPa; (e) $\tau_c = 24.9$ MPa

12.3-13 (a) $t_{min} = 0.675$ in.; (b) $t_{min} = 0.338$ in.

12.3-14 (a) $\sigma_1 = 93.3$ MPa, $\sigma_2 = 46.7$ MPa; (b) $\tau_1 = 23.2$ MPa, $\tau_2 = 46.7$ MPa; (c) $\epsilon_1 = 3.97 \times 10^{-4}$, $\epsilon_2 = 9.33 \times 10^{-5}$; (d) $\theta = 35°$, $\sigma_{x_1} = 62.0$ MPa, $\sigma_{y1} = 78.0$ MPa, $\tau_{x1y1} = 21.9$ MPa;

12.3-15 (a) $\sigma_1 = 7015$ psi, $\sigma_2 = 3508$ psi; (b) $\tau_1 = 1754$ psi, $\tau_2 = 3508$ psi; (c) $\epsilon_1 = 1.988 \times 10^{-4}$, $\epsilon_2 = 4.68 \times 10^{-5}$; (d) $\theta = 28°$, $\sigma_{x1} = 4281$ psi, $\sigma_{y1} = 6242$ psi, $\tau_{x1y1} = 1454$ psi;

12.4-1 (a) $\sigma_1 = 4.96$ ksi, $\sigma_2 = -3.99$ ksi, $\tau_{max} = 4.48$ ksi; (b) $\sigma_1 = 17.08$ ksi, $\sigma_2 = -0.95$ ksi, $\tau_{max} = 9.01$ ksi

12.4-2 (a) $\sigma_1 = 56.7$ MPa, $\sigma_2 = -0.968$ MPa, $\tau_{max} = 28.8$ MPa; (b) $\sigma_1 = 13.18$ MPa, $\sigma_2 = -6.42$ MPa, $\tau_{max} = 9.80$ MPa

12.4-3 $\sigma_1 = 1586$ psi, $\sigma_2 = -2.02$ psi, $\tau_{max} = 794$ psi

12.4-4 $\sigma_1 = 42.8$ MPa, $\sigma_2 = -42.8$ MPa, $\tau_{max} = 42.8$ MPa

12.4-5 $t_{min} = 0.125$ in.

12.4-6 $p_{max} = 9.60$ MPa

12.4-7 (a) $\sigma_{max} = \sigma_1 = 11.09$ ksi, $\tau_{max} = 3.21$ ksi; (b) $T_{max} = 178$ kip-ft; (c) $t_{min} = 0.519$ in.

12.4-8 (a) $P_{max} = 461$ kN; (b) $p_{max} = 8.32$ MPa

12.4-9 $\sigma_t = 10{,}680$ psi: No compressive stresses. $\tau_{max} = 5340$ psi

12.4-10 $\phi_{max} = 0.552$ rad $= 31.6°$

12.4-11 $\sigma_t = 3963$ psi, $\sigma_c = -8791$ psi, $\tau_{max} = 6377$ psi

12.4-12 $\sigma_t = 16.43$ MPa, $\sigma_c = -41.4$ MPa, $\tau_{max} = 28.9$ MPa

12.4-13 $P = 194.2$ kips

12.4-14 (a) $\sigma_{max} = \sigma_1 = 35.8$ MPa, $\tau_{max} = 18.05$ MPa; (b) $P_{max} = 6.73$ kN

12.4-15 (a) $\sigma_x = 0$, $\sigma_y = 6145$ psi, $\tau_{xy} = 345$ psi; (b) $\sigma_1 = 6164$ psi, $\sigma_2 = -19.30$ psi, $\tau_{max} = 3092$ psi

12.4-16 $\tau_A = 76.0$ MPa, $\tau_B = 19.94$ MPa, $\tau_C = 23.7$ MPa

12.4-17 $\sigma_1 = 2689$ psi, $\sigma_2 = 0$, $\tau_{max} = 1344$ psi

12.4-18 $\sigma_1 = 0$, $\sigma_2 = -21.7$ MPa, $\tau_{max} = 10.87$ MPa

12.4-19 $\sigma_1 = 1094$ psi (max. tensile stress at base of pole), $\sigma_2 = -7184$ psi (max. compressive stress at base of pole), $\tau_{max} = 3731$ psi (max. shear stress at base of pole)

12.4-20 $d_{min} = 48.4$ mm

12.4-21 $\sigma_t = 39{,}950$ psi, $\sigma_c = -2226$ psi, $\tau_{max} = 21{,}090$ psi

12.4-22 (a) $\sigma_x = 0$, $\sigma_y = 2.23$ MPa, $\tau_{xy} = 0$; (b) $\tau_{max} = 1.116$ MPa

12.4-23 (a) $\sigma_t = 29.15\dfrac{qR^2}{d^3}, \sigma_c = -8.78\dfrac{qR^2}{d^3}, \tau_{max} = 18.97\dfrac{qR^2}{d^3};$

(b) $\sigma_t = 14.04\dfrac{qR^2}{d^3}, \sigma_c = -2.41\dfrac{qR^2}{d^3}, \tau_{max} = 8.22\dfrac{qR^2}{d^3}$

12.4-24 (a) $\sigma_x = 0, \sigma_y = 2.34$ MPa, $\tau_{xy} = -0.1912$ MPa;
(b) $\sigma_1 = 2.35$ MPa, $\sigma_2 = -0.0155$ MPa;
(c) $\tau_{max} = 1.185$ MPa

12.4-25 $\sigma_t = 4320$ psi, $\sigma_c = -1870$ psi, $\tau_{max} = 3100$ psi

12.4-26 Pure shear: $\tau_{max} = 0.804$ MPa

12.4-27 (a) $d_{min} = 1.65$ in.; (b) $P_{max} = 19.25$ lb

12.4-28 (a) $\sigma_1 = 29.3$ MPa, $\sigma_2 = -175.9$ MPa,
$\tau_{max} = 102.6$ MPa; (b) $\sigma_1 = 156.1$ MPa,
$\sigma_2 = -33$ MPa, $\tau_{max} = 94.5$ MPa

12.4-29 (a) $\sigma_1 = 0, \sigma_2 = -20,730$ psi, $\tau_{max} = 10,365$ psi;
(b) $\sigma_1 = 988$ psi, $\sigma_2 = -21,719$ psi, $\tau_{max} = 11,354$ psi

12.4-30 Maximum: $\sigma_t = 18.35$ MPa, $\sigma_C = -18.35$ MPa,
$\tau_{max} = 9.42$ MPa

12.4-31 Top of beam: $\sigma_1 = 8591$ psi, $\sigma_2 = 0$, $\tau_{max} = 4295$ psi

12.4-32 (a) $d_{AL} = 26.3$ mm; (b) $d_{Ti} = 21.4$ mm

12.4-33 $\sigma_y = 0$, $\sigma_x = \dfrac{(FL)\dfrac{d_2}{2}}{\dfrac{I_p}{2}} = 1943$ psi,

$\tau_{xy} = \dfrac{-T\dfrac{d_2}{2}}{I_p} = -547$ psi, $\sigma_1 = 2087$ psi,

$\sigma_2 = -143.2$ psi, $\tau_{max} = 1115$ psi

12.4-34 (a) $\sigma_1 = 0, \sigma_2 = \sigma_x = -108.4$ MPa,
$\tau_{max} = \dfrac{\sigma_x}{2} = -54.2$ MPa;
(b) $\sigma_1 = 0.703$ MPa, $\sigma_2 = -1.153$ MPa,
$\tau_{max} = 0.928$ MPa; (c) $P_{max} = 348$ N

12.4-35 $\sigma_x = -18.6$ ksi, $\sigma_y = 0, \tau_{xy} = 4.45$ ksi,
$\sigma_1 = 1.012$ ksi, $\sigma_2 = -19.62$ ksi, $\tau_{max} = 10.31$ ksi

CHAPTER 13

13.2-1 Moment M_0 applied at $x = L$

13.2-2 Uniform load

13.2-3 $q = q_0x/L$; Triangular load, acting downward

13.2-4 (a) $q = q_0 \sin \pi x/L$, Sinusoidal load;
(b) $R_A = R_B = q_0L/\pi$; (c) $M_{max} = q_0L^2/\pi^2$

13.2-5 $q = q_0(1 - x/L)$; Triangular load, acting downward

13.2-6 (a) $q = q_0(L^2 - x^2)/L^2$; Parabolic load, acting
downward;
(b) $R_A = 2q_0L/3, M_A = -q_0L^2/4$

13.3-1 $\delta_{max} = 0.0349$ in., $\theta_A = 7.93 \times 10^{-4}$ rad

13.3-2 (a) $q = 1.438$ kN/m; (b) $r = 14.74$ mm

13.3-3 $\delta_{max} = 0.182$ in., $\theta = 0.199°$

13.3-4 $h = 96$ mm

13.3-5 $L = 120$ in. $= 10$ ft

13.3-6 $\delta_{max} = 15.4$ mm

13.3-7 $\delta/L = 1/400$

13.3-8 $E_g = 80.0$ GPa

13.3-9 Let $\beta = a/L : \dfrac{\delta_C}{\delta_{max}}$

$\dfrac{\delta_C}{\delta_{max}} = \dfrac{(3\sqrt{3})(-1 + 8\beta - 4\beta^2)}{16(2\beta - \beta^2)^{3/2}}$

The deflection at the midpoint is close to the maximum
deflection. The maximum difference is only 2.6%.

13.3-10 $\dfrac{\delta_B}{M} = 3.2 \times 10^4 \left[\dfrac{1}{N}\right], \dfrac{\theta_B}{M} = 200 \left[\dfrac{1}{N \cdot \mu m}\right]$

13.3-11 (a) $\delta_B = 0.992$ in., $\delta_C = 1.999$ in.;
(b) $\delta_B = 0.992$ in., $\delta_C = 2.03$ in.

13.3-15 (a) $q_0 = 83.3$ kips/ft (b) $\theta_B = 1.562 \times 10^{-3}$ rad

13.3-16 W360 $\times$ 79

13.3-17 $v(x) = -mx^2(3L - x)/6EI, \delta_B = mL^3/3EI$,
$\theta_B = mL^2/2EI$

13.3-18 $v(x) = -\dfrac{q}{48EI}(2x^4 - 12x^2L^2 + 11L^4), \delta_B = \dfrac{qL^4}{48EI}$

13.3-19 See Table E-2, Case 13.

13.3-20 See Table E-1, Case 2.

13.3-21 $v(x) = \dfrac{q_0L}{24EI}(x^3 - 2Lx^2)$ for $0 \le x \le \dfrac{L}{2}$;

$v(x) = \dfrac{-q_0}{960LEI}(-160L^2x^3 + 160L^3x^2$
$+ 80Lx^4 - 16x^5 - 25L^4x + 3L^5)$

for $\dfrac{L}{2} \le x \le L$;

$\delta_B = \dfrac{7}{160}\dfrac{q_0L^4}{EI}, \delta_C = \dfrac{1}{64}\dfrac{q_0L^4}{EI}$

13.3-22 $v(x) = \dfrac{q_0x}{5760LEI}(200x^2L^2 - 240x^3L$
$+ 96x^4 - 53L^4)$ for $0 \le x \le \dfrac{L}{2}$;

$v(x) = \dfrac{-q_0L}{5760EI}(40x^3 - 120Lx^2$
$+ 83L^2x - 3L^3)$ for $\dfrac{L}{2} \le x \le L$;

$\delta_C = \dfrac{3q_0L^4}{1280EI}$

13.3-23 $v(x) = -\dfrac{PL}{10,368EI}(-4104x^2 + 3565L^2)$

for $0 \le x \le \dfrac{L}{3}$; $v(x) = -\dfrac{P}{1152EI}$

$(-648Lx^2 + 192x^3 + 64L^2x + 389L^3)$

for $\dfrac{L}{3} \le x \le \dfrac{L}{2}$; $v(x) = -\dfrac{P}{144EIL}$
$(-72L^2x^2 + 12Lx^3 + 6x^4 + 5L^3x + 49L^4)$

for $\dfrac{L}{2} \le x \le L$;
$\delta_A = \dfrac{3565PL^3}{10,368EI}, \delta_C = \dfrac{3109PL^3}{10,368EI}$

13.4-3 $v = -M_0x(L - x)^2/2LEI$, $\delta_{max} = 2M_0L^2/27EI$
(downward)

13.4-4 $v(x) = -\dfrac{q}{48EI}(2x^4 - 12x^2L^2 + 11L^4)$, $\theta_B = -\dfrac{qL^3}{3EI}$

13.4-5 See Table E-1, Case 10.

13.4-6 $v = -q_0x^2(45L^4 - 40L^3x + 15L^2x^2 - x^4)/360L^2EI$,
$\delta_B = 19q_0L^4/360EI$, $\theta_B = q_0L^3/15EI$

13.4-7 $v = -q_0x(3L^5 - 5L^3x^2 + 3Lx^4 - x^5)/90L^2EI$,
$\delta_{max} = 61q_0L^4/5760EI$

13.4-8 $v(x) = \dfrac{q_0}{120EIL}(x^5 - 5Lx^4 + 20L^3x^2 - 16L^5)$,
$\delta_{max} = \dfrac{2q_0L^4}{15EI}$

13.4-9 $v(x) = -\dfrac{qL^2}{16EI}(x^2 - L^2)$ for $0 \le x \le L$,

$v(x) = -\dfrac{q}{48EI}(-20L^3x + 27L^2x^2 -$

$12Lx^3 + 2x^4 + 3L^4)$ for $L \le x \le \dfrac{3L}{2}$,

$\delta_C = \dfrac{9qL^4}{128EI}$, $\theta_C = \dfrac{7qL^3}{48EI}$

13.4-10 $v(x) = -\dfrac{q_0L^2}{480EI}(-20x^2 + 19L^2)$ for $0 \le x \le \dfrac{L}{2}$;

$v(x) = -\dfrac{q_0}{960EIL}(80Lx^4 - 16x^5 -$
$120L^2x^3 + 40L^3x^2 - 25L^4x + 41L^5)$

for $\dfrac{L}{2} \le x \le L$; $\delta_A = \dfrac{19q_0L^4}{480EI}$, $\theta_B = -\dfrac{13q_0L^3}{192EI}$,

$\delta_C = \dfrac{7q_0L^4}{240EI}$

13.5-1 $b = 3.44$ in., $h = 6.88$ in.

13.5-2 $\delta_{max} = 0.78$ mm

13.5-3 $P = 3.94$ lb

13.5-4 $\delta_C = 6.36$ mm

13.5-5 $\delta_B = 1.187 \times 10^{-3}$ in., $\theta_B = 4.94 \times 10^{-5}$ rad

13.5-6 $\delta_B = 0.764$ mm, $\theta_B = 4.02 \times 10^{-4}$ rad

13.5-7 $\theta_B = 7PL^2/9EI$, $\delta_B = 5PL^3/9EI$

13.5-8 (a) $\delta_1 = 11PL^3/144EI$; (b) $\delta_2 = 25PL^3/384EI$;
(c) $\delta_1/\delta_2 = 88/75 = 1.173$

13.5-9 (a) $a/L = 2/3$; (b) $a/L = 1/2$

13.5-10 (a) $\delta_C = 6.25$ mm (upward);
(b) $\delta_C = 18.36$ mm (downward)

13.5-11 $y = Px^2(L - x)^2/3LEI$

13.5-12 $\theta_B = 7qL^3/162EI$, $\delta_B = 23qL^4/648EI$

13.5-13 $\delta_C = 0.0905$ in., $\delta_B = 0.293$ in.

13.5-14 (a) $M = PL/2$;
(b) $M = 5PL/24$, $\theta_B = PL^2/12EI$;
(c) $M = PL/8$, $\delta_B = -PL^3/24EI$

13.5-15 $\theta_A = \dfrac{PL^2}{9EI}$ (CW), $\theta_B = \theta_A$ but (CCW)

$\delta_{max} = \dfrac{23}{648}\dfrac{PL^3}{EI}$ at $x = \dfrac{L}{2}$

13.5-16 $\theta_A = \theta_B = \dfrac{PL^2}{81EI}$ (CW), $\delta_{max} = \dfrac{2\sqrt{6}}{2187}\dfrac{PL^3}{EI}$

13.5-17 $M = (19/180)q_0L^2$

13.5-18 (a) $\delta_A = PL^2(10L - 9a)/324EI$ (positive upward);
(b) Upward when $a/L < 10/9$, downward when
$a/L > 10/9$

13.5-19 (a) $\delta_C = PH^2(L + H)/3EI$;
(b) $\delta_{max} = PHL^2/9\sqrt{3}EI$

13.5-20 $\delta_C = 3.5$ mm

13.5-21 $\theta_B = q_0L^3/10EI$, $\delta_B = 13q_0L^4/180\ EI$

13.5-22 $\theta_A = q(L^3 - 6La^2 + 4a^3)/24EI$,
$\delta_{max} = q(5L^4 - 24L^2a^2 + 16a^4)/384EI$

13.5-23 (a) $P/Q = 9a/4L$;
(b) $P/Q = 8a(3L + a)/9L^2$;
(c) $P/qa = 9a/8L$ for $\delta_B = 0$,
$P/qa = a(4L + a)/3L^2$ for $\delta_D = 0$

13.5-24 $\delta = 19WL^3/31,104EI$

13.5-25 $k = 3.33$ lb/in.

13.5-26 $M_1 = 7800$ N·m, $M_2 = 4200$ N·m

13.5-27 $\delta = \dfrac{6Pb^3}{EI}$

13.5-28 $\delta_E = \dfrac{47Pb^3}{12EI}$

13.5-29 $\delta_C = 0.120$ in.

13.5-30 $\delta_C = 2.05$ mm (downward)

13.5-31 $\delta_B = 0.554$ in., $\theta_A = 0.539°$

13.5-32 $q = 16cEI/7L^4$

13.5-33 $\delta_h = Pcb^2/2EI$, $\delta_v = Pc^2(c + 3b)/3EI$

13.5-34 $\delta = PL^2(2L + 3a)/3EI$

13.5-35 (a) $H_B = 0, V_B = \dfrac{M}{L}, V_C = -V_B$;

(b) $\theta_A = \dfrac{5ML}{6EI}$, $\theta_B = \dfrac{ML}{3EI}$, $\theta_C = \dfrac{-ML}{6EI}$,

$\theta_D = \theta_C$; (c) $\delta_A = (7/24)ML^2/EI$ (to the left),

$\delta_D = (1/12)ML^2/EI$ (to the right);

(d) $L_{CD} = \dfrac{\sqrt{14}}{2}L = 1.871L$

13.5-36 (a) $H_B = 0$, $V_B = \dfrac{P}{3}$, $V_C = \dfrac{2P}{3}$;

(b) $\theta_A = \left(\dfrac{-4}{81}\right)\dfrac{ML}{EI}$, $\theta_B = \theta_A$,

$\theta_C = \left(\dfrac{5}{81}\right)\dfrac{ML}{EI}$, $\theta_D = \theta_C$;

(c) $\delta_A = -\theta_B\left(\dfrac{L}{2}\right) = \dfrac{2L^2M}{81EI}$ (to the right),

$\delta_D = \theta_C\left(\dfrac{L}{2}\right) = \dfrac{5L^2M}{162EI}$ (to the left);

(d) $L_{CD} = \dfrac{2\sqrt{5}L}{5} = 0.894L$

13.5-37 (a) $b/L = 0.403$; (b) $\delta_C = 0.00287qL^4/EI$

13.5-38 $\alpha = 22.5°, 112.5°, -67.5°,$ or $-157.5°$

13.5-39 $\delta_D = \dfrac{2}{9}\dfrac{PL^3}{EI} + \dfrac{wL^4}{8EI}$, $\theta_D = \dfrac{5}{18}\dfrac{PL^2}{EI} + \dfrac{wL^3}{6EI}$

13.5-40 $\delta_B = 113.6$ mm, $\theta_B = 0.0238$ rad

13.5-41 $M_A = \dfrac{14}{81}qL^2$, $\theta_A = \dfrac{2}{81}\dfrac{qL^3}{EI}$ (CCW)

13.5-42 $M_A = \dfrac{16}{81}qL^2$, $\theta_A = \dfrac{qL^3}{27EI}$ (CCW)

13.5-43 $d = 0.375\,L$, $\theta_A = 0.0635\dfrac{qL^3}{EI}$ (CCW)

13.5-44 (a) $\delta_C = \dfrac{121}{1920}\dfrac{q_2L^4}{EI} - \dfrac{13}{960}\dfrac{q_1L^4}{EI}$ (upward);

(b) $q_2 = \dfrac{26}{121}q_1$; (c) $\delta_C = \dfrac{19}{384}\dfrac{q_0L^4}{EI}$ (upward)

13.6-1 (a) $R_A = 0.252$ kips, $R_B = 0.151$ kips,
$M_A = 0.579$ kip-ft; (b) $x_0 = 2.88$ ft;
(c) $\delta_{max} = 6.7 \times 10^{-4}$ in., $\theta_B = 1.868 \times 10^{-5}$ rad

13.6-2 (a) $h = 257$ mm; (b) $\delta_0 = 1$ mm

13.6-3 $R_A = -R_B = 3M_0/2L$, $M_A = M_0/2$,
$v = -M_0x^2(L - x)/4LEI$

13.6-4 $R_A = R_B = qL/2$, $M_A = M_B = qL^2/12$,
$v = -qx^2(L - x)^2/24EI$

13.6-5 $R_A = R_B = 3EI\delta_B/L^3$, $M_A = 3EI\delta_B/L^2$,
$v = -\delta_Bx^2(3L - x)/2L^3$

13.6-6 $\theta_B = \dfrac{qL^3}{6(k_RL - EI)}$,

$\delta_B = \left(\dfrac{1}{EI}\right)\left(-\dfrac{1}{8}qL^4 + \dfrac{k_RqL^5}{12(k_RL - EI)}\right)$

13.6-7 $R_A = V(0) = \dfrac{9}{40}q_0L$, $R_B = -V(L) = \dfrac{11}{40}q_0L$,

$M_A = \dfrac{7}{120}q_0L^2$

13.6-8 (a) $R_A = V(0) = \dfrac{7}{60}q_0L$, $R_B = -V(L) = \dfrac{13}{60}q_0L$,

$M_A = \dfrac{1}{30}q_0L^2$, $v = \dfrac{q_0}{360L^2EI}(-x^6 + 7L^3x^3 - 6q_0L^4x^2)$;

(b) $R_A = V(0) = 0.31q_0L$

$= \left(\dfrac{2}{\pi} - 6\dfrac{\pi^2 - 4\pi + 8}{\pi^4}\right)q_0L$,

$R_B = -V(L) = 0.327q_0L$

$= \left(6\dfrac{\pi^2 - 4\pi + 8}{\pi^4}\right)q_0L$,

$M_A = -2q_0L^2\dfrac{\pi^2 - 12\pi + 24}{\pi^4}$,

$v = \dfrac{1}{EI} \times$

$\left[\begin{array}{l} -q_0\left(\dfrac{2L}{\pi}\right)^4\sin\left(\dfrac{\pi x}{2L}\right) - 6q_0L\dfrac{\pi^2 - 4\pi + 8}{\pi^4}\dfrac{x^3}{6} \\ +2q_0L^2\dfrac{\pi^2 - 12\pi + 24}{\pi^4}\dfrac{x^2}{2} + q_0\left(\dfrac{2L}{\pi}\right)^3x \end{array}\right]$

13.6-9 (a) $R_A = \dfrac{61Lq_0}{120}$,

$M_A = \dfrac{11L^2q_0}{120}$,

$R_B = \dfrac{19Lq_0}{120}$,

$v(x) = -\dfrac{q_0x^2(33L^4 - 61L^3x + 30L^2x^2 - 2x^4)}{720EIL^2}$;

(b) $R_A = \dfrac{48Lq_0}{\pi^4}$,

$R_B = \displaystyle\int_0^L q(x)dx - R_A = \dfrac{2Lq_0}{\pi} - \dfrac{48Lq_0}{\pi^4}$,

$M_A = \displaystyle\int_0^L q(x)x\,dx - R_BL = \dfrac{2L^2q_0(\pi - 2)}{\pi^2}$

$- L\left(\dfrac{2Lq_0}{\pi} - \dfrac{48Lq_0}{\pi^4}\right)$,

$v(x) = \dfrac{16L^4q_0 - 24L^2q_0x^2 + 8Lq_0x^3 - 16L^4q_0\cos\left(\frac{\pi x}{2L}\right)}{\pi^4EI}$

13.6-10 (a) $R_A = V(0) = \dfrac{24}{\pi^4}q_0L$,

$R_B = -V(L) = -\dfrac{24}{\pi^4}q_0L$,

$M_A = \left(\dfrac{12}{\pi^4} - \dfrac{1}{\pi^2}\right)q_0L^2$ (CCW),

$M_B = \left(\dfrac{12}{\pi^4} - \dfrac{1}{\pi^2}\right)q_0L^2$ (CCW),

$v = \dfrac{1}{\pi^4 EI}\left[-q_0L^4\cos\left(\dfrac{\pi x}{L}\right)\right.$

$\left. + 4q_0Lx^3 - 6q_0L^2x^2 + q_0L^4\right]$;

(b) $R_A = R_B = q_0L/\pi$, $M_A = M_B = 2q_0L^2/\pi^3$,

$v = -q_0L^2(L^2\sin\pi x/L + \pi x^2 - \pi Lx)/\pi^4 EI$

13.6-11 (a) $R_A = V(0) = \dfrac{48(4-\pi)}{\pi^4}q_0L$,

$R_B = -V(L) = \left(\dfrac{2}{\pi} - \dfrac{48(4-\pi)}{\pi^4}\right)q_0L$,

$M_A = -q_0\left(\dfrac{2L}{\pi}\right)^2 + \dfrac{16(6-\pi)}{\pi^4}q_0L^2$,

$M_B = -\dfrac{32(\pi-3)}{\pi^4}q_0L^2$,

$v = \dfrac{1}{\pi^4 EI}\left[-16q_0L^4\cos\left(\dfrac{\pi x}{2L}\right)\right.$

$+ 8(4-\pi)q_0Lx^3$

$\left. - 8(6-\pi)q_0L^2x^2 + 16q_0L^4\right]$;

(b) $R_A = V(0) = \dfrac{13}{30}q_0L$,

$R_B = -V(L) = \dfrac{7}{30}q_0L$,

$M_A = \dfrac{1}{15}q_0L^2$ (CCW), $M_B = -\dfrac{1}{20}q_0L^2$ (CCW),

$v = \dfrac{q_0}{360L^2 EI}[x^6 - 15L^2x^4 + 26L^3x^3$

$- 12L^4x^2]$

13.6-12 $R_A = V(0) = \dfrac{3}{20}q_0L$,

$R_B = -V(L) = \dfrac{7}{20}q_0L$,

$M_A = \dfrac{1}{30}q_0L^2$,

$v = \dfrac{1}{120LEI}(-q_0x^5 + 3q_0L^2x^3 - 2q_0L^3x^2)$

13.6-13 $R_A = -R_B = 3M_0/2L$,
$M_A = -M_B = M_0/4$,
$v = -M_0x^2(L - 2x)/8LEI$ for $0 \le x \le L/2$

13.6-14 $R_B = -\dfrac{9}{8}\dfrac{M_0}{L}$, $R_A = \dfrac{9}{8}\dfrac{M_0}{L}$, $M_A = \dfrac{1}{8}M_0$,

$v = \dfrac{1}{EI}\left(\dfrac{9M_0}{48L}x^3 - \dfrac{M_0}{16}x^2\right)\left(0 \le x \le \dfrac{L}{2}\right)$,

$v = \dfrac{1}{EI} \times$

$\left(\dfrac{9M_0}{48L}x^3 - \dfrac{9M_0}{16}x^2 + \dfrac{M_0L}{2}x - \dfrac{M_0L^2}{8}\right)$

$\left(\dfrac{L}{2} \le x \le L\right)$

13.6-15 $q = 0.385$ kips/ft

13.6-16 $q = 2.95$ kN/m

13.6-17 (a) $R_A = R_C = 60$ lb, $R_B = 200$ lb;
(b) $\delta_{AB} = -2.54 \times 10^{-3}$ in.

13.6-18 $R_A = 1.76$ kN, $M_A = 2.4$ kN·m,
$R_B = 3.24$ kN, $M_B = 3.6$ kN·m, $\delta_C = 0.0674$ mm

13.6-19 (a) $R_A = 0.505$ kips, $M_A = 1.154$ kip-ft,
$R_B = 0.1701$ kips, $M_B = 0.668$ kip-ft;
(b) $\delta_{max} = 0.277$ in. at $x = 0.466L = 6.99$ ft;
(c) $R_A = R_B = 0.562$ kips,
$M_A = -M_B = 1.406$ kip-ft

13.6-20 $R_A = 375$ N, $M_A = 225$ N·m, $R_B = 225$ N

13.6-21 $R_A = Pb(3L^2 - b^2)/2L^3$, $R_B = Pa^2(3L - a)/2L^3$,
$M_A = Pab(L + b)/2L^2$

13.6-22 $R_A = qL$, $M_A = \dfrac{qL^2}{3}$, $M_B = \dfrac{qL^2}{6}$

13.6-23 $R_A = -\dfrac{1}{8}qL$, $R_B = \dfrac{17}{8}qL$, $M_A = -\dfrac{1}{8}qL^2$

13.6-24 (a) $R_A = M_0/3L$, $H_A = -4M_0/3L$,
$R_B = -R_A$, $R_C = -H_A$;
(b) $\theta_A = -M_0L/18EI$, $\theta_B = M_0L/9EI$,
$\theta_C = \theta_A$; (c) $L_{BC} = 2L$

13.6-25 (a) $R_A = \dfrac{4M_0}{3L}$, $H_A = \dfrac{2M_0}{3L}$, $R_B = -\dfrac{4M_0}{3L}$,
$R_C = -\dfrac{2M_0}{3L}$; (b) $\theta_A = \dfrac{5}{18}\dfrac{M_0L}{EI}$, $\theta_B = \dfrac{-M_0L}{18EI}$,
$\theta_C = \dfrac{M_0L}{36EI}$; (c) $L_{AB} = 2.088L$

13.6-26 (a) $R_A = \dfrac{M_0}{L} + \dfrac{M_0k_R}{2(3EI + Lk_R)}$,

$R_B = -R_A$, $M_B = \dfrac{LM_0k_R}{6EI + 2Lk_R}$ (CCW);

(b) $\theta_A = \dfrac{LM_0}{4EI} + \dfrac{LM_0}{4(3EI + Lk_R)}$

For k_R goes to zero:

$\theta_A = \dfrac{LM_0}{4EI} + \dfrac{LM_0}{4(3EI)} = \dfrac{LM_0}{3EI}$

For k_R goes to infinity: $\theta_A = \dfrac{M_0L}{4EI}$

For k_R goes to $6EI/L$:

$\theta_A = \dfrac{LM_0}{4EI} + \dfrac{LM_0}{4\left[3EI + L\left(\dfrac{6EI}{L}\right)\right]} = \dfrac{5LM_0}{18EI}$

13.6-27 (a) $H_A = \dfrac{3}{2}\dfrac{M_0}{L}$,

$\quad\quad H_B = 0, V_B = 0, V_C = 0, H_D = -H_A$;

(b) $\theta_A = \dfrac{-M_0 L}{16 EI}, \theta_D = -\theta_A$,

$\quad\quad \theta_B = \dfrac{M_0 L}{8 EI}, \theta_C = -\theta_B$;

(c) $H_A = \dfrac{M_0}{L}, H_B = -2\dfrac{M_0}{L}, V_B = \dfrac{M_0}{L}$,

$\quad\quad V_C = \dfrac{-M_0}{L}, H_D = H_A, \theta_A = \dfrac{-M_0 L}{24 EI}$,

$\quad\quad \theta_D = \theta_A, \theta_B = \dfrac{M_0 L}{12 EI}, \theta_C = \theta_B$

13.6-28 $t_{AB}/t_{CD} = L_{AB}/L_{CD}$

13.6-29 $R_A = \dfrac{7}{12}qL, R_B = \dfrac{17}{12}qL, M_A = \dfrac{7}{12}qL^2$

13.6-30 $R_A = 2qL, M_B = \dfrac{7}{12}qL^2$

13.6-31 $R_A = R_B = q_0 L/4, M_A = M_B = 5q_0 L^2/96$

13.6-32 $R_A = qL/8, R_B = 33qL/16, R_C = 13qL/16$

13.6-33 $R_A = 1100$ lb (downward), $R_B = 2800$ lb (upward), $M_A = 30{,}000$ lb-in. (clockwise)

13.6-34 $R_B = 6.44$ kN

13.6-35 $R_A = 40$ lb, $R_B = 10$ lb, $M_A = 66.7$ lb-ft

13.6-36 (a) $R_A = R_B = 45$ kN, $M_A = -M_B = 35$ kN·m;
(b) $h = 167.1$ mm

13.6-37 $R_A = -35.7$ lb, $R_B = -154.3$ lb, $M_A = -262$ lb-ft, $M_B = -506$ lb-ft

13.6-38 $R_A = 33$ kN, $R_B = 17$ kN, $M_A = 25.3$ kN·m

13.6-39 (a) The tension force in the tie rod $=$ $R_D = 604$ lb; (b) $R_A = 796$ lb, $M_A = 1308$ lb-ft $= 1.567 \times 10^4$ lb-in.

13.6-40 $R_A = 31qL/48, R_B = 17qL/48, M_A = 7qL^2/48$

13.6-41 (a) $R_A = -23P/17, R_D = R_E = 20P/17, M_A = 3PL/17$;
(b) $M_{max} = PL/2$

13.6-42 $R_A = R_D = 2qL/5, R_B = R_C = 11qL/10$

13.6-43 $M_B(q) = (-800q)$ lb-in. for $q < 250$ lb/in.,
$M_B(q) = (-200q - 150{,}000)$ lb-in. for
$q \ge 250$ lb/in.

13.6-44 $R_A = \dfrac{19}{128}q_0 L, R_C = -\dfrac{19}{128}q_0 L$,

$\quad\quad M_A = \dfrac{3}{128}q_0 L^2, V_{max} = \dfrac{19}{128}q_0 L$,

$\quad\quad M_{max} = 0.0147 q_0 L^2$

13.6-45 $R_A = 25.1$ lb, $R_B = 4.88$ lb, $M_A = 61.5$ lb-ft

13.6-46 $R_A = -R_B = 6M_0 ab/L^3$,
$\quad\quad M_A = M_0 b(3a - L)/L^2$,
$\quad\quad M_B = -M_0 a(3b - L)/L^2$

13.6-47 $\sigma = 509$ psi

13.6-48 $(M_{AB})_{max} = 121qL^2/2048 = 6.05$ kN·m;
$\quad\quad (M_{CD})_{max} = 5qL^2/64 = 8.0$ kN·m

13.6-49 $F = 3160$ lb, $M_{AB} = 18{,}960$ lb-ft,
$\quad\quad M_{DE} = 7320$ lb-ft

13.6-50 $k = 48EI(6 + 5\sqrt{2})/7L^3 = 89.63EI/L^3$

13.6-51 (a) $V_A = V_C = 3P/32, H_A = P, M_A = 13PL/32$;
(b) $M_{max} = 13PL/32$

13.6-52 $H_A = \dfrac{-29}{64}P, H_C = \dfrac{-35}{64}P, M_{max} = \dfrac{29}{128}PL$

13.6-53 $R_A = R_B = 3000$ lb, $R_C = 0$

13.6-54 (a) $M_A = M_B = qb(3L^2 - b^2)/24L$;
(b) $b/L = 1.0, M_A = qL^2/12$;
(c) For $a = b = L/3, (M_{max})_{pos} = 19qL^2/648$

13.6-55 (a) $d_2/d_1 = \sqrt[4]{8} = 1.682$;
(b) $M_{max} = qL^2(3 - 2\sqrt{2})/2 = 0.08579qL^2$;
(c) Point C is below points A and B by the amount $0.01307qL^4/EI$.

13.6-56 $M_{max} = 19q_0 L^2/256, \sigma_{max} = 13.4$ MPa,
$\quad\quad \delta_{max} = 19q_0 L^4/7680EI = 0.00891$ mm

13.6-57 $M_A = \dfrac{2}{27}qL^2$

13.6-58 $M_A = \dfrac{4}{81}qL^2$

13.6-59 (a) $V_{max} = \dfrac{4}{3}P, M_{max} = -\dfrac{PL}{3}$;

$\quad\quad$ (b) $V_{max} = \dfrac{-17}{27}P, M_{max} = -\dfrac{10}{81}PL$

CHAPTER 14

14.2-1 $P_{cr} = \beta_R/L$

14.2-2 (a) $P_{cr} = \dfrac{\beta a^2 + \beta_R}{L}$; (b) $P_{cr} = \dfrac{\beta a^2 + 2\beta_R}{L}$

14.2-3 $P_{cr} = 2.08$ kips

14.2-4 $P_{cr} = 10$ kN

14.2-5 $P_{cr} = 6\beta_R/L$

14.2-6 $P_{cr} = 1.688$ kN

14.2-7 $P_{cr} = 1.042$ kips

14.2-8 (a) $P_{cr} = \dfrac{(L - a)(\beta a^2 + \beta_R)}{aL}$;

$\quad\quad$ (b) $P_{cr} = \dfrac{\beta L^2 + 20\beta_R}{4L}$

14.2-9 $P_{cr} = \dfrac{3\beta_R}{L}$

14.2-10 $P_{cr} = \dfrac{3}{5}\beta L$

14.2-11 $P_{cr} = \dfrac{7}{4}\beta L$

14.2-12 $P_{cr} = \dfrac{9}{8}\beta L$

14.2-13 $P_{cr} = \dfrac{9}{5}\dfrac{\beta_R}{L}$

14.2-14 $P_{cr1} = 30.2$ kN, $P_{cr2} = 84.6$ kN

14.2-15 $P_{cr1} = 54$ kips, $P_{cr2} = 207$ kips

14.2-16 $P_{cr1} = \beta L$, $P_{cr2} = \dfrac{9}{4}\beta L$

14.3-1 $P_{allow} = 135.3$ kips

14.3-2 W200 × 31.3

14.3-3 (a) $P_{cr} = 453$ kips; (b) $P_{cr} = 152$ kips

14.3-4 (a) $P_{cr} = 2803$ kN; (b) $P_{cr} = 953$ kN

14.3-5 (a) $P_{cr} = 650$ kips; (b) $P_{cr} = 140$ kips

14.3-6 $M_{allow} = 1143$ kN·m

14.3-7 $P_{allow} = 21.2$ kips

14.3-8 $P_{crY} = 254$ kN

14.3-9 (a) $P_{cr} = 26.1$ kips; (b) $d = 4.13$ in.

14.3-10 (a) $P_{cr} = 61.4$ kN; (b) $d = 115.8$ mm

14.3-11 $Q_{allow} = 23.8$ kips

14.3-12 (a) $Q_{cr} = \dfrac{\pi^2 EI}{L^2}$; (b) $Q_{cr} = \dfrac{2\pi^2 EI}{9L^2}$

14.3-13 (a) $Q_{cr} = \dfrac{2\pi^2 EI}{L^2}$; (b) $M_{cr} = \dfrac{3d\pi^2 EI}{L^2}$

14.3-14 $\Delta T = \pi^2 I / \alpha A L^2$

14.3-15 $h/b = 2$

14.3-16 (a) $P_{cr} = 3\pi^3 Er^4/4L^2$; (b) $P_{cr} = 11\pi^3 Er^4/4L^2$

14.3-17 $P_1 : P_2 : P_3 = 1.000 : 1.047 : 1.209$

14.3-18 $P_{allow} = 604$ kN

14.3-19 $F_{allow} = 54.4$ kips

14.3-20 $W_{max} = 124$ kN

14.3-21 $t_{min} = 0.165$ in.

14.3-22 $P_{cr} = 497$ kN

14.3-23 $W_{cr} = 51.9$ kips

14.3-24 $\theta = \arctan 0.5 = 26.57°$

14.3-25 (a) $q_{max} = 142.4$ lb/ft; (b) $I_{b,min} = 38.5$ in⁴; (c) $s = 0.264$ ft, 2.42 ft

14.3-26 $P_{cr} = 3.56$ kN

14.3-27 $P_{cr} = 16.28$ kips

14.4-1 $P_{cr} = 153.3$ kips

14.4-2 $t_{min} = 4.98$ mm

14.4-3 $P_{cr} = 235$ kips, 58.7 kips, 480 kips, 939 kips

14.4-4 $P_{cr} = 62.2$ kN, 15.6 kN, 127 kN, 249 kN

14.4-5 $P_{allow} = 253$ kips, 63.2 kips, 517 kips, 1011 kips

14.4-6 $P_{allow} = 678$ kN, 169.5 kN, 1387 kN, 2712 kN

14.4-7 $P_{cr} = 229$ kips

14.4-8 $T_{allow} = 18.1$ kN

14.4-9 (a) $Q_{cr} = 4575$ lb; (b) $Q_{cr} = 10,065$ lb, $a = 0$ in.

14.4-10 $P_{cr} = 447$ kN, 875 kN, 54.7 kN, 219 kN

14.4-11 $P_{cr} = 4\pi^2 EI/L^2$, $v = \delta(1 - \cos 2\pi x/L)/2$

14.4-12 $L_{max} = 3.61$ m

14.4-13 $P_{max} = 42.3$ kips

14.4-14 (a) $P_{cr} = 10.24$ kN, (b) $d = 71.1$ mm

14.4-15 $t_{min} = 0.284$ in.

14.4-16 $t_{min} = 10.0$ mm

14.4-17 (b) $P_{cr} = 13.89EI/L^2$

14.5-1 $\delta = 0.112$ in., $M_{max} = 1710$ lb-in.

14.5-2 $\delta = 8.87$ mm, $M_{max} = 2.03$ kN·m

14.5-3 $L = 16.33$ ft

14.5-4 $b = 11.07$ mm, $h = 16.61$ mm

14.5-5 For $P = 0.3P_{cr}$: $M/Pe = 1.162(\sin 1.721 x/L) + \cos 1.721x/L$

14.5-6 $P = 583.33\{\arccos[5/(5 + \delta)]\}^2$, in which $P =$ kN and $\delta =$ mm; $P = 884$ kN when $\delta = 10$ mm

14.5-7 $P = 125.58\{\arccos[0.2/(0.2 + \delta)]\}^2$, in which $P =$ kips and $\delta =$ in.; $P = 190$ kips when $\delta = 0.4$ in.

14.5-8 $P_{allow} = 49.9$ kN

14.5-9 $L_{max} = 150.5$ in. = 12.5 ft

14.5-10 $L_{max} = 3.14$ m

14.5-11 $\delta = e(\sec kL - 1)$, $M_{max} = Pe \sec kL$

14.5-12 $L_{max} = 2.21$ m

14.5-13 $L_{max} = 130.3$ in. = 10.9 ft

14.5-14 $T_{max} = 8.29$ kN

14.5-15 (a) $q_0 = 2230$ lb/ft = 186 lb/in.; (b) $M_{max} = 37.7$ kip-in., ratio = 0.47

14.6-1 (a) $\sigma_{max} = 17.3$ ksi; (b) $L_{max} = 46.2$ in.

14.6-2 $P_{allow} = 37.2$ kN

14.6-3 $b_{min} = 4.10$ in.

14.6-4   (a) $\sigma_{max} = 38.8$ MPa; (b) $L_{max} = 5.03$ m

14.6-5   (a) $\sigma_{max} = 9.65$ ksi; (b) $P_{allow} = 3.59$ kips

14.6-6 $d_2 = 131$ mm

14.6-7 $L = 12.5$ ft

14.6-8 $\sigma_{max} = 263$ MPa

14.6-9   (a) $\sigma_{max} = 10.9$ ksi; (b) $P_{allow} = 160$ kips

14.6-10   (a) $\sigma_{max} = 104.5$ MPa; (b) $L_{max} = 3.66$ m

14.6-11   (a) $\sigma_{max} = 9.60$ ksi; (b) $P_{allow} = 53.6$ kips

14.6-12   (a) $\sigma_{max} = 47.6$ MPa; (b) $n = 2.30$

14.6-13   (a) $\sigma_{max} = 13.4$ ksi; (b) $n = 2.61$

14.6-14   (a) $\sigma_{max} = 120.4$ MPa; (b) $P_2 = 348$ kN

14.6-15   (a) $\sigma_{max} = 17.6$ ksi; (b) $n = 1.89$

14.6-16   (a) $\sigma_{max} = 115.2$ MPa; (b) $P_2 = 149.7$ kN

Table F-1

Weights and Mass Densities

Material	Weight Density γ		Mass Density ρ	
	lb/ft³	kN/m³	slugs/ft³	kg/m³
Aluminum alloys	160–180	26–28	5.2–5.4	2600–2800
2014-T6, 7075-T6	175	28	5.4	2800
6061-T6	170	26	5.2	2700
Brass	520–540	82–85	16–17	8400–8600
Bronze	510–550	80–86	16–17	8200–8800
Cast iron	435–460	68–72	13–14	7000–7400
Concrete				
Plain	145	23	4.5	2300
Reinforced	150	24	4.7	2400
Lightweight	70–115	11–18	2.2–3.6	1100–1800
Copper	556	87	17	8900
Glass	150–180	24–28	4.7–5.4	2400–2800
Magnesium alloys	110–114	17–18	3.4–3.5	1760–1830
Monel (67% Ni, 30% Cu)	550	87	17	8800
Nickel	550	87	17	8800
Plastics				
Nylon	55–70	8.6–11	1.7–2.2	880–1100
Polyethylene	60–90	9.4–14	1.9–2.8	960–1400
Rock				
Granite, marble, quartz	165–180	26–28	5.1–5.6	2600–2900
Limestone, sandstone	125–180	20–28	3.9–5.6	2000–2900
Rubber	60–80	9–13	1.9–2.5	960–1300
Sand, soil, gravel	75–135	12–21	2.3–4.2	1200–2200
Steel	490	77.0	15.2	7850
Titanium	280	44	8.7	4500
Tungsten	1200	190	37	1900
Water, fresh	62.4	9.81	1.94	1000
sea	63.8	10.0	1.98	1020
Wood (air dry)				
Douglas fir	30–35	4.7–5.5	0.9–1.1	480–560
Oak	40–45	6.3–7.1	1.2–1.4	640–720
Southern pine	35–40	5.5–6.3	1.1–1.2	560–640